Extinction & Biogeography of Tropical Pacific Birds

Extinction & Biogeography of Tropical Pacific Birds

David W. Steadman

The University of Chicago Press
Chicago and London

David W. Steadman is the curator of birds at the Florida Museum of Natural History at the University of Florida.

The University of Chicago Press, Chicago 60637
The University of Chicago Press, Ltd., London

Printed in the United States of America

15 14 13 12 11 10 09 08 07 06   1 2 3 4 5

ISBN: 0-226-77141-5 (cloth)
ISBN: 0-226-77142-3 (paper)

Library of Congress Cataloging-in-Publication Data

Steadman, David W.
Extinction and biogeography of tropical Pacific birds / David W. Steadman.
p. cm.
Includes bibliographical references (p.) and index.
ISBN 0-226-77141-5 (cloth : alk. paper)—ISBN 0-226-77142-3 (pbk. : alk. paper)
1. Birds—Oceania. 2. Extinction (Biology)—Oceania. 3. Birds—Oceania—Geographical distribution. I. Title.
QL694.O25S73 2006
598.1752099–dc22   2005030534

⊗ The paper used in this publication meets the minimum requirements of the American National Standard for Information Sciences—Permanence of Paper for Printed Library Materials, ANSI Z39.48-2.

For Will and Sarah

Contents

Preface

Most books about island biology, as well as many journal articles, begin by trying to hook the reader on one or both of two themes. The first theme is that islands are wonderful "natural laboratories" where biological processes can be analyzed with unusual clarity in simplified settings. The second theme, related to the first, pays homage to the great names in island biology, who are among the most influential organismal biologists of all time.

This book will begin a little differently. I will not portray islands as outstanding natural laboratories because not many of them are natural any more. Human activities have irreparably altered island biotas, so that the plant and animal communities we have been able to study over the past two centuries have an artificial composition characterized by losses of indigenous species and additions of nonnative species. The full extent of the losses often is poorly known, making it difficult to gauge just how "unnatural" any given island biota really is.

My unwillingness to praise the major figures in island biology is not out of disrespect. Please don't worry; in due course I will give these scientists the credit they so deserve, for anyone studying island biology today stands to some extent on the foundations they have laid. Nevertheless, one of my goals in this book is to nudge island biology in a new direction. All fields of study need a good kick in the pants once in a while. In my opinion, this kick is long overdue in island biology, especially in the area of island biogeography. Despite more than three decades of criticism, the fundamentals of equilibrium theory in island biogeography continue to have widespread support. Data that fail to support an "elegant" model are often regarded as noise or the exception that proves the rule. Elegant models made by deified people die hard.

I will use a historical approach, grounded in carefully collected empirical data, to walk island biogeography down a different path. Unafflicted by physics envy, I am willing to sacrifice elegance for reality, knowing just how muddy reality can be. I see little value in reducing a complex biological situation to an equation whose assumptions have stripped the situation of its integral properties. I am more interested in conceptual progress than in theoretical exercises that are electronically feasible but may have little bearing on understanding how real species exist on real islands. Throughout this book, especially in Parts II and III, I will present and interpret an abundance of data that anyone is welcome to manipulate by whatever methods they choose. I hope that "baring my bones" will stimulate creative new discussions of how island faunas are put together and taken apart through time.

The focus of this study is landbirds on tropical Pacific islands, especially those from Fiji eastward to Easter Island. Many of the patterns and processes that I will describe may be very different for other groups of organisms on the same islands, or for landbirds from other island groups, whether in the Pacific or another ocean. The field of island biogeography is fraught with overgeneralization. The multitudinous groups of plants and animals that live on oceanic islands are highly variable in evolutionary history, taxonomic diversity, dispersal, colonization, population size, habitat preference, reproduction, nourishment, and any other attribute. Therefore, I see little reason why the biology of disparate insular organisms should be governed by any particular set of rules. I will evaluate each situation, in this case each set of landbirds in each island group, on its own merit.

Does this mean, for example, that the distribution of birds on oceanic islands in the South Pacific has little conceptual relationship to that of mammals on mountaintops in the Great Basin of North America? Probably so, other than that neither situation can be understood based only on what exists today.

Since the 1960s, the feverish search for general patterns in nature has, in my opinion, led many fine biologists to look for statistically defensible similarities at the expense of all manner of genuine and interesting variation. Depending on your perspective, the glass is either half full or half empty. Patterns of some sort usually emerge (at least in a statistical sense) if you look hard enough, especially if you transform your data to enhance their graphic or statistical presentation. I will attempt to be unbiased toward fullness or emptiness. I also will interpret my data with little or no transformation.

Two more points are worth making before going on. The first is that the Polynesian islands that form the backbone of this book are true oceanic islands surrounded by sea,which is an absolutely sterile environment for landbirds. These islands were never connected to continents by glacially induced sea-level lowering. They always have lacked indigenous nonvolant mammals. The second point is that while my focus is biological, I will draw regularly on geological and archaeological information to place the biogeography of South Pacific birds in a longer timeframe than is standard for biologists. I apologize to geological and archaeological colleagues if my treatment of their important disciplines seems at times to be cursory. My goal here is to draw from geology and archaeology in ways that enhance the biological information, knowing full well that the positive relationships among these branches of science are reciprocal.

Please let me outline, briefly, how this book is organized. Part I introduces the islands of Oceania (Chapter 1), their biotas (Chapter 2), their cultural histories (Chapter 3), and the field and laboratory methods used to generate my data (Chapter 4); these data are to be presented in Part II. A fairly thorough exposure to the physical and political geography of Oceania is required because the region is so complex and has had such a long, confusing history of nomenclature, whether for island groups or for individual islands. Even the use of the three primary cultural subdivisions of Oceania (Melanesia, Polynesia, Micronesia) is falling into disuse among many scholars. As is the case throughout this book, Part I features numerous maps so that the reader can visualize the juxtaposition of islands and island groups across the vast Pacific Ocean.

Part II (Chapters 5–8) describes the modern and prehistoric avifaunas of each island group in Oceania except the Hawaiian Islands and New Zealand, two major archipelagos with avifaunal histories largely independent of those elsewhere in Oceania. Chapters 5–7 slowly sail eastward from Melanesia through Polynesia. Chapters 6 (West Polynesia) and 7 (East Polynesia) are the empirical core of this book because most of my field and museum studies have been in these island groups. Birds from Melanesia (Chapter 5) and Micronesia and the Remote Central Pacific Islands (Chapter 8) are reviewed primarily to provide perspective for the Polynesian avifaunas in Chapters 6 and 7. The level of detail in Part II varies considerably with the quality and quantity of data; for certain island groups, such as Tonga (Chapter 6) or the Cook Islands and Pitcairn Islands (Chapter 7), both the modern and prehistoric birds have received much attention. As the fossil record grows, I hope that a future edition of this book will treat all of Oceania at least this comprehensively.

Each of Chapters 5–8 has two levels of organization, featuring both a review of the avifauna of each island group and comparisons of avifaunas among the island groups within the chapter. The tables in Part II list species of birds by their scientific names only, with family-level headings that use common names. The appendix compiles the scientific and common names for all families and species of birds mentioned in the text or tables.

Part III (Chapters 9–15) reviews the distribution of birds in Oceania from a taxonomic standpoint. These family-by-family accounts will serve as a sort of avian atlas where biogeographic patterns can be discerned and compared. Because of their broad distribution, taxonomic diversity, and relatively rich fossil record, four families will receive the most thorough treatment—the megapodes (Megapodiidae), rails (Rallidae), pigeons and doves (Columbidae), and parrots (Psittacidae). As in Part II, these chapters will focus on Polynesia (Fiji to Easter Island) but include biogeographic perspective from elsewhere in Oceania. The final chapter in Part III briefly reviews South Pacific seabirds, namely albatrosses, petrels, shearwaters, storm-petrels, tropicbirds, boobies, cormorants, frigatebirds, gulls, and terns. Like the landbirds, seabirds have a rich evolutionary history in Polynesia that has been damaged by people.

Part IV (Chapters 16–22) synthesizes the information from previous chapters to learn how data from Polynesian birds influence major biogeographic topics. Extinction (Chapter 16) permeates all aspects of the biogeography of Polynesian birds, although it is essential to distinguish natural or background extinction from that caused by humans. The extinction that I will analyze is dead-end, long-term extinction at the levels of population/subspecies and species, not the temporary absence of a species that many biogeographers regard as an extinction in studies of faunal turnover. Under natural conditions, true extinction does not occur nearly as often on islands as one might guess from equilibrium theory.

Among the most speculative issues in Part IV are those relating to colonization and faunal attenuation (Chapter 17). In spite of a rapidly growing fossil record, we still know very little about when most birds colonized any given set of islands or about the timing (and, in many cases, the extent) of evolutionary changes subsequent to colonization. The fossil record generally is not old enough to document these critical aspects of island biogeography. Faunal attenuation refers to the decrease in species richness, as well as in any other measure of taxonomic or ecological diversity, as one moves eastward across the South Pacific from New Guinea. Unlike previous studies of faunal attenuation, I will analyze the west-to-east changes in island avifaunas in full consideration of anthropogenic extinction, which attenuates faunas as effectively as (and more permanently than) geographic isolation.

Turnover (Chapter 18) is faunal change through time, typically expressed as colonizations versus extinctions. "Colonization," like "extinction" (although to a lesser extent) has broad, vague definitions in biogeography. I will examine turnover on longer timescales than most ecologists and will develop more stringent definitions of what constitutes a colonization or an extinction. Turnover is the process that is believed to result in an equilibrium state of species richness on islands. I will question whether the concept of equilibrium has any relevance on oceanic islands with differentiated taxa.

Species-area relationships (Chapter 19) are rightly regarded as one of the most fundamental components of the equilibrium theory of island biogeography. I will argue, however, that we do not yet have accurate values for species richness on most islands in Oceania even for landbirds, which are widely believed to be among the best known of all groups of organisms. Because of the pervasive anthropogenic extinction that has swept across the Pacific in recent millennia, our descriptive knowledge of the distributions of Polynesian landbirds has been so incomplete as to be deceptive. Even values for species richness that incorporate the species lost to human impact are underestimates of the true number of species that an island would sustain today under natural conditions.

In Chapter 20, I examine the structure of avian communities on individual islands, especially from a standpoint of feeding guilds and bird-plant interactions. Many niches not represented in modern Polynesian landbird communities were once occupied by species that would still be feeding and breeding today if not for people. Analyses of avian community structure that consider only extant populations would benefit from realizing that the fragmented modern communities developed under conditions very different from those we see today. We have underestimated the biotic potential of islands.

Chapter 21 attempts to bridge the gaps among past, present, and future bird communities by suggesting ways that data on prehistoric birds can be useful in conservation biology. Two related topics will be developed. The first is that prehistoric distributions can help to guide efforts to re-stock selected islands with some of the species that once lived there. The second is that autecological or synecological information useful to conservation biologists should be evaluated in light of the manifest changes that have occurred in the plant and animal communities on any island.

To conclude, Chapter 22 attempts to summarize and intertwine the major findings of all previous chapters. I also will suggest topics for future research, stressing major empirical gaps and some prominent areas of conceptual uncertainty.

That brings me to an overdue statement of humility. I do not pretend to have the last word on anything regarding island birds. They have fascinated me since youth, and I feel a professional as well as personal obligation to share with the reader what I've learned in the field and museum during the past two and a half decades. As in any field of research, the questions far outnumber the answers. By presenting and interpreting so much data on the distribution of birds in Oceania, I can guarantee some degree of improvement in our understanding of island biogeography. If I also can convey to you some of the thrill and curiosity that have fueled my travels to wonderful places that used to be even more wonderful, my mission will be accomplished.

Acknowledgments

First I would like to thank the many people with whom I have worked in the field in Oceania: Melinda Sue Allen, Kennery Alvea, Pia Anderson, Susan Antón, Tokoa Arokapiti, Kurt Auffenburg, Connie Bodner, Leslie Bolick, Henri Bos, Steve Brown, David Burley, Virginia Butler, Scarlet Chiu, Claudio Cristino, Ronald Crombie, Willy Dauron, Scott Derrickson, Bill and Jackie Dickinson, Donald Drake, Maile Drake, Joanna Ellison, Patricia Fall, Christopher Filardi, Jeannie Fitzpatrick, Janet Franklin, Holly Freifeld, Jean-Christophe Galipaud, Tuara George, John Groves, Mark Hafe, Heidi Hirsch, Patrick Kirch, Jeremy Kirchman, Roger Kolomule, Andrew Kratter, David Lee, Chief Pierre Leon, David Luders, William Manehage, Ann Marshall, Paul Martin, Sepeti Matararaba, Claudia Matavalea, Lofia Matavalea, Tuaiva Mautairi, Jim Mead, Peter Ngatokorua, Susan Ngatokorua, Tiria Ngatokorua, Diana Ngu, Ma'ara Ngu, Patrick Nunn, Patrick O'Day, Sharyn O'Day, Gregory Pregill, Rhonda Quinn, Ralph Regenvanu, Dillon Ripley, Gordon Rodda, Austel Ruumana, Jeff Sailer, Catherine Smith, Darren Smith, Tomo Solomona, the late Carol Spaw, Thomas Stafford, John Starmer, William Steadman, Anne Stokes, Ruth Stokes, Sam Stull, Sandy Swift, Atingakau Tangatakino, Julie Endicott Taomia, Sonny Taomia, Pierot Tavouiruja, Alan Tuara, the late George Tuara, Kilisimasi Tupou, Patricia Vargas, Vasa Vi, Price Webb, Karen Weinstein, Art Whistler, Gary Wiles, and Chris Wood.

For research permits and other sorts of cooperation, I thank government officials of the Commonwealth of the Northern Mariana Islands (Calistro Falig, Michael Fleming, Paul Palmer), Cook Islands (the late Stuart Kingan, William Tamarua, Tony Utanga), Easter Island (Claudio Cristino), Fiji (the late President Ratu Sir Kamisese Mara, Patrick Nunn, Nacanieli Bola Rowaico, Tarisi Sorovi-Vunidilo, Randy Thaman), Guam (Tino Aguon, Vic April, Gerald Deutscher, Gary Wiles), Kingdom of Tonga (HRH 'Ulukalala Lavaka 'Ata, Susana Faletau Fotu, Halaevalu Palu), Samoa (Easter Galuvao, Ray Sailimalo Pati Vi), Solomon Islands (Moses Biliki, Joseph Horako, Audrey Rusa), and Vanuatu (Ernest Bani, Donna Kalfatak, Ralph Regenvanu).

For help in the laboratory, I thank Joseph Bopp, Holly Freifeld, Jeannie Justice, Jeremy Kirchman, Andrew Kratter, Sharyn O'Day, Dominique Pahlavan, Gay Petri, Matthew Reetz, Jeff Sailer, Sue Schubel, Terry Taylor, Markus Tellkamp, Annamaria van Doorn, Matthew Williams, and Marie Zarriello. For help in putting this book together, especially drafting figures and formatting tables, I thank Jeremy Kirchman, Matthew Reetz, and Matthew Williams. I always will be in awe of your computer abilities.

The following people have facilitated my museum studies in one way or another primarily by providing access to collections under their care: Allison Andors, Phil Angle, George Barrowclough, Chris Blake, Carla Cicero, James Dean, Gary Graves, Shannon Hackett, Toni Han, Gene Hess, Janet Hinshaw, Ned Johnson, Carla Kishinami, Andrew Kratter, Mary LeCroy, Barbara Mushlitz, Storrs Olson, Robert Payne, Sievert Rohwer, Yosihiko Sinoto, Terry Taylor, Phil Unitt, Francois Vuilleumier, Tom Webber, David Willard, and Chris Wood.

I am very grateful to the following scientists (mainly archaeologists) who have kindly sent me bird bones to identify from their excavations: Jim Allen, Melinda Sue Allen, William Ayres, Stuart Bedford, David Burley, Geoffrey Clark, Eric Conte, John Craib, Tom Dye, Jean Christophe Galipaud, Roger Green, Terry Hunt, Rosalind Hunter-Anderson, Michiko Intoh, Patrick Kirch, Lisa Nagaoka, Maeva Navarro, Patrick and Sharyn O'Day, Pierre Ottino, Gustav Paulay, Barry Rolett, Richard Shutler, Yosi Sinoto, Matthew Spriggs, Christopher Stevenson, Jan Takayama, Julie Endicott Taomia, Marshall Weisler, David Welch, Peter White, Stephen Wickler, and Trevor Worthy.

So many people have discussed issues of island biology with me over the years that listing them all is not possible. Nevertheless, I would like to single out the following scholars for sharing many hours if not days, weeks, or months of informative and very enjoyable conversation and correspondence—Ronald Crombie, Jared Diamond, Janet Franklin, Helen James, Patrick Kirch, Andrew Kratter, Paul Martin, Brian McNab, Storrs Olson, Gregory Pregill, and Gary Wiles. I also thank the many students who have taken my island biogeography class for keeping me on my toes.

My research on Pacific island birds has been funded by the National Geographic Society (grants 2088, 4001-89,

5123-93), the National Science Foundation (grants BSR-8607535, BNS-9020750, EAR-9714819), the Smithsonian Institution, the U.S. Fish and Wildlife Service, the University of California (Berkeley), the University of Florida Division of Sponsored Research (grants RDA 1-23-95-96, ROF U001), and the University of Florida Foundation (Coffey Ornithology Endowment).

Finally, I give my love and gratitude to Anne, Will, and Sarah for tolerating and sometimes sharing my passion for islands and birds.

Part I

Chapter 1 Geography and Geology

STRETCHING ABOUT 15,500 km north to south and 19,500 km east to west, the Pacific Ocean is frighteningly large. No place on earth is so geographically perplexing as the 25,000 islands sprinkled across the Pacific that make up Oceania. Ranging from mere rocks to New Britain at 35,742 km^2, these islands lie in clusters over 13,000+ km from Palau to Easter Island (Figure 1-1). The clusters (= island groups or archipelagos) vary geometrically from chains to seemingly random scatters.

In the eastern Pacific, the islands close (<1500 km) to Central or South America sustain primarily or exclusively Neotropical biotas and thus are not part of Oceania. These include Juan Fernandez, Galápagos, Cocos, Clipperton, Revillagigedos, and many less isolated islands, none of which harbors a biota of Old World origin or has evidence of prehistoric habitation by Oceanic peoples. I do not consider the massive island of New Guinea to be part of Oceania in spite of its many biotic ties with the Bismarcks and Solomons. From my nesophilic perspective, New Guinea, which was joined to the Australian continent as recently as 12,000 years ago, is more of a continent.

Modern political boundaries in Oceania may not agree with those based on geology, biogeography, or ethnicity. Further confusion ensues because of European-induced name changes that are now obsolete, such as for Tonga (the Friendly Islands), Samoa (Navigator's Group), or the Cook Islands (the Hervey Group). Palau, often called "Pelew" by non-Palauans until the mid-1900s, is officially the Republic of Belau. For the island of Emirau in the Bismarcks, the names Emira, Squally, Storm, Sturminsel, Keruë, and Hunter can be found in 20th-century writing (Steadman & Kirch 1998).

Inconsistency occurs as well in definitions of archipelagos, such as Mangareva being regarded as the largest of the Gambier Islands, although it also has been classified as the only eroded volcanic island in the Tuamotu Group. Both statements are correct if you consider (as I do) the Gambier Islands as part of the atoll-dominated Tuamotu Group. Tuamotu, which is East Polynesian for "many (*tua*) islands (*motu*)," was called the Puamotu Group or Low Archipelago in the 19th and early 20th centuries.

Dividing Oceania into Polynesia ("many islands"), Micronesia ("small islands"), and Melanesia ("black islands") is useful but imperfect for culture (Clark 2003) or biogeography. Fiji, for example, can be pigeonholed culturally into neither Melanesia nor Polynesia (Frost 1979). The birdlife of Fiji also has multiple affinities, although tends to resemble that of Tonga or Samoa (West Polynesia) more than that of Melanesian islands to the west (Steadman 1993a). Cultural diversity in Melanesia far exceeds that in Polynesia or Micronesia (Spriggs 1997, Kirch 2000). Within both Micronesia and Melanesia are remote, usually small islands that were colonized prehistorically, and often remain inhabited today, by seafaring Polynesians (the "Polynesian outliers"; Kirch 1984b, 2000).

A useful distinction is that between Near Oceania and Remote Oceania. Near Oceania consists of islands near New Guinea that were colonized by people ca. 35,000 to 30,000 years ago, whereas Remote Oceania comprises all other Pacific island groups, in which people first arrived

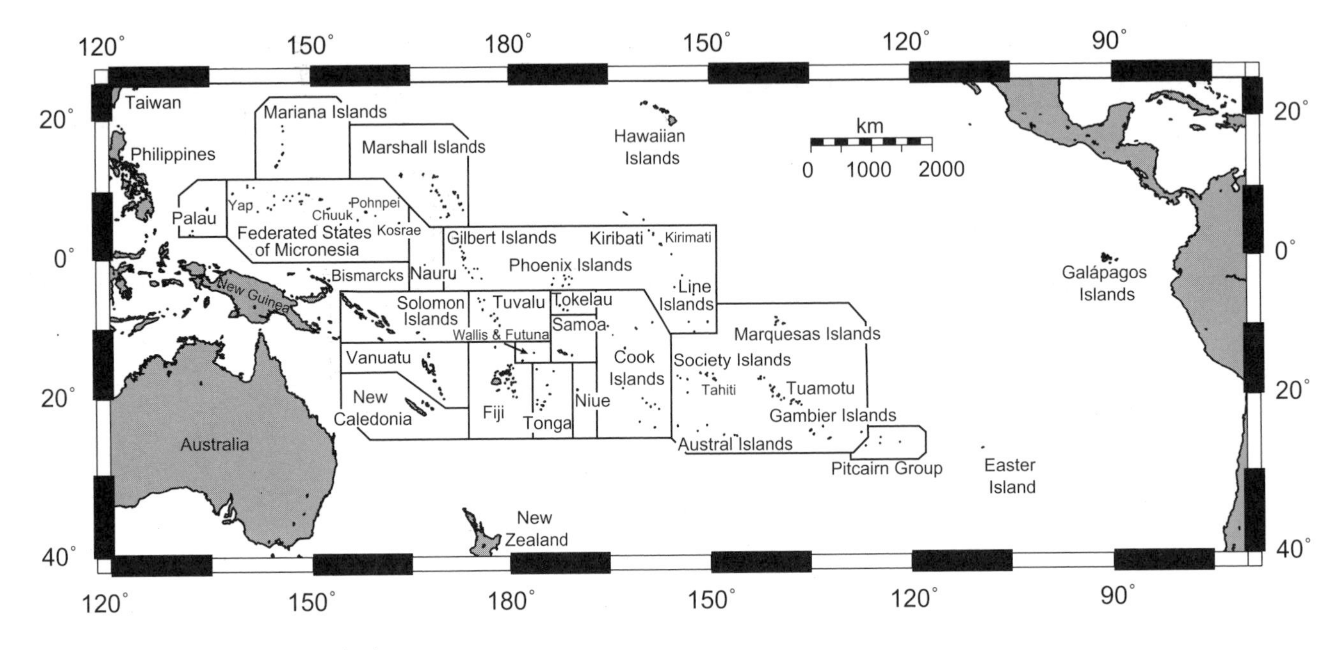

Figure 1-1 Oceania, showing major island groups.

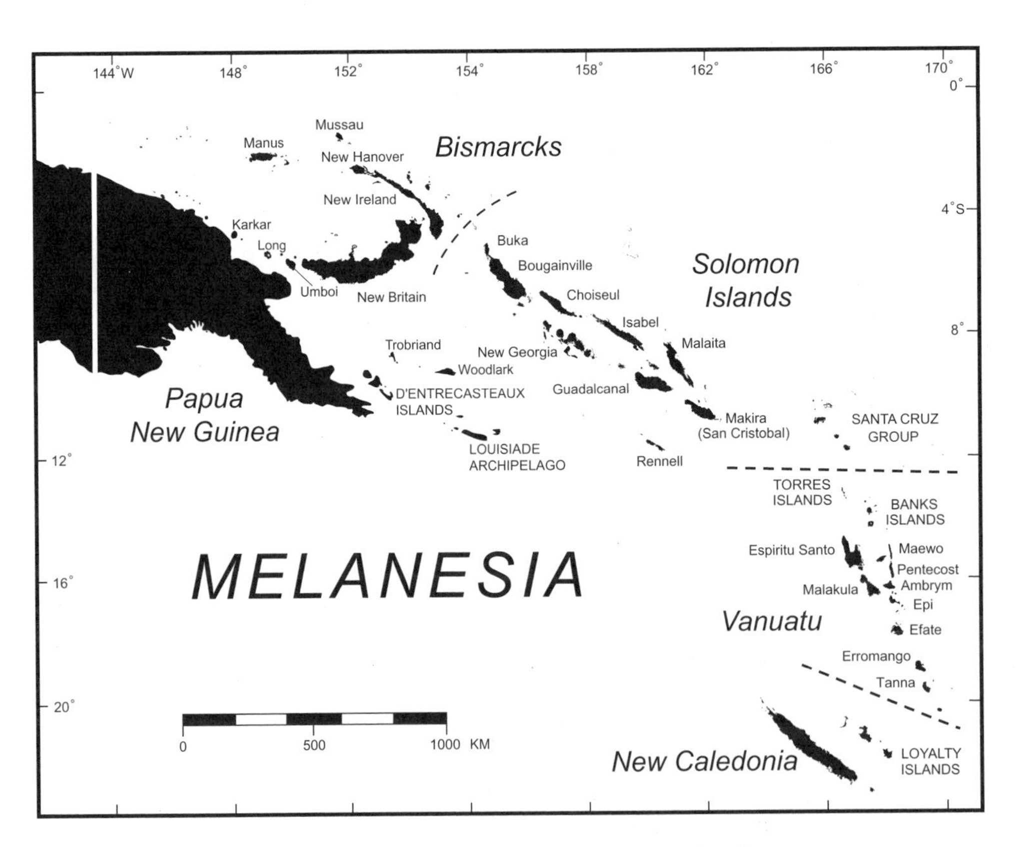

Figure 1-2 Melanesia. Redrawn from New Zealand Department of Lands and Survey (1986).

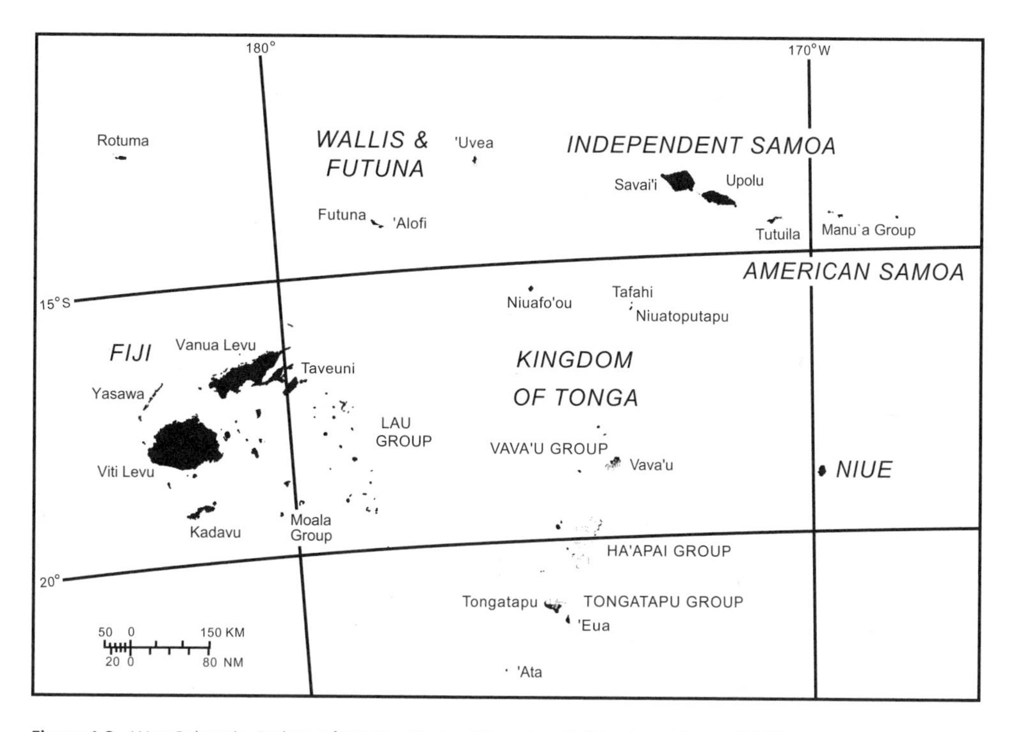

Figure 1-3 West Polynesia. Redrawn from New Zealand Department of Lands and Survey (1986).

(also from the west) less than 3500 years ago (Green 1991). As detailed in Chapters 2, 5, and 9–14, these two culturally derived terms also are useful in biogeography because so many terrestrial taxa are unknown on Pacific islands north or east of Near Oceania, which consists of the Bismarck Archipelago and the Solomon Island Main Chain (Buka through Makira; Figure 1-2). Outlying islands that are politically part of the Solomon Islands today (Rennell, Bellona, Ontong Java, Sikaiana, and the Santa Cruz Group) are part of Remote Oceania.

Islands and island groups on Papua New Guinea's continental shelf, such as the D'Entrecasteaux Islands and Louisiade Archipelago, are not part of Oceania. Neither are other islands near New Guinea, such as Karkar and Bagabag (volcanos that are not part of the Bismarcks) and Japen and the Aru Islands, which were connected to New Guinea ("land-bridge" islands) during times of glacially lowered sea levels. My coverage of birdlife in Oceania excludes the immense continental island of New Guinea (currently divided politically into the independent Papua New Guinea in the east and West Papua or Irian Jaya in the west, the latter controlled by Indonesia), except for pertinent comparisons, especially in Chapter 17. The Bismarcks through New Caledonia can be called Island Melanesia to distinguish them from New Guinea (Spriggs 1997).

This book has many maps because familiarity with the shapes, sizes, and location of islands and archipelagos is essential for critical thinking in biogeography. The geographic sequence by which I will cover most topics is west to east in the southern hemisphere from Melanesia to West Polynesia to East Polynesia (Figures 1-2 through 1-4), then back to the far western equatorial and northern hemisphere part of Oceania in Micronesia (Figure 1-5). Except for occasional faunal comparisons, I will not cover the Hawaiian Islands or New Zealand, two isolated Pacific island groups that are faunistically largely independent from the rest of Oceania.

I will provide the land area and elevation for every island where I tabulate data on the past and present distributions of birds. Data for land areas and elevations can be inconsistent among different sources, with variation exceeding ±5% in some cases. I have tried to choose what seem to be the most reliable sources. In some cases I have measured island area directly from maps.

Important geographic references for Pacific islands include Kennedy (1966), Douglas (1969), Bunge & Cooke (1984), Carter (1984), Menard (1986), Motteler (1986), New Zealand Government (1986), Nunn (1994), Ridgell (1995), Lobban & Schefter (1997), Rapaport (1999), and Lal & Fortune (2000). I will draw regularly from these sources, citing more detailed studies as needed, especially for geology. My geographic and geological narratives are influenced as well by my own observations on 100+ Pacific islands.

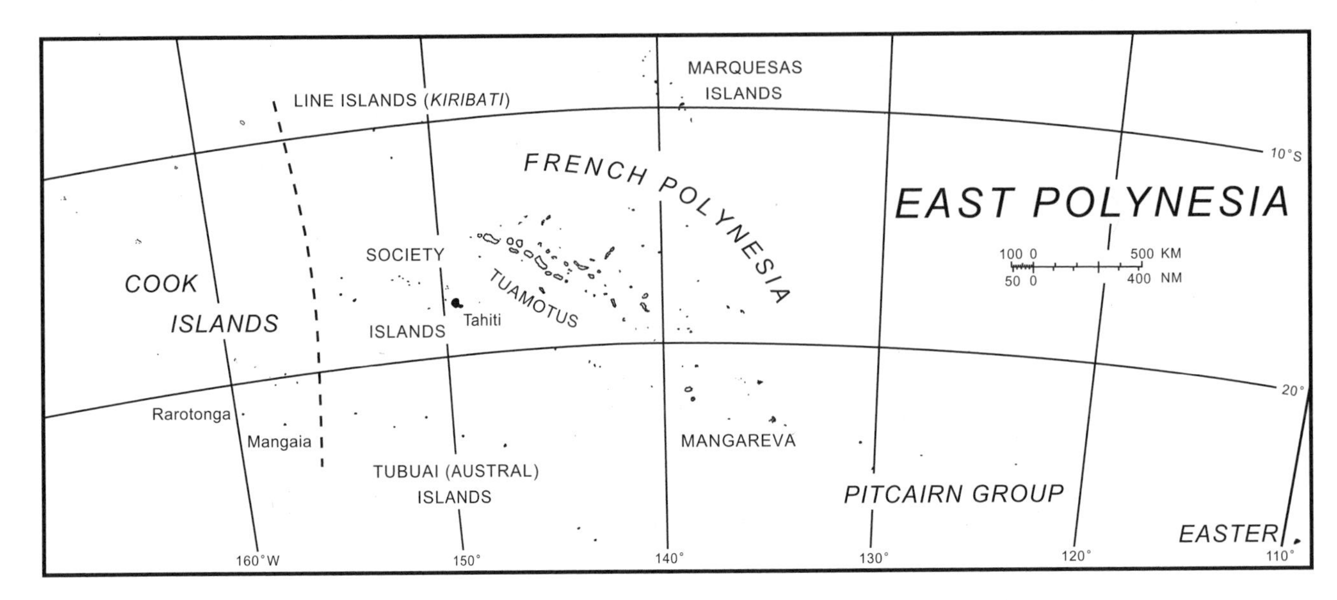

Figure 1-4 East Polynesia. The Line Islands are not part of East Polynesia. Redrawn from New Zealand Department of Lands and Survey (1986).

Regional Geology

To facilitate comprehension by nongeologists, I will minimize the use of geological jargon, with apologies to geologists. A partial geological time scale (Figure 1-6) also should be helpful. I must begin with geochronological abbreviations. A million years is "My," whereas million years ago or a million years old is "Ma." Similarly, a thousand years is "ky" and a thousand years ago or a thousand years old is "ka." Thus an island that is 10,000,000 years old (10 Ma) may have had an eruptive phase that lasted 2,000,000 years (2 My) from 10,000,000 to 8,000,000 years ago (10–8 Ma). A low sea-level stand from 22,000 to 19,000 years ago (22–19 ka) lasted 3000 years (3 ky).

The geochemical (isotopic) dating methods most pertinent to Pacific geology are potassium-argon and argon-argon (K-Ar, Ar-Ar) for volcanic rocks, uranium series (U-series, sometimes involving uranium-thorium or U-Th) for carbonate rocks, and radiocarbon (^{14}C) for carbonate rocks and organic materials. The potential effective age range is ca. 2–5 ka to 4000+ Ma (several thousand years to 4+ billion years) for K-Ar and Ar-Ar dating, ca. 5 to

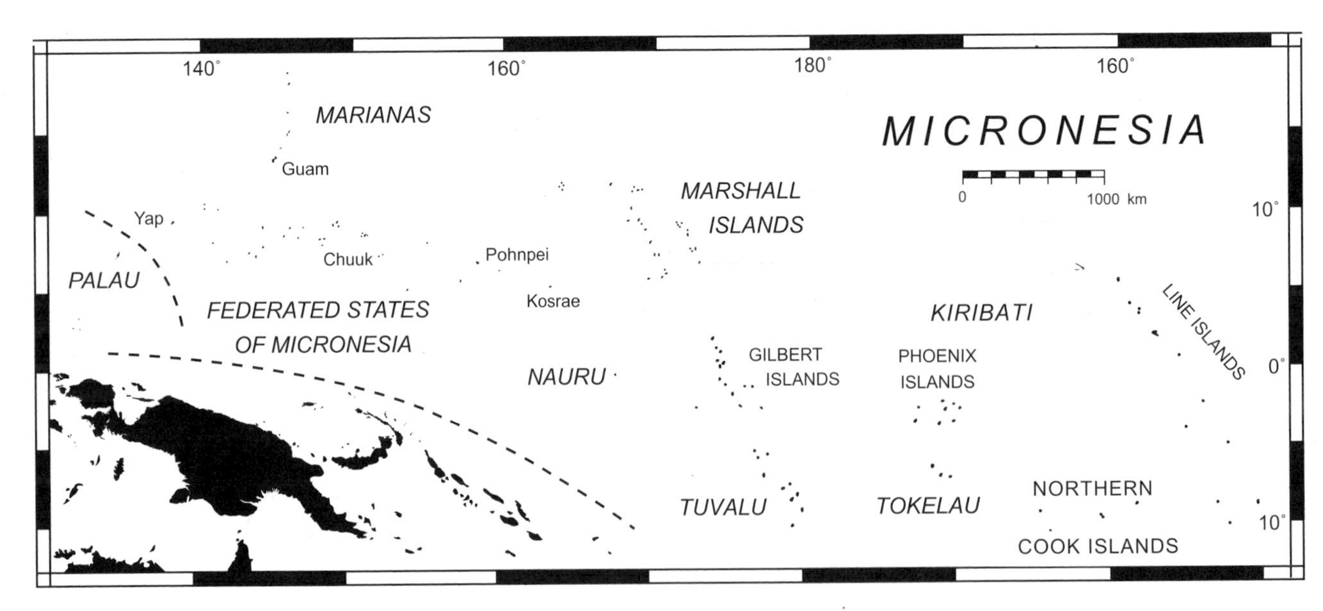

Figure 1-5 Micronesia and Remote Central Pacific islands. Tuvalu and Tokelau are culturally West Polynesian. The Northern Cook Islands are primarily East Polynesian. Redrawn from New Zealand Department of Lands and Survey (1986).

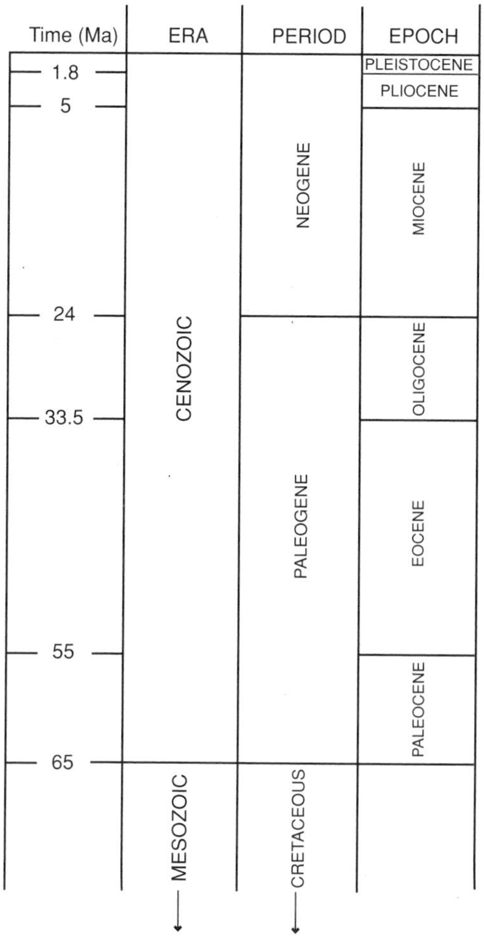

Time (Ma)	ERA	PERIOD	EPOCH
1.8	CENOZOIC	NEOGENE	PLEISTOCENE
5			PLIOCENE
24			MIOCENE
33.5		PALEOGENE	OLIGOCENE
55			EOCENE
65			PALEOCENE
	MESOZOIC	CRETACEOUS	

Figure 1-6 Partial geological time scale, including only the time intervals most pertinent to terrestrial biogeography in Oceania. The Holocene Epoch (past 12,000 years) is too short to depict at this scale. Adapted from Berggren al. (1995).

350 ka for U-series dating, and ca. 0.2 to 45 ka for ^{14}C dating. Terminology specific to ^{14}C dates will be explained in Chapter 4.

Most islands in Oceania lie either on the Pacific Plate or Indo-Australia Plate (Drummond et al. 2000a, b; Figure 1-7 herein). Exceptions are Easter Island (Rapanui), which is on the Nazca Plate just east of the small Easter Plate or Easter Microplate (Handschumacher et al. 1981, Woollard & Kulm 1981), and the westernmost Micronesian islands (Palau, Marianas, and part of Yap) on the Philippine Sea Plate (Dickinson & Shutler 2000). Islands on the Pacific and Nazca Plates typically are made of dark, heavy volcanic rocks characteristic of thin oceanic crust, such as basalts (Calmant & Cazenave 1987, Dostal et al. 1998, Helffrich & Wood 2001). Islands on the Indo-Australia Plate (= Australia-India Plate, India Plate, Australia Plate, etc.) and the Philippine Sea Plate are more geologically varied; in spite of localized plutonic rocks (those formed in the earth's mantle) or oceanic-type volcanic rocks, they generally consist of rocks that are lighter in weight and color, with "andesitic" mineral suites more characteristic of continental crust (Polhemus 1996, Dickinson 2001). Reflecting this fundamental geological difference, the boundary between the Pacific Plate and the Indo-Australian and Philippine Sea Plates is called "the Andesite Line" (see Tarling 1965).

Islands on the Pacific Plate average younger than those on the Indo-Australia Plate. Nearly all currently emergent islands on the Pacific Plate are <5 Ma, with most dating to <3 Ma. This does not mean that the Pacific Plate was island-free >5 Ma; most older islands are now submerged. Land was available for terrestrial life in Remote Oceania long before the subaerial age of any individual island.

Except for New Caledonia and New Zealand, evidence is lacking that Pacific islands, even those in Near Oceania, ever were connected to Australia or any other continent (Dickinson & Shutler 2000, Dickinson 2001). New Caledonia and New Zealand are Gondwanan fragments originally associated with Australia and Antarctica, although their separation from continental landmasses occurred during dinosaur times in the Cretaceous (Kroenke 1996). Therefore, I believe that most of the vertebrate fauna (species still living and those that became extinct since human arrival) of New Caledonia and New Zealand can be explained by Cenozoic over-water dispersal rather than by late Mesozoic or early Cenozoic vicariance (rifting landmasses). Because the

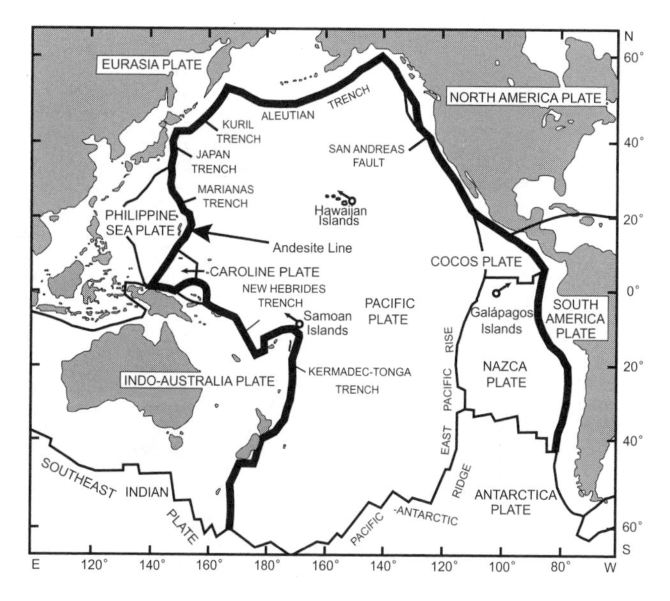

Figure 1-7 Tectonic plates of the Pacific region. ♂= direction of plate motion. Redrawn from Mueller-Dombois & Fosberg (1998: Figure 1.3).

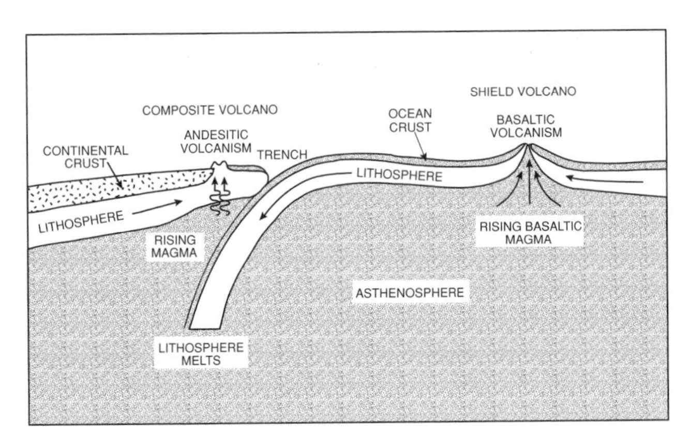

Figure 1-8 Schematic cross-section of a convergent plate boundary. Redrawn from Muller-Dombois & Fosberg (1998: Figure 1.4).

islands in the rest of Oceania always have been islands, overwater dispersal must have been involved at some point in their terrestrial biogeographic history.

Ocean depths average greater, and subsurface contours around islands average steeper, on the Pacific Plate than the Indo-Australia Plate (Summerhayes 1967). Deep trenches may occur at convergent plate boundaries where the Pacific Plate's thinner, heavier crust subducts beneath the lighter crust of the Indo-Australia or Philippine Sea Plates. Dramatic examples are the 10,882 m deep Tonga Trench (Scholl et al. 1985, Pelletier & Louat 1989) and the 11,022 m deep Marianas Trench (Fryer 1995). Island formation behind these trenches features uplifted limestone islands on the forearc belt close to the trench, and active volcanoes slightly farther from the trench (volcanic arc) that largely represent melting of subducted oceanic crust (Dickinson et al. 1999; Figure 1-8 herein). Divergent plate boundaries, also known as spreading centers, usually are not areas of island formation (Mayes et al. 1990).

Island Types

Pacific islands can be classified into seven major types, each with substantial variation. The first type is the active volcano, such as Tofua or Kao in Tonga (Figures 1-9, 1-10). Active volcanoes can be found along island arcs (often trending north to south) associated with subduction zones at convergent plate boundaries (Taylor 1992, Clift et al. 1995) or along linear chains (often trending east-southeast to west-northwest) associated with midplate hotspots (Okal & Batiza 1987, Keating 1992, Dickinson 1998b, Hieronymus & Bercovici 1999). Active volcanic islands vary from large ones such as Savai'i in Samoa (1821 km^2, elev. 1858 m) to tiny, ephemeral ones such as Metis Shoal (Late Iki) in Tonga (Sutherland 2000) or Kavachi in the Solomon Islands (anonymous 2000, Holden 2000). These last two islands

Figure 1-9 The active volcanic islands of Tofua (left) and Kao (right) in the Ha'apai Group, Tonga, with the small, uninhabited raised limestone island of Niniva in the foreground. Photo taken from a boat by DWS, 15 July 1996.

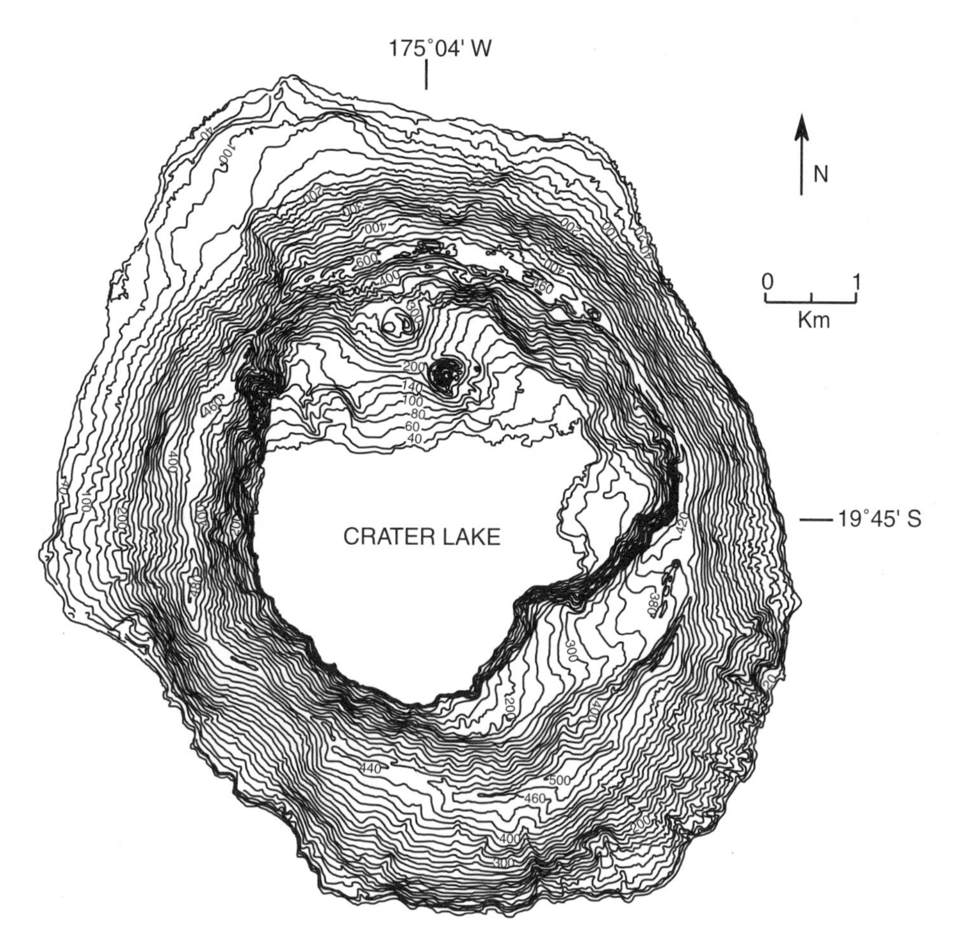

Figure 1-10 Topography of Tofua, an active volcanic island in the Ha'apai Group of Tonga. Contour intervals 20 m. Redrawn from Kingdom of Tonga (1975b).

still come and go above the ocean's surface as their lava domes expand and contract. When Lister (1891) visited Metis Shoal in 1889/90, it was 46 m in elevation and had been above water for about a decade. When I flew low over Metis Shoal (July 1995) it was ca. 120 × 100 × 20 m high and had been above water for only weeks. Since then it has submerged and risen again. Most active volcanic islands date to <2 Ma, often <1 Ma. Their shorelines often are cliffy and bouldery with little or no beach development. Subsurface contours also are steep, with deep nearshore waters that generally lack coral reefs.

The next type is the eroded volcanic island. Usually 1–8 Ma and no longer volcanically active, the bedrock is basaltic to andesitic. Eroded volcanic islands can have steep, knife-edge topography (especially in young stages as in Tahiti, Mo'orea, and many Marquesan islands) or gentler slopes at later stages such as Babeldaob in Palau. A coastal plain of variable width and continuity usually occurs, with some beach development (Gillie 1997a). The beach sands may be mineral (bedrock-derived), calcareous (biogenic), or combinations thereof. Eroded volcanic islands can be large such as Tahiti in the Society Islands (1042 km^2, elev. 2237 m) or very small such as Morotiri in the Tubuai Islands (0.3 km^2, elev. 105 m). The three islands in the Manu'a Group of American Samoa are small, very steep, and high (Figure 1-11). The nearby, larger eroded volcanic island of Tutuila (145 km^2, elev. 652 m) also is steep but not so precipitous as to inhibit forest growth over most of its surface (Figure 1-12). Fringing reefs often develop around eroded volcanic islands. Gaps in the reef may be found near stream mouths when terrigenous sediment inhibits coral growth (Richmond 1997b).

Raised limestone islands, sometimes just called limestone islands, are based in Oceania on emergent coralline limestones, i.e., uplifted coral reefs dominated by calcium carbonate ($CaCO_3$). The uplifted reefs are tens to hundreds of meters thick and cap a core of volcanic rock (Menard 1986, Spencer et al. 1987). These islands range in size from tiny rocks less than 10 m across to such major islands as Rennell (676 km^2, elev. 154 m), in the Solomons, or Niue

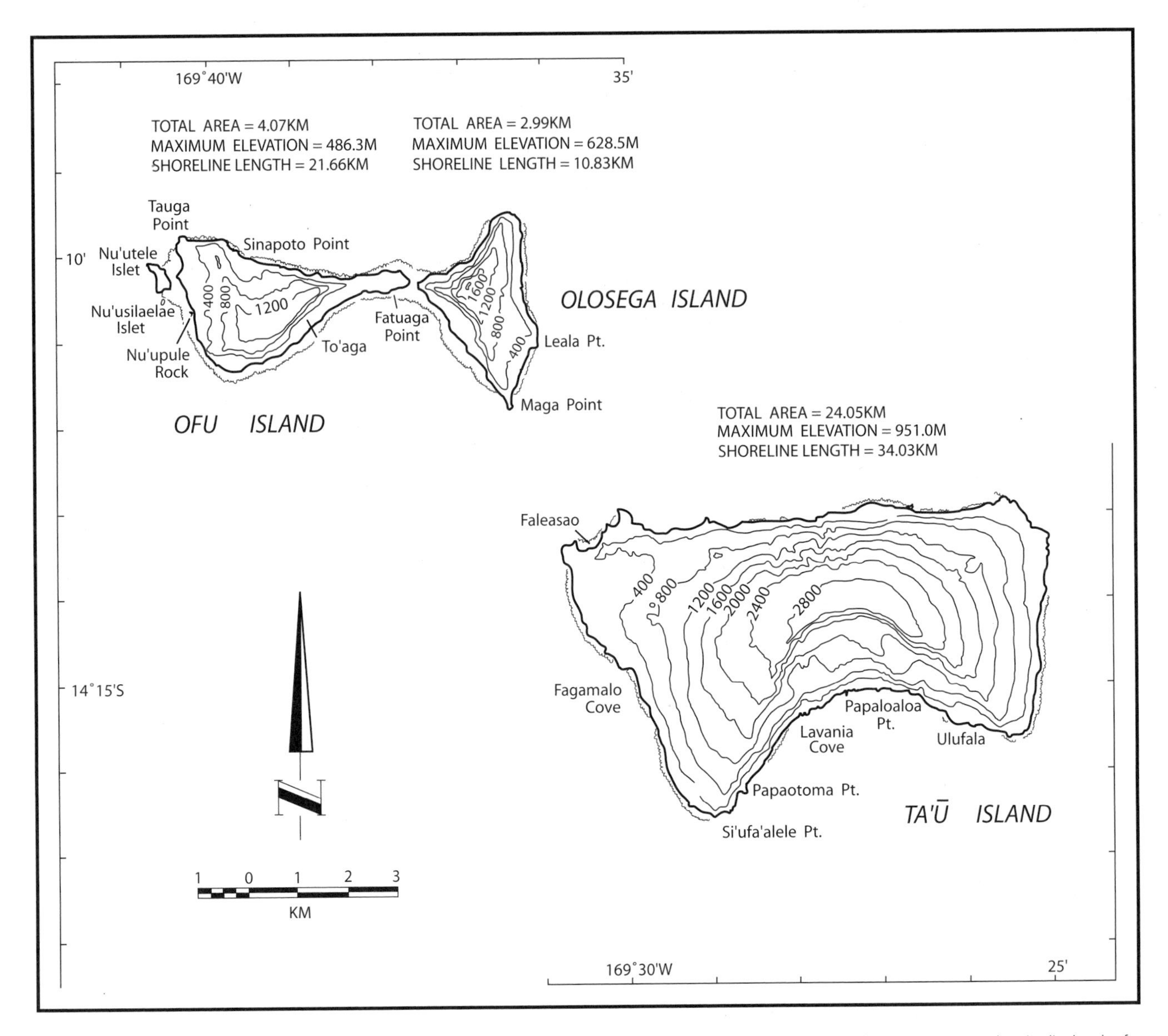

Figure 1-11 Topography of three eroded volcanic islands in the Manu'a Group of American Samoa. Contour intervals 400 feet. Note separate longitudinal scales for Ta'u vs. Ofu and Olosega. Redrawn from Amerson et al. (1982a).

(259 km^2, elev. 68 m), lying between Tonga and the Cook Islands. Although some raised limestone islands are found in most island groups, major concentrations of these islands include the Lau Group in Fiji, Tonga (Figures 1-13 to 1-15), the southern Cook Islands, Tubuai, Palau, and the southern half of the Mariana Islands. Raised limestone islands can have a wedding-cake profile (sets of terraces and cliffs) that represents multiple phases of reef formation and uplift. While the volcanic foundations of raised limestone islands are highly variable in age and can date from only several to occasionally as much as 20 to 30 Ma, these islands became totally submerged (typically through subsidence after hotspot or plate-boundary volcanism; Dickinson 1998b, Dickinson et al. 1999) before being uplifted. Emergence of many raised limestone islands has occurred only during the past 1-2 My or less.

The shorelines of raised limestone islands may be cliffy, gentle with beaches (the sand usually calcareous), or combinations thereof (Dickinson 2004b; Figures 1-13 to 1-15 herein). Limestone islands usually have fringing reefs or occur in clusters more or less surrounded by a barrier reef, as in Palau or the Vava'u Group of Tonga. Because limestone is so prone to develop caves and rockshelters (areas under projecting rock ledges; Farrand 1985) with high-pH, low-energy sediments, this type of island is an excellent source for prehistoric bird bones (Chapters 3, 4). Some raised

Figure 1-12 The eroded volcanic island of Tutuila, American Samoa. Photo by DWS, 29 April 1999.

Figure 1-13 The raised limestone island of Nuapapu, Tonga. Photo by DWS, 28 July 1995.

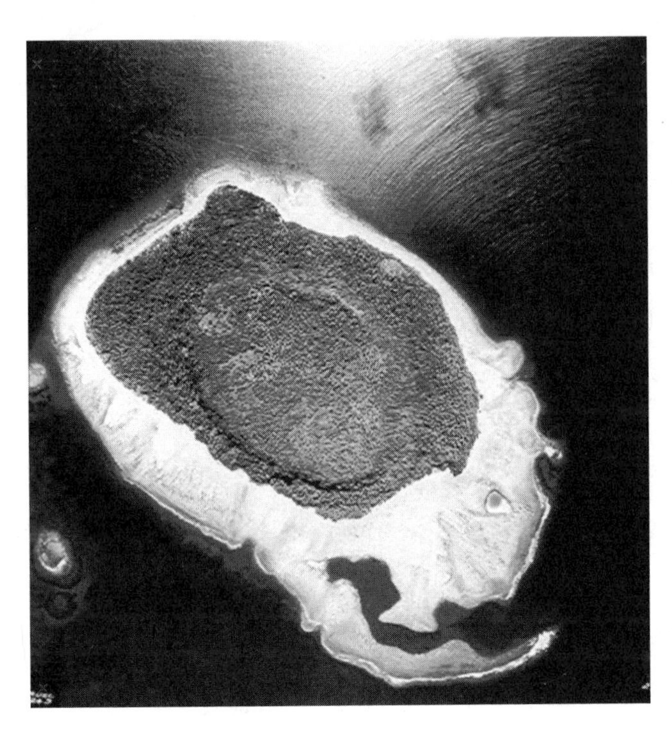

Figure 1-14 The raised limestone island of 'Euakafa, Vava'u Group, Tonga. Aerial photo by the Australian Air Force, 1990.

limestone islands (and atolls; see below) also are covered with phosphate deposits [$Ca_3(PO_4)_2$], often in the form of clayey phosphorite (typically $Ca_{10}(PO_4CO_3)_6F_{2-3}$; Allaby & Allaby 1999:410). Phosphorite deposits are believed to be derived from ancient seabird guano (Rodgers 1948, Stoddart & Scoffin 1983). These deposits can persist even on submerged atolls, known as guyots (Cullen & Burnett 1987).

Atolls typically are a set of low, sandy islets encircling a central lagoon and surrounded by a fringing reef (Bryan 1953, Stone 1953, Wiens 1962, Newhouse 1980, Devaney et al. 1987). No volcanic rock is exposed on atolls (Figures 1-16, 1-17), where calcareous sand overlies the subsurface limestone (which can be as much as 1300 m thick; Leopold 1969) that caps the volcanic rock. On some atolls, reefal limestone or indurated calcareous rubble are exposed intertidally or just above the tideline (Dickinson 1999a). Richmond (1997a) classified the 16 islands in the Gilbert Group of Kiribati into four types: (1) "reef island," for those with exposed, low-lying clastic (sands and gravels) carbonate buildup and little if any lagoon; (2) "atoll, deep lagoon," for those with a lagoon >10 m deep and the land separated by channels into islets; (3) "atoll, shallow lagoon," differing from the last only in depth (<10 m) of lagoon; and (4) "atoll, enclosed lagoon" for those with an essentially continuous landmass or reef surrounding the lagoon. His "reef island" more or less is what I call a "sand cay" below. A shallow, doubly convex layer of fresh water, known as the Ghyben-Herzberg lens, often develops on an atoll's larger islets (Arnow 1954, Wiens 1962, Niering 1963), greatly enhancing the potential for agriculture and therefore human habitation. As current global warming (Wigley & Raper 2001) melts ice at high latitudes and altitudes (Kindler & Hearty 2000, Bradley 2001) and therefore increases sea level (Cabanes et al. 2001, Church 2001, Arendt et al. 2002), many Pacific atolls are threatened with salinization of water supplies and terrestrial habitats, if not downright inundation, in upcoming decades or centuries.

The maximum land area and elevation of a Pacific atoll are 321 km^2 and 11 m for Kirimati (Christmas Island) in Kiribati, which differs from most atolls in having much of the central lagoon filled with calcareous sand. Atolls generally have land areas <15 km^{2}and elevations <5 m, often even <5 km^2 and <3 m. More land is usually exposed on the windward than leeward side, reflecting the greater coral growth (i.e., more marine productivity) and increased sedimentation of a relatively high-energy environment (Maragos et al. 1973, Nunn 1994:245). Another important process for islet growth is lateral spit progradation, principally at the ends of islet chains or at the lagoonal margins of

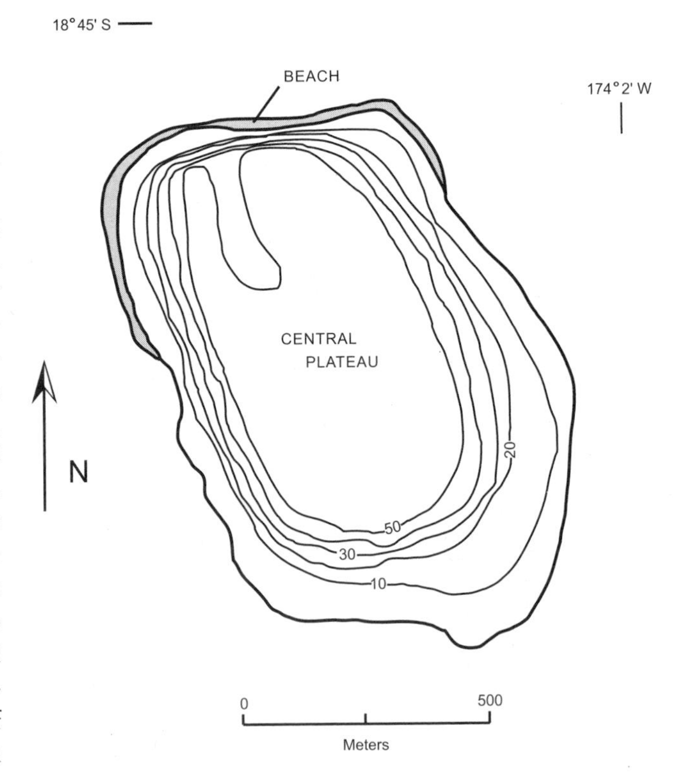

Figure 1-15 Topography of 'Eukafa, Vava'u Group, Tonga. Contour intervals 10 m. Redrawn from Kingdom of Tonga (1975a).

Figure 1-16 The atoll of Palmerston, Cook Islands. Photo by DWS, 4 March 1985.

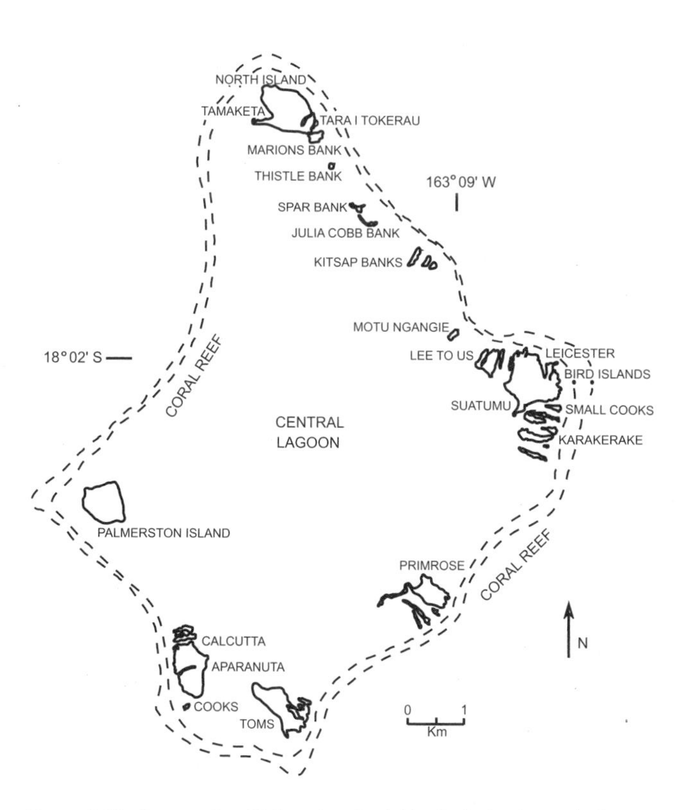

Figure 1-17 Topography of Palmerston, Cook Islands. No land exceeds 5 m elevation. Redrawn from New Zealand Department of Lands and Survey (1986).

inter-islet channels, fueled by abundant calcareous sand from biologically productive lagoons (Paulay 1997, Richmond 1997a). Even large atolls are more susceptible than higher islands to physical perturbations, such as from typhoons or tsunamis. Although more common along the margins of the Pacific Ocean than in most of Oceania, tsunamis have been recorded on central Pacific atolls (Soloviev et al. 1992). Typhoons will be discussed later in this chapter.

During glacially lowered sea levels, atolls become raised limestone islands. Assuming no subsidence from lithospheric flexure (Dickinson 1998b, 2004a), most atolls were drowned during the sea-level highstand at ca. 120 ka (see below). Cores drilled beneath the lagoons of Pukapuka and Rakahanga (Cook Islands) suggest that these atolls have been submerged on five occasions in the past 600 ky, making up 8 to 15% of this time (Gray et al. 1992). Many atolls may have drowned through much of the Holocene, when a rapidly rising sea level outpaced vertical growth of the encircling coral (Marshall & Jacobsen 1985, Gray et al. 1992, Dickinson 2003). Reef growth yields vertical carbonate sediment accumulation from 0.3 to 2.8 m/ky (Richmond 1997b). Some atolls probably remained drowned until the end of the mid-Holocene highstand at only 3 ka (Pirazzoli & Montaggioni 1988). As one would expect, atolls drowned in the Holocene sustain a depauperate, "trampy" terrestrial biota with little or no endemism.

Figure 1-18 The almost-atoll of Aitutaki, Cook Islands, looking seaward from the summit (124 m elevation). Photo by DWS, 26 November 1987.

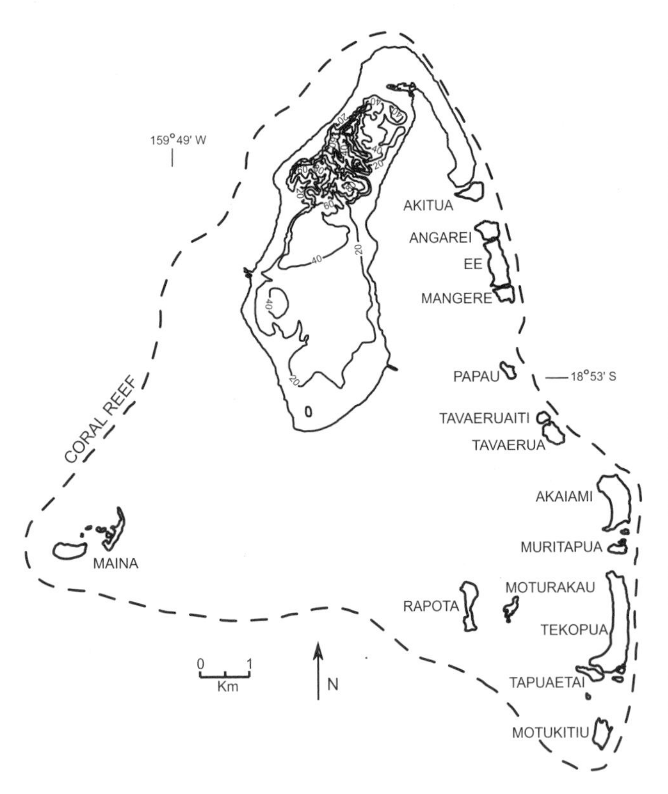

Figure 1-19 Topography of Aitutaki, Cook Islands. Contour intervals 20 m. Redrawn from Aitutaki (1983).

An "almost-atoll" is an island believed to represent the developmental stage between eroded volcanic islands and atolls. Good examples are Aitutaki (Cook Islands; Stoddart & Gibbs 1975, Allen 1997; Figure 1-18 herein), Mangareva (Tuamotu Group; Munschy et al. 1996), and Chuuk (= Truk; Caroline Islands; Keating et al. 1987). Almost-atolls have a lagoon and fringing reef but differ from true atolls in that volcanic bedrock is exposed, up to 124 m elevation on Aitutaki's main island and 10 m elevation on its two islets with exposed volcanic rock, Moturakau and Rapota (Figure 1-19). Mangareva is presumably at an earlier stage of atoll development than Aitutaki because its main island (14 km^2) is higher (482 m elev.) and 10 of its islets, which unlike Aitutaki are not concentrated on the perimeter as in an atoll, have exposed volcanic rock (Steadman & Justice 1998). The eroded volcanic "islets" of Chuuk are as large as 33 km^2 and as high as 446 m; like Mangareva, they lie within the lagoon rather than on the perimeter.

Sand cays (keys) are small individual islands that lack exposed bedrock but are not part of an atoll, i.e., they lack a central lagoon. Typically <1 km^2 in area and <5 m in elevation, sand cays may lack a freshwater lens. The difference between a sand cay and an atoll can be subtle (Richmond 1997a). Similarly, the difference between a sand cay and a low raised limestone island is gradational. For example, the

Figure 1-20 The sand cay of 'Uoleva, Ha'apai Group, Tonga. Photo by DWS, 25 July 1996.

rather large Tongan sand cay of 'Uoleva (Figures 1-20, 1-21; area 2.0 km^2, elev. 8 m) is built on the same limestone bank as the nearby raised limestone islands of Ha'ano, Foa, Lifuka, and 'Uiha (area 5.3-13.3 km^2, elev. 11–20 m). 'Uoleva lacks exposed bedrock today, but this distinction would disappear if sea level dropped by only several meters.

The composite island is an artificial category for any island with multiple types of exposed bedrock. Composite islands with outcrops of limestone and volcanic rock include Mangaia (Cook Islands; Stoddart et al. 1985; Figures 1-22, 1-23) and Rurutu (Tubuai Islands; Stoddart & Spencer 1987, Chauvel et al. 1997) on the Pacific Plate, Lakeba (Fiji; Nunn 1994) on the Indo-Australia Plate, and Guam (Mariana Islands; Tracey et al. 1964, Reagan & Meijer 1984) on the Philippine Sea Plate. On Mangaia, ^{14}C and U-series dating of corals on the modern reef flat and uplifted limestone terraces document interglacial sea-level highstands of +1.5 m at ca. 5 to 2.5 ka, +12 to 15 m at ca. 120 ka, and +27 m at ca. 400 ka (Stoddart et al. 1985, Yonekura et al. 1988, Spencer et al. 1987, Woodroffe et al. 1990, 1991, Dickinson 1998b). Mangaian limestones date from the early to middle Miocene through Holocene (Ward et al. 1971, Stoddart et al. 1985) and reach 73 m elevation. Even though Mangaia's eroded volcanic center also dates to the early Miocene and attains a modern elevation of 169 m, the island's current subaerial period may be only 1–2 Ma.

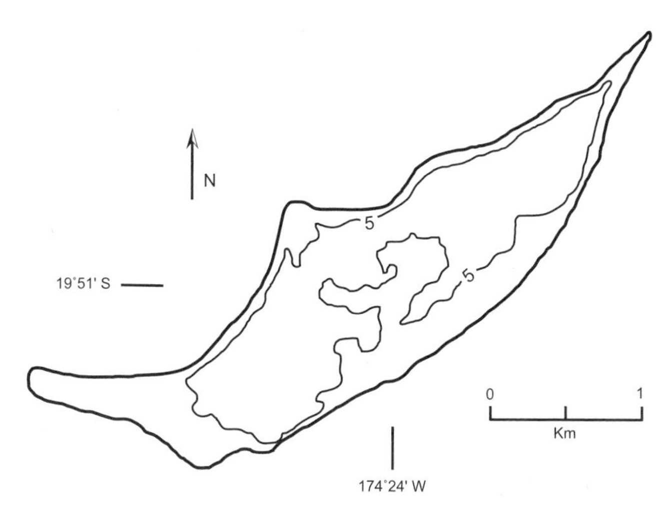

Figure 1-21 Topography of 'Uoleva. Maximum elevation is 8 m. Redrawn from Kingdom of Tonga (1976).

A very different, more geologically complex type of composite island would be those on or near the Indo-Australia Plate margin, such as Viti Levu (Fiji; Stratford & Rodda 2000), Santo or Malakula (Vanuatu; Taylor et al. 1985, Falvey et al. 1991), or large islands in the Solomon Main Chain (Coulson & Vedder 1986, Roy 1997b; Figures 1-24, 1-25 herein). These composite islands are typically older than the eroded volcanic-raised limestone islands just discussed. Their exposed bedrock can consist of oceanic

Figure 1-22 The interior limestone cliff at Lake Tiriara, Mangaia, Cook Islands. Photo by DWS, 20 June 2001.

(± basaltic) and continental (± andesitic) volcanic rocks, igneous rocks, marine and terrestrial sedimentary rocks (limestone, carbonate mudstones, sandstones, siltstones, etc.), and metamorphic rocks. Maximum elevations usually range from 500 to 2000 m. Because of their large size and geological complexity, these islands can impart the feeling of a continental rather than insular landscape. A special type of composite island is Grande Terre (New Caledonia), much of which is a rifted Gondwanan fragment (Kroenke 1996; also see below).

Brief Geographic and Geologic Review of Pacific Island Groups

This section presents a few major geographic and geological features for each island group, maps of which are in Chapters 5–8, where the distributions of birds are analyzed. For those wanting more information, I provide some key references for each island group. Most data for the human populations and sea areas ("200 mile exclusive economic zones"; see Center for Pacific Island Studies 1987) are from Lal & Fortune (2000). Mention of cultural affinities involves only generalities of who was present at European contact.

Like biology, geology is inexact. Models of plate motions, past positions of islands, and their times of emergence and submergence may be informative, plausible, visually pleasing, and based on careful research, but are not as precise chronologically or geographically as depicted in publications. Isotopic dating methods, while exceedingly useful and geochemically defensible, have margins of error and other limitations. Biologists who want to learn how long a certain island has been above the sea, and where it was when it first emerged, may not get a precise answer from geologists. This is nobody's fault. We all are doing the best we can in these challenging fields of inquiry. I must note here that the outstanding papers by William Dickinson in recent years (many of them cited above and below) have made the regional and local geology of Pacific islands much more accessible to biologists and archaeologists than ever before.

Melanesia (Figure 1-2)

The Bismarck Archipelago (land area 48,724 km^2, sea area ca. 800,000 km^2; Figure 5-1) comprises 41 islands or island clusters ranging from tiny islets to New Britain, at 35,742 km^2 and 2439 m elevation the largest and one of the highest islands in tropical Oceania (Mayr & Diamond

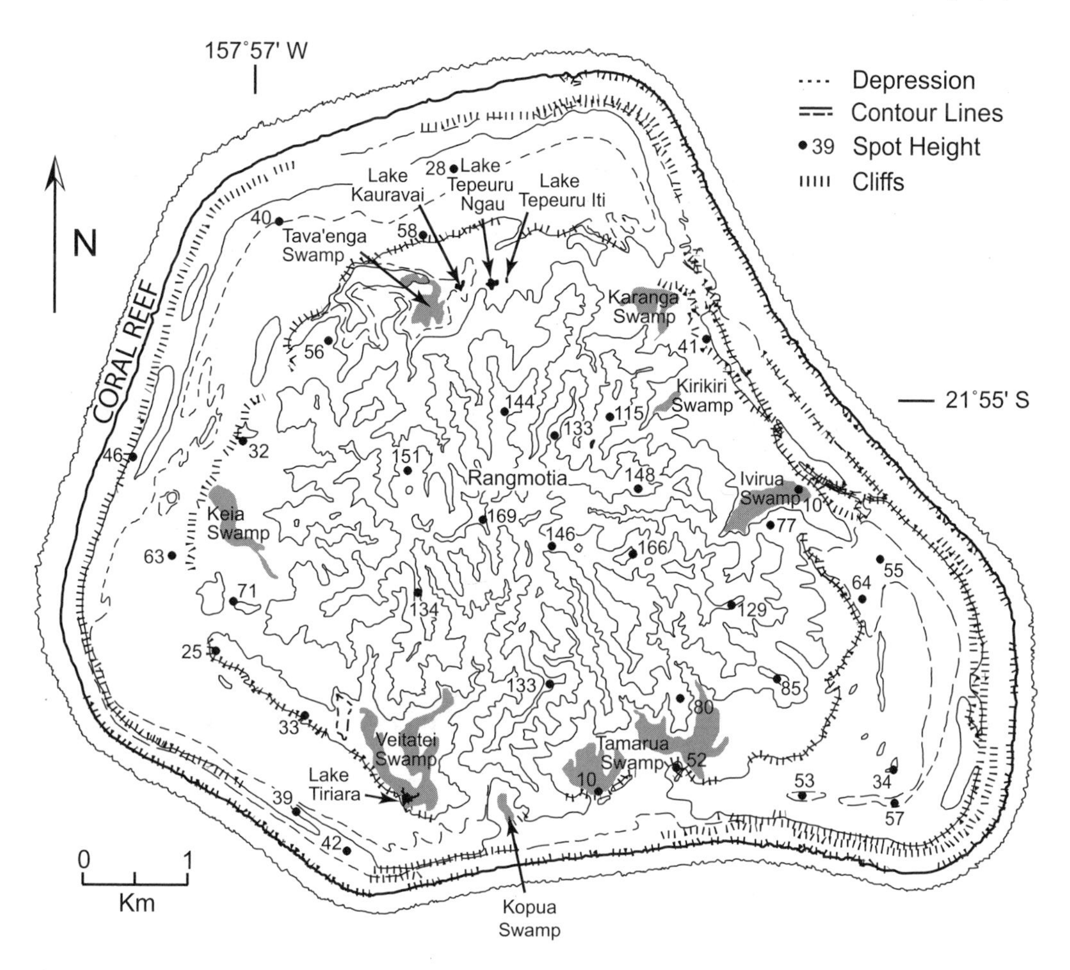

Figure 1-23 Topography of Mangaia, Cook Islands. Contour intervals 20 m. Redrawn from Mangaia (1986).

2001). The northwestern part of the Bismarcks (Manus and nearby smaller islands) is called the Admiralty Group. Politically part of Papua New Guinea today, the Bismarcks are inhabited by diverse Melanesian peoples (population ca. 435,000).

Lower sea levels of glacial intervals did not connect most major islands in the Bismarcks to each other. Although it is the closest very large island in Oceania to New Guinea, neither New Britain nor the rest of the Bismarcks ever has been joined to New Guinea but instead has been getting closer to it with time (Wells et al. 1999). Similarities in the pre-Pliocene volcanic and igneous bedrock of New Britain, New Ireland, New Hanover, and Manus suggest that these islands are displaced segments of the same ancestral island arc (Exon & Marlow 1988, Martinez & Taylor 1996, Dickinson 1998a). Miocene limestones flank and cap the Paleogene volcanic rocks on Manus, New Ireland, and New Britain. Seafloor spreading in the Bismarck Sea carried Manus off to the north. Mussau is younger and tectonically independent of other major islands in the Bismarcks. Neogene volcanism accounts for Watom and the islands that stretch westward from New Britain (Dickinson 1998a). New Britain itself has active volcanoes, especially at Rabaul on its eastern end. Extinct Plio-Pleistocene volcanoes form the Tabar-Lihir-Tanga-Feni island chain northeast of New Ireland.

The Solomon Islands (population ca. 461,000; Figures 5-2 and 5-3) is an independent constitutional monarchy. Buka and Bougainville are politically part of Papua New Guinea today but belong to the Solomons by any geologic, biogeographic, or cultural measure. Another colonial artifact is political inclusion in the Solomon Islands of remote Polynesian outliers such as the northern atolls of Sikaiana and Ontong Java and the small volcanic islands of Tikopia and Anuta in the Santa Cruz Group. The 51 islands or island clusters in the Solomons make up a land area of 35,634 km^2 (including Bougainville and Buka but excluding the Santa Cruz Group, which is related geologically to Vanuatu; see below). Seven islands exceed 2000 km^2, the largest and highest being Bougainville (8591 km^2, 2591 m elev.; Mayr

Figure 1-24 The composite island of Isabel, Solomon Islands, as seen from Fera, a small island off Isabel's northern coast. Photo by DWS, 2 July 1997.

& Diamond 2001). The Solomons lie within a sea area of ca. 1,440,000 km^2. The major islands from Buka southeast through Makira (San Cristobal) are the Main Chain, which, unlike the rest of the group, is part of Near Oceania. Glacially lowered sea levels united these islands from Buka, Bougainville, and Choiseul to Isabel and perhaps New Georgia (but not Malaita, Guadalcanal, or Makira) into a massive island of ca. 46,400 km^2 called "Greater Bougainville" or "Greater Bukida" (Spriggs 1997, Mayr & Diamond 2001).

Lying 170 km south of the Main Chain are the uplifted Neogene limestone islands of Rennell (676 km^2, 110 m elev.) and Bellona (19.5 km^2, 79 m elev.). They are separated from the Main Chain by deep water and have a fairly independent geological, biotic, and cultural history (Grover 1960, Diamond 1984, Filardi et al. 1999). Rennell is the world's largest uplifted former atoll, with relatively little relief over most of its surface but entirely surrounded by subaerial cliffs.

Exposed bedrock in the Solomons runs the gamut of volcanic, igneous, metamorphic, and terrestrial and marine sedimentary rocks (Coulson & Vedder 1986, Falvey et al. 1991: Figure 5; Figure 1-25 herein) because the islands lie at a very complex part of the boundary between the Pacific and Indo-Australia Plates. The Pacific Plate is represented here by basalts of the massive (1,280,000 km^2) Ontong Java Plateau, formed by two phases of volcanism in the Cretaceous (ca. 122 and 90 Ma), when this region was several thousand kilometers farther east (Kroenke 1996, Neal et al. 1997). On most large islands in the Solomons, island-arc magmatism and volcanism on the Indo-Australia Plate account for the varied volcanic, igneous, and metamorphic rocks (Petterson et al. 1997). Malaita, however, lies on the Pacific Plate where high-pressure minerals (xenoliths) derived from 400 to 670 km deep in the earth's mantle occur along with Cretaceous basalts, Miocene limestones, and Neogene mudstones, together forming a dramatic example of a composite island (Collerson et al. 2000). Vanuatu (land area 12,195 km^2, sea area 680,000 km^2; Figure 5-4) is an independent republic once called New Hebrides. Oriented mainly north-south, Vanuatu has 12 islands >270 km^2 and many smaller ones. The 70 inhabited islands sustain 189,300 persons. Primarily Melanesian at European contact and today, a few small islands in Vanuatu are Polynesian outliers. The islands range from active volcanoes (e.g., Tanna, Ambrym) to raised limestone islands (e.g., Aniwa, Futuna) to composite islands with emergent oceanic basement rocks, ocean-floor sediments, volcanics, volcaniclastics, and reefal limestones (e.g., Efate, Malakula, Santo, the last being the largest and highest island at 3900 km^2 and 1879 m elev.) (MacFarlane et al. 1988, Nunn 1994:38–40, Cabioch & Ayliffe 2001).

Vanuatu lies just east of the New Hebrides Trench, a feature on the Indo-Australia Plate where relatively heavy crust

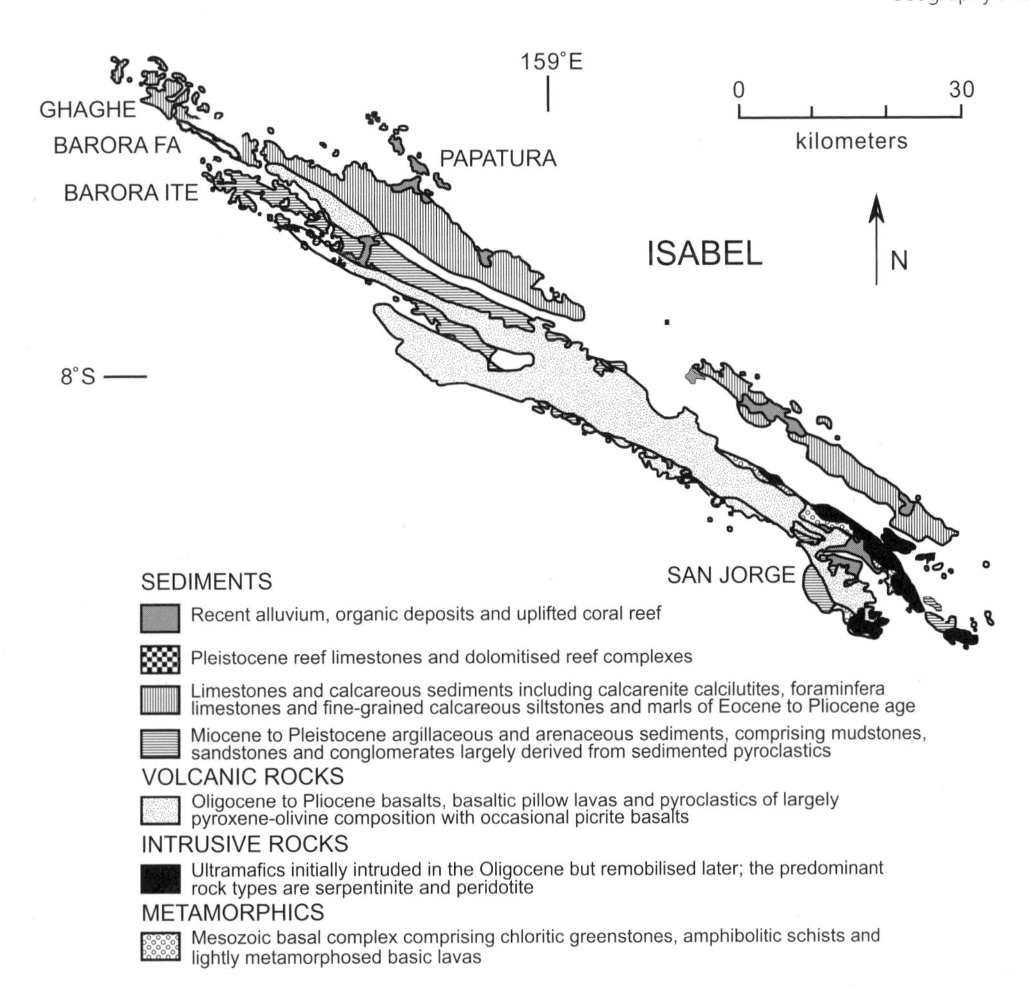

Figure 1-25 Generalized bedrock geology of Isabel. Ca. 10% of the land is >400 m; <1 of the land is >800 m; maximum elevation is 1250 m. Redrawn from Solomon Islands (1976).

is being subducted eastward (Carney et al. 1985, Dickinson & Shutler 2000: Figure 6). The composite islands of Vanuatu form the forearc just behind (east) of the trench. The volcanic rocks are mostly andesitic but range from basalts to dacite (MacFarlane et al. 1988, Dickinson 2001). On many islands the volcanic or volcaniclastic rocks are flanked or capped with uplifted Neogene limestones (Taylor et al. 1985, 1990). The main (central) arc of Neogene volcanic islands extends the entire length of the chain from Matthew and Hunter through the Banks and Torres Islands to the Santa Cruz Group (Tinakula, Nendo, Vanikolo, Reef Islands, Tikopia, Anuta, Taumako), which is politically part of the Solomon Islands (Duncan 1985, Musgrave & Firth 1999, Dickinson & Shutler 2000). Vanuatu has been rotating southwestward since the late Miocene (ca. 7.5 Ma), getting farther from Fiji and Tonga and closer to New Caledonia (Dickinson 2001: Figure 3). The Vanuatu arc also is lengthening northward and southward through volcanism.

New Caledonia (19,103 km^2 of land in 1,740,000 km^2 of sea; Figure 5-5) lies on the Indo-Australia Plate and is an Overseas Territory of France. Grande Terre is Oceania's second largest island (16,890 km^2) and supports most of New Caledonia's 212,800 persons. Much of Grande Terre (1628 m elev.) is an old (Mesozoic), rifted Gondwanan/Australian fragment with ultrabasic igneous rocks (the source of rich nickel deposits; Dupon 1986) as well as volcanic tuffs and lavas, conglomerates, sandstones, shales, limestones, and metamorphic rocks (Paris 1981, Kroenke 1996, Dickinson 2002: Figure 2). The Belep Islands and Ile des Pins are related geologically to Grande Terre; all are inhabited by Melanesians. The Polynesian-inhabited Loyalty Islands to the northeast are limestone, their uplift associated with the forebulge of the New Hebrides Trench, where seafloor is being subducted eastward beneath Vanuatu (Nunn 1994:117–119, Dickinson & Shutler 2000: Figure 6).

West Polynesia (Figure 1-3)

Fiji (Figures 6-1, 6-2) is an independent republic that is more diverse geologically (volcanic, metamorphic, limestone, and

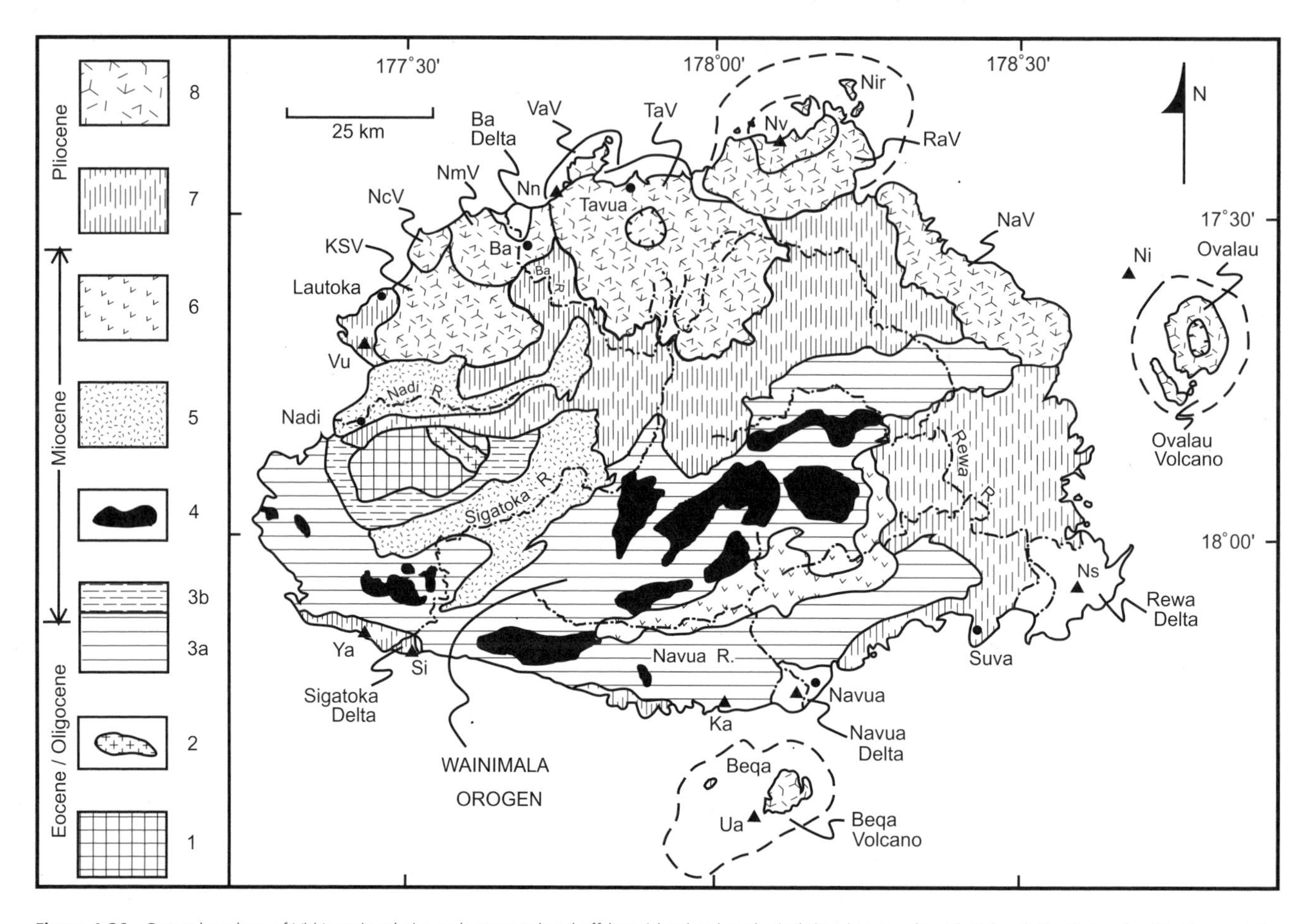

Figure 1-26 General geology of Viti Levu in relation to key coastal and offshore island archaeological sites (Ka, Karobo; Ni, Naigani; Nn, Natunuku; Nir, Nananu-I-Ra; Ns, Nasilai; Nv, Navatu; Si, Sigatoka; Ua, Ugaga; Vu, Vuda; Ya, Yanuca). Key to legend: (1) Eocene volcanic-volcaniclastic Yavuna Group (oldest remnant of ancestral Vitiaz arc system); (2) Eocene-Oligocene (30–35 Ma) Yavuna stock; (3) mid-Oligocene to mid-Miocene volcanic-volcaniclastic Wainimala Group (3a, arc axis; 3b, forearc flank); (4) Middle to Upper Miocene (12.5–7.5 Ma) Colo plutons (dioritic-gabbroic to granitic); (5) Upper Miocene (8–6 Ma) sedimentary strata (Tuva-Navosa-Nadi Groups) unconformable (post-Colo) on Wainimala orogen; (6) Upper Miocene (6.0–5.5 Ma) volcanic-volcaniclastic Namosi Andesite and associated volcaniclastic Medrausucu Group (terminal Vitiaz arc activity); (7) Pliocene (5.5–3.5 Ma) sedimentary strata; (8) Pliocene (5.5–2.5 Ma) postarc volcanic-volcaniclastic rocks (volcanoes along north coast of Viti Levu: KSV, Koroyanitu-Sabeto; Nav, Naimasi; NcV, Namosau; RaV, Rakiraki, TaV, Tavua; VaV, Vatia). Hachured lines denote caldera margins of Tavua and Ovalau (Lovoni caldera) volcanoes, and outer limit of altered and intruded core of eroded Rakiraki volcano. From Dickinson (2001: Figure 8).

other sedimentary rocks), has more islands (>300 that are >0.1 km^2), has more land (18,272 km^2), and sustains more persons (population 801,540) than any South Pacific island group to its east. Viti Levu (10,384 km^2) and Vanua Levu (5535 km^2) are geological composites that make up 87% of Fiji's land area, although five other Fijian islands are larger than any in most Polynesian and all Micronesian island groups. Fiji's sea area is 1,290,000 km^2. Fijian islands lie on the Fiji Platform (the large western islands), Lau Ridge (the smaller eastern islands, or Lau Group), and the intervening South Fiji Basin (Moala, Matuku, Totoya) of the Indo-Australia Plate (Stratford & Rodda 2000, Dickinson 2001). Viti Levu consists of Eocene to late Miocene plutonic rocks and a variety of younger (late Miocene through Quaternary) volcanic rocks (basalts, andesites, volcaniclastics) and sedimentary rocks (conglomerates, sandstones, siltstones, mudstones, marls, limestones; Gill 1987, Stratford & Rodda 2000, Dickinson 2001; Figure 1-26 herein). Other western Fijian islands are less complex geologically and dominated by Neogene volcanics, volcaniclastics, and limestones. The Lau Ridge is an inactive remnant arc of the Tongan Trench-Arc System (see below). Islands in the Lau Group are either eroded volcanic (primarily andesitic), raised limestone, or composites thereof (Nunn 1994:125–128). The Lau Backarc Basin, which separates the Lau Ridge from the Tonga Ridge, is getting wider (Smith et al. 2001, Martinez & Taylor 2002), which means that Fiji and Tonga are getting farther apart. Finally, the remote, eroded volcanic island of Rotuma (42 km^2) is a Polynesian outlier that is politically part of Fiji.

The Kingdom of Tonga (land area 747 km^2, sea area 700,000 km^2; Figures 6-4, 6-5, 6-9) is an independent constitutional monarchy. Tonga features 160 raised limestone islands (0.1 to 259 km^2) in three groups (Tongatapu,

Ha'apai, and Vava'u). The 10 active or probably active (Quaternary) volcanic islands along its western and northern margins range from Tofua (46 km^2) down to the ephemerally emergent Fonuafo'ou and Late Iki. The dominant island is Tongatapu (259 km^2), where 67,000 of the 105,000 Tongans live; 35 other islands are inhabited. Tongatapu is a flat island of uplifted Pliocene through Quaternary limestones dated by U-series from >400 ka to the Holocene (Roy 1997, Spennemann 1997). Like most of Tonga's raised limestone islands, Tongatapu has rich soils developed from volcanic ashes (andesitic tephra; Cowie 1980, Wilson & Beecroft 1983). Other population centers are on nearby 'Eua (87 km^2, elev. 300 m; primarily raised limestone), the much lower (elev. <20 m) limestone islands of Lifuka (11.4 km^2) and Foa (13.3 km^2) in the Ha'apai Group, and the raised limestone 'Uta Vava'u (96 km^2, 215 m elev.) in the Vava'u Group.

Tonga lies on the Indo-Australia Plate just west of the Tonga Trench, where the Pacific Plate subducts westward ca. 9 to 24 cm/year (90 to 240 km/My) to create one of the world's most active regions seismically and volcanically (Isacks et al. 1969, Chen & Brudzinski 2001, Green 2001, Hall & Kincaid 2001, Smith et al. 2001, Tibi et al. 2003). Tonga is part of the Tonga-Kermadec Trench-Arc System, which extends from 15° S to 40° S (Ballance et al. 1999). The Tongan Forearc Belt or Tonga Ridge is divided by east-west transverse faults into at least 12 major structural blocks (Taylor & Bloom 1977, Herzer & Exon 1985, Tappin 1993). Raised limestone islands lie atop five of these forearc blocks (Figure 1-27), each with a related but partially independent history of uplift, subsidence, and tilting (Dickinson et al. 1999). For example, a greater rate or duration of uplift in Vava'u than in Tongatapu has led to higher elevations and a more advanced state of karstic weathering in the former, with better developed solutional cliffs and a lack of primary depositional features in the reefal limestones (Roy 1997). The young volcanic islands of Tonga lie just west of the forearc blocks, with the Tofua Trough between (Bourdon et al. 1999, Dickinson et al. 1999).

The oldest exposed rocks in Tonga are on 'Eua, where late Eocene ages (ca. 42–38 Ma) are estimated for limestones through biochronology of foraminerans and for basalts by K-Ar and Ar-Ar dating (Hoffmeister 1932, Cunningham & Anscombe 1985, Duncan et al. 1985, Chaproniere 1994). This does not mean that 'Eua has been subaerial (and thus inhabitable by terrestrial plants and animals) continuously since the Eocene; it has been submerged for much of its history (Tappin & Ballance 1994). Nobody really knows how long any particular Tongan island has been above the ocean, although 5 Ma or less (Plio-Pleistocene) might be a reasonable estimate even for an "old" island such as 'Eua. I would speculate, nevertheless, that at least some land has been exposed in Tonga since the late Miocene (ca. 10-8 Ma) or perhaps even earlier in the Neogene.

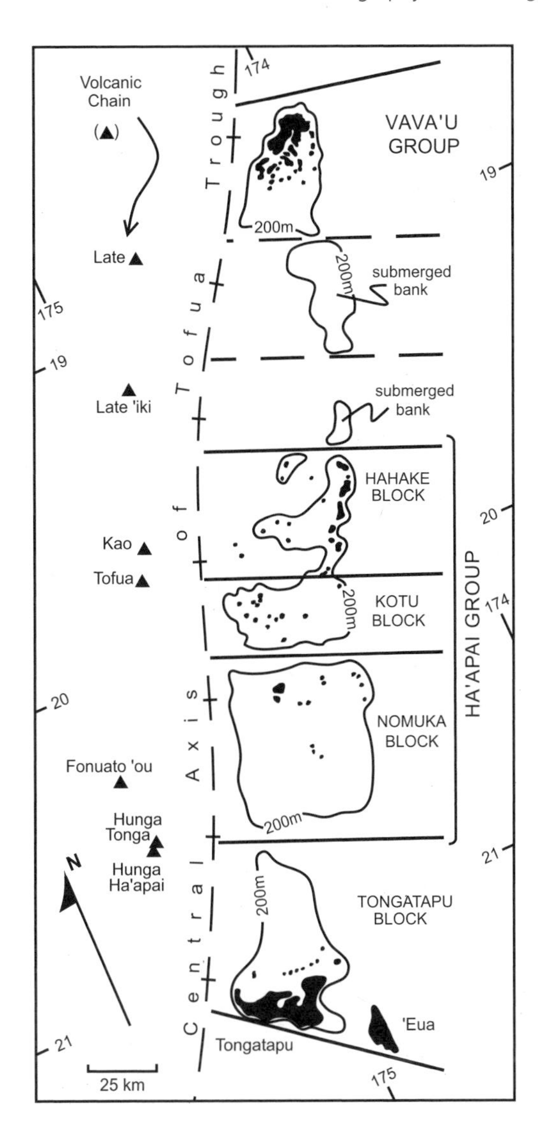

Figure 1-27 "Piano key" arrangement of structural blocks on the Tonga Ridge. Transverse fault-controlled discontinuities (heavy lines) delineate separate forearc structural blocks (dashed lines denote subordinate intrablock structures) along the crest of the Tonga Ridge. 200 m isobaths show approximate shapes of block crests dotted with islands and banks. Solid triangles denote Quaternary island volcanoes along the active volcanic chain west of the Tofua Forearc Trough. From Dickinson et al. (1999: Figure 3).

Samoa (Figure 6-15) is divided politically today. The young volcanic islands of Upolu (1100 km^2), Savai'i (1821 km^2), and their offshore islands make up the independent country of Western Samoa (population 167,990, land area 2935 km^2, sea area 120,000 km^2; now often called Independent Samoa or simply Samoa). To the east, American Samoa (an unincorporated U.S. territory; population 63,330; land area 199 km^2; sea area 390,000 km^2) consists of two atolls and five eroded volcanic islands, the largest being Tutuila at 145 km^2.

Samoan islands are composed of basalt-rich rocks that include late Quaternary volcanism on Savai'i, Upolu, Tutuila, and Ta'u (Simkin et al. 1981). The oldest ages (K-Ar) of Samoan lavas are ca. 2.8 ± 0.20 Ma and are found on Upolu (Natland & Turner 1985). Rocks from Savai'i are normally magnetized, whereas those from Upolu are mixed, and most rocks from Tutuila have reversed magnetic polarity (Keating 1985). This suggests that Savai'i may be slightly older than Upolu (up to 3.4 Ma).

Although the oceanic shield volcanoes that make up Samoa lie on the Pacific Plate, the Samoan terrestrial biota is much more closely related to those of Fiji and Tonga than to any in East Polynesia (Steadman 1993a). Samoan volcanism probably is related to lateral flexure of the Pacific Plate as it passes by the northern part of the Tonga Subduction Zone and the Lau Backarc Basin (Smith et al. 2001). The largest Samoan islands, Upolu and Savai'i, consist of late Pliocene through Holocene volcanic rocks (generally basaltic) along an ESE to WNW alignment of late Quaternary cinder cones that is superimposed on older (Neogene) volcanics that do not follow the steady age progression predicted in a hotspot-island chain model (Kear & Wood 1959, Keating 1992, McDougall 1985, Natland & Turner 1985, Dickinson & Green 1998). Eruptions are unknown in historic times on Upolu but have occurred on Savai'i as recently as AD 1911 and on Ta'u in American Samoa in AD 1866 (Nunn 1994:46). Where sufficient time has elapsed and the land is not too steep, Samoan lavas have weathered into rich soils (Wright 1963).

The isolated islands of Wallis & Futuna (population 14,375, land area 145 km^2, sea area 300,000 km^2; Figure 6-16) are an overseas territory of France. Faunally related to Fiji-Tonga-Samoa, the group consists of the three inhabited islands of Wallis (= 'Uvea), Futuna, and Alofi, and several offshore islets (Kirch 1995). The Polynesians who live here are related to Tongans ('Uvea) or Samoans (Futuna, Alofi). 'Uvea (80 km^2, 151 m elev.) is an eroded volcanic island on the Pacific Plate that is aligned with the Samoan hotspot chain (Nunn 1994:46). Futuna (45 km^2, 524 m elev.) and Alofi (18 km^2, 417 m elev.), formerly called the Horne Islands, lie 250 km southwest of 'Uvea on the Indo-Australia Plate, separated from 'Uvea by the Vitiaz Paleotrench, a westward extension of the Tonga Trench (Dickinson 2001). Futuna and Alofi are composites of eroded volcanics and raised limestones. They are built on late Pliocene volcanic rocks ranging from basalts to andesites, whereas the volcanic bedrock of 'Uvea is mainly basaltic. Uplifted limestone terraces are best developed on Alofi.

Niue is a large (258 km^2), isolated raised limestone island (Figure 6-17) that is self-governing in free association with New Zealand. Niue (population 2040, sea area 390,000 km^2) lies on the Pacific Plate but has a terrestrial biota most similar to that of Tonga (Paulay & Spencer 1992, Steadman et al. 2000). The somewhat phosphatic central plateau (the lagoon of an uplifted Pliocene atoll) is depressed as low as 34 m whereas the island's highest elevation (68 m) is on the annular limestone that represents the former atoll's fringing reef (Schofield & Nelson 1978, Aharon et al. 1987, Paulay & Spencer 1992). The entire coastline of Niue is cliffy, weathered limestone. The Polynesians living on Niue are closely related to Tongans.

East Polynesia (Figure 1-4)

The Cook Islands (population 16,770, land area 238 km^2, sea area 1,830,000 km^2; Figure 7-1) include six northern atolls and a southern group with an eroded volcanic island (Rarotonga, 67 km^2, 652 m elev.; Thompson et al. 1998), four composites of raised limestone and eroded volcanics (Mangaia at 52 km^2 being the largest), the almost-atoll of Aitutaki, two atolls (Manu'ae, Palmerston), and the small limestone island of Takutea (Wood 1967, Wood & Hay 1970, Stoddart & Gibbs 1975, Stoddart et al. 1985, 1990, Gray et al. 1992, Dickinson 1998b). Within the fringing reef of Rarotonga are four very small islands, of which Oneroa and Koromiri are sand cays, Motutapu has exposed reefal limestone, and Ta'akoka has exposed basalt (Stoddart 1972). The Cook Islands are self-governing in free association with New Zealand. The southern Cook Islands are related geologically to the Tubuai (Austral) Islands as part of the Cook-Austral Island-Seamount Chain (CAISC), a linear set of Neogene volcanos (Figure 1-28). Unlike the classic model of age progression of islands and seamounts along a linear chain, the CAISC has had repetitive episodes of volcanism at multiple sites, nonmonotonic subsidence of islands after hotspot volcanism, uplift of certain islands long after initial subsidence, and multiple alignments of volcanic edifices (Turner & Jarrard 1982, Dickinson 1998b). Postimmersion uplift has produced six limestone (or limestone/eroded volcanic composite) islands in the CAISC (Atiu, Mitiaro, Ma'uke, Mangaia, Rimatara, Rurutu; Stoddart & Spencer 1987). In spite of karstic weathering, the current elevations of emergent fringing reefs are good indicators of net uplift on these islands (Stoddart et al. 1985, Dickinson 1998b).

Tubuai (= Austral Islands; Figure 7-7) consists of seven islands (147 km^2) that are related tectonically to the southern Cook Islands as part of CAISC (Dalrymple et al. 1975, Dickinson 1998b; Figure 1-28). Like the Society Islands, Tuamotu Archipelago, and Marquesas Islands, Tubuai (population ca. 6600) is politically part of French Polynesia, an overseas territory of France. Tubuai consists of four eroded

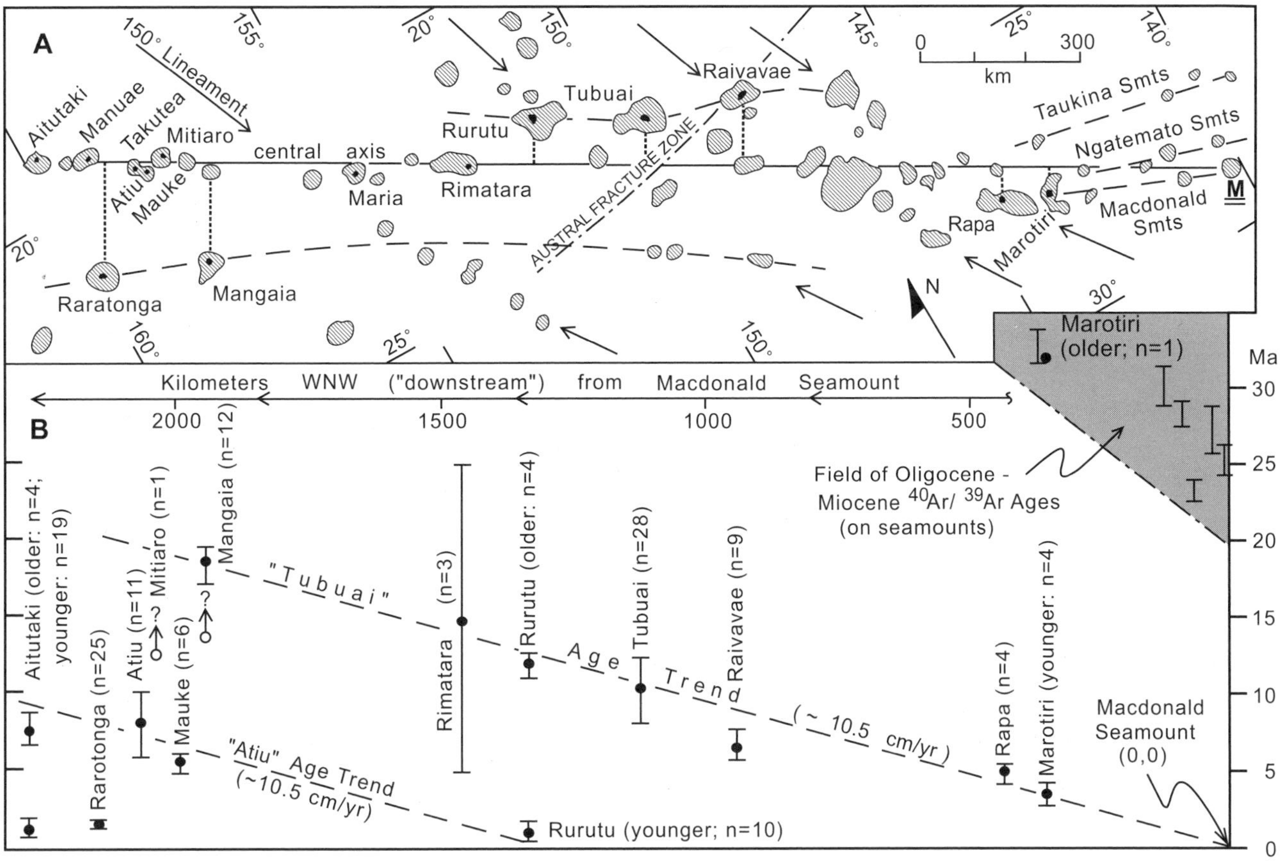

Figure 1-28 Cook-Austral Island-Seamount Chain. A: Geographic configuration (with submerged volcanic edifices indicated by diagonal rules and capping islands, each designated by name, in black) with segments of the compound chain indicated by dashed trend lines, net orientation indicated by a solid trend line (drawn at 120° or S60° E from Aitutaki on the far northwest to MacDonald Seamount at M, the current site of active hotspot volcanism), oblique seamount alignments denoted by arrows ("150° lineament," referring to azimuthal orientation rather than longitude), and dotted lines denoting projection of dated off-trend islands to the net trend of the chain. B: Time-distance plot of Cook-Austral volcanism (mean ages plotted as solid circles with full observed ranges indicated by brackets and minimum ages plotted as open circles), with constituent age trends denoted separately, and anomalous Late Oligocene to Early Miocene seamount ages delimited by shading. From Dickinson (1998b: Figure 2).

volcanic islands (Tubuai, Raivavae, Rapa, Marotiri), two raised limestone/eroded volcanic composites (Rimatara, Rurutu; Stoddart & Spencer 1987), and the small atoll of Maria. Marotiri and Maria are uninhabited.

Aside from several atolls and small, highly eroded volcanic islands, the Society Islands (1400+ km^2; Figure 7-8) consist of six large eroded volcanic islands, dominated by Tahiti (the largest, highest island in East Polynesia at 1042 km^2 and 2241 m elev.) and nearby Mo‘orea. Tahiti (population ca. 150,000) is the main commercial center for the Society Islands (pop. ca. 200,000) and all of French Polynesia (pop. 228,785, sea area 5,030,000 km^2). Tahiti and Mo‘orea are the main islands of the Windward Group of the Society Islands. Bora Bora, Huahine, Ra‘iatea, and Taha‘a are the main islands of the Leeward Group.

Lying between and parallel to the CAISC and Tuamotu Group (Hekinian et al. 1991, Dickinson 1998b: Figure 1), the Society Islands formed through hotspot volcanism and have a fairly consistent age progression from the oldest island in the west (Maupiti) to the youngest island/seamounts of Mehetia and Moua Pihaa in the east (Nunn 1994:44–45; Figure 1-29 herein). With the oldest dated rocks at only 4–5 Ma, the age/distance relationship among the Society Islands yields an inferred rate of Pacific Plate motion of ca. 11 cm/year to the west-northwest (Dickinson 1998b). Immense Tahiti comprises two basaltic volcanoes dated from 1.7 to 0.3 Ma (Duncan et al. 1994).

The Tuamotu Islands (land area 752 km^2, including the Gambier Islands) are also part of French Polynesia. They consist of the raised limestone islands of Makatea (Montaggioni et al. 1985), Niau, and Tepoto, the somewhat outlying almost-atoll of Mangareva, and 72 atolls (Figure 7-10). The Tuamotuan population of ca. 15,400 is scattered across many of these islands. Oriented roughly parallel to

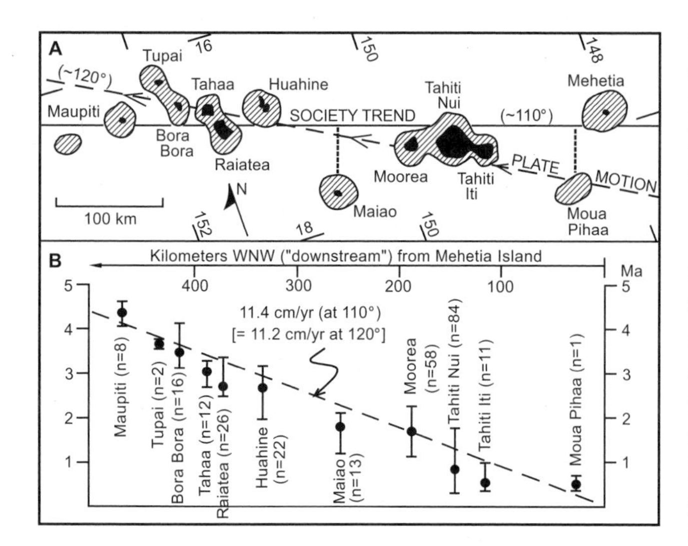

Figure 1-29 Society Islands hotspot track. A: Geographic configuration (with submerged volcanic edifices indicated by diagonal rules and capping islands in black), with net trend (at 110° or S70° E) slightly oblique to plate motion (at 120° or S60° E). B: Time-distance plot of Society Islands volcanism (mean ages plotted as circles with full observed ranges indicated by brackets) based on K-Ar ages. From Dickinson (1998b: Figure 3).

the Society Islands, the Tuamotus are built on Neogene volcanic basement rocks (primarily late Miocene or younger; Gillot et al. 1992, Guillou et al. 1993, Munschy et al. 1996). Many atolls in the Tuamotus may have been inundated as recently as 1.2 ka (Dickinson 2003).

The Marquesas Islands (population ca. 8140) consist of 10 high, eroded volcanic islands (1060 km^2; Figure 3-14) that are rather close to each other but are isolated from the rest of French Polynesia. These islands are developed on the Marquesan Platform, a thickened area of oceanic crust lying at anomalously shallow depths (<4000 m) for the Pacific Plate (Wolfe et al. 1994, Caress et al. 1995). The crude lineation of islands and seamounts here is not parallel to inferred movement of the Pacific Plate, nor is the age progression of islands strictly linear, with Eiao, Hatutu, and Ua Pou having the oldest volcanic rocks, dated at ca. 5 Ma (Dickinson 1998b; Figure 1-30 herein). The northern and southern clusters have five islands each, the former dominated by Nuku Hiva (337 km^2) and the latter by Hiva Oa (324 km^2).

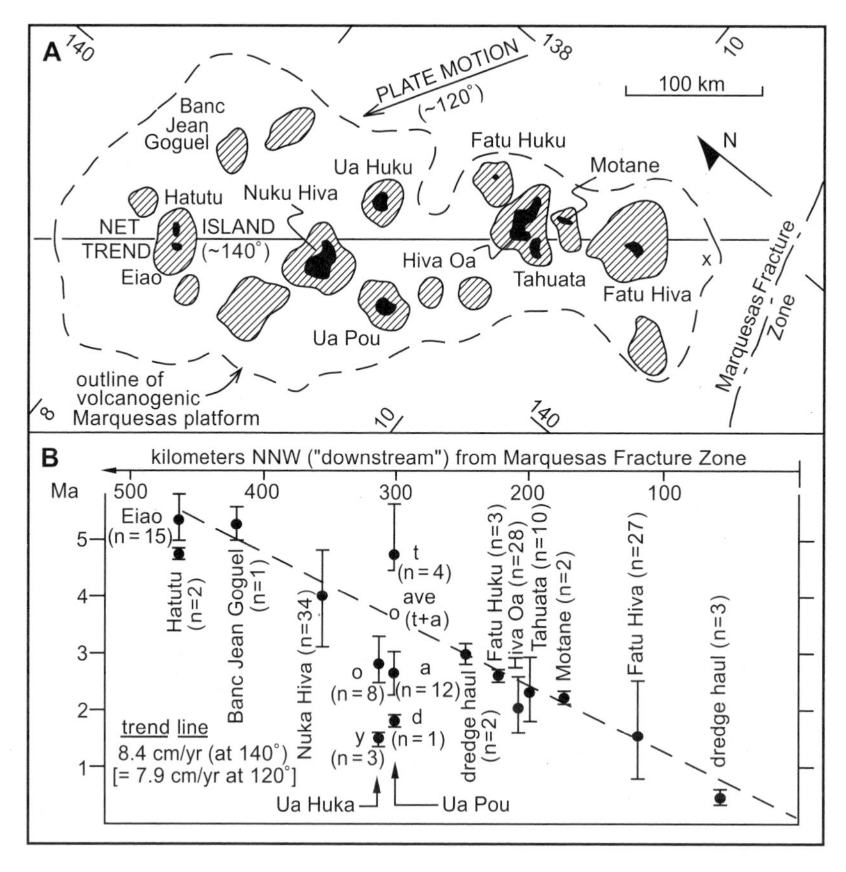

Figure 1-30 Marquesas Islands hotspot track. A: Submerged volcanic edifices indicated by diagonal rules and capping islands in black, with net trend (at 140° or S40° E) oblique to inferred plate motion (at 120°). B: Time-distance plot of Marquesas Islands volcanism (mean ages plotted as circles with full observed ranges indicated by brackets) based on K-Ar ages. From Dickinson (1998b: Figure 4).

The remote Pitcairn Group (land area 45.2 km^2, sea area 800,000 km^2; Figure 7-12) consists of the eroded volcanic Pitcairn (6.6 km^2, 347 m elev.), the raised limestone Henderson (37.2 km^2, 34 m elev.), and the small atolls of Ducie and Oeno. Governed through the British High Commission in New Zealand, only Pitcairn itself is inhabited (population 47). These four islands are the easternmost exposed land along a 1900-km-long hotspot lineation developed over thin oceanic crust in the far eastern Pacific Plate (Spencer 1995). The Gambier Group of the southern Tuamotus is part of this lineation. The age of Pitcairn Island is ca. 1 Ma (Duncan et al. 1974). The basement rocks of Oeno, Henderson, and Ducie are older but not precisely dated (Spencer 1995). Henderson is an uplifted Pleistocene atoll (Pandolfi 1995). U-series dating of fossil corals in Henderson's limestones has yielded ages mainly from 404 to 225 ka, with the island estimated to have been emergent above the sea since ca. 380 ka (Blake 1995).

The most isolated island in Oceania is Rapanui or Easter Island (161 km^2, elev. 507 m, sea area 355,000 km^2; Figure 3-17), which is a province of Chile. Easter Island (population 3000+) comprises three coalesced shield volcanoes with oceanic basalts, trachytes, obsidian, and tuff (Bandy 1937, Baker 1967, Baker et al. 1974, Beardsley et al. 1991). It lies on the Nazca Plate, uniquely in Oceania. Just to the west is the Easter Microplate, an islandless piece of lithosphere sandwiched between the westward-moving Pacific Plate and eastward-moving Nazca Plate (Hanan & Schilling 1989). This divergent plate boundary (where new oceanic crust is produced) is called the East Pacific Rise or East Pacific Ridge (Herron 1972). The oldest dated (K-Ar) rocks on Easter Island are ca. 2.5 Ma (Clark & Dymond 1977). Lying 375 km to the northeast is the small (0.15 km^2, elev. 10 m), uninhabitable seabird island of Sala y Gomez, which is the barely subaerial peak of a large seamount more or less aligned with Easter Island on the Sala y Gomez Ridge (Clark & Dymond 1977, Harrison & Jehl 1988).

Micronesia (Figure 1-5)

Moving back now to the far western Pacific, Micronesia is more geologically diverse in the west (Palau, Yap, Marianas) than in its atoll-dominated east (Figure 1-31). Palau, the Marianas, and most of Yap lie on the Philippine Sea Plate, whereas the rest of Micronesia lies on the Pacific Plate or Caroline Plate (Figure 1-7).

Palau (Republic of Belau; Figure 8-2) is self-governing in free association with the United States. These ca. 340 islands (land area ca. 494 km^2, sea area ca. 629,000 km^2) are just west of the Palau Trench, where the Pacific Plate subducts beneath the Philippine Sea Plate. All but nine Palauan islands are uninhabited; most of the 19,200 people live on the causeway-connected islands of Babelthuap (Babeldaob), Koror (Oreor), Arakabesan (Ngerekebesang), and Malakal (Ngemelachel). Except for the Southwest Islands (outlying atolls 300–600 km toward the Moluccas) and Angaur, the

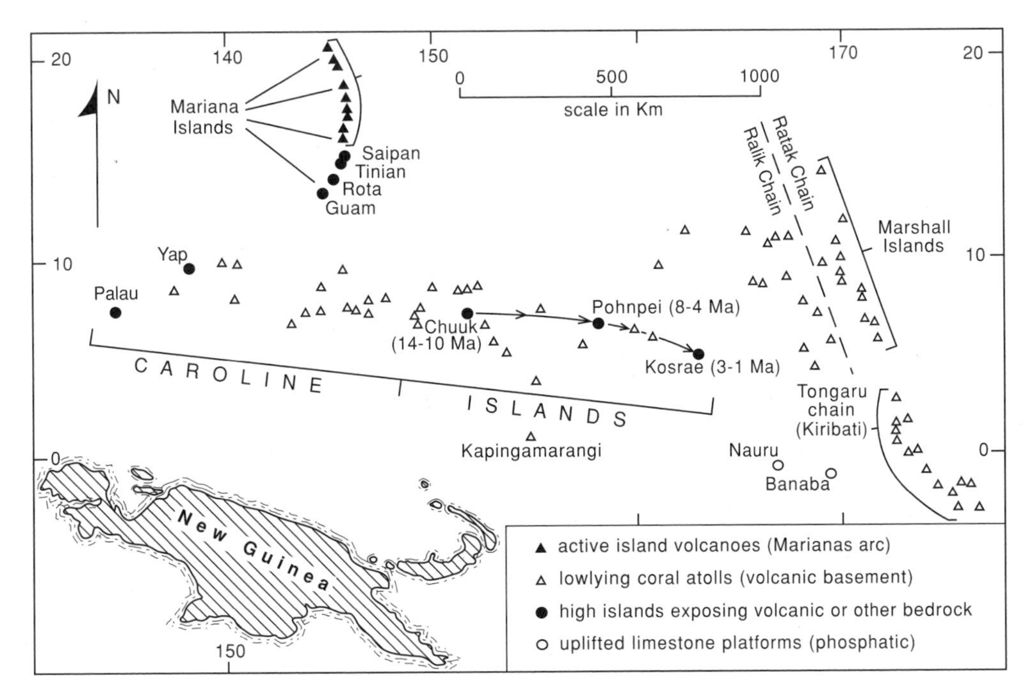

Figure 1-31 Regional geology of Micronesia. Redrawn from Dickinson (2000b: Figure 1).

islands in Palau are more or less encircled by a massive barrier reef (Kayanne et al. 2002).

The four causeway-connected islands are composites of Paleogene/early Neogene volcanic rocks and Neogene raised limestones. Babeldaob (333 km^2) is by far the largest island (Rogers & Legge 1992). Most other Palauan islands consist entirely of raised Neogene limestone. Known locally as the Rock Islands, they are cliffy, difficult of access, mostly forested, and riddled with karst features such as caves, rockshelters, and crevices (Rogers & Legge 1992, Pregill & Steadman 2000).

The Mariana Islands (Figure 8-4) is a north-south island arc with five major raised limestone islands or limestone-volcanic composites in the south (Guam north to Saipan) and 10 volcanic islands (eight still active) in the north (Farallon de Medinilla northward to Farallon de Pajaros). All of the Mariana Islands (1012 km^2 of land in a sea area of 995,000 km^2) are affiliated with the United States, Guam as an unincorporated non-self-governing U.S. territory and the 14 other islands as the more independent Commonwealth of the Northern Mariana Islands (population 72,780). Guam (population 149,640) is the largest island at 541 km^2. The Micronesians who inhabit the Marianas, known as Chamorros, occupy Guam northward to Saipan (except uninhabited Aguiguan) and a few of the northern volcanic islands.

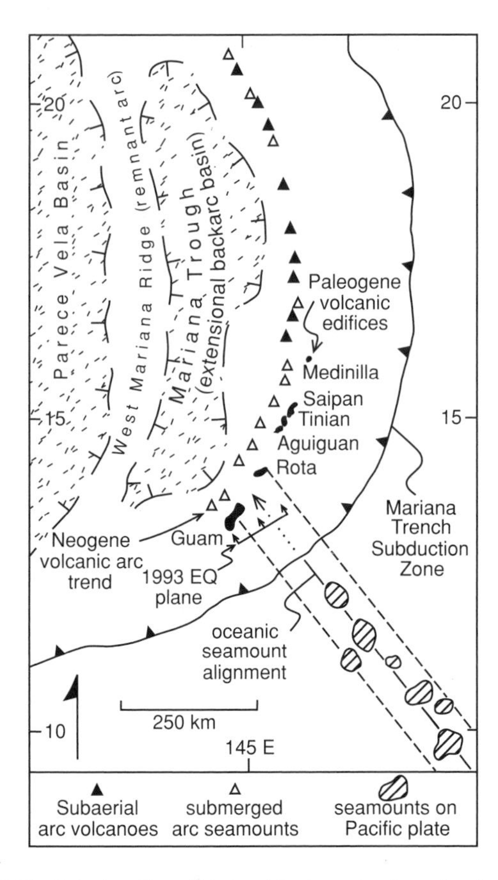

Figure 1-32 Tectonic setting of Mariana island arc, showing active chain of Neogene island volcanoes and submerged seamounts, ancestral chain Paleogene volcanic edifices (black), and a key alignment of oceanic seamounts being subducted with Pacific seafloor at the Mariana trench. Up-dip projection of the seismic plane of 1993 Guam earthquake would intersect at the Mariana Trench. Redrawn from Dickinson (2000b: Figure 4).

The Marianas consist of two parallel, curvilinear chains of islands and seamounts lying west of the Mariana Trench Subduction Zone, where the westward-moving Pacific Plate is subsumed beneath the Philippine Sea Plate (Taylor 1992, Fryer 1995, Dickinson 2000b; Figure 1-32 herein). The western chain is an active volcanic arc of Pleistocene and perhaps Pliocene (ca. 3–0 Ma) islands and seamounts. The eastern chain (Guam to Saipan) is a shorter forearc belt whose islands began with late Eocene to early Oligocene (ca. 45–30 Ma) volcanism, followed by minor volcanism in the middle Miocene (ca. 15–12 Ma). Paleomagnetic data from Guam suggest a stable latitudinal location but a clockwise rotation of the island arc since the Miocene (Larson et al. 1975). Guam has extensive raised limestone as well as eroded volcanic rocks (Tracey et al. 1964, Reagan & Meijer 1984), whereas Rota, Aguiguan, Tinian, and Saipan are capped and flanked almost exclusively by corraline limestones formed from the Miocene through Holocene (ca. 15–0 Ma). The Miocene limestones reach elevations of nearly 500 m on Rota and Saipan, whereas most of the lower-elevation limestones date to the Quaternary (Bell & Siegrist 1991, Siegrist & Randall 1992).

Lying to the northeast, north, and northwest of the Marianas are isolated, primarily volcanic islands belonging to Japan (Daito Islands, Parece Vela, Bonin Islands, Volcano Islands, Minami Tori Shima; see Center for Pacific Island Studies 1987). These subtropical islands will not be considered further.

The Federated States of Micronesia (FSM; 701 km^2, sea area 2,978,000 km^2, population 116,410) is self-governing in free association with the United States. The FSM has four states–Yap, Chuuk, Pohnpei, and Kosrae. Yap (121 km^2, population ca. 9000, Figure 8-3) consists of four high islands separated by narrow, mangrove-lined channels and surrounded by a fringing reef (Yap Proper), the outlying raised limestone island of Fais (Rodgers 1948), a large atoll (Ulithi), and 13 other atolls. Only Yap Proper and Ngulu Atoll lie on the Philippine Sea Plate. Yap Proper consists of metamorphosed and uplifted fragments of seafloor intruded by isolated granitic plutons and their weathering products (Dickinson & Shutler 2000). This results in unusual bedrock associations for Oceania, such as volcanic breccias and conglomerates, amphibolite, green chlorite, and talc schists (often folded, faulted and highly eroded)

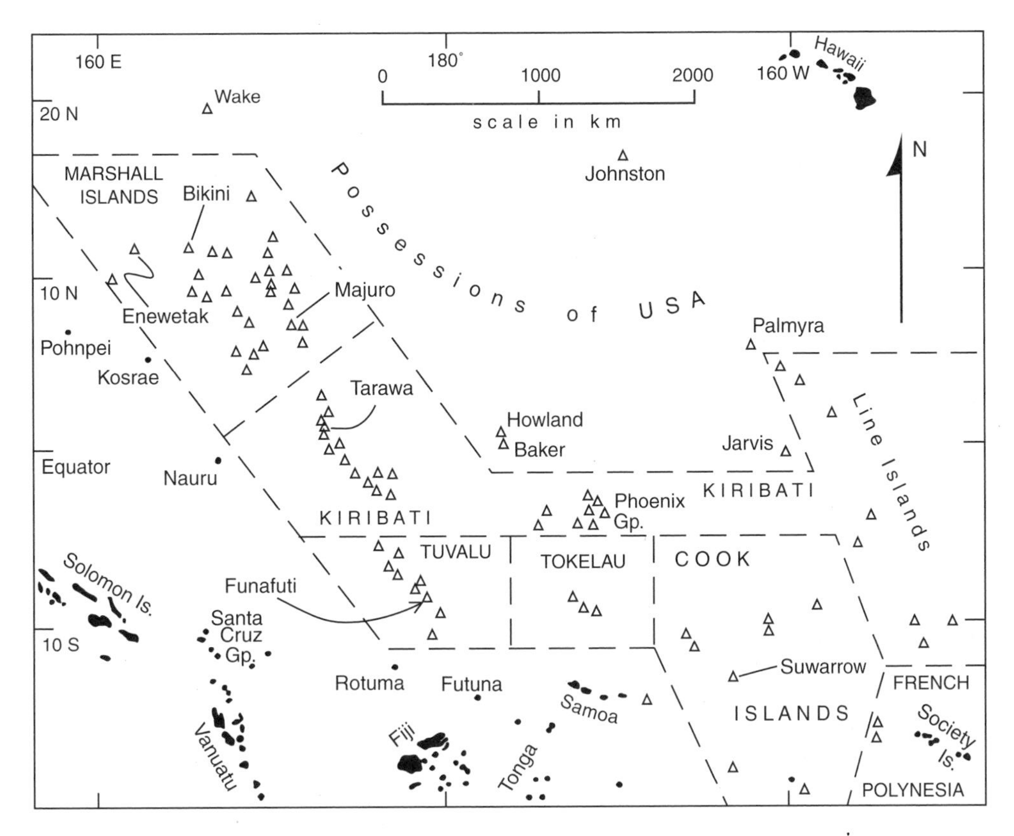

Figure 1-33 Islands (mainly atolls) of "Remote Central Pacific Island Groups" and adjacent areas. Δ, atoll; ●, high islands. Redrawn from Dickinson (1999a: Figure 1).

as well as diverse soil types (Rogers & Legge 1992, Dodson & Intoh 1999).

Chuuk (129 km^2, population ca. 41,000; Figure 8-6.) is also known as Truk. Aside from 10 atolls scattered across 540 km of ocean, Chuuk consists of a large almost-atoll of 15 eroded volcanic islands (up to 33 km^2 and 446 m elev.) within a barrier reef enclosing a lagoon of nearly 2000 km^2. The predominant bedrock on Chuuk is olivine basalt with lesser amounts of siliceous basalt and other volcanic rocks dating perhaps as old as 14–10 Ma (Mattey 1982, Keating et al. 1984, Rogers & Legge 1992).

Pohnpei (345 km^2, population ca. 24,000) is a large and high (337 km^2, elev. 772 m) eroded shield volcano (Figure 8-7). Associated with Pohnpei as a state in the FSM are eight atolls including the Polynesian outliers of Nukuoro and Kapingamarangi (Niering 1963). Uniquely in Micronesia, Pohnpei and some of its atolls to the west and southwest lie on the Ontong Java Plateau, a massive Cretaceous basalt flow (see Solomon Islands account above). The bedrock on Pohnpei is dominated by olivine-rich basalts perhaps as old as 7–5 Ma (Mattey 1982, Rogers & Legge 1992).

Kosrae (109 km^2, elev. 629 m, population ca. 600, Figure 8-8) is the most eastern eroded volcanic island in Micronesia. Reflecting its young age (ca. 1 Ma; Mattey 1982), Kosrae is steeper than Pohnpei and has a much more limited coastal plain, i.e., it is in a less advanced stage of erosion. Unlike the three other states in the FSM, no atolls are affiliated with Kosrae.

Remote Central Pacific Island Groups (RCPIG; Figures 1-5, 1-33) is a geologically, biogeographically, and culturally mixed set of isolated atolls and atoll clusters (Bryan 1953, Douglas 1969) with an occasional raised limestone island or sand cay. Consisting of Wake, the Marshall Islands, Nauru, Kiribati, Tuvalu, Tokelau, and various U.S. possessions, these islands lie on the Pacific Plate from 160° E to 150° W and 20° N to 10° S, a vast region of ca. 10,000,000 km^2 of ocean. The six atolls of the Northern Cook Islands (at ca. 10° S between Tokelau and the southern Line Islands) also could be considered part of RCPIG.

Wake is a single, large atoll (7.4 km^2, elev. 6 m) with three islets. Owned by the United States, Wake is highly isolated (nearest land is ca. 600 km south-southeast in the Marshall Islands) that was uninhabited until a damaging military presence during and after World War II (Bryan 1959, Douglas 1969, Urwin 1997). Higher than most atolls and known to sustain endemic species of insects and birds (Bryan 1926, Greenway 1958), Wake was not inundated during the Holocene.

The Republic of the Marshall Islands (population 63,225, land area 181 km^2, sea area 2,131,000 km^2) is an independent country with 29 atolls, five sand cays, and many barely exposed or unexposed reefs. The islands consist of unconsolidated carbonate sand, gravel, cobbles, and boulders as well as indurated limestone debris and beachrock (Arnow 1954, Fosberg et al. 1965, Fosberg 1990). These surface features typically rest upon hundreds to 1000+ meters of limestone, underlain by basalts (Emery et al. 1954, Leopold 1969). The Marshall Islands are culturally Micronesian and include atolls as large as Kwajalein, with 90 islets (mostly uninhabited), a land area of 16 km^2, and a massive lagoon of 2335 km^2. The highest elevation in the Marshall Islands is only 10 m, with most land <5 m.

Nauru (population 11,360) is an independent republic consisting of one isolated raised limestone island (22 km^2, 65 m elev.) within a sea area of 320,000 km^2. The rich phosphate deposits on Nauru have largely been mined away and exported for fertilizer, leaving the Micronesians who live there with enhanced finances but a degraded landscape with unnatural surface contours (Manner et al. 1985, Thaman 1992).

The Republic of Kiribati (population 88,550) is an independent country inhabited primarily by Micronesians. The 811 km^2 of land scattered over 3,550,000 km^2 of ocean consist of three sets of islands—the Gilbert Islands (16 atolls and the raised limestone island of Banaba), the Phoenix Islands (eight atolls), and Line Islands far to the east (seven atolls, including Kirimati or Christmas Island, with a land area of 388 km^2 the world's largest atoll, and the raised limestone island of Malden). Atolls of the Gilberts vary in land area from 5.2 to 49.2 km^2 and in lagoonal area from 0 to 370 km^2 (Richmond 1997a). At European contact, the Phoenix Islands and Line Islands were uninhabited.

Tuvalu, formerly called the Ellice Islands, is an independent constitutional monarchy with a population of 9600. Polynesian at European contact and still primarily so, the five atolls and four sand cays of Tuvalu make up 26 km^2 of remote land lying in 900,000 km^2 of ocean north of Rotuma and south of the Gilbert Group of Kiribati. As is probably true for atolls everywhere, the larger and more stable atolls in Tuvalu are those where the major islets are underpinned with reefal limestone and where indurated Holocene reef-flat deposits (cemented coral rubble, beachrock) are exposed in the intertidal and supratidal zones (Dickinson 1999a).

Tokelau, a non-self-governing territory under New Zealand administration, consists of three atolls ('Atafu, Nukunonu, Faka'ofo), a mere 12.2 km^2 of land in 290,000 km^2 of ocean. Polynesians (total population 1500) live on one or two leeward islets on each atoll (New Zealand Dept. Lands & Survey 1969). No land in Tokelau is more than 5 m in elevation.

Howland, Baker, Jarvis, Palmyra, and Johnston are U.S. possessions. Each of these atolls was uninhabited at European contact (ca. 200 years ago) and remains so except for permanent or occasional government personnel. Howland and Baker are isolated north of the Phoenix Group. Palmyra (see Dawson 1959) is geologically the northernmost of the Line Islands, with Jarvis lying west of this lineation. Johnston, the most isolated, consists of two natural islets of 0.21 and 0.05 km^2 and two artificial islets made for military purposes (Amerson & Shelton 1976).

Climate

A brief review of climate in Oceania will continue to set the physical stage for biogeographic information (Chapters 2, 5-20) and for changes in sea level (covered later in this chapter). I will begin with modern climate, followed by paleoclimate.

Ocean Currents and General Atmospheric Circulation

Tropical Pacific climates are influenced by circulation of the ocean and the air above it. The winds and ocean currents are complex and vary in direction and speed on annual, seasonal, and shorter timescales. Nevertheless, from the Bismarcks eastward clear to Easter Island, the predominant wind and current direction is from the eastern quarter (northeast to southeast; Figure 1-34).

The general atmospheric circulation in tropical and subtropical Oceania has six main zones (Wyrtki & Meyers 1976, Thompson 1986a, b, Sturman & McGowan 1999, Talley et al. 1999). From south to north, the first is the Subtropical High Pressure Zone (SHPZ), a belt spanning the Pacific at ca. 25-30° S latitude. Within this zone in the eastern Pacific is a semipermanent high-pressure region near 90–100° W longitude, which is east of any islands in Oceania.

Next is the Southern Trade Wind Zone (STWZ), an extensive belt on the northern side of the SHPZ where the winds blow fairly consistently from the easterly quarter but in the western South Pacific may have a more southerly component. Surface wind speeds are normally moderate but may reach 50 km/hr. Clouds are often aligned in bands parallel to the wind direction.

The South Pacific Convergence Zone (SPCZ) is a region of semipermanent clouds where equatorial easterly winds converge with the southeasterly trades of the STWZ. The weather in places like Fiji and Tonga eastward at least to the

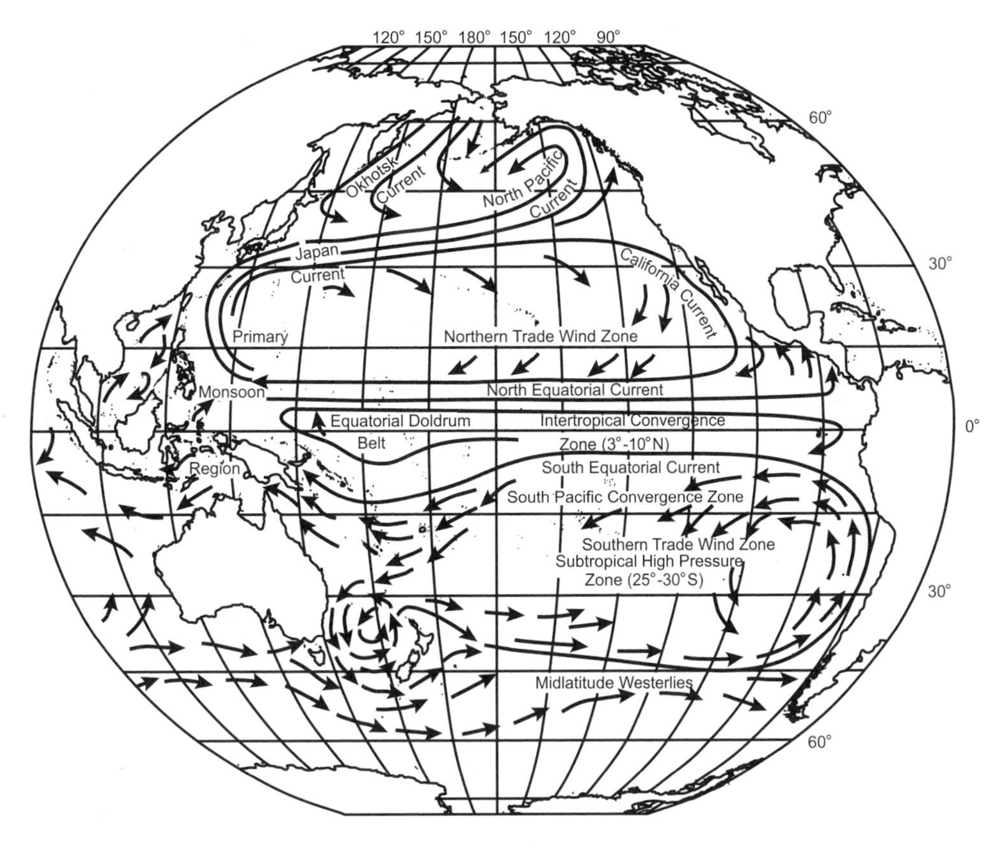

Figure 1-34 Dominant wind and current directions in the Pacific Ocean. Redrawn and partially relabeled from Kirch (2000: Map 4).

Pitcairn Group depends largely on the position and intensity of the SPCZ, which can vary from month to month. During the austral summer or wet season (November to April) the SPCZ becomes an extension of the low pressure "monsoon trough" of northern Australia; then it can bring unsettled weather, heavy rain, and even cyclones (see below). During the dry season or winter (May to October), the SPCZ generally is positioned to the north, and these island groups are affected mainly by the drier southeast trade winds. Variation in the location of the SPCZ is associated with the Southern Oscillation, which represents the pressure differential between the western and eastern regions of the tropical Pacific Ocean (see section below on El Niño/Southern Oscillation).

The Equatorial Doldrum Belt (EDB) is a region of relatively light winds that is present all year in the western Pacific Ocean. Lying within about 5° of the equator, the EDB has high rainfall with seasonal variability. During the austral summer, when the EDB is furthest south, the monsoon trough usually extends from northern Australia into the Coral Sea. During the austral winter, the EDB lies principally in the northern hemisphere, and the monsoon trough is absent.

The Intertropical Convergence Zone (ITCZ) is where the North Pacific trade winds meet the South Pacific trades. Latitudinally similar to the EDB, the ITCZ lies from ca. 3–10° N in the eastern central Pacific, and from ca. 5–15° N near the Americas. The ITCZ is an area of cloud and showers due to a variably intense ascent of air (Augstein 1984). North-south movement of the ITCZ follows the annual latitudinal passage of the sun, with a lag time of about three to four months.

The Northern Trade Wind Zone (NTWZ) extends across most of the Pacific from ca. 10° to 25° N. These trade winds blow fairly reliably from the northeast. Except for the Hawaiian Islands, which are not part of our study area, little land is influenced directly by the NTWZ, which tends to be drier than the ITCZ.

Precipitation and Temperature

Most of this account is taken from Thompson (1986a, b) and Mueller-Dombois & Fosberg (1998). Hail is extremely rare and snow is essentially unknown in Oceania south of the Hawaiian Islands, north of New Zealand, and east of New Guinea. Thus "rainfall" and "precipitation" are

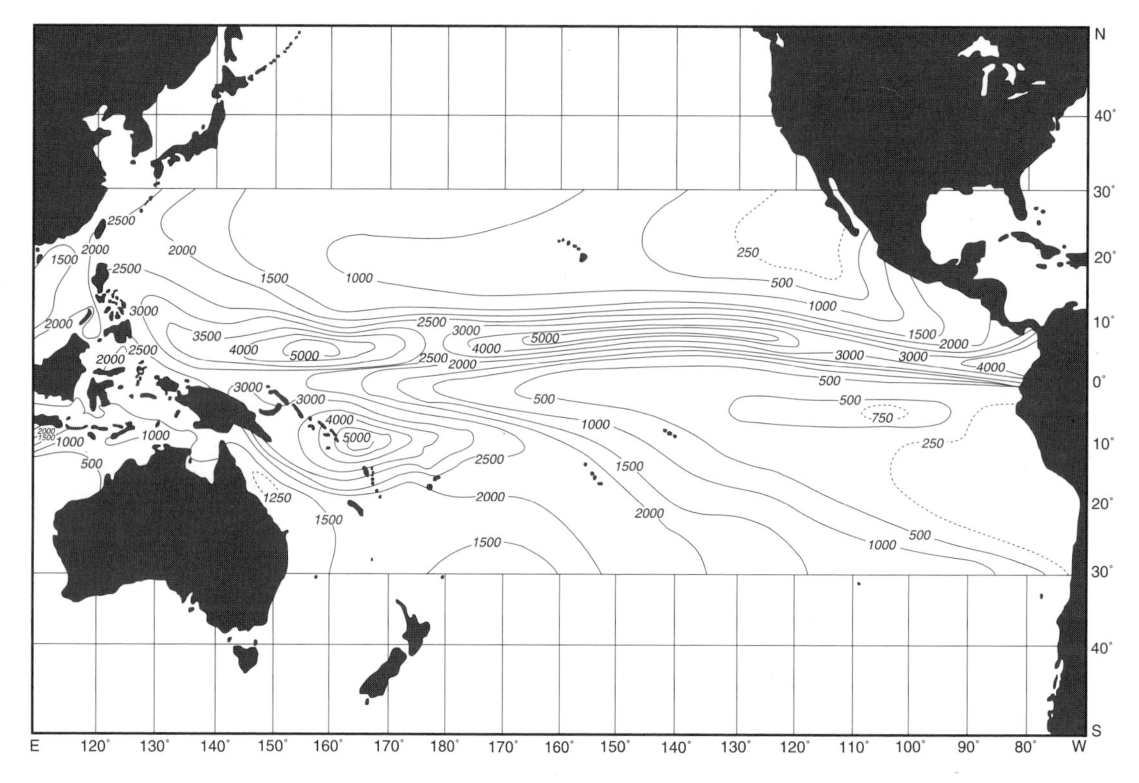

Figure 1-35 Mean annual rainfall at sea-level in tropical Oceania. Redrawn from Taylor (1973: Figure 1).

synonymous. Oceania has some of earth's wettest places. Three very wet regions, with 3500 to 5500 mm of mean annual rainfall at sea level, are: (1) the Caroline Islands from Palau eastward to Kosrae, and then into the southern Marshalls; (2) the northern Line Islands, from Kingman Reef to Fanning Island; and (3) the Solomon Main Chain, the Santa Cruz Group, and northern Vanuatu (Figure 1-35). The south coast or "weather coast" of Guadalcanal (Solomons) receives an astonishing 8000 mm of mean annual rainfall. The western Pacific has no arid zones; among Melanesian islands only New Caledonia (the southernmost) is in an area of moderate rainfall.

Most rainfall in the eastern Pacific occurs in a narrow belt north of the equator. A drier zone lies to the west around the Phoenix Islands, southern Gilberts, and northern Tuvalu, with another wet zone west of this. Particularly over the eastern Pacific, convectional winds precipitate most of their substantial moisture as they rise over the equatorial tropics at the ITCZ. They diverge at higher elevations as anti-tradewinds that form nearly symmetrical patterns in the northern and southern subtropics. Here the winds descend, although little of the remaining moisture is available for precipitation because of adiabatic warming, leaving the eastern Pacific subtropics (SHPZ) relatively dry. From the subtropics the winds are drawn back to the equatorial low-pressure zone as trade winds blowing at low elevations. Because of the Coriolis force, the winds move toward the ITCZ as northeast trades in the northern hemisphere and as southeast trades in the southern hemisphere. As the trade winds blow toward the ITCZ, they pick up moisture from the ocean's warm surface (Figure 1-36). This moisture eventually precipitates, with higher islands receiving heavier, more reliable rainfall than the lower islands. Such a montane (orographic) effect is evident, for example, at Samoa, Rarotonga, and Tahiti, especially on windward slopes.

Seasonality is an important part of Pacific rainfall regimes, with less mean annual precipitation often associated with greater seasonality, i.e., more severe dry seasons. The warm season (approximately November to April in the southern hemisphere, and May to October in the northern hemisphere) usually has more precipitation than the cooler half of the year. In the Cook Islands, for example, roughly two-thirds of the rain falls in the warm season (Table 1-1, Figures 1-37, 1-38). Seasonality can be much more subtle in some parts of Oceania. For example, more than 150 mm of rain may fall in a single "dry season" event in an overall hot, wet place like the Solomons, Samoa, or Palau. When combined with nighttime temperatures in excess of 25°C, terms such as "winter" or "cool season" or "dry season" become

Table 1-1 Seasonal distribution of rainfall (in mm) for seven islands (nine locations) in the Cook Islands and five islands in Tonga, 1951–1980. Wet season is November through April. Dry season is May through October. The increase in precipitation on Rarotonga from the airport to Aroa to Turoa is due to windward effects. Modified from Thompson (1986a: Table 15, 1986b: Table 7).

	Season			% Annual Total	
	Wet	Dry	Total	Wet	Dry
COOK ISLANDS					
Palmerston (18°03′S)	1337	638	1975	68	32
Aitutaki (18°52′S)	1263	617	1880	67	33
Mitiaro (19°51′S)	1185	641	1826	65	35
Atiu (19°59′S)	1336	634	1970	68	32
Ma'uke (20°09′S)	1030	578	1608	64	36
Rarotonga: airport (21°12′S)	1292	729	2021	64	36
Rarotonga: Aroa (21°15′S)	1662	921	2583	64	36
Rarotonga: Turoa (21°16′S)	1706	1094	2800	61	39
Mangaia (21°55′S)	1230	737	1967	63	37
TONGA					
Niuafo'ou (15°34′S)	1575	947	2522	62	38
Niuatoputapu (15°57′S)	1529	822	2351	65	35
'Uta Vava'u, Vava'u (18°39′S)	1555	757	2312	67	33
Lifuka, Ha'apai (19°48′S)	1214	591	1805	67	33
Nuku'alofa, Tongatapu (21°08′S)	1138	676	1814	63	37

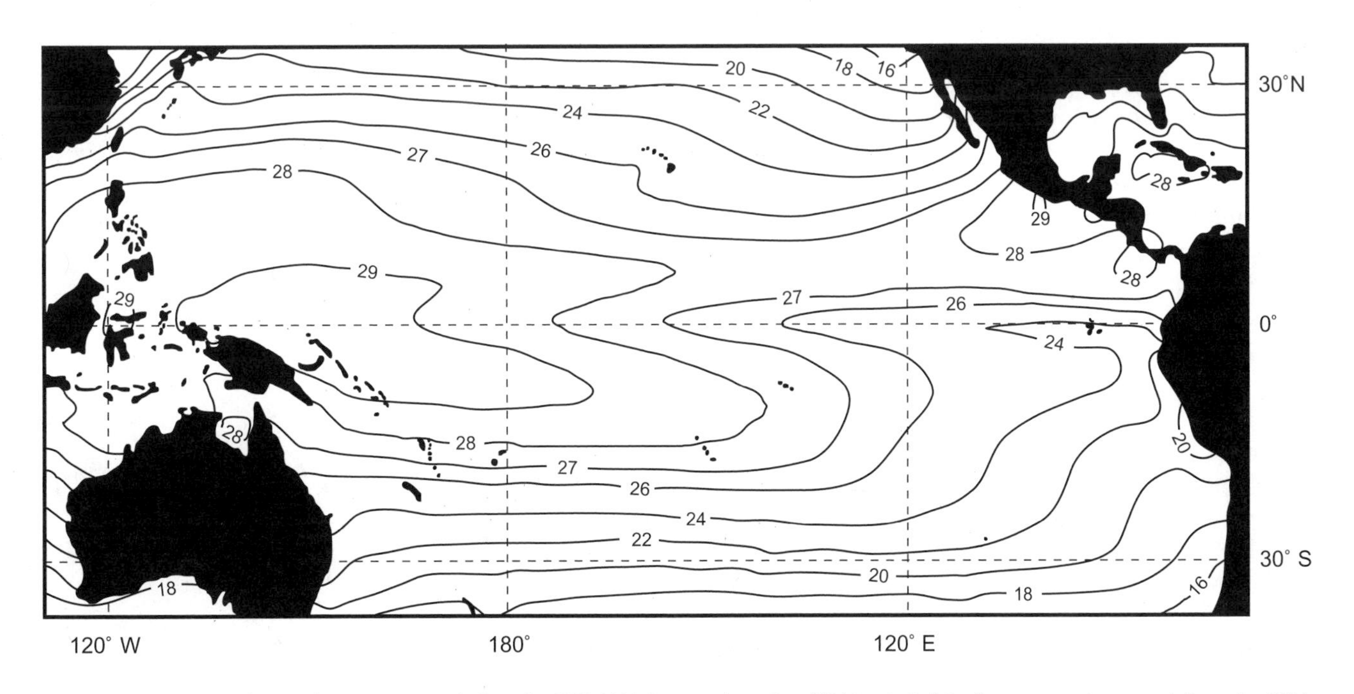

Figure 1-36 Mean annual sea-surface temperatures in Oceania, 1971–2000. Image redrawn from NOAA website (http://www.cpc.ncep.noaa.gov), December 2004.

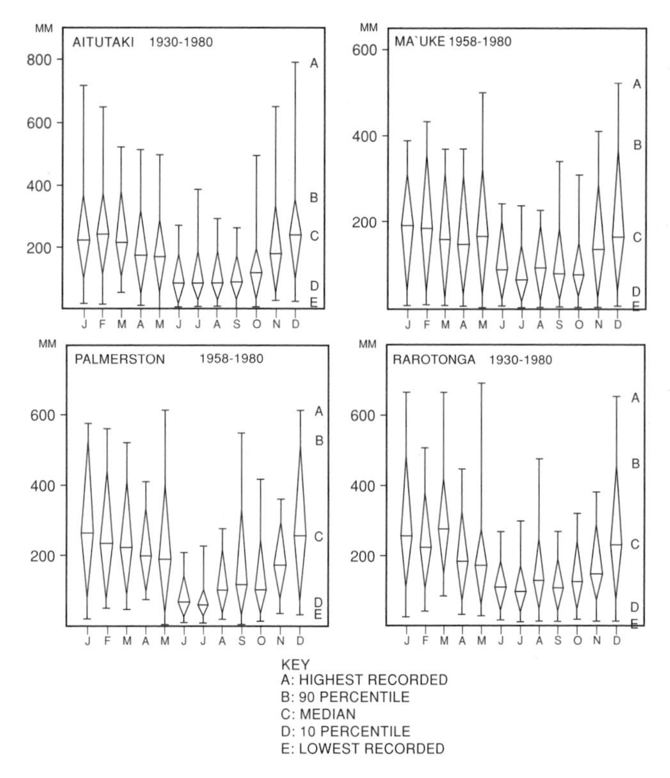

Figure 1-37 Seasonal distribution and variability of rainfall in the Cook Islands. Redrawn from Thompson (1986a: Figure 17).

horrible misnomers to short-term visitors from temperate climates.

Mean annual air temperatures generally increase toward the equator (Table 1-2). Except in equatorial regions with strong rainfall gradients (Figure 1-35), many places in Oceania with the warmest and most aseasonal air temperatures are also the wettest. The interplay of mean annual temperature and rainfall yields a prediction of forest types in Oceania (Table 1-2). These general patterns, based on potential evapotranspiration (Holdridge 1967), are influenced in the real world by seasonality, winds, and edaphic conditions. Chapter 2 will review forest types in Oceania.

Cyclones

Tropical cyclones (also known as hurricanes or typhoons) are severe storms that regularly occur in much of Oceania, especially the western portion (Gray 1984). They usually form between 10° and 25° latitude (north or south) when surface temperatures of the ocean exceed 26°C (Chang 1972). Thus most Pacific cyclones occur in the hot, wet season. These storms tend to track east to west with counterclockwise winds in the northern hemisphere and clockwise winds in the southern hemisphere (Visher 1925). Hurricanes have wind speeds ≥118 km/hr whereas tropical storms have winds of 88-117 km/hr (Kerr 1976). On average worldwide, 45 tropical storms reach hurricane intensity each year, with 30% of them in the western North Pacific (Bengtsson 2001). Island groups particularly susceptible to hurricanes are the Marianas north of the equator, and Fiji, Samoa, and Cook Islands south of the equator.

Cyclones produce short-term and long-term landscape change on Pacific islands and reefs, depending on their strength and duration, proximity to land, and rainfall totals and intensities (Fitchett 1987, Scoffin 1993, Terry & Raj 1999, Nott & Hayne 2001, Terry et al. 2001, pers. obs.). The most dramatic landscape effects of hurricane-force winds, intense rainfall, and temporary storm surges are flooding, landslides, forest blowdown and mortality (snapped stems, uprooting), defoliation, defructation, and coastal degradation or progradation (Lessa 1962,

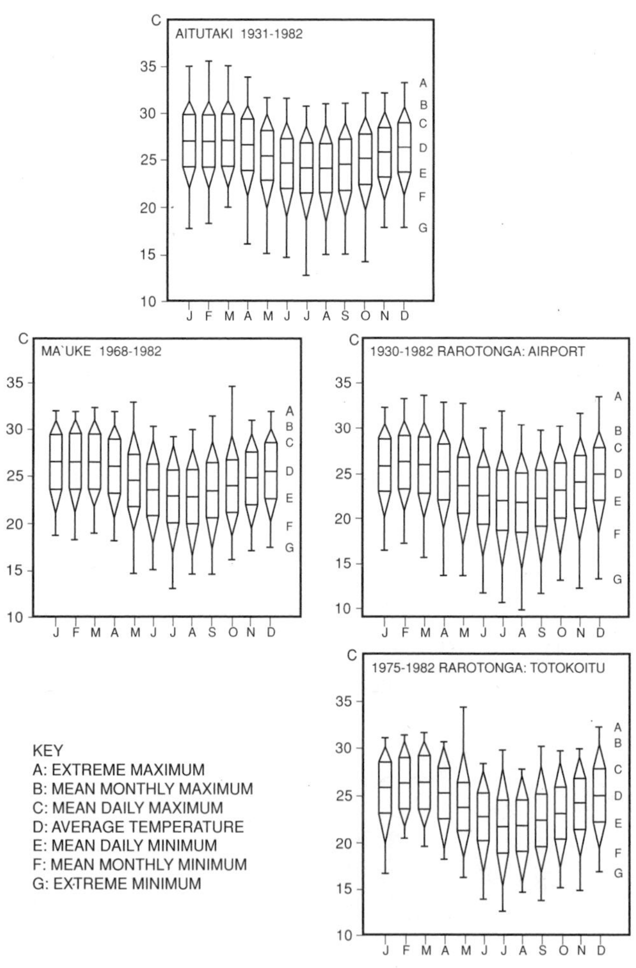

Figure 1-38 Seasonal distribution and variability of rainfall and air temperature in the Cook Islands. Redrawn from Thompson (1986a: Figure 20).

Table 1-2 General climatic conditions, predicted Potential Evapotranspiration (PET), and general vegetation types (Holdridge life zones) in Oceania, oriented north to south. Latitudes and longitudes are rounded to the nearest degree. Expanded and otherwise modified from Mueller-Dombois & Fosberg (1998: Table 1.1), using data from Blumenstock & Rex (1960), Taylor (1973), Walter et al. (1975), Carter (1984), Thompson (1986a, b), Stanley (1989), Karolle (1993), and Spencer (1995).

Thermal belt/latitude	Island, island group	Mean annual temperature (°C)	Mean annual rainfall (mm)	PET[a]	Holdridge life zone[b]
Subequatorial tropics 19°N, 167°E	Wake Island	26.6	1047	2347	dry forest
Equatorial tropics 14°N, 145°E	Guam, Marianas	26.0	2560	917	moist forest
Equatorial tropics 11°N, 162°E	Eniwetok, Marshall Islands	27.9	1470	1839	dry forest
Equatorial tropics 9°N, 138°E	Yap, western Carolines	27.5	3103	846	moist forest
Equatorial tropics 9°N, 167°E	Kwajalein, Marshall Islands	27.6	2578	1026	dry-moist forest
Equatorial tropics 7°N, 134°E	Koror, Palau	27.5	3741	702	moist forest
Equatorial tropics 7°N, 158°E	Pohnpei, Carolines	27.0	4818	525	wet forest
Equatorial tropics 7°N, 171°E	Majuro, Marshall Islands	27.2	3393	757	moist forest
Equatorial tropics 3°S, 172°E	Canton, Phoenix Islands	28.7	509	5620	very dry forest
Equatorial tropics 6°S, 160°E	Kieta, Bougainville, Solomon Islands	27.2	3035	847	moist forest
Equatorial tropics 10°S, 139°W	Hiva Oa, Marquesas	~25	1215	~1786	dry forest
Equatorial tropics 12°S, 167°E	Vanikolo, Santa Cruz	~27	5664	~447	wet forest
Equatorial tropics 14°S, 172°W	Apia, Upolu, Samoa	26.2	2870	831	moist forest
Equatorial tropics 18°S, 177°E	Nadi, Viti Levu, Fiji	25.4	1921	1166	dry forest
Equatorial tropics 18°S, 178°E	Suva, Viti Levu, Fiji	25.1	3027	723	moist forest
Subequatorial tropics 19°S, 174°W	Vava'u, Tonga	26.1	2312	1023	moist forest
Subequatorial tropics 18°S, 168°E	Port Vila, Efate, Vanuatu	24.5	2293	909	moist forest
Subequatorial tropics 21°S, 175°W	Tongatapu, Tonga	25.0	1814	1196	dry forest
Subequatorial tropics 22°S, 166°E	Noumea, New Caledonia	23.4	1064	1787	dry forest
Subequatorial tropics 17°S, 150°W	Papeete, Tahiti, Society Islands	25.0	3101	700	moist forest
Subequatorial tropics 22°S, 160°W	Rarotonga, Cook Islands	~24	2021	~990	moist forest
Subtropical 24°S, 130°W	Pitcairn, Pitcairn Group	~23	1716	~1070	dry-moist forest
Subtropical 27°S, 109°W	Easter Island	20.0	1365	1018	dry-moist forest

[a]PET = Potential Evapotranspiration, calculated according to the formula (mean annual temperature × 58.93)^2/mean annual precipitation (Holdridge 1967).
[b]Prediction of life zones (plant formations) based on Holdridge (1967: Figure 1).

Bayliss-Smith 1988, Gillie 1997b; Franklin et al. 2004, McConkey et al. 2004b).

El Niño/Southern Oscillation

The western tropical Pacific Ocean is normally warmer and wetter than the eastern Pacific because winds and currents build up warm water in the west. During an "El Niño" or "El Niño/Southern Oscillation" (ENSO) event, the eastern equatorial Pacific is abnormally warm and wet (low pressure) while the western tropical Pacific is cooler and drier than normal (high pressure) with a shift in the dominant wind and current from the eastern to western quarter (Henderson-Sellers & Robinson 1986, Thompson 1986a, Meehl 1987). ENSO conditions develop, on average, every four to six years, with varying periodicity, intensity, and duration (see detailed accounts in Hastings 1990, Sturman & McGowan 1999). Planktonic analyses from deep-sea cores in the equatorial Indian and Pacific Oceans suggest that ENSO or ENSO-like phenomena have operated for at least the past 250 ky (Beaufort et al. 2001).

Parts of the Southwest Pacific with highly seasonal rainfall have droughts during ENSO events. Aitutaki (Cook Islands) had only 637 mm of rain from May 1982 through April 1983, compared to an annual mean of 1890 mm (Thompson 1986b). During the 1997–98 ENSO event, after working under cool and dry conditions in the Cook Islands and Tonga, I experienced extreme heat, humidity, heavy rains, and huge populations of nonnative ants and wasps in the normally arid Galápagos Islands (far eastern equatorial Pacific; see Lea 2002, Koutavas et al. 2002). My favorable assessment of the weather on Mangaia (Cook Islands) in May-June 1997 was not shared by the Mangaians themselves, who found the evenings much too cold (15°–18°C). More importantly, Mangaians' taro swamps had dried up, leading to crop failure of their favorite food.

Paleoclimate

Worldwide, climates are believed to have been tropical in the late Cretaceous and much of the Paleocene and Eocene, related to high atmospheric CO_2 levels (Kump 2001, Pearson et al. 2001, Zachos et al. 2001, Pearson & Palmer 2002). In Fiji and Tonga, analyses of fossil foraminiferans suggest tropical marine conditions since at least the late Eocene (ca. 38 Ma), with a slight reduction in ocean temperatures (leading to those of today) beginning in the late Miocene (ca. 10-8 Ma; Chaproniere 1994b, c). Circulation in the equatorial Pacific must have changed in the Pliocene (ca. 5–2 Ma) from formation of the Isthmus of Panama, thereby separating equatorial waters of the Atlantic and Pacific, and constriction (narrowing and shallowing) of the "Indonesian valve" that allows warm Pacific water to flow into the Indian Ocean (Coates et al. 1992, Reynolds et al. 1999, Cane & Molnar 2001, Wright 2001, Gordon et al. 2003). Nevertheless, it seems safe to say that the central and western Pacific Ocean between 20° N and 20° S has been tropical throughout the Cenozoic.

About 22 glacial-interglacial cycles have occurred during the Quaternary, or the past ca. 1.8 My, with the past 0.6 My especially well studied (Figure 1-39). What causes these cycles and their periodicity is a source of much research, speculation, and debate (Crowley 2002, Elkibbi & Rial 2002). Within the tiny part of Earth's history that can be measured by radiocarbon dating (the past 0.04–0.05 My or 40–50 ky), the most profound and best studied interval of climate change is the transition from glacial (late Pleistocene) to interglacial (Holocene) conditions that took place from about 18 to 8 ka.

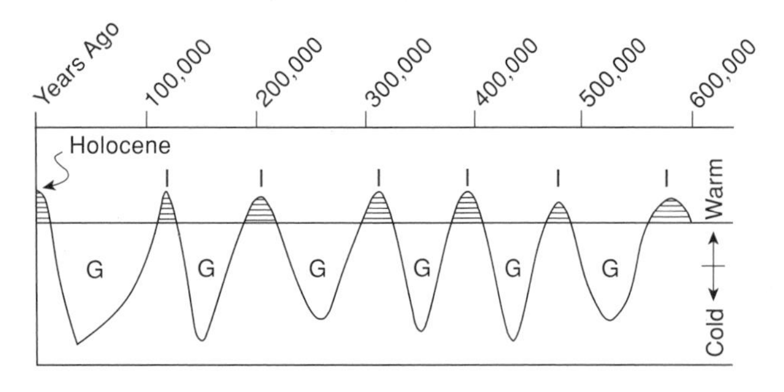

Figure 1-39 Generalized global climatic fluctuations of the past 600,000 years, illustrating glacial (G) versus interglacial (I) cycles. The Holocene represents only the past 12,000 years. Redrawn from Dickinson (2000a: Figure 5).

A very useful method for reconstructing late Quaternary terrestrial climates and vegetation is pollen analysis of well-stratified sediments from lakes or other freshwater wetlands. A limitation of tropical pollen records is that they are biased toward taxa with windblown pollen and against the large number of taxa, including many trees, that are pollinated by animals. In spite of this, pollen analyses have yielded important data on climatic and vegetational changes in Oceania. Holocene (<10 ka) pollen records from Fiji, Tonga, and the Cook Islands relate directly to modern plant communities and will be reviewed in Chapter 2.

Three deposits in tropical Oceania have produced pollen-rich sediments of late Pleistocene age (>10 ka). One is Lac Suprin (elev. 230 m) on Grande Terre, New Caledonia (Hope 1996). The record extends from ca. 60 to 10 ka, with the coldest (no tropical storms) and driest (increased fire frequency) conditions during the height of northern hemisphere glaciation (24 to 18 ka). *Nothofagus* (southern beech) forests were replaced to varying degrees by mixed forests (diverse angiosperms and gymnosperms) during times of low fire frequency. The other two late Pleistocene pollen data sets are from Taveuni and Viti Levu, Fiji (Hope 1996). Neither discloses evidence for a drier late Pleistocene climate, although temperatures may have averaged a few °C cooler.

From four areas adjacent to tropical Oceania, late Pleistocene pollen data help to reconstruct regional vegetation patterns and terrestrial climates. First, on mainland New Guinea during full-glacial times (20–18 ka), alpine snowlines were depressed ca. 1000-1100 m compared to today (Hope & Tulip 1994, Hope & Golson 1995, Hope 1996). Similarly, the transition from subalpine and upper-montane forest was depressed ca. 900 m, and that from lower-montane to lowland forest by ca. 600–780 m. These elevational changes in vegetation types suggest that temperatures on mainland New Guinea at 18 ka were 4–7°C cooler than modern, with the amount of change increasing with elevation. Second, from warm-temperate New Zealand (North Island; 34°–37° S) at 18 ka, the far north remained primarily forested (as it would be today if not for people) whereas the southern half of the island (at that time connected to the South Island by lowered sea levels) was a grassland and shrubland in a region generally forested in the Holocene. This argues for relatively dry conditions with mean temperatures 4–5°C cooler than present (Dodson et al. 1988, McGlone 1988, Markgraf et al. 1992).

Subtropical Easter Island (27° S) is the southeastern corner of Oceania. Except at high elevations, Easter Island had a palm-dominated forest with a shrub layer of up to eight taxa in both the late Pleistocene and Holocene (until human arrival; see Chapters 2, 3, 7), although a more open (less shrubby) lowland forest from ca. 21 to 12 ka may reflect cooler and drier conditions (Flenley & King 1984, Flenley et al. 1991). Finally, from the Hawaiian Islands (18–22° N, just north of our study area), vegetation zones were depressed elevationally in the late Pleistocene (Burney et al. 1995, 2001, Athens 1997, Hotchkiss & Juvik 1999). The climate was probably cooler than today, and drier in montane regions but wetter in the lowlands.

The pollen records from New Zealand, Easter Island, and the Hawaiian Islands lack indication of human presence until just the past 1.6 to 1.0 ky, whereas evidence of people (such as increased charcoal influx and declines in pollen from forest taxa) is found in New Guinea at least as early as 32.5 ka (Hope 1998). Prehistoric human-caused changes in vegetation will be covered in Chapters 2, 3, 16, and 21.

Although Quaternary temperature fluctuations tended to be less in the tropics than at higher latitudes (Clark et al. 1999, Vimeux et al. 1999), tropical Oceania was cooler and probably drier during the late Pleistocene than during the Holocene. Marine data also argue that the low-latitude Pacific Ocean was cooler than present during glacial intervals but remained tropical (de Garidel-Thoron et al. 2005). Marine data should be a decent proxy for terrestrial conditions because >95% of the earth's surface here is ocean rather than land; regional climate is controlled primarily by the hydrosphere.

Glacial-interglacial differences in moisture regimes (precipitation/evaporation) are more difficult to track with marine proxy indicators than glacial-interglacial changes in temperature. Proxy data, such as ratios of isotopes or elements that are known to vary today with temperature, can be used as paleothermometers, although more research is needed to increase resolution and to understand the limitations of these seemingly sound methods (Cohen et al. 2002). The ratios include $^{18}O/^{16}O$ and Mg/Ca in the shells of planktonic foraminifera, and $^{18}O/^{16}O$, Sr/Ca, Mg/Ca, and U/Ca in the skeletal aragonite of corals, which can be dated using either ^{14}C or U-series (Beck et al. 1997, Gagan et al. 2000, Lea et al. 2000, Steig 2001, Tudhope et al. 2001). Periods of time where $^{18}O/^{16}O$ data indicate relatively cool or relatively warm conditions are classified as a "Marine Isotope Stage" or MIS, numbered consecutively back in geological time with MIS 1 being the Holocene.

Estimating sea surface temperatures (SSTs) is crucial in understanding past climates. In low-latitude sediment cores in the far eastern (90° W) and western (160° W) Pacific,

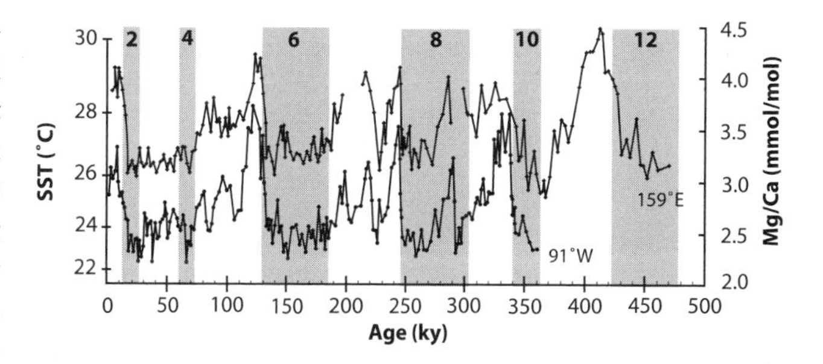

Figure 1-40 Equatorial sea-surface temperature estimates based on oxygen isotopes of the planktonic foraminiferan *Globigerina ruber*. Shaded, even-numbered intervals are Marine Isotope Stages (see text and Figure 1-41). Redrawn from Lea et al. (2000: Figure 3).

$^{18}O/^{16}O$ and Mg/Ca data from planktonic foraminifera suggest that SSTs were 2–3°C cooler in the Last Glacial Maximum (LGM; 24 to 18 ka) than at present (Lea et al. 2000, Stott et al. 2002), with similarly cool SSTs during each of the six glacial intervals since 500 ka (MIS 12, 10, 8, 6, 4, and 2; see Figure 1-40). That the cooling or warming of equatorial surface water preceded the formation and demise of continental ice sheets by 3–5 ky hints at a causal link (Nürnberg 2000, Seltzer et al. 2002).

Analyses of pollen, spores, organic matter, and lithology of a sediment core at 134° E, 3° N suggest cooler, dryer conditions during glacial intervals in the far western Pacific (Kawahata et al. 2002). Sr/Ca and U/Ca ratios in fossil corals, and Mg/Ca ratios in fossil foraminifera from sites in Papua New Guinea, Australian Great Barrier Reef, Vanuatu, and Tahiti suggest that SSTs were depressed by 4–6°C during deglaciation (20–12 ka), then warmed to roughly modern temperatures by 10 to 6 ka, and were even 1°C warmer than modern at 5.5 ka (Barash & Kuptsov 1997, Beck et al. 1997, Gagan et al. 1998, 2000, Visser et al. 2003). Even during glacial times, the low-latitude western Pacific remained warm enough to sustain reef-building corals and other tropical marine invertebrates (Cabioch et al. 2003). By inference, the late Pleistocene terrestrial conditions also were tropical, even if somewhat cooler, at least at low elevations.

Long-term Changes in Sea Level

Variation in sea level has had a profound effect on the shape, size, elevation, and even existence of Pacific islands. Low sea levels can expose seamounts to become islands, coalesce clusters of nearby islands to become a single large island, change atolls to raised limestone islands, or expose reef flats to convert an eroded volcanic island into a composite island with a girdle of coralline limestone. Higher sea levels can drown atolls and sand cays to become shallow reefs rich in marine life but devoid of terrestrial plants or animals. From the perspective of an individual island, sea level is relative and is due, at any given time, to: (1) global changes in sea level related to how much of Earth's water is tied up in ice; (2) regional geological processes that may cause uplift (emergence) or sinking (subsidence) of an island or island group (see, for example, Spencer et al. 1987); (3) local geological and biological processes (reef-building corals, beach progradation, upland erosion, etc.); and (4) thermal expansion of seawater. I will discuss these processes now, citing well-studied examples from Oceania.

An ever-changing sea level is one of the most dramatic and important outcomes of the climate change associated with Quaternary glacial-interglacial cycles. Sea levels go down as ice builds up during glacial intervals; sea levels rise when the ice melts (Cuffey & Marshall 2000, Hvidberg 2000). Polar ice caps are the main reservoirs of ice, so their waxing and waning have the most profound effect on sea levels (Clark et al. 2002, Sabadini 2002). Alpine

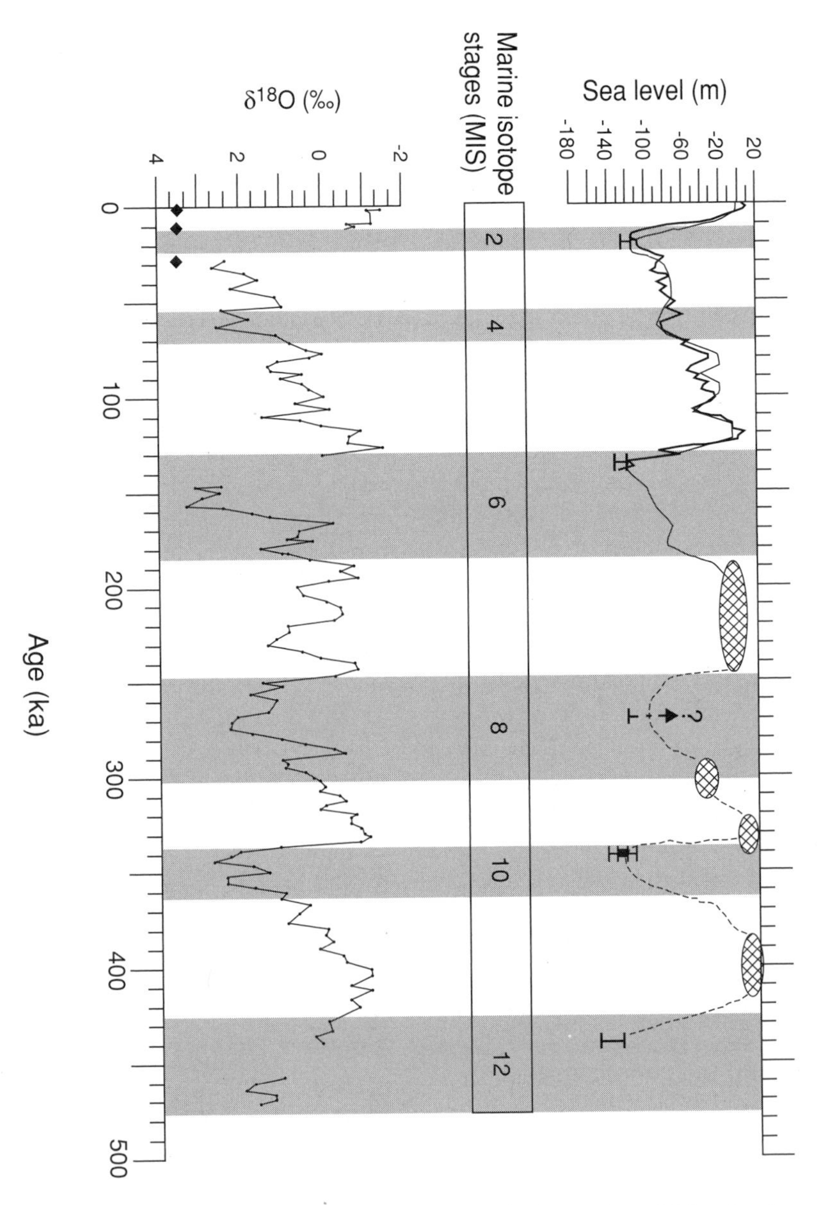

Figure 1-41 Global sea-levels for the past 500,000 years. Marine Isotope Stages (MIS) are based on changing $^{18}O/^{16}O$ ratios of foraminiferan shells in marine sediments, beginning with MIS 1 (the Holocene) and continuing numerically back into time. The even-numbered MISs indicate glacial conditions; the odd-numbered ones portray interglacials. Redrawn from Rohling et al. (1998: Figure 1).

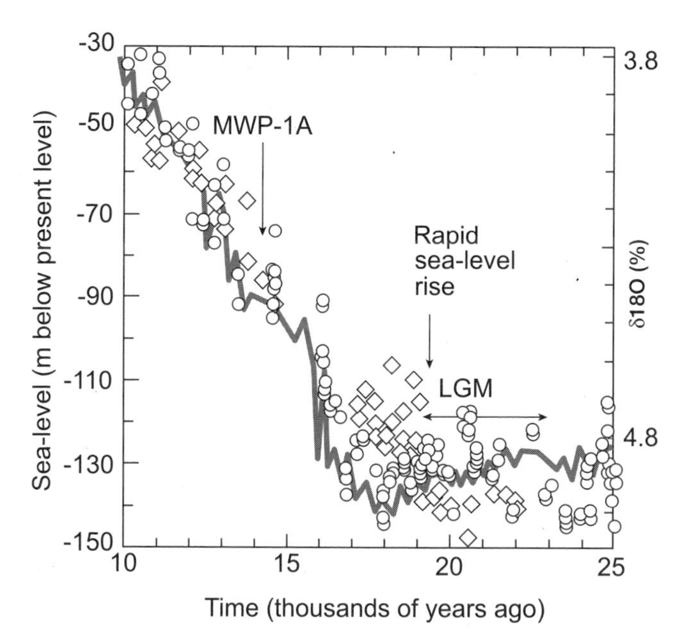

Figure 1-42 Sea-level estimates for 25–10 ka compared to marine oxygen isotope and atmospheric CO_2 records. Each circle or diamond represents a dated coral. Redrawn from Clark & Mix (2000: Figure 1).

glaciers also contribute, but to a much lesser extent. On a global scale, the buildup of polar and alpine ice is roughly in synch.

Over the past 800 ky, an average glacial-interglacial cycle consumed 100,000 years (Lambeck et al. 2002), with highly variable but relatively cool glacial conditions (such as extensive ice sheets over Canada and northern Europe) occupying as much as 90% of that time, and interglacial conditions (such as those we have today) taking up the remaining time (Peltier 1994). The younger the glacial or interglacial interval, the more intensively it can be studied because of multiple dating methods, better proxy indicators, and more sites available.

During the past 500 ky, the coldest glacial event peaked at ca. 450 ka (MIS 12) with an estimated drop in sea level of 140 m below modern (Chappell 1998, Rohling et al. 1998). Other sea-level minima (in the range of −120 to −100 m) occurred at ca. 350 ka, 270 ka, 85 ka, and 20 ka (Figure 1-41). Conversely, sea-level maxima (interglacial peaks) happened at ca. 400 ka, 330 ka, 220 ka, 120 ka, and present. The highest, at 400 ka (MIS 11), is estimated to have reached +20 m (Kindler & Hearty 2000). Assuming little or no uplift or subsidence from geological or biological processes, the MIS 11 highstand would have inundated hundreds of Pacific islands.

Because it is well studied and so pertinent to biogeography, let us focus on the last interglacial-glacial-interglacial triplet. A sea level of ca. +4 to +6 m at 120 ka is expressed in limestones of Oceania as emergent reef flats and shoreline notches, the current elevation of which must be evaluated in terms of local and regional tectonics (Stoddart et al. 1985, Pirazzoli & Salvat 1992, Dickinson 1998b, McCulloch et al. 1999). Also, just as ice loading depresses the earth's crust at high latitudes during glacial times (Milne et al. 2001), the increased depth of seawater during interglacials (whether 120 ka or today) depresses crust below the oceans (Peltier 1994, Dickinson 1998b). In spite of these and other sources of imprecision, a number of studies worldwide has produced evidence of the 120 ka highstand, now known as substage 5e of MIS 5 (Smart & Richards 1992, Rohling et al. 1998, Hvidberg 2000). Higher than any sea level of the Holocene (MIS 1), the MIS 5e highstand undoubtedly drowned many Pacific atolls, although not as many as during the even warmer 400 ka highstand (MIS 11).

Our next focus is the Late Glacial Maximum (LGM), the period (24 to 18 ka, with maximum intensity at 22 to 19 ka) during the last glacial interval with greatest global ice volume and lowest sea level (Yokoyama et al. 2000). This corresponds to MIS 2, which featured overall glacial conditions until ca. 12-10 ka. (I note here that the last glacial interval, which began with sudden cooling at ca. 115 ka, consists of MIS 5d-5a, 4, 3, and 2, all cooler than present

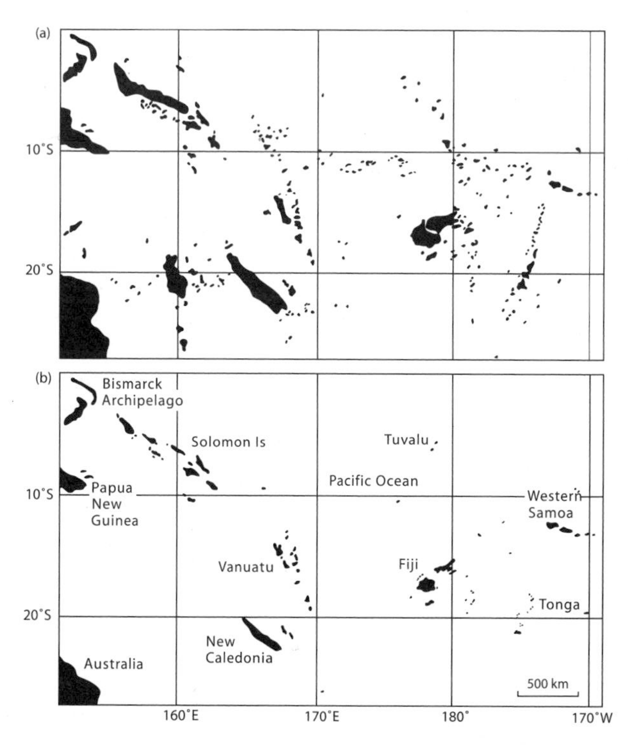

Figure 1-43 A rough comparison of southwest Pacific islands at late glacial maximum (ca. 22–19 ka; a) versus today (b). Redrawn from Gibbons & Clunie (1986: Figure 3).

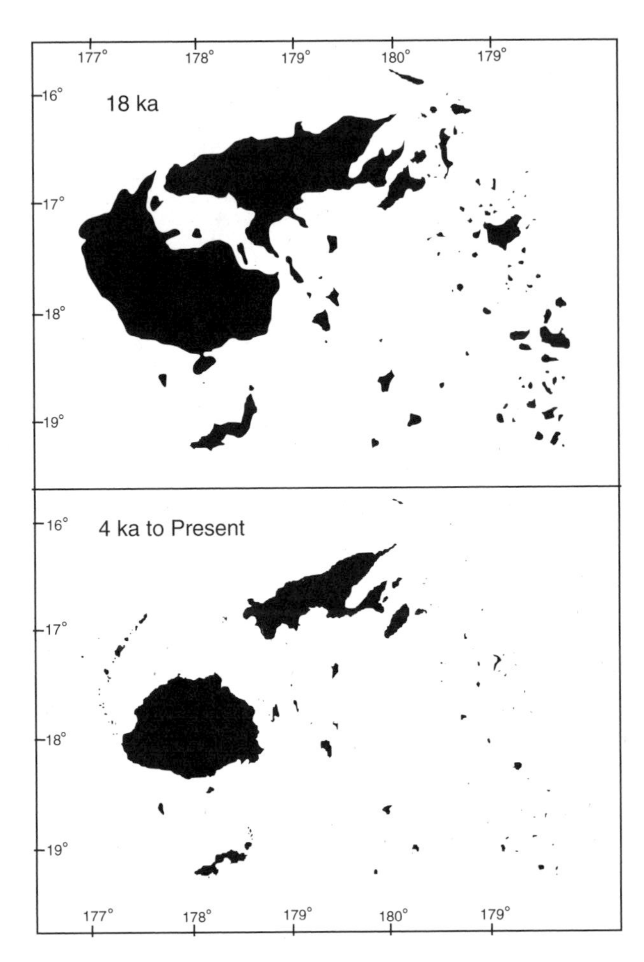

Figure 1-44 A rough comparison of Fiji at late glacial maximum (ca. 22–19 ka) versus today. Redrawn from Gibbons & Clunie (1986: Figure 2).

although MIS 4 (ca. 70–55 ka) and MIS 2 (ca. 25–12/10 ka) were the coldest (Rohling et al. 1998). MIS 3 was not warm enough to be classified as an interglacial. This and other somewhat warmer periods within glacial intervals are called "interstadials."

Sea level at the LGM was as low as -120 to -140 m (Clark & Mix 2000, 2002, Hanebuth et al. 2000, Yokoyama et al. 2000; Figure 1-42 herein), with major impacts in Oceania such as joining some islands in the Solomons and western Fiji, exposing much land now inundated as reefs between northern Australia and New Caledonia, coalescing all 58 islands in the Vava'u Group of Tonga into a single island, and converting hundreds of atolls into raised limestone islands (Gibbons & Clunie 1986, Nunn 1994, Spriggs 1997; Figures 1-43, 1-44 herein). On the other hand, steep subsurface contours and a narrow or absent fringing reef meant that changes in island area and isolation were minor for many active volcanic islands, eroded volcanic islands, and raised limestone islands, especially those on the Pacific Plate. Even a sea level 120–140 m lower than today would not bring drastic changes, for example, to island areas or inter-island distances in the southern Cook Islands or the Marquesas, an exception being that Tahuata would probably join nearby Hiva Oa (Figure 7-23).

After starting to rise at 19 ka, sea level reached −100 m at 15 ka, −80 m at 13 ka, −60 m at 12 ka, and −40 m at 10 ka (Figure 1-42). Sea level reached that of today by 6 ka and then rose another meter or so by 5.5–5 ka (Cabioch et al. 1999; Figure 1-45 herein). This highstand (variously estimated at from +1 to +2.6 m; Dickinson 1999a, 2003) was maintained until ca. 2.5 ka, since which time sea level has been fairly constant except for a small rise in the past century. The rising sea level in the early to mid-Holocene probably outpaced the vertical rate of coral growth on many atolls, meaning that these islands would have been inundated for at least several millennia (Neumann & MacIntyre 1985, Paulay 1997, Cabioch et al. 1999). Terrestrial life on such atolls has colonized only within the past few thousand years, which explains the lack of endemic terrestrial species.

The mid-Holocene (5.5 to 2.5 ka) high sea-level stand is much more studied because: (1) its indicators (emergent reef flats, microatolls, beachrock, and shoreline notches; see Dickinson 1998b, Grossman et al. 1998) are relatively unweathered compared to those of earlier highstands;

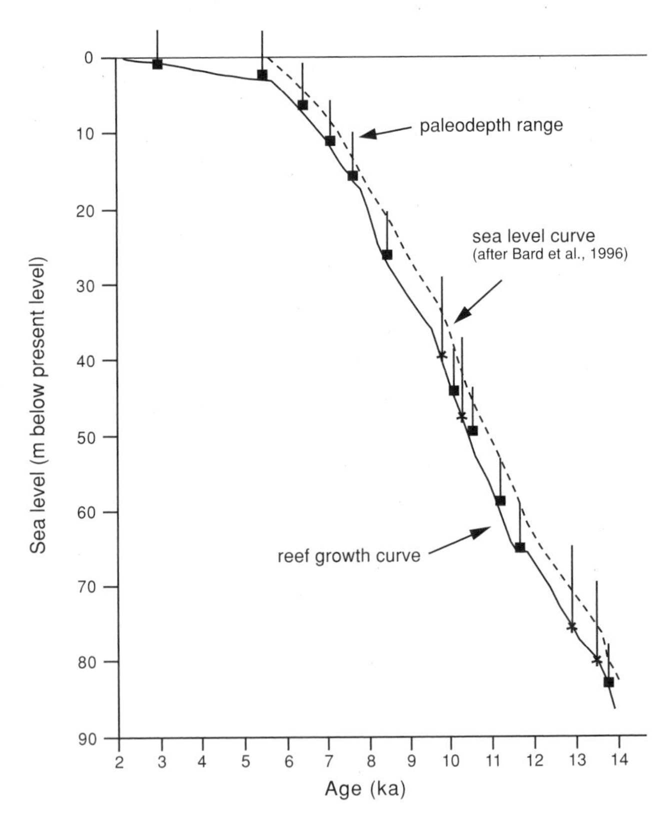

Figure 1-45 Sea-level estimates for 14 to 2 ka. Redrawn from Cabioch et al. (1999: Figure 5).

(2) radiocarbon dating can be used to determine ages of emergent corals in life position; (3) the relatively small amount of elapsed time minimizes the potential effects of lithospheric flexure (regional/local emergence or subsidence); and (4) in much of Remote Oceania, this is when people first arrived (Dickinson 1998a, 2003, Dickinson et al. 1999). An island settled at 3 ka is more likely to preserve archaeological evidence of this event if the island has been tectonically stable or uplifted. On islands with a subsiding Holocene history, such as most of them on the Pacific Plate, early archaeological sites may be drowned (Kirch 1986, Dickinson & Green 1998, but see Dickinson 1998b).

Given that "first contact" archaeological sites in Oceania often contain abundant bones of extinct vertebrates (especially birds) of great biogeographical and evolutionary interest, a mere mention of the complex interplay among changing sea levels, carbonate geomorphology, archaeology, comparative osteology, and vertebrate extinction seems like a great place to end this chapter. Having already covered the first two topics, archaeological and comparative osteology will get some attention in Chapters 3 and 4, after we put some plants and animals on the islands in Chapter 2. The fifth topic, extinction, will infiltrate if not dominate the last 18 chapters.

Chapter 2 Terrestrial Flora and Fauna

The birds of Oceania live alongside fascinating assemblages of plants and animals. On tropical Pacific islands, woody plants provide, directly or indirectly, most food and nest sites for indigenous landbirds, as well as the structure of upland terrestrial habitats. Because birds dominate the rest of the book, I will focus the zoological part of this chapter instead on nonavian terrestrial vertebrates, with a cursory look at a few of the better known groups of nonmarine invertebrates.

The marine environment—such as coral reefs uplifted to become limestones that sustain terrestrial plants and animals, or sea level rising to obliterate terrestrial life on a low-lying island, or the marine control of climates across Oceania (Chapter 1)—profoundly affects terrestrial life. Because of the marine environment, there are marine gastropods that have become secondarily aquatic (Meyer et al. 2000) and seabirds that feed on marine life but nest on land, where they contribute nutrients (Chapter 15). Nevertheless, sticking to my landlubber tendencies, I will not review marine biogeography of the Pacific, an enormously interesting field loaded with important new discoveries (Whatley & Jones 1999, Bouchet et al. 2002, Paulay 2003, Paulay & Meyer 2003).

Isolation

Dispersal has played a far larger role than vicariance in the distribution of terrestrial life in Oceania, where only New Caledonia, New Zealand, and tiny Lord Howe Island have any former continental (Gondwanan) connections. All other tropical Pacific islands have always been islands, and thus require dispersal for biotic enrichment. (Remember that, as a nesophile, I consider New Guinea a continent.) Terrestrial biotas in Oceania are influenced by the island's isolation from source areas (usually other islands), size, age, climate, and geological and edaphic environment. Of these attributes, perhaps none is more important than isolation. Evidence for this is found in the most outstanding feature in Oceania's terrestrial biogeography—the overwhelming dominance of paleotropical (Old World) rather than Neotropical (New World) species.

Pure isolation is an effective barrier to colonization. Even though prevailing winds and currents in the tropical Pacific are from the east (Figure 1-34), the Neotropical element in most groups of plants and animals in Oceania, including birds, is nil. This reflects how much closer the islands are to New Guinea or Australia than the New World tropics (Figure 1-1). The thousands of kilometers of deep ocean in the tropical eastern Pacific, at most latitudes unbroken by islands, have been an effective isolating agent even for many groups of marine organisms with pelagic larvae (Kay 1999).

Interisland and interarchipelago distances change as sea level fluctuates. Even with sea level rising 120-130 m since the last glacial maximum (LGM) at ca. 22-18 ka (Chapter 1), however, many changes in interisland and interarchipelago distances were rather minor. Over-water distances no greater than 198 km need to be traversed today to reach all major islands in Near Oceania (Bismarcks including Mussau and the Admiralty Group, and the Solomon

Main Chain) from New Guinea (Table 2-1). This distance would be reduced only slightly to 190 km at the LGM. In either case, colonization could begin with a relatively short (44–46 km) dispersal from New Guinea to Umboi. If we exclude Mussau and the Admiralties, and plot a more direct course through the Bismarcks involving Umboi, New Britain, New Ireland, and nearby islands, then islands in the Bismarcks can be reached by crossing just 58 km (today) or 57 km (LGM) of ocean. Getting to the Solomon Main Chain, by island-hopping from New Ireland, would require a crossing of at least 157 km even at the LGM. Going from the Solomons to Vanuatu and New Caledonia (via the Santa Cruz Group) requires crossing at least 395 km of coean today and nearly as much (365 km) during the LGM.

The distance between Vanuatu to Fiji was 310 km less in the LGM than today, but still involved 500+ km of ocean. Fiji's Lau Group provides stepping-stones between the large islands of western Fiji and those of Tonga. Samoa is rather isolated from Fiji (at least 500 km at LGM) but is fairly close to Tonga when the northern Tongan volcanic outliers (themselves 230 km from the Vava'u Group of Tonga) are considered.

Two places where interarchipelago distances were greatly reduced during the LGM are New Caledonia from Australia (1280 vs. 110 km) and the southern Cooks from Tonga (1520/1160 vs. 700/520 km). The former explains the Australian influence in the New Caledonian avifauna (Chapter 5) but the latter, which represents the break between West and East Polynesia, was not enough to result in strong Tongan affinities in Cook Island birds (Chapter 7). Even at LGM, interarchipelago distances were substantial within East Polynesia, varying from no less than 280 km (Pitcairn from Tuamotu) to as much as 1280 (Easter from Pitcairn). Distances required to reach all islands within a group also average greater in East Polynesia than in West Polynesia or Melanesia. This is especially so when only high islands are considered.

The general eastward trend through Melanesia, West Polynesia, and East Polynesia is one of reduced floral and faunal diversity at all taxonomic levels (Paulay 1994). Although the west-to-east decrease in diversity in most higher taxa would argue that this has been the general direction for dispersal, there is no reason why east-to-west, north-to-south, or south-to-north colonization cannot occur. The circular winds of a typhoon can move plants or animals in any direction, and we should recall that the winds blow mainly from the east. To name just one example of modern distribution unlikely to be due only to west-to-east dispersal, the Pacific's most widespread species of flying fox (fruit bat), *Pteropus tonganus*, occurs from Karkar and Koil (off New Guinea's northern coast) eastward to the Cook Islands (Koopman & Steadman 1995, Flannery 1995a:295–296, Bonaccorso 1998:148). It probably originated in the Fiji-Tonga-Samoa region, from which it island-hopped ca. 1200 km to the east and up to 4000 km to the west, with individual ocean crossings of at least 530 km and probably much greater.

The closest Micronesian island group to continental areas is Palau. Depending on which of its islands are included, Palau now is 280 to 845 km away from the Philippines, Moluccas, or New Guinea (Table 2-2), the three areas suggested as source areas for the Palauan herpetofauna (Crombie & Pregill 1999). These distances are reduced by only 15–20 km during the LGM. Yap is nearly 300 km from Palau regardless of sea level. Although island-hopping from the northern continental island of Japan (via the Ryukyu and Volcano Islands) is geographically feasible in the Marianas and may account for the regular arrival there of Asian migratory ducks and shorebirds, the resident biota of the Marianas is oceanic rather than continental. Moving eastward, the high islands of Chuuk, Pohnpei, and Kosrae are very isolated from each other as well as from the Marianas and Yap. This high level of interarchipelago isolation continues throughout the Remote Central Pacific Islands, which are dominated by atolls and where within-group interisland distances are extremely variable.

On time scales much greater than that associated with the last glacial interval, tectonic movements in Oceania (Chapter 1) have affected biotic distributions. Briefly, islands on or near the Indo-Australia Plate are classified as part as the Inner Melanesian Arc (IMA) or the Outer Melanesian Arc (OMA). IMA islands consist wholly or in part of rifted continental fragments of Gondwana (Kroenke 1996, Dickinson & Shutler 2000). The IMA includes New Caledonia, eastern New Guinea, Lord Howe Island, and the Norfolk ridge (and island) south to New Zealand (Figures 1-1, 1-7). The OMA, believed to have developed primarily or only since the mid-Tertiary, includes the Bismarcks, Solomons, Vanuatu, Fiji, and Tonga, regions that always have been islands (Dickinson & Shutler 2000, Dickinson 2001). Floral and faunal similarity is high within the OMA, although species richness and taxonomic diversity generally decrease from west to east, with the Bismarcks and Solomons more rich and diverse than Vanuatu, Fiji, or Tonga (see Keast & Miller 1996, and the rest of this chapter).

Biogeographic Regions

The distribution of plants and animals in Oceania has fascinated biogeographers for nearly two centuries. Numerous

Table 2-1 Interarchipelago and interisland distances (in km) in Oceania (Southern Hemisphere only) under modern conditions and during the late-glacial maximum (LGM) at ca. 22–18 ka. These values vary in accuracy and resolution; many are rounded to the nearest 5 or 10 km. Values for LGM consider land then exposed but now submerged. Data measured from these maps and hydrographic charts: Kingdom of Tonga (1969, no date), Mammerickx et al. (1973), Archipel des Nouvelles-Hébrides (1976), Île Mangareva (1976), Tahiti (1977), Bier (1980, 1995), Fiji Islands (1980), Archipel de la Société (1988), Tonga Islands (1990), Santa Cruz Islands (1993), Vanuatu (1994), Bismarck Archipelago and Solomon Islands (1995), Îsles Marquises (1995), Wuvulu Island to Kaniet Islands (1995), Ysabel Channel (1995), Long Island to the Tami Islands (1996), Manus Island and Approaches (1996), Solomon Islands (1996), Western Samoa (1996), Asia West Pacific (no date), Papua New Guinea (no date).

	Minimum interarchipelago distance		Minimum interisland distance to reach all major islands in the archipelago	
Archipelago	Modern	LGM	Modern	LGM
Bismarcks[1] (from New Guinea)	46	44	198	190
Bismarcks[2] (from New Guinea)	46	44	198	190
Solomon Main Chain (from Bismarcks)[1]	174	157	88	41
Rennell & Bellona (from Solomon Main Chain)	280	278	36	35
Santa Cruz[3] (from Solomon Main Chain)	395	365	98	96
Vanuatu (from Santa Cruz)	150	120	102	100
New Caledonia (from Vanuatu)	235	233	106	98
New Caledonia (from Australia)	1280	110	106	98
Fiji[4] (from Vanuatu)	840	530	73	40
Fiji[5] (from Vanuatu)	840	530	105	100
Rotuma (from Viti Levu, Fiji)	540	430	—	—
Tonga[6] (from Fiji)	330	325	87	59
Tonga[7] (from Fiji)	330	325	234	233
Wallis & Futuna (from Fiji)	250	245	230	229
Samoa[8] (from Fiji)	760	500	92	91
Samoa[8] (from Tonga)[7]	275	220	92	91
Samoa[9] (from Fiji)	760	500	290	288
Samoa[9] (from Tonga)[7]	275	220	290	288
Niue (from Tonga)[6]	425	422	—	—
Southern Cooks[10] (from Tonga)[6]	1520	700	205	204
Southern Cooks[10] (from Niue)	1060	700	205	204
Southern Cooks[11] (from Tonga)[6]	1160	520	350	348
Southern Cooks[11] (from Niue)	710	520	350	348
Tubuai (from Southern Cooks)[11]	320	318	550	370
Society[12] (from Southern Cooks)[11]	700	698	134	132
Society[13] (from Southern Cooks)[11]	510	508	180	178
Tuamotu (from Society)[12]	180	178	186	184
Tuamotu (from Society)[13]	180	178	186	184
Marquesas (from Tuamotu)	480	478	84	81
Pitcairn (from Tuamotu)	380	280	335	334
Easter (from Pitcairn)	1280	1280	—	—

[1]Without Mussau or the Admiralty Group.
[2]With Mussau and the Admiralty Group.
[3]Without Tikopia, Anuta, or Fatutaka.
[4]Without Rotuma or Ono-i-lau.
[5]With One-i-lau but not Rotuma.
[6]Without volcanic outliers of Niuatoputapu, Tafahi, Niuafo'ou, or 'Ata.
[7]With all volcanic outliers.
[8]Without Swains or Rose Atoll.
[9]With Swains and Rose Atoll.
[10]Without Palmerston Atoll.
[11]With Palmerston Atoll.
[12]Without Mehetia, Maiao, Motu One, Manuae, or Maupiha'a.
[13]With Mehetia, Maiao, Motu One, Manuae, and Maupiha'a.

Table 2-2 Interarchipelago and interisland distances (in km) in Oceania (mainly Micronesia, mainly Northern Hemisphere) under modern conditions and during the late-glacial maximum (LGM) at ca. 22–18 ka. Tuvalu, Tokelau, and the Northern Cook Islands are Polynesian but are considered alongside Micronesia in Chapter 8 as "Remote Central Pacific Islands." Many values rounded to nearest 5 or 10 km. Values for LGM consider land then exposed but now submerged. Data measured from these maps and hydrographic charts: Tokelau Islands (1969), Mariana Islands (1975), Tol (1983), Trust Territory of the Pacific Islands (1985), Mindanau to Palau Islands (1989), Gilbert Islands to Tuvalu Islands (1990), Kosrae Island to Ngatik Atoll (1990), Bier (1995), Palau Islands (1996), Asia West Pacific (1996), Kanton Island (1998).

	Minimum interarchipelago distance		Minimum interisland distance to reach all major islands in the archipelago	
Archipelago	Modern	LGM	Modern	LGM
Palau[1] (from Philippines)	845	825	32	9
Palau[1] (from Moluccas)	765	745	32	9
Palau[1] (from New Guinea)	810	795	32	9
Palau[2] (from Philippines)	670	655	264	261
Palau[2] (from Moluccas)	280	265	264	261
Palau[2] (from New Guinea)	375	355	264	261
Yap[3] (from Palau)[1]	298	279	176	175
Marianas (from Yap)[3]	570	505	82	63
Marianas (from Volcano Islands)	640	610	82	63
Chuuk[4] (from Marianas)	895	545	3	0
Chuuk[4] (from Yap)[3]	321	314	3	0
Chuuk[5] (from Guam)	730	555	105	103
Chuuk[5] (from Yapese atolls)	182	48	105	103
Pohnpei[6] (from Chuuk)[4]	660	655	—	—
Pohnpei[6] (from Chuuk)[5]	490	485	—	—
Pohnpei[7] (from Chuuk)[4]	345	207	315	313
Pohnpei[7] (from Chuuk)[5]	270	207	315	313
Kosrae (from Pohnpei)[6]	525	523	—	—
Kosrae (from Pohnpei)[7]	257	251	—	—
Marshalls (from Kosrae)	490	480	299	289
Marshalls (from Tungaru)	535	532	299	289
Wake (from Marshalls)	560	545	—	—
Nauru (from Kosrae)	775	774	—	—
Nauru (from Banaba)	292	291	—	—
Tungaru[8] (Gilberts) from Nauru	700	685	122	116
Tungaru[9] (Gilberts) from Tuvalu	340	336	452	448
Phoenix (from Tokelau)	415	413	220	218
Howland & Baker (from Phoenix)	500	498	80	78
Northern Line Islands (from Jarvis)	495	493	300	298
Northern Line Islands (from Southern Line Islands)	720	718	300	298
Southern Line Islands (from Society)	520	516	620	618
Tuvalu (from Rotuma)	300	290	153	149
Tokelau (from Samoa)[11]	450	420	89	87
Tokelau (from Samoa)[10]	180	175	89	87
Northern Cooks (from Samoa)[11]	540	537	370	368
Northern Cooks (from Samoa)[9]	450	447	370	368

[1]Without Southwest Islands.
[2]With Southwest Islands.
[3]Yap Proper, Fais, and Ngulu, Ulithi, and Sorol atolls; eastern Yapese atolls (Eauripik through Pikelot) are biogeographically Trukese.
[4]Neoch only.
[5]All Trukese islands.
[6]Pohnpei island only.
[7]Pohnpei and all atolls in Pohnpei State.
[8]Without Banaba.
[9]With Banaba
[10]Without Swains or Rose Atoll.
[11]With Swains or Rose Atoll.

attempts have been made to classify Pacific islands and island groups into biogeographic regions based on floral or faunal similarities and differences. These classifications were discussed in detail by Stoddart (1992), from whom I now present four maps (Figures 2-1 to 2-4) of biotic regions in Oceania. These four schemes have much in common but differ in details, which in large part reflect distributional patterns in organisms studied by the scheme's developer.

Gressitt (1956, 1961), an entomologist, lumped all of tropical Oceania except the Bismarcks and Solomons (Near Oceania) into one major biogeographic region (Figure 2-1). Thorne (1963), a botanist, proposed a similar classification but subdivided Near Oceania, New Guinea, and northeastern Australia from each other (Table 2-3, Figure 2-2). The vertebrate zoologist Udvardy (1975) split tropical Oceania into more regions (Figure 2-3), which seems reasonable for birds although I would segregate the Bismarcks and Solomons from New Guinea, and relate Fiji as much or more with Tonga and Samoa than Santa Cruz and Vanuatu (Chapter 6). The marine biologist Dahl (1979, 1980) subdivided Oceania even more finely (Figure 2-4), a proposal that would seem to be oversplit for many groups of terrestrial organisms.

Except as noted above, the distribution of landbirds in Oceania is compatible with all four schemes, although I favor Figures 2-3 and 2-4 in excluding tropical adjacent Australia from Oceania. It is important to stress that none of these regions represents an absolute break from another region. Biogeographical classifications depend on taxonomic categories, with higher categories yielding higher levels of similarity. Among birds, for example, a genus or family is much more likely to be shared among island groups than a species. Finally, we must realize that there is much interarchipelago biotic variation within each region, and even quite a bit among islands within a group. Documenting and trying to explain this variation for birds will be a major task in Chapters 5-21.

Table 2-3 Explanation of biogeographic regions of the Pacific as depicted in Figure 2-3. From Stoddart (1992: Table 1), which is based on Thorne (1963).

- I. Oriental Region
  - IA. Papuan Subregion
    - IAi. Papuan province
    - IAii. Torresian province
    - IAiii. Bismarckian province
      - Bismarckian district
      - Solomonian district
  - IB. Polynesian Subregion
    - IBi. Fijian province
      - IBia. New Hebridean district [includes Santa Cruz]
      - IBia. Fijian district [includes Tonga and Samoa]
    - IBii. Polynesian province
      - IBiia. Micronesian district
      - IBiib. Polynesian district
      - IBiic. Hawaiian district
  - IC. Neocaledonian Subregion
    - ICi. Neocaledonian province [includes Loyalties]
- II. Australian Region
  - IIA. Australian Subregion
  - IIB. Neozeylandic Subregion
    - IIBi. Kermadecian province
      - IIBia. Lord Howean district
      - IIBib. Norfolkian district
      - IIBic. Kermadecian district
    - IIBii. Neozeylandic province
- III. Neotropical Region
  - IIIA. Chilean Subregion
    - IIIAi. Fernandezian province
  - IIIB. Peruvian Subregion
    - IIIBi. Galapagean province
- IV. Holarctic Region
  - IVA. Nearctic Subregion
    - IVAi. Caribbean province
      - IVAia. Mexican district [includes Clipperton]
    - IVAii. Sonoran province
    - IVAiii. California province

Tertiary Plant Fossils

Throughout this and the next section, family names will be given at the first mention of a genus. I am unaware of a pre-Quaternary fossil record of plants in Near Oceania, although the sparse Tertiary fossil record of plants in New Guinea indicates that *Nothofagus* (Fagaceae), Proteaceae, Myrtaceae, and Casuarinaceae, each characteristic of the Papuan and Australian region today, were present by the early Eocene (Morley 2000:180, 182). Fossil pollen from the late Miocene to Pleistocene of lowland New Guinea also features Myrtaceae, *Casuarina* (Casuarinaceae), Rubiaceae, *Nothofagus*, Palmae, and various mangroves (Morley 2000:204, 205).

Living genera of angiosperms characteristic of upland habitats have occupied islands in Remote Oceania since at least the Miocene. A landmark study in this regard is that of Leopold (1969), the most important points of which I

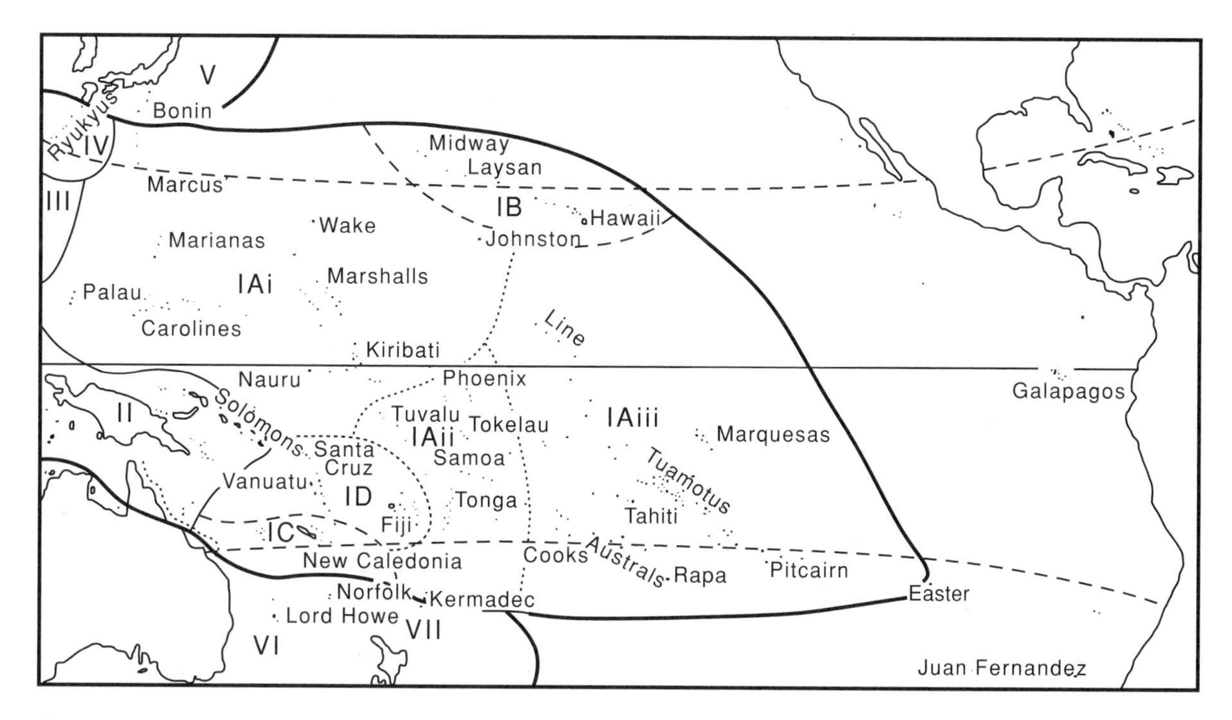

Figure 2-1 Biogeographic regions of the Pacific based on Gressitt (1956, 1961). Region I more or less corresponds with Remote Oceania, whereas II comprises Near Oceania plus New Guinea and part of Australia. The other regions are not part of tropical Oceania. Redrawn from Stoddart (1992: Figure 1).

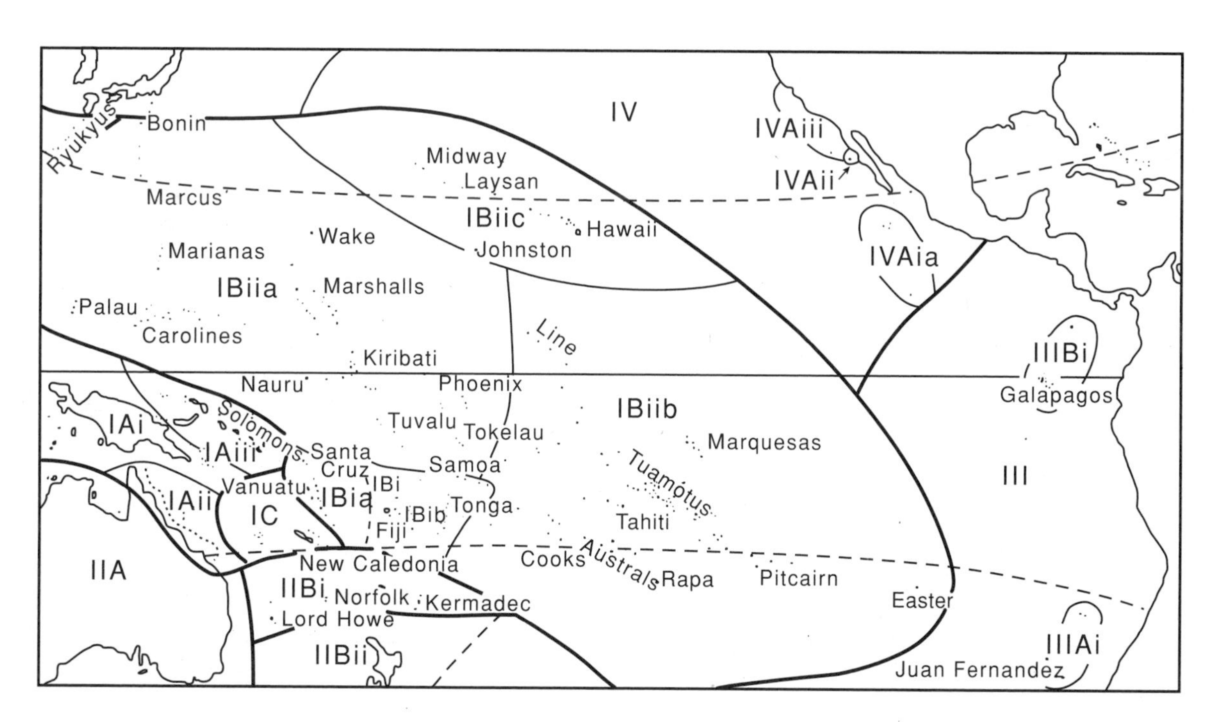

Figure 2-2 Biogeographic regions of the Pacific based on Thorne (1963). Region I comprises tropical Oceania as well as New Guinea westward to Asia, and part of northeastern Australia. For explanation see Table 2-3. Redrawn from Stoddart (1992: Figure 2).

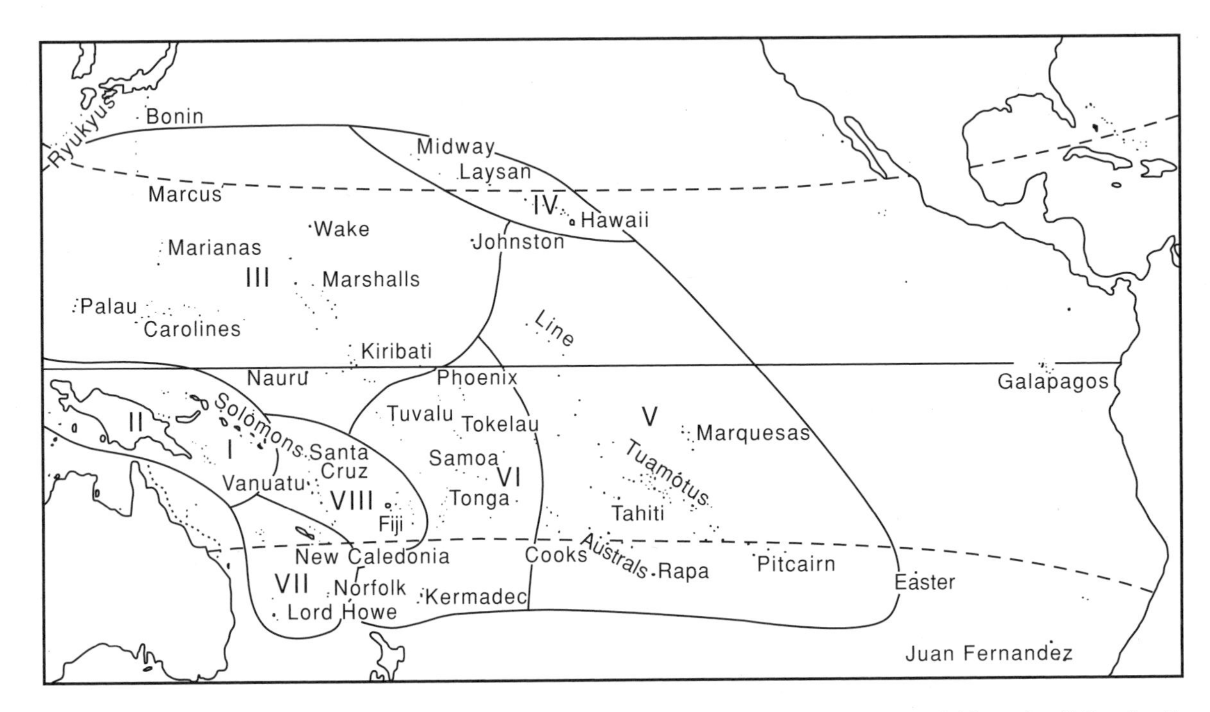

Figure 2-3 Biogeographic regions of the Pacific based on Udvardy (1975). I: Papuan rainforest. II: Papuan savanna. III: Micronesian. IV: Hawaiian. V: Southeast Polynesian. VI: Central Polynesian. VII: New Caledonian. VIII: East Melanesia. Redrawn from Stoddart (1992: Table 3, Figure 4).

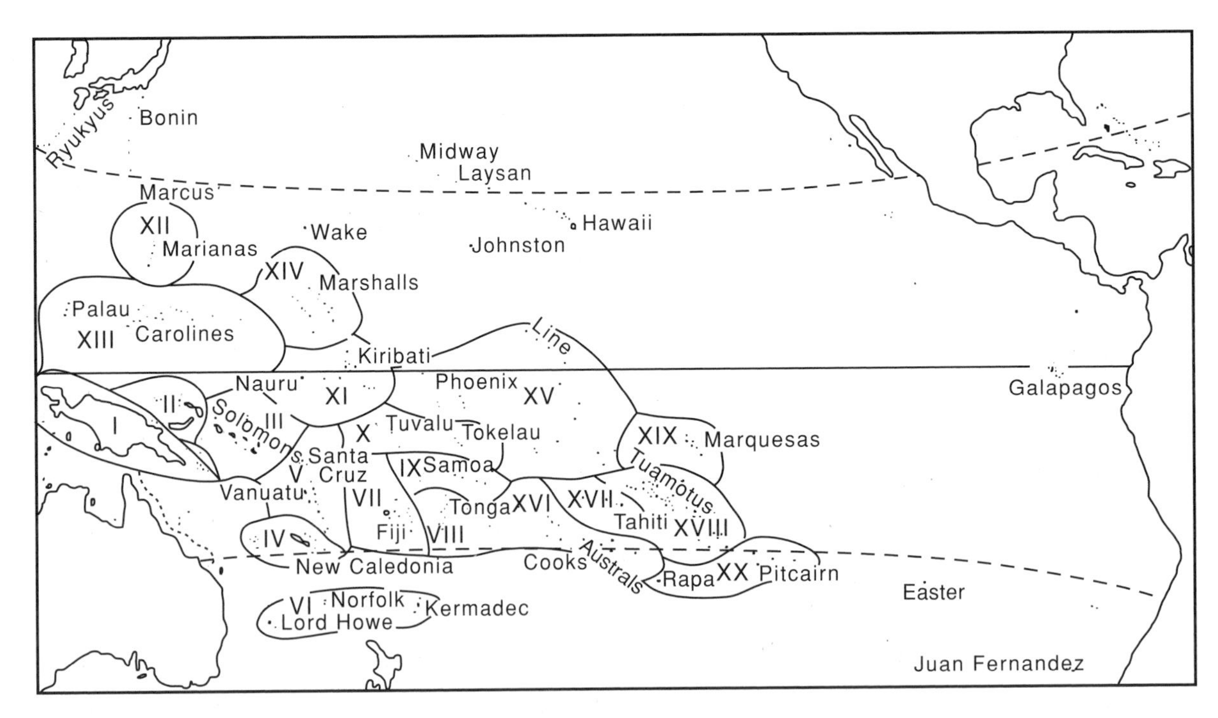

Figure 2-4 Biogeographic regions of the Pacific based on Dahl (1979, 1980). I: New Guinea. II: Bismarck Archipelago. III: Solomon Islands. IV New Caledonia and Loyalty Islands. V: Vanuatu and Santa Cruz Islands. VI: Norfolk, Lord Howe, and Kermadec Islands. VII: Fiji. VIII: Tonga and Niue. IX: Samoa, Wallis, Futuna. X: Tuvalu and Tokelau. XI: Kiribati and Nauru. XII: Marianas. XIII: Caroline Islands. XIV: Marshall Islands. XV: Phoenix, Line, and northern Cook Islands. XVI: Southern Cook Islands and Austral Islands. XVII: Society Islands. XVIII: Tuamotu Archipelago. XIX: Marquesas. XX: Pitcairn, Rapa, Gambier Islands. Redrawn from Stoddart (1992: Table 4, Figure 5).

summarize in the next three paragraphs. Drill holes at the large atolls of Eniwetok and Bikini (Marshall Islands, Micronesia) yielded ca. 1200 m of carbonate sediments underlain by basalt. The stratigraphic sequence is composed mainly of coral and shallow-water sediments. Although Eniwetok, Bikini, and probably the other atolls in the Marshall Islands are cylinders of coral growing slowly in a region of overall subsidence, Leopold's sediment core has three deeply weathered, pollen-free zones from at least three prolonged periods of emergence of Eniwetok and Bikini–one between the late Eocene and Miocene, one during the Miocene, and one in the post-Miocene.

Of 17 angiosperm genera identified by Miocene pollen, the only four native to Eniwetok today are *Pandanus* (Pandanaceae), *Pisonia* (Nyctaginaceae), *Tournefortia* (Boraginaceae), and *Cordia* (Boraginaceae). Six others now inhabit the southern Marshalls but not Eniwetok, namely *Sonneratia* (Sonneratаceae), two Rhizophoraceae (*Rhizophora*, *Bruguiera*), *Lumnitzera* (Combretaceae), and two Rubiaceae (*Morinda*, *Randia*). The five genera occurring in western or central (but not eastern) Micronesia are *Acalypha* (Euphorbiaceae), *Ceriops* (Rhizophoraceae), *Terminalia* (Combretaceae), *Avicennia* (Verbenaceae), and *Ixora* (Rubiaceae). Finally, *Gardenia* (Rubiaceae) occurs in Borneo, New Guinea, and Polynesia but not in Micronesia. Miocene pollen of the tree *Thespesia* (Malvaceae) from Eniwetok is more similar to that of an upland Malaysian species (*T. lampas*) than that of *T. populnea*, a modern strand species from Oceania. The primary affinity of Eniwetok's Miocene flora is with that of high islands now in western Micronesia, such as Palau, Yap, and the Marianas.

Five habitats can be inferred for Eniwetok in the Miocene: intertidal beach rock, based on the presence of blue-green algae; mangrove swamps with abundant *Rhizophora* pollen and occasional pollen of *Sonneratia*, *Lumnitzera*, *Ceriops*, and *Avicennia*; mangrove depressions on beach ridges, based on *Bruguiera*; beach ridges with strand genera (*Morinda*, *Pandanus*, *Terminalia*, and *Cordia*); and upland forest on nonsaline soils, with such woody genera as *Ixora*, *Acalypha*, *Randia*, *Gardenia*, and *Thespesia*. Of these five habitats, only the intertidal beach rock and beach ridges still exist on Eniwetok, a relatively dry atoll with a typically depauperate flora (Blumenstock & Rex 1960, Lane 1960). The Miocene pollen rain on Eniwetok suggests a denser vegetation on the same land area or an increased land area. The presence of high-island forms in the plant assemblage corroborates evidence from fossil endodontoid land snails (Solem 1983) that, at times during the Miocene, Eniwetok was a raised limestone island that supported upland forest communities.

Tertiary plant fossils from elsewhere in Remote Oceania are begging for more research. A pollen and spore assemblage from the Miocene Airai Clay on Babeldaob (Palau) is believed to represent nearly 100 upland and aquatic/estuarine taxa (Tayama 1939). Miocene pollen floras of Guam and Fiji record major range extensions for woody plants. That from Guam has *Anacolosa* (Olacaceae) and probable *Wikstroemia* (Thymeliaceae), two Indo-Malayan genera unknown in Micronesia today. Others certainly or probably occur on Guam today in mangrove or strand (*Rhizophora*, *Pisonia*, *Pandanus*, *Combretum*/*Terminalia*) or the uplands (*Ixora*). Miocene pollen from Viti Levu (Fiji) features the mangrove *Sonneratia* and cf. *Miliusa* (Annonaceae), neither in Fiji today (Ladd 1965, Leopold 1969). The Miocene assemblage from Fiji also has certain or possible coastal taxa such as *Rhizophora*, *Morinda*, cf. *Timonius* (Rubiaceae), *Terminalia* or *Combretum*, Palmae, and various ferns. Finally, undated (presumably Neogene) lignite beds on Rapa have yielded pollen of *Coprosma* (Rubiaceae), a shrub or small tree found in East Polynesia only on high islands (Cranwell 1962, 1964, Ellison 1994a). Other pollen identifications from Rapa's lignite include Myrtaceae, Piperaceae, Sapindaceae, Liliaceae, and Taccaceae, the first three representing woody species.

Modern Plant Communities

Primarily from a floristic rather than phylogenetic view, Mueller-Dombois & Fosberg (1998) surveyed the modern distribution of woody plants in Oceania. My review of Pacific plant life is derived in large part from this important book, supplemented by other references and my own observations. I have chosen 10 island groups that cover all major regions of Oceania and vary from having rich floras to very depauperate ones. For each island group, climate data and predicted "life zones" or plant formations (following Holdridge 1967) were presented in Table 1-2. My prejudice toward woody vegetation ignores some important groups of plants that are excellent long-distance dispersers, such as ferns, grasses, sedges, and composites (Carlquist 1965, Tryon 1970). Stressing inland rather than coastal vegetation is appropriate because inland forests provide much better habitat for landbirds than coastal forests, which lack most of the bird-dispersed tree and shrub taxa that exist in Oceania's noncoastal forests.

Solomon Islands

The Solomon Islands (Figures 5-2, 5-3) are part of Near Oceania and are large, hilly to mountainous, geologically diverse, and wet. The flora is more diverse than on any island group to the east but less than that of the Bismarcks. The Dipterocarpaceae, Fagaceae, and *Eucalyptus* (Myrtaceae) are important in the Bismarcks but absent in the Solomons (Whitmore 1969). In terms of origin, 31% of the Solomon genera are Malesian (a biogeographic region comprising Malaysia, Indonesia, the Philippines, and New Guinea; van Balgooy 1976), 29% are Paleotropical (some with ranges extending to Africa), 29% are Cosmopolitan, and 3% are Tropical Pacific (van Balgooy 1971). The last group includes *Metrosideros* (Myrtaceae), *Fagraea* (Loganiaceae), and *Lepina solomonensis* (Apocynaceae; a genus also known from Pohnpei and Tahiti). Less than 1% of the genera are of probable Australian origin, including *Casuarina*, *Acacia* (Fabaceae), and *Pittosporum* (Pittosporaceae) (Good 1974, van Balgooy 1971). While rich, the Solomon Islands tree flora is dominated at many forest sites by only a dozen or so large species and has low endemism (Whitmore 1969). *Ficus* (Moraceae) is a notable exception, having 63 species in the Solomons, 23 (37%) of which are endemic (Corner 1967).

Mueller-Dombois & Fosberg (1998:70–79) recognize eight woody vegetation types for the Solomon Main Chain—coastal strand forest, mangrove forest, freshwater swamp forest, lowland rain forest on well-drained soil, low-diversity rain forest, seasonally dry forest, montane rain forest, and anthropogenically modified vegetation. The coastal strand forest is dominated by widespread species and differs little from that of adjacent island groups. Mangrove forests and freshwater swamp forests are extensive in the Solomons along river mouths and sheltered coastlines.

Lowland rain forest on well-drained soil is the most widespread type of vegetation in the Solomons, with tall canopies of 25 to 35 m and about 60 common species of trees. Of these, 12 are very common (in the Anacardiaceae, Chrysobalanaceae, Clusiaceae, Combretaceae, Cunoniaceae, Dilleniaceae, Euphorbiaceae, Sapinidaceae, Tiliaceae, and Verbenaceae). On Kolombangara, 22 forest plots (each 100 × 60 m) had 172 species with a girth of ≥10 cm (Whitmore 1969, 1974). Some lowland rain forest areas have low diversity (dominated by one to several species of trees) because of unusual soils or natural disturbance, as in meandering floodplains (Kratter et al. 2001a).

Seasonally dry evergreen forest is developed on the northern side of Guadalcanal, the only island in the Solomons with a major windward/leeward rain shadow (8000 vs. 2000 mm mean annual precipitation). While it might seem odd to regard 2000 mm of rain as "dry," Guadalcanal's leeward forests are much less rich than typical Solomons lowland rain forest. Montane rain forest occurs above 700–1000 m elevation on many islands. The change from lowland rain forest to montane forest is gradual. Myrtaceae is the best represented family in montane forest. The canopy height is short (15–20 m). Finally, human-modified vegetation is due to disturbance mostly in what otherwise would be coastal strand forest or lowland rain forest on well-drained soil.

New Caledonia

With its great habitat differentiation and old geological age (Chapter 1), New Caledonia sustains a rich and highly endemic biota that can be accounted for by dispersal and Gondwanan vicariance (Bauer & Sadlier 1993, 2000, Morat 1993, Mueller-Dombois & Fosberg 1998:136–160). The main island of Grande Terre (16,650 km^2) became isolated from Australia by about 65 Ma (end of the Cretaceous) and from New Zealand by the mid-Miocene or perhaps slightly earlier (Kroenke 1996). New Caledonia also includes Ile des Pins, the Loyalty Islands, and smaller nearby islands. Although it now lies at the southern limit of tropical climates at 20–23° S, New Caledonia has been moving north with time, lying at ca. 45° S in the Paleocene (ca. 55 Ma) and not reaching 25° S until the late Miocene, ca. 7 Ma (Yan & Kroenke 1993). Thus it has occupied temperate and subtropical latitudes for most of its history.

Grande Terre has Oceania's richest gymnosperm flora with Araucariaceae, Cupressaceae, and Podocarpaceae well represented. Although believed ultimately to be Gondwanan in origin, the distribution of conifers across Oceania (de Laubenfels 1996) also reflects substantial dispersal. Angiosperm genera in New Caledonia that may be Gondwanan are in the Casuarinaceae, Epacridaceae, Fagaceae, Myrtaceae, Oncanthaceae, Proteaceae, Strasburgiaceae, and Winteraceae, many with closest relatives in Australia. *Eucalyptus* is absent from New Caledonia. The floristic affinities of New Caledonia's rain forest genera are 35% Malesian, 30% Gondwanan, 28% Tropical Pacific, and 7% with New Zealand/Norfolk Island/Lord Howe Island (Morat et al. 1984).

New Caledonia's major types of vegetation are mangrove/strand/littoral forest, lowland swamp vegetation, lowland (mesic) rain forest, limestone forest, montane rain forest and cloud forest, dry sclerophyll forest, savanna, maquis (scrub) vegetation.

Mangroves, strand, and littoral forest are widely dispersed but not extensive. Lowland swamp vegetation is found in a few places dominated by *Melaleuca quinquenervia* (Myrtaceae). The two rain forest types (lowland and montane) occupy about 20–25% of Grande Terre along the windward (eastern) slopes and ridges of the central mountain range. Lowland (mesic) rain forest occurs from

0 to 800/1000 m elevation in areas with from 1500 to 2000+ mm annual rainfall. Canopy height can exceed 30 m. These forests are typically a blend of araucariad and podocarpic gymnosperms with a great variety of angiosperms. Occasionally they are replaced by monodominant stands of *Araucaria* (Araucariaceae), *Callistemon* (Myrtaceae), or *Nothofagus* (Fagaceae). New Caledonia's five species of *Nothofagus* are the only representation of southern beech in tropical Oceania except for two species on New Britain (R. S. Hill 1996). Limestone forest, an edaphic variation of lowland rain forest, dominates the Loyalty Islands and Ile des Pins but is local on Grande Terre. These forests of ca. 20 m canopy are dominated by Malesian and Tropical Pacific angiosperms; the only gymnosperm is the immense (up to 40 m tall) *Araucaria columnaris*.

Montane/cloud forest occurs above 600/1000 m elevation with annual rainfall up to 3500–4000 mm. The windy, cool conditions can even include frost. Canopies up to 10 m high are interrupted by gymnosperms emergent to 30 m. These temperate/subtropical forests are rich in Gondwanan genera of gymnosperms and angiosperms. Seasonally dry vegetation, mostly savanna with pockets of sclerophyll forest, covers 45–50% of Grande Terre, largely on the western side, and three islands off its northwest end. This low (canopy 5–7 m) forest occupies areas of highly seasonal rainfall (1000–1100 mm annually). *Acacia spirorbis* is dominant, along with other angiosperms and the cycad *Cycas circinalis* (Cycadaceae). The sclerophyll forest gives way to a savanna landscape varying from woodland to parkland to pure grassland. The only native trees are *Melaleuca quinquenervia*, *Acacia spirorbis*, and *Casuarina collina*. Fire is important in maintaining these savannas, which are at least partly anthropogenic.

The maquis on ultramafic soils occupies ca. 30% of New Caledonia's surface. Rich in Gondwanan taxa, this "scrub" or shrubland has a 1 to 5 m canopy with highly variable communities of shrubs (woody plants branched at the base), small trees (woody plants branched from a single trunk), and lignified, semi-woody herbaceous plants. The annual rainfall and elevation are remarkably variable at from 900 to 4000+ mm and from 0 to 1600 m, respectively. Ultramafic bedrock is the common factor, resulting in soils that are poor in phosphorus, potassium, and calcium but with a very high content of nickel, magnesium, chromium, cobalt, and other metals (Manner et al. 1999).

Fiji

Fiji is regarded as East Melanesian by Mueller-Dombois & Fosberg (1998) and as West Polynesian by me (Chapter 6). The distinction is arbitrary; whether with plants, birds, or most other groups of organisms, the Fijian biota has considerable endemism but close ties with Vanuatu to the west and Tonga/Samoa to the east. Although Fiji has 500+ named islands, the western part is dominated by the huge, old, geologically complex islands of Viti Levu (10,388 km^2, 1323 m elev.) and Vanua Levu (5535 km^2, 1032 m). The eastern part (Lau Group) comprises volcanic islands, raised limestone islands, and composites thereof (Chapter 1) and has poorer, less endemic floras.

Among Fiji's nine principal vegetation types, three (mangrove forest and scrub, coastal strand vegetation, and freshwater wetland vegetation) are not highly distinct from those of adjacent island groups (Mueller-Dombois & Fosberg 1998, Thaman et al. 2005). The six others (lowland rain forest, upland rain forest, cloud forest, dry forest, *talasiqa*, and smaller island vegetation) grade into one another (Keppel 2005) and merit further description.

Lowland rain forest is very widespread, although heavily cut over in most places. Canopy heights average 20 to 30 m. The elevation is from 0 to 400/600 m, with rainfall >2250 m. Fijian lowland rain forest has more species than those in Tonga, Samoa, and perhaps even the less studied Vanuatu. Of Fiji's 1769 native species of vascular plants, ca. 1350 (76%) occur in lowland rain forests, including ca. 270 canopy trees and 400 smaller trees and shrubs (Ash 1992). Three endemic species of gymnosperms (in nonendemic genera) occur in both lowland and upland rain forest, namely *Agathis vitiensis* (Araucariaceae), *Dacrydium elatum* (Cupressaceae), and *Podocarpus vitiensis* (Podocarpaceae). The first two represent the eastern limit of their families in the Pacific; podocarps reach their eastern limit in nearby Tonga. Fijian lowland rain forest trees are dominated, however, by angiosperms in families such as Burseraceae, Chrysobalanaceae, Clusiaceae, Euphorbiaceae, Lauraceae, Meliaceae, Myristicaceae, Myrtaceae, Sapotaceae, and Sterculiaceae (Mueller-Dombois & Fosberg 1998, Keppel et al. 2005). A given stand usually has ca. 40-50 species of canopy trees.

Montane rain forest occurs from ca. 400/600 m to 600/1000 m elevation, with the lower values on wetter, windward sides of islands. Rainfall is from 2000 to 3750+ mm. The canopy is shorter than in lowland rain forest. Important conifers are *Agathis* and *Podocarpus* in wetter situations and *Agathis* and *Dacrydium* in drier ones. Angiosperm trees are often dominant and consist mainly of species also found in lowland rain forest, although *Metrosideros collina* is more common in this cooler, wetter environment.

Cloud forest comprises stunted (canopy height 5–10 m), moss-covered trees growing above 600/1000 m elevation. Stunting is related to cool temperatures, much wind, low solar radiation, and waterlogged soils, and sometimes to steep slopes. The cloud forest studied by Ash (1988) at 1210 m on Taveuni had an amazing annual rainfall of 9970 mm. Most of the common trees in lowland and upland rain forest are not found in cloud forest, especially at the highest and wettest sites. Distinctive trees in Taveuni's cloud forest are *Weinmannia* sp. (Cunoniaceae), *Dysoxylum lenticellare* (Meliaceae), *Fagraea vitiensis*, and *Ardina brackenridgii* (Myrsinaceae).

Dry forest occurs in the leeward lowlands of Fiji's large, high islands. Rainfall is very seasonal and averages 1750 to 2250 mm. Being so easy to burn in the dry season, most dry forest has been destroyed. Characteristic trees are the conifers *Dacrydium* and *Podocarpus* and the angiosperms *Dysoxylum*, *Fagraea*, *Gymnostoma* (Casuarinaceae), *Myristica* (Myristicaceae), *Parinari* (Chrysobalanaceae), *Intsia* (Fabaceae), *Syzygium* (Myrtaceae), and *Santalum* (Santalaceae).

Talasiqa is a Fijian word for the plant community that now covers huge areas degraded by fire. The ground cover has ferns (*Pteridium*, *Dicranopteris*) and grasses (*Miscanthus*, *Pennisetum*, *Sporobolus*). Trees typically are sparse and feature mainly *Pandanus tectorius* and *Casuarina equisetifolia*. Soil erosion is prevalent in *talasiqa*, much of which would be dry forest in a less altered state. Nonnative *Pinus caribaea* (Pinaceae) and *Psidium guajava* (Myrtaceae) are locally abundant today, the former planted by people and the latter spread by nonnative birds and mammals.

"Smaller island vegetation" refers to lowland rain forests in the Lau Group and islands <1100 km^2 elsewhere in Fiji. Mangrove, strand, and *talasiqa* vegetation types on these islands are similar to those on larger islands. Where not highly modified by people, "lowland" forest types occur more or less across these smaller islands, which seldom exceed 300 m elevation. The species richness of canopy trees is much lower than on large Fijian islands and, especially in the Lau Group, often is similar to that in Tonga (see below). Gymnosperms are absent except *Podocarpus*. Among the common genera of canopy trees are *Geissois* (Cunoniaceae), *Cyathocalyx* (Annonaceae), *Ficus*, *Litsea* (Lauraceae), *Dysoxylum*, *Cryptocarya* (Lauraceae), *Maniltoa* (Fabaceae), *Ellatostachys* (Sapindaceae), and *Planchonella* (Sapotaceae).

Rain forests on Lakeba, Aiwa Levu, and Nayau in the Lau Group have begun to be surveyed by Janet Franklin and colleagues (unpub. data), in concert with landbird surveys (Steadman & Franklin 2000). Especially interesting is the small (1.2 km^2), uninhabited, 100% forested, raised limestone island of Aiwa Levu (Figure 2-5), where

Figure 2-5 Aiwa Levu, Lau Group, Fiji. This small island (1.2 km^2) is uninhabited and sustains a forest comprising only indigenous species of woody plants. Photo by DWS, 3 March 2000.

late-successional rain forest has many common species of large canopy trees (diameter 40-100 cm, height 15 to 25 m) representing *Xylosma*, *Planchonella*, *Maniltoa*, *Pisonia*, *Pongamia* (Fabaceae), *Heritiera* (Sterculiaceae), *Burckella* (Sapotaceae), *Harpullia* (Sapindaceae), *Garuga* (Burseraceae), and *Ficus*.

Tonga

Lying east of Fiji, the Kingdom of Tonga is the easternmost land on the Indo-Australia Plate and consists mostly of raised limestone islands with active volcanos on its western and northern margins (Dickinson et al. 1999; see Chapter 1). Tonga is the Pacific's eastern limit for conifers and cycads, with a single species each of *Podocarpus* and *Cycas* (de Laubenfels 1996, K. D. Hill 1996). Tonga and Samoa form the modern eastern limit of mangroves in the South Pacific, although *Rhizophora* was in the Cook Islands as recently as ca. 5 ka (Ellison 1991, 1994, Stoddart 1992).

The mangrove and strand vegetation of Tonga is similar to that in Fiji but less rich in species. Moving eastward from Tonga, the woody plant communities along shorelines continue to become less rich at a low rate, with little endemism. From this point forward, I will describe only the upland forests, they being far more important to landbirds than coastal vegetation, which is dominated by species adapted for over-water dispersal.

Tongan islands are too low for montane vegetation types. Rain forest would occur, if not for people, at all altitudes inland from the mangrove or strand. My description of Tongan rain forests is from Drake et al. (1996) on 'Eua and Franklin et al. (1999, 2004) and Steadman et al. (1999a) on Vava'u, influenced as well by Sykes (1981), Whistler (1992), and Wiser et al. (2002).

Drake et al. (1996) recognized four forest types on the large, raised limestone island of 'Eua. The first is *Maniltoa-Pleiogynium* lowland rain forest, which begins at the landward edge of the littoral forest in northwestern 'Eua and continues inland to 60 m elevation on slopes ranging from 11–20°. *Myristica hypargyraea* and *Neisosperma oppositifolium* (Apocynaceae), so common in lowland rain forests elsewhere on 'Eua (see below), are absent. The dominant species are *Maniltoa grandiflora* and *Pleiogynium timoriense* (Anacardiaceae). Other large trees (>35 cm dbh) include *Sapindus vitiensis* (Sapindaceae) and the nonnative *Aleurites moluccana* (Euphorbiaceae). The subcanopy contains *Chionanthus vitiensis* (Oleaceae), *Xylosma simulans*, *Vavaea amicorum* (Meliaceae), and *Diospyros samoensis* (Ebenaceae).

The second type is *Myristica* lowland rain forest (Figure 2-6), which also begins where coastal sand or exposed limestone gives way to volcanic soils overlying limestone, and continues inland to 110 m elevation on slopes up to 33+°. *Myristica hypargyraea*, reaching 25 m high and 97 cm dbh, is very dominant. A few large (>40 cm dbh) individuals of *Pleiogynium timoriense*, *Maniltoa grandiflora*, *Guettarda speciosa* (Rubiaceae), *Canarium harveyi* (Burseraceae), *Diospyros samoensis*, *Planchonella grayana*, and *Calophyllum neo-ebudicum* (Clusiaceae) also occur. The subcanopy has *Neisosperma oppositifolium* and *Xylosma simulans*. Where this forest appears to be recovering from disturbance, a dominant tree is the dreaded skin irritant *Dendrocnide harveyi* (Urticaceae) with *Myristica hypargyraea* as a codominant.

The third type is *Calophyllum* mixed upland rain forest, which occupies volcanic soils overlying limestone on certain terraces, ridges, and ravines from 100 to 180/240 m elevation on slopes of 2–45°. No single species covers more than 12% of the basal area. The dominant large tree is *Calophyllum neo-ebudicum*, found in all plots and reaching 35 m in height. *Dysoxylum tongense* is a codominant. Other common canopy species (>50 cm dbh) are *Elattostachys falcata*, *Canarium harveyi*, *Myristica hypargyraea*, *Alphitonia zizyphoides* (Rhamnaceae), *Maniltoa grandiflora*, *Neonauclea forsteri* (Rubiaceae), *Semecarpus vitiensis* (Anacardiaceae), *Litsea mellifera*, and *Dysoxylum forsteri*. The subcanopy features *Diospyros samoensis*, *Cryptocarya hornei*, *Garcinia myrtifolia* (Clusiaceae), and *Citronella samoensis* (Icacinaceae). Where this forest seems to be recovering from disturbance, the leading dominants are either *Dendrocnide harveyi* with *Bischofia javanica* (Euphorbiaceae) or *Rhus taitensis* (Anacardiaceae) alone.

The fourth type on 'Eua is *Calophyllum-Garcinia* upland rain forest, occupying volcanic soils on limestone at 190–300 m elevation on slopes 5–40°. It grows upslope from *Calophyllum* mixed forest, mainly on steep slopes near the summit of the eastern ridge. The canopy is more dominated by *Calophyllum neo-ebudicum* with other common large trees (>40 cm dbh) including *Neonauclea forsteri*, *Homalium whitmeeanum* (Flacourtiaceae), *Canarium harveyi*, *Podocarpus pallidus*, *Dysoxylum tongense*, *Elaeocarpus graeffei* (Elaeocarpaceae), *Hernandia moerenhoutiana* (Hernandiaceae), *Myristica hypargyraea*, and *Rhus taitensis*. The subcanopy is almost completely *Garcinia myrtifolia*. Where this forest type appears to be recovering from disturbance, the dominants are *Alphitonia zizyphoides* and *Elattostachys falcata*.

The Vava'u Group of Tonga comprises ca. 60 limestone islands on a single submarine platform, with rich soils derived from tephra deposited by nearby volcanos. Franklin et al. (1999) surveyed late-successional forests as well as stands in earlier stages of succession following agricultural

Figure 2-6 *Myristica* lowland rain forest on the eastern side of 'Eua, Tonga. This forest is dominated by the nutmeg *M. hypargyraea*. Photo by J. G. Stull, November 1988.

abandonment. Unlike on 'Eua, species composition and forest structure in Vava'u are related almost exclusively to successional turnover of species rather than slope or elevation. That secondary forest begins to resemble late-successional forest in 30 to 50 years has positive implications for landbirds (Steadman et al. 1999a).

The first of three types of lowland rain forest in Vava'u is *Maniltoa grandiflora–Pleiogynium timoriense–Planchonella grayana* late-successional rain forest (Figure 2-7), found on large and small islands from 5 to 180 m elevation. This forest is considered late-successional because introduced or cultivated early-successional species are rare or absent, and air photos taken in 1968 show no clearing. Two subtypes of late-successional forest can be distinguished from species dominance. One is *Maniltoa grandiflora–Garuga floribunda–Chionanthus vitiensis* late-successional forest, found on steep slopes. The other is *Elattostachys falcata–Cryptocarya turbinata–Zanthoxylum pinnatum–Maniltoa grandiflora* disturbed mid- and late-successional forest, found in plots with evidence of disturbance, such as having *Rhus taitensis* or *Alphitonia zizyphoides* (early-successional species).

Another type is *Cryptocarya turbinata–Rhus taitensis–Alphitonia zizyphoides* mid-successional rain forest. *Elaeocarpus tonganus*, *Zanthoxylum pinnatum* (Rutaceae), *Pleiogynium timoriense*, *Elattostachys falcata*, and *Cocos nucifera* (Arecaceae) also are common in this forest type. The third type is *Rhus taitensis–Alphitonia zizyphoides* early-successional rain forest. These two native species have been identified previously as early-successional (Whistler 1980, Palmer 1988, Drake et al. 1996), which is supported in Vava'u by their presence among cultivated species, by subjective assessment of site characteristics, and in one case by air photos showing that the land was cleared in 1968.

The Cook Islands

"There is no more dramatic biogeographic boundary in the Pacific than that between the southern Cooks and

Figure 2-7 Aerial photo of the Mo'ungalafa region of 'Uta Vava'u island, Vava'u Group, Tonga. The relatively flat terraces are cleared in places and sustain *Maniltoa-Pleiogunium-Planchonella* late-successional lowland rain forest where not cleared and on steep areas between terraces. Also see Figure 1-14 ('Euakafa Island). Photo by Australian Air Force, 1990.

the southern Tongan islands" (Stoddart 1992:288). This statement applies to resident landbirds, where species-level overlap between West Polynesia (Chapter 6) and East Polynesia (Chapter 7) is restricted to four trampy species (Chapter 17). East Polynesian upland forests, exemplified here by the Cook Islands, also share few species with Tonga and have fewer species overall. Isolated Niue lies between Tonga and the Cook Islands but is closer to the former (Table 2-1) and has a depauperate terrestrial biota with more Tongan than Cook Island affinities (Steadman et al. 2000b).

Forests of the Cook Islands are well described on the eroded volcanic island of Rarotonga (Merlin 1985) and on the composite (raised limestone–eroded volcanic) islands of Atiu, Ma'uke, Mitiaro, and Mangaia (Franklin & Steadman 1991, Merlin 1991, Franklin & Merlin 1992). No native forest remains on the almost-atoll of Aitutaki (Stoddart 1975a, Steadman 1991b). Essentially all native lowland forest has been disturbed or removed on Rarotonga, where three forest types are distinguished on the generally steep land above 50 to 200 m elevation. The first is *Homalium* forest, a closed-canopy (height 8–20 m) assemblage at 50 to 400 m elevation, dominated by *H. acuminatum* but also with *Canthium barbatum* (Rubiaceae), *Elaeocarpus tonganus*, *Ixora bracteata* (Rubiaceae), and others. The second is *Fagraea-Fitchia* forest, which prefers knife-edge ridges from 100 to 400 m elevation (Figure 2-8). The short, broken canopy (6–15 m) is dominated by *Fagraea berteroana* and *Fitchia speciosa* (Asteraceae), but also includes *Homalium*, *Canthium*, *Alyxia* (Apocynaceae), *Coprosma* (Rubiaceae), *Meryta* (Araliaceae), and *Metrosideros*. On wet, often cloudy ridges and peaks above 400 m on Rarotonga is *Metrosideros* forest, with a low (6–8 m), broken canopy dominated by *M. collina*. Other common trees are *Ascarina diffusa* (Cloranthaceae), *Elaeocarpus tonganus*, *Weinmannia samoensis*, and *Pittosporum arborescens*.

On Atiu, Ma'uke, Mitiaro, and Mangaia, some tracts of predominantly native inland forest still exist on rugged limestone terrain. Dominant trees are *Elaeocarpus tonganus* and *Hernandia moerênhoutiana*, with *Pandanus tectorius*, *Cocos nucifera* (native but propagated by people), *Guettarda speciosa*, *Ficus prolixa*, *F. tinctoria*, *Calophyllum inophyllum*, and nonnative *Aleurites moluccana* locally common. The canopy is 6–18 m high and may be continuous or broken. *Hibiscus tiliaceus* (Malvaceae) is abundant in disturbed areas in the Cook Islands and most of Remote Oceania.

Lake sediments provide a history of plant communities in the Cook Islands. On Atiu (Parkes 1997), beginning at ca. 9 ka, the relative abundance of tree pollen increased through the Holocene until an abrupt drop at ca. 2 ka, when an increase in pollen of grasses and sedges and fern spores and the first appearance of charcoal represent human arrival. Coconut (*Cocos nucifera*) pollen occurs throughout, indicating that this economically important palm is native. Pollen of another palm (*Pritchardia* sp.; Arecaceae) is from a population that died out after people arrived.

On Mangaia, Kirch et al. (1991, 1995) and Ellison (1994a) analyzed pollen, spores, and sediments in 21 cores from the island's seven wetlands. Although nonwoody plants are also present, the pollen represents a number of trees and shrubs (*Ficus* through *Erythrina*, *Hernandia*, *Rhizophora*, *Cocos*; Figure 2-9a) until ca. 2.5 ka, when overall tree pollen declines, accompanied the first appearance of charcoal, and an increase in disturbance indicators such as the ferns *Dicranopteris* and *Cyclosorus*; Figure 2-9b). Taxa recorded in the sediment cores but lost from Mangaia since human arrival include the tree *Weinmannia* and the palm *Pritchardia*. The pollen data also show that *Cocos nucifera* and *Morinda citrifolia* are indigenous, but *Casuarina equisetifolia* is not.

Society Islands

I will describe the woody vegetation (see Mueller-Dombois & Fosberg 1998:407–433) from upland Tahiti. Gone is the

Figure 2-8 *Fagraea-Fitchia* forest on Rarotonga, Cook Islands. Photo by DWS, 5 December 1987.

undescribed lowland forest that once must have covered much of Tahiti and other eroded volcanic islands in the group. Native and quasi-native forests are confined now to areas too steep to cultivate. Even most of these places are infiltrated today with nonnative woody plants. Nevertheless, indigenous *Freycinetia* (Pandanaceae), *Pandanus*, *Fagraea*, *Canthium*, *Crossostylis* (Rhizophoraceae), *Wikstroemia*, *Dodonaea* (Sapindaceae), *Alyxia*, and tree ferns can still be common in patches of short, montane rain forest. Above the montane rain forest, at >1000 m elevation, is cloud forest or mossy forest. Canopy heights vary from 10 to 20+ m, being higher on flatter land. The cool, wet conditions sustain epiphyte-laden forests dominated by *Weinmannia parviflora*, *Alstonia costata* (Apocynaceae), and tree ferns.

I should add that in the Marquesas, as in the Society Islands, the original patterns of vegetation are obscured nearly everywhere by deforestation and replacement with nonnative species (Florence & Lorence 1997, Mueller-Dombois & Fosberg 1998:444–459).

Pitcairn Group (Henderson Island)

The raised limestone island of Henderson is uninhabited and in relatively good environmental condition, even if not pristine (Steadman & Olson 1985, Wragg & Weisler 1994, Weisler 1995). Henderson is covered by low forest (Fosberg et al. 1983). Two vegetation types that are neither littoral or exclusively scrubby are the "*Thespesia* woodland community" and the "plateau forest and scrub" (Paulay & Spencer 1989). The former corresponds to the "embayment forests" of Waldren et al. (1995), who divide "plateau forest and scrub" into "open limestone scrub" and "plateau forest." *Thespesia* woodland is dominated by the small tree *T. populnea*, which tends to grow in more exclusively coastal situations on larger, higher islands through much of Oceania. The plateau forest and scrub is more diverse, with these small trees generally or locally common: *Celtis* sp. (Ulmaceae), *Geniostoma hendersonensis* (Loganiaceae), *Glochidion pitcairnensis* (Euphorbiaceae), *Guettarda speciosa*, *Hernandia stokesii*, *Nesoluma st-johnianum*

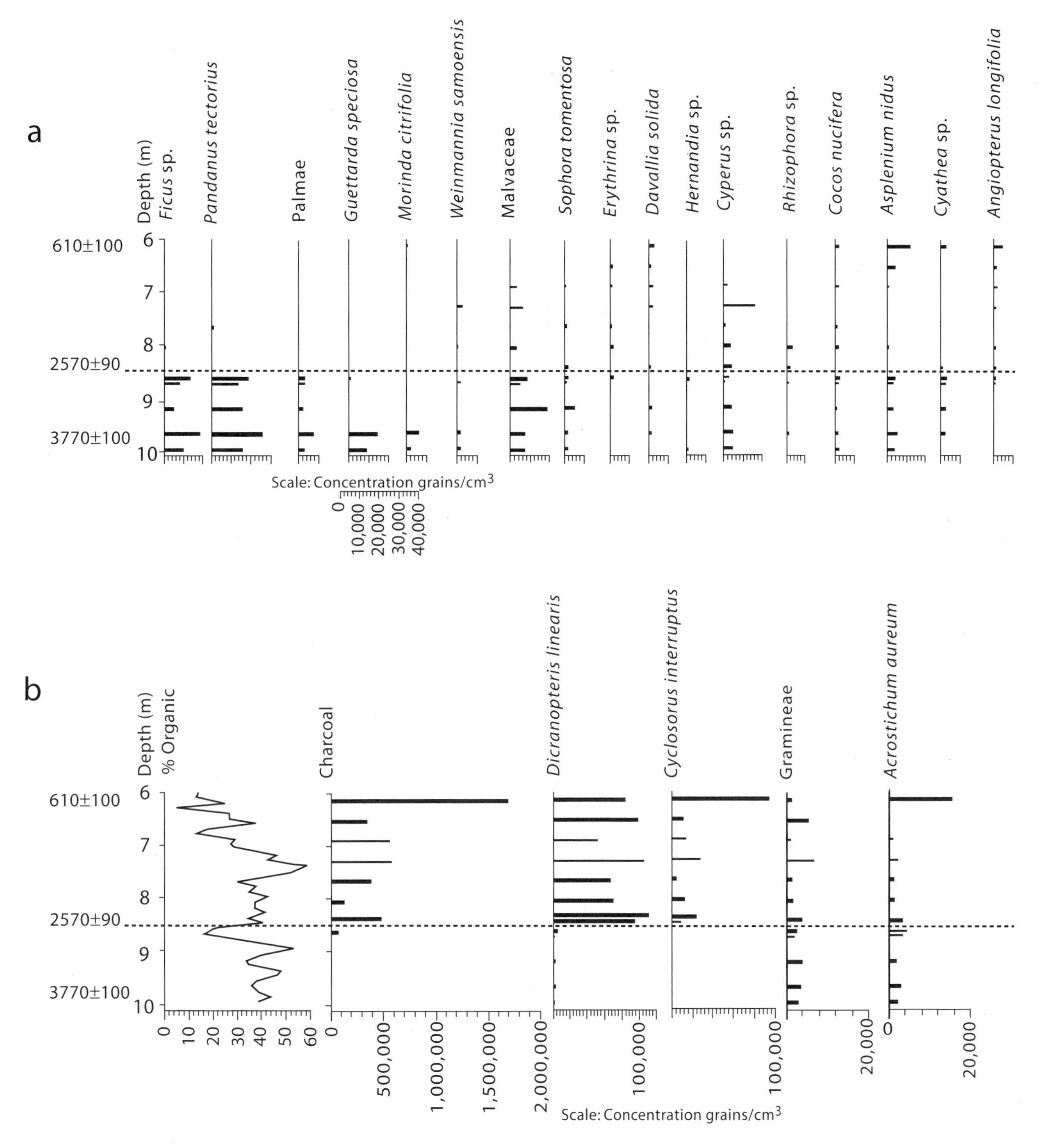

Figure 2-9 Summary pollen diagram from lake sediments on Mangaia, Cook Islands. At human arrival (dashed line), note the changes in pollen and spore influx, % organic material, and charcoal concentration. Radiocarbon dates are given along the left margin. Redrawn from Ellison (1994a: Figure 5).

(Sapotaceae), *Pandanus tectorius*, *Pisonia grandis*, and *Xylosma suaveolens*. Many trees here are superficially shrubby in appearance (Florence et al. 1995).

Easter Island (Rapa Nui)

This East Polynesian outpost is an eroded volcanic island with a depauperate woody flora reflecting its extreme isolation. The native flora of Easter Island is best represented by nonwoody groups with outstanding dispersal abilities, such as grasses, sedges, and ferns. Essentially no native woody vegetation remains, although its former condition has been estimated through pollen from sediment cores in three crater lakes (Flenley et al. 1991, Flenley & King 1984). At first human contact, ca. 1.5 ka, most of Easter Island was a forest or woodland of the palm *Jubaea* sp. (Arecaceae), with several taxa of trees or shrubs such as *Sophora toromiro* (Fabaceae), *Triumfetta* (Tiliaceae; perhaps not woody), Myrtaceae sp., *Coprosma*, another Rubiaceae, *Acalypha*, *Capparis* (Capparidaceae), *Trema* (Ulmaceae), a probable woody composite, and a probable tree-fern. The palm is South American in origin (Dransfield et al. 1984). Each of these taxa became rare or extinct by 0.5 ka, at which point the pollen/spore spectra are dominated by grasses, sedges, and ferns.

Mariana Islands

For brief coverage of Micronesia, I have chosen a group of high islands in the far west (Marianas) and some atolls in the east (Marshall Islands). The Marianas consists of active volcanoes in the north and raised limestone islands or raised limestone/eroded volcanic composite islands in the south. My descriptions of its plant communities will focus on a remnant primary tract of mixed-species limestone forest on Saipan (Craig 1992a) and forests on limestone and volcanic soils on Guam (Fosberg 1960, Mueller-Dombois & Fosberg 1998:270–275). The limestone forest on Saipan had 27 species of canopy trees and 22 species of understory trees in 10 transects that sampled 27.5 ha. The dominant trees were *Pisonia grandis*, *Dendrocnide latifolia*, and *Cynometra ramiflora* (Fabaceae), with *Intsia bijuga*, *Erythrina variegata* (Fabaceae), *Premna serratifolia* (Verbenaceae), *Ficus prolixa*, and *F. tinctoria* also common.

The five types of limestone forest recognized on Guam have broken to closed canopies of 10 to 45 m. *Artocarpus-Ficus* forest is dominated by *A. mariannensis* (Moraceae) and *F. prolixa* but includes many other genera of canopy trees. *Mammea* forest is dominated by *M. odorata* (Clusiaceae) and *Guettarda*, *Cynometra*, and *Ochrosia* (Apocynaceae). *Cordia* forest is dominated by *C. subcordata* but has many other genera. *Merrilliodendron-Ficus* forest is dominated by *M. megacarpum* (Icacinaceae) and *F. prolixa*. *Pandanus* forest features the widespread screw pine *P. tectorius*, with several other genera commonly associated. Forest patches on volcanic soil on Guam are degraded and often heavily infused with nonnative species and native early successional species. The more abundant trees include *Hibiscus tiliaceus*, *Pandanus tectorius*, *P. dubius*, *Ficus prolixa*, *Glochidion mariannensis*, and *Premna serratifolia*. Feral pigs and cattle affect regeneration.

Marshall Islands

These atolls lack land above 5 m elevation. The vegetation has been altered by people to some extent throughout the Marshalls. A strong rainfall gradient (drier in the north, wetter in the south; Table 1-2) also influences the plant communities (Manner et al. 1999). I will describe six types of nonmangrove woody vegetation, with "forest" referring to canopied types and "stands" referring to short, noncanopied vegetation structure (Mueller-Dombois & Fosberg 1998:297–309). The mixed broadleaf forest of the Marshall Islands is similar to the littoral forest on sandy coastlines of larger, higher islands in Remote Oceania. Common trees, all widespread in Oceania, are *Tournefortia argentea*, *Guettarda speciosa*, *Pisonia grandis*, *Pandanus tectorius*, *Allophylus timoriensis* (Sapindaceae), *Cordia subcordata*, *Hernandia nymphaefolia*, and *Thespesia populnea*.

The remaining five types of forests or stands in the Marshalls are overwhelmingly populated by a single woody species. The first is *Neisosperma* forest, dominated by *N. oppositifolium*. The second is *Pisonia grandis* forest, often with enormous trees up to 30 m tall and 2 m in diameter. Next are *Tournefortia argentea* stands, a successional stage leading to mixed broadleaf forest. Fourth are *Pemphis* stands, with low and dense growth of *P. acidula* (Lythraceae) on limestone platforms. Finally, *Suriana* stands, with open to dense growth of *S. maritima* (Simaroubaceae) up to 4 m tall, inhabit certain sandy shores.

Plant Radiations

Adaptive radiation is "the evolution of ecological and phenotypic diversity within a rapidly multiplying lineage. It occurs when a single ancestor diverges into a host of species that use a variety of environments and that differ in traits used to exploit those environments" (Schluter 2000:2). I assume that "rapidly" is from an evolutionary rather than ecological aspect. The adaptive radiation of plants in Oceania is an active field of research. Much more than elsewhere in Oceania, molecular and morphological phylogenetic analyses have yielded sophisticated hypotheses of dispersal and evolution in many Hawaiian groups of plants (Wagner & Funk 1995). These analyses also benefit from the geological ages of individual islands being relatively well known from considerable K-Ar dating of volcanic rocks.

Metrosideros (Myrtaceae) comprises ~26 species of trees and shrubs that occur on the Gondwanan landmasses of New Zealand and New Caledonia, as well as on volcanic islands from Melanesia to Polynesia and the Bonin Islands (Wilson 1996). A phylogenetic analysis of nuclear ribosomal DNA from all named species proposes that *M. umbellata* of New Zealand is basal in the subgenus *Metrosideros*, with the remaining species falling into three monophyletic clades (Wright et al. 2000). One clade includes the seven New Caledonian species and three daughters in western Oceania (Bonin Islands, Solomon Islands, Fiji) that probably dispersed during the mid/late Tertiary. A second contains six species in eastern Melanesia and Samoa that may also have arisen from mid/late Tertiary dispersal, in this instance from New Zealand. The third clade includes three New Zealand endemics along with all species of *Metrosideros* in the Kermadec, Cook, Society, and Hawaiian Islands. Because the rDNA sequences of the non–New Zealand species are either identical to that of *M. excelsa* (New Zealand) or differ by a single nucleotide change, Wright et al. (2000) suggested that all are derived from Pleistocene dispersal from New Zealand. The fossil record of *Metrosideros* includes pollen from the late Paleocene/early Eocene and early Miocene of New Zealand (Mildenhall 1980, Pole 1993) and thus is compatible with these proposals.

Each of New Caledonia's 37 species of palms is endemic, and 13 of these are regarded as vulnerable, endangered, or critically endangered (Pintaud et al. 1999). New Caledonia also has more endemic genera of palms, including the most monotypic genera, than any other island in the world (Johnson 1996). New Caledonia's 43 species of conifers are endemic and represent considerable in situ differentiation on this large, geologically and climatically diverse island (de Laubenfels 1988). On the other hand, New Caledonia's radiation of *Elaeocarpus* (29 species, more than in any other island group in Oceania; Smith 1953) may not reflect a Gondwanan heritage because this genus is most diverse in Indonesia and New Guinea and occurs in much of Oceania.

An old angiosperm family with minimal radiation is the Degeneriaceae (order Magnoliales), endemic to Fiji with *Degeneria vitiensis* on Viti Levu and *D. roseiflora* on Vanua Levu and Taveuni (Miller 1989). These trees reach 35 m in height and nearly 1 m in diameter. Other primitive angiosperms (Magnoliales, Laurales) in New Caledonia are various genera of Winteraceae, Austrobaileyaceae, and the monotypic Amborellaceae (Smith 1970, Soltis et al. 2000), and the much more widespread nutmegs (Myristicaceae), which occur as far east as Tonga and Samoa. Molecular analyses suggest that *Amborella trichopoda*, a shrub endemic to New Caledonia, may be the sister group to all other living angiosperms (Soltis et al. 1999, 2000, 2001).

Invertebrates

Landsnails

Three (Partulidae, Achatinellidae, Amastridae) of six families in the most primitive order (Orthurethra) of pulmonate land snails, as well as the anatomically most generalized family (Endodontidae) of higher landsnails (Sigmurethra), are restricted to oceanic islands (Solem 1976, 1979, Paulay 1994, Goodacre & Wade 2001). In most island groups, endemic species and often endemic genera abound (Solem 1983), although prehistoric and historic extinctions have reduced most modern landsnail faunas in Oceania, just like so many avifaunas, to vestiges of their former richness (Christensen & Kirch 1986, Cowie 1992, Hopper & Smith 1992, Hadfield et al. 1993, Coote et al. 1999, Bouchet & Abdou 2001). On Rota in the Marianas, for example, 23 of the 38 native species of landsnails recorded by Bauman (1996) were found only as prehistoric or historic shells in caves and rockshelters, not as living snails. About one-half of the 23 species are undescribed. Comparable situations exist through much of Oceania, where modern surveys are badly needed and landsnail faunas are increasingly contaminated with nonnative species (Cowie & Rundell 2002).

Insects

The state of systematic and biogeographic information on Pacific insects is poor (Miller 1996), a situation made worse by nonnative species replacing native ones (Jourdan 1997) and the lack of a fossil record outside of Hawai'i and New Zealand. Native ants may occur only as far east as Samoa (Wilson & Taylor 1967b, Wetterer 2002, Wetterer & Vargo 2003). One spectacular radiation consists of 67+ endemic, flightless species of curculionid weevils in the genus *Miocalles* on the remote, eroded volcanic island of Rapa in the Tubuai Group (Paulay 1985). The entire geographic range of many of these species is <1 km^2. Another dramatic radiation of coleopterans comprises 70+ species of *Mecyclothorax* (Carabidae) endemic to the large but young

Table 2-4 Families, genera, and species of frogs on mainland New Guinea versus Oceania. New Britain, New Ireland, and the Admiralty Islands together make up the Bismarcks. Modified from Allison (1996), Allison & Bigilale (2001).

		Near Oceania				Remote Oceania	
	New Guinea	New Britain	New Ireland	Admiralty Islands	Solomon Islands	Fiji	Palau
HYLIDAE							
Genera	2	1	1	1	1	—	—
Species	76	3	2	2	2	—	—
MICROHYLIDAE							
Genera	16	2	—	—	—	—	—
Species	104	2	—	—	—	—	—
MYOBATRACHIDAE							
Genera	5	—	—	—	—	—	—
Species	7	—	—	—	—	—	—
RANIDAE							
Genera	3	3	2	2	6	1/2	1
Species	14	11	5	2	24	3	1
TOTAL							
Families	4	3	2	2	2	1	1
Genera	26	6	3	3	7	1/2	1
Species	201	16	7	4	26	3	1

eroded volcanic island of Tahiti (Perrault 1992). To generalize, species-level radiations of insects in Remote Oceania are often much more extensive on eroded volcanic islands than on raised limestone islands (Paulay 1991).

Nonavian Vertebrates

Amphibians

The distribution and diversity of indigenous amphibians in Oceania are limited. Salamanders, sirenians, and caecilians are absent altogether. Frogs (Anura) have multiple families only in Near Oceania, and even those anuran faunas are much less diverse than in New Guinea (Table 2-4). The only island groups in Remote Oceania with native frogs are Fiji and Palau, in each case with one to several species of the ranid *Platymantis* (Allison 1996, Crombie & Pregill 1999, Morrison et al. 2004; Figure 2-10 herein), although a giant, undescribed species of *Platymantis* has been reported from late Quaternary fossils on Viti Levu, Fiji (Worthy et al. 1999, Worthy 2001a). Anurans lack a fossil record elsewhere in Remote Oceania except for the extant *P. pelewensis* from several late Quaternary sites within its modern range in Palau (Pregill & Steadman 2000).

Reptiles

Freshwater turtles are absent from Oceania today (Crombie & Pregill 1999) even though three families (Carretochelidae, Trionychidae, Chelidae), five genera, and nine species of freshwater turtles inhabit New Guinea (Allison 1996). Perhaps better fossil records or modern surveys would change this situation. Also absent in Oceania are true tortoises (Testudinidae), which do occur in Island Southeast Asia and the Galápagos Islands. Until it was wiped out

Figure 2-10 The frog *Platymantis pelewensis*, Ngerekebesang, Palau. Photo by G. K. Pregill, 9 June 1994.

Figure 2-11 The skink *Emoia trossula* (snout-vent length ca. 80 mm), Aiwa Levu, Lau Group, Fiji. Photo by DWS, 22 February 2000.

prehistorically by people (Balouet 1987), New Caledonia (including the Loyalty Islands and Walpole) was inhabited by the immense "horned" cryptodiran turtle *Meiolania mackayi* (Meiolaniidae, an extinct family also known from Quaternary fossils in Australia and Lord Howe Island; Gaffney 1981, Gaffney et al. 1984, Bauer & Sadlier 2000, Sand 2002).

Also eliminated by humans were three species of high-snouted, terrestrial crocodiles—*Mekosuchus inexpectatus* from New Caledonia and Ile des Pins (Buffetaut 1983, Balouet & Buffetaut 1987, Balouet 1991), *M. kalpokasi* from Efate, Vanuatu (Mead et al. 2002), and an undescribed form from Viti Levu, Fiji (Worthy et al. 1999). Among the three island groups where Quaternary mekosuchines have been found, New Caledonia is closest to Australia, where such fossil crocodiles occur from the Paleogene to late Quaternary (Willis 1997). Vicariance could account for mekosuchine crocodiles in Oceania only on New Caledonia and New Zealand, where they are known from the Miocene (Molnar & Pole 1997). As rifting separated New Caledonia and New Zealand from Australia in the latest Cretaceous (Kroenke 1996), ancestral mekosuchines may have occupied the rafting continental-crust islands. Dispersal is needed to explain the former presence of these crocodiles in Vanuatu and Fiji (Mead et al. 2002).

Aside from mekosuchines, crocodilians are represented in Oceania by only one species, the estuarine or saltwater crocodile *Crocodylus porosus*. It has localized resident populations in the Bismarcks, Solomons, Vanuatu, and Palau, although isolated individuals have been recorded in more remote parts of Oceania such as New Caledonia, Fiji, Pohnpei, and Nauru (King & Burke 1997, Ross 1998, Bauer & Sadlier 2000).

The number of genera of indigenous lizards in Oceania decreases dramatically from west to east (Table 2-5). The prehistory of oceanic lizards is well developed, however, only in Tonga and the Marianas (Pregill 1993, 1998), with some data as well from Vanuatu, Fiji, Samoa and the Cook Islands (Worthy et al. 1999, Pregill & Worthy 2003, G. K. Pregill pers. comm.). Fossils can establish which lizard taxa are indigenous and, in the case of nonnative species, whether the human-assisted dispersal occurred in prehistoric or historic times. Endemic species of lizards are absent from most island groups on the Pacific Plate except Samoa (Steadman & Pregill 2004). Nearby Tonga, on the Indo-Australia Plate, has a number of indigenous and endemic species of skinks and geckos (Ineich & Zug 1993, Pregill 1993).

Although just one of New Guinea's families of lizards (Dibamidae) does not live in Oceania, skinks (Scincidae) and geckos (Gekkonidae s.l.) are the only widespread and taxonomically diverse lizard families in Oceania (Figures 2-11, 2-12). New species of both continue to be described, in part because of the discovery of genetically distinct but morphologically cryptic species (Bruna et al. 1995, 1996, Zug & Moon 1995, Radtkey et al. 1996) and because rigorous modern surveys add new specimen-based records that lead to taxonomic revisions, even on islands in Remote Oceania (Crombie & Steadman 1986, Crombie & Pregill 1999, Bauer & Sadlier 2000, Zug et al. 2003).

Iguanas (subfamily Iguaninae) have one of biology's most inexplicable ranges, being confined to the New World except for two genera in Fiji and Tonga. *Brachylophus* includes both living species of Pacific iguanas, the relatively small, arboreal *B. fasciatus* and *B. vitiensis*, the former widespread but local in Fiji and Tonga and the latter on a

Figure 2-12 The gecko *Nactus pelagicus* (snout-vent length 48 mm), 'Uta Vava'u, Vava'u Group, Tonga. Photo by DWS, 15 July 1995.

Table 2-5 Distribution of lizards on mainland New Guinea versus Oceania. M, modern record; P, prehistoric record; ^{e} endemic; ?, questionable record, or residency not established. Records in brackets indicate that the genus certainly or probably is a prehistoric or historic introduction rather than indigenous. Gekkonidae includes Diplodactylidae. *Hemidactylus* is dispersing rapidly today with human assistance; it may occur on many island groups in addition to those designated here. Only high island groups are included for Micronesia. There is no evidence of possibly indigenous lizard genera on Micronesian atolls or on the more remote island groups of East Polynesia. From McCoy (1980), Crombie & Steadman (1986), Bauer & Vindum (1990), Zug (1991), Pregill (1993, 1998), Allison (1996), Crombie & Pregill (1999), Steadman et al. (1999a), Bauer & Sadlier (2000), Allison & Bigilale (2001), Pregill & Worthy (2003), Pregill & Steadman (2004), Steadman & Pregill (2004), G. K. Pregill (unpub. data).

	New Guinea	Bismarcks	Solomons	Vanuatu	New Caledonia	Fiji	Tonga	Samoa	West Polynesian Outliers	Cook Islands	Tubuai	Society Islands	Palau	Yap	Marianas	Chuuk	Pohnpei	Kosrae
AGAMIDAE																		
Chlamydosaurus	M	—	—	—	—	—	—	—	—	—	—	—	—	—	—	—	—	—
Diporiphora	M	—	—	—	—	—	—	—	—	—	—	—	—	—	—	—	—	—
Hydrosaurus	M	—	—	—	—	—	—	—	—	—	—	—	—	—	—	—	—	—
Hypsilurus	M	M	M	—	—	—	—	—	—	—	—	—	M?	—	—	—	—	—
Lophognathus	M	—	—	—	—	—	—	—	—	—	—	—	—	—	—	—	—	—
Physignathus	M	—	—	—	—	—	—	—	—	—	—	—	—	—	—	—	—	—
DIBAMIDAE																		
Dibamus	M	—	—	—	—	—	—	—	—	—	—	—	—	—	—	—	—	—
GEKKONIDAE																		
Bavayia	—	—	—	—	M^{e}	—	—	—	—	—	—	—	—	—	—	—	—	—
Cosymbotus	M	—	—	—	—	—	—	—	—	—	—	—	—	—	—	—	—	—
Cyrtodactylus	M	—	M	—	—	—	—	—	—	—	—	—	—	—	—	—	—	—
Eurydactylodes	—	—	—	—	M^{e}	—	—	—	—	—	—	—	—	—	—	—	—	—
Gehyra	M	M	M	M	M	M, P	M, P	M, P	M	[M, P]	[M]	[M]	M	M	M, P	M	M	M
Gekko	M	M	M	—	—	—	—	—	—	—	—	—	M	—	—	—	—	—
Hemidactylus	M	M	M	—	[M]	[M]	[M]	[M]	—	[M]	[M]	—	[M]	—	M, P	—	M	—
Hemiphyllodactylus	M	M	M	—	M	M	—	—	—	[M]	—	[M]	M	—	—	—	—	—
Lepidodactylus	M	M	M	M	[M]	M	M	M	[M]	[M]	[M]	[M]	M	M	M, P	M	M	M
Nactus	M	M	M	M	M	M	M	M	—	—	—	[M]	M	M	M, P	—	M	—
Perochirus	—	—	—	M	—	—	P	—	—	—	—	—	M	—	M, P	M	M	M
Rhacodactylus	—	—	—	—	M^{e}	—	—	—	—	—	—	—	—	—	—	—	—	—
Genus uncertain	—	—	—	—	—	—	—	—	—	—	—	—	—	—	P	—	—	—

PYGOPODIDAE																		
Lialis	M	M	—	—	—	—	—	—	—	—	—	—	—	—	—	—	—	—
SCINCIDAE																		
Caledoniscincus	—	—	—	—	M[e]	—	—	—	—	—	—	—	—	—	—	—	—	—
Carlia	M	M	M	—	—	—	—	—	—	—	—	—	[M]	—	[M]	—	—	—
Corucia	—	—	M	—	—	—	—	—	—	—	—	—	—	—	—	—	—	—
Cryptoblepharus	M	M	M	M	M	M	M	M	M	[M]	[M]	[M]	M	—	M, P	—	—	M
Ctenotus	M	—	—	—	—	—	—	—	—	—	—	—	—	—	—	—	—	—
Egernia	M	—	—	—	—	—	—	—	—	—	—	—	—	—	—	—	—	—
Emoia	M	M	M	M	M	M, P	M, P	M, P	M	M, P	[M]	[M]	M	M	M, P	M	M	M
Eugongylus	M	M	M	—	—	—	P?	—	—	—	—	—	M	—	—	M	M	—
Fojia	M	—	—	—	—	—	—	—	—	—	—	—	—	—	—	—	—	—
Geomyersia	—	M	M	—	—	—	—	—	—	—	—	—	—	—	—	—	—	—
Geoscincus	—	—	—	—	M[e]	—	—	—	—	—	—	—	—	—	—	—	—	—
Graciliscincus	—	—	—	—	M[e]	—	—	—	—	—	—	—	—	—	—	—	—	—
Lacertoides	—	—	—	—	M[e]	—	—	—	—	—	—	—	—	—	—	—	—	—
Lamprolepis	M	M	M	—	—	—	—	—	—	—	—	—	M	[M]	—	—	M	M
Leiolopisma	—	—	—	M	—	M	—	—	—	—	—	—	—	—	—	—	—	—
Lioscincus	—	—	—	—	M[e]	—	—	—	—	—	—	—	—	—	—	—	—	—
Lipinia	M	M	M	M	—	M	M	M	M	[M]	[M]	[M]	M	M	M, P	M	M	M
Lobulia	M	—	—	—	—	—	—	—	—	—	—	—	—	—	—	—	—	—
Lygisaurus	M	—	—	—	—	—	—	—	—	—	—	—	—	—	—	—	—	—
Lygosoma[1]	—	—	—	—	M	—	—	—	—	—	—	—	—	—	—	—	—	—
Mabyua	M	—	—	—	—	—	—	—	—	—	—	—	M	M?	—	—	—	—

(continued)

Table 2-5 (continued)

	New Guinea	Bismarcks	Solomons	Vanuatu	New Caledonia	Fiji	Tonga	Samoa	West Polynesian Outliers	Cook Islands	Tubuai	Society Islands	Palau	Yap	Marianas	Chuuk	Pohnpei	Kosrae
Marmorosphax	—	—	—	—	M^{e}	—	—	—	—	—	—	—	—	—	—	—	—	—
Nannoscincus	—	—	—	—	M	—	—	—	—	—	—	—	—	—	—	—	—	—
Papuascincus	M	—	—	—	—	—	—	—	—	—	—	—	—	—	—	—	—	—
Phoboscincus	—	—	—	—	M^{e}	—	—	—	—	—	—	—	—	—	—	—	—	—
Prasinohaema	M	—	M	—	—	—	—	—	—	—	—	—	—	—	—	—	—	—
Sigaloseps	—	—	—	—	M^{e}	—	—	—	—	—	—	—	—	—	—	—	—	—
Simiscincus	—	—	—	—	M^{e}	—	—	—	—	—	—	—	—	—	—	—	—	—
Sphenomorphus	M	M	M	—	—	—	—	—	—	—	—	—	M	—	—	—	—	—
Tachygia	—	—	—	—	—	—	M^{e}	—	—	—	—	—	—	—	—	—	—	—
Tiliqua	M	M	—	—	—	—	—	—	—	—	—	—	—	—	—	—	—	—
Tribolonotus	M	M	M	—	—	—	—	—	—	—	—	—	—	—	—	—	—	—
Tropidoscincus	—	—	—	—	M^{e}	—	—	—	—	—	—	—	—	—	—	—	—	—
VARANIDAE																		
Varanus	M	M	M	—	P	—	—	—	—	—	—	—	M	M	[M]	[M]	[M]	[M]
IGUANIDAE																		
Brachylophus	—	—	—	—	—	M, P	M, P	—	—	—	—	—	—	—	—	—	—	—
Lapitiguana	—	—	—	—	—	P^{e}	—	—	—	—	—	—	—	—	—	—	—	—
TOTAL																		
Families	6	5	4	2	3	3	3	2	2	1	0	0	4	3	2	2	2	2
Genera (M)	34	19	20	8	20	9	8	6	4	1	0	0	14–15	6–7	8	6	9	7
Genera (M + [M])	34	19	20	8	22	10	9	7	5	7	6	7	16–17	7–8	10	7	10	8
Genera (M + [M] + P)	34	19	20	8	23	11	10–11	7	5	7	6	7	16–17	7–8	11	7	10	8

1. Validity uncertain.

Figure 2-13 The iguana *Brachylophus fasciatus* (snout-vent length ca. 195 mm), Aiwa Levu, Lau Group, Fiji. Photo by DWS, 23 February 2000.

few islands in northwestern Fiji (Gibbons & Watkins 1982, Zug 1991, Harlow & Biciloa 2001). The failure to find bones of *B. fasciatus* in any prehistoric site in Tonga suggests that this small iguana was introduced to Tonga from Fiji, particularly from the Lau Group, where it is recorded archaeologically (Pregill 1993, Pregill & Steadman 2004).

Tonga was not, however, without iguanas at human contact. A large, ground-dwelling form (*Brachylophus gibbonsi*) has been found in archaeological sites in Ha'apai (five islands) and on Tongatapu (Pregill & Dye 1989, Pregill & Steadman 2004). This extinct species was lost within a century or two of human arrival in Tonga ca. 2.8 ka (Steadman et al. 2002a). *Brachylophus fasciatus* (Figure 2-13) still exists on a number of Fijian islands, including Viti Levu, where precultural fossils have been found of an even larger, extinct ground-dwelling iguana *Lapitiguana impensa* (Pregill & Worthy 2003). Even though the distribution and taxonomic diversity of Pacific iguanas were much greater at human arrival than today, the fossil record has failed to shed light on their evolutionary origins. Two decades ago I suspected that bones of extinct iguanas would be found in Oceania on islands east of Fiji and Tonga, thereby helping to bridge the huge distance between the New World and these two island groups on the Indo-Australia Plate. No such bones have materialized. It remains today as big a mystery as ever how iguanines could be found in the Americas, Fiji, and Tonga but nowhere else in Oceania.

New Caledonia has an exceptionally rich lizard fauna (Bauer 1999, Bauer & Sadlier 2000). The diplodactyline geckos of New Caledonia (*Bavayia*, *Eurydactylodes*, *Rhacodactylus*) have close affinities with carphodactyline geckos of Australia and New Zealand (Bauer & Russell 1986), suggesting a possible Gondwanan origin. New Caledonia is in a class of its own in generic-level endemism of lizards (Table 2-5). At the species level, Grande Terre shares less than 12% of its modern lizard fauna with Vanuatu, Fiji, Tonga, and Samoa, whereas the nearby Loyalty Islands share >30% of their species with these island groups to the north and east. Shared species on different oceanic island groups certainly reflects dispersal. Grande Terre is the only island in Oceania with multiple congeneric endemic species of lizards, this being the case in three endemic genera of geckos and seven endemic genera of skinks (Bauer & Sadlier 2000). Very large species (a turtle, crocodile, and moniter lizard) are the only part of the New Caledonian fossil reptile fauna that has been studied (Balouet 1991, Pregill 2001). If associated bird fossils are any indication (Balouet & Olson 1989; Chapter 5 herein), the thousands of small reptile fossils already excavated are certain to augment New Caledonia's impressive reptilian diversity once they are examined.

Only two island groups in Oceania have extensive fossil records of small lizards. In Tonga, fossils from precultural strata (>3 ka; some even >60–80 ka) on 'Eua feature the gecko *Perochirus* sp. (a genus now absent east of Vanuatu), a skink (*Emoia trossula* or *Emoia* undescribed sp.), and the smaller, much more widespread skink *E. cyanura* (Pregill 1993). Deposits that postdate human arrival (<3 ka) yielded bones of five species that still exist on 'Eua (geckos *Gehyra oceanica*, *Lepidodactylus lugubris*, and *Nactus pelagicus*, and skinks *Cryptoblepharus poecilopterus* and *Lipinia noctua*), thus showing that these widespread species arrived on 'Eua prehistorically, even though they may not be indigenous. Finally from 'Eua, a single bone of a very large skink (probably *Emoia* or *Eugongylus*; certainly different from all species above) was found in deposits that postdate human arrival; this species is extinct and probably indigenous.

Oceania's other substantial fossil lizard fauna is from prehistoric sites in the Northern Marianas, where hundreds of bones represent 12 species, including all six species of geckos and at least four of the skinks known in modern times from the Marianas (Pregill 1998). Five species from Aguiguan and one from Rota are unrecorded on these islands today. Each island also yielded bones of an undescribed species of gecko in a genus (undetermined) unknown in Micronesia today. Every species of lizard possibly native to the Marianas was recorded in a prehistoric context except the geckos *Gehyra oceanica* and *G. mutilata* and the monitor *Varanus indicus*. The gecko *Perochirus ateles* and skink *Emoia slevini*, both now extirpated or very rare throughout the Marianas, are the two most commonly represented species.

Table 2-6 Distribution of indigenous snakes on mainland New Guinea versus Oceania. M, modern record; P, prehistoric record. Sea snakes are not included. From McCoy (1980), Allison (1996), Crombie & Pregill (1999), Bauer & Sadlier (2000), Allison & Bigilale (2001), Steadman & Pregill (2004).

	New Guinea	Bismarcks	Solomons	Vanuatu	New Caledonia	Fiji	Tonga	Samoa	West Polynesian Outliers	Palau	Marianas
ACROCHORDIDAE											
Acrochordus	M	—	M	—	—	—	—	—	—	—	—
COLUBRIDAE											
Boiga	M	M	M	—	—	—	—	—	—	—	—
Cantoria	M	—	—	—	—	—	—	—	—	—	—
Cerberus	M	—	—	—	—	—	—	—	—	M	—
Dendrelaphis	M	M	M	—	—	—	—	—	—	M, P	—
Enhydris	M	—	—	—	—	—	—	—	—	—	—
Fordonia	M	—	—	—	—	—	—	—	—	—	—
Myron	M	—	—	—	—	—	—	—	—	—	—
Stegonotus	M	M	—	—	—	—	—	—	—	—	—
Tropidonophis	M	M	—	—	—	—	—	—	—	—	—
ELAPIDAE											
Acanthophis	M	—	—	—	—	—	—	—	—	—	—
Aspidomorphus	M	M	—	—	—	—	—	—	—	—	—
Demansia	M	—	—	—	—	—	—	—	—	—	—
Glyphodon	M	—	—	—	—	—	—	—	—	—	—
Loveridgelaps	—	—	M	—	—	—	—	—	—	—	—
Micropechis	M	—	—	—	—	—	—	—	—	—	—
Ogmodon	—	—	—	—	—	M	—	—	—	—	—
Oxyuranus	M	—	—	—	—	—	—	—	—	—	—
Parapistocalamus	—	—	M	—	—	—	—	—	—	—	—
Pseudechis	M	—	—	—	—	—	—	—	—	—	—
Pseudonaja	M	—	—	—	—	—	—	—	—	—	—
Salomonelaps	—	—	M	—	—	—	—	—	—	—	—
Toxicocalamus	M	—	—	—	—	—	—	—	—	—	—
Unechis	M	—	—	—	—	—	—	—	—	—	—
BOIDAE											
Candoia	M	M	M	M	M	M, P	P	M, P	M, P	M, P	—
PYTHONIDAE											
Bothrochilus	M	M	—	—	—	—	—	—	—	—	—
Morelia	M	M	M	—	—	—	—	—	—	—	—
TYPHLOPIDAE											
Acutotyphlops	M	M	M	—	—	—	—	—	—	—	—
Ramphotyphlops	M	M	M	M	M	—	—	—	—	M	M, P
Typhlops	M	M	—	—	—	—	—	—	—	—	—
TOTAL											
Families	6	5	6	2	2	2	1	1	1	3	1
Genera	26	11	10	2	2	2	1	1	1	4	1

Figure 2-14 The Palau boa *Candoia carinata* (adult ♀, snout-vent length ca. 250 mm), Malakal, Palau. Photo by G. K. Pregill, 5 January 1997.

The taxonomic diversity of nonmarine snakes in Oceania (Table 2-6) is much less than that of lizards. Although all six families of snakes in New Guinea occur in Near Oceania, generic diversity drops dramatically in colubrids and elapids. Achrochordids and pythons are absent in Remote Oceania, where colubrids, elapids, and typhlopids all have fewer genera than in Near Oceania. Native snakes are unknown in East Polynesia and in Micronesia outside of Palau (four species; Figure 2-14) and the Marianas (one species). I would caution, however, that snakes have no pre-Quaternary fossil record anywhere in Oceania and a very paltry if informative late Quaternary record. In the Marianas, fossils from Tinian indicate that the blind snake *Ramphotyphlops* cf. *braminus*, thought to be a human introduction, is indigenous (Pregill 1998). Our archaeological excavations in August 2001 on Tongatapu (Tonga) yielded a vertebra of the Pacific boa *Candoia bibronii*, which is recorded today from Fiji and Samoa but not Tonga. In Samoa, *C. bibronii* exists today only in the Manu'a Group of American Samoa. Its bones are common in the single fossil site on Tutuila (Steadman & Pregill 2004). This boa probably once lived throughout Samoa and Tonga.

Mammals

Native mammal faunas on Pacific islands are dominated by bats, which can fly. (Unlike in birds, there is no evidence of flightlessness in insular bats.) The only other indigenous mammals in Oceania are rodents. Missing altogether, as far as we know, are the monotremes and marsupials so conspicuous now and as fossils in Australia and New Guinea (Flannery 1995b, Roberts et al. 2001, Long et al. 2002). Some marsupials live today in the Bismarcks and Solomon Main Chain, but these populations are thought to represent prehistoric introductions (Flannery 1995a,b).

Bats are divided into the suborders Megachiroptera (family Pteropodidae, generally large, frugivorous or nectarivorous "flying foxes" or "fruit bats"; see Rainey & Pierson 1992) and Microchiroptera (various families, generally small and insectivorous). Both groups decrease in all aspects of taxonomic diversity from west to east in Oceania, with a substantial drop between Near and Remote Oceania (Tables 2-7, 2-8). Being volant, however, bats do not decline in diversity from mainland New Guinea to the Bismarcks

Table 2-7 Number of species of bats on mainland Papua New Guinea (PNG) versus Oceania. Bismarcks include the Admiralty Islands. Vanuatu includes the Santa Cruz Group, which politically is part of the Solomon Islands. Data from Flannery (1995a,b), Koopman & Steadman (1995), and Bonaccorso (1998). In parentheses are the number of species shared with mainland Papua New Guinea.

	Suborder	
	Megachiroptera	Microchiroptera
Mainland PNG	19	52
Bismarcks	14 (6)	29 (27)
Solomons	12 (3)	20 (18)
Vanuatu	7 (0)	7 (4)
New Caledonia	4 (0)	5 (0)
Fiji	4 (0)	2 (1)
Tonga	3 (0)	2 (1)
Samoa	2 (0)	1 (0)
West Polynesian Outliers	1 (0)	1 (0)
Cook Islands	1 (0)	—
Palau	2 (0)	1 (0)
Yap	2 (0)	—
Marianas	2 (0)	1 (0)
Chuuk	2 (0)	1 (0)
Pohnpei	1 (0)	1 (0)
Kosrae	1 (0)	—

Table 2-8 Distribution of genera of bats on mainland Papua New Guinea (PNG) versus Oceania. Bismarcks includes the Admiralty Islands. Vanuatu includes the Santa Cruz Group, which politically is part of the Solomon Islands. H, historic record; M, modern record; P, prehistoric record. From Flannery (1995a,b), Koopman & Steadman (1995), Bonaccorso (1998), and DWS unpub. data. Abundant late Quaternary fossils of bats are unstudied from various Melanesian sites.

	Mainland PNG	Bismarcks	Solomons	Vanuatu	New Caledonia	Fiji	Tonga	Samoa	West Polynesian Outliers	Cook Islands	Palau	Yap	Marianas	Chuuk	Pohnpei	Kosrae
MEGACHIROPTERA																
PTEROPODIDAE																
Aproteles	M, P	—	—	—	—	—	—	—	—	—	—	—	—	—	—	—
Dobsonia	M, P	M	M	—	—	—	—	—	—	—	—	—	—	—	—	—
Pteralopex	—	—	M	—	—	M	—	—	—	—	—	—	—	—	—	—
Pteropus	M	M, P	M, P	M, P	M, P	M, P	M, P	M	M, P	M, P	M, P	M	M, P	M	M	M
Rousettus	M	M	M	—	—	—	—	—	—	—	—	—	—	—	—	—
Notopteris	—	—	—	M, P	M	M	P	—	—	—	—	—	—	—	—	—
Nyctimene	M	M	M	H	—	—	—	—	—	—	—	—	—	—	—	—
Paranyctimene	M	—	—	—	—	—	—	—	—	—	—	—	—	—	—	—
Macroglossus	M	M	M	—	—	—	—	—	—	—	—	—	—	—	—	—
Melonycteris	—	M	M	—	—	—	—	—	—	—	—	—	—	—	—	—
Syconycteris	M	M	—	—	—	—	—	—	—	—	—	—	—	—	—	—
TOTAL																
M	8	7	7	2	2	3	1	1	1	1	1	1	1	1	1	1
M + H + P	8	7	7	3	2	3	2	1	1	1	1	1	1	1	1	1
MICROCHIROPTERA																
EMBALLONURIDAE																
Emballonura	M	M	M	—	—	M, P	M, P	M, P	M	—	M	M	M, P	M	M	—
Mosia	M	M	M	—	—	—	—	—	—	—	—	—	—	—	—	—
Saccolaimus	M	M	M	—	—	—	—	—	—	—	—	—	—	—	—	—
Taphozous	M	—	—	—	—	—	—	—	—	—	—	—	—	—	—	—

HIPPOSIDERIDAE																
Anthops	—	—	M	—	—	—	—	—	—	—	—	—	—	—	—	—
Aselliscus	M	M	M	M	—	—	—	—	—	—	—	—	—	—	—	—
Hipposideros	M	M	M	M	—	—	—	—	—	—	—	—	—	—	—	—
RHINOLOPHIDAE																
Rhinolophus	M	M	—	—	—	—	—	—	—	—	—	—	—	—	—	—
VESPERTILIONIDAE																
Chalinolobus	M	—	—	—	M	—	—	—	—	—	—	—	—	—	—	—
Myotis	M	M	M	—	M	—	—	—	—	—	—	—	—	—	—	—
Philetor	M	M	—	—	—	—	—	—	—	—	—	—	—	—	—	—
Pipistrellus	M	M	M	—	—	—	—	—	—	—	—	—	—	—	—	—
Scotorepens	M	—	—	—	—	—	—	—	—	—	—	—	—	—	—	—
Murina	M	M	—	—	—	—	—	—	—	—	—	—	—	—	—	—
Nyctophilus	M	M	—	M	—	—	—	—	—	—	—	—	—	—	—	—
Pharotis	M	—	—	—	—	—	—	—	—	—	—	—	—	—	—	—
Kerivoula	M	M	—	—	—	—	—	—	—	—	—	—	—	—	—	—
Phoniscus	M	M	—	—	—	—	—	—	—	—	—	—	—	—	—	—
Miniopterus	M	M	M	M	M	—	—	—	—	—	—	—	—	—	—	—
MOLOSSIDAE																
Chaerephon	M	M	M	M	M	M	P	—	—	—	—	—	—	—	—	—
Mormopterus	M	M	—	—	—	—	—	—	—	—	—	—	—	—	—	—
Otomops	M	—	—	—	—	—	—	—	—	—	—	—	—	—	—	—
Tadarida	M	—	—	—	—	—	—	—	—	—	—	—	—	—	—	—
TOTAL																
M	22	16	10	5	4	2	1	1	1	0	1	1	1	1	1	0
M + H + P	22	16	10	5	4	2	2	1	1	0	1	1	1	1	1	0

Table 2-9 Distribution of genera, and numbers of species, of indigenous rodents (Muridae) on mainland New Guinea versus Oceania. Compiled from Flannery (1995a,b).

	Mainland New Guinea	Bismarck Archipelago	Solomon Island Main Chain
HYDROMYINI			
Crossomys	1	—	—
Hydromys	4	1	—
Leptomys	3	—	—
Mayeromys	1	—	—
Microhydromys	2	—	—
Neohydromys	1	—	—
Parahydromys	1	—	—
Paraleptomys	2	—	—
Pseudohydromys	2	—	—
CONILURINI			
Conilurus	1	—	—
Pseudomys	1	—	—
UROMYINI			
Abeomelomys	1	—	—
Anisomys	1	—	—
Chiruromys	3	—	—
Coccymys	2	—	—
Hyomys	2	—	—
Lorentzimys	1	—	—
Macruromys	2	—	—
Mallomys	4	—	—
Pogonomelomys	2	—	—
Melomys	13	4	1
Pogonomys	4	1	—
Solomys	—	—	4
Uromys	2	1	3
Xenuromys	1	—	—
MURINAE			
Rattus	6	—	—
Stenomys	5	—	—
TOTAL genera/species	26/68	4/7	3/8

as dramatically as rodents, marsupials, lizards, or snakes. Unlike in lizards and birds, New Caledonia is not especially rich in bats.

Pteropodids do not occur east of the Cook Islands, where *Pteropus tonganus* (Figure 2-15) is known today or as a fossil from Rarotonga, Mangaia, and Aitutaki (Steadman 1991b). The easternmost microchiropterans are the localized populations of sheath-tailed bat *Emballonura sèmicaudata* in Samoa. This widespread species is also the only small bat in Micronesia, where it is a common fossil in the Marianas, including islands without modern records (Steadman 1999a). Another widespread small species is the free-tailed bat *Chaerephon* (formerly *Tadarida*) *jobiensis*, which lives from New Guinea to Fiji and has been found in

Figure 2-15 The flying fox *Pteropus tonganus*, kept as a pet by Tuaiva Mautairi, Mangaia, Cook Islands. Photo by DWS, 17 April 1984.

bone deposits in Tonga, where it is absent today (Koopman & Steadman 1995). The substantial unstudied collections of fossil bats from the Bismarcks, Vanuatu, New Caledonia, and Fiji undoubtedly include species, extant or extinct, no longer found in these island groups or on the individual islands involved.

In the Pacific, native rodents are confined to Near Oceania (Flannery 1995a). Only the family Muridae is represented, with many fewer genera and species in the Bismarcks and Solomons than on mainland New Guinea (Table 2-9). Four genera but no species of native rodents are shared between mainland New Guinea and Near Oceania, within which New Britain has more genera (at least four) and species (at least six) than any other island. The only islands in the Solomons known to have multiple native rats are Bougainville, Choiseul, and Guadalcanal, each with one or two genera and three species (Flannery 1995a). The native rodents of Near Oceania (all murids but not true murines) are poorly known from any standpoint. Rich opportunities await the mammalogist willing to study rodents in the highly malarial Bismarcks or Solomons. For example, a recent survey in New Britain confirmed the formerly doubted presence of *Pogonomys macrourus* and also secured specimens of an undescribed species in an undetermined genus (Anthony 2001). Rodent fossils from Near Oceania are likely to augment the taxonomic diversity and expand the ranges of extant species.

Various species of *Rattus* (true murines) are indigenous through most of the Old World, including New Guinea (Taylor et al. 1982; Table 2-9). Only with the spread of ocean-voyaging peoples did species of *Rattus* colonize Oceania (Flannery et al. 1988, Flannery & Wickler 1990, Matisoo-Smith et al. 1998, White et al. 2000). The two most widespread species on Pacific islands today are the prehistorically dispersed *R. exulans* (the small Pacific rat or Polynesian rat) and the historically dispersed *R. rattus* (medium-sized, indigenous to Eurasia, called the black rat,

Figure 2-16 Two species of *Rattus* trapped on Vaka'eitu, Vava'u Group, Tonga. Left, the Pacific rat *R. exulans*. Right, the black rat *R. rattus*. Photo by DWS, 4 July 1995.

roof rat, or ship rat; Figure 2-16). Another medium-sized rat taken to islands by prehistoric people is *R. praetor*, indigenous to mainland New Guinea (Flannery 1995b). Bones of *R. praetor* have been found at archaeological sites in the Bismarcks, Solomons, Santa Cruz Group, Vanuatu, Fiji, and perhaps Palau (Flannery & White 1991, DWS unpub. data). The negative influence of nonnative rodents on native insular biotas is covered in Chapters 16 and 21.

Among terrestrial plants or vertebrates, nearly any measure of diversity decreases from west to east in Oceania. Although isolation clearly has had an enormous filtering effect on faunas (Chapter 17), no biogeographic treatment of Oceania can ignore people's profound influence on the distribution of indigenous plants and animals. In this spirit, it is time now to look at the human colonization and occupation of Pacific islands.

Chapter 3 Human History

The scientific allure of Oceania has attracted not only biologists and geologists. Some of the premier names in anthropology, such as Sir Raymond Firth, Patrick Kirch, Bronislaw Malinowski, Margaret Mead, and Marshall Sahlins, did much or all of their field research on tropical Pacific islands. Archaeology is a branch of anthropology that studies past peoples, especially through the survey, excavation, and analysis of sites where they once lived. The prehistoric peopling of Oceania is of great biogeographic importance because it initiated huge environmental changes (Kirch & Hunt 1997, Hope et al. 1999) including the extinction of birds, a topic that will dominate this book.

Archaeology in Oceania benefits from the closely related disciplines of physical anthropology (Howells 1970, Turner 1986, Houghton 1996), pathobiology (Clark & Kelley 1993), molecular biology (Serjeantson & Gao 1995, Matisoo-Smith et al. 1998, Cann 2001, Hagelberg 2001, Oppenheimer & Richards 2001, Matisoo-Smith & Robins 2004), cultural anthropology (Kirch 1994, Sillitoe 1998), and linguistics (Pawley & Ross 1995, Lynch 1998, Gray & Jordan 2000). Archaeology also draws heavily from geology (Dickinson 2001, Kirch et al. 2004), pedology (Vitousek et al. 2004), hydrology (Stuck et al. 2003), and biology (Steadman 1995a) in assessing the landscapes, floras, and faunas that were encountered and altered by prehistoric peoples.

Much of this brief, simplified overview of Oceanic archaeology is derived from several excellent books (Irwin 1992, Bellwood et al. 1995, Kirch 1997, 2000, Spriggs 1997, Galipaud & Lilley 1999). After describing patterns of human settlement, I will report on some archaeological sites with well-studied stratigraphy, chronology, material culture (artifacts), and biological materials (bones and, in some cases, plants and mollusks). I do this mostly for biologists, who seldom are exposed to archaeology and who will appreciate prehistoric bird bones (the heart of this book) much better by understanding their cultural and chronological context. I apologize to archaeologists for my ornithocentric view of prehistory.

Before diving into archaeology, a quick note on Pacific languages is needed. For its land area and population size, Melanesia has more languages than anywhere else on earth (Lynch 1998). By contrast, Polynesia and Micronesia have only a few languages (Table 3-1), almost none confined to a single island. This difference is related to the voyaging skills of people who colonized Polynesia and Micronesia, as well as the more tribal, xenophobic nature of most Melanesian societies.

Two major language groups occur in Oceania, with Austronesian languages being widespread (Bellwood et al. 1995, Blust 1999, Diamond 2000) and Papuan languages confined to New Guinea, the Bismarcks, and Solomons (Lynch 1998; Table 3-1 herein). On New Guinea itself, Austronesian languages are generally spoken in the lowlands and Papuan languages in the highlands, with the logical assumption that the Papuan languages may pertain to the original settlement of New Guinea and Near Oceania >30 ka, and the Austronesian languages may relate to the more coastally oriented Lapita colonizations of only about

Table 3-1 Pacific languages by region and country. Papua New Guinea includes the Bismarcks, Buka, and Bougainville. The single continental island of New Guinea is divided politically into Irian Jaya (part of Indonesia) and Papua New Guinea (independent). Solomon Islands includes the Santa Cruz Group. From Lynch (1998: Tables 1, 3).

	Number of languages		
REGION, country	Papuan	Austronesian	Total
MELANESIA (1151+ total)			
Irian Jaya	160+	45+	205+
Papua New Guinea	540+	210+	750+
Solomon Islands	7	56	63
Vanuatu	—	105	105
New Caledonia	—	28	28
FIJI AND POLYNESIA (22 total)			
Fiji, including Rotuma	—	3	3
Tonga	—	2	2
Niue	—	1	1
Samoa	—	1	1
Tuvalu	—	1	1
Tokelau	—	1	1
Wallis & Futuna	—	2	2
Cook Islands	—	3	3
Hawai'i	—	1	1
French Polynesia	—	5	5
Easter Island	—	1	1
New Zealand	—	1	1
MICRONESIA (16 total)			
Belau	—	1	1
Northern Marianas and Guam	—	2	2
Marshall Islands	—	1	1
Kiribati	—	1	1
Nauru	—	1	1
Federated States of Micronesia	—	11	11
TOTAL	707+	482+	1189+

3.6 to 3 ka (Kirch 1997, Sillitoe 1998). Each Polynesian language (summarized in Lynch 1998: Figure 7) has much in common with the others; most are mutually intelligible or, at most, as different as Spanish and Portuguese or Spanish and Italian. The Polynesian outliers are islands, often relatively small and remote, that are geographically part of Melanesia or Micronesia but were settled prehistorically by Polynesians who speak a Samoic dialect (Kirch 1984b).

Human Settlement

Near Oceania

Virtually all islands in Oceania (Figure 3-1) were inhabited at one time or another in prehistory. As defined by Green (1991), Near Oceania refers to those islands relatively near New Guinea that were settled by people in the late Pleistocene. Near Oceania consists of the Bismarck Archipelago

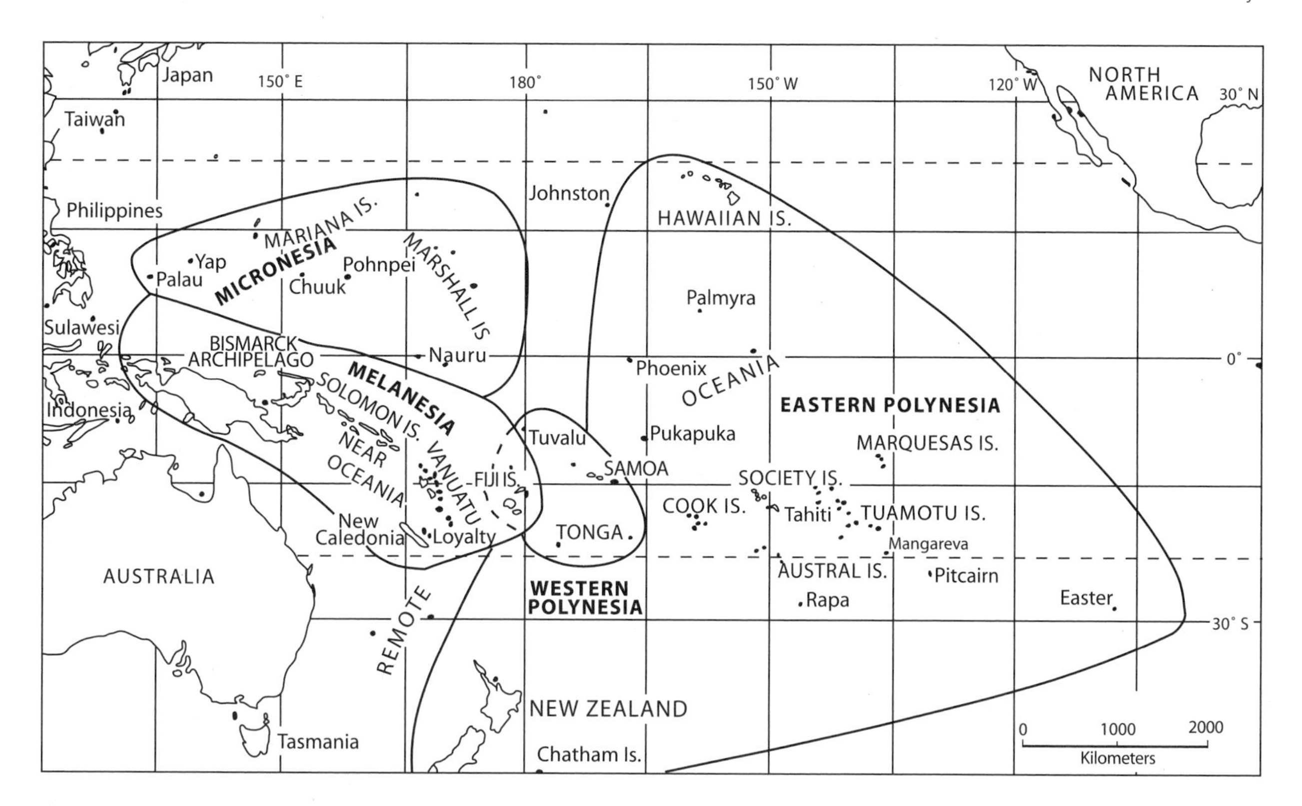

Figure 3-1 Near Oceania vs. Remote Oceania. Redrawn from Kirch (2001: Map 1).

(dominated by the large islands of New Britain and New Ireland, but also including the Admiralty Group and numerous smaller islands) and the Main Chain of the Solomon Islands (the cluster of islands from Buka through Makira; Figure 1-2).

People arrived in Near Oceania as early as 33,000 yr BP (radiocarbon years before present; Chapter 4 explains ^{14}C dates) on New Britain (Pavlides & Gosden 1994), 32,000 yr BP on New Ireland (Allen et al. 1989), and 29,000 yr BP on Buka (Wickler & Spriggs 1988, Wickler 2001). The first colonizers of Near Oceania presumably came from New Guinea, which for much of the late Pleistocene was joined to Australia as the continent of Sahul (Spriggs 1997). Late Pleistocene sea levels were as much as 130 m below modern levels from 25,000 to 18,000 yr BP, although most of the time from 40,000 to 10,000 yr BP the sea levels were from 70 to 40 m below modern (Hanebuth et al. 2000, Yokoyama et al. 2000; Chapter 1 herein). This means that one could have seen New Britain from New Guinea's Huon Peninsula, with New Ireland subsequently visible from northeastern New Britain. From New Ireland, getting to Buka in the northern Solomons would have involved crossing >100 km of ocean, including several tens of km with no land in sight (Spriggs 1997). That such voyages did take place at least 29 ka is clear from the cultural deposits at Kilu Cave on Buka (Wickler & Spriggs 1988, Wickler 2001). For much of the late Pleistocene, islands in the Solomon Main Chain were either connected or very close to each other (Chapter 5), further facilitating their suspected (but as yet undocumented beyond Buka) human colonization at this time.

The first agriculture in the region developed in highland New Guinea in the early Holocene, with evidence of plant exploitation and some cultivation on wetland margins at ca. 10,000 cal BP (Denham et al. 2003). (Holocene radiocarbon dates can be calibrated to an absolute calendric chronology based primarily on tree rings; such dates are reported as "cal BP" rather than "yr BP"; these ^{14}C dates also can be expressed on a BC/AD timescale; see Chapter 4.) By ca. 7000 cal BP, bananas (*Musa* spp.) and taro (*Colocasia esculenta*) were intensively cultivated.

From ca. 3.6 to 3 ka, pottery-making people known as the Lapita Cultural Complex moved from some part of greater Island Southeast Asia (Taiwan, Philippines, and Greater Sundas through the Moluccas) across Near Oceania (as in the Mussau sites to be discussed below) and into Remote Oceania. Lapita-age archaeological sites are rather common in the Bismarcks, especially on small or outlying islands, but rare on the larger islands and throughout the Solomon Main Chain (Kirch 2001).

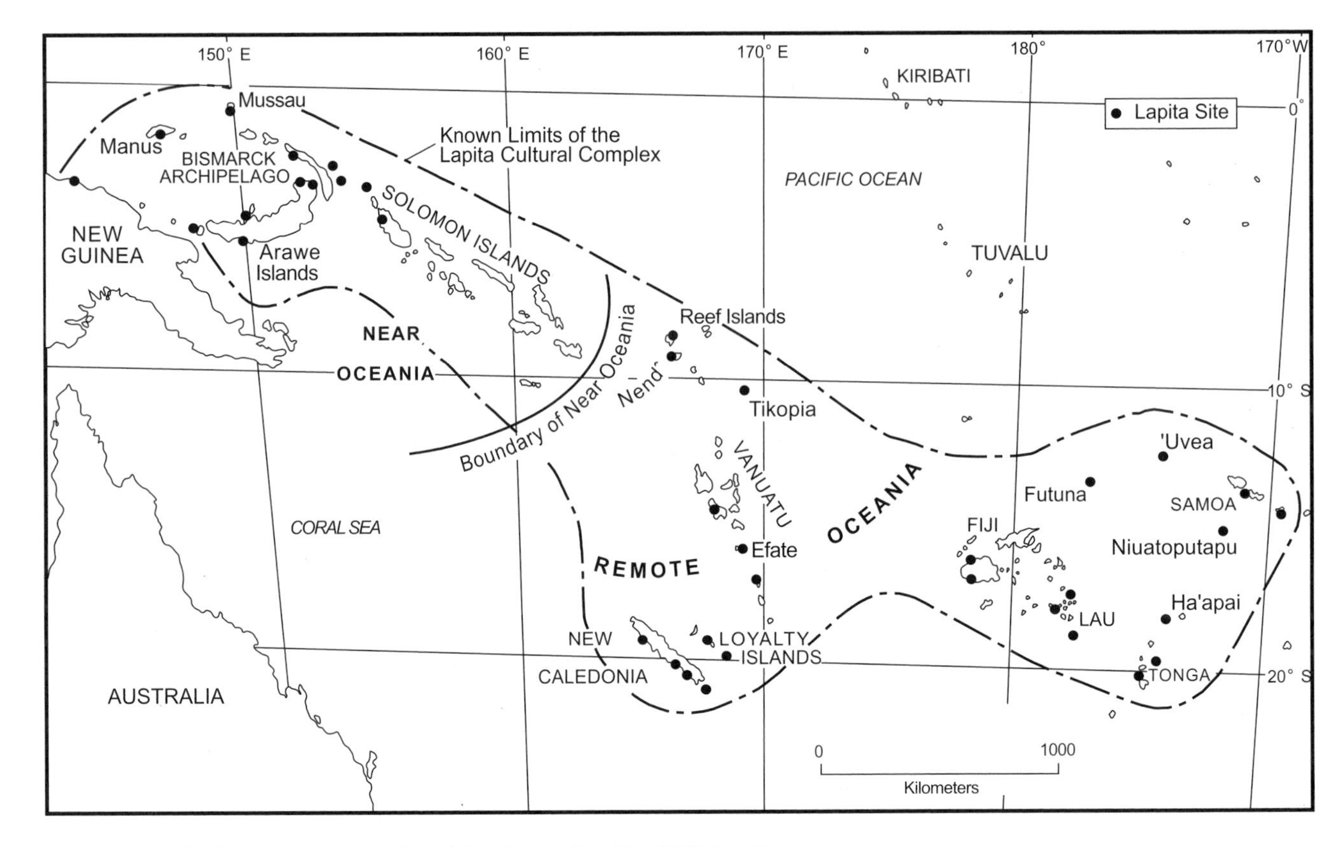

Figure 3-2 The distribution of Lapita archaeological sites. Redrawn from Kirch (2000: Map 7).

Remote Oceania

Remote Oceania refers to the rest of Oceania, i.e., Melanesia east of the Solomon Main Chain and all of Polynesia and Micronesia (Figures 3-1, 3-2). This region was not colonized by people until the late Holocene. In the southern hemisphere, estimated dates of human arrival are 3000 to 2800 cal BP for Vanuatu, New Caledonia, Fiji, Tonga, and Samoa (Kirch & Hunt 1988, Burley et al. 1995, Kirch 1997, 2000, Anderson & Clark 1999, Steadman et al. 2002a), and 2500 to 1000 cal BP for East Polynesia (Kirch 1986, 2000, Burney & Burney 2003, Conte & Anderson 2003).

As far as we know, the Lapita sailors were the first people to colonize the Santa Cruz Group, Vanuatu, New Caledonia, Fiji, Tonga, and Samoa (Irwin 1992, Kirch 1997, 2000, Anderson et al. 2001, Oppenheimer & Richards 2001; Figure 3-2 herein). Lapita pottery bears distinctive if variable decorative patterns produced mainly by dentate stamping in the wet clay, perhaps using modified epidermal scutes of the hawksbill turtle *Eretmochelys imbricata* (Ambrose 1997). Lapita pottery is unknown, and indeed any sort of ceramics is essentially unknown, on Niue and throughout East Polynesia. The Lapita peoples were outstanding seafarers as well as agriculturalists, hunters, and fishers who exploited a wide range of marine and terrestrial animals (Nagaoka 1988, Kirch 1989, Dye & Steadman 1990, V. L. Butler 1994). Their settlements were primarily along coastlines rather than the interiors of islands. The name "Lapita" derives from the Lapita site (WKO-013) on New Caledonia, where the distinctive pottery that is the trademark of the Lapita Cultural Complex was first discovered (Gifford & Shutler 1956, Kirch et al. 1997, Sand 2001a). Based in part on variation in ceramic styles, the Lapita Cultural Complex has been divided into three provinces–Far Western (Bismarcks), Western (Solomons, Vanuatu, New Caledonia), and Eastern (Fiji, Tonga, Samoa; Summerhayes 2001a).

The chronology of human arrival in East Polynesia (the Cook Islands eastward) has been debated for decades (Emory & Sinoto 1965, Sinoto 1970, Kirch 1984a, 1986, Spriggs & Anderson 1993, Kirch & Ellison 1994, Conte & Anderson 2003). The best review of this complex topic is Kirch (2000:230–245). To generalize, archaeological sites older than 1000 cal BP are scarce or absent in East Polynesia, although sedimentological and paleobotanical information suggests human presence in at least some of the island groups from 500 to 1500 years earlier than 1000 cal BP. In spite of these discrepancies in chronology of first arrival, there is a consistent pattern of heavy exploitation of native birds early in the archaeologically documented cultural sequence in East Polynesia (which typically begins at ca. 1000 cal BP), followed by an increased dependence on domesticated and commensal species after the extinction of most species of birds (Dye & Steadman 1990, Kirch et al. 1995, Steadman & Rolett 1996).

The peopling of Micronesia is another challenging issue. The region has 3000+ islands and islets (primarily atolls) scattered across 5000+ km of ocean. Linguistic evidence argues that Palau and the Marianas were colonized separately from Island Southeast Asia, whereas Austronesian languages of the Oceanic subgroup are spoken in the Carolines, Marshalls, and Kiribati, this suggesting (along with archaeological data) that human arrival in these island groups might be related to Lapita and post-Lapita expansion into Remote Oceania nearly 3 ka (Intoh 1997, Kirch 2000:167). The earliest archaeological sites in the Marianas date to ca. 3600–3400 cal BP (Craib 1993, 1999, Hunter-Anderson & Butler 1995, Amesbury et al. 1996). No sites in Palau are older than ca. 2200 cal BP (Liston et al. 1998a,b), although I would caution that the early archaeology (as well as paleontology) of Palau is not well known (Pregill & Steadman 2000). East of Palau and the Marianas, the earliest cultural sites from various island groups date from ca. 2200 to 1000 cal BP (Kirch 2000: Table 6.1).

The avian losses in parts of Remote Oceania are reminiscent of the "blitzkrieg" model of extinction (Mosimann & Martin 1975, Martin 1990; Chapter 16 herein): a rapidly dispersing people with high population growth who hunted intensively wherever they went, wiped out many species, and then moved on to richer hunting grounds. Not everyone, of course, moved on. Polynesians and Micronesians arrived with a productive set of domesticated, nonnative animals (Allen et al. 2001b, Matisoo-Smith & Allen 2001), and on many islands a portion of the founding population remained to establish long-term settlements based largely on horticulture of nonnative plants (Kirch & Lepofsky 1993) and fishing (Allen 1992, 2002). While the prehistoric loss of birdlife on most islands probably was blitzkrieg-like (Steadman et al. 2002a), avian extinction on islands such as Mangaia (Cook Islands) and those in the Marianas required a millennium or more or human presence (Steadman 1995a, 1999a; Chapters 7, 8, and 16 herein).

Selected Sites in Near Oceania

New Ireland

Late Pleistocene cultural sites are found in limestone caves and rockshelters on New Ireland (Figure 3-3). The bird bones from six sites (Panakiwuk, Balof 1, Balof 2, Buang Merabak, Matenkupkum, Matenbek), reported by Steadman et al. (1999b), are summarized in Table 5-4.

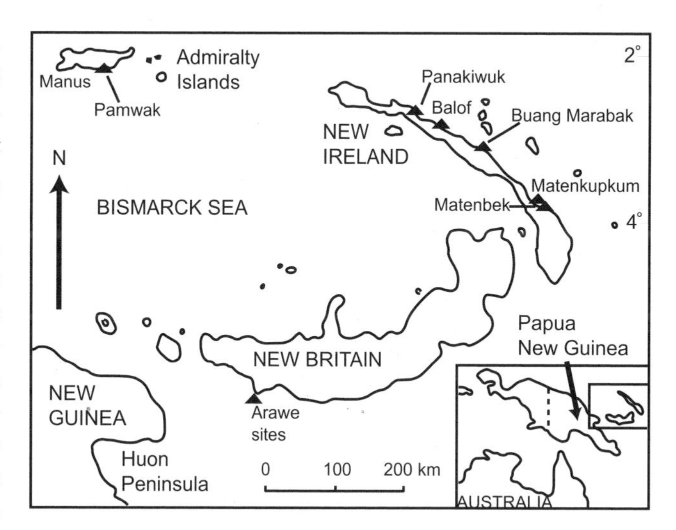

Figure 3-3 Selected archaeological sites in the Bismarck Archipelago.

Panakiwuk (Figure 3-4) is a rockshelter at ca. 150 m elevation and 4 km from either coast (Marshall & Allen 1991). Cultural deposits reach an average depth of 1.45 m in the major excavation (3 m^2). Panakiwuk was first occupied at 15,000–14,000 yr BP, although the richest cultural deposits date from 10,000 to 8000 yr BP. The site was abandoned for most of the Holocene, with minor reoccupation after 16,000 yr BP.

The Balof site consists of two rockshelters formed beneath the edges of a sinkhole 2.7 km from New Ireland's eastern coast. The cultural deposits at Balof 1 are ca. 1 m deep and substantially reworked by oven pits dug in recent centuries, whereas those at Balof 2 are ca. 2 m deep and relatively undisturbed (Downie & White 1978, Allen et al. 1989, White et al. 1991). Both Balof sites are as old as 14,000 yr BP but were occupied only intermittently.

Buang Merabak is a limestone cave ca. 200 m from the eastern coast at 50 m elevation (Leavesley & Allen 1998). A 1 × 1 m excavation, reaching bedrock at 1.6 m depth, discloses early, sparse human occupation beginning at 32,000 yr BP, periodic abandonments, and increased occupation at the end of the Pleistocene (14,000–10,000 yr BP).

Matenkupkum is a large limestone cave in southern New Ireland, ca. 30 m inland and 15 m elevation (Allen et al. 1989, Gosden & Robertson 1991). Although sea level was ca. 40 to 130 m lower during the site's late Pleistocene occupation, the cave's horizontal distance from the sea would not have increased much because of the steep coral reefs. Cultural deposits indicate intermittent occupation from 35,000 to 20,000 yr BP, then abandonment until 16,000 yr BP. The most intensive occupation was from 14,000 to 10,000 yr BP.

Figure 3-4 Matenbek Rockshelter, New Ireland. Scanned from Kirch (2000: Figure 3.6), based on a photo by J. Allen.

Matenbek (Figure 3-4) is a limestone cave 70 m south of Matenkupkum (Allen et al. 1989). The bones are from two test pits (80 × 80 cm) that sampled 90 cm of cultural deposits sandwiched beneath 45 cm of inwashed soils and above 30 cm of sterile sands. Two phases of human occupation at Matenbek are dated at 20,000 to 18,000 yr BP, and 9000 to 6000 yr BP.

The stone artifacts from New Ireland consist mainly of unretouched flakes that, on average, are much smaller at Panakiwuk and Balof than at Matenkupkum and Matenbek, where flaked river cobbles predominate (Figure 3-5). Shell tools also occur throughout the sequences. Obsidian imported from New Britain first appears in the sites as early as 19,000 yr BP (Matenbek) and as late as 7,000 yr BP (Balof 2). Potsherds occur in small numbers only in deposits <3000 yr BP.

Although most of the New Ireland sites have noncultural deposits at their bases, these strata contain only bones of indigenous rodents. Fish bones occur throughout the cultural sequence. Bones in all strata (Analytic Units A to D) are mostly of indigenous rats and bats (Table 3-2). A nonnative marsupial, the phalanger *Phalanger orientalis*, appears after 19,000 yr BP. Across New Ireland, other nonnative mammals (the cuscus *Spilocuscus maculatus*, wallaby *Thylogale brunii*, rats *Rattus praetor* and *R. exulans*, pig *Sus scrofa*, and dog *Canis familiaris*) are found only in Holocene contexts (Flannery & White 1991, Flannery 1995a).

Mussau

The Mussau Islands lie 80 km NNW of Lavongai (New Hanover), which is west of New Ireland. The two major islands are Mussau or St. Matthias (415 km^2, elev. 650 m) and Emirau (32 km^2, elev. <10 m). Just off southern Mussau are at least 10 small, low islets, the largest of which are Eloaua (ca. 5 km^2) and Emanaus (ca. 4 km^2). Of the many archaeological sites on these islets and adjacent Mussau, three (Talepakemalai, Etakosara, and Epakapaka; Figure 3-6) have produced bird bones (Steadman & Kirch 1998). Several Mussau sites were stilt villages at the edges of shallow, calm lagoons, leading to excellent preservation of delicate plant and animal remains in anoxic fine silts. The

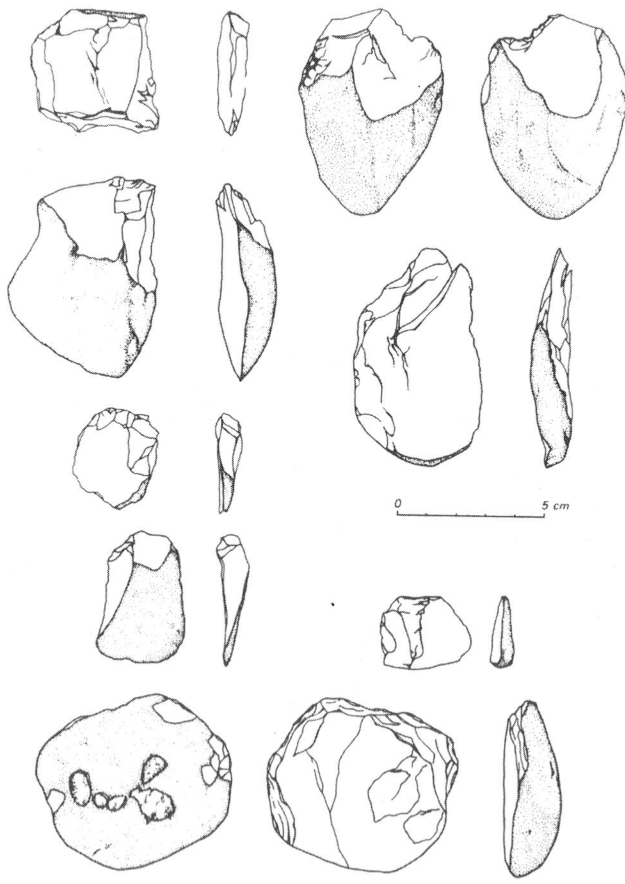

Figure 3-5 Flaked stone artifacts from the Matenkupkum site, New Ireland. Stippling represents river pebble cortex. Scanned from Kirch (2000: Figure 3.7), based on Allen et al. (1989: Figure 3).

Table 3-2 Number of bones from Analytic Units A (youngest) through D (oldest), Panakiwuk site, New Ireland. i, introduced species. Condensed and corrected from Marshall & Allen (1991: Table 14). Birds are identified in Table 5.4.

	Analytic Unit				
	A	B	C	D	Total
FISH					
Osteichthyes	18	17	1	1	37
AMPHIBIANS					
Anura	28	198	2	—	228
REPTILES					
Boidae	14	2	—	—	16
? Elapidae	10	7	—	2	19
Serpentes	—	1	—	—	1
? *Crocodylus*	2	5	—	—	7
? *Chelonia*	6	6	—	1	13
Varanus	11	15	—	—	26
? *Tiliqua*	2	7	4	—	13
Gekkonidae, Scincidae	—	8	1	2	11
BIRDS					
Identified bird	4	20	2	—	26
MAMMALS					
Homo sapiens (i)	—	2	—	—	2
Phalanger orientalis (i)	30	52	1	—	83
Thylogale brunii (i)	6	—	—	—	6
Pteropus neohibernicus	11	8	—	—	19
Dobsonia moluccensis	193	209	5	15	422
Megachiroptera	54	389	11	7	461
Microchiroptera	8	30	—	2	40
Chiroptera	321	900	23	21	1265
Melomys rufescens	1373	7383	314	106	9176
Rattus sp. (i)	406	1186	25	29	1646
Rattus exulans (i)	27	2	1	—	30
TOTAL	2524	10,447	390	186	13,547

Talepakemalai site in particular has an abundant variety of Lapita pottery (Figure 3-7), artifacts of stone, shell, teeth, and bone, and food remains (midden) represented by bone, shell, and plant (Kirch 1988, 1989, 2001, Kirch & Hunt 1988, Kirch et al. 1989, V. L. Butler 1994). Radiocarbon dates on wood charcoal or marine shell from the three sites range from ca. 3600 to 2200 cal BP. The relative scarcity of bones from extirpated species of birds may be because much extinction already had taken place at the hands of hypothesized pre-Lapita inhabitants, although no pre-Lapita sites have been found yet on Mussau (Chapter 5).

Selected Sites in Remote Oceania

The pursuit of unexploited avifaunas, not to mention pristine fishing and shelling grounds, may explain why the Lapita people, and later colonizers of East Polynesia, moved so rapidly across the Pacific. Once beyond Near Oceania, abundant, tame birds and previously unfished reefs awaited on each new island. Navigation toward unknown islands, and subsequent navigation between distant islands already known, was aided by seabirds, whose daily and seasonal movements would have been known to early voyagers

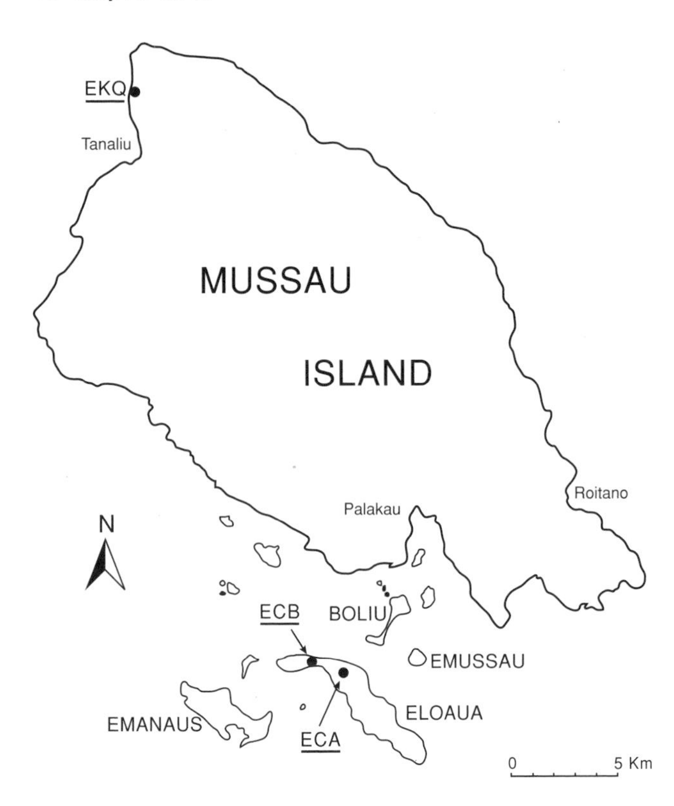

Figure 3-6 Archaeological sites on or near Mussau, Bismarck Archipelago. ECA, Talepakemalai; ECB, Etakosara; EKQ, Epakapaka. Redrawn from Steadman & Kirch (1998: Figure 2).

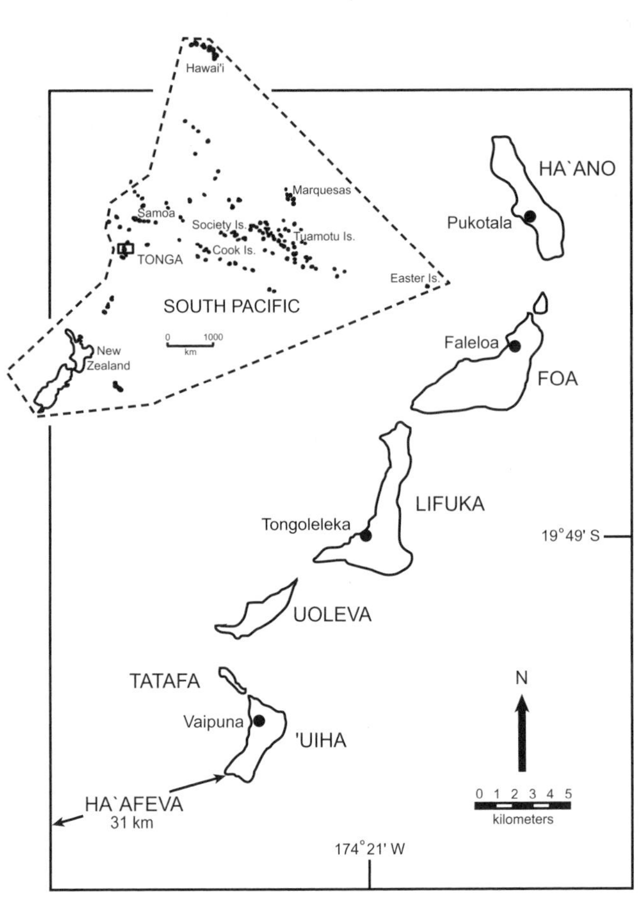

Figure 3-8 Lapita archaeological sites in the Ha'apai Group of Tonga. From Pregill & Steadman (2004).

Figure 3-7 Lapita pottery, Talepakemalai site, Mussau. Scanned from Kirch (2000: Figure 4.3), based on a photo by Thérèse Babineau.

(Chapter 4). Feeding seabirds also helped sailors to locate fish both nearshore and offshore. The eventual loss of most populations of shearwaters, petrels, boobies, terns, and other seabirds must have diminished the importance of seabirds as navigational and fishing aids by late prehistoric times (Steadman 1997b).

The tameness of birds in Remote Oceania at first human contact must have been remarkable, the species having evolved in settings free of people or other mammalian predators. Obtaining birds may have resembled our concept of gathering more than hunting. Native birds furnished easily exploited fat and protein while colonists worked to establish stocks of cultivated plants and domesticated animals. Birds also provided people with pets, feathers for decoration, and bones for tools and ornaments (Chapter 4). These activities must have decreased or become more specialized as one species after another was lost.

Now I describe five examples of archaeological or paleontological sites in Remote Oceania that have produced major samples of bird bones—one in West Polynesia, three in East Polynesia, and one in Micronesia.

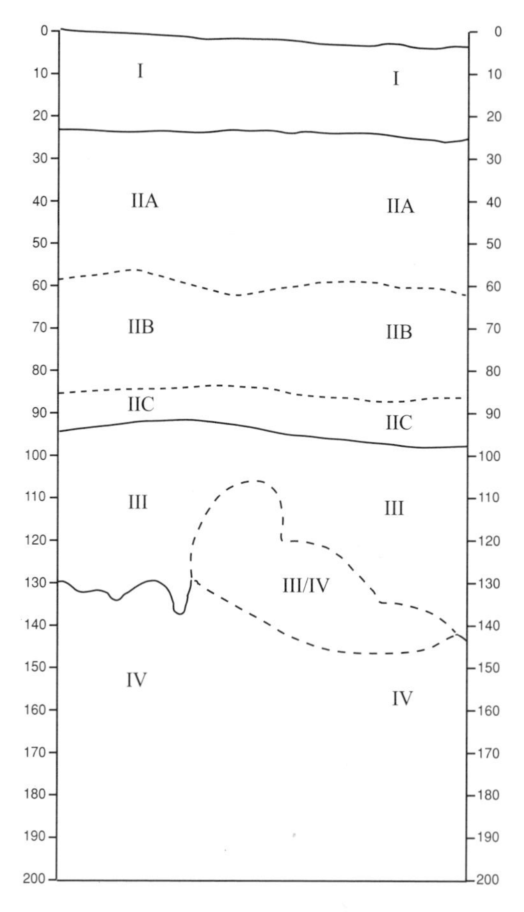

Figure 3-9 Stratigraphic profile, unit 10, Tongoleleka site, Lifuka, Ha'apai Group, Tonga. Vertical scale in cm. From Steadman et al. (2002a).

Tongoleleka Site, Lifuka, Ha'apai Group, Tonga

Lying on the western (leeward) coast of Lifuka (Figure 3-8), Tongoleleka is an accumulation of pottery, marine mollusks, and nonhuman bones (Steadman 1989c, Dye 1990, Dye & Steadman 1990, Dickinson et al. 1994, Burley et al. 1995, Steadman et al. 2002a). Layers I–III (Figure 3-9) have abundant cultural materials, although they often are degraded in Layer I, a dark, clayey, silty topsoil. Layer II (subdivided into IIA, IIB, and IIC because of minor textural and color changes) is a slightly sandy silt lighter in color and less compacted than Layer I. Layer III is even less compacted, lighter, and much sandier. Layer IV is culturally sterile calcareous beach sand. Layer III/IV is a mixture of silty sand (Layer III) and beach sand (Layer IV), the former especially rich in bones that probably represent the first habitation of Lifuka.

The pottery at Tongoleleka includes Lapita-style dentate-stamped bowls, plates, and jars. Other artifacts are of shell (beads, bracelets) and bone (awls, needles). The midden bones are mainly of fish but also sea turtles, small lizards, iguanas, birds, flying foxes, rats, dogs, pigs, and porpoises.

The 2 m deep sediments at Tongoleleka record dramatic changes in artifacts and vertebrates after human arrival 2850 cal BP (Figure 3-10). The lower strata contain decorated Lapita-style pottery and bones of an extinct iguana (*Brachylophus gibbonsi*; Pregill & Steadman 2004) and many species of extinct birds (Chapter 6). These materials are rare or absent in Layers I and II, where "Polynesian plainware" pottery and bones of extant vertebrates are found. Twenty ^{14}C dates on individual bones of the iguana, an extinct megapode (*Megapodius alimentum*), and the nonnative chicken (*Gallus gallus*) suggest that loss of the first two species and establishment of the last occurred over a time period too short (a century or less) to be resolved by ^{14}C dating (Steadman et al. 2002a; Tables 3-3, 3-4 herein).

Tangatatau Rockshelter, Mangaia, Cook Islands

Mangaia (Figures 1-22, 1-23, 3-11) is a composite (limestone and eroded volcanic) island with many archaeological

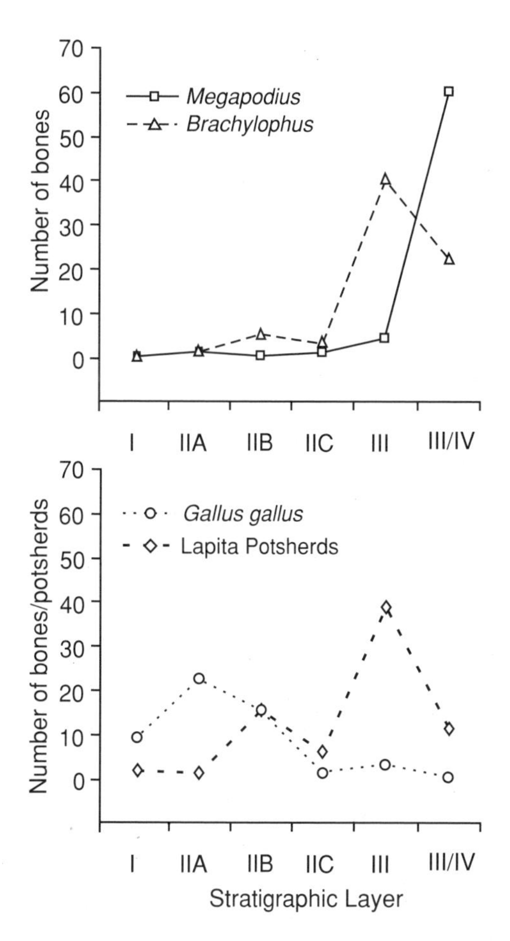

Figure 3-10 Vertical distribution of Lapita (decorated) potsherds and bones of extinct iguanas (*Brachylophus gibbonsi*), extinct megapodes (*Megapodius alimentum*), and introduced chickens (*Gallus gallus*) at Unit 10, Tongoleleka site, Lifuka, Tonga. From Steadman et al. (2002a).

Table 3-3 AMS (accelerator-mass spectrometer) ^{14}C ages on individual bones of chicken (*Gallus gallus*), extinct megapode (*Megapodius alimentum*), and extinct Tongan iguana (*Brachylophus gibbonsi*) from excavation unit 10, Tongoleleka (site Li7), Lifuka, Ha'apai Group, Tonga. Stratum reflects the natural sedimentary structure of the site. Levels are arbitrary within strata, continuous from top to bottom, and average 5 cm in thickness, except for the two samples (Beta-135249, 135250) from adjoining excavation unit 11, where the levels (*) do not correspond to those from unit 10; the stratum and depth of these two samples, however, correlate with those from unit 10. Fea. = Feature. The uncorrected and corrected ^{14}C ages are reported in radiocarbon years before present (yr BP, with AD 1950 being present). The calibrated ^{14}C ages (95% C.I.) are calibrated calendrically for secular variation in atmospheric ^{14}C, reported as Cal BP following Stuiver et al. (1998). From Steadman et al. (2002a).

Laboratory number	Species	Skeletal element	Stratum	Level	Depth (cm below datum)	Conventional ^{14}C age	Calibrated ^{14}C age
Beta–134573	*Gallus gallus*	humerus	IIA	5	25/30–33/35	2730 ± 50	2935–2755
Beta–134574	*Gallus gallus*	coracoid	IIA	8	44/47–52/53	2670 ± 40	2845–2745
Beta–134575	*Gallus gallus*	tarsometatarsus	IIB	13	70/75–76/79	2640 ± 40	2790–2735
Beta–134576	*Gallus gallus*	pedal phalanx	IIC	19a	99/100–102/122	2610 ± 60	2795–2710
							2585–2510
Beta–134577	*Gallus gallus*	sternum	III	22	106/108–111/113	2590 ± 40	2770–2720
Beta–135249	*Gallus gallus*	carpometacarpus	III	*	120–130	2670 ± 40	2845–2745
Beta–134580	*Megapodius alimentum*	carpometacarpus	IIC	19b	118–124 (Fea. 2)	2790 ± 40	2970–2785
Beta–135250	*Megapodius alimentum*	pedal phalanx	III	*	140–160	2750 ± 40	2935–2770
Beta–134581	*Megapodius alimentum*	pedal phalanx	III/IV	33	145/147–148/151	2910 ± 50	3220–2890
Beta–134582	*Megapodius alimentum*	scapula	III/IV	34	146/148–152/159	2770 ± 40	2950–2775
Beta–134583	*Megapodius alimentum*	pedal phalanx	III/IV	35	152/159–165/169	2770 ± 50	2970–2770
Beta–134586	*Megapodius alimentum*	humerus	III/IV	40	178–193 (Fea. 5)	2740 ± 50	2945–2760
Beta–134584	*Megapodius alimentum*	tarsometatarsus	III/IV	42	186/190–197/199 (Fea. 5)	2780 ± 40	2960–2780
Beta–134585	*Megapodius alimentum*	ulna	III/IV	45	202 (Fea. 5)	2760 ± 50	2960–2765
Beta–134587	*Brachylophus gibbonsi*	pedal phalanx	IIB	12	67/68–70/75	2780 ± 40	2960–2780
Beta–134588	*Brachylophus gibbonsi*	femur	IIC	18	93/97–95/99	2630 ± 40	2785–2735
Beta–134589	*Brachylophus gibbonsi*	femur	III	23	111/113–116/119	2660 ± 40	2835–2740
Beta–134590	*Brachylophus gibbonsi*	thoracic vertebra	III	27	128/131–132/136	2730 ± 40	2890–2760
Beta–134591	*Brachylophus gibbonsi*	thoracic vertebra	III/IV	31	141/144–146/149	2700 ± 40	2865–2755
Beta–134592	*Brachylophus gibbonsi*	thoracic vertebra	III/IV	34	146/148–152/159	2680 ± 50	2865–2740

Table 3-4 Stratigraphic and taxonomic means (rounded to nearest 10^1) of 20 AMS ^{14}C dates from excavation units 10 and 11, Tongoleleka (site Li7), Lifuka, Ha'apai Group, Tonga. The calibrated ^{14}C ages are presented at 95% C.I. i, introduced species. Based on data in Table 3-3. From Steadman et al. (2002a: Table 2).

Stratum/Species	Number of AMS ^{14}C dates	Weighted mean average conventional ^{14}C age (yr BP)	Weighted mean average calibrated ^{14}C age (cal BP)
IIA	2	2690 ± 30	2850–2760
IIB/IIC	5	2700 ± 20	2840–2760
III	5	2680 ± 20	2850–2750
III/IV	8	2760 ± 20	2880–2790
Gallus gallus (i)	6	2650 ± 20	2780–2750
Megapodius alimentum	8	2780 ± 20	2950–2780
Brachylophus gibbonsi	6	2700 ± 20	2840–2760

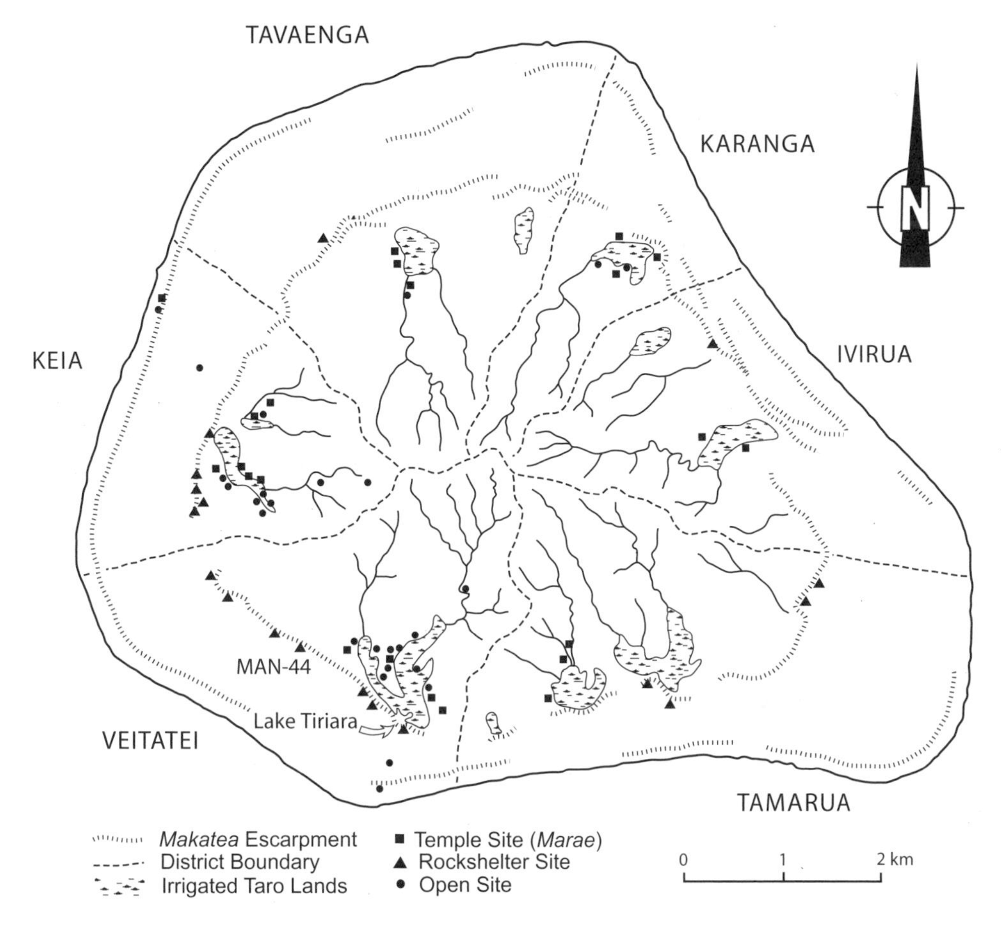

Figure 3-11 Archaeological sites on Mangaia, Cook Islands. Redrawn from Kirch (2000: Figure 8.2).

and paleontological sites (Chapters 4, 7). Tangatatau Rockshelter (site MAN-44) has produced one of East Polynesia's best chrono-stratigraphic sequences of artifacts, bone, shell, and plant materials (Steadman & Kirch 1990, Kirch et al. 1992, 1995, Kirch 1996). The cultural strata at MAN-44 (Figure 3-12) are ashy middens that range in age from ca. 1000 cal BP (Analytic Zone 2) to 200 cal BP (Zone 15). The underlying Zone 1 is culturally sterile but has many avian fossils. The artifacts consist of 570 basalt adzes, shell fishhooks, stone pounders, scrapers, files, bone needles, and beads, not to mention innumerable basalt flakes (Figure 3-13). Conforming to the situation across East Polynesia (Leach 1982, Kirch et al. 1988), pottery is absent, even in the earliest zones. Plant macrofossils from MAN-44 reflect many of the major food crops of prehistoric East Polynesians, such as coconuts (*Cocos nucifera*), ti (*Cordyline terminalis*), sugarcane (*Saccharum* sp.), banana (*Musa* sp.), sweet potato (*Ipomoea batatas*), taro (*Colocasia esculenta*), and giant swamp taro (*Cyrtosperma chamissonis*).

Figure 3-12 Well-stratified sediments at Tangatatau Rockshelter, Mangaia, Cook Islands. Scanned from Kirch (2000: Figure 8.3).

MAN-44 may not reflect the earliest occupation of Mangaia because bones of nonnative species (rat, dog, pig,

Table 3-5 Number of bones from Analytic Units 1A (oldest) through 19 (youngest), Tangatatau Rockshelter, Mangaia. i, introduced species. "Rat" is primarily *Rattus exulans* but includes some *R. rattus* in the upper zones. Condensed and corrected from Kirch et al. (1995: Table 4).

Species	1A	1B	2	3	4	5	6	7	8	9	10	11	13	14	15	17	18	19	Total
FISH	53	1098	4535	4578	4940	2937	809	1280	6763	285	426	418	162	47	1450	276	1039	505	31,601
REPTILES																			
Sea turtle	—	—	—	—	2	—	—	—	1	—	—	—	—	—	—	—	—	—	3
BIRDS	55	153	124	48	125	11	3	13	63	1	5	1	8	7	78	26	7	24	752
MAMMALS																			
Pig (i)	—	—	22	67	14	8	12	9	62	—	4	—	—	—	1	16	6	16	237
Dog (i)	—	—	23	12	—	—	—	—	5	—	2	—	—	—	—	4	1	3	50
Rat (i)	1	62	506	438	195	23	40	15	330	22	131	10	7	29	352	83	47	69	2360
TOTAL	109	1313	5210	5143	5276	2979	864	1317	7224	308	568	429	177	83	1881	405	1100	617	35,003

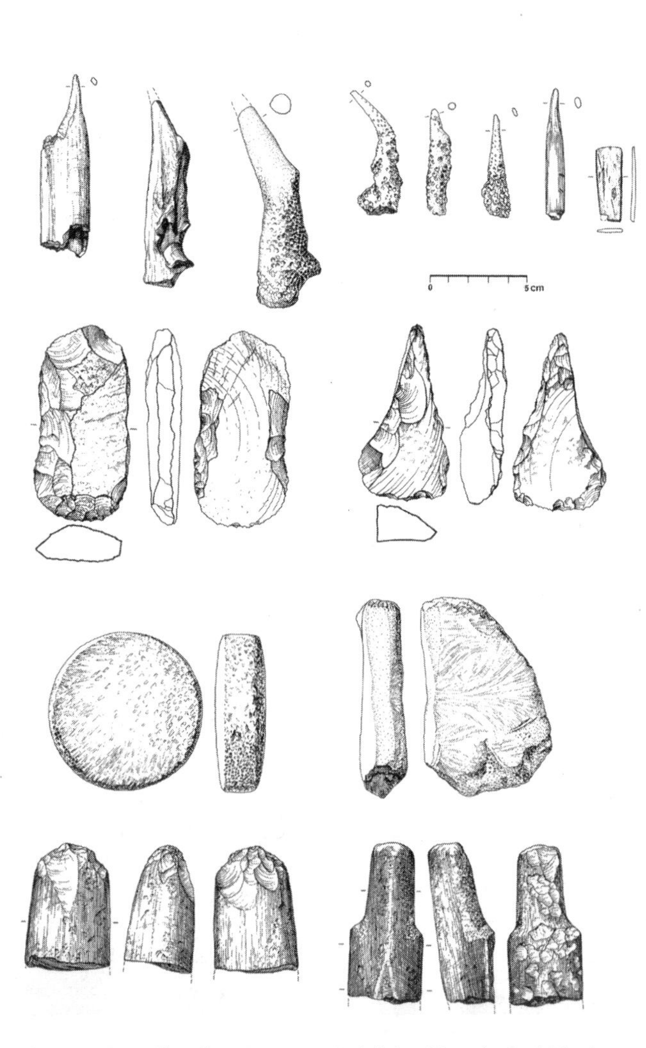

Figure 3-13 Artifacts from Tangatatau Rockshelter, Mangaia, Cook Islands, including bone awls, coral and sea urchin abraders, and a bone tattoo needle (top row), flaked basalts scrapers or knives (second row), a coral gaming stone and abrader (third row), and basalts adzes (bottom row). Scanned from Kirch (2000: Figure 7.19), based on a drawing by Judith Ogden.

chicken) occur regularly in Zone 2 (Table 3-5). One would expect bones of these species to be scarce or absent in Zone 2 if it represented first contact. Also, Zone 2 has little evidence for heavy exploitation of seabirds, in contrast to the case in early sites elsewhere in East Polynesia (Steadman 1995b). Zone 2 of MAN-44 may represent the first long-term occupation of Mangaia's interior, with earlier human habitation of the island intermittent and restricted to the coast.

Consumption of vertebrates changed considerably while MAN-44 was occupied. Fish bones dominate all cultural strata (Butler 2001, Kirch et al. 1995). The relative abundance of pig, dog, rat, and chicken bones increases in the middle levels (Zones 8–10). All of these except rat decline in upper prehistoric levels, perhaps because resource stress forced people to eat lower on the food chain. Birds at MAN-44 are rich in species only in early levels (ca. AD 1000 to 1300; Zones 2–4). The dramatic decline in native species of birds between Zone 4 (22 species) and Zone 5 (five species) may reflect the felling of interior forests to create an agricultural landscape to feed a growing human population (Kirch 1996, Steadman 1997b). By ca. AD 1300, most species of birds were either gone or too rare to be sampled archeologically.

Hanamiai Dune Site, Tahuata, Marquesas Islands

Lying in a coastal valley on Tahuata in the southern Marquesas (Figure 3-14), the Hanamiai Dune site has yielded artifacts, cultural features (hearths, postholes, stone structures, etc.), and faunal materials (Table 3-6) spanning an 800-year period (Rolett 1998). In 1984–1985, Barry Rolett and assistants excavated a 21 m^2 area (Figure 3-15) only 30 m from the coast, revealing cultural strata (Layers A–H;

Table 3-6 Number of bones from Zone G/H (oldest) through AB (youngest), Hanamiai site, Tahuata, Marquesas Islands, i, introduced species. Condensed from Rolett (1998: Table 5.2).

Phase	I	II	III	IV	V	
Zone	G/H	G	EF	CD	AB	Total
FISH	2292	2216	2001	4401	501	11,411
REPTILES						
Sea turtle	20	10	7	9	3	49
BIRDS						
Native seabirds	448	88	13	14	4	567
Native landbirds	55	5	5	5	—	70
Chicken (*Gallus gallus*) (i)	3	9	4	6	—	22
Unidentifiable	26	10	2	4	—	42
MAMMALS						
Pig (*Sus scrofa*) (i)	4	22	65	251	105	447
Dog (*Canis familiaris*) (i)	2	26	20	50	11	109
Rat (*Rattus exulans*) (i)	60	269	165	506	28	1028
Small whale (Odontoceti)	28	17	94	166	14	319
Medium whale (Odontoceti)	10	4	—	2	—	16
Human (*Homo sapiens*) (i)	—	1	3	2	2	8
TOTAL	2948	2677	2379	5416	668	14,088

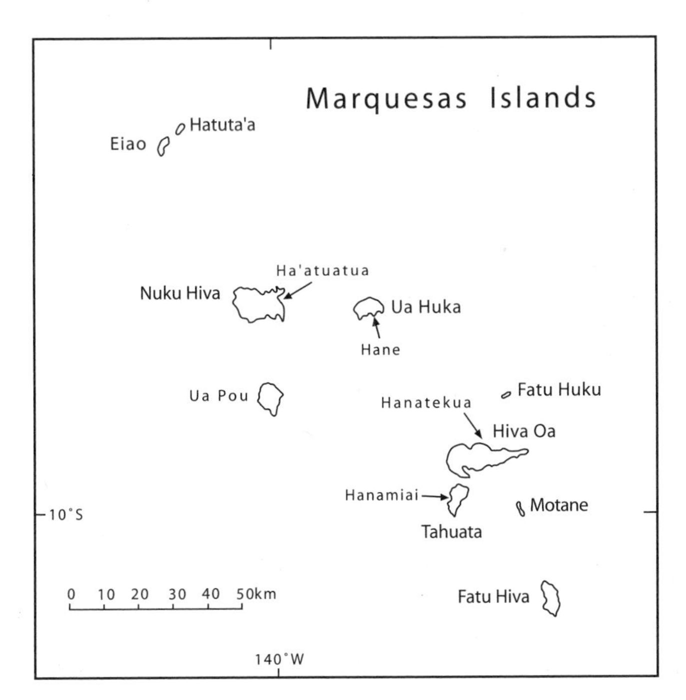

Figure 3-14 The Marquesas Islands, showing the location of Hanamiai and other major archaeological sites. Hatuta'a is an alternative name for Hatutu. Redrawn from Rolett (1988: Figure 2.2).

Figure 3-16) to depths of 1.6 m. The earliest deposits (Layers G/H and H) represent Phase I of the Hanamiai cultural sequence, dated at AD 1025–1300 (= 925–650 cal BP). Layers GH and H contain hearths and manufacturing areas for whalebone tools, pearl shell fishhooks, coconut graters, and ornaments, and basalt adzes. Layer G, a mixed calcareous and volcanic sand representing Phase II, has a massive stone pavement that probably was the veranda of a pole and thatch structure defined by postholes. The site remained a habitation area, with artifactual evidence for butchering, woodworking, and tool manufacture. Radiocarbon dates from Phase II are similar to those from Phase I.

In spite of some change in fauna and artifacts, cultural Phases III and IV (Layers C–F, the latter ^{14}C dated at ca. AD 1300–1450) reflect continued use of the dune for habitation with low stone platforms (presumably for houses), hearths, and materials from reworking adzes, animal butchering, and other domestic activities. There are no ^{14}C dates for Phase IV (Layers C, D). Historic artifacts in Layers A and B place the terminal Hanamiai occupation, Phase V, in the early 19th century.

As is always true in Marquesan sites, fish dominates the bone assemblage from Hanamiai (Table 3-6). Indigenous landbirds are represented by 70 bones of 10 species,

Figure 3-15 Excavation is progress, Hanamiai site, Tahuata, Marquesas Islands. Scanned from Kirch (2000: Fgiure 8.8), based on a photo by Barry Rolett.

including two rails, a pigeon, a dove, and three parrots no longer on Tahuata (Steadman & Rolett 1996; Table 7-21 herein). Most of the ca. 660 identifiable bird bones at Hanamiai are from extinct or extirpated species of seabirds (Rolett 1998; Table 15-3 herein). Exploitation of indigenous birds was most intense during Phase I (ca. AD 1025-1300). By AD 1450 or before, the seven extirpated species of landbirds (and probably others not recorded among only 70 bones) either had been eliminated or had become rare enough to elude archaeological sampling.

Ahu Naunau (Anakena), Easter Island

The stratigraphy, chronology, artifacts, and nonbird bones from an excavation at Ahu Naunau (Figures 3-17, 3-18) are detailed in Steadman et al. (1994). A 1 × 4 m test trench exposed three cultural strata (Layers I, II, and III). Layer I (with subunits Ia and Ib) is a rather uniform calcareous sand, separated from the underlying Layer II by a band of charcoal-stained sand. Layer II is a similar sand, although the lowermost part is interbedded with sandy, silty clays from weathered volcanic rock. Layer III is a dark reddish brown bouldery, pebbly, silty clay derived from basaltic and scoriaceous colluvium from a nearby hillside and perhaps alluvium from a nearby stream. The contact of Layers II and III seems to represent the earliest living surface at the site. Four ^{14}C dates on wood charcoal range from 1220–1430 cal AD (contact of Layers I/II) to 980–1280 cal AD at the sand-clay contact (base of Layer II).

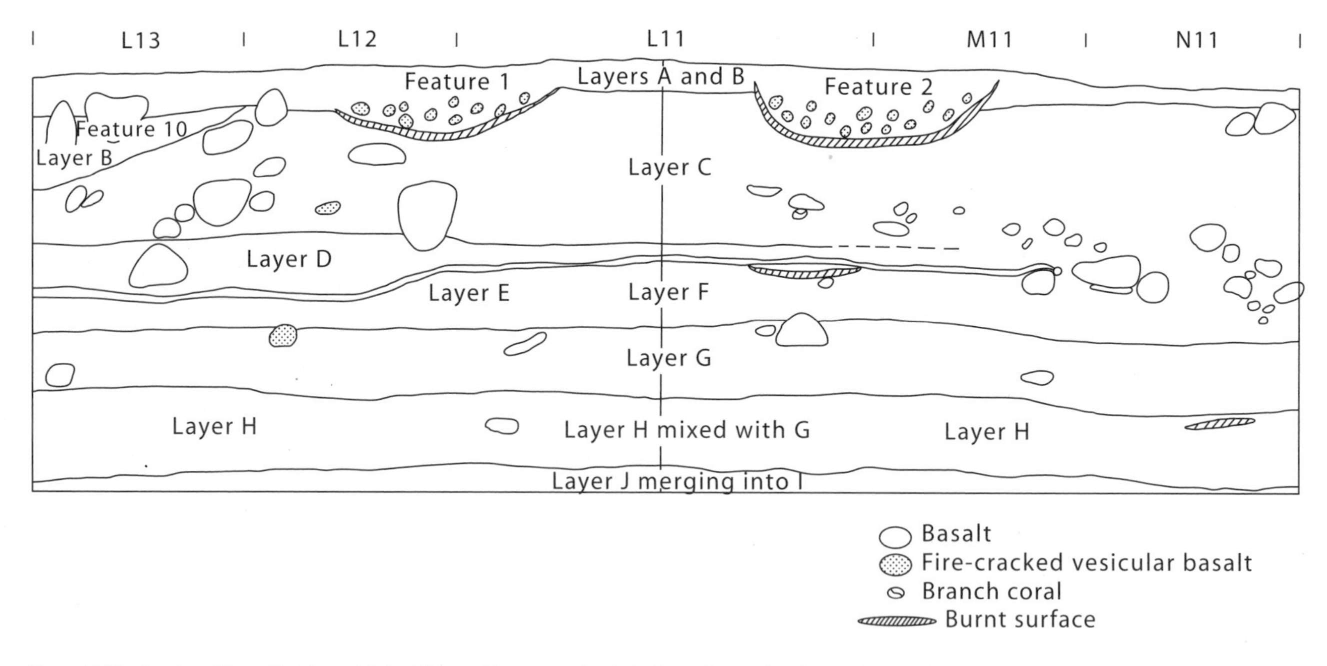

Figure 3-16 Stratigraphic profile, Hanamiai site, Tahuata, Marquesas Islands. Redrawn from Rolett (1998: Figure 4.5).

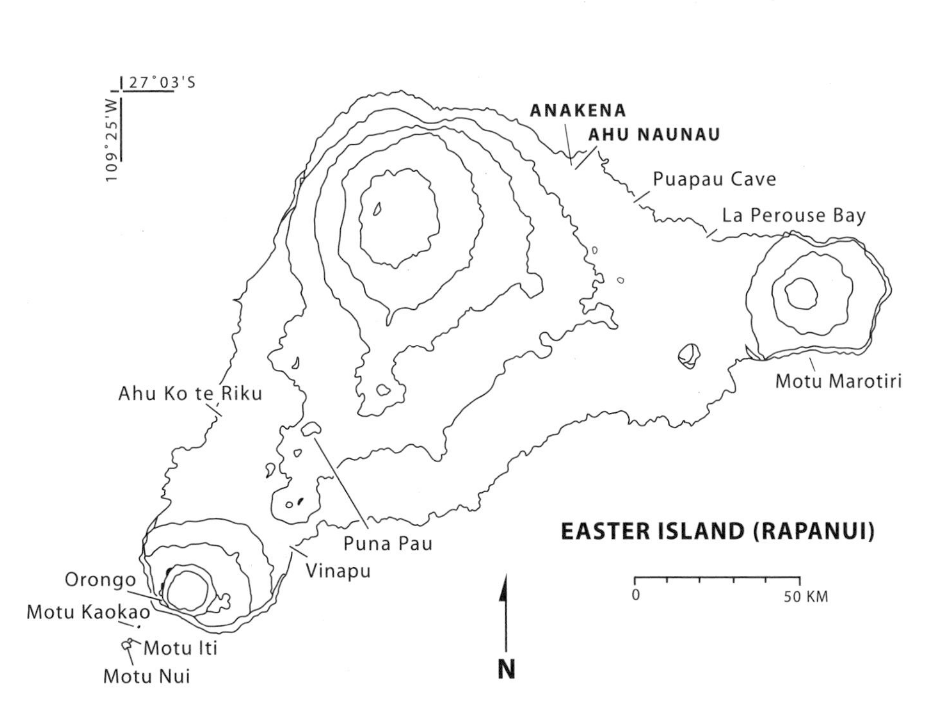

Figure 3-17 Location of Ahu Naunau archaeological site, Easter Island. From Steadman et al. (1994: Figure 1).

Figure 3-18 The excavation in progress at Ahu Naunau, Easter Island. Photo by DWS, 17 July 1991.

Table 3-7 Number of bones from the surface (youngest) to >120 cm depth (oldest), Ahu Naunau site, Anakena, Easter Island. i, introduced species. From Steadman et al. (1994: Table 5).

	Depth (cm)								
	Surface	0–20	20–40	40–60	60–80	80–100	100–120	>120	Total
FISH	—	100	248	168	87	98	205	689	1595
NATIVE BIRD	10	19	78	41	15	5	21	162	351
CHICKEN (i)	3	11	12	1	—	—	—	2	29
MAMMALS									
Rat (i)	—	252	480	616	196	44	19	536	2143
Dolphin	6	530	563	337	285	26	28	537	2312
Pinniped	1	—	1	—	—	1	—	—	3
TOTAL	20	912	1382	1163	583	174	273	1926	6433

About 90% of the 1433 lithic artifacts recovered at Ahu Naunau are obsidian (unretouched flakes, retouched flakes, cores, spearpoints, and drills). The rest are of basalt (flakes, unpolished adze preforms or scrapers, polished adze flakes, grinding stone, and pounders) or red scoria (a probable iris for the eye of a stone statue). The only other artifacts are two needles made from radii of the extirpated Murphy's petrel (*Pterodroma ultima*) and a cut and polished dolphin bone tab that probably was a net gauge.

The bones from Ahu Naunau are distinctive for Polynesia in lacking sea turtles, lizards, pigs, and dogs, and having only a single human bone. The 6433 identifiable bones from units 1–3 are dominated by the common dolphin (*Delphinus delphinus*; 39%), Pacific rat (*Rattus exulans*; 32%), and fish (23%) (Table 3-7). The Ahu Naunau bones differ from younger (post–AD 1450) faunal assemblages from Easter Island (Ayres 1979, 1985) in that marine mammals, seabirds, and native landbirds are much more

Table 3-8 Number of bones from Layer I (youngest) to V (oldest), Pisonia Rockshelter, Aguiguan, Northern Mariana Islands. i, introduced species. From Steadman (1999a: Table 8).

	Layers						
Bone	I	II	III	IV	IV–V	V	Total
FISH	8	31	105	1404	95	126	1769
LIZARD	2	2	6	121	12	42	185
BIRDS							
Identified	8	5	21	748	132	80	994
Bird sp.	—	—	12	451	50	69	582
MAMMALS							
Pteropus sp.	—	—	1	5	—	—	6
Rattus exulans (i)	10	2	19	—	—	—	31
Homo sapiens (i)	—	—	1	14	1	10	26
Sus scrofa (i)	2	—	—	—	—	—	2
Capra hircus (i)	1	—	—	—	—	—	1
Large mammal (i)	—	—	—	1	1	15	17
TOTAL	31	40	165	2780	254	344	3613

common in the former, whereas fish, humans, and chickens are rarer (Steadman 1995a). The bird bones from Ahu Naunau will be summarized in Tables 7-22 and 7-23.

Polynesians hunted marine mammals with harpoons from seaworthy canoes (Dye 1990), manufacture of which required a supply of large trees. Deforestation of Easter Island was essentially complete by about AD 1400 (Flenley et al. 1991, Bahn & Flenley 1992), after which seaworthy canoes would not have been possible to make. While fish are the third most common category of vertebrate, their relative abundance at Ahu Naunau is much less than in other major sites from tropical Polynesia (Steadman & Kirch 1990, Allen & Steadman 1990). This may reflect the difficulty of going to sea from Easter Island's rugged, unprotected coastline, especially during the frequent periods of foul weather.

Pisonia Rockshelter, Aguiguan, Northern Mariana Islands

This site lies beneath an undercut limestone cliff on the small (7 km^2) island of Aguiguan (Steadman 1999a). I discovered and excavated Pisonia Rockshelter in 1994, about 250 m inland from the north-central coast. The sediments (2+ m deep) are well stratified ashy silts with angular to subangular limestone pebbles and cobbles. Layer I is compacted, heavily burnt, and historic in age, as evidenced by pig bones and lack of pottery. Layers II–IV are gray to pinkish gray, pottery-rich, cultural deposits that range in age from AD 1290 to 1450 (Layer II) to AD 80 to 420 (Layer IV; Table 8-5). Layer V is a pale brown, very pebbly, sandy silt that is primarily precultural, based on a sharp decline in pottery content, lack of charcoal, and scarcity of burned bones. The small amount of pottery and bone in Layer V probably is not in situ but rather is from sedimentary mixing during deposition of Layer IV. The pottery from Pisonia Rockshelter, being studied by Connie Bodner, includes at least four major styles.

Bones generally are rare in Layers I–III, with fish and rat being the most common (Table 3-8). In Layer IV, fish and bird bones become abundant while those of rats disappear from the record. Layer IV yielded 2779 (77%) of the 3613 total bones dry-sieved from Test Pits 1–3. In wet-sieved samples from Test Pit 1, Layer IV yielded 25 of 27 total bones (93%). The lower percentage of bones from Layer IV in dry-sieved Test Pits 1–3 is because many bones actually from Layer IV are subsumed in the category "IV–V" in Table 3-8, these being from a loose area of the excavation where Layers IV and V had slumped together. Nearly all nonhuman bones from Layer IV are burned, providing clear evidence of human cookery. The bones identified as human or "large mammal" from Layers IV–V may all be from a prehistorically disturbed human burial.

Having tersely outlined the peopling of the Pacific and described a few exemplar archaeological sites, I will explain in Chapter 4 how prehistoric bones can generate data on the past distribution and diversity of birds. Such data have strengths and weaknesses, just as with any sort of neontological data. There is much common ground, however, and in this spirit I will use both past and present information to build an improved biogeography of tropical Pacific birds.

Chapter 4 Birds Living and Dead, on Islands and in Museums

UNTIL TWO DECADES ago, and sometimes even today, the biogeography of birds in Oceania was studied with two assumptions. One was that modern distributions were well known, especially because of the Whitney South Sea Expedition in the 1920s and 1930s. The other was that whatever extinction had occurred took place in historic times. Prehistory was not considered. For example, Adamson (1939:64) noted that "the birds of the Marquesas and most other parts of the central Pacific have been adequately collected and systematically studied. Few, if any, species remain to be discovered in the Marquesas. It is possible that some may have been exterminated recently by introduced cats and pigs, since several species once known to occur on many islands are now much restricted in distribution." Greenway (1958:95), echoing ideas of prominent colleagues such as Ernst Mayr (1941c), stated that "no form [of bird] is known to have been extirpated from the Marquesas. Bird populations were never large, the birds confiding and the habitat somewhat limited."

We now know that many species of birds became extinct prehistorically in the Marquesas and just about everywhere else in Oceania (Steadman 1995a; Chapters 5–8 herein). This is evident in Marquesan archaeological sites such as Hane and Hanamiai, where bones of extinct birds are confined to lower cultural strata (Kirch 1973, 1983, Dye & Steadman 1990, Steadman & Rolett 1996, Rolett 1998). For some species it may not be possible to determine whether extinction occurred during prehistoric (before AD 1800) or historic times. In other cases, such as five sites in the Ha'apai Group of Tonga (Chapter 6), bones of numerous extinct species occur in the earliest cultural strata (dated from 2850 to 2750 cal BP), but not in sediments only a century younger.

Even if such convincing data from prehistoric bird bones were lacking, it would be illogical to believe that all extinction in Polynesia was confined to the past two centuries of European influence. Nevertheless, scientists before Patrick Kirch assumed that the many centuries of prehistoric human occupation of tropical islands had no influence on indigenous birdlife. This Rousseauian "noble savage" outlook now has been shown, on island after island, to be wrong. The avifaunas seen by the first European explorers of the Marquesas, Tonga, and everywhere else in Oceania already had been damaged irreparably. In attempting to estimate the natural distributions of birds in the Pacific, we must realize that most volant species survive on only a subset of the islands where they lived at first human contact. Furthermore, many species are extinct, with no evidence of their former existence except bones from prehistoric sites. The extinct species include many volant species as well as flightless forms (especially rails) that typically were single-island endemics (see Chapters 9–16).

In part because of the extinction it wrought, the peopling of Oceania (Chapter 3) was of great biological importance. The surviving bird communities on Pacific islands are residuals of ones that were much richer at human arrival, so we must look to the past to estimate natural distribution and diversity. The past that concerns us, however, is modern in a geological or evolutionary sense because it is measured only in centuries, millennia, or, at most, a few tens of millennia.

The 3000 years since people arrived in Remote Oceania would be instantaneous (not discernible) by any radiometric dating method that could measure the millions or tens of millions of years that characterize most studies of fossil vertebrates. For further chronological perspective, a species of bird lost on an East Polynesian island 650 years ago represents an extinction that occurred five orders of magnitude (100,000 times) more recently than that of dinosaurs 65 million years ago. Nevertheless, just as for dinosaurs, we must use modern methods of geochronology, stratigraphy, and comparative osteology to study extinct Polynesian birds.

The extinct birds of Oceania are not some archaic assemblage of species that was destined by nature to go the way of the dinosaurs. If not for people, virtually all of the extinct species and populations of birds known from Oceania would be alive today. This claim is bolstered by the fact that avian extinctions in Remote Oceania occurred during the past 3000 years, well after the major changes in climate and sea-level associated with the Pleistocene-Holocene (glacial-interglacial) transition (Martin & Steadman 1999; Chapter 1 herein).

My goal in this chapter is to provide the background for interpreting data in Chapters 5-15, much of which is based on ancient bird bones. The prehistory of island birds features extinct species, which of course no longer can walk, fly, perch, preen, molt, sing, call, feed, or breed. This obviously restricts what we can learn about them. Here I would like to explain some of the methods, strengths, and limitations of prehistoric faunal data, relating these attributes whenever possible to living birds.

People vs. Birds

Birds were important to Oceanic peoples. The first colonizers of Remote Oceania arrived on one pristine island after another, each inhabited by wonderful birds but, once beyond New Caledonia, Vanuatu, Fiji, Tonga, and Samoa, few if any species of reptiles or mammals. The importance of birds, however, did not save them. Birds were killed for their fat, protein, bones, and colorful feathers by island colonists, who reduced most bird populations further by clearing forests, cultivating crops, and introducing exotic animals (Steadman 1997b; Chapter 16 herein). The resulting losses of birds include: (1) extinction (loss of all populations of a species); (2) extirpation (loss of a species on an entire island, with one or more conspecific populations surviving elsewhere); and (3) reduced population on an island. The last category is difficult to document through fossils, but is of concern because it can lead to extirpation or extinction. I must note here that human-created habitat changes may have increased the populations of some native species that are not forest-obligates, assuming an ability to withstand the effects of disease and nonnative predators.

Each category of loss can be exemplified on Mangaia, Cook Islands (details in Chapter 7). The conquered lorikeet (*Vini vidivici*) is *extinct* because it is gone from Mangaia as well as every other island where it ever occurred. The Society Islands pigeon (*Ducula aurorae*) is *extirpated* on Mangaia but survives on Makatea (Tuamotu Group) and Tahiti (Society Islands). Audubon's shearwater (*Puffinus lherminieri*), a pantropical seabird, has what must be a *reduced population* on Mangaia today, surviving in numbers<100, whereas archaeological and ethnographic evidence suggests that once it was common and widespread on the island.

The losses of landbirds have been greatest for megapodes, rails, columbids (= pigeons and doves), and parrots (Chapters 9–12), although no widespread family of Pacific landbird has been spared human-caused reductions in distribution and diversity (Chapters 13–14). The heaviest losses among seabirds in Oceania have been for petrels and shearwaters, with albatrosses, storm-petrels, tropicbirds, frigatebirds, boobies, terns, and gulls also affected (Chapter 15).

Seabird bones usually are much more common than those of landbirds in sites from Remote Oceania (especially Polynesia), whereas the opposite is true in Near Oceania, Vanuatu, and the Mariana Islands (Table 4-1). Among landbirds, late Pleistocene (30,000 to 10,000 yr BP) bone assemblages from Near Oceania (such as on New Ireland and Buka; Tables 5-3, 5-9) resemble those from late Holocene (<3000 yr BP) sites as distant as the Marquesas (Steadman 1992b, 1993a, 1995a) in being dominated by columbids and rails. This has been devastating to both groups but especially to rails, with all but a few of the many endemic, flightless species now extinct (Steadman 1995a; Chapter 10 herein). Bones of megapodes, parrots, and passerines (songbirds) also are recovered regularly and in good numbers in sites across most of the tropical Pacific.

Only one nonnative species of bird, the chicken or red junglefowl (*Gallus gallus*), is found prehistorically in Oceania. Feral or domesticated populations of *G. gallus* occur today nearly throughout Oceania. Chickens arrived with Lapita peoples in Vanuatu, Fiji, Tonga, and presumably Samoa (Chapters 5, 6), and are unrecorded before Lapita times in Near Oceania. Bones of *G. gallus* have been found in East Polynesian archaeological sites of all ages except on Henderson Island and New Zealand (Anderson 1989, Schubel & Steadman 1989; Chapter 7 herein). In Micronesia, prehistoric chicken bones have not been identified in the Marianas but occur elsewhere (Chapter 8).

Table 4-1 Family-level summary of bird bones (% of identified specimens), comparing data from Near Oceania with those from Remote Oceania. The nonnative species (*Gallus gallus*) is excluded. All passerine families are combined. Values are rounded to the nearest 1% except for Tahuata and Aguiguan (0.1%) and Ua Huka (0.01%). NISP = number of identified specimens.

	Near Oceania		Remote Oceania						
	Bismarcks	Solomons	Vanuatu		Santa Cruz	Society	Marquesas		Marianas
	New Ireland[1]	Buka[2]	Malakula[3]	Efate[4]	Tikopia[5]	Huahine[6]	Tahuata[7]	Ua Huka[8]	Aguiguan[9]
SEABIRDS									
Procellariidae (petrels)	1	—	<1	—	9	15	77	80	—
Oceanitidae (storm-petrels)	—	—	—	—	—	—	0.5	0.01	—
Phaethontidae (tropicbirds)	—	—	—	—	6	9	1.5	0.07	0.1
Sulidae (boobies)	—	—	—	1	7	14	3.1	2.8	—
Fregatidae (frigatebirds)	—	—	—	—	6	10	0.7	0.45	0.1
Laridae [Sterninae] (terns)	—	5	—	—	60	15	4.5	1.18	2.2
LANDBIRDS									
Ardeidae (herons)	1	1	—	—	—	5	—	0.02	—
Anatidae (ducks)	—	1	—	1	—	—	—	—	—
Accipitridae (hawks)	4	5	—	3	—	—	—	—	—
Megapodiidae (megapodes)	4	2	2	1	4	—	—	—	1.0
Phasianidae (quail)	2	—	—	—	—	—	—	—	—
Rallidae (rails)	25	44	23	78	4	16	5.3	0.25	67.0
Charadriidae (plovers)	—	—	—	—	1	1	—	0.01	2.8
Scolopacidae (sandpipers)	—	—	—	—	2	2	—	0.11	—
Columbidae (pigeons, doves)	28	38	12	10	—	9	1.9	14.8	24.3
Psittacidae (parrots)	5	1	2	2	—	2	1.5	2.2	—
Cuculidae (cuckoos)	1	—	1	1	—	—	—	—	—
Tytonidae (barn-owls)	6	—	23	1	—	—	—	—	—
Strigidae (typical owls)	1	1	—	—	—	—	—	—	—
Caprimulgidae (nightjars)	<1	—	—	—	—	—	—	—	—
Apodidae (swifts)	4	—	1	—	—	—	—	—	—
Alcedinidae (kingfishers)	<1	—	3	1	—	1	0.1	—	—
Bucerotidae (hornbills)	9	—	—	—	—	—	—	—	—
Passeriformes (songbirds)	11	—	35	2	—	1	0.3	0.04	2.4
Total NISP	241	81	190	145	178	324	632	11,828	994
% NISP seabirds	1	5	1	1	88	63	91	82	6
% NISP landbirds	99	95	99	99	12	37	9	18	94
Seabird families	1	1	1	1	5	5	6	6	3
Nonpasserine landbird families	14	8	8	9	4	7	4	6	4

1. Five sites (Steadman et al. 1999a; Chapter 5 herein).
2. One site (Steadman et al. 1999a; Chapter 5 herein).
3. Four sites (Chapter 5).
4. Arapus and Mangaasi sites (Chapter 5).
5. Kiki site (Steadman et al. 1990b; Chapter 5 herein).
6. Fa'ahia site (Steadman & Pahlavan 1992; Chapter 7 herein).
7. Hanamiai site (Steadman & Rolett 1996, Rolett 1998; Chapter 7 herein).
8. Hane site (Early Phase only) (Chapter 7).
9. Pisonia Rockshelter (Steadman 1999a; Chapter 8 herein).

Figure 4-1 Boys from Rennell, Solomon Islands, who were very accurate with slingshots at shooting small birds, especially the honeyeater *Myzomela cardinalis*. Photo by DWS, 25 June 1997.

Traditional Pacific islanders are skilled bird hunters, an art that is being lost to westernization (Steadman 1997). Boys learn to hunt birds at an early age (Figure 4-1). For most boys, even on traditional islands, the amount of time spent bird hunting wanes in the teenage years, although some continue to hunt regularly as adults. Their profound knowledge has no evolutionary context, yet the best native hunters in Oceania know more about their local birdlife than any visiting scientist. Men with whom I have worked, such as Mark Hafe of Isabel, Mart Taieha of Rennell, and Pierot Tavouiruja of Santo, dazzled me daily with their intimate knowledge of the plumage, vocalizations, seasonal and daily movements, feeding and nesting habits, and gastronomy-oriented internal anatomy of the birds on their islands. All of this unwritten knowledge (see Diamond 2001) was learned through watching, hunting, butchering, and eating birds, and then having the time to discuss these topics with other hunters. As with avid ornithologists or bird-watchers, these islanders always are conscious of the bird sounds around them, no matter what else they are doing.

I relate strongly to these men. According to my parents, my first word was "bird," uttered while pointing at a housefly circling our kitchen light one wintery night in 1952. Embarrassed by this misidentification, I worked hard at developing skills to approach birds closely. As a youth in rural western Pennsylvania, I avidly hunted birds with an air rifle (BB gun) and slingshot on our dairy farm, specializing in the nonnative sparrows (*Passer domesticus*), starlings (*Sturnus vulgaris*), and pigeons (*Columba livia*) that lived around the barn and corn crib. At age 12, allowed to purchase a hunting license and pursue game species, I was more interested in birds (grouse, pheasant, woodcock, doves, ducks) than mammals. I dissected and ate anything I shot. My interactions with native nongame birds were limited to binoculars, which made them a little less interesting at the time. All of this is to say that I appreciate the innate urge to hold a bird in your hand and learn as much as you can about it. In spite of growing up in the country (a clear advantage for naturalists; Odum 2002) and having an intense interest in birds, both free-flying and in the hand, I never will be nearly as good a hunter as the three islanders just mentioned. Another of my shortcomings is that no bones from the birds I hunted on our Pennsylvania farm will survive long enough to be of archaeological interest, they being deposited on moist, acidic, clayey soil overlying Devonian shale, rather than on sweet silts and sands overlying Cenozoic limestones.

Vertebrate Paleontology and Zooarchaeology

The backbone of my research in Oceania is vertebrate paleontology (the study of bones from noncultural sites) and

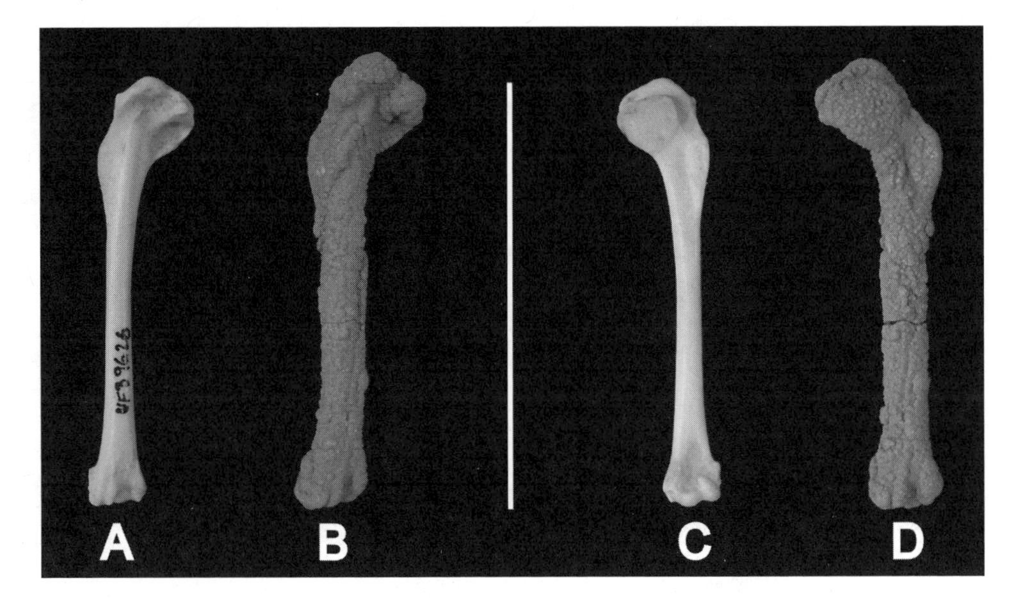

Figure 4-2 Humerus of the tropicbird *Phaethon lepturus*. A, C: Modern specimen (UF 39626, male), Tutuila, American Samoa. B, D: Calcite-encrusted fossil (UF 60269), Te Rua Rere Cave, Mangaia, Cook Islands. Scale = 80 mm. Photo by Jeremy Kirchman.

zooarchaeology (the study of bones from cultural sites). The sites vary from hundreds to thousands or tens of thousands of years old. As already noted, these time frames are young from a geological standpoint. Some sites, especially in caves and rockshelters, have bone-bearing cultural deposits directly overlying bone-bearing noncultural sediments (James et al. 1987, Steadman 1993a, Steadman et al. 2000a). Most bird bones from paleontological sites on tropical islands were deposited by "pitfall" (natural trap) activity (Hearty et al. 2004) or by avian predators such as owls, falcons, or herons. Bird bones from archaeological sites typically are the remains of species that were hunted (and usually consumed) by prehistoric people. These bones occur alongside typically much more abundant fish bones, which are not known to include extinct species but nevertheless are evidence of human impact on marine ecosystems (Butler 2001, Allen 2002, 2003).

I use the terms "bones," "ancient bones," "prehistoric bones," and "fossils" interchangeably for the osseous remains of fish, amphibians, reptiles, birds, and mammals from paleontological or archaeological sites. The extent of mineralization of these bones varies enormously in different sedimentary settings (Figure 4-2). A bone need not be mineralized to be a fossil (Mead et al. 1986, Emslie 1987, Baird & Rowley 1990), nor is a minimum age required. Because age and mineralization do not always go hand in hand, I avoid the vague and almost condescending term "subfossil" that often is used to imply that the specimen in question is less than completely mineralized or is Holocene in age (the last 10,000 radiocarbon years, with true fossils being older).

I have studied 25,000+ bird bones from >60 Pacific islands from the Bismarcks and Palau eastward to Easter Island. Most of the samples have been small (<100 identified specimens per island), although the number of identified prehistoric bird bones now exceeds 500 on 19 islands in eight island groups (Table 4-2). Any fossil record is subject to sampling effects, including a bias toward common species (Signor & Lipps 1982, McKinney et al. 1997). Regardless of the distribution of relative abundance, a richer species pool should require a larger sample of bones for comparable completeness in species representation. This is especially true given the unevenness among species in relative abundance and intra-island distribution, two traits of island avifaunas corroborated by modern censuses and presumably operative in the past as well.

This leads to a major limitation of my data: on no Pacific island is it likely that the bones I or others have analyzed represent all of the species that existed there at first human contact. A possible exception is Henderson Island (Chapter 7), where prehistoric representation probably is >90% and may even be 100%. I believe that faunal completeness is at least being approached (>80% of all species of landbirds) on four other Polynesian islands with >1000 identified bird bones (Ua Huka in the Marquesas, Mangaia in the Cook Islands, and Lifuka and 'Eua in Tonga; Chapters 6 and 7 herein). On most of the 60+ islands, however, the available fossil record certainly or probably represents a smaller portion of the avifauna.

Avian prehistory is very well studied in two other Polynesian island groups, the Hawaiian Islands (Olson & James

Table 4-2 Geographic coverage of paleontological and archaeological sites with bird bones in Oceania. A, archaeological site; P, paleontological site. "Number of identified bird bones" includes only those bones identified to at least the family level. *, data incomplete or imprecise.

ISLAND GROUP Island	Number of sites	Approx. age range of sites (cal BP)	Number of identified bird bones (all taxa/native resident landbirds)	References
BISMARCKS				
New Ireland	5A	35,000–1000	241/239	Steadman et al. 1999b; Chapters 5, 15 herein
Mussau	3A	3500–2000	60/13	Steadman & Kirch 1998; Chapters 5, 15 herein
Arawe	3A	4200–present	4/1	Chapters 5, 15
SOLOMON ISLANDS				
Buka	A	28,000–20,000; 9000–6500; <2500	79/75	Wickler & Spriggs 1988; Chapters 5, 15 herein
Tikopia	8A	3000–present	468/55	Steadman et al. 1990a; Chapters 5, 15 herein
Anuta	A	3000–present	299/0	Steadman et al. 1990a; Chapters 5, 15 herein
VANUATU				
Woga	A	?	22/22	Chapter 5
Toga	A	?	28/20	Chapter 5
Santo	A	?	7/7	Chapter 5
Malakula	10A	2900–present	196/190	Chapters 5, 15
Efate	A	2500–present	183/145	Chapters 5, 15
Erromango	5A, P	2900–present	279/217	Chapters 5, 15
NEW CALEDONIA				
Grande Terre	A, 2P	>10,000?–2000	5324/5313	Balouet & Olson 1989; Chapters 5, 15 herein
Ile de Pins	A, P	>10,000?–2000	999/999	Balouet & Olson 1989; Chapter 5 herein
'Uvea	A	2800–?	10/2	Chapters 5, 15
Lifou	A	2800–?	17/11	Chapter 5
Mare	A	2800–?	147/128	Chapter 5
FIJI				
*Viti Levu	2A, P	Precultural	?/?	Worthy et al. 1999, Worthy 2000, 2001b; Chapter 6 herein
*Naigani	A	2900–2700	?/?	Worthy 2000, Anderson et al. 2001; Chapter 6 herein
Waya	A	2800–2400	12/7	Hunt et al. 1999; Chapters 6, 15 herein
Lakeba	4A, 2P	2800–present	342/254	Steadman 1989a, Worthy 2000, 2001b; Chapters 6, 15 herein
Aiwa Levu	5A	>2800?–present	168/136	Chapters 6, 15
Aiwa Lailai	A	2300–1300	27/19	Chapter 6
Nayau	5A	700–present	155/77	O'Day et al. 2004; Chapters 6, 15 herein
TONGA				
Niuatoputapu	A	2800–2000	14/0	Kirch 1978; Chapter 15 herein
Ha'ano	A	2850–2500	274/146	Chapters 6, 15
Foa	A	2850–2500	352/215	Chapters 6, 15
Lifuka	3A	2850–500	808/607	Steadman 1989b, 1995a, 1999c; Steadman et al. 2002a,b; Chapters 6, 15 herein
'Uiha	A	2850–2500	426/187	Chapters 6, 15
Ha'afeva	A	2850–2500	641/293	Chapters 6, 15

(continued)

Table 4-2 (continued)

ISLAND GROUP Island	Number of sites	Approx. age range of sites (cal BP)	Number of identified bird bones (all taxa/native resident landbirds)	References
Mango	A	undated	2/0	Chapters 6, 15
Tongatapu	2A	2850–2000	473/437	Chapters 6, 15
'Eua	6A, 14P	>80,000–present	2032/1319	Steadman 1989a, 1993a, 1995a; Chapters 6, 15 herein
SAMOA				
Upolu	A	2500–?	12/2	Chapter 6
Tutuila	P	1500	73/50	Steadman & Pregill 2004; Chapters 6, 15 herein
Ofu	A	2800–1900	74/6	Steadman 1993b; Chapters 6, 15 herein
WALLIS & FUTUNA				
'Uvea (Wallis)	2A	2500–2000	4/4	Balouet & Olson 1987; Chapter 6 herein
NIUE	7A, 3P	4500–present	645+/521	Worthy et al. 1998, Steadman et al. 2000b; Chapters 6, 15 herein
COOK ISLANDS				
Aitutaki	6A	1000–200	149/43	Steadman 1991b; Chapters 7, 15 herein
Atiu	4P	undated	41/41	Steadman 1991b; Chapters 7, 15 herein
Mangaia	18A, 6P	>5000–present	3237/1674	Steadman 1985, 1987, 1992b, 1995a,b, 1997b, Steadman & Zarriello 1987, Steadman & Kirch 1990, Kirch et al. 1992, 1995; Chapters 7, 15 herein
TUBUAI				
Tubuai	A	1000–500	11/1	Chapters 7, 15
Rurutu	P	undated	2/0	Chapters 7, 15
SOCIETY ISLANDS				
Huahine	A	1250–750	336/110	Steadman & Pahlavan 1992; Chapters 7, 15 herein
TUAMOTU ARCHIPELAGO				
Mangareva	5A	800–200	215/4	Steadman & Justice 1998; Chapters 7, 15 herein
PITCAIRN ISLANDS				
Henderson	12A	1000–400	16,041/1612	Steadman & Olson 1985, Schubel & Steadman 1989, Wragg 1995; Chapters 7, 15 herein
MARQUESAS ISLANDS				
*Nuku Hiva	A	1000–500	~900/28	Steadman 1989a; Chapters 7, 15 herein
Ua Huka	A	1000–500	12,486/2187	Steadman 1989a; Chapters 7, 15 herein
*Ua Pou	A	?	~200/0	Steadman 1989a; Chapters 7, 15 herein
Hiva Oa	3A	1000–500	1024/168	Steadman 1989a; Chapters 7, 15 herein
Tahuata	A	900–100	716/70	Steadman 1989a, Steadman & Rolett 1996, Rolett 1998; Chapters 7, 15 herein
*EASTER ISLAND	A	900–600	~800/6	Steadman et al. 1994, Steadman 1995a; Chapters 7, 15 herein
PALAU				
Ngerduais	A	undated	1/0	Pregill & Steadman 2000; Chapters 8, 15 herein
Ulebsechel	4A, 4P	undated	62/0	Pregill & Steadman 2000; Chapters 8, 15 herein
Ulong	P	undated	10/5	Pregill & Steadman 2000; Chapters 8, 15 herein

Table 4-2 (continued)

ISLAND GROUP Island	Number of sites	Approx. age range of sites (cal BP)	Number of identified bird bones (all taxa/native resident landbirds)	References
YAP				
Fais	4A	1800–400	197/7	Steadman & Intoh 1994; Chapters 8, 15 herein
MARIANA ISLANDS				
Guam	4A, 3P	undated	28/27	Chapters 8, 15
Rota	3A, 5P	>1000–present	712/498	Steadman 1992a, 1999a; Chapters 8, 15 herein
Aguiguan	A	1800–600	994/943	Steadman 1999a; Chapters 8, 15 herein
Tinian	2A, P	2600–1800	663/618	Steadman 1999a; Chapters 8, 15 herein
Saipan	A	3000–2000	14/12	Chapters 8, 15
CAROLINE ISLANDS				
Pohnpei	5A	2000–200	46/3	Steadman 1999d; Chapters 8, 15 herein
MARSHALL ISLANDS				
Majuro	2A	?	2/1	Weisler 2000; Chapters 8, 15 herein
TUVALU				
Vaitupu	A	?	13/0	Chapter 15
LINE ISLANDS				
Fanning	3A	?	9/0	Chapter 15

1984, 1991, James & Olson 1991, James & Burney 1997) and New Zealand (Holdaway et al. 2001, Worthy & Holdaway 2002). These isolated archipelagos north and south of the rest of Oceania have avifaunas with largely independent evolutionary histories. Thus, I will compare their birdlife with that of our study area (the rest of Oceania) only when it is taxonomically or conceptually relevant, such as for rails (Chapter 10), extinction (Chapter 16), equilibrium and turnover (Chapter 18), or conservation biology (Chapter 21).

To study the timing and extent of prehistoric extinction requires identifying bones from dated contexts. To gather this evidence one must find sites with good bone preservation, determine the age of the sites, obtain samples of bones as large as possible, and identify the bones to species. Accelerator-mass spectrometer (AMS) radiocarbon (^{14}C) dating has revolutionized the ability to build sound chronologies for extinction (Steadman et al. 1991, Kirch et al. 1995, Steadman et al. 2002a). AMS ^{14}C dating requires only 0.5-1 mg of purified carbon, enabling as little as 50 mg of a single, chemically well-preserved bone to be dated (Stafford et al. 1988, 1990, 1999).

Without AMS ^{14}C dating, 100 to 200 grams of bone typically are needed for a ^{14}C date. For perspective, the entire skeleton of a 40.1 g kingfisher (*Halcyon chloris*, immature ♂) weighs 2.81 g (7.0% of body weight). The same measurements for a large flightless rail (*Gallirallus* [*woodfordi*] *immaculatus*, adult ♀) are 569 g, 37.5 g, and 6.6%, and for a large pigeon (*Ducula rubricera*, ad. ♂) are 612 g, 29.31 g, and 4.8%. The thousandfold decrease in the amount of bone needed for AMS vs. conventional ^{14}C dating means that, with good collagen preservation, you can obtain a reliable ^{14}C date from one matchstick-like limb bone of a kingfisher, or a fragmentary megapode, rail, or pigeon bone (Figure 4-3). Conventional ^{14}C dating would require at least 40 entire kingfisher skeletons for a single age determination. Dating an individual bone also avoids problems of time averaging, i.e., a ^{14}C date on one bone represents when that individual bird existed, whereas a ^{14}C date on multiple bones represents the weighted mean age of all the bones, which may represent individuals that did not live at the same time.

Radiocarbon dates are reported in "radiocarbon years before present" (yr BP). If calibrated calendrically for fluctuations of atmospheric ^{14}C, the dates are reported as cal BP or, if using an AD/BC system, as cal AD or cal BC. These calibrations are used routinely for the past 11,000 radiocarbon years (= ~13,000 calendar years) based on tree-ring

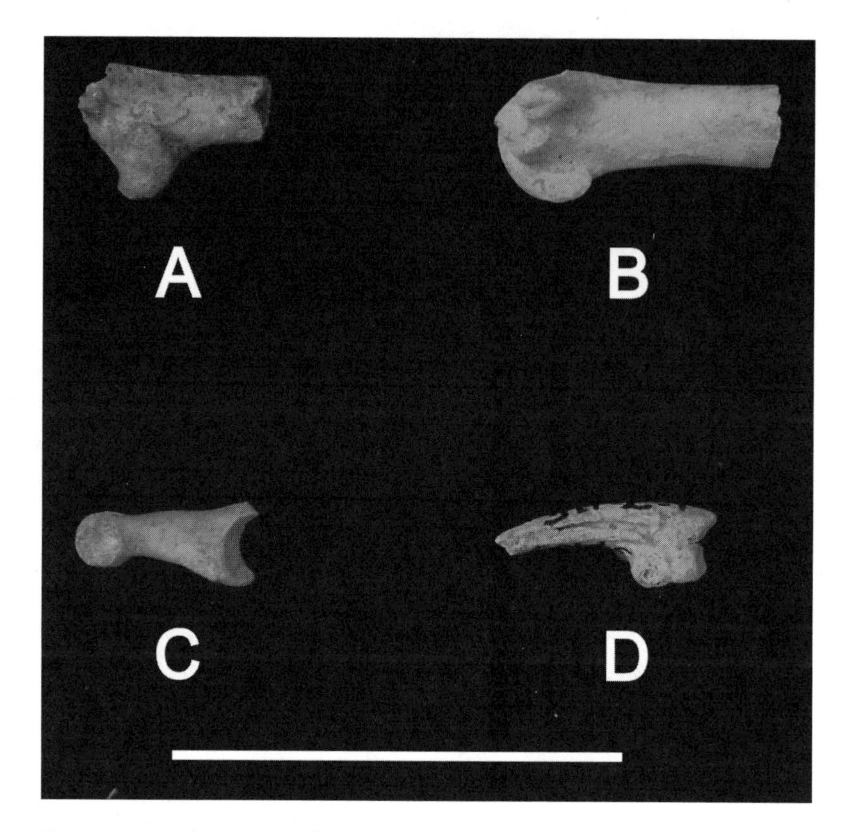

Figure 4-3 Examples of bones (Tongoleleka site, Lifuka, Tonga) of the extinct megapode *Megapodius alimentum*. A: Scapula, UF 58335. B: Ulna, UF 58557. C: Pedal phalanx, UF 57904. D: Ungual phalanx, UF 57903. Similar bones were the source of the AMS ^{14}C dates of ca. 2800 cal BP in Table 3-4. Scale = 30 mm. Photo by Jeremy Kirchman.

dating; sometimes they are extended to 45,000 years by counting dated, annually laminated (varved) sediment, and then cross-dating (with U-series; see Chapter 1) with shallow-water tropical corals and stalagmites (Stuiver et al. 1998, Bard 2001, Beck et al. 2001). Fortunately for Pacific archaeology, the late Holocene time interval that represents Lapita colonization (ca. 3600 to 2700 cal BP) is very well calibrated for atmospheric variation in ^{14}C (Kromer et al. 2001, Manning et al. 2001, Reimer 2001). "Present" in ^{14}C dating is AD 1950. Thus an age of 50 cal AD equals 1900 cal BP. Radiocarbon dates are statistical age estimates with variable precision. A ^{14}C date of 1000 ± 40 yr BP has a 68% (1 σ) probability of being from 1040 to 960 yr BP, and a 95% (2 σ) probability of being from 1080 to 920 yr BP. Calendric calibration can expand or contract the confidence intervals depending on past variation in atmospheric ^{14}C.

Most bone-bearing sites in Oceania are in rockshelters (Farrand 1985), caves, or calcareous beach sands. These sites provide a geochemically favorable setting for preserving bones (and associated shells, lithics, ceramics, etc.) because of high-pH sediments with minerals such as calcite, apatite, and gypsum (Bull 1983). Caves and rockshelters are common on raised limestone islands (Steadman 1985, 1993a, 1999a) and young volcanic islands (Metraux 1940, James et al. 1987, Coello et al. 1999). Many caves and rockshelters have fine-grained, "low-energy" sediments, i.e., have been transported relatively gently by wind, water, and gravity, resulting in excellent preservation of small, fragile bones.

Just as many species of island birds are extinct or endangered, many of the prehistoric sites that potentially could produce faunal histories of islands have been destroyed or damaged by human activities, especially during the past 200 years. A major problem is sand mining, which removes coastal sand deposits for use in construction cement (Lewis et al. 1997). Other sorts of damage, whether on Pacific or other tropical islands, includes removing cave sediments for fertilizer, bulldozing for roads, digging ditches, etc. On island after island, the physical evidence of faunal and cultural prehistory is being sacrificed to the drum beat of economic development.

Fieldwork

Careful fieldwork is essential. Nothing ever can compensate for what is done poorly in the field. My scientific travels in Oceania (100+ islands; Table 4-3) began in 1984, on the heels of 13 months of fieldwork in the West Indies and Galápagos Islands during the previous seven years. The work has included survey and excavation of archaeological and paleontological sites, as well as research on living birds (sight/sound surveys along transects, specimen collections, habitat analyses and mapping, feeding ecology, and blood parasites).

Diverse fieldwork is needed to interpret how island avifaunas have changed through time. I designed my education to be broad. After a B.S. in biology at Edinboro State College (Pennsylvania) with a focus on vertebrate ecology, my M.S. in zoology at the University of Florida concentrated on the osteology, paleontology, and evolution of vertebrates. My Ph.D. in geosciences (University of Arizona) focused on Quaternary paleoecology (stratigraphy, geochronology, paleoclimate, paleobotany, archaeology, vertebrate paleontology) with a minor in Ecology and Evolutionary Biology. Pre- and postdoctoral fellowships at the Division of Birds, Smithsonian Institution, provided great opportunities for travel to oceanic islands and for research in the world's best skeleton collection of birds.

My fieldwork has been with zoologists, botanists, archaeologists, geologists, and geographers. The methods and concepts that we learn from each other are a never-ending source of stimulation. Eventually, in one way or another, our data all feed into island biogeography, a branch of biology begun in the 19th century by systematists and evolutionary biologists. In the 1960s, island biogeography was swept off its feet by two publications by Robert MacArthur and Edward Wilson (MacArthur & Wilson 1963, 1967). From then on, ecologists have been vital players in island biogeography, although not to the exclusion of other biologists (e.g., Pregill & Olson 1981, Whittaker 1998) and nonbiologists (e.g., Kirch & Hunt 1997, Dickinson 2000a).

First, I will discuss the prehistoric component of my fieldwork. Much of my exploration has targeted limestone islands because they are likely to have caves and rockshelters. Calcareous sand deposits can be found on many island types (see Chapter 1) in protected situations on bays or leeward coasts. Once a potential site is located, the sediments must be sampled to evaluate their contents. Often this is done with a soil auger, which is capable of quickly bringing up sediment samples to a depth of 1–2 m depending on compaction and rockiness. In other cases, a test excavation (usually 1 × 1 m in area) is done with a trowel, providing a much larger, more controlled, and more time-consuming first look at a site's potential.

Figure 4-4 Excavation in progress at Tangatatau Rockshelter, Mangaia, Cook Islands. Photo by Patrick Kirch, 24 July 1989.

After initial testing, most places are judged either not to be a paleontological or archaeological site at all, or to contain so little of interest (bones, artifacts, etc.) as to be not worth the effort. Excavation is expanded beyond the initial testing for sites believed to be important. The stratigraphic provenience of bones (and associated artifacts, charcoal, shell, etc.) must be recorded during excavation (Figure 4-4). Unless fine-mesh sieves (3 mm or less) are used, bones of small species of birds (and other vertebrates) are unlikely to be recovered, resulting in an unnecessary loss of information. Equally unfortunate is when archaeologists recover small, fragile bones in the field but then do not package them carefully for shipment. Such irresponsibility or lack of common sense has led to thousands of perfectly fine bird bones being broken beyond species-level recognition at some point between exhumation in the field and arrival

Table 4-3 The author's fieldwork in Oceania, excluding the Hawaiian Islands and New Zealand. Airport layovers are not included unless I could break away long enough to take notes on birds or prehistoric sites.

Island Group and Island	Dates
Solomon Islands Guadalcanal, Isabel, Rennell	15 Jun–29 Jul 1997
Vanuatu Efate, Iririki, Eretoka, Lelepa, Santo, Aore	28 Jul–8 Aug 1996, 29 Jul–6 Aug 1997, 3–18 Jun 2003
Fiji Aiwa Lailai, Aiwa Levu, Lakeba, Nayau, Viti Levu, Yanuca	25–26 Feb 1985, 28–29 Jun 1995, 4–16 May 1999, 17 Feb–7 Mar 2000, 16–27 Oct 2001
Tonga 'Ata, Tongatapu (including Alakipeau, 'Atata, 'Eueiki, Fafa, Makaha'a, Malinoa, Pangaimotu, Polo'a, Tufaka, Toketoke), 'Eua, Ha'apai (Fetoa, Foa, Ha'afeva, Ha'ano, Lifuka, Matuku, Nomuka, Nukunamo, Teaupa, Tofua, Tungua, 'Uiha, Uoleva), Vava'u (A'a, 'Euakafa, 'Euaiki, Foeata, Foelifuka, Hunga, Kapa, Kenutu, Kulo, Mafana, Nuku, Ofu, Pangaimotu, Taunga, 'Uta Vava'u, 'Utungake, Vaka'eitu)	28 Feb–2 Mar 1985, 18 Nov–19 Dec 1988, 21 Nov–19 Dec 1989, 29 Jun–31 Jul 1995, 3–27 Jul 1996, 5–14 Jun 1997, 31 Jul–14 Aug 1999, 3–16 Aug 2001
Samoa Apolima, 'Aunu'u, Fanuatapu, Manono, Namu'a, Nu'ulopa, Nu'usafe'e, Nu'utele, Tutuila, 'Upolu	2–3 Mar 1985, 5 Apr–3 May 1999
Cook Islands Aitutaki (including Maina, Moturakau, Rapota), Atiu, Ma'uke, Mangaia, Mitiaro, Palmerston, Rarotonga (including Motutapu, Oneroa, Ta'akoka)	13 Mar–23 Apr 1984, 4–5 Mar, 28 May–14 Jun 1985, 15 Sep–7 Dec 1987, 14 Jul–11 Aug 1989, 3 Jun–2 Jul 1991, 8 May–4 Jun 1997, 16–25 Dec 1997, 10 Jun–3 Jul 2001
Society Islands Bora Bora, Mo'orea, Tahiti	12–13 Mar 1984, 7–10 Mar 1985, 14–15 Sep, 8–11 Dec 1987, 3–5 Jul 1991
Easter Island	6–26 Jul 1991
Palau Arakabesan, Babelthuap, Bailechesengel, Dmasech, Iilblau, Koror, Malakal, Ngelsibel, Ngerduais, Ngis, Ulong, Ulebsechel, Urukthabel	1–21 Jan 1995, 10–19 Jan 1997
Yap O'Keefe, Tomil, Yap	8–10 Jan 1997
Mariana Islands Aguiguan, Guam, Naftan Rock, Rota, Saipan, Tinian	27 Jan–6 Feb 1990, 21 Jun–24 Jul 1994, 19 Jan–2 Feb 1997, 29 Dec 2004–9 Jan 2005
Caroline Islands Chuuk (Truk), Kosrae, Pohnpei	29 Dec 1994–1 Jan 1995
Marshall Islands Kwajalein, Majuro	29 Dec 1994
Johnston Atoll	28 Dec 1994

at my laboratory. Most archaeologists now take very good care of the bones they excavate.

Studying living birds also is important. Whether based on specimens or sight/sound records, learning the modern distribution and status of island birds is a never-ending task. Birds are touted, perhaps rightly, as the world's best known organisms. Birds on Pacific islands, however, are horribly less studied than those in Britain or the United States, the two countries with the most ornithologists and birdwatchers. Thomas Lovejoy (1997:7) may have exaggerated only slightly when he said, "Indeed I believe that there are so many naturalists in Britain that it is impossible for a bird to lay an egg without three people, including at least one cleric, recording it." Birds on tropical Pacific islands lead more private lives, seldom if ever being in a binocular field of view. The negative side of this obscurity is that many species of birds in Oceania are not represented by modern specimens with high-quality data, or in standardized censuses needed to assess long-term population trends and habitat relationships.

The bird surveys I have done in Oceania have covered islands that vary considerably in area, elevation, isolation, and human impact (Steadman 1998, Steadman & Freifeld 1998, Steadman et al. 1999a, Steadman & Franklin 2000, Freifeld et al. 2001). Comparisons with prehistoric records, on the same or nearby islands, allow me to evaluate long-term changes in distribution. Assessing relative abundance and habitat relationships also helps to define the conservation status of species (Chapter 21), supplemented by data on blood-borne parasitic infections (Steadman et al. 1990a, Frank et al. MS submitted).

Whether with birds living or dead, fieldwork in Oceania seems exotic to an armchair traveler. Pacific islands indeed are some of the world's prettiest and most alluring places. Realistically, however, field research in Oceania boils down to hard, dirty labor under difficult conditions. The labor itself is enjoyable, thought-provoking, and rewarding, but logistical problems often arise with travel, communication, access to sites, research permits, weather, personnel, and health. With experience one can minimize some of these problems on most trips, but with no guarantee of success. Logistical difficulties are seldom acknowledged in journal articles, Whitmore (1989) being an admirable exception. The dearth of basic field research in most of Oceania should be a clue as to how tough it can be.

Like anywhere else, Oceania is not paradise. Just ask anyone who has stayed in filthy, rat-infested hotels in Pago Pago, American Samoa, where the stench of tuna canneries complements the unfriendliness of the staff. The Bismarcks, Solomons, and Vanuatu have some of the world's highest rates of chloroquine-resistant malaria. It is also difficult there to avoid ulcerated sores, especially on your legs, which can lead to systemic bacterial infections. The forests of Oceania are no place for arachnophobes; many islands have high densities of spider webs almost anywhere you walk. Inter-island boat trips can be difficult and dangerous. A visiting scientist unwittingly can become a target for financial exploitation by rival families, villages, or tribes.

While field research on tropical islands seldom resembles the sugar-coated stuff of television or the movies, I must say that many of my most rewarding and enjoyable experiences have taken place while working in Oceania. I cannot describe the field research truthfully, however, without offering a few cautionary tales about these islands of great contrast, where satisfying scientific discoveries and reasonable personal comfort are not guaranteed.

The capitals of some countries are noisy and have rush-hour traffic jams, often with nonnative roosters (*Gallus gallus*), pigeons (*Columba livia*), bulbuls (*Pycnonotus cafer*), and mynas (*Acridotheres tristis* or *A. fuscus*) contributing to the background din. Ethnic and economic tensions in Papua New Guinea, the Solomon Islands, Fiji, and occasionally elsewhere have led to violence and political instability in recent years. Most of the South Pacific has little violent crime, but the abundant dogs can keep you on edge. In July 1996, David Burley and I were warned by many Tongans that we would not enjoy Nomuka, an island lacking an airstrip or electricity and therefore enticing. Driven by our interests, we visited Nomuka with Bill and Jackie Dickinson. On the third day, while walking alone on a public dirt road (that connects Nomuka's only village to agricultural plots and accommodates the island's single motor vehicle), a pack of five angry dogs knocked me down in a sudden, multi-angled attack. They lacerated my arms and legs while I struggled unsuccessfully to regain footing. The Nomukan man who owned the dogs watched all this, unconcerned and perhaps entertained as I desperately tried to keep the biting dogs away from vital parts. Finally, after I was thoroughly cut and bruised, and one of his curs was mortally wounded from the one good swing I could muster from my machete, the man called off his dogs.

Aside from their threats to human health (Langley 1992, Mather & Fielding 2001) and detrimental effects on native birds (Hunt et al. 1996), island dogs are prone to barking all night, usually right outside your window. Peace and tranquility make great copy for South Pacific tourist brochures, which invariably fail to mention the roosters, dogs, and pigs that serenade you through the evening. Camping on uninhabited islands usually eliminates these problems, although rats are often a nuisance.

In spite of these problems, I truly love Oceania. This love is fueled by personal as well as professional fascination and

Table 4-4 Growth in the prehistoric record of birds on Mangaia, Cook Islands. The 1987 trip was by G. K. Pregill. DP, data pending; MS, migratory shorebirds; RL, resident landbirds; SE, seabirds. Species richness values for RL do not include the introduced chicken *Gallus gallus*. Data compiled from Steadman (1985, 1987, 1995a, herein), Steadman & Kirch (1990), Steadman et al. (2000a), and Steadman (unpub.).

Dates of fieldwork	Number of sites sampled	New sites	Running total of sites	Number of bird bones collected	Running total of bird bones	Running total of landbird bones	Number of species recorded			Running total of species		
							SE	RL	MS	SE	RL	MS
26 Mar–18 Apr 1984	6	6	6	530	530	85	7	10	0	7	10	0
1–11 Jun 1985	3	1	7	275	805	154	4	8	1	8	12	1
4–14 Nov 1987	2	0	7	6	811	154	2	0	0	8	12	1
18 Jun–10 Aug 1989	6	4	11	381	1192	332	11	14	1	12	14	1
7 Jun–2 Jul 1991	16	11	22	802	1994	834	14	17	2	17	17	2
9 May–2 Jun 1997	7	2	24	1243	3237	1674	13	17	2	18	19	2
11–30 Jun 2001	8	6	30	DP	DP	DP	DP	DP	DP	DP	DP	DP

curiosity. My fieldwork is done with the utmost respect for islanders and their customs. Permission to do research is typically a three-tiered process involving the national government, a provincial (island) government, and local chiefs and landowners. When what I am trying to accomplish does not sit well with the islanders, I leave. I have had many very good trips, a few very bad ones, and some in between. There are many islands that I cannot wait to visit again. Mangaia, 'Euakafa, 'Euaiki, Lakeba, Aiwa Levu, Aiwa Lailai, and Aguiguan come to mind immediately. Others I probably will avoid. In order to get a firsthand look at its modern birds and terrestrial habitats, I will visit any island at least once. If the paleontological or zooarchaeological potential is good, I will try to visit it another time or two, even if the logistics are difficult.

My most special island always will be Mangaia (Cook Islands). I have gone there off and on since 1984. I know many Mangaians very well. We have watched each other get older. Some of my best Mangaian friends have died, while others have become adults and had babies. Our mutual respect is from being content with who we are—I will never be a Mangaian and would never pretend to be. It does not bother me that there are big parts of Mangaian culture that I will never understand.

Mangaia also is a treasure chest for my research. Its many caves and rockshelters never cease to produce new evidence of the birdlife that once inhabited this 52 km^2, half-limestone, half-volcanic island (Figures 1-22, 1-23, 3-11, 3-12, 4-4). Even though part of me would love to visit every Pacific island before a coconut falls on my head, a bigger part carries me back to Mangaia time after time.

Mangaia has become a test case for what it takes to saturate the fossil record on a single island. In other words, how many times must I visit Mangaia, how many sites must I excavate there, and how many bird bones must I identify, until I add no more species to its prehistoric avifauna? Evaluating diminishing returns, like any other aspect of negative evidence in the fossil record, is a tricky business. Mathematics helps, but cannot account for all variables (see below). For now, it seems that I do not yet know the entire avifauna that existed there at first human contact, because each of my last four trips to Mangaia added species of birds that I had not found on earlier visits (Table 4-4). Because of its extraordinary fossil record, Mangaia now is tied with Henderson Island (Pitcairn Group) for second place among East Polynesian islands in seabird richness at 19 (Easter Island is first with 25) and is first in species richness (20) of landbirds. The combined total of 39 resident species is unsurpassed in East Polynesia, a reflection of Mangaia's high-quality fossil record rather its potential to sustain birds.

As mentioned, the bones recovered and identified from any prehistoric site represent only a subset of the species that lived on the island at the time (Steadman 1995a). Attempting to reach a point of diminishing returns in the prehistoric species richness of birds is not just a matter of increasing the number of identified bones. At least four other variables influence how completely the species lists derived from bone assemblages reflect an island's actual prehistoric avifauna: (1) how thoroughly predators (human or otherwise) sampled the available range of species while accumulating the bone deposit(s); (2) the ages of the bone samples (by late prehistoric times, most species of birds already were gone from most islands so that even a large collection of bird bones from a late context is unlikely to be rich in species); (3) the field collection methods, which should recover (through the use of fine-mesh sieves) even the minute, delicate bones of small species; and (4) the accuracy and resolution of the identifications, which depend on the skill of the researcher, preservational qualities of the fossils, and taxonomic representation of the comparative collection of modern skeletons. To this I must add that many if not most natural communities are believed to have more rare than common species (Magurran & Henderson 2003), this obviously making complete fossil sampling more difficult.

On Mangaia, bone-bearing prehistoric sites now include five of the island's six districts (Figure 3-11). This broad intra-island coverage increases the chances that rare or localized species would be sampled in at least one site. The highly eroded volcanic uplands are practically devoid of such sites but are within 1–2 km (a short walk or flight) of them. Mangaia's coastline also has not yielded prehistoric bird bones, but again is within the predator radius of bone-bearing caves and rockshelters found in the rugged, cliffy limestone that encircles the island (Figures 1-22, 1-23).

Accumulating species is not the only goal of paleontology. Another is to estimate the time range of each species, a topic to be discussed further in Chapters 16–18. In estimating the chronology of colonization and extinction, it is important to realize that an absence of fossil representation may reflect inadequate sampling rather than absence of a species (Signor & Lipps 1982, Peters & Foote 2002, Jablonski 2003). An East Polynesian pigeon or parrot with bones recorded commonly in strata older than 700 cal BP, but never in younger strata, may not have become extinct 700 years ago, but may only have become too rare to show up in the bone assemblages. The actual date of extinction may have been 600 or 500 or even 400 years ago. In one instance (the extinct swamphen *Porphyrio paepae* from the Marquesas; see below), a species that dropped out of the prehistoric record centuries before European contact has been found probably to have survived to within the past 100 years.

Bird Bones

Identification

At the risk of losing readers already suffering from an osteo-overdose, please allow me to sing the praises of bird bones. The distributional data in Chapters 5–15 are from bird bones that I have identified, my surveys of living birds, and the writings of other scientists. To identify and describe bird bones at the species level is an uncommon skill. Competence in avian comparative osteology is essential to generate high-quality data, but requires years of meticulous, hands-on study of skeletal specimens in museums. Those with short attention spans need not apply. You learn the comparative osteology of birds by studying real bones, not by pointing and clicking. Zooarchaeological analyses can suffer from a lack of standards that leads to great variation in the identifications of bones (Gobalet 2001). I simply do not trust species-level identifications of bird bones by certain zooarchaeologists. Fortunately in the Pacific, this problem is minor because the primary workers (Helen James and Storrs Olson in the Hawaiian Islands, Richard Holdaway and Trevor Worthy in New Zealand and Fiji, and yours truly in most other island groups) are skilled in avian osteology.

As nice as it would be, I cannot claim that all of my identifications of bird bones are error-free. With experience, however, my willingness to try to identify very fragmentary specimens has decreased through time. Having higher standards for preservational attributes, especially completeness, of the bones that I study improves the quality of my identifications, which are based on modern and prehistoric specimens catalogued in museums and thus verifiable. With great difficulty, I also have amassed a steadily improving set of modern comparative skeletons. For example, my list of taxa from prehistoric sites on New Ireland (Table 5-4) is improved over that in Steadman et al. (1999a) because of examining bird skeletons collected in the Upper Fly River Drainage of Papua New Guinea by A. W. Kratter and J. K. Sailer in June 2000. These same skeletons also allowed me to finish identifying the bones from Kilu Cave on Buka (Table 5-8), which until then I had analyzed only partially using specimens collected in 1997 and 1998 on Isabel and Rennell, Solomon Islands. As in any identification process, increasing the specimen sample for each species helps to characterize the species more thoroughly and to decrease the odds of misidentification (Hayek et al. 2001).

A difficult thing for fledgling paleontologists to learn is which specimens are diagnostic (and therefore worth studying) and which are undiagnostic and thus not worth your time. The adage that a paleontologist can reconstruct an entire beast from one bone fragment is not true. I may be able to determine that a single whole tibiotarsus from an archaeological site represents an extinct, undescribed species of cuckoo-dove (such as *Macropygia arevarevauupa* from Huahine; Steadman 1992b), but that leg bone by itself tells me nothing about the head or wing of the Society Island cuckoo-dove.

In most families of birds, the diagnostic skeletal elements are, from head to toe, the upper and lower bill (rostrum and mandible), pectoral girdle (coracoid, scapula, sternum), and major long bones of the wing (humerus, ulna, radius, carpometacarpus) and leg (femur, tibiotarsus, tarsometatarsus). Other skeletal elements sometimes can be identified as well, such as the quadrate or furcula (wishbone). Even

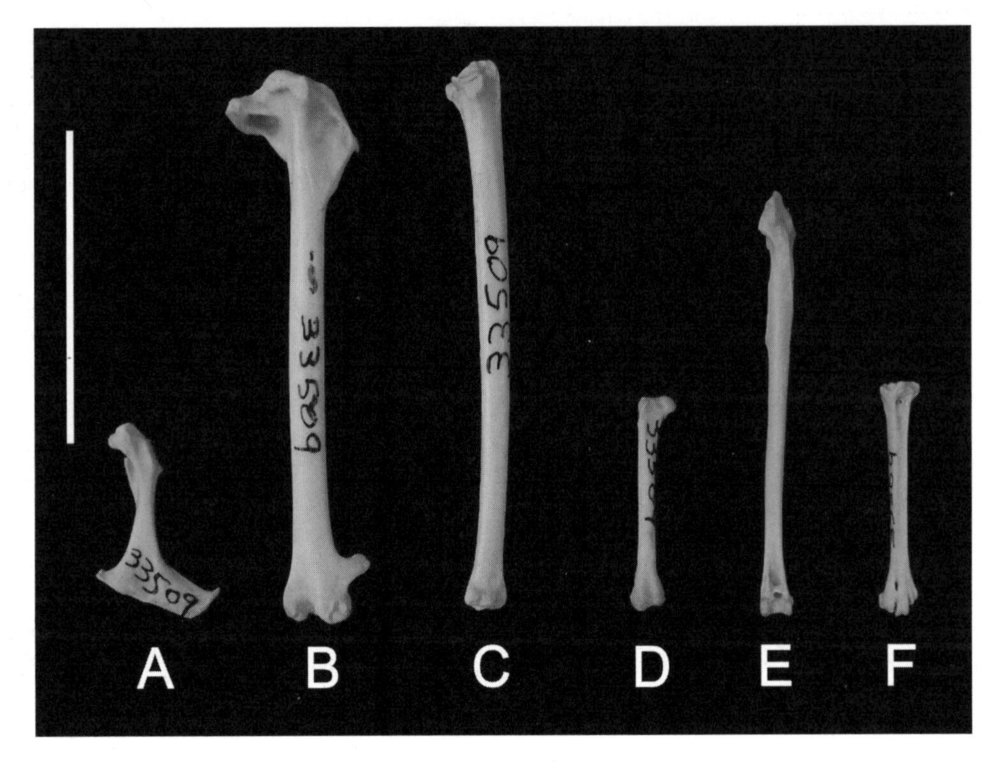

Figure 4-5 Major skeletal elements of a Cook's petrel, *Pterodroma cookii* (UF 33658, adult unsexed from New Zealand). A: Coracoid. B: Humerus. C: Ulna. D: Femur. E: Tibiotarsus. F: Tarsometatarsus. Scale = 60 mm. Photo by Jeremy Kirchman.

the toes (pedal phalanges) can be surprisingly diagnostic at least to the level of genus in many nonpasserine families. Overall, probably the five most identifiable skeletal elements are the rostrum, tarsometatarsus, humerus, coracoid, and tibiotarsus, not necessarily in that order for each family. The osteology of passerines is generally more uniform, and therefore more difficult to resolve taxonomically, than that of nonpasserines.

Entire bones, as in Figures 4-5 to 4-7, are easier to identify (though still often difficult) than the incomplete ones often found in prehistoric contexts. Within limits, paleontologists and zooarchaeologists must be able to identify bone fragments. At least with the more readily identifiable skeletal elements just mentioned, fragmentary specimens often are identifiable to species as long as they retain one articular end. Bones lacking both articular ends ("shafts") typically cannot be identified, which is why careful packaging to prevent breakage during shipment is essential.

In spite of tender care in both the field and lab, many prehistoric bird bones still are undiagnostic. From the carefully excavated Railhunter Rockshelter on Tinian (Mariana Islands), for example, I could not identify 31% of the bird bones beyond being from a bird rather than a fish, reptile, or mammal (Table 8-7). Quantifying unidentifiable bird bones can be helpful to zooarchaeologists interested in the relative contribution of birds to the diets of prehistoric peoples. The patterns of breakage in bird bones also are useful in understanding taphonomic processes (Steadman et al. 2002b). Very often, however, such undiagnostic bones are not tabulated because they add nothing to the species assemblage. My own treatment of unidentifiable bird bones is variable; sometimes I count and tabulate them to remind myself (and others) how common they are. More often, I do no more with these fragments than be sure that they stay associated with their labels.

Analyzing ancient DNA (aDNA) is an important new line of research. Given the expense involved, as well as the limits of organic preservation, aDNA never will replace comparative osteology as the primary means of identifying prehistoric bird bones. Nevertheless, aDNA already has addressed research problems in at least three areas. One is in distinguishing among species that are difficult or impossible to characterize osteologically. This has been done for Hawaiian fossils of ducks, geese, and eagles up to several thousand years old (Cooper et al. 1996, Sorenson et al. 1999, Paxinos et al. 2002), and for salmon bones from Washington and Oregon up to 9000 years old (Butler & Bowers 1998). Another is for phylogenetic hypotheses, such as with New Zealand's moas (Cooper 1997, Cooper et al. 2001). The third is to study genetic changes through time, as has been

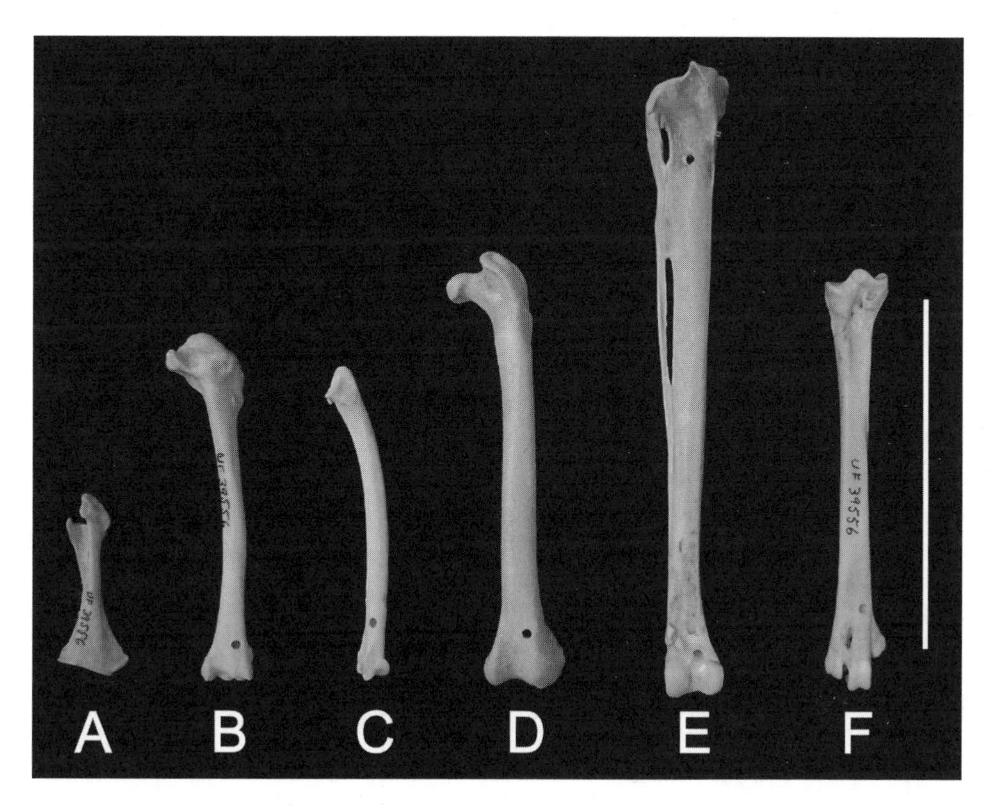

Figure 4-6 Major skeletal elements of the flightless rail, *Gallirallus* [*woodfordi*] *immaculatus* (UF 39556, adult female from Isabel, Solomon Islands). A: Coracoid. B: Humerus. C: Ulna. D: Femur. E: Tibiotarsus. F: Tarsometatarsus. Scale = 50 mm. Photo by Jeremy Kirchman.

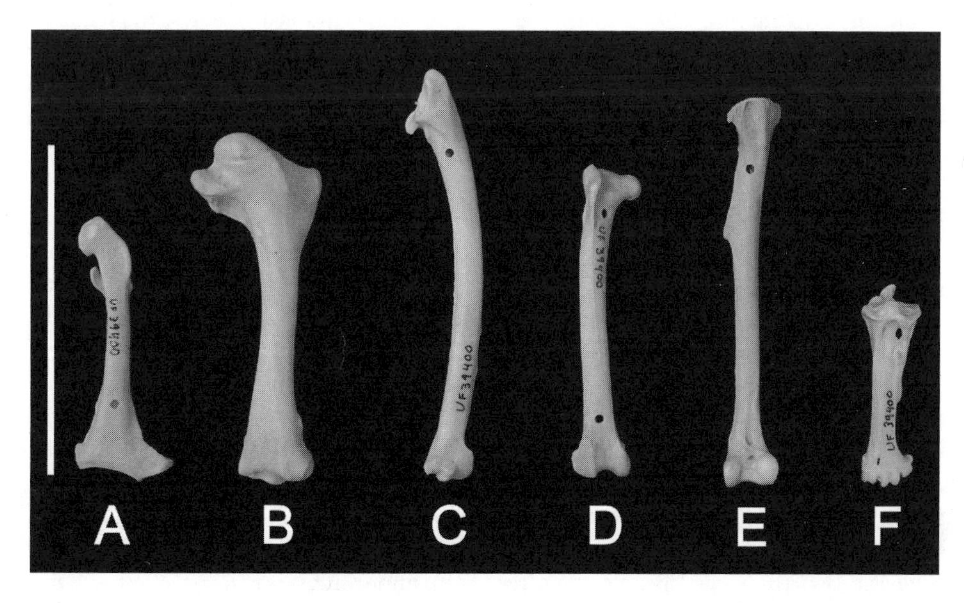

Figure 4-7 Major skeletal elements of the red-knobbed pigeon, *Ducula rubricera* (UF 39400, adult male from Isabel, Solomon Islands). A: Coracoid. B: Humerus. C: Ulna. D: Femur. E: Tibiotarsus. F: Tarsometatarsus. Scale = 60 mm. Photo by Jeremy Kirchman.

done for small mammals in Wyoming (Hadly 1999), bears in Beringia (Barnes et al. 2002), and penguins in Antarctica (Lambert et al. 2002). These last three places are much colder and drier than tropical Oceania, leading to better long-term preservation of DNA. Nevertheless, especially in light of the Hawaiian results, I expect to see great advances in aDNA research on islands over the coming years.

Taphonomy

Taphonomy is the study of what happens to body parts (in this case, bones) from when it was in a living organism until it is recovered in a fossil context. The bones recovered from archaeological sites, whether birds in Oceania or large mammals on a continent, reflect the killing, butchering, and eating practices of prehistoric peoples, as well as the bone's ability to withstand the ravages of time (Grayson & Delpech 1998, Steadman et al. 2002b). Because most Pacific island birds were captured with snares, nets, bird lime (sticky gum on branches), or hands (Steadman 1997a), few bones were broken during hunting except for dispatching the bird through damage to the head or neck. Butchering as well may have contributed little to bone breakage because oceanic peoples tend to cook birds whole (pers. obs.). Most breakage probably occurred during and subsequent to eating. Related to enjoyment of fatty or greasy foods (Kirch & O'Day 2002), Oceanic peoples tend to chew through bird bones much more than westerners, leaving behind less that might ever be identified (pers. obs.). Once discarded, most of the remaining bones are damaged further (often beyond recognition) or completely destroyed from trampling by people, chewing by rats, dogs, or pigs, or being deposited in a geochemically unfavorable setting. Thus, the bird bones recovered from archaeological sites represent an exceedingly tiny but unknowable fraction of the birds that have been killed and eaten on any island.

All else being equal, larger and denser (thicker walled) bird bones are more likely to survive through time than smaller, more delicate ones (Livingston 1989). In seabirds, wing bones generally are larger and stronger than leg bones, which are underrepresented as fossils (Table 4-5, Figure 4-5). To eat seabirds, the longest wing bones (humerus, ulna, radius) were stripped of their meat with no breakage or with clean breaks near one or both ends. (For perspective, how many bones must you break to eat a chicken wing? I should repeat, however, that traditional peoples tend to crush many bones when eating, especially of landbirds such as chickens, columbids, and passerines). Among leg bones of seabirds, the small, weak femur is surrounded by meat and usually was destroyed during or after eating. The somewhat stronger (but still weak when compared to those in a running bird) tibiotarsus supports the "drumstick" and usually was broken at one or both ends, the latter likely to make the bone unidentifiable to species. The highly diagnostic tarsometatarsus, surrounded by little if any meat, often remained unbroken.

Table 4-5 Distribution of skeletal elements, expressed as % of all bones, of seabirds vs. landbirds in the bone assemblages from archaeological sites on three islands: Mussau, Bismarck Archipelago (Steadman & Kirch 1998; Chapter 5 herein); Huahine, Society Islands (Steadman & Pahlavan 1992; Chapter 7 herein); and Fais, Yap (Steadman & Intoh 1994; Chapter 8 herein). Wing = humerus, ulna, radius, carpometacarpus, manus phalanges. Body = axial skeletal plus coracoid, furcula, scapula. Leg = femur, tibiotarsus, tarsometatarsus, pedal phalanges. Seabirds = petrels, shearwaters, frigatebirds, tropicbirds, boobies, gulls, terns. Landbirds = all other families. NISP = number of identified specimens.

	Wing	Body	Leg	NISP
SEABIRDS				
Mussau	50	30	20	30
Huahine	60	35	5	203
Fais	82	13	5	75
Nonweighted mean	64	26	10	—
LANDBIRDS				
Mussau	20	30	50	30
Huahine	22	40	38	133
Fais	17	36	47	122
Nonweighted mean	20	35	45	—

In most landbird families, the leg elements are sturdier than those of seabirds of a similar size and thus are better able to withstand the rigors of being from a bird that has been killed, butchered, eaten, perhaps scavenged by dogs, rats, or pigs, and then interred in or near someone's living quarters. Exceptions are small-legged, strong-flying landbirds such as swifts or kingfishers. Leg bones are particularly strong in ground-dwelling birds such as megapodes, chickens, and rails (Figure 4-6), where all three major elements are strong enough not to always be broken while the bird was being eaten. If they were broken, a clean break of the shaft near either end typically does not prevent identification to species as long as one articulating end survives.

Avian Ethnobiology

Being light and hollow but hard and strong, bird bones were important raw materials for tools, toys, and ornaments (Steadman 1997a). Because bones are readily preserved in prehistoric contexts, our ethnobiological knowledge of bones, unlike that of feathers (Dove 1999), covers the entire period of human history in Oceania. Sewing needles were usually made of the radius (less often, the humerus or ulna)

Table 4-6 Bone needles from Lapita and Polynesian Plainware archaeological sites on Foa, Lifuka, Ha'ano, 'Uiha, Ha'afeva, and Tongatapu, Tonga.

Skeletal element	Procellariidae (shearwater, petrel)	*Fregata* spp. (frigatebird)	*Pandion haliaetus* (osprey)	Bird (family indeterminate)	Pteropodidae (fruit bat)	TOTAL
Humerus	2	—	—	2	—	4
Ulna	22	2	1	—	—	25
Radius	94	—	—	2	4	100
Manus phalanx	—	—	—	—	6	6
TOTAL	118	2	1	4	10	135

by cutting both ends perpendicular to the long axis, drilling a small hole in the shaft near one end, and then filing the other end to form a sharp point. Bone needles probably were used to stitch together *tapa* (bark cloth). They are found commonly in prehistoric sites on Easter Island (Metraux 1940:213, Heyerdahl 1961:412), where they are called *ivi tia nua*. Two such needles from Anakena, Easter Island (ca. 800 to 600 cal BP; Steadman et al. 1994) were made from the radii of Murphy's petrel (*Pterodroma ultima*), a tropical seabird that no longer lives on Easter Island. From sites on six islands in Tonga dating to the Lapita and Polynesian Plainware periods (ca. 2850 to 2200 cal BP), David Burley and I have recovered 100+ sewing needles made of bird bones (Table 4-6, Figure 4-8). As on Easter Island, the preferred bone for Tongan needles was the radius of shearwaters and petrels, although wing bones of frigatebirds (*Fregata* spp.) and fruit bats (*Pteropus* spp.) were used as well.

Prehistoric whistles were made from thin-walled bones of large seabirds such as frigatebirds or boobies. From 'Anatu, 'Eua, Tonga, I found a still functional whistle (48 mm long) made from the radius of a lesser frigatebird (*Fregata ariel*). From Hanatekua Shelter No. 2 site (MH-11), Hiva Oa, Marquesas, I have identified the cut proximal 44 mm of another radius of *F. ariel*. Presumably, the adjoining piece (not recovered) would have finished the whistle. No bird-bone whistles in Oceania are as musically sophisticated as those made from the ulnae of cranes (*Grus japonensis*) during the early Neolithic (ca. 9000–7700 cal BP) in China (Zhang et al. 1999).

Tattooing needles were made of bird bone on Easter Island (skeletal element and species unrecorded; Metraux 1940:237–238, 241), Cook Islands (details unrecorded; Hiroa 1944:128), and Tikopia (humerus, ulna, and radius of frigatebirds; Steadman et al. 1990b). Mammal bones also were used as tattooing needles on Mangaia (Kirch et al. 1995) and Tonga (pers. obs.). The humerus, ulna, and tibiotarsus of chickens and large native birds were cut in short sections to make beads. From strata dated ca. 600 cal BP at Tangatatau Rockshelter on Mangaia, Pat Kirch and I found an entire chicken sternum with two drilled holes, apparently a pendant.

Bird bones from archaeological sites can corroborate the former existence of species otherwise known only by stories or linguistic deduction. For example, cognates of *malau*, the Tongan and proto-Polynesian word for megapode (*Megapodius* spp.), are found in many places in Melanesia and Indonesia (Clark 1982). Because megapodes no longer live in Fiji, through which proto-Polynesian speakers must have passed on their way to Tonga, use of the word *malau* in Tonga (for *M. pritchardii* on Niuafo'ou) suggests that megapodes once existed in Fiji. Prehistoric bones now indicate that various megapodes existed at human arrival not only in Fiji, but also in Samoa, Niue, and many islands in Tonga, which sustained four species of megapodes, three of them extinct (Steadman 1989b, 1993a, 1993b, 1999d; Chapters 6, 9 herein).

Clark (1982) believed that a bird similar to the purple swamphen (*Porphyrio porphyrio*, proto-Polynesian name *kalae*) must have existed in the Marquesas or Society Islands, which have been named as possible source areas of the Hawaiian people (Sinoto 1965, Kirch 1984a), because the Hawaiian cognate word *'alae* also refers to large rails (the gallinule *Gallinula chloropus* and coot *Fulica americana*). No rails in the genera *Porphyrio*, *Gallinula*, or *Fulica*, however, had ever been found in East Polynesia, the nearest occurrence being that of *P. porphyrio* in Tonga, Samoa, and Niue (Pratt et al. 1987), where it is still known as *kalae*. Raynal (1980-1981) noted that the Marquesan name *koau* refers to a flightless bird with bluish purple plumage that existed early in the 20th century on Hiva Oa. With foresight he proposed that the *koau* was related to the large flightless forms of *Porphyrio* of New Zealand (*P. hochstetteri* on the North Island, *P. mantelli* on the South Island; Trewick 1996). In 1986–1987, I discovered 19 bones of an extinct, undescribed species of swamphen subsequently named *Porphyrio paepae* from archaeological sites on Hiva Oa and

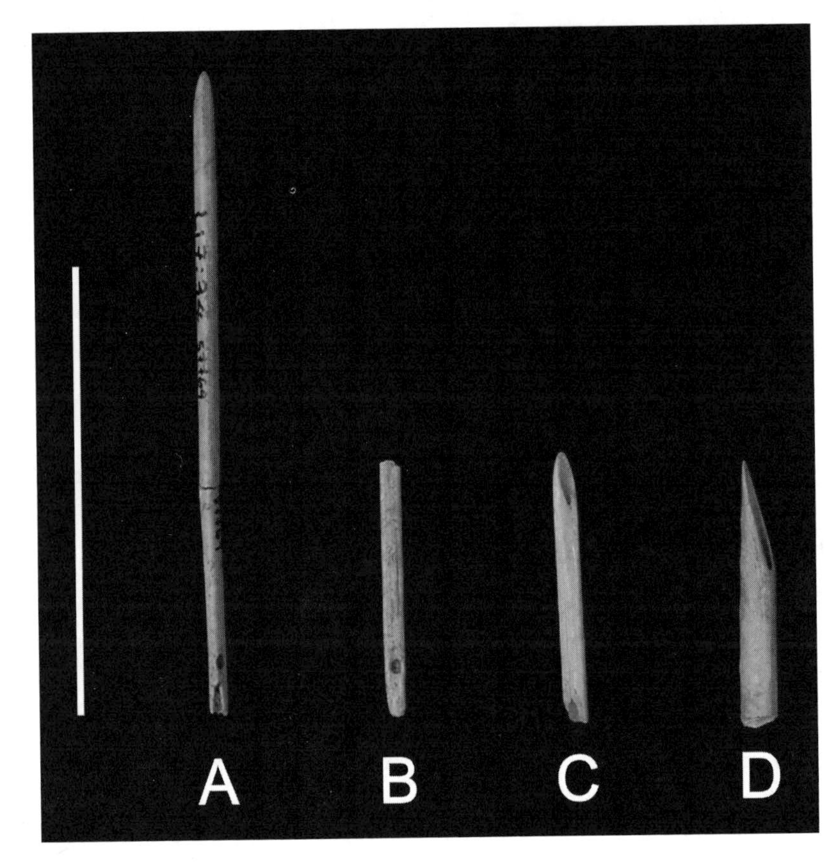

Figure 4-8 Prehistoric bird bone needles made from the radius (A–C: UF 57769, 58603, 57772) and ulna (D: UF 57774) of procellariids (shearwaters or petrels), Tongoleleka Site, Lifuka, Ha'apai Group, Tonga. Note the drilled eyes in needles A and B, and the sharpened points in needles A, C, and D. Only needle A is complete. Scale = 50 mm. Photo by Jeremy Kirchman.

Tahuata (Steadman 1988). While supporting the proposals of Clark (1982) and Raynal (1980-1981), this discovery does not solve the linguistic discrepancy between *kalae* and *koau*. The bones from Tahuata are ca. 700 to 800 years old (Steadman & Rolett 1996, Rolett 1998), and those from Hiva Oa are probably of a similar age.

Raynal & Dethier (1990) suggested that a "cryptozoological" search be made on Hiva Oa to see if *Porphyrio paepae*, the Marquesan swamphen, still exists. I agree that the precipitous terrain of Hiva Oa deserves more ornithological exploration than it has received. Although I doubt that *P. paepae* still exists, I would be the first to cheer if it did. My chances of cheering were increased recently with the discovery that a swamphen, presumably *P. paepae*, may have survived on Hiva Oa until at least AD 1902, the year that it apparently was depicted in a painting done there by Paul Gaugin (Raynal 2002). A successful search for living *P. paepae* thus may require it to have survived for only 100 years beyond its last documentation, rather than six or seven centuries.

Bird imagery is widespread in Oceania. Fishermen in the Solomon Islands carve frigatebirds (*Fregata* spp.) on wooden (*Hibiscus tiliaceus*) fishing floats. Easter Islanders made wooden carvings and petroglyphs of stylized birds and bird-men, the latter having the head of a bird (often *Fregata minor*) and a human body (Metraux 1940:256–259, 270–272, Lee 1986). Bird glyphs also appear in Easter Island's *rongorongo* wooden tablets (Metraux 1940:389–411). The "bird cult" of Easter Island involves sooty terns (*Sterna fuscata*) that nest on the offshore islet of Motu Nui (Metraux 1940:331–341, McCoy 1978, Van Tilburg 1992:57). I believe that the bird cult originated or at least increased in significance in late prehistoric times because of the rarity (and therefore value) of any seabird or its eggs. All but one of the 25 species of seabirds that formerly occupied Easter Island itself had been exterminated (Steadman 1995a; Chapter 7 herein).

Seabirds helped traditional sailors to navigate (Hilder 1963:90, Lewis 1964:364, 1972:162–173, Sharp 1964:40, Gladwin 1970:180–200, Finney 1979:334). One way to detect nearness to an island was to keep an eye out for the species of seabirds that seldom stray more than 20 to 100 km from the islands where they roost at night. Useful in this regard were boobies (*Sula* spp.) and terns (*Procelsterna cerulea*, *Anous stolidus*, *A. minutus*, *Gygis candida*). Tropicbirds (*Phaethon* spp.), frigatebirds (*Fregata* spp.), and the tern *Sterna fuscata* also were helpful in locating islands, although their wanderings made them less reliable. Inhabited islands often have far fewer seabirds than uninhabited ones. Dening (1963:114) noted that "birds in great numbers became accepted . . . as the sign of an uninhabited island. In this we might find an explanation of why almost every uninhabited island in the Pacific gives signs of having been visited by the Polynesians. Lost voyagers would be easily attracted by the sign of birds." In addition to true seabirds, the melodic whistle of the migratory plover *Pluvialis fulva*, heard both day and night, may indicate nearness of land.

Observing birds also helped to locate fish. In the Caroline Islands, nearshore surface feeding by terns (*Anous* spp., *Gygis candida*) would trap "the little fish in their frenzy between enemies above and below. It is the birds which signal to the fishermen that the [larger] fish are running in a school. All the canoes turn and plunge toward the birds" (Gladwin 1970:30). Mangaian fishermen have told me that *Gygis candida* is trustworthy and often reveals productive fishing grounds, whereas *Anous stolidus* is a "cunning" bird likely to lead fishermen to sterile waters. Fishermen with whom I have sailed in the Ha'apai and Vava'u Groups of Tonga keep an eye out for feeding flocks of boobies (*Sula leucogaster*) and terns (*Sterna fuscata*, *Anous stolidus*, and *A. minutus*).

Anous stolidus, *A. minutus*, and *Gygis candida* have been more resistant to overexploitation than other tropical seabirds and thus remain useful in locating islands and schooling fish in areas where semi-traditional sailing and deep-water fishing still occur. The loss, however, of most other seabirds on most islands must have diminished the importance of seabirds in prehistoric navigation and fishing. Shearwaters and petrels have been eliminated or depleted on island after island across Oceania (Chapter 15). Especially outside of the nesting season, most species of shearwaters and petrels are highly pelagic. Their former role in blue-water navigation is uncertain (Hilder 1963:83, 84, Sharp 1964:42, 43, 47), although their feeding activities far offshore may have helped fishermen to locate pelagic fish and marine mammals. While the role of migrant landbirds in facilitating long-distance voyaging is speculative (Lewis 1972:172, 173), those who doubt the potential utility of pelagic seabirds to sailors (i.e., Sharp 1964:59, 61) should bear in mind that a full appreciation of this matter is impossible today because: (1) seabirds are so reduced in range and numbers (the total number of individual resident seabirds in the tropical Pacific today may be from one hundredth to one thousandth of what it was 3000 years ago); and (2) those now studying traditional navigation almost certainly do not understand Pacific seabirds as well as prehistoric sailors.

Before their populations were reduced or eliminated, seabirds also may have been a food supplement for sailors, although methods of luring them near enough to be captured are unrecorded. A fire at night would attract shearwaters, petrels, storm-petrels, and tropicbirds to large double-hulled canoes, but might not be feasible on smaller wooden canoes. Accounts of long-distance voyaging seldom mention taking seabirds at sea; the only reports I have found are of native missionaries from Aitutaki who killed a "few sea-birds" during five months at sea in the 1820s (Dening 1963:138) and of several frigatebirds (*Fregata minor*) caught after landing in the rigging of a European ship off Nassau, Cook Islands, in the mid-1800s (Gill 1885:31).

Extinction

It may sound silly, but nothing is as effective as extinction in decreasing interest in a species of bird. The depletion of insular seabirds and landbirds that followed human arrival narrowed the range of species that people could use, whether as navigational or fishing aids, in legends or imagery, for feathers or bones, as pets, or as food. Thus the importance of birds in Oceanic societies, while substantial at European contact, must have been even greater before so many of the species were lost. Taboos on killing certain species were implemented locally in late prehistoric and historic times for religious or economic reasons (Steadman 1997a, Valeri 2000), although such restrictions may have done little to increase populations of rare species, whose rarity probably was related to other factors in addition to human predation (Chapter 16).

Although archaeological sites tell us that people killed and consumed birds across Oceania, it is difficult from this evidence to assess whether human predation contributed to avian extinction as much as habitat loss, disease, or predation from rats, dogs, and pigs. Undoubtedly the importance of each factor varied from island to island, and from species to species. Regardless of the details, human presence on tropical islands has been called, rightly and without prejudice, an environmental catastrophe (Olson 1989). While this statement might rile certain anthropologists, few birds or biologists would argue against it. I say this without criticizing any person, living or deceased, who has killed birds. We all have a little blood on our hands; even vegetarian birdwatchers occasionally hit a bird on the highway or eat cereal made from grain grown in massive fields where bird-friendly hedgerows have been removed.

Whatever the precise cause, extinction has rearranged bird communities across Oceania. I will document these losses now from a geographic standpoint in Chapters 5–8, followed by a taxonomic perspective in Chapters 9–15.

Part II

Chapter 5 Melanesia

FEW PLACES are as interesting culturally or biogeographically as Melanesia (Figure 1-2). People colonized the Bismarck Archipelago and Solomon Islands (Near Oceania) by 35,000 to 30,000 yr BP (Wickler & Spriggs 1988, Pavlides & Gosden 1994; Chapter 3 herein). Unlike in Remote Oceania, where human arrival in the late Holocene was devastating to birds (Steadman 1995a), the Lapita peoples who moved across Near Oceania from 3600 to 3000 cal BP found islands that already had had tens of millennia of human activity (Kirch 1996, Steadman & Kirch 1998, Steadman et al. 1999b). While Polynesian islands lost most of their species of birds within the past 3000 years, avian extinction in Near Oceania probably was a slower process.

The rich avifauna of the immense, continental island of New Guinea has little evidence of extinction in historic times (Coates 1985, Beehler et al. 1986) and essentially no grounding in avian prehistory. Aside from a few unidentified bones from several sites (Bulmer 1975), the only avian fossils known from New Guinea are of three extant species (the cassowary *Casuarius bennetti*, eagle *Harpyopsis novaeguinae*, and megapode *Aepypodius arfakianus*) at the Nombe Rockshelter in the highlands of Simbu Province (Gillieson & Mountain 1983). The prehistoric void for New Guinean landbirds will require an enormous research investment to fill.

New Guinea's late Pleistocene fossil mammals include extinct marsupials that probably were wiped out by people (Flannery 1995b). Although the extent of prehistoric avian extinction in New Guinea is speculative, major losses of birds did occur on New Ireland, a large island near New Guinea with Oceania's second richest modern landbird fauna (105–110 species; see below). Because such a large, mostly forested island close to New Guinea experienced substantial losses of landbirds following human arrival, New Guinea itself probably was not spared the great loss of birdlife that accompanied the spread of humans across the Pacific. Although renowned since its "discovery" by Europeans for being so wild and forested, the New Guinean environment was far from pristine at European contact (Groube 1989, Gosden & Webb 1994).

New Guinea was connected to Australia during lowered sea levels of the late Pleistocene to form the continent of Sahul (Thorne & Raymond 1989, Spriggs 1997). People colonized New Guinea and Australia about 50 to 60 ka (Lourandos 1997, Miller et al. 1999), a frustrating time interval that is difficult to resolve because ^{14}C dating typically is effective only with samples younger than 40 to 45 thousand years old. (With older samples, so much ^{14}C has decayed that it becomes impossible to distinguish the sample's remaining radioactivity from background levels.) Various large birds were lost in Australia after human arrival, including the immense flightless relatives of waterfowl known as dromornithids or "mihirungs" (Murray & Megirian 1998, Miller et al. 1999) and large megapodes *Progura* spp. (Baird 1991). These conspicuous species may be the tip of the avian extinction iceberg in Australia and, by association, New Guinea as well. I also

Table 5-1 Taxonomic comparison of the landbird faunas of lowland (<1000 m) New Guinea vs. Near Oceania. From Mayr and Diamond (2001); Tables 5-2, 5-3 to 5-8 herein.

	Lowland New Guinea	Bismarcks	Solomon Main Chain
FAMILIES			
Nonpasserines	34	28	23
Pittas	1	1	1
Passerida	11	9	8
Vagile Corvida	8	5	5
Sedentary Corvida	10	3	3
Total	64	46	40
GENERA			
Nonpasserines	114	64	58
Pittas	1	1	1
Passerida	23	15	14
Vagile Corvida	14	7	7
Sedentary Corvida	55	5	6
Total	207	92	86
SPECIES			
Nonpasserines	204	109	89
Pittas	3	3	1
Passerida	42	26	32
Vagile Corvida	84	22	25
Sedentary Corvida	96	14	13
Total	429	174	160

suspect that New Guinea formerly was inhabited by other extinct landbirds in the same families (such as rails, pigeons, and parrots) that have suffered so much extinction in Oceania.

"Island Melanesia" includes all island groups covered in this chapter and excludes New Guinea (Spriggs 1997). Island Melanesia can be divided into "Near Melanesia" (Bismarcks, Solomon Islands Main Chain) and "Remote Melanesia" (outyling islands in the Solomons such as Rennell, Bellona, Santa Cruz Group, Sikaiana, and Ontong Java, and all of Vanuatu and New Caledonia; Figure 1-2). Near Melanesia is the same as Near Oceania. Although more families (47 vs. 37) and genera (102 vs. 81) of landbirds have been recorded in Near Melanesia than in Remote Melanesia, the landbird taxa of the latter region are not merely a subset of those in the former; six families and 24 genera in Remote Melanesia are unknown in Near Melanesia, largely because of New Caledonia's distinctive avifauna. As detailed below, prehistoric bird bones have been identified from only five islands in Near Melanesia and 12 in Remote Melanesia. These assemblages exceed 100 bones on one island in the former region and eight islands in the latter.

At any taxonomic level, the biggest distinction between the landbird faunas of New Guinea and Near Oceania is among 12 passerine families known as the sedentary Corvida (Table 5-1 and the appendix; also see Mayr & Diamond 2001:51). Much of the difference in nonpasserine diversity between the Bismarcks and Solomons is due to New Guinean taxa that reach the Bismarcks but not the Solomons. By contrast, passerines (divided into Pittas, Passerida, vagile Corvida, and sedentary Corvida) are as or more diverse in the Solomons than the Bismarcks at the level of genus or species. More speciation (species per genus) has occurred in the Solomons, especially in sylviid warblers, white-eyes, starlings, and fantails.

Genus-level endemism is rare in Near Oceania. Among 72 nonpasserine genera in Near Malanesia, only five are endemic (three columbids, an owl, and a frogmouth; Table 5-2). For passerines, these values are 33 and three, although all three supposed endemics are meliphagids (honeyeaters) of uncertain generic classification (see Mayr & Diamond 2001:398, 399, plate 9). Each of the eight certain or possible endemic genera occurs in the Solomons but not the Bismarcks.

Bismarck Archipelago

From an ornithological perspective, the Bismarck Archipelago (Figure 5-1) includes the islands of Crown, Long, Tolokiwa, Umboi, and Sakar but not Karkar and Bagabag (Diamond & LeCroy 1979, Diamond 1976a, 1981, Mayr & Diamond 2001). Landbirds of the Bismarcks (and Solomons) have played a major role in the conceptual development of community ecology and island biogeography, primarily when Diamond (1975b) interpreted the distributional data on modern landbirds in terms of "incidence functions" and "assembly rules." I will discuss these topics in Chapter 20, including updated considerations from Mayr & Diamond (2001). Here I will review the past and present distribution of landbirds in the Bismarcks, which have the most taxonomically diverse landbird fauna in Oceania, with 93 genera, 65 of which are nonpasserine (Table 5-2). A total of 14 genera (11 nonpasserines, three passerines) is found in Oceania only in the Bismarcks. The richest (New Britain, 127–132 species) and second richest (New Ireland, 107–111 species) modern landbird faunas on single islands

Table 5-2 Genera of resident landbirds in Melanesia. [e], endemic genus; H, historic record, now certainly or probably extirpated; M, modern record; M/H, a definite historic record with inadequate evidence of current survival; P, prehistoric record; †, extinct genus; ?, questionable record, or residency not established. Excludes *Pelecanus*, in which *P. conspicillatus* occasionally wanders to Melanesia and breeds. Genus-level systematics of certain Melanesian honeyeaters are unresolved. Tentatively, I recognize *Stresemannia, Guadalcanaria*, and *Meliarchus*; see Mayr & Diamond (2001:398, 399) for a brief discussion of possible synonyms. From Mayr (1945), Coates (1985, 1990), Balouet & Olson (1989), Bregulla (1992), Filardi et al. (1999), Steadman et al. (1990b, 1999b), Kratter et al. (2001a), Mayr & Diamond (2001), and herein.

	Near Melanesia		Remote Melanesia			
	Bismarck Archipelago	Solomon Island Main Chain	Rennell & Bellona	Santa Cruz	Vanuatu	New Caledonia
CASSOWARIES						
Casuarius	M, P	—	—	—	—	—
GREBES						
Tachybaptus	M	M	M	—	M	M
CORMORANTS						
Phalacrocorax	M	M	M	M	—	M
HERONS						
Egretta	M, P	M	M	M	M	M
Butorides	M	M	—	M	M	M
Nycticorax	M, P	M,P	—	—	—	M
Ixobrychus	M	M	M	—	—	—
Botaurus	—	—	—	—	—	M
IBISES						
Threskiornis	—	—	M	—	—	—
DUCKS						
Dendrocygna	M	—	—	—	—	H
Anas	M	M, P	M	M, P	M	M, P
Aythya	—	—	—	—	M	M
OSPREYS						
Pandion	M	M	M	—	—	M
HAWKS						
Aviceda	M	M	—	—	—	—
Henicopernis	M	—	—	—	—	—
Haliastur	M, P	M, P	—	—	—	M
Haliaeetus	M	M, P	—	—	—	—
Circus	—	—	—	—	M	M
Accipiter	M, P	M, P	M	M	M, P	M, P
FALCONS						
Falco	M	M	—	—	M	M, P
MEGAPODES						
Megapodius	M, P	M, P	—	P	M, P	P
†*Sylviornis*[e]	—	—	—	—	—	P
QUAILS						
Coturnix	M, P	—	—	—	—	—
BUTTON-QUAILS						
Turnix	M	M	—	—	—	H, P

(*continued*)

Table 5-2 (continued)

	Near Melanesia		Remote Melanesia			
	Bismarck Archipelago	Solomon Island Main Chain	Rennell & Bellona	Santa Cruz	Vanuatu	New Caledonia
RAILS						
Rallina	M, P	—	—	—	—	—
Gallirallus[1]	M, P	M, P	—	P	M, P	M/H, P
Gymnocrex	M	—	—	—	—	—
Porzana	M, P	M	M/H	M	M, P	M, P
Poliolimnas	M, P	M	—	—	M	M
Amaurornis	M, P	M	—	—	—	—
Gallinula	—	—	—	—	—	M, P
Pareudiastes	—	M, P	—	—	—	—
Porphyrio	M, P	M, P	M	M, P	M, P	M, P
KAGU						
Rhynochetos[e]	—	—	—	—	—	M, P
JACANAS						
Irediparra	M	—	—	—	—	—
STILTS						
Himantopus	M	—	—	—	—	—
THICK-KNEES						
Esacus	M	M	—	—	—	M
PLOVERS						
Charadrius	M	—	—	—	—	—
SNIPES						
Coenocorypha?	—	—	—	—	—	P
PIGEONS, DOVES						
Columba	M, P	M	—	—	M, P	M, P
Macropygia	M, P	M	M	M	M, P	—
Reinwardtoena	M	M	—	—	—	—
Ptilinopus	M, P	M	M	M	M, P	M
Drepanoptila[e]	—	—	—	—	—	M, P
Ducula	M, P	M	M	M, P	M, P	M, P
Gymnophaps	M	M	—	—	—	—
Chalcophaps	M, P	M	—	M	M, P	M, P
Henicophaps	M	—	—	—	—	—
Caloenas	M, P	M, P	M/H	—	—	P
Gallicolumba	M, P	M, P	M	M	H	P
†*Microgoura*[e]	—	H	—	—	—	—
†*Extinct genus A*[e]	—	P	—	—	—	—
†*Extinct genus B*[e]	—	P	—	—	—	—
PARROTS						
Cacatua	M, P	M	—	—	—	P
Chalcopsitta	M	M, P	—	—	—	—
Trichoglossus	M	M	—	M	M, P	M
Lorius	M, P	M	M	—	—	—

Table 5-2 (continued)

	Near Melanesia		Remote Melanesia			
	Bismarck Archipelago	Solomon Island Main Chain	Rennell & Bellona	Santa Cruz	Vanuatu	New Caledonia
Charmosyna	M, P	M	—	M	M, P	H
Micropsitta	M	M	M	—	—	—
Eunymphicus[e]	—	—	—	—	—	M
Cyanoramphus	—	—	—	—	—	M
Geoffroyus	M	M	M	—	—	—
Eclectus	M	M	—	—	P	—
Loriculus	M	—	—	—	—	—
CUCKOOS						
Cacomantis	M, P	M	—	—	M, P	M
Chrysococcyx	—	—	M	M?	M, P	M
Eudynamys	M	M	—	—	—	—
Centropus	M, P	M	—	—	—	—
BARN-OWLS						
Tyto	M, P	M	M/H	M	M, P	M, P
STRIGID OWLS						
Ninox	M, P	M	—	—	—	P
Nesasio[e]	—	M, P	—	—	—	—
OWLET-NIGHTJARS						
Aegotheles	—	—	—	—	—	H, P
FROGMOUTHS						
cf. *Podargus*	M	—	—	—	—	—
New genus[e]	—	M	—	—	—	—
NIGHTJARS						
Eurostopodus	—	M	—	—	—	M/H
Caprimulgus	M, P	—	—	—	—	—
CRESTED SWIFTS						
Hemiprocne	M	M	M	—	—	—
SWIFTS						
Collocalia	M, P	M	M	M	M, P	M, P
KINGFISHERS						
Alcedo	M, P	M	—	—	—	—
Ceyx	M	M	—	—	—	—
Halcyon	M	M	M	M	M, P	M, P
Actenoides	—	M	—	—	—	—
Tanysiptera	M	—	—	—	—	—
BEE-EATERS						
Merops	M	—	—	—	—	—
ROLLERS						
Eurystomus	M	M	—	—	—	—
HORNBILLS						
Aceros	M, P	M	—	—	—	P

(continued)

Table 5-2 (continued)

	Near Melanesia		Remote Melanesia			
	Bismarck Archipelago	Solomon Island Main Chain	Rennell & Bellona	Santa Cruz	Vanuatu	New Caledonia
PITTAS						
Pitta	M, P	M	—	—	—	—
SWALLOWS						
Hirundo	M	M	—	M	M, P	M
THRUSHES						
Zoothera	M	M	—	—	—	—
Turdus	M, P	M	M	M	M, P	M
CISTICOLAS						
Cisticola	M	—	—	—	—	—
WARBLERS						
Cettia	—	M	—	—	—	—
Acrocephalus	M	M	—	—	—	—
Phylloscopus	M	M	—	—	—	—
Megalurulus	—	—	—	—	—	M
Cichlornis	M	M	—	—	M	—
OLD WORLD FLYCATCHERS						
Saxicola	M	—	—	—	—	—
SUNBIRDS						
Nectarinia	M	M	—	—	—	—
FLOWERPECKERS						
Dicaeum	M	M	—	—	—	—
WHITE-EYES						
Zosterops	M	M	M	M	M, P	M
Woodfordia[e]	—	—	M	M	—	—
STARLINGS						
Aplonis	M, P	M	M	M	M, P	M
Mino	M, P	M	—	—	—	—
ESTRILDID FINCHES						
Erythrura	M	M	—	—	M	M
Lonchura	M	M	—	—	—	—
CUCKOO-SHRIKES, TRILLERS						
Coracina	M, P	M	M	—	M, P	M
Lalage	M	M	—	M	M	M
GERYGONES						
Gerygone	—	—	M	—	M, P	M
ROBINS						
Monachella	M	—	—	—	—	—
Petroica	—	M	—	—	M	—
Eopsaltria	—	—	—	—	—	M

Table 5-2 (continued)

	Near Melanesia		Remote Melanesia			
	Bismarck Archipelago	Solomon Island Main Chain	Rennell & Bellona	Santa Cruz	Vanuatu	New Caledonia
WHISTLERS						
Pachycephala	M	M	M	M	M, P	M
FANTAILS						
Rhipidura	M	M	M	M	M	M
MONARCHS						
Mayrornis	—	—	—	M	—	—
Neolalage[e]	—	—	—	—	M	—
Clytorhynchus	—	—	M	M	M	M
Monarcha	M	M	—	—	—	—
Myiagra	M	M	M	M	M	M
HONEYEATERS						
Stresemannia[e?]	—	M	—	—	—	—
Guadalcanaria[e?]	—	M	—	—	M	M
Melidectes	M	—	—	—	—	—
Meliarchus[e?]	—	M	—	—	—	—
Lichmera	—	—	—	—	M, P	M
Myzomela	M	M	M	M	M, P	M
Philemon	M	—	—	—	—	M
Gymnomyza	—	—	—	—	—	M
Phylidonyris	—	—	—	—	M	M
DRONGOS						
Dicrurus	M, P	M	—	—	—	—
WOOD-SWALLOWS						
Artamus	—	—	—	—	M	M
CROWS						
Corvus	M, P	M	—	—	—	M
TOTAL NONPASSERINES						
M	65	54	23–25	17–18	25	32–34
M + H	65	55	25	17–18	26	36
M + H + P	65	58	25	19–20	27	44
TOTAL PASSERINES						
M	28	28	11	12	19	21
M + H	28	28	11	12	19	21
M + H + P	28	28	11	12	19	21
TOTAL (ALL GENERA)						
M	93	82	34–36	29–30	44	53–55
M + H	93	83	36	29–30	45	57
M + H + P	93	86	36	31–32	46	65
Number of landbird bones	254+	75	0	67+	601	6353+

1. Includes *Nesoclopeus* (see Chapter 10).

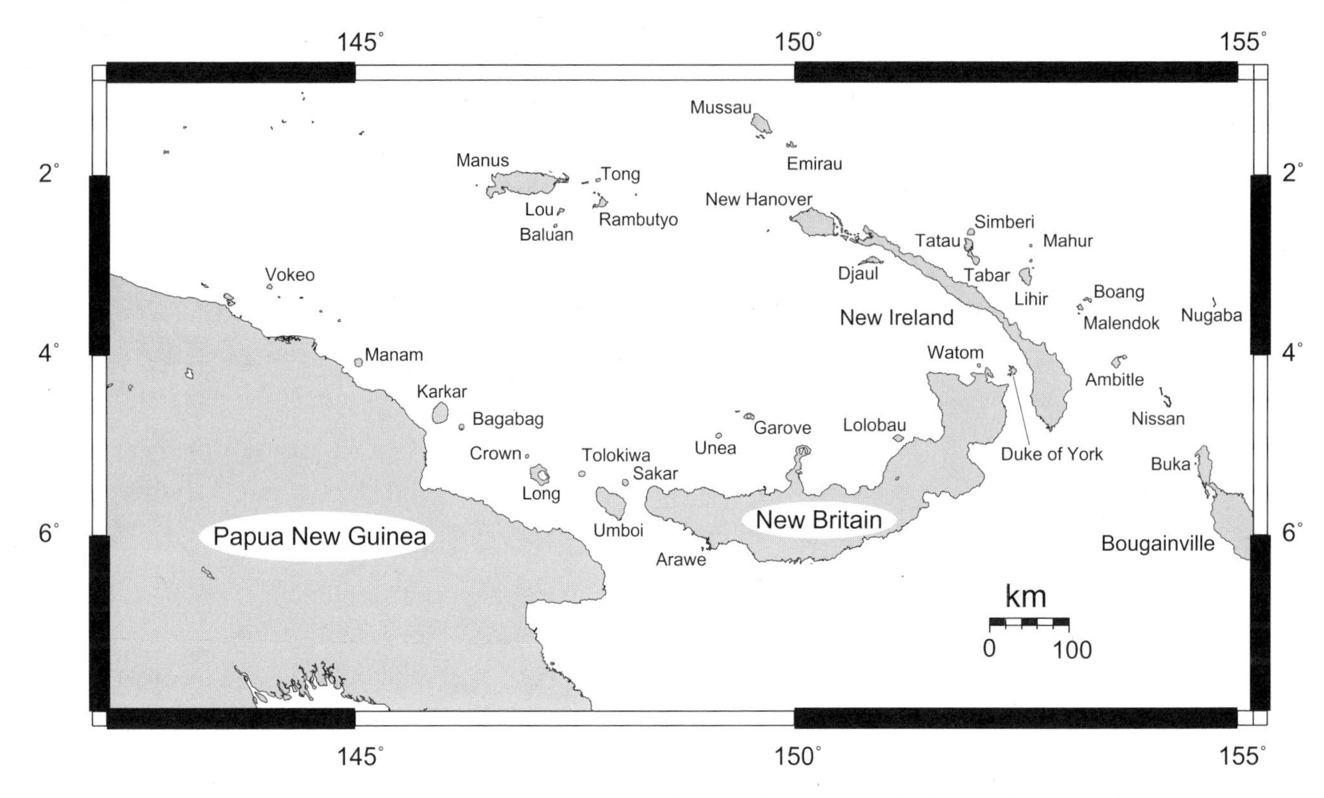

Figure 5-1 The Bismarck Archipelago and adjacent Papua New Guinea. Buka and Bougainville are biogeographically part of the Solomon Islands.

in Oceania are in the Bismarcks, a clear reflection of their nearness to New Guinea (Table 5-3).

No islands in the Bismarcks ever were connected to New Guinea. Although the 120 m lower sea level at 18,000 yr BP would have made all islands in the Bismarcks closer to each other, none of the six largest (New Britain, New Ireland, Manus, Lavongai, Umboi, Mussau) would have been connected. The greater isolation of Mussau and Manus probably accounts for their lower species richness of landbirds. If fossils were available, I predict that any of the species found on New Ireland, past or present, would show up on the nearby but much smaller Lavongai (New Hanover). Similarly, I suspect that Umboi once supported many of the species now confined (at least in the Bismarcks) to New Britain or to New Britain and New Ireland.

New Ireland

The second largest island in the Bismarcks, New Ireland lies 30 km north and east of New Britain (Figure 5-1). The resident landbird fauna now consists of about 107–111 species (103 species listed in Mayr & Diamond 2001), of which 70-74 are nonpasserines (including an amazing 18 pigeons and 11 parrots) and 37 are passerines (Table 5-3). The modern avifauna of New Ireland is most similar to, and largely a subset of, that of New Britain, which is larger, less isolated from New Guinea, and supports about 127–132 species of landbirds. Most of the 30–32 species found on New Britain but not on New Ireland either are endemic to New Britain (as far as we know; see Mayr & Diamond 2001:227–230) or occur as well on mainland New Guinea but are found in Oceania only on New Britain.

Bird bones have been found on New Ireland in six cultural sites as old as 32,000 years (Steadman et al. 1999b; Chapter 3 herein). Of 52 species of birds represented by only 241 bones, at least 11 (petrel, megapode, quail, four rails, cockatoo, two owls, and crow) were not known previously from New Ireland (Table 5-4). The megapode, two of the rails, and probably the cockatoo are extinct, undescribed species.

With the conservative assumption that all landbird taxa marked "#" in Table 5-4 represent species that still live on New Ireland, bones of 40 (~38%) of the 105–110 currently resident species were recovered from the archaeological sites, in addition to 10 species of landbirds no longer occurring on the island. Thus a complete record of New Ireland's late Pleistocene/early Holocene landbirds probably would include at least 25–30 species that are absent today, yielding a landbird fauna of at least 130–140 species. In part because most of New Ireland's extinct/extirpated

Table 5-3 Resident landbirds of the Bismarck Archipelago. M, modern record; P, prehistoric record. Taxa entirely in brackets may represent a species listed more precisely and thus are excluded from totals. Species-level systematics are often unresolved, especially in passerines. Excludes species of uncertain breeding status, such as *Pelecanus conspicillatus, Ardea alba, Egretta intermedia, Scythrops novaehollandiae*, and *Halcyon sancta*. *, generic relationships uncertain; possibly *Cichlornis, Megalurulus*, or *Trichocichla*; ?, questionable record, or residency not established; "most small islands" = fifteen or more islands other than the six listed separately (see Coates 1985, 1990, Mayr & Diamond 2001). Compiled with difficulty and uncertainty from Hartert (1925), Diamond (1970a, b, 1971, 1976, 1981), Orenstein (1976), Beehler (1978), Coates (1985, 1990), Dutson & Newman (1991), Fry et al. (1992), Chantler & Driessen (1995), Cleere (1998), Juniper & Parr (1998), Taylor (1998), Steadman & Kirch (1998), Steadman et al. (1999b), Williams (1999), Beehler et al. (2001), and Mayr & Diamond (2001). Island areas and elevations are from Mayr & Diamond (2001).

Species	Umboi	New Britain	New Ireland	Lavongai	Manus	Mussau	Smaller islands
CASSOWARIES							
Casuarius cf. *casuarius*	—	—	—	—	—	—	P (Arawe)
Casuarius bennetti	—	M	—	—	—	—	—
GREBES							
Tachybaptus ruficollis	M	M	M	—	P?	—	Lolobau, Long, Los Negros, Witu
CORMORANTS							
Phalacrocorax melanoleucos	—	M	—	—	—	—	—
HERONS							
Ardea alba	—	M	—	—	—	—	—
Egretta sacra	M	M	M	M	M	M	most small islands
Butorides striatus	—	M	M	M	—	—	—
Nycticorax caledonicus	M	M	M, P	M	M, P?	—	Duke of York, Feni, Lihir, Lolobau, Long, Rambutyo, Sakar, Tabar, Tanga, Tolokiwa, Watom
Ixobrychus flavicollis	M	M	M	M	M	M	Duke of York, Dyaul, Emirau, Feni, Hermits, Lihir, Lou, Ninigos, Sakar, Tabar, Watom, Witu
DUCKS							
Dendrocygna guttata	M	M	—	—	—	—	—
Dendrocygna arcuata	M	M	—	—	—	—	—
Anas superciliosa	M	M	M	M	M	—	Feni, Lihir, Long, Rambutyo, Sakar, Tabar
OSPREYS							
Pandion haliaetus	M	M	M	M	M	M	most small islands
HAWKS							
Aviceda subcristata	—	M	M	M	M	—	Dyaul, Lolobau, San Miguel, Tabar
Henicopernis infuscata	—	M	—	—	—	—	Lolobau
Haliastur indus	M	M	M, P	M	M	M	most small islands
Haliaeetus leucogaster	M	M	M	M	M	M	most small islands
Accipiter novaehollandiae	M	M	M, P	M	M	M	Duke of York, Dyaul, Feni, Lihir, Lolobau, Nauna, Rambutyo, San Miguel, Tabar?, Tanga
Accipiter albogularis	—	—	—	—	—	—	Feni
Accipiter luteoschistaceus	M	M	—	—	—	—	—
Accipiter princeps	—	M	—	—	—	—	—
Accipiter brachyurus	—	M	M	—	—	—	—
Accipiter meyerianus	M	M	M	—	—	—	Long?, Watom?
[*Accipiter* sp. 1]	—	—	P	—	—	—	—
[*Accipiter* sp. 2]	—	—	P	—	—	—	—
FALCONS							
Falco severus	—	M	M	—	—	—	Duke of York, Tanga, Watom
Falco berigora	—	—	—	—	—	—	Long

(*continued*)

Table 5-3 (continued)

Species	Umboi	New Britain	New Ireland	Lavongai	Manus	Mussau	Smaller islands
Falco peregrinus	M	M	—	—	—	—	Feni, Lihir, Long, Witu
MEGAPODES							
Megapodius eremita	M	M	M, P	M	M	M	most small islands
Megapodius undescribed sp. A	—	—	P	—	—	—	—
QUAILS							
Coturnix cf. *ypsilophorus*	—	—	P	—	—	—	—
Coturnix chinensis	—	M	M, P	M	—	—	Duke of York, Lihir, Tabar, Tanga, Watom
BUTTON-QUAILS							
Turnix maculosa	—	M	—	—	—	—	Duke of York
RAILS							
Rallina tricolor	—	—	M, P	M	—	M	—
Gallirallus philippensis	—	M	M, P	M	M	—	most small islands
Gallirallus insignis	—	M	—	—	—	—	
Gallirallus undescribed sp. A	—	—	P	—	—	—	—
Gymnocrex plumbeiventris	—	—	M?	—	—	—	—
Porzana tabuensis	—	—	P	—	—	—	Watom, Wuvulu
Poliolimnas cinereus	M	M	M	M	—	M	Duke of York, Emirau, Lihir
Amaurornis moluccanus	M	M	M, P	M	—	—	Duke of York, Lihir, Long, Sakar, Tabar, Tanga, Tolokiwa, Watom, Witu
Porphyrio porphyrio	M	M	P	M	M, P?	—	Lou, Rambutyo, Tabar, Watom
Porphyrio undescribed sp. A	—	—	P	—	—	—	—
JACANAS							
Irediparra gallinacea	—	M	—	—	—	—	—
STILTS							
Himantopus himantopus	—	M	—	—	—	—	Long
THICK-KNEES							
Esacus magnirostris	M	M	M	M	M	M	most small islands
PLOVERS							
Charadrius dubius	—	M	M	M?	—	—	—
PIGEONS, DOVES							
Columba vitiensis	—	M?	M	—	—	—	—
Columba pallidiceps	—	M	M	—	—	—	Duke of York?
[*Columba sp.*]	—	—	P	—	—	—	—
Macropygia amboinensis	M	M	M, P	M	M	—	Duke of York, Feni, Lihir, Lolobau, Long, Sakar, Tabar, Tanga, Tolokiwa, Watom
Macropygia nigrirostris	—	M	M, P	M	—	—	Duke of York, Tabar
Macropygia mackinlayi	M	M?	M?	—	M	M	Crown, Emirau, Lihir, Long, Nauna, Rambutyo, Sakar, San Miguel, Tench, Tolokiwa, Watom, Witu
Reinwardtoena browni	M	M	M	M	M	—	Duke of York, Dyaul, Lihir, Lolobau, Nauna, Rambutyo, Tabar, Watom?
Ptilinopus superbus	M	M	M	M	M	M	Duke of York?, Feni, Lihir, Lolobau, Rambutyo, Tabar, Tanga, Tolokiwa, Watom?

Table 5-3 (continued)

Species	Umboi	New Britain	New Ireland	Lavongai	Manus	Mussau	Smaller islands
Ptilinopus rivoli	M	M	M, P	M	—	—	Duke of York, Dyaul, Lihir, Lolobau, Tabar, Tanga
Ptilinopus solomonensis	M	M?	M?	M	M	M	most small islands
Ptilinopus viridis	—	—	—	—	M	—	Lihir
Ptilinopus insolitus	M	M	M	M	—	M	Crown, Duke of York, Emirau, Feni, Lihir, Lolobau, Long, Sakar, Tabar, Tolokiwa, Watom
[*Ptilinopus* sp.]	—	—	P	—	—	—	—
Ducula pacifica	—	—	—	—	—	M?	Anchorites, Hermits, Ninigos, Tench, Wuvulu
Ducula rubricera	M	M	M	M	—	—	Credner, Duke of York, Dyaul, Feni, Lihir, Lolobau, Sakar, Tabar, Tanga, Watom
Ducula finschii	M	M	M	—	—	—	Dyaul, Watom?
Ducula pistrinaria	M	M	M	M	M	M	most small islands
Ducula melanochroa	M	M	M	—	—	—	Duke of York?, Watom?
Ducula spilorrhoa	M	M	M	M	M	—	Credner, Duke of York?, Hein, Lolobau, Long, Lou, Malai, Nauna, Sakar, Tolokiwa, Watom?
[*Ducula* sp.]	—	—	P	—	—	—	—
Gymnophaps albertisii	—	M	M	—	—	—	—
Chalcophaps stephani	M	M	M, P	M	M	M	most small islands
Henicophaps foersteri	M	M	—	—	—	—	Lolobau
Caloenas nicobarica	M	M	M, P	M	M	M	most small islands
Gallicolumba jobiensis	M	M	M, P	M?	—	—	Duke of York?, Feni, Hermits, Lihir, Lolobau, Sakar, Tabar, Tanga, Tong, Watom
Gallicolumba beccarii	M	M	M, P	M	M	M	Bali, Crown, Duke of York, Emirau, Feni, Lihir, Lolobau, Long, Sakar, Tabar?, Tanga, Tolokiwa, Witu
PARROTS							
Cacatua [*galerita*] *ophthalmica*	—	M	—	—	—	—	—
Cacatua [*galerita*] undescribed species/subspecies	—	—	P	—	—	P	—
Chalcopsitta cardinalis	—	—	—	M?	—	—	Feni, Lihir, Tabar, Tanga
Trichoglossus haematodus	M	M	M	M	M	M	most small islands
Lorius hypoinochrous	M	M	M	M	—	—	Dyaul, Lihir, Lolobau, Long, Sakar, Tabar, Watom?, Witu
Lorius albidinuchus	—	—	M	—	—	—	—
[*Lorius* sp.]	—	—	P	—	—	—	—
Charmosyna rubrigularis	—	M	M	M?	—	—	—
Charmosyna placentis	M	M	M	M	M?	—	most small islands
[*Charmosyna* sp.]	—	—	P	—	—	—	—
Micropsitta pusio	M	M	M?	—	—	—	Bali, Duke of York, Lolobau, Sakar, Tolokiwa, Unea, Watom, Witu
Micropsitta bruijnii	—	M	M	—	—	—	—
Micropsitta meeki	—	—	—	—	M	M	Emirau, Lou?, Rambutyo

(*continued*)

Table 5-3 (continued)

Species	Umboi	New Britain	New Ireland	Lavongai	Manus	Mussau	Smaller islands
Micropsitta finschii	—	—	M	M	—	—	Dyaul, Lihir, Tabar
Geoffroyus heteroclitus	M	M	M	M	—	—	Duke of York, Dyaul, Lihir, Lolobau, Tabar, Witu
Eclectus roratus	M	M	M	M	M	—	Bali, Duke of York, Dyaul, Feni, Lihir, Lolobau, Rambutyo, Tabar, Tanga, Unea, Watom, Witu
Loriculus [*aurantiifrons*] *tener*	—	M	M	M	—	—	Duke of York
CUCKOOS							
Cacomantis variolosus	M	M	M, P	M	M	—	most small islands
Eudynamys scolopacea	M	M	M	—	—	—	Crown, Duke of York, Lolobau, Long, Sakar, Tolokiwa, Watom
Centropus violaceus	—	M	M	—	—	—	—
Centropus ateralbus	M	M	M	—	—	—	Dyaul, Lolobau
[*Centropus* sp.]	—	—	P	—	—	—	—
BARN–OWLS							
Tyto novaehollandiae	—	—	—	—	M	—	—
Tyto aurantia	—	M	—	—	—	—	—
Tyto alba	—	M?	—	—	—	—	Long, Tanga
[*Tyto* sp. 1]	—	—	P	—	—	—	—
[*Tyto* sp. 2]	—	—	P	—	—	P	—
STRIGID OWLS							
Ninox meeki	—	—	—	—	M	—	Los Negros
Ninox variegata	—	M?	M, P	M	—	—	—
Ninox odiosa	—	M	—	—	—	—	Watom
FROGMOUTHS							
cf. *Podargus ocellatus*	—	M	—	—	—	—	—
NIGHTJARS							
Caprimulgus macrurus	M	M	M, P	M	—	—	Lihir, Lolobau, Long, Tabar, Tolokiwa, Watom
CRESTED SWIFTS							
Hemiprocne mystacea	M	M	M	M	M	M	Duke of York, Dyaul, Emirau, Feni, Lihir, Lolobau, Long, Los Negros, Mahur, Rambutyo, Tabar, Tanga, Watom
SWIFTS							
Collocalia esculenta	—	M	M, P	M	M	—	most small islands
Collocalia spodiopygia	—	M	M	M	M	M	Emirau, Horno, Lihir, Long, Los Negros, Lou, Pak, Rambutyo, Tabar, Watom
Collocalia vanikorensis	M	M	M	M	M	M	most small islands
Collocalia whiteheadi	—	—	M	—	—	—	—
KINGFISHERS							
Alcedo atthis	M	M	M, P	M	M	M	Duke of York, Dyaul, Emirau, Feni, Lihir, Long, Sakar, Tabar, Tanga, Tolokiwa, Watom
Ceyx websteri	M	M	M	M	—	—	Lihir, Watom?
Ceyx pusillus	M	M	M	M	—	—	Dyaul, Tabar

Table 5-3 (continued)

Species	Umboi	New Britain	New Ireland	Lavongai	Manus	Mussau	Smaller islands
Ceyx lepidus	M	M	M	M	M	—	Lihir, Lolobau, Tabar, Watom
Halcyon albonotata	—	M	—	—	—	—	—
Halcyon chloris	M	M	M	M	—	M	Crown, Dyaul, Emirau, Feni, Lihir, Long, Tabar, Tanga, Tingwon, Tolokiwa, Unea, Watom, Witu, Wuvulu?
Halcyon saurophaga	M	M	M	M	M	M	most small islands
Tanysiptera [sylvia] nigriceps	M	M	—	—	—	—	Duke of York, Lolobau, Watom?
BEE-EATERS							
Merops philippinus	M	M	—	—	—	—	Long, Sakar
ROLLERS							
Eurystomus orientalis	M	M	M	M	—	M	Bali, Duke of York, Feni, Lihir, Sakar, Tabar, Tolokiwa, Unea, Watom?, Witu
HORNBILLS							
Aceros plicatus	—	M	M, P	M	—	—	—
PITTAS							
Pitta sordida	—	—	—	—	—	—	Crown, Long, Tolokiwa
Pitta superba	—	—	—	—	M	—	—
Pitta erythrogaster	M	M	M, P	M	—	—	Duke of York?, Dyaul, Lolobau, Tabar, Tolokiwa
SWALLOWS							
Hirundo tahitica	M	M	M	M	M	—	Crown, Duke of York, Feni, Lihir, Long, Malai, Sakar, Tabar, Tanga, Tolokiwa, Watom, Witu
THRUSHES							
Zoothera heinei	—	—	—	—	—	M	Emirau
Zoothera talasea	M	M	—	—	—	—	—
Turdus poliocephalus	—	M	M, P	—	—	M	Tolokiwa
CISTICOLAS							
Cisticola exilis	M	M	M	M	—	—	Duke of York, Lihir, Long, Masahet, Sakar, Tabar, Watom
OLD WORLD WARBLERS							
Acrocephalus [australis] stentoreus	M	M	M	—	—	—	Long
Phylloscopus poliocephalus	M	M	M	—	—	M	—
Megalurus timoriensis	—	M	M	M	—	—	Tolokiwa, Watom
Cichlornis grosvenori	—	M	—	—	—	—	—
Cichlornis rubiginosus	—	M	—	—	—	—	—
OLD WORLD FLYCATCHERS							
Saxicola caprata	—	M	M	—	—	—	Long, Watom
SUNBIRDS							
Nectarinia [sericea] aspasia	M	M	M	M	—	—	Bali, Credner, Duke of York, Dyaul, Feni, Lihir, Lolobau, Sakar, Tabar, Unea, Watom
Nectarinia jugularis	M	M	M	M	M	M	most small islands
FLOWERPECKERS							
Dicaeum eximium	—	M	M	M	—	—	Dyaul, Lihir, Lolobau, Watom

(*continued*)

Table 5-3 (continued)

Species	Umboi	New Britain	New Ireland	Lavongai	Manus	Mussau	Smaller islands
WHITE-EYES							
Zosterops hypoxanthus	M	M	M	M	M	—	Watom
Zosterops griseotinctus	—	—	—	—	M?	—	Crown, Long, Nauna, Tolokiwa
STARLINGS							
Aplonis metallica	M	M	M, P	M	M	M	most small islands
Aplonis cantoroides	M	M	M, P	M	M	M	most small islands
Aplonis feadensis	—	—	—	—	—	—	Hermits, Ninigos, Tench, Wuvulu
Mino dumontii	M	M	M, P	M	—	—	Lolobau, Duke of York?, Tanga
ESTRILDID FINCHES							
Erythrura trichroa	M	M	M	—	—	M	Crown, Eloaua, Emirau, Feni, Long, Tolokiwa,
Lonchura spectabilis	M	M	—	—	—	—	Long, Tolokiwa, Watom
Lonchura hunsteini	—	—	M	M	—	—	—
Lonchura forbesi	—	—	M	—	—	—	—
Lonchura meleana	—	M	M	—	—	—	—
CUCKOO-SHRIKES							
Coracina lineata	—	M	M	—	—	—	—
Coracina papuensis	M	M	M	M	M	—	Lolobau, Watom, Duke of York
Coracina tenuirostris	M	M	M	M	M	M	Long, Sakar, Lolobau, Duke of York, Watom, Dyaul, Feni, Tabar, Lihir, Tanga, Emirau
[*Coracina* sp.]	—	—	P	—	—	—	—
Lalage leucomela	M	M	M	M	—	M	Duke of York, Lolobau, Dyaul, Tabar, Lihir
ROBINS							
Monachella muelleriana	—	M	—	—	—	—	—
WHISTLERS							
Pachycephala pectoralis	M	M	M	M	M	M	Dyaul, Feni, Lihir, Tabar, Tolokiwa
Pachycephala melanura	M?, islets	islets	islets	islets	—	—	Credner, Duke of York, Lihir, Long, Ritter, Tanga?, Tingwon, Tolokiwa?, Witu, Watom
FANTAILS							
Rhipidura leucophrys	M	M	M	M	—	M	most small islands
Rhipidura rufiventris	—	M	M	M	M	M	Duke of York, Dyaul, Lolobau, Rambutyo, Watom, Tabar, Lihir, Boang, Tanga
Rhipidura matthiae	—	—	—	—	—	M	—
Rhipidura dahli	M	M	M	—	—	—	—
Rhipidura [*rufifrons*] *semirubra*	—	—	—	—	M	—	San Miguel, Tong
MONARCHS							
Monarcha cinerascens	M?	islets	islets	islets	M?	M	most small islands
Monarcha infelix	—	—	—	—	M	—	Lou, Rambutyo, Tong
Monarcha menckei	—	—	—	—	—	M	—
Monarcha verticalis	M	M	M	M	—	—	Duke of York, Dyaul
Monarcha ateralba	—	—	—	—	—	—	Dyaul
Monarcha chrysomela	—	—	M	M	—	—	Dyaul, Lihir, Tabar
Myiagra cyanoleuca	M	M	—	—	—	—	Lihir

Table 5-3 (continued)

Species	Umboi	New Britain	New Ireland	Lavongai	Manus	Mussau	Smaller islands
Myiagra alecto	M	M	M	M	M	—	Credner, Duke of York, Dyaul, Fenu, Lolobau, Lou, Pak, Rambutyo, Sakar, Tabar, Tanga, Unea, Watom
Myiagra hebetior	—	M	M	M	—	M	Dyaul, Watom
HONEYEATERS							
Melidectes whitemanensis	—	M	—	—	—	—	—
Myzomela cineracea	M	M	—	—	—	—	—
Myzomela cruentata	—	M	M	M	—	—	Duke of York?, Dyaul, Tabar
Myzomela pulchella	—	—	M	—	—	—	—
Myzomela sclateri	—	islets	—	—	—	—	Bali, Credner, Crown, Long, Talele, Tamunnianim, Tolokiwa, Unea, Watom, Witu
Myzomela pammelaena	islets	—	islets	islets	M	M	most small islands
Myzomela erythromelas	—	M	—	—	—	—	—
Philemon albitorques	—	—	—	—	M	—	Los Negros
Philemon cockerelli	M	M	—	—	—	—	Duke of York?
Philemon eichhorni	—	—	M	—	—	—	—
DRONGOS							
Dicrurus megarhynchus	—	—	M	—	—	—	—
Dicrurus hottentotus	M	M	—	—	—	—	Lolobau
[*Dicrurus* sp.]	—	—	P	—	—	—	—
WOOD-SWALLOWS							
Artamus insignis	—	M	M	—	—	—	Watom?
CROWS							
Corvus orru	M	M	M	M	—	—	Duke of York, Dyaul, Lolobau, Malai, Sakar, Watom, Witu
Corvus sp.	—	—	P	—	—	—	—
TOTAL NONPASSERINES							
M	57	85–90	69–73	53–57	37–38	27–28	76–78
M + P	57	85–90	78–82	53–57	38–39	29–30	77–79
TOTAL PASSERINES							
M	27–29	42	37	24	15–17	17	42–44
M + P	27–29	42	39	24	15–17	17	42–44
TOTAL (ALL TAXA)							
M	84–86	127–132	106–110	77–81	52–55	44–45	118–122
M + P	84–86	127–132	117–121	77–81	53–56	46–47	119–123
Area (km^2)	816	35,742	7174	1186	1834	414	—
Elevation (m)	1655	2439	2399	960	718	651	—
Number of landbird bones	0	0	239	0	?	13	1 (Arawe)

Table 5-4 Birds from archaeological sites, New Ireland, Bismarck Archipelago. Numbers represent NISP (number of identified specimens). BAL, Balof 1 (17 bones) + Balof 2 (102 bones); BUA, Buang Merabak; MBE, Matenbek; MKK, Matenkupkum; PAN, Panakiwuk; †, extinct species; *, extant species, not previously recorded on New Ireland; #, identification not precise enough to determine current status on New Ireland (assumed to be extant); ##, not necessarily different from a taxon identified more precisely. No species of *Accipiter* are counted for "Total */† bones" or % † bones." cf. *Scythrops* is assumed to be a migrant. Updated from Steadman et al. (1999b).

	PAN	BAL	BUA	MKK	MBE	Total
PETRELS						
* *Pseudobulweria* sp.	—	—	—	—	2	2
HERONS						
Egretta sacra	—	—	1	—	—	1
Nycticorax caledonicus	—	—	1	—	—	1
HAWKS						
Haliastur indus	—	1	—	—	—	1
Accipiter novaehollandiae	—	1	—	—	—	1
#/*/† *Accipiter* sp. 2	—	3	—	—	—	3
#/*/† *Accipiter* sp. 3	—	—	—	2	2	4
MEGAPODES						
Megapodius eremita	—	7	—	—	—	7
†*Megapodius* undescribed sp. A	1	1	—	—	—	2
QUAILS						
**Coturnix* cf. *ypsilophorus*	—	—	—	4	—	4
Coturnix chinensis	—	1	—	—	—	1
RAILS						
cf. *Rallina tricolor*	—	1	—	—	—	1
# *Rallus/Gallirallus* sp.	1	—	—	—	—	1
Gallirallus philippensis	2	1	—	7	—	10
†*Gallirallus* undescribed sp. A	1	10	—	1	3	15
**Porzana* cf. *tabuensis*	—	—	—	2	—	2
Poliolimnas cinereus	1	—	—	—	—	1
Amaurornis moluccanus	1	1	—	—	—	2
**Porphyrio porphyrio*	—	—	—	1	—	1
†*Porphyrio* undescribed sp. A	1	16	—	1	—	18
##Rallidae sp.	1	7	—	1	—	9
PIGEONS, DOVES						
# *Columba* sp.	—	2	—	—	—	2
Macropygia amboinensis	2	1	—	—	—	3
Macropygia nigrirostris	—	5	—	1	—	6
Ptilinopus rivoli	2	—	—	2	—	4
# *Ptilinopus* sp. 1	6	1	—	4	—	11
# *Ducula* sp.	—	1	—	1	—	2
Chalcophaps stephani	1	—	—	—	—	1
Caloenas nicobarica	—	6	—	1	—	7
Gallicolumba jobiensis	2	—	—	—	—	2
Gallicolumba beccarii	—	3	—	—	—	3
##Columbidae sp. (small)	8	—	—	14	—	22
##Columbidae sp. (medium/large)	2	—	—	—	1	3
PARROTS						
*/† *Cacatua* undescribed sp./subsp.	—	5	—	—	—	5
#*Lorius* sp.	3	—	—	—	—	3
#*Charmosyna* sp.	—	1	—	—	—	1
##Loriinae sp.	3	—	—	—	—	3
CUCKOOS						
Cacomantis variolosus	1	—	—	—	—	1
cf. *Scythrops novaehollandiae*	—	1	—	—	—	1
#*Centropus* sp.	—	1	—	—	—	1
BARN-OWLS						
*/† *Tyto* sp. A (small)	—	—	—	4	—	4
*/† *Tyto* sp. B (large)	—	3	—	6	—	9
OWLS						
Ninox cf. *solomonis*	—	1	—	—	1	2
NIGHTJARS						
Caprimulgus macrurus	—	—	—	1	—	1
SWIFTS						
Collocalia esculenta	1	6	—	1	—	8
KINGFISHERS						
Alcedo atthis	—	1	—	—	—	1
HORNBILLS						
Aceros plicatus	—	21	1	—	—	22
PITTAS						
Pitta erythrogaster	1	2	—	—	—	3
THRUSHES						
Turdus poliocephalus	—	1	—	—	—	1
STARLINGS, MYNAS						
Aplonis metallica	2	—	—	—	—	2
Aplonis cantoroides	1	—	—	—	—	1
Mino dumontii	—	3	—	—	—	3
CUCKOO-SHRIKES						
Coracina sp.	—	—	—	2	—	2
DRONGOS						
Dicrurus sp.	—	—	—	1	—	1
CROWS						
*/† *Corvus* sp.	—	2	—	—	—	2
UNIDENTIFIED PASSERINES						
Passeriformes sp.	8	2	—	—	1	11
TOTAL SPECIES	19	32	3	18	6	50
*/†	3	6	0	7	2	11
% */†	16	19	0	39	33	22
TOTAL BONES	52	119	3	57	10	241
*/†	3	37	0	18	5	63
% */†	6	31	0	32	50	26

species belong to families known to be quite vulnerable to people, I favor human impact over postglacial "faunal relaxation" (Diamond 1972b, 1976b) as the primary cause of the late Quaternary reduction in species richness. I also believe that very few of New Ireland's currently resident species of landbirds have colonized the island only within the past several millennia. This is especially so for forest obligates.

The petrel *Pseudobulweria* sp., which may represent the Tahiti petrel (*P. rostrata*), breeds today only in East Polynesia but is recorded offshore in the Bismarcks. The osteology of Melanesian species of *Accipiter* is too poorly known (LeCroy et al. 2001) to evaluate two of the three species of *Accipiter* recorded on New Ireland by bones. *Megapodius* undescribed sp. A was a large megapode in the approximate size range of *M. molistructor* from New Caledonia and Tonga (Balouet & Olson 1989). The quail *Coturnix ypsilophorus* occurs on mainland New Guinea but not in the Bismarcks. It prefers grasslands and thus may never have been abundant in New Ireland.

The medium-sized flightless rail (*Gallirallus* undescribed sp. A) probably was endemic to New Ireland. The huge flightless swamphen (*Porphyrio* undescribed sp. A) was larger than the extinct *P. paepae* of the Marquesas or *P. kukwiedei* of New Caledonia (Steadman 1988, Balouet & Olson 1989), and taller but less stout than *P. hochstetteri* or *P. mantelli* of New Zealand. Two other rails not recorded previously on New Ireland, *Porzana tabuensis* and *Porphyrio porphyrio*, are extant and widespread in Melanesia.

The cockatoo (*Cacatua* undescribed sp./subsp.) was much larger than any other psittacid on New Ireland. It was a member of the *C. galerita* species group, the nearest of which is *C. [galerita] ophthalmica* of New Britain (Coates 1985, Hartert 1926). The New Ireland specimens are stouter than in *C. [g.] ophthalmica* or the Australian *C.[g.]galerita*. No species of barn-owls (*Tyto* spp.) have been recorded previously from New Ireland. The bones of *Tyto* sp. A are about the size of those in *T. alba* (New Guinea and some offshore islands), *T. aurantia* (New Britain), or perhaps *T. longimembris* (local in New Guinea, Australia, and New Caledonia). The bones of *Tyto* sp. B are large as in *T. novaehollandiae* or perhaps *T. tenebricosa* of New Guinea, Japen, and Australia (Rich et al. 1978). In Papua New Guinea today, *T. novaehollandiae* occurs only in lowlands of the southern Trans-Fly region and on Manus Island (Coates 1985:370–371), the latter population regarded as the endemic *T. manusi* by König et al. (1999:202). A large species of *Tyto* has been recorded as well from a bone on Mussau (see below). *Tyto novaehollandiae* s.l. once may have inhabited much of the Bismarcks.

The bones of *Corvus* sp. are larger than those of *C. orru* (New Ireland's only crow), *C. woodfordi* (Solomons), *C. bennetti* (Australia), *C. tristis* (New Guinea), *C. coronoides* (Australia), or *C. macrorhynchos* (Philippines). The extant myna *Mino dumontii* was the only other passerine identified by Steadman et al. (1999b), who found the 21 remaining passerine bones to represent four smaller sizes of songbirds. With more comparative skeletons at hand, I recently restudied these 21 bones, identifying 10 of them in six extant taxa as follows: a mandible, carpometacarpus, and tarsometatarsus of the pitta *Pitta erythrogaster*; a femur and tarsometatarsus of a large cuckoo-shrike *Coracina* sp. (~size of *C. caledonica*); a coracoid of the thrush *Turdus poliocephalus*, which lives today on New Ireland only above 1500 m elevation (Ripley 1977, Diamond 1989); the starlings *Aplonis metallica* (two humeri) and *A. cantoroides* (humerus); and a drongo *Dicrurus* sp. (ulna). Most of these passerines are tasty frugivores, which may explain their presence in cultural deposits.

Bone assemblages from Polynesia typically reveal the loss of 50–90% of an island's species of native landbirds (Olson & James 1991, Steadman 1995a; Chapters 6 and 7 herein). The lower proportion (20+%) of species lost from New Ireland may be related to the presence of an indigenous murid rodent (*Melomys rufescens*). Remote Oceania lacks native mammals other than bats, leading to a naïveté among indigenous birds to ground-based predators and thus vulnerability to predation when humans and associated nonnative mammals arrive (Atkinson 1985). Evolving alongside native rodents, New Ireland's birds had been exposed to potential predation from terrestrial mammals in prehuman times. Nevertheless, the prehistoric introduction to New Ireland of seven species of mammals (Flannery et al. 1988, White et al. 1991) undoubtedly had a negative impact on the local birds, which previously had not experienced marsupials, dogs, pigs, or any species of *Rattus*.

Strata dated to the Pleistocene (>10,000 yr BP) account for 56% of New Ireland's bird bones. Unlike the record at some sites in Remote Oceania with continuous and rich deposition of bones (Chapters 6–8), that of New Ireland is too spotty to say precisely when a species was lost. Only 2% of the bird bones are older than ca. 15,000 yr BP, at which time people already had lived on New Ireland for 20,000 years. Which species may have been lost during those first 1000 human generations is largely unknown.

Extinct/extirpated species make up 31% of Pleistocene and 21% of Holocene bones. Of the three sites with >50 bird bones, Panakiwuk has the fewest extinct species, and dates primarily to ca. 10,000-8000 yr BP, suggesting that much extinction had taken place by that time. For flightless rails (*Gallirallus* undesc. sp. A, *Porphyrio* undesc. sp.

Table 5-5 Latest records of extinct or extirpated species of birds on New Ireland. These are the latest occurrences among the available, small bone samples, not necessarily the time of extinction. †, extinct species or subspecies. Updated from Steadman et al. (1999b).

Species	Latest record (yr BP)	Number of sites/bones
PETRELS		
Pseudobulweria sp.	6000	1/2
MEGAPODES		
†*Megapodius* undescribed sp. A	14,000–10,000	2/2
QUAILS		
Coturnix cf. *ypsilophorus*	>10,000	1/4
RAILS		
†*Gallirallus* undescribed sp. A	<2000	4/15
Porzana cf. *tabuensis*	>10,000	1/2
Porphyrio porphyrio	>10,000	1/1
†*Porphyrio* undescribed sp. A	<1600	3/18
PARROTS		
†*Cacatua* undescribed sp./subsp.	5000–1000	1/5
BARN-OWLS		
†*Tyto* sp. A (small)	>10,000	1/4
Tyto sp. A (large)	10,000–5000	2/9
CROWS		
†*Corvus* sp.	5000–1000	1/2

A), 31 of 33 bones are from Pleistocene strata, hinting that the two isolated bones may be out of stratigraphic context. Three other extinct/extirpated species, however, are recorded at or after 6000 yr BP (Table 5-5). Thus the loss of birds on large, topographically rugged New Ireland was a protracted process, in contrast to the very rapid extinction on many Polynesian islands (Chapters 6, 7). As an avian fossil record develops on New Guinea, we might expect delayed human impacts there as well. This would match some of New Guinea's pollen records, where strong evidence that people disturbed the vegetation may not be apparent until tens of millennia after human arrival (Hope & Tulip 1994).

Mussau

Three archaeological sites on or near Mussau (Talepakemalai, Etakosarai, Epakapaka) have produced bird bones (Steadman & Kirch 1998, Kirch 2001; Chapter 3 herein). The modern resident landbird fauna of Mussau consists of 27–28 species of nonpasserines and 17 species of passerines, of which 7 and 11, respectively, are endemic species or subspecies (Table 5-6). Mussau today has 18 of the 19 species found on smaller and more isolated Emirau (Hartert 1924a–c). Please keep in mind, however, that these islands lack rigorous bird surveys. Unlike other major islands in the region, Mussau was not visited by the Whitney South Seas Expedition in 1928–1929 and 1932–1935 (M. LeCroy pers. comm.).

The prehistoric avifauna of Mussau is based on too few specimens (only eight bones of resident landbirds) to estimate past species richness, but still is important among bird assemblages from Lapita sites (3600 to 2700 cal BP) for being among the best dated, and associated with extensive cultural and environmental information (Steadman & Kirch 1998). Two seabirds (Chapter 15) and two landbirds identified from the sites had not been recorded previously from Mussau. The tibiotarsus of a cockatoo (cf. *Cacatua* sp.) probably represents an undescribed species, but is too fragmentary to identify more precisely. It is shorter than that of *C.* [*galerita*] *triton* (New Guinea), *C.* [*g.*] *ophthalmica* (New Britain), *C. sulphurea*, *C. moluccensis*, *C. alba*, or *C. tenuirostris*, longer than that of *C. haematuropygia*, and about the same length but much stouter than that of *C. leadbeateri*, *C. sanguinea*, or *C. ducorpsii*. The nearest and only population of any cockatoo in the Bismarcks is *C.* [*g.*] *ophthalmica* on New Britain (Hartert 1926, Coates 1985:339). Along with the bones of *Cacatua* from New Ireland (see above), the Mussau specimen suggests a former radiation of cockatoos in the Bismarcks.

No species of barn-owls (*Tyto* spp.) have been recorded previously from Mussau. The fossil ungual phalanx is the size expected in *T. novaehollandiae*, the region's largest tytonid owl, confined in Papua New Guinea today to the southern Trans-Fly region and Manus Island. Bones of a large species of *Tyto* also occur on New Ireland (see above). Especially since *T. novaehollandiae* s.l. (= *T. manusi*) is known on Manus (although not recorded during a three-week survey in 1990; Dutson & Newman 1991), it is likely that such an owl once occupied much of the Bismarcks.

The bone sample from Mussau is very different from those of New Ireland (above) or Buka (below), which are dominated by landbirds; instead it resembles those from other Lapita sites as far away as Tonga (Chapter 6) in featuring chickens (*Gallus gallus*), shorebirds, and a variety of both seabirds and landbirds, the latter with many columbids. The lack of rail bones from Mussau is unusual and may be a sampling effect. The percentage of extinct/extirpated species from Mussau is much lower for both seabirds or landbirds than in Lapita sites from Remote Oceania. This could be related to the presence of malaria, which may limit human populations. Another possibility is that earlier, pre-Lapita peoples had lived on Mussau and already had extirpated various species of birds. The human

Table 5-6 Prehistoric birds and modern birds of Mussau, Bismarck Archipelago. Data for modern birds are from Heinroth (1902), Martens (1922), Hartert (1924b), Rothschild & Hartert (1924), Baker (1948:22), Salomonsen (1964, 1972), Coates (1973), Diamond (1975b:421, 1989), Silva (1975), Anonymous (1985), Coates (1985, 1990), and Mayr & Diamond (2001). e, endemic subspecies; ee, endemic species; [e], endemism shared with Emirau and, in some cases, Tench; i, introduced species; m, migratory species; *, presumably extirpated on Mussau; ?, questionable record, or residency not established. The prehistoric records (number of identified bones) are from the Talepakemalai, Etakosarai, and Epakapaka archaeological sites. "Mussau" refers to both Mussau and nearby Eloaua. Taxa in brackets are likely to represent either a species identified more precisely from bones or a species listed among the modern birds of Mussau. From Steadman & Kirch (1998), except for updated modern records from Mayr & Diamond (2001).

	Modern record	Prehistoric record
SEABIRDS		
PETRELS		
* *Pterodroma/Pseudobulweria* sp.	—	2
BOOBIES		
* *Sula leucogaster*	—	7
TERNS		
Sterna hirundo (m)	X	1
Sterna fuscata	X	3
Anous minutus	X	6
Anous stolidus	X	10
[Sterninae sp.]	—	1
LANDBIRDS		
HERONS		
Egretta sacra sacra	X	—
Ixobrychus flavicollis gouldi	X	—
OSPREYS		
Pandion haliaetus melvillensis	X	1
HAWKS		
Haliastur indus girrenera	X	—
Haliaeetus leucogaster	X	—
Accipiter novaehollandiae matthiae (e)	X	—
MEGAPODES		
Megapodius [*freycinet*] *eremita*	X	—
CHICKENS		
Gallus gallus (i)	X	14
RAILS		
Rallina tricolor laeta (e)	X	—
Poliolimnas cinereus meeki (e)	X	—
THICK-KNEES		
Esacus magnirostris	X	—
PLOVERS		
Pluvialis dominica (m)	X	2
SANDPIPERS		
Numenius sp. (m)	X	1
PIGEONS, DOVES		
Macropygia mackinlayi arossi	X	—
Ptilinopus superbus superbus	X	—
Ptilinopus solomonensis johannis	X	—
Ptilinopus insolitus inferior [e]	X	—
[*Ptilinopus* sp. 1]	—	1
[*Ptilinopus* sp. 2]	—	1
Ducula pacifica tarrali	X?	—
Ducula pistrinaria rhodinolaema	X	—
[*Ducula* sp.]	—	1
Chalcophaps stephani stephani	X	—
Caloenas nicobarica nicobarica	X	2
[*Caloenas nicobarica* or *Ducula* sp.]	—	2
Gallicolumba beccarii eichhorni [e]	X	—
PARROTS		
* cf. *Cacatua* sp.	—	1
Trichoglossus haematodus flavicans	X	—
Micropsitta meeki proxima [e]	X	—
CUCKOOS		
Scythrops novaehollandiae (m)	X	—
BARN-OWLS		
* *Tyto* sp. (large)	—	1
CRESTED SWIFTS		
Hemiprocne mystacea mystacea	X	—
SWIFTS		
Collocalia spodiopygia eichhorni	X	—
Collocalia vanikorensis lihirensis	X	—
KINGFISHERS		
Alcedo atthis hispidoides	X	—
Halcyon chloris matthiae [e]	X	—
Halcyon saurophaga saurophaga	X	—
ROLLERS		
Eurystomus orientalis crassirostris	X	—
THRUSHES		
Zoothera heinei eichhorni (e)	X	—
Turdus poliocephalus heinrothi (e)	X	—
WARBLERS		
Phylloscopus poliocephalus matthiae (e)	X	—
SUNBIRDS		
Nectarinia jugularis flavigaster	X	—
STARLINGS		
Aplonis metallica nitida	X	3

(*continued*)

Table 5-6 (continued)

	Modern record	Prehistoric record
Aplonis cantoroides	X	—
ESTRILDID FINCHES		
Erythrura trichroa eichhorni [e]	X	—
CUCKOO-SHRIKES, TRILLERS		
Coracina tenuirostris matthiae [e]	X	—
Lalage leucomela conjuncta (e)	X	—
FANTAILS		
Rhipidura leucophrys melaleuca	X	—
Rhipidura rufiventris mussaui (e)	X	—
Rhipidura matthiae (ee)	X	—
MONARCHS		
Monarcha cinerascens perpallidus	X	—
Monarcha menckei (ee)	X	—
Myiagra hebetior hebetior (e)	X	—
WHISTLERS		
Pachycephala pectoralis sexuvaria	X	—
HONEYEATERS		
Myzomela pammelaena hades (e)	X	—
TOTAL RESIDENT LANDBIRDS	All species	Endemic species/subspecies
All species	44–45	18 (38–40%)
Nonpasserines	27–28	7 (23–25%)
Passerines	17	11 (65%)
TOTAL BONES		
All species	60	
All species (without i, m)	42	
Resident seabirds	29	
*seabirds	9	
Resident landbirds	13	
*landbirds	2	
TOTAL SPECIES		
All species	17	
All species (without i, m)	13	
Resident seabirds	5	
*seabirds	2	
Resident landbirds	8	
*landbirds	2	

occupation of nearby, less isolated New Britain and New Ireland began 30,000 yr BP and required crossing the ocean from New Guinea (Allen et al. 1988, Pavlides & Gosden 1994, Rosenfeld 1997). That Mussau was occupied in the late Pleistocene seems likely given that Pamwak Rockshelter on Manus (below) dates to at least 13,000 yr BP (Frederickson et al. 1993, Spriggs 1997).

Manus

From Pamwak Rockshelter, Williams (1999) reported an unknown number of bird bones thought to represent a grebe (Podicipedidae; listed in Table 5-3 as possibly *Tachybaptus ruficollis*), heron (cf. *Nycticorax caledonicus*), and swamphen (cf. *Porphyrio porphyrio*). These identifications should be verified.

Watom

One bone of the extant swamphen *Porphyrio porphyrio* was reported (as *P. poliocephalus*) from the Lapita-age Rakival site on the small island of Watom, 6 km off the north coast of New Britain (Specht 1968; also see Green & Anson 1991).

Arawe Islands

From 13 archaeological sites excavated in the small, low Arawe Islands (1 to 8 km off southwestern New Britain) from 1985 to 1992 (Gosden et al. 1989, Gosden 1991, Gosden & Webb 1994, Spriggs 1997, Summerhayes 2001b), I have identified a total of four bird bones from the Apalo site (ca. 4200 to 2000 yr BP) on Kumbun Island, the Maklo site (ca. 2000 yr BP to present) on Maklo Island, and the Winkapiplo site (ca. 1000 yr BP to present) on Kauptimete Island. This has not been reported previously, so I will provide some detail.

A fragmentary tarsometatarsus from Apalo is referred to *Casuarius* cf. *casuarius* because it is larger than that of *C. unappendiculatus* and much larger than that of *C. bennetti*. Among the three living species of cassowaries, only *C. bennetti* occurs on New Britain today, where it has been suspected not to be indigenous, especially since it often is kept as a pet in New Guinea (Diamond 1972a, White 1976, Coates 1985, Beehler et al. 1986). The bone of *C.* cf. *casuarius* suggests that two species of cassowary may have lived on New Britain prehistorically; both probably were taken there by people.

The other three bird bones from the Arawe sites represent seabirds. A tibiotarsus from Winkapiplo is from the shearwater *Puffinus pacificus*, which is found locally in the New Guinea/Bismarck region (Coates 1985, Beehler et al.

1986). The nearest records are at sea off the eastern coast of New Britain and off the Madang area of Papua New Guinea. The nearest nesting areas are off the southern coast of Papua New Guinea. A sternum from Maklo represents a booby (*Sula sula* or *S. leucogaster*). The nearest nesting colony of *S. sula* is on Raine Island off northern Queensland; the nearest nonbreeding record is off Woodlark Island. *Sula leucogaster* is widespread in the New Guinea/Bismarcks region as a nonbreeding bird. The closest nesting colonies are in the Torres Strait region and perhaps in the Nuguria Islands. Finally, a humerus from Winkapiplo represents the frigatebird *Fregata ariel*, which is widespread in the region but very local as a breeder.

Interpretation of the few bird bones from the Arawe sites is limited by how poorly documented the modern fauna is from these islands as well as adjacent West New Britain (Diamond 1971, Coates 1985, Beehler et al. 1986). I assume that the shearwater *Puffinus pacificus* no longer occurs there, but that the booby *Sula* sp. and frigatebird *Fregata ariel* are still found near these islands as nonbreeding birds. That the cassowary *Casuarius* cf. *casuarius* might have been indigenous on the Arawe Islands is intriguing but unlikely. More remarkable about the single cassowary bone from the Apalo site is that the species represented is not *C. bennetti*, the only cassowary that lives on New Britain today. The Arawe archaeological sites resemble those on Mussau (see above) in having relatively few birds bones in the midden deposits. As proposed for Mussau, a likely explanation is that people already had occupied, or at least visited, these islands for many millennia before the Lapita potters arrived. The most vulnerable species of birds may have been lost during those earlier occupations.

Solomon Islands (Main Chain)

Although modern political geography places Bougainville and Buka in Papua New Guinea, avian biogeography has these two islands as the northwestern end of the Solomon Main Chain (Figure 5-2). The Pleistocene archaeology of

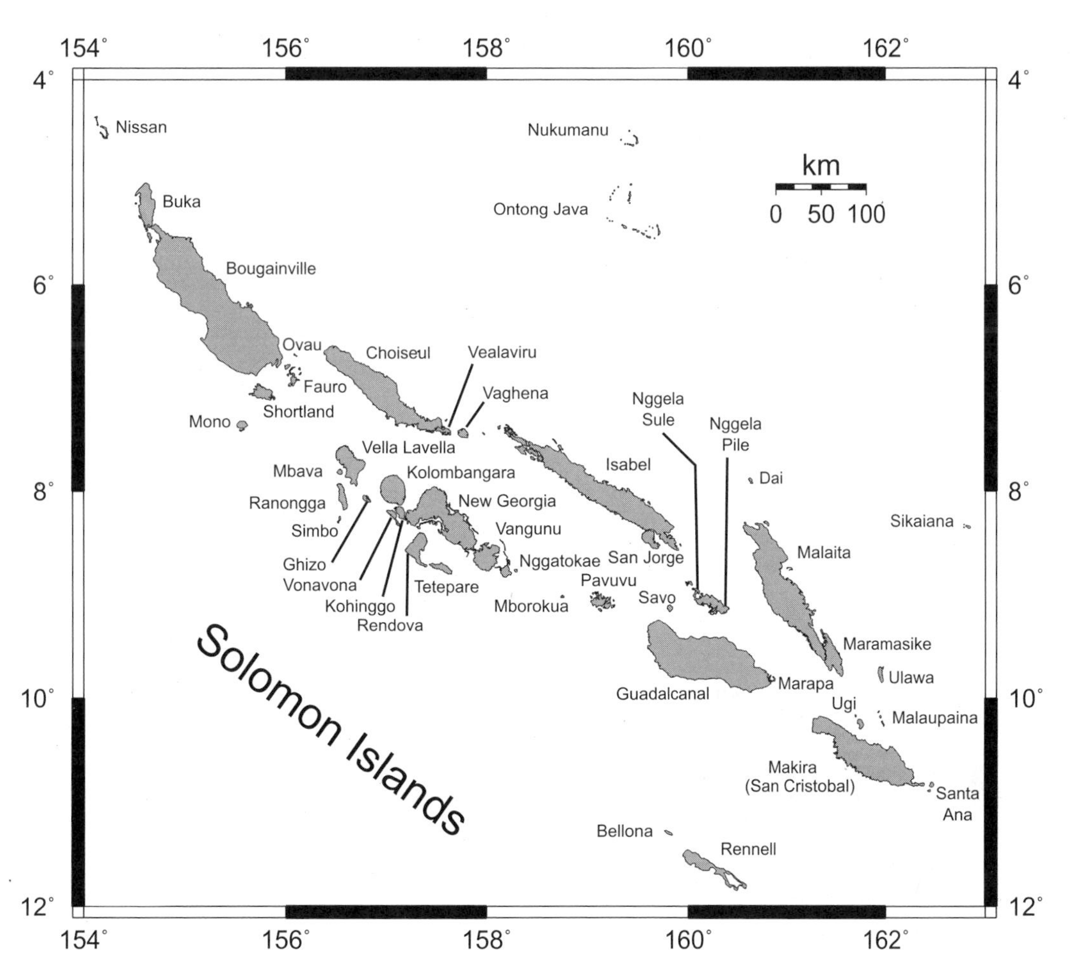

Figure 5-2 The Solomon Islands.

the Main Chain is unknown except on Buka, but the evidence there is enough to suspect that most or all of the Main Chain has been occupied by humans for 30,000 years. Late Pleistocene lowering of sea level would have connected Buka, Bougainville, Choiseul, Isabel, and the Nggela (Florida) Group into a single, immense land mass of ca. 46,400 km^2 known as "Greater Bougainville" or "Greater Bukida" (Wickler & Spriggs 1988, Spriggs 1997: Figure 2.2, Mayr & Diamond 2001; Figure 1-43 herein). The large islands of New Georgia (expanded to 5600 km^2), Malaita, Guadalcanal, and Makira (San Cristobal) would have been within sight of Greater Bougainville but not connected to it.

The modern landbirds of Makira strongly reflect not being connected to Greater Bougainville. Makira has 13 "endemic" species, some of which also occur on adjacent small islands but not on any other large islands of the Main Chain (Mayr & Diamond 2001:248). The eight endemic species in the New Georgia Group include five species of white-eyes (*Zosterops* spp.). By contrast, only three endemic species occur on Malaita, probably because it was nearly connected to Greater Bougainville. Guadalcanal sustains four species not found elsewhere in the Main Chain (table 5-7), three of these endemic to the island. Isabel, on the other hand, was part of Greater Bougainville and has only one possibly endemic species (the flightless rail *Gallirallus* [*woodfordi*] *immaculatus*).

As in the Bismarcks, nonpasserines dominate the landbirds of the Solomons, with a great variety of hawks, rails, columbids, parrots, and kingfishers. Many genera of passerines show species-level inter-island variation, an uncommon situation among nonpasserines. Substantial subspecies-level inter-island variation occurs in nonpasserines and especially passerines. A strict application of the phylogenetic species concept probably would more than double the number of species listed in Table 5-7. I suspect that most absences of widespread species on one to several islands in the Main Chain, as across Oceania, are due to human impact.

Buka

All but one or two of the 67–68 species of landbirds on Buka (611 km^2) are among the 102 species recorded on adjacent Bougainville (8591 km^2), which was connected to it in the late Pleistocene (Hadden 1981, Mayr & Diamond 2001). The oldest cultural site in the Solomons is Kilu Cave on Buka's eastern coast, with 2.2 m of stratified cultural sediments dominated by mammal, lizard, and fish bone, marine shell, and flaked stone (Wickler & Spriggs 1988, Flannery & Wickler 1990, Wickler 1990, 2001). Four charcoal ^{14}C dates from Layer I at Kilu Cave range from 9430 ± 150 to 6670 ± 80 yr BP (= 11,200–10,250 and 7670–7420 cal BP at 95.4% C.I.). The underlying Layer II, which lacks charcoal but is anthropogenic given its content of bone, shell, and flaked stone, has three ^{14}C dates on marine shell that range from 28,740 ± 280 to 20,140 ± 300 yr BP. These ^{14}C dates are uncorrected for the marine reservoir effect, which probably would push the age several hundred years younger.

The 77 bird bones from Kilu Cave have not been described until now. They represent 18 species of landbirds, 11 of which no longer occur on Buka (Table 5-8). Seven of the 11 species are extinct and undescribed. Remarkably, three of the seven represent extinct genera (see below). Except for the heron *Nycticorax* undescribed sp. A and owl *Nesasio solomonensis*, the extirpated/extinct species from Buka are from the three families (megapodes, rails, and columbids) with the most extinction across Oceania, thus suggesting that people caused their losses on Buka. All but one of the 14 landbird bones from Layer II (Pleistocene) represents an extinct species. In Layer I (Holocene), 71% of the bones are from extinct or extirpated species, but these include 10 of the 11 such species from the site.

The night-heron from Buka (*Nycticorax* undescribed sp. A) is represented by a medium-sized tarsometatarsus. *Megapodius* undescribed sp. B was large, in the size range of *M. molistructor* of New Caledonia and Tonga. The extinct rail cf. *Pareudiastes* undescribed sp. is known only from the distal end of a tarsometatarsus that compares favorably with an illustration of the same element in *P. sylvestris* from Makira (Olson 1975a). This is the first evidence of *Pareudiastes* in the Solomons outside of Makira. *Gallirallus* undescribed sp. B was flightless like so many other of its congeneric species. The seven bones that I refer to à larger flightless rail, *Gallirallus* [*woodfordi*] *tertius*, resemble modern skeletons of *G.* [*w.*] *immaculatus* from Isabel; I refer them to *G.* [*w.*] *tertius* on geographic grounds, this being the form found today on Bougainville. Members of this species complex were formerly classified in *Nesoclopeus* (see Chapter 10). *Porphyrio* undescribed sp. B was a massive flightless species, most similar to another undescribed species of swamphen from New Ireland.

The columbid bones from Kilu Cave represent two species that still live on Buka, two living species unrecorded there, and two extinct genera. The surviving species are canopy frugivores whereas the four lost species are understory/ground-dwelling forms, suggesting that hunting and nonnative predators may have been involved in their demise. *Caloenas nicobarica* occurs on Bougainville but has not been recorded on Buka (Hadden 1981). *Gallicolumba rufigula* lives today on mainland New Guinea but nowhere in the Bismarcks or Solomons. "Undescribed genus

Table 5-7 Landbirds of major islands in the Main Chain, Solomon Islands. "New Georgia" = New Georgia itself and Vella Lavella, Mbava, Ranongga, Simbo, Ghizo, Kolombangara, Vonavona, Kohinggo, Rendova, Tetepare, Vangana, Nggatokae, and adjacent smaller islands, these corresponding to islands 15–27 in Mayr & Diamond (2001:366). "Small islands" = any small island within the main chain not listed here; they are not outliers. For Isabel, G = recorded at the Garanga River site (see text). *Nesoclopeus* merged into *Gallirallus* (see Chapter 10). Species-level systematics often unresolved, especially in passerines. Excludes species of uncertain breeding status, such as *Pelecanus conspicillatus, Ardea alba, Egretta intermedia, Platalea regia, Scythrops novaehollandiae*, and *Halcyon sancta*. M, modern record; P, prehistoric record; †, extinct species; ?, questionable record, or residency not established. [a]Area of New Georgia Island only; [b]elevation of Kolombangara, the highest island in the New Georgia Group. Compiled with difficulty and uncertainty from Mayr (1945), Sibley (1951), Cain & Galbraith (1956), Galbraith & Galbraith (1962), Diamond (1975b, 1991), Hadden (1981), Coates (1985, 1990), Blaber (1990), Fry et al. (1992), Webb (1992), Juniper & Parr (1998), Taylor (1998), Kratter et al. (2001a, b), Mayr & Diamond (2001), and herein. Island areas and elevations from Mayr & Diamond (2001).

Species	Buka	Bougainville	Choiseul	Isabel	New Georgia	Guadalcanal	Malaita	Makira	Small islands
GREBES									
Tachybaptus ruficollis	—	M	—	—	—	—	—	—	—
CORMORANTS									
Phalacrocorax melanoleucos	—	M	M	—	M	M	—	M	M
HERONS									
Egretta sacra	M	M	M	M, G	M	M	M	M	M
Butorides striatus	M	M	M	M, G	M	M	M	M	M
Nycticorax caledonicus	M	M	M	M, G	M	M	M	M	M
†*Nycticorax* undescribed sp. A	P	—	—	—	—	—	—	—	—
Ixobrychus sinensis	—	M	—	—	—	—	—	—	—
Ixobrychus flavicollis	—	M	M	M, G	M	M	—	—	—
DUCKS									
Anas superciliosa	M, P	M	M	M, G	M	M	M	M	M
OSPREYS									
Pandion haliaetus	M	M	M	M, G	M	M	M	M	M
HAWKS									
Aviceda subcristata	M	M	M	M, G	M	M	M	M	M
Haliastur indus	M, P	M	M	M, G	M	M	M	M	M
Haliaeetus sanfordi	M, P	M	M	M, G	M	M	M	M	M
Accipiter novaehollandiae	M, P	M	M	M, G	M	M	M	—	M
Accipiter albogularis	M	M	M	—	M	M	M	M	M
Accipiter imitator	—	M	M	M, G	—	—	—	—	—
Accipiter meyerianus	M?	—	—	—	M	M	—	—	—
FALCONS									
Falco severus	M	M	M	M, G	M	M	—	M	M?
MEGAPODES									
Megapodius eremita	M	M	M	M, G	M	M	M	M	M
†*Megapodius* undescribed sp. B	P	—	—	—	—	—	—	—	—
BUTTON-QUAILS									
Turnix maculosa	—	—	—	—	—	M	—	—	—
RAILS									
Gallirallus philippensis	—	M	M	—	—	M	—	M	M
Gallirallus [w.] *woodfordi*	—	—	—	—	—	M	—	—	—
Gallirallus [w.] *immaculatus*	—	—	M?	M, G	—	—	—	—	—
Gallirallus [w.] *tertius*	P	M	—	—	—	—	—	—	—

(*continued*)

Table 5-7 (continued)

Species	Buka	Bougainville	Choiseul	Isabel	New Georgia	Guadalcanal	Malaita	Makira	Small islands
Gallirallus rovianae	—	—	—	—	M	—	—	—	—
†*Gallirallus* undescribed sp. B	P	—	—	—	—	—	—	—	—
Porzana tabuensis	—	—	—	—	—	M	—	M	—
Poliolimnas cinereus	M	M	M?	—	—	M	M?	M	M
Amaurornis moluccanus	M	M	—	M, G	M	—	M	M	M
(†?) *Pareudiastes silvestris*	—	—	—	—	—	—	—	M	—
†*Pareudiastes* undescribed sp.	P	—	—	—	—	—	—	—	—
Porphyrio porphyrio	M	M	M	M, G	M	M	M	M	M
†*Porphyrio* undescribed sp. B	P	—	—	—	—	—	—	—	—
THICK-KNEES									
Esacus magnirostris	—	M	M?	M	M	M?	M?	—	M?
PIGEONS, DOVES									
Columba vitiensis	—	M	—	M?	M	M	M	M	—
Columba pallidiceps	—	M	M	—	M	M	—	M	M
Macropygia mackinlayi	M	M	M	M, G	M	M	M	M	M
Reinwardtoena crassirostris	—	M	M	M?	M	M	M	M	M
Ptilinopus superbus	M	M	M	M, G	M	M	M	M	M
Ptilinopus richardsii	—	—	—	—	—	—	—	—	M
Ptilinopus greyii	—	—	—	—	—	—	—	—	M
Ptilinopus solomonensis	M	M	—	—	M	M	M	M	M
Ptilinopus viridis	M	M	M	M, G	M	M	M	M	M
Ducula pacifica	—	—	—	—	—	—	M?	—	M
Ducula rubricera	M, P	M	M	M, G	M	M	M	M	M
Ducula pistrinaria	M, P	M	M	M	M	M	M	M	M
Ducula brenchleyi	—	—	—	—	—	M	M	M	—
Gymnophaps solomonensis	—	M	—	—	M	M	M	—	—
Chalcophaps stephani	M	M	M	M, G	M	M	M	M	M
Caloenas nicobarica	P	M	M	M	M	M	M	—	M
Gallicolumba rufigula	P	—	—	—	—	—	—	—	—
Gallicolumba jobiensis	—	—	—	—	M	M	—	M	—
†*Gallicolumba salamonis*	—	—	—	—	—	—	M	M	M
Gallicolumba beccarii	—	M	M?	—	M	M	—	M	M
†*Microgoura meeki*	—	—	M	—	—	—	—	—	—
†Undescribed genus & sp. A	P	—	—	—	—	—	—	—	—
†Undescribed genus & sp. B	P	—	—	—	—	—	—	—	—
PARROTS									
Cacatua ducorpsii	M	M	M	M, G	M	M	M	—	M
Chalcopsitta cardinalis	M, P	M	M	M, G	M	M	M	M	M
Trichoglossus haematodus	M	M	M	M, G	M	M	M	M	M
Lorius chlorocercus	—	—	—	—	—	M	M	M	M
Charmosyna meeki	—	M	—	M	M	M	M	—	—

Table 5-7 (continued)

Species	Buka	Bougainville	Choiseul	Isabel	New Georgia	Guadalcanal	Malaita	Makira	Small islands
Charmosyna placentis	M	M	—	—	—	—	—	—	—
Charmosyna margarethae	—	M	M?	M	M	M	M	M	M
Micropsitta bruijnii	—	M	—	—	M	M	—	—	—
Micropsitta finschii	M	M	M	M, G	M	M	M	M	M
Eclectus roratus	M	M	M	M, G	M	M	M	M	M
Geoffroyus heteroclitus	M	M	M	M, G	M	M	M	M	M
CUCKOOS									
Cacomantis variolosus	M	M	—	M, G	M	M	M	M	M
Eudynamys scolopacea	—	M	M	M, G	M	M	M	M	M
Centropus milo	—	—	—	—	M	M	—	—	M
BARN-OWLS									
Tyto alba	M	M	—	M	M	M	M	M	M
STRIGID OWLS									
Ninox jacquinoti	M	M	M	M, G	—	M	M	M	M
Nesasio solomonensis	P	M	M	M	—	—	—	—	—
FROGMOUTHS									
Undescribed genus, *inexpectatus*	—	M	M	M, G	—	—	—	—	—
NIGHTJARS									
Eurostopodus mysticalis	—	M	—	M	M	M	—	—	M
CRESTED SWIFTS									
Hemiprocne mystacea	M	M	M	M, G	M	M	M	M	M
SWIFTS									
Collocalia esculenta	M	M	M	M, G	M	M	M	M	M
Collocalia spodiopygia	M	M	M	M, G	M	M	M	M	M
Collocalia orientalis	—	—	—	—	—	M	—	—	—
Collocalia vanikorensis	M	M	M	M, G	M	M	M	M	M
KINGFISHERS									
Alcedo atthis	M	M	M	M, G	M	M	M	M	M
Ceyx pusillus	M	M	M	M, G	M	M	M	—	M
Ceyx lepidus	M	M	M	M, G	M	M	M	M	—
Halcyon leucopygia	M	M	M	M, G	M	M	—	—	M
Halcyon chloris	M	M	M	M, G	M	M	M	M	M
Halcyon saurophaga	M	M	M	M	M	M	M	M	M
Actenoides bougainvillei	—	M	—	—	—	M	—	—	—
ROLLERS									
Eurystomus crientalis	M	M	M	M, G	M	M	M	M	M
HORNBILLS									
Aceros plicatus	M	M	M	M, G	M	M	M	—	M
PITTAS									
Pitta anerythra	—	M	M	M, G	—	—	—	—	—

(continued)

Table 5-7 (continued)

Species	Buka	Bougainville	Choiseul	Isabel	New Georgia	Guadalcanal	Malaita	Makira	Small islands
SWALLOWS									
Hirundo tahitica	M	M	M	M, G	M	M	M	M	M
THRUSHES									
Zoothera heinei	—	—	M	—	—	—	—	—	—
Zoothera talasea	—	M	—	—	—	—	—	—	—
Zoothera margaretae	—	—	—	—	—	M	—	M	—
Turdus poliocephalus	—	M	—	—	M	M	—	—	—
WARBLERS									
Cettia parens	—	—	—	—	—	—	—	M	—
Acrocephalus [*australis*] *stentoreus*	M	M	—	M	—	M	—	—	M
Phylloscopus poliocephalus	—	M	—	M	M	M	M	—	—
Phylloscopus makirensis	—	—	—	—	—	—	—	M	—
Phylloscopus amoenus	—	—	—	—	M	—	—	—	—
Cichlornis llaneae	—	M	—	—	—	—	—	—	—
Cichlornis whitneyi	—	—	—	—	—	M	—	—	—
SUNBIRDS									
Nectarinia jugularis	M	M	M	M, G	M	M	M	—	M
FLOWERPECKERS									
Dicaeum aeneum	M	M	M	M, G	—	M	M	—	M
Dicaeum tristrami	—	—	—	—	—	—	—	M	—
WHITE-EYES									
Zosterops vellalavella	—	—	—	—	M	—	—	—	M
Zosterops splendidus	—	—	—	—	M	—	—	—	M
Zosterops luteirostris	—	—	—	—	M	—	—	—	M
Zosterops kulambangrae	—	—	—	—	M	—	—	—	M
Zosterops murphyi	—	—	—	—	M	—	—	—	M
Zosterops metcalfii	M	M	M	M, G	—	—	—	—	M
Zosterops rendovae	—	M	—	—	M	—	—	—	M
Zosterops ugiensis	—	M	—	—	—	M	—	M	—
Zosterops stresemanni	—	—	—	—	—	—	M	—	—
STARLINGS									
Aplonis metallica	M	M	M	M, G	M	M	M	M	M
Aplonis cantoroides	M	M	M	M, G	M	M	M	M	M
Aplonis feadensis	—	—	—	—	—	—	—	—	M
Aplonis brunneicapilla	—	M	M	—	M	M	—	—	M
Aplonis grandis	M	M	M	M, G	M	M	M	—	—
Aplonis dichroa	—	—	—	—	—	—	—	M	—
Mino dumontii	M	M	M	M, G	M	M	M	—	M
ESTRILDID FINCHES									
Erythrura trichroa	—	M	—	—	M	M	—	—	—
Lonchura melaena	M	—	—	—	—	—	—	—	—

Table 5-7 (continued)

Species	Buka	Bougainville	Choiseul	Isabel	New Georgia	Guadalcanal	Malaita	Makira	Small islands
CUCKOO-SHRIKES									
Coracina caledonica	—	M	—	M, G	M	—	—	—	M
Coracina novaehollandiae	—	—	—	—	—	—	—	—	M?
Coracina lineata	M	M	M	M, G	M	M	M	M	M
Coracina papuensis	M	M	M	M, G	M	M	M	—	M
Coracina tenuirostris	M	M	M	M, G	M	M	M	M	M
Coracina holopolia	M	M	M	M, G	M	M	M	—	—
Lalage leucopyga	—	—	—	—	—	—	—	M	M
ROBINS									
Petroica multicolor	—	M	—	—	M	M	—	M	—
WHISTLERS									
Pachycephala pectoralis	M	M	M	M, G	M	M	M	M	M
Pachycephala melanura	islets	islets	—	—	—	—	—	—	M
Pachycephala implicata	—	M	—	—	—	M	—	—	—
FANTAILS									
Rhipidura leucophrys	M	M	M	M, G	M	M	M	M	M
Rhipidura cockerelli	M	M	M	M, G	M	M	M	—	M
Rhipidura drownei	—	M	—	—	—	M	—	—	—
Rhipidura tenebrosa	—	—	—	—	—	—	—	M	—
Rhipidura fuliginosa	—	—	—	—	—	—	—	M	—
Rhipidura malaitae	—	—	—	—	—	—	M	—	—
Rhipidura rufifrons	M	M	M	M, G	M	M	M	M	M
MONARCHS									
Monarcha cinerascens	islets	islets	islets	islets	—	—	—	—	M
Monarcha castaneiventris	M	M	M	M, G	—	M	M	M	M
Monarcha richardsii	—	—	—	—	M	—	—	—	—
Monarcha browni	—	—	—	—	M	—	—	—	M
Monarcha viduus	—	—	—	—	—	—	—	M	M
Monarcha barbatus	M	M	M	M, G	—	M	M	—	M
Myiagra ferrocyanea	M	M	M	M, G	M	M	M	—	M
Myiagra cervinicauda	—	—	—	—	—	—	—	M	M
HONEYEATERS									
Stresemannia [*Melilestes?*] *bougainvillei*	—	M	—	—	—	—	—	—	—
Guadalcanaria [*Meliphaga?*] *inexpectata*	—	—	—	—	—	M	—	—	—
Meliarchus [*Melidectes?*] *sclateri*	—	—	—	—	—	—	—	M	—
Myzomela cardinalis	—	—	—	—	—	—	—	M	M
Myzomela lafargei	M	M	M	M, G	—	—	—	—	M
Myzomela eichhorni	—	—	—	—	M	—	—	—	—
Myzomela malaitae	—	—	—	—	—	—	M	—	—
Myzomela melanocephala	—	—	—	—	—	M	—	—	M
Myzomela tristrami	—	—	—	—	—	—	—	M	M

(continued)

Table 5-7 (continued)

Species	Buka	Bougainville	Choiseul	Isabel	New Georgia	Guadalcanal	Malaita	Makira	Small islands
DRONGOS									
Dicrurus bracteatus	—	—	—	—	—	M	—	M	—
CROWS									
Corvus woodfordi	—	—	M	M, G	—	M	—	—	M
Corvus meeki	M	M	—	—	—	—	—	—	M
TOTAL NONPASSERINES									
M	44–45	66	52–54	51–53	59	66–67	50–53	52	56–58
M + P	55–56	66	52–54	57–53	59	66–67	50–53	52	56–58
TOTAL PASSERINES									
M	23	36	24	25	31	33	22	25	39–40
M + P	23	36	24	25	31	33	22	25	39–40
TOTAL (ALL TAXA)									
M	67–68	102	76–78	76–78	90	99–100	72–75	77	95–98
M + P	78–79	102	76–78	76–78	90	99–100	72–75	77	95–98
Area (km^2)	611	8591	2966	4095	2044[a]	5281	4307	3090	—
Elevation (m)	402	2591	970	1250	1768[b]	2448	1280	1040	—
Number of landbird bones	75	0	0	0	0	0	0	0	0

& species A" is a large ground-dwelling columbid about the size of *C. nicobarica* but with shorter wings. "Undescribed genus & species B" is another, much larger ground-dwelling columbid with hind limbs nearly the size of those in *Goura* spp. (the New Guinea crested pigeons) but with shorter wings. Neither of these undescribed columbids is very similar to *Microgoura meeki*, a large ground-dwelling pigeon known only from Choiseul but which must have lived on other islands in the Solomon Main Chain, at least those that once were part of Greater Bougainville. The hindlimb osteology of *M. meeki* resembles that in *Gallicolumba* more than in *Goura* or *Caloenas*. Lastly, an ungual phalanx from Kilu Cave represents the owl *Nesasio solomonensis*, unknown today on Buka (Hadden 1981).

The prehistoric avifauna of Buka has 61% extinct/extirpated species vs. 22% on New Ireland. A factor that could account for this is island area, with avifaunas on smaller islands, on average, being more severely depleted by human impact (Chapters 16, 19). How much of Buka's higher level of extinction was due to human impact vs. "faunal relaxation" is unknown although I favor the former as more important. Another possible factor may be that the bones from Kilu Cave better represent the initial human occupation of Buka (Table 5-5). This is not the case, however, because bones of only one of the 11 extinct or extirpated species from Kilu Cave are confined to Layer II. Thus, as on New Ireland, the loss of landbirds following human arrival was slow from an ecological or cultural perspective, albeit rapid in a geological sense.

Buka lacks endemic landbirds today, whereas Bougainville has 14 endemic genera, species, or subspecies (Mayr & Diamond 2001: Table 25.1, Figure 32.1). I believe that each of these taxa once inhabited Buka as well. I also believe that each of the extinct species found as fossils on Buka used to live on Bougainville.

Isabel

Prehistoric bone deposits are badly needed from any Main Chain island southeast of Buka. In the meantime, community-level analyses of modern birds hint at possible prehistoric losses. Based on two dry-season visits (9-23 July 1997, 22–29 June 1998) to a forest site stretching for 4–5 km^2 along the lower Garanga River on Isabel (4095 km^2), Kratter et al. (2001a, b) recorded 65 resident and six migrant species of birds, using specimen and observational data to document relative abundance, habitat preferences, and foraging guilds. Nonnative species of birds are absent. The Garanga River site sustains all but 11 to 13 of the 76 to 78 species of landbirds known from Isabel (Table 5-7).

Table 5-8 Prehistoric birds from Kilu Cave, Buka, Northern Solomon Islands, Papua New Guinea. †, extinct species; *, extant species but extirpated on Buka. Taxa in brackets are not necessarily different from those identified more precisely. Modern status of species from Hadden (1981:97–99). Chronology of layers (from Wickler & Spriggs 1988, Wickler 2001) discussed in text.

	Layer		
Species	I	II	Total
SEABIRDS			
TERNS			
Anous stolidus	4	—	4
LANDBIRDS			
HERONS			
†*Nycticorax* undescribed sp. A	1	—	1
DUCKS			
Anas superciliosa	1	—	1
HAWKS, EAGLES			
Haliastur indus	1	—	1
Haliaeetus sanfordi	2	—	2
Accipiter novaehollandiae	1	—	1
MEGAPODES			
†*Megapodius* undescribed sp. B	2	—	2
RAILS			
**Gallirallus* [*woodfordi*] *tertius*	6	1	7
†*Gallirallus* undescribed sp. B	9	2	11
†cf. *Pareudiastes* undescribed sp.	1	—	1
†*Porphyrio* undescribed sp. B	8	2	10
[Rallidae sp.]	5	—	5
PIGEONS, DOVES			
Ducula rubricera	6	—	6
Ducula pistrinaria	—	1	1
**Caloenas nicobarica*	2	—	2
**Gallicolumba rufigula*	—	1	1
†Undescribed genus & sp. A	4	1	5
†Undescribed genus & sp. B	11	6	17
[Columbidae sp.]	1	—	1
PARROTS			
Chalcopsitta cardinalis	1	—	1
OWLS			
**Nesasio solomonensis*	1	—	1
TOTAL LANDBIRD BONES	63	14	77
†/*	42	13	58
% †/*	67	93	—
TOTAL LANDBIRD SPECIES	16	7	18
†/*	10	6	11
% †/*	62	86	—

Of the missing species, four are small island/beach specialists (*Esacus magnirostris*, *Ducula pistrinaria*, *Caloenas nicobarica*, *Halcyon saurophaga*), three are mainly or only montane (*Charmosyna meeki*, *Nesasio solomonensis*, *Phylloscopus [trivirgatus] poliocephalus*), and six are of uncertain status on Isabel (*Columba vitiensis*, *Reinwardtoena crassirostris*, *Charmosyna margarethae*, *Tyto alba*, *Eurostopodus mysticalis*, *Acrocephalus stentoreus*). Habitat heterogeneity, maintained by floodplain dynamics, is a major contributor to avian diversity at the site.

As is typical for Near Oceania, the avifauna along the Garanga River is dominated by nonpasserines, especially parrots, pigeons, kingfishers, and hawks. The flightless rail *Gallirallus* [*woodfordi*] *immaculatus*, regarded as rare and threatened with extinction (Taylor 1996, 1998), was common. We recorded the heron *Ixobrychus flavicollis*, falcon *Falco severus*, and cuckoo *Eudynamys scolopacea* for the first time on Isabel. We also found four species in the lowlands that previously were thought to be confined on Isabel to upper elevations (*Micropsitta finschii*, *Collocalia spodiopygia*, *Coracina caledonica*, and *Pachycephala pectoralis*). Thus the distinction between lowland and montane bird communities is not as clear-cut as had been thought. I suspect that this also may be true elsewhere in Near Oceania. In fact, the tiny parrot *M. finschii* may have been the most common species of forest bird along the lower Garanga River.

The understory avifauna of the Garanga River site is depauperate, which suggests that anthropogenic extinctions (as yet undocumented by bones) have taken place on Isabel. Among the species that may have lived once on Isabel are the extinct heron, megapode, rails (flightless species of *Pareudiastes*, *Porphyrio*, and perhaps another *Gallirallus*), and the various ground-dwelling pigeons recorded at Kilu Cave on Buka. (Both Isabel and Buka were part of "Greater Bougainville.") Other possible losses on Isabel would include ground-dwelling pigeon (*Microgoura meeki*), canopy or mid-level columbids (*Gymnophaps solomonensis*, *Columba pallidiceps*, *Reinwardtoena crassirostris*), the thrushes *Zoothera* sp. and *Turdus poliocephalus*, and the finch *Erythrura trichroa*. Some of these taxa may yet be found alive on Isabel. As shown for New Ireland (Steadman et al. 1999b; see above), even large, well-forested islands in Near Oceania do not sustain "intact" avifaunas.

Rennell and Bellona

These intriguing raised limestone islands lie south of the Solomon Main Chain, with Rennell separated from Guadalcanal and Makira by 190 and 170 km of deep ocean.

Table 5-9 Landbirds of Solomon Island outliers. Santa Cruz: Main Islands = Ndende, Utupua, Vanikolo, Tevai, Tinakula, Lomlom, Fenualoa, and the Duff Islands. E, endemic species; e, endemic subspecies; H, historic record, no longer occurs; M, modern record; P, prehistoric record; ?, questionable record, or residency not established. Rennell and Bellona are considered jointly for endemism. From Mayr (1945), Bayliss-Smith (1972), Diamond (1984), Steadman & Kirch (1990), Filardi et al. (1999), and herein, [a]species absent from Solomon Main Chain but with conspecific or closely related congener in Australia, New Caledonia, or Vanuatu; [b]area of Ndende only; [c]elevation of Vanikolo, highest island in Santa Cruz.

	Rennell	Bellona	Santa Cruz: Main Is.	Tikopia	Anuta	Ontong Java	Sikaiana
GREBES							
Tachybaptus novaehollandiae	M (e)	—	—	—	—	—	—
CORMORANTS							
Phalacrocorax carbo[a]	M	—	—	—	—	—	—
Phalacrocorax melanoleucos	M (e)	—	—	M	—	—	—
HERONS							
Egretta sacra	M	M	M	M	—	M	M
Butorides striatus	—	—	M	—	—	—	
Ixobrychus flavicollis	M	H	—	—	—	—	—
IBISES							
Threskiornis molucca[a]	M (e)	M (e)	—	—	—	—	—
DUCKS							
Anas gracilis[a]	H (e)	—	—	—	—	—	—
Anas superciliosa	H	—	—	M, P	—	—	—
OSPREYS							
Pandion haliaetus	M	M	—	—	—	—	—
HAWKS							
Accipiter fasciatus[a]	M	M	—	—	—	—	—
Accipiter albogularis	—	—	M (e)	—	—	—	—
MEGAPODES							
Megapodius eremita	—	—	—	—	—	M	M
Megapodius layardi	—	—	P	P	—	—	—
RAILS							
Gallirallus philippensis	—	—	—	P	—	—	—
Porzana tabuensis	H/M	H	M	—	—	—	—
Porphyrio porphyrio	M	M	M, P	M, P	—	M?	—
PIGEONS, DOVES							
Ptilinopus richardsii	M (e)	M (e)	—	—	—	—	—
Ptilinopus greyii[a]	—	—	M	—	—	—	—
Ducula pacifica	M	M	M, P	M, P	M	M	M
Macropygia mackinlayi	M	M?	M	—	—	—	—
Chalcophaps indica	—	—	M	—	—	—	—
Caloenas nicobarica	H/M	H/M	—	—	—	M	M
Gallicolumba sanctaecrucis	—	—	M (E)	—	—	—	—
Gallicolumba beccarii	M	M	—	—	—	—	—
PARROTS							
Trichoglossus haematodus	—	—	M	M	—	—	—
Lorius chlorocercus	M	—	—	—	—	—	—

Table 5-9 (continued)

	Rennell	Bellona	Santa Cruz: Main Is.	Tikopia	Anuta	Ontong Java	Sikaiana
Charmosyna palmarum	—	—	M	—	—	—	—
Micropsitta finschii	M	—	—	—	—	—	—
Geoffroyus heteroclitus	M (e)	—	—	—	—	—	—
CUCKOOS							
Chrysococcyx lucidus[a]	M (e)	M	M?	—	—	—	—
BARN-OWLS							
Tyto alba	H/M	M	M	—	—	—	—
CRESTED SWIFTS							
Hemiprocne mystacea	M	—	—	—	—	—	—
SWIFTS							
Collocalia esculenta	M (e)	M (e)	M	—	—	—	—
Collocalia spodiopygia	—	—	M (e)	—	—	—	—
Collocalia vanikorensis	M	M	M	M	—	—	—
KINGFISHERS							
Halcyon chloris	M (e)	M (e)	M (e)	M	—	M	M
SWALLOWS							
Hirundo tahitica	—	—	M	—	—	—	—
THRUSHES							
Turdus poliocephalus	M (e)	—	M (e)	—	—	—	—
WHITE-EYES							
Zosterops rennellianus	M (E)	—	—	—	—	—	—
Zosterops sanctaecrucis	—	—	M (E)	—	—	—	—
Woodfordia superciliosa	M (E)	—	—	—	—	—	—
Woodfordia lacertosa	—	—	M (E)	—	—	—	—
STARLINGS							
Aplonis cantoroides	M	—	—	—	—	—	—
Aplonis insularis/feadensis	M (e)	—	—	—	—	M	—
Aplonis zelandica[a]	—	—	M (e)	—	—	—	—
Aplonis tabuensis	—	—	M (e)	M (e), P	—	—	—
CUCKOO-SHRIKES							
Coracina lineata	M (e)	M (e)	—	—	—	—	—
Lalage maculosa[a]	—	—	M (e)	—	—	—	—
WHISTLERS							
Pachycephala pectoralis	M (e)	—	M (e)	—	—	—	—
FANTAILS							
Rhipidura rennelliana	M (E)	—	—	—	—	—	—
Rhipidura rufifrons	—	—	M (e)	—	—	—	—
MONARCHS							
Mayrornis schistaceus	—	—	M (E)	—	—	—	—
Clytorhynchus hamlini[a]	M (E)	—	—	—	—	—	—
Clytorhynchus nigrogularis[a]	—	—	M (e)	—	—	—	—

(continued)

Table 5-9 (continued)

	Rennell	Bellona	Santa Cruz: Main Is.	Tikopia	Anuta	Ontong Java	Sikaiana
Monarcha cinerascens	—	—	—	—	—	M	M
Myiagra caledonica[a]	M (e)	—	—	—	—	—	—
Myiagra vanikorensis	—	—	M (e)	—	—	—	—
HONEYEATERS							
Myzomela cardinalis	M (e)	—	M	M (e)	—	—	—
GERYGONES							
Gerygone flavolateralis[a]	M (e)	—	—	—	—	—	—
TOTAL							
M	34–36	14–16	30–31	10	1	7–8	6
M + H	38	17–18	30–31	10	1	7–8	6
M + H + P	38	17–18	31–32	12	1	7–8	6
Area (km^2)	676	19.5	438[a]	4.6	0.4	9.6	1.3
Elevation (m)	154	76	923[b]	360	80	<10	<10
Number of landbird bones	0	0	12+	55	0	0	0

Culturally Polynesian rather than Melanesian, Rennell (676 km^2, elev. 154 m) and Bellona (19.5 km^2, elev. 76 m) are politically part of the Solomon Islands today. Most of the genera and species of landbirds in the Solomon Main Chain are not found on either island (Tables 5-2, 5-7, 5-9). The avifaunas of Bellona and especially of Rennell are instead renowned for species- and subspecies-level endemism, as well as having affinities with certain birds from Australia, New Caledonia, or Vanuatu rather than the Solomon Main Chain (Diamond 1984, Mayr & Diamond 2001:250–255).

The modern landbird fauna of Bellona is a subset (18 of 40) of that on Rennell (Table 5-9). Although Rennell's area is 35 times that of Bellona, I believe that Bellona once sustained most of the landbirds still found on Rennell except the species confined to Rennell's large lake (grebes, cormorants, ducks). Birds would be unlikely to colonize Rennell, either from the Solomon Main Chain or from Australia, without seeing Bellona. At 19.5 km^2, Bellona is small by Solomon Island standards but still a substantial forested chunk of limestone. A forest passerine requiring a hectare for a territory, for example, could have a population of >1000 pairs on Bellona, which lacks surface drainage, is surrounded by cliffs, and therefore probably was forested entirely before human arrival.

During a much-too-brief visit (20–30 June 1997), we observed three species previously unrecorded on Rennell: a putative caprimulgid species, swallow *Hirundo nigricans*, and cuckoo-shrike *Coracina novaehollandiae* (Filardi et al. 1999). While none of these species may breed on Rennell, two others reported initially as vagrants, the cormorant *Phalacrocorax carbo* (aquatic) and starling *Aplonis cantoroides* (forest edge), have colonized Rennell and increased in abundance over the past several decades. The establishment of the cormorant and starling portends the significance of vagrant nonforest species on islands in or close to Near Oceania that have substantial human disturbance (Mayr & Diamond 2001:402–405). Furthermore, Rennell's geographical position seems to make it well suited for avian propagules directly to or from Australia, New Guinea, New Caledonia, and Vanuatu.

Although Rennell remains mostly forested, it lacks some expected taxa of landbirds, which would suggest that prehistoric extinctions have occurred. Prehistory on Rennell is any time more than a century ago, it being one of the last islands in Oceania to have contact with westerners (Woodford 1916, Wolff 1958). Among the taxa that I believe once occurred on Rennell would be any hawk other than *Accipiter fasciatus*, a megapode (*Megapodius*), flightless rails (perhaps *Gallirallus, Pareudiastes, Porphyrio*), a pigeon (*Ducula*) other than *D. pacifica*, lorikeets (*Trichoglossus, Charmosyna*), a cockatoo (*Cacatua*), and parrot (*Eclectus*).

Santa Cruz Group

Politically part of the Solomon Islands, the Santa Cruz Group consists of the eroded volcanic islands of Ndende

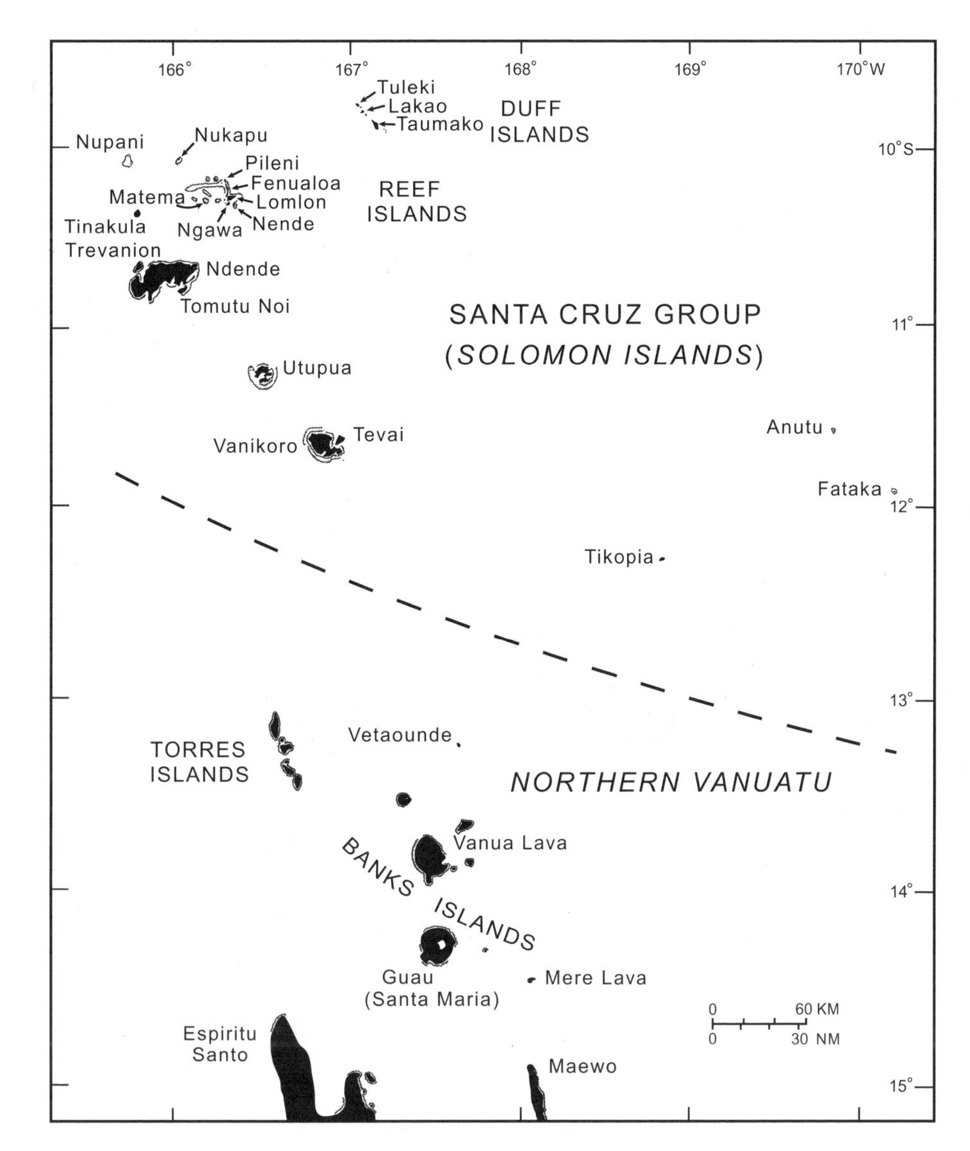

Figure 5-3 The Santa Cruz Group (Solomon Islands), in relation to northern Vanuatu. Redrawn from New Zealand Department of Lands and Survey (1986).

(Nendo), Utupua, Vanikolo (Vanikoro), and Tevai, the Reef Islands (volcanic Tinakula and raised limestone Lomlom, Fenualoa, and islets), the Duff Islands (Tuleki, Taumako, and Bass Islands), and the tiny, isolated, eroded volcanic islands of Tikopia, Anuta, and Fataka (Figure 5-3). The Santa Cruz Group is related geologically to Vanuatu rather than the Solomon Main Chain (Chapter 1). Its modern avifauna, although not well studied, is less rich than that of nearby Vanuatu or the Solomon Main Chain (Tables 5-2, 5-7, 5-9, 5-11). Mayr and Diamond (2001) mentioned the Santa Cruz landbird fauna several times but provided no detailed analyses.

Prehistoric data on birds are lacking for Santa Cruz except for Tikopia and Anuta (see below) and a Lapita site on Lomlom (Green 1976), the latter's record consisting of *Megapodius* cf. *layardi* (extirpated; Chapter 9 herein), the extant *Gallirallus philippensis*, *Porphyrio porphyrio*, and *Ducula pacifica*, and nonnative *Gallus gallus*. I suspect that a substantial fossil record from Santa Cruz would reveal many more losses.

The very small, remote, malaria-free islands of Tikopia (4.6 km^2, elev. 360 m) and Anuta (0.4 km^2, elev. 80 m) lie as close to northern Vanuatu as to large islands in Santa Cruz. The nearest major islands to Tikopia are Vanikolo (215 km WNW) and Valua (185 km SW; politically part of Vanuatu). Vanikolo is 310 km W and Valua 315 km SW of Anuta. Renowned in anthropology (Firth 1936), Tikopia and Anuta are 137 km from each other and are occupied by

Table 5-10 Birds of Tikopia and Anuta, Polynesian outliers in the Santa Cruz Group, Solomon Islands. i, introduced species; m, migratory species; M, modern record; P, prehistoric record; ?, questionable record, or residency not established. i, m are excluded from totals. From Steadman et al. (1990b).

	Tikopia	Anuta
SEABIRDS		
PETRELS, SHEARWATERS		
Pseudobulweria rostrata	M, P	—
Puffinus pacificus	M?, P	P
Puffinus lherminieri	P	P
TROPICBIRDS		
Phaethon rubricauda	M, P	—
Phaethon lepturus	M, P	M, P
BOOBIES		
Papasula abbotti	P	—
Sula dactylatra	M, P	—
Sula sula	P	P
Sula leucogaster	M, P	M, P
FRIGATEBIRDS		
Fregata minor	M?, P	M, P
Fregata ariel	M?, P	P
TERNS		
Sterna lunata	M	—
Sterna fuscata	P	P
Anous minutus	M, P	M, P
Anous stolidus	M, P	M, P
Gygis candida	M	M
LANDBIRDS		
CORMORANTS		
Phalacrocorax melanoleucos	M	—
HERONS		
Egretta sacra	M	—
DUCKS		
Anas superciliosa	M, P	—
MEGAPODES		
Megapodius layardi	P	—
CHICKENS		
Gallus gallus (i)	M, P	M, P
RAILS		
Gallirallus philippensis	P	—
Porphyrio porphyrio	M, P	—
PLOVERS		
Charadrius mongolus (m)	M	—
Pluvialis fulva (m)	M, P	M, P
SANDPIPERS		
Limosa lapponica (m)	M	—
Numenius phaeopus (m)	M?	—
Numenius tahitiensis (m)	M, P	—
Heteroscelus incanus (m)	M, P	—
Arenaria interpes (m)	M	M, P
PIGEONS		
Ducula pacifica	M, P	M
PARROTS		
Trichoglossus haematodus	M	—
CUCKOOS		
Eudynamy taitensis (m)	M, P	M, P
SWIFTS		
Collocalia vanikorensis	M	—
KINGFISHERS		
Halcyon chloris	M	—
STARLINGS		
Aplonis tabuensis	M, P	—
HONEYEATERS		
Myzomela cardinalis	M	—
TOTAL SPECIES (M)		
Resident seabirds	9–12	6
Resident landbirds	10	1
All	19–22	7
TOTAL SPECIES (M + P)		
Resident seabirds	16	11
Resident landbirds	12	1
All	28	12
TOTAL NISP		
Resident seabirds	323	275
Resident landbirds	55	0
All	378	275

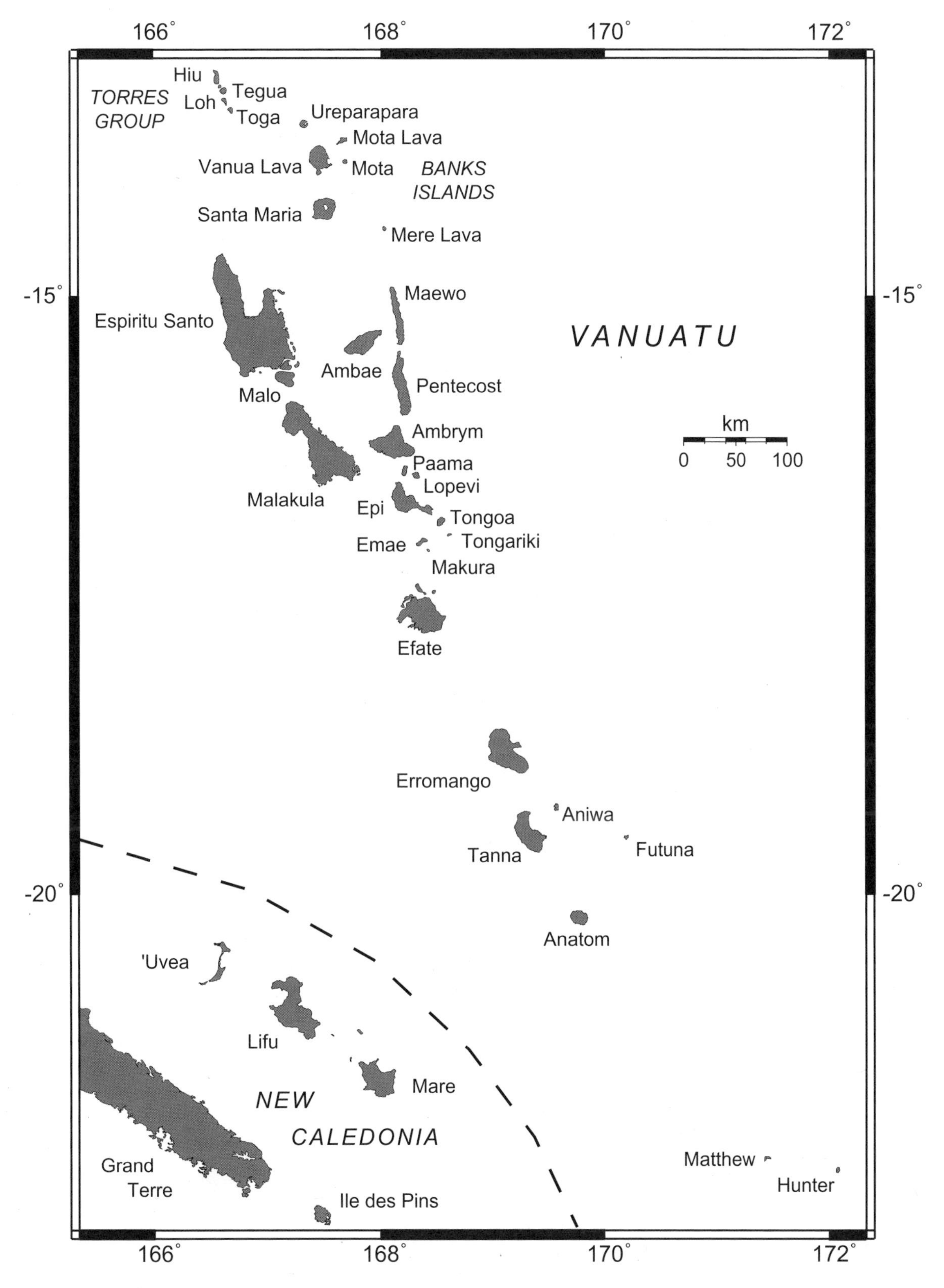

Figure 5-4 Vanuatu.

Table 5-11 Landbirds of most major islands, Vanuatu. Not listed are fifty-nine islands (land areas <60 km^2) because their landbird faunas are too poorly known or unknown. E, endemic to Vanuatu; e, endemic to Vanuatu and Santa Cruz Islands or to Vanuatu and New Caledonia; M, modern record; P, prehistoric record; †, extinct species; ?, questionable record, or residency not established. From Medway & Marshall (1975), Bregulla (1992:27–29, 71–75), Eakle (1997), Kratter et al. (in press), and Steadman (unpub. data). Island area for Torres Group combines the six largest islands; the largest single island (Hiu) is 50 km^2.

	Torres Group	Ureparapara	Mota Lava	Vanua Lava	Gaua	Mere Lava	Santo	Malo	Maewo	Ambae	Pentecost	Malakula
GREBES												
Tachybaptus novaehollandiae	—	—	—	—	M	—	M	M	M	M	—	—
HERONS												
Egretta sacra	M	M	M	M	M	M	M	M	M	M	M	M
Butorides striatus	M	M	—	M	M	—	M	—	—	—	—	M
DUCKS												
Anas gibberifrons	—	—	—	—	—	—	—	—	—	—	—	—
Anas superciliosa	—	—	—	—	M	—	M	—	—	M	—	—
Aythya australis	—	—	—	—	M	—	M	M	—	M	—	—
HAWKS												
Circus approximans	M	M	—	M	M	—	M	M	M	M	M	M
Accipiter fasciatus	—	—	—	—	—	—	—	—	—	—	—	—
FALCONS												
Falco peregrinus	—	—	—	—	M	—	M	M	—	—	—	M
MEGAPODES												
Megapodius layardi (e)	M	M	M	M	M	—	M	M	M	M	M	M, P
†*Megapodius* undesc. sp. C (E)	—	—	—	—	—	—	—	—	—	—	—	—
RAILS												
Gallirallus philippensis	—	—	—	M	—	—	M	M	M	M	M	M, P
†*Gallirallus* undescribed sp. C (E)	P	—	—	—	—	—	—	—	—	—	—	—
Porzana tabuensis	—	—	—	—	—	—	—	—	—	—	—	P
†*Porzana* undescribed sp. A (E)	—	—	—	—	—	—	—	—	—	—	—	P
Poliolimnas cinereus	—	—	—	—	M	—	—	—	—	—	—	—
Porphyrio porphyrio	P	—	—	—	—	—	M	M	M	M	M	M, P
PIGEONS, DOVES												
Columba vitiensis	M	M	—	M	—	M	M	M	M	M	M	M, P
Macropygia mackinlayi	—	M	M	M	M	M	M	M	M	M	M	M, P
Ptilinopus tannensis (E)	—	—	M	M	M	M	M, P	M	M	M	M	M, P
Ptilinopus greyii	M, P	M	M	M	M	M	M	M	M	M	M	M, P
Ducula pacifica	M, P	M	M	M	M	M	M	M	M	M	M	M
Ducula bakeri (E)	—	M	M	M	M	—	M	—	M	M	M	—
Chalcophaps indica	M	M	M	M	M	M	M, P	M	M	M	M	M, P
Gallicolumba sanctaecrucis (e)	—	—	—	—	—	—	M	—	—	—	—	—
†*Gallicolumba ferruginea* (E)	—	—	—	—	—	—	—	—	—	—		
PARROTS												
Trichoglossus haematodus	M	M	M	M	M	M	M	M	M	M	M	M
Charmosyna palmarum	—	M	M	M	M	M	M, P	M	M	M	M	M
†*Eclectus* undescribed sp. (?E/e)	—	—	—	—	—	—	—	—	—	—	—	P
CUCKOOS												
Cacomantis flabelliformis	—	M	M	M	M	—	M, P	M	—	M	—	M
Chrysococcyx lucidus	—	—	—	M	M	—	M	—	—	—	—	M, P

Ambrym	Paama	Lopevi	Epi	Tongoa	Tongariki	Emae	Makura	Nguna	Emau	Efate	Erromango	Aniwa	Tanna	Futuna	Aneityum	Total
—	—	—	—	—	—	—	—	—	—	M	—	—	—	—	—	6
M	M	M	M	M	M	M	M	M	M	M	M, P	M	M	M	M	28
—	—	—	—	—	—	—	—	—	—	—	M	M	—	—	—	8
—	—	—	—	—	—	—	—	—	—	M	—	—	M?	—	—	1–2
M	—	—	—	—	—	—	—	—	—	M, P	M	—	M	—	M	8
—	—	—	—	—	—	—	—	—	—	M	M	—	M	—	—	7
M	M	M	M	M	—	M	M	M	M	M	M	M	M	M	M	25
—	—	—	—	—	—	—	—	—	—	P	—	—	M	—	M	3
—	—	—	—	M	—	—	—	M	—	M	M	M	M	M	M	12
M	M	M	M	M	M	M	—	M	M	M, P	—	—	M	—	—	22
—	—	—	—	—	—	—	—	—	—	P	—	—	—	—	—	1
M	—	—	M	M	—	M	—	M	M	M, P	M, P	—	M	—	M	17
—	—	—	—	—	—	—	—	—	—	—	—	—	—	—	—	1
—	—	—	—	—	—	—	—	—	—	M, P	M	—	M	—	M	5
—	—	—	—	—	—	—	—	—	—	—	—	—	—	—	—	1
—	—	—	—	—	—	—	—	—	—	—	M	—	M	—	—	3
M	—	—	M	M	M	M	M	M	M	M, P	M, P	—	M	M	M	20
M	M	M	M	M	—	M	M	M	M	M, P	M, P	M	M	M	M	25
M	M	M	M	M	M	M	M	M	M	M	M, P	—	M	M	M	26
M	M	M	M	M	M	M	—	M	M	M	M	—	M	—	M	23
M	M	M	M	M	M	M	M	M	M	M, P	M, P	M	M	M	M	28
M	—	—	M	M	—	M	—	M	M	M	M, P	M	M	M	M	24
M	—	—	—	—	—	—	—	—	—	—	—	—	—	—	—	9
M	M	M	M	M	M	M	M	M	M	M	M	M	M	M	M	28
—	—	—	—	—	—	—	—	—	—	—	—	—	—	—	—	1
M	M	—	M	M	M	M	M	M	M	M, P	M, P	M	M	—	M	26
M	M	M	M	M	M	M	—	M	M	M	M	—	M	M	M	25
—	—	—	—	—	—	—	—	—	—	—	—	—	—	—	—	1
M	M	M	M	—	—	M	—	M	—	M	M	M	M	M	—	19
M	—	M	M	—	—	—	—	M	M	M	M	M	—	M	—	13

(*continued*)

Table 5-11 (continued)

	Torres Group	Ureparapara	Mota Lava	Vanua Lava	Gaua	Mere Lava	Santo	Malo	Maewo	Ambae	Pentecost	Malakula
BARN-OWLS												
Tyto alba	—	M	—	M	—	M	M	M	M	M	M	M, P
SWIFTS												
Collocalia esculenta	M, P	M	M	M	M	M	M	M	M	M	M	M, P
Collocalia spodiopygia	—	M	—	—	—	—	M	M	—	—	—	M
Collocalia vanikorensis	M	—	—	M	M	—	M	M	M	M	M	M
KINGFISHERS												
Halcyon farquhari (E)	—	—	—	—	—	—	M	M	—	—	—	M
Halcyon chloris	M, P	M	M	M	M	M	M	M	M	M	M	M, P
SWALLOWS												
Hirundo tahitica	M, P	—	—	—	—	M	M	M	M	M	M	M
THRUSHES												
Turdus poliocephalus	—	M	—	M	M	M	M	M	M	M	M	M, P
WARBLERS												
Cichlornis whitneyi	—	—	—	—	—	—	M	—	—	—	—	—
WHITE-EYES												
Zosterops lateralis	M, P	M	M	M	M	M	M	M	M	M	M	M, P
Zosterops flavifrons (E)	M	—	—	M	M	M	M	M	M	M	M	M, P
STARLINGS												
Aplonis zelandica (e)	—	M	—	—	M	M	M	M	—	M	M	M
Aplonis santovestris (E)	—	—	—	—	—	—	M	—	—	—	—	—
†*Aplonis* undescribed sp. (E/e)	—	—	—	—	—	—	—	—	—	—	—	—
ESTRILDID FINCHES												
Erythrura trichroa	—	—	—	—	M	—	—	—	—	M	—	—
Erythrura cyaneovirens	—	—	—	—	M	—	M	—	—	M	M	M
CUCKOO-SHRIKES, TRILLERS												
Coracina caledonica	—	—	—	—	—	—	M	M	—	—	—	M, P
Lalage maculosa	—	—	—	—	—	—	M	M	—	—	—	M
Lalage leucopyga	M	M	M	M	M	M	M	M	M	M	M	M
GERYGONES												
Gerygone flavolateralis	—	—	—	M	M	—	M	M	M	M	M	M, P
ROBINS												
Petroica multicolor	—	—	—	—	M	M	M	M	M	M	—	M
WHISTLERS												
Pachycephala pectoralis	—	M	—	M	M	—	M	M	M	M	M	M, P
FANTAILS												
Rhipidura fuliginosa	M	M	M	M	M	M	M	M	M	M	M	M
Rhipidura spilodera	—	—	—	M	M	—	M	M	M	M	M	M
MONARCHS												
Neolalage banksiana (E)	—	—	—	M	—	—	M	M	M	M	M	M
Clytorhynchus pachycephaloides (e)	M	M	M	M	M	M	M	M	M	M	M	M
Myiagra caledonica	M	M	M	M	M	M	M	M	M	M	M	M
HONEYEATERS												
Lichmera incana (e)	—	—	—	—	—	—	—	—	—	—	—	M, P

Ambrym	Paama	Lopevi	Epi	Tongoa	Tongariki	Emae	Makura	Nguna	Emau	Efate	Erromango	Aniwa	Tanna	Futuna	Aneityum	Total
M	—	—	M	M	—	M	—	M	M	M, P	M, P	M	M	M	M	21
M	M	M	M	M	—	M	—	M	M	M	M, P	M	M	M	M	26
M	—	—	M	—	—	M	—	—	—	M	M	—	M	—	M	11
M	—	—	M	M	—	M	M	M	M	M	M, P	—	M	—	M	20
—	—	—	—	—	—	—	—	—	—	—	—	—	—	—	—	3
M	M	M	M	M	M	M	M	M	M	M, P	M, P	M	M	M	M	28
M	M	M	M	M	M	M	—	M	—	M	M	M	M	—	M	21
M	M	M	M	—	—	M	—	M	M	M, P	M, P	—	M	M	—	21
—	—	—	—	—	—	—	—	—	—	—	—	—	—	—	—	1
M	M	M	M	M	M	M	M	M	M	M	M	M	M	—	—	26
M	M	M	M	M	M	M	M	M	M	M	M	M	M	M	M	26
M	M	M	M	—	—	—	—	—	—	M	—	—	—	—	—	13
—	—	—	—	—	—	—	—	—	—	—	—	—	—	—	—	1
—	—	—	—	—	—	—	—	—	—	—	P	—	—	—	—	1
M	—	M	—	—	—	—	—	M	M	M	M	—	M	—	M	10
M	M	M	M	M	M	M	—	—	—	M	—	—	—	—	M	14
—	—	—	—	—	—	—	—	—	—	—	M	—	—	—	—	4
—	M	—	M	M	—	M	—	M	M	M	—	—	—	—	—	10
M	M	M	M	M	—	M	—	M	M	M	M	M	M	—	M	25
M	—	M	M	—	—	M	—	—	—	—	—	—	—	—	—	12
M	M	M	M	M	M	M	—	M	M	M, P	M	—	M	—	M	20
M	M	M	M	M	M	M	—	M	M	M	M	—	—	—	M	21
M	M	M	M	M	M	M	—	M	M	M	M	—	—	M	M	25
M	M	—	M	M	M	M	—	M	M	M	—	—	—	—	—	17
M	—	—	M	—	—	—	—	—	—	M	—	—	—	—	—	10
—	M	M	M	M	—	M	—	—	—	M	M	—	—	—	—	19
M	—	M	M	M	—	M	M	M	M	M	M	M	M	M	M	26
M	M	M	M	M	M	M	M	M	M	M	M	—	—	—	—	13

(continued)

Table 5-11 (continued)

	Torres Group	Ureparapara	Mota Lava	Vanua Lava	Gaua	Mere Lava	Santo	Malo	Maewo	Ambae	Pentecost	Malakula
Myzomela cardinalis	M	M	M	M	M	M	M	M	M	M	M	M, P
Phylidonyris notabilis (E)	—	M	—	M	—	—	M	—	M	M	M	M, P
WOOD-SWALLOWS												
Artamus leucorhynchus	—	—	—	—	M	—	M	M	M	M	M	M
TOTAL												
M	20	27	19	33	38	23	50	41	35	41	35	44
M + P	22	27	19	33	38	23	50	41	35	41	35	47
Island area (km^2)	120	39	31	331	310	15	3900	180	270	400	439	2030
Elevation (m)	366	764	411	946	797	883	1879	326	811	1496	946	879
Isolation (from nearest island > 100 km^2)	85	20	14	26	26	50	3	3	6	12	6	15
Number of bird bones	43	0	0	0	0	0	9	0	0	0	6	15
Number of landbird bones	43	0	0	0	0	0	7	0	0	0	0	195

Polynesians. Fataka (Fatutaka, Patutaka, or Mitre Island) is steep and uninhabited, although visited by Anutans for fishing and hunting birds (Feinberg 1981, 1988). Tragically, both Tikopia and Anuta were decimated by the extremely strong (winds to 305 km/hr) Cyclone Zoe on 28 December 2002, killing one person and destroying all villages and crops.

Steadman et al. (1990b) analyzed bird bones from eight archaeological sites on Tikopia and Anuta that have been described in great detail (Kirch & Rosendahl 1973, 1976, Kirch 1982, Kirch & Yen 1982). Among 468 bird bones from Tikopia are those of six species (four seabirds, two landbirds) unknown there in modern times: the shearwater *Puffinus lherminieri*, boobies *Papasula abbotti* and *Sula sula*, tern *Sterna fuscata*, megapode *Megapodius layardi*, and rail *Gallirallus philippensis* (Table 5-10). Among the 299 bird bones from one site on Anuta are those of four seabirds not otherwise recorded there: the shearwaters *Puffinus pacificus* and *P. lherminieri*, booby *Sula sula*, and tern *Sterna fuscata*.

The bone samples from Tikopia and Anuta are similar in size and chronology (the past ca. 3000 years, with most bones from strata containing Lapita pottery). Most of their differences in species composition probably are due to: (1) sampling (sets of bird bones numbering in the hundreds probably are too small to represent the entire avifauna, even on islands as small as Tikopia or Anuta; Chapter 4); and (2) the extremely small land area of Anuta. Combined with its great isolation, Anuta is so tiny that any populations of resident landbirds undoubtedly would have been small (<500 individuals), highly vulnerable to environmental disruptions (whether or not of human origin) and difficult to sustain over long time scales. Tikopia, on the other hand, is 11 times as large in land area and, in spite of its isolation, sustains a substantial resident landbird fauna, including endemic subspecies. Whatever landbirds were on Anuta at first human contact probably perished so rapidly as to elude archaeological sampling. Thorough biotic surveys of both islands are needed, especially to evaluate the status of native woody plants and birds after the recent hurricane.

Vanuatu

This north-south-oriented archipelago is geologically complex (Chapter 1, Figure 5-4). By my inspection of South Pacific Maps (1994), at least 57 islands in Vanuatu exceed 1 km^2 in land area. The modern distribution of landbirds on 28 of these islands is known well enough (based mainly on Bregulla 1992) to be compiled (Table 5-11), which reveals many illogical absences of widespread species. These absences have been treated as natural gaps in "checkerboard distributions" without consideration of human impact (but see Chapter 18). The modern landbird fauna of Vanuatu consists of 56 species, with as many as 50 on a single island (Santo, the largest and highest island). As elsewhere in Oceania, I believe that much of Santo's greater species richness can be attributed to differential human impact. Under natural conditions, I suspect that variation in species richness values of landbirds in Vanuatu was minimal among islands >50 km^2, perhaps even >10 km^2. Adequate fossils to test this idea have not been found.

Ambrym	Paama	Lopevi	Epi	Tongoa	Tongariki	Emae	Makura	Nguna	Emau	Efate	Erromango	Aniwa	Tanna	Futuna	Aneityum	Total
M	M	M	M	M	M	M	M	M	M	M	M	M	M	M	M	28
M	M	—	M	—	—	—	—	—	—	—	—	—	—	—	—	10
M	—	M	M	M	—	M	M	M	M	M	M	—	M	—	M	19
41	29	30	40	33	20	36	16	35	32	45	40	20	36–37	20	32	57
41	29	30	40	33	20	36	16	35	32	47	41	20	36–37	20	32	61
665	33	30	445	42	6	33	1.7	25	8	915	900	8	572	11	160	—
1270	544	1413	833	487	521	644	297	593	448	647	886	42	1084	666	859	—
11	7	14	22	7	24	22	32	6	6	81	37	25	37	76	66	—
0	0	0	0	0	0	0	0	0	0	137	218	0	0	0	0	0
0	0	0	0	0	0	0	0	0	0	135	217	0	0	0	0	0

Vanuatu's modern avifauna has much less endemism than do those of the Solomon Main Chain and New Caledonia. A single genus, the monarch *Neolalage*, is endemic to Vanuatu. Seven extant species are endemic, with another five endemic to Vanuatu plus the Santa Cruz Group and/or New Caledonia. Fossils add five more presumably endemic species (see below).

The prehistory of birds in Vanuatu is just now being developed (Table 5-12), based on bones (under study by Jeremy Kirchman and myself) from archaeological sites excavated by Stuart Bedford, Jean-Christophe Sand, and Matthew Spriggs. We have not yet been able to identify to the species level the bones of starlings (*Aplonis*) from these sites because skeletal specimens of *A. santovestris* and *A. zelandica* are not available. In fact, no museum in the world has a skeletal specimen of either starling (Wood & Schnell 1986).

Vanuatu's avian fossils are dominated by extant species of landbirds, but also include five extinct ones and several others extirpated from individual islands. The poor showing of extinct or extirpated species of landbirds (only 45 of 601 identified bones; Table 5-12) suggests that strata from first human contact have not been heavily sampled. As on New Ireland, Buka, and New Caledonia, nearly all ancient bird bones from Vanuatu are of landbirds, not seabirds. Nevertheless, the mere three seabird bones represent three extirpated populations of shearwaters and booby.

From Toga Island in the Torres Group, a small set of bird bones represents four species, all of which still occur there except a flightless rail (*Gallirallus* undescribed sp. C) and perhaps the swamphen *Porphyrio porphyrio*. From nine sites on Malakula (see Bedford 2001), 190 bird bones represent 31 species of resident landbirds, of which the volant rail *Porzana tabuensis*, an extinct, flightless rail (*Porzana* undescribed sp. A), and an extinct parrot (*Eclectus* undescribed sp.) are gone from the island. Known on Malakula from an ulna and tibiotarsus, this parrot also was found prehistorically in Tonga (Steadman in press; Chapters 6, 12 herein).

From Efate, 145 landbird bones from two sites (Arapus, Mangaasi) represent 15 species, of which a hawk (*Accipiter* cf. *fasciatus*) and megapode (*Megapodius* undescribed sp. C) no longer occur there (Table 5-12). This hawk is found today on two other islands in Vanuatu (Eakle 1997) as well as New Caledonia, Rennell, Bellona, and Australia. The megapode is an extinct species similar to *M. alimentum* of Tonga and Fiji (Steadman 1989b, 1999d; Chapter 9 herein). Both species are from Arapus, a site that, by Vanuatu standards, has high representation (5 of 27 bones, or 18%) of extinct or extirpated species of birds. Arapus (see Spriggs & Bedford 2001) also has yielded bones of the extinct terrestrial crocodile *Mekosuchus kalpokasi* (Mead et al. 2002).

Lapita and post-Lapita archaeological sites on Erromango (reviewed by Bedford 1999, Spriggs 1999) have produced 217 landbird bones from 15 species, 14 of which still live on the island. The exception is *Aplonis* undescribed sp., a large starling in the size range of *A. grandis* of the Solomon Islands.

I believe that each of Vanuatu's extinct or extirpated landbird populations known thus far was part of a

Table 5-12 Preliminary summary of prehistoric records of birds from Vanuatu, based on the number of bones excavated through the 2001 field season. TOG = Toga; WOG = Woga; MAL = Malososaba; WOP = Woplamplam; YAL = Yalu; NAV = Navaprah; MSM = minor sites of Malakula (Navapule, Waprap, Woapraf, Ndavru, Wambraf, Malua Bay Cave); ARA = Arapus; MAN = Mangaasi; PON = Ponamla; IFO = Ifo; MSE = minor sites of Erromango (Raowalai, Velemendi, Ilpin, Velivo). i, introduced species; m, migratory species; *, species extirpated on this island; †, extinct species. Seabirds, i, and m are excluded from totals. [a]includes only extinct species. Taxa in brackets are not necessarily different from those identified more precisely. From Steadman (unpub. data).

	Torres Group		Santo	Malakula				Efate		Erromango			
	TOG	WOG	MAL	WOP	YAL	NAV	MSM	ARA	MAN	PON	IFO	MSE	Total
SEABIRDS													
SHEARWATERS													
Puffinus pacificus	—	—	—	—	—	—	—	—	—	1*	—	—	1
Puffinus cf. *gavia*	—	—	—	1*	—	—	—	—	—	—	—	—	1
BOOBIES													
Papasula abbotti	—	—	—	—	—	—	—	—	1*	—	—	—	1
LANDBIRDS													
HERONS													
Egretta sacra	—	—	—	—	—	—	—	—	—	1	—	—	1
DUCKS													
Anas superciliosa	—	—	—	—	—	—	—	—	1	—	—	—	1
HAWKS													
Accipiter cf. *fasciatus*	—	—	—	—	—	—	—	4*	—	—	—	—	4
MEGAPODES													
Megapodius layardi	—	—	—	3	—	—	—	—	1	—	—	—	4
†*Megapodius* undescribed sp. C	—	—	—	—	—	—	—	1*	—	—	—	—	1
CHICKENS													
Gallus gallus (i)	13	—	1	—	—	2	2	13	23	33	23	5	105
RAILS													
Gallirallus philippensis	—	—	—	2	8	1	2	1	17	20	—	—	51
†*Gallirallus* undescribed sp. C	5*	—	—	—	—	—	—	—	—	—	—	—	5
Porzana tabuensis	—	—	—	3*	3*	12*	—	—	3	1	—	—	22
†*Porzana* undescribed sp. A	—	—	—	—	2*	1*	—	—	—	—	—	—	3
Porphyrio porphyrio	11 (* ?)	—	—	—	4	1	5	11	81	65	8	2	188
SHOREBIRDS													
Heteroscelus incanus (m)	—	—	1	—	—	—	—	—	—	—	—	—	1
PIGEONS, DOVES													
Columba vitiensis	—	—	—	—	1	—	—	1	2	20	1	1	26
Macropygia mackinlayi	—	—	—	—	—	1	—	—	—	1	—	—	2
Ptilinopus cf. *tannensis*	—	—	1	2	—	2	—	—	—	—	—	—	5
Ptilinopus greyii	1	4	—	2	1	6	—	—	2	—	1	—	17
Ducula pacifica	2	4	—	—	1	—	1	3	—	15	17	—	43
[*Ducula* sp.]	2	—	—	—	1	—	—	1	—	7	—	—	11
Chalcophaps indica	—	—	1	—	4	1	1	—	—	—	—	—	7
[Columbidae sp.]	—	—	—	—	1	—	—	4	1	12	—	—	19

Table 5-12 (continued)

	Torres Group		Santo	Malakula				Efate		Erromango			
	TOG	WOG	MAL	WOP	YAL	NAV	MSM	ARA	MAN	PON	IFO	MSE	Total
PARROTS													
Trichoglossus haematodus	—	—	—	—	—	1	—	—	3	1	—	—	5
Charmosyna palmarum	—	—	2	—	—	—	—	—	—	—	—	—	2
†*Eclectus* undescribed sp.	—	—	—	—	—	—	2*	—	—	—	—	—	2
CUCKOOS													
Cacomantis flabelliformis	—	—	1	—	—	—	—	—	—	—	—	—	1
Chrysococcyx lucidus	—	—	—	—	1	—	—	—	—	—	—	—	1
Eudynamys taitensis (m)	—	1	—	—	—	1	—	—	1	—	—	—	3
BARN-OWLS													
Tyto alba	—	3	—	2	6	35	—	—	2	6	1	—	55
SWIFTS													
Collocalia esculenta	—	6	—	—	1	—	—	—	—	—	—	15	22
Collocalia vanikorensis	—	—	—	—	—	—	—	—	—	—	—	1	1
[*Collocalia* sp.]	—	—	1	—	—	1	—	—	—	—	—	10	12
KINGFISHERS													
Halcyon chloris	—	1	—	—	6	—	—	—	2	1	—	—	10
SWALLOWS													
Hirundo tahitica	—	1	—	—	—	—	—	—	—	—	—	—	1
THRUSHES													
Turdus poliocephalus	—	—	—	1	10	5	—	—	—	1	—	—	17
WHITE-EYES													
Zosterops lateralis	—	1	—	3	—	—	—	—	—	—	—	—	4
Zosterops flavifrons	—	—	—	—	2	—	—	—	—	—	—	—	2
[*Zosterops* cf. *lateralis*]	—	—	—	—	—	1	—	—	—	—	—	—	1
[*Zosterops* sp.]	—	—	—	—	—	—	—	—	—	—	—	1	1
STARLINGS													
Aplonis sp. (small)	—	—	—	—	3	13	—	—	—	—	—	—	16
†*Aplonis* undescribed sp.	—	—	—	—	—	—	—	—	—	1*	—	—	1
ESTRILDID FINCHES													
Erythrura sp.	—	—	—	—	1	—	—	—	—	—	—	—	1
CUCKOO-SHRIKES, TRILLERS													
Coracina caledonica	—	—	—	—	—	1	—	—	—	—	—	—	1
Lalage sp.	—	—	—	—	—	2	—	—	—	—	—	—	2
GERYGONES													
Gerygone flavolateralis	—	—	—	—	1	1	—	—	—	—	—	—	2
ROBINS													
Petroica multicolor	—	—	—	—	—	—	—	1	—	—	—	—	1
WHISTLERS													
Pachycephala pectoralis	—	—	—	1	1	—	—	—	—	—	—	—	2

(continued)

Table 5-12 (continued)

	Torres Group		Santo	Malakula				Efate		Erromango			
	TOG	WOG	MAL	WOP	YAL	NAV	MSM	ARA	MAN	PON	IFO	MSE	Total
FANTAILS													
Rhipidura sp.	—	—	—	—	1	—	—	—	—	—	—	—	1
MONARCHS													
Myiagra/Neolalage	—	—	—	1	—	—	—	—	—	—	—	—	1
HONEYEATERS													
Lichmera incana	—	—	—	—	—	2	—	—	—	—	—	—	2
Myzomela cardinalis	—	—	—	—	—	1	—	—	—	—	—	—	1
Phylidonyris notabilis	—	—	—	—	1	1	—	—	—	—	—	—	2
[Passeriformes sp.]	—	—	1	2	5	7	—	—	2	—	—	7	24
SITE TOTALS													
Species	4	7	6	10	20	20	5	7	11	12	5	6	42
†/* species	1–2	0	0	1	2	2	1	2	0	1	0	0	5[a]
Number of bones	22	20	7	22	61	96	11	27	118	152	28	37	601
Number of †/*bones	5–16	0	0	3	5	13	2	5	0	1	0	0	45
ISLAND TOTALS													
Species	4	7	6	31				15		15			42
†/* species	1–2	0	0	3				2		1			5[a]
Number of bones	21	20	7	190				145		217			600
Number of †/*bones	5–16	0	0	23				5		1			45

widespread species/superspecies before human arrival. I suspect as well that the fossil record in hand provides just a glimpse at the species that once lived in these islands. The mother lode of bones of extinct birds, if it exists, has not been discovered. If Lapita peoples were the first to settle Vanuatu, then either they did not wipe out nearly as many species as in New Caledonia, Fiji, or Tonga, or archaeologists have yet to sample first-contact bone deposits in any volume. Another possibility is that pre-Lapita peoples inhabited Vanuatu and already had eliminated the most vulnerable species. The small likelihood of pre-Lapita habitation in Vanuatu (and New Caledonia) has been reviewed by Spriggs (1997:40–42). I agree with his assessment, which can be tested by more excavations, followed by careful osteological studies.

New Caledonia

The name "New Caledonia" often is used to refer both to the largest island (Grande Terre) and to the entire group (Figure 5-5). Grande Terre is huge (16,750 km^2), high (1628 m), geologically old and complex (Jaffre & Veillon 1990, Hope 1996, Kroenke 1996), and yet lacks native mammals except bats (Nicholson & Warner 1953, Flannery 1995a). Off Grande Terre are numerous raised limestone islands, the largest being Isle of Pines (Ile des Pins). The small Belep Islands off the northern tip of Grande Terre are, as far as I can determine, ornithologically unknown. The Loyalty Islands ('Uvea, Lifu, Mare, and associated islets) lie only 210–230 km from Tanna and Anatom in southern Vanuatu.

The modern New Caledonian flora has been attributed to vicariant and dispersal events throughout at least the Cenozoic (past 65 million years; R. S. Hill 1996, van Balgooy et al. 1996, Weston & Crisp 1996, Wilson 1996). To what extent vicariance vs. dispersal accounts for New Caledonia's avifauna is speculative, although I favor dispersal for most if not all groups. Of course the most distinctive forms would be the best candidates for vicariance, although even flightless forms such as *Sylviornis neocaledoniae*, *Rhynochetos jubatus*, and *R. orarius* may have lost their ability to fly after colonization by a volant ancestor. Considering past changes

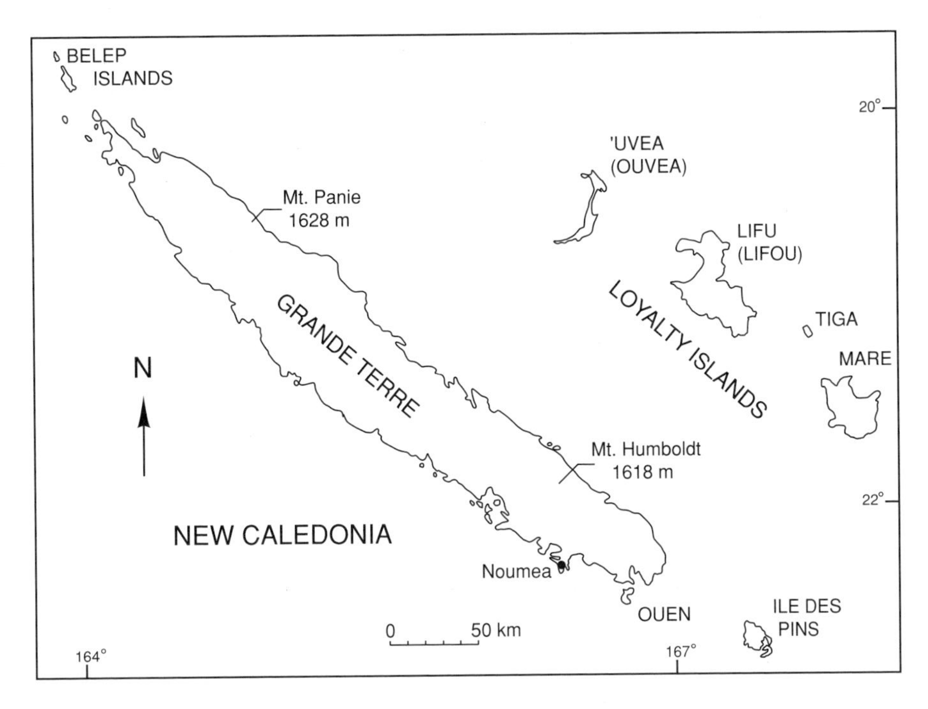

Figure 5-5 New Caledonia.

in sea level is essential when evaluating the biogeography of New Caledonia. Most importantly, its modern isolation from Australia (ca. 1350 km) was reduced greatly during the last glacial maximum (22 to 19 ka), when a series of small to very large islands in the Chesterfield Reef region allowed island hopping between Australia and New Caledonia without crossing a water gap larger than ca. 100 km. The deep ocean between New Caledonia and Vanuatu, on the other hand, has been a substantial barrier to dispersal regardless of sea level.

Highly endemic at both the generic and species levels, the New Caledonian avifauna is the most taxonomically diverse in Remote Oceania (Table 5-13). Kagus (*Rhynochetos jubatus*, *R. orarius*) are the only endemic family of birds in Oceania outside of New Zealand. Resident species of buttonquail (*Turnix*), thick-knees (*Esacus*), snipes (?*Coenocorypha*), cockatoos (*Cacatua*), strigid owls (*Ninox*), owlet-nightjars (*Aegotheles*), nightjars (*Eurostopodus*), hornbills (*Aceros*), and crows (*Corvus*) are or were found in Remote Melanesia only in New Caledonia (Tables 5-2, 5-13).

Bird bones from three caves on Grande Terre and Ile des Pins represent 31 species of nonpasserine landbirds, 14 of which are either extinct or no longer occur on these islands (Balouet & Olson 1989). The abundant passerine fossils have not been studied. Nonpasserine fossils from the three mainly or exclusively precultural sites consist of 96–100% extinct or extirpated species (Table 5-14). I have examined much smaller bone assemblages from the Lapita site (the "type" site for Lapita-style pottery) and immediately post-Lapita archaeological sites on Grande Terre, Ile des Pins, 'Uvea, Lifu, and Mare (Steadman 1997a; Table 5-14 herein). The chronology and ceramic sequence of these and similar New Caledonian sites were reviewed by Sand (1999, 2001a, b, c, Sand et al. 2001, 2002). The timing of human arrival is independently substantiated by deforestation signals between 3000 and 2500 cal BP in sediment and pollen data from two lakes on Grande Terre (Stevenson 1999).

Avian fossils disclose much more extinction in New Caledonia than Vanuatu. Renowned for its endemic modern avifauna (Mayr 1945, Delacour 1966), the birdlife of New Caledonia had even more endemic species at human contact. As more fossils are studied, I believe that the avifaunas of Grande Terre, Ile de Pins, and the Loyalty Islands will be found to have been much more similar in the past than today, because of higher levels of extinction on the smaller islands.

The two extinct species of *Accipiter* mean that New Caledonia once had four sympatric congeneric species of hawks, unmatched in Oceania except on New Britain with five (Table 5-3). All four species of New Caledonian megapodes are extinct (Table 5-13, Chapter 9), as are both flightless rails (*Gallirallus lafresnayanus*, *Porphyrio kukwiedei*), although the former survived into historic times. Grande Terre and Ile des Pins shared three of the four megapodes. With an improved fossil record, all four species may prove to have lived on both islands. The Loyalty Islands probably

Table 5-13 Landbirds of New Caledonia. The Loyalty Islands are 'Uvea, Lifu, Tiga, and Mare. H, historic record; M, modern record; P, prehistoric record; †, extinct species; e, endemic subspecies; E, endemic species; EE, endemic genus; ?, questionable record, or residency not established. Passerine fossils from New Caledonia have not been studied. The multiple conspecific subspecies in the Loyalty Islands are allopatric. The data for modern birds on Ile des Pins are highly suspect. I have been unable to find a good list of the species occurring there; most authors have not consistently distinguished Ile des Pins from Grande Terre. The range maps in Doughty et al. (1999) are unlabeled, extremely small, and when interpretable, often inaccurate. [a]approximate combined area of Lifu (1150 km^2), Mare (650 km^2), Uvea (160 km^2), and Tiga (10 km^2). From Mayr (1940b, 1945), Hannecart & Letocart (1980, 1983), Ross (1988), Balouet & Olson (1989), and Steadman (1997a, unpub. data).

	Grande Terre	Ile des Pins	Loyalty Islands
GREBES			
Tachybaptus novaehollandiae leucosternos	M	—	—
CORMORANTS			
Phalacrocorax m. melanoleucos	M	—	—
HERONS			
Egretta novaehollandiae nana (e)	M	—	M
Egretta sacra albolineata (e)	M, P	M	M
Butorides striatus macrorhynchos	M	—	M
Nycticorax c. caledonicus (e)	M	—	—
Botaurus poiciloptilus	M	—	M
DUCKS			
Dendrocygna arcuata australis	H	—	—
Anas gracilis subsp.?	M, P	—	—
Anas superciliosa pelewensis	M	M?	M
Aythya australis australis	M	—	—
OSPREYS			
Pandion haliaetus melvillensis	M	M?	M
HAWKS			
Haliastur sphenurus	M	—	—
Circus a. approximans	M	M?	M
Accipiter fasciatus vigilax	M	—	M
Accipiter haplochrous (E)	M, P	—	—
†*Accipiter quartus* (E)	P	—	—
†*Accipiter efficax* (E)	P	—	—
FALCONS			
Falco peregrinus nesiotes	M, P	M?	M
MEGAPODES			
†*Megapodius molistructor*	P	P	—
†*Megapodius* undescribed sp. D (E)	P	P	—
†*Megapodius* undescribed sp. E (E)	—	—	P
†*Sylviornis neocaledoniae* (EE)	P	P	—
BUTTON-QUAILS			
Turnix [varia] novaehollandiae (E/e)	M, P	P	—
RAILS			
Gallirallus philippensis swindellsi (e)	M, P	M	M, P
†*Gallirallus lafresnayanus* (E)	H, P	P	—
Porzana t. tabuensis	M, P	—	M, P
Poliolimnas cinereus	M	—	M
Porphyrio porphyrio samoensis	M, P	M?	M, P
†*Porphyrio kukwiedei* (E)	P	P	—
Gallinula tenebrosa subsp.?	M, P	—	—
KAGUS			
Rhynochetos jubatus (EE)	M	—	—
†*Rhynochetos orarius* (EE)	P	P	—
THICK-KNEES			
Esacus magnirostris	M	—	—
SNIPES			
†*Coenocorypha*? undescribed sp. (E)	P	—	—
PIGEONS, DOVES			
Columba vitiensis hypoenochroa (e)	M, P	M	M, P
Ptilinopus greyii	M	M	M, P
Drepanoptila holosericea (EE)	M, P	M, P	—
Ducula p. pacifica	P	M?, P	M, P
Ducula goliath (E)	M, P	M	P
Chalcophaps indica chrysochlora/sandwichensis	M, P	M?, P	M
†*Caloenas canacorum*	P	—	—
Gallicolumba cf. *beccarii*	—	—	P
†*Gallicolumba longitarsus* (E)	P	P	—
PARROTS			
†*Cacatua* undescribed sp. (E)	P	—	—
Trichoglossus haematodus deplanchii (e)	M	M	M
†*Charmosyna diadema* (E)	H	—	—
Eunymphicus c. cornutus (EE)	M	—	—
Eunymphicus cornutus uveaensis (EE)	—	—	M
Cyanoramphus novaezelandiae saisseti (e)	M	—	—
CUCKOOS			
Cacomantis flabelliformis pyrrhophanus (e)	M	M?	M
Chrysococcyx lucidus layardi	M	M?	M
BARN-OWLS			
Tyto longimembris longimembris/oustaleti	M	—	—
Tyto alba lulu	M, P	M	M, P
†*Tyto? letocarti*	P	—	—
STRIGID OWLS			
Ninox cf. *novaeseelandiae* (e?)	P	—	—

Table 5-13 (continued)

	Grande Terre	Ile des Pins	Loyalty Islands
OWLET-NIGHTJARS			
†*Aegotheles savesi* (E)	H, P	—	—
NIGHTJARS			
Eurostopodus mysticalis	H	—	—
SWIFTS			
Collocalia esculenta albidior (e)	M, P	M?	M
Collocalia spodiopygia leucopygia (e)	M, P	M?	M
KINGFISHERS			
Halcyon sancta canacorum (e)	M, P	M?	—
Halcyon sancta macmilliani (e)	—	—	M
HORNBILLS			
†*Aceros* undescribed sp. (E)	—	—	P
SWALLOWS			
Hirundo tahitica subfusca	M	M?	M
THRUSHES			
Turdus poliocephalus xanthops (e)	M	—	—
Turdus poliocephalus mareensis (e)	—	—	M
Turdus poliocephalus pritzbueri (e)	—	—	M
WARBLERS			
Cichlornis mariei (E)	M	—	—
WHITE-EYES			
Zosterops inornatus (E)	—	—	M
Zosterops xanthochroa (E)	M	M?	M
Zosterops minutus (E)	—	—	M
Zosterops lateralis griseonota (e)	M	—	—
Zosterops lateralis nigrescens (e)	—	—	M
Zosterops lateralis melanops (e)	—	—	M
STARLINGS			
Aplonis striata striata (e)	M	M?	—
Aplonis striata atronitens (e)	—	—	M
ESTRILDID FINCHES			
Erythrura trichroa cyaneifrons	—	—	M
Erythrura psittacea (E)	M	M?	—
CUCKOO-SHRIKES, TRILLERS			
Coracina caledonica caledonica (e)	M	M?	—
Coracina caledonica lifuensis (e)	—	—	M
Coracina analis (E)	M	—	—
Lalage leucopyga montrosieri (e)	M	M?	—
Lalage leucopyga simillima	—	—	M
GERYGONES			
Gerygone f. flavolateralis (e)	M	M?	M
ROBINS			
Eopsaltria flaviventris (E)	M	M?	—
WHISTLERS			
Pachycephala pectoralis littayei (e)	—	—	M
Pachycephala caledonica (E)	M	M	—
Pachycephala rufiventris xanthetraea (e)	M	M?	—
FANTAILS			
Rhipidura fuliginosa bulgeri (e)	M	M?	M
Rhipidura spilodera verreauxi (e)	M	M?	M
MONARCHS			
Clytorhynchus p. pachycephaloides (e)	M	—	—
Myiagra caledonica caledonica (e)	M	M?	—
Myiagra caledonica melanura	—	—	M
Myiagra caledonica viridinitens (e)	—	—	M
HONEYEATERS			
Lichmera incana incana (e)	M	M?	—
Lichmera incana poliotis (e)	—	—	M
Lichmera incana mareensis (e)	—	—	M
Myzomela caledonica (e)	M	M	—
Myzomela cardinalis lifuensis (e)	—	—	M
Philemon diemenensis (E)	M	M?	M
Gymnomyza aubryana (E)	M	—	—
Phylidonyris undulata (E)	M	M?	—
WOOD-SWALLOWS			
Artamus leucorhynchus melaleucus (e)	M	M?	M
CROWS			
Corvus moneduloides (E)	M	M?	M?
TOTAL			
M	64	10–39	46–47
M + H	68	10–39	46–47
P	33	11	11
M + H + P	82	20–47	50–51
Area (km^2)	16,750	134	1970[a]
Elevation (m)	1628	266	188
Number of bones (nonpasserines only)	5313	999	141

Table 5-14 Nonpasserine landbirds from prehistoric sites in New Caledonia. In the Loyalty Islands, sites LTA037 and LTA042 are on Mare, sites LWT008 (Hnajoisisi) and LWT054 (Keny) are on Lifu, and sites LUV029 and LUV030 are on Uvea (Sand 2002). †, extinct species; *, extirpated population. Data for Pindai Cave, Gilles Cave, and Kanumera are from Balouet & Olson (1989). Data for the Lapita site are from Steadman (1997a). Data for misc. sites" and the six Loyalty Islands sites are from Steadman (unpub. data).

	Grande Terre					Ile des Pins		Loyalty Islands		
	Pindai Cave	Upper Gilles Cave	Lower Gilles Cave	Lapita	Misc. sites	Kanumera	Vatcha Chi	LTA037 LTA042	LWT008 LWT054	LUV029 LUV030
HERONS										
Egretta sacra	—	—	—	—	2	—	—	—	—	—
DUCKS										
Anas gracilis	17	—	—	—	—	—	—	—	—	—
HAWKS										
Accipiter haplochrous	—	—	—	—	2	—	—	—	—	—
†*Accipiter quartus*	3	—	—	—	—	—	—	—	—	—
†*Accipiter efficax*	19	—	—	—	—	—	—	—	—	—
FALCONS										
Falco peregrinus	2	—	—	—	—	—	—	—	—	—
MEGAPODES										
†*Megapodius molistructor*	6	—	—	—	—	9	3	—	—	—
†*Megapodius* undescribed sp. D	—	—	—	1	—	—	22	—	—	—
†*Megapodius* undescribed sp. E	—	—	—	—	—	—	—	3	1	—
†*Sylviornis neocaledoniae*	4263	45	22	1	—	617	—	—	—	—
BUTTON–QUAILS										
Turnix [varia] novaehollandiae	7*	4*	84*	—	—	1*	—	—	—	—
RAILS										
†*Gallirallus lafresnayanus*	13	3	1	—	—	238	—	—	—	—
Gallirallus philippensis	3	—	81	—	6	—	—	32	—	—
Porzana tabuensis	—	—	5	—	—	—	—	1	—	—
Gallinula tenebrosa	3	—	—	—	—	—	—	—	—	—
Porphyrio porphyrio	15	—	—	—	19	—	—	79	—	2
†*Porphyrio kukwiedei*	15	—	1	—	—	12	—	—	—	—
KAGU										
†*Rhynochetos orarius*	216	—	—	—	—	70	—	—	—	—
SNIPES										
†*Coenocorypha?* undescribed sp.	—	3	—	—	—	—	—	—	—	—
PIGEONS, DOVES										
Columba vitiensis	1	—	5	1	—	—	—	3	3	—
Ptilinopus greyii	—	—	—	—	—	—	—	—	1	—
Drepanoptila holosericea	1	—	—	—	—	—	20	—	—	—
Ducula pacifica	—	—	—	1*	—	—	2	2	1	—
Ducula goliath	3	—	1	2	—	—	1	—	2*	—
Chalcophaps indica	—	—	1	1	—	—	1	—	—	—
†*Caloenas canacorum*	8	—	—	—	—	—	—	—	—	—
Gallicolumba cf. *beccarii*	—	—	—	—	—	—	—	—	1*	—
†*Gallicolumba longitarsus*	6	—	—	—	—	—	3	—	—	—
PARROTS										
†*Cacatua* undescribed sp.	—	—	—	1	—	—	—	—	—	—

Table 5-14 (continued)

	Grande Terre					Ile des Pins		Loyalty Islands		
	Pindai Cave	Upper Gilles Cave	Lower Gilles Cave	Lapita	Misc. sites	Kanumera	Vatcha Chi	LTA037 LTA042	LWT008 LWT054	LUV029 LUV030
BARN-OWLS										
Tyto alba	4	—	186	—	8	—	—	8	—	—
†*Tyto? letocarti*	—	7	—	—	—	—	—	—	—	—
STRIGID OWLS										
Ninox cf. *novaeseelandiae*	3*	—	1*	—	—	—	—	—	—	—
OWLET-NIGHTJARS										
†*Aegotheles savesi*	—	3	—	—	—	—	—	—	—	—
SWIFTS										
Collocalia esculenta	3	—	—	—	—	—	—	—	—	—
Collocalia spodiopygia	202	—	—	—	—	—	—	—	—	—
KINGFISHERS										
Halcyon sancta	—	—	2	—	—	—	—	—	—	—
HORNBILLS										
†*Aceros* undescribed sp.	—	—	—	—	—	—	—	—	2	—
TOTAL BONES										
All species	4813	65	390	8	37	947	52	128	11	2
†/*	4559	65	109	4	0	947	28	3	6	0
%†/*	95	100	28	50	0	100	54	2	55	0
TOTAL SPECIES										
All	22	6	12	7	5	6	7	7	7	1
†/*	11	6	5	4	0	6	3	1	4	0
%†/*	50	100	42	57	0	100	43	14	57	0

sustained more than one species of megapode. The extinct kagu *Rhynochetos orarius* was larger than its living congener (Balouet & Olson 1989; Chapter 13 herein). The extinct snipe (cf. *Coenocorypha* undescribed sp.) is poorly known.

Abundant bones of pigeons and doves on Grande Terre feature extinct species of *Caloenas* and *Gallicolumba*. A tibiotarsus of *Cacatua* undescribed sp. from the Lapita site on Grande Terre is the only record of a cockatoo east of the Solomon Main Chain. The tibiotarsus of the Grande Terre species is larger than that of *C. ducorpsi* (Solomons) but similar in size to that of *C.* [*galerita*] *ophthalmica* of New Britain, the two extant species of *Cacatua* in Oceania. I suspect that a species of *Eclectus* also lived in New Caledonia, given the former presence of these parrots in Vanuatu (above) and Tonga (Chapters 6, 12).

New Caledonia once sustained the most diverse set of nightbirds (Tytonidae, Strigidae, Caprimulgidae, Aegothelidae) in Remote Oceania, although few of these survive. While having three species of *Tyto* is unparalleled in Oceania, the barn-owl *T. alba* may have colonized New Caledonia only after people and rats arrived (Balouet & Olson 1989). The extinct hornbill (*Aceros* undescribed sp.) from Lifu (Loyalty Islands) must have lived also on Grande Terre and Ile de Pins. The only hornbill now in Oceania is the much larger *Aceros* (*Rhyticeros*) *plicatus* in the Bismarcks and Solomon Main Chain. We should look for prehistoric bones of hornbills in Vanuatu as well.

Finally, from the isolated raised limestone island of Walpole (135 km SE of Mare, 140 km E of Ile des Pins), a number of unidentified bird bones were recovered from archaeological sites by Sand (2002).

Chapter 6 West Polynesia

West Polynesia includes Fiji, Tonga, Samoa, and outlying Rotuma, Wallis & Futuna, and Niue (Figure 1-3). Tokelau and Tuvalu are remote groups of atolls that are culturally West Polynesian but have very depauperate avifaunas that are covered with those of other atolls in Chapter 8. Fiji, which is culturally intermediate between Polynesia and Melanesia (Chapter 4), has a landbird fauna more similar to those of Tonga and Samoa than to those of any Melanesian island group to the west.

The West Polynesian avifauna is defined by the current or former presence on most islands of these living species (* = endemic to the region): the rails *Gallirallus philippensis, Porzana tabuensis*, and *Porphyrio porphyrio*, columbids **Gallicolumba stairi*,* *Ptilinopus perousii, P. porphyraceus, Ducula pacifica*, and **D. latrans,* parrots **Vini solitarius* and **V. australis*, barn-owl *Tyto alba,* swift *Collocalia spodiopygia*, kingfisher *Halcyon chloris*, triller **Lalage maculosa*, shrikebill **Clytorhynchus vitiensis*, honeyeater **Foulehaio carunculata*, and starling *Aplonis tabuensis.* Nine of these species inhabit islands west of West Polynesia, but only two of them (*Porzana tabuensis, Ducula pacifica*) live to the east. In fact, the only landbirds shared by West and East Polynesia, today or in the past, are four trampy species, the heron *Egretta sacra*, duck *Anas superciliosa*, rail *Porzana tabuensis*, and pigeon *Ducula pacifica.* The first three of these are primarily aquatic species rather than forest dwellers.

The distinction between West Polynesian and East Polynesian landbirds is reinforced when prehistoric species are included (Steadman 1993a); not a single extinct species is shared by the two regions. The break in landbird faunas between West and East Polynesia is abrupt at higher taxonomic levels as well, with 38 of West Polynesia's 54 genera (70%) and 14 of its 25 families (56%) not occurring in East Polynesia (Table 6-1). This distinctiveness is not confined to the large, western islands of Fiji; among the taxa unknown in East Polynesia, only 10 of the 38 genera and two of the 14 families (grebes, woodswallows) occur in West Polynesia solely in Fiji. Just three of these 10 genera are found as well in Vanuatu (Chapter 5). Only two (*Hirundo, Aplonis*) of West Polynesia's 20 passerine genera occur in East Polynesia.

Fiji

The largest archipelago in West Polynesia is Fiji (Figure 6-1), with a land area of 18,272 km^2 and ca. 106 islands >1 km^2 by my assessment of Derrick (1965) and Fiji Islands (1980). Derrick (1965:3) listed 520 named Fijian islands and islets, 322 large enough for habitation and 106 actually inhabited. Fiji is also the most geologically complex island group in West Polynesia, having various combinations of sedimentary, metamorphic, igneous, and volcanic rocks (Rodda 1994; Chapter 1 herein). Viti Levu, the region's oldest island, may date to the late Eocene (Hathway & Colley 1994). Land probably has been continuously emergent above the sea in Fiji for longer than in any other island group in West Polynesia, East Polynesia, or Micronesia. Only the Gondwanan fragments in Oceania (New Caledonia, New Zealand) are older.

Table 6-1 Distribution of genera of landbirds in West Polynesia (Fiji, Tonga, Samoa, and outliers). H, historic record, now extirpated; M, modern record; P, prehistoric record; e, endemic to West Polynesia; *, genus absent as residents in East Polynesia; **, family absent as residents in East Polynesia. Outliers = Rotuma, Wallis & Futuna, and Niue. Modern records are mainly from Watling (2001). Prehistoric records are from many references cited in the text as well as my previously unpublished data; fossil passerines are unstudied in Fiji.

	Fiji	Tonga	Samoa	Outliers
**GREBES				
*Tachybaptus	P	—	—	—
HERONS				
Egretta	M, P	M, P	M	M
Butorides	M, P	—	—	—
Nycticorax	—	P	—	P
DUCKS				
Dendrocygna	H	—	—	—
Anas	M, P	M, P	M	M
**OSPREYS				
*Pandion	—	P	—	—
**HAWKS				
*Circus	M	M	—	—
*Accipiter	M	P	—	—
**FALCONS				
*Falco	M	—	M	—
**MEGAPODES				
*Megapodius	P	M, P	P	P
*Megavitiornis (e)	P	—	—	—
RAILS				
Gallirallus	M, P	M, P	M, P	M, P
*Vitirallus (e)	P	—	—	—
Porzana	M, P	M, P	M, P	M, P
*Poliolimnas	M, P	—	M	—
*Pareudiastes	—	—	H	—
Porphyrio	M, P	M, P	M	M
PIGEONS				
*Columba	M, P	—	M	—
Ptilinopus	M, P	M, P	M	M
Ducula	M, P	M, P	M	M, P
*Undescribed genus C (e)	—	P	—	—
*Caloenas	—	P	—	—
*Didunculus (e)	—	P	M	—
Gallicolumba	M, P	M, P	M, P	M
*Natunaornis (e)	P	—	—	—
PARROTS				
Vini	M	M, P	M	M
*Charmosyna	M	—	—	—
*Prosopeia (e)	M	—	—	—
*Eclectus	—	P	—	—
**CUCKOOS				
*Cacomantis	M	P	—	—
**BARN-OWLS				
*Tyto	M, P	M, P	M, P	M, P
SWIFTS				
Collocalia	M, P	M, P	M, P	M, P
KINGFISHERS				
Halcyon	M, P	M, P	M	M
SWALLOWS				
Hirundo	M	M	—	—
**THRUSHES				
*Turdus	M	P	M	—
WARBLERS				
*Cettia	M	P	—	—
*Trichocichla (e)	M	—	—	—
**WHITE-EYES				
*Zosterops	M	—	M	—
*Undescribed genus (e)	—	P	—	—
STARLINGS				
Aplonis	M, P	M, P	M, P	M, P
**ESTRILDID FINCHES				
*Erythrura	M	—	M	—
**TRILLERS				
*Lalage	M	M, P	M	M
**ROBINS				
*Petroica	M	P	M	—
WHISTLERS				
*Pachycephala	M	M, P	M	—
FANTAILS				
*Rhipidura	M	—	M	—
MONARCHS				
*Mayrornis	M	—	—	—
*Clytorhynchus	M	M, P	M	M
*Myiagra	M	P	M	—
*Lamprolia (e)	M	—	—	—
**HONEYEATERS				
*Myzomela	M	P	M	M
*Foulehaio (e)	M	M, P	M, P	M
*Gymnomyza	M	—	M	—
**WOOD-SWALLOWS				
*Artamus	M	—	—	—
TOTAL				
M	40	20	29	17
M + H	41	20	30	17
M + H + P	46	34	31	19

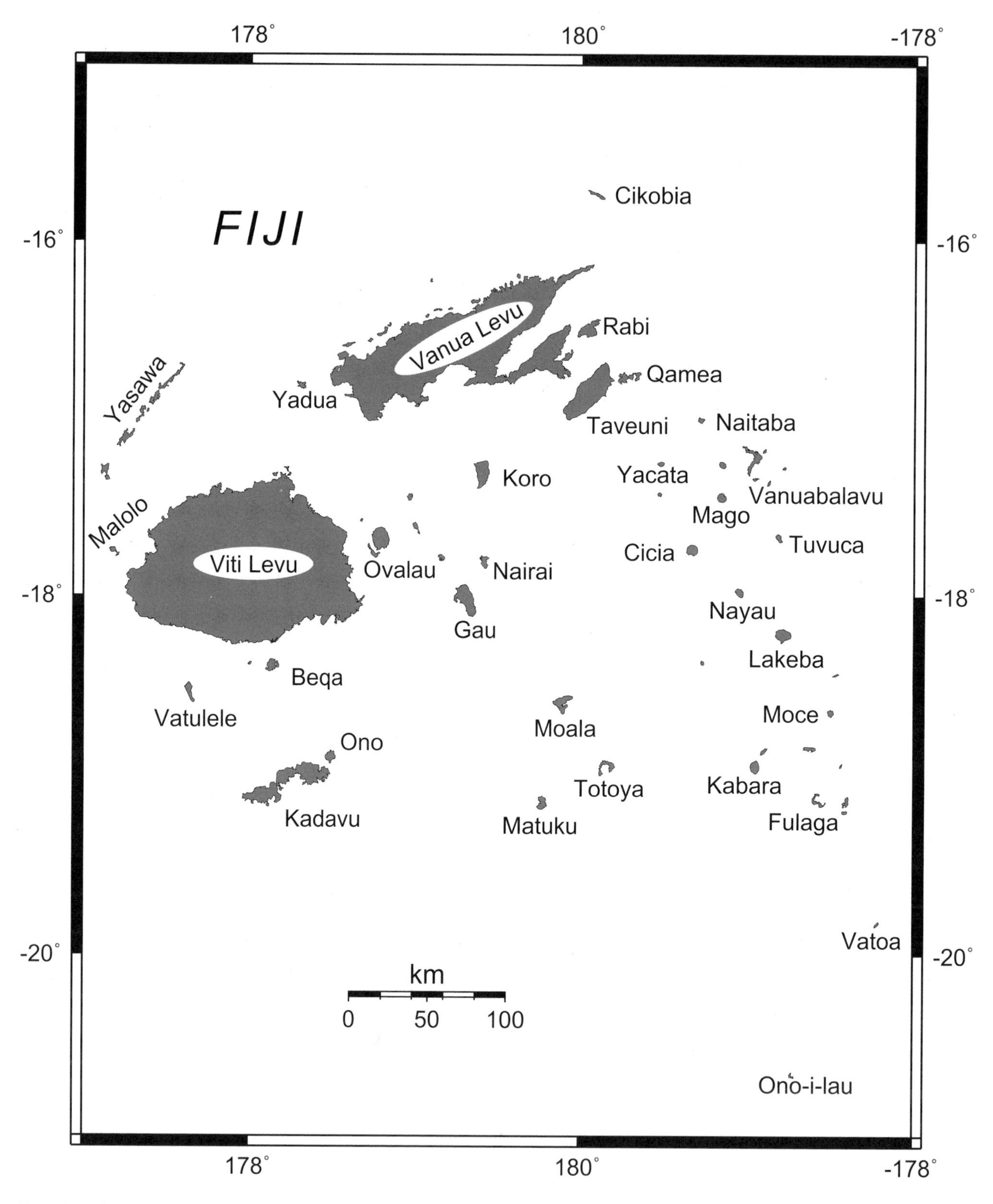

Figure 6-1 Fiji.

Table 6-2 Distribution of landbirds on major islands and island clusters in Fiji. [e], endemic to Fiji; H, historic record of extirpated population; M, modern record; P, prehistoric record; † extinct species; *54 islands combined; ?, questionable record, or residency not established. Fossil passerines are unstudied. From Watling (1978, 1982, 2001), Clunie (1984), Pratt et al. (1987), Worthy et al. (1999), Worthy (2000, 2001a,b), Steadman & Franklin (2000), Steadman (herein).

	Viti Levu	Vanua Levu	Taveuni	Kadavu	Gau	Ovalau	Lau Group*	Rotuma
GREBES								
Tachybaptus novaehollandiae	P	—	—	—	—	—	—	—
HERONS								
Egretta novaehollandiae	M	—	—	—	—	—	—	—
Egretta sacra	M, P	M	M	M	M	M	M, P	M?
Butorides striatus	M	M	M	M	M	M	M, P	—
DUCKS								
Dendrocygna arcuata	H	—	—	—	—	—	—	—
Anas superciliosa	M, P	M	M	M	M	M	M, P	—
HAWKS								
Circus approximans	M	M	M	M	M	M	M	—
Accipiter rufitorques	M	M	M	M	M	M	—	—
FALCONS								
Falco peregrinus	M	M	M	—	M	M	M	—
MEGAPODES								
[e]†*Megapodius alimentum*	—	—	—	—	—	—	P	—
[e]†*Megapodius amissus*	P	—	—	—	—	—	—	—
[e]†*Megapodius amissus/molistructor*	—	—	—	—	—	—	P	—
[e]†*Megavitiornis altirostris*	P	—	—	—	—	—	—	—
RAILS								
Gallirallus philippensis	H, P	H	M	M	M	M	M, P	M
[e]†*Gallirallus poecilopterus*	H	—	—	—	—	H	—	—
[e]†*Gallirallus* undescribed sp. D	—	—	—	—	—	—	P	—
[e]†*Vitirallus watlingi*	P	—	—	—	—	—	—	—
Porzana tabuensis	M, P	M	M	M	M	M	M, P	—
[e]†*Porzana* undescribed sp. B	—	—	—	—	—	—	P	—
Poliolimnas cinereus	M	M	M	M	M	M	P	—
Porphyrio porphyrio	H, P	H	M	M	M	M	M, P	M
PIGEONS, DOVES								
Columba vitiensis	M	M	M	M	M	M	M, P	M
Ptilinopus perousii	M	M	M	M	M	M	M, P	—
Ptilinopus porphyraceus	—	—	—	—	—	—	M, P	M
[e]*Ptilinopus victor*	—	M	M	—	—	—	—	—
[e]*Ptilinopus luteovirens*	M	—	—	—	M	M	—	—
[e]*Ptilinopus layardi*	—	—	—	M	—	—	—	—
Ducula pacifica	—	—	—	—	M	M	M, P	M
[e]†*Ducula lakeba*	P	—	—	—	—	—	P	—
Ducula latrans	M	M	M	M	M	M	M, P	—
Gallicolumba stairi	M	M	M	M	M	M	M, P	—
[e]†*Natunaornis gigoura*	P	—	—	—	—	—	—	—

(*continued*)

Table 6-2 (continued)

	Viti Levu	Vanua Levu	Taveuni	Kadavu	Gau	Ovalau	Lau Group*	Rotuma
PARROTS								
Vini solitarius	M	M	M	M	M	M	M	—
Vini australis	—	—	—	—	—	—	M	—
[e]*Charmosyna amabilis*	M	M	M	—	—	H	—	—
[e]*Prosopeia tabuensis*	—	M	M	M	M	—	—	—
[e]*Prosopeia personata*	M	—	—	—	—	—	—	—
CUCKOOS								
Cacomantis flabelliformis	M	M	M	M	M	M	—	—
BARN-OWLS								
Tyto longimembris	H	—	—	—	—	—	—	—
Tyto alba	M, P	M	M	M	M	M	M, P	M
SWIFTS								
Collocalia spodiopygia	M	M	M	M	M	M	M, P	—
KINGFISHERS								
Halcyon chloris	M	M	M	M	M	M	M, P	—
SWALLOWS								
Hirundo tahitica	M	M	M	M	M	M	M	—
THRUSHES								
Turdus poliocephalus	M	M	M	M	M	M	—	—
WARBLERS								
[e]*Cettia ruficapilla*	M	M	M	M	—	—	—	—
[e]*Trichocichla rufa*	M	M	—	—	—	—	—	—
WHITE-EYES								
[e]*Zosterops explorator*	M	M	M	M	—	M	—	—
Zosterops lateralis	M	M	M	M	M	M	—	—
STARLINGS								
Aplonis tabuensis	M	M	M	M	M	M, P	M	M
ESTRILDID FINCHES								
Erythrura cyaneovirens	M	M	M	M	—	—	—	—
[e]*Erythrura kleinschmidti*	M	—	—	—	—	—	—	—
TRILLERS								
Lalage maculosa	M	M	M	M	M	M	M	M
ROBINS								
Petroica multicolor	M	M	M	M	—	—	—	—
WHISTLERS								
Pachycephala pectoralis	M	M	M	M	M	M	M	—
FANTAILS								
Rhipidura spilodera	M	M	M	—	—	M	—	—
[e]*Rhipidura personata*	—	—	—	M	—	—	—	—

Table 6-2 (continued)

	Viti Levu	Vanua Levu	Taveuni	Kadavu	Gau	Ovalau	Lau Group*	Rotuma
MONARCHS								
[e]*Mayrornis versicolor*	—	—	—	—	—	—	M	—
[e]*Mayrornis lessoni*	M	M	M	M	M	M	M	—
Clytorhynchus vitiensis	M	M	M	M	M	M	M	M
[e]*Clytorhynchus nigrogularis*	M	M	M	M	—	M	—	—
Myiagra vanikorensis	M	M	M	M	M	M	M	—
[e]*Myiagra azureocapilla*	M	M	M	—	—	—	—	—
[e]*Lamprolia victoriae*	—	M	M	—	—	—	—	—
HONEYEATERS								
[e]*Myzomela chermesina*	—	—	—	—	—	—	—	M
[e]*Myzomela jugularis*	M	M	M	M	M	M	M	—
Foulehaio carunculata	M	M	M	—	M	M	M	—
[e]*Foulehaio provocator*[1]	—	—	—	M	—	—	—	—
[e]*Gymnomyza viridis*	M	M	M	—	—	—	—	—
WOOD-SWALLOWS								
Artamus leucorhynchus[2]	M	M	M	—	M	—	—	—
TOTAL								
M	42	42	43	37	34	35	29	10–11
M + H	47	44	43	37	34	37	29	10–11
M + H + P	52	44	43	37	34	37	35	10–11
Area (km^2)	10,384	5535	435	407	140	100	376*	43
Elevation	1323	1032	1241	805	715	626	314	262
Total landbird bones	?	0	0	0	0	0	338	0

1. Listed as the endemic *A. mentalis* in Watling (2001).
2. Listed in the endemic genus *Xanthotis* in Watling (2001).

The Fijian landbird fauna (Table 6-2) is West Polynesia's richest. As the fossil record grows, the increase in species richness may be substantial. More species (47) have been recorded in historic times on huge Viti Levu (10,387 km^2) than on any other island in West Polynesia. At least five of these species (a duck, three rails, and an owl) have not withstood the past 200 years of habitat loss and predation from rats, cats, dogs, mongoose, pigs, and people. Six other species (a grebe, two megapodes, a rail, and two pigeons) are known on Viti Levu only from prehistoric bones (Worthy et al. 1999; Worthy 2000, 2001b; Table 6-2 herein). These are some of Oceania's most spectacular extinct birds, such as the large megapode *Megavitiornis altirostris* (also recorded from the nearby small island of Naigani), the rail *Vitirallus watlingi*, and the giant pigeon, *Natunaornis gigoura*. Knowing whether or not these distinctive species (or congeners) occurred outside of Viti Levu awaits improvement of Fiji's fossil record. Either way, Viti Levu's substantial geological age at least partly explains why such distinctive forms could have evolved there.

Seven of the 10 genera believed to be endemic to West Polynesia are restricted to Fiji (Table 6-1). The other three are pigeons (*Didunculus* from Tonga and Samoa and an undescribed Tongan genus) and an undescribed genus of Tongan white-eye. Their absences from Fiji may only reflect inadequate sampling; Fijian fossil birds are known thus far only on Viti Levu, Beqa, Waya, Naigani, and four islands in Lau. Excluding flightless rails, most of the genera and species endemic to Fiji are confined, as far as we know, to one or more of the six largest islands in western Fiji (Viti Levu through Ovalau in Table 6-2), sometimes also including nearby smaller islands. Except for apparent superspecies such as the doves *Ptilinopus* (*Chrysoenas*) *victor-luteovirens-layardi*, parrots *Prosopeia tabuensis-personata*,

Table 6-3 Bird bones from post-Lapita archaeological deposits at Sigatoka Sand Dunes (Area A), Viti Levu. Column headings represent depths in cm below surface. Excavations by D. V. Burley, June 2000. Identifications by DWS.

	Surface	0–10	10–20	20–30	30–40	40–50	TOTAL
GREBES							
Tachybaptus novaehollandiae	—	1	—	—	—	—	1
HERONS							
Egretta sacra	1	—	—	—	—	—	1
DUCKS							
Anas superciliosa	—	1	—	1	1	—	3
RAILS							
Porzana tabuensis	—	—	—	—	—	1	1
Gallirallus philippensis	1	—	—	—	—	—	1
Porphyrio porphyrio	—	1	—	1	—	—	2
BARN-OWLS							
Tyto alba	—	1	—	—	—	—	1
TOTAL	2	4	0	2	1	1	10

fantails *Rhipidura spilodera-personata*, and honeyeaters *Foulehaio carunculata-provocator*, I believe that many absences of volant species on any of the six largest islands are due to human impact. Examples would be the absence of *Charmosyna amabilis* on Kadavu and Gau, and of *Cettia ruficapilla* on Gau and Ovalau. As the fossil data improve, more of the volant western Fijian endemic species probably will be found to have lived in the Lau Group as well.

Fiji also sustains Remote Oceania's only surviving sympatric, congeneric species triplet, namely the fruit-doves *Ptilinopus perousii*, *P. porphyraceus*, and *P. victor*), on the small (12.2 km^2), high (880 m) island of Laucala, off Taveuni (Watling 1989). Fiji also has or had more sympatric, congeneric species pairs than elsewhere in Polynesia, involving megapodes (*Megapodius*), pigeons (*Ducula*), fruit-doves (*Ptilinopus*), barn-owls (*Tyto*), monarchs (*Clytorhynchus*, *Myiagra*), white-eyes (*Zosterops*), and parrotfinches (*Erythrura*).

From post-Lapita cultural deposits excavated by David Burley in the Sigatoka Sand Dunes, southwestern Viti Levu, I identified bones of seven species of birds (Table 6-3). Among these is a tibiotarsus of the grebe *Tachybaptus novaehollandiae*, an inhabitant of freshwater lakes and marshes that occurs locally today only as far east as Vanuatu. Bones of a large, extinct species of *Ducula*, similar in size to *D. lakeba* of the Lau Group (see below, and Chapter 11), have been found on Viti Levu and nearby Beqa (Worthy 2001b).

From Qara ni Cagi Rockshelter on the small island of Waya in the Yasawa Group (Hunt et al. 1999), I identified 12 bird bones from Lapita contexts that represent the tern *Sterna lunata* (1), rails *Gallirallus philippensis* (1) and *Porzana tabuensis* (1), pigeons *Columba vitiensis* (1) and *Ducula latrans* (4), and nonnative chicken *Gallus gallus* (4). Watling (1985, 2001) mentions no modern records from Waya for *P. tabuensis*, *C. vitiensis*, or *D. latrans*.

Excluding the eroded volcanic trio of Moala, Totoya, and Matuku (geologically independent from other Lauan islands), the Lau Group in Eastern Fiji consists of 54 named islands (35 of them >1 km^2) totaling 376 km^2 (Tables 6-4, 6-5). Today, Lau (Figure 6-1, 6-2) lacks most of the species of landbirds found on the six largest, western Fijian islands (Watling 1982, 2001; Table 6-2 herein). Only two (*Mayrornis lessoni*, *Myzomela jugularis*) of the 21 species still found on Lakeba, the largest island in Lau (55.9 km^2) and the 10th largest in Fiji, are endemic to Fiji. These values compare with 14 of 47 modern species on Viti Levu and 12 of 43 on Taveuni (Table 6-2). Some of the species formerly regarded as Fijian endemics, such as *Accipiter rufitorques*, *Ducula latrans*, and *Vini solitarius*, have been found prehistoricaly in Tonga (see below). *Mayrornis versicolor*, a monarch confined to Ogea Levu, is the only extant species of landbird endemic to Lau.

Birds of the Lau Group have been, until recently, perhaps the most poorly known in West Polynesia. Mentioned in a very general way in review works such as Watling (1982, 1985, 2001), Clunie (1984), and Pratt et al. (1987), nothing had been reported about their relative abundance or habitat relationships until Janet Franklin and I began point-count surveys on Lakeba in May 1999 (Steadman &

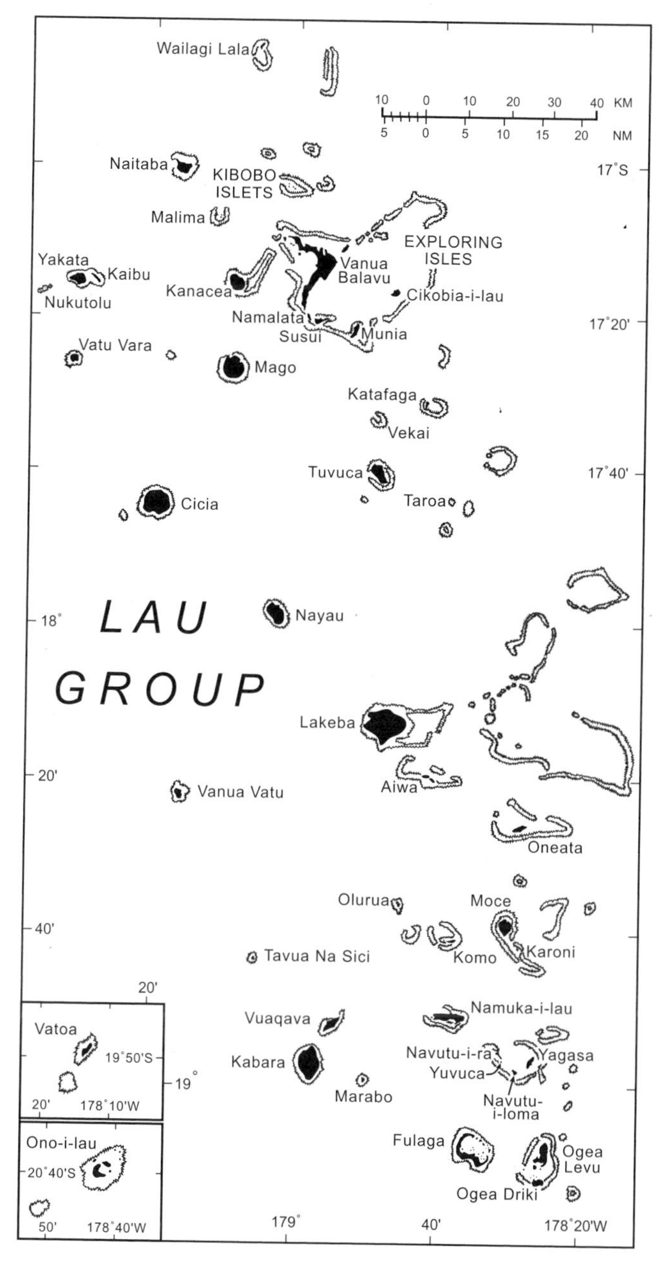

Figure 6-2 Islands of the Lau Group, Fiji. Redrawn from New Zealand Department of Lands and Survey (1986).

Franklin 2000). During February and March 2000 and October 2001, we did more point-counts on Lakeba as well as Nayau and Aiwa Levu (Figure 6-2). I also visited Aiwa Lailai briefly, but was unable to conduct point-counts.

The relative abundance and species richness of birds on Lakeba are greatest in submature/mature limestone forest, the habitat least affected by people (Table 6-6). This is due largely to increased numbers of four species of columbids (*Gallicolumba stairi*, *Ptilinopus perousii*, *P. porphyraceus*, *Ducula pacifica*) and two passerine insectivores (*Mayrornis lessoni*, *Myiagra vanikorensis*). By contrast, the relative abundance of passerine nectarivores (*Myzomela jugularis*, *Foulehaio carunculata*) is greatest in disturbed habitats with many coconut trees.

Based on our point-count and other survey data from Lakeba, Nayau, Aiwa Levu, and Aiwa Lailai, it is clear that habitat quality, even within one habitat category, is more important than island area in structuring landbird communities (Table 6-7). Lakeba is mostly deforested and has 2500 people as well as chickens, Pacific rats, black rats, cats, dogs, horses, pigs, and cows. Aiwa Levu is uninhabited, 100% forested, and lacks nonnative mammals except Pacific rats and goats. Columbids are more abundant per unit area on Aiwa Levu than on Lakeba, especially when seasonality is considered. Two of the noncolumbid species on Aiwa Levu (*Falco peregrinus*, *Clytorhynchus vitiensis*) seem to be extirpated on Lakeba; *Myzomela jugularis* is the only species found in Lakeba's forest but not on Aiwa Levu (Table 6-7). Nayau is mostly deforested and has 430 people as well as chickens, rats (both spp.?), cats, dogs, horses, pigs, and cows. From our brief visit in October 2001, the most distinctive aspects of Nayau's current landbird fauna are the presence of the pigeon *Ducula latrans* and an absence or extreme rarity (island-wide) of the passerines *Lalage maculosa*, *Mayrornis lessoni*, *Clytorhynchus vitiensis*, and *Myiagra vanikorensis*.

Most species of landbirds in Lau also occur in Tonga, where I have surveyed birds on about 30 islands (Steadman 1998, Steadman & Freifeld 1998, Steadman et al. 1999a; also see below). Many of these species are found as well in American Samoa, where Freifeld (1999) did monthly surveys for four years. Looking at all data from Lau, Tonga, and Samoa, the relative abundance and habitat association of widespread West Polynesian species varies sometimes dramatically and sometimes very little, whether between or within island groups. For example, the pigeon *Ducula pacifica* is widely distributed but generally absent on islands lacking secondary or mid-/late-successional native forest; it tends to increase in abundance with greater maturity of the forest. The fruit-dove *Ptilinopus porphyraceus* occurs on more islands than any columbid in the region. It also increases in abundance as forests mature, although is tolerant of secondary forest in abandoned plantations. The ground-dove *Gallicolumba stairi* is a forest obligate that is absent or very rare on inhabited islands. In contrast, three very widespread species, the barn-owl *Tyto alba*, swift *Collocalia spodiopygia*, and kingfisher *Halcyon chloris*, show no clear inter- or intra-island trends in habitat preference or relative abundance.

Table 6-4 Modern distribution of landbirds in Northern Lau Group, Fiji. [e], endemic to Fiji; W, recorded by Whitney South Sea Expedition in 1924–25 (Watling 1985); ?, questionable record, or residency not established.

	Vanuabalavu	Cicia	Mago	Tuvuca	Kanacea	Yacata	Naitaba	Munia	Vatuvara	Cikobiailau	Yavea	Katavaga	Wailagilala	Kibobo	Sovu	Yaroua	Vekai	Total
Egretta sacra	W	W	W	—	—	—	—	—	—	W	W	—	—	—	—	—	—	5
Butorides striatus	W	—	—	—	—	—	—	—	—	—	—	—	—	—	—	—	—	1
Anas superciliosa	—	W	W	—	W	—	—	—	—	—	—	—	—	—	—	—	—	3
Circus approximans	W	—	W	—	W	W	W	W	—	W	—	—	—	—	—	—	—	7
Accipiter rufitorques	W	W	—	—	—	—	—	—	—	—	W	—	—	—	—	—	—	3
Gallirallus philippensis	W	—	W	—	—	—	—	—	—	—	—	—	—	—	—	—	—	2
Columba vitiensis	W	W	W	W	W	W	W	W	W	W	W	W	—	W	—	—	—	13
Ptilinopus perousii	W	—	W	—	W	W	W	W	—	W	—	W	W	—	—	—	—	9
Ptilinopus porphyraceus	—	W	W	W	W	W	W	W	W	W	W	W	W	—	—	—	—	12
Ducula pacifica	—	—	W	W	—	W	W	—	W	W	W	—	—	W	W	—	—	9
Ducula latrans	W	W	W	W	—	W	W	—	W	W	W	—	—	—	—	—	—	9
Gallicolumba stairi	W	—	—	W	—	W	W	—	W	W	—	—	W	—	—	—	—	7
Vini solitarius	—	—	—	W	—	W	W	—	W	—	—	—	—	—	—	—	—	4
Tyto alba	W	—	—	—	—	—	—	—	—	—	—	—	—	—	W	—	—	2
Collocalia spodiopygia	W	W	W	W	W	W	—	W	—	W	W	W	—	W	W	—	—	12

Halcyon chloris	W	W	W	W	W	W	W	W	W	W	W	W	—	W	W	—	—	14
Hirundo tahitica	—	W	W	W	W	—	W	W	—	W	—	—	—	—	—	—	—	7
Aplonis tabuensis	W	W	W	W	W	W	W	W	W	W	W	W	W	W	W	—	—	15
Lalage maculosa	—	—	—	—	—	W	—	—	W	—	—	—	—	—	—	—	—	2
Pachycephala pectoralis	—	—	—	—	—	—	—	—	W	—	—	—	—	—	—	—	—	1
[e]*Mayrornis lessoni*	W	W	W	W	—	—	W	W	W	W	W	—	—	—	W	—	—	10
Clytorhynchus vitiensis	—	—	—	W	—	W	—	—	W	—	—	—	—	—	—	—	—	3
Myiagra vanikorensis	W	W	W	—	W	W	W	W	W	—	W	—	W	W	W	—	—	12
[e]*Myzomela jugularis*	W	W	W	—	W	—	—	W	—	W	W	W	—	—	W	—	—	9
Foulehaio carunculata	W	—	—	W	—	W	—	—	W	—	—	—	—	—	—	—	—	4
Artamus leucorhynchus	W	W	—	—	—	—	—	—	—	—	W	—	—	—	—	—	—	3
TOTAL (W)	18	14	16	13	11	15	13	11	14	14	13	7	5	6	8	0	0	26
Island area (km^2)	53.20	34.60	21.89	12.51	12.48	~9	8.81	4.49	3.84	2.98	2.16	0.81	0.30	0.17	0.17	~0.1	0.08	—
Inhabited	+	+	+	+	?	+	–	–	–	–	+	–	–	–	–	–	–	6–7
Field effort (days)	4	3	2	3	2	1	1	1	1	1	1	1	2	2	1	2	1	29
Field effort (person-days)	8	6	4	6	4	2	2	2	2	2	2	2	4	4	2	4	2	58

Table 6-5 Modern distribution of landbirds in Southern Lau Group, Fiji. [e], endemic to Fiji; S, recorded by D. W. Steadman et al. in 1999–2001; W, recorded by Whitney South Sea Expedition in 1924–1925 (Watling 1985).

	Lakeba	Kabara	Vulaga	Nayau	Ogea Levu	Namula	Moce	Vuaqava	Ogea Driki	Vatoa	Onoilau	Vanuavatu
Egretta sacra	S, W	W	W	S, W	—	W	W	—	—	W	—	W
Anas superciliosa	S	—	—	S	—	—	W	—	—	—	W	—
Circus approximans	S, W	W	W	S, W	—	W	W	—	—	—	—	—
Falco peregrinus	—	—	—	—	—	—	—	—	—	—	—	—
Gallirallus philippensis	S	W	—	S	W	—	—	—	—	—	—	—
Porphyrio porphyrio	S	—	—	S	—	—	—	—	—	—	W	—
Columba vitiensis	S, W	W	W	S	W	W	W	—	—	—	—	W
Ptilinopus perousii	S, W	W	W	S, W	W	W	W	W	—	W	—	W
Ptilinopus porphyraceus	S, W	W	W	S, W	W	W	W	W	—	—	—	W
Ducula pacifica	S, W	W	W	S, W	W	W	—	W	W	W	—	—
Ducula latrans	—	—	—	S, W	—	—	—	—	—	—	—	—
Gallicolumbia stairi	S	—	—	—	—	W	—	—	—	—	—	—
Vini solitarius	S, W	—	—	—	—	—	—	—	—	—	—	—
Vini australis	—	—	W	—	W	—	W	—	—	W	W	—
Tyto alba	S	—	—	S	—	—	—	—	—	—	W	—
Collocalia spodiopygia	S, W	W	W	S, W	W	W	W	—	—	W	—	W
Halcyon chloris	S, W	W	W	S, W	W	W	W	W	—	—	—	W
Hirundo tahitica	S, W	—	—	S, W	W	—	W	—	—	—	—	—
Aplonis tabuensis	S, W	—	W	S, W	W	W	W	W	—	W	W	W
Lalage maculosa	S, W	W	W	—	W	W	W	W	—	W	W	—
Pachycephala pectoralis	—	—	W	—	W	—	—	W	—	—	—	—
[e]*Mayrornis versicolor*	—	—	—	—	W	—	—	—	—	—	—	—
[e]*Mayrornis lessoni*	S	W	—	—	W	W	W	—	—	—	—	W
Clytorhynchus vitiensis	—	W	W	—	W	W	—	W	—	—	—	—
Myiagra vanikorensis	S, W	W	W	—	W	W	W	W	—	—	—	W
[e]*Myzomela jugularis*	S, W	W	W	S	W	W	W	W	—	W	—	W
Foulehaio carunculata	S, W	—	W	S, W	W	—	W	—	—	—	W	—
TOTAL												
W	15	14	16	11	18	15	16	10	1	8	7	10
S	21	—	—	17	—	—	—	—	—	—	—	—
W + S	21	14	16	17	18	15	16	10	1	8	7	10
Island area (km^2)	55.94	31.24	18.52	18.44	13.29	12.80	10.79	8.13	5.23	4.46	4.46	4.10
Inhabited	+	+	+	+	+	+	+	–	–	+	+	+
Field effort (days) W	3	6	4	1	10	4	3	1	1	3	5	1
Field effort (person-days) W	6	12	8	2	20	8	6	2	2	6	10	2
Field effort (days) S	20	—	—	10	—	—	—	—	—	—	—	—
Field effort (person-days) S	40	—	—	13	—	—	—	—	—	—	—	—
Total person-days	46	12	8	15	20	8	6	2	2	6	10	2

Table 6-5 (continued)

Oneata	Yagasalevu	Komo	Marabo	Aiwa Levu	Aiwa Lailai	Yuvuca	Tavunasici	Karoni	Olorua	Vanuamasi	Lateitoga	Lateiviti	Reid Reef	Total
—	W	W	W	S	S	—	—	W	—	—	—	W	—	15
W	—	—	—	—	—	—	—	W	—	—	—	—	—	6
W	—	W	—	S, W	S	—	—	—	—	—	—	—	—	10
—	—	—	—	S	S	—	—	—	—	—	—	—	—	2
W	—	—	—	S, W	—	W	—	—	W	—	—	—	—	8
—	—	—	—	—	—	—	—	—	—	—	—	—	—	3
—	—	—	W	S, W	S	—	—	—	—	—	—	—	—	11
W	W	—	—	S, W	S	—	—	W	W	—	—	—	—	16
W	W	W	W	S, W	S	—	—	W	—	—	—	—	—	16
W	W	—	W	S, W	S	—	W	—	W	—	—	—	—	16
W	—	—	—	—	—	—	—	—	—	—	—	—	—	2
W	W	—	—	S, W	S	—	—	—	W	—	—	—	—	7
—	—	—	—	—	—	—	—	—	—	—	—	—	—	1
W	—	—	—	—	—	—	—	—	—	—	—	—	—	6
—	—	—	—	S, W	—	—	—	—	W	—	—	—	—	5
W	W	—	W	S, W	S	—	—	—	—	—	—	—	—	14
W	W	W	W	S, W	S	—	W	—	W	—	—	—	—	17
—	—	W	—	—	—	—	—	—	—	—	—	—	—	5
W	W	W	W	S, W	S	—	W	—	W	W	—	—	—	19
W	W	—	W	S, W	S	—	W	—	W	—	—	—	—	16
—	—	W.	—	—	—	—	—	—	—	—	—	—	—	4
—	—	—	—	—	—	—	—	—	—	—	—	—	—	1
W	W	—	W	S, W	S	—	—	W	—	—	—	—	—	12
W	W	—	—	S, W	S	—	—	—	—	—	—	—	—	10
W	W	W	—	S, W	S	—	—	—	—	—	—	—	—	13
W	W	W	—	—	—	—	W	W	W	—	—	—	—	16
W	—	—	W	S, W	S	—	—	—	—	—	—	—	—	10
18	13	9	10	16	—	1	5	6	9	1	0	1	0	26
—	—	—	—	18	16	—	—	—	—	—	—	—	—	23
18	13	9	10	18	16	1	5	6	9	1	0	1	0	27
4.03	2.59	1.53	1.29	1.21	∼1	0.47	0.40	0.40	0.16	0.15	∼0.1	∼0.1	∼0.1	—
+	–	–	—	—	—	—	—	—	—	—	—	—	—	11
4	4	2	1	4	—	1	1	1	2	1	1	1	1	66
8	8	4	2	8	—	2	2	2	4	2	2	2	2	132
—	—	—	—	7	1	—	—	—	—	—	—	—	—	28
—	—	—	—	14	2	—	—	—	—	—	—	—	—	56
8	8	4	2	22	2	2	2	2	4	2	2	2	2	188

Table 6-6 Relative abundance of forest birds during the wet season (18 February–6 March 2000) and early dry season (10–15 May 1999) on Lakeba, Lau Group, Fiji. Data are expressed as mean individuals seen/heard per point-count, rounded to the nearest 0.1. The number of point-counts is given in parentheses for each time interval in each habitat type. *Collocalia* is excluded from total relative abundance but not total species (see "Methods" in Steadman & Franklin 2000). *Gallus* (i, introduced) is excluded from all totals. *, recorded during point-counts but only at distances >50 m. Updated from Steadman & Franklin (2000).

	Coastal coconut plantation		Secondary forest on carbonate bedrock		Mid-/late-successional forest on limestone		Mixed pine woodland		Pure pine woodland	
	Feb–Mar (8)	May (7)	Feb–Mar (12)	May (9)	Feb–Mar (4)	May (7)	Feb–Mar (10)	May (9)	Feb–Mar (8)	May (10)
Egretta sacra	—	—	*	—	—	—	—	—	—	—
Circus approximans	*	—	0.1	0.2		—	*	—	—	—
Gallus gallus (i)	*	0.3	*	*	—	—	*	—	—	*
Gallirallus philippensis	—	—	—	—	—	—	0.1	—	—	—
Porphyrio porphyrio	—	—	—	—	—	—	0.1	—	—	—
Columba vitiensis	*	—	*	*	*	—	0.2	*	0.1	0.1
Ptilinopus perousii	—	—	—	—	—	0.6	—	—	—	—
Ptilinopus porphyraceus	—	*	0.2	0.4	0.2	1.0	0.1	*	*	—
Ducula pacifica	*	—	0.2	0.1	1.0	0.9	—	—	—	—
Gallicolumba stairi	—	—	—	—	—	0.1	—	—	—	—
Vini solitarius	0.5	—	—	—	—	—	—	—	—	—
Tyto alba	—	—	—	—	—	0.1	—	—	—	—
Collocalia spodiopygia	3.1	0.6	0.8	0.9	1.5	0.1	1.3	3.4	0.6	0.9
Halcyon chloris	0.4	0.3	0.3	0.4	1.0	0.4	0.5	0.4	0.1	0.1
Aplonis tabuensis	0.4	0.9	0.9	1.0	1.2	1.0	1.1	0.9	0.6	0.5
Lalage maculosa	—	—	—	—	—	—	0.1	0.4	—	—
Mayrornis lessoni	0.1	—	1.1	0.7	2.5	3.3	0.4	1.1	—	—
Myiagra vanikorensis	—	0.3	0.8	1.4	2.0	2.1	0.8	1.0	0.1	0.1
Myzomela jugularis	3.4	2.1	1.2	1.0	—	0.3	1.8	1.6	1.6	1.3
Foulehaio carunculata	2.9	2.4	0.8	1.8	1.0	0.7	1.2	1.0	0.9	0.3
TOTAL										
Relative abundance	7.7	6.0	5.6	7.0	8.9	10.5	6.4	6.4	3.4	2.4
Species (without *)	7	6	10	10	8	12	12	8	7	7
Species (all)	10	7	12	11	9	12	13	10	8	7
Columbid abundance	***	*	0.4	0.5	1.2	2.6	0.3	**	0.1	0.1
Nectarivore abundance	6.8	4.5	1.9	2.8	1.0	1.0	3.0	2.6	2.5	1.6

Table 6-7 Distribution and relative abundance of landbirds in mid-/late-successional limestone forest on Lakeba and Aiwa Levu, Lau Group, Fiji. Data are expressed as mean individuals seen/heard per point-count, rounded to the nearest 0.1. The number of point-counts is given in parentheses for each island. *Collocalia* is excluded from total abundance but not total species (see "Methods" in Steadman & Franklin 2000). *, recorded during point-counts but only at distances >50 m; **, occurs in this forest type but not recorded during point-counts.

	Lakeba Feb–Mar (4)	Lakeba May (7)	Aiwa Levu Feb–Mar (15)
Circus approximans	—	—	*
Falco peregrinus	—	—	**
Gallirallus philippensis	—	—	0.1
Columba vitiensis	*	—	0.1
Ptilinopus perousii	**	0.6	0.1
Ptilinopus porphyraceus	0.2	1.0	0.5
Ducula pacifica	1.0	0.9	1.9
Gallicolumba stairi	**	0.1	0.4
Tyto alba	**	0.1	0.1
Collocalia spodiopygia	1.5	0.1	0.1
Halcyon chloris	1.0	0.4	1.2
Aplonis tabuensis	1.2	1.0	1.7
Lalage maculosa	—	—	0.1
Mayrornis lessoni	2.5	3.3	2.3
Clytorhynchus vitiensis	—	—	0.9
Myiagra vanikorensis	2.0	2.1	1.5
Myzomela jugularis	—	0.3	—
Foulehaio carunculata	1.0	0.7	1.8
TOTAL			
Relative abundance	8.9	10.5	12.7
Species (without *, **)	8	12	15
Species (with *, **)	12	12	17
Columbid abundance	1.2	2.6	3.0
Nectarivore abundance	1.0	1.0	1.8
Island area (km^2)	55.94		1.21
Forest area (km^2)	~1.0		1.21

The triller *Lalage maculosa* is enigmatic in its distribution and relative abundance. Of 13 islands surveyed in Ha'apai, it was absent from five, uncommon on seven, and common on only one (Steadman 1998). In Vava'u, *L. maculosa* occurred on all 16 islands surveyed, with roughly similar relative abundance in all habitat categories (Steadman & Freifeld 1998). Absent, rare, or uncommon in the Lau Group, this habitat generalist is common in Western Samoa but absent in American Samoa. The starling *Aplonis tabuensis* is a habitat generalist although, like the three columbids mentioned above, it increases in relative abundance with greater forest cover in Tonga and Samoa. In Lau, we detected no such habitat-related trend. I found the honeyeater *Foulehaio carunculata* on eight of 13 islands in Ha'apai, although it was common only where *Lalage maculosa* was rare or absent, and was most abundant in the magnificent forests of Tofua (Figure 6-3). In Vava'u, *F. carunculata* occurred on 11 of 16 islands and was more abundant with increasing forest maturity, as also was true in American Samoa. On Lakeba, however, *F. carunculata* was abundant in coconut plantations, reminiscent of the situation on two small islands (Ha'afeva, Tungua) in Ha'apai.

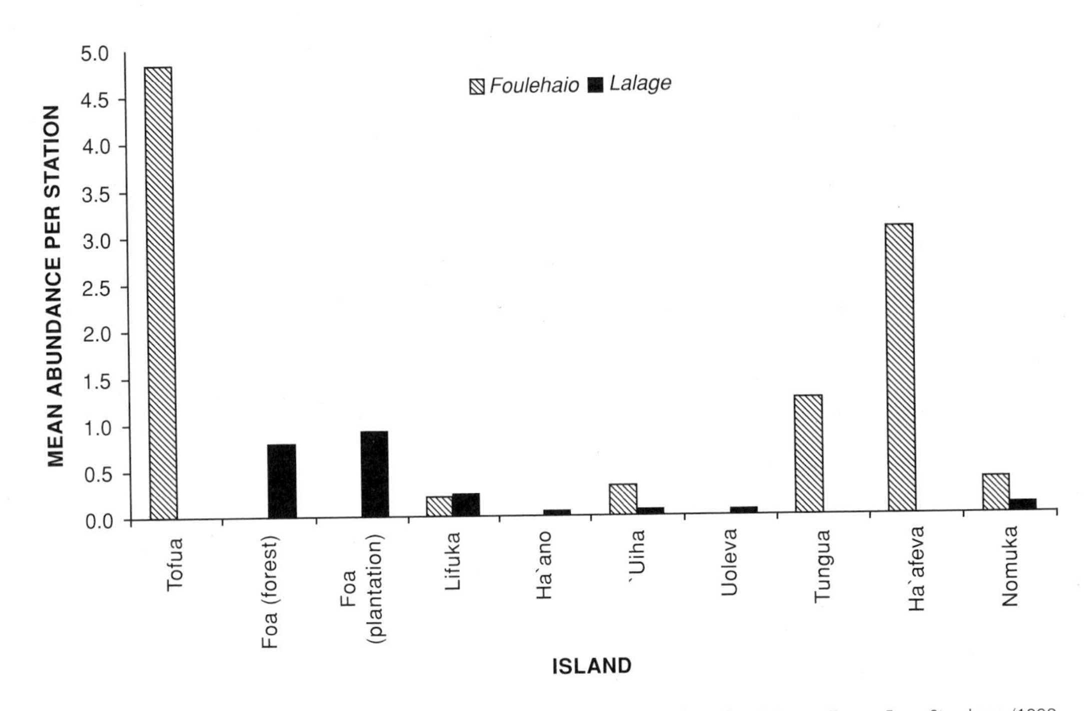

Figure 6-3 Relative abundance of *Lalage maculosa* and *Foulehaio carunculata* in the Ha`apai Group, Tonga. From Steadman (1998: Figure 4).

Over the past several years, Sean Connaughton, Sepeti Matararaba, Patrick O'Day, Sharyn O'Day, Gregory Pregill, and I have begun to excavate prehistoric sites on the same four islands in Lau where Janet Franklin and I have surveyed modern birds (Tables 6-8, 6-9). Bird bones also were excavated from Qara ni Puqa Rockshelter on Lakeba in the 1970s (Best 1984, 2002), initially studied by me but published (the megapodes and pigeons) mainly by Worthy (2000, 2001b). Because skeletons of living Fijian passerines still are not well represented in museum collections, I have not yet identified the passerine fossils from Lau.

Another qualification for Lau's fossil record is that only on Lakeba and Aiwa Levu are any of the bones old enough to be associated with Lapita pottery (ca. 2800–2700 cal BP) and thus likely to be from extinct or extirpated species; as far as I can say now, many of the bird bones from Lakeba and Aiwa Levu, most from Aiwa Lailai, and all from Nayau are much younger, typically <2000 cal BP ('O'Day et al. 2004, Steadman et al. unpub. data). This probably explains why we have not found extinct species of megapodes, rails, and pigeons thus far on Nayau or Aiwa Lailai. On no island in Lau is the currently developed fossil record as thorough as that from any one of the seven Tongan islands described below.

While the seabirds of Lau never have been surveyed, from my observations (1999–2001) I would say that the bones identified thus far represent at least three extirpated populations—the petrel *Pterodroma nigripennis* and storm-petrel *Nesofregetta fuliginosa* on Aiwa Levu, and the tern *Sterna sumatrana* on Aiwa Lailai. I am uncertain of the status of *Anous minutus*, a tern whose bones are found on all four islands but which may not nest on any of them.

From Lakeba (56 km^2), one site (Qara ni Puqa, or site 197) has yielded bird bones and Lapita pottery (Cassells 1984, Best 1984, 2002, Steadman 1989a, 1999d, Worthy 2000, 2001b). Extinct species recognized at this site are the megapode *Megapodius alimentum* (also reported from Mago Island in Lau; Worthy 2000), the flightless rail *Gallirallus* undescribed sp. D, and the pigeon *Ducula lakeba*. The megapode also occurred in Tonga (Steadman 1989b, 1999d). The rail is not fully studied but presumably was endemic to Lakeba and nearby Aiwa Levu and Aiwa Lailai, which were joined as one large island (ca. 1000 km^2) in the late Pleistocene. *Ducula lakeba* may have been endemic to Fiji. Without benefit of direct comparison with appropriate specimens, Steadman (1989a) tentatively referred the large pigeon bones from Qara ni Puqa to the similarly sized, extinct *D. david*, which was described from Wallis & Futuna (Balouet & Olson 1987). The three living but extirpated species recorded from Lakeba are the small heron *Butorides striatus* and two volant rails, *Porzana tabuensis* and *Poliolimnas cinereus*. Fossils increase Lakeba's landbird fauna from 21 to 27 species (Table 6-9).

From Nayau (18 km^2), 80 bird bones from post-Lapita cultural and noncultural (barn-owl roost) contexts

Table 6-8 Prehistoric records of birds from the Lau Group, Fiji. Passerine bones are unstudied. i, introduced species; m, migratory species (i, m not included in totals); NISP, number of identified specimens. Lakeba: QNP, Qara ni Pusi; GYR, Gyrocarpus Rockshelter; QSE, Qara Selesele; QNQ, Qara ni Puqa Rockshelter (site 197); 197OR, site 197 Owl Roost; OSOC, Osonabukete Cave; OSOR, Osonabukete Rockshelter; WAI, Wainiabia Cave; VAR, Vagadra Rockshelter; LVR, Lost Village Rockshelter; WaiTW, Waituruturu West; WaiTE, Waituruturu East; KV2, Korovatu Rockshelter 2; UluNK, Ulunikoro; DKT, Daku ni Tuba; ALC1, Aiwa Levu Cave 1; ALC2, Aiwa Levu Cave 2; ALR1, Aiwa Levu Rockshelter 1; GRS, Goat Rockshelter. Aiwa Lailai: DAUR, Dau Rockshelter. Data on *Megapodius alimentum* and *Ducula lakeba* from QNQ are from Worthy (2000, 2001b); all other data are from Steadman (unpublished); QNQ also has unidentified, and therefore untabulated, bones of parrots and additional columbids.

	Lakeba										Nayau					Aiwa Levu				Aiwa Lailai
	QNP	GYR	QSE	QNQ	1970R	OSOC	OSOR	WAI	VAR	LVR	WaiTE	WaiTW	KV2	UluNK	DKT	ALC1	ALC2	ALR1	GRS	DAUR
SEABIRDS																				
Pterodroma nigripennis	—	—	—	—	—	—	—	—	—	—	—	—	—	—	—	1	—	—	—	—
Nesofregetta fuliginosa	—	—	—	—	—	—	—	—	—	—	—	—	—	—	—	—	—	1	—	—
Phaethon lepturus	—	—	21	5	—	—	—	51	—	—	—	—	1	—	—	—	—	1	—	—
Fregata ariel	—	—	—	1	—	—	—	—	—	—	—	—	—	—	—	—	—	—	—	—
Sterna sumatrana	—	—	—	—	—	—	—	—	—	—	—	—	—	—	—	—	—	—	—	1
Anous stolidus	—	—	—	—	—	—	—	—	—	—	—	—	—	—	—	—	—	6	—	3
Anous minutus	—	—	2	—	—	—	—	—	—	—	1	—	—	—	—	—	—	7	—	1
Gygis candida	—	—	—	—	—	1	—	—	—	—	—	—	—	—	—	—	—	1	—	—
LANDBIRDS																				
Egretta sacra	—	—	6	1	—	—	—	—	—	—	—	—	—	—	—	—	—	—	—	—
Butorides striatus	—	—	—	1	—	—	—	—	—	—	—	—	—	—	—	—	—	—	—	—
Anas superciliosa	—	—	—	—	—	—	—	—	—	—	—	—	—	—	—	—	—	1	—	—
†*Megapodius alimentum*	—	—	—	49	—	—	—	—	—	—	—	—	—	—	—	—	—	3	—	—
†*Megapodius amissus/molistructor*	—	—	—	—	—	—	—	—	—	—	—	—	—	—	—	—	—	1	—	—
Gallus gallus (i)	1	3	1	1	—	—	1	—	—	—	—	1	—	—	2	—	7	8	—	3
Gallirallus philippensis	2	—	—	2	2	—	—	—	—	—	—	1	—	—	1	10	1	7	2	—
[e]†*Gallirallus* undesc. sp. D	—	—	—	7	—	—	—	—	—	—	—	—	—	—	—	—	—	3	—	—

(continued)

Table 6-8 (continued)

	Lakeba										Nayau					Aiwa Levu				Aiwa Lailai
	QNP	GYR	QSE	QNQ	1970R	OSOC	OSOR	WAI	VAR	LVR	WaiTE	WaiTW	KV2	UluNK	DKT	ALC1	ALC2	ALR1	GRS	DAUR
Porzana tabuensis	—	—	—	—	—	—	1	1	—	1	—	—	—	2	—	—	—	—	—	—
[e]† *Porzana* undesc. sp. B	—	—	—	—	—	—	—	—	—	—	—	—	—	—	—	—	—	1	—	—
Poliolimnas cinereus	—	—	—	—	1	—	—	—	—	—	—	—	—	—	—	—	—	—	—	—
Porphyrio porphyrio	—	—	—	7	—	—	—	—	—	—	—	—	—	—	—	1	3	1	—	—
Pluvialis fulva (m)	—	—	—	1	2	—	—	—	—	—	1	—	—	—	—	—	—	—	—	—
Columba vitiensis	1	—	—	1	—	—	—	—	—	—	—	—	—	3	—	—	1	5	—	—
Ptilinopus perousii	—	—	1	—	—	—	—	—	—	—	4	30	—	2	—	—	—	1	—	1
Ptilinopus porphyraceus	—	—	—	—	8	—	—	—	—	—	1	—	—	2	—	—	—	—	—	—
Ducula pacifica	—	—	—	—	—	—	—	—	—	—	—	—	—	—	—	—	—	10	—	14
Ducula latrans	—	—	—	—	—	—	—	—	—	—	—	—	—	—	—	—	—	5	—	1
[e]†*Ducula lakeba*	—	—	—	92	—	—	—	—	—	—	—	—	—	—	—	—	1	—	—	—
Gallicolumba stairi	—	—	—	—	—	—	—	—	—	1	—	3	—	—	—	3	—	3	—	2
Vini solitarius	—	—	—	—	—	—	—	—	—	—	—	—	—	—	—	—	—	—	—	—
Tyto alba	—	—	1	2	1	5	—	—	—	—	—	—	—	2	—	—	1	—	—	—
Collocalia spodiopygia	—	—	—	—	—	—	—	—	—	—	—	8	—	—	—	—	4	1	—	—
Halcyon chloris	1	—	4	—	—	—	—	—	—	—	3	16	—	4	—	2	—	3	—	—
Passeriformes	21	—	5	1	5	—	5	3	13	11	5	53	—	15	—	5	2	58	—	1
Total resident birds																				
NISP	26	—	34	166	16	6	6	55	13	13	14	111	1	28	1	22	13	119	2	24
Species	>3	—	>5	>11	>3	2	>1	>2	>0	>2	>4	>5	1	>5	1	>5	>6	>18	1	>7
Total nonpasserine landbirds																				
NISP	4	—	6	157	11	5	—	1	—	2	8	58	—	13	1	16	11	45	2	18
Species	3	—	3	9	3	1	—	1	—	2	3	5	—	5	1	4	6	13	1	4

Table 6-9 Modern and prehistoric landbirds from the four islands with prehistoric data in the Southern Lau Group, Fiji. [e], endemic to Fiji; M, modern record; P, prehistoric record; †, extinct species. Passerine bones are unstudied. Based on data in Tables 6-5 and 6-8.

	Lakeba	Nayau	Aiwa Levu	Aiwa Lailai
Egretta sacra	M, P	M	M	M
Butorides striatus	P	—	—	—
Anas superciliosa	M	M	P	—
Circus approximans	M	M	M	M
Falco peregrinus	—	—	M	M
†*Megapodius alimentum*	P	—	P	—
†*Megapodius amissus/molistructor*	—	—	P	—
Gallirallus philippensis	M, P	M, P	M, P	—
[e]†*Gallirallus* undesc. sp. D	P	—	P	—
Porzana tabuensis	P	P	—	—
[e]†*Porzana* undesc. sp. B	—	—	P	—
Poliolimnas cinereus	P	—	—	—
Porphyrio porphyrio	M, P	M	P	—
Columba vitiensis	M, P	M, P	M, P	M
Ptilinopus perousii	M, P	M, P	M, P	M, P
Ptilinopus porphyraceus	M, P	M, P	M	M
Ducula pacifica	M	M	M, P	M, P
Ducula latrans	—	M	P	P
[e]†*Ducula lakeba*	P	—	P	—
Gallicolumba stairi	M, P	P	M, P	M, P
Vini solitarius	M	—	—	—
Tyto alba	M, P	M, P	M, P	—
Collocalia spodiopygia	M	M, P	M, P	M
Halcyon chloris	M, P	M, P	M, P	M
Hirundo tahitica	M	M	—	—
Aplonis tabuensis	M	M	M	M
Lalage maculosa	M	—	M	M
[e]*Mayrornis lessoni*	M	—	M	M
Clytorhynchus vitiensis	—	—	M	M
Myiagra vanikorensis	M	—	M	M
[e]*Myzomela jugularis*	M	M	—	—
Foulehaio carunculata	M	M	M	M
TOTAL				
M	21	17	18	16
M + P	27	19	26	17
Island area (km^2)	55.94	18.44	1.21	~1.0
Inhabited	+	+	–	–
Total person-days (M)	46	15	22	2
Total nonpasserine landbird bones (P)	186	80	74	18

represent nine extant species of nonpasserine landbirds (Table 6-8). Two of these are extirpated (the volant rail *Porzana tabuensis* and ground-dove *Gallicolumba stairi*), increasing Nayau's landbird fauna to 19 species (Table 6-9). The bird bones recovered thus far on Nayau are <1000 cal BP ('O'Day et al. 2004).

From small (1.21 km^2) Aiwa Levu, I have identified 73 nonpasserine landbird bones from post-Lapita through pre-Lapita contexts (Steadman et al. unpub. data). This small but rich set of bones represents 16 species, including five extinct (two megapodes, two rails, and a pigeon; Table 6-8) and three that are extirpated (a duck, rail, and pigeon). Most of these losses are documented in one site, the deep and well-stratified Aiwa Levu Rockshelter 1, 470–280 cal BP in upper strata to 2370–2150 cal BP in lower strata, with some bird bones certainly older. The bones from Aiwa Levu Cave 2 are from strata dated from 960–760 cal BP (stratum I) to 2710–2330 cal BP (stratum II). The mere 73 bones increase Aiwa Levu's landbird fauna from 18 to 26 species, just one fewer than recorded from Lakeba. Along with the Tongan island of Ha'afeva (1.8 km^2), Aiwa Levu plays a key role in learning how much species richness a very small West Polynesian island can sustain (see below in this chapter, and Chapter 19).

Finally, the neighboring island of Aiwa Lailai (~1.0 km^2) has yielded one bone deposit, a cultural site called Dau Rockshelter. Its 18 nonpasserine landbird bones represent four species of columbids, among which only the extant *Ducula latrans* does not live today on Aiwa Lailai. Much of the difference in species richness between Aiwa Levu and nearby Aiwa Lailai (Table 6-9) must be due to the poorer sampling of the latter's modern and prehistoric birds.

A priority of future research in Lau will be to expand the fossil records on Aiwa Levu and Aiwa Lailai. Among the species expected to be found there eventually, or on any island in Lau, are four that live today in western Fiji and that also have been found as fossils in Tonga, namely the hawk *Accipiter rufitorques*, cuckoo *Cacomantis flabelliformis*, thrush *Turdus poliocephalus*, and robin *Petroica multicolor*.

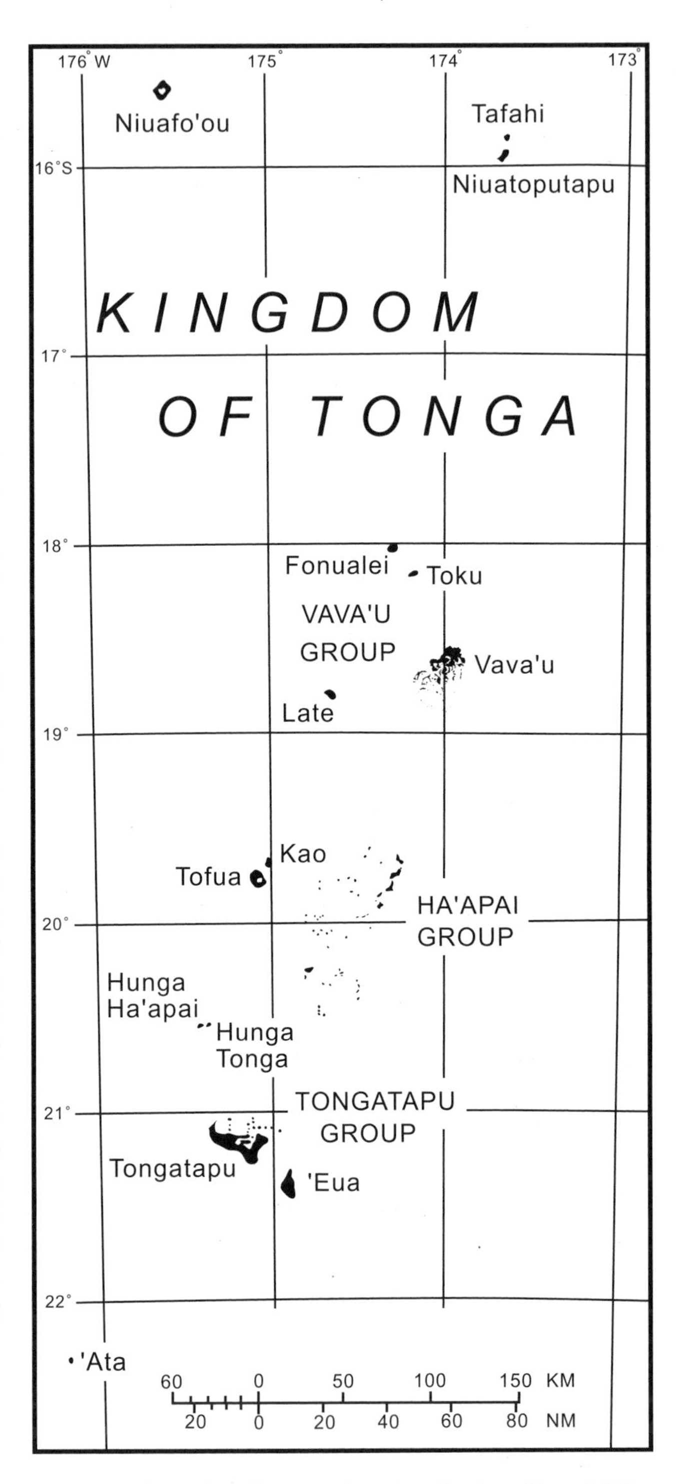

Figure 6-4 The Kingdom of Tonga. Redrawn from Kingdom of Tonga (1969).

Tonga

The Kingdom of Tonga (Figure 6-4) lies 320 km east of the Lau Group. The most eastern islands on the Indo-Australia Plate, Tonga consists primarily of raised limestone islands, with active volcanic islands lying to the north and west. Some of the volcanos have estimated ages of ca. 60 ka or less (Turner & Hawkesworth 1997, Bourdon et al. 1999). The raised limestone islands tend to be older, although probably none has been continuously emergent longer than the late Miocene or Pliocene (Stearns 1971, Taylor & Bloom 1977, Chaproniere 1994a, Tappin & Balance 1994, Roy 1997a, Dickinson et al. 1999).

The modern inter-island distribution of Tongan birds is fairly well-documented for most major islands or island clusters (Table 6-10). The extant, indigenous landbirds

Table 6-10 Distribution of landbirds in Tonga and Niue. [e], endemic to Tonga; [en], endemic to Niue; H, historic record of extirpated population; M, modern record; P, prehistoric record; † extinct species; *, recent arrival, not included in totals; ?, questionable record, or residency not established. Volcanic outliers are north to south, Niuafo`ou, Tafahi, Niuatoputapu, Fonualei, Toku, Late, Kao, Tofua, Hunga Tonga, Hunga Ha`apai, and `Ata. From Kinsky & Yaldwyn (1981), Pratt et al. (1987), Rinke (1987), Steadman (1993a, 1998, herein), Steadman & Freifeld (1998), Worthy et al. (1998), and Steadman et al. (2000b, 2002b).

	Vava`u	Ha`apai	Tongatapu	`Eua	Volcanic outliers	Niue
HERONS						
**Egretta novaehollandiae*	—	—	M	—	—	—
Egretta sacra	M	M, P	M	M, P	M	M
[en]†*Nycticorax kalavikai*	—	—	—	—	—	P
[e]†*Nycticorax* undescribed sp. B	—	P	—	P	—	—
DUCKS						
Anas superciliosa	M	M, P	M	M	M	—
OSPREYS						
Pandion haliaetus	—	P	P	—	—	—
HAWKS						
Circus approximans	—	M	M	—	M	—
Accipiter rufitorques	—	—	—	P	—	—
MEGAPODES						
Megapodius pritchardii	—	P	—	P	M	P
[e]†*Megapodius alimentum*	—	P	P	P	—	—
†*Megapodius molistructor*	—	P	P	—	—	—
[e]†*Megapodius* undescribed sp. F	—	—	—	P	—	—
†Genus uncertain	—	—	P	—	—	—
RAILS						
Gallirallus philippensis	M	M, P	M, P	M, P	M	M, P
[en]†*Gallirallus huiatua*	—	—	—	—	—	P
[e]†*Gallirallus* undescribed sp. E	—	P	—	—	—	—
[e]†*Gallirallus* undescribed sp. F	—	P	—	—	—	—
[e]†*Gallirallus* undescribed sp. G	—	—	P	—	—	—
[e]†*Gallirallus* undescribed sp. H	—	—	—	P	—	—
Porzana tabuensis	H	P	M, P	P	—	M, P
Porphyrio porphyrio	M	M, P	M, P	M, P	M	M, P
PIGEONS, DOVES						
Ptilinopus perousii	H	M, P	M, P	M, P	M	—
Ptilinopus porphyraceus	M	M, P	M, P	M, P	M	M
Ducula pacifica	M	M, P	M, P	M, P	M	M, P
Ducula latrans	—	P	P	P	—	—
[e]†*Ducula* undescribed sp.	—	P	—	P	—	—
[e]†Undescribed genus C	—	P	P	P	—	—
†*Caloenas canacorum*	—	P	—	—	—	—
[e]†*Didunculus* undescribed sp.	—	P	P	P	—	—
Gallicolumba stairi	M	M, P	H, P	P	M	—

(continued)

Table 6-10 (continued)

	Vava`u	Ha`apai	Tongatapu	`Eua	Volcanic outliers	Niue
PARROTS						
Vini solitarius	—	P	—	P	—	—
Vini australis	H	M	H	P	M	M
†*Eclectus* undescribed sp.	—	P	—	P	—	—
CUCKOOS						
Cacomantis flabelliformis	—	P	P	—	—	—
BARN-OWLS						
Tyto alba	M	M, P	M	M, P	M	M, P
SWIFTS						
Collocalia spodiopygia	M	M	M	M, P	M	M, P
KINGFISHERS						
Halcyon chloris	M	M, P	M	M, P	M	—
SWALLOWS						
Hirundo tahitica	—	M	M	M?	—	—
THRUSHES						
Turdus poliocephalus	—	P	P	P	—	—
WARBLERS						
[e]† cf. *Cettia* sp.	—	—	—	P	—	—
WHITE-EYES						
[e]†Genus unknown	—	—	P	P	—	—
STARLINGS						
Aplonis tabuensis	M	M, P	M, P	M, P	M	M, P
TRILLERS						
Lalage maculosa	M	M, P	M, P	M	M	M
[e]†cf. *Lalage* sp.	—	—	P	P	—	—
ROBINS						
Petroica cf. *multicolor*	—	P	—	—	—	—
WHISTLERS						
Pachycephala jacquinoti	M	P	P	—	—	—
MONARCHS						
Clytorhynchus vitiensis	H	M, P	H, P	H, P	M	—
Myiagra sp.	—	—	—	P	—	—
HONEYEATERS						
Myzomela cf. *cardinalis/jugularis*	—	P	—	P	—	—
Foulehaio carunculata	M	M, P	M, P	M, P	M	—
TOTAL						
M	14	18	16	13–14	18	11
M + H	18	18	19	14–15	18	11
M + H + P	18	38	32	35–36	18	14
NISP landbirds	0	1448	437	1319	0	482
NISP passerines	0	216	48	282	0	34

represent 20 genera (none endemic; Table 6-1) and 21 species, two of which are believed to be endemic—the megapode *Megapodius pritchardii* and whistler *Pachycephala jacquinoti*, although the former once lived on Niue (Steadman et al. 2000b). As I will detail shortly, at least 26 additional species of landbirds, most of them extinct, occurred in Tonga when people arrived ca. 2800 years ago. These include two extinct, undescribed, possibly endemic genera (a pigeon and white-eye).

A brief review of Tonga's modern landbirds will place the prehistoric avifauna in context. Based on fieldwork in 1995 and 1996, my colleagues and I assessed the distribution, relative abundance, and habitat preferences of forest trees, lizards, and birds in the Vava'u Group (Steadman & Freifeld 1998, Franklin et al. 1999, Steadman et al. 1999a). The 16 raised limestone islands that we surveyed were all conjoined in the late Pleistocene, but vary today in habitat composition, land area (0.02–96 km^2), elevation (20–215 m), and distance (0–10.1 km) from the largest island of 'Uta Vava'u (Table 6-11, Figure 6-5). The two major types of nonlittoral mature forest are similar in species composition to forest communities described on 'Eua in southern Tonga (Drake et al. 1996) and lowland Samoa (Whistler 1983, 1992), but with different patterns of species dominance. Littoral forests are dominated by trees with water-dispersed seeds and, unlike inland forests, provide poor-quality habitat for

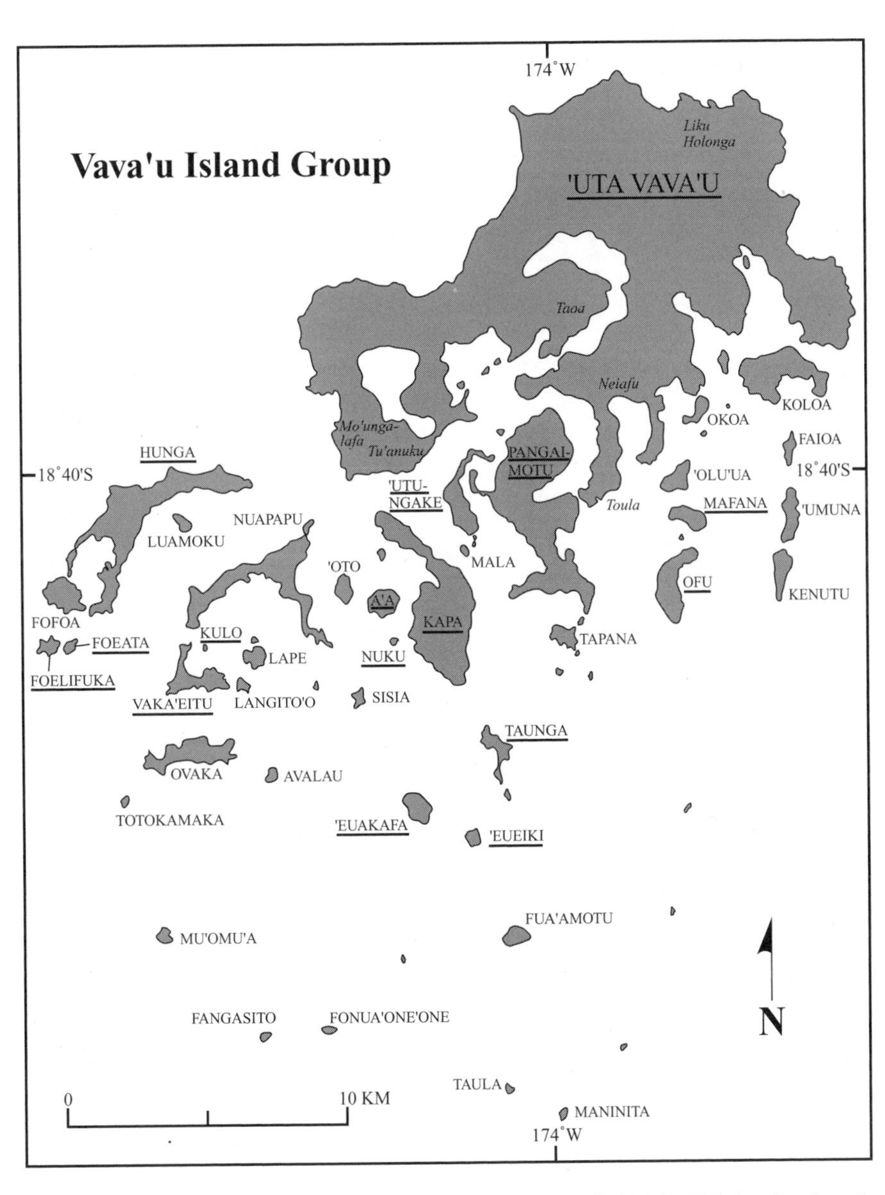

Figure 6-5 The Vava`u Group, Tonga. Underlined islands were surveyed for birds in 1995. From Steadman & Freifeld (1998: Figure 2).

Table 6-11 Characteristics of islands surveyed in the Vava`u Group, Tonga. Most elevations are rounded to the nearest 5 m or 10 m. *Transect surveys were not conducted on Nuku or Kulo, although habitat assessments and counts of all birds were made; **includes Pangaimotu (connected by a causeway). Habitat types: 1, village; 2, open plantation; 3, wooded plantation/early successional forest; 4, submature/disturbed mature forest; 5, mature forest. Habitat types are defined in detail in Steadman & Freifeld (1998), Franklin et al. (1999), Steadman et al. (1999a).

				Number of stations in each habitat type					
Island	Area (km^2)	Max. elev. (m)	Distance to `Uta Vava`u (km)**	1	2	3	4	5	Dates surveyed
`Uta Vava`u	95.95	215	0	18	2	41	37	41	2, 8, 9, 14, 21–25, 29 July 1995, 7 July 1996
Pangaimotu	9.24	70	0	—	7	12	—	—	7 July 1996
Kapa	6.06	100	1.2	—	—	10	6	—	26 July 1995
Hunga	5.34	90	2.8	—	—	20	—	—	11–12 July 1995
Ofu	1.32	30	2.1	—	—	16	—	—	1 August 1995
`Utungake	1.09	80	0.2	6	—	—	6	—	8 July 1996
Vaka`eitu	0.85	45	7.0	—	—	—	5	6	4 July 1995
`Euakafa	0.70	62	6.8	—	—	—	7	15	19 July 1995
Taunga	0.56	40	4.0	2	—	23	—	—	20 July 1995
A`a	0.54	45	3.4	—	—	—	2	10	7 July 1995
Mafana	0.46	25	1.5	—	—	8	8	—	31 July 1995
`Eueiki	0.24	35	6.6	—	—	—	3	—	18 July 1995
Foelifuka	0.21	40	10.1	—	—	14	—	—	28 July 1995
Foeata	0.06	30	9.8	—	—	4	—	—	28 July 1995
Nuku	0.04	20	3.8	—	—	—	—	*	30 July 1995
Kulo	0.02	30	6.8	—	—	—	*	—	6 July 1995
TOTAL	—	—	—	26	9	148	74	72	—

Table 6-12 Relative abundance (mean number of birds per station) for each island or locality sampled in Vava`u, Tonga. UV, `Uta Vava`u localities. Totals are calculated without *Collocalia spodiopygia*, an aerial feeder. Numbers of point-counts are in parentheses. No point-counts were done on Nuku or Kulo (see Tables 6-11, 6-13). From Steadman & Freifeld (1998: Table 3).

	UV-Taoa (16)	UV-Liku Holonga (7)	UV-Mo`ungalafa (59)	UV-Neiafu (30)	UV-Tu`anuku (24)	UV-Toula (3)	Pangaimotu (19)	Kapa (16)
Gallirallus philippensis	0.13	—	0.05	0.13	—	0.7	0.05	—
Porphyrio porphyrio	0.06	—	—	0.03	0.46	—	0.05	—
Gallicolumba stairi	—	—	0.05	—	—	—	—	—
Ptilinopus porphyraceus	1.88	1.3	1.76	0.60	1.33	1.7	0.53	1.31
Ducula pacifica	0.69	1.7	1.80	—	0.25	—	0.26	0.56
Collocalia spodiopygia	1.38	0.1	0.15	0.70	1.54	1	1.78	0.38
Halcyon chloris	0.75	0.4	0.69	0.07	0.21	0.7	1.05	0.19
Aplonis tabuensis	3.69	0.9	1.41	1.33	1.08	1.7	1.26	1.38
Lalage maculosa	1.88	1.3	0.81	1.2	1.5	1.7	0.74	1.25
Pachycephala jacquinoti	0.56	2.1	2.12	0.07	0.67	1	0.47	0.69
Foulehaio carunculata	1.38	1	1.75	0.43	0.38	—	0.05	0.31
TOTAL	11.02	8.7	10.44	3.86	5.88	7.5	4.47	5.69

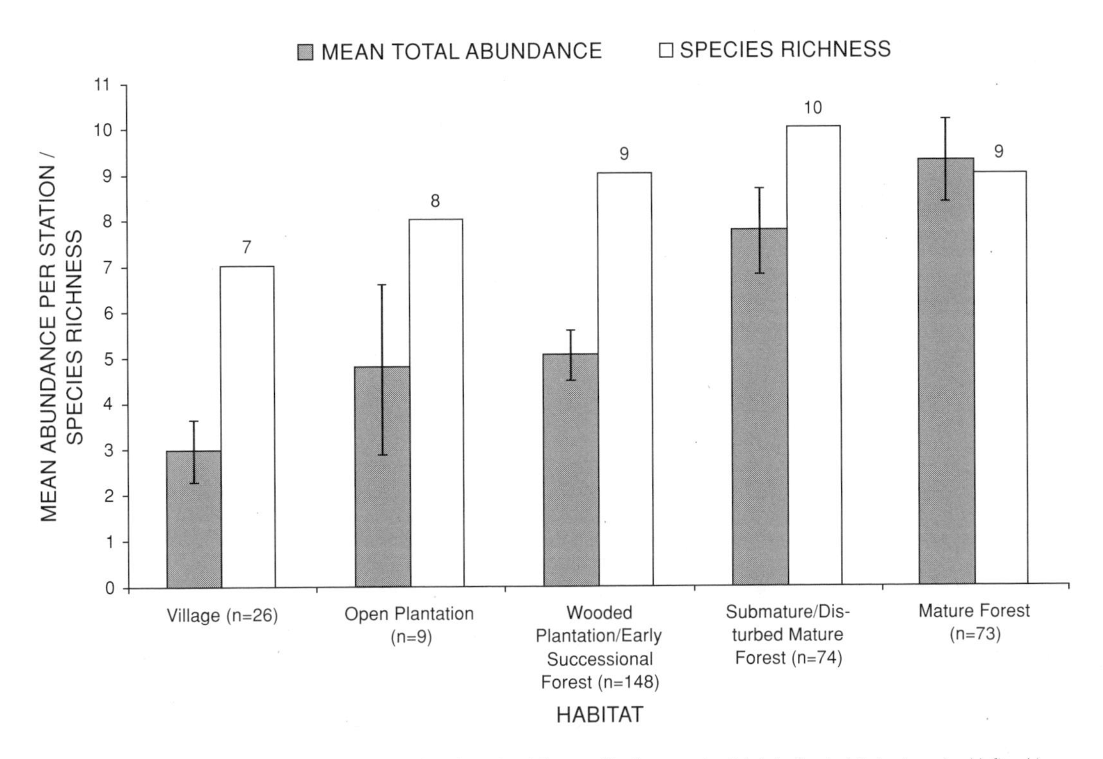

Figure 6-6 Total relative abundance (mean ± SE) and species richness of indigenous landbirds in five habitat categories (defined in Steadman & Freifeld 1998), Vava`u, Tonga. From Steadman & Freifeld (1998: Figure 4).

Hunga (20)	Ofu (16)	`Utungake (12)	Vaka`eitu (11)	`Euakafa (22)	Taunga (25)	A`a (12)	Mafana (16)	`Eueilki (3)	Foelifuka (14)	Foeata (4)
—	—	—	—	—	—	—	0.31	—	—	—
—	—	—	—	0.09	—	—	—	—	—	—
—	—	—	—	—	—	—	—	—	—	—
1.1	1	1.33	1	0.5	0.28	2.25	0.81	0.7	0.57	0.5
0.65	0.38	1	2.09	0.59	0.04	2.25	0.69	1	0.14	—
0.35	0.81	0.67	0.27	0.08	0.37	0.25	0.69	0.1	0.29	—
0.35	0.56	0.67	0.27	0.64	0.04	0.33	0.44	—	0.57	0.2
1.25	2.13	1.17	0.73	0.82	1.16	0.5	1.63	1.3	0.43	—
0.95	1	1.42	1.55	0.36	1.16	0.83	1.75	0.7	1.5	1.5
0.35	—	0.67	0.91	1.23	—	0.25	—	1	0.64	0.5
0.25	—	0.08	0.18	2.18	0.2	0.17	0.06	2	—	—
4.9	5.07	6.34	6.73	6.41	2.88	6.58	5.69	6.7	3.85	2.7

Table 6-13 Distribution of landbirds in the Vava`u Group, Tonga. X, recorded July 1995 or July 1996; —, not recorded in 1995/1996 and presumably truly absent; A, not recorded in 1995/1996 but likely occurs; * breeding unlikely; i, introduced species (not included in vertical totals). From Steadman & Freifeld (1998: Table 4).

	`Uta Vava`u	Pangaimotu	Kapa	Hunga	Ofu	`Utungake	Vaka`eitu	`Euakafa	A`a	Taunga	Mafana	`Eueiki	Foelifuka	Foeata	Nuku	Kulo	Totals X	Totals X + A
Gallus gallus (i)	X	X	X	X	X	X	X	A	A	X	X	X	X	X	—	—	12	14
Gallirallus philippensis	X	X	A	A	A	A	—	—	A	A	X	—	—	—	—	—	3	9
Porphyrio porphyrio	X	X	A	A	—	A	A	X	A	—	—	A	—	—	—	—	3	9
Ptilinopus porphyraceus	X	X	X	X	X	X	X	X	X	X	X	X	X	X	A	X	15	16
Ducula pacifica	X	X	X	X	X	X	X	X	X	X*	X	X	X*	A*	A*	X*	14	16
Gallicolumba stairi	X	—	—	—	—	—	—	—	—	—	—	—	—	—	—	—	1	1
Tyto alba	A	A	A	X	A	A	A	X	A	A	A	A	X	X	A	A	4	16
Collocalia spodiopygia	X	X	X	X	X*	X	X	X	X	X*	X*	X	X*	A*	A*	A*	13	16
Halcyon chloris	X	X	X	X	X	X	X	X	X	X	X	A	X	X	A	A	13	16
Pycnonotus cafer (i)	X	—	—	—	—	—	—	—	—	—	—	—	—	—	—	—	1	1
Aplonis tabuensis	X	X	X	X	X	X	X	X	X	X	X	X	X	A	A	X	14	16
Lalage maculosa	X	X	X	X	X	X	X	X	X	X	X	X	X	X	X	X	16	16
Pachycephala jacquinoti	X	X	X	X	—	X	X	X	X	—	—	X	X	X	—	—	11	11
Foulehaio carunculata	X	X	X	X	—	X	X	X	X	X	X	X	—	—	—	—	11	11
TOTAL																		
X	11	10	8	9	6	8	8	10	8	7	8	7	8	5	1	4	14	—
X + A	12	11	11	11	8	11	10	10	11	9	9	10	8	8	7	7	—	14

landbirds. Our surveys were confined therefore to inland forests. Mature inland forest persists in Vava'u mainly in areas too steep for cultivation and covers about 10% of the land area.

Among modern native landbirds in Vava'u, 13 species are widespread and at least locally common, one (the ground-dove *Gallicolumba stairi*) is extremely rare, and four have been extirpated in the past century or so (Tables 6-10, 6-12, 6-13). To evaluate the habitat association of landbirds, we classified the vegetation at each point-count station into one of five categories (Table 6-11). Total relative abundance of landbirds increased steadily with forest maturity (Figure 6-6). This trend held regardless of island area (Figure 6-7). Columbids (*Ptilinopus*, *Ducula*, *Gallicolumba*), whistlers (*Pachycephala*), and honeyeaters (*Foulehaio*) increased in abundance with forest maturity, whereas no such trend is apparent in the six other species (Figure 6-8). The species richness and relative abundance of both indigenous plants and vertebrates in Vava'u have been affected more by deforestation than by the three most often used physical variables of island biogeography—area, elevation, or isolation.

The Ha'apai Group of Tonga, lying south of Vava'u, consists of 57 raised limestone islands and five volcanic outliers (Kao, Tofua, Falcon, Hunga Tonga, Hunga Ha'apai). On average, the islands in Ha'apai (Figure 6-9, which does not depict many small islands) are more isolated from each other than those of Vava'u. Based on their tectonic setting,

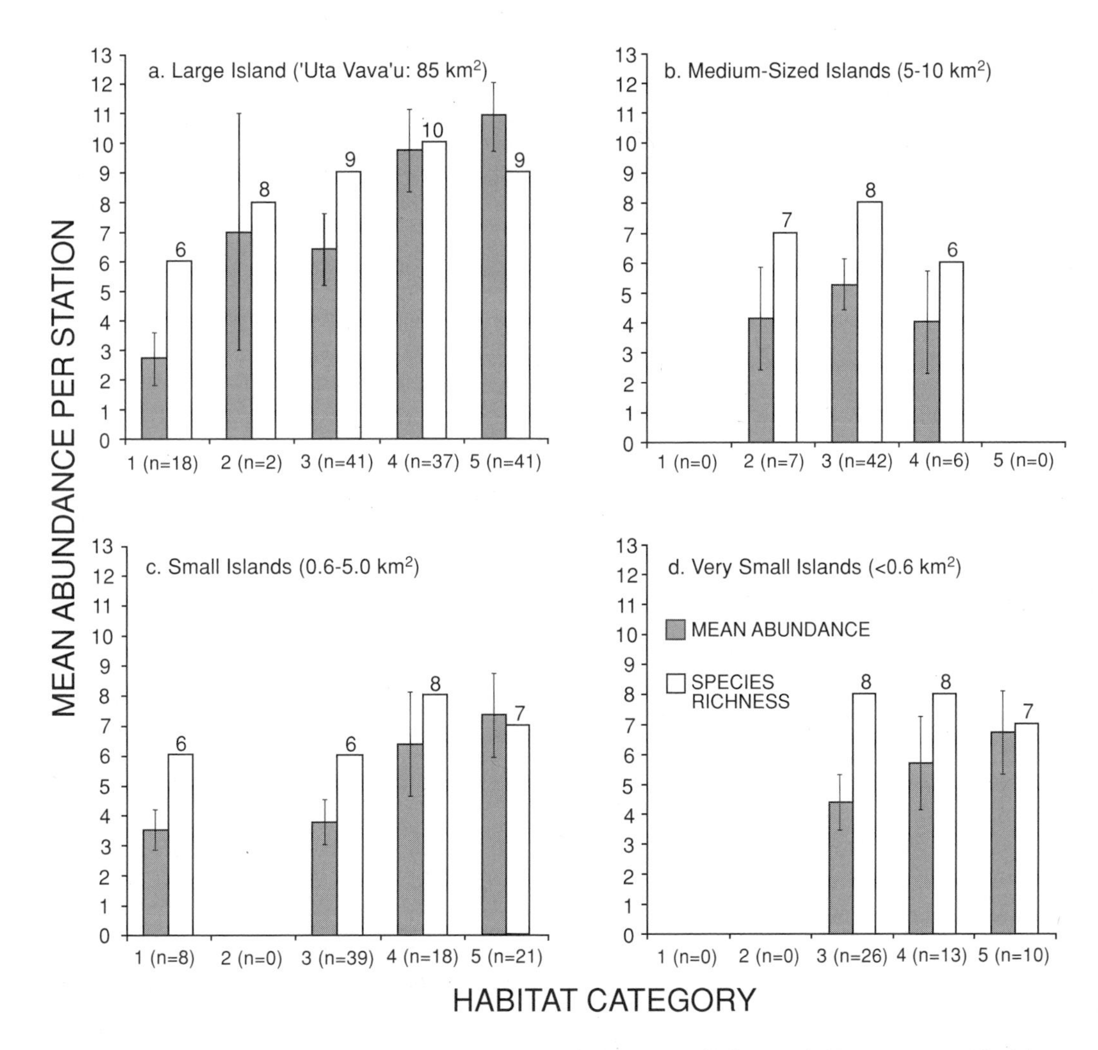

Figure 6-7 Relative abundance (mean ± SE) and species richness of indigenous landbirds among habitat categories on islands in different size classes, Vava`u, Tonga. Numbers in parentheses following habitat category refer to the number of stations in that habitat for that island size class. The habitat categories are (1) village, (2) open plantation, (3) wooded plantation/early successional forest, (4) submature/disturbed mature forest, and (5) mature forest. Large island (a) is `Uta Vava`u. Medium-sized islands (b) are Pangaimotu, Hunga, and Kapa. Small islands (c) are Ofu, `Euakafa, and Taunga. Very small islands (d) are Mafana, A`a, `Eueiki, Foelifuka, and Foeata. From Steadman & Freifeld (1998: Figure 5).

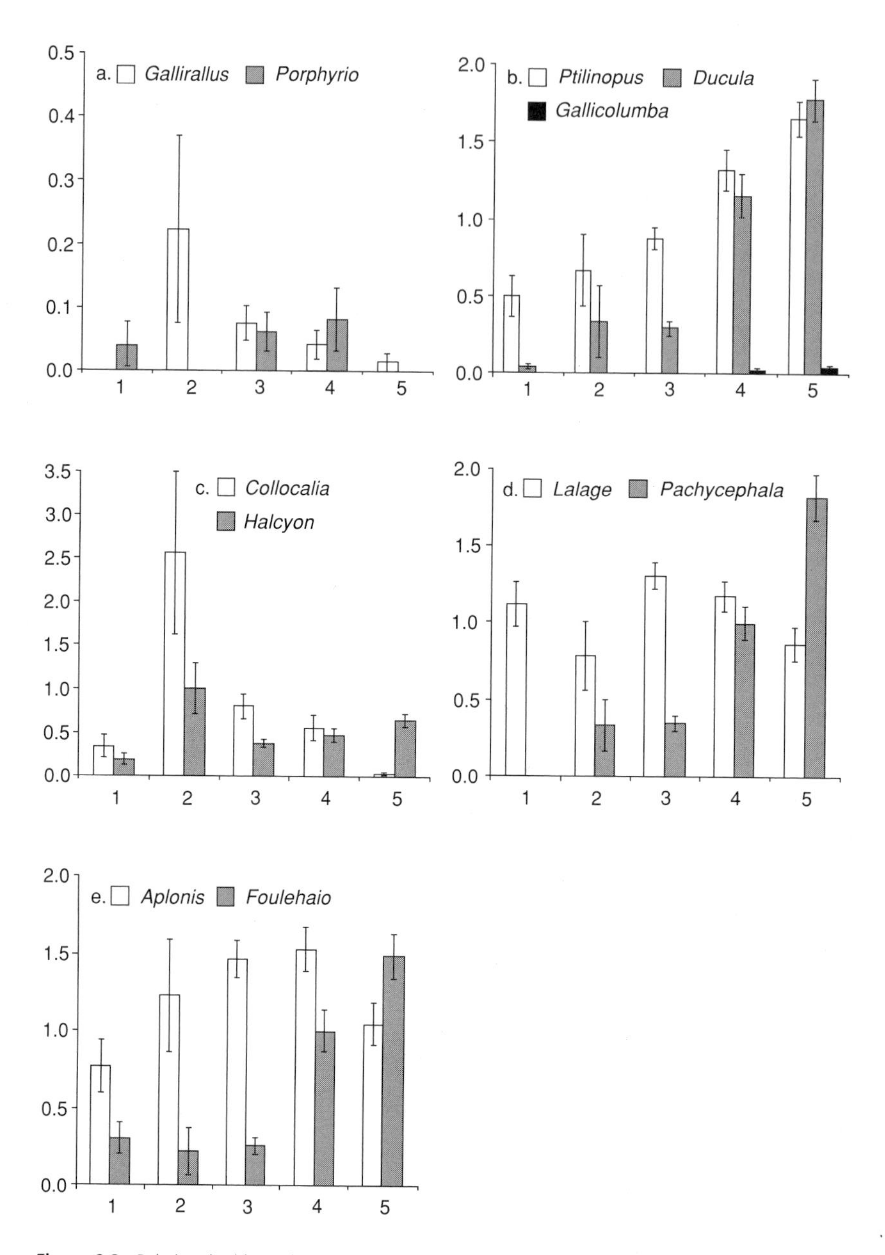

Figure 6-8 Relative abundance (mean ± SE) of indigenous landbirds in five habitat categories, Vava`u, Tonga. 1, village; 2, open plantation; 3, wooded plantation/early successional forest; 4, submature/disturbed mature forest; 5, mature forest. From Steadman & Freifeld (1998: Figure 6).

the 57 nonvolcanic islands in Ha'apai (raised limestone islands or sand cays; conjoined as just six islands in the late Pleistocene) are divided into three subgroups (Dickinson et al. 1994, 1999). The largest island in Ha'apai is volcanic Tofua (46.6 km^2) whereas the highest is nearby volcanic Kao (1046 m). The largest nonvolcanic islands are Foa and Lifuka (Table 6-14), with 10 others (six that I surveyed for birds) >1 km^2. Most of them are <20 m in elevation, although several are higher, such as Mango (38 m), Nomuka (45 m), and Fotuha'a (60 m). About 9000 persons lived in Ha'apai in 1993 (Christopher 1994), with about one half of them on Foa and Lifuka.

Among the 13 islands I surveyed for birds in Ha'apai in July 1995 and July 1996 (Table 6-14), mature inland forest exists only on Tofua (Steadman 1998). Vegetation on the 12 smaller (0.15–13.3 km^2), lower (6–45 m) islands is a mosaic of active and abandoned agricultural plots, nearly all with an overstory of coconut trees. The minimal topographic relief and intensive land use on these 12 islands lead to a simpler habitat classification than in Vava'u. Cultivation prevents inland vegetation from reverting to secondary forest that is more than ca. 40 years old. Unlike in Vava'u, I did not survey birds in villages in Ha'apai, where "plantation" habitat is equivalent to both "open plantation" and

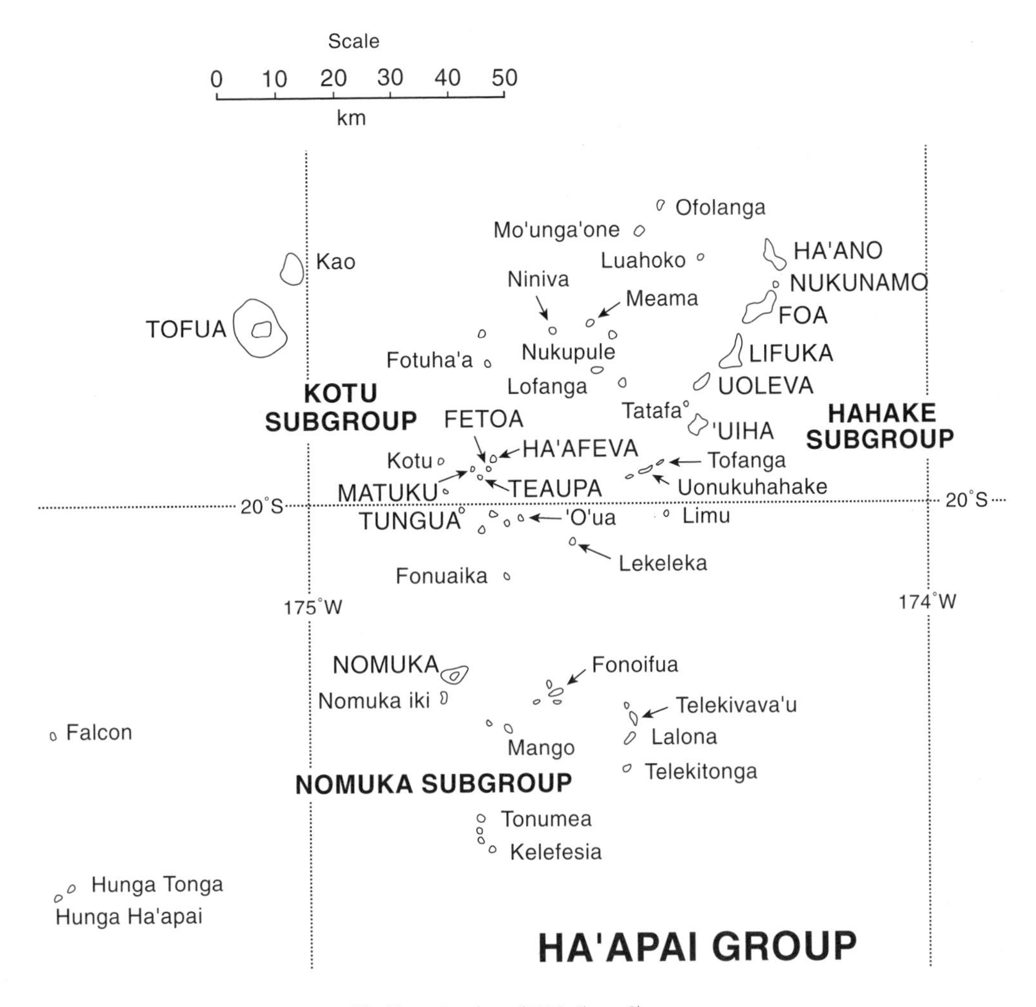

Figure 6-9 Ha`apai Group, Tonga. Modified from Steadman (1998: Figure 2).

"wooded plantation/early successional forest" in Vava'u. "Secondary forest" in Ha'apai is equivalent to "early successional forest" in Vava'u. The mature forest categories are similar, although the trees are larger, taller, and more widely spaced on Tofua than in Vava'u.

Of the 16 or 17 resident species of landbirds in Ha'apai, nine are widespread and at least locally common, although only four (*Gallirallus philippensis*, *Ptilinopus porphyraceus*, *Halcyon chloris*, *Aplonis tabuensis*) certainly or probably occur now on all 13 surveyed islands (Table 6-15). Three (*Gallicolumba stairi*, *Ptilinopus perousii*, *Clytorhynchus vitiensis*) are extirpated or extremely rare except on Tofua. On the seven limestone islands where I did point-counts, each dominated by plantation habitat, only three species were found on every island (Table 6-16). The overall species richness and abundance of landbirds are by far greatest on Tofua, even when only forested habitats are considered (Tables 6-16, 6-17, Figure 6-10). I believe that this difference is due more to Tofua's higher-quality forests than its greater area and elevation. The low, flat islands of Ha'apai no longer have mature forest, which persists on Tofua because of a steep, roadless interior and a cliffy coastline that restricts boat access. Except on Tofua, the overall abundance of birds in Ha'apai is less than in Vava'u.

Tongatapu is a fairly low, flat, raised limestone island of Pliocene and Quaternary age (Roy 1997a). It is Tonga's commercial center and by far its largest and most populous island (259 km^2, ca. 80,000 persons). Little native forest remains on Tongatapu, an island that has been densely populated for the past three millennia. More than 90% of the land area is occupied by active or recently abandoned agricultural plots or villages (Hau'ofa 1977, Wiser et al. 2002, pers. obs. 1985–2001). I did point-counts of landbirds at 54 localities across Tongatapu during August 1999 and 2001 (Table 6-18). Unlike on Vava'u and Ha'apai, four introduced species of birds (the chicken *Gallus gallus*, pigeon *Columba livia*, bulbul *Pycnonotus cafer*, and starling *Sturnus vulgaris*) are a major component of Tongatapu's

Table 6-14 Characteristics of islands surveyed in the Ha`apai Group, Tonga. Human population values are from the 1986 Tongan census. Habitat types: 1, plantation; 2, secondary forest; 3, mature forest. [a] includes two stations mixed with freshwater wetlands; [b] includes thirteen stations mixed with secondary forest, seven stations mixed with secondary coastal forest, and two stations at the edge of a village. From Steadman (1998: Table 1).

				Number of stations in each habitat type			
Island	Area (km^2)	Max. elev. (m)	Human population	1	2	3	Dates visited
VOLCANIC SUBGROUP							
Tofua	46.6	558	89	—	—	6	15–17 Jul 1995
HAHAKE SUBGROUP							
Foa	13.3	20	1409	25	19[a]	—	11, 23, 24 Jul 1996
Lifuka	11.4	16	2840	51	—	—	9–15, 23–26 Jul 1996
Ha`ano	6.6	12	728	20	—	—	12 Jul 1996
`Uiha	5.3	11	913	16	—	—	13 Jul 1996
Uoleva	2.0	8	0	20	—	—	25 Jul 1996
Nukunamo	0.25	6	0	—	—	—	12 Jul 1996
KOTU SUBGROUP							
Ha`afeva	1.8	10	449	42	—	—	17–22 Jul 1996
Tungua	1.5	17	305	27	—	—	19 Jul 1996
Matuku	0.3	11	142	—	—	—	20 Jul 1996
Fetoa	0.15	16	0	—	—	—	20 Jul 1996
Teaupa	0.15	17	0	—	—	—	20 Jul 1996
NOMUKA SUBGROUP							
Nomuka	7.0	45	687	53[b]	—	—	15–17 Jul 1996

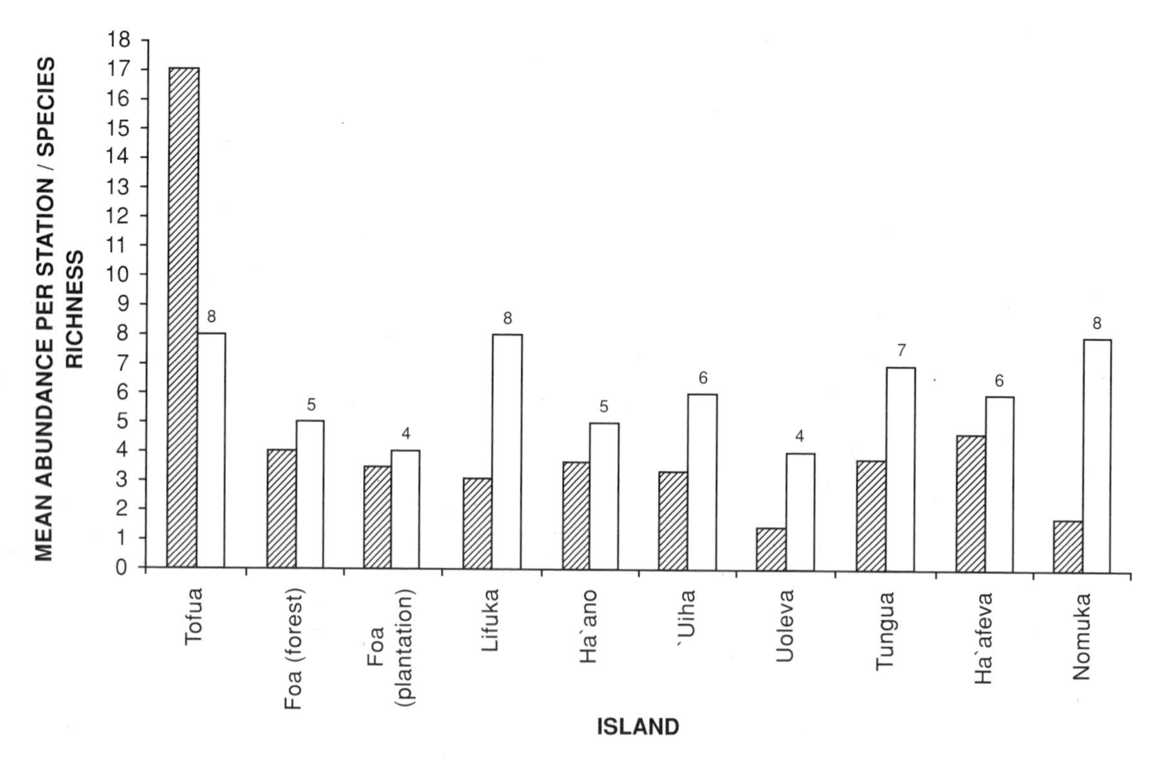

Figure 6-10 Relative abundance and species richness of landbirds, Ha`apai, Tonga. Based on transect data in Tables 6-15, 6-16. See Table 6-17 for expanded species richness values. Data from Tofua are all from forest; data from Foa are from both forest and plantation; data from the other seven islands are from plantation. From Steadman (1998: Figure 3).

Table 6-15 Distribution of landbirds, Ha`apai Group, Tonga. X, recorded July 1995 or July 1996; A, not recorded in 1995/96 but likely occurs; B, not recorded in 1995/96 but possibly occurs; i, introduced species; m, migrant species; *, breeding unlikely. i, m, X*, A* are not included in column totals. From Steadman (1998: Table 10), except for *Egretta sacra* and *Anas superciliosa*.

	Tofua	Foa	Lifuka	Ha`ano	`Uiha	Uoleva	Nukunamo	Ha`afeva	Tungua	Matuku	Fetoa	Teaupa	Nomuka	Totals	
														(X)	(X + A)
Egretta sacra	B	X	X	X	X	X	X	X	X	X	X	X	X	13	3
Anas superciliosa	X	X	X	B	B	B	B	A	B	B	B	B	X	4	5
Circus approximans	A	X	A*	A*	A*	B*	B*	B*	B*	—	—	—	X	2	3
Gallus gallus (i)	A	X	X	X	X	X	X	X	X	X	—	—	X	10	11
Gallirallus philippensis	A	X	X	X	X	A	A	X	X	A	A	A	A	6	13
Porphyrio porphyrio	A	A	X	A	A	A	A	X	A	X	B	B	X	4	11
Ptilinopus perousii	X	B	X	B	B	—	—	—	—	—	—	—	—	2	2
Ptilinopus porphyraceus	X	X	X	X	X	X	X	X	X	X	A	A	X	11	13
Ducula pacifica	X	A*	X*	A*	A*	A*	A*	B*	B*	B*	B*	B*	B*	1	1
Gallicolumba stairi	A	—	—	—	—	—	—	—	—	—	—	—	—	0	1
Vini australis	A	X	X	A	A	A	B	X	X	X	A	X	A	6	12
Eudynamys taitensis (m)	A*	A*	X*	X*	A*	X*	A*	A*	A*	A*	A*	A*	A*	—	—
Tyto alba	X	A	A	A	A	A	A	A	X	A	B	B	X	3	11
Collocalia spodiopygia	X	—	—	—	—	—	—	—	—	—	—	—	—	1	1
Halcyon chloris	X	X	X	X	X	X	X	X	X	X	A	X	X	12	13
Aplonis tabuensis	X	X	X	X	X	X	X	X	X	X	X	X	X	13	13
Lalage maculosa	X	X	X	X	X	X	A	B	B	B	—	—	X	7	8
Clytorhynchus vitiensis	X	—	—	—	B	—	—	B	—	—	—	—	—	1	3
Foulehaio carunculata	X	A	X	A	X	B	B	X	X	X	A	X	X	8	11
TOTAL															
X	11	9	11	6	7	5	4	8	8	7	2	5	10	16	—
X + A	16	12	12	10	10	9	7	10	9	9	7	7	12	—	17

Table 6-16 Relative abundance of birds (mean birds per station) in plantation transects, Ha`apai Group, Tonga. i, introduced species; m, migrant species. i, m are not included in totals. Modified from Steadman (1998: Table 5).

	Foa	Lifuka	Ha`ano	`Uiha	Uoleva	Ha`afeva	Tungua	Nomuka	Mean
Circus approximans	—	—	—	—	—	—	—	0.02	<0.01
Gallus gallus (i)	—	0.24	0.25	0.38	0.15	0.33	0.04	0.32	.21
Gallirallus philippensis	—	0.18	0.10	0.50	—	0.07	0.07	—	.12
Porphyrio porphyrio	—	0.04	—	—	—	0.02	—	0.02	.01
Ptilinopus porphyraceus	0.68	0.43	0.80	0.56	0.20	0.62	0.33	0.23	.48
Ducula pacifica	—	0.02	—	—	—	—	—	—	<0.01
Vini australis	—	—	—	—	—	0.50	0.52	—	.13
Eudynamys taitensis (m)	—	0.04	0.05	—	0.05	—	—	—	.02
Halcyon chloris	0.44	0.26	0.65	0.50	0.20	0.38	0.26	0.30	.37
Hirundo tahitica	—	—	—	—	—	—	—	0.07	.01
Aplonis tabuensis	0.92	0.92	0.95	0.82	0.75	0.64	0.26	0.40	.71
Foulehaio carunculata	—	0.16	—	0.31	—	0.88	1.22	0.36	.36
Lalage maculosa	0.84	0.20	0.05	0.06	0.05	—	—	0.08	.16
Total species	4	8	5	6	4	7	6	8	6.0
Total mean pairs	2.88	2.21	2.55	2.75	1.20	3.11	2.66	1.48	2.33

Table 6-17 Relative abundance of landbirds in forest transects, Ha`apai Group, Tonga. The habitat on Tofua is mature forest; on Foa it is secondary forest. From Steadman (1998: Table 3).

	Mean pairs per station		Mean birds per station	
	Tofua	Foa	Tofua	Foa
Circus approximans	—	0.05	—	0.05
Ptilinopus perousii	0.17	—	0.17	—
Ptilinopus porphyraceus	1.00	1.58	1.00	1.58
Ducula pacifica	2.33	—	4.33	—
Collocalia spodiopygia	0.50	—	0.83	—
Halcyon chloris	0.67	0.42	0.83	0.42
Aplonis tabuensis	1.17	0.84	2.67	1.16
Lalage maculosa	—	0.53	—	0.79
Clytorhynchus vitiensis	1.17	—	2.33	—
Foulehaio carunculata	2.17	—	4.83	—
Total species	8	5	8	5
Total mean pairs/birds	9.18	3.42	16.99	4.00

modern avifauna in three of the five habitat categories (Figure 6-11). Contamination of Tongatapu's bird communities by nonnative species has increased since my visits to the island in 1985, 1988, and 1989, although I have no point-count data from these years. The chicken, of course, is a prehistoric introduction whereas the pigeon, bulbul, and starling arrived on Tongatapu within the past century and continue to spread to other Tongan islands.

Among Tongatapu's native landbirds, five resident species (the swiftlet *Collocalia spodiopygia*, kingfisher *Halcyon chloris*, triller *Lalage maculosa*, honeyeater *Foulehaio carunculata*, and starling *Aplonis tabuensis*) are widespread and common, being found in all five habitats surveyed (Table 6-18). The other five recorded on point-counts are more localized, occurring mainly in mangroves (the heron *Egretta sacra*, duck *Anas superciliosa*, and rails *Gallirallus philippensis* and *Porphyrio porphyrio*) or forests (fruit-dove *Ptilinopus porphyraceus*). Similar trends (relative abundance of columbids and nectarivores increasing with forest maturity) were evident on Tongatapu as in Vava'u (Figures 6-8, 6-12), although on Tongatapu the pigeon *Ducula pacifica* is rare, the forest-obligate whistler *Pachycephala jacquinoti* is absent, and the fruit-dove *Ptilinopus perousii* still exists but is highly localized and depends on the large *Ficus obliqua* trees growing mainly in or near villages.

Table 6-18 Relative abundance of birds on Tongatapu, Tonga, 1–14 August 1999 and 13–14 August 2001. Values are mean individuals seen/heard per point-count, rounded to the nearest 0.1. The number of point-counts is given in parentheses for each habitat type. i, introduced; m, migratory shorebird (m not included in totals); *, heard only at distances >50 m. Early successional/submature/disturbed forest is intermediate between categories 3 and 4 in Vava`u and 2 and 3 in Ha`apai (Tables 6-11, 6-14).

	Urban/ suburban (9)	Cultivated coconut plantation (13)	Partially wooded coconut plantation (14)	Early successional/ submature/disturbed forest (9)	Mangrove forest (9)
Egretta novaehollandiae	—	—	—	—	0.2
Egretta sacra	—	—	—	—	0.9
Anas superciliosa	—	—	—	—	0.9
Gallirallus philippensis	—	—	—	—	0.4
Porphyrio porphyrio	—	—	—	—	0.6
Pluvialis fulva (m)	—	—	—	—	0.3
Heteroscelus incanus (m)	—	—	—	—	0.4
Ptilinopus perousii	—	—	—	—	0.1
Ptilinopus porphyraceus	—	—	0.1	1.8	0.1
Ducula pacifica	—	—	—	0.1	—
Collocalia spodiopygia	1.1	2.1	2.6	0.1	0.9
Halcyon chloris	0.2	0.4	0.8	0.8	1.0
Aplonis tabuensis	1.4	0.8	1.3	1.0	0.4
Lalage maculosa	0.3	0.8	0.4	0.4	1.6
Foulehaio carunculata	1.6	2.2	3.1	4.1	2.2
Gallus gallus (i)	2.1	0.3	0.2	*	—
Columba livia (i)	1.4	0.2	0.4	—	—
Pycnonotus cafer (i)	3.6	1.2	0.2	—	—
Sturnus vulgaris (i)	5.0	2.4	0.8	0.4	—
TOTAL INDIVIDUALS					
Native species, with *Collocalia*	4.6	6.3	8.3	8.3	9.3
Native species, without *Collocalia*	3.5	4.2	5.7	8.2	8.4
Introduced species	12.1	4.1	1.6	0.4	0
% native individuals (with *Collocalia*)	28	61	84	95	100
% native individuals (without *Collocalia*)	22	51	78	94	100
TOTAL SPECIES					
Native	5	5	6	7	12
Nonnative	4	4	4	2	0

'Eua (Figure 6-13) is a large, high island 18 km southeast of Tongatapu. Most of 'Eua consists of terraced, uplifted limestone (Taylor & Bloom 1977) that sustained substantial tracts of rain forest in the late 1980s and early 1990s (Drake et al. 1996, DWS pers. obs.). Although quantified surveys of landbirds have not been done on 'Eua, Rinke (1987) and Steadman (1993a) have described the modern birdlife from fieldwork in the 1970s and 1980s. With 20-20 hindsight, how I wish that I would have spent two or three extra weeks on 'Eua in 1988 and 1989 to do point-counts of living birds as well as unearth prehistoric ones. Of 13–14 native species of landbirds on 'Eua today (Table 6-10), each is common and widespread except the swallow *Hirundo tahitica*. Nonnative species present in the 1980s were chickens (*Gallus gallus*, both feral and domesticated), a parrot (*Prosopeia tabuensis*), starling (*Sturnus*

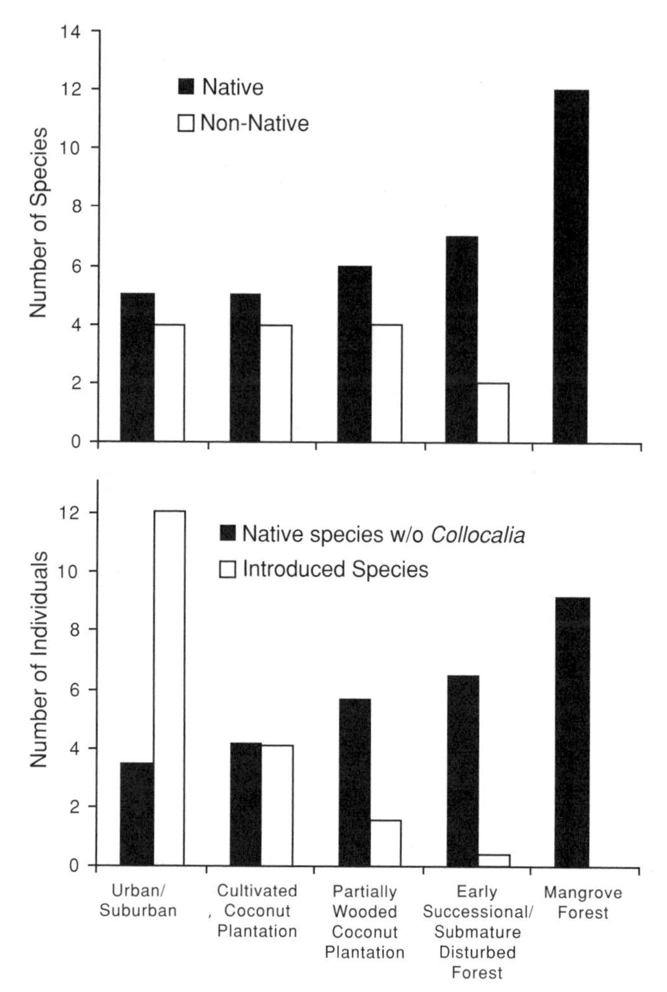

Figure 6-11 Native vs. nonnative landbirds in five habitat categories on Tongatapu, Tonga. Based on data in Table 6-18.

vulgaris), and a small colony of mynas (*Acridotheres fuscus*).

The Ha'apai Group, Tongatapu, and 'Eua all have excellent data on ancient birdlife. On five islands in Ha'apai (Figures 3-8, 6-9), David Burley and colleagues (occasionally including me) have excavated five Lapita sites (2850–2700 cal BP) as follows: Faleloa on Foa, Tongoleleka on Lifuka, Pukotalo on Ha'ano, Vaipuna on 'Uiha, and Mele Havea on Ha'afeva (Burley et al. 1995, Burley et al. 2001, Steadman et al. 2002b). The stratigraphy and chronology of human arrival in Ha'apai, and its sudden impact on native species, have been studied intensively at Tongoleleka (site Li7) on Lifuka (Steadman et al. 2002a; Chapter 3 herein). Based on 20 AMS ^{14}C dates on individual bones, the well-stratified profile at Li7 portrays a short (less than two centuries) period of human occupation during which major cultural and faunal changes took place, such as the loss of Lapita pottery, the introduction of chickens, and the extinction or extirpation of iguanas and megapodes (Tables 3-3, 3-4, Figures 3-9, 3-10), not to mention the loss of seabirds (Chapter 15) and other landbirds.

I have identified all 1448 bones of native landbirds from the five Lapita sites in Ha'apai, although have not yet tabulated all of them stratigraphically. Prehistoric species richness increased with the number of identified bones (Table 6-19). If 600 landbird bones were available from every island, as for Lifuka, I believe that 31 ± 2 species would be identified on each. For example, a flightless rail was sampled only on the two islands (Lifuka, Ha'afeva) with the most identified bones. Inter-island variation in species richness that might be genuine is the fewer species of passerines on Ha'afeva, a small island that holds its own, however, for nonpasserines. Many more than the 21 passerine bones currently available from Ha'afeva are needed to test this.

Except for two flightless rails (*Gallirallus* undescribed spp. G, H) and the robin *Petroica* cf. *multicolor*, each of the 21 extinct or extirpated species from Ha'apai has been found as well on Tongatapu or 'Eua. Five of the 21 lost species have been found on only one island—the three just mentioned, the thrush *Turdus poliocephalus* and honeyeater *Myzomela cardinalis* (Table 6-19). The row totals for NISP in Table 6-19, just like the column totals discussed in the last paragraph, reveal strong sampling effects. For example, the three extinct/extirpated species where NISP ≥60 are recorded on all five islands, as are five of six species where NISP ≥28.

The small, low limestone islands of Ha'apai sustained a diverse avifauna at first human contact. At least 11 species are extinct (a heron, two megapodes, two rails, five columbids, and a parrot), and 10 others occur on none of these five islands today (an osprey, rail, columbid, parrot, cuckoo, thrush, robin, whistler, and honeyeater). One can only imagine the magnificent forests that presumably grew on these islands when as many as nine species of pigeons and doves co-occurred, not to mention three megapodes and six or more passerines. The ecological implications of such rich landbird faunas on small islands will be discussed in Chapters 19-21.

The prehistory of birds on Tongatapu is based on excavations by David Burley and me in 1999 and 2001. Major archaeological projects took place on Tongatapu in the past but yielded little in the way of bird bones (Poulsen 1987, Spennemann 1989, 1997). The most productive site for prehistoric birds is the Lapita occupation at Ha'ateiho, excavated by David Burley, Andrew Barton, Jim Mead, Sandy Swift, and me. Several smaller sites have yielded a few bones of species found as well at Ha'ateiho. Nearly all Lapita pottery at Ha'ateiho is from stratum III, which also contains most of the bird bones. Each of the extinct or

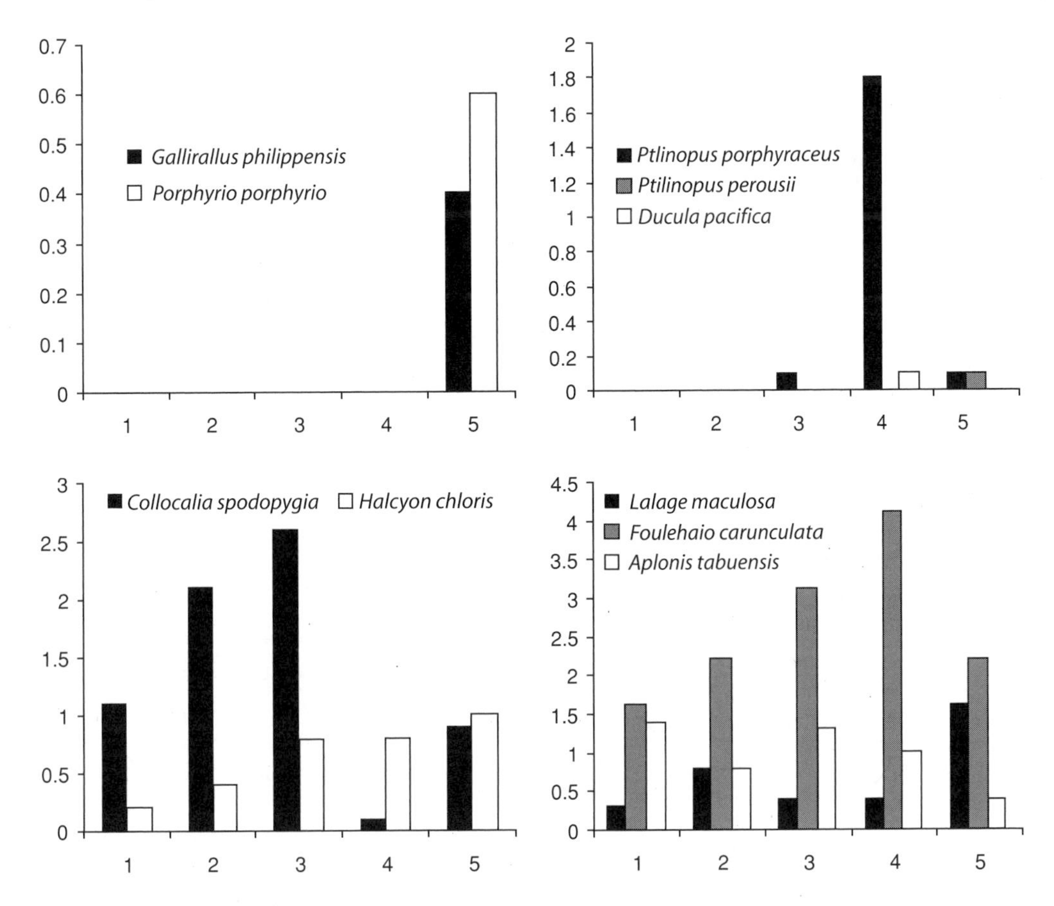

Figure 6-12 Relative abundance (mean ± SE) of native landbirds in five habitat categories on Tongatapu, Tonga. 1, urban/suburban; 2, cultivated coconut plantation; 3, partially wooded coconut plantation; 4, early successional/submature disturbed forest; 5, mangrove forest. Based on data in Table 6-18.

extirpated species found at Ha'ateiho has been recorded on other Tongan islands except an enigmatic large megapode, which may not belong in the genus *Megapodius*. The difference in combined species richness of landbirds between Tongatapu (31) and 'Eua (37–38; Table 6-10) is probably a sampling artifact (NISP = 437 for Tongatapu vs. 1319 for 'Eua).

The prehistoric record of vertebrates on 'Eua is rich with extinct or extirpated species of lizards, birds, and bats (Steadman 1993a, 1995a, Pregill 1993, Koopman & Steadman 1995). Unlike the vertebrate history of Ha'apai or Tongatapu, that of 'Eua features hundreds of precultural fossils as well as those from cultural sites. At least 27 species of landbirds have been identified from precultural contexts (Tables 6-20, 6-21), including six of the 13 or 14 species surviving on the island. Of the other survivors, four (*Egretta sacra*, *Anas superciliosa*, *Collocalia spodiopygia*, *Hirundo tahitica*) probably inhabited 'Eua in prehuman times but were not in the fossil sample; the others (volant rails *Gallirallus philippensis*, *Porzana tabuensis*, and *Porphyrio porphyrio*, barn-owl *Tyto alba*) may have colonized 'Eua since human arrival ca. 2800 years ago (Chapters 10, 13). The three volant species of extinct or extirpated landbirds recorded from 'Eua but nowhere else in Tonga (a hawk, warbler, and monarch; Table 6-10) are all rare as fossils (NISP = 3–5; Table 6-21) and probably await discovery on other islands.

The prehuman avifauna of 'Eua was similar in species composition to that which naïvely welcomed human settlers. I believe that most if not all species not yet recorded from cultural strata on 'Eua (Table 6-20, Figure 6-14) were in fact there at human arrival. This suggests low rates of species turnover in landbirds during prehuman times, a topic developed further in Chapter 18. It certainly is safe to say that the arrival of humans has influenced the Tongan avifauna more than any other climatic, tectonic, or biological event of the past ~100,000 years (Steadman 1993a).

Of the 24 species of landbirds recorded as fossils from anywhere in Tonga but no longer living in the island group (Table 6-10), the nearest geographic occurrences of conspecifics or most closely related congeners are from

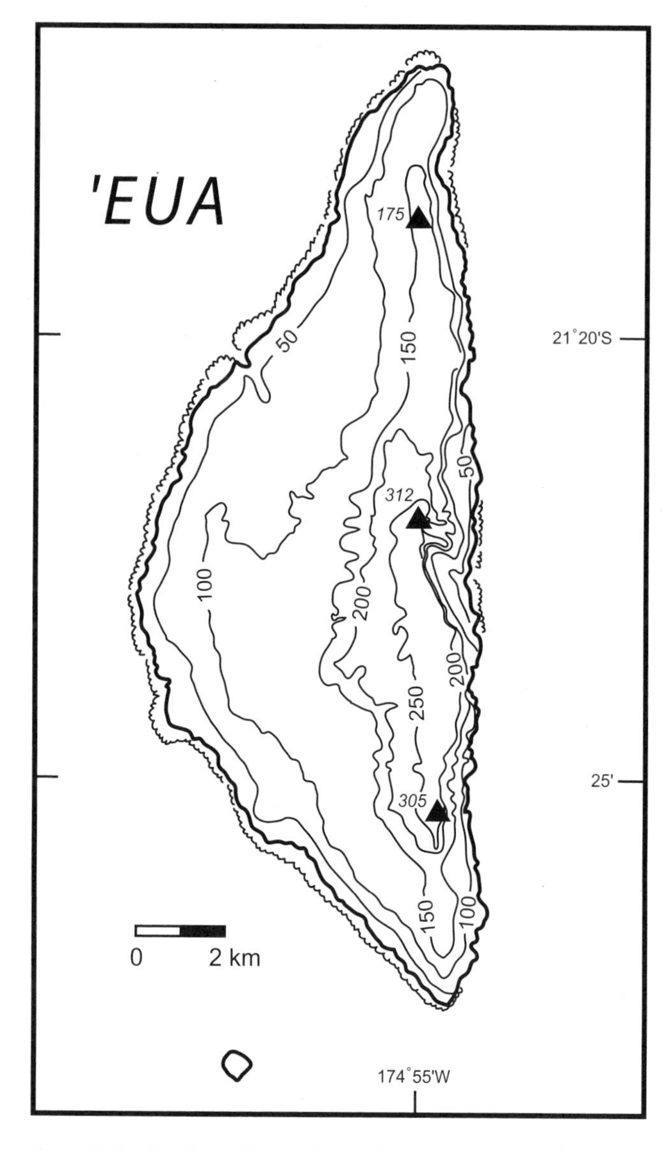

Figure 6-13 `Eua, Tonga. Contour intervals in m. Redrawn from Rinke (1987: Figure 1).

the Solomon Islands (1 sp.), Vanuatu (1 sp.) New Caledonia (1 sp.), Fiji and/or Samoa (11 spp.), or unknown (10 spp.). The large number of unknowns is a gauge of how poorly we understand the phylogenies of landbirds in Oceania.

Samoa

Samoa (Figure 6-15) consists of Independent (formerly Western) Samoa, with the large islands of Upolu and Savai'i and their small offshore islands, and American Samoa with Tutuila, Aunu'u, the Manu'a Group (Ta'u, Ofu, Olosega), and the atolls of Swains and Rose. The island group is dominated by active and eroded volcanic islands, mainly Quaternary but no older than late Pliocene (Chapter 1).

Today, more species of landbirds live on Upolu and Savai'i than elsewhere in Samoa (Table 6-22). The distribution and relative abundance of landbirds in American Samoa are well documented (Amerson et al. 1982a, 1982b, Banks 1984, Engbring & Ramsey 1989, Freifeld 1999, Freifeld et al. 2004), but the species composition is obviously artificial. Missing are 13 species (in 12 genera) known from Independent Samoa (Table 6-23), consisting of two rails, two columbids, and nine passerines (mostly understory insectivores). Six of the 12 missing genera (*Didunculus* and five passerines) occur today or formerly in Tonga. Nearly all of the landbirds shared by the two Samoas fall into five categories: aquatic/semiaquatic/grassland species (ducks, rails), predators (barn-owl), aerial insectivores (swiftlet), canopy frugivores (pigeons, doves, starlings), and canopy nectarivores (lorikeets, honeyeaters). The only exception is the ground-dove *Gallicolumba stairi*, a species extirpated or very rare throughout Samoa.

The species that now inhabit Independent Samoa but not American Samoa probably used to live in both places. From Upolu it is easy to see Tutuila 70 km to the east. That all islands in American Samoa are smaller than Savai'i and Upolu may not account for the difference in species richness. Evidence that habitat alteration might be a factor in American Samoa's much poorer avifauna comes from our bird surveys on small islands off Upolu in April 1999 (Freifeld et al. 2001; Table 6-23; Figure 6-15 herein). Nu'utele, the only one of these islands with extensive mature forest (Whistler 1983), has the richest avifauna. Apolima, which has the same land area as Nu'utele, supports fewer species of landbirds because it is inhabited (one village of ca. 100 persons) and substantially deforested. Five of the six species of landbirds found on Nu'utele but no other offshore islands are forest dwellers that are more common in relatively mature forest than in more degraded habitats on Upolu and Savai'i (Bellingham & Davis 1988, Evans et al. 1992). Just as in Fiji and Tonga (Steadman & Freifeld 1998; Tables 6-5 to 6-9, 6-12, 6-13 herein), forest-obligate columbids and passerines can occur on very small Samoan islands as long as some forest is intact.

Ironically, the four largest islands in American Samoa (Tutuila, Ta'u, Ofu, Olosega) have a higher percentage of forest cover today (Whistler 1980) than the much larger Upolu and Savai'i. The major causes of forest damage in Samoa are recent hurricanes (Elmqvist et al. 1994) and people. Once prized for extensive tracts of forest (Whistler 1992), Independent Samoa now has little forest left. Furthermore, unlike on limestone islands in Tonga (Franklin et al. 1999), secondary succession in Independent Samoa is dominated

Table 6-19 Prehistoric and modern landbirds on five limestone islands in the Ha`apai Group, Tonga. M, modern record (A or X from Table 6-15); M?, uncertain modern record (X*, A*, B, B*, from Table 6-15); P, prehistoric record, based on a single Lapita archaeological site on each island; NISP, number of identified specimens.

	Foa	Lifuka	Ha`ano	`Uiha	Ha`afeva	Total (M)	Total (M + P)	NISP
HERONS								
Egretta sacra	M	M, P	M	M	M	5	5	1
†*Nycticorax* undescribed sp. B	P	P	P	—	P	0	4	6
DUCKS								
Anas superciliosa	M, P	M	—	—	M, P	3	3	11
OSPREYS								
Pandion haliaetus	—	—	P	P	—	0	2	2
HAWKS								
Circus approximans	M	M?	M?	M?	M?	1–5	1–5	0
MEGAPODES								
Megapodius pritchardii	P	P	—	P	P	0	4	8
†*Megapodius alimentum*	P	P	P	P	P	0	5	369
†*Megapodius molistructor*	P	P	P	P	P	0	5	29
RAILS								
Gallirallus philippensis	M, P	M, P	M, P	M, P	M, P	5	5	165
†*Gallirallus* undescribed sp. E	—	P	—	—	—	0	1	4
†*Gallirallus* undescribed sp. F	—	—	—	—	P	0	1	6
Porzana tabuensis	P	P	P	P	P	0	5	28
Porphyrio porphyrio	M, P	M, P	M, P	M, P	M, P	5	5	180
PIGEONS, DOVES								
Ptilinopus perousii	P	M, P	—	P	P	1	4	24
Ptilinopus porphyraceus	M, P	M, P	M, P	M, P	M, P	5	5	28
Ducula pacifica	M, P	M, P	M?, P	M?, P	M?, P	2–5	5	60
†*Ducula latrans*	P	P	P	P	P	0	5	60
†*Ducula* undescribed sp.	P	P	—	—	P	0	3	25
†Undescribed genus C	P	P	—	P	P	0	4	35
†*Caloenas canacorum*	—	P	P	—	P	0	3	15
†*Didunculus* undescribed sp.	—	P	P	P	P	0	4	12
Gallicolumba stairi	P	P	P	P	P	0	5	65
PARROTS								
Vini solitarius	—	P	—	P	—	0	2	3
Vini australis	M	M	M	M	M	5	5	0
†*Eclectus* undescribed sp.	—	P	—	P	—	0	2	10
CUCKOOS								
Cacomantis cf. *flabelliformis*	—	P	—	P	—	0	2	2
BARN-OWLS								
Tyto alba	M, P	M	M, P	M	M	5	5	3
KINGFISHERS								
Halcyon chloris	M, P	M, P	M, P	M, P	M, P	5	5	111

(continued)

Table 6-19 (continued)

	Foa	Lifuka	Ha`ano	`Uiha	Ha`afeva	Total (M)	Total (M + P)	NISP
SWALLOWS								
Hirundo tahitica	—	—	—	—	M	1	1	0
THRUSHES								
Turdus poliocephalus	—	P	—	—	—	0	1	1
STARLINGS								
Aplonis tabuensis	M, P	M, P	M, P	M, P	M, P	5	5	132
TRILLERS								
Lalage maculosa	M, P	M, P	M, P	M	M?, P	4–5	5	6
ROBINS								
Petroica cf. *multicolor*	P	—	—	—	—	0	1	1
WHISTLERS								
Pachycephala jacquinoti	P	P	P	—	—	0	3	5
MONARCHS								
Clytorhynchus vitiensis	P	P	P	M?, P	—	0–1	4	18
HONEYEATERS								
Myzomela cf. *cardinalis/ jugularis*	—	—	P	—	—	0	1	1
Foulehaio carunculata	M, P	M, P	M, P	M, P	M, P	5	5	27
TOTAL SPECIES								
M	13	13–14	10–12	10–13	11–14	15–16	—	—
M + P	26	31–32	23–24	24–26	26–27	—	37	—
TOTAL NISP								
All landbirds	215	607	146	187	293	—	—	1449
Passerines	60	61	23	26	21	—	—	191
								—
Area (km^2)	13.3	11.4	6.6	5.3	1.8	—	—	—
Elevation (m)	20	16	12	11	10	—	—	—
Isolation (km from nearest island >10 km^2)	0.6 to Lifuka	0.6 to Foa	2.6 to Foa	7.5 to Lifuka	38 to Lifuka	—	—	—

by nonnative trees, such as *Adenanthera pavonina*, *Castilla elastica*, *Cananga odorata*, *Funtumia elastica*, and *Psidium guajava* (pers. obs.).

Four factors probably account for American Samoa being more forested than Independent Samoa. First, its steepness makes deforestation more difficult. Second, less subsistence agriculture is practiced today in American Samoa, especially on Tutuila, where most food is purchased in grocery stores rather than grown. In the past, it is likely that much forest on steep slopes in American Samoa was cleared for plantations. Third, Independent Samoa has been hit by six severe hurricanes in the 1960s and 1990s (Elmqvist et al. 2001, Hjerpe et al. 2001). Finally, major tracts of forest are protected on Tutuila and Ta'u as the National Park of American Samoa. As long as American Samoa has enough cash to offset the need for extensive agriculture, the future of its forests may be bright. Simplistically, American Samoa has the forests while Independent Samoa has the birds. What a difference 70 km of ocean can make, especially as reflected in the history of land use. Based in part on the age of a bone deposit on Tutuila (below) and the extinction data from Tonga (above), I would speculate that most of the hypothesized avian extinction in American Samoa occurred more than 2000 years ago.

Table 6-20 Chronology of resident landbirds from `Eua, Tonga. Prehuman record = >3000 yr BP (= strata II, II/III, and III of Anatú). Archaeological record = 3000-200 yr BP (15 sites). i, introduced by humans; not included in totals; *, extirpated species; †, extinct species. Updated from Steadman (1993a).

	Prehuman record	Archaeological record	Historic record	Extant in 1988
HERONS				
Egretta sacra	—	X	X	X
†*Nycticorax* undescribed sp. B	X	—	—	—
DUCKS				
Anas superciliosa	—	—	X	X
HAWKS				
**Accipiter* cf. *rufitorques*	X	—	—	—
MEGAPODES				
**Megapodius pritchardi*	X	—	—	—
†*Megapodius alimentum*	X	X	—	—
†*Megapodius* undescribed sp. F	X	X	—	—
CHICKENS				
Gallus gallus (i)	—	X	X	X
RAILS				
Gallirallus philippensis	—	X	—	X
†*Gallirallus* undescribed sp. H	X	—	—	—
**Porzana tabuensis*	—	X	X	—
Porphyrio porphyrio	—	X	X	X
PIGEONS AND DOVES				
Ptilinopus perousii	X	X	—	X
Ptilinopus porphyraceus	X	X	X	X
Ducula pacifica	—	X	X	X
†*Ducula latrans*	X	X	—	—
†*Ducula* undescribed sp.	X	X	—	—
†Undescribed genus C	X	—	—	—
†*Didunculus* undescribed sp.	X	X	—	—
**Gallicolumba stairi*	X	X	—	—
PARROTS				
**Vini solitarius*	X	—	—	—
**Vini australis*	X	X	X	—
Prosopeia tabuensis (i)	—	X	X	X
†*Eclectus* undescribed sp.	X	X	—	—
BARN-OWLS				
Tyto alba	—	X	X	X
SWIFTS				
Collocalia spodiopygia	—	X	X	X
KINGFISHERS				
Halcyon chloris	X	X	X	X
THRUSHES				
**Turdus poliocephalus*	X	X	—	—

(continued)

Table 6-20 (continued)

	Prehuman record	Archaeological record	Historic record	Extant in 1988
WARBLERS				
*/†cf. *Cettia* sp.	X	—	—	—
WHITE-EYES				
†Zosteropidae undescribed sp.	X	—	—	—
STARLINGS				
Aplonis tabuensis	X	X	X	X
Acridotheres fuscus (i)	—	—	—	X
Sturnus vulgaris (i)	—	—	—	X
TRILLERS				
Lalage maculosa	X	X	X	X
*/†cf. *Lalage* sp.	X	X	—	—
MONARCHS				
**Clytorhynchus vitiensis*	X	X	X	—
**Myiagra sp.*	X	X	—	—
HONEYEATERS				
**Myzomela cardinalis*	X	X	—	—
Foulehaio carunculata	X	X	X	X
TOTAL				
Species	27	26	14	13
*/†species	21	14	3	0
# sites / # landbird bones	1/401	14/918	—	—

The archaeology of Upolu and Savai'i has received relatively little attention since Roger Green and Janet Davidson's surveys and excavations from 1957 to 1967 (Green & Davidson 1969, 1974). Their sites yielded few bones overall (mostly fish, dog, pig, or human) and no bird bones that they identified to species, although in the Auckland Museum (in 1988) I identified 12 bones from the Lotofaga site on Upolu, as follows: the chicken *Gallus gallus* (9), rail *Gallirallus philippensis* (2), and tern *Anous stolidus* (1). Each species still exists on Upolu. Subsequent excavations there in 1974 yielded no bird bones (Jennings et al. 1976).

The wet volcanic soils of Upolu and Savai'i are poorly suited to preserve bones. Even lava tubes, which can house rich bone deposits in the Hawaiian Islands (James et al. 1987, James & Olson 1991, Olson & James 1991) and the Galápagos Islands (Steadman 1986, Steadman & Burke 1999), have not been productive in Samoa. Streams run through many Samoan lava tubes (Thomson 1921, Kear & Wood 1959, Ollier & Zarriello 1979), creating an unfavorable setting for bone deposition and preservation that includes stagnant air saturated with humidity (Hoch & Asche 1988). Archaeological excavations at the relatively dry entrances of the Falemaunga Caves on Upolu are not known to have yielded bird bones (Freeman 1944).

During my 15 days of exploring lava tubes on Upolu in April 1999, the only bird bones found were a few surface remains of the swiftlet *Collocalia spodiopygia*, which still nests in the same caves. Thus we know nothing about prehistoric birds on Upolu and Savai'i two of the largest, highest, most verdant (although now largely deforested) and species-rich islands in Polynesia. This may be the most glaring and challenging void in the historic biogeography of Polynesian birds.

The situation is only slightly better in American Samoa, with just two small sets of bird bones from two islands. The Toaga site on Ofu (6.56 km^2) in the Manu'a Group produced rich cultural materials dating from ca. 2800 to 1900 yr BP (Kirch & Hunt 1993). The 74 bird bones identified from Toaga (Steadman 1990, 1993b) represent 10 species of seabirds (Chapter 15) and five species of landbirds that include the easternmost record of a megapode (*Megapodius* sp., larger than *M. pritchardii*, smaller than *M. alimentum*;

Table 6-21 Stratigraphic summary of landbird bones from `Anatu, `Eua, Tonga.•, recorded from a posthuman context at another site on `Eua;••, recorded from a posthuman site not on `Eua but elsewhere in Tonga. Totals do not include introduced species (i). Updated from Steadman (1993a).

	Stratum						Stratum				
	I	II	II/III	III	Total		I	II	II/III	III	Total
HERONS						KINGFISHERS					
†Nycticorax undescribed sp. B	—	3	—	—	3	Halcyon chloris	18	5	1	3	27
HAWKS						THRUSHES					
*Accipiter cf. rufitorques	—	—	1	2	3	*Turdus poliocephalus	1	16	5	14	36
MEGAPODES						WARBLERS					
*Megapodius pritchardi	••	—	—	1	1	*/† cf. Cettia sp.	—	3	—	2	5
†Megapodius alimentum	••	20	14	9	43	WHITE-EYES					
†Megapodius undescribed sp. F	•	—	—	2	2	†Zosteropidae undescribed sp.	••	3	—	—	3
CHICKENS						STARLINGS					
Gallus gallus (i)	25	—	—	—	25	Aplonis tabuensis	40	33	9	6	88
RAILS						TRILLERS					
Gallirallus philippensis	18	—	—	—	18	Lalage maculosa	5	1	1	1	8
†Gallirallus undescribed sp. H	—	39	20	31	90	*/†cf. Lalage sp.	1	1	—	—	2
*Porzana tabuensis	85	—	—	—	85	MONARCHS					
Porphyrio porphyrio	38	—	—	—	38	*Clytorhynchus vitiensis	26	3	2	—	31
PIGEONS AND DOVES						*Myiagra sp.	•	3	—	1	4
Ptilinopus perousii	3	3	—	—	6	HONEYEATERS					
Ptilinopus porphyraceus	•	3	—	2	5	*Myzomela cardinalis	•	11	—	2	13
Ducula pacifica	10	—	—	—	10	Foulehaio carunculata	80	10	2	4	96
†Ducula latrans	•	5	3	4	12	BIRD BONES (all species)	333	208	76	116	759
†Undescribed genus C	••	5	2	2	9	*/† bones	119	153	63	101	436
†Didunculus undescribed sp.	3	11	5	6	25	% bones */†	36	74	83	87	58
*Gallicolumba stairi	•	27	10	18	55	SPECIES	17	22	14	20	32
PARROTS						*/† species	7	16	10	15	21
*Vini solitarius	••	—	1	—	1	% */† species	41	74	71	75	66
*Vini australis	1	1	—	1	3	SPECIES (including•,••)	28				
*Eclectus undescribed sp.	2	2	—	6	10	*/† species	17				
SWIFTS						% */†	61				
Collocalia spodiopygia	2	—	—	—	2						

Chapter 9). Known from an ulna and femur, this megapode is the Toaga site's only extirpated landbird.

On Tutuila (142 km^2), a small cave named Ana Pe'ape'a is the only source of prehistoric bird bones (Steadman & Pregill 2004). Ana Pe'ape'a is on the south-central coast only ca. 20 m inland from, and ca. 10 m above, the sea. The primary bone deposit is a small pocket of dry sediment perched on a ledge on the cave's northern wall. I believe that the barn-owl *Tyto alba* is responsible for the bones because: (1) it is the most commonly represented bird, including bones of nestlings; (2) all other bones recovered are of reptiles, birds, and mammals small enough to be eaten by *T. alba*; (3) 95+% of the 13,700 bones are of *Rattus exulans* (Table 6-24), the preferred prey of *T. alba* in Oceania when available (pers. obs.); and (4) no cultural evidence (artifacts, charcoal, features, etc.) was found. Based on the chronology of human arrival in West Polynesia (Kirch & Hunt 1993, Anderson & Clark 1999, Burley et al. 1999),

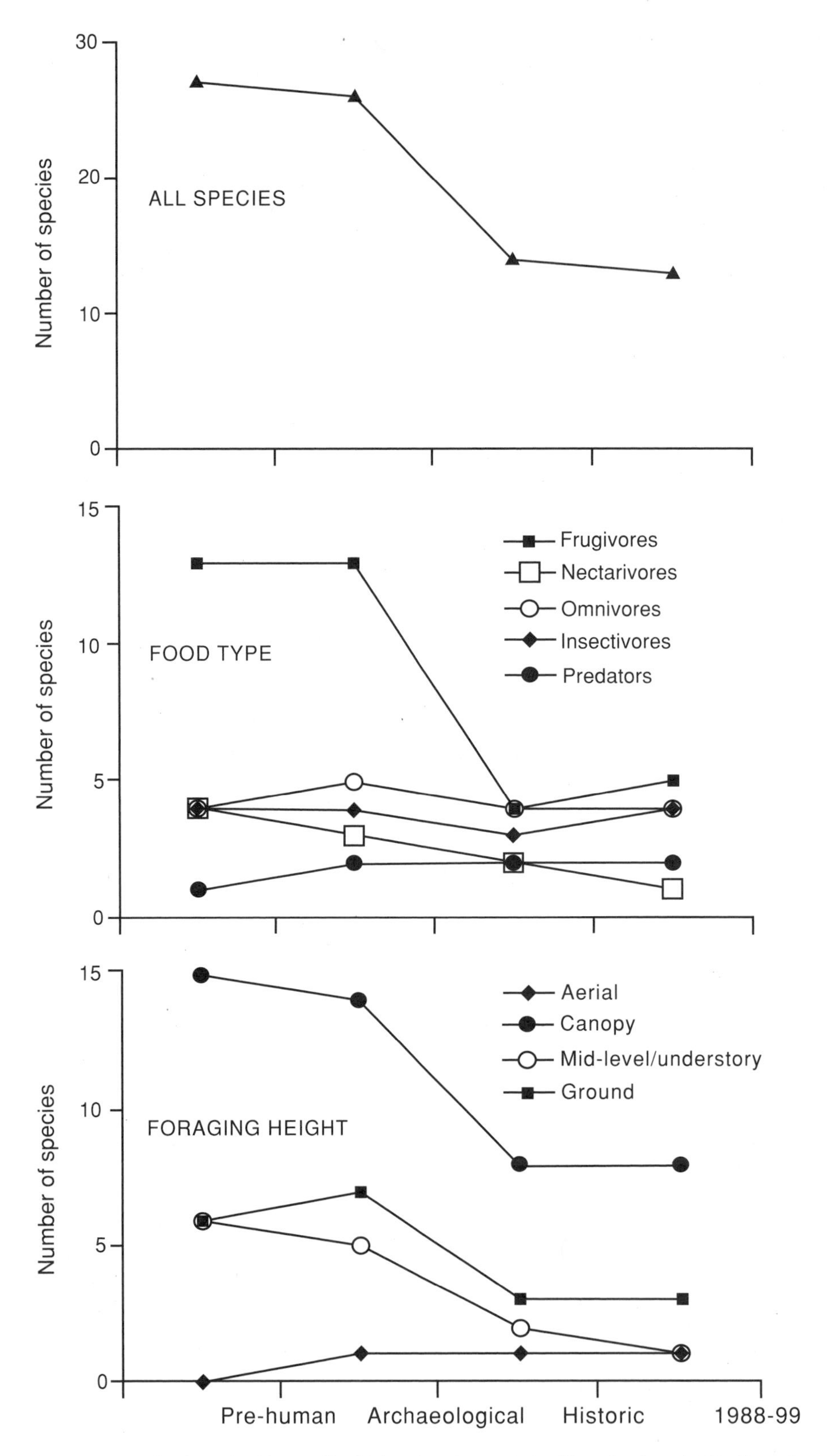

Figure 6-14 Decline in species richness of landbirds on `Eua, Tonga. Updated from Steadman (1995a: Figure 4), based on data in Table 6-20.

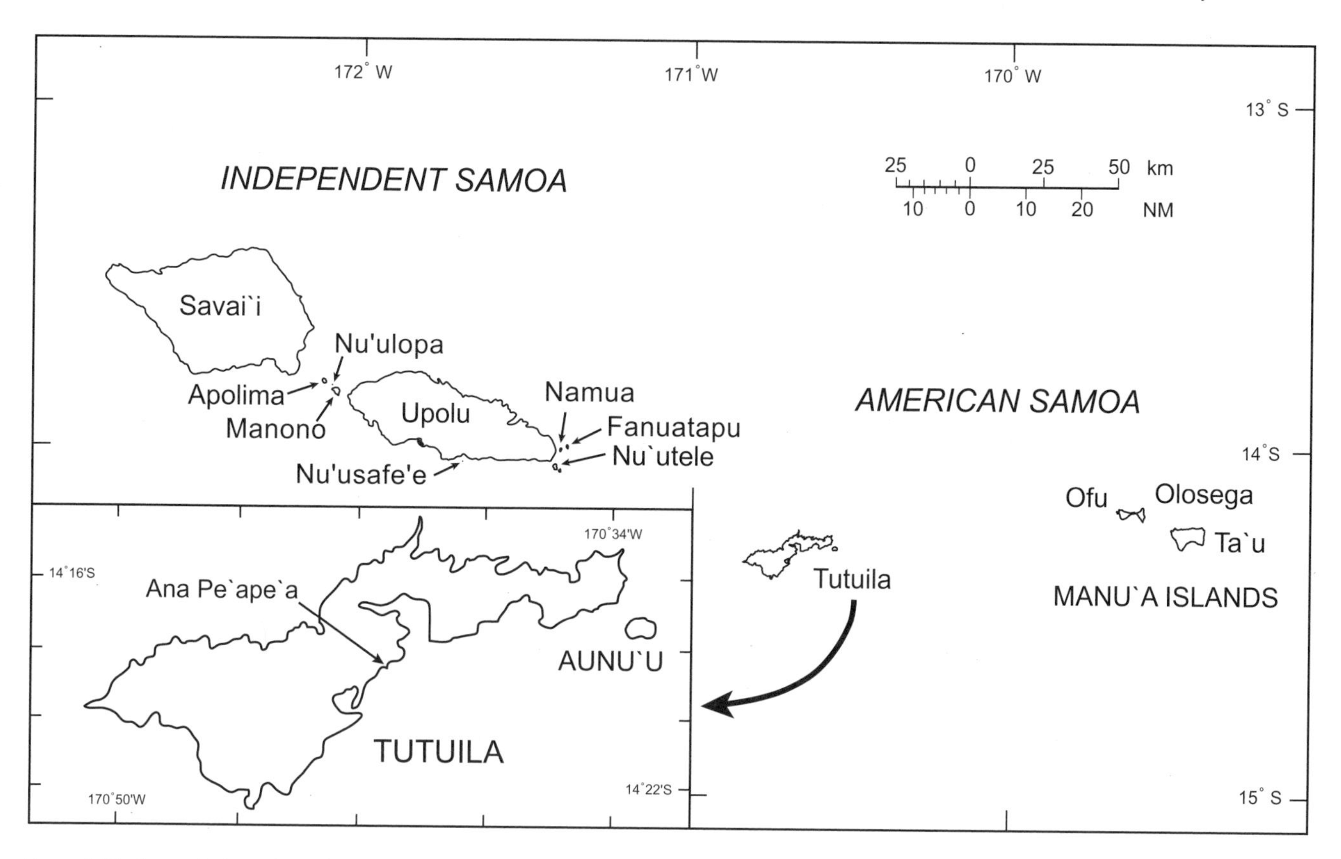

Figure 6-15 Samoa. Redrawn from Bier (1980) and Western Samoa (1996).

the presence of *R. exulans* bones throughout the strata at Ana Pe'ape'a indicates an age less than 2800 yr BP. An AMS ^{14}C date (Beta-134572) on the femur of a chicken (*Gallus gallus*) is cal AD 445 to 640 (1505 to 1310 cal BP), which is reasonable because none of the site's nine species of birds is extinct and only one (the rail *Porzana tabuensis*) is extirpated on Tutuila (Table 6-25). On other Polynesian islands where it no longer occurs, *P. tabuensis* often survived into late prehistoric or historic times rather than being lost during Lapita times (Steadman 1993a; Chapter 10 herein).

West Polynesian Outliers

Three isolated islands or island groups are part of the West Polynesian faunal region. One is Rotuma, lying 560 km north of Viti Levu, Fiji. The next is Wallis & Futuna, three islands that are 350 to 530 km west of Savai'i in Samoa, and 500 to 650 km east of Rotuma. The last is Niue, 440 km east of Vava'u, Tonga. The landbird faunas of these outliers are West Polynesian in origin but depauperate from their isolation. In no West Polynesian outlier is the prehistoric landbird record good enough to estimate species richness at human contact.

Rotuma

Rotuma (43 km^2, 256 m elev.; Figure 1-3) is politically part of Fiji today, although Rotumans speak a Proto Central Pacific language more similar to Samoan than Fijian (Lynch 1998). Rotuma is an eroded volcanic island with 11 offshore islets, one ('Uea) being nearly 1 km^2 and as high as 262 m. Thus 'Uea may sustain landbirds, although it is unexplored scientifically.

Ornithological investigation of Rotuma is in its infancy, with just three published visits: J. Stanley Gardiner (September to December 1896; Gadow 1898); the Whitney South Sea Expedition (18 to 25 May 1925; Watling 1985); and Fergus Clunie (30 November to 5 December 1985; Clunie 1985). Each of its 10 or 11 species of landbirds (Table 6-2) is widely distributed in West Polynesia except the endemic honeyeater *Myzomela chermesina* (listed as *M. jugularis* by Clunie 1985), which is related to the widespread *M. cardinalis* (Koopman 1957). The Rotuman landbird fauna resembles that of American Samoa in the prevalence of supertramps, with columbids being restricted to canopy frugivores and passerines being dominated by starlings and honeyeaters. Rotuma lacks parrots, swifts, kingfishers, and many West Polynesian passerine families, not to mention having a much reduced columbid assemblage.

Table 6-22 Distribution of landbirds in Independent Samoa (Savai`i, Upolu, offshore islands), American Samoa (Tutuila, Manu`a), and Wallis & Futuna. H, historic record of extirpated population; M, modern record; P, prehistoric record; † extinct species; *, multiple islands considered. The Samoan offshore islands are listed in Table 6-23. The island-by-island distribution of landbirds in Wallis & Futuna is given in Table 6-26. From Amerson et al. (1982a, b), Balouet & Olson (1987), Pratt et al. (1987), Steadman (1994), Freifeld et al. (2001), and Steadman & Pregill (2004).

	Savai`i	Upolu	Offshore islands	Tutuila	Manu`a	Wallis & Futuna
HERONS						
Egretta sacra	M	M	M	M	M, P	M?
DUCKS						
Anas superciliosa	M	M	M	M	M	—
FALCONS						
Falco peregrinus	M?	—	—	—	—	—
MEGAPODES						
Megapodius sp.	—	—	—	—	P	—
RAILS						
Gallirallus philippensis	M	M, P	M	M, P	M	M
Porzana tabuensis	M	—	—	P	M	—
Poliolimnas cinereus	M	M	—	—	—	—
†*Pareudiastes pacificus*	H	—	—	—	—	—
Porphyrio porphyrio	M	M	M	M	M	—
PIGEONS, DOVES						
Columba vitiensis	M	M	—	—	—	—
Ptilinopus perousii	M	M	M	M	—	—
Ptilinopus pophyraceus	M	M	M	M	M	M
Ducula pacifica	M	M	M	M	M	M, P
†*Ducula david*	—	—	—	—	—	P
Didunculus strigirostris	M	M	M	—	—	—
Gallicolumba stairi	M	M	M	H	M, P	M
PARROTS						
Vini australis	M	M	—	M	M	—
BARN-OWLS						
Tyto alba	M	M	—	M, P	M	M
SWIFTS						
Collocalia spodiopygia	M	M	M	M	M	M
KINGFISHERS						
Halcyon recurvirostris	M	M	M	—	—	—
Halcyon chloris	—	—	—	M, P	M	—
THRUSHES						
Turdus poliocephalus	M	M	—	—	—	—
WHITE-EYES						
Zosterops samoensis	M	—	—	—	—	—
STARLINGS						
Aplonis tabuensis	M	M	M	M, P	M	M
Aplonis atrifusca	M	M	M	M, P	M	—

Table 6-22 (continued)

	Savai`i	Upolu	Offshore islands	Tutuila	Manu`a	Wallis & Futuna
ESTRILDID FINCHES						
Erythrura cyaneovirens	M	M	—	—	—	—
TRILLERS						
Lalage sharpei	M	M	—	—	—	—
Lalage maculosa	M	M	M	—	—	M
ROBINS						
Petroica multicolor	M	M	—	—	—	—
WHISTLERS						
Pachycephala flavifrons	M	M	M	—	—	—
MONARCHS						
Rhipidura nebulosa	M	M	M	—	—	—
MONARCHS						
Clytorhynchus vitiensis	—	—	—	—	M	—
Myiagra albiventris	M	M	M	—	—	—
HONEYEATERS						
Myzomela cardinalis	M	M	M	M	M	—
Foulehaio carunculata	M	M	M	M, P	M	M
Gymnomyza samoensis	M	M	M	H	—	—
TOTAL						
M	30–31	28	20	15	17	9–10
M + H	31–32	28	20	17	17	9–10
M + H + P	31–32	28	20	18	18	10–11
Area (km^2)	1820	1100	5.3*	142	49*	211*
Elevation (m)	1858	1113	200	653	965*	480*
Total landbird bones	0	2	0	50	6	5

Rotuma has no avian fossil data with which to evaluate these absences, any of which could be due to human impact rather than to isolation. Uncommon bird bones from the Maka Bay archaeological site, dated to ca. 600–1000 cal AD, are unidentified (Allen et al. 2001a). If a prehistoric record of birds is developed on Rotuma, any of the genera already found as fossils in Tonga, Samoa, or other West Polynesian outliers might be expected. Because of its remoteness, I would hesitate to predict how well represented, if at all, would be Fijian endemic genera such as *Charmosyna*, *Prosopeia*, *Trichocichla*, and *Lamprolia*.

Wallis & Futuna

These West Polynesian outliers are a French Territory that, in spite of their collective name, consists of three main islands—'Uvea (= Wallis), Futuna, and 'Alofi (Figure 6-16). Each island consists of eroded volcanic rocks as well as uplifted limestone (Kirch 1994). Futuna (45.7 km^2, 524 m elev.) and 'Alofi (17.5 km^2, 417 m) lie ca. 240 km southwest of 'Uvea (95 km^2, 142 m) but only 2.5 km from each other.

The avifauna of Wallis & Futuna is poor in species (Tables 6-22, 6-26). As in American Samoa, the few surviving passerines are canopy frugivores or nectarivores. Only six or seven species of landbirds are found on all three islands. The negative species-area relationship among these three islands is probably related to anthropogenic extinction, before which they probably had very similar landbird faunas.

The only evidence of prehistoric birds from Wallis & Futuna comes from 'Uvea, where two sites (see Frimigacci

Table 6-23 Landbirds recorded on offshore islands in Samoa, 10–17 April 1999. In parentheses below island names is the number of person-hours spent surveying each island. V, possible visitor from a nearby island. Plant communities scale from 1 (mainly indigenous species) to 5 (mainly nonnative/cultivated species). *, Nu`ulopa lies 0.7 km from Manono; **, distance from Savai`i. From Freifeld et al. (2001).

	Nu`usafe`e (1.5)	Nu`ulo pa (2.25)	Fanuatapu (1.5)	Namu`a (5.25)	Nu`utele (6.5)	Apolima (2.5)	Manono (4.15)	Total islands
RAILS								
Gallirallus philippensis	—	—	1	—	—	4	11	3
Porphyrio porphyrio	—	—	—	—	—	—	1	1
PIGEONS, DOVES								
Ptilinopus perousii	—	—	—	—	—	—	1, V	1
Ptilinopus porphyraceus	—	—	—	1, V	22	1, V	7	4
Ducula pacifica	—	—	—	—	12	—	—	1
Didunculus strigirostris	—	—	—	—	2	—	—	1
Gallicolumba stairi	—	—	—	—	1	—	—	1
SWIFTS								
Collocalia spodiopygia	—	—	—	—	5	—	—	1
KINGFISHERS								
Halcyon recurvirostris	—	—	1	1	—	—	2	3
STARLINGS								
Aplonis tabuensis	1, V	—	—	1, V	1	—	5	4
Aplonis atrifusca	2, V	2, V	2, V	16–28	32	30	41	7
TRILLERS								
Lalage maculosa	—	2, V	—	2, V	27	8	16	5
WHISTLERS								
Pachycephala flavifrons	—	—	—	—	18	—	—	1
FANTAILS								
Rhipidura nebulosa	—	—	—	1, V	—	—	—	1
MONARCHS								
Myiagra albiventris	—	—	—	—	7	—	—	1
HONEYEATERS								
Myzomela cardinalis	—	—	—	—	3	1	13	3
Foulehaio carunculata	1, V	2, V	11	30–49	72	36	112	7
Total species	3	3	4	7	12	6	10	
Land area (km^2)	.02	.02	.08	.18	1.05	1.08	2.9	
Elevation (m)	3–4	30	50	70	200	165	105	
Distance from Upolu (km)	1.3	6.6*	2.5	.8	1.3	8.0**	3.7	
Inside vs. outside of reef	inside	inside	inside	inside	outside	outside	outside	
Plant communities	1	1	1	2	1	4	4	
Upland forest	—	+	–	+	+	+	+	5
Human habitation	–	–	–	+	–	+	+	3
Gallus gallus	–	–	–	+	–	+	+	3

Table 6-24 The bone assemblage from excavation units 1 and 2, Ana Pe`ape`a, Tutuila, American Samoa, expressed in numbers of identified specimens (NISP). From Steadman & Pregill (2004).

	Level 1	Level 2	Total	~ % NISP
REPTILES				
Gehyra oceanica	25	9	34	0.25
Nactus pelagicus	1	4	5	0.04
Emoia spp.	3	—	3	0.02
Lizard sp.	9	2	11	0.08
Candoia bibroni	317	157	474	3.47
BIRDS	57	57	114	0.81
MAMMALS				
Emballonura semicaudata	105	35	140	1.02
Rattus exulans	9481+	3413+	12,894+	94.3
TOTAL	10,009+	3666+	13,675+	100

Table 6-25 Bones of birds from excavation units 1 and 2, Ana Pe`ape`a, Tutuila, American Samoa. i, introduced species; *, extirpated species. From Steadman & Pregill (2004).

	Level 1	Level 2	Total
CHICKENS			
Gallus gallus (i)	2	2	4
RAILS			
Gallirallus philippensis	1	—	1
**Porzana tabuensis*	10	5	15
TERNS			
Procelsterna cerulea	11	8	19
BARN-OWLS			
Tyto alba	14	10	24
KINGFISHERS			
Halcyon chloris	2	1	3
STARLINGS			
Aplonis tabuensis	1	—	1
Aplonis atrifusca	1	4	5
HONEYEATERS			
Foulehaio carunculata	—	1	1
Unidentified bird	26	15	41
TOTAL			
Bones	68	46	114
Bones of native landbirds	29	21	50
Species	8	7	9
Native landbird species	6	5	7

Table 6-26 Distribution of landbirds in Wallis & Futuna. M, modern record; P, prehistoric record. From Balouet & Olson (1987), Thibault & Guyot (1987), and Gill (1995).

	`Uvea	Futuna	`Alofi
HERONS			
Egretta sacra	M	M	M
DUCKS			
Anas superciliosa	M	—	—
RAILS			
Gallirallus philippensis	M	M	M
Porphyrio porphyrio	M	M?	M
PIGEONS, DOVES			
Ptilinopus porphyraceus	M	M	M
Ducula pacifica	M, P	M	M
Ducula david	P	—	—
Gallicolumba stairi	—	—	M
PARROTS			
Vini australis	—	M	M
BARN—OWLS			
Tyto alba	M	M	M
SWIFTS			
Collocalia spodiopygia	—	M	M
KINGFISHERS			
Halcyon chloris	—	M	M
STARLINGS			
Aplonis tabuensis	M	M	M
TRILLERS			
Lalage maculosa	—	M	M
MONARCHS			
Clytorhynchus vitiensis	—	M	M
HONEYEATERS			
Foulehaio carunculata	—	M	M
TOTAL			
M	8	12–13	14
M + P	9	12–13	14
Area (km^2)	96	80	35
Elevation (m)	155	480	370
Number of landbird bones	8	0	0

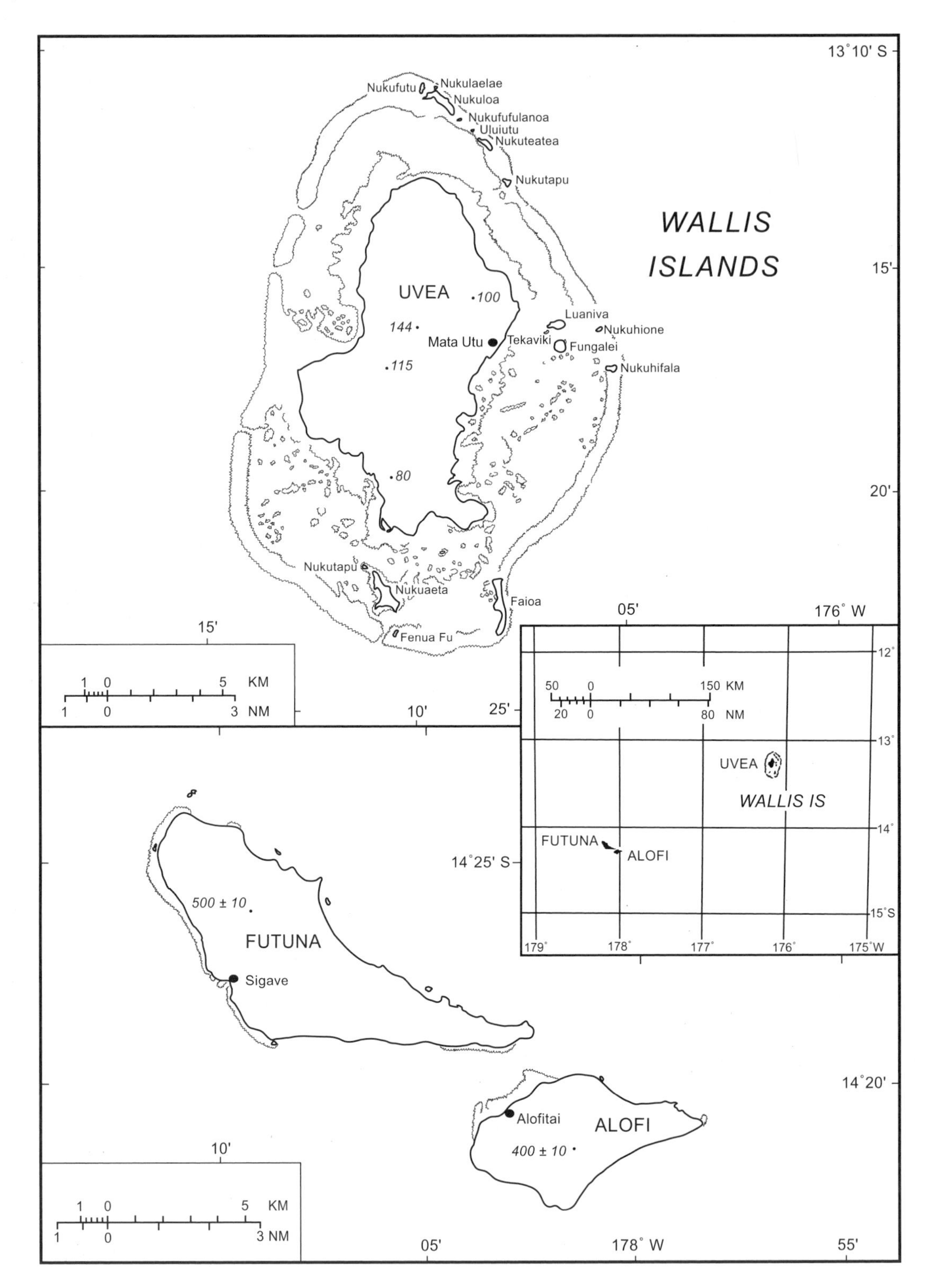

Figure 6-16 Wallis & Futuna. Redrawn from New Zealand Department of Lands and Survey (1986).

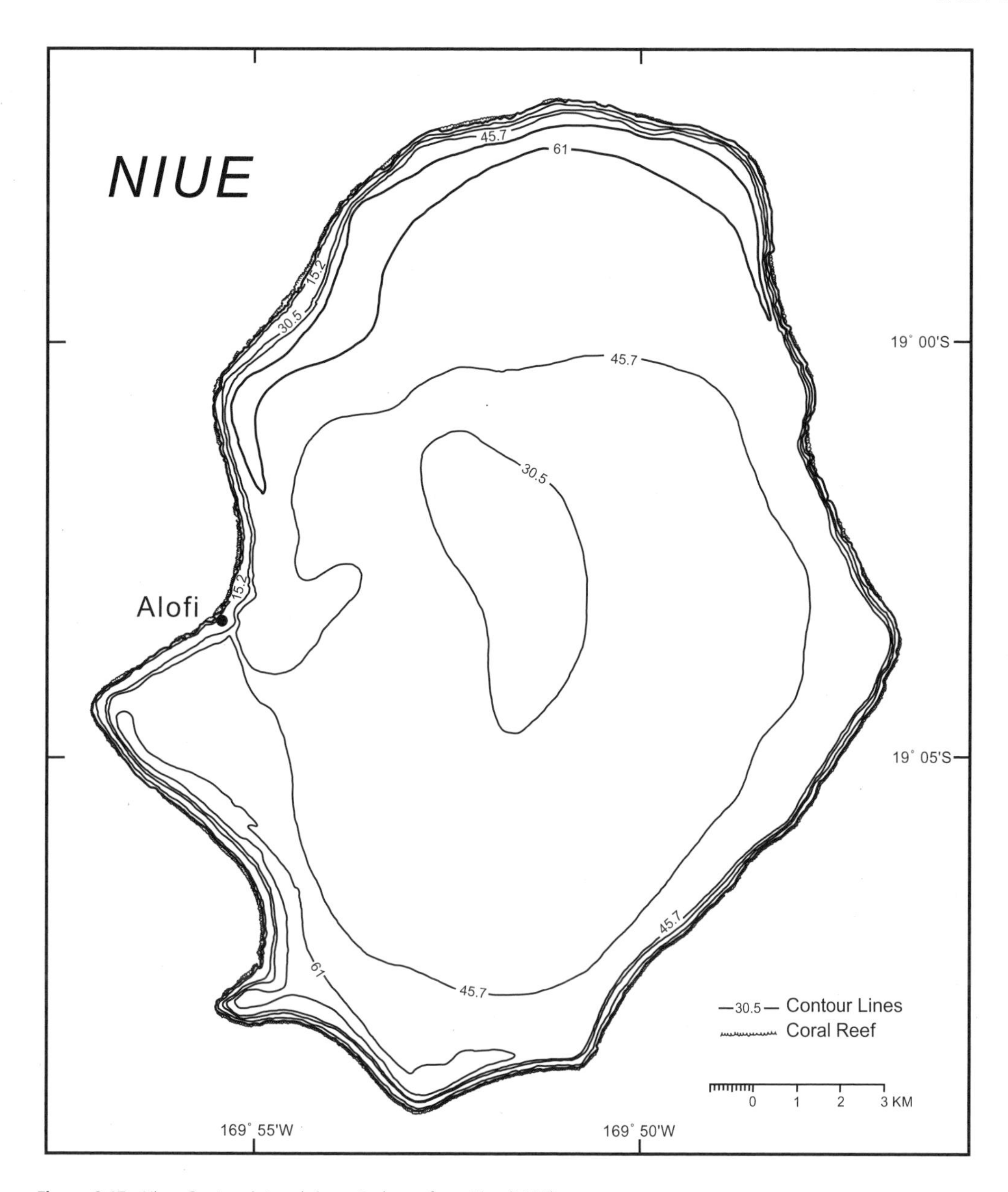

Figure 6-17 Niue. Contour intervals in m. Redrawn from Niue (1985).

et al. 1987) produced three bones of the large, extinct pigeon *Ducula david* associated with Lapita pottery (Balouet & Olson 1987). Subsequently, *D. david* was believed to have been widespread in West Polynesia before human impact (Steadman 1989a, 1993a). With much more extensive fossil samples of large pigeons now available from Fiji (Worthy 2000) and Tonga (Chapter 11 herein), it seems that the large species of *Ducula* in these two island groups were distinct from *D. david* and from each other, although probably closely related.

Niue

Niue (Figure 6-17) is isolated between the Cook Islands (800 km to the east), Samoa (560 km north-northwest), and Tonga (440 km west; Figure 1-3). A large island (259 km^2, the same as Tongatapu), Niue attains an elevation of 68 m in the perimeter ring of coralline limestone that surrounds a gently depressed (min. elev. 34 m) basin (Niue 1985). Niue is an uplifted Pliocene atoll (Aharon et al. 1987, Wheeler & Aharon 1991, Paulay & Spencer 1992).

The modern Niuean landbird fauna comprises 11 species (Townsend & Wetmore 1919, Wodzicki 1971, Kinsky & Yaldwyn 1981, Hay & Powlesland 1998, Powlesland & Hay 1998; Table 6-10 herein), three of which are endemic subspecies (the dove *Ptilinopus porphyraceus whitmeei*, triller *Lalage maculosa whitmeei*, and starling *Aplonis tabuensis brunnescens*). Each surviving species is widespread in West Polynesia. Only the trampy heron *Egretta sacra*, rail *Porzana tabuensis*, and pigeon *Ducula pacifica* occur on islands east of Niue.

Noncultural Holocene fossils of three extinct or extirpated species of landbirds have been reported from the Niuean cave Anakuli (Steadman et al. 2000b). The night-heron *Nycticorax kalavikai*, known only from Niue, is closely related to an undescribed species from Tonga. Next is the "Niuafo'ou" megapode *Megapodius pritchardii*, historically only from Niuafo'ou (Tonga) but found in prehistoric sites elsewhere in Tonga (see above). Finally, *Gallirallus huiatua* is a flightless rail presumably endemic to Niue. The megapode reinforces the West Polynesian rather than East Polynesian affinities of Niue's avifauna. *Gallirallus huiatua* is very distinct from the nearest known extinct, flightless, congeneric species, namely *G. ripleyi* from Mangaia (Cook Islands) and *G.* undescribed sp. H from 'Eua (Tonga). Extinct night-herons also are found in both Tonga and the Cook Islands (Chapter 13).

A better fossil record probably would fill in some of Niue's missing taxa. Glaring among these absences are three taxa found both to the west (Tonga) and east (Cook Islands): the ground-dove *Gallicolumba*, a pigeon (*Ducula*) larger than *D. pacifica*, and kingfisher *Halcyon*. Other West Polynesian genera likely to have occurred on Niue before human impact, because they once inhabited both Tonga and Samoa, are the pigeon *Didunculus*, thrush *Turdus*, shrikebill *Clytorhynchus*, and honeyeaters *Myzomela* and *Foulehaio*.

Chapter 7 East Polynesia

In this vast area lie islands that are recognized worldwide, such as Tahiti and Easter Island, and others of great obscurity, such as either named Maria (one in Tubuai, one in the Tuamotus). The 135 major islands of East Polynesia are a collection of eroded volcanic islands, raised limestone islands, atolls, almost-atolls, and composites. Active volcanoes are lacking. The seven island groups in East Polynesia form a natural avifaunal region (Figure 1-4). All but Easter Island are on the Pacific Plate. Although some island groups have similar tectonic histories (Dickinson 1998b), none was ever connected to any of the others, even during late Pleistocene low-sea-level stands. Furthermore, with few exceptions, the individual islands never have been connected, being surface remnants of separate volcanoes, surrounded by deep water.

Culturally, "East Polynesia" also includes the Hawaiian Islands far to the north and New Zealand far to the southwest. Throughout this chapter, East Polynesia will refer to the avifaunal rather than cultural region.

East Polynesia has been explored for birds since the three voyages of Captain James Cook from 1772 to 1780 (Holyoak & Thibault 1984). The East Polynesian avifauna differs from that of West Polynesia (Chapter 6) by the presence in most island groups of resident sandpipers (*Prosobonia* spp.), multiple sympatric species of ground-doves (*Gallicolumba* spp.), one or two species of fruit-doves in the *Ptilinopus purpuratus* species group, the often sympatric pigeons *Ducula aurorae* and *D. galeata*, a radiation of lorikeets (*Vini* spp.; at least six species total, with up to three of these sympatric), kingfishers of the *Halcyon tuta* species-group, and parapatric radiations of monarchs (*Pomarea* spp.) and reed-warblers (*Acrocephalus* spp.). Flightless rails are represented in East Polynesia by three genera (*Gallirallus, Porzana, Porphyrio*).

Easter Island is included in the East Polynesian avifaunal region by default. No landbirds survived to historic times on Easter Island, where all we know about them is based on six fragmentary prehistoric bones (described below). Only when Easter Island's landbirds are much better documented will it be possible to say whether they may have differed from all others in Oceania by including a Neotropical element.

East Polynesia has half as many landbird genera as West Polynesia (23 vs. 46; Tables 6-1, 7-1). Only *Prosobonia* and *Pomarea* are endemic to East Polynesia. The other 21 are widespread in Oceania, although three (*Macropygia, Cyanoramphus, Acrocephalus*) have not been recorded in West Polynesia but occur in Melanesia or Micronesia. Each of East Polynesia's 19 nonpasserine and four passerine genera is represented by prehistoric bones in at least one island group. Five of these genera no longer occur in East Polynesia, the nearest modern record being in West Polynesia (*Nycticorax, Dendrocygna, Gallirallus, Porphyrio*) or Vanuatu (*Macropygia*). The first two of these have been found in East Polynesia only in the Cook Islands.

Many East Polynesian prehistoric bird assemblages are biased toward larger species because they were not collected with small-mesh screens. Thus swifts, kingfishers, and passerines are absent or underrepresented in many bone samples. To the west in Tonga, passerines underwent losses

Table 7-1 Distribution of genera of landbirds in East Polynesia. M, modern (includes historic); P, prehistoric; †, extinct genus; *, extant genus but extirpated in East Polynesia; ?, questionable record, or residency not established. From Pratt et al. (1987), Steadman & Pahlavan (1992), Wragg & Weisler (1994), and Steadman (1995a, herein).

	Cook Islands	Tubuai Islands	Society Islands	Tuamotu Archipel.	Pitcairn Group	Marquesas Islands	Easter Island	TOTAL
HERONS								
Egretta	M, P	M	M, P	M, P	M	M	P?	6–7
Butorides	—	—	M, P	—	—	—	—	1
**Nycticorax*	P	—	—	—	—	—	—	1
DUCKS								
**Dendrocygna*	P	—	—	—	—	—	—	1
Anas	M, P	M	M	—	—	—	—	3
RAILS								
**Gallirallus*	P	—	M, P	—	—	P	P?	3–4
Porzana	M, P	M	M, P	M	M, P	M, P	P	7
**Porphyrio*	P	—	—	—	—	P	—	1–2
SANDPIPERS								
Prosobonia	P	—	M	M	P	P	—	5
PIGEONS, DOVES								
**Macropygia*	—	—	P	—	—	P	—	2
Ptilinopus	M, P	M, P	M, P	M	M, P	M, P	—	6
Ducula	M, P	—	M, P	M	P	M, P	—	5
†Undescribed genus?	—	—	—	—	P	—	—	1
Gallicolumba	P	—	M, P	M, P	P	M, P	—	5
PARROTS								
Vini	P	M	M, P	—	M, P	M, P	—	5
**Cyanoramphus*	—	—	M	—	—	—	—	1
†Genus uncertain	—	—	—	—	—	—	P	1
SWIFTS								
Collocalia	M, P	—	M	—	—	M	—	3
KINGFISHERS								
Halcyon	M, P	—	M, P	M	—	M, P	—	4
SWALLOWS								
Hirundo	—	—	M	—	P	—	—	2
WARBLERS								
Acrocephalus	M, P	M	M, P	M	M, P	M	—	6
STARLINGS								
Aplonis	M	—	M, P	—	—	—	—	2
MONARCHS								
Pomarea	M, P	—	M	—	—	M	—	3
TOTAL								
M	10	6	17	8	5	10	0	17
P	16	1	12	2	9	10	5	24
M + P	17	6	18	8	10	14	5	24
Number of bones		1	110	4	1612	2330	6	

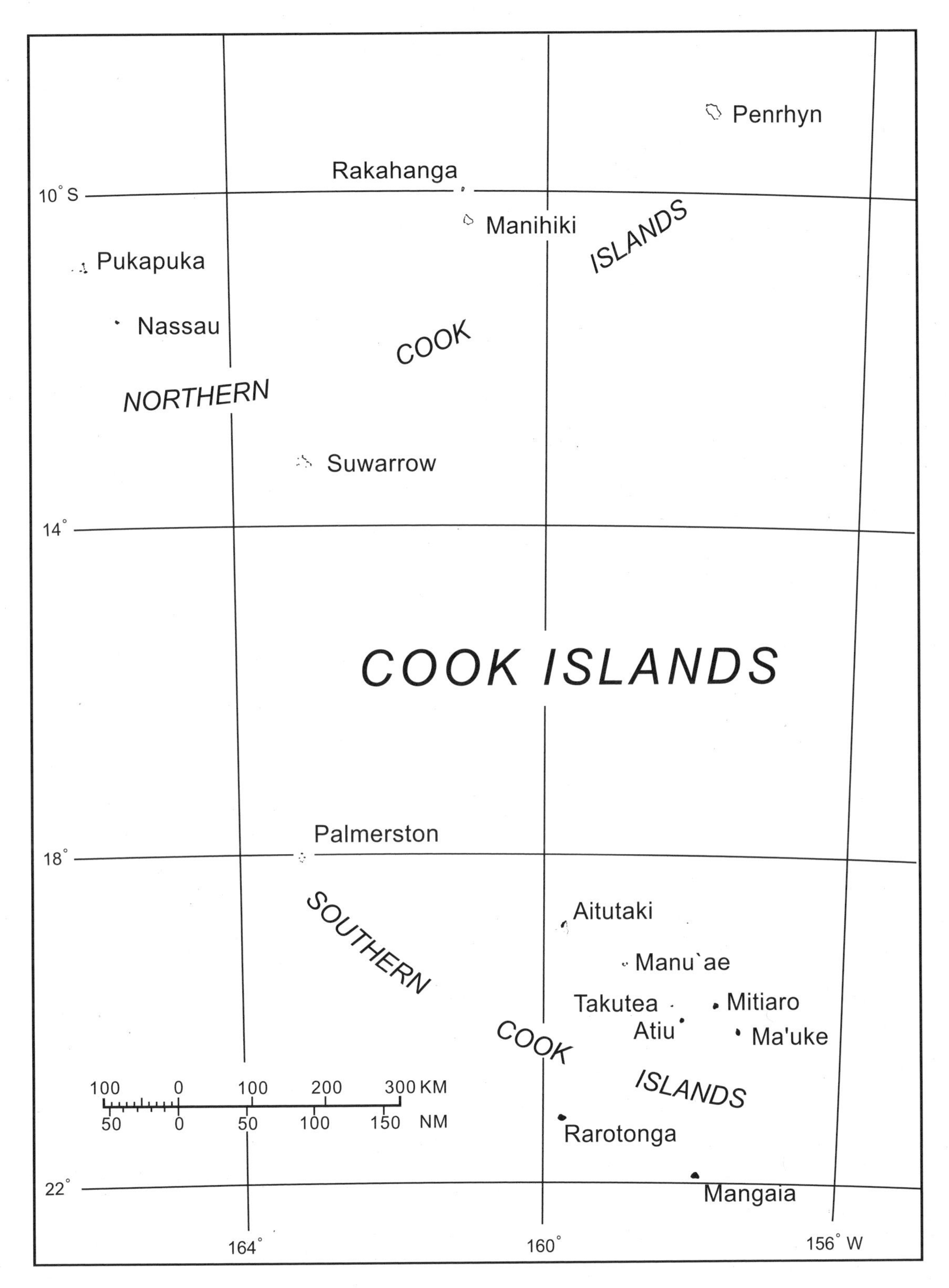

Figure 7-1 The Cook Islands. Redrawn from New Zealand Department of Lands and Survey (1986).

Table 7-2 Landbirds of main islands in the Southern Cook Islands. H, early historic records, no longer survives; M, modern record, still survives; P, prehistoric record; †, extinct species; ?, questionable record, or residency not established. From Burland (1964), Holyoak (1980), Steadman (1985, 1989a, 1991b, 1995a, 2002b, herein), and Olson (1986a).

	Rarotonga	Mangaia	Atiu	Mitiaro	Aitutaki	Ma`uke	Palmerston	Manu`ae	Total
HERONS									
Egretta sacra	M	M, P	M	M	M, P	M	M	M	8
†*Nycticorax* undescribed sp. C	—	P	—	—	—	—	—	—	1
DUCKS									
†cf. *Dendrocygna* sp.	—	—	—	—	P	—	—	—	1
Anas superciliosa	H	M, P	M	M	M, P	H	—	—	6
RAILS									
†*Gallirallus ripleyi*	—	P	—	—	—	—	—	—	1
Porzana tabuensis	—	M, P	M, P	M	P	H?	—	—	4–5
†*Porzana rua*	—	P	—	—	—	—	—	—	1
†*Porzana* undescribed sp. C	—	P	—	—	—	—	—	—	1
†*Porphyrio* undesc. sp. C	—	P	—	—	—	—	—	—	1
SANDPIPERS									
†*Prosobonia* undescribed sp. A	—	P	—	—	—	—	—	—	1
PIGEONS, DOVES									
Ptilinopus rarotongensis	M	P	M	—	H	H	—	—	5
Ducula pacifica	M	—	M	M	H	M	M	M	7
Ducula aurorae	—	P	P	—	—	—	—	—	2
Ducula galeata	—	P	—	—	—	—	—	—	1
Gallicolumba erythroptera	—	P	P	—	—	—	—	—	2
†*Gallicolumba nui*	—	P	—	—	—	—	—	—	1
PARROTS									
†*Vini sinotoi*	—	P	—	—	—	—	—	—	1
†*Vini vidivici*	—	P	—	—	—	—	—	—	1
Vini kuhlii	H	P	H, P	—	P	H	—	—	5
SWIFTS									
Collocalia sawtelli	—	—	M, P	H	—	—	—	—	2
†*Collocalia manuoi*	—	P	—	—	—	—	—	—	1
KINGFISHERS									
Halcyon mangaia	—	M, P	—	—	—	—	—	—	1
Halcyon tuta	H	—	M, P	H	—	M	—	—	4
WARBLERS									
Acrocephalus kerearako	—	M, P	—	—	—	—	—	—	1
Acrocephalus kaoko	—	—	—	M	—	—	—	—	1
STARLINGS									
Aplonis cinerascens	M	—	—	—	—	—	—	—	1
†*Aplonis mavornata*	—	—	—	—	—	H	—	—	1
MONARCHS									
Pomarea dimidiata	M	—	—	—	—	—	—	—	1

Table 7-2 (continued)

	Rarotonga	Mangaia	Atiu	Mitiaro	Aitutaki	Ma`uke	Palmerston	Manu`ae	Total
TOTAL SPECIES									
M	5	5	7	5	2	3	2	2	12
M + H	8	5	8	7	4	7–8	2	2	14
M + H + P	8	20	10	7	7	7–8	2	2	28
TOTAL BONES									
All species	0	3237+	41	0	151	0	0	0	3429+
Landbirds	0	1674+	41	0	43	0	0	0	1758+
Island area (km^2)	67	52	27	22	18	18	6	2.6	213
Elevation (m)	652	169	72	15	124	29	5	5	—
Distance (km) to nearest island	204	204	50	50	102	59	95	350	—

comparable to those of larger landbirds (Steadman 1993a; Chapter 6 herein). In East Polynesia, screens with mesh sizes of 1/8″ have been used at some sites in the Cook Islands, Pitcairn Islands, Easter Island, and Marquesas (see below). Given the variation in fossil sampling, the generic level diversity of landbirds is almost certainly underestimated in Tubuai, the Tuamotus, and Easter Island and is best estimated in the Cook and Pitcairn Islands. Nevertheless, the great difference in generic level diversity of passerine birds between West Polynesia (20) and East Polynesia (4) is unlikely to decrease much by improving the fossil record. In East Polynesia, only on Mangaia (Cook Islands) and Henderson (Pitcairn Islands) have multiple sites with abundant bird bones from small species been sampled carefully. In both cases we find extinct small species (a sandpiper, swift, and tiny rail on Mangaia, and a sandpiper and swallow on Henderson), although not to the extent seen in Tonga.

Cook Islands

The Cook Islands (Figure 7-1) are divided into the geologically and faunally more diverse Southern Group (eight major islands) and the Northern Group, consisting of six scattered atolls lying closer to Samoa, Tokelau, and the Line Islands than to the Southern Group. The heron *Egretta sacra* and pigeon *Ducula pacifica* are the only landbirds today in the Northern Group or on the Southern Group's two atolls (Manu'ae, Palmerston), where no prehistoric bones have been studied (Beaglehole & Beaglehole 1938, Burland 1964, Clapp 1977, Holyoak 1980). Although the Cook Islands are the westernmost island group in East Polynesia, *D. pacifica* is the only landbird it shares with West Polynesia that is found nowhere else in East Polynesia.

The species-level distribution of birds in the southern Cooks has many instances of extirpation (Table 7-2). The obvious null hypothesis to test with yet-to-be-discovered prehistoric bones is that one or more species in each genus in Table 7-2 once occurred on each of the eight islands except perhaps the atolls of Palmerston and Manuae. Extinction on any island in the southern Cooks is not difficult to imagine. Most of the vegetation cover is anthropogenic, although native forest exists in some rugged areas on the eroded volcanic island of Rarotonga (Wilder 1931, Merlin 1985, McCormack & Künzlé 1995, Gill & Sykes 1996) and the mainly limestone islands of Mangaia, Atiu, Mitiaro, and Ma'uke (Merlin 1985, 1991, Franklin & Steadman 1991, Franklin & Merlin 1992). These islands also have rats, cats, pigs, and dogs.

The prehistoric record of birds in the southern Cooks is very rich on Mangaia, quite limited on Atiu and Aitutaki, and lacking on the other islands (Table 7-2). On Atiu, several limestone caves have yielded 41 undated bones of six species of birds, three of which no longer exist there or elsewhere in the group (the ground-dove *Gallicolumba erythroptera*, pigeon *Ducula aurorae*, and parrot *Vini kuhlii*; Steadman 1991b). The limestone portion of Atiu retains some native forest that supported substantial populations of the pigeon *Ducula pacifica*, dove *Ptilinopus rarotongensis*, swift *Collocalia sawtelli*, and kingfisher *Halcyon tuta* in 1987 (Franklin & Steadman 1991), although *C. sawtelli* nests and roosts in just two caves (Tarburton 1990, Steadman 1991b, Fullard et al. 1993, Gill 1996).

Six archaeological sites up to 1000 cal BP have been excavated on the almost-atoll of Aitutaki (Allen & Schubel

Table 7-3 Site summary of prehistoric birds from Mangaia, Cook Islands. Additional data on these sites are presented in Table 4-4. Data updated from Steadman (1985, 1995a), Steadman & Kirch (1990), Kirch et al. (1995). i, introduced species; m, migrant species; †, extinct species; *, extirpated on Mangaia; **, data incomplete. i, m not included in landbird totals. Numbers for certain taxa at MAN-84 exceed those in Table 7-7 because they include bones from the 1991 excavations in addition to the 1997 excavations. Taxa in brackets are not necessarily different from those identified more specifically.

	Te Rua Rere MAN-24	Tangatatau Rockshelter MAN-44	Te Ana Manuku MAN-84	Te Ana Tautua MAN-54	Te Ana Tuatini MAN-41	Te Ana Toroa	Te Ana Toruapuru MAN-57	Te Ana Kakaia MAN-55	Unnamed cave ~100 m W of Lake Tiriara MAN-60	Small shelter ~100 m NE of To'uri Cave	Horseshoe Cave (~50 m NE of MAN-44)
SEABIRDS											
SHEARWATERS, PETRELS											
**Pterodroma inexpectata*	—	—	1	—	—	—	—	—	—	—	—
**Pterodroma cookii*	30	—	24	—	—	—	—	—	2	—	—
**Pterodroma nigripennis*	507	104	92	1	2	—	13	—	2	4	5
[**Pterodroma cookii/nigripennis*]	8	—	75	2	—	1	—	—	—	—	—
**Pterodroma* sp. (large)	—	—	—	1	—	—	—	—	—	—	—
**Pachyptila* sp.	—	—	2	—	—	—	—	—	—	—	—
**Bulweria* sp.	—	—	5	—	—	—	—	—	—	—	—
**Puffinus* cf. *gavia*	—	—	5	—	4	1	—	—	1	—	—
Puffinus lherminieri	37	17	14	1	—	—	5	—	1	—	1
[Procellariidae sp.]	—	1	109	—	—	—	—	—	1	—	—
STORM-PETRELS											
**Nesofregetta fuliginosa*	1	8	44	—	—	1	3	—	—	—	1
TROPICBIRDS											
Phaethon rubricauda	—	1	—	2	—	—	—	—	—	—	—
Phaethon lepturus	3	43	18	—	—	—	2	—	—	—	—
BOOBIES											
**Sula sula*	—	1	—	—	—	5	—	—	—	—	—
FRIGATEBIRDS											
Fregata minor	—	5	—	—	—	—	—	—	—	—	—
Fregata ariel	—	4	—	—	—	—	—	—	—	—	—
TERNS											
**Anous minutus*	—	—	1	—	—	—	—	—	—	—	—
Anous stolidus	—	17	—	—	—	—	—	—	—	—	—
Procelsterna cerulea	—	1	7	—	1	—	—	—	—	—	—
Gygis candida	44	15	—	119	1	—	—	6	—	—	8
**Gygis microrhyncha*	—	13	1	—	—	—	—	—	—	—	—
LANDBIRDS											
HERONS											
Egretta sacra	—	—	—	—	—	—	—	—	—	—	—
†*Nycticorax* undescribed sp. C	—	—	6	—	—	—	—	—	—	—	—
DUCKS											
Anas superciliosa	—	30	—	2	—	—	—	—	—	—	—
CHICKENS											
Gallus gallus (i)	—	107	2	—	—	—	—	—	—	—	—

Fissure Cave in LS block NE of MAN-44 MAN-62	Ana Tuara	Tiny cave in LS block NE of MAN-44 MAN-61	Piri Te Umeume	MAN-81	MAN-83	MAN-87	MAN-82	MAN-63	MAN-95	Te Rua o Ngauru	Te Ana Tapukeu	**Te Ana o Ivi Kanapanapa	Total sites/ total bones
—	—	—	—	—	—	—	—	—	—	—	—	—	1/1
—	—	—	—	—	—	—	—	—	—	—	—	—	3/56
—	2	—	—	4	—	1	1	—	1	—	—	10+	15/749+
2	—	2	1	—	—	—	—	—	—	—	—	10+	8/101+
—	—	—	—	—	—	—	—	—	—	—	—	—	1/1
—	—	—	—	—	—	—	—	—	—	—	—	—	1/2
—	—	—	—	—	—	—	—	—	—	—	—	—	1/5
—	—	—	—	—	—	—	—	—	—	—	—	—	4/11
—	—	—	—	—	—	—	—	—	1	—	—	2	9/79
—	—	—	—	—	—	—	—	—	—	—	—	—	3/11
—	—	—	—	—	—	—	—	—	—	—	—	47	7/105
—	—	—	—	—	—	—	—	—	—	—	—	—	2/3
—	—	—	—	1	1	—	—	—	—	1	—	—	7/69
—	—	—	—	—	—	—	—	—	—	—	—	—	2/6
—	—	—	—	—	2	—	—	—	—	—	—	—	2/7
—	—	—	—	—	—	—	—	—	—	—	—	—	1/4
—	—	—	—	—	—	—	—	—	—	—	—	—	1/1
—	—	—	—	1	—	—	—	—	—	—	—	—	2/18
—	—	—	—	1	—	—	—	—	—	—	—	—	4/10
—	—	—	—	—	—	—	—	—	—	1	—	—	7/194
—	—	—	—	—	—	—	—	—	—	—	—	—	2/14
—	—	—	—	—	—	—	—	1	—	—	—	—	1/1
—	—	—	—	—	—	—	—	—	—	—	—	—	1/6
—	—	—	—	—	—	1	1	—	—	—	—	—	4/34
—	—	—	—	1	7	3	1	—	—	—	—	—	6/121

(continued)

Table 7-3 (continued)

	Te Rua Rere MAN-24	Tangatatau Rockshelter MAN-44	Te Ana Manuku MAN-84	Te Ana Tautua MAN-54	Te Ana Tuatini MAN-41	Te Ana Toroa	Te Ana Toruapuru MAN-57	Te Ana Kakaia MAN-55	Unnamed cave ~100 m W of Lake Tiriara MAN-60	Small shelter ~100 m NE of To'uri Cave	Horseshoe Cave (~50 m NE (of MAN-44)
RAILS											
†*Gallirallus ripleyi*	6	45	67	—	—	—	1	—	—	—	—
Porzana tabuensis	7	6	9	—	—	—	32	—	—	—	—
†*Porzana rua*	30	120	164	—	2	—	2	—	—	—	—
†*Porzana* undescribed sp. C	—	4	10	—	—	—	—	—	—	—	—
[*Porzana* sp.]	—	9	25	—	—	—	—	—	—	—	—
†cf. *Porphyrio* undesc. sp. C	—	2	—	—	—	—	—	—	—	—	—
PLOVERS											
Pluvialis fulva (m)	—	2	1	—	—	—	—	1	—	—	—
SANDPIPERS											
Numenius tahitiensis (m)	—	11	2	—	1	—	—	—	—	—	—
†*Prosobonia* undescribed sp. A	—	4	1	—	—	—	—	—	—	—	—
PIGEONS, DOVES											
**Gallicolumba erythroptera*	16	17	418	—	3	—	6	—	—	—	—
†*Gallicolumba nui*	6	10	3	—	3	—	—	—	—	—	—
**Ptilinopus rarotongensis*	6	13	8	1	—	—	—	—	—	—	—
**Ducula aurorae*	—	8	1	—	2	—	—	—	—	—	—
**Ducula galeata*	2	15	1	—	—	—	—	—	—	—	—
[Columbidae sp.]	—	—	11	—	—	—	—	—	—	—	—
PARROTS											
†*Vini sinotoi*	—	—	3	—	—	—	—	—	—	—	—
†*Vini vidivici*	—	65	2	—	—	—	—	—	—	—	—
**Vini kuhlii*	1	94	4	—	—	—	—	—	—	—	—
[†/* *Vini vidivici/kuhlii*]	—	9	—	—	—	—	—	—	—	—	—
CUCKOOS											
Eudynamys taitensis (m)	—	—	1	—	—	—	—	—	—	—	—
SWIFTS											
†*Collocalia manuoi*	—	—	6	—	—	—	—	—	—	—	—
KINGFISHERS											
Halcyon mangaia	22	9	5	—	2	—	—	—	—	—	—
WARBLERS											
Acrocephalus kerearako	5	16	82	—	—	—	1	—	—	—	—
TOTAL SPECIES	17	34	37	8	10	4	9	2	5	1	4
Seabirds	7	13	14	6	4	4	4	1	4	1	4
†/* seabirds	4	4	10	3	2	4	2	0	3	1	2
Landbirds	10	19	21	2	5	0	5	0	1	0	0
†/* landbirds	7	13	14	1	4	0	3	0	1	0	0
TOTAL BONES	731	826	1230	129	21	8	65	7	8	4	15
Seabirds	630	230	398	126	8	8	23	6	7	4	15
Landbirds	101	583	829	3	12	0	42	0	1	0	0

Fissure Cave in LS block NE of MAN-44 MAN-62	Ana Tuara	Tiny cave in LS block NE of MAN-44 MAN-61	Piri Te Umeume	MAN-81	MAN-83	MAN-87	MAN-82	MAN-63	MAN-95	Te Rua o Ngauru	Te Ana Tapukeu	**Te Ana o Ivi Kanapanapa	Total sites/ total bones
—	—	—	—	—	—	—	—	—	—	—	—	—	4/119
—	—	—	—	—	—	—	—	—	—	—	—	—	4/54
—	1	—	—	—	—	—	—	—	—	—	2	—	7/321
—	—	—	—	—	—	—	—	—	—	—	—	—	2/14
—	—	—	—	—	—	—	—	—	—	—	—	—	2/34
—	—	—	—	—	—	—	—	—	—	—	—	—	1/2
—	—	—	—	—	1	—	—	—	—	—	—	—	4/5
—	—	—	—	—	1	—	—	—	—	—	—	—	4/15
—	—	—	—	—	—	—	—	—	—	—	—	—	2/5
—	—	—	—	1	—	—	—	—	—	—	—	—	6/461
—	—	—	—	—	—	—	—	—	—	—	—	—	5/23
—	—	—	—	—	—	—	—	—	1	—	—	—	5/29
—	—	—	—	—	—	—	—	—	—	—	—	—	3/11
—	—	—	—	—	—	—	—	—	—	—	—	—	3/18
—	—	—	—	—	—	—	—	—	—	—	—	—	1/11
—	—	—	—	—	—	—	—	—	—	—	—	—	1/3
—	—	—	—	—	1	—	—	—	—	—	—	—	3/68
—	—	—	—	—	—	—	—	—	—	—	—	—	3/99
—	—	—	—	—	—	—	—	—	—	—	—	—	1/9
—	—	—	—	—	—	—	—	—	—	—	—	—	1/1
—	—	—	—	—	—	—	—	—	—	—	—	—	1/6
—	—	—	—	—	—	—	—	—	—	—	—	—	4/38
—	—	—	1	—	—	—	—	—	—	—	—	3	6/108
1	2	1	2	5	—	—	—	—	—	—	—	—	—
1	1	1	1	4	2	1	1	0	2	2	0	3	19
1	1	1	1	1	0	1	1	0	1	0	0	2	11
0	1	0	1	1	1	1	1	1	1	0	1	1	20
0	1	0	0	1	1	0	0	0	1	0	1	0	15
2	3	2	2	8	13	5	3	1	3	1	2	72+	3018+
2	2	2	1	7	3	1	1	0	2	1	0	69+	1533+
0	1	0	1	1	1	1	1	1	1	0	2	3	1356

1990, Allen & Steadman 1990, Allen 1992). These sites have yielded 48 bird bones from 15 species (Steadman 1991b), of which two seabirds and three landbirds no longer occur on the island (the petrel *Pterodroma rostrata*, booby *Sula sula*, whistling-duck cf. *Dendrocygna* undescribed sp., rail *Porzana tabuensis*, and parrot *Vini kuhlii*). Only the booby and rail survive anywhere in the Cook Islands. The nearest record of any species of *Dendrocygna* is in Fiji, where the much smaller *D. arcuata* is extirpated (Chapter 13). Other than the aquatic, trampy heron *Egretta sacra* and duck *Anas superciliosa*, the only possibly native, resident landbird on Aitutaki today is the parrot *Vini peruviana*, which probably was introduced from Tahiti, however, since all prehistoric bones of small species of *Vini* from the Cook Islands are of *V. kuhlii*. The columbids *Ducula pacifica* and *Ptilinopus rarotongensis* also occurred until the 1940s or 1950s on Aitutaki, where indigenous forest no longer exists (Steadman 1991b).

Mangaia (Figures 1-22, 1-23, 3-11) has one of Oceania's best records of prehistoric birds, based on many archaeological and paleontological sites (Tables 4-4, 7-3). Bones have increased Mangaia's landbird fauna from five to 20 species, more than double that of any other island in the Cooks. One indication of the relative thoroughness of Mangaia's avian prehistory is that every species still living on the island, whether seabird or landbird, has been recovered as well in one or more bone deposits. Only Henderson Island (Pitcairn Group) can also make this claim. Of the 16 genera recorded from the Cook Islands (Table 7-1), only three (cf. *Dendrocygna*, *Pomarea*, *Aplonis*) have not been found yet on Mangaia. The former is represented by just one bone from Aitutaki (see above), whereas *Pomarea* and *Aplonis* survive only on Rarotonga, the latter also known historically on Ma'uke (Holyoak 1980, Olson 1986a).

Mangaia's uplifted limestone is riddled with caves and rockshelters (Gill 1894, Hiroa 1934, Steadman 1985, Stoddart et al. 1985, Ellison 1994b, Kirch 1996). Among the 30 sites that have produced prehistoric bird bones (Table 7-3), three merit special mention. The first is Te Rua Rere (site MAN-24; Figure 7-2), a 600+ m–long cave where surface collections of bird bones are dominated by seabirds, especially the petrel *Pterodroma nigripennis* (Table 7-4). I have visited MAN-24 on each trip to Mangaia (Table 4-3). Te Rua Rere has a largely horizontal floor, a walk-in entrance (after climbing down a 7 m crevice), and several skylights (Ellison 1994b), thus providing an ideal physical setting for deposition of bones. Furthermore, much of the passage in MAN-24 is dominated by secondary carbonates precipitated from coral limestone, an ideal geochemical situation for long-term preservation of bones.

Aside from its many fossils of birds, Te Rua Rere has 26 human burial sites (mostly prehistoric) representing at least 66 persons (Antón & Steadman 1998, 2003). The oldest ^{14}C date on a human bone from MAN-24 is 1275–1410 cal AD (95.4% probability). The bird bones lying on the floor of this cave have no obvious cultural association. They are not

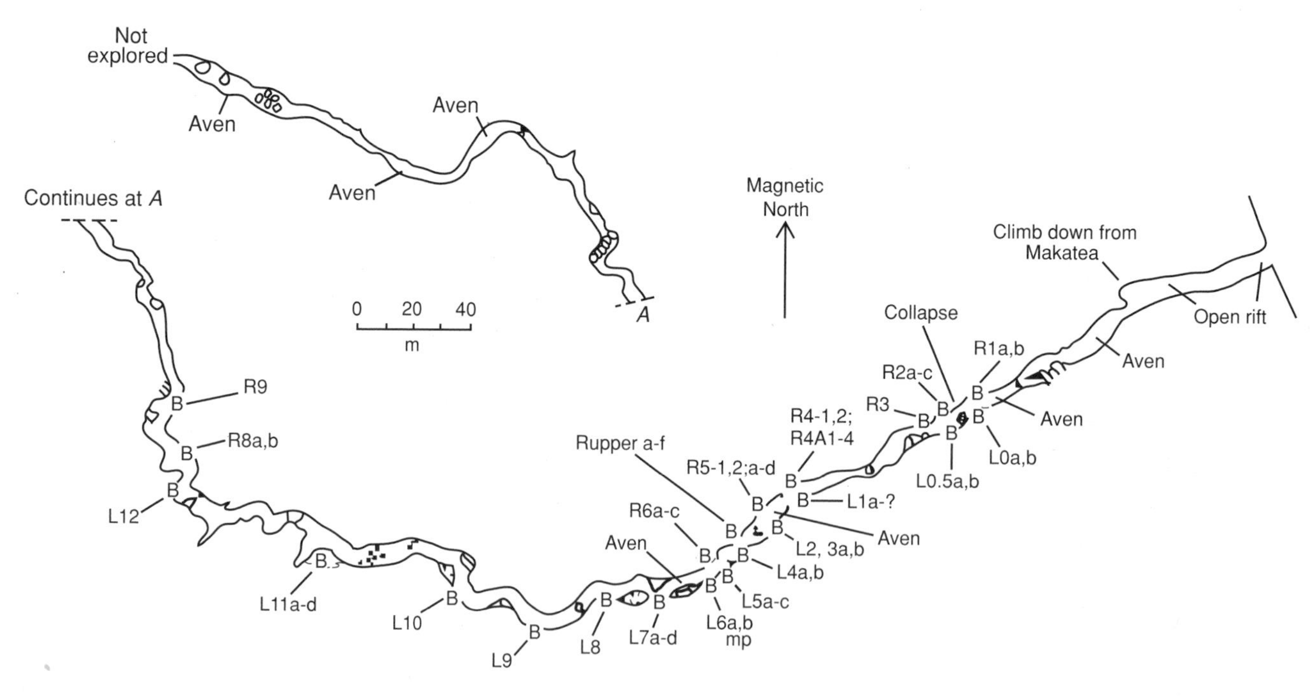

Figure 7-2 Plan view of Te Rua Rere Cave (MAN-24), Mangaia, Cook Islands. Redrawn from Anton & Steadman (2003: Figure 4).

Table 7-4 Bone summary from Te Rua Rere (MAN-24), Mangaia. See Table 4-3 for itinerary of these surface collections, the earliest of which were reported in Steadman (1985). Data for human bones from Antón & Steadman (2003). Taxa in brackets are not necessarily different from those identified more specifically.

	Number of bones	Minimum number of individuals
Fish	9	2
Lizard	3	3
Pacific rat (*Rattus exulans*)	310	58
Black rat (*Rattus rattus*)	17	2
Fruit bat (*Pteropus tonganus*)	6	4
Human (*Homo sapiens*)	1000+	66
SEABIRDS		
SHEARWATERS, PETRELS		
**Pterodroma cookii*	30	6
**Pterodroma nigripennis*	507	61
**Pterodroma cookii/nigripennis*	8	5
Puffinus lherminieri	37	7
STORM-PETRELS		
**Nesofregetta fuliginosa*	1	1
TROPICBIRDS		
Phaethon lepturus	3	2
TERNS		
Gygis candida	44	14
LANDBIRDS		
RAILS		
†Gallirallus ripleyi	6	5
Porzana tabuensis	7	6
†Porzana rua	30	16
PIGEONS AND DOVES		
**Ptilinopus rarotongensis*	6	5
**Ducula galeata*	2	2
**Gallicolumba erythroptera*	16	6
†Gallicolumba nui	6	4
PARROTS		
**Vini kuhlii*	1	1
KINGFISHERS		
Halcyon mangaia	22	4
WARBLERS		
Acrocephalus kerearako	5	5
TOTAL		
Seabirds	630	96
Landbirds	101	54
All species	731	150

in a midden context nor are they clustered near the human skeletons as "grave goods." There is almost no fish bone at Te Rua Rere; nearly all nonhuman bones are from birds or the Pacific rat *Rattus exulans* (Table 7-4). I believe that these bones represent animals that died naturally in the cave over the millennia. The avens (skylights) that pierce the roof of Te Rua Rere have allowed the occasional bird, rat, lizard, or bat to fly or fall into the cave and suffer injury or disorientation in the darkness that prevented it from leaving. The few fish bones may be the food remains from seabirds (petrels, shearwaters, tropicbirds, terns) that died here. Another indication that the nonhuman bones at Te Rua Rere are not cultural in origin is that they often are found as partially associated skeletons, perhaps somewhat scattered, broken, or chewed by scavenging rats or crabs, but unburnt. Bones from midden deposits tend to be fragmentary, disassociated, and burnt (Steadman et al. 2000b).

Te Rua Rere is the type locality of the flightless rails *Gallirallus ripleyi* and *Porzana rua*, the latter named after the site (Steadman 1987). Among landbirds, the high proportion of rail bones at MAN-24 is further evidence that one-way trips through avens may have been the primary means of bone deposition. Aside from rails, the remains of five other species of extinct or extirpated landbirds have been found at Te Rua Rere (Table 7-4). During my initial visit to the cave in 1984, these bones were the first evidence that pigeons, doves, and parrots once lived on Mangaia. They furnished the impetus to work with archaeologist Patrick Kirch to excavate and study bone deposits from Mangaia's cultural sites, to confirm that these and other extirpated species had coexisted with prehistoric people.

As a result of this collaboration, the prehistory of Mangaian birds has been examined in chronostratigraphic detail at many sites, but especially at Tangatatau Rockshelter (MAN-44; Figures 3-12, 3-13, 7-3). The age of MAN-44 ranges from older than ca. 1000 cal AD (the largely noncultural Zones 1A–1B) to ca. 1000-1250 cal AD (Zones 2–3, the earliest cultural strata) to ca. 1650-1750 cal AD (the late prehistoric Zone 15). The 2100+ bones from Zones 17–19 are not included in Table 7-5 because they are a mixture of late prehistoric and historic materials. Fish bones dominate all purely cultural levels at MAN-44. The human consumption of nonfish vertebrates underwent dramatic changes from Zones 2–3 to 15, such as the domesticated dog *Canis familiaris*, pig *Sus scrofa*, and chicken *Gallus gallus* becoming rare (Table 7-5), perhaps from overconsumption by a large human population.

The number of bones from native birds is high in Zones 1–4, low in the middle zones, and then fairly high again in Zone 15. The species composition, however, is very different between the early and late zones. Extinct and extirpated

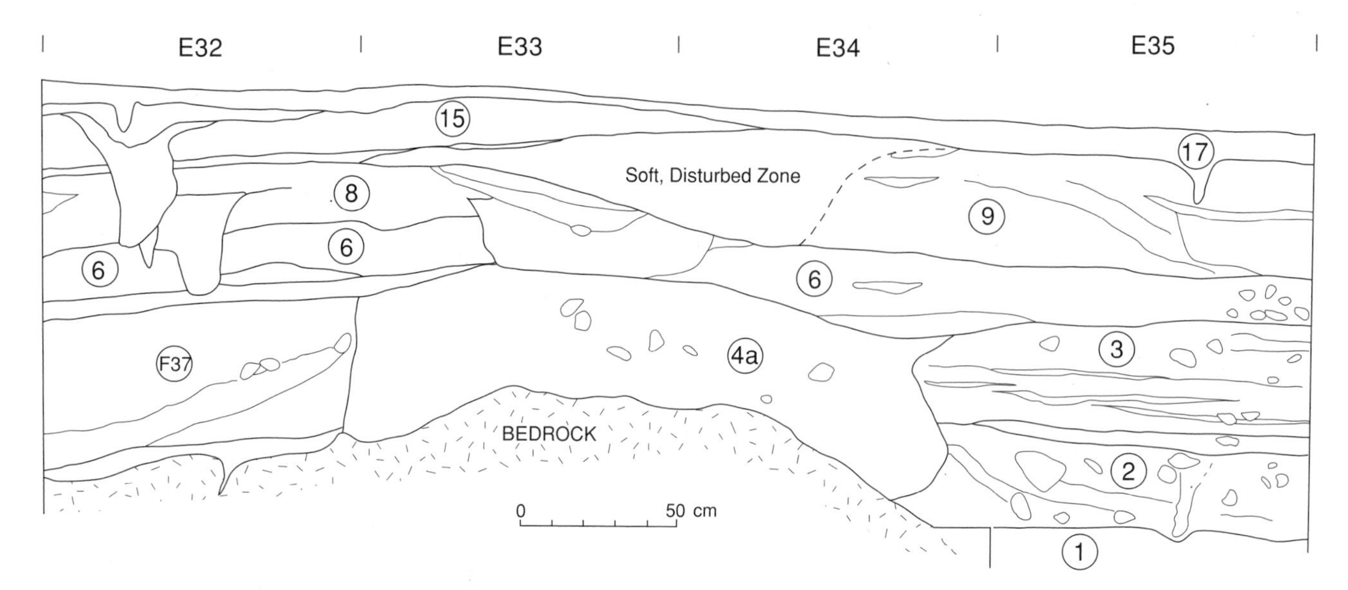

Figure 7-3 Stratigraphic profile, Tangatatau Rockshelter (MAN-44), Mangaia, Cook Islands. Circled numbers are "analytic zones" (chronostratigraphic units) that correspond with Tables 7-5 and 7-6. Many of the 17 analytic zones are not represented in this profile. Redrawn from Kirch et al. (1995: Figure 4).

species of landbirds dominate Zones 1–4, whereas seabirds account for the late increase in exploitation of birds (Table 7-6). This differs from other East Polynesian sites, where bones of seabirds are much more abundant than those of landbirds in the oldest cultural levels (Steadman 1989a, 1995a, Dye & Steadman 1990, Steadman et al. 1994; Chapter 15 herein). Seabirds may have been able to survive longer on Mangaia because its precipitous, creviced limestone cliffs furnished an extensive band of relatively rat-free nesting habitat.

As elsewhere in East Polynesia, most species of extinct or extirpated landbirds from MAN-44 are rails, sandpipers, pigeons, doves, and parrots. At least three of the extinct rails were flightless; volancy is uncertain for *Porphyrio* undescribed sp. C, a small swamphen known only from a scapula and femur. Based on bone counts, predation on landbirds at MAN-44 tapered off to practically nil by the end of Zone 4 (Table 7-6). Each extinct or extirpated species of bird first discovered at Te Rua Rere has been found in cultural association at MAN-44.

Table 7-5 Faunal summary, main excavation block, Tangatatau Rockshelter (site MAN-44), Mangaia, by analytic zone. Zone 1A is almost entirely noncultural. The bones from zone 1B are both cultural and noncultural in origin. Zone 2 is the oldest purely cultural zone. Zone 15 is the youngest purely prehistoric context at MAN-44. Zones that are closely related, based on stratigraphy, chronology, and artifact styles, are combined. Modified from Kirch et al. (1995: Tables 4-7).

	Analytic Zone								
	1A	1B	2–3	4	5–7	8	9–14	15	TOTAL
Fish	53	1098	9113	4940	5026	6763	1338	1450	29,781
Human	0	0	7	1	2	3	0	4	17
Rat	1	62	944	195	78	330	199	352	2161
Dog	0	0	35	0	0	5	2	0	42
Pig	0	0	89	14	29	62	4	1	199
Chicken	0	0	22	20	27	16	12	1	98
Native bird	105	143	127	119	15	40	45	78	672
TOTAL	159	1303	10, 337	5289	5177	7219	1600	1886	32,970

Table 7-6 Bird bones from the main excavation block, Tangatatau Rockshelter (site MAN-44), Mangaia, Cook Islands. †, extinct species; *, extant species extirpated on Mangaia; i, introduced species. Updated from Steadman (1995a). Taxa in brackets are not necessarily different from those identified more specifically.

	Analytic Zone									
	1A	1B	2–3	4	5–7	8	9–14	15	17–19	TOTAL
SEABIRDS										
SHEARWATERS, PETRELS										
**Pterodroma nigripennis*	—	—	—	—	—	19	23	47	11	100
Puffinus lherminieri	—	—	—	—	—	2	2	10	3	17
STORM-PETRELS										
**Nesofregetta fuliginosa*	—	1	—	2	3	—	1	1	—	8
TROPICBIRDS										
Phaethon rubricauda	—	1	—	—	—	—	—	—	—	1
Phaethon lepturus	4	2	19	12	2	—	1	1	—	41
BOOBIES										
Sula sula	—	—	—	1	—	—	—	—	—	1
FRIGATEBIRDS										
Fregata minor	—	—	1	—	—	4	—	—	—	5
Fregata ariel	—	—	—	—	—	5	—	2	—	7
TERNS										
Anous stolidus	1	1	—	—	1	5	3	3	1	15
Procelsterna cerulea	—	—	—	1	—	—	—	—	—	1
Gygis candida	2	1	6	2	1	—	—	—	—	12
**Gygis microrhyncha*	—	4	7	1	—	—	—	—	—	12
LANDBIRDS										
DUCKS										
Anas superciliosa	—	—	3	1	2	2	8	11	3	30
CHICKENS										
Gallus gallus (i)	—	—	22	20	27	16	12	1	2	100
RAILS										
†*Gallirallus ripleyi*	10	5	12	16	1	—	—	—	—	44
Porzana tabuensis	—	—	2	1	—	1	—	1	1	6
†*Porzana rua*	11	44	41	21	2	1	—	—	—	120
†*Porzana* undescribed sp. C	—	2	1	1	—	—	—	—	—	4
†*Porphyrio* undesc. sp. C	—	1	—	1	—	—	—	—	—	2
SANDPIPERS										
†*Prosobonia* undescribed sp. A	1	1	1	1	—	—	—	—	—	4
PIGEONS, DOVES										
**Ptilinopus rarotongensis*	1	4	4	2	—	—	—	1	—	12
**Ducula aurorae*	1	2	1	1	—	—	—	—	—	5
**Ducula galeata*	3	1	5	2	—	1	—	—	—	12
**Gallicolumba erythroptera*	3	5	3	5	—	—	—	—	—	16
†*Gallicolumba nui*	3	3	—	3	—	—	—	—	—	9

(continued)

Table 7-6 (continued)

	Analytic Zone									
	1A	1B	2–3	4	5–7	8	9–14	15	17–19	TOTAL
PARROTS										
†*Vini vidivici*	42	14	3	14	2	—	—	—	—	75
**Vini kuhlii*	14	41	15	23	1	—	—	—	—	94
[†/* *Vini vidivici/kuhlii*]	2	2	—	5	—	—	—	—	—	9
KINGFISHERS										
Halcyon mangaia	1	2	2	2	—	—	—	—	1	8
WARBLERS										
Acrocephalus kerearako	6	6	1	2	—	—	—	—	—	15
TOTAL SPECIES										
All	15	20	19	22	10	10	7	10	7	29
All native	15	20	18	21	9	9	6	9	6	28
Seabirds	3	6	4	6	4	5	5	6	3	12
Native landbirds	12	14	14	15	5	4	1	3	3	16
†/* native landbirds	10	12	10	11	4	2	0	1	0	12
TOTAL BONES										
All species	105	143	149	140	42	56	50	78	22	785
All native species	105	143	127	120	15	40	38	77	19	685
Seabirds	7	10	33	19	7	35	20	64	15	220
Native landbirds	98	133	94	101	8	5	8	13	5	465
†/*landbirds	91	125	86	95	6	2	0	1	0	416
†/* landbird bones as a percentage of those of all birds and mammals	72	57	5	21	4	0.4	0	0.1	0	11
all birds	87	87	58	68	14	4	0	1	0	53
all native landbirds	93	94	92	94	75	40	0	8	0	88

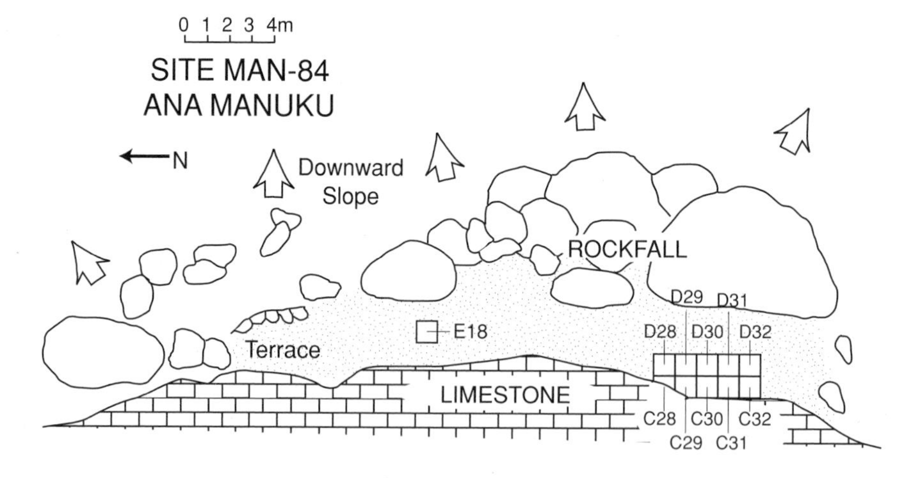

Figure 7-4 Plan view of Ana Manuku (MAN-84), Mangaia, Cook Islands. The 11 excavation squares (units) are labeled in a number/letter grid. From Steadman et al. (2000a: Figure 2).

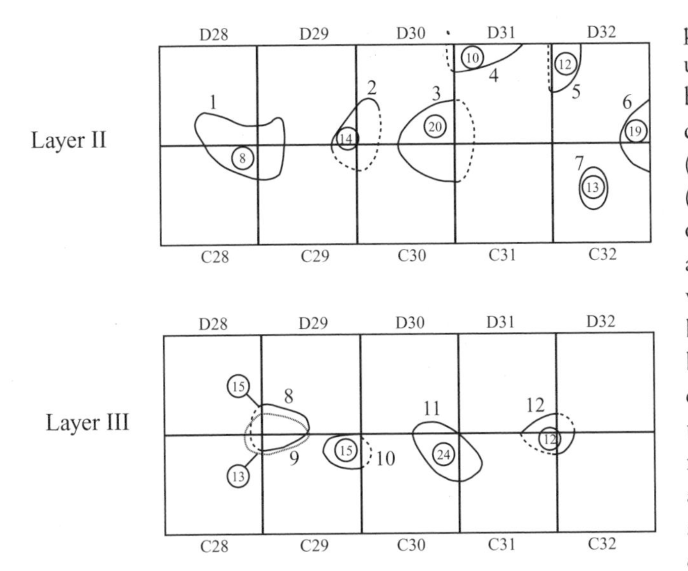

Figure 7-5 Plan view showing oven features from the cultural strata (Layers II and III), Ana Manuku (MAN-84), Mangaia, Cook Islands. From Steadman et al. (2000a: Figure 4). The thickness of each oven (in cm) is given in circles.

The Mangaian rockshelter called Ana Manuku (MAN-84) was excavated by Patrick Kirch in 1991, and by Susan Antón and myself in 1997 (Steadman et al. 2000a). The faunal and artifactual composition of MAN-84 is utterly unlike that of any other Polynesian site. Two cultural strata (Layers II, III) with 12 earth ovens (Figures 7-4 to 7-6) are rich in charcoal and burned bones of humans, fish, rats, and birds (Table 7-7). Essentially all of the rat and fish bones, and even most (84%) of the human bones are burned. The precultural Layer IV is dominated (96% of all bones) by unburned bones of extirpated species of birds (discussed below). Layers II and III differ from other major cultural deposits on Mangaia, particularly Tangatatau Rockshelter (MAN-44), in these attributes: short period of occupation (probably less than a century, based on nine AMS ^{14}C dates on human bone that cluster at 1390–1470 cal AD); very few artifacts, including no adzes or one-piece fishhooks; shellfish very rare; fish bone relatively scarce; pig, dog, and chicken bones absent or rare; and human bone abundant, mostly burned and broken, representing >40 persons of diverse age classes, and all in a midden rather than burial context. These unusual characteristics, as well as the location of MAN-84 near several temples (*marae*) of great religious, political, and ceremonial importance, suggest that Ana Manuku was a ritualistic or special-use site, where human bodies were cooked, rather than a typical habitation site.

Because the cultural strata at MAN-84 represent such a short interval of time, this site is not nearly as good as MAN-44 for documenting chrono-stratigraphic changes in birdlife after human arrival. MAN-84 excels, however, in its rich assemblage of precultural bird bones from Layer IV (Table 7-7). Most bird bones from Layer III probably are precultural as well, given that they are unburned (Table 7-8) and of identical preservation to those from Layer IV.

Among the 398 seabird bones from MAN-84, only 12 (from two species) are from Layer II. About 83% of the seabird bones are from Layer IV, with nine of the 12 species no longer living on Mangaia. The landbirds include four extinct species (the night-heron *Nycticorax* undescribed sp. C,

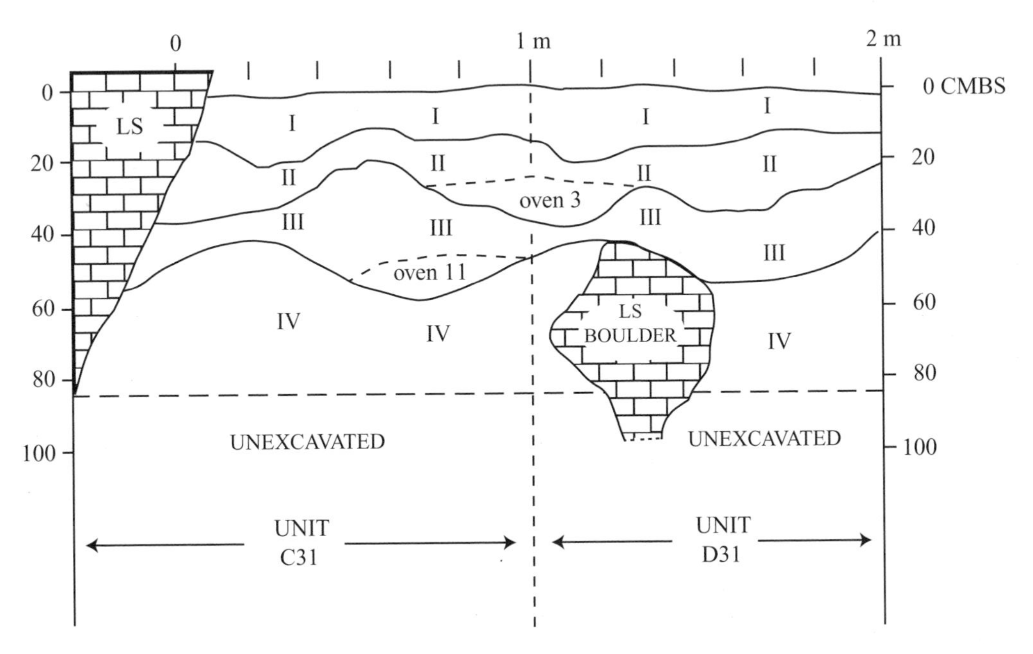

Figure 7-6 Stratigraphic profile, Ana Manuku (MAN-84), Mangaia, Cook Islands. From Steadman et al. (2000a: Figure 3).

Table 7-7 Stratigraphic summary of bird bones from the 1997 excavations at Ana Manuku (site MAN-84), Mangaia, Cook Islands. i, introduced species; m, migratory species; †, extinct species; *, extant species, extirpated on Mangaia. i, m not included in totals. Taxa in brackets are not necessarily different from those identified more precisely.

	Layer			
	II	III	IV	TOTAL
SEABIRDS				
SHEARWATERS, PETRELS				
**Pterodroma inexpectata*	—	—	1	1
**Pterodroma cookii*	—	3	21	24
**Pterodroma nigripennis*	—	11	81	92
[**Pterodroma nigripennis/cookii*]	—	7	68	75
**Pachyptila* sp.	—	—	2	2
**Bulweria* sp.	—	3	2	5
**Puffinus* cf. *gavia*	—	1	4	5
Puffinus lherminieri	5	2	7	14
[Procellariidae sp.]	3	9	97	109
STORM-PETRELS				
**Nesofregetta fuliginosa*	4	5	35	44
TROPICBIRDS				
Phaethon lepturus	—	12	6	18
TERNS				
**Anous minutus*	—	1	—	1
Procelsterna caerulea	—	1	6	7
**Gygis microrhyncha*	—	—	1	1
LANDBIRDS				
HERONS				
†*Nycticorax* undescribed sp. C	—	—	6	6
CHICKENS				
Gallus gallus (i)	2	—	—	2
RAILS				
†*Gallirallus ripleyi*	—	13	54	67
Porzana tabuensis	6	3	—	9
†*Porzana rua*	—	18	146	164
†*Porzana* undescribed sp. C	6	—	4	10
[*Porzana* sp.]	3	5	17	25
PLOVERS				
Pluvialis fulva (m)	—	—	1	1
SANDPIPERS				
Numenius tahitiensis (m)	—	1	1	2
†*Prosobonia* undescribed sp. A	—	—	1	1
PIGEONS, DOVES				
**Ptilinopus rarotongensis*	—	1	7	8
**Ducula aurorae*	—	—	1	1
**Ducula galeata*	—	—	1	1
**Gallicolumba erythroptera*	4	51	359	414
†*Gallicolumba nui*	—	—	3	3
[Columbidae sp.]	—	1	10	11
PARROTS				
†*Vini sinotoi*	—	3	—	3
†*Vini vidivici*	—	—	2	2
**Vini kuhlii*	—	—	4	4
CUCKOOS				
Eudynamys taitensis (m)	—	—	1	1
SWIFTS				
†*Collocalia manuoi*	—	2	4	6
KINGFISHERS				
Halcyon mangaia	—	1	4	5
WARBLERS				
Acrocephalus kerearako	—	12	70	82
SEABIRD TOTALS				
Bones	12	55	331	398
Species	2	9	11	12
†/*species	1	6	8	9
LANDBIRD TOTALS				
Bones	19	110	695	819
Species	3	9	15	17
†/*species	2	6	13	14

Table 7-8 Major categories of vertebrates, Ana Manuku (site MAN-84), Mangaia, Cook Islands. Layers II and III are cultural; Layer IV is precultural. Based on recovery in situ and dry-screening through 3.3 mm mesh. Numbers are NISP (number of identified specimens). From Steadman et al. (2000a).

	Layer			
	II	III	IV	TOTAL
Fish	352	275	11	638
Lizard	1	5	0	6
Bird	28	181	688	897
Fruit bat	1	1	0	2
Human	316	1475	7	1798
Rat	185	694	10	889
Dog	1	0	0	1
TOTAL	884	2631	716	4231
Bird bones: % burned	69	28	1	—

a tiny rail *Porzana* undescribed sp. C, parrot *Vini sinotoi*, and swift *Collocalia manuoi*) that had not been found before on Mangaia. The precultural Layer IV includes 15 of the 17 species of landbirds recorded from the site, and 13 of the 14 extinct or extirpated landbird species. Ground-dwellers (rails, ground-doves) dominate the MAN-84 landbird assemblage, suggesting that the precultural predator that accumulated the bones hunted primarily on the ground, where nesting seabirds also would have been available. The identity of this predator is unknown although a prime suspect, on this island where no hawks, falcons, or owls are recorded today or in the past, is the extinct night-heron.

Lacking artifactual, sedimentary, or faunal evidence of cultural activities, the lowest levels of both MAN-84 (Layer IV) and MAN-44 (Zones 1a–1b) were deposited before people occupied these rockshelters. The species composition of their precultural landbirds is fairly similar, although parrots are common at MAN-44 but rare at MAN-84, where the only landbirds found in good numbers are rails, the ground-dove *Gallicolumba erythroptera*, and warbler *Acrocephalus kerearako*. Among seabirds, procellariids are absent at MAN-44 but abundant at MAN-84, including eight species now gone from Mangaia (Table 7-9). These dissimilarities suggest different modes of precultural bone accumulation at the two sites.

The overall prehistory of birds on Mangaia (Table 7-4) portrays an avifauna that has been decimated over the past 1000 years. Late prehistoric archaeological sites on Mangaia, such as those dating after 1600 cal AD reported by Taomia (2000), are nearly or absolutely devoid of bird bones. Deforestation (Figure 2-9), as well as predation by humans and rats, has eliminated all Mangaian populations of the pigeons, doves, parrots, and flightless rails (see Chapter 16). The resident sandpiper, while probably not flightless, undoubtedly was tame and easily captured. Pigeons, doves, and parrots are tasty and favored as food by Polynesians (Steadman 1997b). Parrots, and sometimes pigeons and doves, were used also for their brightly colored feathers.

The five surviving species of landbirds on Mangaia (Tables 7-2, 7-4) all can withstand substantial forest clearance. Three are widespread, aquatic species. The heron *Egretta sacra* feeds in the intertidal zone and freshwater habitats. Because they prefer marshes or dense growths of ferns, grasses, and sedges, the duck *Anas superciliosa* and the rail *Porzana tabuensis* probably have benefited from prehistoric landscape changes that promote irrigated cultivation of taro (Kirch 1996). The endemic kingfisher *Halcyon mangaia* tolerates moderate forest clearance, but requires tree cavities for nesting (Rowe & Empson 1996). Its current scarcity on Mangaia may be related to nest-site competition with the abundant and aggressive myna *Acridotheres tristis*, introduced early in the 20th century. The endemic reed-warbler *Acrocephalus kerearako* inhabits forests, thickets, and shrublands of varying disturbance and maturity. The warbler is widespread and common, although its population is dwarfed by that of the raucous myna.

Tubuai Group (Austral Islands)

This irregularly linear chain of seven islands (Figure 7-7) is related geologically to the southern Cook Islands (Dickinson 1998b). In fact Mangaia lies closer to Maria (westernmost island in the Tubuai Group) than to Rarotonga, its closest neighbor in the Cook Islands. The modern landbirds of Tubuai are few in species and very patchily distributed among the islands (Table 7-10). The prehistoric data needed to estimate the past distribution of birds have yet to be gathered, however, except for very limited information from Rurutu and Tubuai. If the Tubuai Group had more avian fossils, any of the genera recorded from nearby Mangaia might be expected.

From Rurutu, which is nearly lacking in landbirds today, Gustav Paulay collected 12 associated bones of a petrel in a noncultural context (beneath a calcite layer) from Coconut Cave in August 1983. It is uncertain whether Coconut Cave is the same Rurutan cave of cultural significance mentioned by Emory (1927, 1932) and Teimaore (1927). The 12 bones are from a small species of *Pterodroma* in the size range of *P. nigripennis* or *P. cookii*, the former the most commonly recorded prehistoric seabird on Mangaia.

Table 7-9 Comparison of birds from precultural strata on Mangaia, Cook Islands: Zones 1A–1B at Tangatatau Rockshelter (site MAN-44) versus Layer IV at Ana Manuku (site MAN-84). m, migratory species; †, extinct species; *, extant species, extirpated on Mangaia. Taxa in brackets are not necessarily different from those identified more precisely.

	Number of bones		% of total bird bones	
	Zones 1A–1B MAN-44	Layer IV MAN-84	Zones 1A–1B MAN-44	Layer IV MAN-84
SEABIRDS				
SHEARWATERS, PETRELS				
**Pterodroma inexpectata*	—	1	—	0.1
**Pterodroma cookii*	—	21	—	2.1
**Pterodroma nigripennis*	—	81	—	8.0
[**Pterodroma cookii/nigripennis*]	—	68	—	6.7
**Pachyptila* sp.	—	2	—	0.2
**Bulweria* sp.	—	2	—	0.2
**Puffinus* cf. *gavia*	—	4	—	0.4
Puffinus lherminieri	—	7	—	0.7
[Procellariidae sp.]	—	97	—	9.5
STORM-PETRELS				
**Nesofregetta fuliginosa*	1	35	0.4	3.4
TROPICBIRDS				
Phaethon rubricauda	1	—	0.4	—
Phaethon lepturus	6	6	2.4	0.6
TERNS				
**Anous stolidus*	2	—	0.8	—
Procelsterna cerulea	—	6	—	0.6
Gygis candida	3	—	1.2	—
**Gygis microrhyncha*	4	1	1.6	0.1
LANDBIRDS				
HERONS				
†*Nycticorax* undescribed sp. C	—	5	—	0.5
RAILS				
†*Gallirallus ripleyi*	15	54	6.0	5.3
†*Porzana rua*	55	146	22.2	14.4
†*Porzana* undescribed sp. C	2	4	0.8	0.4
[*Porzana* sp.]	—	17	—	1.7
†*Porphyrio* undesc. sp. C	1	—	0.4	—
PLOVERS				
Pluvialis fulva (m)	—	1	—	0.1
SANDPIPERS				
Numenius tahitiensis (m)	—	1	—	0.1
†*Prosobonia* undescribed sp. A	2	1	0.8	0.1

Table 7-9 (continued)

	Number of bones		% of total bird bones	
	Zones 1A–1B MAN-44	Layer IV MAN-84	Zones 1A–1B MAN-44	Layer IV MAN-84
PIGEONS, DOVES				
**Ptilinopus rarotongensis*	5	7	2.0	0.7
**Ducula aurorae*	3	1	1.2	0.1
**Ducula galeata*	4	1	1.6	0.1
**Gallicolumba erythroptera*	8	359	3.2	35.2
†*Gallicolumba nui*	6	3	2.4	0.3
[Columbidae sp.]	—	10	—	
PARROTS				
†*Vini vidivici*	56	2	22.6	0.2
**Vini kuhlii*	55	4	22.2	0.4
[†/**Vini vidivici/kuhlii*]	4	—	1.6	—
CUCKOOS				
Eudynamys taitensis (m)	—	1	—	0.1
SWIFTS				
†*Collocalia manuoi*	—	4	—	0.4
KINGFISHERS				
Halcyon mangaia	3	4	1.2	0.4
WARBLERS				
Acrocephalus kerearako	12	70	4.8	6.9
TOTAL BONES				
All species	248	1016	100.0	101.0
Landbirds	231	685	93.2	67.4
TOTAL SPECIES				
Seabirds	6	11		
†/*seabirds	2	8		
Landbirds	14	15		
†/*landbirds	12	13		

The modern native landbirds on Tubuai Island consist of three aquatic tramps (Table 7-10). During a survey in February 1980, introduced chickens (*Gallus gallus*) and mynas (*Acridotheres tristis*) were the only species found in Tubuai's forests (Ehrhardt 1980). Richard Shutler excavated a small sample of the cultural sediments at the Atiahara site on Tubuai in 1995. Atiahara is an open, calcareous sand site with an Early East Polynesian artifact assemblage (see Kirch 1986, 1996, Weisler 1994) that probably dates from 1000 to 600 cal BP. The bones from Atiahara are dominated by fish (Table 7-11). The scarcity of bird bones, and the fact that pig bones are nearly as frequent as rat bones, would argue that the recovery of small bones was less than optimal. Nevertheless, the 11 bird bones (Table 7-12) provide a tantalizing glimpse of Tubuai's prehistoric avifauna. The presence of the prion *Pachyptila* would ally the prehistoric seabird fauna with that of the Cook Islands, Marquesas, and Easter Island.

The fruit-dove (*Ptilinopus* undescribed sp.) from Atiahara is represented by a nearly complete tibiotarsus that is very different from those of the geographically nearest congeneric species, *P. rarotongensis* to the west (Cook Islands)

Table 7-10 Distribution of landbirds in the Tubuai (Austral) Islands. H, historic record, no longer survives; M, modern record; P, prehistoric record. From Petitot & Petitot (1975), Ehrhardt (1980), Holyoak & Thibault (1984), Pratt et al. (1987), Seitre & Seitre (1991), and herein.

	Tubuai	Rurutu	Rapa	Raivavae	Rimatara	Maria	Marotiri	TOTAL
HERONS								
Egretta sacra	M	M	M	M	M	M	—	6
DUCKS								
Anas superciliosa	M	—	M	M	M	—	—	4
RAILS								
Porzana tabuensis	M/H	—	M	—	—	—	—	2
DOVES								
Ptilinopus huttoni	—	—	M	—	—	—	—	1
†*Ptilinopus* undescribed sp.	P	—	—	—	—	—	—	1
PARROTS								
Vini kuhlii	—	—	—	—	M	—	—	1
WARBLERS								
Acrocephalus vaughani	—	—	—	M	M	—	—	2
TOTAL SPECIES	4	1	4	3	4	1	0	—
Number of bird bones	11	12	0	0	0	0	0	23
Number of native landbird bones	1	0	0	0	0	0	0	1
Island area (km^2)	50	32	22	21	18	1.3	0.3	—
Elevation (m)	399	390	633	437	95	5	105	—
Isolation (km to nearest high island)	180	145	505	180	145	200	90	—

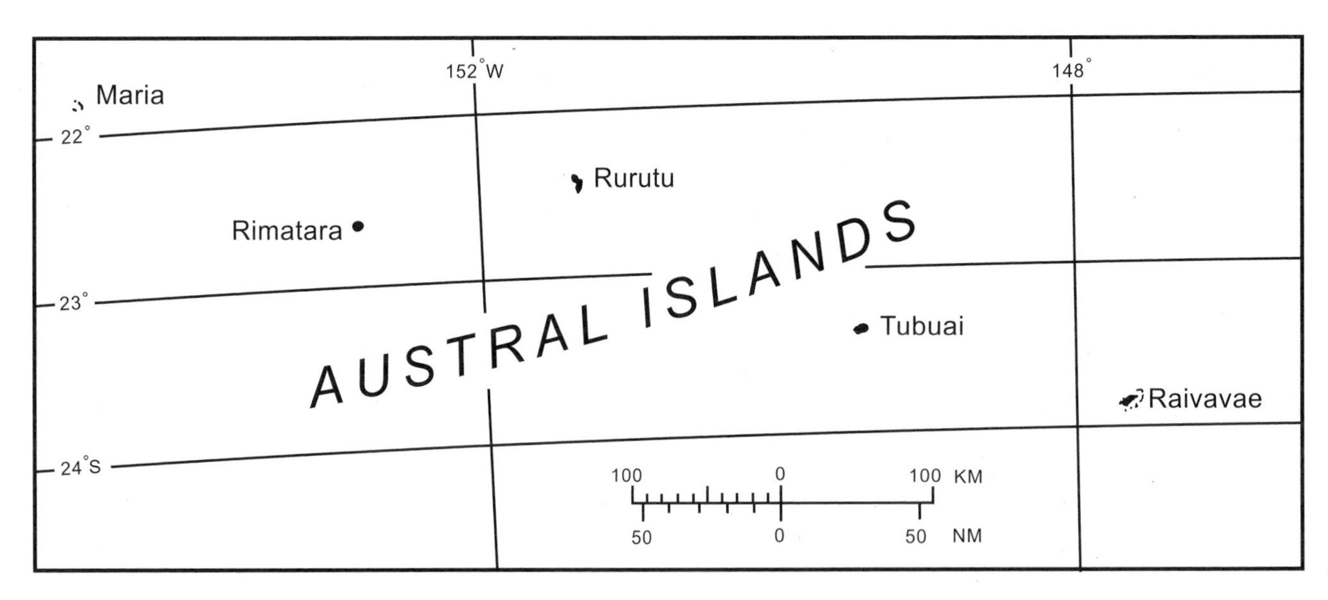

Figure 7-7 The Tubuai Islands, or Austral Islands. Redrawn from New Zealand Department of Lands and Survey (1986).

Table 7-11 Vertebrates (number of bones) from the Atiahara site, Tubuai, Tubuai (Austral) Islands.

	cm below surface					
	0–10	10–20	20–30	30–40	>40	TOTAL
Fish	230	639	940	185	104	2098
Sea turtle	1	14	7	2	—	24
Rat	3	9	64	—	—	76
Pig	37	9	13	4	1	64
Bird	2	2	3	3	1	11
TOTAL	273	673	1027	194	106	2273

and *P. huttoni* to the east (Rapa). The only two species known in the Tubuai Group (*Ptilinopus* undescribed sp. from Tubuai, *P. huttoni* from Rapa) are highly distinct from other East Polynesian species and from each other. The former is extinct, whereas fewer than 300 individuals of *P. huttoni* were estimated to have existed in 1989-1990, confined to Rapa's last patches of forest (Thibault & Varney 1991a).

Society Islands

The six largest of the Society Islands (Figure 7-8) are high, eroded volcanic islands with extensive fringing reefs. The remainder are small eroded volcanic islands or atolls. Raised limestone islands are lacking. Tahiti, Huahine, and the other major islands date to only the past 4–5 million years (Hoffman 1991, Dickinson 1998b; Chapter 1 herein). The lush greenery of the Society Islands is beautiful, but the coastal plain, and nearly any forested slope visible from the sea, has been dominated for a century or more by nonnative species of plants and early successional native species (Setchell 1926, Whittier 1976, Meyer 1996). Contamination of the flora with invasive species is approaching the horrendous levels seen in the Hawaiian Islands.

Tahiti is the largest island in East Polynesia and dominates the Society Islands politically, economically, and scientifically. Far more species (19) of landbirds have been recorded alive on Tahiti over the past 230 years (since European contact) than anywhere else in East Polynesia. This may relate to several factors. First, explorers and scientists have spent much more time on Tahiti than on other islands. Second, Tahiti's extreme elevation, steepness, and large size probably have retarded the pace of extinction since first human arrival, an event that has not been disclosed archaeologically. Finally, the Society islands may have had a somewhat richer avifauna than elsewhere in East Polynesia, a topic to be developed shortly.

No more than 12 of the 19 species recorded from Tahiti still survive there. Among the species certainly lost, both flightless rails (*Gallirallus pacificus*, *Porzana nigra*) are known only by paintings from Captain Cook's visits in the 1770s (Lysaght 1959, Olson 1973b, Medway 1979, Walters 1988, 1989). The future is not bright for some of the survivors, such as the heron *Butorides striatus*, pigeon *Ducula aurorae*, swift *Collocalia leucophaea*, kingfisher *Halcyon tuta*, and monarch *Pomarea nigra* (Monnet et al. 1993b). Nonnative species of landbirds have been more common than native species since the late 19th century (Townsend & Wetmore 1919, King 1958, Holyoak 1974, Monnet et al. 1993b, Lockwood et al. 1999).

Two genera (the heron *Butorides* and parrot *Cyanoramphus*) are known in East Polynesia only in the Society Islands (Table 7-1). The Society Islands are further unique in East Polynesia in having sympatric species of kingfishers (*Halcyon tuta*, *H. venerata*), both of which have been recorded on Tahiti and Bora Bora (Wilson 1907, Steadman & Pahlavan 1992).

Whether evaluated by genus or species, modern birds have highly fragmented ranges in the Society Islands (Table 7-13). That the more nonsensical distributional gaps are due

Table 7-12 Bird bones from Atiahara site, Tubuai, Austral Islands, i, introduced species; †, extinct species; *, extant species, extirpated on Tubuai.

	cm below surface					
	0–10	10–20	20–30	30–40	>40	TOTAL
SHEARWATERS, PETRELS						
†/**Pachyptila* sp.	—	—	1	—	—	1
Puffinus lherminieri	—	—	—	1	—	1
TROPICBIRDS						
Phaethon lepturus	—	1	1	1	—	3
TERNS						
Anous stolidus	1	—	—	—	—	1
Gygis candida	1	—	—	—	—	1
CHICKENS						
Gallus gallus (i)	—	1	1	1	—	3
DOVES						
†*Ptilinopus* undescribed sp.	—	—	—	—	1	1
TOTAL	2	2	3	3	1	11

Table 7-13 Resident landbirds of the Society Islands. From Wilson (1907), Townsend & Wetmore (1919), Holyoak (1974), Thibault (1976), Holyoak & Thibault (1984), Walters (1988, 1989), Seitre & Seitre (1991), and Steadman & Pahlavan (1992). H, historic record, no longer survives; M, modern record; P, prehistoric record; †, extinct species; ?, questionable record, or residency not established.

	Tahiti	Ra`iatea	Mo`orea	Taha`a	Huahine	Bora Bora	Tupai	Maiao	Tetiaroa	Maupiti	Mopelia	Mehetia	Scilly	Bellingshausen	TOTAL
HERONS															
Egretta sacra	M	M	M	M	M, P	M	M	M	M	M	M?	—	M	M	12–13
Butorides striatus	M	—	—	—	P	—	—	—	—	—	—	—	—	—	2
DUCKS															
Anas superciliosa	M	M	M	M	M	—	—	M	H	—	—	—	—	—	7
RAILS															
†*Gallirallus pacificus*	H	—	—	—	—	—	—	—	—	—	—	—	—	—	1
†*Gallirallus* undescribed sp. H	—	—	—	—	P	—	—	—	—	—	—	—	—	—	1
Porzana tabuensis	M	M	M	—	H, P	—	—	—	—	—	—	—	—	—	4
†*Porzane nigra*	H	—	—	—	—	—	—	—	—	—	—	—	—	—	1
†*Porphyrio* undescribed sp. D	—	—	—	—	P	—	—	—	—	—	—	—	—	—	1
SANDPIPERS															
†*Prosobonia leucoptera*	H	—	—	—	—	—	—	—	—	—	—	—	—	—	1
†*Prosobonia ellisi*	—	—	H	—	—	—	—	—	—	—	—	—	—	—	1
PIGEONS, DOVES															
†*Macropygia arevarevauupa*	—	—	—	—	P	—	—	—	—	—	—	—	—	—	1
Ptilinopus purpuratus	M	M	M	M	M, P	M	—	—	—	M	—	—	—	—	7
Ducula aurorae	M	—	H	—	P	—	—	—	—	—	—	—	—	—	3
Ducula galeata	H	—	—	—	P	—	—	—	—	—	—	—	—	—	2
Gallicolumba erythroptera	H	—	H	—	P	—	—	—	—	—	—	—	—	—	3
†*Gallicolumba nui*	—	—	—	—	P	—	—	—	—	—	—	—	—	—	1
PARROTS															
†*Vini sinotoi*	—	—	—	—	P	—	—	—	—	—	—	—	—	—	1
†*Vini vidivici*	—	—	—	—	P	—	—	—	—	—	—	—	—	—	1
Vini peruviana	H	H	H	H	H	H	—	—	H	H	H	—	M	M	11
†*Cyanoramphus zealandicus*	H	—	—	—	—	—	—	—	—	—	—	—	—	—	1
†*Cyanoramphus ulietanus*	—	H	—	—	—	—	—	—	—	—	—	—	—	—	1

SWIFTS															
Collocalia leucophaea	M	—	M	—	H	—	—	—	—	—	—	—	—	—	3
KINGFISHERS															
Halcyon venerata	M	—	M	—	—	H	—	—	—	—	—	—	—	—	3
Halcyon tuta	M	M	—	M	M, P	M	H	—	—	M	—	—	—	—	7
SWALLOWS															
Hirundo tahitica	M	—	M	—	—	—	—	—	—	—	—	—	—	—	2
WARBLERS															
Acrocephalus caffer	M	H	M	—	H, P	—	—	—	—	—	—	—	—	—	4
STARLINGS															
†*Aplonis diluvialis*	—	—	—	—	P	—	—	—	—	—	—	—	—	—	1
MONARCHS															
Pomarea nigra	M	—	—	—	—	—	—	—	—	—	—	—	—	—	1
†*Pomarea pomarea*	—	—	—	—	—	—	—	—	—	H	—	—	—	—	1
TOTAL SPECIES															
M	12	5	8	4	4	3	1	2	1	3	0–1	0	2	2	—
M + H	19	8	12	5	8	5	2	2	3	5	1–2	0	2	2	—
M + H + P	19	8	12	5	19	5	2	2	3	5	1–2	0	2	2	—
TOTAL BONES	0	0	0	0	110	0	0	0	0	0	0	0	0	0	110
Island area (km^2)	1042	238	133	99	78	38	21	9	6.5	5	2.6	<2.4	<2.4	—	
Elevation (m)	2241	1017	1207	590	669	727	5	170	5	380	5	433	5	5	—
Isolation (km to nearest high island)	16	4.3	16	4.3	34	15	16	73	49	12	175	110	243	246	—

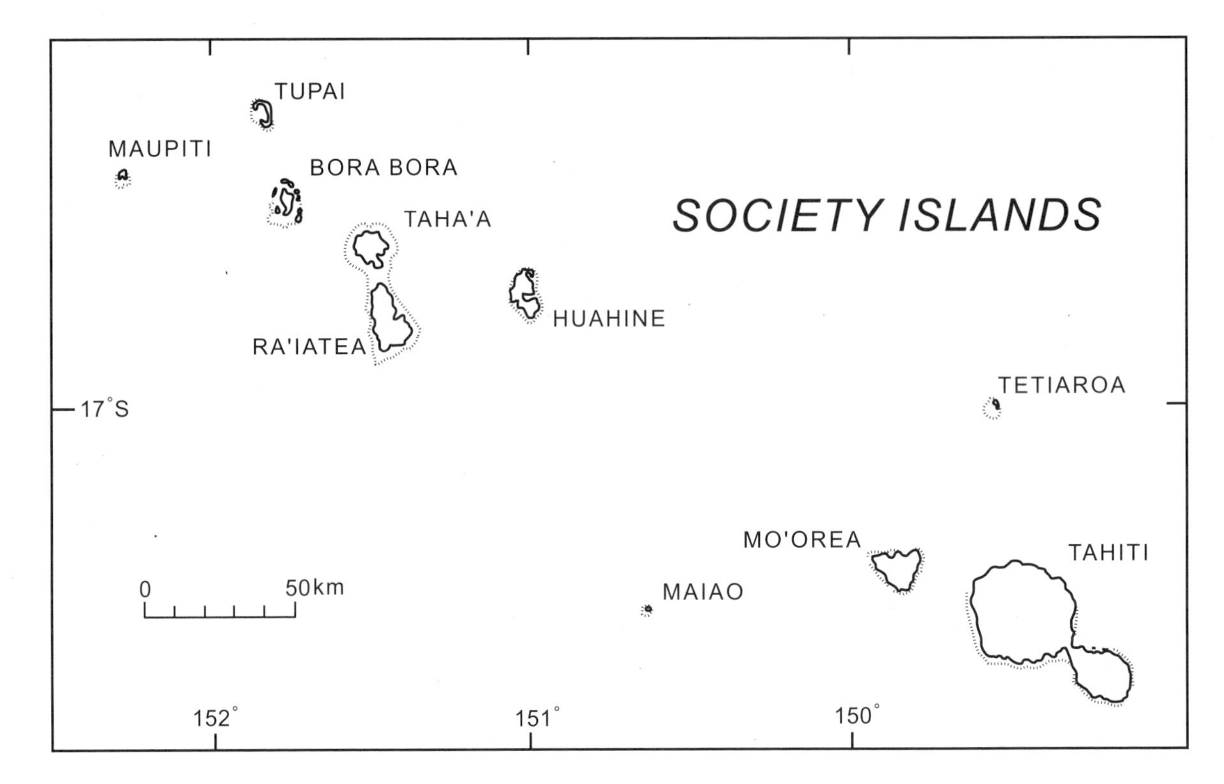

Figure 7-8 The Society Islands. Redrawn from Steadman and Pahlavan (1992: Figure 2).

to anthropogenic extinction is indicated by the archipelago's only prehistoric evidence of birds, from the coastal, waterlogged site of Fa'ahia on medium-sized Huahine (Figure 7-9). The Fa'ahia site is sometimes called Vaito'otia because it straddles the boundary of these two land districts (Sinoto 1988). The spectacular artifacts from Fa'ahia include adzes, fishhooks, shell ornaments, and items of wood and bone exceptionally preserved under anoxic conditions (Sinoto & McCoy 1975, Sinoto 1979, 1983, 1988, Pigeot 1985, 1986). These artifacts are stylistically typical of Early East Polynesian sites, which is supported by radiocarbon dates with an age range of ca. 700 to 1150 cal AD (Sinoto 1983, 1988, Kirch 2000).

The avifauna from the Fa'ahia site is exceptionally rich in species considering that it involves only 335 identified bird bones, more than 60% of which are of seabirds (Steadman & Pahlavan 1992). The 110 bones of native landbirds represent 15 species and boost Huahine's species richness from eight to 19, the same number as recorded from immense Tahiti, which has no fossil record (Tables 7-13, 7-14). While similar numbers of species are known from Mangaia, Cook Islands (20) and Ua Huka, Marquesas (17), both Mangaia and Ua Huka have much larger samples of landbird bones than Huahine. Thus the Society Islands probably had a richer landbird fauna at human arrival than other East Polynesian island groups.

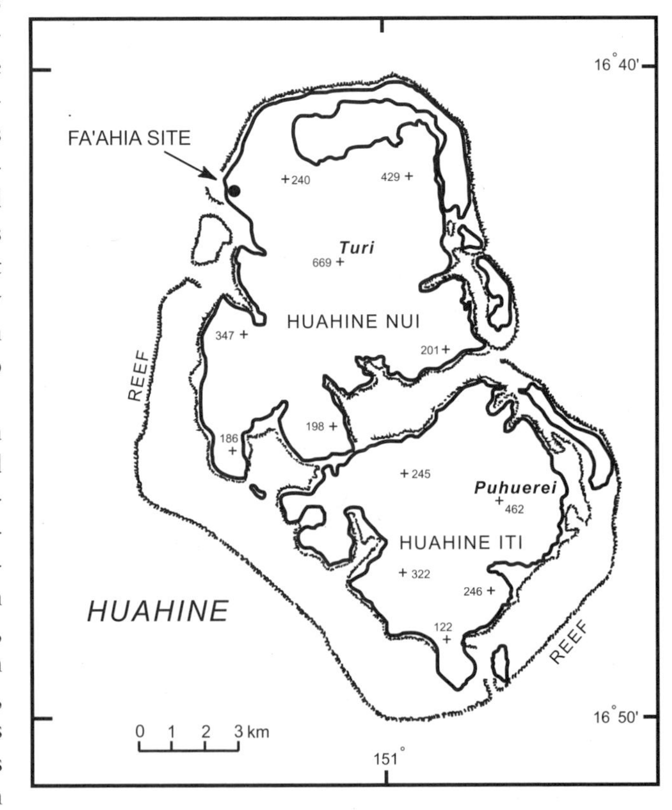

Figure 7-9 Huahine. Redrawn from Steadman and Pahlavan (1992: Figure 3).

Table 7-14 Prehistoric and modern birds from Huahine, Society Islands. Updated from Steadman & Pahlavan (1992). †, extinct species; e, recorded in 19th century but how extirpated; i, introduced species; m, migratory species; x, certainly or probably still exists on Huahine; i, m not included in totals for native landbirds. Taxa in brackets are not necessarily different from those identified more specifically.

	Bones from Fa`ahia site	Modern record
SEABIRDS		
SHEARWATERS, PETRELS		
Pseudobulweria rostrata	5	—
Pterodroma alba	3	—
Pterodroma heraldica	4	—
[*Pterodroma* sp.]	3	—
Puffinus pacificus	5	—
Puffinus nativitatis	1	—
Puffinus lherminieri	11	—
[Procellariidae sp.]	17	—
TROPICBIRDS		
Phaethon lepturus	30	x
BOOBIES		
Sula sula	23	—
Sula leucogaster	9	—
[*Sula* sp.]	12	—
FRIGATEBIRDS		
Fregata minor	5	—
Fregata ariel	17	—
[*Fregata* sp.]	10	—
GULLS, TERNS		
†*Larus utunui*	12	—
Anous minutus	2	x
Anous stolidus	13	x
Gygis candida	17	x
[Sterninae sp.]	4	—
LANDBIRDS		
HERONS		
Egretta sacra	11	x
Butorides striatus	5	—
DUCKS		
Anas superciliosa	—	x
CHICKENS		
Gallus gallus (i)	12	x
RAILS		
†*Gallirallus* undescribed sp. H	47	—
Porzana tabuensis	3	—
†*Porphyrio* undescribed sp. D	3	—
PLOVERS		
Pluvialis fulva (m)	4	x
SANDPIPERS		
Numenius tahitiensis (m)	5	x
Heteroscelus incanus (m)	1	x
PIGEONS, DOVES		
†*Macropygia arevarevauupa*	2	—
Ptilinopus purpuratus	10	x
Ducula aurorae	2	—
Ducula galeata	3	—
Gallicolumba erythroptera	9	—
†*Gallicolumba nui*	3	—
[Columbidae sp.]	1	—
PARROTS		
†*Vini sinotoi*	2	—
†*Vini vidivici*	3	—
Vini peruviana	—	e
SWIFTS		
Collocalia leucophaea	—	e
KINGFISHERS		
Halcyon cf. *tuta*	3	x
WARBLERS		
Acrocephalus caffer	2	e
STARLINGS		
†*Aplonis diluvialis*	1	—
TOTAL, SPECIES	34	15
Seabirds	15	4
Seabirds (columns combined)	15	—
Native landbirds	16	4
Native landbirds (columns combined)	19	—
Migratory landbirds	3	3
TOTAL BONES		
All species	335	
Seabirds	203	
Native landbirds	110	
Migratory landbirds	10	
Chicken	12	

Of the 15 species of seabirds recorded from Fa'ahia, only three (the petrel *Pterodroma heraldica*, extinct gull *Larus utunui*, and tern *Gygis candida*) do not occur as well at prehistoric sites in the Marquesas (see below). The species-level similarity among landbirds is much less striking, with only six of 18 species shared. As is typical for East Polynesian sites, however, bones of rails, pigeons, doves, and parrots dominate the landbird assemblage at Fa'ahia.

Had Huahine already been permanently inhabited for centuries before the Fa'ahia site was occupied, it seems unlikely that such a variety of seabirds and landbirds could have been brought to the site. The high species richness at Fa'ahia, with a flightless rail as the most common species, might argue that it is a "first-contact" overkill site, reminiscent of the Lapita sites in Tonga (Chapter 6). The chronology of first human arrival in East Polynesia, however, is still debated, with estimates ranging from 2500 to 1000 cal BP (see Kirch 1986, 2000, Anderson et al. 1994). From elsewhere in the Society Islands, husks of cultivar-type coconuts from Mo'orea have been ^{14}C dated as early as 654–775 cal AD (68% probability; Lepofsky 1995, Lepofsky et al. 1996), which may be a century or two older than the earliest ^{14}C dates from Huahine. Dated lake sediments on Mo'orea suggest that deforestation and soil erosion began at 658–977 cal AD near Lake Vaihiria and 428–651 cal AD near Lake Temae (Parkes et al. 1992). Pending more archaeological and paleoecological research, I cannot suggest confidently whether the bird fossils on Huahine represent first-contact overkill or a Mangaia-like situation with evidence for an earlier, perhaps temporary human presence but no habitation sites older than 1000 cal AD (see above).

Mo'orea is the only Society island aside from Huahine with studied bird bones. In 1996, I identified three diagnostic seabird bones recovered from two archaeological

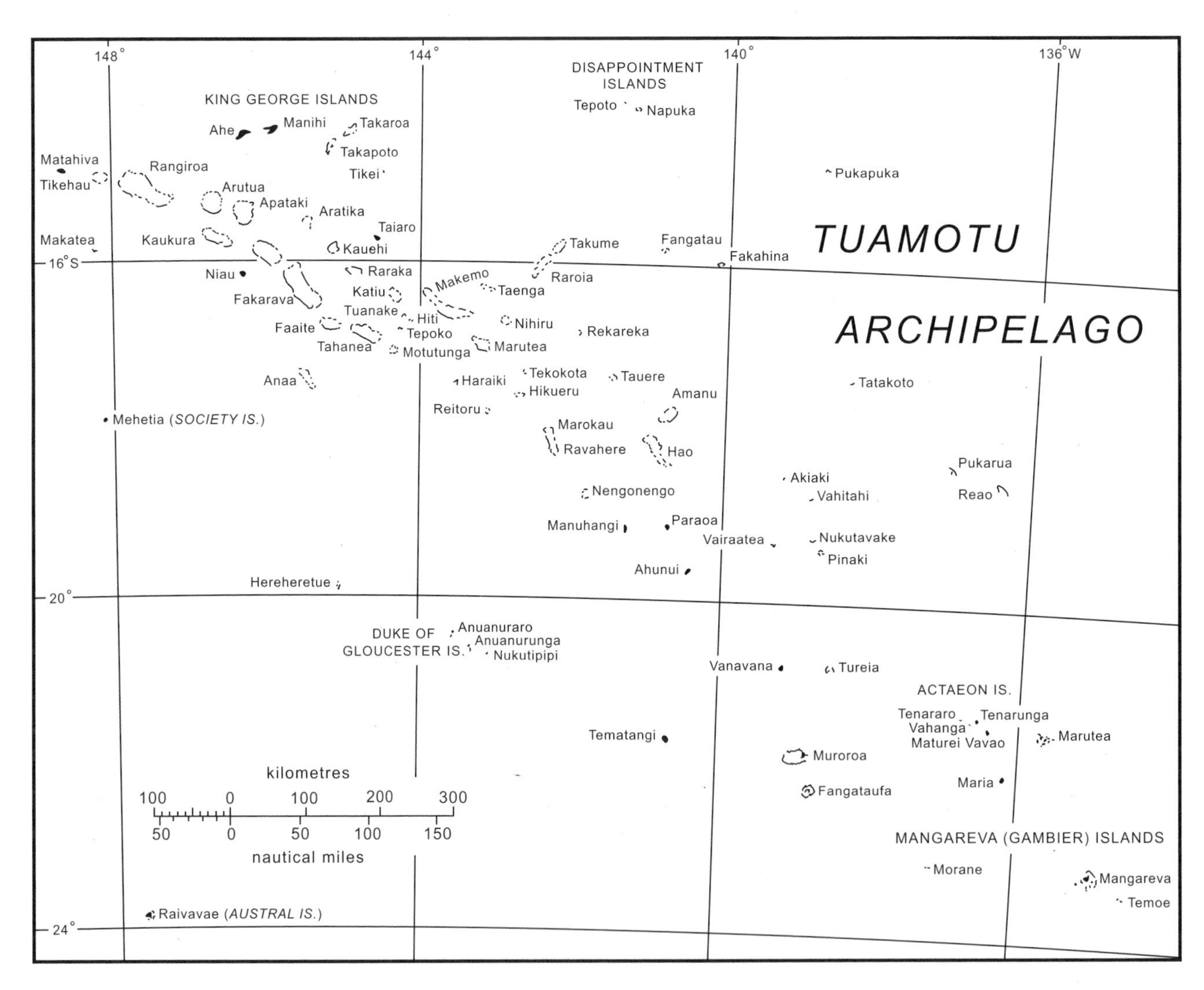

Figure 7-10 The Tuamotu Archipelago. Redrawn from New Zealand Department of Lands and Survey (1986).

localities (Te Amaama, site ScMf-5; Mata'i Taria, site ScMf-6) on 'Opunohu Bay excavated in 1960-1962 (Green et al. 1967). They consist of a shearwater *Puffinus lherminieri* (ulna), petrel *Pterodroma* sp. (large tarsometatarsus, as in *P. lessoni*), and frigatebird *Fregata minor* (radius). The first species has been seen at sea near Mo'orea and the last is a widespread visitor in the Society Islands (Thibault 1974). The petrel no longer occurs on Mo'orea or in the region.

Table 7-15 Resident landbirds of the Tuamotu Archipelago. Makatea is a relatively large raised limestone island. "Atolls" number seventy-three and, for convenience in this table, include the two small raised limestone islands of Tepoto North and Mataiva. "Mangareva" includes all islands in the Gambier Group (the almost-atoll of Mangareva and Temoe Atoll). Data from duPont (1976), Holyoak & Thibault (1984), Thibault & Guyot (1987), and Steadman & Justice (1998). H, historic record, no longer survives; M, modern record; P, prehistoric record; †, extinct species. For atolls, the number in parentheses is the number of atolls where the species has been recorded during the twentieth century, primarily by the Whitney South Sea Expedition in 1922.

	Makatea	Atolls	Mangareva
HERONS			
Egretta sacra	M	M (35)	M, P
RAILS			
Porzana tabuensis	—	M (8)	M/H
SANDPIPERS			
Prosobonia cancellata	—	M (21)	H
PIGEONS, DOVES			
Ptilinopus chalcurus	M	—	—
Ptilinopus coralensis	—	M (26)	—
Ducula aurorae	M	—	—
Gallicolumba erythroptera	—	M (5)	P
†*Gallicolumba nui*	—	—	P
PARROTS			
Vini peruviana	—	M (5)	—
KINGFISHERS			
Halcyon gambieri	—	M (1)	H
WARBLERS			
Acrocephalus atyphus	M	M (38)	H
TOTAL SPECIES			
M	4	8	1–2
M + H	4	8	5
M + H + P	4	8	7
Number of landbird bones	0	0	4

Tuamotu Archipelago

A celestial sprinkle of 73 atolls and three raised limestone islands makes up the main group of the Tuamotu Archipelago (Figure 7-10). The lack of bird bones from this region is a void in the avian prehistory of Polynesia that rivals that of Upolu and Savai'i, the two largest islands in Samoa (Chapter 6). Within the Tuamotus, ancient bird bones have been recovered and analyzed only in the outlying Gambier Islands, which consist of the almost-atoll of Mangareva and Temoe Atoll (Steadman & Justice 1998). Sailing upwind (ESE) from the Society Islands through Tuamotu (a Polynesian word meaning "a number of islands"), you will traverse a gap of 2500 km (from Huahine to Mangareva), across which appears island after island where no scientist ever has carefully recovered the delicate bones of prehistoric birds.

Modern birds of the Tuamotus also are not well studied. Eight species of landbirds still exist, although only four species are widespread (Table 7-15). Individual atolls are inhabited by only one to four species of landbirds today. I would speculate that the raised limestone islands and the atolls that survived Holocene inundation (Chapter 1) were inhabited at first human contact by 12 to 15 species of landbirds in at least these genera: *Egretta*, *Gallirallus*, *Porzana*, *Porphyrio*, *Prosobonia*, *Gallicolumba*, *Ptilinopus*, *Ducula*, *Vini*, *Halcyon*, *Acrocephalus*, and *Pomarea*. No bone records yet exist to test this proposal, nor are there paleovegetational analyses to estimate how the forests changed on atolls subsequent to human arrival.

Makatea is the largest raised limestone island in the Tuamotus. In spite of habitat damage from phosphate mining, it sustains what may be the only viable population of the pigeon *Ducula aurorae* (Thibault & Guyot 1987). Makatea also is inhabited by an endemic fruit-dove *Ptilinopus chalcurus* and endemic subspecies of warbler *Acrocephalus atypha eremus*.

The most thorough bird survey of the Tuamotus was by the Whitney South Sea Expedition in 1922. Most islands still are accessible only by sea and seldom visited by scientists. Thus it is difficult to say how the distribution of landbirds has changed over the last 80+ years. In recent decades, the sandpiper *Prosobonia parvirostris* (*cancellatus*) has been found on Anuanu, Morane, Tenararo, Rangiroa, and Tahanea, the ground-dove *Gallicolumba erythroptera* on Tenararo, the fruit-dove *Ptilinopus coralensis* on Manihi, Niau, Tenararo and Rangiroa, the lorikeet *Vini peruviana* on Apataki, Niau, and Rangiroa, and the warbler *Acrocephalus atypha* on Anuanu, Hereheretue, Hiti, Manihi, Motutunga, Nukutipipi, Tepoto, Tuanake, and Rangiroa (Petitot and Petitot 1975, anonymous 1989, Salvat & Salvat 1991, Seitre & Seitre 1992, Salvat et al. 1993, Blanvillain et al. 2002). Aside from an extirpated

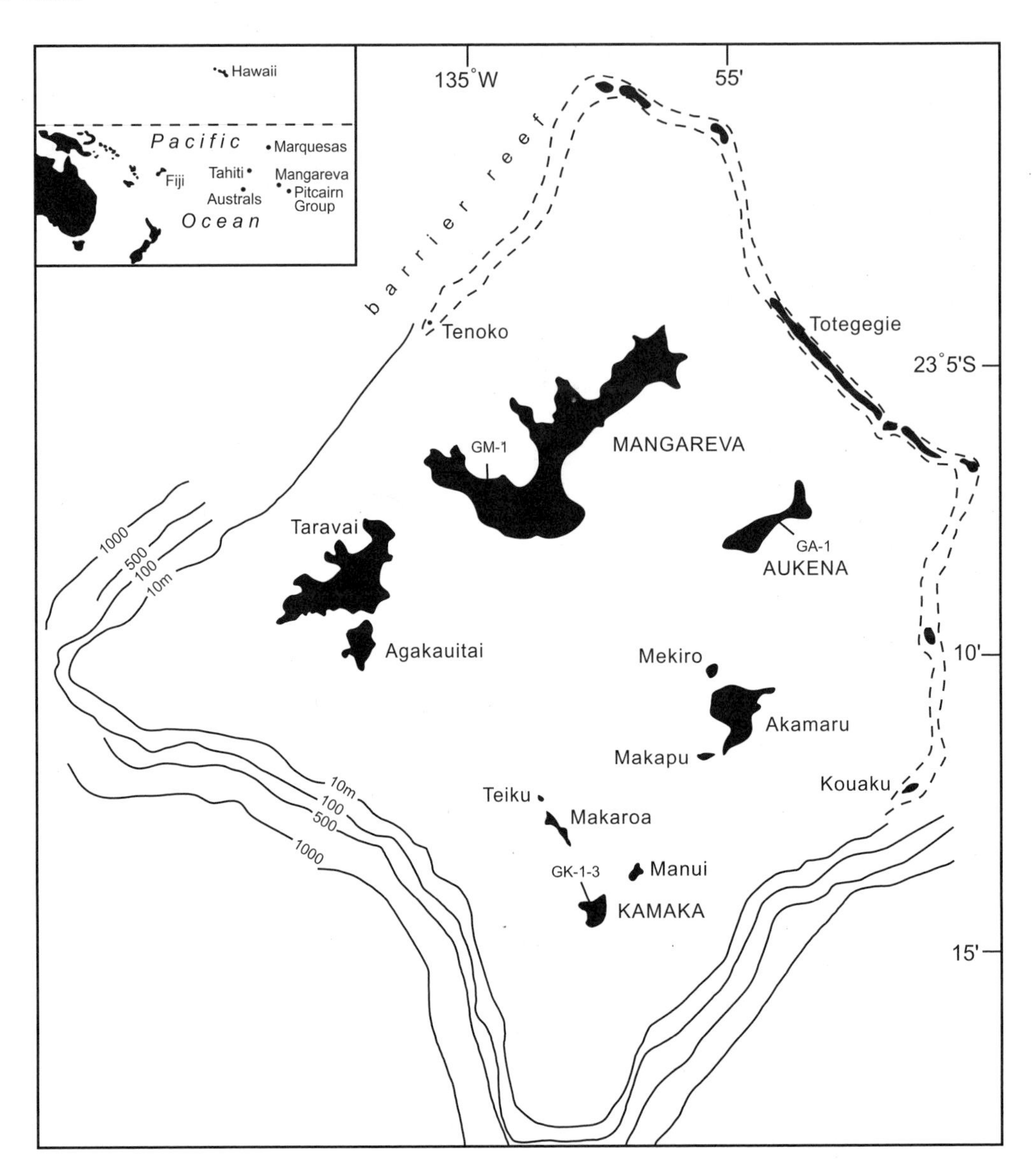

Figure 7-11 The almost-atoll of Mangareva, showing bathymetric contours and the location of five archaeological sites (GA-1, GK-1-3, GM-1). From Steadman and Justice (1998: Figure 1).

population on Mangareva, the kingfisher *Halcyon gambieri* is known only from Niau Atoll (Holyoak & Thibault 1977).

In the Gambier Islands (sometimes referred to simply as "Mangareva"), Steadman and Justice (1998) identified 215 bird bones from five archaeological sites on Aukena, Kamaka, and Mangareva (Figure 7-11). Two to seven of the 22 species of seabirds are extirpated (Table 7-16). The mere four bones of indigenous resident landbirds represent the heron *Egretta sacra* and two extirpated ground-doves *Gallicolumba erythroptera* and *G. nui*, all from the tiny islet of Kamaka. This bone sample is much too small to estimate the true extent of Mangareva's prehistoric avifauna. From a biogeographic standpoint, the most important result is showing that sympatric species of *Gallicolumba* occurred this far southeast in Polynesia.

Pitcairn Group

Four remote islands, all uninhabited at European contact, make up the Pitcairn Group (Figure 7-12; Spencer 1995). A rich record of ancient birds has been developed on the raised limestone island of Henderson, whereas nothing is known about past birdlife on the three other islands—the eroded volcanic Pitcairn (inhabited by descendants of the *Bounty* mutineers) and the small, uninhabited atolls of Ducie and

Table 7-16 The distribution of modern and prehistoric birds in the Gambier Islands. B, breeds; E, extirpated; H, historic record, no longer survives; i, introduced; m, migrant; U, status uncertain; ?, tentative identification, bones listed as *Pterodroma* sp.; *, certainly or probably now extirpated; †, extinct species. Modern distributions are from Lacan & Mougin (1974), Mougin & Naurois (1981), and Holyoak & Thibault (1984), supplemented by Pratt et al. (1987) and Steadman & Pahlavan (1992). Prehistoric data (number of bones from archaeological sites) from Steadman & Justice (1998). Taxa in brackets are not necessarily different from others identified more precisely.

	MODERN	PREHISTORIC		
	Gambier Islands	Aukena Islet	Kamaka Islet	Mangareva Island
SEABIRDS				
SHEARWATERS, PETRELS				
Pterodroma lessonii	U	—	—	—
Pterodroma ultima	B	?	?	—
Pterodroma heraldica	U	?	?	—
[*Pterodroma* sp.]	—	7	34	—
Bulweria cf. *B. fallax*	E	1	—	—
Puffinus pacificus	B	—	5	—
Puffinus tenuirostris	E	2	4	—
Puffinus nativitatis	B	9	44	—
Puffinus lherminieri	B	1	16	—
[*Puffinus* sp.]	—	—	1	—
[Procellariidae sp.]	—	—	11	—
STORM-PETRELS				
Nesofregetta fuliginosa	B	—	4	—
TROPICBIRDS				
Phaethon rubricauda	B	4	15	—
Phaethon lepturus	B	—	13	—
BOOBIES				
Sula dactylatra	B	—	—	—
Sula sula	U	—	—	—
Sula leucogaster	B	—	—	—
FRIGATEBIRDS				
Fregata minor	U	—	11	—
Fregata ariel	U	2	—	—
TERNS				
Sterna bergii	B	—	—	—
Sterna fuscata	B	—	—	—
Anous minutus	B	—	—	—
Anous stolidus	B	—	6	—
Procelsterna cerulea	B	—	1	—
Gygis candida	B	—	14	—

	MODERN	PREHISTORIC		
	Gambier Islands	Aukena Islet	Kamaka Islet	Mangareva Island
LANDBIRDS				
HERONS				
Egretta sacra	B	—	1	—
CHICKENS				
Callus grallus (i)	B	—	3	1
RAILS				
Porzana tabuensis	H	—	—	—
PLOVERS				
Pluvialis fulva	m	—	—	—
SANDPIPERS				
Numenius tahitiensis	m	—	2	—
Heteroscelus incanus	m	—	—	—
Prosobonia cancellata	H	—	—	—
Calidris alba	m	—	—	—
PIGEONS, DOVES				
Gallicolumba erythroptera	E	—	1	—
†*Gallicolumba nui*	E	—	2	—
CUCKOOS				
Eudynamys taitensis	m	—	—	—
KINGFISHERS				
Halcyon gambieri	H	—	—	—
WARBLERS				
Acrocephalus atyphus	H	—	—	—
TOTAL SPECIES	17B, 4H, 5m, 4E, 5U	7	17	1
TOTAL BONES				
All species	—	26	188	1
Seabirds	—	26	179	0
Native landbirds	—	0	4	0
Migratory landbirds	—	0	2	0
Chicken	—	0	3	1

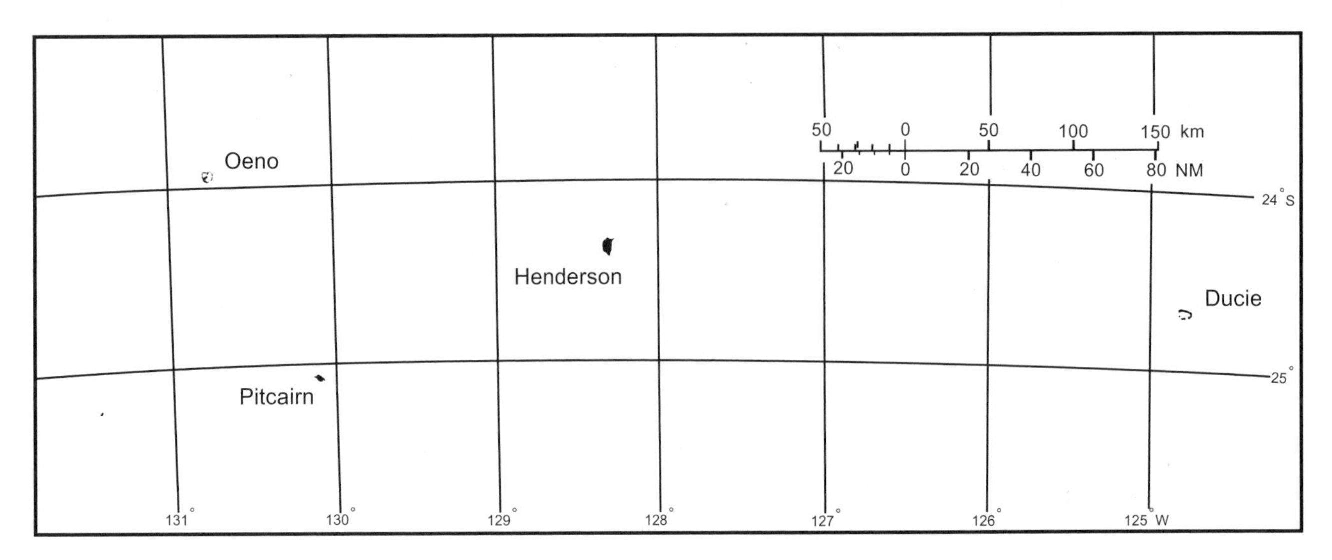

Figure 7-12 The Pitcairn Group. Redrawn from New Zealand Department of Lands and Survey (1986).

Oeno. Lying 335 km east of Henderson, Ducie is the last speck (0.8 km^2) of land until Easter Island, another 1280 km to the east.

The modern landbird fauna of the Pitcairn Group is depauperate (Table 7-17). Ignoring the trampy heron *Egretta sacra*, only Henderson sustains multiple species of landbirds. Except for this heron and the rail *Porzana tabuensis*, all other surviving species of landbirds are endemic to the island group, although representing radiations within widespread East Polynesian species-groups. If ancient landbirds were known on Pitcairn Island, which like Henderson was inhabited prehistorically by Polynesians (Weisler 1995), I believe that each species known from Henderson (or closely related congeners) would be found to have lived there. I would speculate further that some of these species or species-groups once occupied Oeno and Ducie as well.

Henderson (Figure 7-13) is one of the most intensely studied islands in Oceania (Benton & Spencer 1995). Uninhabited when found by Europeans in AD 1606 and ever since, Henderson had been regarded, uniquely among substantial islands in Oceania, as "pristine" (Fosberg et al. 1983). Henderson's coastline, however, has a series of archaeological sites (rockshelters) that range in age from ca. 1000-1100 to 1600 cal AD (Sinoto 1983, Weisler 1994). Imported raw materials (pearlshell, volcanic oven stones, basalt for adzes, volcanic glass) indicate that the Polynesians who lived on Henderson interacted with people from Pitcairn and Mangareva (Weisler 1994, 1995). These interactions decreased or ceased about 1400 to 1450 cal AD, a time of reduced long-distance voyaging across East Polynesia (Irwin 1992).

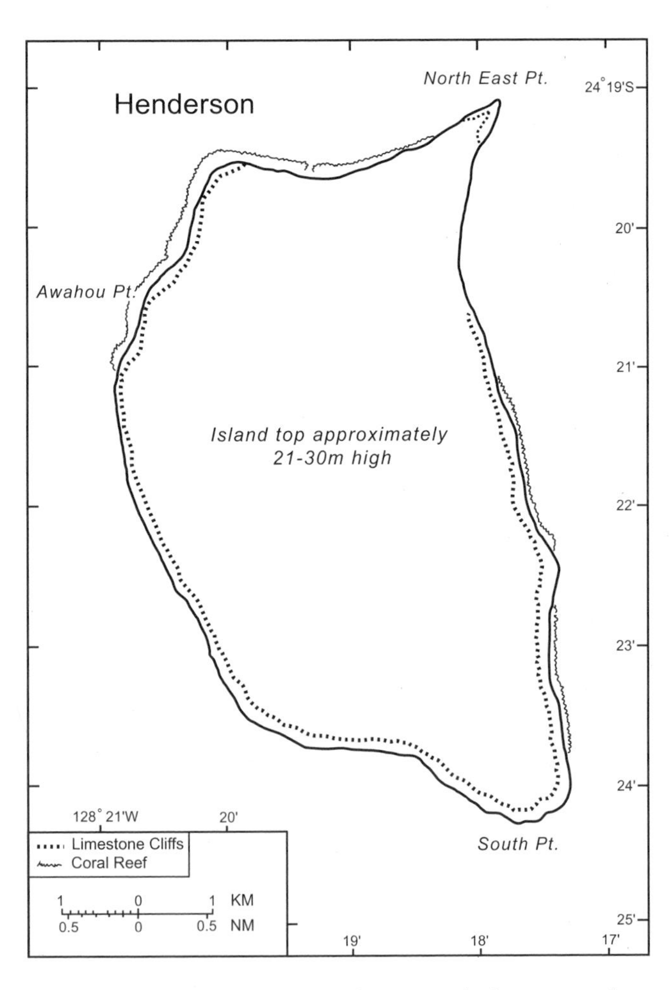

Figure 7-13 Henderson Island. Redrawn from New Zealand Department of Lands and Survey (1986).

Table 7-17 Resident landbirds of the Pitcairn Group. M, modern record; m, modern record, likely not a permanent resident; P, prehistoric record; †, extinct species; *, extirpated species. Data from Williams (1960), Steadman & Olson (1985), Schubel & Steadman (1989), Wragg & Weisler (1994), and Worthy & Wragg (2003). *Ducula* sp. was listed as *Ducula* cf. *aurorae* and *D.* cf. *galeata* by Steadman & Olson (1985) and as *Ducula* new sp. by Wragg & Weisler (1994).

	Henderson	Pitcairn	Oeno	Ducie
HERONS				
Egretta sacra	m	—	m	—
RAILS				
Porzana tabuensis	—	—	M	—
Porzana atra	M, P	—	—	—
SANDPIPERS				
†*Prosobonia* undescribed sp. B	P	—	—	—
PIGEONS & DOVES				
Ptilinopus insularis	M, P	—	—	—
†/**Ducula* sp.	P	—	—	—
†Undescribed genus	P	—	—	—
†*Gallicolumba leonpascoi*	P	—	—	—
PARROTS				
Vini stepheni	M, P	—	—	—
SWALLOWS				
†/**Hirundo* sp.	P	—	—	—
WARBLERS				
Acrocephalus vaughani	M, P	M	—	—
TOTAL SPECIES				
M + m	5	1	2	0
M + m + P	10	1	2	0
Island area (km^2)	29.8	4.8	0.6	0.8
Elevation (m)	~30	304	4	4
Isolation from nearest island (km)	185	130	130	335

The avian prehistory of Henderson Island is based on 15,000+ bird bones from at least 12 archaeological sites (Steadman & Olson 1985, Schubel & Steadman 1989, Weisler 1994, Wragg & Weisler 1994, Wragg 1995, Worthy & Wragg 2003; Table 7-18 herein). Each of the seabirds and landbirds recorded this century on Henderson is represented, a gauge of the thoroughness of this prehistoric record, which comprises more specimens than from any other island covered in this book.

Each bird bone assemblage from Henderson is dominated by seabirds (88.0 to 97.6%), especially procellariids (Chapter 15). The landbird bones reveal the former presence of a resident sandpiper *Prosobonia* sp., ground-dove *Gallicolumba leonpascoi*, pigeon *Ducula* sp., an undescribed genus of pigeon (not otherwise known from East Polynesia), and a swallow *Hirundo* sp. (either *H. tahitica* or an undescribed species). Henderson's four living species of landbirds, including the only surviving flightless species of *Porzana*, have withstood 500 to 600 years of prehistoric Polynesian presence and an even longer period (900 to 1000 years) of cohabitation with *Rattus exulans*. One factor in the persistence of these species is the extremely rugged, pinnacled karst that renders most of Henderson's interior uninhabitable to humans. Unlike on Easter Island (below), geomorphology limited the extent of inhabitable and arable land on Henderson, and therefore stifled the size of its human population.

Isolation has affected the Henderson landbird data. Missing are multiple species of ground-dove (*Gallicolumba*) or parrot (*Vini*), or any species of swiftlet (*Collocalia*), kingfisher (*Halcyon*), or large flightless rail (*Gallirallus*). The 1500+ landbird bones from Henderson account for only nine species. By contrast, on Huahine in the Society Islands, only 110 bones of resident landbirds represent 15 species (Table 7-14). As is typical nevertheless for East Polynesia, columbids dominate both faunas in species richness (four on Henderson, six on Huahine).

Marquesas Islands

Lying about 450 km north of the Tuamotus, the Marquesas Islands (10 eroded volcanic islands plus offshore islets; Figure 3-14) are high, rugged, and surrounded by a deep ocean unfettered by major reefs. Except on tiny Hatutu, nonnative species and successional native species dominate the Marquesan vegetation today, with semi-natural rainforest confined to patches at high elevations (Decker 1973, Halle 1978). The vegetation of inhabited valleys is almost wholly anthropogenic.

The Marquesas Islands are begging for modern bird surveys. For example, the fruit-dove *Ptilinopus mercierii* may now be extinct (Seitre & Seitre 1992), although thorough bird surveys either have not been undertaken or have not been published. While concern for the future of all endemic species of Marquesan birds is justified, the precise status of most species is not well understood.

Four genera of landbirds are known in the Marquesas only from prehistoric bones–the rails *Gallirallus* and *Porphyrio*, sandpiper *Prosobonia*, and pigeon *Macropygia* (Tables 7-1, 7-19). Nearly one-half of the 23 Marquesan species of landbirds also are known from bones only. At human contact, I believe that each species or species-group

Table 7-18 Prehistoric birds from Henderson Island. Data of Steadman & Olson (1985) are based on bones excavated by Y. H. and A. Sinoto in 1971 (Sinoto 1983). Data of Schubel & Steadman (1989) are based on bones excavated by S. E. Schubel in 1987. Data of Weisler (1994) combine four sites (HEN-3, 5, 10, and 11) excavated in 1991–1992. Data of Wragg (1995) combine twelve sites excavated in 1991–1992, including those represented by Weisler (1994). NISP, number of identified specimens; m, migrant; †, extinct species; *, extirpated species; ?, questionable record, or residency not established. *Pterodroma nigripennis* was listed as *Pterodroma* sp. (small) in Schubel & Steadman (1989). The numerous bones assigned to "*Pterodroma* sp. (medium)" by Weisler (1994) and Wragg (1995) probably represent, in large part, *P. alba* and *P. externa*. *Puffinus assimilis/lherminieri* was listed as *Puffinus* sp. (small) in Weisler (1994). *Anous stolidus* was listed as *Anous stolidus* and *A. minutus* in Wragg (1995). Schubel & Steadman (1989) reported 2795 bird bones rather than 2762, the difference being 33 bones they listed as "Aves sp." Taxa in brackets are not necessarily different from others identified more precisely.

	Steadman & Olson (1985)		Schubel & Steadman (1989)		Weisler (1994)		Wragg (1995)	
	NISP	%	NISP	%	NISP	%	NISP	%
SEABIRDS								
ALBATROSSES								
**Diomedea epomophora* (m?)	—	—	—	—	3	0.10	3	0.02
SHEARWATERS, PETRELS								
Pterodroma alba	165	54.4	1961	71.0	—	—	?	—
Pterodroma externa	—	—	2	0.07	—	—	?	—
**Pterodroma nigripennis*	—	—	4	0.14	—	—	228	1.8
Pterodroma sp. (medium)	—	—	—	—	1852	61.8	6894	53.1
**Bulweria bulwerii*	—	—	—	—	—	—	2	0.02
Puffinus pacificus	—	—	—	—	—	—	7	0.05
Puffinus nativitatis	8	2.6	1	0.04	163	5.4	344	2.6
Puffinus assimilis/lherminieri	—	—	—	—	28	0.93	34	0.26
[Procellariidae sp.]	—	—	457	16.5	—	—	—	
STORM-PETRELS								
Fregetta tropica/grallaria	—	—	—	—	—	—	1	0.01
Nesofregetta fuliginosa	3	1.0	1	0.04	—	—	113	0.87
TROPICBIRDS								
Phaethon rubricauda	20	6.6	104	3.8	206	6.9	907	7.0
Phethon lepturus	—	—	—	—	—	—	1	0.01
BOOBIES								
Sula dactylatra	—	—	—	—	3	0.10	20	0.15
Sula sula	3	1.0	—	—	—	—	16	0.12
FRIGATEBIRDS								
Fregata minor	9	3.0	2	0.07	6	0.20	73	0.56
TERNS								
Sterna fuscata	—	—	—	—	—	—	3	0.02
Anous stolidus	4	1.3	12	0.43	88	2.9	265	2.0
Procelsterna cerulea	—	—	—	—	—	—	11	0.08
Gygis candida	70	23.1	151	5.5	427	14.2	2502	19.3

Table 7-18 (continued)

	Steadman & Olson (1985)		Schubel & Steadman (1989)		Weisler (1994)		Wragg (1995)	
	NISP	%	NISP	%	NISP	%	NISP	%
LANDBIRDS								
PLOVERS								
Pluvialis fulva (m)	—	—	—	—	—	—	6	0.05
SANDPIPERS								
Numenius tahitiensis (m)	—	—	2	0.07	—	—	16	0.12
Heteroscelus incanus (m)	—	—	1	0.04	—	—	1	0.01
†*Prosobonia* undescribed sp.	—	—	—	—	13	0.43	69	0.53
RAILS								
Porzana atra	8	2.6	6	0.2	70	2.3	569	4.4
PIGEONS, DOVES								
Ptilinopus insularis	1	0.3	54	2.0	90	3.0	280	2.2
†/**Ducula* sp.	12	4.0	—	—	1	0.03	218	1.7
†Undescribed genus	—	—	—	—	—	—	16	0.12
†*Gallicolumba leonpascoi*	—	—	1	0.04	41	1.4	210	1.6
PARROTS								
Vini stepheni	—	—	2	0.07	4	0.13	9	0.07
CUCKOOS								
Eudynamys taitensis (m)	—	—	—	—	—	—	2	0.02
SWALLOWS								
†/**Hirundo tahitica* or undesc. sp.	—	—	—	—	—	—	92	0.71
WARBLERS								
Acrocephalus vaughani taiti	—	—	1	0.04	3	0.10	64	0.49
TOTAL SPECIES								
All	11		16		16		30–33	
Seabirds	8		9		9		17–19	
Resident landbirds	3		5		7		9	
Migrant landbirds	0		2		0		4	
TOTAL BONES								
All species	303	100	2762	100	2998	100	12,976	100
Seabirds	282	93.1	2695	97.6	2776	92.6	11,424	88.0
Resident landbirds	21	6.9	64	2.3	222	7.4	1527	11.8
Migrant landbirds	0	0	3	0.1	0	0	25	0.19

Figure 7-14 The Hanamiai valley, Tahuata, Marquesas Islands. Photo by B. V. Rolett, November 1985.

occurred throughout the island group. In other words, at least one species in each genus listed in Table 7-19 probably was found on each major Marquesan island. The only species confined to a single island may have been flightless rails. Support for my proposed archipelago-wide distribution of species or species-groups is bolstered by illogically patchy modern ranges, and by the small or isolated islands of Eiao, Hatutu, Motane, and Fatu Huku still sustaining populations of the ground-dove *Gallicolumba rubescens*, fruit-dove *Ptilinopus dupetithouarsii*, warbler *Acrocephalus mendanae*, and monarchs *Pomarea iphis* and *P. mendoza*.

The Marquesas is the only East Polynesian archipelago where multiple islands have extensive data on prehistoric birds (ca. 15,000 bones from five islands (Table 7-19). The ancient bones are from sand dunes or rockshelters at the mouths of valleys. Human habitation in the Marquesas has been, and still is, restricted to valleys because the uplands are too steep. Although each bone assemblage is dominated by seabirds (Chapter 15), landbirds are represented as well except on Ua Pou (Steadman 1989a, 1995a, 1997b).

The five Marquesan bone samples with large prehistoric records vary in the number of sites per island, the sedimentary context of the site, the volume of excavated sediment, the number of bones per volume of sediment, the methods of excavation and bone recovery, and perhaps the time span represented. That no landbird bones were recovered at the Anapua site on Ua Pou (excavated by Pierre Ottino in 1992) suggests that people occupied this coastal cave only after most species of landbirds had been lost on Ua Pou. Except for excavations on Tahuata by Barry Rolett in the 1980s and those on Nuku Hiva by Rolett and Eric Conte in the 1990s, the bone samples are biased toward large and medium-sized species because the sediments were sifted through screens of 1/4″ or 1/2″ mesh rather than 1/8″ or 1/16″ mesh. Therefore, very few bones of small Marquesan birds such as kingfishers and passerines have been recovered. For example, the immense set of bird bones from the Hane site, while well collected for its time (1960s), includes only three bones of passerines and none of kingfishers. The lack of swift bones, on the other hand, may not be due to insufficient bone-recovery methods; bones of swifts in Oceania generally are confined to noncultural deposits (Steadman 2002b).

From Ua Huka (Table 7-20), the Hane site (MUH 1) was excavated by Yosihiko Sinoto and colleagues in the mid-1960s (Sinoto 1966, 1968). Hane has yielded 12,000+ bird bones, nearly 10,000 of which represent at least seven species of shearwaters and petrels, readily obtained from their nesting burrows or perhaps attracted to campfires at night. The current seabird community on Ua Huka is depleted from what it was at first human contact in numbers of individuals (by perhaps two or three orders of magnitude) and in numbers of species. Most species of seabirds believed to exist today on Ua Huka probably no longer nest on Ua Huka itself, but only on tiny offshore islets (Holyoak & Thibault 1984, Seitre & Seitre 1992). Although Marquesan seabirds have not been surveyed in detail, the situation probably resembles that on Rapa, where most surviving species exist either only on offshore islets or on steep, seaward-facing cliffs on the main island (Thibault & Varney 1991b).

Pigeons and doves are the most common family of landbirds at the Hane site both in number of either species or bones (Table 7-20). Only one of Hane's six columbids survives on Ua Huka (Steadman 1989a, 1992b). Parrots are the next most common, represented by the extinct, widespread East Polynesian *Vini sinotoi* and *V. vidivici* and the smaller Marquesan endemic *V. ultramarina* (Steadman & Zarriello 1987). Rails consist of two flightless, undescribed species of *Gallirallus* and *Porzana*.

The age of Hane and other early Marquesan sites has been disputed for decades (Sinoto 1979, Kirch 1986, Rolett 1993, 1998, Spriggs & Anderson 1993, Anderson et al. 1994, Dye 1996). Initial human occupation of the Marquesas, such as represented by earliest levels at Hane and the Ha'atuatua site on Nuku Hiva (Rolett & Conte 1995), may have occurred no earlier than ca. 900 to 1000 cal AD. This is about when East Polynesians expanded to very remote places such as Easter Island, the Hawaiian Islands, and New Zealand (Irwin 1992, Green 1999, Kirch 2000). While there may have been a time lag of several centuries between first arrival in the Marquesas and permanent occupation of all major valleys (Rolett 1998), the many extinct and extirpated species of landbirds at the Hane site (and the early sites on Nuku Hiva, Hiva Oa, and Tahuata) seem to reflect a "first contact" avifauna just as may be the case at the Fa'ahia site on Huahine. The likely similarity in age between the Fa'ahia and early Marquesan sites is supported also by the close stylistic resemblance of fishhooks, adzes, and other artifacts (Sinoto 1983, Kirch 1986, Rolett 1998).

Regardless of when people arrived, the stratigraphic pattern of avian exploitation at Hane and other early Marquesan sites is clear, with bird bones being common in lower strata but rare or absent in upper strata (Dye & Steadman 1990). The abundance of bird bones, especially of seabirds, in the early levels at Hane was noted first by Kirch (1973). These 12,500 bird bones remained unidentified, however, until I began to sort, clean, catalogue, and identify them in 1986, resulting (Table 7-20) in 41 species (20 seabirds, 4 migrant shorebirds, chicken, 16 native landbirds) from cultural phases I and II (Early; ca. 900/1000 to 1300 cal AD) compared to only 160 bones of 9 or 10 species from cultural phases III and IV (Late; ca. 1300 to 1800 cal AD). Because 135 of the 160 "Late" bird bones are from mixed Early and Late strata, the true Late period avifauna was even more depauperate than indicated in Table 7-20, where the similar family-level composition in the Early and Late columns suggests further that most "Late" bones may actually be from Early levels. The chronostratigraphic resolution at Hane is not adequate to pinpoint just when the severe depletion of birdlife occurred on Ua Huka, except that it preceded the onset of cultural phase III (defined by artifact typology; Sinoto 1966), at which point most species of birds either were extirpated or so rare that they escaped archaeological sampling.

I have identified ca. 900 bird bones, >90% of them seabirds, from Ha'atuatua (MN 1) and lesser sites on Nuku Hiva. Site MN 1 was excavated in 1956–1957 (Suggs 1961), 1965 (Sinoto 1966), and 1992–1993 (Rolett & Conte 1995). The 28 landbird bones represent nine species, of which two rails, two columbids, and three parrots no longer occur on Nuku Hiva (Table 7-19). Each of these species was found as well at the Hane site, except for the endemic flightless species of *Gallirallus* and *Porzana*.

From Hiva Oa, nine archaeological sites have yielded 1024 identifiable bird bones. The richest site by far is Hanatekua Shelter No. 2 (MH 3-12). The ages of the Hiva Oa sites are not well resolved, although their early strata may be coeval with early sites on Ua Huka and Nuku Hiva (Kirch 1986). The seabird assemblage from MH 3–12 is unusual in having relatively few procellariids and many frigatebirds (*Fregata* spp.). The landbird bones consist mainly of two large columbids, the extirpated pigeon *Ducula galeata* and the extinct ground-dove *Gallicolumba nui*. No bones of swifts, kingfishers, passerines, small rails, or small parrots were recovered from Hiva Oa, suggesting that the sediment was not passed through screens with mesh finer than 1/2″. The current situation with native birds on Hiva Oa is tragic, the avifauna having been depleted over the past 70 years by predation from the introduced owl *Bubo virginianus* (Seitre & Seitre 1992).

On Tahuata (Figure 7-14), the Hanamiai site (MT 1) produced 716 identifiable bird bones (Steadman & Rolett 1996). Nearly all bones of extinct birds are confined to the deepest strata (Figures 3-15, 3-16), which have a maximum age of ca. 1000 cal AD (Rolett 1998). The 70 landbird bones represent 10 species, seven of them extinct or extirpated (Table 7-21). That fewer extinct species of landbirds were found on Tahuata than on Ua Huka or Hiva Oa probably is a sampling effect because the carefully controlled excavations at MT 1 covered 21 m^2, a much smaller area than on the other islands. The Hanamiai bone assemblage is especially important because the sediments were sieved with 1/8″ mesh and the bones are associated with well-described artifacts, high-quality stratigraphic data, and defensible ^{14}C dates (Steadman & Rolett 1996, Rolett 1998). The vertical sequence of bird bones at Hanamiai reflects intense exploitation only in cultural phase I (ca. 1025-1300 cal AD), even though several of the extinct or extirpated species survived into phases II (same age as phase I) or III (ca. 1300–1450 cal AD). The only species shared uniquely between Tahuata and nearby Hiva Oa is the extinct swamphen *Porphyrio paepae* (Steadman 1988).

All modern species of Marquesas landbirds except the heron *Egretta sacra* and the rail *Porzana tabuensis* are regarded as endemic to the island group (Bruner 1972, duPont 1976, Holyoak & Thibault 1984, Pratt et al. 1987). This was not so, however, among the volant extinct species. The large ground-dove *Gallicolumba nui*, for example, is known from prehistoric sites on Huahine and Mangaia as well as Ua Huka and Hiva Oa (Tables 7-3, 7-13, 7-19; Chapter 11). Two parrots, *Vini sinotoi* and *V. vidivici*, were described

Table 7-19 Native landbirds of the Marquesas Islands. H, historic record, no longer survives; M, modern record, certainly or probably survives; P, prehistoric record; R, reintroduced population; †, extinct species; ?, questionable record, or residency not established. Modified from Fisher & Wetmore (1931), Holyoak & Thibault (1984), Steadman (1989a), Seitre & Seitre (1991, 1992), and Kuehler et al. (1997).

	Nuku Hiva	Hiva Oa	Ua Pou	Ua Huka	Fatu Hiva	Tahuata	Eiao	Hatutu	Motane	Fatu Huku	Total
HERONS											
Egretta sacra	M	M	M	M, P	M	M	M	M	M	M	10
RAILS											
†*Gallirallus* undescribed spp. I–L	P	P	—	P	—	P	—	—	—	—	4
Porzana tabuensis	M	M	—	—	M	M	—	M	—	—	5
†*Porzana* undescribed sp. D	P	—	—	—	—	—	—	—	—	—	1
†*Porzana* undescribed sp. E	—	—	—	P	—	—	—	—	—	—	1
†*Porzana* undescribed sp. F	—	—	—	P	—	—	—	—	—	—	1
†*Porphyrio paepae*	—	P	—	—	—	P	—	—	—	—	2
SANDPIPERS											
Prosobonia cf. *cancellata*	—	—	—	P	—	—	—	—	—	—	1
PIGEONS, DOVES											
†*Macropygia heana*	P	—	—	P	—	—	—	—	—	—	2
Ptilinopus dupetithouarsii	M, P	M	M	M, P	M	M, P	—	—	M	—	7
†*Ptilinopus mercierii*	M	M	—	P	M?	—	—	—	—	—	3–4
Ducula galeata	M, P	P	—	P	—	P	—	—	—	—	4
Gallicolumba rubescens	H, P	P	—	P	—	P	—	M	—	M	6
†*Gallicolumba nui*	—	P	—	P	—	P	—	—	—	—	3
PARROTS											
†*Vini sinotoi*	P	P	—	P	—	P	—	—	—	—	4
†*Vini vidivici*	P	P	—	P	—	P	—	—	—	—	4
Vini ultramarina	M/H, P	—	M	P, R	R	P	—	—	—	—	5

SWIFTS											
Collocalia ocista	M	M	M	M	M?	M	M	M?	M	M	8–10
KINGFISHERS											
Halcyon godeffroyi	—	M	—	P	H	M, P	—	—	—	—	4
WARBLERS											
Acrocephalus mendanae	M	M	M	M, P	M	M	M	M	M	—	9
MONARCHS											
Pomarea iphis	—	—	—	M, P	—	—	M	—	—	—	2
Pomarea mendozae	M/H	M/H	M	—	—	M	—	—	M	—	5
Pomarea whitneyi	—	—	—	—	M	—	—	—	—	—	1
TOTAL SPECIES											
M	7–9	7–8	6	5	5–7	7	4	4–5	5	3	
M + H	10	8	6	5	6–8	7	4	4–5	5	3	
M + H + P	15	15	6	18	6–8	15	4	4–5	5	3	
TOTAL BONES											
All species	~900	1024	~200	12,486	0	716	0	0	0	0	~15,326
Landbirds	27	146	0	2187	0	70	0	0	0	0	2330
Island area (km^2)	337	324	104	78	77	53	52	18	16	1	
Elevation (m)	1185	1190	1232	855	960	1000	576	428	520	361	
Distance (km) to nearest island	19	3	19	21	21	3	4	4	12	14	
Distance (km) from Nuku Hiva or Hiva Oa	0	0	19	21	38	3	47	50	12	14	

Table 7-20 Birds of Ua Huka, Marquesas Islands. i, introduced; m, migrant (i, m not included in resident totals); x, one or more bones present; †, extinct species or subspecies; *, certainly or probably extirpated; ?, questionable record, or residency not established. Following Sinoto (1966), "Early" = cultural phases I, II (ca. 300–1200 AD), "Late" = cultural phases III, IV (ca. 1200–1800 AD) of the Hane site (MUH 1). See text for an alternative chronology, where "Early" is ca. 900/1000 to 1300 AD, and "Late" is ca. 1300 to 1800 AD. The Early vs. Late comparison is presence/absence rather than quantified bone for bone because many bones of shearwaters and petrels have not been identified to species. Number of identified specimens (NISP) are provided at the family level. The 12,486 Early phase bones include 484 of uncertain but probable assignment to the Early phase. *Vini ultramarina* exists today on Ua Huka because of a twentieth-century reintroduction; it had been extirpated. Data for "may exist today on Ua Huka" includes offshore islets and is from Fisher & Wetmore (1931), Holyoak & Thibault (1984), Seitre & Seitre (1992). Updated from Steadman (1991a).

	Bones from Hane site			
	Early	Late	Total	May exist today on Ua Huka
SEABIRDS	10,153	143	10,296	
SHEARWATERS, PETRELS	9598	141	9739	
**Pseudobulweria rostrata*	x	x	x	—
**Pterodroma* cf. *alba*	x	x	x	x
**Pterodroma* small sp.	x	x	x	—
Bulweria cf. *bulwerii*	x	—	x	x
Puffinus pacificus	x	x	x	x
**Puffinus nativitatis*	x	—	x	—
**Puffinus lherminieri*	x	x	x	x
STORM-PETRELS	2	0	2	
**Fregetta grallaria*	x	—	x	—
Nesofregetta fuliginosa	x	—	x	x
TROPICBIRDS	8	0	8	
Phaethon lepturus	x	—	x	x
BOOBIES	340	2	342	
†*Papasula abbottii costelloi*	x	—	x	—
**Sula dactylatra*	x	—	x	x
Sula sula	x	?	x	x
Sula leucogaster	x	—	x	x
FRIGATEBIRDS	54	0	54	
Fregata minor	x	—	x	x
Fregata ariel	x	—	x	x
TERNS	151	0	151	
Sterna lunata	—	—	—	x
Sterna fuscata	x	—	x	x
Anous minutus	x	—	x	x
Anous stolidus	x	—	x	x
Procelsterna cerulea	—	—	x	x
Gygis microrhyncha	x	—	x	x
LANDBIRDS	2133	14	2147	
HERONS	2	0	2	
Egretta sacra	x	—	x	x

Table 7-20 (continued)

	Bones from Hane site			
	Early	Late	Total	May exist today on Ua Huka
PLOVERS	2	0	2	
Pluvialis fulva (m)	x	—	x	x
SANDPIPERS	13	0	13	
Numenius tahitiensis (m)	x	—	x	x
Heteroscelus incanus (m)	x	—	x	x
Arenaria interpres (m)	x	—	x	—
* Prosobonia cf. *cancellata*	x	—	x	—
CHICKENS	38	3	41	
Gallus gallus (i)	x	x	x	x
RAILS	37	0	37	
†*Gallirallus* undescribed sp. J	x	—	x	—
†*Porzana* undescribed sp. E	x	—	x	—
†*Porzana* undescribed sp. F	x	—	x	—
PIGEONS, DOVES	1808	13	1821	
†*Macropygia heana*	x	—	x	—
Ptilinopus dupetithouarsii	x	x	x	x
**Ptilinopus mercierii*	x	—	x	—
**Ducula galeata*	x	—	x	—
**Gallicolumba rubescens*	x	—	x	—
†*Gallicolumba nui*	x	x	x	—
PARROTS	270	1	271	
†*Vini sinotoi*	x	x	x	—
†*Vini vidivici*	x	—	x	—
Vini ultramarina	x	—	x	—
CUCKOOS	0	0	0	
Eudynamys taitensis (m)	—	—	—	x
SWIFTS	0	0	0	
Collocalia ocista	—	—	—	?
KINGFISHERS	0	0	0	
**Halcyon godeffroyi*	—	—	—	—
WARBLERS	0	0	0	
Acrocephalus mendanae	—	—	—	x
MONARCHS	3	0	3	
Pomarea iphis	x	—	x	x
TOTAL SPECIES				
Seabirds	20	5–6	—	17
Seabirds (combining columns)	22			
Resident landbirds	16	3	16	4–5
Resident landbirds (combining columns)	16–17			

Table 7-21 Stratigraphic summary of bird bones from the Hanamiai site, Tahuata, Marquesas. †, extinct species; *, extant species extirpated on Tahuata; #, interpreted as a precultural deposit. Zone = Stratigraphic sequence; phase = cultural sequence. Modified from Steadman & Rolett (1996) and Rolett (1998). Taxa in brackets are not necessarily different from others identified more precisely.

Zone	AB	CD	EF	G	G/H	I#	
Phase	V	IV	III	II	I		TOTAL
SEABIRDS	4	14	13	88	448	—	567
NATIVE LANDBIRDS							
RAILS							
†*Gallirallus* undescribed sp. L	—	—	2	3	19	—	24
†*Porphyrio paepae*	—	—	1	1	9	3	14
[†*Gallirallus/Porphyrio*]	—	—	—	—	1	—	1
PIGEONS, DOVES							
Ptilinopus dupetithouarsii	—	—	1	—	1	—	2
**Ducula galeata*	—	—	—	—	6	—	6
**Gallicolumba rubescens*	—	—	—	—	5	—	5
[Columbidae sp.]	—	—	—	—	1	—	1
PARROTS							
†*Vini sinotoi*	—	—	—	—	3	—	3
†*Vini vidivici*	—	—	—	—	4	—	4
**Vini ultramarina*	—	3	1	—	—	1	5
[†*Vini* sp.]	—	—	—	—	2	—	2
KINGFISHERS							
Halcyon godeffroyi	—	—	—	1	—	—	1
WARBLERS							
Acrocephalus mendanae	—	2	—	—	—	—	2
Native landbird totals							
All bones	0	5	5	5	51	4	70
Bones (†/*only)	0	3	4	4	49	4	64
All species	0	2	4	3	7	2	10
†/*species only	0	1	3	2	6	2	7

from bones on Ua Huka, Hiva Oa, and Tahuata (Steadman & Zarriello 1987), but have been found since on Nuku Hiva, Huahine, and Mangaia (Tables 7-3, 7-13, 7-19; Chapter 12). *Ducula galeata*, the largest extant pigeon in Oceania and regarded as endemic to Nuku Hiva, once lived on Ua Huka, Hiva Oa, and Tahuata. Furthermore, bones or early historic records of *D. galeata* have been discovered on Huahine and Tahiti (Society Islands), Mangaia (Cook Islands), and perhaps Henderson Island (Pitcairn Group) (Steadman 1989a, 1997c). Thus a huge pigeon, now seen as endemic to a single island, actually had a range that spanned most of East Polynesia (thousands of kilometers) before people caused its nearly total demise. The modern ranges of other volant Marquesan landbirds, such as *Gallicolumba rubescens* or *Vini ultramarina,* also have been reduced (Table 7-19).

Easter Island (Rapa Nui)

Three shield volcanos coalesced during the Quaternary to form Easter Island (Bandy 1937, Baker 1967, Isaacson & Heinrichs 1976), which has rugged coastlines but a gently contoured interior (Figures 3-17, 7-15). Easter is the most southern and eastern island with prehistoric Polynesian habitation. The nearest land to the west is tiny Ducie

Figure 7-15 Horses grazing in the interior of Easter Island, now largely a grassland dominated by nonnative species of plants and animals. Photo by DWS, 8 July 1991.

Atoll in the Pitcairn Group, 1280 km of deep ocean away. To the east, the nearest potentially inhabitable islands are Mas Afuera of the Juan Fernandez Group (33°45″S, 3300 km away) and Isla San Felix (26°15″S, 3250 km away). These very isolated islands have floras and faunas of Neotropical origin (Skottsberg 1956, Stuessy et al. 1984) and lack evidence of human prehistory (Haberle 2003). South America is 4000 km east of Easter Island.

Easter Island is unusual biologically and culturally. It is too cold to cultivate coconuts, breadfruit, or other truly tropical plants. Being so far east in the huge Pacific Ocean, the depauperate native flora of Easter Island has a Neotropical influence that is lacking elsewhere in Oceania (Skottsberg 1956). This begs the question of possible Neotropical affinities as well in its birdlife. The resident landbirds of the Pitcairn Group (closest islands to the west) and all other non-Hawaiian island groups in Oceania are all of Old World (mainly Papuan) origin. No landbirds have survived, however, on Easter Island. Johnson et al. (1970:532) assumed that it simply was too isolated to have been colonized by landbirds, noting, "Whether there were any land-birds on the island when the first settlers arrived is extremely doubtful, but cannot be proved one way or the other." We now know that the first people on Easter Island did find landbirds, although the evidence is limited.

Polynesians colonized Easter Island at ca. 900–1000 cal AD, perhaps from Mangareva (Green 1999, Finney 2001). Extreme isolation did not serve the Easter Islanders well. There is no evidence for long-distance exchange of goods and people after initial arrival (Van Tilburg 1994). They had no terrestrial resources other than what the island itself could offer (native plants and birds) and the sweet potato, taro, sugar cane, chickens, and rats that they had brought with them. (Unlike in most of Polynesia, pigs and dogs are not evident prehistorically on Easter Island). Isolation prevented the Rapa Nui people from replenishing depleted stocks of plants or animals through trading or raiding with other islands.

In 1991, I excavated a 1 × 4 m trench at the coastal site of Ahu Naunau at Anakena on Easter Island's leeward coast (Steadman et al. 1994; Martinsson-Wallin & Crockford 2002 also describe this site). The cultural deposits consist of a basal silty clay overlain by 1.3 m of calcareous sand (Figures 3-18, 7-16). Four ^{14}C dates on wood charcoal range from 1000–1280 to 1220–1430 cal AD. More than 1000 obsidian artifacts were recovered, including two of the distinctive spearpoints (*mata'a*) that are so common in the island's later prehistoric contexts. The Ahu Naunau vertebrate fauna differs from those of the younger prehistoric sites in that bones of marine mammals and seabirds

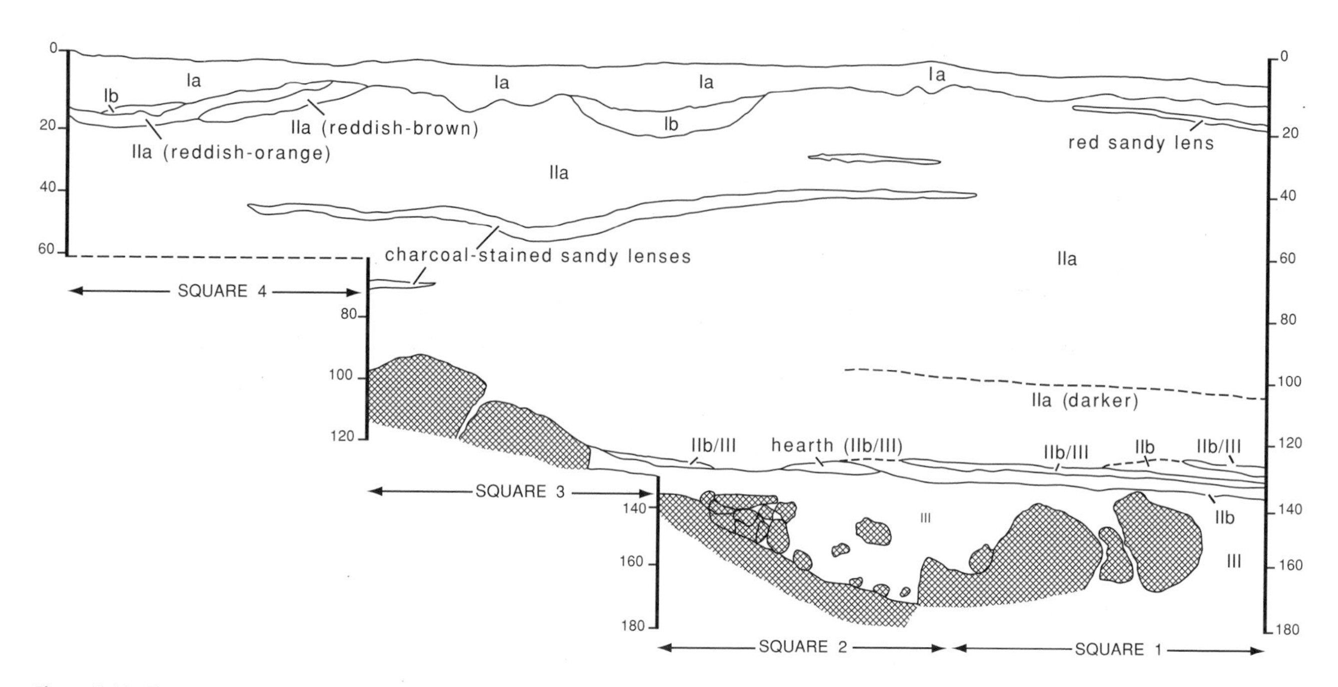

Figure 7-16 The 1991 excavations at Ahu Naunau, Anakena, Easter Island. Stratigraphic profile of west wall of units 1–4. From Steadman et al. (1994).

are common, whereas bones of fish, humans, and chickens are rare (pers. obs.). Vertebrates at Ahu Naunau are dominated by the common dolphin (*Delphinus delphis*), Pacific rat (*Rattus exulans*), and various fish (Table 7-22). Bones of fish are much rarer, and those of dolphins much more abundant, than in midden assemblages from similarly-aged sites in the more tropical parts of East Polynesia, including any site covered in this chapter.

The prehistoric Rapa Nui must have had seaworthy canoes for harpooning dolphins offshore during the time interval represented at Ahu Naunau. Pollen data from Easter Island's three lakes provide clear evidence of deforestation beginning at ca. 900 cal AD (Flenley et al. 1991, Bahn & Flenley 1992). For the island's principal woody species of plants (a palm *Jubaea* sp., a legume *Sophora toromiro*, a tiliaceous *Triumfetta* sp., a rubiaceous *Coprosma* sp., and a probably woody composite), pollen virtually if not absolutely disappears from the lake sediments by ca. 1400 cal AD, which is 400 to 500 years after colonization. Deforestation eliminated the trees, especially the palms, needed to make canoes, eliminating dolphins from the diet (Steadman et al. 1994).

Table 7-22 Stratigraphic summary of bones from Squares 1–3 of the 1991 excavations at Ahu Naunau, Anakena, Easter Island. From Steadman et al. (1994). Surface bones were not sieved. Square 4 is not included because it was not excavated completely (see Figure 7-23).

				cm below surface					
	Surface	0–20	20–40	40–60	60–80	80–100	100–120	>120	TOTAL
Fish	—	100	248	168	87	98	205	689	1595
Rat	—	252	480	616	196	44	19	536	2143
Dolphin	6	530	563	337	285	26	28	537	2312
Pinniped	1	—	1	—	—	1	—	—	3
Chicken	3	11	12	1	—	—	—	2	29
Native bird	10	19	78	41	15	5	21	162	351
TOTAL	20	912	1382	1163	583	174	273	1926	6433

Figure 7-17 Statues (*moai*) carved from welded tuff at Rano Raraku, Easter Island. Photo by DWS, 7 July 1991.

Humans caused deforestation not only by clearing land for agriculture, but also (especially for palms) by using trees as rollers to move the massive stone statues, called *moai* (Bahn & Flenley 1992; Figure 7-17 herein). The average *moai* was 4.05 m tall and weighed 12.5 metric tons (Van Tilburg 1994). An astonishing 992 *moai* have been discovered, including 397 that never left the statue quarry at Rano Raraku and hundreds of others moved up to 15 km away to be set up on stone platforms such as Ahu Naunau (Bahn & Flenley 1992, Van Tilburg 1994).

Bones of extinct and extirpated seabirds and landbirds occur throughout the deposit at Ahu Naunau but are more common in deeper strata (Steadman et al. 1994), a trend opposite that of chicken bones, which never exceed 0.1% NISP in strata deeper than 20–40 cm (Figure 7-18). The 25 species of seabirds include 12 species of procellariids (Table 7-23). All but three of the seabirds have been found archaeologically, 14 of them exclusively so. Hundreds of seabird bones in my lab remain sorted only to family or genus. Only one species of seabird (the tropicbird *Phaethon rubricauda*) still nests on Easter Island itself (Johnson et al. 1970); eight to 10 others certainly or perhaps breed on the steep offshore islets of Motu Nui, Motu Iti, and Motu Kaokao. The seabird fauna of Easter Island probably exceeded 30 resident species when Polynesians arrived, making it the richest seabird island in the world (Chapter 15).

Indigenous landbirds are known on Easter Island only from six broken bones that I recovered at Ahu Naunau (Table 7-23). These fragments seem to represent five ex-

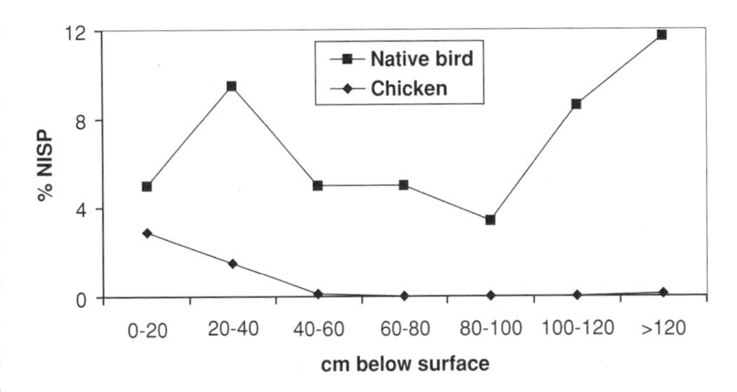

Figure 7-18 Relative abundance (% NISP, number of identified specimens) of chicken versus native bird at Ahu Naunau, Anakena, Easter Island. Based on data in Table 7-22. Because of variation in extent of breakage, bones of dolphin (*Delphinus delphis*) are not considered in the calculations.

Table 7-23 Resident native birds from Easter Island (27°S). Prehistoric record categories: x, recorded from one or more archaeological sites; -, no record. Modern status categories: B, breeds today on Easter Island; b, breeds today on one or two offshore islets but not on Easter Island itself; E, extirpated on Easter Island and all offshore islets; ?, questionable record, or residency not established. Modified from Steadman (1995a).

	Prehistoric record	Modern status	Modern breeding range (°S)
SEABIRDS			
ALBATROSSES			
Diomedea sp.	x	E	01–51
SHEARWATERS, PETRELS			
Fulmarus glacialoides	x	E	53–68
Pterodroma macroptera/lessoni	x	E	34–50
Pterodroma ultima	x	E	15–27
Pterodroma neglecta	—	b	20–33
Pterodroma heraldica	—	b	08–27
Pterodroma externa	x	E	24–33
Pachyptila vittata	x	E	36–65
Procellaria sp.	x	E	36–50
Puffinus carneipes	x	E	31–42
Puffinus griseus	x	E	32–50
Puffinus nativitatis	x	b	00–27
†Procellariidae undescribed sp.	x	E	—
STORM-PETRELS			
Nesofregetta fuliginosa	x	E	00–27
TROPICBIRDS			
Phaethon rubricauda	x	B	00–27
Phaethon lepturus	x	E	00–24
FRIGATEBIRDS			
Fregata minor	x	b/E	00–27
BOOBIES			
Sula dactylatra	x	b	00–27
TERNS			
Sterna paradisaea	x	E?	—
Sterna lunata	—	b/E	00–27
Sterna fuscata	x	b	00–27
Anous stolidus	x	b	00–33
Procelsterna cerulea	x	b	00–33
Gygis candida	x	b	00–27
Gygis microrhyncha	x	E	08–10
LANDBIRDS			
HERONS			
†cf. Ardeidae undescribed sp.	x	E	
RAILS			
Porzana undescribed sp. G	x	E	
†cf. Rallidae undescribed sp.	x	E	
PARROTS			
†cf. Psittacidae undescribed sp. 1	x	E	
†cf. Psittacidae undescribed sp. 2	x	E	
TOTAL SPECIES			
Seabirds (B)	1		
Seabirds (B + b)	9–11		
Seabirds (B + b + E)	25		
Landbirds (all E)	5		

tinct, undescribed species (a heron, two rails, and two parrots). The bone originally reported as "cf. Tytonidae sp." in Steadman (1995a) in fact represents an indeterminate procellariid. The only bone confidently assigned to genus is the tibiotarsus of a rail (*Porzana* undescribed sp. G) that is smaller than the widespread *P. tabuensis*. Although species of *Porzana* are characteristic of East Polynesia, the genus also occurs in the Neotropics, as does each of the landbird families recorded from Ahu Naunau. The other extinct, undescribed rail is much larger but the two bones (a synsacral fragment and a complete hallux) are not diagnostic other than differing from available specimens of *Porzana*, *Gallirallus*, *Porphyrio*, or *Gallinula*. The synsacral fragment of a heron resembles that in *Egretta* much more than in other likely genera (*Butorides*, *Nycticorax*, *Ixobrychus*). Parrots are represented by a partial quadrate of a very large species (larger than in *Nestor*, *Prosopeia*, *Eclectus*, or any lorikeet; dissimilar from that in Neotropical parrots) and digit I, phalanx 2 of the wing (larger than in *Vini* or *Cyanoramphus*, smaller than in *Nestor* or *Eclectus*; ca. the size in *Prosopeia*). Except for the tibiotarsus of *Porzana*, these specimens are as frustrating as they are informative; I anxiously await another opportunity to excavate small bones on Easter Island.

Having covered avian prehistory in the last three chapters on a generally ESE track from the Bismarcks through Easter Island, let us now return in Chapter 8 to the far western Pacific, but in the northern hemisphere.

Chapter 8 Micronesia and Remote Central Pacific Islands

THE 3000+ Micronesian islands and islets combined consist of only ca. 2700 km^2 of land, which is less than a number of individual Melanesian or West Polynesian islands. Most of Micronesia lies in the northern hemisphere (Figure 1-5). The term "Remote Central Pacific Islands" refers to the very isolated, equatorial, atoll-dominated island groups lying northeast to southeast of Pohnpei and Kosrae. Some of these, such as the Marshall Islands, Nauru, and much of Kiribati, are culturally Micronesian. Others, such as Tuvalu and Tokelau, are Polynesian. Many of the most remote atolls, such as the Phoenix Islands, Line Islands, and Wake, are inhabited sporadically or not at all.

The nomenclature and geography of islands in Micronesia can be confusing. For example, the term "Caroline Islands" has been used to cover all Micronesian islands (except the Marianas) from Kosrae westward either to Palau or to Yap or to the atolls lying just west of Chuuk. Today, the high islands of Yap, Chuuk, Pohnpei, and Kosrae, along with associated atolls, are united politically as the Federated States of Micronesia (Figure 8-1). I will attempt to use the names for islands or island groups most often used by islanders, identifying common synonyms as needed.

Avian prehistory is well developed only in the Mariana Islands, with 23 genera of landbirds recorded thus far from ca. 3000 bones (Table 8-1). The fossil record in the Marianas differs from that of Polynesia, or elsewhere in Micronesia, in that bones of landbirds outnumber those of seabirds. At Pisonia Rockshelter on Aguiguan, for example, 94% of the ca. 1000 identifiable bird bones are from resident landbirds, especially rails. I do not know if this is due to a genuine prehistoric scarcity of seabirds in the Marianas, or to a cultural preference for hunting landbirds.

Palau today supports more genera of landbirds than any other Micronesian island group, including eight unknown elsewhere in the region. Two are endemic to Palau (the owl *Pyrroglaux* and white-eye *Megazosterops*). Each of the others occurs in one or more of the nearest continental island regions (Philippines, Moluccas, New Guinea). In the Marianas, fossils increase the genera of landbirds from 20 to 26, just three fewer than in Palau, where a very limited fossil record (see below) adds nothing to its landbird diversity. Four of the five genera of landbirds endemic to Micronesia are passerines. At the level of genus or species, more endemism occurs today in white-eyes (Zosteropidae sensu lato) than any other family of Micronesian birds. If not for anthropogenic extinction, rails would exceed white-eyes in species-level endemism.

Among the five trampy species so widely distributed in Polynesia and Melanesia (the heron *Egretta sacra*, duck *Anas superciliosa*, and rails *Gallirallus philippensis*, *Porzana tabuensis*, and *Porphyrio porphyrio*), only the heron is widespread in Micronesia. The others either do not occur in Micronesia or do so only in Palau.

Palau (Republic of Belau)

With a land area of 493.8 km^2 (Karolle 1993) and 12 islands >1 km^2 (Figure 8-2), Palau consists of the large, eroded volcanic island of Babeldaob (365 km^2), raised limestone islands, mixtures thereof, and seven outlying atolls or low

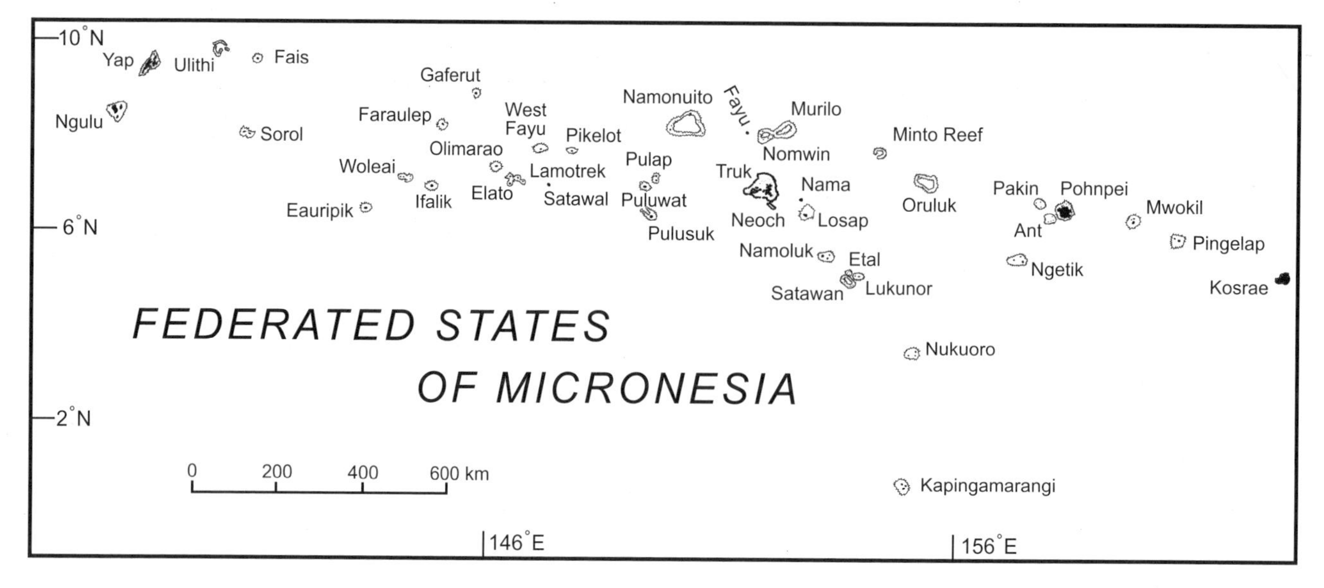

Figure 8-1 The Federated States of Micronesia. Redrawn from UNEP/IUCN (1988).

coral islands. Closer to potential source areas for landbirds (Philippines, Moluccas, and New Guinea) than any other Micronesian island group, Palau sustains the region's richest modern landbird fauna (Table 8-2). This probably is related as well to Palau having more mature forest than elsewhere in Micronesia. Most of Palau's raised limestone islands (known locally as the Rock Islands) are uninhabited, largely or entirely forested, and difficult to explore because of cliffs, crevices, and pinnacle karst. Unfortunately, deforestation on the large island of Babeldaob is proceeding along a new road to the island's northern half, previously accessible by boat only.

Nine of Palau's 31 resident species of landbirds are endemic. The high species richness value begs the question of how much, if any, prehistoric extinction has taken place in these lush islands. Fossils have shed no light on this issue. In spite of 60 days (1993 to 1997) of searching in Palau for bone deposits in caves and rockshelters, we found no sites with abundant bones of terrestrial vertebrates (Pregill & Steadman 2000). We located and sampled only 10 very small bone deposits in caves and rockshelters on four islands (Ulebsechel, Ngeruktabel, Ngerduais, Ulong). None of these sites yielded bones of extirpated species, and only five landbird bones were recovered. In addition, chicken (*Gallus gallus*) and seabird bones were reported from an archaeological site on Koror (Kaneko & Abe 1979). The failure to find extinct species of birds distinguishes Palau from other groups of limestone islands that are well explored paleontologically.

Learning about prehistoric birds on Palau has been difficult for three reasons. First, wartime activities destroyed the natural integrity of most of Palau's caves and rockshelters and therefore, presumably, most of Palau's vertebrate fossils (Pregill & Steadman 2000). The caves most likely to have bone deposits were used by the Japanese military for more than a decade prior to the American invasion in 1944 (Hough 1990, Ross 1991). To convert them to storage facilities, communication stations, gun emplacements, command posts, or troop quarters, the caves were excavated, their walls scraped and tunneled, and the floors were trampled or cemented over (Osborne 1966, 1979, Rogers & Legge 1992). Throughout Palau but especially on Ngeaur (Angaur) and Beliliou (Peleliu), caves almost invariably contain evidence of the war: rusted canteens and ammunition casings, gerry cans, broken saki bottles, melted glass, and occasionally human bone. Caves were occupied and disturbed even on remote islands with no apparent strategic value. Nonhorizontal floors (and therefore limited sedimentation) and damp microclimates have compromised our ability to discover bone deposits even in the least modified caves and rockshelters. A reflection of Palau's high rainfall, the perpetual dampness promotes decay rather than preservation of bones.

Does Palau's limited fossil record mean that this island group, with a rich avifauna by Micronesian standards, has been spared the extinction suffered elsewhere in Remote Oceania? I doubt it. Rats occur on all islands and probably have done so for the entire two millennia of human presence. Also, some conspicuous absences hint at what might be found if substantial fossil deposits ever materialize. These include flightless rails (*Gallirallus, Porzana, Porphyrio*), any sort of hawk or parrot, a reed-warbler (*Acrocephalus*), and

Table 8-1 Distribution of resident landbird genera in Micronesia. e, genus endemic to Micronesia; H, historic record of extirpated species; M, modern record; P, prehistoric record. Remote Islands are Nauru, Marshall Islands, Kiribati (including Gilbert, Phoenix, and most Line Islands), Tuvalu, Tokelau, and U.S. possessions (Wake, Johnston, Howland, Baker, Jarvis, Palmyra). Data for Yap, Chuuk, Pohnpei, and Kosrae are for the high islands, not the atolls. From Pyle & Engbring (1985), Pratt et al. (1987), Steadman & Intoh (1994), Steadman (1999a), Pregill & Steadman (2000).

	Palau	Yap	Marianas	Chuuk	Pohnpei	Kosrae	Remote Islands	TOTAL
HERONS								
Egretta	M	M	M, P	M	M	M	M	7
Nycticorax	M	—	—	M	—	—	—	2
Ixobrychus	M	M	M	M	—	—	—	4
DUCKS								
Anas	M	—	H, P	M	—	—	H	4
FALCONS								
Falco	—	—	P	—	—	—	—	1
MEGAPODES								
Megapodius	M, P	—	M, P	—	P	—	—	3
RAILS								
Rallina	M	—	—	—	—	—	—	1
Gallirallus	M	—	M, P	—	—	—	H	3
Porzana	—	—	P	—	—	H	—	2
Poliolimnas	M	M, P	H, P	M	M	—	—	5
Gallinula	M	—	M, P	—	—	—	—	2
Porphyrio	M	—	P	—	—	—	—	2
SANDPIPERS								
Prosobonia	—	—	—	—	—	—	H	1
PIGEONS, DOVES								
Ptilinopus	M	—	M, P	M	M	M	H	6
Ducula	M	M, P	P	M	M	M	M	7
Caloenas	M	—	—	—	—	—	—	1
Gallicolumba	M, P	M, P	M, P	M	M	—	—	5
PARROTS								
Trichoglossus	—	—	—	—	M	—	—	1
other parrot	—	—	P	—	—	—	—	1
OWLS								
Pyrroglaux (e)	M	—	—	—	—	—	—	1
Asio	—	—	—	—	M	—	—	1
NIGHTJARS								
Caprimulgus	M	—	—	—	—	—	—	1
SWIFTS								
Collocalia	M	H	M, P	M	M	M	—	6
KINGFISHERS								
Halcyon	M	M?	M, P	—	M	—	—	3–4
WARBLERS								
Cettia	M	—	—	—	—	—	—	1
Acrocephalus	—	M?	M, P	M	M	M	M, P	5–6

(continued)

Table 8-1 (continued)

	Palau	Yap	Marianas	Chuuk	Pohnpei	Kosrae	Remote Islands	TOTAL
WHITE-EYES								
Zosterops	M	M	M, P	M	M	M	—	6
Rukia (e)	—	M	—	M	M	—	—	3
Cleptornis (e)	—	—	M, P	—	—	—	—	1
Megazosterops (e)	M	—	—	—	—	—	—	1
STARLINGS								
Aplonis	M	M, P	M, P	M	M	M	—	6
ESTRILDID FINCHES								
Erythrura	M	—	P	M	M	M	—	5
CUCKOO-SHRIKES								
Coracina	M	M	—	—	M	—	—	3
WHISTLERS								
Colluricincla[1]	M	—	—	—	—	—	—	1
FANTAILS								
Rhipidura	M	M	M, P	—	M	—	—	4
MONARCHS								
Monarcha	—	M	M	—	—	—	—	2
Metabolus (e)	—	—	—	M	—	—	—	1
Myiagra	M	—	H, P	M	M	—	—	4
HONEYEATERS								
Myzomela	M	M	M, P	M	M	M	—	6
WOOD-SWALLOWS								
Artamus	M	—	—	—	—	—	—	1
CROWS								
Corvus	—	—	M	—	—	—	—	1
TOTAL GENERA								
M	29	12–14	17	17	18	9	3	38
M + H	29	13–15	20	17	18	10	7	38
P	2	4	23	0	1	0	1	23
M + H + P	29	13–15	26	17	19	10	7	40
Total resident landbird bones	5	7	~3000	0	3	0	1	~3000+

[1]Sometimes classified in the genus *Pitohui*.

a monarch (*Monarcha*). Given its nearness to continental islands, Palau also may have sustained genera otherwise unknown in Micronesia.

Yap

Yap (Figures 8-1, 8-3) consists of a cluster of four islands ("Yap Proper") made of gently contoured, eroded volcanic and metamorphic rocks (Yap, Maap, Rumung, and Tamil) and two main outlying islands—Ulithi Atoll (Lessa 1966) and the raised limestone Fais. Seventeen other, more remote atolls are also included in Yap State of the Federated States of Micronesia (Nicholson 1969). The four islands of Yap Proper (99.9 km^2 of Yap's 118.9 km^2 of land; Karolle 1993) are separated from each other only by narrow, mangrove-lined waterways.

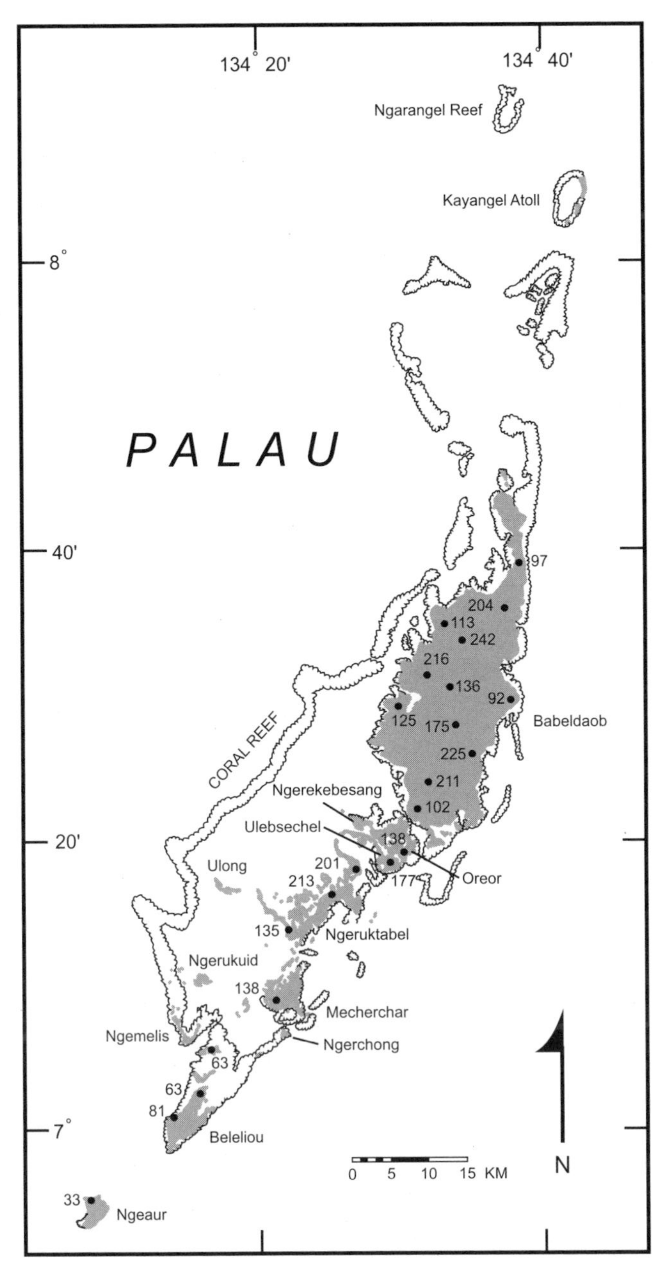

Figure 8-2 Palau, not including the Southwest Islands (320 km southwest of Ngeaur). Redrawn from Palau Islands (1996).

The Yapese avifauna (see Fisher 1950, Pratt et al. 1977) is not "intact" as suggested by Engbring & Pratt (1985). Missing from Yap Proper, for example, are species in these genera that occur in Palau to the southwest or in Micronesian island groups to the east or north: *Nycticorax*, *Megapodius*, *Gallirallus*, *Porphyrio*, *Gallinula*, *Ptilinopus*, *Halcyon*, *Acrocephalus*, *Myiagra*, and *Erythrura* (Table 8-1). Yap features three endemic species (all passerines) compared to nine in Palau (Table 8-2). The landbirds of Palau outnumber those of Yap in total species by 31 to 12. While some of this difference probably is due to Yap's greater isolation, I believe that even more of it is due to human impact, especially deforestation. As far as Steadman and Intoh (1994) could determine from conflicting and confusing sources, the only resident landbirds on Ulithi may be the heron *Egretta sacra*, rail (*Poliolimnas cinereus*?), kingfisher *Halcyon* cf. *cinnamomina*, warbler *Acrocephalus syrinx*, and starling *Aplonis opaca*.

During my brief visit (three days) to Yap and Tamil islands in 1997, I did not see a single patch of mature, undisturbed upland forest. Yapese upland vegetation can be categorized into four types (Falanruw et al. 1987, Falanruw 1994, Dodson & Intoh 1999): cultivated areas (gardens and taro patches); a pyrophytic *Pandanus* savanna with a ground cover of ferns; young successional forests and shrublands; and late successional, degraded forests. The deforestation is not just modern; pollen analyses of wetland sediment on Yap and Tamil islands show that "a major period of forest destruction, accompanied by fire, took place about 3300 yr BP" (Dodson & Intoh 1999:17). Each modern type of upland vegetation is wholly or partially anthropogenic.

Archaeological sites in Yap are no older than ca. 2000 yr BP. Evidence of past birdlife is limited to chicken bones from Yap Proper (Gifford & Gifford 1959) and Ngulu Atoll (Takayama 1982), and one set of bones from small, outlying Fais (2.8 km^2, elev. ca. 20 m; Steadman & Intoh 1994). We have no boney evidence at all of the birds (or lizards or bats) that were lost during the first 1300 years of human influence.

Four archaeological sites on the southern coast of Fais (Intoh 1991, 1997) yielded nearly 200 identifiable bird bones from strata dating from ca. 1800 to 400 yr BP (Table 8-3). Because the modern birds of Fais are not well studied, it is difficult to determine which of the 14 species of seabirds and four species of landbirds from the four sites still occur on Fais. Based on modern distributions of birds from Yap Proper, Ulithi, and adjacent island groups (Baker 1948, Pratt et al. 1977), the modern environment of Fais (Fosberg & Evans 1969), and what is known about the vulnerability of individual species, it is likely that about nine of the seabirds (two petrels, two boobies, and five terns) and three of the landbirds (the rail *Poliolimnas cinereus* and the columbids *Gallicolumba* cf. *xanthonura* and *Ducula oceanica*) no longer live on Fais (Steadman & Intoh 1994). Each of the landbirds has been recorded on Yap Proper (Pratt et al. 1987).

Mariana Islands

This island arc has a Southern Group of raised limestone islands (Guam to Saipan) and a Northern Group of

Table 8-2 Landbirds of Palau and Yap. e, endemic species; H, historic record, no longer survives; M, modern record; P, prehistoric record. The distribution of species on selected individual islands is from Pratt et al. (1980), Engbring (1983a, 1992), and Wiles & Conry (1990). Also from Baker (1951), Pratt et al. (1987), Steadman & Intoh (1994), Pregill & Steadman (2001).

	Palau									
	Babeldaob	Ngerekebesang	Koror	Ngeruktabel	Eil Malk	Ngerukewid	Beliliou	Ngeaur	Southwest islands	Yap
HERONS										
Egretta sacra	M	M	M	M	M	M	M	M	M	M
Nycticorax caledonicus	M	M	M	M	M	M	M	M	—	—
Ixobrychus sinensis	M	M	M	M	M	—	M	M	—	M
DUCKS										
Anas superciliosa	M	—	H	—	—	—	M	M	—	—
MEGAPODES										
Megapodius laperouse	M	M	M	M	M	M	M	M	—	—
RAILS										
Rallina eurizonoides	M	M	M	M	M	—	M	M	—	—
Gallirallus philippensis	M	M	M	—	—	—	M	M	—	—
Poliolimnas cinereus	M	—	M	—	—	—	M	M	—	M, P
Gallinula chloropus	H	—	H	—	—	—	M	M	—	—
Porphyrio porphyrio	M	—	M	—	—	—	M	M	—	—
PIGEONS, DOVES										
Ptilinopus pelewensis (e)	M	M	M	M	M	M	M	M	—	—
Ducula oceanica	M	M	M	M	M	M	M	—	—	M, P
Caloenas nicobarica	M	M	M	M	M	M	M	—	—	—
Gallicolumba xanthonura	—	—	—	—	—	—	—	—	—	M, P
Gallicolumba canifrons (e)	M	M	M	M	M	—	M	M	—	—
OWLS										
Pyrroglaux podargina (e)	M	M	M	M	M	M	M	M	—	—
NIGHTJARS										
Caprimulgus indicus	M	M	M	M	M	—	M	—	—	—
SWIFTS										
Collocalia vanikorensis	M	M	M	M	M	M	M	—	—	H
KINGFISHERS										
Halcyon cinnamomina	M	M	M	M	M	M	M	—	—	—
Halcyon chloris	M	M	M	M	M	M	M	M	M	—
WARBLERS										
Cettia annae (e)	M	M	M	M	M	—	M	—	—	—
WHITE-EYES										
Zosterops hypolais (e)	—	—	—	—	—	—	—	—	—	M
Zosterops semperi	M	M	M	M	M	M	M	—	—	—
Zosterops finschii (e)	M	M	M	M	M	M	M	—	—	—
Rukia oleaginea (e)	—	—	—	—	—	—	—	—	—	M
Megazosterops palauensis (e)	M?	—	—	M	—	—	M	—	—	—

Table 8-2 (continued)

	Palau									
	Babeldaob	Ngerekebesang	Koror	Ngeruktabel	Eil Malk	Ngerukewid	Beliliou	Ngeaur	Southwest islands	Yap
STARLINGS										
Aplonis opaca	M	M	M	M	M	M	M	M	—	M, P
ESTRILDID FINCHES										
Erythrura trichroa	M	M	M	M	M	—	—	—	—	—
CUCKOO-SHRIKES										
Coracina tenuirostris	M	M	M	M	M	M	M	—	—	M
WHISTLERS										
Colluricincla tenebrosa[1] (e)	M	M	M	M	M	M	M	—	—	—
FANTAILS										
Rhipidura lepida (e)	M	M	M	M	M	M	M	—	—	—
Rhipidura rufifrons	—	—	—	—	—	—	—	—	—	M
MONARCHS										
Monarcha godeffroyi (e)	—	—	—	—	—	—	—	—	—	M
Myiagra erythrops (e)	M	M	M	M	M	M	M	—	—	—
HONEYEATERS										
Myzomela rubrata	M	M	M	M	M	M	M	M	—	M
WOOD-SWALLOWS										
Artamus leucorhynchus	M	—	M	—	—	—	—	—	—	—
TOTAL SPECIES										
M	29–30	25	28	25	24	18	29	16	2	12
M + H	30–31	25	30	25	24	18	29	16	2	13
M + H + P	30–31	25	30	25	24	18	29	16	2	13
TOTAL BONES										
All species	0	0	0	0	0	0	0	0	0	197
Resident landbirds	0	0	0	0	0	0	0	0	0	7

[1]Sometimes classified in the genus *Pitohui*.

10 young volcanic islands (Farallon de Medinilla to Farallon de Pajaros; Figures 1-32, 8-4). Guam is the largest island, making up 554 km^2 of the archipelago's total land area of 1032 km^2 (Karolle 1993). Guam is distinguished further by being a separate political entity today from the other major islands in the Marianas, which are known as the Commonwealth of the Northern Mariana Islands.

Compared to most Pacific island groups, the modern distribution of birds is well known in the Marianas (Table 8-4 and references therein). Lying more northwest than other Micronesian groups, the Marianas regularly receive migratory non-charadriiform landbirds from Asia (such as herons, ducks, hawks, falcons, and passerines; Wiles et al. 1993, 2004, Stinson et al. 1997, Kessler 1999). The resident avifauna features certain or probable species-level endemism in ducks, rails, columbids, parrots, warblers, monarchs, white-eyes, crows, and estrildid finches, including seven of the 10 species of passerines, such as Micronesia's only species of crow (*Corvus kubaryi*). *Cleptornis* is the only endemic genus of landbird. Variously regarded as a honeyeater (Meliphagidae) or a white-eye (Zosteropidae), a recent analysis of mitochondrial DNA (Slikas et al. 2000) suggests that *Cleptornis* is the sister group to all other white-eyes, in spite of differing in having a large 10th primary remex. I tentatively list *Cleptornis* in the Zosteropidae.

The prehistoric record of birds in the Marianas is very incomplete but by far the best in Micronesia. Bird bones have been found on all five major limestone islands—Guam,

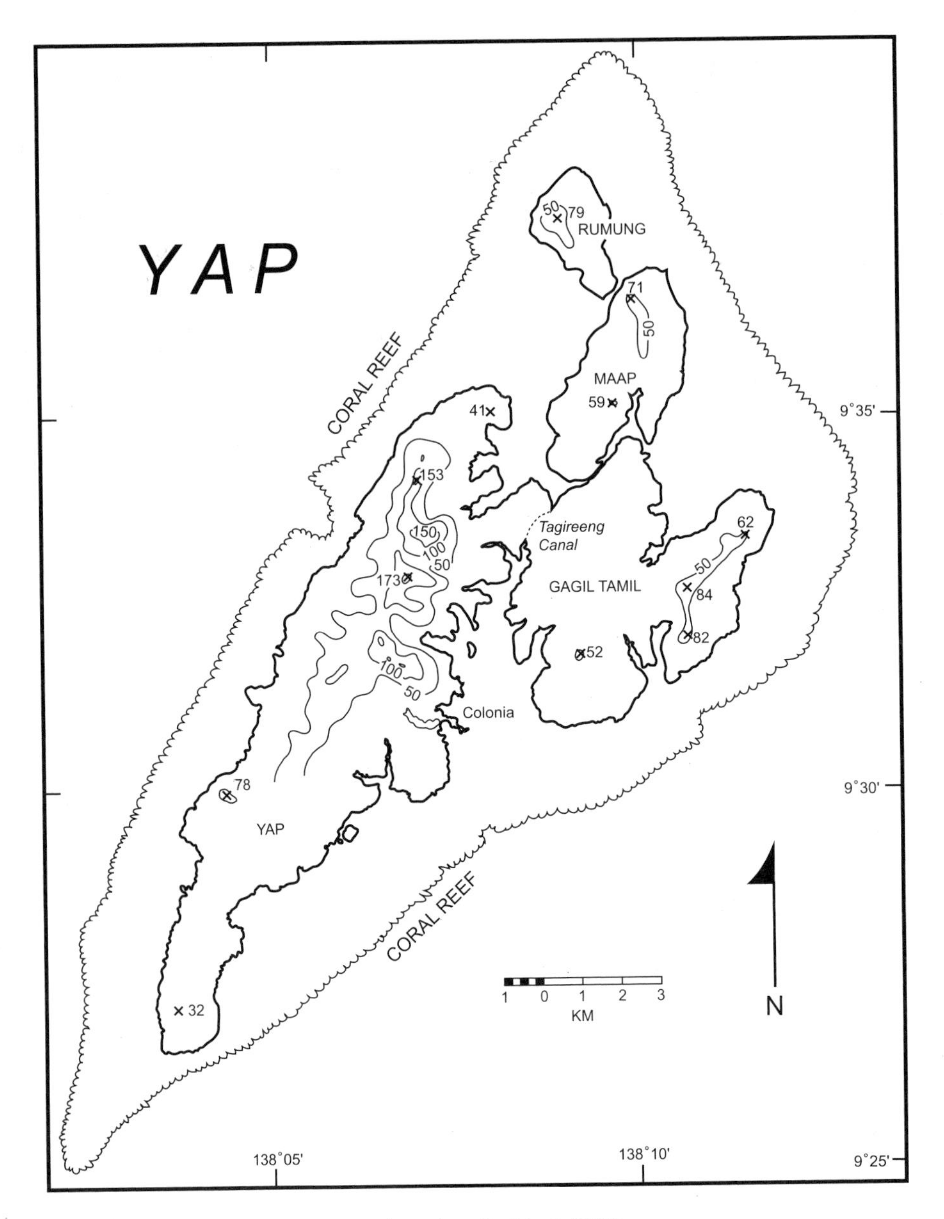

Figure 8-3 Yap. Contour intervals in m. Redrawn from Yap Islands (1996).

Rota, Aguiguan, Tinian, and Saipan (Figure 8-4). Fossils are fewest on Guam and Saipan. Even on Rota, Aguiguan, and Tinian, new fossil material would increase the number of species of birds known from each island.

In 1990 and 1994, I evaluated caves and rockshelters in the Marianas for bone deposits, eventually excavating nine sites (four on Rota, two on Aguiguan, and three on Tinian). The abundant caves and rockshelters provide ideal physical and geochemical settings for bones. Unfortunately, caves in the Marianas have fared poorly since their descriptions in 1887–1889 by Antoine-Alfred Marche (Marche 1982). As in Palau, activities associated with World War II have ruined the prehistoric potential of most caves and rockshelters in the Mariana Islands. On Rota in 1990, for example, I examined nearly 50 caves and rockshelters before finding one (Payapai Cave; see below) with undisturbed, bone-rich sediments. Some of the effects of World War II on modern birdlife can be gleaned from Marshall (1949).

The ages of the cultural sites on Rota, Aguiguan, and Tinian do not exceed ca. 2500 cal BP (Steadman 1999a; Table 8-5 herein), whereas those in calcareous sand deposits on Rota, Tinian, and Saipan are as old as ca. 3400 cal BP (Craib 1993, 1999, Butler 1994, Amesbury et al. 1996), and span most of the prehistoric ceramic sequence in the Marianas (Dickinson et al. 2001). Only on Tinian have the very early cultural sites produced abundant bird bones, which

Table 8-3 Bird bones from prehistoric sites on Fais, Yap. FSFA-1, Faligochol-1 site; FSFA-2, Faligochol-2 site; FSPO-1, Powag-1 site, FSPO-2, Powag-2 site. i, introduced species; m, migratory species; *, certainly or probably extirpated. Taxa in brackets are not necessarily different from those identified more specifically. From Steadman & Intoh (1994).

	FSFA-1	FSFA-2	FSPO-1	FSPO-2	TOTAL
SEABIRDS					
PETRELS					
**Pterodroma* sp.	—	1	—	—	1
**Bulweria bulwerii*	—	—	1	—	1
TROPICBIRDS					
Phaethon lepturus	1	—	—	—	1
BOOBIES					
**Sula dactylatra*	1	—	—	—	1
**Sula sula*	—	2	—	—	2
[**Sula sula* or *leucogaster*]	2	—	—	—	2
FRIGATEBIRDS					
Fregata minor	1	—	1	2	4
Fregata ariel	1	1	—	—	2
[*Fregata* sp.]	—	—	1	—	1
TERNS					
**Sterna sumatrana*	1	1	—	—	2
**Sterna lunata*	4	1	1	2	8
**Sterna fuscata*	4	2	—	—	6
[*Sterna* sp.]	1	—	—	—	1
**Anous minutus*	2	1	—	—	3
Anous stolidus	13	1	2	2	18
**Procelsterna cerulea*	1	1	1	—	3
Gygis candida	5	3	—	—	8
[Sterninae sp.]	6	5	—	—	11
LANDBIRDS					
CHICKENS					
Gallus gallus (i)	56	26	7	13	102
PLOVERS					
Pluvialis fulva (m)	1	1	—	—	2
Charadrius leschenaultii (m)	—	1	—	—	1
SANDPIPERS					
Numenius phaeopus (m)	3	1	—	1	5
Tringa erythropus (m)	1	1	—	—	2
Heteroscelus sp. (m)	1	—	—	—	1
[Scolopacidae sp.] (m)	—	2	—	—	2
RAILS					
**Poliolimnas cinereus*	—	1	—	—	1

(continued)

Table 8-3 (continued)

	FSFA-1	FSFA-2	FSPO-1	FSPO-2	TOTAL
PIGEONS, DOVES					
**Ducula oceanica*	1	—	—	—	1
**Gallicolumba* cf. *xanthonura*	1	1	—	—	2
STARLINGS					
Aplonis opaca	—	3	—	—	3
TOTAL BONES					
Seabirds	43	19	7	6	75
Migrant shorebirds	6	6	0	1	13
Native landbirds	2	5	0	0	7
Chicken	56	26	7	13	102
All species	107	56	14	20	197
TOTAL SPECIES					
All native species	17	18	5	4	23
Extirpated native species	8	9	3	1	12

remain incompletely studied (see below). Thus our evidence for which species of birds were hunted by the first inhabitants is limited.

Guam

This is Micronesia's largest, most populous, and most commercially developed island. Guam is sadly famous for the avifaunal devastation wrought in recent decades by the introduced brown tree snake *Boiga irregularis* (Savidge 1987, Jaffe 1994, Rodda et al. 1997, Fritts & Rodda 1998). Continued spread of *B. irregularis* now threatens the birdlife on any human-inhabited island in the Marianas, and potentially any Micronesian island with an airport or wharf.

In historic times, more species of landbirds (18) have been recorded on Guam than on other major islands in the Marianas (Table 8-4). Only three of these species survive on Guam, in reduced numbers and range. The crow *Corvus kubaryi* persisted into the late 1990s (Wiles 1998) but now is gone. Most of the extinct species or extirpated populations were recorded last in the 1980s, a period of great abundance for *Boiga irregularis*. Although the causes of rarity and extinction in Marianas landbirds are diverse (Engbring & Ramsey 1984, Stinson et al. 1991, Reichel et al. 1992, Savidge et al. 1992), predation by the snake has been by far the leading cause on Guam since World War II. How much extinction occurred prehistorically is unknown because Guam's avian fossil record (just 28 bones) represents only species recorded in historic times. Minimally, any species or species-group listed in Table 8-4 might be expected if more fossils were available.

Rota

Unlike on Guam, the extensive losses of birds on Rota are documented mainly through prehistoric bones rather than historic specimens and observations. Because of its fossils, Rota boasts more species of landbirds (25) than any other island in the Marianas (Table 8-4). Only 10 of those species still live on Rota, an island with moderate forest cover (though very little mature, undisturbed upland forest) and considerable development pressure (pers. obs. 1990, 1994).

Rota's ancient birdlife first was glimpsed with a report of the moorhen *Gallinula chloropus* from the Borja and Maratita archaeological sites (Becker & Butler 1988). The most important site for prehistoric birds on Rota is Payapai Cave, excavated in 1990 and 1994 (Steadman 1992a, 1999a). Bird bones are abundant in the contiguous Units 1–3 (Table 8-6) but rare in Unit 4 (Table 8-7), located closer to the cave's immense entrance. Although the wood charcoal in Units 1–3 probably is from human-set fires, the excavations yielded no human artifacts, not even a potsherd, which is typically the most abundant type of prehistoric artifact in the Marianas. The fish bones at Payapai Cave are not burned and probably represent the prey of seabirds and herons that nested or roosted in the cave, rather than fish caught by people. The ^{14}C dates on wood charcoal from

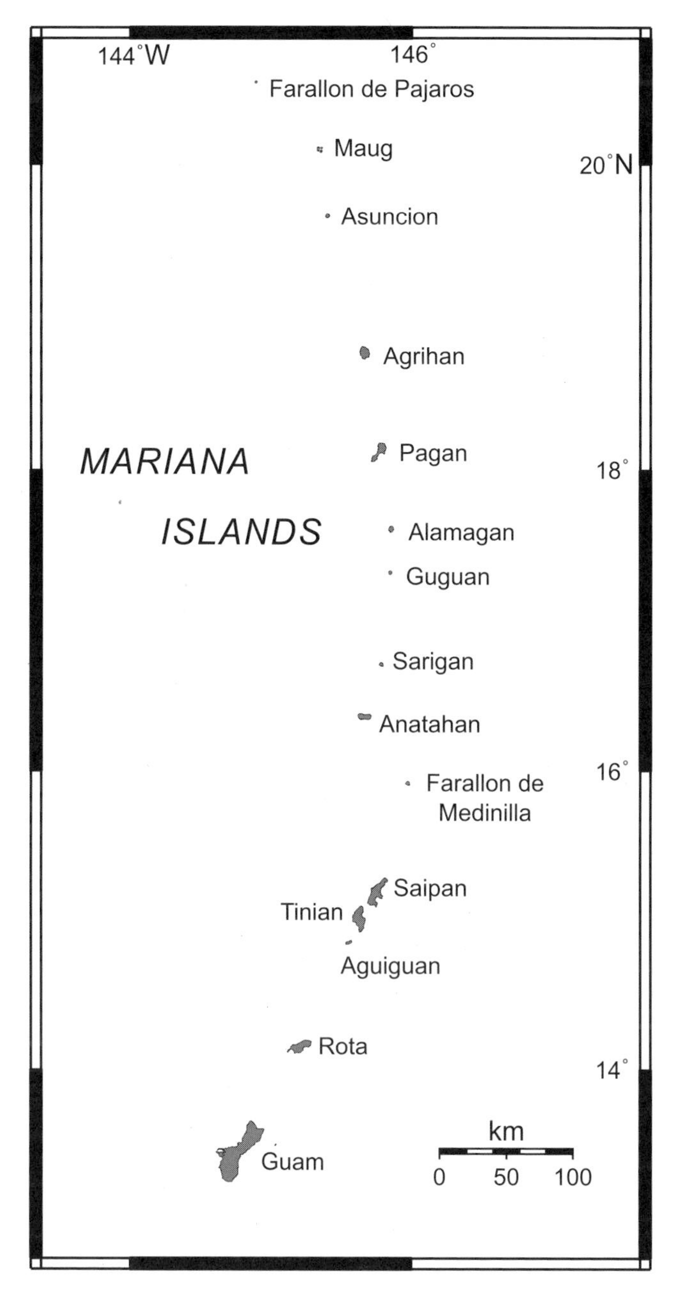

Figure 8-4 The Mariana Islands.

Payapai Cave are reversed; that from Unit 3 (Layer III, Level 4, at the contact with Layer II) is 530–310 cal BP, whereas that from Unit 2 (Layer II, Level 2) is 960–690 cal BP (Table 8-5). This suggests that the massive wood charcoal deposit (Layer II) in Units 1–3 has been mixed, or perhaps that heartwood from a very old tree was the source of charcoal for the older radiocarbon date. Finding bones of the introduced rat *Rattus exulans* in all strata of Units 1–3 (Table 8-6) corroborates the late prehistoric age (<1000 cal BP) of this bone deposit.

The abundance of bones of *Aplonis opaca* at Payapai Cave may reflect that this starling will nest on cliffsides, especially in or near cave entrances (Jenkins 1983, pers. obs.). Of 17 resident species of landbirds from Payapai Cave, 11 no longer occur on Rota. Two (the megapode *Megapodius laperouse* and the swiftlet *Collocalia vanikorensis*) were lost in historic times; the nine others are known only from bones. Six of them have been recorded from other sites as well, whereas Payapai Cave has furnished Rota's only evidence of a small flightless duck (Anatidae undescribed sp.), falcon *Falco peregrinus*, and pigeon *Ducula oceanica*.

The seven sets of bird bones from sites other than Units 1–3 of Payapai Cave account for 15 species, nine of which no longer occur on Rota (Table 8-7). Small accumulations of sediment near the rear of Payapai Cave produced 1300+ additional bones, mostly of lizards (Pregill 1998, Steadman 1999a). The noncultural, lizard-dominated faunal assemblage in Crevice 150m N of Payapai Cave is similar to that from the rear of Payapai Cave. Aside from lizards, this site yielded many bones of the extirpated sheath-tailed bat *Emballonura semicaudata* and an important set of passerine bones that includes Rota's only record of the monarch flycatcher *Myiagra* cf. *freycineti*.

On Rota's northeast coast, As Matmos Cliffside Cave has small pockets of noncultural, bone-rich sediment. Perched just above the short, windward forest canopy, this site as well as Payapai Cave and Crevice 150m N of Payapai Cave would be an ideal roost for an avian predator. Among raptorial birds (hawks, falcons, owls), the only bone found on Rota or anywhere else in the Marianas is a single pedal phalanx of the falcon *Falco peregrinus* from Payapai Cave. This identification, made subsequent to Steadman (1999a), suggests that the ornithophagous *F. peregrinus* (Vice & Vice 2004) may have accumulated bones in caves in the Marianas, perhaps especially in seaward-facing caves that could be used as nests or plucking perches (see White et al. 2002). The falcon bone from Payapai Cave is from an adult, thus shedding no light on whether *F. peregrinus* was formerly a resident or a migrant in the Marianas. Six of the 11 species of landbirds from As Matmos Cliffside Cave are extinct or extirpated, including Rota's only record of *Cleptornis marchei*. The only bones of *Gallirallus* undescribed sp. L found thus far on Rota are from Mochong, an open archaeological site excavated by John Craib.

Aguiguan

Smallest of the five major raised limestone islands in the Marianas, Aguiguan is encircled by cliffs and lacks a beach or anchorage. It lies 9 km south of Tinian, from which it is accessible by helicopter or by jumping ashore during calm

Table 8-4 Distribution of resident landbirds in the Mariana Islands. e, endemic species; H, historic record, now extirpated; M, modern record; P, prehistoric record; †, extinct species; *, survives in captivity. Modern distribution from Baker (1951), Jenkins (1983), Engbring & Pratt (1985), Engbring et al. (1986), Pratt et al. (1987), Reichel (1991), Reichel & Glass (1991), Craig et al. (1992a), Stinson (1994), and Stinson et al. (1997). Prehistoric distribution updated from Steadman (1999a). The precise distribution of birds in the remote Northern Group (the ten volcanic islands north of Saipan) is too poorly known to compile except on Sarigan, which is largely deforested (Fancy et al. 1999), and Maug, which consists of three steep islets with a depauperate flora (Eldredge et al. 1977). The two different bone totals from Tinian reflect that the 900+ bones from the Unai Chulu site are not completely studied.

	Guam	Saipan	Tinian	Rota	Aguiguan	Sarigan	Maug
HERONS							
Egretta sacra	M, P	M	M	M, P	—	—	M
Ixobrychus sinensis	M	M	M	M	M	—	—
DUCKS							
†*Anas oustaletii* (e)	H, P	—	—	—	—	—	—
†Anatidae undescribed sp. (e)	—	—	—	P	—	—	—
FALCONS							
Falco peregrinus	—	—	—	P	—	—	—
MEGAPODES							
Megapodius laperouse	H, P	M	M, P	H, P	M, P	M	M
RAILS							
Gallirallus owstoni (e)	H*,P	—	—	—	—	—	—
†*Gallirallus* undescribed sp. L (e)	—	—	—	P	—	—	—
†*Gallirallus* undescribed sp. M (e)	—	—	—	—	P	—	—
†*Gallirallus* undescribed sp. N (e)	—	—	P	—	—	—	—
†*Gallirallus* undescribed sp. O (e)	—	P	—	—	—	—	—
†/* *Porzana* undescribed spp. G–I (e)	—	—	P	P	P	—	—
Poliolimnas cinereus	H, P	—	P	P	—	—	—
Gallinula chloropus	H, P	M	M	P	—	—	—
†*Porphyrio* undescribed sp. E (e)	—	—	P	P	—	—	—
PIGEONS, DOVES							
Ptilinopus roseicapilla (e)	H	M	M	M, P	M, P	—	—
Ducula oceanica	—	—	P	P	—	—	—
†*Gallicolumba* undescribed sp. (e)	—	—	—	P	—	—	—
Gallicolumba xanthonura	H, P	M, P	M, P	M, P	M, P	M	—
PARROTS							
†Psittacidae sp. (e)	—	—	—	P	—	—	—
SWIFTS							
Collocalia vanikorensis	H, P	M	P	H, P	M	—	—
KINGFISHERS							
Halcyon cinnamomina	H*	—	—	—	—	—	—
Halcyon chloris	—	M	M	M	—	M	M
WARBLERS							
Acrocephalus luscinia (e)	H	M	P	—	M	—	—
WHITE-EYES							
Zosterops conspicillatus (e)	H	M	M, P	M, P	M, P	—	—
Cleptornis marchei (e)	—	M	P	P	M, P	—	—

Table 8-4 (continued)

	Guam	Saipan	Tinian	Rota	Aguiguan	Sarigan	Maug
STARLINGS							
Aplonis opaca	M, P	M, P	M, P	M, P	M, P	M	M
ESTRILDID FINCHES							
†*Erythrura* undescribed sp. (e)	—	—	—	P	—	—	—
FANTAILS							
Rhipidura rufifrons	H	M	M, P	M	M	—	—
MONARCHS							
Monarcha takatsukasae (e)	—	H	M	—	—	—	—
†*Myiagra freycineti* (e)	H, P	—	—	P	—	—	—
HONEYEATERS							
Myzomela rubrata	H, P	M	M, P	M, P	M, P	M	—
CROWS							
Corvus kubaryi (e)	H	—	—	M	—	—	—
TOTAL SPECIES							
M	3	14	12	10	11	5	4
M + H	18	15	12	12	11	5	4
M + H + P	18	16	20	25	13	5	4
Island area (km^2)	554	122	102	85	7.2	4.6	2.1
Elevation (m)	406	466	170	491	166	538	227
Number of resident landbird bones	28	12	647 (~1600)	475	944	0	0

Table 8-5 Radiocarbon dates from caves and rockshelters with prehistoric bird bones in the Mariana Islands. Lab abbreviations: Beta, Beta Analytic, Inc.; CAMS, Center for Accelerator Mass Spectrometry, Lawrence Livermore National Laboratory. Provenience: Roman numerals are Layers (natural stratigraphy); arabic numerals are Levels (arbitrary subdivisions). Radiocarbon age is corrected for $^{13}C/^{12}C$ and presented in years before present $\pm 1\sigma$. Calibrated age is given at $\pm 2\sigma$ (95.4% confidence) using OxCal v3.3 (1999). Updated from Steadman (1999a).

Lab number	Mass of carbon	Island	Site	Provenience	Radiocarbon age (yr BP)	Calibrated age (cal BP)
Beta-36074	~12 gm charcoal	Rota	Payapai Cave	Unit 2: II, 2	930 ± 70	960–690
CAMS-17397	1.31 mg charcoal	Rota	Payapai Cave	Unit 3: III, 4	400 ± 60	530–310
CAMS-17392	1.11 mg charcoal	Aguiguan	Pisonia Rockshelter	Unit 1: II, 5	540 ± 60	660–500
CAMS-32973	1.17 mg bone *Rattus exulans*	Aguiguan	Pisonia Rockshelter	Unit 2: IIIB, 6	790 ± 50	800–650
CAMS-17390	1.09 mg charcoal	Aguiguan	Pisonia Rockshelter	Unit 1: IIIC, 15	1480 ± 60	1520–1290
CAMS-17403	0.82 mg charcoal	Aguiguan	Pisonia Rockshelter	Unit 1: IV, 16	1520 ± 60	1530–1300
CAMS-17393	0.97 mg charcoal	Aguiguan	Pisonia Rockshelter	Unit 1: IV, 16	1780 ± 70	1870–1530
CAMS-32971	1.55 mg bone *Gallirallus* undesc. sp. N	Tinian	Railhunter Rockshelter	Unit 6: I, 2	1880 ± 50	1930–1700
CAMS-32972	1.54 mg bone *Gallirallus* undesc. sp. N	Tinian	Railhunter Rockshelter	Unit 6: IIA, 5	2110 ± 60	2310–2230, 2210–1920
CAMS-17394	1.00 mg charcoal	Tinian	Railhunter Rockshelter	Unit 4: IIB, 7	2200 ± 60	2350–2040
CAMS-17389	0.91 mg charcoal	Tinian	Railhunter Rockshelter	Unit 4: IIC, 8	2460 ± 60	2720–2350
CAMS-31063	0.87 mg bone *Porzana* undesc. sp. I	Tinian	Railhunter Rockshelter	Unit 4: III, 11	2420 ± 60	2720–2340

Table 8-6 Stratigraphic bone summary, Units 1–3, Payapai Cave, Rota. Dry-sieving in the field (1.6 mm mesh). i, introduced species; m, migratory species; *, extirpated; †, extinct. Taxa in brackets are not necessarily different from others identified more specifically. Totals do not include "Bird sp." Updated from Steadman (1992a, 1999a).

	Layer					
	Surface	I	II	III	Mixed	TOTAL
MAMMALS						
Rattus exulans (i)	—	30	16	7	1	54
SEABIRDS						
SHEARWATERS						
**Puffinus lherminieri*	—	12	11	36	4	63
TROPICBIRDS						
Phaethon rubricauda	1	—	—	—	—	1
Phaethon lepturus	41	22	1	1	5	70
TERNS						
**Procelsterna cerulea*	—	2	1	18	2	23
Gygis candida	1	5	—	1	—	7
LANDBIRDS						
HERONS						
Egretta sacra	4	60	9	—	3	76
DUCKS						
*Anatidae undescribed sp.	—	—	1	—	—	1
FALCONS						
**Falco peregrinus*	—	—	—	1	—	1
MEGAPODES						
**Megapodius laperouse*	—	3	2	9	—	14
RAILS						
**Poliolimnas cinereus*	—	—	1	1	—	2
*cf. *Porphyrio* undescribed sp. E	—	—	—	—	1	1
PLOVERS						
Pluvialis fulva (m)	—	—	—	1	—	1
SANDPIPERS						
Heteroscelus sp. (m)	—	—	—	1	—	1
PIGEONS, DOVES						
Ptilinopus roseicapilla	—	1	1	8	—	10
**Ducula oceanica*	—	1	—	2	—	3
Gallicolumba xanthonura	—	3	2	14	4	23
†*Gallicolumba* undescribed sp.	—	—	—	5	1	6
[Columbidae sp.]	—	—	—	3		3
PARROTS						
*Psittacidae sp.	—	—	1	—	—	1
SWIFTS						
**Collocalia vanikorensis*	—	—	4	—	1	5

Table 8-6 (continued)

	Layer					
	Surface	I	II	III	Mixed	TOTAL
WHITE-EYES						
Zosterops conspicillatus	—	2	—	7	—	9
STARLINGS						
Aplonis opaca	—	68	19	114	16	217
ESTRILDID FINCHES						
†*Erythrura* undescribed sp.	—	—	1	1	—	2
MONARCHS						
**Myiagra* cf. *freycineti*	—	—	—	2	—	2
HONEYEATERS						
Myzomela rubrata	—	—	1	4	—	5
[Passeriformes sp.]	—	1	4	13	—	18
[Bird sp.]	—	187	45	193	—	425
TOTAL BIRD BONES						
All species	47	180	59	242	37	565
Resident landbirds	4	139	46	184	26	399
†/*resident landbirds	0	4	10	21	3	38
TOTAL SPECIES						
Seabirds	3	4	3	4	3	5
*Seabirds	0	2	2	2	2	2
Resident landbirds	1	7	11	12	6	17
†/*resident landbirds	0	2	6	7	3	11

seas from a boat at one place on the leeward side. Previous archaeological work on this uninhabited island was limited to a surface survey in 1990 (Butler [1991 or later]). In 1994, I excavated Pisonia Rockshelter, a human-habitation site with well-stratified sediments 2.2 m deep (Chapter 3). The upper stratum (Layer I) is historic in age, whereas Layers II–IV are pottery-rich, prehistoric cultural deposits with ^{14}C dates that range from 660–500 cal BP (Layer II) to 1870-1530 cal BP in Layer IV, the primary bone-bearing stratum (Table 8-8). Layer V may have been deposited in part before people occupied the site, based on a sharp decline in pottery (which may be intrusive from Layer IV), a lack of charcoal, and a scarcity of burned bones.

Considering all taxa, 3600+ bones were recovered at Pisonia Rockshelter (Steadman 1999a). Each category of bones is scarce in Layers I–III, with those of fish and rat most frequent. In Layer IV, bones of fish and birds become abundant. Among bird bones, Layer IV yielded 748 (75%) of the 994 identified specimens dry-sieved from Units 1–3 (Table 8-8). This value is conservative because many bird bones from Layer IV are subsumed in the category "IV–V" in Table 8-8, these being from a partially collapsed area of the excavation wall that had to be salvaged in such a way that Layers IV and V could not be distinguished. Essentially all nonhuman bones from Layer IV are burned. Bones of *Rattus exulans* do not occur in Layers IV and V. They drop out within Layer III, being present in its upper portion (IIIA, IIIB) but absent in IIIC. The ^{14}C dates (Table 8-5) suggest that rats arrived on Aguiguan from ca. 1300 and 800 cal BP, or roughly 1000 years ago.

The nearly 1000 identified landbird bones at Pisonia Rockshelter represent just nine species. I believe that this low richness value reflects the specialized hunting of ground-dwelling birds (especially rails) by prehistoric people rather than a genuinely depauperate assemblage of forest birds on Aguiguan. That snares were the primary way to capture birds is suggested by the abundance of bones from ground-dwelling species; rails and the ground-dove

Table 8-7 Bone summary (rodents, birds only) from sites on Rota other than Units 1–3 of Payapai Cave. U4PC, Unit 4 at Payapai Cave (dry-sieving in the field, 1.6 mm mesh); PCSB, Scott Bauman samples 1 and 2 at rear of Payapai Cave (wet-sieving in the lab, 0.60 mm mesh); CNPC, Crevice 150 m N of Payapai Cave (dry-sieving in the field, 1.6 mm mesh); AR, Alaguan Rockshelter (dry-sieving in the field, 1.6 mm mesh); AMCC, As Matmos Cliffside Cave (dry-sieving in the field, 1.6 mm mesh); OS, other sites (unnamed caves 1–3 of Steadman 1992a); MO, Mochong site (excavated by J. L. Craib). i, introduced species; m, migratory species; ND, no data; *, extirpated; †, extinct. Updated from Steadman (1999a).

	Site							
	U4PC	PCSB	CNPC	AR	AMCC	OS	MO	TOTAL
RODENTS								
Rattus exulans (i)	—	10	11	—	7	—	ND	28
Rattus rattus (i)	1	—	—	1	—	12	ND	14
SEABIRDS								
SHEARWATERS								
**Puffinus lherminieri*	—	—	—	—	1	—	—	1
TROPICBIRDS								
Phaethon lepturus	10	—	2	1	—	—	17	30
BOOBIES								
Sula dactylatra	—	—	—	—	—	—	2	2
TERNS								
Anous stolidus	—	—	—	1	—	—	14	15
LANDBIRDS								
HERONS								
Egretta sacra	1	—	—	2	—	—	—	3
MEGAPODES								
**Megapodius laperouse*	—	—	—	—	3	—	13	16
RAILS								
†*Gallirallus* undescribed sp. L	—	—	—	—	—	—	4	4
†*Porzana* undescribed sp. G	—	—	—	1	1	—	—	2
**Gallinula chloropus*	—	—	—	—	—	1	—	1
PIGEONS, DOVES								
Ptilinopus roseicapilla	—	—	—	—	2	—	—	2
†*Gallicolumba* undescribed sp.	—	—	—	—	1	—	—	1
Gallicolumba xanthonura	—	—	1	—	2	—	4	7
SWIFTS								
**Collocalia vanikorensis*	2	—	—	—	1	—	—	3
WHITE-EYES								
Zosterops conspicillatus	1	1	—	—	2	—	—	4
**Cleptornis marchei*	—	—	—	—	2	—	—	2
STARLINGS								
Aplonis opaca	3	—	3	—	18	3	—	27
ESTRILDID FINCHES								
†*Erythrura* undescribed sp.	—	—	1	—	1	—	—	2
MONARCHS								
†*Myiagra* cf. *M. freycineti*	—	—	2	—	—	—	—	2

Table 8-7 (continued)

	Site							
	U4PC	PCSB	CNPC	AR	AMCC	OS	MO	TOTAL
HONEYEATERS								
Myzomela rubrata	—	—	3	—	3	—	—	6
[Passeriformes sp.]	2	4	3	—	8	—	—	17
TOTAL BONES								
All birds	19	5	15	5	45	4	54	147
Landbirds	9	5	13	3	44	4	21	99
†/*landbirds	2	0	3	1	9	1	17	33
TOTAL SPECIES								
Seabirds	1	0	1	2	1	0	3	4
Landbirds	4	1	5	2	11	2	3	15
†/*landbirds	1	0	2	1	6	1	2	9

Gallicolumba xanthonura make up >85% of all landbird bones at the site. Rails also dominate the bird assemblages at other early cultural sites throughout the Marianas, whether in open sites such as Mochong (Rota), Unai Chulu (Tinian), and Chalan Piao (Saipan), or in rockshelters such as Railhunter Rockshelter (Tinian). If I had been able to locate and excavate noncultural sites on Aguiguan, such as the various cliffside caves from Rota where avian predators accumulated bones of small landbirds, a broader range of landbirds probably would have been sampled on Aguiguan.

Nevertheless, the nine species of landbirds from Pisonia Rockshelter (based on 943 landbird bones) do not compare favorably with 14 species of landbirds based on 587 bones from Railhunter Rockshelter on Tinian (Tables 8-8, 8-9). One more factor makes me reluctant, however, to attribute the difference in species richness to island size (Aguiguan at 7.2 km^2 vs. Tinian at 102 km^2). This is that the bone deposit at Pisonia Rockshelter (Layer IV) is at least several hundred years younger than that at Railhunter Rockshelter (Layer II; Table 8-5). Assuming that these two islands were colonized at about the same time (3400 to 2800 cal BP based on the Unai Chulu site and sediment cores from Lake Hagoi on Tinian; Craib 1993, 1999, Athens & Ward 1998), then we have no evidence of the birds that may have been lost on Aguiguan during the first 1000 to 1500 years of human influence. Aguiguan's smaller size would have facilitated extinction.

The only extinct birds at Pisonia Rockshelter are flightless rails—*Gallirallus* undescribed sp. M (similar to the Guam Rail) and the much smaller *Porzana* undescribed sp. G. More study is needed to learn if the highly varied bones of *Porzana* from Pisonia Rockshelter represent a single, very polymorphic species or two different species, at least one of which was flightless.

Tinian

Signs of horrific wartime activities still exist in nearly every cave or rockshelter on Tinian in the form of gas masks, munitions, etc. The prehistoric potential of most sites was destroyed in the 1940s. An exception is Railhunter Rockshelter, excavated by International Archaeological Research Institute, Inc., in 1992 and by myself in 1994, yielding 600 identifiable bird bones (Steadman 1999a). Another major source of bird bones on Tinian is Unai Chulu, an open archaeological site excavated by Paul H. Rosendahl, Ph.D. Inc. in 1994. I have sorted 900+ bird bones from the fish-rich bone assemblage at Unai Chulu. These bones are not completely studied and thus are not tabulated here. As at Railhunter Rockshelter, the bird bones from Unai Chulu are dominated by rails.

Layer I at Railhunter Rockshelter is poorly stratified, gray organic silt with limestone pebbles and cobbles. Layer II is similar but more stratified and ashy, and less organic. Layer III is unstratified, light brownish orange silt with abundant landsnails and angular limestone pebbles, cobbles, and boulders. Minor variation in color and texture subdivide Layer II into IIA, IIB, and IIC. Layers I and especially II yielded abundant prehistoric pottery but no other prehistoric artifacts. Layer II, which dates from ca. 2600 to

Table 8-8 Stratigraphic bone summary (rodents, birds only), Units 1–3, Pisonia Rockshelter, Aguiguan. Dry-sieving in the field (1.6 mm mesh). i, introduced species; m, migratory species; *, extirpated; †, extinct. Taxa in brackets are not necessarily different from those identified more specifically. Updated from Steadman (1999a).

	Layer						
	I	II	III	IV	IV–V	V	TOTAL
RODENTS							
Rattus exulans (i)	10	2	19	—	—	—	31
SEABIRDS							
TROPICBIRDS							
Phaethon lepturus	—	—	—	—	—	1	1
FRIGATEBIRDS							
Fregata ariel	—	—	—	1	—	—	1
TERNS							
Anous stolidus	2	—	—	12	1	1	16
Gygis candida	—	—	—	5	—	1	6
LANDBIRDS							
PLOVERS							
Pluvialis fulva (m)	—	—	—	4	22	2	28
MEGAPODES							
Megapodius laperouse	—	—	—	10	—	—	10
RAILS							
†*Gallirallus* undescribed sp. M	—	—	5	197	21	16	239
†*Porzana* undescribed sp. H	1	4	9	149	19	17	199
†/*[Rallidae sp.]	—	—	4	156	54	14	228
PIGEONS, DOVES							
Ptilinopus roseicapilla	—	—	—	9	1	—	10
Gallicolumba xanthonura	5	—	2	105	11	14	137
[Columbidae sp.]	—	—	—	82	1	12	95
WHITE-EYES							
Zosterops conspicillatus	—	—	—	2	—	1	3
Cleptornis marchei	—	—	—	—	1	—	1
STARLINGS							
Aplonis opaca	—	—	1	11	—	1	13
HONEYEATERS							
Myzomela rubrata	—	—	—	1	—	—	1
[Passeriformes sp.]	—	1	—	4	1	—	6
[Bird sp.]	—	—	12	451	50	69	582
TOTAL BONES							
All birds	8	5	33	1199	182	149	1576
All birds without Bird sp.	8	5	21	748	132	80	994
Resident landbirds	6	5	21	726	109	75	943
†/*resident landbirds	1	4	18	502	94	47	666
TOTAL SPECIES OF BIRDS							
All species	3	2	4	12	7	9	14
Seabirds	1	0	0	3	1	3	4
Resident landbirds	2	2	4	8	5	5	9
†/*resident landbirds	1	1	2	2	2	2	2

1. Two species may be represented.

Table 8-9 Stratigraphic bone summary (rodents, birds only), Units 4–6 combined, Railhunter Rockshelter (site 131), Tinian. Dry-sieving in the field (1.6 mm mesh) and wet-sieving in the field (1.6 mm mesh) and laboratory (0.60 mm mesh) combined. Layer I is youngest (uppermost); Layer III is oldest (lowermost). i, introduced species; m, migratory species; *, extirpated; †, extinct; i not included in species totals. Taxa in brackets are not necessarily different from those identified more specifically. Updated from Steadman (1999a).

	Layer					
	I	IIA	IIB	IIC	III	TOTAL
RODENTS						
Rattus exulans (i)	15	—	—	—	—	15
Rattus rattus (i)	3	—	—	—	—	3
SEABIRDS						
TERNS						
Sterna sp.	—	—	1	—	—	1
Anous minutus	—	1	2	—	—	3
Anous stolidus	—	2	3	5	—	10
**Procelsterna cerulea*	1	2	1	—	—	4
Gygis candida	1	11	4	4	—	20
**Gygis microrhyncha*	—	—	2	—	—	2
LANDBIRDS						
PLOVERS						
Pluvialis fulva (m)	—	1	3	—	—	4
MEGAPODES						
Megapodius laperouse	—	1	—	1	—	2
CHICKENS						
Gallus gallus (i)	1	—	—	—	—	1
RAILS						
†*Gallirallus* undescribed sp. N	13	49	82	58	1	203
†*Porzana* undescribed sp. I	7	33	92	115	5	252
**Poliolimnas cinereus*	1	3	4	1	—	9
†/*[Rallidae sp.]	4	3	3	3	1	14
PIGEONS, DOVES						
Ptilinopus roseicapilla	1	3	2	1	1	8
*cf. *Ducula oceanica*	—	1	—	—	—	1
Gallicolumba xanthonura	4	14	16	20	—	54
SWIFTS						
**Collocalia vanikorensis*	1	—	—	—	—	1
WARBLERS						
**Acrocephalus luscinia*	—	—	3	2	—	5
WHITE-EYES						
Zosterops conspicillatus	1	1	—	—	1	3
**Cleptornis marchei*	1	—	—	—	—	1
STARLINGS						
Aplonis opaca	1	1	7	7	—	16

(continued)

Table 8-9 (continued)

	Layer					
	I	IIA	IIB	IIC	III	TOTAL
FANTAILS						
Rhipidura rufifrons	1	—	—	—	—	1
HONEYEATERS						
Myzomela rubrata	—	—	1	—	—	1
[Passeriformes sp.]	1	6	4	4	1	16
[Bird sp.]	16	75	106	75	9	281
TOTAL BONES						
All birds	55	207	336	296	19	913
All birds without Bird sp.	39	132	230	221	10	632
Resident landbirds	36	115	214	212	10	587
†/*resident landbirds	27	89	184	179	7	486
TOTAL SPECIES						
All birds	12	14	15	10	4	21
Seabirds	2	4	6	2	0	6
Resident landbirds	10	9	8	8	4	14
†/*resident landbirds	5	4	4	4	2	7

2000 cal BP (Table 8-5), produced 92% (583 of 632) of the site's bird bones. Major faunal changes, such as the absence of rats and an increase in bones of extinct birds, further distinguish Layer II from the overlying Layer I (Table 8-9). Unlike Layers II or III, Layer I also contained historic (20th-century) glass, gas masks, rusted iron, etc. Layer III lacked ash, charcoal, or any other cultural materials. Because the single ^{14}C date on bone from Layer III (2720–2340 cal BP) is virtually identical to that from Layer IIC (Table 8-5), it is likely that the few bird bones from Layer III were deposited when people first occupied Railhunter Rockshelter.

Most bones in Layer II, the primary cultural stratum, are burned. Railhunter Rockshelter is unique among archaeological sites in Oceania in that, using careful techniques of fine-mesh sieving, more bones were recovered of birds than any other class of vertebrates. Fish are typically the most abundant bone category in culturally derived faunal assemblages across Remote Oceania, including Tinian's coastal site of Unai Chulu. As at another inland site, Pisonia Rockshelter on Aguiguan, the people who lived at Railhunter Rockshelter hunted mostly ground-dwelling birds, especially rails.

At Railhunter Rockshelter bones of nonnative vertebrates are confined to Layer I (the prehistorically introduced Pacific rat *Rattus exulans* and the historically introduced toad *Bufo marinus*, black rat *Rattus rattus*, dog *Canis familiaris*, and chicken *Gallus gallus*). No records of pig (*Sus scrofa*) or goat (*Capra hircus*) from Tinian or elsewhere in the Marianas are from a securely prehistoric context. The only nonnative, nonvolant mammal with solid evidence of being in the Marianas prehistorically is *Rattus exulans*. The absence of nonnative mammals at European contact is supported by this statement by Fray Juan Pobre, who resided on Rota in AD 1602: "There are so many rats in these islands, however, that they destroy more than half of what is planted. There are no other animals and very few birds on any of the islands" (Driver 1989:12).

The arrival of *Rattus exulans* in the Marianas did not occur until >2000 years after human colonization. Rat bones are absent from the concentrations of extinct birds on both Tinian (Layer II of Railhunter Rockshelter) and Aguiguan (Layer IV of Pisonia Rockshelter). The absence of *R. exulans* for most of prehistory, and the absence of dog, pig, and chicken throughout prehistory, probably explains the large time lag in the Marianas between human arrival (ca. 3400 cal BP) and the extinction of most flightless rails and other landbirds. I will develop this idea further in Chapter 16.

The birds from Railhunter Rockshelter consist of six species of terns, one migrant shorebird, and 14 resident landbirds (Table 8-9), including seven of Tinian's eight extinct or extirpated species. (The other is the swamphen *Porphyrio* undescribed sp. E from Unai Chulu). The rails *Gallirallus* undescribed sp. N and *Porzana* undescribed sp. I each have >200 bones at Railhunter Rockshelter. The other five extirpated species (the rail *Poliolimnas cinereus*, pigeon cf. *Ducula oceanica*, swift *Collocalia vanikorensis*, warbler *Acrocephalus luscinia*, and white-eye *Cleptornis marchei*) are represented by only 17 bones altogether.

Saipan

The only bird bones from Saipan are 14 specimens from Chalan Piao, an open site excavated by Micronesian Archaeological Research Services in 1995. Seven are leg or foot bones of the Saipan form of the flightless Guam-like rail (*Gallirallus* undescribed sp. O). Four bones are of the ground-dove *Gallicolumba xanthonura*, and one is of the starling *Aplonis opaca*. Finally, there are two bones of migratory curlews—a carpometacarpus referred only to the genus *Numenius*, and an ulna of *N. arquata*. The earliest reliable ^{14}C dates for people on Saipan (Chalan Piao and Achugao sites) are ca. 3400 cal BP (Butler 1994, Craib 1999). Saipan has little mature native forest today but had, up until at least a decade ago, very high overall densities of landbirds (Craig 1996). Because the brown tree snake has been found regularly beginning in the 1990s (Rodda et al. 1997), Saipan probably is facing a Guam-like extinction crisis during the next few decades.

Biogeography and Extinction in the Marianas

Of the 54 prehistoric island records of landbirds, 29 represent lost species or populations (Table 8-10). The richest sites on Aguiguan and Tinian are primarily cultural whereas most sites on Rota are noncultural. The cultural sites typically have more bones of fish, fruit bats, seabirds, and ground-dwelling birds (megapodes, rails, columbids), whereas the noncultural sites feature more remains of lizards, small bats, and passerines. Rota, Aguiguan, and Tinian each have at least one site with hundreds of bird bones. The prehistoric bird records of Guam and Saipan are very limited.

Vertical distribution of bones from extinct or extirpated species is strikingly different in the three most stratified sites. Such bones are uncommon (ca. 20% or less) at Payapai Cave, a relatively young noncultural site with many bones of herons, doves, and starlings but few of rails (Figure 8-5). A similar uncommoness is found in Layer I of Pisonia Rockshelter, where, however, the percentage of bones from extinct or extirpated species skyrockets to 80% in Layer II (ca. same age as Payapai Cave; Table 8-5) and remains high through Layers III, IV, and V as well. At Railhunter Rockshelter, the value is consistently 70 to 80%. Unlike at the other two sites, no strata at Railhunter Rockshelter postdate the extinction or extreme rarity of rails, which commenced only after *Rattus* arrived between 1200 and 800 cal BP.

The prehistoric landbirds of the Marianas include several extinct species and many range extensions of locally extirpated species. A flightless duck (Anatidae undescribed sp.) is known only from one juvenile coracoid from Rota. Bones of the megapode *Megapodius laperouse* have been found on four islands. This species is extirpated on Guam and Rota, is nearly so on Tinian, and survives on Aguiguan and Saipan. Each of the five limestone islands was inhabited by an endemic Guam-like rail (*Gallirallus* spp. L–O) closely related to the Guam Rail *Gallirallus owstoni*, which thrives in captivity but has been gone from the wild since the mid-1980s. At least five species of rails (in the genera *Gallirallus*, *Porzana*, *Poliolimnas*, *Porphyrio*, *Gallinula*) have been lost on Rota (Steadman 1992a, 1999a). If the fossil record were complete, each island probably would be found to have sustained a species in each of these genera. The species-level systematics of *Gallirallus*, *Porzana*, and *Porphyrio* in the Marianas is under study but has not yet been reconciled.

The large species of *Gallicolumba* from Rota is the first extinct species of ground-dove from Micronesia, although others are known from New Caledonia and East Polynesia (Balouet & Olson 1989, Steadman 1992a; Chapters 5, 7, 11 herein). Bones of the pigeon *Ducula oceanica* from Tinian and Rota suggest that this canopy frugivore, which is widespread in Micronesia (Amadon 1943) but now absent from the Marianas, once lived through much of the island group. The extinct large parrot from Rota is known only from one tibiotarsus that may represent *Cacatua* or *Eclectus* (Steadman 1992a). The swiftlet *Collocalia vanikorensis* was lost on Tinian and Rota within recent decades (Engbring et al. 1986). Across the Pacific, bones of *Collocalia* occur regularly in noncultural cave deposits but are scarce or absent in cultural sites (Steadman 2002b), which might explain their absence from sites on Aguiguan and Saipan.

Fossils also document many range extensions for living species of passerines, none of which is likely to be endemic to a single island in the Marianas. Losses of island populations, in historic and prehistoric times, have led to one distributional artifact after another. For example, the monarch *Monarcha takatsukasae* is not endemic to Tinian (contra Balis-Larsen & Sutterfield 1997) but lived historically on Saipan (Peters 1996). If Aguiguan had more fossil passerines, I would bet that *M. takatsukasae* would be found there as well. Similarly, the flycatcher *Myiagra freycineti* is not endemic to Guam, as traditionally believed, but once lived

Table 8-10 Prehistoric records of native resident landbirds from the five major raised limestone islands in the Mariana Islands. Data from Steadman (1992a, 1999a, herein). Modern status is from Pratt et al. (1987), supplemented by Steadman (1999a). E, extinct species; e, extirpated species (survives elsewhere); P, prehistoric record of locally extant species.

	Guam	Rota	Aguiguan	Tinian	Saipan
HERONS					
Egretta sacra	P	P	—	—	—
DUCKS					
Anatidae undescribed sp.	—	E	—	—	—
FALCONS					
Falco peregrinus	—	e	—	—	—
MEGAPODES					
Megapodius laperouse	e	e	P	P	—
RAILS					
Gallirallus owstoni	e	—	—	—	—
Gallirallus undescribed spp. L–O	—	E	E	E	E
Porzana undescribed spp. G–I	—	E	E	E	—
Poliolimnas cinereus	—	e	—	e	—
Porphyrio undescribed sp. E	—	E/e	—	E/e	—
Gallinula chloropus	—	e	—	—	—
PIGEONS, DOVES					
Gallicolumba undescribed sp.	—	E	—	—	—
Gallicolumba xanthonura	—	P	P	P	P
Ptilinopus roseicapilla	—	P	P	P	—
Ducula oceanica	—	e	—	e	—
PARROTS					
Psittacidae undescribed sp.	—	E	—	—	—
SWIFTS					
Collocalia vanikorensis	e	e	—	e	—
KINGFISHERS					
Halcyon sp.	—	—	—	P	—
WARBLERS					
Acrocephalus luscinia	—	—	—	e	—
WHITE-EYES					
Zosterops conspicillatus	—	P	P	P	—
Cleptornis marchei	—	e	P	e	—
STARLINGS					
Aplonis opaca	P	P	P	P	P
ESTRILDID FINCHES					
Erythrura undescribed sp.	—	E	—	—	—
FANTAILS					
Rhipidura rufifrons	—	—	—	P	—
MONARCHS					
Myiagra cf. *M. freycineti*	—	E	—	—	—
HONEYEATERS					
Myzomela rubrata	—	P	P	P	—
TOTAL SPECIES					
P	2	6	7	8	2
E + e	3	15	2	8	1
P + E + e	5	21	9	16	3
Number of identified bones	23	475	944	647 (~1600)	12

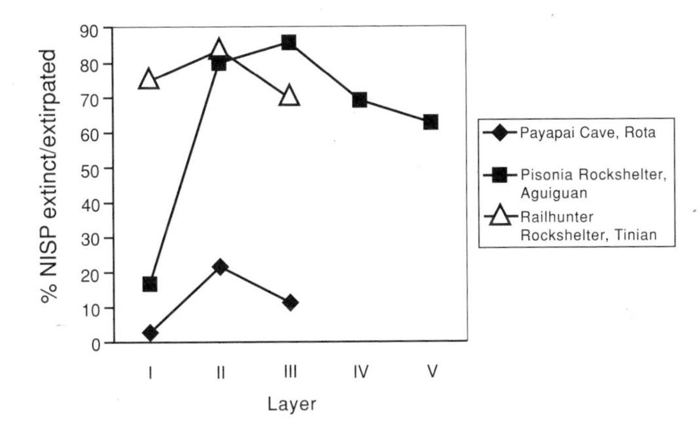

Figure 8-5 Stratigraphic comparison of the percentage of landbird bones (NISP, number of identified specimens) that represent extinct/extirpated species from three sites in the Mariana Islands.

on Rota. The reed-warbler *Acrocephalus luscinia*, known in the nonvolcanic Marianas from Guam, Aguiguan, and Saipan (Reichel et al. 1992), occurred prehistorically on Tinian. The enigmatic passerine *Cleptornis marchei* has been thought to be endemic to Saipan and Aguiguan, with an illogical absence on Tinian. I have identified bones of *C. marchei*, however, on Tinian as well as Rota. Lastly, the large, extinct parrotfinch (*Erythrura* undescribed sp.) from Rota undoubtedly lived as well on other islands. The nearest congeneric species is *E. trichroa*, with populations on Palau, Chuuk, Pohnpei, and Kosrae (Pratt et al. 1987:316).

The sites on Rota have produced more species of landbirds than any other island in the Marianas, including all but four of the species recorded in modern times. Until less than 1000 years ago, Rota was inhabited by at least 15 species of landbirds that no longer live there. If its fossil record were complete, Rota's landbirds probably would number 30 to 35 species, compared to the 10 resident species that persist today. I do not believe that Rota had an unusually rich avifauna or that it has been subjected to more extinction than nearby islands; it simply has a better fossil record. Before human impact, representatives of most if not all of the volant species or superspecies of Mariana Island landbirds probably occurred on all five major limestone islands (Guam north through Saipan). Flightless species would have been endemic to single islands, with closely related congeners replacing each other geographically.

On each island ranging in size from Guam (554 km^2) down to Aguiguan (7.2 km^2), I believe therefore that species richness values were very similar before human impact (Chapter 19) and included each genus known today or prehistorically from the group (Tables 8-1, 8-4). This includes 10 genera and species of passerines. This is worth considering when interpreting the modern foraging strategies of passerines (Craig 1989, 1990, 1992b, Craig et al. 1992b) in terms of potential competitors in similar feeding guilds. Ecologically similar species pairs of passerines that once were sympatric have not been so for centuries because of human-caused extirpations. Examples can be gleaned from Table 8-4. How these losses influenced the foraging (and nesting) ecology that we can observe today is a potentially rich source of speculation.

As more fossils are found in the Marianas, they are likely also to represent genera now living elsewhere in Micronesia (Palau, Yap, Chuuk, or Pohnpei). Possibilities include *Nycticorax* (night-herons), *Rallina* (crakes), *Caloenas* ("Nicobar" pigeons), *Trichoglossus* (lorikeets), *Otus* [*Pyrroglaux*] (small scops-like owls), *Coracina* (cuckoo-shrikes), *Cettia* (bush-warblers), *Colluricincla* (morningbirds), and *Rukia* or *Megazosterops* (white-eyes).

Chuuk (Truk)

The almost-atoll (11 eroded vocanic islands surrounded by a single massive reef) of Chuuk (127.4 km^2; formerly called Truk; Figure 8-6) has been a household word for six decades because of World War II. The 60 sunken ships in Truk Lagoon still attract many divers each year. (The 17 species of landbirds in Chuuk lure almost no visitors). The largest and highest island in Truk Lagoon is Tol (34.2 km^2, 439 m elev.). Also included in Chuuk State (Federated States of Micronesia; Figure 8-1) are 10 outlying atolls (Nicholson 1969).

Four genera of landbirds (*Nycticorax*, *Ixobrychus*, *Anas*, *Metabolus*) occur in Chuuk but not to the east in Pohnpei or Kosrae (Table 8-11). Whether this is due to the greater isolation of Pohnpei and Kosrae, or is an artifact of human impact, awaits a fossil record of landbirds, which now is lacking on Kosrae, limited on Chuuk to 13 unidentified bird bones from Moen Island (King & Parker 1984:220), some chicken (*Gallus gallus*) bones on Tol Island (Takayama & Seki 1973, Takayama & Intoh 1978), and restricted on Pohnpei to five bones (see below). Chuuk lies closer than Pohnpei or Kosrae to Palau and the Marianas, which have much richer avifaunas. Pohnpei is ca. 780 km east of Chuuk yet has a somewhat richer modern avifauna. These two facts together suggest that considerable avian extinction awaits discovery on Chuuk, an island group that retains little if any mature upland forest. Among the genera now missing from Chuuk, but to be expected there as fossils, are *Megapodius*, *Gallirallus*, *Porzana*, *Trichoglossus*, *Asio*, *Halcyon*, *Coracina*, and *Rhipidura*. I also believe that a second species of starling (*Aplonis*) lived in Chuuk, as is or was the case on Pohnpei and Kosrae.

Three species of passerines are endemic to Chuuk (Table 8-11). The monarch flycatcher *Myiagra oceanica* and

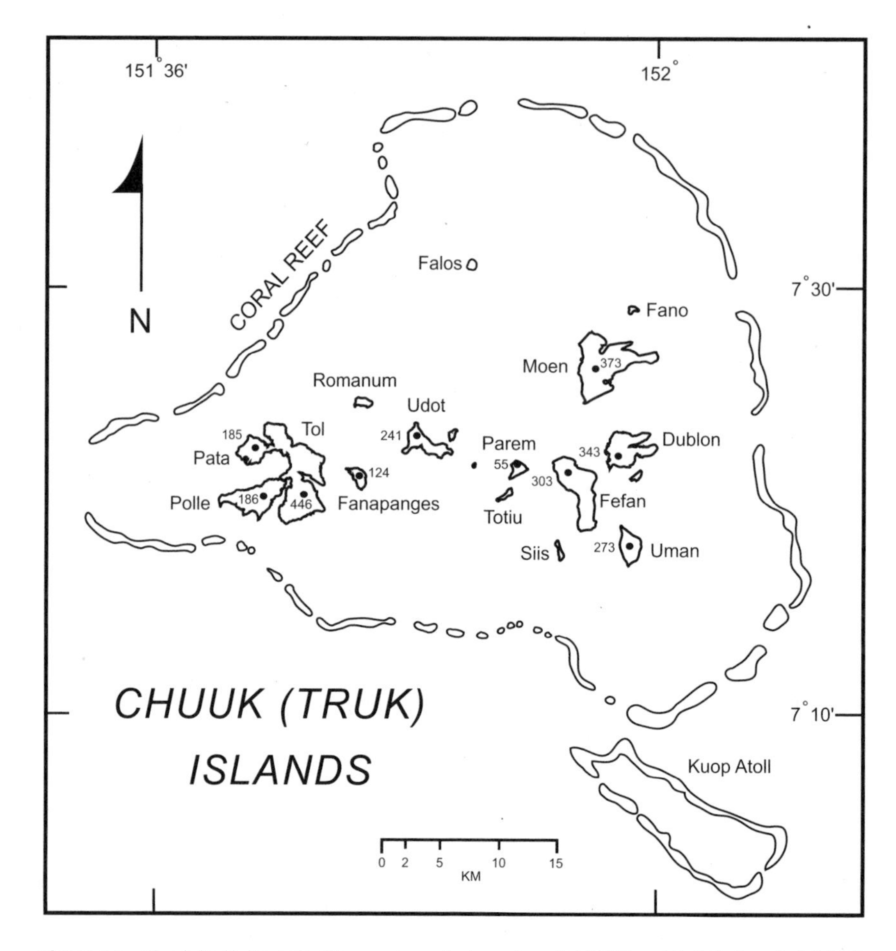

Figure 8-6 Chuuk (Truk). Spot elevations are in m. Redrawn from Tol (1983) and Truk-Eastern Part (1996).

white-eye *Rukia ruki* (see Pyle & Engbring 1988) are part of widespread radiations in Micronesia. The monarch *Metabolus rugensis* is regarded as an endemic genus, although it seems to me, based only on skins, to be closely related to or congeneric with *Monarcha*, which occurs in Yap and the Marianas (Table 8-1). The only atoll in Chuuk State with respectable data on modern landbird distribution is Namoluk, which has a small subset of the species found on Chuuk proper (Marshall 1977; Table 8-11 herein).

Pohnpei (Ponape)

This young, wet volcanic island (Gulick 1858, Merlin & Juvik 1995) has a land area (337 km^2) exceeded in Micronesia only by Guam (Marianas) and Babeldaob (Palau). Pohnpei (Figures 8-1, 8-7) also is Micronesia's second highest (772 m) island, exceeded by volcanic Agrihan (965 m) in the Northern Marianas. Pohnpei State of the Federated States of Micronesia includes eight atolls ranging from the nearby Ant and Pakin (Buden 1996b, 1996c) to the very remote Polynesian outliers of Nukuoro and Kapingamarangi to the south (Niering 1963, Merlin et al. 1992, Buden 1998).

The survival of a relatively rich landbird fauna on Pohnpei (Table 8-11) is due in part to its large size, mountainous interior, and having been spared battle in World War II. The aquatic/semiaquatic avifauna is poorly developed, with only the heron *Egretta sacra* and the rail *Poliolimnas cinereus*. The much richer upland avifauna features four endemic passerines, including the rare starling *Aplonis pelzelni* (Buden 1996a, 2000). The parrot *Trichoglossus rubiginosus* is confined to Pohnpei and nearby atolls. Five genera are on Pohnpei but not Chuuk or Kosrae—*Trichoglossus*, the owl *Asio* (see Mayr 1933), kingfisher *Halcyon*, cuckoo-shrike *Coracina*, and fantail *Rhipidura*.

Pohnpei's scant avian fossil record features 45 bones from cultural sites in or near Nan Madol, excavated by Williams Ayres and colleagues in the 1980s and 1990s. The overall fauna from these sites was reported by Kataoka (1990). Nan Madol, on the eastern coast, is famous for its spectacular prehistoric structures made of columnar basalt pillars (Morgan 1988). The nonnative chicken *Gallus gallus*

Table 8-11 The distribution of landbirds on Chuuk (Truk), Pohnpei (Ponape), Kosrae (Kusaie), and nearby well-surveyed atolls. e, endemic species; H, historic record, now extirpated; M, modern record; P, prehistoric record; †, extinct species; *, extirpated species; ?, questionable record, or residency not established. From Baker (1951), Marshall (1977), Pratt et al. (1987), Engbring et al. (1990), Buden (1996b, 1996c, 2000), and Steadman (1999a).

	Chuuk	Namoluk Atoll	Pohnpei	Ant Atoll	Pakin Atoll	Kosrae
HERONS						
Egretta sacra	M	M	M, P	M	M	M
Nycticorax caledonicus	M	—	—	—	—	—
Ixobrychus sinensis	M	H?	—	—	—	—
DUCKS						
Anas superciliosa	M	M	—	—	—	—
MEGAPODES						
**Megapodius* cf. *laperouse*	—	—	P	—	—	—
RAILS						
†*Porzana monasa* (e)	—	—	—	—	—	H
Poliolimnas cinereus	M	—	M	—	—	—
PIGEONS, DOVES						
Ptilinopus porphyraceus	M	—	M	—	—	M
Ducula oceanica	M	H	M	M	M	M
Gallicolumba kubaryi (e)	M	—	M	—	—	—
PARROTS						
Trichoglossus rubiginosus (e)	—	H	M, P	M	—	—
OWLS						
Asio flammeus	—	—	M	—	—	—
SWIFTS						
Collocalia vanikorensis	M	—	M	—	—	M
KINGFISHERS						
Halcyon cinnamomina	—	—	M	M	—	—
WARBLERS						
Acrocephalus syrinx (e)	M	M	M	M	M	H
WHITE-EYES						
Zosterops semperi	M	—	M	—	—	—
Zosterops cinereus	—	—	M	—	—	M
Rukia ruki (e)	M	—	—	—	—	—
Rukia longirostra (e)	—	—	M	—	—	—
STARLINGS						
Aplonis opaca	M	M	M	M	M	M
Aplonis pelzelni (e)	—	—	M	—	—	—
†*Aplonis corvina* (e)	—	—	—	—	—	H
ESTRILDID FINCHES						
Erythrura trichroa	M	—	M	—	—	M
CUCKOO-SHRIKES						
Coracina tenuirostris	—	—	M	—	—	—

(continued)

Table 8-11 (continued)

	Chuuk	Namoluk Atoll	Pohnpei	Ant Atoll	Pakin Atoll	Kosrae
FANTAILS						
Rhipidura kubaryi (e)	—	—	M	—	—	—
MONARCHS						
Metabolus rugensis (e)	M	—	—	—	—	—
Myiagra pluto (e)	—	—	M	—	—	—
Myiagra oceanica (e)	M	—	—	—	—	—
HONEYEATERS						
Myzomela rubrata	M	M	M	M	M	M
TOTAL SPECIES						
M	17	5	20	7	5	8
M + H	17	7–8	20	7	5	11
M + H + P	17	7–8	21	7	5	11
TOTAL BONES						
All species	0	0	48	0	0	0
Resident landbirds	0	0	5	0	0	0

is by far the most common species (30 bones). Ten other bones are of seabirds (Chapter 15). The only landbird bones, from Nan Madol itself, are of the heron *Egretta sacra* (ulna, tibiotarsus), megapode *Megapodius* cf. *laperouse* (tarsometatarsus; Steadman 1999a), and parrot *Trichoglossus rubiginosus* (coracoid, humerus). With its presence today in Palau and the Marianas, the record of *M.* cf. *laperouse* from Pohnpei means that a megapode probably once inhabited Yap and Chuuk as well. Bath (1984) reported several chicken bones and a columbid bone from the Sapwtakai site on Pohnpei's southwestern coast.

Recent surveys by Buden (1996b,c) hint at the potential of Pohnpei's atolls to sustain landbirds. Ant Atoll (12 islets, land area 1.86 km^2) lies 18.5 km southwest of Pohnpei. Pakin Atoll (21 islets, 1.09 km^2) lies 33 km west of Pohnpei. Ant and Pakin support seven and five species of landbirds (Table 8-11). These subsets of the Pohnpei landbird fauna include four genera (the parrot *Trichoglossus*, kingfisher *Halcyon*, honeyeater *Myzomela*, and starling *Aplonis*) unknown on atolls to the east in the Remote Central Pacific (Table 8-12). The closeness of Ant and Pakin atolls to Pohnpei might explain their relatively rich landbird faunas, given that two other atolls (Kapingamarangi 770 km south-southwest of Pohnpei, and Oroluk 325 km west-northwest of Pohnpei) are inhabited only by the heron *Egretta sacra* and starling *Aplonis opaca* (Niering 1963, Buden 1999). This notion needs to be tested, however, with bird bones currently lacking on all four atolls. Arguing against nearness to Pohnpei as the sole reason for enhanced species richness on Ant and Pakin atolls is the landbird fauna of Namoluk Atoll in Chuuk State (Marshall 1971; Table 8-11 herein), where seven or eight species are known from 0.83 km^2 of land that is 210 km from Chuuk itself (the nearest high islands).

Kosrae (Kusaie)

The easternmost and thus most isolated of Micronesia's eroded volcanic islands is Kosrae (109 km^2, elev. 629 m; Figure 8-8). Human settlement on Kosrae is restricted to the narrow coastal perimeter; the interior is extremely wet, steep, green, and lush (Merlin et al. 1993). Also included in Kosrae State of the Federated States of Micronesia (Figure 8-1) are Mokil and Pingelap atolls (Nicholson 1969).

No endemic species survive on Kosrae, which is famous nevertheless for the former presence of two such species—the rail *Porzana monasa* and starling *Aplonis corvina*, neither of which seems to have survived into the 20th century (Finsch 1881, Baker 1951, Greenway 1958, Engbring et al. 1990). *Porzana monasa* was flightless and had the same limb proportions as the extinct *P. rua* of Mangaia (Steadman 1987). It is the only modern record of *Porzana* in Micronesia, where the trampy *P. tabuensis*, so widely distributed in Melanesia and Polynesia (Taylor 1998), is absent.

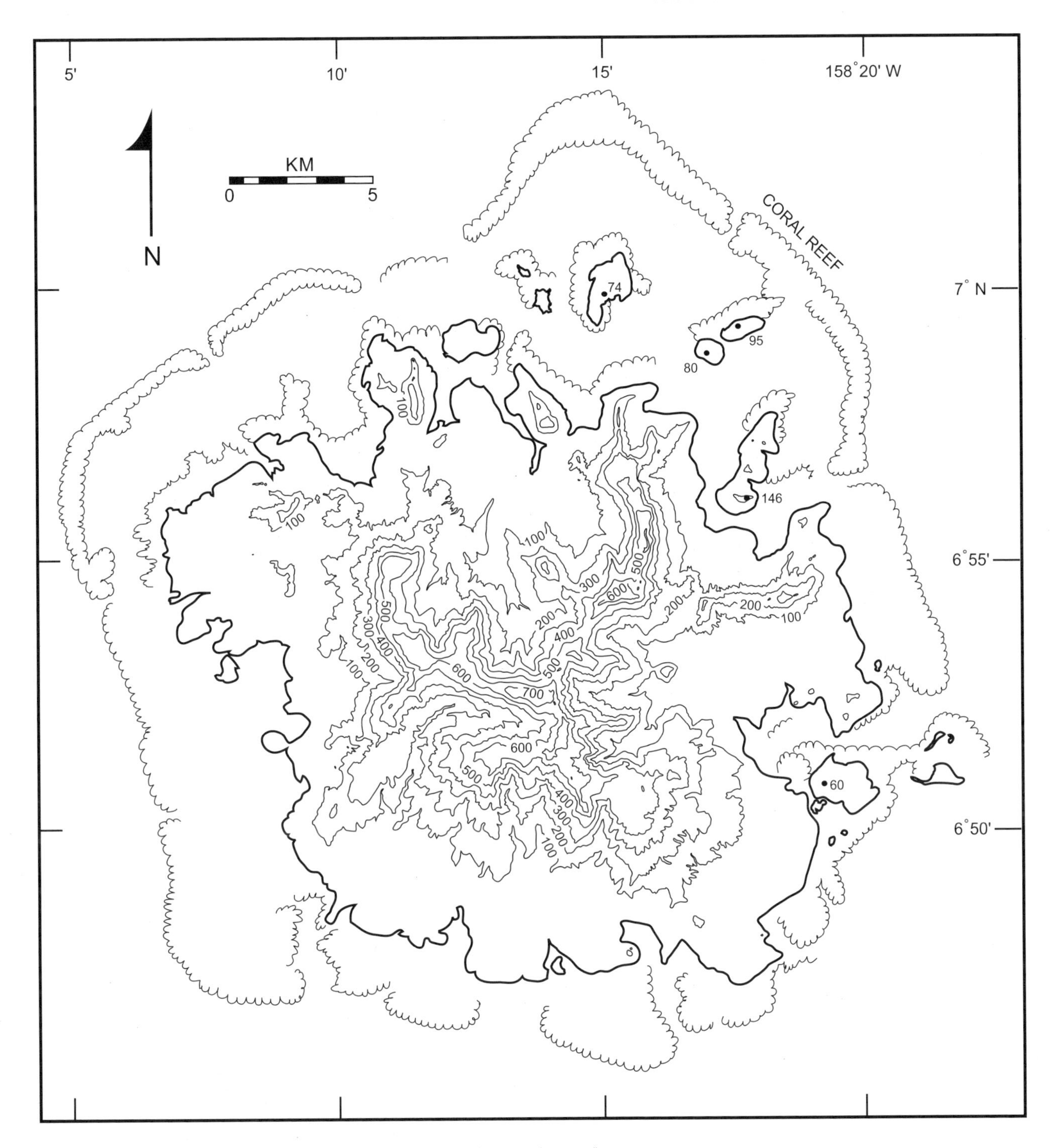

Figure 8-7 Pohnpei. Contour intervals are in m. Redrawn from Island of Pohnpei (2001a–d).

All surviving species of landbirds on Kosrae occur as well on Pohnpei and usually Chuuk (Table 8-11). Missing are 10 genera found today or in the past on Pohnpei. A fossil record from Kosrae would be fascinating to test how much of its avifaunal attenuation is due to isolation versus human impact.

Remote Central Pacific Islands

These widely scattered islands and island groups consist of numerous atolls and a few raised limestone islands (Dickinson 1999a; Figures 1-5, 1-33 herein). Coconuts dominate the vegetation on many atolls, in part reflecting historic

Table 8-12 Landbirds on Remote Central Pacific Islands. e, endemic species; H, historic record, now extirpated; M, modern record; P, prehistoric record; †, extinct species or subspecies. The Gilbert Islands, Phoenix Islands, and most of the Line Islands are united politically today as Kiribati. From Baker (1951), Wodzicki & Laird (1970), Pratt et al. (1987).

	Wake	Marshall Islands	Nauru	Gilbert Islands	Phoenix Islands	Line Islands	Tuvalu	Tokelau	TOTAL
HERONS									
Egretta sacra	—	M	M	M	M	M	M	M	7
DUCKS									
†*Anas strepera couesi*	—	—	—	—	—	H	—	—	1
RAILS									
Gallirallus philippensis	—	—	—	—	—	—	M	—	1
†*Gallirallus wakensis* (e)	H	—	—	—	—	—	—	—	1
SANDPIPERS									
†*Prosobonia cancellatus*	—	—	—	—	—	H	—	—	1
PIGEONS, DOVES									
Ptilinopus porphyraceus	—	H	—	—	—	—	—	—	1
Ducula pacifica	—	—	—	—	M	—	M	M	3
Ducula oceanica	—	M	M	M	—	—	—	—	3
WARBLERS									
Acrocephalus rehsei (e)	—	—	M	—	—	—	—	—	1
Acrocephalus aequinoctialis (e)	—	—	—	—	—	M	—	—	1
Acrocephalus sp. (e?)	—	P	—	—	—	—	—	—	1
TOTAL SPECIES									
M	0	2	3	2	2	2	3	2	5
M + H	1	3	3	2	2	4	3	2	8
M + H + P	1	4	3	2	2	4	3	2	9
Total landbird bones	0	1	0	0	0	0	0	0	1

copra plantations. The prehistoric inhabitants of these islands also cultivated coconuts as well as breadfruit, pandanus, taro, and sugar cane, a much more limited selection of crops than on high islands (Thaman 1990, Kirch 2000). The climate is influenced strongly by seasonal movements of the intertropical convergence zone and equatorial doldrum belt (Burgess 1987; Chapter 1 herein). Annual precipitation decreases to the east and north, where the atolls become increasingly prone to drought.

The only widespread landbird in the Remote Central Pacific Islands is the heron *Egretta sacra* (Table 8-12). Determining the original landbird faunas is a challenge in this vast, atoll-dominated region. No more than three species of landbirds occur today on any single island. Most islands have one or two species, or even none. For example, no resident landbirds are known from Johnston Atoll (Amerson & Shelton 1976), the most isolated island in this chapter (1440 km northwest of Palmyra in the Line Islands). Prehistory has shed little light on the landbird faunas at first human contact. One bone from the Marshall Islands (see below) is the entire fossil record of resident landbirds.

Wake

The endemic rail *Gallirallus wakensis* is the only landbird known from Wake, a large, isolated, high atoll (7.4 km^2, 6 m elev.) with three islets (Bryan 1959, Nicholson 1969, Pratt et al. 1987). This flightless rail did not survive military activities during World War II (Greenway 1958). The former existence of *G. wakensis* on Wake, similar to those of flightless species of *Porzana* on the atolls of Laysan (3.4 km^2) and Lisianski (1.7 km^2) atolls in the leeward Hawaiian Islands (Olson & Ziegler 1995), demonstrates that even

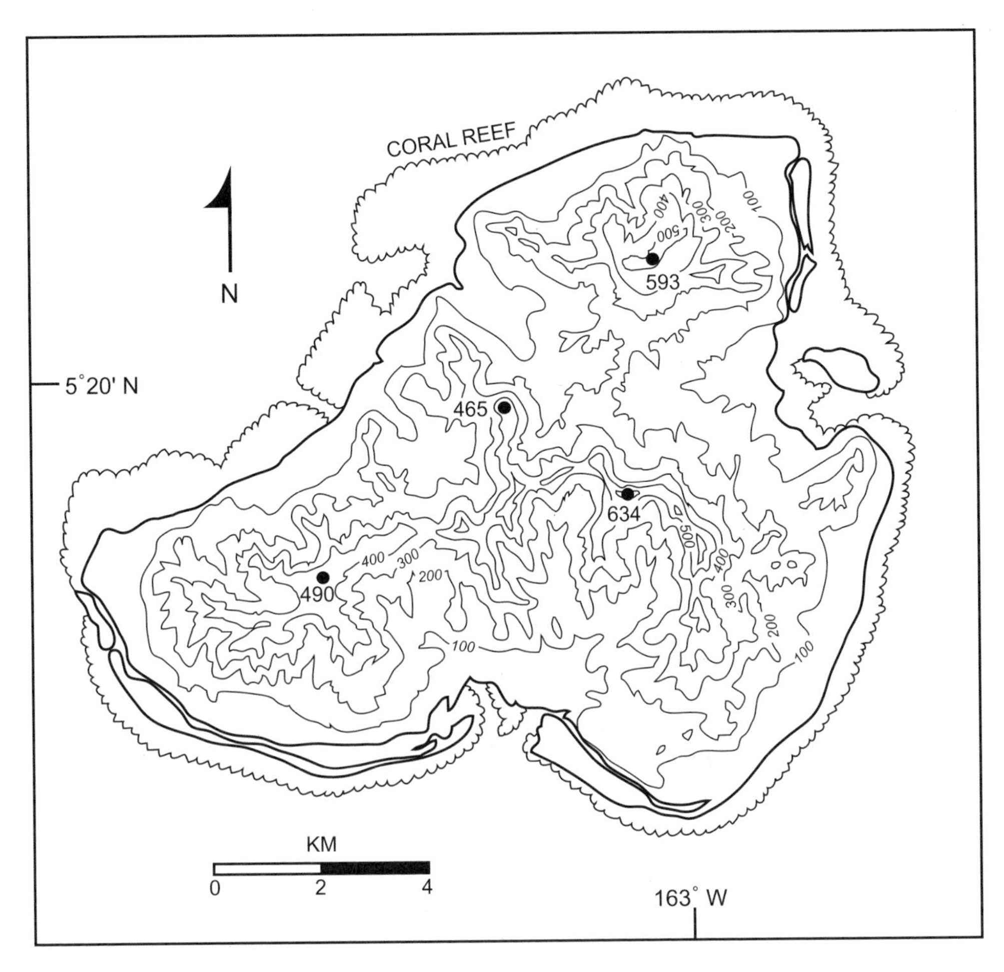

Figure 8-8 Kosrae. Contour intervals are in m. Redrawn from Jane's Oceania web page (www.janesoceania.com/micronesia_prehistory/index.htm) and Bier (1995).

atolls were capable of sustaining endemic flightless species of rails in at least two genera. The implications of this for evolution, extinction, and species richness will be explored in Chapters 10, 16, and 17. An important factor in sustaining endemic species is that Wake, unlike most atolls, was not inundated during the rapid early Holocene rise in sea level (Chapter 1).

Marshall Islands

A vast set of 29 atolls and five very small islands or "table-reefs" make up the Republic of the Marshall Islands (Nicholson 1969, Fosberg 1990, Merlin et al. 1994). With 2000+ individual islets, the archipelago consists of two clusters of islands, the eastern Radak Chain and western Ralik Chain. The largest Marshallese atoll is Kwajalein (16.3 km^2), which, like so much of Micronesia, was badly damaged by military activities before, during, or after World War II. Nearly all land in the Marshall Islands is below 5 m elevation. In this fragile setting live ca. 60,000 people, a five-fold increase in population since 1940 (Merlin et al. 1994). Not surprisingly, little if any unaltered forest exists in the Marshall Islands today (Fosberg 1953, 1990).

The heron *Egretta sacra* is widespread in the Marshalls and typically the only landbird present (Fosberg 1966, Pearson & Knudsen 1967). The other resident landbirds recorded are two widespread (but local) frugivorous columbids. The pigeon *Ducula oceanica* occurs or did occur on four Marshallese atolls (*D. o. oceanica* on Jaluit and Ailinglaplap, and the endemic *D. o. ratakensis* on Arno and Wotje; Finsch 1880a, 1880b, Peters 1937, Baker 1951, Amerson 1969, Pratt et al. 1987). The fruit-dove *Ptilinopus porphyraceus* formerly inhabited Ebon Atoll, based on a specimen collected in 1859 (Ripley & Birckhead 1942, Baker 1951, Amerson 1969:315). Fosberg (1990) recorded the native fig *Ficus tinctoria* from three atolls in the Marshalls, thus improving the likelihood that a fruit-dove once may have lived here. I agree with Baker (1951:184) that the fruit-dove from Ebon may have been "the final vestige of a formerly well-distributed population in the Marshall Islands." The same may be said of *D. oceanica.* I tentatively follow Amerson (1969:309) in regarding a single record of the rail *Poliolimnas cinereus* from Bikini Atoll as a vagrant. Similarly, a record (July 1973) of the owl *Asio flammeus* from Enewetak (Eniwetok) probably does not represent a resident population (Johnson & Kienholz 1975, Berger 1987).

A coracoid from the Laura archaeological site on Majuro Atoll (excavated by Marshall Weisler in 1995) represents a warbler (*Acrocephalus*), probably an undescribed species. Species of *Acrocephalus* are known from Pohnpei, Kosrae, Nauru, and Kirimati, so the past occurrence of a warbler in the Marshalls is expected.

Nauru

A raised limestone island (21.2 km^2, 65 m elev.), independent Nauru is fabled for rich phosphate deposits that have brought financial wealth at an environmental price (Manner et al. 1984, 1985, Thaman 1992, Thaman et al. 1994). Although phosphate mining has removed the soil and upper bedrock from about two-thirds of the island's surface, the pigeon *Ducula oceanica* and the endemic warbler *Acrocephalus rehsei* survive in unknown numbers on Nauru. Vertebrate fossils have not been found; it is likely that much of the island's fossil potential has been mined away.

Kiribati

The country of Kiribati consists of three equatorial island groups formerly known as the Gilbert Islands, Phoenix Islands, and Line Islands (in part). The Gilberts are Micronesian and consist of 16 atolls (land area 295 km^2) with a depauperate flora lacking endemic species (Thaman 1992). The raised limestone island of Banaba (or Ocean Island; 6.5 km^2, 81 m elev.) lies 400 km west of the Gilberts. Like that of Nauru 300 km to the west, most of Banaba's surface has been removed for phosphate. The largest Gilbertese atoll is Tabiteuea at 49 km^2.

The Phoenix Islands (8 atolls, land area 30 km^2) lie 1000 km east of Arorae, the easternmost island in the Gilberts. Uninhabited at European contact and generally today as well, many of these atolls have been mined for guano but retain substantial coastal forest (Nicholson 1969). Howland (1.9 km^2) and Baker (1.7 km^2) atolls belong to the United States and lie 500 km northwest of the Phoenix Islands.

Also uninhabited at European contact, the Line Islands consist of a southern group (5 atolls, 56.8 km^2) and northern group (6 atolls, 367 km^2), including the world's largest atoll, Kirimati or Christmas Island (321 km^2), renowned for extensive though highly variable seabird populations (Schreiber & Schreiber 1984). Sand dunes on Kirimati reach 11 m elevation, extraordinarily high for an atoll. Three atolls in the northern group (Jarvis, Palmyra, and Kingman Reef) belong to the United States. The vegetation varies from being dominated by coconut plantations and damaged by guano mining to being well forested (Dawson 1959). Polynesian archaeological sites in both the Phoenix and Line groups (Bellwood 1979:352) indicate that these remote atolls felt prehistoric human impact. Probably none of them was spared.

Landbirds in Kiribati are restricted to the heron *Egretta sacra* in all three island groups, an extirpated population of duck *Anas strepera couesi* on Teraina (Washington) Atoll in the Line Islands, an extinct sandpiper *Prosobonia cancellata* on Kirimati, the pigeon *Ducula pacifica* on unspecified islands in the Phoenix Group (Pratt et al. 1987), another pigeon *D. oceanica* extirpated from Kuria and Aranuka atolls in the Gilberts (Amerson 1969), and the endemic warbler *Acrocephalus aequinoctialis* on Kirimati, Tabuaeran (Fanning), and Teraina (Kirby 1925). Kirimati is the only island known to have supported as many as three species of landbirds. Like Wake (above), Kirimati was not inundated in the Holocene.

The extirpated duck (*Anas strepera couesi*) on Teraina was a small, female-plumaged, resident form of the gadwall *A. strepera*, which otherwise breeds in the temperate northern hemisphere. The Teraina population is known from two specimens, believed to be immatures, taken in 1874 (Streets 1876, Wetmore 1925, Greenway 1958, Weller 1980:9–10). Vagrant records for other north-temperate species of dabbling ducks (*Anas* spp.) are scattered through the Remote Central Pacific Islands (Baker 1951, Amerson 1969, Pratt et al. 1987). Whether any species other than *A. strepera* ever formed resident island populations can be determined only through fossils.

From Makin Atoll in the Gilbert Group, I identified 44 bird bones from the prehistoric Utiroa site, excavated by Jun Takayama in the 1980s. Dominated by seabirds (Chapter 15) and the nonnative chicken *Gallus gallus*, these bones include none of native landbirds.

Before human arrival I suspect that some or most islands in Kiribati sustained, aside from *Egretta sacra*, species of *Anas* (ducks), *Gallirallus* and *Porzana* (flightless rails), *Prosobonia* (sandpipers), *Ducula* (pigeons), *Halcyon* (kingfishers), *Acrocephalus* (warblers), and *Aplonis* (starlings). Given the high vulnerability of atolls to human impact, the loss of these hypothesized populations/species probably occurred shortly after people landed, thereby decreasing the chances of finding their bones.

Tuvalu

Formerly called the Ellice Islands, Tuvalu's nine small atolls (land area 28.5 km^2) stretch over 650 km of ocean. Tuvalu is inhabited by Polynesians. Coconuts dominate the woody vegetation on most atolls although some native forest is present as well (Nicholson 1969, Woodroffe 1985). The northernmost atoll (Nanumea) is 350 km south of Arorae Atoll in the Gilberts (Kiribati). The southernmost atoll in Tuvalu (Niulakita) lies 280 km northeast of Rotuma (a Fijian outlier; Chapter 6).

Two resident landbirds, the heron *Egretta sacra* and pigeon *Ducula pacifica*, are widespread in Tuvalu (Child 1960). The report of *D. pistrinaria* (North 1898) is an error. The rail *Gallirallus philippensis* colonized Niulakita in 1972 and established a breeding population (McQuarrie 1991). Thus three trampy species inhabit Tuvalu today. My guess at which landbirds lived in Tuvlau before human arrival resembles that for Kiribati (above), although I would add *Ptilinopus* (fruit-doves) to the list.

Arguing against a diverse landbird fauna, however, is the proposal by Dickinson (1999a) for a mid-Holocene (ca. 4000–3000 cal BP) sea level in Tuvalu that was 2.2–2.4 m higher than modern, due to both tectonic and eustatic causes. This would have eliminated the fresh water lens on most or all atolls, thus removing the terrestrial flora or reducing it to only the most salt-tolerant species. Such an atoll would be unsuited for columbids or starlings. Again, a fossil record could test this proposal. Bird bones from Tuvalu consist of 13 specimens I identified from the Temei site on Vaitupu, excavated by Jun Takayama in the 1980s. Only seabirds (Chapter 15) and chickens were found.

Tokelau

These three Polynesian atolls (land area 10.1 km^2) lie 800 km east of Tuvalu and 450 km north of Savai'i (Samoa). The vegetation is coconut plantation and coastal forest (Nicholson 1969). No resident landbirds are recorded from Tokelau in Pratt et al. (1987), although the heron *Egretta sacra* and pigeon *Ducula pacifica* occur on all three atolls (Amadon 1943, Wodzicki & Laird 1970). Before human impact, any of the taxa proposed for Tuvalu may have existed as well in Tokelau.

Part III

Chapter 9 Megapodes

THE MEGAPODES (also called mound-builders, incubatorbirds, and scrubfowl) are unique in that the parents do not directly incubate their eggs, which instead are warmed through burial in soils heated by decaying vegetation, volcanic activity, or the sun. Only the male attends the incubation chamber, which often is a conspicuous mound of soil and leaf litter. Uniquely among galliforms, megapodes (Figure 9-1) disperse well over the ocean. They colonized all of Near Oceania and much of Remote Oceania, although much of their distribution has been lost to prehistoric human activities.

Extant megapodes (seven genera and 22 species; Jones et al. 1995) are most diverse in eastern Indonesia, New Guinea, and Australia, with species of *Megapodius* on certain continental islands to the west (the Nicobars) and north (Philippines), and on oceanic islands north to Micronesia and locally eastward to Tonga (Figure 9-1). Olson (1980) proposed that the Greater Sundas (Sumatra, Borneo, Java) and mainland Asia lack megapodes because of competition with phasianids (pheasants, quail, partridges), whereas Dekker (1989) and Jones et al. (1995:25) suggested that predation by placental carnivores explains why megapodes have such a limited range west of Wallace's Line. I lean toward predation, especially given my lack of faith in competition as controlling species assemblages of megapodes (Chapter 20).

No matter what controls their western limit in island Southeast Asia, the natural distribution and diversity of megapodes have been poorly understood until recently at their eastern limit in Oceania, a region without phasianids or placental carnivores (Steadman 1999b). A growing fossil record is filling many puzzling gaps in the modern distribution of living species of megapodes; it also exposes their much greater taxonomic diversity in Oceania before people wiped out so many populations and species.

The fossil record of megapode-like birds begins with *Quercymegapodius depereti* and *Q. brodkorbi* from the late Eocene (ca. 40-36 Ma) of France (Mourer-Chauviré 1982, 1992, 1995). They and the late Oligocene–early Miocene (ca. 30–20 Ma) *Ameripodius* of France and Brazil portray an early Tertiary distribution drastically different from that of their modern counterparts (Mourer-Chauviré 2000). Although regarded as a distinct family (Quercymegapodiidae) from the Megapodiidae, the French fossils are much more similar to bones of living megapodes than to those of any other living or extinct family. Quercymegapodiids are one of many examples of avian families with early Tertiary fossils from the northern hemisphere but closest living relatives mainly or strictly in the southern hemisphere (Mourer-Chauviré 1995).

True megapodes (Megapodiidae) lived in Australia by the late Oligocene (ca. 28–25 Ma), as evidenced by the extinct genus *Ngawupodius* (Boles & Ivison 1999). No Miocene fossils of megapodes have been discovered from within the modern range of the family. Miocene avifaunas from mainland Asia include phasianids but no megapodes (Cheneval et al. 1991), suggesting (perhaps especially if you subscribe to the competition-with-phasianids theory) that megapodes already were confined to the Indonesian-Papuan-Australian-Oceanic region by that time.

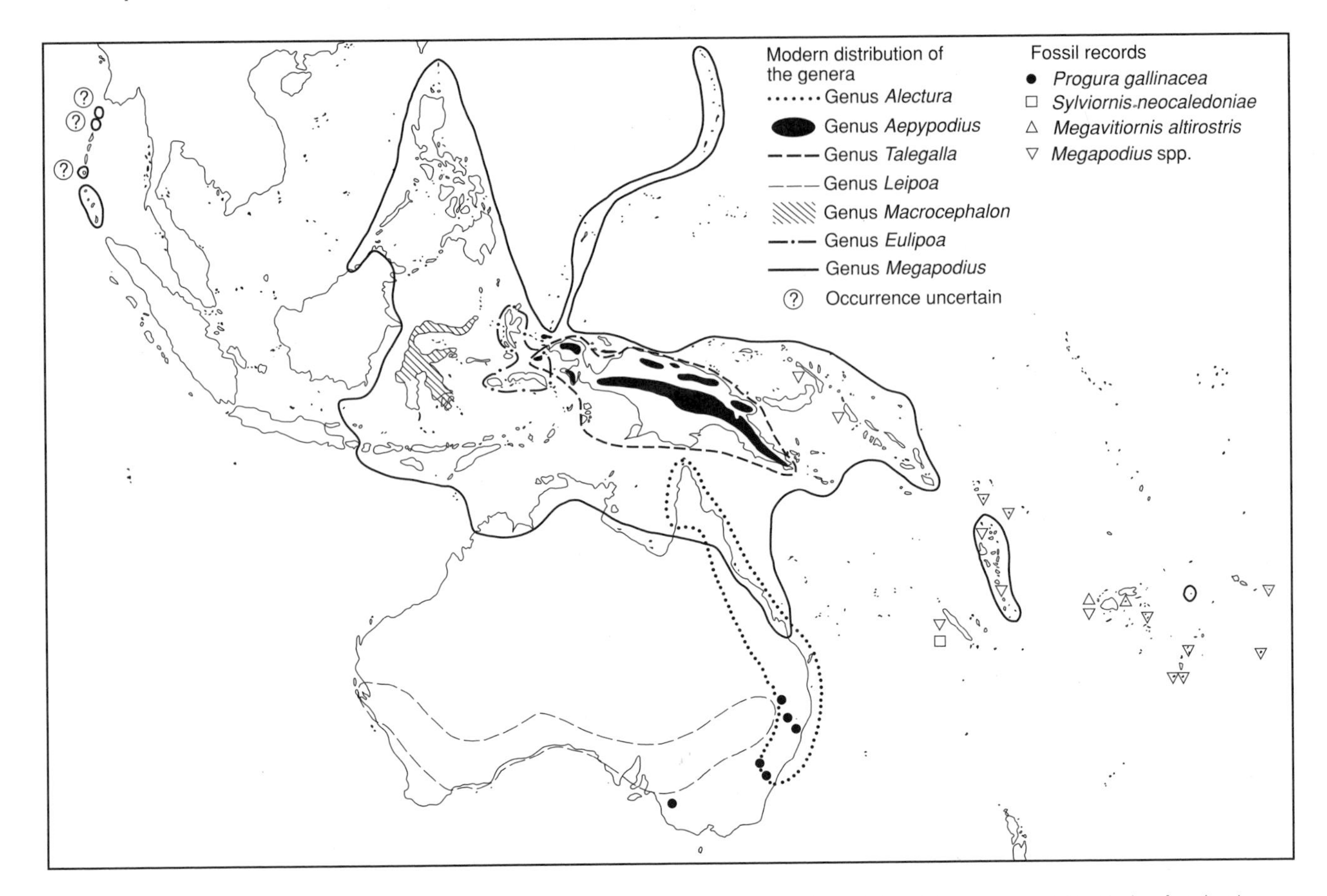

Figure 9-1 The modern and prehistoric (Quaternary) distribution of megapodes. Modern data from Jones et al. (1995: Figure 3.1). Prehistoric data from herein.

Pliocene and Pleistocene fossils of megapodes outside of Oceania are limited to two species of *Progoura* (another extinct genus) from Australia (Olson 1985b, Rich & van Tets & 1990, Boles & Mackness 1994). As the late Pleistocene and Holocene fossil record improves in Indonesia, New Guinea, and Australia, I predict that, as in Oceania, it will show that prehistoric humans affected the distribution and diversity of megapodes in these more continental regions as well.

My current evaluation of the distribution of megapodes in Oceania (Tables 9-1, 9-2, Figure 9-1) is updated substantially from that in Steadman (1999b).

Systematic Review

Megapodius

Following Mayr (1938) and Amadon (1942a), all living forms of *Megapodius* except the two smallest species, *M. laperouse* and *M. pritchardii*, generally had been regarded as subspecies of *M. freycinet*. With improved knowledge of the biology of megapodes, more recent authors have tended to recognize sets of the many allopatric populations of *M. freycinet* s.l. as full species, defined on characters of plumage, soft part colors, size, vocalization, nesting ecology, and behavior (Roselaar 1994, Jones et al. 1995, Birks & Edwards 2002). I follow these authors in recognizing 11 living species of *Megapodius*, four confined to Oceania. Whether to regard the living forms of *M. freycinet* s.l. as subspecies or species is, however, rather subjective, at least pending molecular studies.

I believe that, before human arrival, a representative of the *Megapodius freycinet* species-group occurred throughout Melanesia and West Polynesia. The fossil evidence presented below suggests that, at first human contact, most or all of these islands also were occupied by one or two additional species of *Megapodius*.

One species of megapode in the archaeological sites on New Ireland, Bismarck Archipelago (Chapter 5) is *Megapodius* cf. *eremita*, the extant local form of the *M. freycinet* species-group (see Mayr 1938:12–15). The second is *Megapodius* undescribed sp. A, in the approximate size range of the extinct *M. molistructor* of New Caledonia

Table 9-1 Distribution of megapodes in Oceania by island group. M, modern record; P, prehistoric record; †, extinct species. Solomons = Solomon Main Chain. Based on data in Table 9-2.

	Bismarcks	Solomons	Santa Cruz	Vanuatu	New Caledonia	Fiji	Tonga	Samoa	Niue	Palau	Marianas	Pohnpei
Megapodius eremita	M, P	M, P	—	—	—	—	—	—	—	—	—	—
Megapodius layardi	—	—	P	M, P	—	—	—	—	—	—	—	—
Megapodius pritchardii	—	—	—	—	—	—	M, P	P	P	—	—	—
Megapodius laperouse	—	—	—	—	—	—	—	—	—	M, P	M, P	P
†*Megapodius alimentum*	—	—	—	—	—	P	P	—	—	—	—	—
†*Megapodius amissus*	—	—	—	—	—	P	—	—	—	—	—	—
†*Megapodius molistructor*	—	—	—	—	P	P?	P	—	—	—	—	—
†*Megapodius* undescribed sp. A	P	—	—	—	—	—	—	—	—	—	—	—
†*Megapodius* undescribed sp. B	—	P	—	—	—	—	—	—	—	—	—	—
†*Megapodius* undescribed sp. C	—	—	—	P	—	—	—	—	—	—	—	—
†*Megapodius* undescribed sp. D	—	—	—	—	P	—	—	—	—	—	—	—
†*Megapodius* undescribed sp. E	—	—	—	—	P	—	—	—	—	—	—	—
†*Megapodius* undescribed sp. F	—	—	—	—	—	—	P	—	—	—	—	—
†Genus uncertain, undescribed sp.	—	—	—	—	—	—	P	—	—	—	—	—
†*Megavitiornis altirostris*	—	—	—	—	—	P	—	—	—	—	—	—
†*Sylviornis neocaledoniae*	—	—	—	—	P	—	—	—	—	—	—	—
TOTAL SPECIES												
M	1	1	0	1	0	0	1	0	0	1	1	0
M + P	2	2	1	2	4	3–4	5	1	1	1	1	1

and Tonga. *Megapodius* undescribed sp. A is extinct and known from only two fragmentary late Pleistocene (>10,000 yr BP) specimens, a scapula from Panakiwuk Cave and a tarsometatarsus from Balof 2 Rockshelter (Steadman et al. 1999a).

On Buka (politically in Papua New Guinea but biologically part of the Solomon Islands), the late Pleistocene archaeological site of Kilu Cave (Wickler 2001; Chapter 5 herein) yielded two femora of an extinct species (*Megapodius* undescribed sp. B) that resembles *M. molistructor* in size.

The Kiki and Sinapupu archaeological sites on the small, isolated island of Tikopia produced bones of a medium-sized species of *Megapodius*, most likely in the *M. freycinet* species-group and therefore perhaps referable (based only on geography) to the extant *M. layardi* (Steadman et al. 1990b, Steadman 1999b; Table 5-9, Figure 9-2 herein). Eight of the 10 megapode bones from Tikopia are from the island's earliest cultural strata (Lapita), which are ca. 2900 years old (Kirch & Yen 1982).

From a Lapita site of similar age on Lomlom in the Reef Islands, north of Santa Cruz (Nendo), an unknown number

Figure 9-2 Orange-footed scrubfowl, *Megapodius reinwardt*. Photo by Neil Fifer, Port Douglas, Queensland, October 2003.

Table 9-2 Prehistoric records of megapodes in Oceania. †, extinct species; *, extirpated population. Based on data in Green (1976), Poplin (1980), Kirch & Yen (1982), Poplin et al. (1983), Balouet (1984, 1987), Poplin & Mourer-Chauviré (1985), Balouet & Olson (1989), Steadman (1989a, 1989b, 1991c, 1992a, 1993a,b, 1995a, 1999a, unpub. data), Steadman et al. (1990b, 1999a, 1999b, 2000b), and Worthy (2000). Updated from Steadman (1999b).

ISLAND GROUP / island	Species
BISMARCK ARCHIPELAGO	
New Ireland	*Megapodius* cf. *eremita*
	†*Megapodius* undescribed sp. A
SOLOMON ISLANDS	
Buka	†*Megapodius* undescribed sp. B
Tikopia, Santa Cruz	**Megapodius* cf. *layardi*
Lomlom, Reef Is., Santa Cruz	**Megapodius* cf. *layardi*
VANUATU	
Malakula	*Megapodius layardi*
Efate	*Megapodius layardi*
	†*Megapodius* undescribed sp. C
NEW CALEDONIA	
Grande Terre	†*Megapodius* undescribed sp. D
	†*Megapodius molistructor*
	†*Sylviornis* cf. *neocaledoniae*
Ile des Pins	†*Megapodius* undescribed sp. D
	†*Megapodius molistructor*
	†*Sylviornis neocaledoniae*
Loyalty Islands	†*Megapodius* undescribed sp. E
FIJI	
Viti Levu	†*Megapodius amissus*
	†*Megavitiornis altirostris*
Naigani	†*Megavitiornis altirostris*
Lakeba	†*Megapodius alimentum*
Aiwa Levu	†*Megapodius alimentum*
	†*Megapodius amissus/molistructor*
Mago	†*Megapodius alimentum*
TONGA	
Ha`ano	†*Megapodius alimentum*
	†*Megapodius molistructor*
Foa	**Megapodius pritchardii*
	†*Megapodius alimentum*
	†*Megapodius molistructor*
Lifuka	**Megapodius pritchardii*
	†*Megapodius alimentum*
	†*Megapodius molistructor*
`Uiha	**Megapodius pritchardii*
	†*Megapodius alimentum*
	†*Megapodius molistructor*
Ha`afeva	**Megapodius pritchardii*
	†*Megapodius alimentum*
	†*Megapodius molistructor*
`Eua	†*Megapodius* undescribed species F
	**Megapodius pritchardii*
	†*Megapodius alimentum*
Tongatapu	†*Megapodius alimentum*
	†*Megapodius molistructor*
	†Genus uncertain, undescribed sp.
AMERICAN SAMOA	
Ofu	**Megapodius* cf. *pritchardii*
NIUE	**Megapodius pritchardii*
PALAU	
Ulong	*Megapodius laperouse*
MARIANA ISLANDS	
Guam	**Megapodius laperouse*
Rota	**Megapodius laperouse*
Aguiguan	*Megapodius laperouse*
Tinian	*Megapodius laperouse*
CAROLINE ISLANDS	
Pohnpei	**Megapodius laperouse*

of megapode bones, which I examined briefly in 1982, also can be tentatively referred to *Megapodius layardi* or possibly *M. eremita*. The Reef Islands lie closer than Tikopia to the modern range of *M. eremita*. The two modern populations of megapodes nearest to Tikopia and the Reef Islands are *M. eremita* at its eastern limit on Makira (Solomon Main Chain) and *M. layardi* at its northern limit in the Banks Islands, Vanuatu (Roselaar 1994:33). Other galliform bones from Lomlom, which I examined in 2002, represent only the nonnative chicken *Gallus gallus*.

The absence of megapodes on Rennell and Bellona, two outlying raised limestone islands in the Solomons, is almost certainly due to human impact. No prehistoric bird bones are known from either island.

In Vanuatu, archaeological sites on Malakula and Efate produced five megapode bones (Table 5-11). *Megapodius* undescribed sp. C (Efate) was qualitatively similar to, but larger than, the extinct *M. alimentum* of Fiji and Tonga. The extant *M. layardi* is widespread in Vanuatu today (Bregulla 1992, Bowen 1996). The two species probably were sympatric not just on Efate, but through most of Vanuatu.

Megapode bones are common in late Quaternary sites on New Caledonia and nearby islands (Chapter 5). All four species (three of *Megapodius*, one of *Sylviornis*) are extinct but have been found at cultural sites, showing that they survived until human arrival ca. 2900 cal BP. The extinct, volant *Megapodius molistructor* (Balouet & Olson 1989) was much smaller than *Sylviornis* but much larger than any living species of *Megapodius*. Because bones of *M. molistructor* also have been found in Tonga (see below), this species may have inhabited Fiji, Vanuatu, and perhaps Samoa.

I recently discovered two other extinct species of *Megapodius* from New Caledonia (unpub. data). One is the small- to medium-sized *M.* undescribed sp. D, represented by 22 bones at the Vatcha Chi site on Ile des Pins and tentatively by a coracoid from the Lapita site on Grande Terre. It may have been the New Caledonian form of the *M. freycinet* species-group. The next, *M.* undescribed sp. E, is a short, stout megapode known by four bones from the Loyalty Islands (Tables 5-12, 5-13).

Now megapode-free like New Caledonia, Fiji once sustained at least two species of *Megapodius*. One was *M. alimentum* (Figure 9-3), a rather large extinct species that is common in Tongan sites (see below) and known in Fiji from three islands in the Lau Group, as follows: the Lapita-age Qara-ni-puqa Rockshelter on Lakeba (although the island labeled Lakeba in Worthy 2000 is really Cicia); Rockshelter 1 on Aiwa Levu; and the Lapita-age Votua site on Mago (Steadman 1989a, Worthy 2000, Clark et al. 2001; Chapter 6 herein). A second, larger species of *Megapodius* from Fiji is *M. amissus* (Figure 9-3), known from 31 specimens at two noncultural sites on Viti Levu (Udit Cave, Voli Voli Cave; Worthy 2000). Slightly smaller than *M. molistructor*, *M. amissus* was also exceeded in size by *M.* undescribed spp. A and B of Near Oceania. Uniquely among species of *Megapodius*, the limb proportions of *M. amissus* suggest that it was flightless (Worthy 2000). Also from Aiwa Levu is a single pedal phalanx of a megapode larger than *M. alimentum*, which probably pertains to *M. amissus* or *M. molistructor* (Chapter 6). As Fiji's fossil record grows, I predict that most or all islands will be found to have supported two or three species of *Megapodius* at human contact.

In Tonga, megapodes are known from hundreds of bones from seven islands. The only extant population in Tonga (or anywhere in Polynesia; Steadman 1991c) is that of *Megapodius pritchardii* from the volcanic island of Niuafo'ou. I identified bones of *M. pritchardii* s.l., long regarded as endemic to Niuafo'ou, at Lapita-age sites on four islands in the Ha'apai Group (Chapter 6, Table 9-2), a precultural context on 'Eua (Steadman 1993a), an archaeological site on Ofu, American Samoa (Steadman 1993b; Chapter 6 herein), and a precultural cave deposit on Niue (Steadman et al. 2000b). Rather than being endemic to Niuafo'ou or even to Tonga, *M. pritchardii* merely has been able to survive on this single, remote island. On Niuafo'ou, it nests in volcanically warmed soils near a crater lake (Weir 1973, Todd 1983, Rinke 1991, Rinke et al. 1993, Göth & Vogel 1995, 1997). Before people arrived at ca. 2850 cal BP (Burley et al. 1995, Steadman et al. 2002a), *M. pritchardii* probably occurred across the Tonga-Niue-Samoa region. Possibly 100 or more island populations of *M. pritchardii* have been lost.

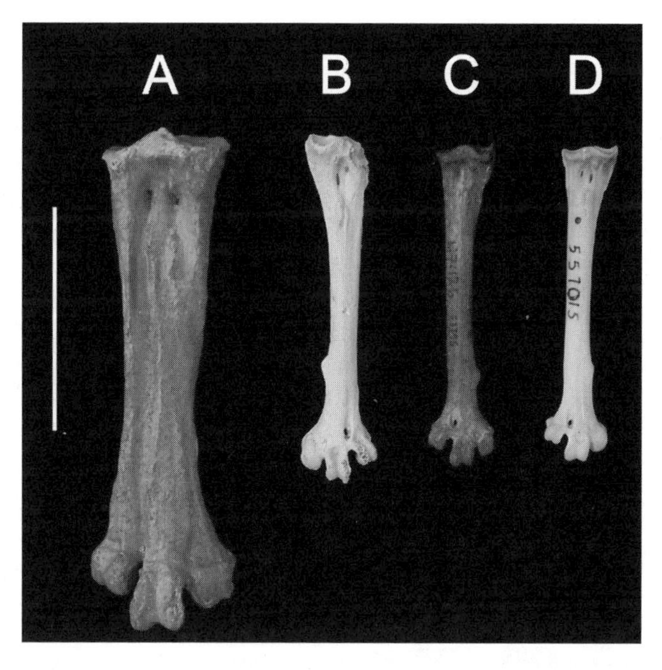

Figure 9-3 The tarsometatarsus of megapodes in acrotarsial aspect. A: *Megavitiornis altirostris*, Viti Levu, Fiji, MNZ S. 37016 (cast). B: *Megapodius amissus*, Viti Levu, Fiji, MNZ S. 37468 (cast). C: *Megapodius alimentum*, Lifuka (Tonga), UF 57795 (fossil). D: *Megapodius freycinet*, Halmahera, Moluccas, USNM 557001 (modern). Scale = 50 mm. Photo by J. J. Kirchman.

The extinct *Megapodius* undescribed sp. F, the smallest megapode known, is from cultural and noncultural contexts on 'Eua (Steadman 1993a). The extinct *M. alimentum* is the megapode most often found on all seven Tongan islands with fossils. *Megapodius alimentum* probably once lived throughout Tonga (and Fiji?). Most bones of *M. alimentum* are fairly similar qualitatively to those in living members of the *M. freycinet* species-group, although averaging larger. The tarsometatarsus and toes, however, are quite distinct (Steadman 1989b, pers. obs.). If it is the Tongan-Fijian representative of the *M. freycinet* species-group, then *M. alimentum* is the largest and perhaps most distinctive species in this "superspecies." The much smaller *M. pritchardii* was broadly sympatric with *M. alimentum* in Tonga, and therefore they cannot be allopatric representatives of a single lineage, i.e., a superspecies. The idea that *M. pritchardii* is part of the *M. freycinet* superspecies (Mayr 1938, Amadon 1942a, Jones et al. 1995) needs to be tested with osteological and molecular data in light of the discovery of so many extinct species and populations of megapodes in West Polynesia.

A fourth Tongan species of *Megapodius* is the larger, extinct *M. molistructor*, known from cultural sites on Ha'ano, Foa, Lifuka, Ha'afeva, and Tongatapu (Steadman 1989b, herein). Described originally from New Caledonia (Balouet & Olson 1989), *M. molistructor* probably once inhabited Vanuatu and Fiji as well. The scarcity of bones of *M. molistructor* in the Ha'apai Group of Tonga, relative to those of *M. alimentum*, might reflect a lower population density that led to its more rapid extinction. As Tonga's largest ground-dwelling bird, *M. molistructor* may have been favored prey of the Lapita colonists (Chapters 6, 16). I should note that *M. molistructor* is somewhat aberrant among *Megapodius* and may, with further study, warrant its own genus.

Another large species of megapode may have existed in Tonga; a humerus and two radii from Tongatapu (Ha'ateiho site), shorter but stouter than in *Megapodius molistructor*, resemble those in extant *Tallegalla* more than in extinct *Megavitiornis* of Fiji or any species of *Megapodius*. An unusual pedal phalanx from Foa (Faleloa site) also may pertain to this species.

Moving to American Samoa, an ulna and femur of a small- to medium-sized species of *Megapodius* from the Toaga site on 'Ofu Island (Manu'a Group) are smaller than in *M. freycinet* s.l. (i.e., *M. eremita* and *M. layardi*) but slightly larger than in modern specimens of *M. pritchardii* (Steadman 1993b, 1999b). Considering that the bones of *M. pritchardii* from Niue (below) also are slightly larger than in modern specimens, I tentatively refer the two bones from 'Ofu as well to *M. pritchardii*. Megapodes undoubtedly inhabited Western Samoa as well.

On Niue, seven bones from a precultural site (Anatuli) are referred to *Megapodius pritchardii* (Steadman et al. 2000b). The Niuean form of *M. pritchardii* was slightly larger than the birds surviving on Niuafo'ou. Along with American Samoa, Niue establishes the eastern limit of megapodes. Given that megapodes have not been found among the abundant fossils from Mangaia (Cook Islands), this limit might be genuine, although negative evidence is never absolute.

In the Mariana Islands, bones of the extant *Megapodius l. laperouse* occur in cultural and noncultural sites on Guam, Rota, Aguiguan, and Tinian (Steadman 1999a; Chapter 8). This small species no longer occurs on Guam or Rota, probably is gone from Tinian, remains common on uninhabited Aguiguan, and is widespread but of unknown population size on the northern volcanic islands (Falanruw 1975, Engbring et al. 1986, Pratt et al. 1987; pers. obs.). Abundant prehistoric bones from the Marianas have yet to reveal a second species of megapode.

From the Nan Madol archaeological site on Pohnpei (see Morgan 1988, Ayres 1990), a tarsometatarsus of *Megapodius laperouse* s.l. recovered by William Ayres is the first record of the species outside of Palau or the Mariana Islands (Steadman 1999b; Chapter 8 herein). This small megapode, or a closely related congeneric species, probably once lived through much of Micronesia, including the high islands of Yap, Chuuk, Pohnpei, and Kosrae, as well as many of the atolls.

Megavitiornis

This flightless Fijian megapode was described from cultural and noncultural sites on Viti Levu (189 bones from Udit Tomo, Qara-ni-vokai, Voli Voli Cave, and Wai-ni-buku Cave) and an archaeological site on nearby Naigani Island (Worthy et al. 1999, Worthy 2000). *Megavitiornis altirostris* was substantially smaller than *Sylviornis* but larger than any living or extinct species of *Megapodius* (Figure 9-3). As suggested by its name, *M. altirostris* had a high, narrow rostrum and matching deep mandible, unique in the Megapodiidae (Worthy 2000). The combined length of the femur, tibiotarsus, and tarsometatarsus in *Megavitiornis* is 459 mm (Worthy 2000). In a modern specimen of *Megapodius [freycinet] eremita* (UF 40180; body mass 0.795 kg) from Isabel, Solomon Islands, this length is 238 mm. Given the scaling relationship of linear measurements vs. body mass, this would suggest a body mass of ca. 6 kg for *Megavitiornis*. The added stoutness of leg elements in *Megavitiornis* probably means that its mass was even greater, perhaps 8–10 kg.

Sylviornis

The endemic, flightless *Sylviornis neocaledoniae* (Figure 9-4) stood ca. 1.2-1.4 m tall and weighed 30–40 kg. This largest of megapodes probably built the soil mounds (diameter 10–50 m, height 0.8–5 m) in New Caledonia known as "tumuli" (Green & Mitchell 1983, Poplin & Mourer-Chauviré 1985, Green 1988). Originally described from two fragmentary bones on Ile des Pins (Poplin 1980), *Sylviornis* was found with study of more fossils to be a giant megapode rather than a paleognathous bird (Poplin et al. 1983, Poplin & Mourer-Chauviré 1985). Much additional material of *Sylviornis* from caves on Grande Terre and Ile des Pins was noted briefly by Balouet (1984, 1987) and Balouet & Olson (1989). A single, eroded distal tibiotarsus

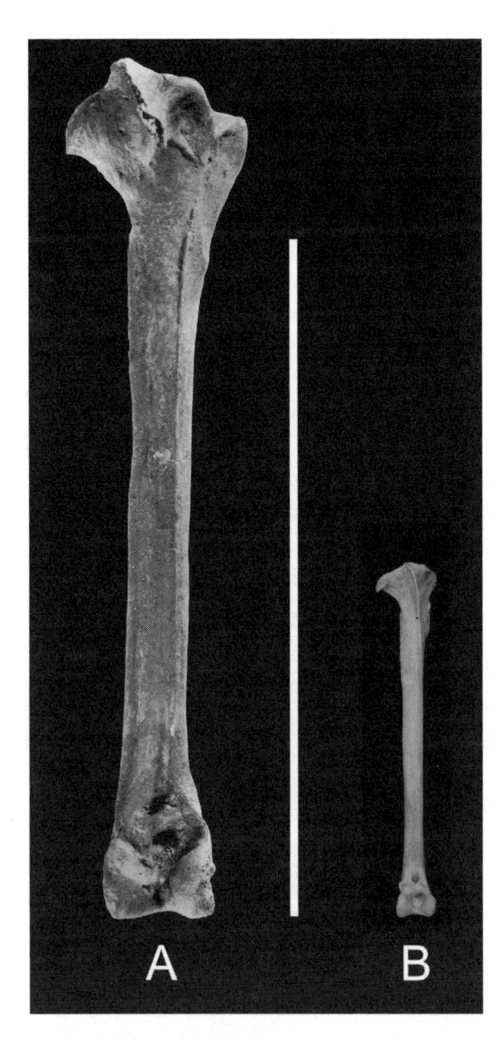

Figure 9-4 Tibiotarsus of *Sylviornis neocaledoniae* (A. extinct, New Caledonia) and *Megapodius eremita* (B. extant, Isabel, Solomon Islands). Photo of *Sylviornis* by Christophe Balouet (courtesy of Cecile Mourer-Chauviré), and of *Megapodius* by J. J. Kirchman. Spliced together by M. J. Reetz. Scale bar = 200 mm.

from the Lapita archaeological site on Grande Terre also is referable to *Sylviornis* (Steadman 1999b). The species-level relationships of *Sylviornis* on Ile des Pins versus Grande Terre are unresolved.

Patterns and Processes

Extinction

Today, 22 species of megapodes are estimated to exist, including four species in Oceania (Jones et al. 1995). At least 12 extinct species of megapodes are known now from late Quaternary fossils in Oceania, bringing the total on Pacific islands to at least 16 species. The only extinct Quaternary megapodes outside of Oceania are two species of *Progura* from Australia, although there is essentially no fossil record of birds in Island Southeast Asia, Indonesia, Philippines, or New Guinea. Maybe 15 or more species of megapodes, beyond those already discovered, have become extinct since human arrival, with about one-half of them from Oceania. In the absence of people, perhaps 45 to 55 species of megapodes would be alive today. How thrilling it must have been for the Lapita colonists of New Caledonia to watch a 40-kg *Sylviornis* building, tending, and defending its huge mound. Had people not wiped out *Sylviornis*, New Caledonia might be as famous today for giant megapodes as Komodo is for its "dragons."

Bones from cultural sites on many islands indicate that megapodes were killed and eaten prehistorically. Much exploitation of megapodes today focuses on eggs, the harvesting of which is potentially sustainable if strictly controlled and if adequate habitat remains (Broome et al. 1984, Lepofsky 1989, Jones et al. 1995, Foster 1999). Megapode eggs probably were taken prehistorically as well, although eggshells have not been found at archaeological sites. Regardless, the combination of habitat loss and predation from people, rats, dogs, and pigs led to the extinction of megapodes. Disease transmitted from a nonnative galliform, the chicken *Gallus gallus*, also may have been involved.

The rapid movement of Lapita peoples to islands across thousands of kilometers of ocean certainly led to catastrophic extinction of megapodes (and other birds) in Remote Oceania (Chapter 16). The radiocarbon and faunal data from Lapita sites on five islands in the Ha'apai Group of Tonga, for example, suggest that at least four species of megapodes were lost on these small (1.8 to 13.3 km^2) islands within a century or two of human contact (Steadman et al. 2002a). More time probably was needed to wipe out megapodes on larger islands.

Biogeography

Megapodes are absent from the fossil record in East Polynesia (Steadman 1995a, herein). American Samoa and Niue, where megapodes lived at human arrival, may be the family's true eastern limit. North and west of Samoa, no megapodes are known among the small bone samples from remote atolls in Tokelau, Tuvalu, Kiribati, and the Marshall Islands. Living forms of *Megapodius* inhabit atolls today in spite of species-poor floras and limited forest structure. Examples include *M. laperouse* on Kayangel in Palau and *M. eremita* on Ontong Java and Sikaiana in the Solomons (Bayliss-Smith 1972, Engbring 1988, Jones et al. 1995). Among high islands in Micronesia, bone records are lacking on Kosrae and Nauru, which, given the former occurrence of *M. laperouse* on Pohnpei, also may have sustained megapodes. To the west in Chuuk and Yap, it seems almost certain that megapodes once inhabited these high islands that lie between Palau, the Marianas, and Pohnpei, where megapodes already are known. Atolls in the Yap-Chuuk-Pohnpei region probably also had megapode populations that were lost soon after human arrival.

The puzzling modern absence of megapodes throughout New Caledonia, Fiji, Tonga (except *M. pritchardii* on outlying Niuafo'ou), and Samoa is shown by bones to be an artifact of human-caused extinction. It is not because megapodes "overshot Efate [Vanuatu] by at least 1600 km" (Diamond & Marshall 1977:719). Equally suspect are modern absences of megapodes on individual islands within the ranges of extant species. For example, human impact can account for the absence of *M. cumingii* on Panay (Philippines), and of *M. eremita* on Vangunu, Nggatukai, and Florida (Solomon Islands), and of *M. layardi* on Mere Lava, Makura, Erromango, Aniwa, Futuna, and Aneityum in Vanuatu (Jones et al. 1985; Table 5-11 herein).

Megapodes incubate their eggs through solar radiation, geothermal activity, and organic decomposition (Frith 1956, Clark 1964, Jones 1989). The method is dictated mainly by available substrates. On Ambrym (Vanuatu), *M. layardi* uses all three methods (Bowen 1996). *Megapodius laperouse* nests today on atolls, limestone islands, eroded volcanic islands, and active volcanic islands in Palau and the Marianas (Pratt et al. 1987). The only population left of *M. pritchardii* nests in burrows in volcanically warmed soil on Niuafo'ou (Göth & Vogel 1995, 1997). On the many limestone islands where it formerly occurred, *M. pritchardii* must have used organic decomposition and perhaps solar radiation to incubate eggs. Geological variation within and among archipelagos warranted intraspecific flexibility in incubation strategies. The bedrock and soil of an island do not limit the distribution of megapodes.

Community Ecology

Island size had little effect on the distribution of megapodes in precultural times. In Tonga, for example, *Megapodius alimentum* lived on islands from at least as small and low as Ha'afeva (1.8 km^2, elev. 10 m) to the two largest islands of 'Eua (87 km^2, elev. 325 m) and Tongatapu (259 km^2, elev. 67 m). An even greater range of island sizes is inhabited by single species of megapodes today. Islands spanning at least five orders of magnitude of area are occupied by *M. eremita* in the Bismarcks and Solomons, by *M. decollatus* in northern New Guinea and offshore islands, and by *M. reinwardt* on southern New Guinea, northern Australia, Lesser Sundas, and intervening islands (Bayliss-Smith 1972, Broome et al. 1984, Jones et al. 1995).

"Species packing" is another issue enlightened by prehistory. No species of *Megapodius* are sympatric today (Mayr 1938). The large, high, landbridge island of Misol (= Misool) off western New Guinea sustains three and perhaps four sympatric species of megapodes, more than anywhere else (Ripley 1960, 1964). They are *Eulipoa wallacei* (breeding on Misol not yet proven; R. W. R. J. Dekker, pers. comm.), *Megapodius freycinet freycinet*, *Aepypodius arfakianus misoliensis*, and *Talegalla cuvieri cuvieri*, the first two being smaller. Even in floristically rich forests of the Lesser Sundas, Moluccas, or New Guinea, it is rare that as many as three species of megapodes occur sympatrically; when they do so occur, none is congeneric (Sinclair 2002).

Before human arrival, however, multiple species of megapodes lived on many if not most individual islands from the Bismarcks through Tonga (Table 9-2). On New Ireland (Bismarcks), Buka (Solomons), and Efate (Vanuatu), the extant *Megapodius eremita* or *M. layardi* was sympatric with a larger, extinct congener. On New Caledonia, *Sylviornis neocaledoniae* was sympatric with *M. molistructor* (Balouet & Olson 1989) and an undescribed species of *Megapodius*. On massive Viti Levu in Fiji, where no megapodes live today, two species are known from bones (*Megapodius amissus*, *Megavitiornis altirostris*). Were Viti Levu's prehistoric avifauna perfectly known, I believe that it would have at least four species of megapodes.

Tonga, near the family's eastern limit, sustained more species of megapodes than any other place in Oceania. Four flat Tongan islands ranging in size from Foa (13.3 km^2, elev. 20 m) down to Ha'afeva (1.8 km^2, elev. 11 m) each had three species of *Megapodius* when people arrived (Table 9-2). The larger, higher island of 'Eua (87 km^2, elev. 325 m) also was inhabited by three species of *Megapodius*, two of them the same as on Foa and Ha'afeva. As the fossil record grows, the past occurrence of two or three sympatric species of *Megapodius* probably will be found to have been the rule across Melanesia and West Polynesia.

Modern distributions of megapodes (and other birds) are residues of human impact. Therefore, "assembly rules" (*sensu* Diamond 1975b) to explain sympatry or allopatry are risky if based on modern distributions alone (Chapter 20). How did three species of *Megapodius* coexist until 2800 years ago on tiny, flat Ha'afeva? The three species differed in overall size and therefore most likely in bill size (the rostrum and mandible of *M. molistructor* are not yet known), leading to a presumption of different food habits. I would speculate, however, that many of the same foods were consumed by *M. pritchardii*, *M. alimentum*, and *M. molistructor*. Their nesting ecology may have differed, perhaps with one or two species preferring coastal sands and the others using interior forest soils. Nevertheless, I doubt that directional selection away from potential competition is needed to account for this sympatry. Two or three of the species in Tonga even may have laid their eggs in the same mounds, a special sort of brood parasitism that has been recorded for *Megapodius reinwardt* and *Talegalla jobiensis* in New Guinea (Dwyer 1981). How wonderfully informative it would be if we could go back 3000 years to observe the multiple megapodes on small Tongan islands.

Chapter 10 Rails

Rails are the most intriguing and tragic family of birds in Oceania if not the world. The intrigue is because, for reasons that even rails may not understand, a few of their species tend to get sprinkled (or sprinkle themselves) across vast distances to oceanic islands. In the prehuman absence of placental carnivores, formerly volant rails evolved into endemic, flightless species on island after island, no matter how isolated. The tragedy is that nearly all of Oceania's many hundreds of endemic species of island rails have become extinct since human arrival. From a standpoint of numbers of species lost, the most dramatic extinction story in the Pacific is that of flightless rails.

Flightlessness, which had been just fine until people and rats arrived, became terminally maladaptive for island rails, which had been the world's most species-rich family of birds. Reversibility of evolution is poorly understood (Teotónio & Rose 2000); if regaining the power of flight is possible in flightless rails once nonnative predators arrive, it does not happen quickly enough to prevent extinction.

Each island in the tropical Pacific with a good prehistoric record of birds has yielded bones of one to four endemic species of flightless rails. At least five genera are involved (nine counting New Zealand species), together providing the most species yet one of the least understood examples of evolutionary radiation in island birds. These rails did not undergo the Darwin's finch sort of radiation with allopatric speciation followed by sympatry (Lack 1947) because flightlessness precluded secondary contact.

Here I will review the very different prehistoric and modern distributions of rails in Oceania. Most of the prehistoric species were flightless, whereas most living species are volant. This fact is instructive in thinking about the dispersal and evolution that led to the vast radiation of flightless rails that greeted human colonists of islands, because many of today's volant populations of rails in Remote Oceania probably did not exist before people arrived in recent millennia. Furthermore, earlier populations of these same volant species may have been the progenitors of many of the extinct flightless congeners.

Flightlessness

More species have become flightless in rails than in any other family of birds (Olson 1973b), although flightlessness has been found in living or recently extinct species in at least 24 other families of birds, namely the ostriches (Struthionidae), rheas (Rheidae), emus and cassowaries (Casuariidae), kiwis (Apterygidae), moas (†Dinornithidae, †Emeidae), elephantbirds (†Aepyornithidae), penguins (Spheniscidae), grebes (Podicipedidae), cormorants (Phalacrocoracidae), herons (Ardeidae), ibises (Threskiornithidae), waterfowl (Anatidae), mihirungs (†Dromornithidae), megapodes (Megapodiidae), phorusrhacids (†Phorusrhacidae), kagus (Rhynochetidae), adzebills (†Apterornithidae), auks (Alcidae), pigeons (Columbidae), parrots (Psittacidae), owls (Strigidae), New Zealand wrens (Acanthisittidae), and emberizine finches (Emberizidae). This is flightlessness for an entire lifetime, not a temporary condition related to extreme fat deposition (Jehl 1997) or the intense molt of flight

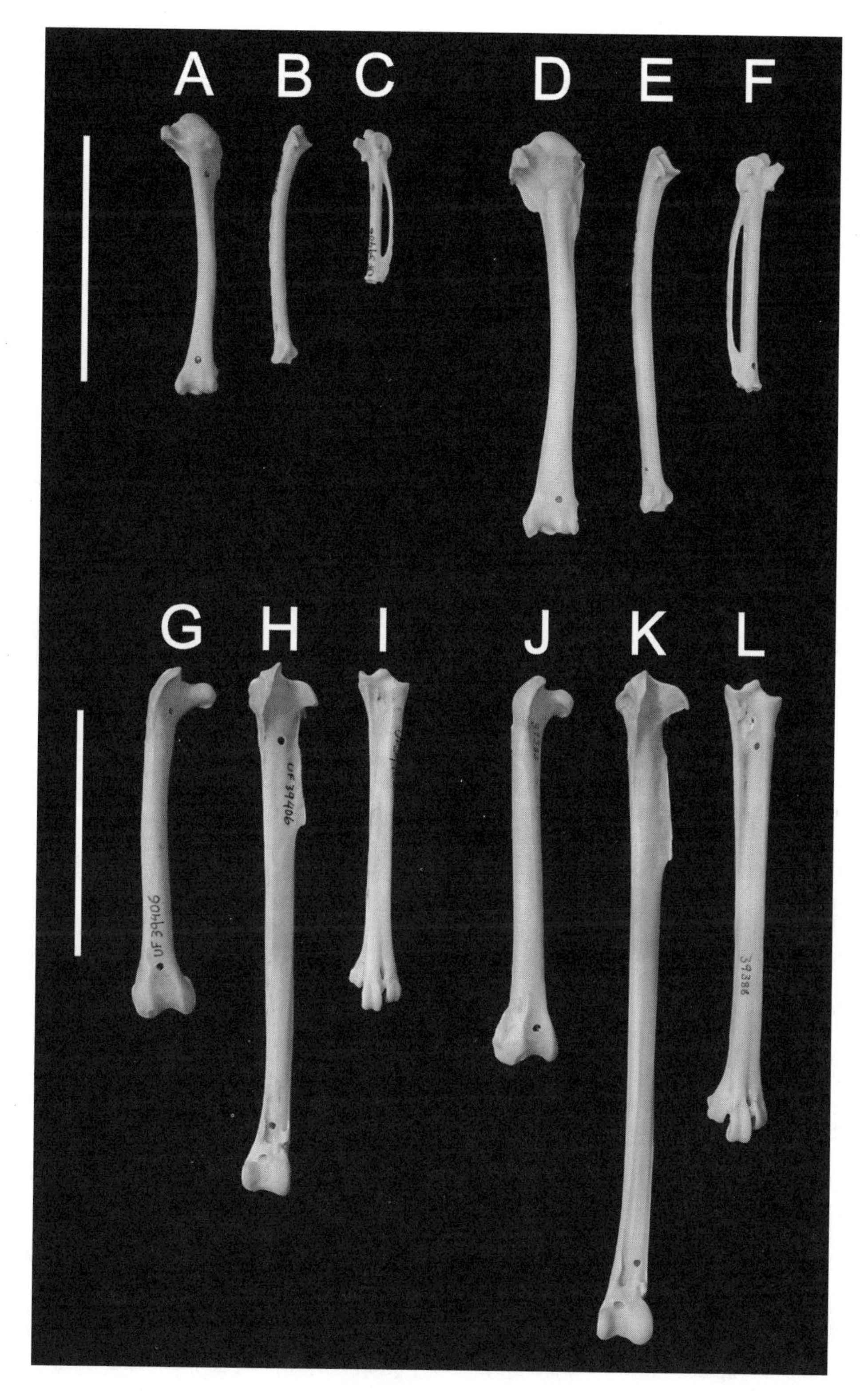

Figure 10-1 Major wing bones (humerus, ulna, carpometacarpus; above) and leg bones (femur, tibiotarsus, tarsometatarsus; below) of a large (571 g) flightless rail *Gallirallus immaculatus* (UF 39406, adult ♀, Isabel, Solomon Islands; A–C, G–I) and a larger (812 g) volant rail *Porphyrio porphyrio samoensis* (UF 39388, adult ♂, Rennell, Solomon Islands; D–F, J–L). Because of breakage, opposite sides are shown for the carpometacarpus. Scale = 50 mm. Photo by J. J. Kirchman.

Table 10-1 Distribution of genera of rails in tropical Oceania. Certain or presumed vagrants are excluded. M, modern record (volant); m, modern record (flightless); P, prehistoric record (volant); p, prehistoric record (flightless); †, extinct genus. From Pratt et al. (1987), Balouet & Olson (1989), Steadman (1987, 1988, 1995a, unpublished), Taylor (1998), Steadman et al. (1999a, 2000b), and Worthy (2004). Sequence of extant genera follows Olson (1973b) except that *Amaurornis* precedes *Porphyrio* rather than *Porzana*; see text for generic synonymies. Note that the totals represent island groups, not individual islands, and that individual species can be represented by both an M and P, or an m and p. Multiple species within an island group are represented by only a single M, P, m, and/or p. See Tables 10-3 through 10-11 for species-level compilations.

	Bismarcks	Solomon Islands	Vanuatu	New Caledonia	Fiji	Tonga	Samoa	Niue	Cook Islands	Tubuai	Society Islands	Tuamotu Arch.	Pitcairn Group
Rallina	M	—	—	—	—	—	—	—	—	—	—	—	—
Gallirallus	MmPp	Mmp	MPp	MmPp	MmPp	MPp	M	MPp	p	—	mp	—	—
†*Vitirallus*	—	—	—	—	p	—	—	—	—	—	—	—	—
Gymnocrex	M	—	—	—	—	—	—	—	—	—	—	—	—
Porzana	MP	M	MPp	MP	MPp	MP	MP	MP	MPp	M	MP	M	Mmp
Poliolimnas	MP	M	MP	M	M	—	M	—	—	—	—		—
Amaurornis	MP	M(m?)	—	—	—	—	—	—	—	—	—	—	—
Gallinula	—	—	—	MP	—	—	—	—	—	—	—	—	—
Pareudiastes	—	mp	—	—	—	—	m	—	—	—	—	—	—
Porphyrio	MPp	Mp	MP	MPp	MP	MP	M	MP	p?	—	—	—	—
TOTAL	7M, m, 5P, 2p	5M, 2–3m, 3p	4M, 4P, 2p	5M, m, 4P, 2p	4M, m, 4P, 3p	3M, 3P, 1–2p	4M, m, P	3M, 3P, p	M, P, 3p	M	M, m, P, p	M	M, m, P

feathers (Humphrey & Clark 1964, Marks 1993). Flightlessness involves more orders and families of birds in New Zealand than anywhere else (Worthy & Holdaway 2002, Steadman 2002c).

Morphological adaptations for flightlessness in rails include shortened and weakened bones of the pectoral girdle (sternum, furcula, coracoid, and scapula) and wing (humerus, ulna, radius, and carpometacarpus) relative to the pelvic and leg elements (Figures 10-1 to 10-4), reduced pectoral and wing muscle mass relative to total body mass, and a weaker aerodynamic function of flight feathers in the tail and especially the wing (Figures 10-5, 10-6). Other characters associated with flightlessness are a reduced area on the sternal keel for attachment of small pectoralis major and supracoracoideus muscles, a small facies articularis humeralis on the coracoid, a more cranially rotated crista deltopectoralis on the humerus (which thus reduces thrust during the wing's downstroke), a straightening and thinning of the ulna (associated with smaller flexor and extensor muscles), and much shortening of the carpometacarpus (which no longer supports aerodynamic primaries). These trends and others (see Olson & Wingate 2000, Livesey 2003; Figures 10-1 to 10-5 herein) also are found among volant vs. flightless species in many other families of birds. The most extreme case is winglessness in moas (Worthy & Holdaway 2002). Freedom from flight also has led to highly asymmetrical wing molt in flightless rails (Kratter et al. 2001a). The lax condition (noninterlocking barbs) of the plumage in some flightless rails extends to body feathers as well as flight feathers (Lowe 1928).

Flightlessness in rails has evolved independently on oceanic islands around the world (Olson 1977). Dispersal to oceanic islands is undoubtedly facilitated by the propensity of some (certainly not all or even most) species of temperate and tropical rails for long-distance movements that may have a seasonal component but often are not a hard-wired, north-south migration (Remsen & Parker 1990). Oceanic islands colonized by rails had no placental carnivores in their prehuman condition, although various indigenous marsupials, primates, insectivores, or rodents occupied islands with flightless rails in the Moluccas, New Guinea, Australia, Bismarcks, Solomons, and West Indies.

The timing of morphological and behavioral development in rails may make them well prepared for flightlessness. In *Porzana carolina* and *Rallus limicola* (volant North American species), the young are covered with down and capable of running one day after hatching, begin to feed themselves on the third day, and are self-sufficient in picking up food by the seventh day, and yet require weeks to develop substantial wings, flight feathers, and effective thermoregulation (Kaufmann 1987).

The time required for a volant species of rail to evolve flightlessness is unknown, although Olson (1973a) speculated that it could occur rapidly (perhaps as few as tens

Marquesas Islands	Easter Island	Palau	Yap	Mariana Islands	Chuuk	Pohnpei	Kosrae	Marshall Islands	Kiribati	Wake Island	TOTAL M	m	P	p
—	—	M	—	—	—	—	—	—	—	—	2	—	—	—
p	p?	M	—	mp	—	—	—	—	—	m	9	7	6	11–12
—	—	—	—	—	—	—	—	—	—	—	—	—	1	—
—	—	—	—	—	—	—	—	—	—	—	1	—	—	—
Mp	p	—	—	Pp	—	—	m	—	—	—	14	3	10	7
—	—	M	M	MP	M	M	—	M	—	—	12	—	3	—
—	—	—	—	—	—	—	—	—	—	—	2	0–1	1	—
—	—	M	—	MP	—	—	—	—	—	—	3	—	2	—
—	—	—	—	—	—	—	—		—	—	—	2	—	1
p	—	M	—	p	—	—	—	—	—	—	9	—	7	6–7
M, 3p	2p	5M	M	2M, m, 3P, 3p	M	M	m	M	—	m	52	12–13	29	25–27

or hundreds of generations) through neoteny once a population was established on a predator-free oceanic island. Reduction of the flight apparatus would yield the physiological benefit of a lower basal metabolic rate and thus lower overall energy requirements (Ryan et al. 1989, McNab 1994a, 1994b). There is, however, no clear trend in the body size of flightless rails, which range from the smallest to largest species in the family.

Systematic Review

Here I will review each genus of rails in Table 10-1, noting synonyms used with some regularity in the 20th century. The most problematic genus is *Gallirallus*, the limits of which are poorly defined, especially in the New Zealand region. Among the five genera of rails represented by flightless species in tropical Oceania (thus excluding New Zealand), three (*Gallirallus, Porzana, Porphyrio*) also include volant species, often excellent dispersers, that live on both islands and continents. The two others, *Vitirallus* and *Pareudiastes*, are known only from flightless species on Pacific islands.

To provide more complete coverage of Oceania's immense radiation of rails, those from the Hawaiian Islands and New Zealand are compiled in Table 10-2. Hawaiian

Table 10-2 Distribution of genera of rails in the Hawaiian Islands and New Zealand. Certain or presumed vagrants are excluded. M, modern record (volant); m, modern record (flightless); P, prehistoric record (volant); p, prehistoric record (flightless); †, extinct genus. From Olson & James (1991), Trewick (1996, 1997), Taylor (1998), Worthy et al. (1998), Gill (1999), and Holdaway et al. (2001). Note that the totals represent island groups, not individual islands, and that individual species can be represented by both an M and P or an m and p.

	Hawaiian Islands	New Zealand	TOTAL M	m	P	p
Gallirallus	—	M, 3m, P, 3p	1	3	1	3
†*Cabalus*	—	p	—	—	—	1
†*Capellirallus*	—	p	—	—	—	1
†*Diaphorapteryx*	—	p	—	—	—	1
Porzana	2m, 10p	M, P	1	2	1	10
Gallinula	M, P	p	1	—	1	1
Porphyrio	—	M, m, P, 2p	1	1	1	2
Fulica	M, P	M, 2p	2	—	1	2
TOTAL	2M, 2m, 2P, 10p	4M, 4m, 3P, 11p	6	6	5	21

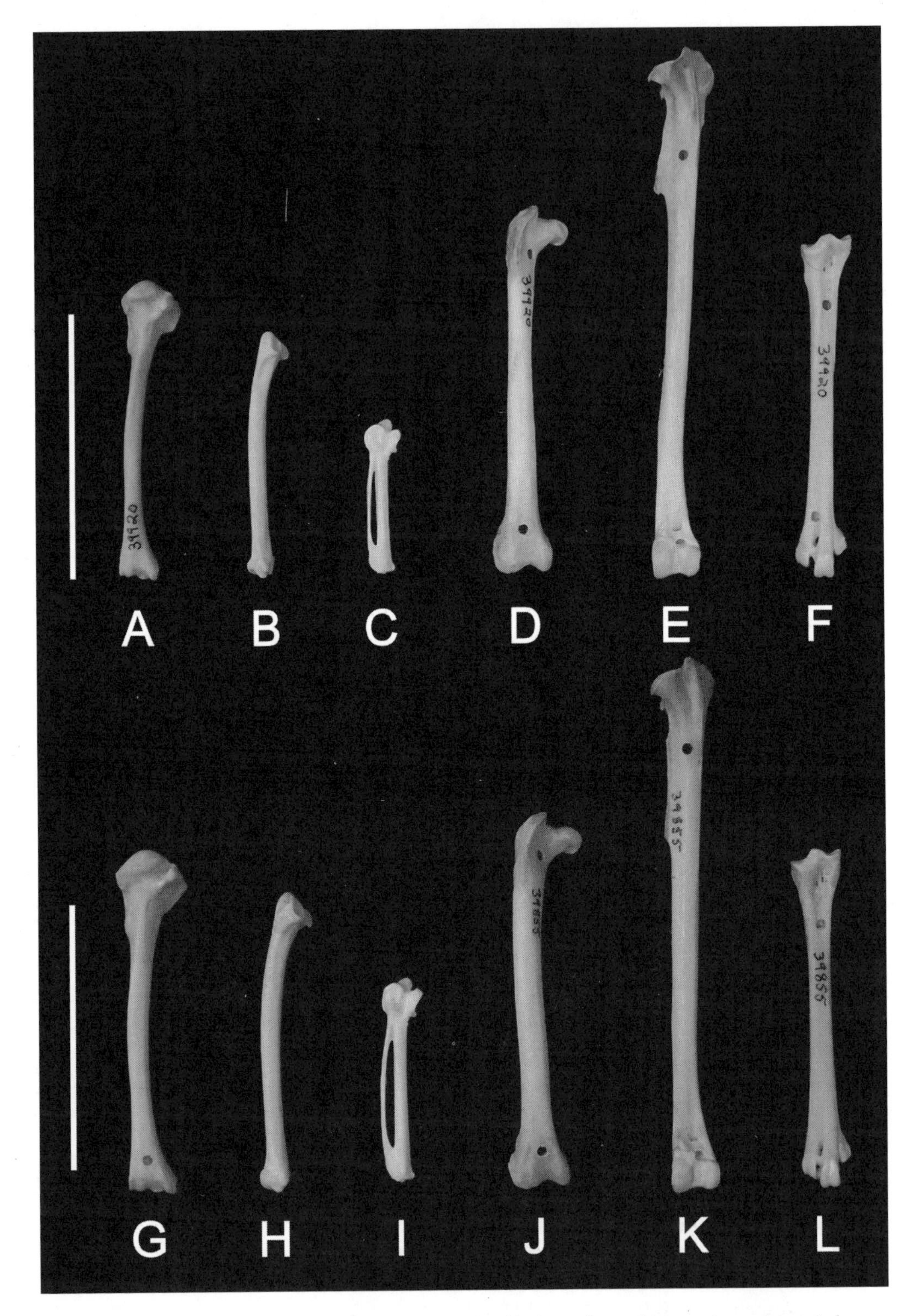

Figure 10-2 Major wing bones (humerus, ulna, carpometacarpus) and leg bones (femur, tibiotarsus, tarsometatarsus) of a medium-sized (245 g) flightless rail *Gallirallus owstoni* (UF 39920, adult ♂, Guam, Mariana Islands; A–F) and a similarly sized (269 g) volant congener *G. philippensis* (UF 39855, adult ♂, Tutuila, American Samoa; G–L). Scale = 40 mm. Photo by J. J. Kirchman.

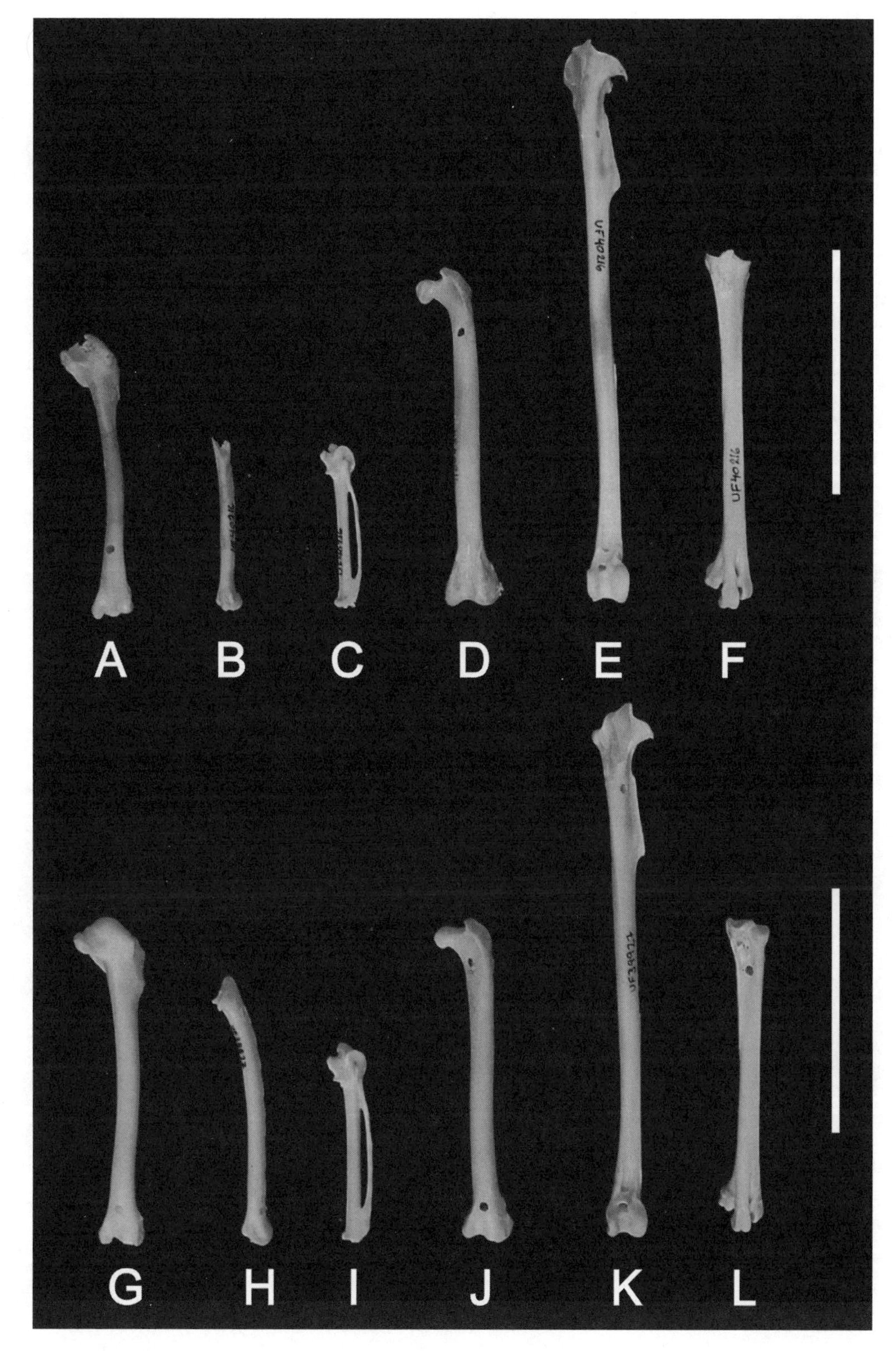

Figure 10-3 Major wing bones (humerus, ulna, carpometacarpus) and leg bones (femur, tibiotarsus, tarsometatarsus) of a medium-sized (229 g), probably flightless rail *Amaurornis moluccanus nigrifrons* (UF 40216, adult ♂, Isabel, Solomon Islands; A–F) and a similarly sized (mass unknown) volant rail *Gallinula chloropus guami* (UF 39927, sex unknown, Guam, Mariana Islands; G–L). Scale = 40 mm. Photo by J. J. Kirchman.

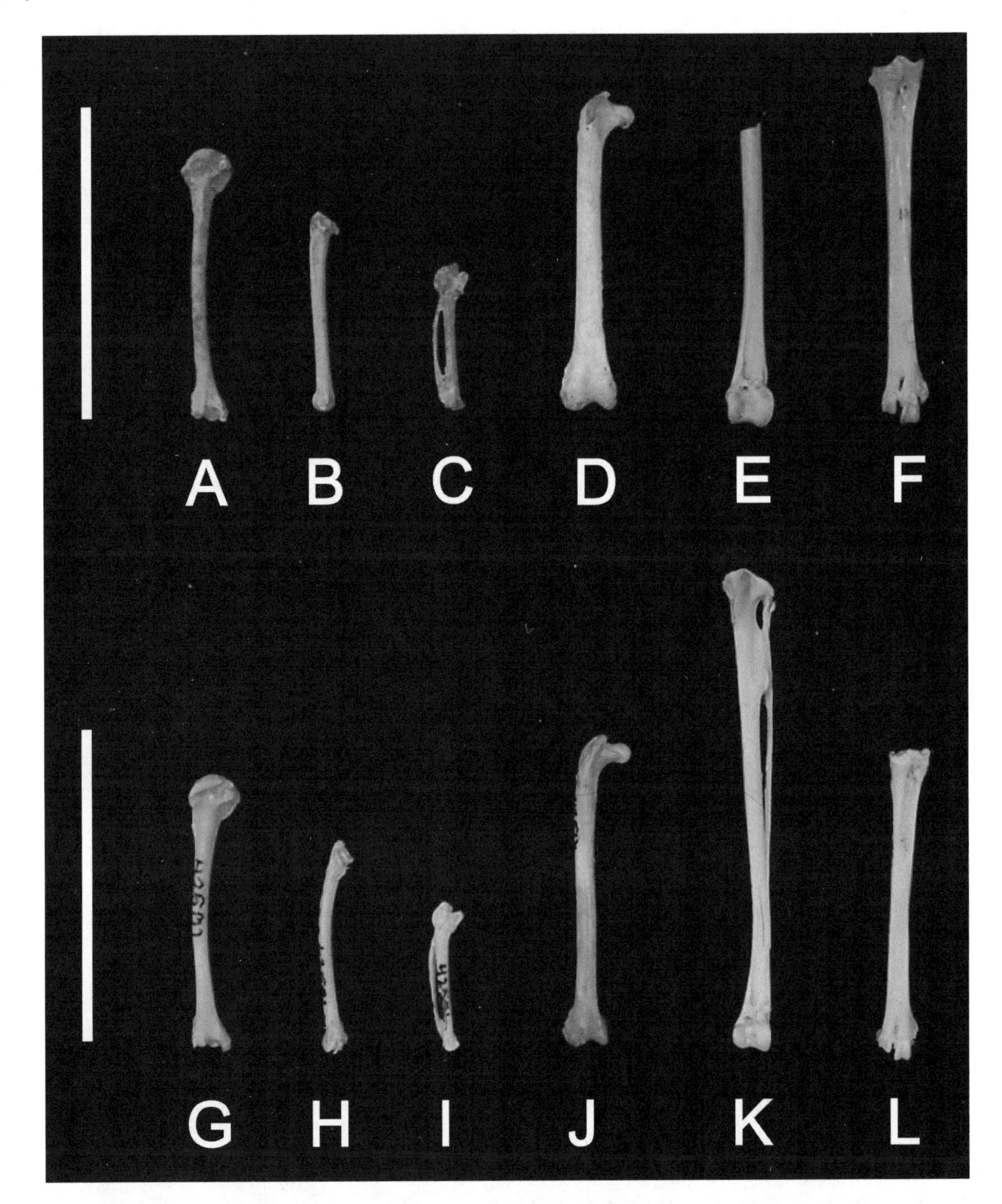

Figure 10-4 Major wing bones (humerus, ulna, carpometacarpus) and leg bones (femur, tibiotarsus, tarsometatarsus) of a small (mass unknown) flightless rail *Porzana rua* (UF 54010, 54905, 55159, 55278, 55287, 59572, representing different unsexed individuals, Mangaia, Cook Islands; A–F) and a small (42 g) volant rail *Porzana tabuensis* (UWBM 42501, imm. ♂, Atiu, Cook Islands; G–L). Scale = 30 mm. Photo by J. J. Kirchman.

Figure 10-5 Spread wing of *Gallirallus [woodfordi] immaculatus* (UF 39547, imm. ♂, 623 g; above) and *Porphyrio porphyrio* cf. *samoensis* (UF 39407, imm. ♂, 918 g; below). Both specimens are from Isabel, Solomon Islands. In spite of heavy wear, note the interlocking barbs on UF 39407, and the lax remiges on UF 39547. Scale = 100 mm. Photo by J. J. Kirchman.

rails are dominated by 12+ flightless species of *Porzana* (Olson & James 1991). In New Zealand and its satellite islands, 13 flightless species (in six to eight genera) were present at human arrival ca. 1000 years ago, with no evidence for prehuman presence of two of the extant, volant species (Olson 1975b, 1977, Trewick 1996, 1997, Worthy 1997a, Worthy & Holdaway 2002). As many as three genera of rails are endemic to New Zealand and its satellite islands (*Cabalus*, *Capellirallus*, and *Diaphorapteryx*, each flightless, extinct, and with single species). Resident forms of coots (*Fulica*) are known in Oceania only in the Hawaiian Islands and New Zealand. The Hawaiian species (*F. alai*) is extant and volant, whereas the two extinct species from New Zealand (*F. chathamensis*, *F. prisca*) were flightless (Millener 1981, Olson 1985b), the world's only known instance of nonvolant coots.

In the accounts of genera that follow, modern distributions of species are from Pratt et al. (1987) and Taylor (1998) unless indicated otherwise.

Rallina

Five of the eight species of *Rallina*, all poorly studied but thought to be volant, live in New Guinea. One of these, *R. tricolor*, occurs in Oceania only in the Bismarcks. The other species in Oceania is *R. eurizonoides*, an uncommon resident in Palau (Engbring 1988) that otherwise occurs in Indonesia, Philippines, and southern Asia. The biology of both species of *Rallina* in Oceania is essentially unknown.

Gallirallus

Gallirallus includes *Habropteryx*, *Habroptila*, *Hypotaenidia*, *Lewinia*, *Nesoclopeus*, *Nesolimnas*, *Tricholimnas*, and possibly *Cabalus* and *Capellirallus*. It is sometimes merged with *Rallus*, but see Olson (1973b).

Because *Nesoclopeus* typically is recognized as a distinct genus, I will begin the *Gallirallus* account with the species often assigned to *Nesoclopeus*, a genus that I cannot define osteologically and with molecular evidence placing it in

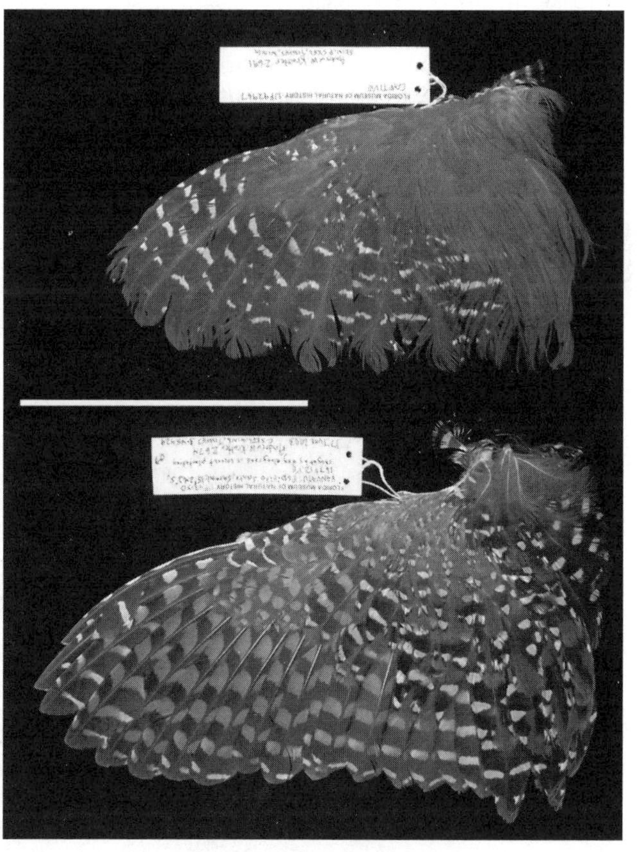

Figure 10-6 Spread wing of *Gallirallus owstoni* (UF 42967, ♀, captive, body mass unknown) and *G. philippensis* (UF 43150, adult ♂, Espiritu Santo, Vanuatu, 210 g). Scale = 100 mm. Photo by J. J. Kirchman.

Table 10-3 The modern (m) and prehistoric (p) distribution of flightless species of *Gallirallus* formerly classified in *Nesoclopeus*. †, extinct species. These species are listed as well in Table 10-4. *Gallirallus rovianae* was described in *Gallirallus*, not *Nesoclopeus* (Diamond 1991), but is included here for the sake of completeness.

									Fiji	
	Buka	Bougain-ville	New Georgia Group	Kolom-bangara	Choiseul	Isabel	Malaita	Guadal-canal	Viti Levu	Ovalau
G. woodfordi	—	—	—	—	—	—	—	m	—	—
G. immaculatus	—	—	—	—	—	m	—	—	—	—
G. tertius	p	m	—	—	—	—	—	—	—	—
G. rovianae	—	—	m	—	—	—	—	—	—	—
G. sp. undetermined	—	—	—	m	m	—	m	—	—	—
†*G. poecilopterus*	—	—	—	—	—	—	—	—	m	m

Gallirallus (B. Slikas, pers. comm.). The three forms of *Nesoclopleus* from the Solomon Islands (Table 10-3) have been regarded as subspecies of a single species (Mayr 1949a), although I consider this to be unresolved. Each is flightless. *Nesoclopeus* (= *Gallirallus*) *woodfordi* is known from Guadalcanal by two specimens, one taken by C. M. Woodford on 6 December 1888 (DWS notes, British Museum, 28 September 1990) and another by G. Dutson in 1998 (Hadden 2002). *Nesoclopeus* (= *Gallirallus*) *immaculatus* (Figure 10-7) of Isabel had been known only from two skins and a set of eggs taken in 1927, and sight records in 1987 and 1988 (Mayr 1949a, Webb 1992), although now it has been found to be common (see below).

Six specimens of *Nesoclopeus* (= *Gallirallus*) *tertius* were taken on Bougainville in 1904, 1908, and 1936, followed by a 60-year absence of records until discovered to be very common in 1999 (Hadden 2002). Bones from Kilu Cave show that *G.* cf. *tertius* once lived on nearby Buka as well (Chapter 5). A sighting of a large rail from New Georgia (Blaber 1990) could pertain to *Gallirallus rovianae* or a closely related undescribed species (Diamond 1991, Taylor 1998). Large, flightless forms of *Gallirallus* s.l. are suspected to exist as well on Choiseul, Kolombangara, and Malaita (Taylor 1998, Diamond 2002, Anonymous 2002, Hadden 2002) and undoubtedly used to inhabit every major island in the Solomon Main Chain (see Figure 5-2).

Figure 10-7 Isabel rail, *Nesoclopeus* (= *Gallirallus*) *immaculatus*, Garanga River, Isabel, Solomon Islands. Photo by DWS, 12 July 1997.

Very little was known about the biology of any species of flightless rail in the Solomons until 1997, when Christopher Filardi, Andrew Kratter, Catherine Smith, Price Webb, and I found *G. immaculatus* to be common in early successional habitats along the Garanga River (Isabel), occurring as well in the adjacent mature forests (Kratter et al. 2001a,b). In spite of its scientific obscurity, this large rail is well known to people on Isabel, who call it *dodo-ili* and loath it for an alleged tendency to damage crops. *Gallirallus immaculatus* had been regarded by scientists as "in low numbers and endangered" (Diamond 1987:79), "obviously scarce" (Blaber 1990:213), a species that "may be close to extinction" (Taylor 1996:161), or "possibly close to extinction" (Taylor 1998:230). On the other hand, Webb (1992) reported *G. immaculatus* to occur locally on Isabel in old gardens and river-edge habitats. Very crudely, I would estimate the population of *G. immaculatus* on Isabel to be no less than 10,000 individuals.

Our specimens from Isabel represent adults and immatures of both sexes. The plumage characters used to distinguish *G. immaculatus* from other species (Mayr 1949a) need to be reevaluated given the extensive variation in our specimens, even within age/sex classes. Males and females of *G. immaculatus* do differ in the color of the hood and its contrast with the breast and mantle. Other characters, such as the extent of white spotting in the wings and tail, do not vary consistently with age or sex (Kratter et al. 2001a). A modern analysis of interisland variation in flightless *Gallirallus* in the Solomons is needed, using data from skins, skeletons, DNA, vocalizations, and life history. Pending new fieldwork, such an analysis would be hampered right now by inadequate specimens and associated data from everywhere but Isabel.

Previous authors described *Gallirallus woodfordi*, *G. tertius*, and *G. immaculatus* as "probably flightless." Men from Isabel told us that the *dodo-ili* never flies, which was confirmed in our specimens, where the distal part of each remix or rectrix is lax, with a weak rachis and barbs that interlock poorly or not at all (Figure 10-5). The nonaerodynamic flight feathers of *G. immaculatus* cannot support the body weight. Skeletal elements of its wing are small relative to those of the leg (Figure 10-1). Also, its sternal keel is much reduced, as is the relative mass of the breast muscles (pectoralis major and supracoracoideus) compared to volant rails of similar body mass.

The only species assigned to "*Nesoclopeus*" outside of the Solomons is "*N.*" (= *Gallirallus*) *poecilopterus* from Viti Levu and nearby Ovalau, Fiji, represented by five specimens collected by Edgar Layard in 1875, another taken before 1864 (DWS notes, British Museum, 28 September 1990), at least eight other 19th-century specimens (Fullagar et al. 1982), and possible sightings in the 1970s (Holyoak 1979). I disagree with the claim (Watling 2001:111) that *G. poecilopterus* was volant. That the Fijian *G. poecilopterus* may be conspecific with *G. woodfordi* s.l. of the Solomons (Ripley 1977, Pratt et al. 1987, Taylor 1996, 1998) is illogical, as these species are flightless, differ considerably in external morphology (Peters 1932, DWS pers. obs.), and are separated by 1100 km of deep ocean. Perhaps these authors took too literally the statement by Mayr (1949b:14) that *G. woodfordi* s.l. "is a geographical representative of *N.* [= *G.*] *poecilopterus* of Fiji." Flightlessness undoubtedly evolved independently from volant ancestors in both Fiji and the Solomons.

Three prehistoric bone fragments collected on Viti Levu by Edward Gifford in 1927, now at the National Museum of Natural History (USNM), were identified by Alexander Wetmore as *Nesoclopeus poecilopterus* (unpublished notes with the specimens). In fact, two of these bones are tibiotarsi of the rail *Porphyrio porphyrio* and the third is from a fish. The only prehistoric evidence of a flightless *Gallirallus* in Fiji is from the Lau Group, where *Gallirallus* undescribed sp. D is known from seven bones from Lakeba and three more from Rockshelter 1 on nearby Aiwa Levu, a small island connected to Lakeba in the late Pleistocene (Chapter 6).

Moving beyond the species formerly regarded as *Nesoclopeus*, one of Oceania's most widespread landbirds is *Gallirallus philippensis* (Figure 10-8), a volant species that prefers nonforested habitats (Steadman 1998, Steadman & Freifeld 1998). Geographic variation in plumage and size of *G. philippensis* within Oceania is minimal (Schodde & de Naurois 1982), suggesting that substantial interisland

Figure 10-8 *Gallirallus philippensis*, captured in a leg-hold snare by boys on `Eua, Tonga. Photo by J. G. Stull, 4 December 1988.

Table 10-4 The distribution of *Gallirallus* in Oceania. Only *G. philippensis* is volant. M, modern record (volant); m, modern record (flightless); P, prehistoric record (volant); p, prehistoric record (flightless); †, extinct species. BA, Bismarck Archipelago; CH, Chuuk; CI, Cook Islands; EI, Easter Island; FI, Fiji; KO, Kosrae; MA, Marquesas Islands; MI, Mariana Islands; NC, New Caledonia; NI, Niue; PA, Palau; PI, Pitcairn Group; PO, Pohnpei; RC, Remote Central Pacific island groups; SA, Samoa; SI, Solomon Islands; SO, Society Islands; TA, Tuamotu Archipelago; TO, Tonga; TU, Tubuai; VA, Vanuatu; YA, Yap.

	BA	SI	VA	NC	FI	TO	SA	NI	CI	TU	SO	TA	PI	EI	MA	PA	YA	MI	CH	PO	KO	RC
G. philippensis	MP	MP	MP	MP	MP	MP	MP	MP	—	—	—	—	—	—	—	M	—	—	—	—	—	M
G. insignis	m	—	—	—	—	—	—	—	—	—	—	—	—	—	—	—	—	—	—	—	—	—
†*G.* undescribed sp. A	p	—	—	—	—	—	—	—	—	—	—	—	—	—	—	—	—	—	—	—	—	—
G. woodfordi	—	m	—	—	—	—	—	—	—	—	—	—	—	—	—	—	—	—	—	—	—	—
G. immaculatus	—	m	—	—	—	—	—	—	—	—	—	—	—	—	—	—	—	—	—	—	—	—
G. tertius	—	mp	—	—	—	—	—	—	—	—	—	—	—	—	—	—	—	—	—	—	—	—
G. sp. undetermined	—	m	—	—	—	—	—	—	—	—	—	—	—	—	—	—	—	—	—	—	—	—
G. rovianae	—	m	—	—	—	—	—	—	—	—	—	—	—	—	—	—	—	—	—	—	—	—
†*G.* undescribed sp. B	—	p	—	—	—	—	—	—	—	—	—	—	—	—	—	—	—	—	—	—	—	—
†*G.* undescribed sp. C	—	—	p	—	—	—	—	—	—	—	—	—	—	—	—	—	—	—	—	—	—	—
†*G. lafresnayanus*	—	—	—	mp	—	—	—	—	—	—	—	—	—	—	—	—	—	—	—	—	—	—
†*G. poecilopterus*	—	—	—	—	m	—	—	—	—	—	—	—	—	—	—	—	—	—	—	—	—	—
†*G.* undescribed sp. D	—	—	—	—	p	—	—	—	—	—	—	—	—	—	—	—	—	—	—	—	—	—
†*G.* undescribed sp. E	—	—	—	—	—	p	—	—	—	—	—	—	—	—	—	—	—	—	—	—	—	—
†*G.* undescribed sp. F	—	—	—	—	—	p	—	—	—	—	—	—	—	—	—	—	—	—	—	—	—	—
†*G.* undescribed sp. G	—	—	—	—	—	p	—	—	—	—	—	—	—	—	—	—	—	—	—	—	—	—
†*G. huiatua*	—	—	—	—	—	—	—	p	—	—	—	—	—	—	—	—	—	—	—	—	—	—
†*G. ripleyi*	—	—	—	—	—	—	—	—	p	—	—	—	—	—	—	—	—	—	—	—	—	—
†*G. pacificus*	—	—	—	—	—	—	—	—	—	—	p	—	—	—	—	—	—	—	—	—	—	—
†*G.* undescribed sp. H	—	—	—	—	—	—	—	—	—	—	p	—	—	—	—	—	—	—	—	—	—	—
†*G.* undescribed sp. I	—	—	—	—	—	—	—	—	—	—	—	—	—	—	P	—	—	—	—	—	—	—
†*G.* undescribed sp. J	—	—	—	—	—	—	—	—	—	—	—	—	—	—	p	—	—	—	—	—	—	—
†*G.* undescribed sp. K	—	—	—	—	—	—	—	—	—	—	—	—	—	—	p	—	—	—	—	—	—	—
†*G.* undescribed sp. L	—	—	—	—	—	—	—	—	—	—	—	—	—	—	p	—	—	—	—	—	—	—
†*G. owstoni*	—	—	—	—	—	—	—	—	—	—	—	—	—	—	—	—	—	m	—	—	—	—
G. undescribed sp. M	—	—	—	—	—	—	—	—	—	—	—	—	—	—	—	—	—	p	—	—	—	—
†*G.* undescribed sp. N	—	—	—	—	—	—	—	—	—	—	—	—	—	—	—	—	—	p	—	—	—	—
†*G.* undescribed sp. O	—	—	—	—	—	—	—	—	—	—	—	—	—	—	—	—	—	p	—	—	—	—
†*G.* undescribed sp. P	—	—	—	—	—	—	—	—	—	—	—	—	—	—	—	—	—	p	—	—	—	—
†*G. wakensis*	—	—	—	—	—	—	—	—	—	—	—	—	—	—	—	—	—	—	—	—	—	m
TOTAL	M, P, m, p	M, P, 5m, 2p	M, P, p	M, P, m, p	M, P, m, p	M, P, 3p	M, P	M, P, p	p	—	2p	—	—	—	4p	M	—	m, 4p	—	—	—	M, m

dispersal may still be taking place. Absent from precultural bone deposits in New Caledonia and Tonga, *G. philippensis* may have colonized much of Remote Oceania (the current eastern limit is Niue) only after human arrival ca. 2800 years ago (Balouet & Olson 1989, Steadman 1993a). The recent (past few decades) establishment of a population of *G. philippensis* on Niulakita Atoll in remote Tuvalu (McQuarrie 1991) indicates that the range of this trampy rail is still expanding.

Contrasting with the volant, poorly differentiated *G. philippensis* is the vast radiation of flightless species of *Gallirallus* that once occupied islands from the Marianas and Bismarcks at least as far east as the Marquesas (Table 10-4). Just west of what generally is regarded as Oceania, other flightless species of *Gallirallus* occur on Okinawa (Ryukyu Islands, Japan; *G. okinawae*) and Halmahera (Moluccas; *G.*["*Habroptila*"] *wallacei*). Because a flightless species of *Gallirallus* even lived on remote Wake Atoll, they probably were widespread in Micronesia at human contact. The absence of flightless species of *Gallirallus* in the rich fossil deposits of the Hawaiian Islands (Olson & James 1991) or Henderson Island (Chapter 7) defines the probable eastern margin of their range in both the northern and southern hemispheres. No species of *Gallirallus* are known from islands in the Indian or Atlantic Oceans (Olson 1986b).

Among the estimated hundreds of species of *Gallirallus* that existed at human contact, the only survivors are *G. owstoni* (Figure 10-9) on Guam (Jenkins 1979; exists today as captive only), *G. insignis* on New Britain, *G. rovianae* s.l. on at least five formerly connected islands in the New Georgia Group (Solomons; Diamond 1991, Taylor 1998), *G. woodfordi, G. immaculatus,* and *G. tertius* elsewhere in the Solomons (see above), *G. sylvestris* on Lord Howe Island (*Tricholimnas conditicius* Peters & Griscom 1928 is a synonym; Olson 1992), and *G. australis* in New Zealand. Conservation programs have helped to sustain *G. owstoni, G. sylvestris,* and *G. australis*; the others are on their own.

Figure 10-9 *Gallirallus owstoni*, captive-bred adult, Rota, Commonwealth of the Northern Mariana Islands. Photo by W. Burt, February 1990.

Endemic flightless species of *Gallirallus* are recorded from 12 island groups in Oceania, with multiple species known from nine of these groups, representing all four major regions of Oceania (Table 10-4). Six of the extinct flightless species have been described; 16 others await description in my lab. In the Marquesas (East Polynesia) and Marianas (Micronesia), each of the four islands with landbird fossils has an endemic species of *Gallirallus*. Among island groups with respectable fossil records from multiple islands, only in Tonga (West Polynesia) have any islands failed to yield evidence for extinct, flightless species of *Gallirallus*. This probably is an artifact of inadequate bone samples (see, for example, Table 6-21). On 'Eua, the one Tongan island with precultural fossils (Steadman 1993a; Chapter 6 herein), an endemic flightless species of *Gallirallus* was one of the most common species of landbirds.

On Buka (Solomons), I have identified two flightless species of *Gallirallus*. One is the size of *G.* [*Nesoclopeus*] *immaculatus* and thus probably referable to *G. tertius*, the form now on Bougainville, which formerly was connected to Buka. The other species is smaller and undescribed. This is the first instance of multiple, sympatric, flightless species of *Gallirallus* in tropical Oceania. Depending on generic-level taxonomy, the only other example of this would be in the chilly Chatham Islands, off New Zealand, where *Cabalus* (= *Gallirallus*?) *modestus* occurred with *G. dieffenbachii* (Trewick 1997, Worthy & Holdaway 2002:571).

The relationships of the many flightless species of *Gallirallus* to extant, volant congeners will be difficult to resolve using osteology alone. Maybe a blend of data from osteology, plumage (where possible), and DNA will be able to untangle at least part of this complex situation, currently under study by Jeremy Kirchman. The volant *G. pectoralis, G. mirificus, G. philippensis, G. striatus,* and *G. torquatus* are prime candidates for being ancestral to endemic flightless species on islands. Olson (1973b) proposed that *G. owstoni, G. wakensis, G. pacificus, G. sylvestris,* and the various flightless New Zealand species were derived from *G. philippensis*, whereas *G. insignis* evolved from *G. torquatus*. Among the five volant candidate species, only *G. philippensis* has resident populations in Oceania.

Table 10-5 The distribution of *Porzana* in Oceania. Only *P. tabuensis* is volant. M, modern record (volant); m, modern record (flightless); P, prehistoric record (volant); p, prehistoric record (flightless). †, extinct species. BA, Bismarck Archipelago; CH, Chuuk; CI, Cook Islands; EI, Easter Island; FI, Fiji; KO, Kosrae; MA, Marquesas Islands; MI, Mariana Islands; NC, New Caledonia; NI, Niue; PA, Palau; PI, Pitcairn Group; PO, Pohnpei; RC, Remote Central Pacific island groups; SA, Samoa; SI, Solomon Islands; SO, Society Islands; TA, Tubuai, TO, Tonga; TU, Tuamotu Archipelago; VA, Vanuatu; YA, Yap.

	BA	SI	VA	NC	FI	TO	SA	NI	CI	TU	SO	TA	PI	MA	EI	PA	YA	MI	CH	PO	KO	RC
P. tabuensis	MP	M	MP	MP	MP	MP	MP	MP	MP	M	MP	M	M	—	M	—	—	—	—	—	—	—
†*P.* undescribed sp. A	—	—	p	—	—	—	—	—	—	—	—	—	—	—	—	—	—	—	—	—	—	—
†*P.* undescribed sp. B	—	—	—	—	p	—	—	—	—	—	—	—	—	—	—	—	—	—	—	—	—	—
†*P. rua*	—	—	—	—	—	—	—	—	p	—	—	—	—	—	—	—	—	—	—	—	—	—
†*P.* undescribed sp. C	—	—	—	—	—	—	—	—	p	—	—	—	—	—	—	—	—	—	—	—	—	—
†*P. nigra*	—	—	—	—	—	—	—	—	—	—	m	—	—	—	—	—	—	—	—	—	—	—
P. atra	—	—	—	—	—	—	—	—	—	—	—	—	mp	—	—	—	—	—	—	—	—	—
†*P.* undescribed sp. D	—	—	—	—	—	—	—	—	—	—	—	—	—	p	—	—	—	—	—	—	—	—
†*P.* undescribed sp. E	—	—	—	—	—	—	—	—	—	—	—	—	—	p	—	—	—	—	—	—	—	
†*P.* undescribed sp. F	—	—	—	—	—	—	—	—	—	—	—	—	—	p	—	—	—	—	—	—	—	—
†*P.* undescribed sp. G	—	—	—	—	—	—	—	—	—	—	—	—	—	—	p	—	—	—	—	—	—	—
†*P.* undescribed sp. H	—	—	—	—	—	—	—	—	—	—	—	—	—	—	—	—	—	p	—	—	—	—
†*P.* undescribed sp. I	—	—	—	—	—	—	—	—	—	—	—	—	—	—	—	—	—	p	—	—	—	—
†*P.* undescribed sp. J	—	—	—	—	—	—	—	—	—	—	—	—	—	—	—	—	—	p	—	—	—	—
†*P. monasa*	—	—	—	—	—	—	—	—	—	—	—	—	—	—	—	—	—	—	—	—	p	—
TOTAL	M, P	M	M, P, p	M, P	M, P, p	M, P	M, P	M, P	M, P, 2p	M	M, P, m	M	M, m, p	3p	p	—	—	3p	—	—	P	—

Vitirallus

As presently understood, the extinct genus *Vitirallus* consists of a single flightless species, *V. watlingi*, known from numerous fossils found in three caves on Viti Levu, Fiji (Worthy et al. 1999, Worthy 2004). Perhaps its most distinctive features are a very long, narrow, curved bill and extremely short wing elements. Worthy (2004) notes that *Vitirallus* could be a highly derived offshoot of *Gallirallus*. Similar forms can be expected from other islands in Fiji.

Gymnocrex

The three species of *Gymnocrex* are poorly studied, forest-dwelling rails confined to Indonesia, New Guinea, and nearby islands (Lambert 1989, 1998a). The only record in Oceania is *G. plumbeiventris* on New Ireland, where it presumably but not certainly is resident. Its apparent absence on New Britain probably is due to inadequate modern or prehistoric sampling.

Porzana

Porzana includes *Aphanolimnas*, *Nesophylax*, *Pennula*, and *Porzanula*. These small, typically dark-plumaged rails (often called "crakes") usually have red eyes as adults. One or more species of *Porzana* once inhabited island groups from Near Oceania eastward to Easter Island, northward to the Mariana and Hawaiian Islands, and south to New Zealand (Table 10-5). The volant *Porzana tabuensis* (Figure 10-10) is the most widespread rail in Oceania today, in spite of many populations extirpated in recent centuries. Occurring locally in wetlands across Melanesia and Polynesia but not Micronesia, *P. tabuensis* can be difficult to detect; rigorous surveys with tape recorders, such as those by Kaufmann (1988) in New Zealand, might find *P. tabuensis* living on a number of islands where previously unknown, especially in Melanesia. As with *Gallirallus philippensis*, the morphological variation in *P. tabuensis* in Oceania lacks obvious geographic patterns (Onley 1982), supporting a hypothesis of recent colonization and continuing interisland gene flow.

Figure 10-10 *Porzana tabuensis*, downy young, Atiu, Cook Islands. Photo by DWS, 24 September 1987.

Endemic flightless species of *Porzana* have not been discovered in Near Oceania, although the fossil evidence is limited. In Remote Oceania outside of the Hawaiian Islands, 14 flightless species (13 of them extinct) are known so far from Vanuatu, Fiji, East Polynesia, and Micronesia. The absence of flightless *Porzana* in Tonga (Table 10-5) may be due to inadequate sampling, although Tonga has good fossil coverage (Chapter 6). The prehistoric lack of flightless species of *Porzana* in New Zealand (Table 10-2) seems to be genuine, given this region's vast fossil record (Holdaway et al. 2001, Worthy & Holdaway 2002). The former presence of flightless species of *Porzana* in the Marianas (*P*. undescribed spp. H, I, J) and on Kosrae (*P. monasa*; Steadman 1987: Figure 5) suggests that other flightless congeners once lived elsewhere in Micronesia.

The islands once inhabited by flightless species of *Porzana* range from as small, low, and isolated as Laysan (1.8 km^2, 12 m elevation) in the Hawaiian Islands (Baldwin 1947) to medium-sized, high islands such as Kosrae (109 km^2, 629 m elevation) or Mangaia (52 km^2, 169 m elev.), to Tahiti (1042 km^2, 2241 m elevation), to the immense (10,458 km^2, 4025 m elev.) island of Hawai'i (Steadman 1987, Olson & Steadman 1987, Walters 1988, 1989, Olson & James 1991, Olson & Ziegler 1995). Similar forms probably once inhabited even atolls of the Remote Central Pacific Islands (Chapter 8), which are distantly bounded on the north (Hawaiian Islands), southeast (Marquesas, Society Islands), south (Cook Islands), and west (Kosrae) by islands where flightless species of *Porzana* once lived.

Porzana provides the most examples of multiple, sympatric, flightless species. On Mangaia (Cook Islands), *P. rua* (Figure 10-4) and *P*. undescribed sp. C are each known from precultural as well as archaeological strata (Steadman 1987, Steadman & Kirch 1990, Kirch et al. 1995). In Mangaia's earliest cultural deposits, both flightless species also are found alongside bones of the volant, extant *P. tabuensis* (Chapter 7). On Ua Huka (Marquesas), a large and a small flightless species of *Porzana* (*P*. undescribed spp. F and G) co-occur at the Hane site, which has the largest bone sample in the Marquesas. The bones of *P. rua* and *P*. undescribed sp. F are approximately the size of those in *P. atra* of Henderson Island, the world's only surviving flightless species of *Porzana*. In these three species, the wing bones are slightly smaller than in *P. tabuensis* and are modified for flightlessness, whereas the leg bones are larger. All bones of

P. undescribed sp. C (Mangaia) and *P*. undescribed sp. G (Ua Huka) are smaller than in any living species of Pacific rail. These tiny rails must have had wing measurements <30 mm and body weights <20 g.

In the Hawaiian Islands, 12 flightless species of *Porzana* (all extinct) have been discovered (Olson & James 1991, Slikas 2003). Endemic small-large pairs of species used to live on Kaua'i and O'ahu, whereas Maui and the island of Hawai'i each had endemic triplets of small-medium-large species. On Maui, the tarsometatarsal lengths were 17.7–23.1 mm ($n = 9$) in *P. keplerorum*, 26.1 mm ($n = 1$) in the undescribed "medium Maui rail," and 34.2–37.9 mm ($n = 6$) in *P. severnsi* (Olson & James 1991). The only two Hawaiian flightless species of *Porzana* that survived to historic times were *P. sandwichensis* on the largest, easternmost island (Hawai'i) and *P. palmeri* on the distant atoll of Laysan. The rail that died out on Lisianski in the early 19th century, before specimens were obtained, probably also was a species of *Porzana* (Olson & Ziegler 1995).

I believe that many, if not most, of the flightless species of *Porzana* from Polynesia and Micronesia were derived from *P. tabuensis*. The dark plumage and red eyes and legs of *P. atra* (Henderson) and *P. monasa* (Kosrae) would support this. In the Hawaiian Islands, on the other hand, some of the flightless species probably were derived from *P. pusilla*, which today inhabits Africa, Eurasia, Australia, and New Zealand (Olson 1973a, Olson & James 1991, Slikas et al. 2002). Outside of the Pacific, a flightless species of *Porzana* is known from St. Helena, South Atlantic (Olson 1973a).

Poliolimnas

Recognition of *Poliolimnas* as distinct from *Porzana* follows Olson (1970) and Steadman (1987). The only Old World species of *Poliolimnas*, the extant and volant *P. cinereus*, is widespread in Melanesia, West Polynesia, Micronesia (Table 10-1), and tropical Australasia. Nevertheless, this long-toed rail is localized, rare (cryptic?), and poorly known. It has been extirpated on some islands in Fiji and the Marianas (Steadman 1999a; Chapters 6, 8 herein) but has not been found prehistorically on any island east of its current range. No flightless forms of *Poliolimnas* are known. The lack of recognized geographic variation in *P. cinereus* in Oceania or elsewhere (Taylor 1998) suggests that it is a relatively recent colonizer of islands or maintains some interisland gene flow.

Amaurornis

The nine species of *Amaurornis* are collectively widespread in the Old World tropics. Olson (1973b) and Taylor (1998) pointed out the similarities of *Amaurornis* to *Porzana*; at least the species of *Amaurornis* that I have examined (*A. olivaceus*, *A. moluccanus*, and *A. phoenicurus*) are similar osteologically to gallinules (*Gallinula*) as well. The only species of *Amaurornis* in Oceania is *A. moluccanus*, recorded from 13 islands in the Bismarcks and six islands in the Solomons (Tables 5-3, 5-7). The prehistoric record of *A. olivaceus* from New Ireland (Steadman et al. 1999a) can be changed to *A. moluccanus* following Lambert (1998b), who restricted the name *A. olivaceus* to Philippine populations.

A specimen of *Amaurornis moluccanus* from Isabel, Solomon Islands (Kratter et al. 2001a,b), has forelimb osteology (Figure 10-3) and flight-feather morphology suggesting that it may be flightless. A reappraisal of the systematics of *A. moluccanus* in Oceania is needed, but would require many more skins, skeletons, and tissue samples than are available now. Our poor understanding of *Amaurornis* is exemplified further by the recent discovery of an undescribed living species in the Talaud Islands, Indonesia (north of Sulawesi and Halmahera, south of Mindanao; Lambert 1998b).

Gallinula (includes *Tribonyx*)

Called moorhens, gallinules, or native-hens, the two species of *Gallinula* in tropical Oceania are extant and volant (Table 10-6). The aquatic, highly dispersive, and nearly cosmopolitan *G. chloropus* occurs in the Hawaiian Islands, Palau, and Marianas, with historic and prehistoric losses of populations in the Marianas (Pratt et al. 1987, Stinson et al. 1991, Worthington 1998, Steadman 1999a, Takano & Haig 2004). The other species is *G. tenebrosa* of Indonesia, New Guinea, Australia, and New Caledonia (Balouet & Olson 1989; New Caledonia not included in the range map of Taylor 1998). The absence of *G. tenebrosa* in freshwater habitats in the Bismarcks, Solomons, and Vanuatu is puzzling;

Table 10-6 The distribution of *Gallinula* and *Pareudiastes* in Oceania. M, modern record (volant); m, modern record (flightless); P, prehistoric record (volant); p, prehistoric record (flightless); †, extinct species. MI, Mariana Islands; NC, New Caledonia; PA, Palau; SA, Samoa; SI, Solomon Islands.

	SI	NC	SA	PA	MI
G. chloropus	—	—	—	M	MP
G. tenebrosa	—	MP	—	—	—
P. sylvestris	m	—	—	—	—
†*P.* undescribed sp.	p	—	—	—	—
†*P. pacificus*	—	—	m	—	—

Figure 10-11 *Pareudiastes pacificus*, unsexed adult, Savai'i, Samoa, in the British Museum (BMNH 1923. 9.7.10, collected by Rev. G. Brown 1874). Photo by DWS, September 1991.

either it has been extirpated, or the New Caledonian population represents colonization directly from Australia.

Flightless species of *Gallinula* from outside of tropical Oceania are *G. mortierii* from mainland Australia (extinct) and Tasmania (extant) and *G. hodgeni* (extinct) from New Zealand (*G. hodgeni*; Olson 1975b, 1977, Baird 1984, 1986). In the South Atlantic are two more flightless species, *G. nesiotis* on Tristan da Cunha and *G. comeri* on Gough Island, both extant (Olson 1973a).

Pareudiastes (includes *Edithornis*)

Recognition of *Pareudiastes* as distinct from *Gallinula* and *Porphyrio* follows Olson (1975a). The documented distribution of *Pareudiastes* is geographically illogical (Table 10-6) and must be incomplete. Both species known from skins have short, flimsy rectrices and remiges as in other flightless rails. One is *P. pacificus* (Figure 10-11), represented by three skins and an egg taken from 1869 to 1873 on Savai'i, Samoa (and possible sight records there as late as 1908; Hartlaub & Finsch 1871, Stair 1897, Olson 1975a, duPont 1976, Pratt et al. 1987, Taylor 1998). It also may have lived on nearby Upolu (Whitmee 1874), although specimens are lacking. The second species is *P. sylvestris*, based on a specimen taken at ca. 600 m elevation on Makira (San Cristobal), Solomon Islands, on 4 December 1929, described by Mayr (1933) as a new genus and species, *Edithornis sylvestris*. Little else is known about *P. sylvestris*. It was not found during an expedition in 1953 although the species was well known to local people (Cain & Galbraith 1956, Galbraith & Galbraith 1962).

The record of "*Gallinula* [= *Pareudiastes*] undescribed sp." from 'Eua, Tonga (Steadman 1993a) is incorrect; it was based on two misidentified bones from the 'Anatu site. Nevertheless, Makira and Savai'i, which lie 2700 km from each other, were not the only islands inhabited by a species of *Pareudiastes*. A distal end of a tarsometatarsus from Kilu Cave, Buka, Solomon Main Chain, agrees overall with the illustration of *P. pacificus* in Olson (1975a) and differs markedly from that in *Gallinula*, *Poryphyrio*, or any other rail. Given the geological history of the Solomons (Chapters 1, 5), the Buka specimen is unlikely to represent the same flightless species as on Makira. I expect that fossils of *Pareudiastes* will be found elsewhere in Melanesia and West Polynesia.

Modern sightings of possible *Pareudiastes* need to be scrutinized; juveniles of *Porphyrio porphyrio* are dark, seldom fly, and easily could be misidentified as a species of *Pareudiastes*. Specimen-based records are essential to learn whether any flightless gallinules, regardless of genus, survive in tropical Oceania. While it is difficult to be optimistic about *P. pacificus*, the survival of *P. sylvestris* is more likely because so little fieldwork has been done in recent decades on logistically challenging Makira.

Porphyrio (includes *Notornis* and *Porphyrula*)

In Oceania, swamphens (sometimes called gallinules) of the genus *Porphyrio* feature an extant, volant species (the highly variable *P. porphyrio*; Figures 10-1, 10-5, 10-12) and several extinct flightless species (Table 10-7). Widespread in

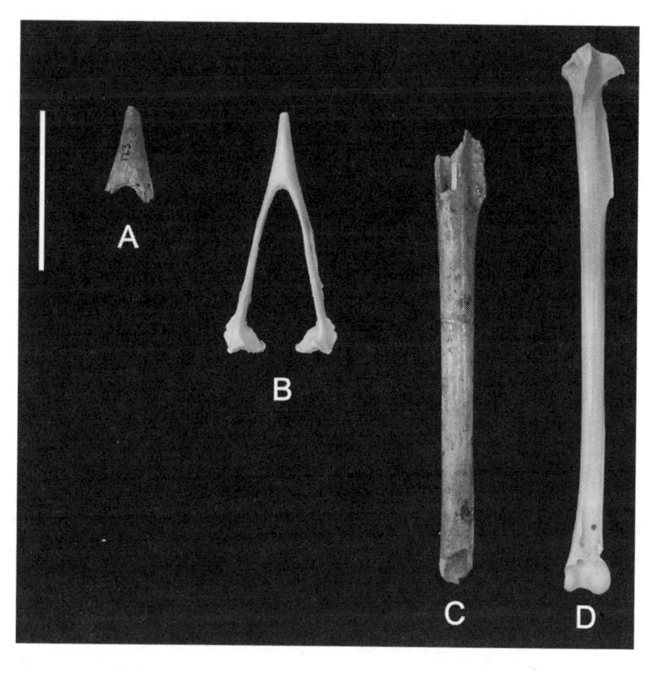

Figure 10-12 Mandible (A, B) and tibiotarsus (C, D) of *Porphyrio* undescribed sp. A (A, C) from the Balof 1 site, New Ireland, compared to those in *P. porphyrio samoensis* (B, D; UF 39388, adult ♂, 812 g Rennell, Solomon Islands). Scale = 40 mm. Photo by J. J. Kirchman.

Table 10-7 The distribution of *Porphyrio* in Oceania. M, modern record (all are volant); P, prehistoric record (volant); p, prehistoric record (flightless); †, extinct species. BA, Bismarck Archipelago; CH, Chuuk; CI, Cook Islands; EI, Easter Island; FI, Fiji; KO, Kosrae; MA, Marquesas Islands; MI, Mariana Islands; NC, New Caledonia; NI, Niue; PA, Palau; PI, Pitcairn Group; PO, Pohnpei; RC, Remote Central Pacific island groups; SA, Samoa; SI, Solomon Islands; SO, Society Islands; TA, Tubuai; TO, Tonga; TU, Tuamotu Archipelago; VA, Vanuatu; YA, Yap.

	BA	SI	VA	NC	FI	TO	SA	NI	CI	TU	SO	TA	PI	EI	MA	PA	YA	MI	CH	PO	KO	RC
P. porphyrio	MP	M	MP	MP	MP	MP	M	MP	—	—	—	—	—	—	—	M	—	—	—	—	—	—
† *P.* undescribed sp. A	p	—	—	—	—	—	—	—	—	—	—	—	—	—	—	—	—	—	—	—	—	—
† *P.* undescribed sp. B	—	p	—	—	—	—	—	—	—	—	—	—	—	—	—	—	—	—	—	—	—	—
† *P. kukwiedei*	—	—	—	p	—	—	—	—	—	—	—	—	—	—	—	—	—	—	—	—	—	—
† *P. paepae*	—	—	—	—	—	—	—	—	—	—	—	—	—	—	p	—	—	—	—	—	—	—
† *P.* undescribed sp. C	—	—	—	—	—	—	—	—	p	—	—	—	—	—	—	—	—	—	—	—	—	—
† *P.* undescribed sp. D	—	—	—	—	—	—	—	—	—	—	p	—	—	—	—	—	—	—	—	—	—	—
† *P.* undescribed sp. E	—	—	—	—	—	—	—	—	—	—	—	—	—	—	—	—	—	p	—	—	—	—

the Old World tropics and subtropics, *P. porphyrio* occurs across Melanesia and West Polynesia, but is absent from East Polynesia and from Micronesia except Palau. All populations in Oceania were regarded as *P. p. samoensis* by Taylor (1998), a judgment compatible with its highly dispersive nature (as in other volant species of *Porphyrio* s.l.; Remsen & Parker 1990). This subspecific evaluation, however, will require much more morphological and molecular data to resolve. Nevertheless, as with *Gallirallus philippensis* and *Porzana tabuensis* (Oceania's other widespread species of volant rails), there are no precultural records of *Porphyrio porphyrio* from New Caledonia or Tonga, suggesting a recent (late Holocene) arrival in Remote Oceania.

The flightless *Porphyrio* undescribed sp. A (New Ireland; Figure 10-12) and *P.* undescribed sp. B (Buka) were much larger than the extinct *P. paepae* of the Marquesas (Steadman 1988), taller but less stout than *P. hochstetteri* (extinct) and *P. mantelli* (barely extant) of New Zealand, and even larger in all dimensions than the extinct *P. kukwiedei* of New Caledonia, which was the largest species of *Porphyrio* when it was described (Balouet & Olson 1989). I list the Marquesan *P. paepae* as flightless although its limb proportions are borderline; it may have been capable of a little flight. Joined as one island in the late Pleistocene, only 3 km of ocean now separate Hiva Oa and Tahuata (Figure 3-14), the two islands where *P. paepae* has been found.

Table 10-8 The distribution of rails in Melanesia. M, modern record (volant); m, modern record (flightless); P, prehistoric record (volant); p, prehistoric record (flightless); †, extinct species. From data in Chapter 5.

	Bismarcks	Solomons	Vanuatu	New Caledonia
Rallina tricolor	MP	—	—	—
Gallirallus philippensis	MP	MP	MP	MP
Gallirallus woodfordi	—	m	—	—
Gallirallus immaculatus	—	m	—	—
Gallirallus tertius	—	mp	—	—
Gallirallus rovianae	—	m	—	—
Gallirallus insignis	m	—	—	—
† *Gallirallus lafresnayanus*	—	—	—	mp
† *Gallirallus* undescribed sp. A	p	—	—	—
† *Gallirallus* undescribed sp. B	—	p	—	—
† *Gallirallus* undescribed sp. C	—	—	p	—
Gyrmnocrex plumbeiventris	M	—	—	—
Porzana tabuensis	MP	M	MP	MP
† *Porzana* undescribed sp. A	—	—	p	—
Poliolimnas cinereus	M	M	M	M
Amaurornis moluccanus	M(m?)	m	—	—
Gallinula tenebrosa	—	—	—	MP
Pareudiastes silvestris	—	m	—	—
Pareudiastes undescribed sp.	—	p	—	
Porphyrio porphyrio	MP	MP	MP	MP
† *Porphyrio kukwiedei*	—	—	—	p
† *Porphyrio* undescribed sp. A	p	—	—	—
† *Porphyrio* undescribed sp. B	—	p	—	—
TOTAL				
M + m	8	10	4	6
M + m + P + p	10	13	6	7

The extinct species of swamphen or gallinule from Mangaia, Cook Islands (*Porphyrio* undescribed sp. C; Table 10-7) is known only by a scapula and femur. The Mangaian swamphen was smaller than any other species of *Porphyrio*, living or extinct. *Porphyrio* undescribed sp. D is known only from three fremora from Huahine, Society Islands. I await discovery of more bones of *Porphyrio* in both the Cook and Society Islands.

Porphyrio undescribed sp. E is represented by 10 bones (six skeletal elements) from the Unai Chulu site on Tinian (Chapter 8). This species was about the size of a robust *P. porphyrio* but had somewhat reduced wings as in *P. paepae*. A fragmentary radius from Rota may pertain to *P.* undescribed sp. E or a related species.

One other species of *Porphyrio*, at the fringe of Oceania, is *P. albus*, a flightless swamphen from Lord Howe Island, 480 km east of Australia. This large white rail became extinct in the early 19th century. One wonders if the white plumage in *P. albus* was due to an allele fixed by a founder effect.

Table 10-9 The distribution of rails in West Polynesia, M, modern record (volant); m, modern record (flightless); P, prehistoric record (volant); p, prehistoric record (flightless); †, extinct species. From data in Chapter 6.

	Fiji	Tonga	Samoa	Niue	Wallis & Futuna
Gallirallus philippensis	MP	MP	MP	MP	M
†*Gallirallus poecilopterus*	m	—	—	—	—
†*Gallirallus* undescribed sp. D	p	—	—	—	—
†*Gallirallus* undescribed sp. E	—	p	—	—	—
†*Gallirallus* undescribed sp. F	—	p	—	—	—
†*Gallirallus* undescribed sp. G	—	p	—	—	—
†*Gallirallus huiatua*	—	—	—	p	—
†*Vitirallus watlingi*	p	—	—	—	—
Porzana tabuensis	MP	MP	MP	MP	—
†*Porzana* undescribed sp. B	p	—	—	—	—
Poliolimnas cinereus	M	—	M	—	—
Pareudiastes pacificus	—	—	m	—	—
Porphyrio porphyrio	MP	MP	M	MP	M
TOTAL					
M + m	5	3	5	3	2
M + m + P + p	8	6	5	4	2

Biogeographic Patterns

Among the 10 genera of rails in tropical Oceania (Table 10-1), only *Gymnocrex* and *Amaurornis* do not occur in Remote Oceania. By contrast, six of Oceania's 17 columbid genera, and five of 11 genera of parrots, are confined to Near Oceania (Tables 11-1, 12-1). This testifies to the impressive dispersal abilities of rails, which consisted of more species on islands than on continents. Nevertheless, the only genus of rail restricted to Remote Oceania (outside of New Zealand) is *Vitirallus* from Fiji (Worthy 2004). Generic-level diversity of rails in Melanesia, West Polynesia, East Polynesia, and Micronesia stands at 9, 6, 3 or 4, and 6 (Tables 10-8 to 10-11).

Flightless rails once inhabited all parts of Oceania, no matter how distant from New Guinea or other continental regions. As currently understood, the rail faunas of Melanesia, West Polynesia, and Micronesia include more species of *Gallirallus* than any other genus, whereas flightless species of *Porzana* slightly outnumber those of *Gallirallus* in East Polynesia (Tables 10-8 to 10-11). This possible trend would be strengthened if the situation of multiple sympatric species of *Porzana*, as discovered in the Cook and Marquesas Islands, proves to be the rule in East Polynesia. Similarly, multiple sympatric species of *Gallirallus* may have been widespread in Near Oceania, as found on Buka (Chapter 5).

The indigenous rodents of the Bismarcks and Solomons have not prevented the evolution of flightless rails in three or four genera in Near Oceania. Even New Guinea, with major radiations of marsupials as well as rodents (Flannery 1995b), has a flightless rail, the spectacular if little-studied *Megacrex ineptus*, whose hindlimb osteology suggests a close relationship with *Amaurornis* (pers. obs.). I predict the eventual discovery of other flightless rails in New Guinea, minimally in *Gallirallus* and *Porphyrio*.

How Many Species of Rails Were There?

Endemic species of flightless rails evolved on Pacific islands of all sizes, shapes, elevations, isolations, ages, and geological compositions. Until the arrival of people, I believe that one or more endemic species of flightless rails inhabited many atolls as well as each high island (active volcanic, eroded volcanic, raised limestone, etc.) across Oceania. While atolls are regarded as providing marginal habitat for most landbirds, flightless species of both *Porzana* and *Gallirallus* have evolved on atolls, such as the endemic species of *Porzana* on Laysan (1.8 km^2, 12 m elev.) and Lisianski (1.7 km^2, 6 m elev.). Both of these atolls are more than 1200 km from the main group of high Hawaiian islands. Similarly, on Wake (7.4 km^2, 6 m elev., 600 km from any

Table 10-10 The distribution of rails in East Polynesia. M, modern record (volant); m, modern record (flightless); P, prehistoric record (volant); p, prehistoric record (flightless); †, extinct species. From data in Chapter 7.

	Cook Islands	Tubuai	Society Islands	Tuamotu Arch.	Marquesas Islands	Pitcairn Group	Easter Island
†*Gallirallus ripleyi*	p	—	—	—	—	—	—
†*Gallirallus pacificus*	—	—	m	—	—	—	—
†*Gallirallus* undescribed sp. H	—	—	p	—	—	—	—
†*Gallirallus* undescribed sp. I	—	—	—	—	p	—	—
†*Gallirallus* undescribed sp. J	—	—	—	—	p	—	—
†*Gallirallus* undescribed sp. K	—	—	—	—	p	—	—
†*Gallirallus* undescribed sp. L	—	—	—	—	p	—	—
Porzana tabuensis	MP	M	MP	M	M	M	—
†*Porzana rua*	p	—	—	—	—	—	—
†*Porzana nigra*	—	—	m	—	—	—	—
Porzana atra	—	—	—	—	—	mp	—
†*Porzana* undescribed sp. C	p	—	—	—	—	—	—
†*Porzana* undescribed sp. D	—	—	—	—	—	—	p
†*Porzana* undescribed sp. E	—	—	—	—	p	—	—
†*Porzana* undescribed sp. F	—	—	—	—	p	—	—
†*Porzana* undescribed sp. G	—	—	—	—	p	—	—
†*Porphyrio/Pareudiastes* undescribed sp.	p	—	—	—	—	—	—
†*Porphyrio* undescribed sp. D	—	—	p	—	—	—	—
†*Porphyrio paepae*	—	—	—	—	p	—	—
†Genus uncertain	—	—	—	—	—	—	p
TOTAL							
M + m	1	1	3	1	1	2	0
M + m + P + p	5	1	4	1	9	2	2

island) in the far north-central Micronesian region, a flightless species of *Gallirallus* evolved. Atolls inundated during part of the Holocene (see Dickinson 2003) would be less likely to have sustained flightless rails at human contact. Flightless species of *Gallirallus* also have been found on limestone islands as small as Ha'afeva (1.8 km^2, 10 m elev.) in Tonga and Aiwa Levu (1.21 km^2, 52 m elev.) in Fiji.

Now I will use the limited but informative fossil record to estimate, admittedly crudely, the distribution and diversity of each genus with flightless rails at first human contact (Table 10-12). These estimates include some Pacific island groups not usually treated in this book but which are known or suspected to have had flightless rails, namely the last nine archipelagos or individual islands in Table 10-12, among which only the Hawaiian Islands and New Zealand have well-developed fossil records. The highly conservative estimate assumes that the genera already recorded as having flightless species will not be found as such in island groups other than where they already are known. It adds up to 442 species, which I believe must be an underestimate of Oceania's species richness of rails when humans arrived. The less conservative estimate assumes that, within reasonable biogeographic parameters, the genera represented by flightless species occupied more island groups than documented by available evidence. Both sets of estimates assume that any island of reasonable size within that island group was occupied by numbers of species comparable to what already has been revealed on the few islands with good fossil records.

Given the vast incompleteness of the fossil data, it is conceivable that even 1579 species is less than the true species richness of rails in Oceania before human intervention. For example, the discovery of flightless species of *Porzana* on

Table 10-11 The distribution of rails in Micronesia. M, modern record (volant); m, modern record (flightless); P, prehistoric record (volant); p, prehistoric record (flightless); †, extinct species. From data in Chapter 8.

	Palau	Yap	Mariana Islands	Chuuk	Pohnpei	Kosrae	Remote Islands
Rallina eurizonoides	M	—	—	—	—	—	—
Gallirallus philippensis	M	MP	—	—	—	—	—
Gallirallus owstoni	—	—	mp	—	—	—	—
†*Gallirallus wakensis*	—	—	—	—	—	—	m
†*Gallirallus* undescribed sp. M	—	—	p	—	—	—	—
†*Gallirallus* undescribed sp. N	—	—	p	—	—	—	—
†*Gallirallus* undescribed sp. O	—	—	p	—	—	—	—
†*Gallirallus* undescribed sp. P	—	—	p	—	—	—	—
†*Porzana monasa*	—	—	—	—	—	m	—
†*Porzana* undescribed sp. H	—	—	p	—	—	—	—
†*Porzana* undescribed sp. I	—	—	p	—	—	—	—
†*Porzana* undescribed sp. J	—	—	p	—	—	—	—
Poliolimnas cinereus	M	MP	MP	M	M	—	—
Gallinula chloropus	M	—	MP	—	—	—	—
Porphyrio porphyrio	M	—	—	—	—	—	—
†*Porphyrio* undescribed sp. E	—	—	p	—	—	—	—
TOTAL							
M + m	5	2	3	1	1	1	1
M + m + P + p	5	2	11	1	1	1	1

islands off New Guinea and in the Moluccas would add tens of species to the overall estimate, as would the discovery of an extinct radiation of flightless *Gallinula* in Micronesia, or any of many other examples of recording a genus with flightless species in an island group where it is currently unknown.

Further fossil sampling will help to refine these estimates, but they will always have broad confidence intervals. Highly simplified estimates of the number of extinct rails (and other birds) that lived in Oceania at human contact, such as in Pimm et al. (1994, 1995) or Curnutt & Pimm (2001), reflect little appreciation of the strengths or the limitations of prehistoric data. Presented in detail in Chapter 4, the most important limitation is that most islands (>90%) have no fossil record, and on those that do, bone-based data seldom if ever represent the complete avifauna because of small sample size, inadequate use of fine-meshed screens, lack of precultural fossils, or poor sampling of bone-rich deposits that represent first human contact.

Bathymetry plays a large role in my estimates. Consulting bathymetric charts for each island group, I count as just one island any set of islands that were connected during the late Pleistocene sea level low stand (120-130 m below modern at 18,000 yr BP; Chapter 1). This adds to the conservative nature of my estimates, which also assume that no new species of flightless rails have evolved on formerly connected islands during the past 18,000 years of generally rising sea levels. In the Hawaiian Islands, single-island endemic species of *Porzana* did occupy adjacent islands that were joined in the late Pleistocene, such as Maui and Moloka'i (Olson & James 1991, Price & Elliott-Fisk 2004). Similarly, the North and South Islands of New Zealand were connected in late glacial times, yet the flightless species *Porphyrio hochstetteri* and *P. mantelli* are distinct (Holdaway et al. 2001). On the other hand, some islands in the Solomons that were joined in the late Pleistocene may have the same species of flightless rails (Diamond 1991) while others may not (see account of the "*Nesoclopeus*" group of *Gallirallus*).

As detailed in Table 10-12, my interpretation of the fossil record and historic specimens roughly suggests that from 500 to 1600 species of flightless rails inhabited Pacific islands at human contact. Rails alone account for most of the

Table 10-12 Estimated number of species of flightless rails at human contact in Greater Oceania (including nine islands or island groups, beginning in this table with the Hawaiian Islands, that typically are not covered in this book but are listed here for sake of completeness). For each island group, the first column represents the number of species actually recorded (AR) thus far, the second column is a "highly conservative" (HC) estimate of the number of species at human contact, and the third column is a "less conservative" (LC) estimate of that number. See text for additional details. †, extinct genus. Note that the Santa Cruz Group is analyzed separately from the rest of the Solomon Islands; "Kiribati" includes all of the Line Islands & adjacent U.S. possessions; "New Zealand" includes its satellite islands and the Kermadecs; totals for this column (*) include flightless species in five genera from New Zealand (†*Cabalus*, †*Capellirallus*, †*Diaphorapteryx*, †*Pleistorallus*, *Fulica*) unrecorded elsewhere in Oceania; "Islands off New Guinea" includes all islands along the entire northern coast of Papua New Guinea (excluding the Bismarcks) eastward through the Louisiade Archipelago, and all islands along both coasts of Irian Jaya from Japen counterclockwise through the Aru Islands, including Misool but not Ceram; "Moluccas" includes Morotai and Halmahera southward through Buru and Ceram. The unknown genus from Easter Island is listed, for convenience, as ?*Gallirallus*.

	Bismarcks			Solomon Islands			Santa Cruz Group			Vanuatu			New Caledonia			Fiji			Tonga		
	AR	HC	LC	AR	HC	LC	AR	HC	LC	AR	HC	LC	AR	HC	LC	AR	HC	LC	AR	HC	LC
Gallirallus	2	22	64	4	19	38	—	6	11	1	24	48	1	5	6	2	12	46	4	13	20
†*Vitirallus*	—	0	0	—	0	0	—	0	6	—	0	24	—	0	4	1	6	12	—	0	3
Porzana	—	0	64	—	0	21	—	0	11	1	12	26	—	0	6	1	10	46	—	10	20
Amaurornis	—	0	23	1	1	17	—	0	0	—	0	12	—	0	4	—	0	0	—	0	0
Gallinula	—	0	0	—	0	0	—	0	0	—	0	0	—	0	0	—	0	0	—	0	0
Pareudiastes	—	0	4	2	5	19	—	0	6	—	0	12	—	0	4	—	0	6	—	0	9
Porphyrio	1	4	23	1	5	17	—	0	6	—	0	12	1	4	6	—	0	6	—	0	9
TOTAL	3	26	178	8	30	112	0	6	40	2	36	134	2	9	30	4	28	116	4	23	61

	Samoa			Wallis & Futuna			Niue			Tuvalu			Tokelau			Cook Islands			Tubuai			Society Islands		
	AR	HC	LC	AR	HC	LC	AR	HC	LC	AR	HC	LC	AR	HC	LC	AR	HC	LC	AR	HC	LC	AR	HC	LC
Gallirallus	—	4	6	—	3	3	1	1	1	—	9	9	—	3	3	1	8	15	—	5	7	2	8	10
†*Vitirallus*	—	0	4	—	0	0	—	0	0	—	0	0	—	0	0	—	0	0	—	0	0	—	0	0
Porzana	—	0	6	—	0	3	—	1	2	—	0	9	—	0	3	2	17	21	—	5	12	1	8	13
Amaurornis	—	0	0	—	0	0	—	0	0	—	0	0	—	0	0	—	0	0	—	0	0	—	0	0
Gallinula	—	0	0	—	0	0	—	0	0	—	0	0	—	0	0	—	0	0	—	0	0	—	0	0
Pareudiastes	1	1	4	—	0	3	—	0	1	—	0	0	—	0	0	—	0	0	—	0	0	—	0	0
Porphyrio	—	0	4	—	0	0	—	0	1	—	0	0	—	0	0	1	2	8	—	0	5	—	0	8
TOTAL	1	5	24	0	3	9	1	2	5	0	9	18	0	3	6	4	27	44	0	10	24	3	16	31

	Tuamotu Arch.			Pitcairn Group			Easter Island			Marquesas Islands			Palau			Yap		
	AR	HC	LC	AR	HC	LC	AR	HC	LC	AR	HC	LC	AR	HC	LC	AR	HC	LC
Gallirallus	—	4	77	—	0	0	1?	1	1	4	8	8	—	2	8	—	3	15
†*Vitirallus*	—	0	0	—	0	0	—	0	0	—	0	0	—	0	0	—	0	0
Porzana	—	38	77	1	3	4	1	1	2	2	12	16	—	2	8	—	3	15
Amaurornis	—	0	0	—	0	0	—	0	0	—	0	0	—	0	2	—	0	0
Gallinula	—	0	0	—	0	0	—	0	0	—	0	0	—	0	0	—	0	0
Pareudiastes	—	0	0	—	0	0	—	0	0	—	0	0	—	0	0	—	0	0
Porphyrio	—	0	4	—	0	0	—	0	1	1	2	8	—	0	2	—	0	3
TOTAL	0	42	158	1	3	4	2	2	4	7	22	32	0	4	20	0	6	33

(continued)

Table 10-12 (continued)

	Mariana Islands			Chuuk			Pohnpei			Kosrae			Marshall Islands			Kiribati			Nauru		
	AR	HC	LC	AR	HC	LC	AR	HC	LC	AR	HC	LC	AR	HC	LC	AR	HC	LC	AR	HC	LC
Gallirallus	4	15	15	—	1	13	—	1	8	—	1	1	—	8	32	—	0	0	—	0	0
†*Vitirallus*	—	0	0	—	0	0	—	0	0	—	0	0	—	0	0	—	2	35	—	1	1
Porzana	3	5	20	—	1	13	—	1	8	1	1	1	—	8	32	—	0	0	—	0	0
Amaurornis	—	0	0	—	0	0	—	0	0	—	0	0	—	0	0	—	2	35	—	1	1
Gallinula	—	0	0	—	0	0	—	0	0	—	0	0	—	0	0	—	0	0	—	0	0
Pareudiastes	—	0	0	—	0	0	—	0	0	—	0	0	—	0	0	—	0	0	—	0	0
Porphyrio	1	1	5	—	0	1	—	0	1	—	0	1	—	0	0	—	0	0	—	0	0
TOTAL	8	21	40	0	2	27	0	2	17	1	2	3	0	16	64	0	4	70	0	2	2

	Wake Island			Hawaiian Islands			New Zealand			Lord Howe			Norfolk			Bonin Islands			Volcano Islands			Ryukyu Islands		
	AR	HC	LC	AR	HC	LC	AR	HC	LC	AR	HC	LC	AR	HC	LC	AR	HC	LC	AR	HC	LC	AR	HC	LC
Gallirallus	1	1	1	—	0	0	6	6	6	1	1	1	—	1	1	—	4	4	—	3	3	1	13	15
†*Vitirallus*	—	0	0	—	0	0	0	0	0	—	0	0	—	0	0	—	0	0	—	0	0	—	0	0
Porzana	—	0	1	12	24	32	0	0	0	—	0	1	—	0	1	—	0	4	—	0	3	—	0	13
Amaurornis	—	0	0	—	0	0	0	0	0	—	0	0	—	0	0	—	0	0	—	0	0	—	0	0
Gallinula	—	0	0	—	0	0	—	0	0	—	0	0	—	0	0	—	0	0	—	0	0	—	0	0
Pareudiastes	—	0	0	—	0	0	1	1	6	—	0	1	—	0	1	—	0	0	—	0	0	—	0	0
Porphyrio	—	0	0	—	0	0	2	3	6	1	1	1	—	0	1	—	0	4	—	0	3	—	0	13
TOTAL	1	1	2	12	24	32	14*	15*	30*	2	2	4	0	1	4	0	4	12	0	3	9	1	13	41

	Islands off New Guinea			Moluccas			TOTAL		
	AR	HC	LC	AR	HC	LC	AR	HC	LC
Gallirallus	—	14	27	1	4	47	36	236	607
†*Vitirallus*	—	0	0	—	0	0	1	6	53
Porzana	—	0	0	—	0	0	24	165	550
Amaurornis	—	0	27	—	0	12	1	1	97
Gallinula	—	0	0	—	0	0	1	1	8
Pareudiastes	—	0	14	—	0	0	3	6	68
Porphyrio	—	0	14	—	0	12	8	22	186
TOTAL	0	14	82	1	4	71	74*	442*	1579*

roughly 1000 to 2000 species of birds that I believe would exist today had people never colonized Oceania (Chapter 16). If not for human impact, more species of rails would be alive today than of any other family of birds.

What a shame that 98 or 99% of these rails are gone. The few that still exist deserve our best efforts to ensure their survival (Chapter 21). On islands where all nonnative mammals can be eliminated, I cannot help but be curious to see if flightlessness might evolve in currently volant populations of rails (Burney et al. 2002). Future biologists would be grateful if we began long-term genetic, morphological, and ecological monitoring of rail populations on islands uncontaminated with rats, cats, dogs, pigs, etc. Even if flightlessness did not evolve, the data would help us to understand the population biology of island rails, a most worthy cause.

Chapter 11 Pigeons and Doves

In spite of all the lost species and populations about to be mentioned, pigeons and doves (order Columbiformes, family Columbidae) remain Oceania's most taxonomically diverse family of landbirds, whether evaluated by genus, species, or subspecies. Columbid populations have been reduced or eliminated on islands by clearing forests, introducing exotic animals, and hunting. Being so palatable, hunting may have been a larger factor in the depletion of columbids than in that of most other landbirds. Ground-dwelling/understory species have been more affected than canopy species.

Prehistoric data on columbids from Polynesia (Steadman 1997c) and Melanesia have grown rapidly in the past few years, whereas prehistoric Micronesian columbids remain poorly known. There is no evidence of columbids on Easter Island, although its fossil landbirds are not well known (Chapter 7). Columbids never colonized the Hawaiian Islands, which have abundant landbird fossils (James 1995). No island in Oceania known to be inhabited by columbids is more than 600 km from another columbid-bearing island, whereas the isolation of the Hawaiian Islands and Easter Island exceeds 1000 km. *Hemiphaga novaeseelandiae* is the only native columbid in New Zealand today. In spite of a rich fossil history of landbirds, no other species of pigeon or dove has been found in New Zealand except the extinct *Gallicolumba norfolciensis* from remote, subtropical Norfolk Island (Worthy & Holdaway 2002:573). Relative isolation, combined with New Zealand's temperate climate and vegetation, prevented most columbid colonization from tropical Oceania or Australia.

Of Melanesia's 17 columbid genera, only six are shared with Polynesia, and just four with Micronesia, which has no endemic genera of columbids(Table 11-1). The only three columbid genera found in Remote Oceania but not in Melanesia are Undescribed genus C, *Didunculus*, and *Natunaornis*, each recorded only from West Polynesia. At the species level(Table 11-2), roughly similar richness values are found in Remote Melanesia (New Caledonia, Vanuatu) and West Polynesia (Fiji, Tonga). This may be, at least in part, an artifact of Tonga's more extensive fossil record.

Melanesia (Figure 1-2)

New Guinea

As in so many families of landbirds, columbid diversity is substantially greater in New Guinea than anywhere in Oceania (Tables 11-1, 11-2). In fact, New Guinea has the world's most extensive modern columbid fauna (14 genera, 39 species) but no prehistoric data with which to evaluate long-term changes. For perspective, only eight genera and 45 species of columbids inhabit all of South America. Borneo, the largest island on the Sunda shelf and comparable in size to New Guinea, has only seven genera and 14 species of pigeons and doves (MacKinnon & Phillips 1993).

The four genera of columbids found in New Guinea but not in Oceania are ground-dwelling forms. Most New Guinean species of columbids inhabit lowland rainforest. An indication of their diversity at a single site can be appreciated from field work in lowland rainforest of near Kiunga (upper Fly River drainage) of Papua New Guinea by Andrew Kratter and Jeff Sailer in May–June 2000. From a single site of only several km^2, they recorded 11 genera and 23 species of columbids, and were informed by locals of another genus and two more species that probably also occur. Among the 23 species, 13 generally inhabit the canopy/subcanopy, three the mid-level/understory, and seven are ground-dwellers (see guild categories in Table 11-1).

Bismarck Archipelago (Figure 5-1)

Although fossils have not augmented the 10 genera and 21 species of pigeons and doves in the Bismarcks(Tables 5-3, 11-3), species-level identification of many columbid bones from New Ireland in Steadman et al. (1999b) was precluded by a dearth of comparative skeletons. I hope to reexamine these columbid bones some day, using skeletons recently collected in the Solomons and New Guinea. Each of the 10 genera in the Bismarcks also occurs in New Guinea, and nine of the 10 (all except *Henicophaps*) occur in the Solomons (Mayr & Diamond 2001).

There are no modern single-site analyses of columbid faunas from lowland forests in the Bismarcks. The closest approximations are from New Britain, with seven genera and 13 species reported from the Talasea Peninsula (Hartert 1926), and six genera and eight species found at Walindi (Putnam & Iova 2001).

Solomon Islands (Figures 5-2, 5-3)

The late Pleistocene site of Kilu Cave on Buka (Chapter 5) has produced bones of two extinct ground-dwelling columbids (Undescribed genera A and B; Table 11-3). Comparing these specimens with skeletons of each genus known from New Guinea or Oceania (Table 11-1), I find them to be substantially different. These two genera must have had a broader distribution, at least within the Solomons.

At lowland rainforest sites, columbids are depauperate today in the Solomons compared to New Guinea or even the Bismarcks. Although nine genera and 13 species are known from Bougainville (Hadden 1981; Table 5-6 herein), most are altitudinally restricted or rare; perhaps no more than seven or eight species of columbids could be found at a single site in lowland forest. We found only four genera and six species of pigeons and doves along the Garanga River, Isabel (Kratter et al. 2001a,b). Missing were most ground-dwelling forms (*Gallicolumba*, *Microgoura*, Undescribed genera A and B), strongly suggesting human impact. On Rennell, isolated from the Solomon Main Chain, only four genera and four species were found at single forest sites (Filardi et al. 1999), with a fifth genus and species (*Caloenas nicobarica*) very rare or an historic extinction. Neither Bougainville, Isabel, or Rennell has a fossil record.

Vanuatu (Figure 5-4)

Up to eight species of columbids in six genera inhabit individual major islands in Vanuatu today (Table 5-10). On Santo, Andrew Kratter, Jeremy Kirchman, and I found five genera and seven species at each of two sites in lower montane forest in 2002 and 2003 (Kratter et al. in press). Fossils have not added extinct species or extirpated populations of columbids in Vanuatu (Table 11-3), although I suspect that endemic *Ducula bakeri*, a large canopy frugivore found now on only nine islands, is represented among prehistoric bone samples from Malakula, Efate, and Erromango (Table 5-11), three islands where it no longer occurs. Vanuatu is today's eastern limit for *Macropygia* and *Chalcophaps*, the former known prehistorically from East Polynesia. The distribution of *Gallicolumba* in Vanuatu is suspect, being represented only by *G. sanctaecrucis* on Santo and the historically extinct *G. ferruginea* on Tanna (Bregulla 1992). Before people arrived, probably every island in the group had at least one species of *Gallicolumba*.

New Caledonia (Figure 5-5)

A canopy frugivore, *Drepanoptila holosericea*, is a New Caledonian endemic genus and species that occurs today on Grande Terre and Ile des Pins, and prehistorically in the Loyalty Islands (Tables 5-12, 5-13). Another surviving endemic species the large, poorly studied canopy frugivore *Ducula goliath*. The large, extinct *Gallicolumba longitarsus* is known only from Grande Terre and Ile des Pins, New Caledonia (Balouet & Olson 1989). *Caloenas canacorum* also was described from New Caledonian fossils (Grande Terre; Balouet & Olson 1989) but has been found as well in Tonga (Steadman 1989b; Chapter 6 herein), suggesting that this large, extinct species also used to live in Vanuatu and Fiji.

West Polynesia (Figure 1-3)

Fiji (Figures 6-1, 6-2)

Fiji shares six of its 11 species of columbids with Tonga and Samoa(Table 11-4). It shares only *Ducula pacifica* with Vanuatu. Large Lauan islands have three to four genera and five to six species of columbids today (Watling 1982,

Table 11-1 The modern (M) and prehistoric (P) distribution of columbid genera in Oceania. BA, Bismarck Archipelago; CH, Chuuk; CI, Cook Islands; FI, Fiji; KO, Kosrae; MA, Marquesas Islands; MI, Mariana Islands; NC, New Caledonia; NG, New Guinea; NI, Niue; PA, Palau; PI, Pitcairn Group; PO, Pohnpei; RC, Remote Central Pacific island groups; SA, Samoa; SI, Solomon Islands; SO, Society Islands; TB, Tubuai; TO, Tonga; TU, Tuamotu Archipelago; VA, Vanuatu; WF, Wallis & Futuna; YA, Yap. Feeding guild categories: CF, canopy/subcanopy frugivore/granivore; MF, mid-level/understory frugivore/granivore; GF, ground-dwelling frugivore/granivore. Feeding guild categories are generalizations; distinctions between CF and MF, and between MF and GF, may be arbitrary. Row totals do not include NG. From data in Chapters 5–8. Polynesian data updated from Steadman (1997c).

	NG	BA	SI	VA	NC	FI	TO	SA	WF	NI	CI	TB	SO	TU
Columba	M	MP	M	MP	MP	MP	—	M	—	—	—	—	—	—
Macropygia	M	MP	M	MP	—	—	—	—	—	—	—	—	P	—
Reinwardtoena	M	M	M	—	—	—	—	—	—	—	—	—	—	—
Ptilinopus	M	MP	M	MP	M	M	MP	M	M	MP	MP	MP	MP	M
Drepanoptila	—	—	—	—	MP	—	—	—	—	—	—	—	—	—
Ducula	M	MP	M	MP	MP	MP	MP	M	MP	MP	MP	—	MP	M
Gymnophaps	M	M	M	—	—	—	—	—	—	—	—	—	—	—
†Undescribed genus C	—	—	—	—	—	—	P	—	—	—	—	—	—	—
Chalcophaps	M	MP	M	MP	MP	—	—	—	—	—	—	—	—	—
Henicophaps	M	M	—	—	—	—	—	—	—	—	—	—	—	—
Geopelia	M	—	—	—	—	—	—	—	—	—	—	—	—	—
Caloenas	M	MP	MP	—	P	—	P	—	—	—	—	—	—	—
Didunculus	—	—	—	—	—	—	P	M	—	—	—	—	—	—
Gallicolumba	M	MP	MP	M	P	MP	MP	MP	M	—	P	—	MP	MP
Microgoura	—	—	M	—	—	—	—	—	—	—	—	—	—	—
Trugon	M	—	—	—	—	—	—	—	—	—	—	—	—	—
Otidiphaps	M	—	—	—	—	—	—	—	—	—	—	—	—	—
†Undescribed genus A	—	—	P	—	—	—	—	—	—	—	—	—	—	—
†Undescribed genus B	—	—	P	—	—	—	—	—	—	—	—	—	—	
Goura	M	—	—	—	—	—	—	—	—	—	—	—	—	—
†*Natunaornis*	—	—	—	—	—	P	—	—	—	—	—	—	—	—
TOTAL GENERA														
M	14	10	10	6	5	4	3	5	3	2	2	1	3	3
P	0	7	4	5	6	4	6	1	1	2	3	1	4	1
M + P	14	10	12	6	7	5	6	5	3	2	3	1	4	3
Number of islands with prehistoric columbid bones	0	2	2	6	4	4	7	1	1	1	2	1	1	1
Presence/absence of feeding guilds (M + P)														
CF	+	+	+	+	+	+	+	+	+	+	+	+	+	+
MF	+	+	+	+	+	+	+	+	−	−	−	−	+	−
GF	+	+	+	+	+	+	+	+	+	−	+	−	+	+

[1]The three Fijian endemics (*P. layardi, P. victor, P. luteovirens*) are CF/MF. The other species of *Ptilinopus* are CF, except *P. coralensis*, which is MF/GF.

MA	PI	PA	YA	MI	CH	PO	KO	RC	TOTAL M	TOTAL P	TOTAL M + P	Feeding guild
—	—	—	—	—	—	—	—	—	6	4	6	CF/MF/GF
P	—	—	—	—	—	—	—	—	3	4	5	MF
—	—	—	—	—	—	—	—	—	2	0	2	CF/MF
MP	MP	M	—	MP	M	M	M	M	21	10	21	CF
—	—	—	—	—	—	—	—	—	1	1	1	CF
MP	P	M	MP	P	M	M	M	M	19	13	21	CF
—	—	—	—	—	—	—	—	—	2	0	2	CF
—	—	—	—	—	—	—	—	—	0	1	1	GF/MF
—	—	—	—	—	—	—	—	—	4	3	4	GF
—	—	—	—	—	—	—	—	—	1	0	1	GF
—	—	—	—	—	—	—	—	—	0	0	0	GF
—	—	M	—	—	—	—	—	—	3	4	5	MF/GF
—	—	—	—	—	—	—	—	—	1	1	2	CF/MF/GF
MP	P	MP	MP	MP	M	M	—	—	15	14	18	GF
—	—	—	—	—	—	—	—	—	1	0	1	GF
—	—	—	—	—	—	—	—	—	0	0	0	GF
—	—	—	—	—	—	—	—	—	0	0	0	GF
—	—	—	—	—	—	—	—	—	0	1	1	GF
—	—	—	—	—	—	—	—	—	0	1	1	GF
—	—	—	—	—	—	—	—	—	0	0	0	GF
—	—	—	—	—	—	—	—	—	0	1	1	GF
3	1	4	2	2	3	3	2	2	77	—	—	—
4	3	1	2	3	0	0	0	0	—	57	—	—
4	3	4	2	3	3	3	2	2	—	—	92	—
4	1	1	1	5	0	0	0	0				
+	+	+	+	+	+	+	+	+				
+	+	+	−	−	−	−	−	−				
+	−	+	+	+	+	+	−	−				

Table 11-2 Number of species of columbids in Oceania. Abbreviations: BA, Bismarck Archipelago; CH, Chuuk; CI, Cook Islands; FI, Fiji; KO, Kosrae; MA, Marquesas Islands; MI, Mariana Islands; NC, New Caledonia; NG, New Guinea; NI, Niue; PA, Palau; PI, Pitcairn Group; PO, Pohnpei; RC, Remote Central Pacific island groups; SA, Samoa; SI, Solomon Islands; SO, Society Islands; TB, Tubuai; TO, Tonga; TU, Tuamotu Archipelago; VA, Vanuatu; WF, Wallis & Futuna; YA, Yap. Feeding guild categories: CF, canopy/subcanopy frugivore/granivore; MF, mid-level/understory frugivore/granivore; GF, ground-dwelling frugivore/granivore. Feeding guild categories are generalizations; distinctions between CF and MF, and between MF and GF, may be arbitrary. In totals, ind. = individual, M = modern, P = prehistoric. *, number increased by prehistoric records; †, extinct genus. Updated from Mayr (1945), Watling (1982), Hannecart & Letocart (1980, 1983), Coates (1985), Beehler et al. (1986), Pratt et al. (1987), Balouet & Olson (1989), Bregulla (1992), Steadman (1995a, 1997c, unpub.), and Worthy et al. (1999).

	NG	BA	SI	VA	NC	FI	TO	SA	WF	NI	CI
Columba	1	2	2	1	1	1	—	1	—	—	—
Macropygia	3	3	1	1	—	—	—	—	—	—	—
Reinwardtoena	1	1	1	—	—	—	—	—	—	—	—
Ptilinopus	12	4	5	2	1	5	2	2	1	1	1
Drepanoptila	—	—	—	—	1	—	—	—	—	—	—
Ducula	7	5	4	2	2	*3	*3	1	*2	1	*3
Gymnophaps	1	1	1	—	—	—	—	—	—	—	—
†Undescribed genus C	—	—	—	—	—	—	*1	—	—	—	—
Chalcophaps	2	1	1	1	1	—	—	—	—	—	—
Henicophaps	1	1	—	—	—	—	—	—	—	—	—
Geopelia	2	—	—	—	—	—	—	—	—	—	—
Caloenas	1	1	1	—	*1	—	*1	—	—	—	—
Didunculus	—	—	—	—	—	—	*1	1	—	—	—
Gallicolumba	3	2	*4	1	*2	1	1	1	—	—	*2
Microgoura	—	—	1	—	—	—	—	—	—	—	—
Trugon	1	—	—	—	—	—	—	—	—	—	—
Otidiphaps	1	—	—	—	—	—	—	—	—	—	—
†Undescribed genus A	—	—	*1	—	—	—	—	—	—	—	—
†Undescribed genus B	—	—	*1	—	—	—	—	—	—	—	—
Goura	3	—	—	—	—	—	—	—	—	—	—
†*Natunaornis*	—	—	—	—	—	*1	—	—	—	—	—
TOTAL GENERA											
M	14	10	10	6	5	4	3	5	2	2	2
M + P	14	10	12	6	7	5	6	5	2	2	3
Individual islands (M + P)	14	10	9	6	7	5	6	5	2	2	3
TOTAL SPECIES											
M	39	21	20	8	6	9	4	6	2	2	2
M + P	39	21	23	8	9	11	9	6	3	2	6
Individual islands (M + P)	39	21	15	8	9	6	9	6	3	2	5
Number of islands with prehistoric columbid bones	0	2	1	3	3	4	7	1	1	1	2
Feeding guild totals (M + P)											
CF	20	10	10	4	4	6	6	3	3	2	4
MF	5	6	4	2	1	3	2	2	—	—	—
GF	14	5	9	2	4	2	2	1	—	—	2

TB	SO	TU	MA	PI	PA	YA	MI	CH	PO	KO	RC	Feeding guild
—	—	—	—	—	—	—	—	—	—	—	—	CF/MF/GF
—	*1	—	*1	—	—	—	—	—	—	—	—	MF
—	—	—	—	—	—	—	—	—	—	—	—	CF/MF
*2	1	2	2	1	1	—	1	1	1	1	1	CF/MF
—	—	—	—	—	—	—	—	—	—	—	—	CF
—	*2	1	1	*1	1	1	*1	1	1	1	1	CF
—	—	—	—	—	—	—	—	—	—	—	—	CF
—	—	—	—	—	—	—	—	—	—	—	—	GF/MF
—	—	—	—	—	—	—	—	—	—	—	—	GF
—	—	—	—	—	—	—	—	—	—	—	—	GF
—	—	—	—	—	—	—	—	—	—	—	—	GF
—	—	—	—	—	1	—	—	—	—	—	—	MF/GF
—	—	—	—	—	—	—	—	—	—	—	—	CF/MF/GF
—	*2	*2	*2	*1	1	1	*2	1	1	—	—	GF
—	—	—	—	—	—	—	—	—	—	—	—	GF
—	—	—	—	—	—	—	—	—	—	—	—	GF
—	—	—	—	—	—	—	—	—	—	—	—	GF
—	—	—	—	—	—	—	—	—	—	—	—	GF
—	—	—	—	—	—	—	—	—	—	—	—	GF
—	—	—	—	—	—	—	—	—	—	—	—	GF
—	—	—	—	—	—	—	—	—	—	—	—	GF
1	3	3	3	1	4	2	2	3	3	2	2	
1	4	3	4	3	4	2	3	3	3	2	2	
1	4	3	4	3	4	2	3	3	3	2	2	
1	3	4	4	1	4	2	2	3	3	2	2	
2	6	5	6	3	4	2	4	3	3	2	2	
1	6	3	6	3	4	2	4	3	3	2	2	
1	1	1	4	1	1	—	4	—	—	—	—	
2	3	2	3	2	2	1	2	2	2	2	2	
—	1	1	1	—	1	—	—	—	—	—	—	
—	2	2	2	1	1	1	2	1	1	—	—	

Table 11-3 The modern (M) and prehistoric (P) distribution of pigeons and doves in Melanesia. †, extinct species. From data in Chapter 5.

	Bismarcks	Solomons	Vanuatu	New Caledonia
Columba vitiensis	M	M	M	MP
Columba pallidiceps	M	M	—	—
Macropygia amboinensis	MP	—	—	—
Macropygia nigrirostris	MP	—	—	—
Macropygia mackinlayi	M	M	M	—
Reinwardtoena browni	M	—	—	—
Reinwardtoena crassirostris	—	M	—	—
Ptilinopus tannensis	—	—	MP	—
Ptilinopus superbus	M	M	—	—
Ptilinopus richardsii	—	M	—	—
Ptilinopus greyii	—	M	MP	MP
Ptilinopus rivoli	MP	—	—	—
Ptilinopus solomonensis	M	M	—	—
Ptilinopus viridis	M	M	—	—
Ptilinopus insolitus	M	—	—	—
Drepanoptila holosericea	—	—	—	MP
Ducula pacifica	M	MP	MP	MP
Ducula rubricera	M	MP	—	—
Ducula finschii	M	—	—	—
Ducula pistrinaria	M	MP	—	—
Ducula brenchleyi	—	M	—	—
Ducula bakeri	—	—	M	—
Ducula goliath	—	—	—	MP
Ducula melanochroa	M	—	—	—
Ducula spilorrhoa	M	—	—	—
Gymnophaps albertisii	M	—	—	—
Gymnophaps solomonensis	—	M	—	—
Chalcophaps indica	—	M	MP	MP
Chalcophaps stephani	MP	M	—	—
Henicophaps foersteri	M	—	—	—
Caloenas nicobarica	MP	MP	—	—
†*Caloenas canacorum*	—	—	—	P
Gallicolumba rufigula	—	P	—	—
Gallicolumba jobiensis	MP	M	—	—
Gallicolumba sanctaecrucis	—	M	M	—
†*Gallicolumba ferruginea*	—	—	M	—
Gallicolumba salamonis	—	M	—	—
Gallicolumba beccarii	MP	M	—	P
†*Gallicolumba longitarsus*	—	—	—	P
†*Microgoura meeki*	—	M	—	—
†Undescribed genus A	—	P	—	—
†Undescribed genus B	—	P	—	—
TOTAL SPECIES				
M	23	22	9	6
M + P	23	25	9	9

1985, Pratt et al. 1987; Table 6-2 herein) compared to five to six genera and six to eight species in Vanuatu (Bregulla 1992). Prehistoric avifaunas, however, are only beginning to be documented in Vanuatu or Fiji. Bones from an archaeological site on Lakeba (Lau Group, eastern Fiji) include a large species of *Ducula* originally believed to be *D. david* (Cassells 1984, Steadman 1989a) but now recognized as another extinct species, *D. lakeba* (Worthy 2001b). From the tiny (1.2 km^2) uninhabited island of Aiwa Levu near Lakeba, five species of columbids survive today (*Columba vitiensis*, *Ptilinopus perousii*, *P. porphyraceus*, *D. pacifica*, *Gallicolumba stairi*), and a limited fossil record includes the extinct *D. lakeba* as well (Steadman & Franklin 2000; Chapter 6 herein). An archaeological site on Waya (22 km^2, Yasawa Group, western Fiji) has yielded bones of *Columba vitiensis* and *D. latrans*, the latter unrecorded on Waya today (Amadon 1943, Watling 1985, Steadman 1997c).

The large Fijian island of Viti Levu was inhabited by Oceania's only flightless species of columbid, *Natunaornis gigoura* (Worthy et al. 1999, Worthy 2001a). Not yet recorded from a cultural context, this immense pigeon undoubtedly was lost early in Viti Levu's human history. The world's only other huge, flightless pigeons, the dodo *Raphus cucullatus* and Rodriguez solitaire *Pezophaps solitaria*, died out within two centuries of people discovering the Mascarene Islands in ca. AD 1500 (Mourer-Chauviré et al. 1999, Roberts & Solow 2003).

Fiji hosts five species of fruit-doves (*Ptilinopus*); nevertheless, no single island has records of more than two species except Laucala (12.2 km^2) near Qamea, Taveuni, and Vanua Levu (Watling 1989), on which *P. perousii*, *P. porphyraceus*, and *P. victor* all occur. *Ptilinopus perousii* is widespread in Fiji whereas *P. porphyraceus* is confined mainly to the Lau Group. The three other Fijian fruit-doves (*P. victor*, *P. luteovirens*, *P. layardi*) have plumages,

Table 11-4 The modern (M) and prehistoric (P) distribution of pigeons and doves in West Polynesia. Abbreviations: FI, Fiji; NI, Niue; SA, Samoa; TO, Tonga; WF, Wallis & Futuna. †, extinct species. Feeding guild categories: CF, canopy/subcanopy frugivore/granivore; MF, mid-level/understory frugivore/granivore; GF, ground-dwelling frugivore/granivore. Feeding guild categories are generalizations; distinctions between CF and MF, and between MF and GF, may be arbitrary. Updated from Steadman (1997c).

						TOTAL			
	FI	TO	SA	WF	NI	M	P	M + P	Feeding guild
Columba vitiensis	MP	—	M	—	—	2	1	2	CF/MF/GF
Ptilinopus perousii	MP	MP	M	—	—	3	2	3	CF
Ptilinopus porphyraceus	MP	MP	M	M	MP	5	3	5	CF
Ptilinopus victor	M	—	—	—	—	1	—	1	CF/MF
Ptilinopus luteovirens	M	—	—	—	—	1	—	1	CF/MF
Ptilinopus layardi	M	—	—	—	—	1	—	1	CF/MF
Ducula pacifica	MP	MP	M	MP	MP	5	4	5	CF
†*Ducula david*	—	—	—	P	—	—	1	1	CF
†*Ducula lakeba*	P	—	—	—	—	—	1	1	CF
†*Ducula* undescribed sp.	—	P	—	—	—	—	1	1	CF
Ducula latrans	MP	P	—	—	—	1	2	2	CF
†Undescribed genus C	—	P	—	—	—	—	1	1	CF/MF/GF
†*Caloenas canacorum*	—	P	—	—	—	—	1	1	MF/GF
Didunculus strigirostris	—	—	M	—	—	1	—	1	CF/MF/GF
†*Didunculus* undescribed sp.	—	P	—	—	—	—	1	1	CF/MF/GF
Gallicolumba stairi	MP	MP	MP	—	—	3	3	3	GF
†*Natunaornis gigoura*	P	—	—	—	—	—	1	1	GF
TOTAL SPECIES									
M	9	4	6	2	2	23	—	—	—
P	8	9	1	2	2	—	22	—	—
M + P	11	9	6	3	2	—	—	31	—
Number of islands with prehistoric columbid bones	5	7	1	1	1				

soft-part colors, and vocalizations that are very distinct from those in other Polynesian congeners (Pratt et al. 1987). These three species (the "golden dove group," for which the generic name *Chrysoenas* is available) are allopatric among themselves on major islands in western Fiji (Amadon 1943), where they are sympatric with *P. perousii* (Table 6-2).

Tonga (Figures 6-4, 6-5, 6-9, 6-13)

Three genera and four species of columbids occur today in Tonga, compared to six genera and nine species at human arrival(Tables 11-4, 11-5). Bones of the three most widespread extant species (*Ptilinopus porphyraceus*, *P. perousii*, *Ducula pacifica*) occur along with those of up to four extinct and two extirpated species in prehistoric sites in the Ha'apai Group (five different islands), Tongatapu, and 'Eua (Chapter 6). All nine species have been found on Lifuka and Ha'afeva (Ha'apai). One species extirpated on all seven islands, *Gallicolumba stairi*, still survives on a few other Tongan islands and locally in Fiji and Samoa (Steadman & Freifeld 1998, Steadman & Franklin 2000). The other, *D. latrans*, had been regarded as a Fijian endemic, but was widespread in Tonga at human arrival.

The large, extinct *Ducula* undescribed sp. has been found on all Tongan islands with landbird fossils. Also widespread was another extinct frugivore, Undescribed genus C, which was the largest volant columbid in Oceania. The extinct

Table 11-5 Prehistoric records of columbids in Tonga. †, extinct species; ?, questionable record, or residency not established. Feeding guild categories: CF, canopy/subcanopy frugivore/granivore; MF, mid-level/understory frugivore/granivore; GF, ground-dwelling frugivore/granivore. Feeding guild categories are generalizations; distinctions between CF and MF, and between MF and GF, may be arbitrary. Details of sites are presented in Chapters 3 and 6. NISP, number of columbid bones identified to species. Modern data from Rinke (1987) and Steadman (1998, unpub.). Prehistoric data from Steadman (1989b, 1993a, 1997c, 1998, unpub.).

	Foa	Lifuka	Ha`ano	`Uiha	Ha`afeva	Tongatapu	`Eua	Feeding guild	NISP
Ptilinopus perousii	P	MP	—	P	P	MP	MP	CF	46
Ptilinopus porphyraceus	MP	MP	MP	MP	MP	MP	MP	CF	35
Ducula pacifica	MP	MP	M?, P	M?, P	M?, P	MP	MP	CF	103
†*Ducula* undescribed sp.	P	P	—	—	P	—	P	CF	32
Ducula latrans	P	P	P	P	P	P	P	CF	98
†Undescribed genus C	P	P	—	P	P	P	P	CF/MF/GF	75
†*Caloenas canacorum*	—	P	P	—	P	—	—	MF/GF	15
†*Didunculus* undescribed sp.	—	P	P	P	P	P	P	CF/MF/GF	47
Gallicolumba stairi	P	P	P	P	P	P	P	GF	142
TOTAL SPECIES									
M	2	3	1–2	1–2	1–2	3	3	—	—
P	7	9	6	7	9	7	8	—	—
M + P	7	9	6	7	9	7	8	—	—
TOTAL NISP	40	108	32	42	104	108	159	—	593

Caloenas canacorum is known in Tonga from Lifuka, Ha'ano, and Ha'afeva; its absence from nearby Foa and 'Uiha may be a sampling effect, whereas the failure to find it on Tongatapu and 'Eua might reflect a genuine absence on these southern islands (see NISP values in Table 11-5). *Didunculus* undescribed sp., originally reported from 'Eua (Steadman 1993a), now has been found on Tongatapu and four islands in Ha'apai (Steadman in press b; Chapter 6 herein). Thus Tonga once had five species (in four genera) that were larger than *Ducula pacifica*, the largest columbid in the island group today.

Fruit-doves are represented by the two widespread West Polynesian species, *Ptilinopus porphyraceus* and *P. perousii*, the first common and found nearly throughout Tonga, and the second gone from many (most?) islands and often rare where it persists (Steadman 1998, Steadman & Freifeld 1998, Steadman et al. 1999a). The only large frugivorous columbid in Tonga today is the widespread, locally common *Ducula pacifica*, which is an excellent disperser of seeds (McConkey et al. 2004a). Two other species of *Ducula* once lived in Tonga. In precultural deposits on 'Eua, *D. pacifica* is absent but *D.* undescribed sp. occurs. In Ha'apai and Tongatapu, these two species are found in the same cultural strata along with *D. latrans*, which survives now only in Fiji. Today, three or more sympatric (and syntopic) species of *Ducula* occur nowhere in Remote Oceania.

Samoa (Figure 6-15)

Five genera and six species of columbids live on the large Samoan islands of Upolu and Savai'i (Beichle 1991; Table 6-26 herein), more than on any other islands east of Fiji (also with up to six species). American Samoa lacks two (*Columba vitiensis*, *Didunculus strigirostris*) of the six species, undoubtedly because of human impact. The only bony evidence of Samoan columbids is *Gallicolumba stairi* from a cultural site on Ofu, Manu'a Group, American Samoa, where it may still exist (Steadman 1993b, H. Freifeld pers. comm.). Each Samoan species of columbid except *Didunculus strigirostris* is shared at least with Tonga or Fiji (Table 11-4). Any of the extinct or extirpated species of columbids discovered in Fiji or Tonga (or their close relatives) may be expected in Samoa if its fossil record improves.

Wallis & Futuna (Figure 6-16)

The only extant columbids here (*Ptilinopus porphyraceus*, *Ducula pacifica*) are widespread in West Polynesia (Table 11-4). The large, extinct *Ducula david* was described from a cultural site on Wallis (= 'Uvea; Balouet & Olson 1987) and since was reported, at least tentatively, from Tongan islands and Lakeba in Fiji (Steadman 1989b, 1993a). As more fossil and modern specimens have accumulated, all of the Tongan and Fijian bones of *Ducula* in this size class have been assigned to other extinct species (see above), which may be part of the same radiation as *D. david*.

Niue (Figure 6-17)

The widespread *Ptilinopus porphyraceus* and *Ducula pacifica* are the only columbids on Niue (Kinsky & Yaldwyn 1981). Limited fossil data have added nothing to this small columbid fauna (Worthy et al. 1998, Steadman et al. 2000b). Given the rich prehistoric columbid fauna in Tonga (480 km to the west), Niue may have been inhabited at human arrival by an additional species each of *Ptilinopus* (perhaps *P. perousii*) and *Ducula* (perhaps *D. latrans* or D.undescribed sp.), one or more species of *Gallicolumba* (*G. stairi* most likely), and maybe a species of *Didunculus*, *Caloenas*, or Undescribed genus C.

East Polynesia (Figure 1-4)

Cook Islands (Figure 7-1)

The Northern Group (six atolls) lacks avian fossils. The only extant columbid here is *Ducula pacifica* on Pukapuka (Holyoak 1980). Seven columbid populations survive in the Southern Group (five of *D. pacifica*, two of *Ptilinopus rarotongensis*), a case of fragmented distributions comparable to that in the Marquesas or Society Islands. At least four other species once inhabited the Cooks(Tables 11-6, 11-7), basedon extensive data from Mangaia (Steadman 1985, 1989a, 1995a, Steadman & Kirch 1990, Kirch et al. 1995; chapter 7 herein) and limited data from Atiu and Aitutaki (Steadman 1991b). Except on remote, very small islands (Palmerston, Manu'ae, and Takutea), all five species (in three genera) recorded from Mangaia (52 km^2, now lacking columbids) probably used to live throughout the Southern Group. The largest (Rarotonga, 67 km^2) and third largest (Atiu, 22 km^2) islands have identical columbid faunas today; finding just five columbid bones on Atiu added two more species. The decline in columbids on Mangaia, from five to zero species over the past 1000 years, can be traced at Tangatatau Rockshelter and Ana Manuku(Table 11-8). The species of *Ducula* (*D. aurorae*, *D. galeata*) and *Gallicolumba* (*G. erythroptera*, *G. nui*) once found on Mangaia were widespread in East Polynesia at human arrival (Table 11-6) but are unknown in West Polynesia.

Ducula pacifica reaches the eastern limit of its immense distribution in the Cook Islands today. It has not been found among Mangaia's 550+ columbid fossils (Table 7-3), nor are there modern specimens or sight records of *D. pacifica* on Mangaia. Thus it is uncertain whether the pigeon called *rupe* in Mangaia stories and place names (Gill 1894, Hiroa 1934) pertains to *D. pacifica* (extant on four other islands in the Southern Group) or to *D. aurorae* (recorded in Mangaian prehistoric sites). Either way, I would suggest a recent colonization (probably <1000 years ago) of the Cook Islands by *D. pacifica*, which is a "supertramp" in Near Oceania (Mayr & Diamond 2001). Perhaps *D. pacifica* never reached Mangaia, easternmost of the Cook Islands. Just east of Mangaia in the Tubuai Group, no species of *Ducula* have been recorded (see below).

Ironically, some rugged areas of Mangaia are still forested largely with native trees (Merlin 1991). Thus prehistoric deforestation, while extensive (Kirch et al. 1992, 1995, Ellison 1994a, Kirch 1996), by itself may not account for Mangaia's columbid losses. Predation from people and other mammals probably also was important.

Tubuai (Figure 7-7)

The only extant columbid in the Tubuai Group is the large fruit-dove *Ptilinopus huttoni* in remnant patches of native forest on Rapa (Seitre & Seitre 1992). I suspect that *P. huttoni*, regarded since its discovery as endemic to Rapa (40 km^2), or a closely related congener once lived on other islands in Tubuai. The only other columbid known from the Tubuai Group is an undescribed, medium-sized species of *Ptilinopus* from an archaeological site on Tubuai Island (Table 7-12). The original columbid fauna of the Tubuai Group probably included one or two species in each of the three genera (*Ptilinopus*, *Ducula*, *Gallicolumba*) that were widespread in East Polynesia. Some of these columbids may prove to be close relatives of, or conspecific with, species already discovered on nearby Mangaia (Cook Islands).

Society Islands (Figures 7-8, 7-9)

Zooarchaeological data on Society Island birds are restricted to Huahine (77 km^2), which has lost five of its six columbid species (Tables 7-13, 11-6). All six (in four genera) probably once lived on each of the group's major islands, although Tahiti (1048 km^2) is the only one that retains even two species (Steadman & Pahlavan 1992). One of these (*Ducula aurorae*) has been rare for at least a century (Monnet et al. 1993b). The other, *Ptilinopus purpuratus*, is the only columbid still living on multiple islands in the Society Group. *Gallicolumba erythroptera*, *G. nui*, and *D. galeata* were widespread in East Polynesia, whereas *Macropygia arevarevauupa* is known only from Huahine.

That two species of columbids survive on Tahiti parallels the situation on Nuku Hiva (at 337 km^2 the largest island in the Marquesas) and Rapa, the second largest island in

Table 11-6 The modern (M) and prehistoric (P) distribution of pigeons and doves in East Polynesia. Abbreviations: CI, Cook Islands; MA, Marquesas Islands; PI, Pitcairn Group; SO, Society Islands; TB, Tubuai; TU, Tuamotu Archipelago. †, extinct species. Feeding guild categories: CF, canopy/subcanopy frugivore/granivore; MF, mid-level/understory frugivore/granivore; GF, ground-dwelling frugivore/granivore. Feeding guild categories are generalizations; distinctions between CF and MF, and between MF and GF, may be arbitrary. Updated from Steadman (1997c).

							TOTAL			
	CI	TB	SO	TU	MA	PI	M	P	M + P	Feeding guild
†*Macropygia arevarevauupa*	—	—	P	—	—	—	—	1	1	MF
†*Macropygia heana*	—	—	—	—	P	—	—	1	1	MF
Ptilinopus rarotongensis	MP	—	—	—	—	—	1	1	1	CF
Ptilinopus purpuratus	—	—	MP	—	—	—	1	1	1	CF
Ptilinopus chalcurus	—	—	—	M	—	—	1	—	1	CF
Ptilinopus coralensis	—	—	—	M	—	—	1	—	1	MF/GF
Ptilinopus huttoni	—	M	—	—	—	—	1	—	1	CF
Ptilinopus dupetithouarsii	—	—	—	—	MP	—	1	1	1	CF
†*Ptilinopus mercierii*	—	—	—	—	MP	—	1	1	1	CF
†*Ptilinopus* undescribed sp.	—	P	—	—	—	—	—	1	1	CF
Ptilinopus insularis	—	—	—	—	—	MP	1	1	1	CF
Ducula pacifica	M	—	—	—	—	—	1	—	1	CF
Ducula aurorae	P	—	MP	M	—	P	2	3	4	CF
Ducula galeata	P	—	MP	—	MP	P	2	4	4	CF
Gallicolumba erythroptera	P	—	MP	M	—	—	2	2	3	GF
†*Gallicolumba leonpascoi*	—	—	—	—	—	P	—	1	1	GF
Gallicolumba rubescens	—	—	—	—	MP	—	1	1	1	GF
†*Gallicolumba nui*	P	—	P	P	P	—	—	4	4	GF
TOTAL SPECIES										
M	2	1	4	4	4	1	16	—	—	
P	5	1	6	1	6	4	—	23	—	
M + P	6	2	6	5	6	4	—	—	29	
Number of islands with columbid bones	2	1	1	1	4	1				
Number of columbid bones	555	1	30	3	1985	724				3298

Tubuai. The rugged interiors of these large islands have less deforestation and hunting, thereby allowing a species or two to persist for centuries longer than on nearby smaller islands.

Tuamotu Islands (Figure 7-10)

Three genera and four species of columbids inhabit the Tuamotus today, with never more than two species on one island. The raised limestone island of Makatea sustains the endemic fruit-dove *Ptilinopus chalcurus* (Murphy 1924b) and one of two extant populations of the formerly widespread pigeon *Ducula aurorae*. Many of the 85 Tuamotuan atolls have not been surveyed for birds in recent decades. Among those that have, the endangered *Gallicolumba erythroptera* has been found on only one, whereas the fruit-dove *P. coralensis* is much more widespread (Pratt et al. 1987, Seitre & Seitre 1992, Monnet et al. 1993a, Blanvillain et al. 2002).

The only prehistoric evidence of columbids from the Tuamotus is from the almost-atoll of Mangareva, where none survives today but bones of *G. erythroptera* and the large, extinct *G. nui* have been found in archaeological contexts (Steadman & Justice 1998). Thus these two ground-doves occurred nearly throughout East Polynesia; many tens if not 100+ populations of each have been lost.

Table 11-7 The modern (M) and prehistoric (P) distribution of pigeons and doves in the Southern Cook Islands. m, modern record, now extirpated, †, extinct species. Feeding guild categories: CF, canopy/subcanopy frugivore/granivore; MF, mid-level/understory frugivore/granivore; GF, ground-dwelling frugivore/granivore. Feeding guild categories are generalizations; distinctions between CF and MF, and between MF and GF, may be arbitrary. Updated from Holyoak (1980), Steadman (1985, 1991b, 1992b, 1995a), Steadman & Kirch (1990), and Kirch et al. (1995).

								TOTAL			
	Rarotonga	Mangaia	Atiu	Mitiaro	Ma`uke	Aitutaki	Palmeston	M	M + m	M + m + P	Feeding guild
Ptilinopus rarotongensis	M	P	M	—	—	m	—	2	3	4	CF
Ducula pacifica	M	—	M	M	M	m	M	5	6	6	CF
Ducula aurorae	—	P	P	—	—	—	—	—	—	2	CF
Ducula galeata	—	P	—	—	—	—	—	—	—	1	CF
Gallicolumba erythroptera	—	P	P	—	—	—	—	—	—	2	GF
†*Gallicolumba nui*	—	P	—	—	—	—	—	—	—	1	GF
TOTAL SPECIES											
M	2	0	2	1	1	0	1	7	—	—	
M + m	2	0	2	1	1	2	1	—	9	—	
M + m + P	2	5	4	1	1	2	1	—	—	16	
Number of columbid bones	0	550	5	0	0	0	0				555
Island area (km^2)	67	52	27	22	18	18	2				
Island elevation	652	169	72	15	29	124	5				
Island isolation (km to nearest island >10 km^2)	204	204	50	50	59	102	370				

Table 11-8 Stratigraphic summary of columbids from two prehistoric sites on Mangaia, Cook Islands, expressed in numbers of identified bones. MAN-44 = Tangatatau Rockshelter (Main Excavation Block only), with Analytic Zone 1 the oldest and 17 the youngest. MAN-84 = Ana Manuku, with Layer IV the oldest and II the youngest. †, extinct species; *, extant species, extirpated on Mangaia. Feeding guild categories: CF, canopy/subcanopy frugivore/granivore; MF, mid-level/understory frugivore/granivore; GF, ground-dwelling frugivore/granivore. Feeding guild categories are generalizations; distinctions between CF and MF, and between MF and GF, may be arbitrary. Modified from Steadman & Kirch (1990), Steadman (1995b, 1997c; Chapters 4 and 7 herein), and Kirch et al. (1995). Bones of "Columbidae sp." not included.

	MAN-44			MAN-84				
	Analytic zones			Layer				
	1–4	5–17	Total	IV	III	II	Total	Feeding guild
**Ptilinopus rarotongensis*	11	1	12	7	1	—	8	CF
**Ducula aurorae*	5	—	5	1	—	—	1	CF
**Ducula galeata*	11	1	12	1	—	—	1	CF
**Gallicolumba erythroptera*	16	—	16	359	51	4	414	GF
†*Gallicolumba nui*	9	—	9	3	—	—	3	GF
TOTAL SPECIES	5	1	5	5	2	1	5	
TOTAL BONES	52	2	54	371	52	4	427	

Table 11-9 The modern (M) and prehistoric (P) distribution of pigeons and doves in the Marquesas Islands. m, modern record, now extirpated; †, extinct species; ?, questionable record, or residency not established. Feeding guild categories: CF, canopy/subcanopy frugivore/granivore; MF, mid-level/understory frugivore/granivore; GF, ground-dwelling frugivore/granivore. Feeding guild categories are generalizations; distinctions between CF and MF, and between MF and GF, may be arbitrary. Modified from Steadman (1997c).

											TOTAL			
	Nuku Hiva	Hiva Oa	Ua Pou	Ua Huka	Fatu Hiva	Tahuata	Eiao	Hatutu	Motane	Fatu Huku	M	M + m	M + m + P	Feeding guild
†*Macropygia heana*	P	—	—	P	—	—	—	—	—	—	0	0	2	MF
Ptilinopus dupetithouarsii	MP	M	M	MP	M	MP	—	—	M	—	7	7	7	CF
†*Ptilinopus mercierii*	m	mP	—	m?P	m?	—	—	—	—	—	0	2–4	3–4	CF
Ducula galeata	M	P	—	P	—	P	—	—	—	—	1	1	4	CF
Gallicolumba rubescens	mP	P	—	P	—	P	—	M	—	M	2	3	6	GF
†*Gallicolumba nui*	—	P	—	P	—	—	—	—	—	—	0	0	2	GF
TOTAL SPECIES														
M	2	1	1	1	1	1	0	1	1	1	10	—	—	
M + m	4	2	1	1–2	1–2	1	0	1	1	1	—	13–15	—	
M + m + P	5	5	1	6	1–2	3	0	1	1	1	—	—	24–25	
TOTAL BONES	6	144	0	1821	0	14	0	0	0	0				1985
Island area (km²)	337	324	104	78	77	53	52	18	16	1				
Island elevation (m)	1185	1190	1232	855	960	1000	576	428	520	361				
Island isolation (km to nearest island)	36	3	33	36	40	3	4	4	22	27				

Marquesas Islands (Figure 3-14)

No more than two species of columbids occur on any Marquesan island today(Table 11-9). If all six species of Marquesan columbids once occurred on at least the nine largest islands in the archipelago, as I believe was the case, then 45 of 54 individual island populations have been lost. This idea could be tested by obtaining columbid bone samples from each Marquesan island that are comparable to the large sample (1821 identified specimens) from Ua Huka, where all six species have been found. I found three species among only six and 14 columbid bones from Nuku Hiva and Tahuata, respectively. The 144 columbid bones from Hiva Oa represent only four species, but this sample is biased because fine-mesh screens were not used. Among the 10 islands listed in Table 11-9, perhaps only Fatu Huku (1 km^2) is too small to have sustained all six Marquesan species of columbids, although it is one of only two islands where the endemic *Gallicolumba rubescens* survives.

The extinct *Gallicolumba nui* was described from Marquesan bones but also lived in the Cook, Society, and Tuamotu groups (Steadman 1992b; Table 11-6 herein). *Ducula galeata*, the largest extant pigeon in Polynesia, had been regarded as endemic to Nuku Hiva (Mayr 1941b, 1976), where it has been rare for decades (Seitre & Seitre 1992). Bones of *D. galeata* have been found, however, on islands in three other East Polynesian island groups (Steadman 1989a). We are fortunate that this pigeon survived on even one island. Many of its columbid companions, such as *Macropygia heana*, *Ptilinopus mercierii*, and *Gallicolumba nui*, have not.

Four species of columbids probably are endemic to the Marquesas, reflecting the island group's relative isolation. The Marquesan cuckoo-dove *Macropygia heana* is part of an East Polynesian radiation in a genus no longer found east of Vanuatu (see Systematic Review later in this chapter). Marquesas is the only East Polynesian island group that had sympatric species of *Ptilinopus*, the endemic *P. dupetithouarsii* and *P. mercierii* (Cain 1957). Another Marquesan endemic is *Gallicolumba rubescens*, Oceania's smallest species of ground-dove.

Pitcairn Group (Figure 7-12)

In this isolated set of four islands, columbids occur today only on Henderson, where one species (*Ptilinopus insularis*) survives (Graves 1992). Henderson also has very large samples of ancient bird bones that document the loss of at least

Table 11-10 The modern (M) and prehistoric (P) distribution of pigeons and doves in Micronesia and Remote Central Pacific Islands (RCPI). †, extinct species. Feeding guild categories: CF, canopy/subcanopy frugivore/granivore; MF, mid-level/understory frugivore/granivore; GF, ground-dwelling frugivore/granivore. Feeding guild categories are generalizations; distinctions between CF and MF, and between MF and GF, may be arbitrary. From data in Chapter 8.

								TOTAL		
	Palau	Yap	Marianas	Chuuk	Pohnpei	Kosrae	RCPI	M	M + P	Feeding guild
Ptilinopus porphyraceus	—	—	—	M	M	M	M	4	4	CF
Ptilinopus pelewensis	M	—	—	—	—	—	—	1	1	CF
Ptilinopus roseicapilla	—	—	MP	—	—	—	—	1	1	CF
Ducula pacifica	—	—	—	—	—	—	M	1	1	CF
Ducula oceanica	M	MP	P	M	M	M	M	6	7	CF
Caloenas nicobarica	M	—	—	—	—	—	—	1	1	MF/GF
Gallicolumba kubaryi	—	—	—	M	M	—	—	2	2	GF
†*Gallicolumba* undescribed sp.	—	—	P	—	—	—	—	—	1	GF
Gallicolumba xanthonura	—	MP	MP	—	—	—	—	2	2	GF
Gallicolumba canifrons	M	—	—	—	—	—	—	1	1	GF
TOTAL SPECIES										
M	4	2	2	3	3	2	3	19	—	
M + P	4	2	4	3	3	2	3	—	21	
Number of islands with prehistoric columbid bones	0	1	5	0	0	0	0	6		

three species of columbids since Polynesians arrived about 1000 years ago (Steadman & Olson 1985, Schubel & Steadman 1989, Wragg & Weisler 1994, Worthy & Wragg 2003; Chapter 7 herein). Taxonomy of the extinct columbids is unresolved, but the prehistoric fauna included at least one species each of *Gallicolumba* and *Ducula*. I list two species of *Ducula* in Henderson in Table 11-6, following Steadman and Olson (1985), although with many more specimens available, the large columbid bones from Henderson were listed as an undetermined species of *Ducula* and an extinct, undescribed genus of large pigeon (Wragg & Weisler 1994, Wragg 1995). Uninhabited since European discovery and formerly regarded as "pristine," Henderson has lost most of its columbid fauna to human impact.

Pitcairn Island has no bone record but probably once sustained more or less the same species as Henderson. One or more species of columbid even may have lived on the small atolls of Ducie and Oeno.

Micronesia (Figure 1-5)

Palau's scant fossil record (Pregill & Steadman 2000) has added nothing to its columbid fauna, which alone in Micronesia includes *Caloenas nicobarica*(Tables 11-1, 11-10). The columbids of Palau otherwise consist of the standard Micronesian set of one species each in *Ptilinopus*, *Ducula*, and *Gallicolumba*.

The absence of any form of *Ptilinopus* in Yap (Table 11-10) must reflect human impact. All nearby island groups have indigenous if not endemic species of *Ptilinopus*.

Fossils in the Marianas (Chapter 8) add two species of columbids to its fauna. Several bones of a large, undescribed species of *Gallicolumba* from Rota provide Micronesia's only known instance of a sympatric, congeneric species pair of columbids. From Rota and Tinian are bones of *Ducula oceanica*, a large frugivore currently lacking in the Marianas (Table 11-10).

Chuuk and Pohnpei each sustain a species of *Ptilinopus*, *Ducula*, and *Gallicolumba* (Table 11-10). Of these, only *D. oceanica* has been recorded on nearby atolls (Buden 1996a,b). Kosrae lacks *Gallicolumba*, a situation unlikely to be natural.

Six of the eight islands or groups included in Remote Central Pacific Islands are inhabited by either *Ducula pacifica* or *D. oceanica*. The only other columbid from these atolls or low coral islands is *Ptilinopus porphyraceus*, extirpated in the Marshall Islands (Table 8-12).

Systematic Review

This section reviews the distributional highlights of each columbid genus in Oceania (Table 11-1). I will not cover *Geopelia*, *Trugon*, *Otidiphaps*, or *Goura*, which inhabit New Guinea but not Oceania. The arboreal frugivores *Ptilinopus*, *Drepanoptila*, *Ducula*, and *Gymnophaps* are similar osteologically and probably are closely related. The ground-dwelling/understory granivores (*Chalcophaps*, *Caloenas*, *Didunculus*, *Gallicolumba*, and *Microgoura*) may be another natural group (Steadman 1992b, Sailer & Steadman unpub. data, but see Johnson & Clayton 2000, Shapiro et al. 2002). The relationships of *Columba*, *Macropygia*, and *Reinwardtoena* are less certain, although they are more similar osteologically to each other and to the arboreal frugivores than to the ground-dwelling/understory granivores.

Figure 11-1 *Macropygia mackinlayi*, Santo, Vanuatu. Photo by DWS, 9 June 2003.

Columba

The only species of *Columba* in Polynesia is *C. vitiensis*, with endemic subspecies at its eastern limit in Samoa (*C. v. castaneiceps*) and Fiji (*C. v. vitiensis*). It also occurs westward through much of Melanesia (five subspecies) and beyond Oceania to the Philippines, Moluccas, and New Guinea (Peters 1937, Amadon 1943). Given its probable long residence in Samoa and Fiji, the absence of *C. vitiensis* in Tongan prehistory is surprising. The other species of *Columba* in Oceania is *C. pallidiceps*, endemic to large islands in the Bismarcks and Solomons (Tables 5-3, 5-7).

Macropygia

Cuckoo-doves occur today in Southeast Asia, Indonesia, the Philippines, Australia, and New Guinea east to Vanuatu. In Oceania, *Macropygia amboinensis* and *M. nigrirostris* are restricted to the Bismarcks, whereas *M. mackinlayi* (Figure 11-1)occurs from the Bismarcks to Vanuatu. All three are recorded from New Britain (Table 5-3; Chapter 20). Given the two large, extinct species from the Marquesas and Society Islands (Steadman 1992b), I suspect that the failure to find bones of *Macropygia* in island groups between Vanuatu and the Society Islands might be due to inadequate sampling, although the Cook Islands and Tonga have extensive fossil records.

Reindwardtoena

These large cuckoo-doves are confined within Oceania to Near Oceania, with *Reinwardtoena browni* endemic to the Bismarcks and *R. crassirostris* endemic to the Solomons (Tables 11-1 through 11-3). They have no fossil record.

Ptilinopus

Fruit-doves (Figure 11-6) are very widespread in Oceania (Tables 11-1, 11-2). The current absence of any *Ptilinopus* from atolls as in Tokelau or Tuvalu may be due to human impact. The presence of *P. coralensis* in the Tuamotus (Blanvillain et al. 2002) and the former occurrence of *P. porphyraceus* in the Marshall Islands (Peters & Griscom 1928; Chapter 8 herein) demonstrate that fruit-doves can inhabit atolls as well as high islands.

On large islands in Near Oceania, three to four (Bismarcks) or two to three (Solomons) species of *Ptilinopus* occur sympatrically (Tables 5-3, 5-6). Nowhere in Remote Oceania do more than two species of *Ptilinopus* coexist except for three species on Laucala, Fiji (Watling 1989). *Ptilinopus greyii* and *P. tannensis* are sympatric in much of Vanuatu, the former also being the only species of *Ptilinopus* in New Caledonia (Tables 5-12, 11-3). *Ptilinopus perousii* (see Cain 1954) and *P. porphyraceus* are sympatric in much of Fiji-Tonga-Samoa, although the latter is replaced on large islands in western Fiji by a member of the *P.* (*Chrysoenas*) *layardi-luteovirens-victor* species-group (Holyoak & Thibault 1978a, Pratt et al. 1987).

All East Polynesian and Micronesian fruit-doves belong to the *Ptilinopus porphyraceus/purpuratus* species group. The only extinct, undescribed species of *Ptilinopus* discovered thus far is from Tubuai (Chapter 7). Two endemic Marquesan species, *P. mercierii* (which may have become extinct in recent decades) and *P. dupetithouarsii*, form the only East Polynesian case of sympatry in *Ptilinopus* (Murphy 1924a, Ripley & Birckhead 1942, Steadman 1989a); these two species are more brightly colored (especially in the head plumage) than other East Polynesian *Ptilinopus*, perhaps from selection for visual species recognition (Cain 1957).

Figs (*Ficus* spp.; Moraceae) are important food for *Ptilinopus* from Australia and New Guinea (Frith et al. 1976, Frith 1982, Innis 1989) to East Polynesia (Franklin & Steadman 1991, Steadman & Freifeld 1999), except for *P. insularis* on fig-free Henderson, Pitcairn Group (Brooke & Jones 1995). That native species of *Ficus* occur across Oceania except the Pitcairn Group and Easter Island (Corner 1963, Smith 1963) supports the idea that fruit-doves once inhabited even many atolls in Tokelau, Tuvalu, the Northern Cook Islands, and Micronesia.

Drepanoptila

Drepanoptila holosericea is an endemic genus and species on Grande Terre and Ile des Pins, New Caledonia (Table 5-13). The frugivorous *Drepanoptila* probably is closely related to *Ptilinopus* (Shapiro et al. 2002, Sailer & Steadman unpub. data).

Ducula

Up to five species of these large frugivores, often called imperial-pigeons, are sympatric on large islands in the Bismarcks (Table 5-3). Whether all five can be syntopic is unknown. Two of them, *Ducula rubricera* and *D. pistrinaria* (Figures 11-7, 11-8) also inhabit the Solomons, where they are sympatric with the endemic *D. brenchleyi* (Table 5-6). In Vanuatu and New Caledonia, the widespread *D. pacifica* is sympatric with endemic *D. bakeri* and *D. goliath*, respectively (Tables 5-10, 5-12). In the Bismarcks and Solomons, *D. pacifica* typically inhabits only small, often remote islands, frequently with *D. pistrinaria* (Diamond 1975b, Mayr & Diamond 2001). In New Caledonia and Vanuatu, *D. pacifica* inhabits both large and small islands. In Fiji, however, *D. pacifica* favors small islands again, whereas *D. latrans* lives on islands of any size. These two species are believed to be allopatric in Fiji except on at least nine islands (none larger than 26 km^2) in the Lomaiviti and Lau groups, and on Laucala off Taveuni (Holyoak & Thibault 1978a, Watling 1982, 1989). On Nayau (22 km^2, Lau Group) in October 2001, I found *D. pacifica* and *D. latrans* together wherever there was forest.

Ducula pacifica (Figure 11-9) is unknown from prehistoric sites in the Cook Islands, where *D. aurorae* and *D. galeata* lived when people arrived. In Tonga, *D. pacifica* coexisted, at least immediately after human arrival, with two larger, extinct or extirpated congeners, *D. latrans* and *D.*undescribed sp., even on small, flat islands in the Ha'apai Group. This congeneric triplet lasted only for a century or two after human colonization at 2850 cal BP, after which only *D. pacifica* survived. The rich precultural bone deposit on 'Eua features *D. latrans* and *D.*undescribed sp. but not *D. pacifica* (Table 6-21). Thus it may be that two rather than three species of *Ducula* lived in Tonga before people arrived. In Fiji, it is possible that *D. latrans* and *D. pacifica* once coexisted with the extinct *D. lakeba*. The three large, extinct West Polynesian species of *Ducula* (*david*, *lakeba*, undescribed sp.) may be closely related.

While widely distributed in East Polynesia when evaluated by island group (Table 11-6), species of *Ducula* are absent today on all but eight of the 100+ individual major islands in this region. *Ducula galeata* (Figure 11-10) survives only on Nuku Hiva (Marquesas), whereas *D. aurorae* lives only on Tahiti and Makatea today. Both species once were widespread in East Polynesia, and were sympatric in at least the Cook Islands and Society Islands. The primary species of *Ducula* in Micronesia is *D. oceanica*, which is widespread today (Amadon 1943, Pratt et al. 1987) but was even more so in the past (Chapter 8, Table 11-10).

The living Polynesian species of *Ducula* feed on the fruits of native trees such as *Myristica* spp., *Syzygium* spp., *Dysoxylum* spp., *Elaeocarpus* spp., *Ficus* spp., *Guettarda speciosa*, and many others (Steadman & Freifeld 1999, McConkey et al. 2004a). The extinct *D. david* of 'Uvea, *D. lakeba* of Fiji, and *D.* undescribed sp. of Tonga were larger than any extant species of *Ducula* and probably could have swallowed whole fruits with diameters up to ca. 5 cm (Meehan et al. 2002). Except for cassowaries on New Britain, no native birds in Oceania today consume and thus potentially disperse fruits this large.

Gymnophaps

Two of the three species of these long-tailed, canopy frugivores are found in Oceania, with *Gymnophaps albertisii* on large islands in the Bismarcks (and mainland New Guinea) and *G. solomonensis* endemic to the Solomon Main Chain (Mayr & Diamond 2001; Tables 5-3, 5-7). There are no fossils of *Gymnophaps*.

Undescribed genus C

Known only from Tonga, where it was widespread at first human contact (Table 11-5), Undescribed genus C represents a large pigeon with skeletal features suggestive of both ground-dwelling and canopy species (Steadman unpub.

Figure 11-2 *Chalcophaps indica*, Santo, Vanuatu. Photo by DWS, 9 June 2003.

data). Its possible relationships to other Oceanic columbids are undetermined.

Chalcophaps

There is no fossil evidence of a larger former distribution of these green-winged ground-doves, represented in Oceania by *Chalcophaps stephani* in the Bismarcks and Solomons and *C. indica*(Figure 11-2) in New Caledonia and Vanuatu (Tables 11-1 through 11-3). Both species occur in New Guinea.

Henicophaps

The sole occurrence of *Henicophaps* in Oceania is that of *H. foersteri* (Tables 11-1 through 11-3), a virtually unstudied species found only on New Britain and nearby Umboi and Lolobau (Mayr & Diamond 2001; Table 5-3 herein).

Caloenas

A single living species, *Caloenas nicobarica*, inhabits islands from the western Indian Ocean to the Solomons and Palau. The larger, extinct *C. canacorum*(Figure 11-3) is known from prehistoric sites in New Caledonia (Balouet & Olson 1989) and Tonga (Steadman 1989b; Chapter 6 herein). Thus *C. canacorum* or a related species probably once lived in Vanuatu and Fiji. As with *C. nicobarica* (Coates 1985, Pratt et al. 1987), *C. canacorum* may have foraged mainly on or near the ground, alongside species of *Gallicolumba*, *Didunculus*, megapodes, and flightless rails.

Didunculus

The living species of *Didunculus*, *D. strigirostris* (Figure 11-5),inhabits Savai'i, Upolu, and Nu'utele in Western Samoa (Beichle 1987, 1991, Freifeld et al. 2001). An extinct, larger species of tooth-billed pigeon, *Didunculus* undescribed sp., has been found on six Tongan islands in both cultural and precultural sites (Steadman 1993a, 1995a, in press b; Chapter 6, Table 11-5 herein). In spite of the unusual size and shape of its bill, *Didunculus* does not warrant placement in a separate subfamily of the Columbidae; it resembles *Gallicolumba*, *Caloenas*, *Chalcophaps*, and other genera of ground-dwelling columbids in postcranial osteology (Steadman 1992b, Sailer & Steadman unpub. data) and clusters with *Goura victoriae*, *Caloenas nicobarica*, and the extinct, flightless *Raphus cucullatus* and *Pezophaps solitaria* of the Mascarene Islands in a single clade based on mitochondrial DNA (Shapiro et al. 2002).

The bill of *Didunculus strigirostris* seems to be well adapted for slicing through the pericarp and extracting the hard seed from the surrounding fleshy pulp in fruits of *Dysoxylum* spp. (Meliaceae), although other fruits as well as tubers are also consumed (Beichle 1987). The mandible

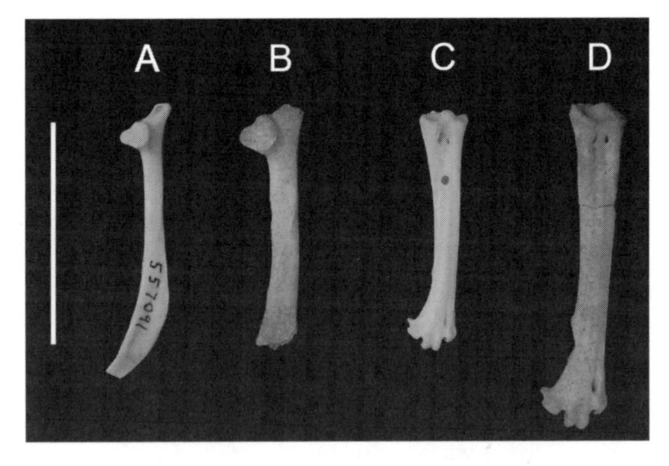

Figure 11-3 Scapulae (A, B) and tarsometatarsi (C, D) of *Caloenas*. A, C: Modern *C. nicobarica*, USNM 557091, adult ♀, Halmahera, *Moluccas. B*, D: Extinct *C. canacorum*, UF 57921 (scapula) and UF 58508/58511 (tarsometatarsus), Lifuka, Tonga. Scale = 40 mm. Photo by J. J. Kirchman.

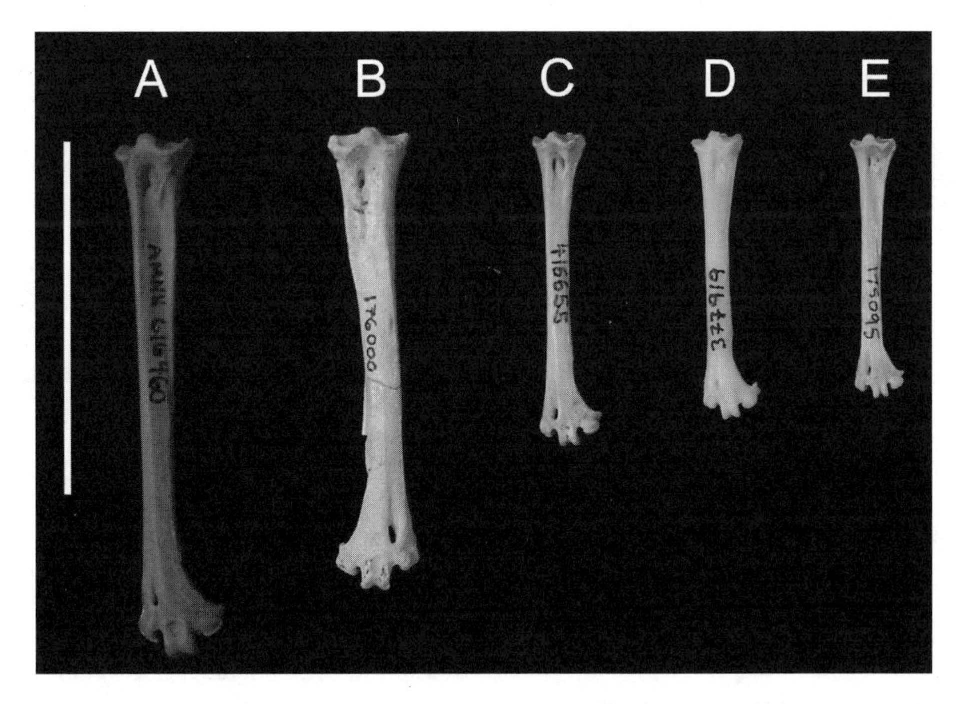

Figure 11-4 Tarsometatarsi of *Microgoura* and *Gallicolumba*. A: Modern (extinct?) *M. meeki*, AMNH 616460 (cast), Choiseul, Solomon Islands. B: Prehistoric (extinct) *G. nui*, BPBM 176000 (holotype; left side), Hane site, Ua Huka, Marquesas. C: Prehistoric (extirpated) *G. erythroptera*, USNM 416655, Te Rua Rere, Mangaia, Cook Islands. D: Modern (extirpated) *G. xanthonura*, USNM 377919, ♂, Guam, Mariana Islands. E: Prehistoric (extirpated) *G. rubescens*, BPBM 175095, Hane site, Ua Huka, Marquesas. Scale = 40 mm. Photo by J. J. Kirchman.

of the extinct Tongan species of *Didunculus* is 18–39% larger in various linear measurements than that of the living Samoan species (Steadman in press b), suggesting that it may have taken larger and harder foods.

Gallicolumba

The modern distribution of species of *Gallicolumba* in Melanesia is illogically patchy, an indication of considerable extirpations. In the Bismarcks, *G. jobiensis* and *G. beccarii* are widespread and often sympatric (Table 5-3). In the Solomons, *G. jobiensis* is confined to New Georgia, Guadalcanal, and Makira (Mayr & Diamond 2001), whereas the more widespread *G. beccarii* is absent from some major islands such as Buka, Isabel, and Malaita (Table 5-6). Furthermore, *G. salamonis* is known only from early 20th-century records on Makira and Ramos (Mayr & Diamond 2001), and *G. rufigula* (an otherwise Papuan species) occurred prehistorically on Buka (Table 5-7). Vanuatu lacks records of *Gallicolumba* except for *G. sanctaecrucis* on Santo and the recently extinct *G. ferruginea* on Tanna (Table 5-11). The only evidence of *Gallicolumba* in New Caledonia is bones of the extinct *G. longitarsus* from Grande Terre and Ile des Pins and the extirpated *G. beccarii* from the Loyalty Islands (Table 5-12).

The only species of *Gallicolumba* from West Polynesia, modern or prehistoric, is *G. stairi* from Fiji, Tonga, and Samoa (Table 11-4), within which it has been extirpated on many islands (Chapter 6). The absence on Wallis & Futuna and Niue of *G. stairi* or a related species probably is not natural.

Nearly all East Polynesian islands now lack *Gallicolumba*, and none has more than a single species. Both living species, *G. erythroptera* and *G. rubescens*, have declined

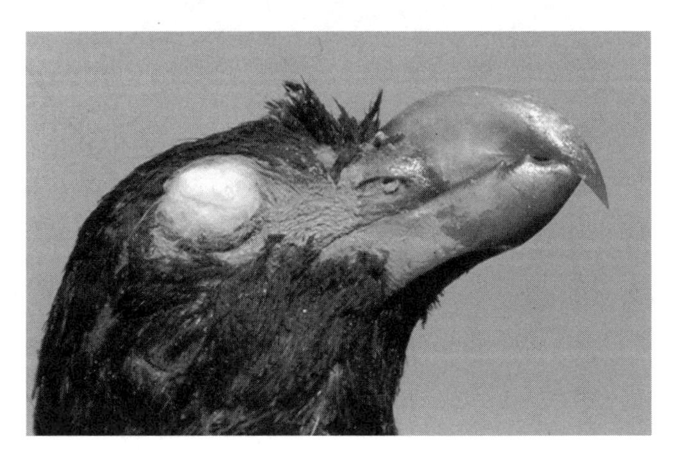

Figure 11-5 *Didunculus strigirostrus*, mounted specimen, British Museum (BMNH). Photo by DWS, September 1990.

Figure 11-6 *Ptilinopus richardsii*, Rennell Island, Solomon Islands. Photo by DWS, 23 June 1997.

Figure 11-8 *Ducula pistrinaria*, Isabel, Solomon Islands (close-up of same individual as in Figure 11-8). Photo by DWS, 7 July 1997.

in range over the past two centuries, surviving now on perhaps as few as one and two islands, respectively (Pratt et al. 1987, Blanvillain et al. 2002). Fossils reveal even greater losses of both species prehistorically (Steadman 1992b, 1995a, 1997c; Tables 11-6 through 11-9 herein). That *G. erythroptera* can live on remote atolls in the Tuamotus suggests that even Tokelau, Tuvalu, the Northern Cook Islands, and other remote atolls once may have sustained a ground-dove. One or two species of *Gallicolumba* used to occupy most if not all East Polynesian island groups, the most widespread being the extinct *G. nui*(Figure 11-4), which was sympatric with a smaller species in the Cook Islands, Society Islands, Tuamotus, and Marquesas. Another extinct species, *G. leonpascoi*, has been discovered on Henderson Island (Schubel & Steadman 1989, Wragg & Weisler 1994, Worthy & Wragg 2003).

A species in the *Gallicolumba canifrons-xanthonura-kubaryi* complex inhabits most high islands in Micronesia and probably lived formerly on the atolls as well. A larger,

Figure 11-7 *Ducula pistrinaria*, Isabel, Solomon Islands, kept as a pet by Joseph Manehage, who captured it on an offshore island. Photo by DWS, 7 July 1997.

Figure 11-9 Pacific pigeon *Ducula Pacifica*, captive adults, Tongatapu, Tonga. Photo by DWS, 30 July 1995.

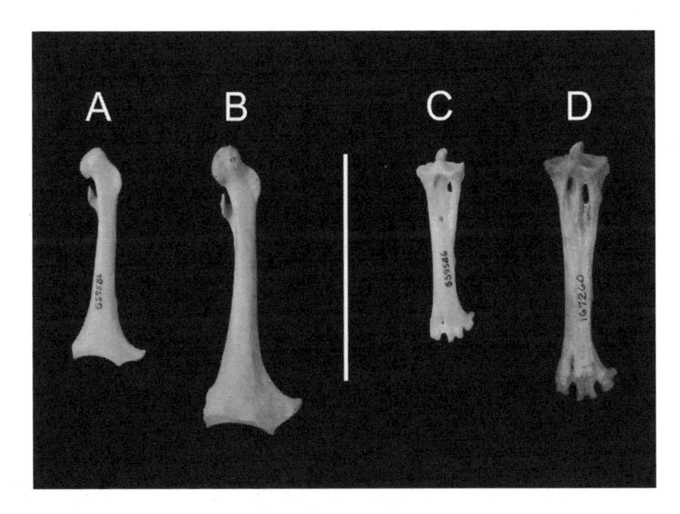

Figure 11-10 Coracoid (A, B) and tarsometatarsus (C, D) of *Ducula*. A, C: Modern *D. pacifica*, USNM 559586, ♂, body mass 401 g, Rarotonga, Cook Islands. B, D: Prehistoric (extirpated) *D. galeata*, BPBM 166056 (coracoid) and BPBM 167260 (tarsometatarsus), Hanatekua site, Hiva Oa, Marquesas. Scale = 40 mm. Photo by J. J. Kirchman.

extinct, undescribed species of *Gallicolumba* was sympatric with *G. xanthonura* in the Marianas (Steadman 1999a).

Microgoura

Microgoura has not been recorded since six specimens were taken in 1903 on Choiseul, Solomon Islands (Rothschild 1904, Parker 1967, 1972). It may be extinct, although Choiseul is not particularly well explored for birds. The single species, *M. meeki*, was a large ground-dwelling pigeon with a distinctive crest and facial markings (see plate 21 of Rothschild 1904, reproduced as plate 25 of Fuller 1987; also see plate 8 of Mayr & Diamond 2001). The hindlimb osteology of *Microgoura* is similar to that in *Gallicolumba* (Sailer & Steadman unpub. data; Figure 11-4 herein).

Undescribed genus A

This very large, ground-dwelling pigeon is known from 17 bones (coracoid, humeri, ulnae, femora, tibiotarsi, tarsometatarsi, and pedal phalanges) from Kilu Cave on Buka, Solomon Islands. Undescribed genus A had leg bones nearly the size of those in *Goura* spp. but proportionately much smaller shoulder and forelimb elements, as in *Otidiphaps*. It probably was widespread at human contact in the Solomon Main Chain, and perhaps the Bismarcks as well.

Undescribed genus B

This fairly large (approximately the size of *Otidiphaps nobilis*) ground-dwelling columbid is known only from five very distinctive bones (three coracoids and two ulnae) from Kilu Cave, Buka, Solomon Islands. Its distribution beyond Buka is speculative, as is that of Undescribed genus A (see above).

Natunaornis

This immense flightless pigeon is based on undated, noncultural fossils from four sites on Viti Levu, Fiji (Worthy et al. 1999, Worthy 2001b). Some features of its wing and leg bones are more similar to those in *Goura* spp. (the New Guinean crowned pigeons) than in other genera (Worthy 2001b). Among columbids anywhere, *Natunaornis gigoura* is exceeded in size only by the recently extinct *Raphus cucullatus* and *Pezophaps solitaria* of the Mascarene Islands, Indian Ocean.

Discussion

Past vs. Present Distribution of Genera

Three genera of columbids (*Ptilinopus*, *Ducula*, *Gallicolumba*) are widespread and species-rich in both Near and Remote Oceania. I believe that at least one species in each of these genera lived at human contact in every tropical Pacific island group except perhaps Easter Island and a few of the most isolated Remote Central Pacific Islands. More research is needed to learn if columbids on atolls ever involved genera other than *Ptilinopus*, *Ducula*, *Gallicolumba*, and occasionally *Caloenas*.

Most New Guinean columbid genera also occur in Near Oceania. All four cases of nonoverlap are ground-dwelling forms. If New Guinea had a fossil record, the three genera in the Solomons but not in New Guinea (or in the Bismarcks) might be found. The large difference in columbid genera between the Solomons (12) and Vanuatu (6) or New Caledonia (7) might be reduced but probably not erased with more fossils.

For each of the six Polynesian island groups with a respectable prehistory of columbids (Fiji, Tonga, Cook Islands, Society Islands, Pitcairn Group, Marquesas), three to six genera have been disclosed, compared to one to four genera today (Table 11-1). Five of the nine genera of Polynesian columbids occur in West but not in East Polynesia. Among these five, *Columba* is found nearly worldwide (Johnston 1962), *Caloenas* is an island specialist from the eastern Indian Ocean to the western Pacific, *Didunculus* may be endemic to West Polynesia, Undescribed genus C is known only from Tonga, and *Natunaornis* is only from Fiji. The former presence of *Macropygia* in East Polynesia and

Melanesia, but not in West Polynesia, is enigmatic and may be only a sampling artifact.

Past vs. Present Distribution of Species

Species-level identification of most columbid fossils from New Ireland was not possible in Steadman et al. (1999a) because comparative skeletons were inadequate at the time. The more recently studied columbid bones from Buka have benefited from availability of modern skeletons of all Papuan-Oceanic genera, including most species that occur in Near Oceania. As a result, I have identified two extinct and one extirpated species that no longer exist on Buka or anywhere in the Solomons, increasing the columbid fauna from seven to 10 species (Tables 5-6, 11-3). Similarly, prehistoric bones increase New Caledonia's columbid fauna from five to seven genera and from six to nine species (Tables 5-12, 11-3).

Precultural bone deposits from 'Eua (Tonga) and Mangaia (Cook Islands) establish that at least seven and five species of columbids lived in these islands at human contact (Chapters 6, 7). Although several genera are widely distributed, *Ducula pacifica* is the only species found in both East Polynesia (Cook Islands) and West Polynesia (all island groups). In much of Melanesia, *D. pacifica* is classified as a supertramp, i.e., an excellent inter-island colonizer and small-island specialist (Diamond 1974, 1975b, 1982, Diamond & Marshall 1977, Mayr & Diamond 2001). *Ducula pacifica* has the lowest basal metabolic rate of the 16 South Pacific columbids studied by McNab (2000), this being a logical adaptation for a highly dispersive species.

Some of the modern range of *Ducula pacifica* probably was colonized after the arrival of people. As in Tonga and the Cook Islands (the current eastern limit of *D. pacifica*; see above), bones of *D. pacifica* are absent from the earliest cultural levels but occur in later ones on Tikopia, a small island in the Santa Cruz Group (Steadman et al. 1990b). That *D. pacifica* is a relatively recent colonizer of Polynesia and perhaps eastern Melanesia is especially tenable given its minimal geographic variation in size and plumage (the same subspecies occurs from the Solomon Islands to the Cook Islands; Amadon 1943). *Ducula pacifica*, like the volant nonforest rails *Porzana tabuensis*, *Gallirallus philippensis*, and *Porphyrio porphyrio*, may have dispersed to Polynesia only after the arrival of humans (Steadman 1993a, 1995a). Partial deforestation created suitable habitat for the rails and may not have been especially disadvantageous for *D. pacifica*, which is well adapted for life among forest patches (Steadman & Freifeld 1998), even though its relative abundance in Tonga and Fiji is greatest in large tracts of mature forest (Steadman 1998, Steadman & Franklin 2000).

Except for five individuals translocated to Ua Huka in June 2000 (Marie-Noel de Visscher pers. comm.), *Ducula galeata* lives only on Nuku Hiva, Marquesas. The apparent endemism of *D. galeata* is an artifact of anthropogenic extinction. The same may be true of the three East Polynesian fruit-doves that now occur on single islands (*Ptilinopus chalcurus* on Makatea, *P. insularis* on Henderson, and *P. huttoni* on Rapa; Table 11-6). Additional range contractions are documented by bones in other East Polynesian species (*Gallicolumba rubescens*, *P. rarotongensis*, *D. aurorae*) that survive on only two or three islands. It also is improbable that the extinct columbids known from single islands, such as *Macropygia arevarevauupa* on Huahine or Undescribed genera A and B from Buka, were really endemic to these islands. A possible exception would be the flightless *Natunaornis gigoura* from Viti Levu, Fiji (Worthy 2001b), although I suspect that other islands in Fiji eventually will be found to have had related flightless pigeons.

Species now confined to one or a few islands, but that had much larger ranges at human contact, can be called "pseudoendemics" (Steadman in press). *Ducula galeata*, for example, is pseudoendemic to Nuku Hiva, an island with no special properties to explain the evolution of Oceania's largest living pigeon. *Ducula galeata* did not originate on Nuku Hiva; it just happens to survive there. The extirpation that leads to pseudoendemism still occurs today. For example, when the small population of *Didunculus strigirostris* on Upolu and nearby Nu'utele (Beichle 1987, 1991, Freifeld et al. 2001) eventually is lost, the species will be confined to Savai'i. Had *D. strigirostris* been discovered by scientists in the year 2044 rather than 1844, it might have been regarded as a single-island endemic. Pseudoendemism can also occur at the generic level. Tongan fossils, for example, render *Didunculus* pseudoendemic to Samoa.

Determining the natural ranges of species has implications beyond improving our understanding of dispersal and endemism. Because *Ducula galeata* once was broadly sympatric with *D. aurorae*, it is no longer appropriate to consider these two as a "superspecies." Given the sympatric species pairs and triplets of *Ducula* in both West and East Polynesia (Tables 11-4 to 11-8), the superspecies concept may not be applicable for any Polynesian species of *Ducula* (contra Mayr 1940b, Amadon 1943, Goodwin 1960, Holyoak & Thibault 1978a, Mayr & Diamond 2001). Pending further testing with fossils, the superspecies concept still may be useful, however, for some East Polynesian species of *Ptilinopus*, where similar but recognizably distinct forms replace each other geographically, and where sympatric species-pairs are known only in the Marquesas.

Conservation

Pigeons and doves are important frugivores in tropical forests on Pacific islands (Franklin & Steadman 1991, Steadman & Freifeld 1999). Whereas deforestation has fueled the loss of columbids, the converse probably also is true; the loss of so many populations and species of columbids probably has affected the forests themselves by limiting inter- and intra-island dispersal of plant propagules (Steadman 1997c). Populations of certain large-fruited species of trees now isolated in forest patches may have lost their means of dispersal because of columbid extinction (Meehan et al. 2002).

The surviving populations of pigeons and doves in Oceania deserve sound conservation programs. More translocations should be considered (Chapter 21), although no program of protecting forests and translocating birds can restore the entire Oceanic columbid fauna. To borrow from Aldo Leopold (1949:110), whatever "intelligent tinkering" we do with Oceanic pigeons does not have the luxury of keeping every "cog and wheel," for many species already are gone. That does not mean we should give up and watch the others slip away. In his lament to the passenger pigeon (*Ectopistes migratorius*), Leopold (1991:213) noted that "The sailor who clubbed the last auk thought of nothing at all. But we, who have lost our pigeons, mourn the loss." The few of us who are interested have a debt to pay while we mourn the extinct pigeons and doves of Oceania. Humans have been altering the distribution of birdlife on Pacific islands in negative ways for thousands of years. We already have quite a history of "playing God" with these birds. It would be nice to use our godly powers constructively.

Chapter 12 Parrots

COLORFUL, entertaining, and tasty, parrots (order Psittaciformes) have been important to Pacific peoples for millennia. This importance has not had much benefit for the parrots themselves, which have undergone considerable anthropogenic extinction and range contraction in Oceania (see below) and other tropical islands (Fuller 1987:142–148, Williams & Steadman 2001).

The losses of parrots that I am about to describe are due to an unknowable blend of habitat loss, hunting, and predation from nonnative mammals. Hunting could be an especially important factor with parrots; brightly colored feathers, such as red ones, were valued by Polynesians and Melanesians, who also ate parrots regularly. Parrots could be caught by any of the standard bird-catching methods (see Steadman 1997b) but are particularly vulnerable at their nest cavities, which can be easy to locate if you spend enough time in the woods. Nestlings can be plucked from cavities in trees and raised by people, who use the feathers and eventually eat the parrot as well.

Among landbird families, parrots rank third (behind rails and pigeons) in geographic range and taxonomic diversity in Oceania. I use the term "parrot" here in the broadest sense, including those with common names such as cockatoo, lory, lorikeet, or parakeet. I regard all living parrots as best classified in a single family defined by many unique morphological and other traits (Smith 1975). I must note that many recent authors recognize cockatoos (Cacatuinae) and lories (Loriinae) as separate families (Cacatuidae, Loriidae) from other parrots (Adams et al. 1984, Sibley & Ahlquist 1990, Christidis et al. 1991, Boles 1993, Brown & Toft 1999). Regardless of taxonomic category, these three groups seem to be natural.

As do pigeons and doves (Chapter 11), parrots have their modern center of diversity in Australia, New Guinea, and Near Oceania. Like columbids further, the absence of parrots in the Hawaiian Islands (Olson & James 1991) is natural rather than due to human impact. All extinct species of parrots discovered thus far in the tropical Pacific were volant; nothing resembling the large flightless *Strigops habroptilus* of New Zealand has been found. Of the three genera of parrots in New Zealand (see McNab & Salisbury 1995), only *Cyanoramphus* is shared with tropical Oceania.

Among the 13 genera of parrots in Oceania, five are restricted to Near Oceania (Bismarcks and Solomons), four are confined to Remote Oceania, and four occur in both regions (Table 12-1). Without fossils, these values would be seven, two, and four. The Bismarcks and Solomons also are much richer in species than the rest of Melanesia (Table 12-2). In West Polynesia, Fiji is richer in species than other island groups in spite of having no parrot fossils (Table 12-3).

While shedding considerable light on late Quaternary extinctions, fossils have done little to elucidate the early evolutionary history of parrots in the Australian-Oceanic region, where the only Tertiary record is that of *Cacatua* sp. from the early to middle Miocene Riversleigh site in

Figure 12-1 Rainbow lorikeet, *Trichoglossus haematodus* (captive). Photo by Trinn (`Pong) Suwannapha.

northwestern Queensland (Boles 1993). As detailed below, prehistoric data have augmented the distribution and taxonomic diversity of Pacific parrots thus far mainly in *Vini*, *Cacatua*, and *Eclectus*. Modern distributions given below are from Juniper & Parr (1998) unless cited otherwise.

Systematic Review

Chalcopsitta

The only species of *Chalcopsitta* in Oceania is the mainly red *C. cardinalis*, a lorikeet found locally in the Bismarcks and widely in the Solomons, where it can be abundant (Kratter et al. 2001a).

Trichoglossus

The rainbow lorikeet, *Trichoglossus haematodus* (Figure 12-1), is Oceania's most widespread species of parrot, living from the Lesser Sundas, New Guinea, and Australia through Melanesia to New Caledonia and Vanuatu. The 20 or 21 subspecies of *T. haematodus* represent more intraspecific geographic variation than in any other Oceanic psittacid, with five of the subspecies confined to Oceania (Peters 1937:150, Cain 1955). This lorikeet is common in much of its range.

The only indigenous parrot in Micronesia today is *Trichoglossus rubiginosus* on Pohnpei and nearby Ant Atoll (Buden 1996b, 2000; Table 8-11 herein). I suspect that other Micronesian islands once supported this or another species of *Trichoglossus*, an idea that awaits recovery of bird bones from the nearest high islands of Chuuk (700 km to the west) and Kosrae (550 km to the east). That *P. rubiginosus* still survives on Ant Atoll (only 18.5 km SW of Pohnpei) suggests that lorikeets may have inhabited atolls as well as high islands, which would mean that their prehistoric losses in Micronesia could have been immense. Minimally, the absence of any species of *Trichoglossus* in Palau, Yap, Marianas, Chuuk, and Kosrae is suspicious.

Lorius

In Oceania, *Lorius* is represented in the Bismarcks by *L. hypoinochrous* and *L. albidinuchus* (sympatric only on New Ireland) and in the Solomons by *L. chlorocercus*.

Vini

While fossils have not augmented the distribution of the four previous genera of nectar-feeding parrots, they have added a great deal for *Vini*. Confined to Polynesia, five of the six extant species of *Vini* occur today on only a fraction

Table 12-1 The modern (M) and prehistoric (P) distribution of genera of parrots in Oceania. Modern distributions from Juniper & Parr (1998). Prehistoric distributions from Chapters 5–8 herein.

	Bismarcks	Solomons	Vanuatu	New Caledonia	Fiji	Tonga	Samoa	West Polynesian Outliers	Cook Islands	Tubuai	Society Islands
Cacatua	M, P	M	—	P	—	—	—	—	—	—	—
Chalcopsitta	M	M, P	—	—	—	—	—	—	—	—	—
Trichoglossus	M	M	M, P	M	—	—	—	—	—	—	—
Lorius	M, P	M	—	—	—	—	—	—	—	—	—
Vini	—	—	—	—	M	M, P	M	M	P	M	M, P
Charmosyna	M, P	M	M, P	M	M	—	—	—	—	—	—
Micropsitta	M	M	—	—	—	—	—	—	—	—	—
Prosopeia	—	—	—	—	M	—	—	—	—	—	—
Eunymphicus	—	—	—	M	—	—	—	—	—	—	—
Cyanoramphus	—	—	—	M	—	—	—	—	—	—	M
Geoffroyus	M	M	—	—	—	—	—	—	—	—	—
Eclectus	M	M	P	—	—	P	—	—	—	—	—
Loriculus	M	—	—	—	—	—	—	—	—	—	—
Genus uncertain	—	—	—	—	—	—	—	—	—	—	—
TOTAL											
M	9	8	2	4	3	1	1	1	0	1	2
M + P	9	8	3	5	3	2	1	1	1	1	2
Number of bones	13	1	9	1	0	26	0	0	180	0	5

of the islands they once occupied (Tables 12-4 and 12-5, which list prehistoric records but not the considerable historic losses). *Vini solitarius* (included here in *Vini* rather than "*Phigys*"; see Steadman & Zarriello 1987) now is restricted to Fiji but once lived in Tonga as well (Steadman 1993a; Chapter 6 herein). *Vini australis* (Figure 12-2) now is gone from a number of islands it once inhabited in Tonga and Samoa (Steadman 1998, Steadman & Freifeld 1998). Certainly in Tonga, and probably as well in the Lau Group of Fiji, *V. solitarius* and *V. australis* were sympatric, unlike today.

Vini kuhlii, confined today to Rimatara Island in Tubuai (McCormack & Künzlé 1996), was widespread prehistorically in the Cook Islands, 480 km to the WNW (Steadman 1991b). Similarly, *V. peruviana* is gone from most islands in the Society Group, as is *V. ultramarina* in the Marquesas (Steadman & Zarriello 1987, Steadman & Pahlavan 1992). Only for *V. stepheni* of Henderson Island do we lack evidence of range contraction since human arrival. Probably every West and East Polynesian island (except perhaps Easter Island) formerly sustained at least one species of *Vini*.

Figure 12-2 Blue-crowned Lorikeet, *Vini australis*, captive nestling, Tongatapu, Tonga. Photo by DWS, 12 August 1999.

Tuamotu	Pitcairn	Easter Island	Marquesas	Palau	Yap	Marianas	Chuuk	Pohnpei	Kosrae	Remote Central Pacific Islands	TOTAL M	TOTAL M + P
—	—	—	—	—	—	—	—	—	—	—	2	3
—	—	—	—	—	—	—	—	—	—	—	2	2
—	—	—	—	—	—	—	—	M	—	—	5	5
—	—	—	—	—	—	—	—	—	—	—	2	2
M	M, P	—	M, P	—	—	—	—	—	—	—	9	10
—	—	—	—	—	—	—	—	—	—	—	5	5
—	—	—	—	—	—	—	—	—	—	—	2	2
—	—	—	—	—	—	—	—	—	—	—	1	1
—	—	—	—	—	—	—	—	—	—	—	1	1
—	—	—	—	—	—	—	—	—	—	—	2	2
—	—	—	—	—	—	—	—	—	—	—	2	2
—	—	—	—	—	—	—	—	—	—	—	2	4
—	—	—	—	—	—	—	—	—	—	—	1	1
—	—	P	—	—	—	P	—	—	—	—	0	2
1	1	0	1	0	0	0	0	1	0	0	36	—
1	1	1	1	0	0	1	0	1	0	0	—	42
0	15	2	300	0	0	1	0	0	0	0	0	553

No more than one species of *Vini* occupies any one island today. In prehistory, three species of *Vini* were sympatric in at least three East Polynesian groups (Cook, Society, and Marquesas; Chapter 7). No indigenous species of *Vini* survive in the Cooks; in the other two island groups, the species that persist (on one to three islands each) are the smallest in the genus, namely *V. peruviana* in the Society Islands and *V. ultramarina* in the Marquesas (Figure 12-3). On Mangaia (Cook Islands), precultural fossils represent *V. kuhlii*, *V. vidivici*, and *V. sinotoi*, thus showing that the sympatry is natural rather than being possibly due to interisland trading by Polynesians. The two large, widespread species (conquered lorikeet *Vini vidivici* and Sinoto's lorikeet *V. sinotoi*) are extinct. In Oceania today, three sympatric species of nectar-feeding parrots (Loriinae) do not occur east of the Solomon Islands. Not even the Bismarcks or Solomons have three sympatric congeneric lorikeets.

The behavior and ecology of species of *Vini* are known only in general terms (Bruner 1972, Montgomery et al. 1980, Watling 1982, Pratt et al. 1987, Steadman 1998). These small, brushy-tongued parrots feed on nectar and fruits from many types of flowering plants, but often prefer

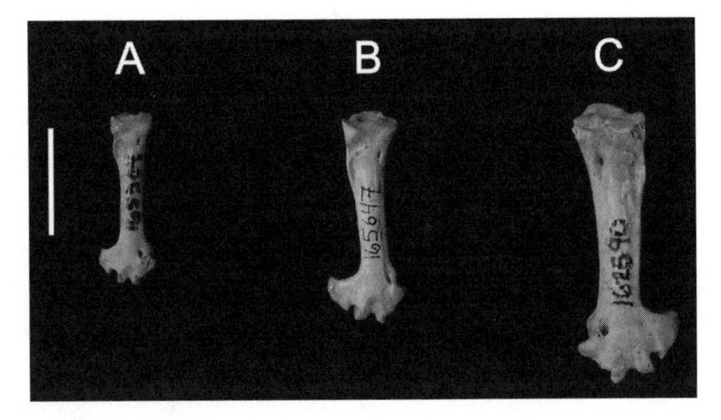

Figure 12-3 Tarsometatarsi of three species of *Vini* from archaeological sites in the Marquesas Islands. A: Extant *V. ultramarina*, BPBM 165504, Hane site, Ua Huka. B: Extinct *V. vidivici*, BPBM 165647 (holotype), Hanatekua site, Hiva Oa. C: Extinct *V. sinotoi*, BPBM 162590 (holotype; right side), Hane site, Ua Huka. Scale = 10 mm. Photo by J. J. Kirchman.

Table 12-2 The modern (M) and prehistoric (P) distribution of parrots in Melanesia. †, extinct species. Taxa in brackets do not necessarily represent different species than those already listed. From data in Chapter 5.

	Bismarcks	Solomons	Vanuatu	New Caledonia	TOTAL M	TOTAL M + P
Cacatua ducorpsii	—	M	—	—	1	1
Cacatua [galerita] ophthalmica	M	—	—	—	1	1
†*Cacatua [galerita]* undescribed sp. A	P	—	—	—	0	1
†*Cacatua* undescribed sp. B	—	—	—	P	0	1
Chalcopsitta cardinalis	M	M, P	—	—	2	2
Trichoglossus haematodus	M	M	M	M	4	4
Lorius hypoinochrous	M	—	—	—	1	1
Lorius albidinuchus	M	—	—	—	1	1
Lorius chlorocercus	—	M	—	—	1	1
[*Lorius* sp.]	P	—	—	—	—	—
Charmosyna palmarum	—	—	M, P	—	1	1
Charmosyna rubrigularis	M	—	—	—	1	1
Charmosyna meeki	—	M	—	—	1	1
Charmosyna placentis	M	M	—	—	2	2
Charmosyna margarethae	—	M	—	—	1	1
†*Charmosyna diadema*	—	—	—	M	1	1
[*Charmosyna* sp.]	P	—	—	—	—	—
Micropsitta pusio	M	—	—	—	1	1
Micropsitta bruijnii	M	M	—	—	2	2
Micropsitta meeki	M	—	—	—	1	1
Micropsitta finschii	M	M	—	—	2	2
Eunymphicus cornutus	—	—	—	M	1	1
Cyanoramphus novaezelandiae	—	—	—	M	1	1
Geoffroyus heteroclitus	M	M	—	—	2	2
Eclectus roratus	M	M	—	—	2	2
†*Eclectus* undescribed sp.	—	—	P	—	0	1
Loriculus [aurantiifrons] tener	M	—	—	—	1	1
TOTAL						
M	14	11	2	4	31	—
M + P	15	11	3	5	—	34
Number of bones	13	1	9	1	24	

Table 12-3 The modern (M) and prehistoric (P) distribution of parrots in West Polynesia. †, extinct species. From data in Chapter 6.

	Fiji	Tonga	Samoa	Niue	Wallis & Futuna	TOTAL M	TOTAL M + P
Vini solitarius	M	P	—	—	—	1	2
Vini australis	M	M, P	M	M	—	4	4
Charmosyna amabilis	M	—	—	—	—	1	1
Prosopeia tabuensis	M	—	—	—	—	1	1
Prosopeia personatus	M	—	—	—	—	1	1
†*Eclectus* undescribed sp.	—	P	—	—	—	0	1
TOTAL							
M	5	1	1	1	0	8	—
M + P	5	3	1	1	0	—	10
Number of bones	0	26	0	0	0	26	

the nectar of coconut palms (*Cocos nucifera*). Macrofossil and pollen records prove that *C. nucifera* is indigenous in Oceania at least as far east as the Cook and Society Islands (Spriggs 1984, Parkes 1997). The species of *Vini* nest mainly in cavities or crevices in trees, such as in coconut palms or *Pandanus*.

Charmosyna

These poorly studied, very small nectar-feeders occur from the Moluccas to Fiji. Amadon (1942b) synonymized *Charmosyna* with *Vini*, although I have found them to be distinct osteologically (Steadman & Zarriello 1987). Single, endemic species of *Charmosyna* inhabit Vanuatu (plus Santa Cruz), New Caledonia, and Fiji. The Fijian endemic *C. amabilis* is now very rare or extinct (Watling 2001). Six species of *Charmosyna* live in New Guinea, whereas the Bismarcks and Solomons have two and three, respectively (Hadden 1981, Coates 1985, Beehler et al. 1986, Mayr & Diamond 2001). These lorikeets are difficult to detect because most species live in the forest canopy, where even large parrots can disappear in the foliage when foraging. Furthermore,

Table 12-4 The modern (M) and prehistoric (P) distribution of *Vini* in Polynesia by island group; †, extinct species. From data in Chapters 6, 7.

	Fiji	Tonga	Samoa	Niue	Cook	Tubuai	Society	Tuamotu	Pitcairn	Marquesas	TOTAL M	TOTAL M + P
†*V. sinotoi*	—	—	—	—	P	—	P	—	—	P	0	3
†*V. vidivici*	—	—	—	—	P	—	P	—	—	P	0	3
V. solitarius	M	P	—	—	—	—	—	—	—	—	1	2
V. australis	M	M, P	M	M	—	—	—	—	—	—	4	4
V. kuhlii	—	—	—	—	P	M	—	—	—	—	1	2
V. stepheni	—	—	—	—	—	—	—	—	M, P	—	1	1
V. peruviana	—	—	—	—	—	—	M	M	—	—	2	2
V. ultramarina	—	—	—	—	—	—	—	—	—	M, P	1	1
TOTAL												
M	2	1	1	1	0	1	1	1	1	1	10	—
M + P	2	2	1	1	3	1	3	1	1	3	—	18
Number of bones	0	6	0	0	180	0	5	0	15	300	506	

Table 12-5 The modern (M) and prehistoric (P) distribution of extirpated populations or extinct species (†) of *Vini* on individual Polynesian islands; CI, Cook Islands; MI, Marquesas Islands; SI, Society Islands; TO, Tonga. From data in Chapters 6, 7.

	Lifuka TO	`Uiha TO	`Eua TO	Atiu CI	Mangaia CI	Huahine SI	Nuku Hiva MI	Ua Huka MI	Hiva Oa MI	Tahuata MI	TOTAL M	TOTAL M + P
†*V. sinotoi*	—	—	—	—	P	P	P	P	P	P	0	6
†*V. vidivici*	—	—	—	—	P	P	P	P	P	P	0	6
V. solitarius	P	P	P	—	—	—	—	—	—	—	0	3
V. australis	M	M	P	—	—	—	—	—	—	—	2	3
V. kuhlii	—	—	—	P	P	—	—	—	—	—	0	2
V. peruviana	—	—	—	—	—	M	—	—	—	—	1	1
V. ultramarina	—	—	—	—	—	—	M, P	P	—	P	1	3
TOTAL												
M	1	1	0	0	0	1	1	0	0	0	4	—
M + P	2	2	2	1	3	3	3	3	2	3	—	24
Number of bones	1	1	4	1	179	5	11	271	4	14	491	

the generally short, high-pitched calls of species of *Charmosyna* are not as easily heard as those of most other parrots.

Cacatua

Cacatua represents large, mainly white parrots known as cockatoos. Locally common on New Britain (*C. galerita ophthalmica*; Coates 1985) and in the Solomon Main Chain (*C. ducorps*; Kratter et al. 2001a), bones of undetermined species of *Cacatua* have been found on New Ireland and Mussau (Steadman & Kirch 1998, Steadman et al. 1999a; Chapter 5 herein), the former classified as *Cacatua* undescribed sp. A. I recently also discovered a burned tibiotarsus of an extinct species (*Cacatua* undescribed sp. B) from the Lapita archaeological site on New Caledonia (Grande Terre), the first record of an indigenous cockatoo in Remote Oceania.

Micropsitta

The six species of pygmy-parrots range from the Moluccas through Near Oceania, with three species on New Guinea. Three species (*Micropsitta pusio*, *M. meeki*, *M. finschii*) live allopatrically in the Bismarcks except that the first and last are both on New Britain. The two species in the Solomons (*M. finschii*, *M. bruijnii*) are sympatric on Bougainville, Kolombangara, and Guadalcanal (Mayr & Diamond 2001). These tiny forest dwellers feed on the lichens and fungi growing on trunks and large limbs. They move along in a sort of nuthatch- or woodpecker-like fashion, aided by stiffened rectrices. In lowland forests of the Solomons, *M. finschii* (Figure 12-4) is common on Rennell and extremely common on Isabel (Filardi et al. 1999, Kratter et al. 2001a).

Geoffroyus

Geoffroyus is confined in Oceania to the Bismarcks and Solomons, where the endemic *G. heteroclitus* occurs. The only recognized variation in this rather low-profile canopy dweller is an endemic subspecies (*G. h. hyacinthus*) on Rennell, where it is common (Filardi et al. 1999).

Eclectus

One species of *Eclectus*, the spectacular *E. roratus* (Figure 12-5), is extant. This cooperative breeder (Heinsohn et al. 1997) inhabits eastern Indonesia (five subspecies), New Guinea and offshore islands (three subspecies), Queensland (*E. r. macgillivrayi*), the Bismarcks (*E. r. goodsoni*), and the Solomons (*E. r. solomonensis*). The plumage of *E. roratus* is unique among birds in that the female is mostly red and the male is mainly green. A larger, extinct species of *Eclectus* has been found in cultural and noncultural sites on four islands in Tonga (2700 km southeast of the present range of *E. roratus*) and on Malakula in Vanuatu (Steadman in press; Chapters 5, 6 herein). I predict,

Figure 12-4 Finsch's pygmy-parrot, *Micropsitta finschii finschii*, Rennell, Solomon Islands. Photo by DWS, 23 June 1997.

Figure 12-6 Masked shinning parrot, *Prosopeia tabuensis* (captive). Photo by Thorston Bönte.

Figure 12-5 Eclectus parrot, *Eclectus roratus* (captive). Photo by Robert Lanciones, 2004.

therefore, that lost populations or species of *Eclectus* await discovery in New Caledonia and Fiji.

Prosopeia

The large, raucous shining parrots are endemic to western Fiji (*Prosopeia tabuensis* on Vanua Levu, Taveuni, Qamea, Koro, Kioa, Kadavu, Gau, and Ono; *P. personata* on Viti Levu; Amadon 1942b, Rinke 1989). Those from Kadavu and Ono were classified by Rinke (1989) as a third species, *P. splendens*, which I do not recognize. *Prosopeia tabuensis* (Figure 12-6) has been introduced on Viti Levu and on Tongatapu, 'Euaiki, and 'Eua in southern Tonga. The Tongan introduction probably was prehistoric, as Captain Cook found it there in the 1770s. Also, *P. tabuensis* is absent among prehuman fossils on 'Eua and from Lapita contexts in Ha'apai and Tongatapu (Chapter 6), where the only large parrot is *Eclectus* undescribed sp.

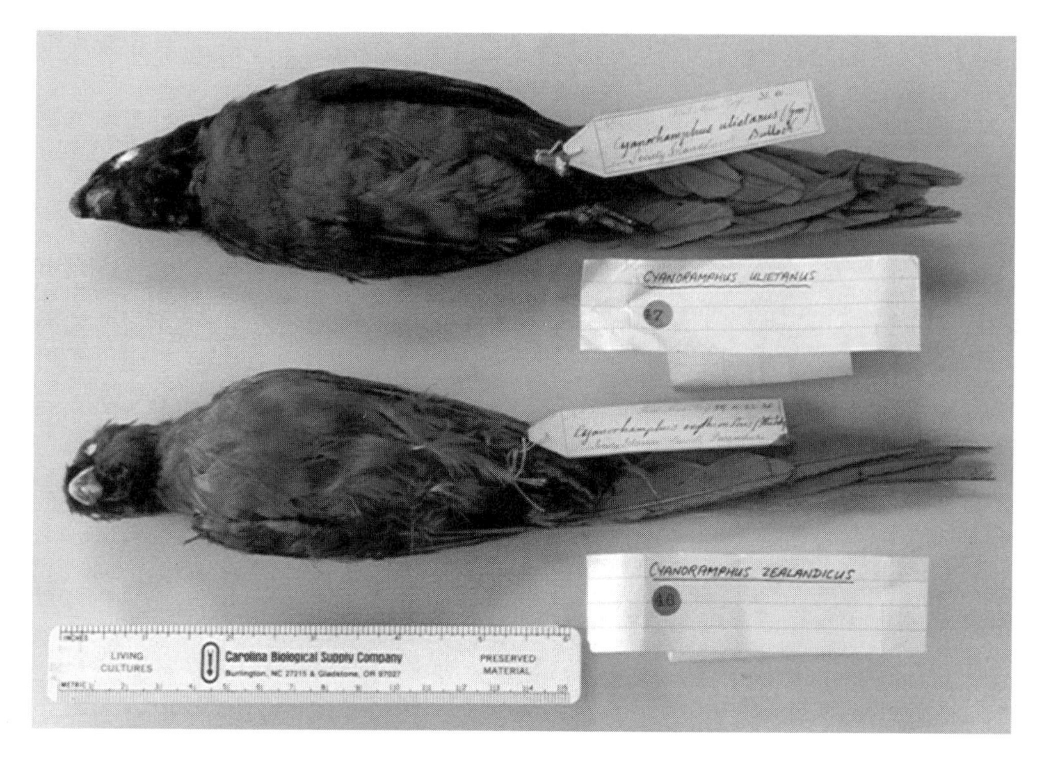

Figure 12-7 British Museum specimens of the Ra`iatea parakeet *Cyanoramphus ulietanus* (above; BMNH 21.R.28.1.84, Ra'iatea, Society Islands) and Tahiti parakeet *Cyanoramphus zealandicus* (below; BMNH 14.R.5.2.84, Tahiti, Society Islands). BMNH 21.R.28.1.84 was collected during the second voyage of Captain James Cook in 1773. BMNH 14.R.5.2.84 was collected during either the first (1769), second (1773), or third (1777) of Cook's voyages. Photo by DWS, September 1990.

Cyanoramphus

Small to medium-sized parakeets with long, narrow tails, *Cyanoramphus* has an enigmatic distribution. Two species that became extinct in the 19th century, both larger than any that survives, are known from the Society Islands (*C. zealandicus* from Tahiti, *C. ulietanus* from Ra'iatea; Greenway 1958:332–337, Schauenberg 1970, Olson 1990, Knox & Walters 1994, Voisin & Mary 1995; Figures 12-7, 12-8 herein). After a *Cyanoramphus*-free gap of 4450 km, *C. novaezelandiae saisseti* is endemic to New Caledonia. Temperate New Zealand and its satellite islands (4100 km SW of the Society Islands) are inhabited by *C. novaezelandiae* (four subspecies), *C. auriceps* (two subspecies), and monotypic *C. unicolor* (Taylor 1985, McNab & Salisbury 1995; see Triggs & Daugherty 1996 for alternative classification). Finally, *C.* [*novaezelandiae*]*cookii* is endemic to Norfolk Island, between Australia and New Zealand.

I once believed that the huge gap in the distribution of *Cyanoramphus* would be reduced by fossils from Fiji, Tonga, or the Cook Islands. The gap, however, has not changed. Fossils of *Cyanoramphus* are found regularly, sometimes abundantly, in New Zealand (Worthy & Holdaway 1994, Worthy 1997b), so I doubt that some unknown factor has prevented their deposition and preservation in more tropical parts of Oceania. While it is true that parrot bones are uncommon or absent in many prehistoric sites, the exceptions (certain sites in Tonga, Mangaia, and the Marquesas) are all in the *Cyanoramphus*-free zone. Unless ancient bones of *Cyanoramphus* begin to fill the void, we must entertain the possibility that the natural distribution of these parrots was inexplicably disjunct.

Eunymphicus

Confined to New Caledonia, *Eunymphicus* comprises *E. c. cornutus* on Grand Terre and *E. c. uveaensis* on 'Uvea in the Loyalty Islands. The 'Uvea population is small (Robinet & Salas 1999), suggesting that these horned parakeets once lived throughout New Caledonia, although there is no fossil record.

Loriculus

The 13 species of hanging-parrots range from Southeast Asia to the Bismarcks. *Loriculus* reaches its eastern limit with *L. tener*, a very poorly known species from New Britain, New Ireland, New Hanover, and Duke of York (Mayr & Diamond 2001; Table 5-3 herein).

Figure 12-8 Profile of the Tahiti parakeet, *Cyanoramphus zealandicus* (same specimen as in Figure 12-7; BMNH 14.R.5.2.84), Tahiti, Society Islands. Photo by DWS, September 1990.

Biogeographic Patterns

Parrots are represented on individual islands in Near Oceania by as many as nine genera and 12 species in the Bismarcks (New Britain, New Ireland; Table 5-3) or eight genera and 10 species in the Solomons (Bougainville, Guadalcanal; Table 5-6). Richness and diversity of parrots diminish from Near to Remote Oceania at comparable rates as in columbids (Chapter 11) but with consistently lower absolute values. Any psittacid comparisons between Near and Remote Oceania are preliminary, however, because parrot fossils are scarce or lacking in the Solomons, Vanuatu, New Caledonia, and Fiji. The only island in Remote Oceania with more than three genera of parrots is Grande Terre (New Caledonia) with five genera and five species.

Among Oceania's five genera of nectar-feeding parrots, only *Vini* is found east of Fiji. Within *Vini*, fossils have disclosed sympatric species triplets in at least three East Polynesian island groups, an situation unparalleled on other Pacific islands. At least two genera previously believed to be confined in Oceania to the Bismarcks and Solomons, *Cacatua* and *Eclectus*, occur as fossils (undescribed species) in Remote Oceania.

Finally, the two scrappy parrot bones from Easter Island are inadequate for generic assignment (Chapter 7). Because parrots occur as well in South America (the Psittacinae but not the Loriinae or Cacatuinae), the specimens from Easter Island cannot even be classified as having Oceanic vs. Neotropical affinities. Either way, they are further evidence of the great dispersal ability in this unusual family of birds.

Chapter 13 Other Nonpasserine Landbirds

RELEGATED to this chapter are 29 nonpasserine families that are less diverse or less widespread in Oceania (Figure 13-1) than the megapodes, rails, pigeons, and parrots covered in Chapters 9-12. Nevertheless, eight of these families (Ardeidae, Anatidae, Accipitridae, Cuculidae, Tytonidae, Strigidae, Apodidae, and Alcedinidae) have substantial representation in Oceania (Table 13-1). Ten of them are not found at all in Remote Oceania, whereas six others are known there only from one or two island groups, often New Caledonia. Just two of the 29 families (Rhynochetidae, Scolopacidae) have resident species in Remote Oceania but not in Near Oceania.

Some of the families consist of aquatic (freshwater or estuarine) species that are questionably called "landbirds." These include grebes, cormorants, jacanas, most herons, and some ducks, which often are little if at all differentiated from populations in New Guinea or Australia. As with aquatic birds worldwide, most of these species are good dispersers that respond opportunistically to changing conditions in lakes, marshes, swamps, and rivers. Thus, while I include aquatic birds for the sake of completeness and because they are not marine species, their distributional ecology (considering habitat preference, dispersal, and human influence) is different from that of forest birds.

Because Table 13-1 treats genera and island groups, not species and individual islands, it is insensitive to prehistoric records of one species from multiple islands within an island group, or of multiple congeneric species within an island group. One more qualifier is this chapter's brevity. While a more detailed review of the biogeography of these families in Oceania is desirable, it is not feasible in part because the systematics, past and present distribution, and ecology of many species are not well documented. For many species I will mention, usually uncritically, their subspecific classification as a rough gauge of differentiation. This usually is based on assessments of external morphology (skins) or osteology; very little research has been done on species- or subspecies-level molecular systematics.

Systematic Review

Casuariidae: Cassowaries

This family of large, flightless birds consists of three extant species—*Casuarius unappendiculatus* in New Guinea, *C. casuarius* in New Guinea, Aru, Ceram, and Queensland, and *C. bennetti* on New Guinea, Japen, and New Britain, the last being the only modern record of cassowaries in Oceania (Mayr 1940a, Coates 1985:58, Beehler et al. 1986:45–46). In the late Pleistocene, *C. bennetti* (= *C. lydekkeri*) occurred in Australia as far south as New South Wales (Miller 1962). Prehistorically, *C.* cf. *casuarius* occurred on the Arawe Islands off New Britain (Chapter 5). Both instances of cassowaries in the Bismarcks may represent human-assisted introductions from New Guinea because cassowaries seem incapable of unaided interisland dispersal.

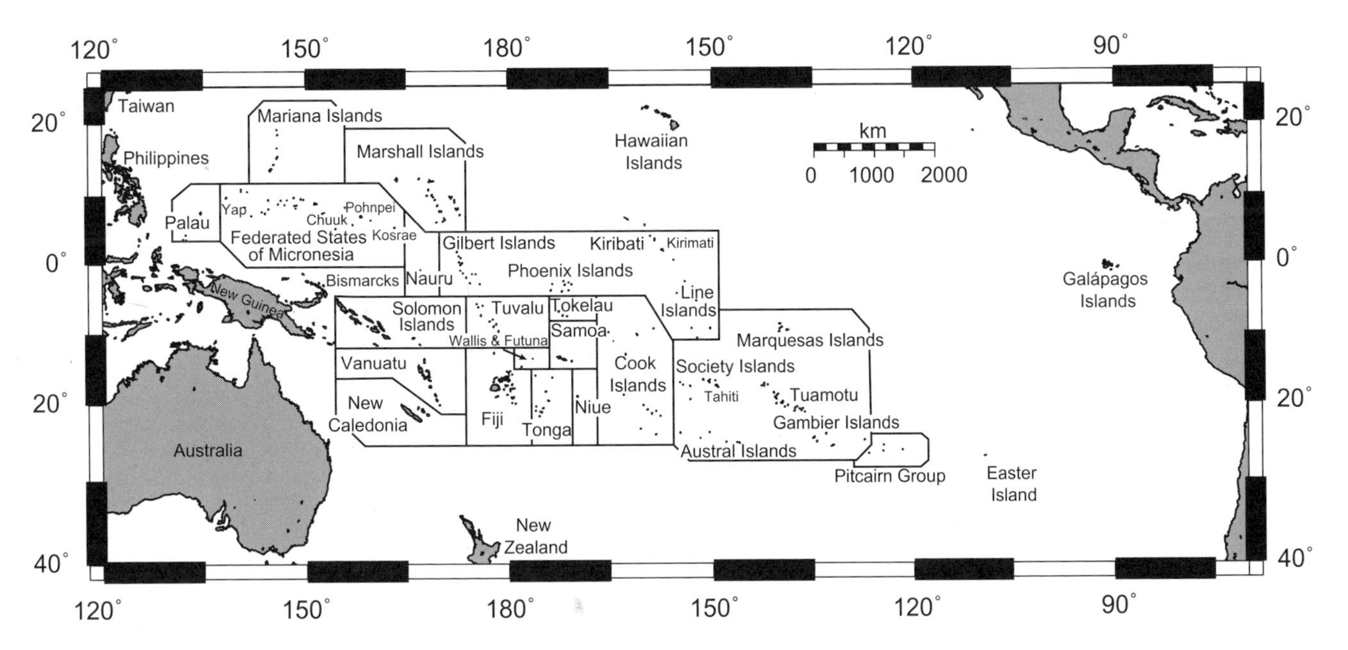

Figure 13-1 Oceania, showing major island groups.

Podicipedidae: Grebes

The distribution of the two species of grebes in Oceania, *Tachybaptus ruficollis* and *T. novaehollandiae*, is probably limited by island isolation and the scarcity of suitable freshwater lakes. Whereas grebe bones often are common in prehistoric contexts in the Americas (Brodkorb 1963, Kent et al. 1999), the only grebe bone from Oceania is a tibiotarsus of *T.* cf. *novaehollandiae* that I recently identified from the Sigatoka Dune archaeological site on Vitu Levu, Fiji. This is ca. 950 km east of its modern range limit in Vanuatu.

Phalacrocoracidae: Cormorants

Two freshwater species of cormorants, *Phalacrocorax melanoleucus* and *P. carbo*, occur locally in Near Oceania and western Remote Oceania. Within Oceania, the former species is differentiated from populations in Australia and New Guinea only on Rennell, Solomon Islands, where the endemic *P. m. brevicauda* (Amadon 1942a) is the world's smallest cormorant. The nearly cosmopolitan *P. carbo* is undifferentiated in Oceania, where it resides only on Rennell, a colonization (directly from Australia?) that probably took place in recent decades (Filardi et al. 1999).

Ardeidae: Herons, Bitterns

Four of the five ardeid genera occur, at least locally, in Remote Oceania as well as Near Oceania. The breeding status of *Ardea alba* in the Bismarcks and Solomons is not well established, so *Ardea* is excluded from Table 13-1. The species of *Egretta* inhabit reef flats, estuaries, lakes, marshes, and rivers. *Egretta novaehollandiae* resides in New Caledonia and is becoming more common in Fiji and Tonga, where resident populations are being established (Watling 2001; DWS pers. obs. 1996–2001). *Egretta sacra* (Figure 13-2) is the most widespread species of "landbird" in Oceania, recorded from every island group except the most isolated parts of East Polynesia (Pitcairn, Easter, and the Marquesas) and certain of the Remote Central Pacific Islands (Tables 8-12, 13-1). Mayr and Amadon (1941) recognize *E. s. albolineata* as endemic to New Caledonia and the Loyalty Islands, with all other populations in Oceania treated as *E. s. sacra*, which also occurs in New Guinea. Geographic variation in the distribution of dark versus white plumage is not well documented or explained (Rohwer 1990, Itoh 1991).

The only extant night-heron in our study area is *Nycticorax caledonicus*, with endemic subspecies in the Bismarcks and Solomons (*N. c. mandibularis*), New Caledonia (*N. c. caledonicus*), and Micronesia (*N. c. pelewensis*; Amadon 1942a). Well east of the modern range of *Nycticorax* in Oceania, three extinct species have been found in Tonga (*N.*undescribed sp. B), Niue (*N. kalavikai*), and the Cook Islands (*N.* undescribed sp. C). Another, *N.* undescribed sp. A, is from Buka (Solomon Islands), within the modern range of *N. caledonicus*. The extinct night-herons may have preyed on crabs, lizards, and birds on the forest floor, rather than on aquatic fauna as their living counterparts do (Steadman

Table 13-1 Distribution of resident nonpasserine genera in Oceania, except the megapodes, rails, columbids, and parrots covered in Chapters 9–12. "Bismarcks" includes the Admiralty Islands. "Remote Central Pacific Islands" consists of the Marshall Islands and Nauru eastward through the Line Islands (see Chapter 8). H, historic; M, modern; P, prehistoric; *, no endemic species known in Oceania. Based on data in Chapters 5–8.

	Bismarcks	Solomons	Vanuatu	New Caledonia	Fiji	Tonga	Samoa	West Polynesian Outliers	Cook Islands	Tubuai	Society Islands
CASSOWARIES											
Casuarius *	M, P	—	—	—	—	—	—	—	—	—	—
GREBES											
Tachybaptus *	M	M	M	M	P	—	—	—	—	—	—
CORMORANTS											
Phalacrocorax *	M	M	M?	M	—	—	—	—	—	—	—
HERONS											
Egretta *	M, P	M	M	M	M	M, P	M	M	M, P	M	M
Butorides *	M	M	M	M	M	—	—	—	—	—	M, P
Nycticorax	M, P	M, P	—	M	—	P	—	P	P	—	—
Ixobrychus *	M	M	—	—	—	—	—	—	—	—	—
Botaurus *	—	—	—	M	—	—	—	—	—	—	—
IBISES											
Threskiornis *	—	M	—	—	—	—	—	—	—	—	—
DUCKS											
Dendrocygna	M	—	—	H	H	—	—	—	P	—	—
Anas	M	M	M	M, P	M	M, P	M	M	M, P	M	M
Aythya *	—	—	M	M	—	—	—	—	—	—	—
OSPREYS											
Pandion *	M	M	—	M	—	P	—	—	—	—	—
HAWKS											
Aviceda *	M	M	—	—	—	—	—	—	—	—	—
Henicopernis	M	—	—	—	—	—	—	—	—	—	—
Haliastur *	M, P	M, P	—	M	—	—	—	—	—	—	—
Haliaeetus	M	M, P	—	—	—	—	—	—	—	—	—
Circus *	—	—	M	M	M	M	—	—	—	—	—
Accipiter	M, P	M	M	M, P	M	P	—	—	—	—	—
FALCONS											
Falco *	M	M	M	M, P	M	—	M	—	—	—	—
QUAILS											
Coturnix *	M, P	—	—	—	—	—	—	—	—	—	—
BUTTON-QUAILS											
Turnix	M	M	—	M, P	—	—	—	—	—	—	—
KAGUS											
Rhynochetos	—	—	—	M, P	—	—	—	—	—	—	—
JACANAS											
Irediparra *	M	—	—	—	—	—	—	—	—	—	—

Tuamotu	Pitcairn	Easter Island	Marquesas	Palau	Yap	Marianas	Chuuk	Pohnpei	Kosrae	Remote Central Pacific Islands
—	—	—	—	—	—	—	—	—	—	—
—	—	—	—	—	—	—	—	—	—	—
—	—	—	—	M	—	—	—	—	—	—
M	—	P?	—	M	M	M, P	M	M	M	M
—	—	—	—	—	—	—	—	—	—	—
—	—	—	—	M	—	—	M	—	—	—
—	—	—	—	M	M	M	M	—	—	—
—	—	—	—	—	—	—	—	—	—	—
—	—	—	—	—	—	—	—	—	—	—
—	—	—	—	—	—	—	—	—	—	—
—	—	—	—	M	—	M	M	—	—	H
—	—	—	—	—	—	—	—	—	—	—
—	—	—	—	—	—	—	—	—	—	—
—	—	—	—	—	—	—	—	—	—	—
—	—	—	—	—	—	—	—	—	—	—
—	—	—	—	—	—	—	—	—	—	—
—	—	—	—	—	—	—	—	—	—	—
—	—	—	—	—	—	—	—	—	—	—
—	—	—	—	—	—	—	—	—	—	—
—	—	—	—	—	—	P	—	—	—	—
—	—	—	—	—	—	—	—	—	—	—
—	—	—	—	—	—	—	—	—	—	—
—	—	—	—	—	—	—	—	—	—	—
—	—	—	—	—	—	—	—	—	—	—

(continued)

Table 13-1 (continued)

	Bismarcks	Solomons	Vanuatu	New Caledonia	Fiji	Tonga	Samoa	West Polynesian Outliers	Cook Islands	Tubuai	Society Islands
THICK-KNEES											
*Esacus**	M	M	—	M	—	—	—	—	—	—	—
STILTS											
*Himantopus**	M	—	—	—	—	—	—	—	—	—	—
PLOVERS											
*Charadrius**	M	—	—	—	—	—	—	—	—	—	—
SANDPIPERS											
Coenocorypha?	—	—	—	P	—	—	—	—	—	—	—
Prosobonia	—	—	—	—	—	—	—	—	P	—	H
CUCKOOS											
*Cacomantis**	M, P	M	M, P	M	M	P	—	—	—	—	—
*Chrysococcyx**	—	M	M	M	—	—	—	—	—	—	—
*Eudynamys**	M	M	—	—	—	—	—	—	—	—	—
Centropus	M, P	M	—	—	—	—	—	—	—	—	—
BARN-OWLS											
Tyto	M, P	M	M, P	M, P	M, P	M, P	M, P	M	—	—	—
OWLS											
Pyrroglaux	—	—	—	—	—	—	—	—	—	—	—
Ninox	M, P	M	—	P		—	—	—	—	—	—
*Asio**	—	—	—	—	—	—	—	—	—	—	—
Nesasio	—	M	—	—	—	—	—	—	—	—	—
NIGHTJARS											
*Eurostopodus**	—	M	—	H/M	—	—	—	—	—	—	—
*Caprimulgus**	M, P	M	—	—	—	—	—	—	—	—	—
OWLET-NIGHTJARS											
Aegotheles	—	—	—	H, P	—	—	—	—	—	—	—
FROGMOUTHS											
Undescribed genus	—	M	—	—	—	—	—	—	—	—	—
CRESTED SWIFTS											
*Hemiprocne**	M	M	—	—	—	—	—	—	—	—	—
SWIFTS											
Collocalia	M, P	M	M, P	M, P	M	M, P	M	M	M, P	—	M
KINGFISHERS											
*Alcedo**	M, P	M	—	—	—	—	—	—	—	—	—
Ceyx	M	M	—	—	—	—	—	—	—	—	—
Halcyon	M	M	M, P	M, P	M	M, P	M	M	M, P	—	M
Actenoides	—	M	—	—	—	—	—	—	—	—	—
*Tanysiptera**	M	—	—	—	—	—	—	—	—	—	—

Tuamotu	Pitcairn	Easter Island	Marquesas	Palau	Yap	Marianas	Chuuk	Pohnpei	Kosrae	Remote Central Pacific Islands
—	—	—	—	—	—	—	—	—	—	—
—	—	—	—	—	—	—	—	—	—	—
—	—	—	—	—	—	—	—	—	—	—
—	—	—	—	—	—	—	—	—	—	—
M	P	—	P	—	—	—	—	—	—	H
—	—	—	—	—	—	—	—	—	—	—
—	—	—	—	—	—	—	—	—	—	—
—	—	—	—	—	—	—	—	—	—	—
—	—	—	—	—	—	—	—	—	—	—
—	—	—	—	—	—	—	—	—	—	—
—	—	—	—	M	—	—	—	—	—	—
—	—	—	—	—	—	—	—	—	—	—
—	—	—	—	—	—	—	—	M	—	—
—	—	—	—	—	—	—	—	—	—	—
—	—	—	—	—	—	—	—	—	—	—
—	—	—	—	M	—	—	—	—	—	—
—	—	—	—	—	—	—	—	—	—	—
—	—	—	—	—	—	—	—	—	—	—
—	—	—	—	—	—	—	—	—	—	—
—	—	—	M	M	H	M	M	M	M	—
—	—	—	—	—	—	—	—	—	—	—
—	—	—	—	—	—	—	—	—	—	—
M	—	—	M	M	—	M	M	M	—	—
—	—	—	—	—	—	—	—	—	—	—
—	—	—	—	—	—	—	—	—	—	—

(continued)

Table 13-1 (continued)

	Bismarcks	Solomons	Vanuatu	New Caledonia	Fiji	Tonga	Samoa	West Polynesian Outliers	Cook Islands	Tubuai	Society Islands
BEE-EATERS											
*Merops**	M	—	—	—	—	—	—	—	—	—	—
ROLLERS											
*Eurystomus**	M	M	—	—	—	—	—	—	—	—	—
HORNBILLS											
Aceros	M, P	M	—	P	—	—	—	—	—	—	—
TOTAL											
M + H	37	34	13–14	24	11	6	6	5	4	2	6
M + H + P	37	34	13–14	27	12	10	6	6	7	2	6

et al. 2000b). Extinct species of *Nycticorax* have been found as well in the Mascarene Islands, Indian Ocean (Mourer-Chauviré et al. 1999).

Butorides (one species in Oceania, the nearly cosmopolitan *B. striatus*) prefers woody vegetation along streams or estuaries. The population in the Solomons and New Ireland (unrecorded elsewhere in the Bismarcks; Mayr & Diamond 2001) is *B. s. solomensis*, quite different from New Guinean birds (Schodde et al. 1980). The large gap in modern distribution of *B. striatus* between Fiji (*B. s. diminutus*, this subspecies also in Santa Cruz and Vanuatu; Mayr 1940a) and Tahiti (the endemic *P. s. patruelis*) is beginning to be filled by prehistoric bones from Huahine, Society Islands (Steadman & Pahlavan 1992) and Lakeba, Lau Group, Fiji (Chapter 6). Especially considering the past and present distribution of mangroves in Oceania (Woodroffe 1987, Ellison 1994a), I expect bones of *B. striatus* to be found in Tonga, Samoa, and the Cook Islands.

Figure 13-2 Pacific reef-heron *Egretta sacra sacra*, dark-phase adult. Photo by DWS, Isabel, Solomon Islands, 10 July 1997.

Tuamotu	Pitcairn	Easter Island	Marquesas	Palau	Yap	Marianas	Chuuk	Pohnpei	Kosrae	Remote Central Pacific Islands
—	—	—	—	—	—	—	—	—	—	—
—	—	—	—	—	—	—	—	—	—	—
—	—	—	—	—	—	—	—	—	—	—
3	0	0	2	9	3	5	6	4	2	3
3	1	1	3	9	3	6	6	4	2	3

The bitterns (*Ixobrychus, Botaurus*) prefer freshwater habitats and are represented by *I.* [*Dupetor*] *flavicollis* in the Bismarcks (nonendemic *I. f. australis*) and Solomons (endemic *I. f. woodfordi*), *I. sinensis* in Micronesia, and *B. poiciloptilus* in New Caledonia, the latter two species undifferentiated from continental populations (Baker 1951:90–93, Martínez-Vilalta & Motis 1992:426–28).

Finally, a single ardeid synsacrum from Easter Island is referred to cf. *Egretta* (certainly not *E. sacra*; Chapter 6). Its affinities are vague until more bones are found.

Threskiornithidae: Ibises

The only resident ibis in Oceania is *Threskiornis molucca pygmaea*, a small endemic subspecies on Rennell and Bellona, Solomon Islands (Mayr 1931). Common in both forested and nonforested habitats on Rennell (Filardi et al. 1999), *T. m. pygmaea* probably represents another instance of Rennell being colonized directly from Australia or New Guinea rather than island hopping through the Bismarcks and Solomons (Diamond 1984). No fossils of flightless ibises have been found in the region, unlike in the West Indies (Olson & Steadman 1977) and Hawaiian Islands (Olson & James 1991).

Anatidae: Ducks

Compared to New Zealand and the Hawaiian Islands, with diverse radiations of endemic anatids (Miller 1937, Weller 1980:27–43, Olson & James 1991, Worthy & Holdaway 1994, 2002, Cooper et al. 1996, James & Burney 1997, Rhymer 2001, McNab 2003), the rest of Oceania is poor in waterfowl. Among whistling-ducks (or tree-ducks), the semiaquatic *Dendrocygna guttata* occurs (undifferentiated) only on Umboi and New Britain in the Bismarcks (Coates 1985:96–98). A more aquatic species (*D. arcuata*) occurs (as *D. a. pygmaea*) locally in the Bismarcks and historically in New Caledonia and Viti Levu, Fiji (Miles 1964, Weller 1980:33, Bolen & Rylander 1983:10, 11, Coates 1985:94–96). The former residency of *D. arcuata* on Viti Levu is well documented. I have examined both specimens (unsexed) collected there in September 1857 by J. D. Macdonald of the H.M.S. *Herald* (British Museum cat. no. 1859.1.10.39 and 1859.1.10.40). In a letter written in Fiji in September 1874, Layard (1875) noted that *D. arcuata* [which he called *D. vagans*] was common in the coastal marshes of western Viti Levu, although it was known as "the Mountain Duck" because it came as well from the island's interior. Additional details are provided by Clunie (1983), who noted that the Fijian name (*gadamu*) for *D. arcuata* literally means "red duck." Other subspecies of *D. arcuata* range from the Philippines and Indonesia to New Guinea and Australia (Carboneras 1992b:576).

A single, distinctive pedal phalanx of a very large species of *Dendrocygna* is known from the Ureia archaeological site, Aitutaki, Cook Islands (Steadman 1991b). This species probably was skilled at walking and perching, as in the neotropical *D. autumnalis* (see Rylander & Bolen 1974) but unlike most other living species of whistling-ducks. I doubt that this extinct, undescribed species was endemic to Aitutaki; its bones are likely to turn up elsewhere in East Polynesia.

By far the most widespread species of duck in Oceania is *Anas superciliosa* (sometimes called *A. poecilorhyncha*; Figure 13-3), which is second only to the heron *Egretta sacra* as Oceania's most widespread landbird. The subspecies

Figure 13-3 Gray duck *Anas superciliosa*. Photo by DWS, Isabel, Solomon Islands, 10 July 1997.

of *A. superciliosa* recognized across tropical Oceania is *A. s. pelewensis*, which may also inhabit northern New Guinea (Amadon 1943, Baker 1951:98–100, Carboneras 1992b:607, Rhymer et al. 2004). Two species of dabbling ducks (*Anas* spp.) have evolved endemic forms on Micronesian islands. The extinct Marianas mallard (*Anas oustaleti*) has been recorded on Guam, Rota, Tinian, and Saipan (Baker 1951, Weller 1980:17–19, Pratt et al. 1987:98). This highly variable form may have been a hybrid swarm of the temperate Mallard (*A. platyrhynchos*) and tropical *A. superciliosa* (Yamashina 1948, Weller 1980:17–19). A single coracoid from Rota represents an extinct flightless duck, perhaps a species of *Anas* (Chapter 8).

Coues Gadwall (*A. strepera couesi*) once lived on Teraina (Washington) Island, Kiribati (Chapter 8). Other resident ducks, currently unknown, probably have been lost since human arrival. Perhaps the northern pintail *Anas acuta*, with more migrant records in Oceania than any other temperate species of duck (Udvardy & Engilis 2001), developed resident populations on some remote islands. Flightless species of *Anas* have evolved on temperate Amsterdam and St. Paul islands, southern Indian Ocean (Olson & Jouventin 1996) and in the New Zealand region (Worthy & Holdaway 2002, McNab 2003).

The teal *Anas gracilis* (often listed as *A. gibberifrons*) occurs in New Guinea, Australia, and New Zealand. In Oceania, *A. g. remissa* was endemic to Rennell but now is extinct (Filardi et al. 1999). The Australian and Papuan race, *A.g. gracilis*, was first recorded on Vanuatu in 1970 and now breeds there (Bregulla 1992:120). *Anas g. gracilis* is found today on New Caledonia as well. Fossils of *A. gracilis* from New Caledonia are smaller, however, than modern bones of *A. g. gracilis* from Australia, perhaps suggesting that a small endemic form has been lost, only to be replaced in modern times by another colonization from Australia (Balouet & Olson 1989).

The only diving duck in Oceania is *Aythya australis*, a mobile, mainly Australian species that occurs locally and undifferentiated in Vanuatu and New Caledonia (Mayr 1940a, Hannecart & Letocart 1980:101, Bregulla 1992:121–122, Carboneras 1992b:616-617). Its absence in the Bismarcks and Solomons, at least as a resident, may be due to a lack of suitable freshwater lakes rather than isolation or human impact.

Pandionidae: Ospreys

Pandion haliaetus resides in Oceania today in the Bismarcks, Solomons, and New Caledonia (*P. h. cristatus*, a subspecies shared with Australia and New Guinea). Thus I was surprised to discover adult and juvenile bones of osprey from several Lapita sites in the Ha'apai Group of Tonga (Chapter 6), 1800 km east of New Caledonia. The former occurrence of *P. haliaetus* in Fiji and Vanuatu is likely, although perhaps not in Samoa, where deep nearshore waters may be poorly suited for a piscivore that hunts in shallow marine, estuarine, and freshwater habitats.

Accipitridae: Hawks, Eagles

Four of Oceania's six genera of hawks and eagles inhabit only Near Oceania. *Aviceda* is represented by one species (*A. subcristata*; Figure 13-4) also found on New Guinea but with endemic subspecies in the Bismarcks (*A. s. coultasi*, *A. s. bismarckii*) and Solomons (*A. s. gurneyi*; Debus 1994). The forms of *Haliastur* in Oceania are *H. sphenurus* on New Caledonia (undifferentiated from Australian and New Guinean populations), *H. indus girrenera* in the Bismarcks (same subspecies in Australia and New Guinea), and *H. i. flavirostris* endemic to the Solomons (Debus 1994).

Henicopernis occurs in Oceania only as *H. infuscatus*, endemic to New Britain but closely related to *H. longicauda* of New Guinea (Coates 1985:108–109). Among eagles, *Haliaeetus leucogaster* occupies coastal and subcoastal habitats in New Guinea and the Bismarcks, whereas *H. sanfordi* is endemic to the Solomons in inland forests and coastal areas (Coates 1985:116–117, Debus 1994, Kratter et al. 2001a).

Only two genera of hawks are known in Remote Oceania. The harriers (*Circus*) are represented by one species, *C. approximans*, which has no fossil record, is undifferentiated, often occurs in open habitats, and may be a relatively recent colonizer from Australia. The other is *Accipiter*, various species of which are or were found in each island group from the Bismarcks to Tonga (Chapters 5, 6), although the distribution, species-level systematics, comparative osteology, and ecology of Oceanic species of *Accipiter* are not well understood (LeCroy et al. 2001). One

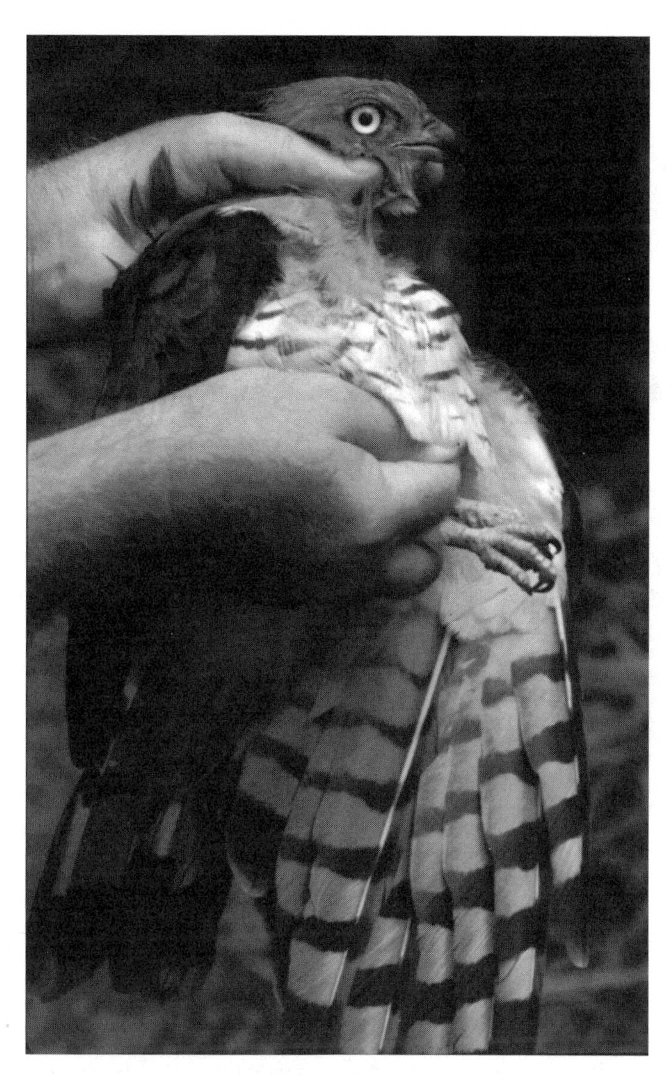

Figure 13-4 Pacific baza *Aviceda subcristata gurneyi*, adult. Photo by DWS, 18 July 1997, Isabel, Solomon Islands.

to five species of *Accipiter* now occupy each major island in the Bismarcks, compared to one or three species in the Solomons (Tables 5-3, 5-7). Two extinct species of *Accipiter* have been described from fossils on Grande Terre, New Caledonia (Balouet & Olson 1989), raising the total there to four species. An endemic species of *Accipiter* probably once occupied Vanuatu as well, where fossils from Efate reveal a broader past distribution of the genus although species-level identification is hampered by limited modern and fossil material (Chapter 5). In Tonga, fossils of *Accipiter* from 'Eua may represent *A. rufitorques* (Steadman 1993a), regarded today as endemic to Fiji, where it is common on most large, western islands (Watling 1982:69, Pratt et al. 1987:112) and must have occurred formerly in the Lau Group as well.

No resident species of hawks or eagles are known, today or prehistorically, from anywhere in East Polynesia or Micronesia. Considering the birdlife that once flourished in these many island groups, the lack of hawks, if genuine, must be due to isolation rather than a lack of sustenance. I suspect, however, that resident forms of *Accipiter* formerly existed in at least Palau, Yap, and the Marianas.

Falconidae: Falcons

Falcons are represented in Oceania by three species. *Falco berigora* of Australia and New Guinea occurs on Long in the Bismarcks (Coates 1985:133). *Falco severus* inhabits the Bismarcks and Solomons, undifferentiated from New Guinea populations. *Falco peregrinus* is a local resident from the Bismarcks (nonendemic *F. p. ernesti*) and Vanuatu to Fiji and perhaps Samoa (the oceanic endemic *F. p. nesiotes*; Mayr 1941b). The only detailed data on the distribution or ecology of *F. peregrinus* in Oceania are for Fiji, where it is widespread with a varied diet dominated by fruit bats (*Pteropus* spp.) and columbids (White et al. 1988). A prehistoric record from Rota, Marianas (Chapter 8) hints at a formerly larger resident range for of *F. peregrinus*.

Phasianidae: Quails

The only species of quail in Oceania today is *Coturnix chinensis* in the Bismarcks, where the endemic subspecies *C. c. lepida* resides (Coates 1985:149). A second species, *C.* cf. *ypsilophorus*, lived on New Ireland prehistorically (Steadman et al. 1999a).

Turnicidae: Buttonquails (Hemipodes)

These cryptic birds are poorly known in Oceania. *Turnix maculosa* lives in the Lesser Sundas, New Guinea, Australia, and has endemic subspecies in the Bismarcks (*T. m. saturata*) and Solomons (*T. m. salomonis*; Mayr 1938, Blaber & Milton 1991). Although regarded as conspecific by many previous authors, *T. novaehollandiae* of New Caledonia is osteologically distinct from *T. varia* of Australia (Balouet & Olson 1989).

Rhynochetidae: Kagus

Endemic to New Caledonia, the flightless kagus consist of the endangered *Rhynochetos jubatus* (Figure 13-5) and extinct *R. orarius* (Balouet & Olson 1989), the latter known only from undated fossil deposits. Balouet & Olson (1989) suggest that the larger *R. orarius* was a lowland species and that *R. jubatus* lived in the highlands of Grande Terre. Kagus are preyed upon by dogs these days (Hunt 1996, Hunt et al. 1996), a good reason for eliminating dogs in New Caledonia.

Figure 13-5 A spread wing and skin of the kagu *Rhynochetos jubatus*, UF 42714, a seventeen-month old ♂ that was born and died in captivity at the San Diego Zoological Park. Scale is in cm. Photo by J. J. Kirchman.

Recurvirostridae: Stilts

Widespread in Australia, New Guinea, and elsewhere, *Himantopus himantopus* nests in Oceania only on New Britain (Coates 1985).

Jacanidae: Jacanas

Irediparra gallinacea is rare, localized, and undifferentiated on New Britain today (Coates 1985:168–169). This is the only evidence of jacanas in Oceania.

Burhinidae: Thick-knees (Stone-curlews)

Esacus magnirostris occurs locally on beaches in the Bismarcks, Solomons, and New Caledonia. It seems to be undifferentiated from populations in New Guinea and Australia.

Charadriidae: Plovers

This family, like the next, is well represented in Oceania by migratory species that breed in the northern hemisphere. The one resident species of plover, *Charadrius dubius*, inhabits much of Eurasia and Africa (Wiersma 1996:426) but does not extend beyond the Bismarcks in Oceania (undifferentiated from New Guinean populations).

Scolopacidae: Sandpipers

Prosobonia (synonym *Aechmorhynchus*; see Zusi & Jehl 1970) is endemic to Remote Oceania. At least seven species, and perhaps hundreds of populations, of *Prosobonia* have been lost from East Polynesia and Remote Central Pacific Islands. The only living species is *P. parvirostris* (usually called *P. cancellatus*, but see Walters 1993), which survives on some atolls in the Tuamotus (van Gils & Wiersma 1996, Wijpkema & Wijpkema 1997). Various extinct species of *Prosobonia* once lived on eroded volcanic islands and raised limestone islands in the Cooks (*P.* undescribed species A), Society Islands (*P. leucoptera* on Tahiti, *P. ellisi* on Mo'orea; Walters 1991), Marquesas (*P.* sp., near *P. parvirostris*), Pitcairn Group (*P.* undescribed species B; Chapter 7), and Kirimati Atoll in Kiribati (*P. cancellata*; Chapter 8).

The distinctive snipes of New Zealand (*Coenocorypha* spp.; van Gils & Wiersma 1996, Worthy 1997b) are represented by an extinct species (*C. miratropica*) from Viti Levu, Fiji (Worthy 2003). An extinct, undescribed species of cf. *Coenocorypha* also has been reported from New Caledonia (Balouet & Olson 1989; Chapter 5 herein). Worthy et al.

(2002b) recognize five or six allopatric species of *Coenocorypha* from New Zealand and its satellite islands.

Cuculidae: Cuckoos

The skin and flesh of cuckoos emit an unpleasant odor (pers. obs.), which might explain their relative rarity in bone deposits. None of Oceania's four resident genera of cuckoos has breeding species east of Tonga or anywhere in Micronesia. The only endemic species of cuckoos in Oceania are *Centropus violaceus* and *C. ateralbus* in the Bismarcks, and the weakly flying *C. milo* in the Solomons (Diamond 2002).

Cacomantis consists in Oceania of *C. variolosus* (three endemic subspecies in the Bismarcks, one in the Solomons), and *C. flabelliformis* (= *C. pyrrophanus*) with an endemic subspecies in each of New Caledonia/Loyalties, Vanuatu, and Fiji (Amadon 1942b, Payne 1997:560–561). Bones from archaeological sites extend the range of *C. flabelliformis* from western Fiji to Lifuka and Tongatapu, Tonga (Chapter 6). Thus it probably occurred widely in Tonga and the Lau Group of Fiji.

The single species of *Chrysococcyx* in Oceania is *C. lucidus*, which breeds in Australia and New Zealand and has endemic subspecies on Rennell and Bellona (*C. l. harterti*) and New Caledonia, Loyalties, Vanuatu, Banks Islands, and Santa Cruz (*C. l. layardi*; Payne 1997:564).

The one resident species of *Eudynamys* in Oceania is *E. scolopacea*, with endemic subspecies in the Bismarcks (*E. s. salvadorii*) and Solomons (*E. s. alberti*; Payne 1997:570). *Eudynamys taitensis* is a widespread migrant in Oceania, especially in West and East Polynesia, but breeds only in New Zealand (Bogert 1937, Payne 1997:571). No other landbird has this migratory pattern. *Scythrops novaehollandiae* is widespread in the Bismarcks but is not known to breed there (Coates 1985:363–364).

Tytonidae: Barn-owls

Although unrecorded in East Polynesia or Micronesia, the nearly cosmopolitan *Tyto alba* is Oceania's only widespread tytonid owl. Geographic variation of *T. alba* in Oceania is poorly understood (König et al. 1999:195). The few specimens of *T. alba* from the Bismarcks, Solomons, Santa Cruz Group, and northern Vanuatu represent various named subspecies (Mayr 1935, 1936, Amadon 1942b) that are difficult to evaluate without more specimens (Galbraith & Galbraith 1962). Mainly nocturnal and calling less conspicuously than most strigid owls, *T. alba* undoubtedly has been overlooked on many islands in Oceania (see Steadman 1998). The specimens of *T. alba* from New Caledonia, southern Vanuatu, Fiji, Tonga, Samoa, and Niue, united as *T. a. lulu* by Mayr (1936), seem to be rather uniform in external characters. Fossils of *T. alba* are lacking in prehuman cave deposits on New Caledonia and 'Eua (Tonga) but are common after human arrival (Balouet & Olson 1989; Chapter 6 herein). This suggests that *T. a. lulu* may have colonized 'Eua (and all of West Polynesia) only after people brought rats (*Rattus* spp.), the current favorite food of insular barn-owls.

Fossils from the Bismarcks (New Ireland and Mussau) indicate the past presence of a larger species of *Tyto*, most likely *T. novaehollandiae* s.l., which is known today only from Manus (the endemic *T. n. manusi*, considered a full species by König et al. 1999:202), southern New Guinea, and Australia (Chapter 5). New Ireland once had at least two species of *Tyto* (Steadman et al. 1999a), a situation that may have been common in the Bismarcks. The smaller species possibly was *T. alba* (unrecorded today on New Ireland) or *T. aurantia* (considered endemic now to New Britain; König et al. 1999:198). The widespread (India to Australia) *T. longimembris* was recorded in the 19th century on Viti Levu (Fiji; listed by Watling 1982:94 as *T. capensis*) and perhaps on New Caledonia or the Loyalty Islands (see Balouet & Olson 1989 for the confused history of the last record). There is no prehistoric evidence to back up these claims, or 20th-century evidence that any *Tyto* other than *T. alba* might survive in either place. The large, extinct *T. letocarti* as well as *T. alba* occur as fossils on New Caledonia, although all six bones of the former were found deeper than those of *T. alba* (Balouet & Olson 1989).

Strigidae: Strigid Owls

The large, powerful *Nesasio solomonensis* is confined to the Solomon Islands, where native mammals consist of bats and rodents but not marsupials (Flannery 1995a,b). The food habits of *N. solomonensis* are poorly understood, although it is said to favor marsupials, especially phalangers (Olsen 1999:242), a habit that must postdate human arrival. *Nesasio* may have occurred in the Bismarcks as well, especially given its past presence on Buka (Chapter 5). *Ninox* is restricted in Oceania today to Near Oceania, with three species endemic to the Bismarcks (*N. meeki* on Manus, *N. variegata* [synonym *solomonis*] on New Britain, New Ireland, and New Hanover, and *N. odiosa* on New Britain) and another (*N. jacquinoti*) endemic to the Solomons (Coates 1985:377–379, König et al. 1999:412–419). *Ninox* occurs as a fossil in New Caledonia (*N.* cf. *novaeseelandiae*, otherwise on Tasmania, New Zealand, Lord Howe, and Norfolk Island; Balouet & Olson 1989). I suspect that a species of *Ninox* also once inhabited Santa Cruz, Vanuatu, and perhaps Fiji.

Figure 13-6 Specimens of the marbled frogmouth, *Podargus ocellatus* (UMMZ 215554, ♂, New Guinea; left) and the Solomon Island frogmouth (undescribed genus; UF 40210, ♀, Isabel, Solomon Islands; right). Scale = 10 cm. Photo by J. J. Kirchman.

The last two genera of strigid owls in Oceania are restricted to Micronesia. *Pyrroglaux* is a small, *Otus*-like owl endemic to Palau, where it is abundant (Engbring 1988, pers. obs. 1995, 1997), not "highly endangered" as claimed by König et al. (1999:281). The nearly cosmopolitan *Asio flammeus* resides on Pohnpei (endemic *A. f. ponapensis*; Mayr 1933) and has been recorded, presumably as vagrants, in Micronesia from the Marianas to the Marshall Islands (Pyle & Engbring 1985, Pratt et al. 1987:216). Other islands or island groups with resident populations of *A. flammeus* are the Hawaiian Islands, Galápagos, Falklands, Cuba, Hispaniola, and Puerto Rico, with the Hawaiian form a poorly differentiated recent colonizer and the others more distinctive (Swarth 1931, Steadman & Zousmer 1988, James 1995, König et al. 1999:430–431).

Caprimulgidae: Nightjars

Outside of Oceania, *Eurostopodus mysticalis* breeds in Australia and winters on New Guinea (Coates 1985:386). The endemic *E. m. nigripennis* occurs in the Solomons and *E. m. exul* is endemic to New Caledonia, where it has been recorded only in 1939 (Cleere & Nurney 1998:176–178). *Caprimulgus macrurus* lives in Oceania only in the Bismarcks, undifferentiated from New Guinea populations (Cleere & Nurney 1998:247–250). *Caprimulgus indicus*, widespread in southern and eastern Asia, resides in Palau as the endemic but little studied *C. i. phaloena*.

Aegothelidae: Owlet-nightjars

As far as is known, *Aegotheles* is confined in tropical Oceania to the recently extinct, endemic *A. savesi* of New Caledonia (Olson et al. 1987, Balouet & Olson 1989). This probably is another instance of colonizing New Caledonia directly from Australia or perhaps New Guinea. Outside of the study area, *A. novaezealandiae* of New Zealand represents an even larger, extinct insular form of owlet-nightjar (Rich & Scarlett 1977, McCulloch 1994, Worthy 1997b).

Podargidae: Frogmouths

The absence of *Podargus* in the Bismarcks seemed especially odd when Oceania's only species (in the Solomon Main Chain) was regarded as *P. ocellatus inexpectatus*, an endemic race of a species otherwise in Australia (*P. o. plumiferus*) and New Guinea (*P. o. ocellatus*; Schodde 1977, Coates 1985:382, Beehler et al. 1986:133; Vanuatu erroneously included in its range by Cleere & Nurney 1998:123). The osteology and plumage of a specimen of frogmouth collected in 1998 on Isabel (Solomons), however, are so different from those of New Guinean *P. ocellatus* that a new genus is required for the Solomon Island form (Cleere et al. MS submitted; Figure 13-6 herein). Regardless of differences between frogmouths in the Solomons vs. New Guinea, I wonder if a systematic search for frogmouths in the Bismarcks, such as that of Smith et al. (1998) in New South Wales, would not reveal their presence.

Hemiprocnidae: Tree-swifts (Crested Swifts)

Hemiprocne mystacea is Oceania's only species of tree-swift, occurring from the Moluccas through New Guinea (*H. m. mystacea*) to the Bismarcks and Solomons (*H. m. woodfordiana*; Coates 1985:398–401). Perching high on snags and feeding aerially at fast speeds (Beehler et al. 1986:136, Chantler & Driessens 1995), *H. mystacea* seems to have been spared the level of predation by raptors or people that would result in a prehistoric record.

Apodidae: Swifts

A single genus of resident swifts, *Collocalia* (synonym *Aerodramus*; Price et al. 2004), is found in Oceania. The earliest evidence of *Collocalia* is an extinct species from the late Oligocene–early Miocene of Queensland, Australia (Boles 2001). Because species of *Collocalia* roost and nest in caves, they have a considerable fossil record in Oceania as far east as the Cook Islands, particularly in noncultural contexts (Chapters 5-8). Three species (*C. esculenta*, *C. spodiopygia*, *C. vanikorensis*) are widespread and sympatric, often even syntopic (Kratter et al. 2001a), as far east as Vanuatu. Locally in the Bismarcks and Solomons, a fourth sympatric species is found (Tables 5-3, 5-7). In Micronesia, *C. vanikorensis* (including *C. bartschi*) is widespread though declining on some islands (Pratt et al. 1987:218–219, Browning 1993, Steadman 1999a). *Collocalia spodiopygia* occurs widely and commonly in West Polynesia (Pratt et al. 1987:218, Steadman & Freifeld 1998).

The remaining four species of swiftlets in Oceania (*Collocalia sawtelli*, *C. manuoi*, *C. ocista*, *C. leucophaea*) are allopatric in East Polynesia (Cook Islands, Society Islands, Marquesas; Steadman 2002b; Figure 13-7 herein). Each is either rare or extinct, and known from only one to several islands.

Alcedinidae: Kingfishers

Four of Oceania's five genera of kingfishers are restricted to Near Oceania. *Alcedo atthis* ranges from Europe and Africa through Near Oceania (Fry et al. 1992:74–75, 219–221).

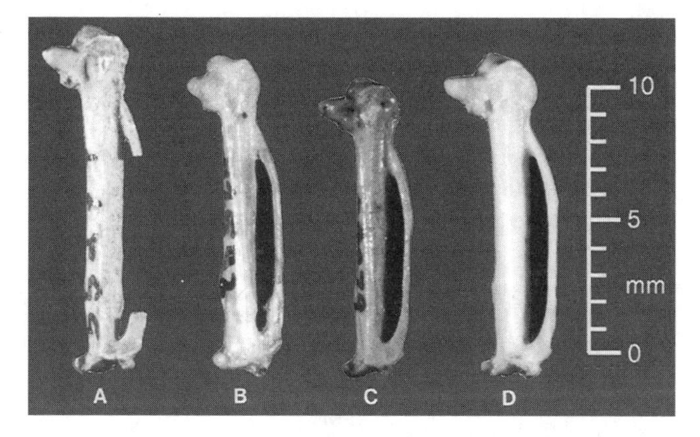

Figure 13-7 The carpometacarpus of *Collocalia*. A: Extinct *C. manuoi*, UF 55277, paratype, Mangaia, Cook Islands. B: *C. sawtelli*, UWBM 42542, Atiu, Cook Islands. C: *C. spodiopygia*, UF 50279, `Eua, Tonga. D: *C. vanikorensis*, UF 39592, Efate, Vanuatu. From Steadman (2002b: Figure 3).

Figure 13-8 Common (river) kingfisher *Alcedo atthis salomonensis*. Photo by DWS, Isabel, Solomon Islands, 10 July 1997.

The nonendemic *A. a. hispidoides* occurs in the Bismarcks, whereas the endemic *A. a. salomonensis* (Figure 13-8) inhabits the Solomons. Three species of *Ceyx* (closely related to and perhaps synonymous with *Alcedo*) inhabit Near Oceania: *C. websteri* endemic to the Bismarcks; *C. pusillus* in the Bismarcks (endemic *C. p. masauji*) and Solomons (three endemic subspecies) as well as New Guinea and Australia; and *C. lepidus* with three endemic subspecies in the Bismarcks, six in the Solomons, and five others from the Philippines to New Guinea (Peters 1945:179–181, 183–184, Fry et al. 1992:66–67, 72–73, 202–204, 215–217).

The only species of *Tanysiptera* in Oceania is *T. sylvia* s.l., with *T. s. nigriceps* endemic to New Britain, Duke of York, Watom, and Lolobau, *T. s. leucura* endemic to Umboi, and two other races in Australia and New Guinea (Fry et al. 1992:118–119, Mayr & Diamond 2001:386). Coates (1985:433–436) regards both Bismarck forms as an endemic species (*T. nigriceps*) rather than a subspecies of *T. sylvia*.

The genus *Actenoides* has a highly discontinuous distribution (Fry et al. 1992:26–29, 40–41, 107–113). The only species in Oceania is *A. bougainvillei* in the Solomons (*A. b. bougainvillei* on Bougainville, *A. b. excelsus* on Guadalcanal). This illogical range, combined with its rarity, would suggest loss of populations on at least the large adjacent islands of the former Greater Bougainville (Chapter 5), such as Choiseul and Isabel. Outside of the Solomons, other

Figure 13-9 Collared kingfisher *Halcyon chloris sacra*. Photo by DWS, Vaka`eitu, Vava`u, Tonga 5 July 1995.

species of *Actenoides* inhabit Sulawesi (two species), the Philippines (two species), and the Greater Sundas and mainland Southeast Asia (one species). The apparent absence of *Actenoides* in the Moluccas, New Guinea, and Bismarcks is perplexing.

Halcyon (synonym *Todirhamphus*) is the only genus of kingfisher that is both widespread and diverse in the Pacific. Endemic to the Bismarcks is *H. albonotata* and to the Solomons is *H. leucopygia*. The large *H. saurophaga* inhabits coastal settings from the Moluccas to the Solomons with three subspecies recognized in the Bismarcks (Fry et al. 1992:60–61, 183–185). The collared kingfisher *H. chloris* (Figure 13-9) has an extraordinary range from the Red Sea across the southern margin of Asia to New Guinea and Australia, then through Melanesia and Polynesia as far east as Tonga and American Samoa, with curious absences in New Caledonia and Western Samoa (Fry et al. 1992:56–59). Of 47 subspecies of *H. chloris*, 31 are endemic to Oceania (Peters 1945:207–213). In Micronesia, *H. chloris* occurs only in the Marianas (Rota, Tinian, Saipan) and Palau. In most of West Polynesia, *H. chloris* has no clear inter- or intraisland trends in habitat preference or relative abundance (Steadman & Franklin 2000).

Four subspecies of *Halcyon cinnamomina* inhabit the Ryuku Islands, Guam, Palau, and Pohnpei. The cooccurrence of *H. cinnamomina* and *H. chloris* in Palau is Micronesia's only instance of sympatric, congeneric kingfishers. *Halcyon sancta* has an unusual distribution (Fry et al. 1992:60–61, 185–188). It is the only resident kingfisher in New Caledonia (*H. s. canacorum*), the Loyalty Islands (*H. s. macmillani*), and islands to the south (New Zealand, Lord Howe, Norfolk, and the Kermadecs; *H. s. vagans*). The Australian race, *H. s. sancta*, migrates north to the Moluccas, New Guinea, and Solomons but also breeds on Guadalcanal and Makira (San Cristobal) in the Solomons.

At least eight other species of *Halcyon* inhabit Oceania. Endemic to Vanuatu is *H. farquhari*, known from just three islands (Santo, Malo, Malakula) where it is sympatric with *H. chloris* (Bregulla 1992:212–213, pers. obs. 2003). The small *H. recurvirostris* is endemic to Western Samoa; it was regarded as a subspecies of *H. sancta* by Fry et al. (1992:60–61, 176, 185–188). I disagree with this, and with the suggestion by Pratt et al. (1987:223) that *H. chloris vitiensis* and *H. c. eximia* of Fiji, and *H. c. regina* of Futuna, are races of *H. sancta*. *Halcyon tuta* (Cook Islands, Society Islands) and its probable close relatives (*H. mangaia* on Mangaia, *H. gertrudae* on Niau, *H. gambieri* on Mangareva, *H. godeffroyi* in the Marquesas) replace *H. chloris* in East Polynesia but have a very discontinuous distribution that must reflect anthropogenic extinction (Chapter 7). *Halcyon venerata* is endemic to the Society Islands, where it is sympatric with *H. tuta*. This is the only case of sympatric species of *Halcyon* in East or West Polynesia (Steadman & Pahlavan 1992).

Meropidae: Bee-eaters

Oceania's single resident species of bee-eater is *Merops [superciliosus] philippinus*, which nests from Southeast Asia to the Bismarcks (Long, Umboi, Sakar, New Britain; Coates 1985:436, Fry et al. 1992:92–93, 273–275, Mayr & Diamond 2001:386–387). Another species, *M. ornatus*, been recorded as a migrant on many islands in the Bismarcks, Solomons, and Palau (Coates 1985:437–439, Pratt et al. 1987:227, Fry et al. 1992:92–93, 273–275).

Coraciidae: Rollers

Oceania's only roller is the dollarbird *Eurystomus orientalis*, which resides in Near Oceania and lacks a fossil record for the same reasons as *Hemiprocne mystacea* (see above). The Bismarcks (*E. o. crassirostris*) and Solomons (*E. o. solomensis*) each have an endemic subspecies (Coates 1985:440, Fry et al. 1992:104–105, 305–308). Vagrants of the Australian *E. o. pacificus* have been recorded in Palau, Yap, and Pohnpei (Engbring 1983b).

Bucerotidae: Hornbills

The species of hornbill (*Aceros plicatus*) in Near Oceania has conspecific populations on New Guinea and eastern Indonesia. The Bismarcks (*A. p. dampieri*) and Solomons (*A. p. harterti, A. p. mendanae*) each have endemic subspecies (Kemp 1995:224–227). This tasty frugivore was the most common species of bird in the late Pleistocene strata at the

Balof site, New Ireland (Steadman et al. 1999a). While this, as well as the current cultural importance of hornbills in the Bismarcks, might suggest human agency in interisland dispersal of *A. plicatus*, its good over-water dispersal ability (Diamond 2002) and its subspecies-level endemism in the Bismarcks and Solomons argue otherwise. An analysis of molecular variation in *A. plicatus* probably would be informative in this regard.

I recently discovered two pedal phalanges of an extinct hornbill (*Aceros* undescribed sp., smaller than *A. plicatus*) from the Hnajoisisi and Keny archaeological sites on Lifou, Loyalty Islands, New Caledonia (Table 5-13). This is the first evidence of hornbills east of the Solomon Main Chain. Bones of *Aceros* may show up, therefore, on Grande Terre and Ile des Pins as well as Vanuatu and the Santa Cruz Group.

Chapter 14 Passerines

PASSERINES OR SONGBIRDS (order Passeriformes) constitute more than 50% of the world's living species of birds, a well-known fact that probably would not be true had people not wiped out so many species of rails and other nonpasserines across Oceania. In the continental regions surrounding Oceania, breeding landbird faunas consist of ca. 50% passerines, such as 48% (82 of 169 species) for the main islands of Japan, 49% (173 of 352 species) for the Philippines, 53% (225 of 426 species) for lowland New Guinea (Table 17-2), and 54% (240 of 441 species) for Queensland, Australia (Storr 1973, Massey et al. 1982, Kennedy et al. 2000). Many of the dominant Australasian passerine families are altogether absent, however, in Oceania.

On relatively well-surveyed islands in Near Oceania, such as New Ireland (Bismarcks) or Isabel (Solomons), passerines constitute only ca. 33% (38 of 115–120, and 25 of 76, respectively) of the resident species of landbirds (Steadman et al. 1999a, Kratter et al. 2001a, Mayr & Diamond 2001). In Remote Oceania, these values are, for example, 44% (22 of 50 species) on Santo (Melanesia; Chapter 5), 26-30% (10–11 of 37–38) on 'Eua (West Polynesia; Chapter 6), 5% (1 of 20) on Mangaia, 11% (2 of 18) on Huahine and Ua Huka, and 22% (2 of 9) on Henderson (East Polynesia; Chapter 7), and 32% (8 of 25) on Rota (Micronesia; Chapter 8). Passerines clearly are a more minor faunal element in East Polynesia than elsewhere in Oceania.

Passerines nevertheless include three of the region's 15 most widespread species of landbirds, the swallow *Hirundo tahitica*, thrush *Turdus poliocephalus*, and honeyeater *Myzomela cardinalis* (Chapter 17). The thrush and honeyeater, along with two other songbirds with ranges nearly as broad (the cicadabird *Coracina tenuirostris* and whistler *Pachycephala pectoralis*), also are strikingly variable geographically in plumage and mensural characters.

As I will document briefly in this chapter, passerines have not been spared the untoward circumstances that followed human arrival in Oceania. What we know about the prehistoric loss of songbirds is in its infancy, however, for two reasons. The first is that many sites have not been sampled thoroughly enough to record the small bones of passerines in respectable quantities, although fine-mesh sieves (see Chapter 4) have been used at a number of sites in Vanuatu, Tonga, the Cook Islands, and the Marianas. The second is that the comparative osteology of passerines, whether on Pacific islands or worldwide, is a difficult topic that few have investigated rigorously. Many prehistoric bones of songbirds are not diagnostic beyond the ordinal or familial level. Those that can be identified to a genus or species typically are so assigned only after considerably more study under the microscope than is needed with most nonpasserine fossils. For any set of landbird bones in my lab, I nearly always identify the nonpasserines first, knowing how difficult and time-consuming it will be to identify the passerines, a process complicated further by an incomplete availability of comparative skeletons of endemic species in many parts of Oceania.

Following Sibley and Ahlquist (1990), I adopt the suborders Passerida and Corvida to accommodate all passerine families in Oceania other than Pittidae (Table 14-1 and the appendix). While most species of passerines in Oceania that occur on multiple islands are polytypic, I have not attempted to list or to evaluate subspecies in this chapter. Very little molecular systematic research has taken place with birds in Oceania, although the worldwide trend in this field has been to recognize more species than in traditional classifications, typically by elevating the level of distinction from subspecies to species. We probably can expect the same trend to apply with passerines on Pacific islands. I hope that interpretations of the molecular data do not obscure the likely fact that most of these insular sets of congeneric allopatric populations represent monophyletic radiations. This notion applies especially to species of passerines currently classified as sets of well-differentiated subspecies, such as in four of the widespread species just mentioned (*Turdus poliocephalus*, *Myzomela cardinalis*, *Coracina tenuirostris*, and *Pachycephala pectoralis*).

Flightless species of passerines have not yet been discovered in the tropical Pacific, even in the hyper-diverse assemblage of cardueline-derived birds known as Hawaiian honeycreepers or, more appropriately, Hawaiian finches (James & Olson 1991, 2003). The only flightless passerines described anywhere are three extinct acanthisittid "wrens" from New Zealand (*Traversia lyalli*, *Pachyplichas yaldwyni*, *Dendroscansor decurvirostris*; Millener 1988, Millener & Worthy 1991, Holdaway et al. 2001) and *Emberiza alcoveri*, an extinct emberizine finch from the Canary Islands (Rando et al. 1999). All four of these species still would hopping or running around today if not for people, cats, and rats. Even on oceanic islands lacking mammalian predators (under natural circumstances), there seems to be a limit to wing reduction among nearly all foliage-gleaning and/or insect-hawking passerines (Steadman 1986, Olson 1994).

Now let us begin to survey the past and present distribution of passerines in tropical Oceania. My treatment of songbirds will differ from that of most nonpasserine families in that I do not include text for all genera, which nevertheless are listed in Table 14-1. The genera to be discussed in any detail are primarily those for which fossils add to our understanding of their biogeography.

Geographic Review

Melanesia

The only areas of Melanesia with substantial, although still small, sets of passerine fossils are New Ireland (Bismarcks) and several islands in Vanuatu (Chapter 5). On New Ireland, six genera and eight species are represented among 29 passerine bones, only 18 of which I have been able to identify to genus or species. Among these is one species no longer found on the island, a large species of crow (*Corvus* sp. indet.). These two fossils provide the first evidence anywhere in Oceania, outside of the Hawaiian Islands (James & Olson 1991), of multiple sympatric species of *Corvus* (Table 14-2).

The 83 passerine bones from five islands in Vanuatu represent 14 species (Table 5-12). The only recorded loss is of a large starling (*Aplonis* undescribed sp.) from Erromango, an island now lacking starlings. The species may be similar to *A. brunneicapillus* or *A. grandis* of the Solomon Main Chain, although refinement of this identification awaits the availability of more skeletons of *Aplonis*. No starling this large lives in Vanuatu today. It is likely that at least one of the three smaller species of *Aplonis* currently in Vanuatu or Santa Cruz (*A. santovestris*, *A. tabuensis*, and especially *A. zelandica*) also once lived on Erromango. Perhaps two or three species of *Aplonis* co-occurred on most islands in Vanuatu at human contact. Nowadays in this region, sympatric species of starlings are found only on Santo (the endemic *A. santovestris* and widespread *A. zelandica*) and parts of the Santa Cruz Group (*A. tabuensis*, *A. zelandica*).

West Polynesia

My study of passerine fossils from the Lau Group of Fiji is hampered by lacking comparative skeletons for many living Fijian species. All 34 passerine fossils from Niue represent the extant starling *Aplonis tabuensis* (Worthy et al. 1998). The only rigorous prehistory of passerines in West Polynesia is from Tonga, where 800+ songbird fossils from seven islands are unparalleled in Oceania (Tables 6-21, 6-23, 14-3). As expected, the largest sample (550 specimens from 'Eua) is the richest, including all three extant species and seven extinct or extirpated ones. Much smaller samples (22 to 75 specimens) from six other Tongan islands yield up to five species of passerines per island. I suspect that most or all of Tonga's 13+ species of passerines occurred across the three main clusters of islands (Vava'u, Ha'apai, Tongatapu) at human contact.

With 20-20 hindsight, the former presence of *Turdus*, *Myiagra*, *Petroica*, and *Myzomela* in Tonga should have been predictable because they occur today in both Fiji and Samoa. I have found no Tongan fossils, however, of *Zosterops*, *Erythrura*, *Rhipidura*, or *Gymnomyza*, four other passerine genera in Fiji and Samoa but not in Tonga. As the Tongan fossil record grows, I expect to find these genera as well.

Table 14-1 The modern (M) and prehistoric (P) distribution of resident genera of Passeriformes in Oceania. "Bismarcks" includes the Admiralty Islands. RCPI = Remote Central Pacific Islands (Marshall Islands and Nauru eastward through the Line Islands; Chapter 8). [e], endemic to Oceania; *, no endemic species known in Oceania (status uncertain for *Hirundo* and *Artamus*). Based on data in Chapters 5–8. SWALLOWS through ESTRILDID FINCHES belong to the suborder Passerida. CUCKOO-SHRIKES through CROWS are in the suborder Corvida.

	Bismarcks	Solomon Main Chain	Solomon Outliers	Vanuatu	New Caledonia	Fiji	Tonga	Samoa	West Polynesian Outliers	Cook Islands
PITTAS										
Pitta	M, P	M	—	—	—	—	—	—	—	—
SWALLOWS										
*Hirundo**	M	M	M	M, P	M	M	M	—	—	—
THRUSHES										
Zoothera	M	M	—	—	—	—	—	—	—	—
*Turdus**	M, P	M	M	M, P	M	M	P	M	—	—
OLD WORLD FLYCATCHERS										
*Saxicola**	M	—	—	—	—	—	—	—	—	—
CISTICOLAS										
Cisticola	M	—	—	—	—	—	—	—	—	—
OLD WORLD WARBLERS										
Cettia	—	M	—	—	—	M	P	—	—	—
Acrocephalus	M	M	—	—	—	—	—	—	—	M, P
Phylloscopus	M	M	—	—	—	—	—	—	—	—
Megalurus	M	—	—	—	—	—	—	—	—	—
Cichlornis	M	M	—	M	M	—	—	—	—	—
Trichocichla[e]	—	—	—	—	—	M	—	—	—	—
SUNBIRDS										
*Nectarinia**	M	M	—	—	—	—	—	—	—	—
FLOWERPECKERS										
Dicaeum	M	M	—	—	—	—		—	—	—
WHITE-EYES										
Zosterops	M	M	M	M, P	M	M	—	M	—	—
Rukia[e]	—	—	—	—	—	—	—	—	—	—
Cleptornis[e]	—	—	—	—	—	—	—	—	—	—
Woodfordia[e]	—	—	M	—	—	—	—	—	—	—
Megazosterops[e]	—	—	—	—	—	—	—	—	—	—
†Zosteropidae undescribed genus[e]	—	—	—	—	—	—	P	—	—	—
STARLINGS										
Aplonis	M, P	M	M, P	M, P	M	M, P	M, P	M, P	M, P	M
*Mino**	M	M	—	—	—	—	—	—	—	—
ESTRILDID FINCHES										
Erythrura	M	M	M	M, P	M	M	—	M	—	—
Lonchura	M	—	—	—	—	—	—	—	—	—
CUCKOO-SHRIKES										
Coracina	M, P	M	M	M, P	M	—	—	—	—	—
Lalage	—	M	M	M, P	M	M	M, P	M	M	—
GERYGONES										
Gerygone	—	—	M	M, P	M	—	—	—	—	—

													TOTAL	
Tubuai	Society Islands	Tuamotu	Pitcairn	Easter Island	Marquesas	Palau	Yap	Marianas	Chuuk	Pohnpei	Kosrae	RCPI	M	M + P
—	—	—	—	—	—	—	—	—	—	—	—	—	2	2
—	M	—	P	—	—	—	—	—	—	—	—	—	8	9
—	—	—	—	—	—	—	—	—	—	—	—	—	2	2
—	—	—	—	—	—	—	—	—	—	—	—	—	7	8
—	—	—	—	—	—	—	—	—	—	—	—	—	1	1
—	—	—	—	—	—	—	—	—	—	—	—	—	1	1
—	—	—	—	—	—	M	—	—	—	—	—	—	3	4
M	M, P	M	M, P	—	M	—	—	M, P	M	M	M	M, P	13	13
—	—	—	—	—	—	—	—	—	—	—	—	—	2	2
—	—	—	—	—	—	—	—	—	—	—	—	—	1	1
—	—	—	—	—	—	—	—	—	—	—	—	—	4	4
—	—	—	—	—	—	—	—	—	—	—	—	—	1	1
—	—	—	—	—	—	—	—	—	—	—	—	—	2	2
—	—	—	—	—	—	—	—	—	—	—	—	—	2	2
—	—	—	—	—	—	M	M	M, P	M	M	M	—	13	13
—	—	—	—	—	—	—	M	—	M	M	—	—	3	3
—	—	—	—	—	—	—	—	M, P	—	—	—	—	1	1
—	—	—	—	—	—	—	—	—	—	—	—	—	1	1
—	—	—	—	—	—	M	—	—	—	—	—	—	1	1
—	—	—	—	—	—	—	—	—	—	—	—	—	1	1
—	P	—	—	—	—	M	M, P	M, P	M	M	M	—	16	17
—	—	—	—	—	—	—	—	—	—	—	—	—	2	2
—	—	—	—	—	—	M	—	P	M	M	M	—	11	12
—	—	—	—	—	—	—	—	—	—	—	—	—	1	1
—	—	—	—	—	—	M	M	—	—	M	—	—	8	8
—	—	—	—	—	—	—	—	—	—	—	—	—	8	8
—	—	—	—	—	—	—	—	—	—	—	—	—	3	3

(continued)

Table 14-1 (continued)

	Bismarcks	Solomon Main Chain	Solomon Outliers	Vanuatu	New Caledonia	Fiji	Tonga	Samoa	West Polynesian Outliers	Cook Islands
AUSTRALIAN ROBINS										
*Monachella**	M	—	—	—	—	—	—	—	—	—
*Petroica**	—	M	—	M, P	—	M	P	M	—	—
Eopsaltria	—	—	—	—	M	—	—	—	—	—
WHISTLERS										
Pachycephala	M	M	M	M, P	M	M	M, P	M	—	—
Colluricincla	—	—	—	—	—	—	—	—	—	—
FANTAILS										
Rhipidura	M	M	M	M, P	M	M	—	M	—	—
MONARCHS										
Pomarea[e]	—	—	—	—	—	—	—	—	—	M
Mayrornis[e]	—	—	M	—	—	M	—	—	—	—
Neolalage[e]	—	—	—	M	—	—	—	—	—	—
Clytorhynchus[e]	—	—	M	M	M	M	M, P	M	M	—
Metabolus[e]	—	—	—	—	—	—	—	—	—	—
Monarcha	M	M	M	—	—	—	—	—	—	—
Myiagra	M	M	M	M	M	M	P	M	—	—
Lamprolia[e]	—	—	—	—	—	M	—	—	—	—
HONEYEATERS										
Stresemannia[e]	—	M	—	—	—	—	—	—	—	—
Guadalcanaria[e]	—	M	—	—	—	—	—	—	—	—
Melidectes[e]	M	M	—	—	—	—	—	—	—	—
Lichmera	—	—	—	M, P	M	—	—	—	—	—
Myzomela	M	M	M	M, P	M	M	P	M	—	—
Foulehaio[e]	—	—	—	—	—	M	M, P	M, P	—	—
Philemon	M	—	—	—	M	—	—	—	—	—
Gymnomyza[e]	—	—	—	—	M	M	—	M	—	—
Phylidonyris	—	—	—	M, P	M	—	—	—	—	—
DRONGOS										
Dicrurus	M, P	M	—	—	—	—	—	—	—	—
WOOD-SWALLOWS										
*Artamus**	M	—	M	M	M	M	—	—	—	—
CROWS										
Corvus	M, P	M	—	—	M	—	—	—	—	—
TOTAL FAMILIES										
M	20	17	12	14	15	13	6	10	3	3
M + P	20	17	12	14	15	13	10	10	3	3
TOTAL GENERA										
M	29	27	17	19	21	19	6	13	3	3
M + P	29	27	17	19	21	19	12	13	3	3
Number of bones	29	0	2	83	0	0	822	7	34	105

Tubuai	Society Islands	Tuamotu	Pitcairn	Easter Island	Marquesas	Palau	Yap	Marianas	Chuuk	Pohnpei	Kosrae	RCPI	TOTAL M	TOTAL M + P
—	—	—	—	—	—	—	—	—	—	—	—	—	1	1
—	—	—	—	—	—	—	—	—	—	—	—	—	4	4
—	—	—	—	—	—	—	—	—	—	—	—	—	1	1
—	—	—	—	—	—	—	—	—	—	—	—	—	8	8
—	—	—	—	—	—	M	—	—	—	—	—	—	1	1
—	—	—	—	—	—	M	M	M, P	—	M	—	—	11	11
—	M	—	—	—	M, P	—	—	—	—	—	—	—	3	3
—	—	—	—	—	—	—	—	—	—	—	—	—	2	2
—	—	—	—	—	—	—	—	—	—	—	—	—	1	1
—	—	—	—	—	—	—	—	—	—	—	—	—	7	7
—	—	—	—	—	—	—	—	—	M	—	—	—	1	1
—	—	—	—	—	—	—	M	M	—	—	—	—	5	5
—	—	—	—	—	—	M	—	M, P	—	M	—	—	10	11
—	—	—	—	—	—	—	—	—	—	—	—	—	1	1
—	—	—	—	—	—	—	—	—	—	—	—	—	1	1
—	—	—	—	—	—	—	—	—	—	—	—	—	1	1
—	—	—	—	—	—	—	—	—	—	—	—	—	2	2
—	—	—	—	—	—	—	—	—	—	—	—	—	2	2
—	—	—	—	—	—	M	M	M, P	M	M	M	—	13	14
—	—	—	—	—	—	—	—	—	—	—	—	—	3	3
—	—	—	—	—	—	—	—	—	—	—	—	—	2	2
—	—	—	—	—	—	—	—	—	—	—	—	—	3	3
—	—	—	—	—	—	—	—	—	—	—	—	—	2	2
—	—	—	—	—	—	—	—	—	—	—	—	—	2	2
—	—	—	—	—	—	M	—	—	—	—	—	—	6	6
—	—	—	—	—	—	—	—	M	—	—	—	—	4	4
1	3	1	1	0	2	10	6	7	6	8	5	1	—	
1	4	1	2	0	2	10	6	8	6	8	5	1	—	
1	3	1	1	0	2	11	7	9	7	9	5	1	—	
1	4	1	2	0	2	11	7	10	7	9	5	1	—	
0	3	0	160	0	3	0	3	380+	0	0	0	2	1633+	

Table 14-2 Distribution of passerines in Oceania with multiple congeneric species within an island group, including fossil data. The first value is the largest number of species on a single island; the second value is the total number of species in the island group.

													TOTAL ISLANDS/ISLAND GROUPS WITH:	
	Bismarcks	Solomon Main Chain	Solomon Outliers	Vanuatu	New Caledonia	Fiji	Samoa	Cook Islands	Marquesas	Palau	Pohnpei	Kosrae	Sympatric congeners	Multiple congeners in the island group
PITTAS														
Pitta	2/3	1/1	—	—	—	—	—	—	—	—	—	—	1	1
THRUSHES														
Zoothera	1/2	1/3	—	—	—	—	—	—	—	—	—	—	0	2
OLD WORLD WARBLERS														
Acrocephalus	1/1	1/1	—	—	—	—	—	1/2	1/1	—	1/1	1/1	0	1
Phylloscopus	1/1	2/3	—	—	—	—	—	—	—	—	—	—	1	1
Cichlornis	1/1	1/2	—	1/1	1/1	—	—	—	—	—	—	—	0	1
SUNBIRDS														
Nectarinia	2/2	1/1	—	—	—	—	—	—	—	—	—	—	1	1
FLOWERPECKERS														
Dicaeum	1/1	1/2	—	—	—	—	—	—	—	—	—	—	0	1
WHITE-EYES														
Zosterops	1–2/2	3/9	1/2	2/2	2/4	2/2	1/1	—	—	2/2	2/2	1/1	6–7	8
Woodfordia	—	—	1/2	—	—	—	—	—	—	—	—	—	0	1
STARLINGS														
Aplonis	2/3	4/6	2/4	2/3	1/1	1/1	2/2	1/2	—	1/1	2/2	2/2	7	8
ESTRILDID FINCHES														
Erythrura	1/1	1/1	—	2/2	1/2	2/2	1/1	—	—	1/1	1/1	1/1	2	3
Lonchura	3/4	1/1	—	—	—	—	—	—	—	—	—	—	1	1

CUCKOO-SHRIKES														
Coracina	3/3	5/6	1/1	1/1	2/2	—	—	—	—	1/1	1/1	—	3	3
Lalage	1/1	1/1	1/1	2/2	1/1	1/1	2/2	—	—	—	—	—	2	2
WHISTLERS														
Pachycephala	2/2	3/3	1/1	1/1	2/3	1/1	1/1	—	—	—	—	—	3	3
FANTAILS														
Rhipidura	3/5	4/7	1/2	2/2	2/2	1/1	1/1	—	—	1/1	1/1	—	4	5
MONARCHS														
Pomarea	—	—	—	—	—	—	—	1/1	1/3	—	—	—	0	1
Mayrornis	—	—	1/1	—	—	1/2	—	—	—	—	—	—	0	1
Clytorhynchus	—	—	1/2	1/1	1/1	2/2	1/1	—	—	—	—	—	1	2
Monarcha	4/6	3/6	1/1	—	—	—	—	—	—	—	—	—	2	2
Myiagra	3/3	1/1	1/2	1/1	1/1	2/2	1/1	—	—	1/1	1/1	—	2	3
HONEYEATERS														
Myzomela	4/6	2/6	1/1	1/1	1/2	1/1	1/1	—	—	1/1	1/1	1/1	2	3
Foulehaio	—	—	—	—	—	1/2	1/1	—	—	—	—	—	0	1
DRONGOS														
Dicrurus	1/2	1/1	—	—	—	—	—	—	—	—	—	—	0	1
CROWS														
Corvus	2/2	1/2	—	—	1/1	—	—	—	—	—	—	—	1	2
TOTAL GENERA WITH:														
Sympatric congeners	11–12	8	1	5	4	4	2	0	0	1	2	1	39–40	—
Multiple congeners in the island group	14	12	6	5	6	6	2	2	1	1	2	1	—	58
Number of bones	29	0	2	83	0	0	7	105	3	0	0	0		229

Table 14-3 The modern (M) and prehistoric (P) distribution of passerines on seven islands in Tonga. H, historic record, now extirpated; †, extinct species, subspecies, or population. From data in Chapter 6.

	Foa	Lifuka	Ha`ano	`Uiha	Ha`afeva	Tongatapu	`Eua	TOTAL ISLANDS		
								M	M + H	M + H + P
SWALLOWS										
Hirundo tahitica	M?	M?	M?	M?	M	M	M?	2–6	2–6	2–6
THRUSHES										
†*Turdus poliocephalus*	—	P	—	—	—	P	P	0	0	3
OLD WORLD WARBLERS										
†*Cettia* sp.	—	—	—	—	—	—	P	0	0	1
WHITE-EYES										
†Undescribed genus	—	—	—	—	—	P	P	0	0	2
STARLINGS										
Aplonis tabuensis	M, P	M, P	M, P	M, P	M, P	M, P	M, P	7	7	7
CUCKOO-SHRIKES										
Lalage maculosa	M, P	M, P	M, P	M	M?, P	M, P	M, P	6–7	6–7	7
†cf. *Lalage* sp.	—	—	—	—	—	P	P	0	0	2
ROBINS										
†*Petroica* sp.	P	—	—	—	—	—	—	0	2–3	6
WHISTLERS										
†*Pachycephala jacquinoti*	P	P	P	—	—	P	—	0	0	1
MONARCHS										
Clytorhynchus vitiensis	P	P	P	H?, P	—	H, P	H, P	0	0	1
†*Myiagra* sp.	—	—	—	—	—	—	P	0	0	4
HONEYEATERS										
†*Myzomela* cf. *cardinalis*	—	—	P	—	—	—	P	0	0	2
Foulehaio carunculata	M?, P	M, P	H, M?, P	M, P	M, P	M, P	M, P	5–7	6–7	7
TOTAL SPECIES										
M	2–4	3–4	2–4	3–4	3–4	4	3–4	4	—	—
M + H	2–4	3–4	3–4	3–5	3–4	5	4–5	—	5	—
M + H + P	6–7	6–7	6–7	4–5	4	9	10–11	—	—	13
Number of bones	70	75	29	29	22	47	550		822	

East Polynesia

Passerine fossils are unknown in three of East Polynesia's seven island groups. Each of the 105 passerine fossils from Mangaia (Cook Islands) is of the extant warbler *Acrocephalus kerearako* (Table 7-3). On Rarotonga, 200 km to the west, a starling (*Aplonis cinerascens*) and monarch (*Pomarea dimidiata*) persist but no warbler. Although I have expected a species of both *Aplonis* and *Pomarea* to appear in Mangaia's fossil record, neither has. On Huahine in the Society Islands, Steadman (1989c) described the extinct *Aplonis diluvialis* from the Fa'ahia archaeological site, this being the easternmost starling in Oceania. Huahine's only other known passerine, the warbler *Acrocephalus caffer*, died out there in historic times (Holyoak & Thibault 1978b, Steadman & Pahlavan 1992; Figure 14-1 herein).

On Henderson Island in the Pitcairn Group, Wragg (1995) reported fossils of the swallow *Hirundo tahitica*, which is widespread today in Melanesia and West Polynesia but occurs in East Polynesia only in the Society Islands.

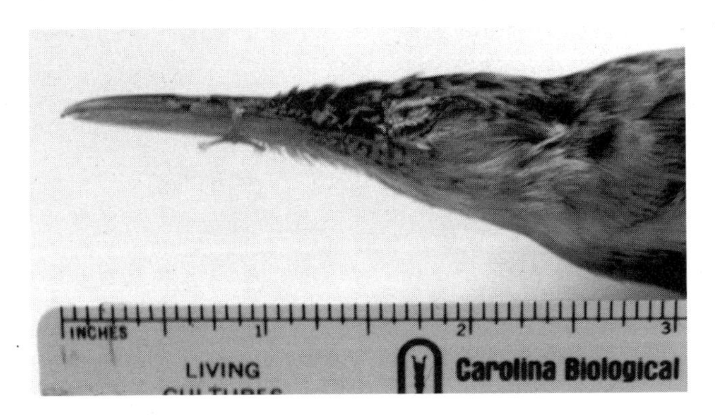

Figure 14-1 *Acrocephalus caffer garretti*, an extinct subspecies of Society Island reed-warbler from Huahine, Society Islands. Holotype in the British Museum (BMNH 98.9.1.2535). Photo by DWS, September 1990.

All three passerine fossils recovered from the Marquesas are of the monarch flycatcher *Pomarea* cf. *iphis* from Ua Huka, where it is rare today. In fact, the entire radiation of *Pomarea* in East Polynesia is close to extinction (Cibois et al. 2004). These specimens were mentioned erroneously by Steadman (1989a) as *Myiagra* sp., a monarch genus widespread in Melanesia, West Polynesia, and Micronesia but unrecorded in East Polynesia.

Micronesia

Passerine fossils have been found in three Micronesian island groups. In Yap, three bones of the extant starling *Aplonis opaca* were identified on Fais (Steadman & Intoh 1994). From Majuro Atoll in the Marshall Islands is a coracoid of an extinct warbler (*Acrocephalus* undescribed sp.) from the Laura archaeological site, excavated by Marshall Weisler.

The Marianas have produced most of Micronesia's passerine fossils (Table 14-4). Small samples (1–24 specimens) from Guam, Aguiguan, and Saipan disclose no species otherwise unrecorded on these three islands. On Tinian (45 specimens), the warbler *Acrocephalus luscinia* and white-eye *Cleptornis marchei* are known only from bones. Both species occur on Aguiguan and Saipan, the islands on either side of Tinian. The large sample of 313 passerine fossils from Rota also includes *C. marchei* and thus suggests that this Marianas endemic once occurred as well on Guam. The extinct monarch flycatcher *Myiagra freycineti* was thought to be endemic to Guam (where it died out in the 1980s; Chapter 8) until fossils were found on Rota (Steadman 1999b). The estrildid finch from Rota, *Erythrura* undescribed sp., is larger than *E. trichroa*, which is widespread in Micronesia but unrecorded in the Marianas. The foraging ecology of living passerines may be better known in the Marianas than anywhere else in tropical Oceania (Craig 1989, 1990, 1992a, 1992b, Craig et al. 1992b). What a shame that Robert Craig and his students will never have an opportunity to study songbirds on an island in the Marianas where the fauna is intact.

Taxonomic Review

Just like the geographic review above, this section provides some detail to flesh out Tables 14-1 and 14-2, stressing genera with fossil records that have augmented what we know about their distributional or taxonomic diversity.

Suborder Passerida

The swallow *Hirundo tahitica* has a fairly continuous range in Melanesia and Fiji. Its distribution in Tonga gets spotty as mangroves and freshwater habitats become more localized than in Fiji or Melanesia. The only extant population of *H. tahitica* east of Tonga is the endemic subspecies *H. t. tahitica* on mangrove-free Tahiti and Mo'orea (Society Islands). The discovery of its fossils on Henderson (Pitcairn Group) opens up the possibility that *H. tahitica* once occurred widely in East Polynesia. The mangrove *Rhizophora*, which now is absent from East Polynesia, lived on Mangaia (Cook Islands) in the mid-Holocene, this suggesting more extensive estuarine development than today (Ellison 1994a). Although swallows are osteologically distinctive among passerines and would be easy to recognize as fossils, the 92 specimens from Henderson (Wragg 1995) are the only swallow fossils known in Oceania.

The thrush *Turdus poliocephalus* (Figure 14-2) has a continuous range from an archipelagal perspective (Table 14-1) but a very spotty distribution within island groups. It occurs on only four islands in the Bismarcks, three in the Solomon Main Chain, one (Rennell) among the Solomon Outliers, 21 in Vanuatu, three in New Caledonia, six in Fiji, and two in Samoa (Chapters 5, 6). Fossils of *T. poliocephalus* have been found on three islands in Tonga, an island group lacking thrushes today (Figure 14-3). I believe that, across its range in Oceania, most populations of *T. poliocephalus* that existed at human contact have been extirpated. Today's patchy distribution is an anthropogenic artifact. The past vs. present distribution of *T. poliocephalus* has other implications for concepts of community ecology (Chapter 20).

Warblers or reed-warblers of the genus *Acrocephalus* (Figures 14-1, 14-4) have a unique modern range, being found (very discontinuously) in Near Oceania, East Polynesia, and Micronesia but nowhere from the Solomon Main Chain through Niue (Bocheński & Kuśnierczyk 2003). Fossils have added species or populations within

Table 14-4 The modern (M) and prehistoric (P) distribution of resident passerines on five islands in the Mariana Islands. H, historic record, now extirpated; †, extinct species. From data in Chapter 8.

	Guam	Rota	Aguiguan	Tinian	Saipan	TOTAL ISLANDS		
						M	M + H	M + H + P
OLD WORLD WARBLERS								
Acrocephalus luscinia (e)	H	—	M	P	M	2	3	4
WHITE-EYES								
Zosterops conspicillatus (e)	H	M, P	M, P	M, P	M	4	5	5
Cleptornis marchei (e)	—	P	M, P	P	M	2	2	4
STARLINGS								
Aplonis opaca	M, P	M, P	M, P	M, P	M, P	5	5	5
ESTRILDID FINCHES								
†*Erythrura* undescribed sp. (e)	—	P	—	—	—	0	0	1
FANTAILS								
Rhipidura rufifrons	H	M	M	M, P	M	4	5	5
MONARCHS								
Monarcha takatsukasae (e)	—	—	—	M	H	1	2	2
†*Myiagra freycineti* (e)	H, P	P	—	—	—	0	1	2
HONEYEATERS								
Myzomela rubrata	H, P	M, P	M, P	M, P	M	4	5	5
CROWS								
Corvus kubaryi (e)	H	M	—	—	—	2	2	2
TOTAL SPECIES								
M	1	5	6	5	6	24	—	—
M + H	7	5	6	5	7	—	30	—
M + H + P	7	8	6	7	7	—	—	35
Number of bones	3	313	24	45	1		386	

Figure 14-2 Island thrush *Turdus poliocephalus rennellianus*. Rennell, Solomon Islands. Photo by DWS, 22 June 1997.

the three major regions of Oceania but have not helped to close the 4700 km gap between *A.* [*australis*] *stentoreus* on Guadalcanal and *A. kaoko* on Mitiaro, Cook Islands. Given the current or former presence of *Acrocephalus* on most high islands in Micronesia as well as such remote outposts as Nauru (*A. rehsii*), Majuro (*A.* undescribed sp.), Kiritimati, Kiribati (*A. bokikokiko*), and the leeward Hawaiian islands of Laysan and Nihoa (*A. familiaris*), I suspect that some sort of warbler once occupied most islands, high or low, in Micronesia and adjacent parts of the central Pacific. This may not have been the case, however, in the Hawaiian Islands, where fossils of *A. familiaris* have not been found on any of the high, main islands (James & Olson 1991).

Typical white-eyes (*Zosterops* spp.; Figure 14-5) are absent from East Polynesia but widespread in Oceania's other

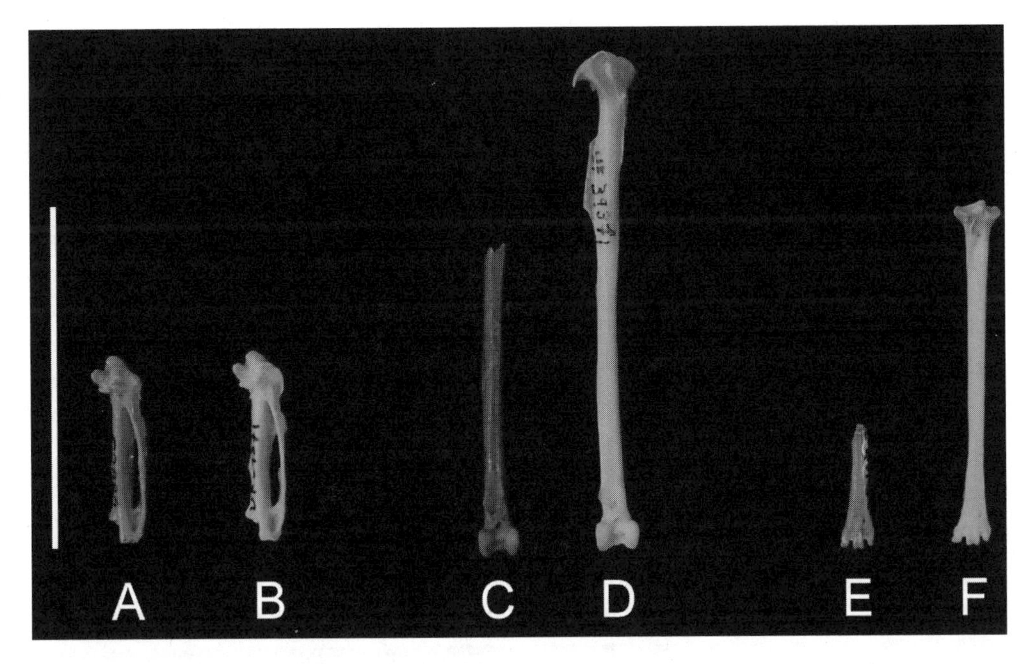

Figure 14-3 Fossils and modern bones of *Turdus poliocephalus*. A, B: carpometacarpus. C, D: Tibiotarsus. E, F: Tarsometatarsus. Fossils (A, UF 52996; C, UF 52611; E, UF 50497) are from an extirpated population at 'Anatu, 'Eua, Tonga. Modern bones (B, D, F, UF 39371) are of *T. p. rennellianus* from Rennell, Solomon Islands. Scale = 30 mm. Photo by J. J. Kirchman.

major regions. Even with better skeletal collections, species-level identification of *Zosterops* fossils would be most difficult. Zosteropidae undescribed genus is recorded only on 'Eua, Tonga. This enigmatic taxon is based on three zosteropid bones that do not match those of *Zosterops* or any other zosteropid genus listed in Table 14-1, such as *Woodfordia* (Figure 14-6), the two species of which are confined to Solomon outliers.

Starlings (*Aplonis*; Figure 14-7) have been recorded in more Pacific island groups, today or prehistorically, than any other passerine genus. Palatability and large size (for a passerine) undoubtedly have contributed to their relatively rich fossil record. The distribution and radiation of starlings in East Polynesia remain poorly understood, however, with just three species described, each from a single island. Two are extinct (*A. mavornata* of Ma'uke, Cook Islands; *A. diluvialis* of Huahine, Society Islands; Olson 1986a, Steadman 1989c), whereas *A. cinerascens* (Rarotonga, Cook Islands) is extant but rare (Figures 14-8, 14-9). *Aplonis* is the only genus of passerine other than *Zosterops* with sympatric species in many regions of Oceania (Table 14-2). In Melanesia, two or three species of starlings are sympatric in much of the Bismarcks, Solomon Main Chain, Solomon Outliers, and on Santo in Vanuatu. In West Polynesia, *A. tabuensis* co-occurs with the larger *A. atrifusca* (Figure 14-9) on most major islands in Samoa. In Micronesia, the widespread *A. opaca* is or was sympatric with single-island endemics on Pohnpei (*A. pelzelni*) and Kosrae (*A. corvina*). I suspect that two species of starlings once inhabited other high islands in Micronesia, such as Chuuk and Yap, and perhaps the Marianas and Palau.

The sole genus of estrildid finch widespread in Oceania is *Erythrura*, the parrotfinches. Given their modern occurrence in Fiji (*E. cyaneovirens*, *E. kleinschmidti*) and Samoa (*E. cyaneovirens*), I expect that fossils of *Erythrura* will be found eventually in Tonga. Similarly, fossils of *E. trichroa* are to be expected from Yap, a conspicuous gap in its current range. A large, stout-billed, extinct species of *Erythrura* from Rota (Marianas) awaits description (Steadman 1999b). Given that *E. trichroa* is the only other species of parrotfinch in Micronesia, the extinct species may prove to be a derivative of *E. trichroa*, which does not live in the Marianas today.

Suborder Corvida

The cuckoo-shrikes or cicadabirds (*Coracina*; Figure 14-10) are widespread in Melanesia and Micronesia, with a considerable radiation in Near Oceania and one highly variable species in Micronesia (*C. tenuirostris*) that I expect will be found some day as a fossil from the Marianas and Chuuk because it still lives in Palau, Yap, and Pohnpei (Table 14-1). Bones of fantails (*Rhipidura*) have not been discovered yet in Tonga although the genus occurs today in Fiji

Figure 14-4 Mitiaro reed-warbler *Acrocephalus kaoko*, nestling, Mitiaro, Cook Islands. Photo by DWS, 29 October 1987.

Figure 14-5 Ranongga white-eye *Zosterops splendidus*, ♂, Ranongga, New Georgia Group, Solomon Islands. Photo by C. E. Filardi, 7 May 2004.

Figure 14-6 Bare-eyed white-eye *Woodfordia superciliosa*, Rennell, Solomon Islands. Photo by DWS, 22 June 1997.

Figure 14-7 Metallic starling *Aplonis metallica*, Isabel, Solmon Islands. Photo by DWS, 15 July 1997.

Figure 14-8 Ma`uke Starling *Aplonis mavornata*, BMNH old vellum catalogue vol. 12. no. 192a. Ma`uke, Cook Islands. Photo by DWS, September 1990.

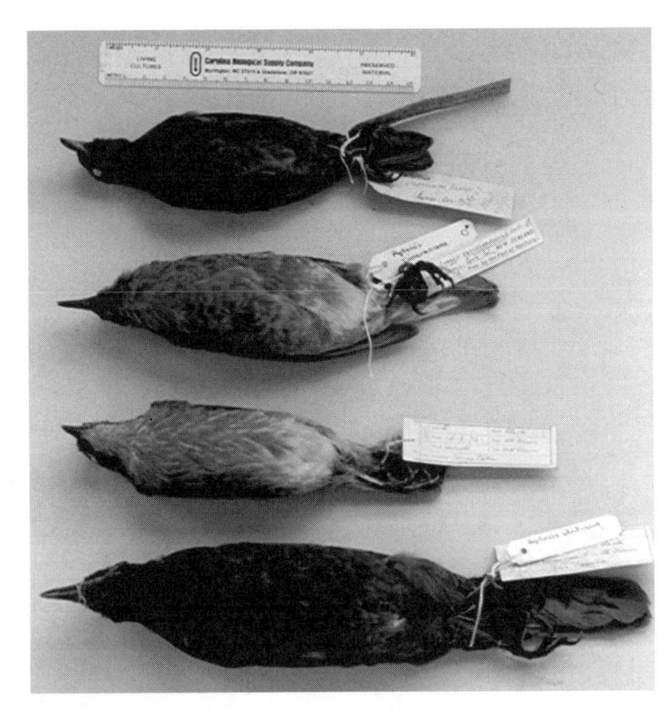

Figure 14-9 Four Polynesian species of starlings in the British Museum. Top to bottom: *Aplonis mavornata*, Ma`uke, Cook Islands; *A. cinerascens*, Rarotonga, Cook Islands; *A. tabuensis*, Fiji; *A. atrifusca*, Samoa. Photo by DWS, September 1990.

Figure 14-10 Yellow-eyed cuckoo-shrike *Coracina lineata*, Rennell, Solomon Islands. Photo by DWS, 25 June 1997.

Figure 14-11 Willie wagtail, *Rhipidura leucophrys*, Isabel, Solomon Islands. Photo by DWS, 20 July 1997.

(*R. spilodera*) and Samoa (*R. nebulosa*), as well as most of Melanesia and high islands in Micronesia (Figure 14-11).

The shrikebills (*Clytorhynchus*; Figure 14-12) consist of four species in eastern Melanesia and West Polynesia, one of which (*C. vitiensis*) is widespread overall but has been lost from many or most islands in its range of Fiji, Tonga, and Samoa. From surveys on 45 islands in Fiji and Tonga, I suspect that deforestation is the primary cause of the many losses, probably assisted by predation from black rats (*Rattus rattus*) and feral cats (*Felis catus*).

The monarchs (*Monarcha*) have a considerable radiation in Near Oceania (Filardi & Smith 2005) and occur in Remote Oceania as a single species each in the Solomon outliers (*M. schistaceus*), Yap (*M. godeffroyi*), and Marianas (*M. takatsukasae*). I suspect that the Chuuk monarch ("*Metabolus*" *rugensis*) may prove to be congeneric with *Monarcha*. The ongoing molecular research by C. E. Filardi and C. E. Smith is certain to change our perspective on evolution and systemics of monarchs on Pacific islands.

Figure 14-12 Rennell shrikebill, *Clytorhynchus hamlini*, Rennell, Solomon Islands. Photo by DWS, 23 June 1997.

Figure 14-13 Melanesian flycatcher *Myiagra caledonica*, Santo, Vanuatu. Photo by DWS, 12 June 2003.

The species of *Myiagra* (Figure 14-13) are usually called broadbills or flycatchers, both of which are confusing names often associated with nonmonarchine passerine families such as the Eurylaimidae, Tyrannidae, and Muscicapidae. Fossils of *Myiagra* (sp. uncertain) have been found on ʻEua in Tonga, an island group where the genus no longer occurs. Species of *Myiagra* still inhabit nearby Fiji (*M. vanikorensis*, *M. azureocapilla*) and Samoa (*M. albiventer*). In the Marianas, bones of extinct *M. freycineti*, confined in historic times to Guam, have been found on Rota.

The robin *Petroica multicolor* occurs today from Australia into Oceania as far east as Fiji and Samoa. In Tonga, it is known from a single fossil at the Faleloa site on Foa (Haʻapai Group). Whistlers (*Pachycephala*) are absent from East Polynesia and Micronesia, whether today or as fossils. The Tongan Whistler *P. jacquinoti* (Figures 14-14, 14-15) is confined today to the Vavaʻu Group and the nearby volcanic island of Late, although its bones have been found in prehistoric sites on Tongatapu and three islands in the Haʻapai Group. Although it has not yet been found among the many passerine fossils from ʻEua, I suspect that *P. jacquinoti* once lived there as well.

Small honeyeaters of the genus *Myzomela* (Figure 14-16) are widespread in Melanesia, West Polynesia, and Micronesia, with sympatric species only in Near Oceania. The current absence of *Myzomela* in Tonga, where it occurs on two islands as a fossil (Steadman 1993a; Tables 6-21, 6-23, 14-3 herein), seems odd given how common *M. jugularis* is today on some islands in the nearby Lau Group of Fiji (Steadman & Franklin 2000). The genus *Foulehaio* is confined to West Polynesia, with *F. carunculata* (Figure 14-17) widespread in Fiji, Tonga, and Samoa and *F. provocator* (sometimes classified as an endemic genus *Xanthotis*) restricted to Kadavu, Ono, and nearby islets in Fiji. The two species are nowhere sympatric. That bones of honeyeaters such as *F. carunculata* and *Myzomela* spp. are found regularly in prehistoric sites may be due in part to the sweet-tasting meat and fat of these nectarivores. When I visited

Figure 14-14 Tongan whistler *Pachycephala jacquinoti*, adult ♀, `Euakafa, Vava`u Group, Tonga. Photo by DWS, 17 July 1995.

Figure 14-15 Tongan whistler *Pachycephala jacquinoti*, adult ♂, `Euakafa, Vava`u Group, Tonga. Photo by DWS, 17 July 1995.

Figure 14-16 Cardinal honeyeater *Myzomela cardinalis*, Rennell, Solomon Islands. Photo by DWS, 22 June 1997.

Rennell (Solomon Islands) in June 1997, young boys daily caught and ate some of the abundant, obese *M. cardinalis* that fed in flowering coconut trees.

Crows (*Corvus*) have a limited range in Oceania and are the sole corvid genus on true oceanic islands. The only outposts for *Corvus* in Remote Oceania (outside of the Hawaiian Islands and New Zealand; see James & Olson 1991, Gill 2003) are New Caledonia (endemic *C. moneduloides*; see Weir et al. 2002) and the Marianas (endemic *C. kubaryi*). In Near Oceania, *C. orru* occurs in the Bismarcks (and New Guinea), whereas *C. meeki* and *C. woodfordi* are endemic to the Solomon Main Chain. The larger crow from New Ireland may be an undescribed species; I have compared its bones (humerus, tibiotarsus) with those of *C. orru* and *C. woodfordi*, as well as *C. tristis* of New Guinea. Pacific island crows may be independently derived in the northern vs. southern hemispheres, with species in the Bismarcks, Solomons, and New Caledonia based on colonizations from Australia and New Guinea, and those in the Marianas and Hawaiian Islands derived from invasions of Asian stock.

Figure 14-17 Wattled honeyeater *Foulehaio carunculata*, `Euakafa, Vava`u Group, Tonga. Photo by DWS, 18 July 1995.

Distributional Trends

Just as with nonpasserine landbirds (Chapters 9-13), the modern ranges of many passerine genera and species have illogical discontinuities that are more likely to be due to anthropogenic extinction than to natural processes such as a failure to colonize, or hypothesized competition-based extinction. These gaps may involve individual islands or entire island groups. Perhaps the most blatant gaps are in *Turdus*, *Cichlornis*, *Cettia*, *Acrocephalus*, *Phylloscopus*, *Aplonis* (especially in East Polynesia), *Coracina* (especially in Micronesia), *Pomarea*, *Clytorhynchus*, *Petroica*, and*Gymnomyza*. Competition-based ideas about which sets of passerines can or cannot coexist are based on anthropogenically depleted distributions and therefore lack empirical rigor (Chapter 20). Holyoak and Thibault (1978b), for example, try to explain through interspecific competition whether species of *Acrocephalus* and *Pomarea* can co-occur on individual islands in East Polynesia. These arguments become moot if, as I believe, any island with *Pomarea* also sustained *Acrocephalus* at human contact.

In the Ha‘apai Group of Tonga in 1995–96, I noticed that the honeyeater *Foulehaio carunculata* was rare or absent on islands where the triller *Lalage maculosa* was common, and vice versa (Steadman 1998; Figure 6-3 herein). I cannot say what caused this inverse relationship, but rather than invoke competition, I would suggest that the answer may lie in the food supply or some other aspect of habitat quality on these degraded islands. On ‘Eua (Tonga), bones of both species are found commonly both before and after human arrival (Chapter 6).

Although I will cover the topic of west-to-east faunal attenuation in detail in Chapter 17, a few comments about passerines seem pertinent here. The two island groups of Near Oceania sustain 35 genera of passerines, compared to 38 genera among 21 island groups in Remote Oceania, with 19 of them shared (Table 14-1). The three most diverse island groups in Remote Oceania (New Caledonia, Vanuatu, and Fiji) each have 19 to 21 genera of passerines, compared to 30 in the Bismarcks and 27 in the Solomon Main Chain. At the opposite end of the scale, passerines are unknown on Easter Island, although very few landbird fossils have been found there. Most island groups in East Polynesia support only one to three genera of passerines, with no instances of sympatric congeners. Unmatched among passerines in their ability to disperse over the ocean, reed-warblers (*Acrocephalus*) have colonized very remote islands in both Micronesia and East Polynesia.

More than 90% of passerine species in Oceania are endemic to the region. The generic level endemism of passerines in Oceania is complicated by the poorly resolved

Table 14-5 The modern (M) and prehistoric (P) distribution of endemic genera of passerines in Oceania. Names in brackets are of unresolved validity. For example, each of the monarch genera except *Lamprolia* might be congeneric with *Monarcha* (C. E. Filardi, pers. comm.).

	Bismarcks	Solomon Main Chain	Solomon Outliers	Vanuatu	New Caledonia	Fiji	Tonga	Samoa	West Polynesian Outliers	Cook Islands	Society Islands	Marquesas	Palau	Yap	Marianas	Chuuk	Pohnpei	TOTAL
OLD WORLD WARBLERS																		
Cichlornis	M	M	—	M	M	—	—	—	—	—	—	—	—	—	—	—	—	4
Trichocichla	—	—	—	—	—	M	—	—	—	—	—	—	—	—	—	—	—	1
WHITE-EYES																		
Rukia	—	—	—	—	—	—	—	—	—	—	—	—	—	M	—	M	M	3
Cleptornis	—	—	—	—	—	—	—	—	—	—	—	—	—	—	M, P	—	—	1
Woodfordia	—	—	M	—	—	—	—	—	—	—	—	—	—	—	—	—	—	1
Megazosterops	—	—	—	—	—	—	—	—	—	—	—	—	M	—	—	—	—	1
†Zosteropidae undescribed genus	—	—	—	—	—	—	P	—	—	—	—	—	—	—	—	—	—	1
MONARCHS																		
[*Pomarea*]	—	—	—	—	—	—	—	—	—	M	M	M, P	—	—	—	—	—	3
[*Mayrornis*]	—	—	M	—	—	M	—	—	—	—	—	—	—	—	—	—	—	2
[*Neolalage*]	—	—	—	M	—	—	—	—	—	—	—	—	—	—	—	—	—	1
[*Clytorhynchus*]	—	—	M	M	M	M	M, P	M	M	—	—	—	—	—	—	—	—	7
[*Metabolus*]	—	—	—	—	—	—	—	—	—	—	—	—	—	—	—	M	—	1
Lamprolia	—	—	—	—	—	M	—	—	—	—	—	—	—	—	—	—	—	1
HONEYEATERS																		
[*Stresemannia*]	—	M	—	—	—	—	—	—	—	—	—	—	—	—	—	—	—	1
[*Guadalcanaria*]	—	M	—	—	—	—	—	—	—	—	—	—	—	—	—	—	—	1
[*Vosea*]	M	—	—	—	—	—	—	—	—	—	—	—	—	—	—	—	—	1
[*Meliarchus*]	—	M	—	—	—	—	—	—	—	—	—	—	—	—	—	—	—	1
Foulehaio	—	—	—	—	—	M	M, P	M	—	—	—	—	—	—	—	—	—	3
Gymnomyza	—	—	—	—	M	M	—	M	—	—	—	—	—	—	—	—	—	3
TOTAL GENERA																		
M	1–2	1–4	1–3	1–3	2–3	4–6	1–2	2–3	0–1	0–1	0–1	0–1	1	1	1	1–2	1	
M + P	1–2	1–4	1–3	1–3	2–3	4–6	2–3	2–3	0–1	0–1	0–1	0–1	1	1	1	1–2	1	
Number of bones	29	0	2	83	0	0	822	7	34	105	3	3	0	3	380+	0	0	

taxonomy in Melanesian honeyeaters, involving *Stresemannia* (? = *Lichmera*, *Melilestes*, or *Meliphaga*), *Guadalcanaria* (? = *Meliphaga*), *Vosea* (? = *Melidectes*), and *Meliarchus* (? = *Melidectes*); see Mayr & Diamond (2001: 398–399). Also unresolved is whether *Metabolus* (for *M. rugensis* on Chuuk) and related genera such as *Neolalage* and *Clytorhynchus* are synonymous with *Monarcha*, the only passerine genus here that has received rigorous molecular research (Filardi & Smith 2005). The generic level classification of non-*Acrocephalus* sylviid warblers is poorly understood as well. Aware of these and other uncertainties, we can estimate roughly 15 to 20 endemic genera of passerines in Oceania (Table 14-5). The only endemic genera of Passerida are among sylviid warblers (Melanesia, West Polynesia) and white-eyes (Melanesia, West Polynesia, and especially Micronesia). Similarly among the Corvida, all genuine or potentially endemic genera are found among just two families, the monarchs (widespread) and the honeyeaters (Melanesia, West Polynesia).

Fiji has six genera of passerines endemic to Oceania (a warbler, three monarchs, and two honeyeaters), more than any other island group. This may be related to Fiji's old age (Chapter 1). Assuming that adequate comparative skeletons become available for study, it will be fascinating to see how much of Fiji's distinctive passerine fauna has been lost since human arrival.

Chapter 15 Seabirds

Up to now, my focus has been on resident landbirds, whose dependence on terrestrial ecosystems makes them ideal for biogeographic analyses. The ocean that separates islands is uninhabitable to landbirds. They may fly above it on occasion, but no landbird feeds solely (if at all) in marine habitats. Seabirds, by contrast, feed on fish and invertebrates in the ocean, and are capable of regular long-distance flights to do so (Shealer 2002). Given the dynamic biological, chemical, and physical conditions of the ocean on various timescales (Murray et al. 1994, Chavez et al. 1999, Rutherford et al. 1999), an ability to disperse hundreds if not thousands of kilometers to find food is essential for seabirds, no matter where they nest. Over much longer (geological) timescales than the fossil record in Oceania, the distributions of families and genera of seabirds have undergone major changes (Olson & Rasmussen 2001, Warheit 2002).

Thus landbirds and seabirds are biogeographic apples and oranges. Dedicating only one chapter to the oranges is unfair, for bones of seabirds far outnumber those of landbirds in most prehistoric sites in Remote Oceania. I seek forgiveness from seabird biologists for the cursory treatment given here to marine birds. Bones of landbirds are my highest priority for study. Nowadays, for lack of time, I often cannot attempt to identify all of the seabird bones from a prehistoric site. The study of past seabird faunas is in its infancy and holds wonderful opportunities for anyone willing to learn their complex comparative osteology. Exhausted from sorting through more than 30,000 of their bones, I present this terse chapter as a *despedida* from seabirds.

The prehistoric records of seabirds are based on many fewer bones from Melanesia or Micronesia than from West Polynesia or especially East Polynesia (Tables 15-1 to 15-4). This uneven situation is related to the relative paucity of prehistoric data on any birds from Melanesia and Micronesia, as well as to the likelihood that these two regions (especially the former) in fact did sustain less rich seabird faunas at human contact. The larger data sets from West and East Polynesia are incompletely analyzed (i.e., many procellariid bones remain unstudied) but provide an overall flavor of the formerly rich seabird faunas in these regions.

Three more qualifications are needed. First, the modern distribution of seabirds in Oceania is imperfectly known (Garnett 1984). Most islands have not had systematic surveys of breeding seabirds. Especially considering nocturnal species, such as many of the procellariids (Reynolds & Ritchotte 1997), a special effort is required to have much certainty about the presence or absence of breeding populations. Badly needed in all island groups are at least a semi-quantified survey such as Thibault (1974) did for the Society Islands. In evaluating whether a prehistoric record represents an extant or an extirpated population, I have been conservative toward extant.

Second, I believe that each of the prehistoric records represents a resident, breeding population. Two lines of evidence support this assumption. One is that on the two islands with good precultural fossil records ('Eua and Mangaia), a very diverse seabird paleofauna is evident at inland sites. For these precultural bones, there is no chance that a predator may have brought the bird to the site from

Table 15-1 Seabirds from prehistoric sites in Melanesia. See Chapter 5 for details of sites. E, extirpated population; X, extant population. SMC, Solomon Main Chain; Sol. Out., Solomon Outliers. Taxa in brackets may not be different from those identified more specifically. From Balouet & Olson (1989), Steadman et al. (1990a, 1999b), Steadman & Kirch (1998), and Steadman (unpub. data).

	Bismarcks			SMC	Sol. Out.		Vanuatu			New Caledonia		TOTAL	
	New Ireland	Mussau	Arawe	Buka	Tikopia	Anuta	Malakula	Efate	Erromango	Grande Terre	`Uvea	All	E
SHEARWATERS, PETRELS													
Pseudobulweria cf. *rostrata*	E	—	—	—	E	—	—	—	—	E	—	3	3
Pterodroma sp.	—	E	—	—	—	—	—	—	—	—	E	2	2
Puffinus pacificus	—	—	E	—	E?	E	—	—	E	—	—	3–4	3–4
Puffinus cf. *gavia*	—	—	—	—	—	—	E	—	—	—	—	1	1
Puffinus lherminieri	—	—	—	—	E	E	—	—	—	—	—	2	2
TROPICBIRDS													
Phaethon rubricauda	—	—	—	—	X	—	—	—	—	—	—	1	0
Phaethon lepturus	—	—	—	—	X	X	—	—	—	—	—	2	0
BOOBIES													
Papasula abbotti abbotti	—	—	—	—	E	—	—	E	—	—	—	2	2
Sula dactylatra	—	—	—	—	X	—	—	—	—	—	—	1	0
Sula sula	—	—	—	—	E	E	—	—	—	—	—	2	2
Sula leucogaster	—	E	—	—	X	X	—	—	—	—	—	3	1
[*Sula* sp.]	—	—	E	—	—	—	—	—	—	—	—	1	1
FRIGATEBIRDS													
Fregata minor	—	—	—	—	X?	X?	—	—	—	—	—	2	0–2
Fregata ariel	—	—	X	—	X?	E	—	—	—	—	—	3	1–2
TERNS													
Sterna bergii	—	—	—	—	—	—	—	—	—	—	X	1	0
Sterna fuscata	—	X	—	—	E	E	—	—	—	—	—	3	2
Anous minutus	—	X	—	—	X	X	—	—	—	—	—	3	0
Anous stolidus	—	X	—	X	X	X	—	—	—	—	—	4	0
TOTAL													
All	1	5	3	1	14	10	1	1	1	1	2	40	—
E	1	2	2	0	5–8	5–6	1	1	1	1	1	—	20–24
Number of bones	2	29	3	4	323	275	1	1	1	11	5	655	

far at sea. The next bit of evidence is the presence in both cultural and precultural sites of juvenile seabird bones (nestling or recently fledged individuals) or long bones with the medullary deposits of laying females. Such bones demonstrate residency of the species on the island.

Third, the comparative osteology of seabirds is not as well understood as that of landbirds. This is especially true of procellariids, a family abounding with unresolved taxonomy (Brooke 2002). My identifications are conservative (i.e., they do not exaggerate the prehistoric species richness); a major effort toward species-level identifications of all procellariid bones would disclose more species, which might require that some identifications of living species being reinterpreted as representing extinct but closely related forms. Two of many potential examples of this would involve the fulmar *Fulmarus glacialoides* from Easter Island (Figure 15-1) or the shearwater *Puffinus griseus* from various islands in Polynesia and Micronesia. These records are based on favorable osteological comparisons, but if more material were available, the prehistoric bones might be interpreted as

Table 15-2 Seabirds from prehistoric sites in West Polynesia. See Chapter 6 for details of sites. *, procellariid bones not fully studied. E, extirpated population; X, extant population; NTT, Niuatoputapu. Taxa in brackets may not be different from those identified more specifically. From Steadman (1989b, 1993b, unpub. data), Worthy et al. (1999).

	Fiji			Tonga									Samoa				TOTAL	
	Waya	Lakeba	Aiwa Levu	NTT	Foa*	Lifuka*	Ha`ano*	`Uiha*	Ha`afeva*	Mango	Tongatapu*	`Eua	Upolu	Tutuila	Ofu	Niue	All	E
SHEARWATERS, PETRELS																		
Pseudobulweria rostrata	—	—	—	—	—	—	—	—	—	—	—	—	—	—	E	—	1	1
Pterodroma macroptera	—	—	—	—	—	—	—	—	—	—	—	E	—	—	—	—	1	1
Pterodroma solandri	—	—	—	—	—	—	—	—	—	—	—	E	—	—	—	—	1	1
†*Pterodroma* undesc. sp.	—	—	—	—	—	—	—	—	—	—	—	E	—	—	—	—	1	1
Pterodroma externa	—	—	—	—	—	—	—	—	—	—	—	E	—	—	—	—	1	1
Pterodroma sp. (med.)	—	—	—	—	—	E	—	E	E	—	E	—	—	—	E	—	5	5
Pterodroma nigripennis	—	—	E	—	—	—	—	—	—	—	—	E	—	—	—	—	2	2
Pachyptila sp.	—	—	—	—	—	E	—	—	—	—	—	—	—	—	—	—	1	1
Puffinus pacificus	—	—	—	—	E	E	E	E	E	E	E	—	—	—	E	—	8	8
Puffinus bulleri	—	—	—	—	—	—	—	—	E	—	—	—	—	—	—	—	1	1
Puffinus griseus	—	—	—	E	—	—	—	—	—	—	—	—	—	—	E	—	2	2
Puffinus lherminieri	—	—	—	—	—	E	E	—	—	—	—	X	—	—	E	—	4	3
†*Puffinus* undesc. sp.	—	—	—	—	—	—	—	—	—	—	—	E	—	—	—	—	1	1
STORM-PETRELS																		
Nesofregetta fuliginosa	—	—	—	—	—	—	—	—	—	—	—	E	—	—	—	—	1	1
TROPICBIRDS																		
Phaethon rubricauda	—	—	—	—	E	—	—	—	—	—	—	—	—	—	—	—	1	1
Phaethon lepturus	—	X	—	—	X	X	—	—	E	—	—	X	—	—	—	X	6	1

BOOBIES																		
Sula dactylatra	—	—	—	—	—	—	—	—	E	—	—	—	—	—	—	—	1	1
Sula sula	—	—	—	E	—	E	E	E	—	—	—	—	—	—	E	—	5	5
Sula leucogaster	—	—	—	—	—	—	—	E	—	—	—	—	—	—	—	—	1	1
FRIGATEBIRDS																		
Fregata minor	—	—	—	—	—	X	X	E	E	—	—	—	—	—	—	—	4	2
Fregata ariel	—	—	—	—	—	—	—	X	X	—	—	X	—	—	—	—	3	0
[*Fregata* sp.]	—	—	—	—	—	—	—	—	—	—	—	—	—	—	X	—	1	0
TERNS																		
Sterna lunata	E?	—	—	—	—	—	—	E	—	—	—	—	—	—	—	—	2	1–2
Sterna anaethetus	—	—	—	—	—	E	—	—	—	—	—	—	—	—	—	—	1	1
Sterna fuscata	—	—	—	—	—	—	E	E	E	—	—	—	—	—	—	—	3	3
Anous minutus	—	X?	—	—	X?	X?	X?	X?	—	—	E	—	—	—	—	—	6	1–6
Anous stolidus	—	—	—	X	X	X	X	X	—	—	X	X	X	—	X	X	10	0
Procelsterna cerulea	—	—	—	—	—	—	—	—	—	—	—	—	—	X	—	—	1	0
Gygis candida	—	X	—	X	X	—	—	X	X	—	—	X	—	—	X	X	8	0
Gygis microrhyncha	—	—	—	—	—	—	—	E	E	—	—	E	—	—	—	—	3	3
TOTAL																		
All	1	3	1	4	6	10	7	12	10	1	4	13	1	1	9	3	86	—
E	0–1	0–1	1	2	2–3	6–7	4–5	8–9	8	1	3	8	0	0	6	0	—	49–55
Number of bones	1	76	1	14	137	201	128	239	348	2	36	713	1	19	51	100	2067	

Table 15-3 Seabirds from prehistoric sites in East Polynesia. See Chapter 7 for details of sites. E, extirpated population; X, extant population; X+, extant population only on offshore islets; SI, Society Islands; TA, Tuamotu Archipelago; *, procellariid bones not fully studied; **, few of the seabird bones from Hiva Oa have been studied; †, extinct species. Taxa in brackets may not be different from those identified more specifically. Updated from Steadman (1985, 1991b, 1995a, 1997b, unpub. data), Steadman & Olson (1985), Schubel & Steadman (1989), Steadman & Pahlavan (1992), Steadman & Justice (1998), Wragg & Weisler (1994), Wragg (1995), and Rolett (1998).

	Cook Islands		Tubuai		SI	TA	Pitcairn	Marquesas						TOTAL	
	Aitutaki	Mangaia	Tubuai	Runutu	Huahine	Mangareva	Henderson	Nuku Hiva*	Ua Huka*	Ua Pou	Hiva Oa**	Tahuata*	Easter Island*	All	E, X+
ALBATROSSES															
Diomedea epomophora	—	—	—	—	—	—	E	—	—	—	—	—	—	1	1
Diomedea sp.	—	—	—	—	—	—	—	—	—	—	—	—	E	1	1
SHEARWATERS, PETRELS															
Macronectes sp.	—	—	—	—	—	—	—	—	—	—	—	—	E	1	1
Fulmarus glacialoides	—	—	—	—	—	—	—	—	—	—	—	—	E	1	1
Pseudobulweria rostrata	E	—	—	—	E	—	—	—	—	—	—	—	—	2	2
Pterodroma macroptera/lessoni	—	—	—	—	—	—	—	—	—	—	—	—	E	1	1
Pterodroma alba	—	—	—	—	E	—	X	—	E	—	—	—	—	3	2
Pterodroma inexpectata	—	E	—	—	—	—	—	—	—	—	—	E	—	2	2
Pterodroma ultima	—	—	—	—	—	—	—	—	—	—	—	E	E	2	2
Pterodroma heraldica	—	—	—	—	E	E	—	—	—	—	—	E	—	3	3
Pterodroma phaeopygia	—	—	—	—	—	—	—	—	—	—	—	—	E	1	1
Pterodroma externa	—	—	—	—	—	—	X			—		E	E	3	2
Pterodroma cookii	—	E	—	—	—	—	—	—	—	—	—	—	—	1	1
Pterodroma nigripennis	—	E	—	E	—	—	X	—	E	—	—	E	—	5	4
[*Pterodroma* sp. (large)]	—	E	—	—	—	—	—	—	—	—	—	—	—	1	1
[*Pterodroma* sp. (medium)]	—	—	—	—	—	—	X	—	—	—	—	—	—	1	0
†Undescribed genus & sp.	—	—	—	—	—	—	—	—	—	—	—	—	E	1	1
Pachyptila sp.	—	E	E	—	—	—	—	—	—	—	—	E	E	4	4
Bulweria bulwerii	—	E	—	—	—	—	E	E	X+	—	—	—	—	4	4
Bulweria cf. *fallax*	—	—	—	—	—	E	—	—	—	—	—	E	—	2	2
Procellaria sp.	—	—	—	—	—	—	—	—	—	—	—	—	E	1	1
Puffinus carneipes	—	—	—	—	—	—	—	—	—	—	—	—	E	1	1
Puffinus pacificus	—	—	—	—	E	X	X	E	X+	—	—	E	—	6	4
Puffinus griseus	—	—	—	—	—	—	—	—	—	—	—	—	E	1	1
Puffinus tenuirostris	—	—	—	—	—	E	—	—	—	—	—	—	—	1	1
Puffinus nativitatis	—	—	—	—	E	X	X	—	E	—	—	E	X+	6	4
Puffinus cf. *gavia*	—	E	—	—	—	—	—	—	—	—	—	—	—	1	1
Puffinus lherminieri	—	X	X	—	E	X	X	E	E	E	—	E	—	9	5

STORM-PETRELS															
Fregetta grallaria/tropica	—	—	—	—	—	—	E	—	E	—	—	E	—	3	3
Nesofregetta fuliginosa	—	E	—	—	—	X	X	X+	X+	X+	—	E	E	8	6
TROPICBIRDS															
Phaethon rubricauda	—	X	—	—	—	X	X	—	—	—	—	—	X	4	0
Phaethon lepturus	—	X	X	—	X	X	X	—	X	—	—	X	E	8	1
BOOBIES															
Papasula abbotti costelloi	—	—	—	—	—	—	—	—	E	—	E	E	—	3	3
Sula dactylatra	—	—	—	—	—	—	—	—	E	—	—	—	X+	2	2
Sula sula	E	E	—	—	E	—	X	—	E	—	—	E	—	6	5
Sula leucogaster	—	—	—	—	E	—	—	—	E	—	—	E	—	3	3
[*Sula* sp.]	—	—	—	—	—	—	—	E	—	E	—	—	—	2	2
FRIGATEBIRDS															
Fregata minor	X	X	—	—	E	X	X	X	X	—	X	X	X+/E	10	2
Fregata ariel	X	X	—	—	E	X	—	—	X	—	—	—	—	5	1
GULLS															
†*Larus utunui*	—	—	—	—	E	—	—	—	—	—	—	—	—	1	1
TERNS															
Sterna lunata	—	—	—	—	—	—	—	—	—	—	—	—	X+/E	1	1
Sterna fuscata	—	—	—	—	—	—	X	—	X+	—	—	E	X+	4	3
[*Sterna* sp.]	—	—	—	—	—	—	—	X	—	—	—	—	—	1	0
Anous minutus	X	E	—	—	E	—	—	X+	X+	—	—	E	E	7	6
Anous stolidus	X	X	X	—	X	X	X	—	X	X	—	X	X+	10	1
Procelsterna cerulea	—	X	—	—	—	X	X	—	—	—	—	E	X+	5	2
Gygis candida	X	X	X	—	X	X	X	—	—	—	—	E	X+	8	2
Gygis microrhyncha	—	E	—	—	—	—	—	X	X+	—	—	X	E	5	3
TOTAL															
All	7	19	5	1	15	14	19	9	19	4	2	23	25	162	—
E, X+	2	11	1	1	12	3	3	6	15	3	1	19	24	—	101
Number of bones	22	1604	7	12	203	205	14,301	~500	10,296	59	~50	567	~800	~28,626	

Table 15-4 Seabirds from prehistoric sites in Micronesia. See Chapter 8 for details of sites. E, extirpated population; X, extant population. From Steadman (1992a, 1999a, unpub. data), Steadman & Intoh 1994, Pregill & Steadman 2000). Taxa in brackets may not be different from those identified more specifically.

	Palau			Yap	Marianas					Marshall		Tuvalu	Line	TOTAL	
	Ngerduais	Ulebsechel	Ulong	Fais	Guam	Rota	Aguiguan	Tinian	Pohnpei	Majuro	Makin	Vaitupu	Fanning	All	E
SHEARWATERS, PETRELS															
Pseudobulweria sp.	—	—	—	—	—	—	—	—	E	—	—	—	—	1	1
Pterodroma sp.	—	—	—	E	—	—	—	—	E	—	—	—	—	2	2
Bulweria bulwerii	—	—	—	E	—	—	—	—	—	—	—	—	—	1	1
Puffinus griseus	—	—	—	—	—	—	—	—	—	E	E	—	—	2	2
Puffinus nativitatis	—	—	—	—	—	—	—	—	E	—	—	—	X	2	1
Puffinus lherminieri	—	X	—	—	—	E	—	—	E	—	—	—	—	3	2
TROPICBIRDS															
Phaethon rubricauda	—	—	—	—	E	X	—	—	—	—	—	—	—	2	1
Phaethon lepturus	X	X	X	X	—	X	X	—	—	—	—	—	X	7	0
BOOBIES															
Sula dactylatra	—	—	—	E	—	—	—	—	—	—	—	—	—	1	1
Sula sula	—	—	—	E	—	—	—	—	—	E	—	—	—	2	2
Sula leucogaster	—	—	—	—	—	—	—	—	—	—	E	—	X	2	1
FRIGATEBIRDS															
Fregata minor	—	—	—	X	—	—	—	—	X	—	—	—	X	3	0
Fregata ariel	—	—	—	X	—	—	E	—	—	—	E	—	—	3	2
TERNS															
Sterna sumatrana	—	X	—	E	—	—	—	—	—	—	—	—	—	2	1
Serna lunata	—	—	—	E	—	—	—	—	—	—	—	—	—	1	1
Sterna fuscata	—	—	—	E	—	—	—	—	—	—	—	—	X	2	1
[*Sterna* sp.]	—	—	—	—	—	—	—	X/E	—	—	—	—	—	1	0–1
Anous minutus	—	—	—	E	—	—	—	X	—	—	—	X	—	3	1
Anous stolidus	—	X	X	X	—	X	X	X	X	X	X	X	—	9	0
Procelsterna cerulea	—	—	—	E	—	E	—	E	—	—	—	—	—	3	3
Gygis candida	—	X	—	X	—	X	X	X	—	—	—	X	—	6	0
Gygis microrhyncha	—	—	—	—	—	—	—	E	—	—	—	—	—	1	1
TOTAL															
All	1	5	2	14	1	6	4	6	6	2	4	3	5	59	—
E	0	0	0	9	1	2	1	2–3	4	2	3	0	0	—	24–25
Number of bones	1	63	5	75	1	179	24	39	14	2	13	4	11	431	

similar but extinct species. *Puffinus griseus* is a subtropical to subantarctic breeder that migrates each year through the tropical Pacific (Spear & Ainley 1999, Kratter & Steadman 2003). The extirpated resident populations on tropical islands may have had different breeding seasons, feeding areas, and migratory patterns.

Seabirds tend to be long-lived and single-brooded and to have long incubation and fledgling periods (Wynne-Edwards 1955, Ashmole 1963). All species of seabirds require mammal-free situations for successful nesting (bats are OK). From this standpoint alone, Remote Oceania would have been prime breeding grounds for seabirds

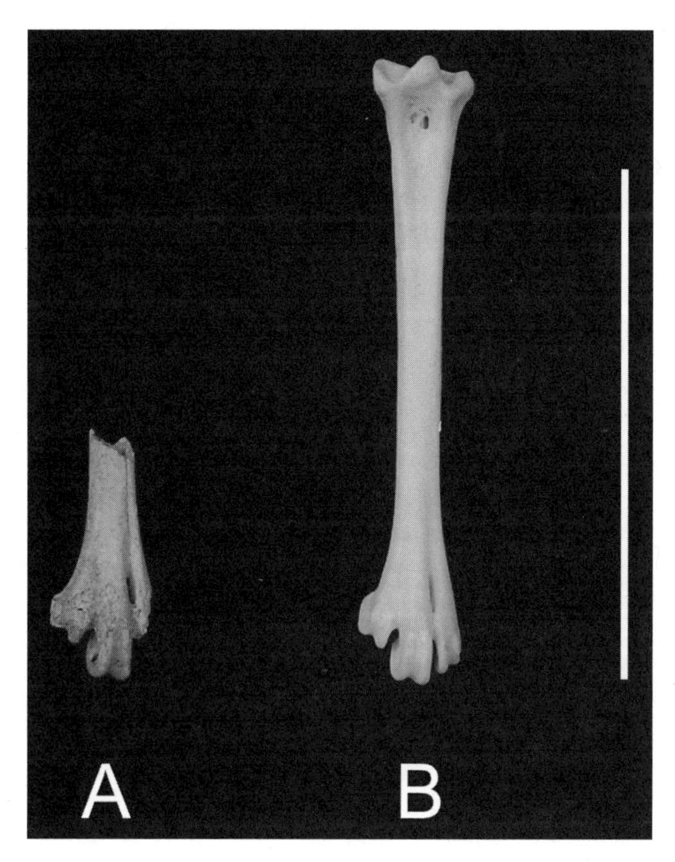

Figure 15-1 The tarsometatarsus of *Fulmarus glacialoides* in acrotarsial aspect. A: From Ahu Naunau site, Easter Island (UF 53189). B: Modern specimen from southern Chile (UF 38940, sex unknown). Scale = 40 mm. Photo by J. J. Kirchman.

before people, rats, dogs, and pigs showed up. Islands in Near Oceania have indigenous rodents (Chapter 2) and therefore may have been marginal for most seabirds even before human arrival. The islands in Remote Oceania with terrestrial mekosuchine crocodiles (New Caledonia, Vanuatu, western Fiji; Mead et al. 2002) may also have been suboptimal for seabirds, a suggestion born out thus far by how few seabird fossils have been found in these island groups. Other potentially significant indigenous predators on seabirds would be the monitor lizard *Varanus indicus* (native in Oceania to the Bismarcks, Solomons, and perhaps Palau and Yap; Chapter 2), the boas *Candoia* spp. (native from the Bismarcks and Solomons locally to Samoa), and the colubrid snakes *Boiga irregularis* and *Dendrelaphis* spp. (both native to the Bismarcks and Solomons, the last in Palau as well; McCoy 1980, Crombie & Pregill 1999).

Human damage to tropical Pacific seabird faunas continues today (Moors & Atkinson 1984, Vermeer & Rankin 1984, Spennemann 1999, Lyver et al. 2000). This chapter demonstrates that the seabird populations found by European explorers of the Pacific already had been depleted by people, a phenomenon not unique to Oceania (Pregill et al. 1994, Mourer-Chauviré 1999, Serjeantson 2001).

Brief Geographic Review

Melanesia (Figure 1-2)

The mere 40 prehistoric records of 17 species of seabirds from Melanesia represent from 20 to 24 extirpated populations, about half of them procellariids (Tables 15-1, 15-5). Nine of the 11 islands involved have 29 or fewer seabird bones. The only "Melanesian" islands with diverse bone assemblage of seabirds are Tikopia and Anuta. More than 90% of the 655 seabird bones in Table 15-1 are from these two tiny, remote islands, which are classified as Melanesian only because of a political affiliation with the Solomon Islands (Chapter 5). Without Tikopia and Anuta, the Melanesian fossil record would consist of 11 species and 16 populations (10 extirpated), and the mean number of seabird bones per island would drop from 60 to just six.

Bones from New Ireland and Buka show that shearwaters, petrels, boobies, frigatebirds, and terns once nested even on islands with native rodents. While the scant fossil record would argue that procellariids once were widespread in Melanesia, nowhere is there evidence that a large Melanesian island once had a wide variety of resident seabirds, even in rodent-free Vanuatu and New Caledonia. As is generally the case across Oceania, most seabird extirpation in Melanesia involved procellariids, whereas tropicbirds and terns had no or relatively few losses. Boobies also were depleted considerably.

West Polynesia (Figure 1-3)

The fossil record of seabirds is much richer (29 species) and more extensive (16 islands, seven with >100 seabird bones) in West Polynesia (Tables 15-2, 15-5) than in Melanesia. Nearly all records of procellariids and boobies represent extirpated populations, whereas most records of tropicbirds and terns do not. Only three of the 13 species of procellariids still have breeding populations anywhere in West Polynesia. 'Eua, with West Polynesia's largest sample of seabird bones (713), had two extinct species (of *Pterodroma* and *Puffinus*) as well as five other extirpated populations of procellariids. The procellariid bones from eight other Tongan islands (incompletely studied) combined represent six species, dominated by the shearwater *Puffinus pacificus*. The most common seabird in the precultural strata at 'Anatu on 'Eua is

Table 15-5 Regional family-level summary of prehistoric records of seabird populations from Oceania. Based on data in Tables 15-1 to 15-4. E, extirpated population; X, extant population; X+, extant population only on offshore islands. *unweighted mean.

	Melanesia		West Polynesia		East Polynesia		Micronesia		Total populations	
Family	All	E	All	E	All	E, X+	All	E	All	E, X+
ALBATROSSES	—	—	—	—	2	2	—	—	0	2
SHEARWATERS, PETRELS	12	11–12	29	28	64	52	11	9	116	100–101
STORM-PETRELS	—	—	1	1	11	9	—	—	12	10
TROPICBIRDS	3	0	7	2	12	1	9	1	31	4
BOOBIES	9	6	7	7	16	15	5	4	37	32
FRIGATEBIRDS	5	1–4	8	2	15	3	6	2	34	8–11
GULLS	—	—	—	—	1	1	—	—	1	1
TERNS	11	2	34	9–15	41	19	28	8–9	114	37–45
Total populations	40	20–24	86	49–55	162	102	59	24–25	345	194–205
% extirpated populations	50–60		57–64		62		41–42		*56–59	
Number of species	17		29		44		21		51	
Number of islands with seabird bones	11		16		13		13		53	
Number of seabird bones	655		2047		~28,626		430		~31,758	
Mean number of seabird bones/island	60		128		~2202		33		—	

the extirpated storm-petrel *Nesofregetta fuliginosa*, which has a very fragmented distribution today (Harrison & Jehl 1988).

There is little evidence of prehistoric seabirds in Fiji, much as in Vanuatu to the west but unlike in Tonga to the east (Tables 15-1, 15-2). More fossil sites are needed to evaluate whether the seabird situation in Fiji and Vanuatu is genuine (and perhaps related to predation by terrestrial crocodiles; see above). The only set of Samoan seabird bones of any size is from Ofu, with six extirpated species (five procellariids, a booby) and three extant ones among only 51 bones. For now it is unclear whether warmer Samoa was inhabited by a less diverse seabird fauna than cooler southern Tonga. Regardless, extensive seabird faunas once did occur on Tongan islands ranging from ʻEua (87 km^2, 325 m elev.) down to Haʻafeva (1.8 km^2, 11 m elev.). The depauperate fossil record of seabirds on Niue must be a sampling artifact, as this raised limestone island lies between Tonga and the Cook Islands, both of which had rich prehistoric seabird faunas.

East Polynesia (Figure 1-4)

This region has 28,600+ seabird bones from 13 islands (Tables 15-3, 15-5). Extirpated populations have been found in all families but especially among procellariids, terns, and boobies. The 44 species of seabirds identified in East Polynesia represent all but seven of the species known as fossils from across Oceania. The 50+% increase in prehistoric species richness from West Polynesia (29) to East Polynesia (44) is due entirely to procellariiforms (albatrosses, shearwaters, petrels, and storm-petrels).

Decent data on prehistoric seabirds are available from six East Polynesian island groups. Procellariids dominate the seabird bones on the four islands with the most fossils (Henderson, Ua Huka, Mangaia, and Easter, each in a different island group). Except on Henderson, which may have the most intact seabird fauna left in East Polynesia, most populations are extirpated. The abundant bones of procellariids from the Marquesas and Easter Island are incompletely studied. More species probably would be found if an attempt were made to identify all of them to the species level. Nevertheless, the former richness (24+ species) and taxonomic diversity (at least nine genera) of East Polynesian procellariids are truly remarkable. Only seven of these 24 species nest anywhere in East Polynesia today. Six to nine species of procellariids have been recorded prehistorically on each of Mangaia (Cook Islands), Huahine (Society Islands), Mangareva (Tuamotu Group), and Henderson (Pitcairn Group). From five Marquesan islands, 12 species of procellariids are known, 10 of these on Tahuata alone. The seabird bones from Nuka Hiva, Ua Huka, and Hiva Oa are less thoroughly analyzed than those from Tahuata, many of which also have not been identified to species.

On Easter Island, at 27° S the coldest island in the region, 12 species of procellariids are recorded by bones, including Oceania's only evidence of resident *Macronectes* sp., *Fulmarus glacialoides*, and *Procellaria* sp., each of which today is a subantarctic breeder. No bones of penguins (Spheniscidae) or diving-petrels (Pelecanoididae) have been found thus far, but I would not be shocked if some day they did. The prehistoric seabird fauna from Easter Island rivals that of Tahuata (9° S) as the richest in Oceania (Steadman 1995a).

Micronesia (Figure 1-5)

The prehistory of seabirds in Micronesia comprises 431 bones from 13 islands, representing 21 species and 59 populations (Tables 15-4, 15-5). These data reveal fewer lost populations (24 or 25) than in West or East Polynesia. This may in part be because most Micronesian seabird bones are from rather late in an island's cultural sequence, after many losses already had taken place. That much more extirpation awaits discovery is suggested by the nine extirpated species of seabirds among only 75 bones from an archaeological site on Fais, Yap (Steadman & Intoh 1994) or on Pohnpei, with four extirpated species of procellariids among only 14 seabird bones. We do not understand the true nature of Micronesian seabird faunas at human contact.

At least six species of procellariids once inhabited the relatively warm waters of western and central Micronesia, including four (*Pseudobulweria* sp., *Pterodroma* sp., *Bulweria bulwerii*, *Puffinus griseus*) that nest nowhere in the region today. Terns make up 40% of the prehistoric records of Micronesian seabirds, more than elsewhere in Oceania. As is true in West and East Polynesia (Tables 15-2, 15-3), the terns identified most often in Micronesia are *Anous stolidus* and *Gygis candida*, both of which can nest in trees and be relatively resilient to human presence. All five Micronesian records of three other terns (*Sterna lunata*, *Procelsterna cerulea*, and *Gygis microrhyncha*) represent extirpated populations.

Brief Systematic Review

Albatrosses (Diomedeidae)

Albatrosses nest nowhere today in our study area. In the northern hemisphere, *Diomedea nigripes* breeds locally in the Hawaiian Islands, Bonin and Ryukyu islands, and formerly in the Marianas, whereas *D. immutabilis* nests locally only in the Hawaiian Islands (Harrison & Seki 1987, Carboneras 1992a). In the equatorial eastern Pacific, *D. irrorata* (= *D. leptorhyncha*; see Steadman & Zousmer 1988) breeds only on in the Galápagos Islands and Isla de la Plata off Ecuador (Douglas 1998). In the southern hemisphere, albatrosses now nest only on subtropical to subantarctic islands. The two prehistoric records of albatrosses from East Polynesia, namely *D. epomophora* on Henderson (24° S) and *Diomedea* sp. on Easter (27° S), suggest a more northerly past breeding range in the southern hemisphere. Today, *D. epomophora* nests from ca. 40 to 53° S on New Zealand and its satellites (Campbell, Auckland, and Chatham Islands; Carboneras 1992a). The single albatross bone (a jugal) from Easter Island is large, in the size range of *D. amsterdamensis*, *D.* (*Thalassarche*) *cauta*, *D.* (*Phoebastria*) *irrorata*, and *D.* (*Phoebastria*) *albatrus*. It is smaller only than in *D. exulans*.

Shearwaters and Petrels (Procellariidae)

This family of seabirds has lost far more populations (99–100 documented by bones thus far) in Oceania than any other (Table 15-5). The losses span Oceania but were greatest in East Polynesia. The past diversity of procellariids increases in the tropical Pacific along a more or less WNW-to-ESE gradient, with no more than four species from a single island in Micronesia or Melanesia, up to seven from single islands in Tonga, eight on Mangaia (Cook Islands), eight on Henderson (Pitcairn Group), 10 on Tahuata (Marquesas), and 12 on Easter Island (Tables 15-1 to 15-4). None of these values is likely to represent the entire procellariid fauna at human contact (see Chapter 4). The astonishing 24 species (in nine genera) from East Polynesia include nine species of *Pterodroma* (petrels; Figure 15-2) and seven of *Puffinus* (shearwaters). Also from East Polynesia are four genera unrecorded elsewhere in Oceania (the giant petrel *Macronectes*, fulmar *Fulmarus*, petrel *Procellaria*, and a petrel-like undescribed genus, each known thus far only from Easter Island).

Following Bretagnolle et al. (1998), I recognize *Pseudobulweria* for *P. rostrata* and its allies such as *P. becki*, although I question that it is more closely related to *Puffinus* than to *Pterodroma*. Species-level systematics of the small "*Cookilaria*" forms of *Pterodroma* are difficult to reconcile even with skins (Falla 1942) much less with isolated bones. A similar statement also pertains to some larger species of *Pterodroma* (Murphy & Pennoyer 1952). I note as well that the generic level systematics of *Pterodroma* is also in flux (Olson 2000).

The prions (*Pachyptila* spp.) comprise four to six (wide-billed) species that breed today on subtropical to subantarctic islands from 36 to 63° S (Carboneras 1992b). Bones of prions (Figure 15-3) have been found across Polynesia in Tonga, the Cook Islands, Tubuai, Marquesas, and

Figure 15-2 Dark-rumped Petrel *Pterodroma phaeopygia* found dead on Santa Cruz, Galápagos Islands. Bones inseparable from those of *P. phaeopygia* (now confined to the Hawaiian and Galápagos Islands) have been found on Easter Island. Photo by DWS, 8 August 1984.

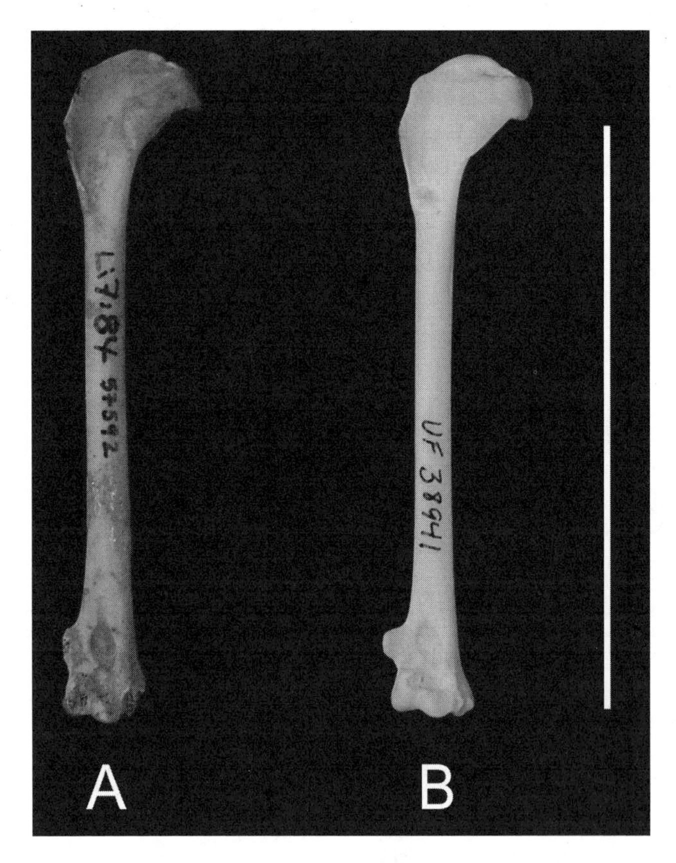

Figure 15-3 The humerus of *Pachyptila* in ventral aspect. A: *Pachyptila* sp., Tongoleleka site, Lifuka, Tonga (UF 57592). B: *P. belcheri*, modern specimen from southern Chile, UF 38941, female. Scale = 50 mm. Photo by J. J. Kirchman.

Easter Island, spanning from 8 to 27° S and 175 to 110° W. The small, delicate petrels of the genus *Bulweria* have two species in Oceania. One is *B. bulwerii* from Mangaia, Henderson, and the Marquesas (Nuku Hiva, Ua Huka). The second, larger species from Mangareva and the Marquesas (Tahuata) may be either the poorly known *B. fallax*, which is believed to breed today only off the northwestern Indian Ocean, more than 15,000 km away (Gallagher et al. 1984, Olson 1985a, Steadman & Justice 1998) or perhaps the extinct *B. bifax*, described from fossils on St. Helena in the Atlantic Ocean (Olson 1975c).

All prehistoric bones of *Puffinus gavia* and *P. huttoni* from Oceania should be compared with specimens of the extinct *P. spelaeus* from New Zealand (Holdaway & Worthy 1994). Similarly, bones referred to *P. pacificus* or *P. bulleri* should be compared with those of the extinct *P. pacificoides* from St. Helena (Olson 1975c).

Storm-petrels (Oceanitidae)

The only widespread species of storm-petrel in Oceania today, *Nesofregetta fuliginosa* (Figure 15-4), had a larger range in Polynesia before human arrival (Tables 15-2, 15-3). It occurs abundantly in precultural faunas on 'Eua (Tonga) and Mangaia (Cook Islands). Prehistoric records of storm-petrels are lacking in Melanesia and Micronesia. Bones of *Fregetta tropica* or *F. grallaria* have been identified from Henderson and the Marquesas. Both species of *Fregetta* are subtropical breeders today; the nearest nesting locality (and only one in Polynesia) is for *F. grallaria* on subtropical Rapa (26° S) in the Tubuai Group (Thibault & Varney 1991a).

Tropicbirds (Phaethontidae)

Two species of tropicbirds, *Phaethon lepturus* and *P. rubricauda* (Figure 15-5), are widespread but local in Oceania, especially Remote Oceania. Fossils from Polynesia and Micronesia have added records from four islands that now lack one of these species. Where birds are still hunted, *P. rubricauda* is favored over *P. lepturus* because it has more meat and its long central tail feathers are red rather than white (pers. obs.).

Figure 15-5 Red-tailed tropicbird *Phaethon rubricauda*, adult, `Ata, Tonga. Photo by DWS, 9 August 2001.

Boobies (Sulidae)

Bones of one or more of Oceania's three resident species of *Sula* are found regularly but usually in low numbers in prehistoric sites from the Bismarcks to Easter Island (Steadman 1995a, Steadman & Kirch 1998). Boobies no longer reside on most of these islands. The tree- and shrub-nesting *S. sula* (Figure 15-6) has been lost from at least 14 islands, compared to four and five islands, respectively, for the ground-nesting *S. dactylatra* and *S. leucogaster* (Figures 15-7, 15-8). It is probably safe to say that hundreds of populations of *Sula* spp. have been lost in Oceania since human arrival.

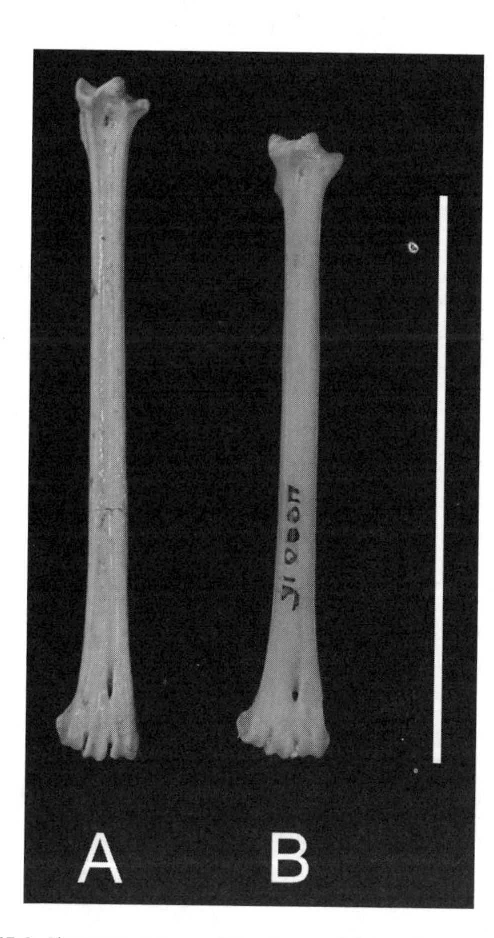

Figure 15-4 The tarsometatarsus of *Nesofregetta fuliginosa* in acrotarsial aspect. A: Fossil from `Anatu, `Eua, Tonga (UF 52194). B: Modern specimen from Kirimati, Kiribati (USNM 498015). Scale = 40 mm. Photo by J. J. Kirchman.

A more dramatic story is that of Abbott's booby, *Papasula abbotti*. Distinctive from *Sula* in many ways including osteology (Olson & Warheit 1988), the tree-nesting *P. a. abbotti* survives today only on Christmas Island in the Indian Ocean, although it was described (Ridgway 1893) from a specimen taken on Assumption Island (also Indian Ocean, but 6300 km west of Christmas Island). Subsequently, *P. a. abbotti* has been extirpated on Assumption (Nelson 1974, Bourne 1976, Prŷs-Jones et al. 1981, Stoddart 1981). I identified its bones from archaeological sites on Tikopia (Santa Cruz Group, Solomon Islands; Steadman et al. 1990) and Efate (Vanuatu; Chapter 5). These islands are 6000+ km east of Christmas Island. Considering the innumerable intervening islands, *P. a. abbotti* undoubtedly has suffered a much more severe population decline than previously recognized.

An extinct subspecies, *Papasula abbotti costelloi*, was described from bones in cultural sites on Tahuata and Hiva Oa, Marquesas (Steadman et al. 1988; Figure 15-9 herein). Since then I have discovered a coracoid of *P. a. costelloi* from Ua Huka as well. *Papasula a. costelloi* differs from *P. a. abbotti* in size and qualitative features. If both forms were alive today, Costello's booby might be regarded as a separate species from Abbott's, especially given the current trend toward taxonomic splitting. *Papasula a. costelloi* is known thus far only from the Marquesas Islands, which are 4800 km east of Tikopia. A tibiotarsus of *Papasula a. costelloi* from Tahuata contains medullary deposits,

Figure 15-6 Red-footed booby, *Sula sula*, pair of adults, Genovesa, Galápagos Islands. Photo by DWS, 10 November 1980.

Figure 15-7 Masked booby, *Sula dactylatra*, adult, Española, Galápagos Islands. Photo by DWS, 7 May 1985.

indicating a female that either was laying eggs or preparing to do so.

Whether *Papasula a. costelloi* occurred outside of the Marquesas is unknown but seems likely. We also do not know where or if the ranges of *P. a. costelloi* and *P. a. abbotti* once met in the Pacific. So far, all booby bones found between the Marquesas and Tikopia/Vanuatu pertain to *Sula* rather than *Papasula*. Sadly, *P. a. abbotti* is at risk on its last outpost, Christmas Island, from deforestation and perhaps as well from a nonnative snake (Reville et al. 1990, Fritts 1993).

Frigatebirds (Fregatidae)

Two species of frigatebirds, *Fregata minor* (Figure 15-10) and *F. ariel*, are widespread in Oceania and found regularly in prehistoric sites. Both are great wanderers that nest nowadays on few islands, although roosting on many islands where they do not nest (Nelson 1976). Thus, you can see frigatebirds today, unlike other seabirds, almost anywhere in Oceania, even many hundreds of kilometers from a nesting colony. This makes it difficult to assess whether prehistoric bones represent breeding populations. I seldom find bones of nestling or fledgling frigatebirds, while I do find such bones of other seabirds; this is in spite of the extended period that these tropical marine scavengers require

Figure 15-8 Brown booby, *Sula leucogaster*, adult, `Ata, Tonga. Photo by DWS, 9 August 2001.

to fledge. Exceptions are ʻUiha and Haʻafeva in Tonga, two islands where frigatebirds do not nest today but with bones of fledglings in their archaeological sites.

Seabirds have been proposed as agents of interisland dispersal for plants with sticky fruits or seeds (Carlquist 1974, Valdebenito et al. 1990). I would suggest that frigatebirds may be especially important in this regard because of their poor site fidelity. Most other seabirds come ashore only to nest and thus are unlikely to land in their lifetimes on more than one island. Because frigatebirds are incapable of landing on the ocean, they are also less likely to have a seed or fruit wash away from their bodies once it adheres to them.

I should note here that the other resident pelecaniform family in Oceania (cormorants, Phalacrocoracidae) is found locally in the Bismarcks, Solomons, New Caledonia, and Palau), where I regard the only two species (*Phalacrocorax carbo*, *P. melanoleuca*) as aquatic (fresh water, estuarine) rather than marine inhabitants (Chapter 13). There is no fossil record of cormorants in Oceania.

Gulls and Terns (Laridae)

Whereas resident species of terns (Sterninae) occur across Oceania, the only resident gulls (Larinae) are *Larus novaehollandiae* in New Caledonia and the extinct *L. utunui* from Huahine, Society Islands (Steadman 2002a; Figure 15-11 herein). The latter probably evolved from an isolated population of *L. novaehollandiae*, which occurs as well in New Zealand and Australia. I expect that other evidence of resident gulls will be disclosed eventually in Oceania's bone deposits.

Four species of terns (*Sterna fuscata*, *Anous stolidus*, *A. minutus*, and *Gygis candida* [= *G. alba*]) are very

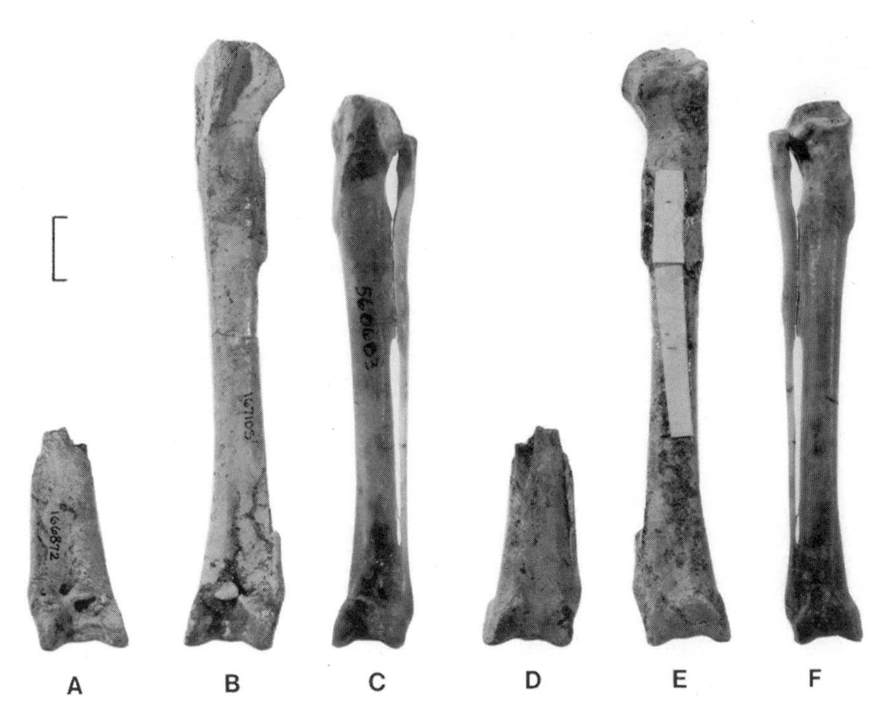

Figure 15-9 The tibiotarsus of *Papasula* in dorsal (A–C) and ventral (D–F) aspects. A, D: *P. a. costelloi*, BPBM 166872, Tahuata, Marquesas. B, E: *P. a. costelloi*, BPBM 167105, holotype, Tahuata, Marquesas. C, F: *P. a. abbotti*, USNM 560683, Christmas Island, Indian Ocean. A, D are from the right side. B, C, E, F are from the left side. Scale bar = 1 cm. From Steadman (1988: Figure 2).

Figure 15-10 Great frigatebird, *Fregata minor*, adult ♂, Genovesa, Galápagos Islands. Photo by DWS, 17 May 1983.

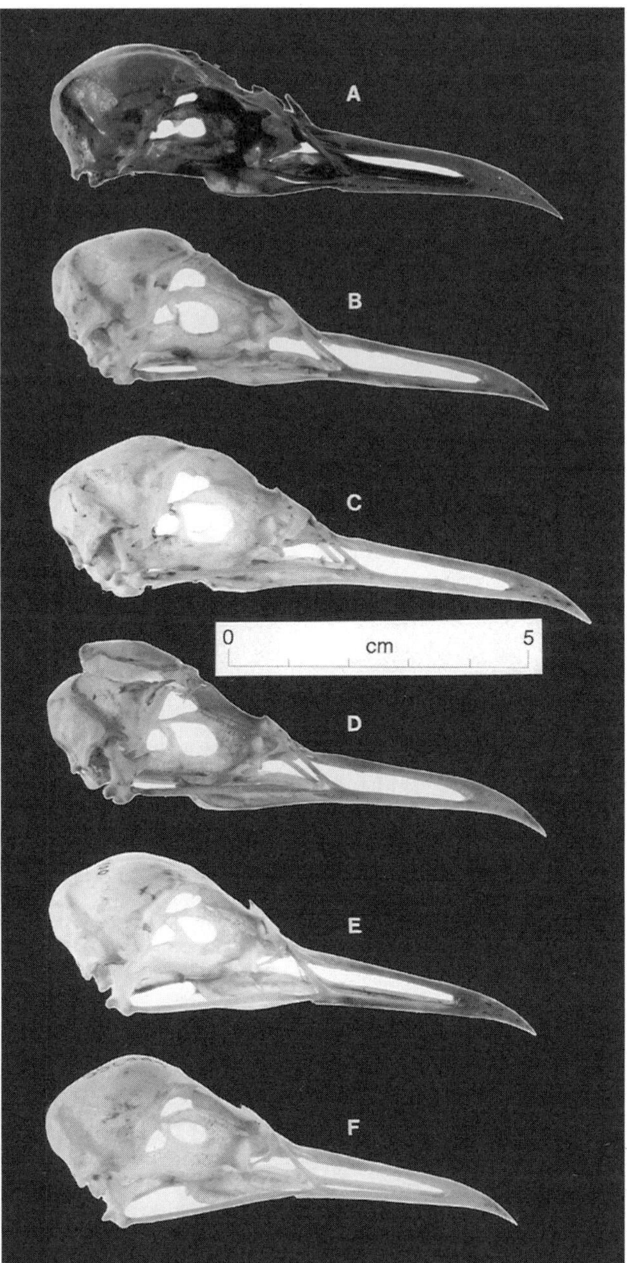

Figure 15-11 The skull of *Larus* in lateral aspect. A: *L. utunui*, DAPT 1, Fa`ahia archaeological site, Huahine, Society Islands. B: *L. novaehollandiae* ♂, USNM 347670, Northern Territory, Australia. C: *L. cirrocephalus* ♂, USNM 227346, Buenos Aires, Argentina. D: *L. atricilla* ♀, NYSM 44, Florida. E: *L. maculipennis* ♂, USNM 343100, Llico, Chile. F: *L. ridibundus* ♂, USNM 556297, Leiden, Holland. From Steadman (2002a: Figure 3).

widespread in Oceania today and were in the past as well. On the other hand, fossils portray large contractions (>1000 km) in breeding range since human arrival for the terns *Procelsterna cerulea* and *G. microrhyncha*. The former occurs in prehistoric sites on three islands in the Marianas (Table 15-4), 2500 km west of the nearest breeding population in the Marshall Islands. A sighting of *P. cerulea* off the west coast of Thailand (Eller 1989) hints that modern range extensions are also possible. *Gygis microrhyncha* is restricted today to the Marquesas but its bones have been found in prehistoric sites on Mangaia (2200 km SW), Easter Island (3300 km SE), ʻUiha, Haʻafeva, and ʻEua in Tonga (3600 km SW), and Tinian in the Marianas (8900 km W; Tables 15-2 to 15-4). *Gygis microrhyncha* was sympatric with *G. candida* (Figure 15-12) on these islands, thus reinforcing the distinctiveness of these two species (Pratt et al. 1987:186-187), which many authors have regarded as conspecific.

Figure 15-12 A chick of the white or fairy tern, *Gygis candida* on Midway Island. Photo by Sheila Conant, June 1999.

Discussion

Nesting Grounds

In Oceania today, most species of seabirds nest mainly on small, uninhabited islands where it is difficult for people to get ashore and where rats may be absent. For example, most of the surviving species of seabirds in the Marquesas, Rapa, or Easter Island breed today only on steep, uninhabited, offshore islets rather than on the major islands (Johnson et al. 1970, Garnett 1984, Thibault & Varney 1991a). These tiny offshore islands probably did not sustain most of the nesting seabirds in Remote Oceania before human arrival. Rather, the modern absence or scarcity of most nesting seabirds on substantial islands is due to long-term predation from people, rats, dogs, and pigs, as well as habitat loss. Overall seabird populations in Remote Oceania must be at least one or two orders of magnitude less than they were at human arrival, with most of the decrease from procellariids.

I am not claiming that each major island in Remote Oceania was covered with breeding seabirds, even though, at least outside of New Caledonia, Vanuatu, and Fiji, this may have been the case. Minimally, the more windward parts of islands, which often have stunted vegetation that is marginal habitat for most landbirds, were prime nesting areas for seabirds before people and rats took over. Many procellariids nest in burrows, and thus any areas with deep soils may have been used. Weathered limestone provides abundant crevices and other solution features that also were well suited for petrels, shearwaters, and storm-petrels, given the boney evidence from 'Eua, Mangaia, and elsewhere. Volcanic tuff also can form cliffs well suited for nesting seabirds. I believe that land area, elevation, and bedrock type were of minimal concern to most species of seabirds before human arrival.

Extinction vs. Extirpation

A major distinction in the anthropogenic decline of birds in Oceania is that in landbirds it features much extinction of species and extirpation of populations, whereas the seabird losses involve populations of extant species much more than extinction of species. In spite of examining tens of thousands of seabird bones from Oceania, the only extinct taxa that I have noticed are *Pterodroma* undescribed sp. and *Puffinus* undescribed sp. ('Eua), Procellariidae undescribed gen. & sp. (Easter Island), *Papasula abbotti costelloi* (Marquesas), and *Larus utunui* (Huahine; Tables 15-2, 15-3).

Further study of the seabird bones already in hand probably would disclose more extinct taxa, at least of procellariids. This immense task could be accomplished only by compiling, in one place, modern skeletal specimens of all extant forms, regardless of their current range. The one museum with holdings approaching this is the National Museum of Natural History, Smithsonian Institution (USNM). By borrowing specimens of the species inadequately represented at USNM, and then spending several months there immersed in modern and prehistoric bones of shearwaters, petrels, prions, and fulmars, one could identify many of the currently nameless prehistoric bones of procellariids. Difficulties, perhaps not insurmountable in most cases, would ensue within *Puffinus* and especially *Pterodroma*. I would admire and support anyone willing to undertake such an important project.

Distribution

The most striking message from Oceania's prehistoric seabird data is that, as with landbirds, modern distributions of most species are subsets of those that existed at human arrival. For seabirds this involves huge east-west range extensions, such as 11,200 km for *Papasula abbotti*, 8900 km for *Gygis microrhyncha*, 2500 km for *Procelsterna cerulea*, and comparable values for many species of Procellariiformes. Given how spotty the fossil coverage is, and how many islands lie within these range extensions, we face the likelihood that thousands of populations of seabirds have been lost. The limited fossil record already has disclosed ca. 200 lost populations (Table 15-5).

Especially in procellariids, the large range extensions involve both latitude and longitude. Many species that breed today only south of 31° S, or 36° S, or even 53° S, certainly or probably nested on Easter Island (27° S; Table 7-23) or Henderson Island (24° S) or, in the case of *Pachyptila* sp.

(not north of 36° S today), as far north as 9° S in the Marquesas. Thus the prehistoric data challenge even our notion of which species make up a tropical vs. subtropical vs. subantarctic seabird fauna. For the South Pacific, Ainley and Boekelheide (1983) defined these faunal regions by mean annual sea surface temperature (SST) and to some extent salinity (SSS), with tropical seabird faunas in areas with SST >22°C, subtropical 14–22°C, and subantarctic 4–14°C. All islands covered in this chapter are tropical except Easter and Rapa, which are subtropical. The modern distributions of species are very clean-cut by SST standards. The prehistoric distributions reveal much more overlap in the species that inhabit each faunal region.

Paleoecology

The expanded prehistoric distributions confound our understanding of the role of seabirds in marine food webs, a poorly understood topic that would benefit from more communication between seabird biologists and fisheries biologists (Croxall 1987, Cairns 1992). In the Pacific and other oceans, marine productivity increases as SST decreases (Longhurst 1998), with primary productivity in the equatorial Pacific thought to be limited by inadequate iron (Behrenfeld & Kolber 1999). The South Pacific gyre, stretching across most of Oceania from ca. 15° to 30° S, has been called "Earth's largest oceanic desert" (Claustre & Maritorena 2003) because of its low primary productivity. As top predators that usually are two to four trophic levels above phytoplankton, modern seabirds roughly track the gradients of temperature, salinity, and productivity in their overall numbers and species richness (Murphy 1936, Ainley & Boekelheide 1983, Brooke 2002), with considerable interannual regional variation related to ENSO (Schreiber & Schreiber 1984, Schreiber 2002; also see Chapter 1 herein). Ashmole (1963) believed that today's numbers of tropical seabirds are limited by food shortages during the nesting season, an idea supported by recent studies of temperate seabirds (Lewis et al. 2001). This concept is difficult to reconcile, however, with the diverse seabird faunas that existed in Earth's largest oceanic desert before human arrival. This dilemma may be fertile grounds for new research.

Furthermore, given that cooler SSTs are associated with increased primary productivity, the seabirds of Oceania probably had an improved food supply during glacial times when tropical SSTs were ca. 3 to 6°C lower than today (Barash & Kuptsov 1997, Beck et al. 1997). Primary productivity in the tropical Pacific was greater during glacial than interglacial intervals over the past 450,000 years, based on analyses in marine sediments of fossil plankton and their chemical components (Paytan et al. 1996, Kawahata et al. 1998). Therefore our modern tropical seabird faunas may be doubly depleted, first by existing during an interglacial interval, and second by being overrun by human impact. The latter is a much worse situation than the former.

A final observation is that seabirds can play a major role in terrestrial food webs by transporting marine-based energy and nutrients to islands (Mizutani & Wada 1988, Polis & Hurd 1996, Anderson & Polis 1999). Stable isotope ratios of carbon ($^{13}C/^{12}C$) and nitrogen ($^{15}N/^{14}N$) are substantially different on seabird vs. nonseabird islands off Baja California (Anderson & Polis 1998, Stapp et al. 1999). I speculate that essentially all islands in Polynesia (and, to a lesser extent, elsewhere in Oceania) once had a measurable input of marine nutrients from nesting seabirds. This idea could be tested on Mangaia, for example, by determining the stable isotope ratios of landbird bones before human arrival, early in human occupation, late in prehistory, and today. If the island formerly sustained abundant seabirds, as I would surmise from fossil evidence, then one would expect the isotopic signature of landbird bones to become increasingly terrestrial with time as Mangaia's seabird populations declined. This would be reflected in lower isotopic rations for both carbon and nitrogen (Fariña et al. 2003).

A decrease in marine nutrients may also have led to less vigorous growth of terrestrial vegetation on Mangaia and many other major Pacific islands. A low, flat island such as Lifuka (Tonga; 11.4 km^2, 16 m elev.) is not expected to sustain lush forest today. When Lapita people arrived 2850 years ago, however, Lifuka had from 27 to 31 species of landbirds (Table 6-19) and its forests probably were impressive. Nutrient input from at least 10 species of seabirds (Table 15-3) may have helped to fuel this forest. Based on my visits to Lifuka in 1996 and 1997, the three or four species of seabirds that still roost or nest there are uncommon and probably have a negligible effect on soil nutrients.

Part IV

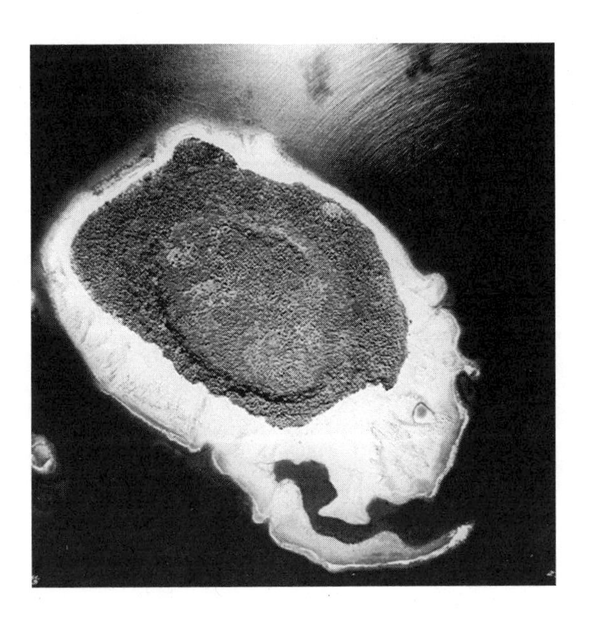

Chapter 16 Extinction

Extinction infiltrates every part of this book. By now you are probably getting sick of extinction. I am. I would give almost anything to experience the birdlife on even one major island in Oceania in its prehuman form. Just imagine a completely forested, rat-free Tongan island with 30+ species of landbirds and a dozen breeding species of seabirds. Time travels in just one direction, however, so our burden is to document and explain the losses that have occurred.

The extinction described in this chapter is genuine; it represents the permanent loss of a species or, in the case of extirpation, the long-term (usually, as far as we know, permanent) loss of an island population. As explained in Chapter 18, my brand of extinction does not include the ephemeral absences of species on continental islands (as in Diamond & May 1977, Simberloff 1983, Cook & Hanski 1995, Maurer 1999:160, and many other studies) that ecologists have called "extinctions."

Over the entire Phanerozoic (the past 543 million years), extinction rates in the marine fossil record correlate with origination rates that are roughly 10 million years later (Erwin 2000, Kirchner & Weil 2000, Jablonski 2003). In other words, an average of 10 million years has been needed to replace the marine biotic diversity lost to extinction, although participation in the recovery is highly variable taxonomically (Jablonski 2002). A roughly similar phenomenon has been suggested for Cenozoic (past 65 million years) mammals in North America (Webb 1989, Webb and Opdyke 1995). No data on such long timescales are available for vertebrates on oceanic islands, so it is difficult to speculate how long it may take for new colonization and evolution to substitute for the taxa lost to human impact in recent millennia. This is probably a moot point, at least in Remote Oceania, where most contemporary floras and faunas are too depleted in potentially island-hopping native species and so contaminated by alien species that it is unrealistic to expect unassisted colonization and evolution to replace the losses wrought by people. The oldest fossil avifaunas from Oceania are only ca. 100,000 years old, and they suggest low turnover rates since human arrival (Chapter 18).

Killing Birds on Pacific Islands

Birds were easily exploited by early human colonists. In Remote Oceania, birds evolved without mammalian predators and thus had little defense against humans and their rats, dogs, and pigs. Tameness is a well-known feature of island birds (Humphrey et al. 1987). Naïveté to human predation can develop even in continental species (Berger et al. 2001, Gittleman & Gompper 2001). Considering that most species of birds would have shown little or no fear of people at first contact, early birding in Oceania probably resembled gathering more than hunting. Megapodes, rails, pigeons, parrots, kingfishers, and passerines were obtained by hand and with snares, nets, bird lime, slings, and thrown rocks (Steadman 1997a). Bird populations were reduced further as indigenous forests were altered through burning, cutting down trees, and the introduction of nonnative plants.

Herons, shorebirds, and terns were hunted along the reefs, beaches, and estuaries. Petrels, shearwaters, storm-petrels, and tropicbirds were plucked from their nesting burrows and crevices. Boobies and frigatebirds were taken from their nests, on the ground or in low shrubs.

Based on bone counts from their sites, the Pleistocene colonists of Near Oceania ate native bats and rodents much more than birds (White et al. 1991). In Remote Oceania, birds were an important source of fat and protein for the first people (Dye & Steadman 1990). An island overflowing with birds must have been a blessing after days or weeks at sea. The need to produce food through agriculture, fishing, and domesticated animals would have increased after indigenous birds were depleted and human populations grew. Aside from supplying easily obtainable sustenance for landing parties and during the early period of human occupation, seabirds probably helped navigators to locate islands in the first place (Lewis 1972, Steadman 1997a). Flocks of seabirds would have alerted sailors to the nearby presence of land, especially during the largely unidirectional flights in the morning or late afternoon of seabirds commuting between nesting islands and feeding areas.

Although people were responsible for depleting indigenous avifaunas, some and perhaps most of the losses were indirect and beyond human control. Rats, for example, prey on eggs, nestlings, and adults of birds regardless of any human wishes to the contrary. Although they killed and ate lots of birds, Oceania peoples probably did not want most of the species to become rare or extinct. Even an islandwide prohibition on hunting certain species may have been in vain because of rats and deforestation.

The birds people killed ranged from the smallest passerines (<10 g) to seabirds such as the booby *Papasula abbotti costelloi* (2000 g), to immense landbirds such as the flightless megapode *Sylviornis* (30–40 kg). The killing of some species may have been *tapu* (forbidden) to commoners on some islands. Nevertheless, based on archaeological and ethnobiological evidence, nearly all birds were eaten except, locally, herons and some small insectivores such as cuckoos, swifts, and certain passerines (Steadman 1997a). Frugivores, granivores, and nectarivores such as megapodes, pigeons, doves, parrots, honeyeaters, and starlings, were especially desirable, as were fat nestlings of seabirds. Even birds of limited palatability might be taken for their feathers or bones.

All of this is compatible with rapid prehistoric extinction, such as in the overkill and blitzkrieg models of Paul Martin (Martin 1958, 1967, 1984, 1990, Mosimann & Martin 1975, Martin & Steadman 1999). These models have intuitive appeal on oceanic islands, where species tend to be more vulnerable than on continents because of small land areas and population sizes, low rates of increase, naïveté to predators, and little resistance to pathogens (Diamond 1985). Since AD 1500, we have witnessed numerous overkill-like events—the extinction of multiple species of birds (and other vertebrates) after the first peopling of an island that was unoccupied prehistorically (Diamond 1984, Olson 1989, Steadman et al. 1991, MacPhee & Flemming 1999).

The differences between overkill and blitzkrieg can be subtle. Overkill simply refers to a species being hunted at a rate beyond replacement, eventually resulting in extinction. The time required for overkill is usually short in a geological sense even if often long (centuries, millennia) in an ecological sense. Blitzkrieg is a rapid form of overkill where an advancing front of people leaves no viable populations of targeted species in its wake. Blitzkrieg also features geometric population growth in people who hunt selected species beyond their need for nourishment (Mosimann & Martin 1975). Both the overkill and blitzkrieg models were developed primarily to explain the loss of most large mammals in North America at the end of the Pleistocene. Modern examples of potential blitzkrieg would be the annual predation rate of ca. 34% (from commercial fishing) for the Pacific leatherback (*Dermochelys coriacea*), leading to predictions of extinction within decades (Spotila et al. 2000), or the overfishing of many species of marine fishes (Hutchings 2000). If extinction occurs, the elapsed time of these events would be instantaneous in radiocarbon time.

Overkill and blitzkrieg remain controversial for continental mammals (e.g., Graham et al. 1996), although both models receive support from diverse authors (e.g., Alroy 1999, Flannery & Roberts 1999, Haynes & Eiselt 1999, Miller et al. 1999, Owen-Smith 1999). Rapid, prehistoric, human-caused extinction is easier to envision on islands (even huge, continental islands such as Madagascar; Burney et al. 2003) than on continents. As I will describe below, the elapsed times for avian extinction on oceanic islands resemble those for continents in varying from geologically instantaneous, blitzkrieg-like events to much slower processes. Teasing apart this variation provides clues about what may have caused the extinction.

Prehistoric Extinction in Oceania

Background Extinction

Large samples of fossils from the Galápagos Islands suggest that the background (prehuman) rate of extinction during the past 8000 years was hundreds of times less than after people arrived several centuries ago (Steadman et al.

1991). Unlike in Oceania, the Galápagos Islands had no people prior to their discovery by Europeans in AD 1535 (Steadman 1986). As a result, human impact there is confined to the past five centuries and seems to have been minor until permanent settlements were established in the 19th century. The Holocene (past 10,000 years) fossil record in the Galápagos comprises about 500,000 bones, more than 90% of which predate human arrival. Bone-based prehistoric faunas (from five islands) reveal the loss of only 0–3 vertebrate populations in 4000–8000 years before human arrival, compared to 21–24 populations lost from these same five islands in the past 150–300 years. Thus, the extinction rate in the Galápagos increased at least 100-fold when people arrived. If undisturbed by people, the natural processes of dispersal, colonization, and evolution result in a very low rate of vertebrate extinction on tropical oceanic islands (Chapters 17, 18).

Unlike in the Galápagos, prehistoric people lived on nearly all islands in Oceania. Human presence over the past 33,000+ years (Near Oceania) or 3000 years (Remote Oceania) has wiped out thousands of populations and species of birds that otherwise would exist today. As in the Galápagos, there is almost no evidence for avian extinction on high islands in Oceania before human arrival (James 1987, 1995, Steadman 1993a, Worthy & Holdaway 2002). On some atolls, Quaternary changes in sea level would drown out or severely deplete the terrestrial biota. Fossil records have yet to document this in Oceania, unlike on Bermuda (Western Atlantic) where an extremely high sea level stand (+21 m), perhaps ca. 400,000 years ago, seems to have obliterated four endemic species of birds (a duck, crane, and two rails; Olson & Wingate 2000, Hearty et al. 2004).

After people arrived, some prehistoric extinctions in Oceania occurred within a century or less. Others required millennia, although either time frame is rapid in a geological sense. Especially devastating to birds were the Lapita people. These first settlers of Remote Oceania moved eastward from the Bismarcks (3500 cal BP) to Tonga and Samoa by 2850 cal BP (Irwin 1992, Kirch 2000, Steadman et al. 2002b; Chapter 3 herein). They were horticulturalists who also killed a wide range of marine and terrestrial animals. Their activities forever changed insular landscapes and biotas (Kirch 1997). The East Polynesian expansion (1500 to 700 cal BP; Kirch 2000) was comparably destructive to indigenous flora and fauna (Steadman & Kirch 1990). The avian losses are believed to be a result of predation by humans and associated nonnative vertebrates (especially rats, dogs, and pigs), habitat changes (loss or alteration of indigenous forests through cutting, burning, and the spread of nonnative plants), and introduced pathogens (Olson & James 1982, Kirch 1983, van Riper et al. 1986, Steadman 1989a).

Prehuman bone assemblages, such as that from ‘Eua, Tonga (Chapter 6), also can help to assess anthropogenic extinction because it is unlikely that any cultural site, even one with abundant bird bones from first human contact, has sampled all of the species that existed on that island at that time (Chapter 4). The most vulnerable species may be lost within decades or a century or two of human arrival, leaving behind few if any bones in archaeological sites. To understand patterns of avian exploitation, one would like to know if the loss required years, decades, centuries, or millennia. From an evolutionary or geological standpoint, knowing the speed of extinction is less critical because all of these timescales are short.

Volcanic eruptions probably are only a minor source of extinction in Oceania. The one known case involves islands very close to the rich New Guinean source area (Diamond 1981). Biotically devastating eruptions have not been documented in Remote Oceania during the past two centuries. In Tonga, for example, the major eruption of Niuafo‘ou in 1946 resulted in temporary human abandonment of the island (Rogers 1986) but there is no evidence that this eruption, or those in 1886, 1929, or 1943 (Simkin et al. 1981) eliminated any bird populations. By far the greatest relative abundance of landbirds in the Ha‘apai Group of Tonga is on Tofua, another active volcano (Steadman 1998). This is not to say that fauna-obliterating Krakatau-like eruptions never have occurred in Remote Oceania; surely they have. A possible example is the massive Kuwae eruption in Vanuatu (ca. AD 1452) that devastated Epi and separated it from Tongoa (Robin et al. 1994). We have not witnessed such an event in the past 200 years, during which many extinctions and extirpations have taken place, each certainly or probably linked to people. Thus, as in the Galápagos Islands, it seems likely that anthropogenic extinction in Remote Oceania, and perhaps even in Near Oceania, exceeds background extinction by at least two orders of magnitude.

To take this a step further, I am unaware of any avian extinction by volcanos, whether in the West Indies, Canary Islands, Mascarenes, Solomons, Vanuatu, Tonga, New Zealand, Galápagos, or the extinction-prone Hawaiian Islands. (Krakatau lies in a continental setting and, as far as is known, harbored no endemic species of birds before it erupted in 1882; Thornton 1996). In Remote Oceania, no eruptions of the past two centuries are known even to have extirpated an island's population of a more widespread landbird. Many eruptions on basaltic islands consist of nonexplosive lava flows. Where major explosive eruptions have occurred, such as on Montserrat since 1995 (Voight et al. 1999), the only endemic species of bird, the oriole *Icterus oberi*, has persisted (several thousand on the northern half of the island; Atkinson & Gibbons 1998). Montserrat also erupted in AD 1646 (Simkin et al. 1981). Bones from an archaeological site show that *I. oberi* was present on Montserrat ca. 2000 years ago (Reis & Steadman 1999), pushing back the time frame for coexistence of volcanism and apparent endemism. I say "apparent" because I would not be surprised to find prehistoric bones of *I. oberi* on nearby Antigua or Barbuda.

Variable Timing of Human-caused Extinction

In Near Oceania, where human history extends to 30,000+ years ago (Chapter 5), what little we know about avian extinction (mainly on New Ireland, Buka, and Mussau) suggests a protracted series of losses over many millennia rather than a single dramatic event. The zooarchaeological data argue against blitzkrieg in Near Oceania, at least on large, topographically complex islands (Steadman et al. 1999b; Table 5-5 herein). What the case may have been on smaller islands in this region is unknown.

In Remote Oceania, human presence begins with the arrival of Lapita people at 3000–2800 cal BP (Kirch 2000). At the eastern edge of Lapita occupation, in the Ha'apai Group of Tonga, the extinction of birds has good chronostratigraphic control. On the small (1.8–13.3 km^2), low (elev. 10–20 m) islands of Foa, Lifuka, Ha'ano, 'Uiha, and Ha'afeva, the Lapita people left behind rich deposits of pottery, other artifacts, charcoal, shell, and bone (Burley et al. 1995, 1999). From these beach sites I have sorted >100,000 bones, dominated by fish, but also including 1448 identifiable landbird bones. At least 23 to 31 species of landbirds existed at human contact on each island, far exceeding the 10 to 14 species per island today and approximating the 30+ species known prehistorically from the much higher, larger Tongan island of 'Eua (Steadman 1993a, 1998; Table 6-21, Figure 6-14 herein).

Bones of extinct birds are abundant in Lapita contexts at the Ha'apai sites but are virtually absent in Polynesian plainware contexts (Steadman et al. 2002a,b; Table 16-1, Figure 3-10 herein). The ^{14}C dates from Polynesian plainware contexts usually overlap (at 1 or 2σ) those from the underlying Lapita strata. This suggests that the loss of most species of landbirds was so rapid as to be within the margin of error in ^{14}C dating, i.e., almost certainly less than 200 years and perhaps less than 100 years (Steadman et al. 2002b). After the short period of intense extinction, the species that survived were essentially the same ones that I surveyed in 1996, reflecting a bird community nearly unchanged in species composition over the past 2600 to 2700 years.

Table 16-1 Indigenous landbirds recorded from archaeological sites on five islands in the Ha'apai Group, Tonga. Based on data in Chapter 6. Lapita sites date to ca. 2900–2700 cal BP. Polynesian plainware and later sites date to <2700 cal BP.

	Lapita contexts	Polynesian plainware and later contexts
All species	34	5
Extinct/extirpated species	21–22	1
Number of identified bones	1448	500+

The islands in Ha'apai are well suited for rapid and extensive extinction. They are small, flat, low, and covered with fertile soils (Dickinson et al. 1994). They lack native mammals except bats. The 100,000+ bones prove that the Lapita colonists hunted and fished indigenous fish, sea turtles, iguanas, birds, and fruit bats, as well as brought chickens, rats, dogs, and pigs to the islands (Steadman et al. 2002b). These people also introduced nonnative plants and cut and burned the original forests (Kirch 2000). They began the overwhelming human presence that continues today. On these small islands, survival of landbirds is more difficult to comprehend than extinction.

In East Polynesia (Easter, Henderson, Marquesas, Society, Cook Islands), most landbird extinction occurred from 1000 to 500 cal BP (Steadman 1995a, Steadman & Rolett 1996). The environmental and cultural contexts for these losses are well studied on Mangaia, Cook Islands (Chapters 3, 7). In a framework of nearly 50 ^{14}C dates, the cultural deposits at Tangatatau Rockshelter depict much of the decline of landbirds on Mangaia from at least 20 species at 1000 cal BP to five species surviving today (a widespread heron, duck, and rail, an endemic kingfisher and warbler). Most extinction on Mangaia was accomplished by 700 to 600 cal BP (Steadman & Kirch 1990, unpub. data, Steadman 1995b, Kirch 1996; Figure 16-1 herein). Another East Polynesian island with good chronostratigraphic control on avian extinction is Tahuata (Marquesas), where seven species of rails, pigeons, and parrots are confined to cultural strata older than 500 cal BP (Steadman & Rolett 1996; Figure 16-2 herein).

How Many Species Were Lost?

At the level of population or species, the anthropogenic loss of birdlife in Oceania may be the largest vertebrate extinction event ever detected. Previously, I estimated that roughly 2000 species of birds, dominated by flightless rails, have been lost in Oceania since its peopling (Steadman 1995a).

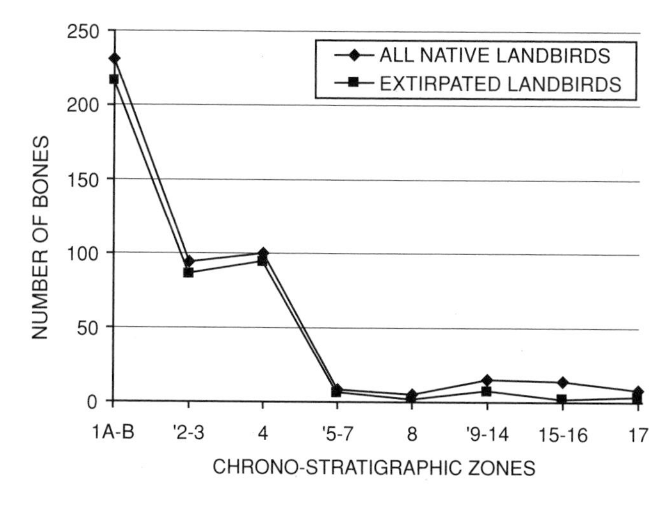

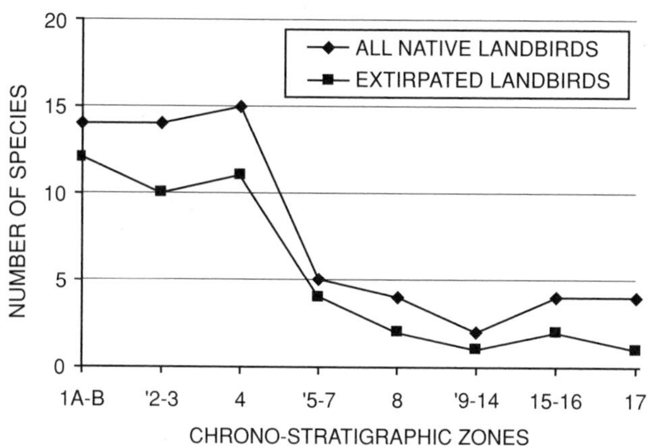

Figure 16-1 Chronostratigraphic summary of landbird decline at Tangatatau Rockshelter (site MAN-44), Mangaia, Cook Islands. From data in Chapter 7.

With more data, I now would estimate, conservatively but still crudely, that 559 to 1696 species of nonpasserine landbirds and 82 species of passerines have been lost since human arrival in Oceania, excluding the Hawaiian Islands and New Zealand (Tables 16-2 and 16-3). Rails are the largest source of uncertainty in the estimate (see Chapter 10). Using different methods, Curnutt and Pimm (2001) estimated ca. 1200 extinct species of Pacific landbirds.

Adding seabirds as well as the species lost from New Zealand, my overall estimate of extinct species of Pacific island birds is 745 to 1882 (Table 16-4). The total number of extirpated populations may be in the neighborhood of 10 times greater, in part because it includes many (most?) extant species as well as all extinct ones. Improved fossil data will continue to refine these estimates. The Hawaiian Islands would add perhaps 70 or 80 extinct species, lifting my estimate of extinct species on Pacific islands to a range from 820 to 1960 species.

Factors That Affect Extinction Rates on Islands

The three sets of factors that influence anthropogenic extinction of birds on oceanic islands were presented as the ABC model by Steadman and Martin (2003). Each factor, as explained here, affects others and may vary among islands (Table 16-5, Figure 16-3). The factors can be evaluated for any oceanic island or island group.

Abiotic Factors

These inherent physical characteristics of an island provide the background for biological and cultural factors.

A1. Island size (land area).

On larger islands it should take longer for human impact to spread across the entire island. In the case of introduced predators (rats, cats, monitor lizards, snakes, etc.) or

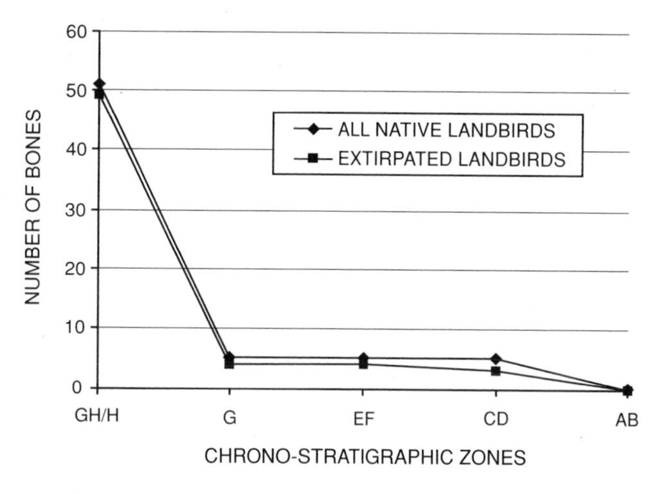

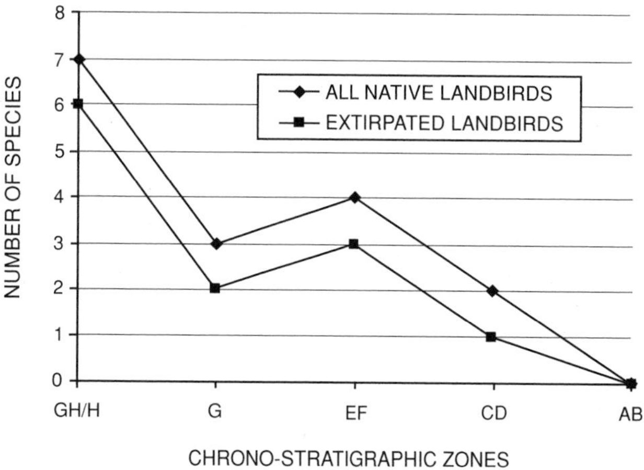

Figure 16-2 Chronostratigraphic summary of landbird decline at the Hanamiai archaeological site, Tahuata, Marquesas. From data in Steadman & Rolett (1996).

Table 16-2 Estimated numbers of living and extinct species of resident nonpasserine landbirds in Oceania, excluding the Hawaiian Islands and New Zealand. The values in columns A and B are based on data in Chapters 5–13. The values in column C are based on my estimate of incompleteness in the fossil record.

Family	A Living species	B Known extinct species	C Estimated total extinct species	D Estimated total species (A + C)
GREBES	2	0	0	2
CORMORANTS	2	0	0	2
HERONS	5	4	8	13
IBISES	1	0	0	1
DUCKS	6	3	6	12
OSPREYS	1	0	0	1
HAWKS	15	2	4	19
FALCONS	3	0	0	3
MEGAPODES	4	11	22	26
QUAILS	2	0	0	2
BUTTON-QUAILS	2	0	0	2
RAILS	15	47	442–1579	457–1594
KAGUS	1	1	1	2
JACANAS	1	0	0	1
STILTS	1	0	0	1
THICK-KNEES	1	0	0	1
PLOVERS	1	0	0	1
SANDPIPERS	1	7	10	11
PIGEONS & DOVES	61	18	36	97
PARROTS	33	9	18	53
CUCKOOS	7	0	0	7
BARN-OWLS	4	1	2	6
TYPICAL OWLS	6	0	2	8
OWLET-NIGHTJARS	0	1	1	1
FROGMOUTHS	1	0	0	1
NIGHTJARS	3	0	0	3
TREE-SWIFTS	1	0	0	1
SWIFTS	7	1	3	10
KINGFISHERS	19	1	2	21
BEE-EATERS	1	0	0	1
ROLLERS	1	0	0	1
HORNBILLS	1	1	2	3
TOTAL SPECIES	209	107	559–1696	770–1907

pathogens (such as avian malaria or pox), the differences in dispersal times across small vs. large islands may be minor (only years vs. decades).

A2. Topography (relief)

Extreme steepness (high relief) may deter extinction from deforestation, cultivation, and hunting. In Oceania, as throughout much of the world, most relatively flat areas no longer sustain mature forest. Topographic constraints on human activities tend to increase with island size simply because large islands are more likely to have more extensive areas that are very steep.

Table 16-3 Estimated numbers of living and extinct species of resident passerine birds in Oceania, excluding the Hawaiian Islands and New Zealand. The values in columns A and B are based on data in Chapters 5–8 and 14. The values for column C, based on my estimate of incompleteness in the fossil record, are even more speculative than in Table 16-2.

Family	A Living species	B Known extinct species	C Estimated total extinct species	D Estimated total species (A + C)
PITTAS	4	0	0	4
SWALLOWS	1	0	0	1
THRUSHES	4	0	5	9
OLD WORLD FLYCATCHERS	1	0	0	1
CISTICOLAS	1	0	0	1
OLD WORLD WARBLERS	22	1	15	37
SUNBIRDS	2	0	0	2
FLOWERPECKERS	3	0	0	3
WHITE-EYES	30	1	15	45
STARLINGS	17	4	15	32
ESTRILDID FINCHES	8	1	5	13
CUCKOO-SHRIKES	12	0	1	13
GERYGONES	1	0	0	1
AUSTRALIAN ROBINS	3	0	0	3
WHISTLERS	7	0	0	7
FANTAILS	14	0	1	15
MONARCHS	44	2	15	59
HONEYEATERS	32	0	5	37
DRONGOS	3	0	0	3
WOOD-SWALLOWS	2	0	0	2
CROWS	5	1	5	10
TOTAL SPECIES	216	10	82	298

Table 16-4 Estimated numbers of living and extinct species of resident birds in Oceania, including all island groups except the Hawaiian Islands. *, number adjusted downward because of species shared between New Zealand and Tropical Oceania. Based on data in Tables 16-2, 16-3, Chapter 15 (tropical Oceania), and Holdaway et al. (2001; New Zealand and associated islands).

	Landbirds				
	Nonpasserines	Passerines	Total	Seabirds	All species
NEW ZEALAND					
A. Living species	55	31	86	90	176
B. Known extinct species	47	15	62	4	66
C. Estimated total extinct species	62	20	82	6	88
D. Estimated total species (A + C)	117	51	168	96	264
TROPICAL OCEANIA					
A. Living species	209	216	425	50	475
B. Known extinct species	105	10	115	6	121
C. Estimated total extinct species	561–1696	82	643–1778	16	659–1794
D. Estimated total species (A + C)	770–1905	298	1068–2203	66	1134–2269
TOTAL					
A. Living species	261*	246*	507*	130*	637*
B. Known extinct species	152	25	177	10	187
C. Estimated total extinct species	623–1758	102	725–1860	22	747–1882
D. Estimated total species (A + C)	884–2019*	348*	1232–2367*	152*	1384–2519*

A3. Bedrock type

This factor influences topography and soil types. Limestone often features pinnacles, crevices, and sinkholes that can deter human activity and thus delay extinction. Cliffs that separate limestone terraces often remain forested, as in the Vava'u Group of Tonga (Franklin et al. 1999, Steadman & Freifeld 1998). Volcanic rocks may weather into knife-edge ridges that cannot be cultivated because of extreme steepness and inadequate soils, as in the Marquesas (Rolett 1998). By contrast, gentle volcanic slopes, such as those on most of Easter Island, encourage soil development and thus human activity.

A4. Soil type

Indigenous plant communities rarely survive where soils are well suited for cultivation. Deforestation of the raised limestone islands in the Ha'apai Group of Tonga, for example, is due to the rich soils created by volcanic ashes rather than the much poorer soils that would otherwise form on limestone (Dickinson et al. 1994). The extremely sandy soils of atolls are unsuitable for most crops (Ayres & Haun 1990).

A5. Isolation

Highly isolated islands tend to have depauperate indigenous plant and animal communities. After people arrive, high levels of isolation also lead to a greater intra-island dependence for natural resources, as in the extreme resource sink on Easter Island (Flenley et al. 1991, Steadman et al. 1994). These factors together facilitate overexploitation of indigenous species. While isolation may delay human colonization, nowhere in the tropical Pacific has the delay been sufficient to preserve a truly intact flora and fauna into the time of scientific exploration.

A6. Climate

Forests on seasonally dry islands are easier to burn than those on very wet islands. Conversely, droughts can restrict agriculture (as on Micronesian atolls or the Marquesas Islands; Kirch 2000) and thus limit human populations.

Indigenous Biological Factors

These factors pertain to the native plant and animal communities that existed on any island at first human contact.

B1. Floristic diversity.

Having more species available, an island with a richer flora might be exploited in a way that spreads human impact across many species of plants. On the other hand, any flora is likely to include naturally rare species that, if subjected

Table 16-5 Summary of factors that influence extinction of vertebrates on oceanic islands after human arrival. See text for further explanation. The human factors are modeled in Figure 16-2. From Steadman and Martin (2003).

	Potentially Promotes Extinction	Potentially Delays Extinction
ABIOTIC FACTORS		
A1. ISLAND SIZE	small	large
A2. TOPOGRAPHY	flat, low	steep, rugged
A3. BEDROCK TYPE	sandy, or non-calcareous sedimentary	limestone or knife-edge volcanics
A4. SOIL TYPE	nutrient-rich	nutrient-poor
A5. ISOLATION	very isolated	many nearby islands
A6. CLIMATE	seasonal aridity	reliably wet
INDIGENOUS BIOLOGICAL FACTORS		
B1. FLORAL DIVERSITY	depauperate	rich (short-term delay only)
B2. FAUNAL DIVERSITY	depauperate	rich (short-term delay only)
B3. TERRESTRIAL MAMMALS	absent	present
B4. MARINE RESOURCES	depauperate; difficult access	rich (temporary delay only); easy access
B5. SPECIES-SPECIFIC ECOLOGICAL, BEHAVIORAL, OR MORPHOLOGICAL TRAITS	ground-dwelling; flightless; large; tame; fatty; good taste; colorful feathers; long & straight bones	canopy-dwelling; volant; small; wary; little fat; bad taste; drab plumage; short & curved bones
CULTURAL FACTORS		
C1. OCCUPATION	permanent	temporary
C2. SETTLEMENT PATTERN	island-wide	restricted (coastal)
C3. POPULATION GROWTH AND DENSITY	rapid growth; high density	slow growth; low density
C4. SUBSISTENCE	farmers as well as h-f-g	h-f-g only, especially if marine oriented
C5. INTRODUCED PLANTS	many species; invasive	few species; noninvasive
C6. INTRODUCED ANIMALS	many species; feral populations	few or no species; no feral populations

to specialized exploitation, would be easy to exterminate. Strict dependencies of a species of bird on a plant depleted by humans could spell trouble for the bird. On an island in the Solomons with 20 species of *Ficus* (figs), for example, the loss of one species of fig would be unlikely eliminate any frugivorous birds, whereas the loss on an atoll of the only indigenous species of *Ficus* could wipe out a population of fruit-doves (*Ptilinopus* sp.). The nonrandom distribution of tropical forest trees (Condit et al. 2000) might facilitate their exploitation. Regardless of the distribution of rarity in an island flora, sustained exploitation of rich plant communities eventually will deplete or eliminate more species than in depauperate ones.

B2. Faunal diversity

As with floral diversity, being part of a relatively rich fauna is not necessarily an advantage or disadvantage for a species once people arrive. The overall richer terrestrial vertebrate faunas in Near Oceania, however, did lead to much more relative exploitation of bats and rodents versus birds than in Remote Oceania. On the other hand, the presence of large iguanas and a variety of edible bats in Fiji and Tonga did little to prevent massive prehistoric avian extinction (Chapter 6). Indigenous birds from islands with low diversities of pathogens and disease vectors might be more vulnerable to disease when nonnative pathogens and biting dipterans arrive with humans (Steadman et al. 1990a). The details of how this plays out across Oceania, however, are unresolved (Frank et al. MS submitted; Chapter 21 herein).

B3. Presence/absence of nonvolant terrestrial mammals

On islands without indigenous non-chiropteran mammals (as in all of Remote Oceania), native birds may have little defense against alien predators (humans, rats, dogs, etc.).

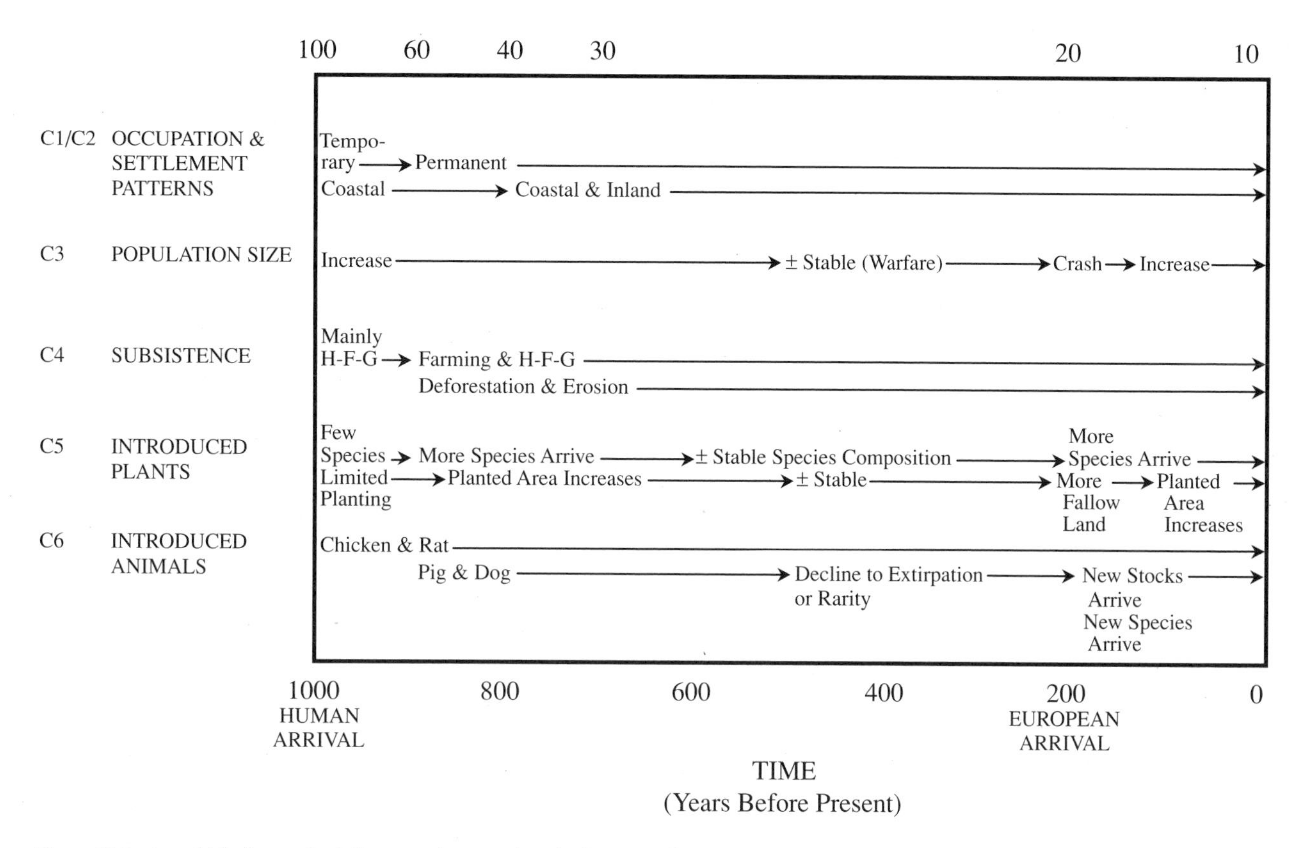

Figure 16-3 A model for factors that influence anthropogenic extinction on Pacific oceanic islands. See text for additional details. Human arrival is arbitrarily set at 1000 years ago, a reasonable estimate for many East Polynesian islands. Earlier arrival dates would apply for most of Melanesia, West Polynesia, and Micronesia. From Steadman and Martin (2003).

In Oceania, only the major islands in the Bismarcks and Solomons (Near Oceania) have native rodents (Flannery 1995a). Native marsupials inhabit mainland New Guinea but not Oceania (Flannery 1995a,b). Thus I suspect, although it is unproven, that the birdlife in Near Oceania averages less naïve to nonnative predators than in Remote Oceania.

B4. Marine resources

Rich marine resources might lessen the short-term dependence on terrestrial resources. In the long run, however, rich marine resources would fuel human population growth, thereby depleting terrestrial resources as well. Furthermore, the bone assemblages from all Lapita sites in Tonga are mainly from marine species (fish, sea turtles) even in the lower levels alongside bones of extinct iguanas, bats, and birds. An emphasis on terrestrial rather than marine resources would result if access to the sea were restricted by intertribal violence (Dye 1990) or unfavorable natural conditions (cliffs, no reefs, windward coasts, storms, rough seas, etc.).

B5. Species-specific ecological, behavioral, or morphological traits

Island birds are renowned for their tameness (Grant & Grant 1979, Barre & Barau 1982, Jehl & Parkes 1983, Humphrey et al. 1987), a trait that does not serve them well after human arrival. Vestiges of this tameness probably remain in the species that have survived. We should keep in mind, however, that the tamest species of island birds are now extinct and that the survivors probably are not as tame as they were at first human contact. One example of how quickly birds can modify their behavior in the presence versus absence of hunting is the extreme tameness of ocellated turkeys (*Meleagris ocellata*) at Tikal National Park, Guatemala, compared to nearby populations that are hunted (Steadman et al. 1979). Birds living in the forest canopy are typically more difficult to hunt than those in the understory or on the ground. Flightlessness is an obvious vulnerability when nonnative predators arrive. Although volant, most species of seabirds nest on the ground and thus are highly vulnerable. Certain birds may be favored (megapodes, pigeons, parrots) or avoided (herons, cuckoos, warblers) by hunters because of their fattiness, taste, plumage color, shape and size of bones (for making tools and ornaments), song, spiritual meaning, or other traits (Burley 1996, Steadman 1997a). As is true elsewhere (Kay 1997, Winterhalder & Lu 1997), the rarity of favored species in Oceania probably did not prevent traditional hunters from killing them.

Cultural Factors

These closely interrelated factors concern the changing socioeconomic structure, demography, and cultural ecology of prehistoric peoples (Figure 16-3). Levels of sociopolitical complexity (see Kirch 1990, 2000) influence these conditions.

C1. Permanent vs. temporary occupation

Permanent settlement of an island may lead to the introduction of more nonnative species of plants and animals than a temporary occupation, such as using an island as an occasional hunting and fishing outpost. While temporary occupation would not preclude introducing nonnative species, or burning forest in the dry season, these activities would tend to take place less than on permanently inhabited islands.

C2. Settlement pattern

The earliest habitation sites in Remote Oceania are coastal (Kirch 1986, 1996, 2000), although the interiors of islands undoubtedly were used for hunting and cultivation fairly early if not immediately after initial settlement. On many islands habitation sites eventually expanded to the island interiors, perhaps as a response to increased human population. Exploitation of natural resources under these circumstances would have been intense island-wide (Kirch & Ellison 1994).

C3. Population growth and density

Islands with extensive settlement data show evidence of population increases with time, even though prehistoric controls on population included resource shortages, disease, warfare, infanticide, and others (Firth 1936:373–374, Kirch 1990). How long human malaria has been endemic in Oceania is an important but unresolved issue in this regard (Tishkoff et al. 2001). Population growth would lead to increased resource consumption, including agriculturally fueled deforestation. An extreme case would be Easter Island, where deforestation was accelerated by the need for timber to move hundreds of massive stone statues (Flenley et al. 1991, Van Tilburg 1994). Evidence of prehistoric deforestation exists across Oceania (Ellison 1994a, Athens 1997). On many islands, however, the most vulnerable species of birds may already have been wiped out before the onset of large-scale agriculture.

C4. Horticulturalists vs. hunter-fisher-gatherers

Lapita and later peoples in Remote Oceania were horticulturalists with much hunting-fishing-gathering (h-f-g) activity as well. People who plant crops might develop larger populations and cut more forest than those with an exclusively h-f-g economy. Continental h-f-g peoples, however, are known to set fires to herd game animals (Forbis 1978, Flood 1983:168, Steele & Baker 1993, Lourandos 1997:95–97). Similar fires may have been used on some islands in Oceania to move megapodes, ground-dwelling columbids, and flightless rails toward hunters with nets, snares, slings, rocks, and quick hands.

C5. Introduced plants

Prehistoric peoples transported many species of useful but alien plants through Oceania (Kirch 1994, 2000). These plants ranged from yams, sweet potatoes, aroids, sugar cane, and bananas to trees such as candlenut, breadfruit, Malay apple, etc. Deforestation, especially through dry-season burning, exposed tillable soils for cultivating these plants (Athens 1997), most of which are of little use to native birds. Similar prehistoric situations have been documented on temperate islands (Hannon & Bradshaw 2000) and continents (Denevan 1992).

C6. Introduced animals

Four species of nonnative vertebrates accompanied people to most of Remote Oceania: chicken (*Gallus gallus*), Pacific rat (*Rattus exulans*), dog (*Canis familiaris*), and pig (*Sus scrofa*). Their prehistoric range in Micronesia is patchy and poorly documented. Each can form feral populations, although only the rat did so consistently. Chickens possibly transmitted disease to native birds (Chapter 21), while the nonnative mammals preyed on birds and disrupted forest succession. New Guinean marsupials were transported prehistorically to Near Oceania (Flannery 1995a,b). A larger species of rat, *R. praetor*, also was taken as far east as Fiji (Flannery et al. 1988, Matisoo-Smith et al. 1998, Matisoo-Smith & Allen 2001).

Some of the past impacts of the six cultural factors are difficult or impossible to quantify. For example, an island's understory vegetation might have been devastated by feral pigs from 600 to 500 years ago, only then to have every pig killed within a decade by a few dedicated pig hunters, followed by recovery of the understory vegetation. Ground-dwelling or shrub-nesting birds may have been extirpated during the pig outbreak on this island, which today might have an understory suitable for these birds if they still existed. The prehistoric recovery of the vegetation would be changed if seed predation by pigs (and rats) influences forest composition, a topic that only is beginning to be investigated (McConkey et al. 2002).

Alien impacts also can be difficult to evaluate over shorter time frames. On 16 January 1995, for example, megapodes (*Megapodius laperouse senex*) were still common on Ulong Island (Palau) in spite of an amazingly high concentration of black rats (*Rattus rattus*). During the first three hours of darkness on that hot, rainy, miserable evening, I killed 29 rats with snap-traps, all within a 15-m radius of my tent, on a forested beach occupied by the megapodes. Because the reproductive success of this megapode population is unknown, I cannot say whether the hyperabundance of rats is sustainable for the megapodes, although it would seem unlikely.

While predation on birds by prehistoric people may result in well-preserved bone deposits, predation on birds by rats has little chance of leaving behind boney remains that could be recovered hundreds or thousands of years later. Thus the intensity of prehistoric predation by rats (or dogs or pigs) cannot be inferred from three-dimensional evidence. The chronology of the arrival of rats versus that of extinction of birds provides, however, some evidence for the relative importance of rat predation in the decline of birdlife. In Tonga, rats (*Rattus exulans*) are recorded at first human contact (Lapita sites, ca. 2850 cal BP), and most extinction occurred within the next century or two. In the Mariana Islands, by contrast, bones of *R. exulans* (or any other rat) are absent from strata older than ca. 1000 cal BP, even though people arrived at ca. 3400 cal BP (Chapters 3, 8). That people occupied the Marianas for more than 2000 years before rats arrived may explain why flightless rails survived so long there (Chapter 21). Athens et al. (2002) suggest that *R. exulans* destroyed native forests and birds on the ʻEwa Plain of Oʻahu (Hawaiian Islands) before people lived in this inhospitable area.

Prehistoric Extinctions: Continents vs. Islands

The differences in landbird faunas between islands and continents are obvious when remote, true oceanic islands are compared with North, Central, or South America, Africa, Eurasia, or Australia. The distinction between island and continent gets fuzzy with continental islands such as Borneo or New Guinea, blessed with rich birdlife in large part because of Pleistocene continental connections. A continuum from the most "insular" of islands to the most "continental" of continents might read like this: Easter, Wake, Mangaia, Tongatapu, Viti Levu, Isabel, New Ireland, New Britain, New Guinea, Borneo, Australia, and mainland Asia. This supports the proposal of Sauer (1969) not to segregate islands and continents in biogeographic theory. I would agree only partially; oceanic islands (such as Easter Island through New Britain in the list above) and continental islands or continents (New Guinea through mainland Asia) have some fundamental differences in geological development, evolutionary histories, and barriers to colonization.

Let us return now to extinction models. To explain the catastrophic extinction of North and South American large mammals at ca. 13,000 cal BP (= 11,000 yr BP), Mosimann & Martin (1975) developed the blitzkrieg model–a rapidly advancing front of skilled hunters who killed preferred prey beyond their actual needs and responded to an unlimited food supply with rapid population growth. The blitzkrieg model and its refinements (Martin 1984, 1990) propose that human presence and megafaunal collapse could have swept across North and South America in as little as several centuries, a period equivalent to only 10 to 20 human generations.

Whether or not you believe that blitzkrieg was involved, two features stand out concerning the late Pleistocene extinction of mammals on continents. One is that, with rare exception, only large (>44 kg) species were affected, pointing to big-game hunters rather than changing climate or plant communities as the primary cause of extinction. Another is that where the losses were greatest (North America, South America, and Australia), megafaunal collapse postdated the first arrival of people in a previously people-free landscape. By contrast, humans and other large mammals evolved side by side in Africa and Eurasia, where late Pleistocene extinctions affected a smaller percentage of the megafauna and were spread out over tens of millennia (Martin & Stuart 1995, Stuart 1999).

The human invasions of North America, South America, and Australia resemble those on oceanic islands in that naïve faunas were subjected without warning to earth's most ingenious predator. Whether large mammals on continents or birds of any size on oceanic islands, most species could not cope with suddenly becoming the prey of people, or at least were unable to survive the cascade of environmental changes that the people caused. Another similarity between vertebrate extinctions on oceanic islands and these three continents is that no other extinction episode of this magnitude has been found earlier in the Cenozoic fossil record (Martin & Steadman 1999). When considered in geological timescales, the coincidence of these losses with human arrival is by itself overwhelming evidence for cause and effect.

The rapid loss of birds on small Tongan islands (Ha'apai Group) fits the blitzkrieg model of extinction, occurring within a century or two of Lapita arrival. Just as the North American large mammal fauna collapsed during the Clovis period rather than during the slightly later Folsom period, the Tongan avifaunas collapsed during the brief time that decorated (Lapita) pottery was made on these islands (Martin & Steadman 1999, Steadman et al. 2002b). By Polynesian plainware times, the species composition of the decimated avifauna was essentially as it is today. For either the Clovis or Lapita colonization, a fauna naïve to humans was decimated by skilled hunters.

On large islands in East Polynesia (>50 km^2, >100 m elev.), estimates of the elapsed time between first human arrival and the loss of most birds range from a few centuries to >1500 years. Most East Polynesian species of birds became extinct from 1000 to 500 cal BP (Steadman 1995a). Variation in estimating the time of extinction arises not so much from dating the species themselves but from differing opinions on when people arrived in the Cook, Society, or Marquesas islands, ranging from as early as 2500 cal BP (Kirch 1986, Kirch & Ellison 1994) to as late as 1000 cal BP (Anderson 1995). The younger estimates of human arrival are compatible with blitzkrieg-type extinction. The older ones require explaining how species could have coexisted with people for so long on a remote island. One such explanation is that human habituation was intermittent from 2500 to 1000 cal BP, with an island such as Mangaia serving as a temporary coastal "fishing camp" without agriculture (Steadman 1995b, Kirch 1996). A similar situation has been proposed for Mayor Island off New Zealand's North Island, with sporadic visits from ca. 2200 to 450 cal BP, after which significant forest decline (and presumed permanent human occupation) took place (Empson et al. 2002).

Here Table 16-5 and Figure 16-3 offer guidance. Acting together, factors A2, A3, C2, C3, and C6 could help to delay extinction on high East Polynesian islands. In other island groups, such as the West Indies (Pregill et al. 1994) or the Mediterranean islands (Alcover et al. 1999), local variation in factors A1, B3, B4, C2, and C5 might further postpone extinctions. In the two large but isolated archipelagos of New Zealand and the Hawaiian Islands, the amount of time between human colonization and extinction of birds varies from species that probably were lost within a century or two to those just now dying out (James & Olson 1991, Olson & James 1991, Tarr & Fleischer 1995, Heather & Robertson 1997, Holdaway 1999, Holdaway & Jacomb 2000). Although these differences in the timing of extinction are of interest to conservation biologists, in an evolutionary or geological sense, any of these losses is instantaneous.

Additional Tests and Uses of the ABC Extinction Model

The ABC model of extinction can be evaluated and refined through careful excavations at stratified, bone-rich sites, especially those believed to represent first human contact. Gathering high-quality data requires laborious bone-retrieval methods (sieving all sediment through 3 mm or finer mesh), a meticulous effort to identify bones to the species level, and extensive ^{14}C dating (Chapter 4). It requires a considerable investment of time, money, cooperation, and expertise.

Some surviving species of landbirds in Oceania (especially Remote Oceania) are threatened with extinction. Conservation strategies (Chapter 21) should focus, when possible, on islands with multiple ABC traits that retard extinction rates.

The Loss of Evolutionary History

Human-caused extinction has overwhelmed landbird faunas in Remote Oceania to the point where many islands in East Polynesia and Micronesia have few if any species left except the heron *Egretta sacra* and, if freshwater wetlands are present, perhaps the duck *Anas superciliosa*. Most single-island endemics are extinct, and most archipelago endemics either are gone or live on a fraction of the islands they once inhabited. The limited prehistoric data from Near Oceania suggest that a smaller but still very substantial proportion of its landbird fauna has been lost to human impact. What do these losses mean in terms of evolutionary history? In other words, to what extent has evolution been halted by the arrival of people across Oceania?

One way to assess the losses from an evolutionary standpoint is to see which taxonomic categories have been affected, with the assumption that random extinction of species affects a smaller number of higher taxa because classifications are hierarchical (Purvis et al. 2000). The extinction of five congeneric species, for example, will not result in the extinction of a genus that has six or more species. The loss of "phylogenetic diversity" increases as the category of the extinct taxon becomes higher (Nee & May 1997). Thus the possible extinction of the kagu *Rhynochetos jubatus* in coming decades would be a great phylogenetic loss because this species is the sole survivor of a genus and family endemic to New Caledonia.

By contrast, the extinction of *Gallirallus ripleyi* on Mangaia 500 years ago was the loss of one species in a genus that once had hundreds of endemic flightless species on islands across the Pacific. The contrast between *Rhynochetos jubatus* and *Gallirallus ripleyi* is not quite so simple, however. Flightless species of *Gallirallus* survive only on several islands today, each of them thousands of kilometers west of Mangaia (Chapter 10). Furthermore, none of the surviving species, nor any of the many extinct forms known thus far, is very similar morphologically to *G. ripleyi* (Steadman 1987, Steadman et al. 2000b). Geographic/morphological outliers may be more than "just another species." From the standpoint of evolutionary history and phylogenetic diversity, the extinction of *G. ripleyi* on Mangaia may be a greater loss than that of any of the closely related species of flightless Guam-like rails in the Marianas (see Chapter 8). Because the four others are gone, however, the sole surviving flightless rail in the Marianas (*G. owstoni*) has become a much more precious evolutionary commodity than if the other members of its species-group were alive.

Examples of extinct geographic or morphological outliers include: the undescribed species of Tongan megapodes, which are much larger or smaller than any living forms of *Megapodius*; the immense flightless swamphens (*Porphyrio* undescribed sp. A and B) from New Ireland and Buka; the large-billed gull *Larus utunui* from Huahine; the large, geographically bizarre species of *Macropygia* in the Society and Marquesas islands; the large East Polynesian lorikeets *Vini vidivici* and *V. sinotoi*; the parrot *Eclectus* undescribed sp. from Tonga and Vanuatu; and the swiftlet *Collocalia manuoi* from Mangaia. Good examples of sole survivors of formerly more widespread species or diverse lineages include the megapode *Megapodius pritchardii* of Tonga, flightless rail *Porzana atra* of Henderson, and ground-dove *Gallicolumba erythroptera* of the Tuamotus.

At the generic level, the fossil record of birds currently in hand has revealed no losses in Micronesia. In East Polynesia, the only extinct genera are the unnamed "Henderson archaic pigeon" (Wragg 1995) and the enigmatic but poorly substantiated heron, large rail, and parrots from Easter Island (Chapter 7). Genus-level losses in West Polynesia include the megapode *Megavitiornis* of Fiji, the rail *Vitirallus* from Fiji, and the pigeons *Natunaornis* from Fiji and Undescribed genus C from Tonga (Worthy 2000, 2001b, Steadman unpub. data; Chapters 6, 9, 11 herein). In Melanesia, the giant megapode *Sylviornis* (New Caledonia) and two ground-dwelling pigeons (Buka) represent extinct genera.

The phylogenetic diversity of landbirds in Remote Oceania cannot recover from the human impact that began 3000 years ago. Most species already are extinct, and the factors that eliminated them continue to operate. Nonnative plants and animals still are moved regularly among islands. This contamination includes highly invasive species of plants (Chapter 21) and an ugly list of nonnative mosquitoes, ants, snakes, rats, cats, dogs, chickens, mynas, etc. The evolutionary history of native birds is winding down in much of Remote Oceania. It is already over on Easter Island, and nearly so on Guam. The resident landbirds recorded on a recent "Christmas count" on northern Guam (26 December 1999) consisted of two native species (23 individuals) and five nonnative species (2114 individuals). Another count on southern Guam (2 January 2000) recorded 28 individuals of three native species versus 777 of six nonnative species (Wiles 2000a,b). Many Polynesian and Micronesian islands have no landbirds other than the trampy heron *Egretta sacra*. On a number of islands in French Polynesia and perhaps even in Fiji, the overall relative abundance of nonnative birds (chickens, certain columbids, bulbuls, estrildid finches, mynas) now exceeds that of native species. A similar situation may be developing on Tongatapu, Tonga (Table 6-18).

In Near Oceania, the situation is less grim. A larger percentage of the original avifauna survives, and contamination by nonnative species generally has been less severe than in Remote Oceania. This is likely to continue. I hope that the difference is not just a matter of scale where what took several millennia to accomplish in Remote Oceania merely will require more time in Near Oceania. In spite of increases in commercial logging and human populations, the short-term future of birdlife in Near Oceania may not be bleak. Keeping these insular avifaunas relatively intact is a worthy goal.

Chapter 17 Dispersal, Colonization, and Faunal Attenuation

DISPERSAL, colonization, and evolution form a logical series of events but I have little to write about the last, for at least two related reasons. The first is a paucity of rigorous phylogenetic information on birds from Oceania. This compromises any attempt to analyze distributions from an evolutionary perspective (or vice versa). A second reason is how little we know about the timescales of dispersal (getting from one place to another) and colonization (establishing a breeding population in the new place). Good phylogenetic information, whether based on morphology (plumages, bones, muscles, etc.), molecules, or both, would provide a framework for estimating routes of dispersal and their timing. Some of these estimates could be tested with fossils. Unfortunately, molecular data for landbirds in Oceania are rare and, when more become available, need to be evaluated as critically as any other scientific information. Morphological data, including my own, are mostly descriptive and seldom have been used to generate phylogenetic hypotheses. There is nothing wrong with this state of affairs; it only means that our understanding of tropical Pacific birds is in an early stage of scholarly pursuit with more description than synthesis or interpretation.

Evolutionary studies of birds in Oceania suffer further because so many species are extinct. In my lab, thousands of bones from 50 undescribed species of herons, megapodes, rails, pigeons, parrots, and passerines remind me daily that the descriptive phase of avian systematics in Oceania, a prerequisite for evolutionary studies, is far from finished. The description of each new species adds morphological data that eventually can be put in a phylogenetic context. Because many of these bones retain collagen, we have the potential for combined morphological and molecular phylogenies that include both living and extinct species, as in the Hawaiian Islands (Fleischer & McIntosh 2001, Fleischer et al. 2001, James 2001, in press, Slikas 2003). In the meantime, we must admit our ignorance about the evolution of birds in most of Oceania. This ignorance stands in great contrast to what we know, for example, about many aspects of the evolution of finches in the Galápagos (Grant 1998, Grant & Grant 1998). This ignorance extends to the interplay of evolution and ecology.

Speciation (see Peterson 1998) is a rich area for research in the Pacific, where the fossil record now shows that intra-archipelago congeneric species of landbirds once were widespread and involved many more genera in Remote Oceania than previously believed. Nevertheless, we lack both the genetic and life history information to know whether some closely related species of birds in Oceania may have, for example, contrasting genetic structures that may have arisen because of differences in ecology and social systems, as has been suggested for certain species pairs on continents (McDonald et al. 1999). Also unexplored in Oceania are possible intra-island, intraspecific differences in morphology and life history (and thus probably genetics) as reported for the blue tit (*Parus caeruleus*) on Corsica in the Mediterranean (Blondel et al. 1999). Morphology-based evidence of hybridization in landbirds from Near Oceania (Mayr & Diamond 2001:179–181) need to be tested genetically. I have found no morphological evidence for hybridization of landbirds in Remote Oceania.

Dispersal and Colonization

Background, Assumptions, More Limitations

Oceania is blessed with all sizes and shapes of islands, not to mention interisland distances from a kilometer or less to hundreds and even thousands of kilometers (Tables 2-1, 2-2, Figure 17-1). All of Oceania lies east of Wallace's Line, yet avian colonization of islands in the Bismarcks lying closest to New Guinea (such as Umboi and New Britain; see Mayr & Diamond 2001:97–111) resembles a continental, Krakatau-like situation more than in Remote Oceania. The principles of dispersal and colonization gleaned from studying the Krakatau fauna (such as a colonization curve that flattens with time; Thornton et al. 1993, Thornton 1996) have little applicability in Oceania outside of the Bismarcks and even there are compromised on all islands except active volcanoes by the largely successional nature of the Krakatau situation and the different source areas (Java and Sumatra on the Sunda Shelf for Krakatau vs. New Guinea for the Bismarcks). I see even less of a conceptual connection to birds in Oceania with biogeographic studies further removed geographically or taxonomically, such as birds on continental islands in northern Europe (Cook & Hanski 1995), or mammals in montane habitats in North America (Brown 1971, Grayson & Madsen 2000).

Biologists often debate whether some current distribution is due to dispersal vs. vicariance (Rieppel 2002). Geology argues against vicariance for most taxa in Oceania because most of the islands have independent origins in the ocean (Chapter 2). Even on Gondwanan New Caledonia, dispersal over the ocean is a larger factor overall.

Over geological time, what would seem to be a highly unlikely dispersal event becomes much more likely. The proposal, for example, by Raxworthy et al. (2002) that chameleons (Chameleonidae) have dispersed several times from Madagascar to Africa has no commonsense argument against it. This does not necessarily exclude vicariance, but does suggest that the modern phylogeography of these lizards carries a larger dispersive than vicariant signal. Following the lead of Voelker (1999), I believe that dispersal and vicariance are not mutually exclusive; both concepts can help to explain the distribution of many taxa, whether on continents or islands.

As a gross generalization, the species richness and taxonomic diversity of birds decrease from west to east from mainland New Guinea through Polynesia in the southern hemisphere and from Palau through the rest of Micronesia in the northern hemisphere (see "Faunal Attenuation," below in this chapter). The isolated archipelagos of New Zealand (including the Kermadecs, Chathams, etc.; see Holdaway et al. 2001, Worthy & Holdaway 2002) and the Hawaiian Islands are, as always, exceptions because of their fantastic, dead-end radiations. By "dead-end" I mean that these island groups are on the way to nowhere. Unlike what

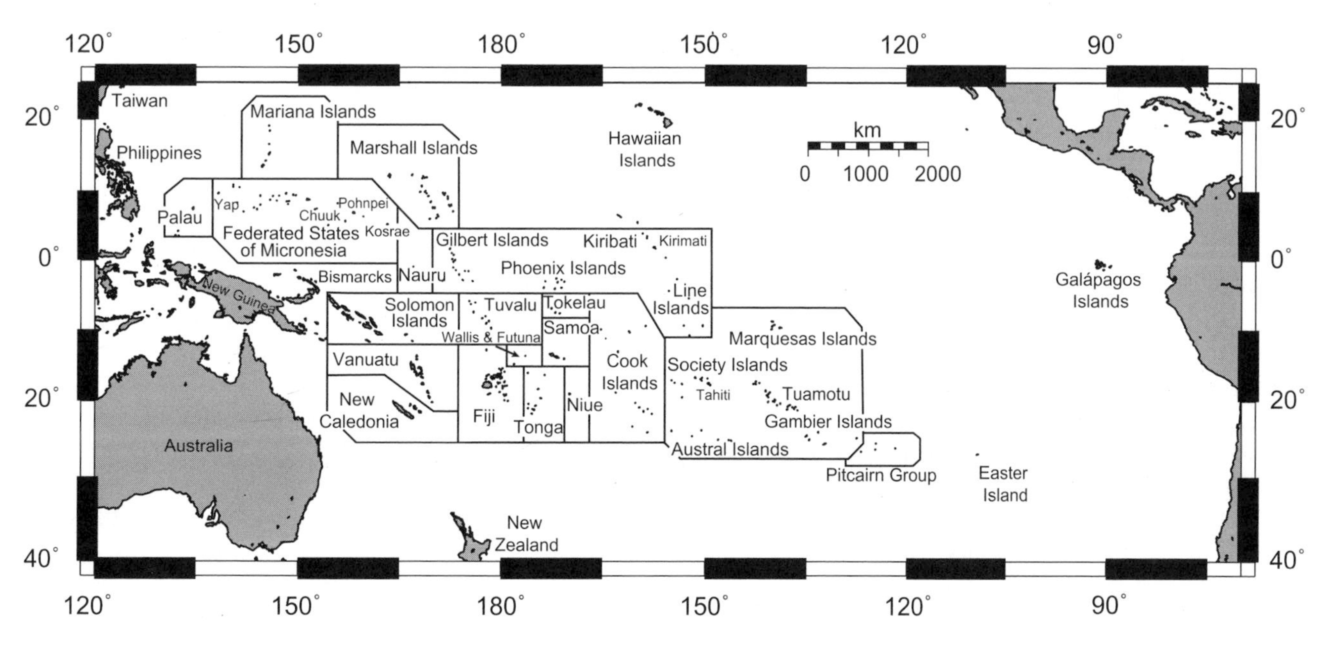

Figure 17-1 Oceania.

Table 17-1 Distribution of the fifteen most widespread species of landbirds in Oceania. From data in Chapters 5–8. Solomon Outliers = Rennell, Bellona, Santa Cruz Group. RCPI = Remote Central Pacific Islands (Chapter 8). Categories of intraspecific variation: 1, minimal (fewer than three subspecies recognized in Oceania); 2, substantial (three or more well-marked subspecies in Oceania). *, allospecies present (does not include flightless rails). Dispersal categories from Mayr & Diamond (2001: Appendix 5), as follows: A, highly vagile; B, intermediate; C, highly sedentary; —, no category assigned. Note that no species is assigned to category C. Row totals exclude Mainland New Guinea.

	Intraspecific variation	Dispersal category	Mainland New Guinea	Bismarcks	Solomon Main Chain	Solomon Outliers	Vanuatu	New Caledonia	Fiji	Tonga	Samoa	Niue	Cooks
Egretta sacra	1	A	X	X	X	X	X	X	X	X	X	X	X
Anas superciliosa	1	A	X	X	X	X	X	X	X	X	X	—	X
Gallirallus philippensis	2	—	X	X	X	X	X	X	X	X	X	X	—
Porzana tabuensis	1	B	X	X	X	X	X	X	X	X	X	X	X
Poliolimnas cinereus	2	B	X	X	X	—	X	X	X	—	X	—	—
Porphyrio porphyrio	1	A	X	X	X	X	X	X	X	X	X	X	—
Columba vitiensis	2	B	X	X	X	—	X	X	X	—	X	—	—
Ducula pacifica	1	A	—	X	X	X	X	X	X	X	X	X	X
Tyto alba	1	B	X	X	X	X	X	X	X	X	X	X	—
Collocalia vanikorensis	2	A	X	X	X	X	X	X	—	—	—	—	—
Collocalia spodiopygia	2	B	—	X	X	X	X	X	X	X	X	X	—
Halcyon chloris	2	A	X	X	X	X	X	X	X	X	X	—	*
Hirundo tahitica	2	A	X	X	X	X	X	X	X	X	—	—	—
Turdus poliocephalus	2	B	X	X	X	X	X	X	X	X	X	—	—
Myzomela cardinalis	2	B	*	*	X	X	X	X	*	X	X	—	—
TOTAL													
X	—	—	12	14	15	13	15	15	13	12	13	7	4
X+*	—	—	13	15	15	13	15	15	14	12	13	7	5

has been proposed regarding the tree *Metrosideros* (Wright et al. 2000; Chapter 2 herein), there is no evidence of colonization by landbirds from New Zealand (or the Hawaiian Islands) to other island groups in Oceania.

The amount of time elapsed since colonization should increase with higher taxonomic categories. In other words, one would expect the time since last gene flow for two confamilial genera to be greater than for congeneric species, which in turn should be greater than for conspecific subspecies, which should be greater than for consubspecific populations. The last two categories, especially the consubspecific populations, may still have interisland gene flow. Organisms that rarely disperse to islands probably are more prone to speciation than those that disperse regularly and for long distances (Wagner 1991, Jablonski & Roy 2003, Paulay & Meyer 2003). As pointed out long ago (Sauer 1969), only the few species of landbirds that disperse regularly among islands, and therefore have differentiated little if at all, behave in a way applicable to the equilibrium theory of island biogeography. Biology is not physics, and the movement of birds is not an example of Brownian motion. I cannot see how to resolve the conflict between interisland differentiation and the equilibrium (or neutral) theory of biogeography, especially given the broad range of dispersal abilities among landbirds in Oceania.

The age of an island sets an absolute maximum limit on the timing of colonization. The quality of geochronological information in Oceania is highly variable but is pretty good in some island groups. Most high islands (non-atolls) have ages in the range of one to 10 million years (see Chapter 1). Some high islands, especially the largest ones from Fiji westward, can have ages that exceed 10 million years, although New Caledonia, a Gondwanan fragment, may be the only one that has been emergent continuously for more than 25 million years.

Atolls present a special case. Considering the growth of coral, and assuming little or no tectonic uplift or subsidence in the late Quaternary, all current atolls probably were submerged beneath the ocean during the +8 to 10 m high sea level stand of the last interglacial (marine isotope stage 5e, which began at 130 ka, peaked at 125 ka, and ended at ca. 116 ka; Kukla 2000; Chapter 1 herein). Thus the entire

	Society												Totals	
Tubuai	Islands	Tuamotu	Pitcairn	Easter Island	Marquesas	Palau	Yap	Marianas	Chuuk	Pohnpei	Kosrae	RCPI	X	X+*
X	X	X	—	—	X	X	X	X	X	X	X	X	21	21
X	X	—	—	—	—	X	—	—	X	—	—	X	14	14
—	—	—	—	—	—	X	—	—	—	—	—	—	10	10
—	X	X	X	—	—	—	—	—	—	—	—	—	13	13
—	—	—	—	—	—	X	X	X	X	X	—	—	11	11
—	—	—	—	—	X	X	—	—	—	—	—	—	11	11
—	—	—	—	—	—	—	—	—	—	—	—	—	6	6
—	—	—	—	—	—	*	*	*	*	*	*	*	10	17
—	—	—	—	—	—	—	—	—	—	—	—	—	9	9
—	—	—	—	—	—	X	X	X	X	X	X	—	11	11
—	—	—	—	—	—	—	—	—	—	—	—	—	9	9
—	*	*	—	—	*	X	—	X	—	—	—	—	10	14
—	X	—	X	—	—	—	—	—	—	—	—	—	9	9
—	—	—	—	—	—	—	—	—	—	—	—	—	8	8
—	—	—	—	—	—	*	*	*	*	*	*	—	6	14
2	4	2	2	0	2	7	3	4	4	3	2	2		
2	5	3	2	0	3	9	5	6	6	5	4	3		

terrestrial biota on most if not all atolls must have developed during only the past ca. 120,000 years. The MIS 5e high sea level stand (along with the much more recent Holocene inundation in some regions; Chapter 1) could account for how little endemism there is today on atolls, which usually are inhabited by trampy, strand species of plants and animals. On the other hand, 100,000+ years might be plenty of time for endemic species to evolve; the problem is that most endemic species of birds on atolls not inundated in the Holocene would have been obliterated almost effortlessly by the first people. The fossil data now available on atolls are not adequate to test this idea.

Within the framework of island age, three lines of evidence can be pursued to estimate the time of colonization for individual species: (1) the fossil record; (2) systematic data; and (3) observations on living populations.

High-quality fossil data on when a species arrived require a chronology of the prehistoric site and careful identification of the fossils. While an island's geological age places a maximum time limit on biotic colonization, the fossil record, because of its incomplete temporal coverage, yields a minimum value on the time of colonization. Most fossil records on Pacific islands are very young; on just a few islands have bone deposits been found that precede human arrival. Thus, for most species of terrestrial vertebrates on most islands, the "bookends" of geological age of the island vs. the oldest fossils do not place tight constraints on the chronology of colonization. Typically we have data such as a Fijian or Tongan island with an estimated age of one to three million years and an oldest fossil deposit three orders of magnitude younger at only 3000 years. Although unsatisfying, this is much better than nothing.

The rare instances of older fossils are informative. The two oldest vertebrate fossil sites in tropical Oceania are at 'Anatu ("ground-dove cave") on 'Eua, Tonga (Steadman 1993a; Chapter 6 herein) and at 'Ulupau Head (Mokapu Point) on O'ahu, Hawaiian Islands (James 1987, 1995). At 'Anatu, a cultural stratum (Layer I) is radiocarbon dated to as old as 2970–2710 Cal BP. Underlying Layer I are two bone-rich strata (Layers II, III) regarded as prehuman because they lack any of the cultural features, charcoal, artifacts, or bones of chickens, rats, pigs, dogs, or humans found in Layer I. Layers II and III are separated by a bed of calcite flowstone with three uranium-series dates of 60–80 ka. Thus Layer II is more than 3 ka but less than 60 to 80 ka, whereas Layer III is more than 60 to 80 ka. Only six of the 21 species of landbirds from Layers II and III still occur on ʻEua. Most or all of the species of birds, bats, and lizards living on ʻEua at human arrival already had been there for more than 60,000 to 80,000 years. They had passed unscathed through the last glacial-interglacial transition (18 to 10 ka) and an unknown number of other such transitions before it. At ʻUlupau Head, 15 or 16 of the 17 species of landbirds recorded from a site at least 120,000 years old (and perhaps even older) still existed on ʻOahu at human contact less than 2000 years ago.

Whereas finding a species in a dated fossil context establishes a minimum age of its presence on an island, negative fossil data also can help to estimate when a species colonized by suggesting its prehistoric absence (James 1987, 1995, Balouet & Olson 1989, Steadman et al. 1991, Olson & James 1991, Steadman 1993a, Holdaway et al. 2001, Worthy & Holdaway 2002). The likelihood that negative data are valid, i.e., that the absence is genuine, is strengthened by increased fossil sampling but never can reach 100%. Nevertheless, the information in hand from several fossil-rich islands argues for recent colonization by some species. Fossils of two widespread species of volant rails (*Porzana tabuensis*, *Porphyrio porphyrio*) are absent in precultural (>3 ka) sites in New Caledonia, Tonga, and New Zealand, but are common in archaeological sites, which are <3 ka in New Caledonia and Tonga, and <1 ka in New Zealand (Chapter 10). These two species probably colonized only after human arrival. The same is true for the volant rail *Gallirallus philippensis* in New Caledonia and Tonga but not in New Zealand, where its fossils have been found in precultural sites (Holdaway et al. 2001). On Mangaia (Cook Islands), bones of *Porzana tabuensis* are common in archaeological sites but absent in bone-rich precultural strata (Chapters 7, 10).

Before people arrived, most rails in New Caledonia, Tonga, Cook Islands, New Zealand, Hawaiian Islands, and presumably across Remote Oceania were flightless species. Recent arrival of the three volant rails is compatible with their poorly defined geographic variation (Chapter 10). The same may be true for the pigeon *Ducula pacifica* and barn-owl *Tyto alba*, which are absent from precultural sites in New Caledonia and Tonga (Balouet & Olson 1989, Steadman 1993a). In the Hawaiian Islands, fossils of the night-heron *Nycticorax nycticorax* and owl *Asio flammeus* are absent from the many precultural sites, again suggesting recent colonization in species that lack morphological evidence of in situ differentiation (Olson & James 1991).

On their own, systematic data lack an absolute chronology although, as already noted, levels of differentiation among taxa should increase with time (Bermingham & Dick 2001, Ricklefs & Bermingham 2001). One can place the differences on an absolute timescale by assuming a uniform and known rate of change, yielding a proxy chronometer or "clock." Purely morphological data are seldom interpreted with artificial clocks, but this is common in molecular studies, where the only characters are the nucleic acids A, C, G, and T. To be accurate and absolute, molecular clocks need to be calibrated by chronometrically precise geological or paleontological data. Assumptions of uniform rates of change, such as the admittedly provisional estimate of 2% sequence divergence per million years between pairs of lineages in mitochondrial DNA of birds and mammals (Avise et al. 1998), are difficult to believe. Estimated divergence times based on DNA generally are older than fossil-based estimates (i.e., Brodkorb 1971, Steadman 1980, Zink & Slowinski 1995, Feduccia 1998, Foote et al. 1999, Strauss 1999).

Taxonomic levels of endemism should be a crude gauge of relative divergence times. In Oceania outside of New Zealand and the Hawaiian Islands, there is little avian endemism above the species level. The Rhynochetidae (kagus; on New Caledonia, which is Gondwanan) is the only endemic family of birds. Only 21 (22%) of the 96 genera of resident, nonpasserine landbirds are endemic to the region. These include two megapodes, two rails, a kagu, a sandpiper, a snipe, seven columbids, four parrots, two owls, and a frogmouth (Chapters 9–13). Among 50–54 genera of resident passerines, 14–18 (28–33%) are endemic to Oceania. These are confined to four families (up to six monarchs, five white-eyes, two to five honeyeaters, and a warbler; Chapter 14).

Observational data on living landbirds can be informative about dispersal over the ocean, the advantages and risks of which were reviewed by Diamond (1985). For Near Oceania, Mayr & Diamond (2001:66–77, 422–430) summarized evidence for overwater dispersal in each species of landbird, based on whether it has been seen flying over the ocean, whether it occurs on offshore islets or Holocene volcanoes, and whether it has colonized new islands in recent decades. In the Vavaʻu and Haʻapai Groups of Tonga, the pigeon *Ducula pacifica* flies between islands more than any other species, although the dove *Ptilinopus porphyraceus* and kingfisher *Halcyon chloris* also are seen regularly over the water (Steadman 1998, unpub. data, Steadman & Freifeld 1998). It is undoubtedly adaptive that the trampy *D. pacifica* has a very low basal metabolic rate for a columbid (McNab 2000). Observational data on interisland dispersal generally correspond with morphological data on intraspecific variation; species seen to fly

Table 17-2 Representation in Oceania of landbirds from the lowlands (<1000 m elev.) of mainland Papua New Guinea (PNG), derived from distributional maps and descriptions of Coates (1985, 1990), with slightly altered classification. m, montane species, included because it is present in Oceania; *, genus occurs, but different species than in PNG.

Family	Genus	Number of lowland PNG species	Number of lowland PNG species in Near Oceania	Number of lowland PNG species in Remote Oceania
CASSOWARIES	*Casuarius*	3	2	—
GREBES	*Tachybaptus*	2	1	1
CORMORANTS	*Phalacrocorax*	3	2	2
HERONS	*Ardea*	2	—	—
	Egretta	3	1	2
	Butorides	1	1	1
	Nycticorax	1	1	1
	Zonerodius	1	—	—
	Ixobrychus	1	1	1
STORKS	*Xenorhynchus*	1	—	—
IBISES	*Plegadis*	1	—	—
	Threskiornis	1	—	1
	Platalea	1	—	—
DUCKS	*Anseranas*	1	—	—
	Dendrocygna	2	2	2
	Tadorna	1	—	—
	Anas	3	1	2
	Aythya	1	—	1
	Nettapus	2	—	—
OSPREYS	*Pandion*	1	1	1
HAWKS	*Aviceda*	1	1	—
	Henicopernis	1	*	—
	Machaerhamphus	1	—	—
	Elanus	1	—	—
	Milvus	1	—	—
	Haliastur	2	1	—
	Haliaaetus	1	*	—
	Harpyopsis	1	—	—
	Circus	1	—	*
	Megatriorchis	1	—	—
	Accipiter	7	2	1
	Aquila	1	—	—
	Hieraaetus	1	—	—
FALCONS	*Falco*	3	2	1
MEGAPODES	*Aepypodius*	1	—	—
	Talegalla	2	—	—
	Megapodius	1	1	1

(*continued*)

Table 17-2 (continued)

Family	Genus	Number of lowland PNG species	Number of lowland PNG species in Near Oceania	Number of lowland PNG species in Remote Oceania
QUAILS	*Coturnix*	2	1	—
BUTTON-QUAILS	*Turnix*	1	1	*
RAILS	*Rallina*	1	1	*
	Gallirallus	2	2	1
	Gymnocrex	1	1	—
	Porzana	2	1	1
	Poliolimnas	1	1	1
	Amaurornis	1	1	—
	Megacrex	1	—	—
	Gallinula	1	*	*
	Porphyrio	1	1	1
CRANES	*Grus*	1	—	—
BUSTARDS	*Otis*	1	—	—
JACANAS	*Irediparra*	1	1	—
THICK-KNEES	*Esacus*	1	1	—
PLOVERS	*Vanellus*	1	—	—
	Charadrius	1	1	—
PIGEONS, DOVES	*Columba*	1	1	1
	Macropygia	2	2	*
	Reinwardtoena	1	*	—
	Ptilinopus	10	3	*
	Ducula	5	1	*
	Gymnophaps	1	1	—
	Chalcophaps	2	1	1
	Henicophaps	1	*	—
	Caloenas	1	1	1
	Geopelia	2	—	—
	Gallicolumba	3	2	*
	Trugon	1	—	—
	Otidiphaps	1	—	—
	Goura	2	—	—
PARROTS	*Cacatua*	2	*	*
	Probosciger	1	—	—
	Chalcopsitta	2	*	—
	Trichoglossus	1	1	1
	Pseudeos	1	—	—
	Psitteuteles	1	—	—
	Lorius	2	1	*
	Charmosyna	5	1	*
	Psittaculirostris	2	—	—

Table 17-2 (continued)

Family	Genus	Number of lowland PNG species	Number of lowland PNG species in Near Oceania	Number of lowland PNG species in Remote Oceania
	Opopsitta	2	—	—
	Micropsitta	3	2	1
	Psittrichas	1	—	—
	Alisterus	1	—	—
	Aprosmictus	1	—	—
	Geoffroyus	2	*	*
	Eclectus	1	1	*
	Loriculus	1	*	—
CUCKOOS	*Cacomantis*	2	1	*
	Rhamphomantis	1	—	—
	Chrysococcyx	2	*	*
	Caliechthrus	1	—	—
	Microdynamis	1	—	—
	Eudynamys	1	1	—
	Scythrops	1	1	—
	Centropus	3	*	—
BARN-OWLS	*Tyto*	3	2	2
STRIGID OWLS	*Uroglaux*	1	—	—
	Ninox	4	*	*
OWLET-NIGHTJARS	*Aegotheles*	5	—	*
FROGMOUTHS	*Podargus*	2	—	—
NIGHTJARS	*Eurostopodus*	1	*	—
	Caprimulgus	1	1	*
SWIFTS	*Collocalia*	5	4	3
	Chaetura	1	—	—
CRESTED SWIFTS	*Hemiprocne*	1	1	1
KINGFISHERS	*Alcedo*	1	1	—
	Ceyx	3	2	—
	Dacelo	3	1	—
	Clytoceyx	1	—	—
	Melidora	1	—	—
	Halcyon	4	3	1
	Syma	2	—	—
	Tanysiptera	5	*	—
BEE-EATERS	*Merops*	2	1	—
ROLLERS	*Eurystomus*	1	1	—
HORNBILLS	*Aceros*	1	1	*
PITTAS	*Pitta*	3	1	—
LARKS	*Mirafra*	1	—	—
SWALLOWS	*Hirundo*	1	1	1
PIPITS	*Anthus*	1	—	—

(*continued*)

Table 17-2 (continued)

Family	Genus	Number of lowland PNG species	Number of lowland PNG species in Near Oceania	Number of lowland PNG species in Remote Oceania
THRUSHES	*Turdus* (m)	1	1	1
	Zoothera	1	*	—
OLD WORLD FLYCATCHERS	*Saxicola*	1	1	1
CISTICOLAS	*Cisticola*	2	1	—
OLD WORLD WARBLERS	*Acrocephalus*	1	1	*
	Phylloscopus	1	1	—
	Megalurus	2	1	—
SUNBIRDS	*Nectarinia*	2	2	—
FLOWERPECKERS	*Dicaeum*	1	*	—
	Toxorhamphus	2	—	—
	Oedistoma	1	—	—
	Melanocharis	4	—	—
	Rhamphocharis	1	—	—
	Oreocharis	1	—	—
WHITE-EYES	*Zosterops*	3	*	*
STARLINGS	*Aplonis*	3	2	1
	Mino	2	1	—
ESTRILDID FINCHES	*Erythrura*	2	1	1
	Lonchura	7	1	—
	Neochima	1	—	—
CUCKOO-SHRIKES	*Coracina*	10	4	1
	Campochaera	1	—	—
	Lalage	3	1	*
ACANTHIZID WARBLERS	*Crateroscelis*	1	—	—
	Sericornis	4	—	—
	Gerygone	6	—	*
SHRIKES	*Lanius*	1	—	—
QUAIL-THRUSHES	*Ptilorrhoa*	3	—	—
	Cinclosoma	1	—	—
AUSTRALIAN BABBLERS	*Pomatostomus*	2	—	—
FAIRY-WRENS	*Sipodotus*	1	—	—
	Malurus	3	—	—
ROBINS	*Monachella*	1	1	—
	Microeca	4	—	—
	Drymodes	1	—	—
	Amalocichla	1	—	—
	Poecilodryas	4	—	—
	Tregellasia	1	—	—
	Eopsaltria	1	—	*
	Peneothello	2	—	—

Table 17-2 (continued)

Family	Genus	Number of lowland PNG species	Number of lowland PNG species in Near Oceania	Number of lowland PNG species in Remote Oceania
WHISTLERS	*Pachycephalopsis*	2	—	—
	Rhagologus	1	—	—
	Pachycare	1	—	—
	Pachycephala	7	1	*
	Colluricincla	2	—	*
	Pitohui	5	—	—
FANTAILS	*Rhipidura*	10	3	1
MONARCHS	*Monarcha*	8	1	*
	Arses	2	—	—
	Grallina	2	—	—
	Myiagra	4	1	*
	Machaerirhynchus	2	—	—
HONEYEATERS	*Melilestes*	1	—	—
	Meliphaga	9	—	—
	Melidectes	2	—	—
	Lichmera	2	—	*
	Myzomela	7	1	*
	Pycnopygius	3	—	—
	Xanthotis	2	—	—
	Lichenostomus	3	—	—
	Glycichaera	1	—	—
	Timeliopsis	1	—	—
	Entomyzon	1	—	—
	Melithreptus	1	—	—
	Conopophila	1	—	—
	Ramsayornis	1	—	—
	Philemon	4	*	*
ORIOLES	*Oriolus*	3	—	—
	Specotheres	1	—	—
DRONGOS	*Dicrurus*	2	1	—
WOOD-SWALLOWS, BUTCHERBIRDS	*Peltops*	2	—	—
	Artamus	3	*	1
	Cracticus	3	—	—
	Gymnorhina	1	—	—
BOWERBIRDS	*Ailuroedus*	2	—	—
	Amblyornis	1	—	—
	Sericulus	2	—	—
	Chlamydera	2	—	—

(*continued*)

Table 17-2 (continued)

Family	Genus	Number of lowland PNG species	Number of lowland PNG species in Near Oceania	Number of lowland PNG species in Remote Oceania
BIRDS-OF-PARADISE	*Melampitta*	1	—	—
	Loboparadisea	1	—	—
	Manucodia	4	—	—
	Epimachus	2	—	—
	Ptiloris	1	—	—
	Seleucidis	1	—	—
	Lophorina	1	—	—
	Parotia	1	—	—
	Cicinnurus	2	—	—
	Paradisea	4	—	—
CROWS	*Corvus*	2	1	*
TOTAL SPECIES				
Nonpasserines		201	71	34
Passerines		223	30	8
All		424	101	42

between islands tend to have little or no detectable interisland variation in plumage or mensural characters. An unexplained exception is the kingfisher *Halcyon chloris*, which is widespread, seen over water, but with well marked interisland variation in plumage (Chapter 13).

For nearly all species of landbirds in Remote Oceania, however, interisland dispersal has been and still is rare. I am aware of no case in East Polynesia over the past 200 years of a landbird population being lost and then recolonizing the same island under its own power. The same may be true for Micronesia except possibly Pelilou (Palau), an island teeming with birds today but was badly bombed and burned during World War II. It is not clear, however, which populations of birds were lost vs. reduced on Pelilou during those tragic times (Marshall 1949, Engbring 1988).

Patterns

First let us look at the 15 most widespread species of landbirds in Oceania, 12 of which also occur on New Guinea (Table 17-1). The three species that inhabit the most island groups today (the heron *Egretta sacra*, duck *Anas superciliosa*, and rail *Porzana tabuensis*) prefer coastal or inland wetlands. Among the nine next most widespread species (each in nine to 11 island groups), the rails *Poliolimnas cinereus* and *Porphyrio porphyrio* and swallow *Hirundo tahitica* also are semiaquatic. The highly dispersive nature of many species of wetland birds is a global phenomenon; nothing is special about the situation in Oceania. I should note that these species forage at least occasionally on or over dry land.

Of the 15 most widespread species only the thrush *Turdus poliocephalus* is a true forest-obligate, although the pigeons *Columba vitiensis* and *Ducula pacifica* certainly prefer forest. The remaining six nonaquatic species (the rail *Gallirallus philippensis*, barn-owl *Tyto alba*, swiftlets *Collocalia vanikorensis* and *C. spodiopygia*, kingfisher *Halcyon chloris*, and honeyeater *Myzomela cardinalis*) are habitat generalists; among these only *G. philippensis* is rare or absent in forested habitats.

Each of the 15 species in Table 17-1 occurs in the Solomons, Vanuatu, and New Caledonia, demonstrating an ability to disperse across at least 395 km of ocean (today) or 365 km of ocean under full-glacial conditions (Table 2-1). All but one or two inhabit the Bismarcks, Fiji, and Samoa, the last island group requiring ocean crossings of at least 840 km (Holocene) or 530 km (full-glacial). Intraspecific morphological variation among the 15 species correlates generally with dispersal ability, as one would expect, a trend that is untested by molecular data. Four of the six species with minimal geographic variation were classified as highly vagile by Mayr and Diamond (2001), compared to only three of the nine species with three or more subspecies in Oceania.

Table 17-3 Summary of the representation in Oceania of species that live today in lowland mainland Papua New Guinea. From data in Table 17-2.

	Family		Genus		Species	
	Number	%	Number	%	Number	%
NONPASSERINES						
Lowland Papua New Guinea	34	100	114	100	201	100
Near Oceania	29	85	66	58	71	35
Remote Oceania	21	62	42	37	33	16
PASSERINES						
Lowland Papua New Guinea	28	100	93	100	225	100
Near Oceania	16	57	28	30	30	13
Remote Oceania	13	46	20	22	7	3
ALL LANDBIRDS						
Lowland Papua New Guinea	62	100	207	100	426	100
Near Oceania	45	73	94	45	101	24
Remote Oceania	34	55	63	30	40	9

Table 17-4 Summary of the representation in Oceania of Passeriformes that live today in lowland mainland Papua New Guinea. From data in Table 17-2.

	Family		Genus		Species	
	Number	%	Number	%	Number	%
PITTAS						
Lowland Papua New Guinea	1	100	1	100	3	100
Near Oceania	1	100	1	100	1	33
Remote Oceania	0	0	0	0	0	0
PASSERIDA						
Lowland Papua New Guinea	10	100	23	100	42	100
Near Oceania	6	60	12	52	14	33
Remote Oceania	4	40	5	22	5	12
VAGILE CORVIDA						
Lowland Papua New Guinea	7	100	14	100	51	100
Near Oceania	5	71	7	50	12	24
Remote Oceania	4	57	6	43	2	4
SEDENTARY CORVIDA						
Lowland Papua New Guinea	10	100	55	100	128	100
Near Oceania	4	40	5	9	3	2
Remote Oceania	5	50	8	15	1	1
TOTAL						
Lowland Papua New Guinea	28	100	93	100	225	100
Near Oceania	16	57	28	30	30	13
Remote Oceania	13	46	20	22	7	3

Finally, the concept that landbirds with widespread mainland (continental) ranges are disproportionately widespread on islands (Graves & Gotelli 1983) may not apply in Oceania, where "mainland" or "continental" probably should refer to New Guinea, Australia, Indonesia, and the Philippines rather than mainland Asia. The range outside of Oceania for the 15 most widespread species varies from nil (three species) to the nearly cosmopolitan barn-owl *Tyto alba*. Other species, not listed in Table 17-1, that are widespread outside of Oceania may occur in only one to several island groups in western Oceania, such as the cormorant *Phalacrocorax carbo*, hawks *Aviceda subcristata* and *Haliastur indus*, jacana *Irediparra gallinacea*, plover *Charadrius dubius*, nightjar *Caprimulgus indus*, and kingfisher *Alcedo atthis*.

Faunal Attenuation

The overall west-to-east decline in taxonomic diversity of landbirds in Oceania has been called "faunal attenuation" (Keast 1996). I will analyze faunal attenuation in three ways. One is from the standpoint of being shared between lowland New Guinea and Near Oceania or Remote Oceania (Tables 17-2 to 17-4). While it is difficult to say in which of these regions each species originated, the overall dispersal trend most likely is eastward and must be west to east for many species that just barely reach Oceania (Mayr & Diamond 2001:93–111). The second perspective is that of species-level representation in genera shared between lowland New Guinea and Oceania, whether or not the species are the same (Table 17-5). This analysis, unlike the last one, begins to accommodate speciation. The third perspective is by island group, at the levels of family, genus, and species (Tables 17-6, 17-7). These analyses more fully account for speciation than Keast (1996) and provide a more refined look at real and potential species richness values across Oceania. Although all tables in this chapter incorporate data from the growing late Quaternary fossil record, the total species richness values still are underestimates in every island group.

About 62 families, 207 genera, and 426 species of resident "landbirds" (including herons, ducks, and other freshwater birds) inhabit the lowlands (<1000 m elevation) of eastern New Guinea (Table 17-2). Determining what is an appropriate source area is difficult in any region (Graves & Gotelli 1983). Australia aside, precisely which part of New Guinea provides the best comparison with Oceania I cannot say. Keast (1996: Table 1), for example, listed 68 families/subfamilies, 232 genera, and 565 species for his New Guinean "source area." Mayr and Diamond (2001: Table 8.1) listed 64 families, an unstated number of genera, and 432 species. Of the New Guinean landbirds that I include in Table 17-2, 73% of the families, 45% of the genera, and 24% of the species are known today or prehistorically from Near Oceania, compared to 55%, 30%, and 9%, respectively, from Remote Oceania (Table 17-3). Nearly half (17 of 39) of the species shared by New Guinea and Remote Oceania are aquatic (grebes, cormorants, herons, ducks, osprey, rails, swallow).

New Guinean passerines are much more poorly represented than nonpasserines in Oceania at all taxonomic levels. Among the 128 species of New Guinean passerines (in 10 families and 55 genera) known as the "sedentary corvida" (see Mayr & Diamond 2001; Chapter 14 herein), only three species reach Near Oceania and just one makes it to Remote Oceania (Table 17-4), this being the woodswallow *Artamus leucorhynchus*, an aerial insectivore of open habitats in New Caledonia, Vanuatu, Fiji, and Palau, although with unresolved species-level systematics (Chapter 6).

Another way to compare the New Guinean avifauna with that of Oceania is in the number of species found in the 102 shared genera (in 50 shared families), whether or not the species are shared (Table 17-5). Only in two genera, the megapode *Megapodius* (Chapter 9) and swamphen *Porphyrio* (Chapter 10), does the number of species on any oceanic island exceed that on New Guinea. In both cases, New Guinea has one species, whereas multiple species are found on oceanic islands when bones are considered. If New Guinea had a decent fossil record, it could be that multiple species of *Megapodius* or *Porphyrio* once lived there as well. The overall message from Table 17-5 is that, even though oceanic islands are regarded as places that promote speciation, almost never does a single island in Oceania have more species in a genus shared with New Guinea.

The number of sympatric congeneric species in Near Oceania vs. Remote Oceania is greater in 62 of the 102 genera, the same in 34 genera, and less in only six genera. The last are cormorants (*Phalacrocorax*) on Rennell (a Solomon outlier; Chapter 13), herons (*Egretta*) in New Caledonia, Fiji, and Tonga (Chapters 5, 6, 13), ducks (*Anas*) on Rennell, New Caledonia, and Vanuatu (Chapters 5, 13), megapodes (*Megapodius*) in Tonga (Chapters 6, 9, 20), crakes (*Porzana*) in the Cook Islands (Chapters 7, 10), and trillers (*Lalage*) on many islands in Vanuatu, Tonga, and Samoa (Chapters 5, 6, 14). Only in the last three genera are species endemic to Oceania involved.

My distributional summary of landbirds in Oceania (Tables 17-6, 17-7) differs from that in Keast (1996: Table 1) in several major ways aside from having a slightly smaller New Guinean source pool of species, and minor differences in taxonomy or geography. First, I name each genus and

Table 17-5 Species richness of landbirds in genera shared by lowland Papua New Guinea and Oceania. These numbers are the highest values for single islands in each island group. Both modern and prehistoric records are considered.

		Species		
Family	Genus	Lowland Papua New Guinea	Near Oceania	Remote Oceania
CASSOWARIES	*Casuarius*	3	2	—
GREBES	*Tachybaptus*	2	1	1
CORMORANTS	*Phalacrocorax*	3	1	2
HERONS	*Egretta*	3	1	2
	Butorides	1	1	1
	Nycticorax	1	1	1
	Ixobrychus	1	1	1
IBISES	*Threskiornis*	1	—	1
DUCKS	*Dendrocygna*	2	2	1
	Anas	3	1	2
	Aythya	1	—	1
OSPREYS	*Pandion*	1	1	1
HAWKS	*Aviceda*	1	1	—
	Henicopernis	1	1	—
	Haliastur	2	1	—
	Haliaeetus	1	1	—
	Circus	1	—	1
	Accipiter	7	4	3
FALCONS	*Falco*	3	2	1
MEGAPODES	*Megapodius*	1	2	3
QUAILS	*Coturnix*	2	1	—
BUTTON-QUAILS	*Turnix*	1	1	1
RAILS	*Rallina*	1	1	1
	Gallirallus	2	2	1
	Gymnocrex	1	1	—
	Porzana	2	1	2
	Poliolimnas	1	1	1
	Amaurornis	1	1	—
	Gallinula	1	1	1
	Porphyrio	1	2	1
JACANAS	*Irediparra*	1	1	—
THICK-KNEES	*Esacus*	1	1	—
PLOVERS	*Charadrius*	1	1	—
PIGEONS, DOVES	*Columba*	1	1	1
	Macropygia	2	2	1
	Reinwardtoena	1	1	—
	Ptilinopus	10	3	2

(continued)

Table 17-5 (continued)

		Species		
Family	Genus	Lowland Papua New Guinea	Near Oceania	Remote Oceania
	Ducula	5	4	2
	Gymnophaps	1	1	—
	Chalcophaps	2	1	1
	Henicophaps	1	1	—
	Gallicolumba	3	2	2
PARROTS	*Cacatua*	2	1	1
	Chalcopsitta	2	1	—
	Trichoglossus	1	1	1
	Lorius	2	1	1
	Charmosyna	5	3	1
	Micropsitta	3	2	1
	Geoffroyus	2	1	1
	Eclectus	1	1	1
	Loriculus	1	1	—
CUCKOOS	*Cacomantis*	2	1	1
	Chrysococcyx	2	1	1
	Eudynamys	1	1	—
	Scythrops	1	1	—
	Centropus	3	1	—
BARN-OWLS	*Tyto*	3	2	2
STRIGID OWLS	*Ninox*	4	1	1
OWLET-NIGHTJARS	*Aegotheles*	5	—	1
NIGHTJARS	*Eurostopodus*	1	1	—
	Caprimulgus	1	1	1
SWIFTS	*Collocalia*	5	4	3
CRESTED SWIFTS	*Hemiprocne*	1	1	1
KINGFISHERS	*Alcedo*	1	1	—
	Ceyx	3	2	—
	Halcyon	4	3	2
	Tanysiptera	5	1	—
BEE-EATERS	*Merops*	2	1	—
ROLLERS	*Eurystomus*	1	1	—
HORNBILLS	*Aceros*	1	1	1
PITTAS	*Pitta*	3	1	—
SWALLOWS	*Hirundo*	1	1	1
OLD WORLD FLYCATCHERS	*Saxicola*	1	1	1
THRUSHES	*Turdus*	1	1	1
	Zoothera	1	1	—

Table 17-5 (continued)

Family	Genus	Species		
		Lowland Papua New Guinea	Near Oceania	Remote Oceania
CISTICOLAS	*Cisticola*	2	1	—
OLD WORLD WARBLERS	*Acrocephalus*	1	1	1
	Phylloscopus	1	1	—
	Megalurus	2	1	—
SUNBIRDS	*Nectarinia*	2	2	—
FLOWERPECKERS	*Dicaeum*	1	1	—
WHITE-EYES	*Zosterops*	3	3	2
STARLINGS	*Aplonis*	3	3	2
	Mino	2	2	—
ESTRILDID FINCHES	*Erythrura*	2	1	1
	Lonchura	7	1	—
CUCKOO-SHRIKES	*Coracina*	10	4	1
	Lalage	3	1	2
ACANTHIZID WARBLERS	*Gerygone*	6	—	1
ROBINS	*Monachella*	1	1	—
	Eopsaltria	1	—	1
WHISTLERS	*Pachycephala*	7	2	1
	Colluricincla	2	—	1
FANTAILS	*Rhipidura*	10	3	1
MONARCHS	*Monarcha*	8	2	1
	Myiagra	4	2	2
HONEYEATERS	*Lichmera*	2	—	1
	Myzomela	7	1	1
	Philemon	4	1	1
DRONGOS	*Dicrurus*	2	1	—
WOOD-SWALLOWS	*Artamus*	3	1	1
CROWS	*Corvus*	2	1	1
TOTAL				
Nonpasserines 29	70	146	92	60
Passerines 21	32	105	42	25
All 50	102	251	134	85

Table 17-6 Number of resident species in genera of nonpasserine landbirds in Oceania. "Bismarcks" includes the Admiralty Islands. "Remote Central Pacific Islands" = the Marshall Islands and Nauru eastward through the Line Islands (see Chapter 8). *, no endemic species known in Oceania; †, extinct species. Based on data in Chapters 5–13. These values represent the maximum number of congeneric species found on a single island in each island group. See text for additional explanation.

	Bismarcks	Solomon Main Chain	Solomon Outliers	Vanuatu	New Caledonia	Fiji	Tonga	Samoa	West Polynesian Outliers	Cook Islands	Society Islands	Marquesas
CASSOWARIES												
*Casuarius**	2	—	—	—	—	—	—	—	—	—	—	—
GREBES												
*Tachybaptus**	1	—	1	1	1	1	—	—	—	—	—	—
CORMORANTS												
*Phalacrocorax**	1	1	2	1	1	—	—	—	—	—	—	—
HERONS												
*Egretta**	1	1	1	1	2	1	1	1	1	1	1	1
*Butorides**	1	1	1	1	1	1	—	—	—	—	1	—
Nycticorax	1	2	—	—	1	—	1	—	1	1	—	—
*Ixobrychus**	1	1	1	—	—	—	—	—	—	—	—	—
*Botaurus**	—	—	—	—	1	—	—	—	—	—	—	—
IBISES												
*Threskiornis**	—	—	1	—	—	—	—	—	—	—	—	—
DUCKS												
Dendrocygna	2	—	—	—	1	1	—	—	—	1	—	—
Anas	1	1	2	2	2	1	1	1	1	1	1	—
*Aythya**	—	—	1	1	1	—	—	—	—	—	—	—
OSPREYS												
*Pandion**	1	1	1	—	1	—	1	—	—	—	—	—
HAWKS												
*Aviceda**	1	1	—	—	—	—	—	—	—	—	—	—
Henicopernis	1	—	—	—	—	—	—	—	—	—	—	—
*Haliastur**	1	1	—	—	1	—	—	—	—	—	—	—
Haliaeetus	1	1	—	—	—	—	—	—	—	—	—	—
*Circus**	—	—	—	1	1	1	1	—	—	—	—	—
Accipiter	4	3	1	1	3	1	1	—	—	—	—	—
FALCONS												
*Falco**	2	1	—	1	1	1	—	1	—	—	—	—
MEGAPODES												
Megapodius	2	2	1	2	2	2	4	1	1	—	—	—
Megavitiornis	—	—	—	—	—	1	—	—	—	—	—	—
Sylviornis	—	—	—	—	1	—	—	—	—	—	—	—
QUAILS												
*Coturnix**	2	—	—	—	—	—	—	—	—	—	—	—
BUTTON-QUAILS												
Turnix	1	1	—	—	1	—	—	—	—	—	—	—

Tuamotu	Tubuai	Pitcairn	Easter Island	Palau	Yap	Marianas	Chuuk	Pohnpei	Kosrae	Remote Central Pacific Islands
—	—	—	—	—	—	—	—	—	—	—
—	—	—	—	—	—	—	—	—	—	—
—	—	—	—	1	—	—	—	—	—	—
1	1	—	1?	1	1	1	1	1	1	1
—	—	—	—	—	—	—	—	—	—	—
—	—	—	—	1	—	—	1	—	—	—
—	—	—	—	1	1	1	1	—	—	—
—	—	—	—	—	—	—	—	—	—	—
—	—	—	—	—	—	—	—	—	—	—
—	—	—	—	—	—	—	—	—	—	—
—	1	—	—	1	—	1	1	—	—	1
—	—	—	—	—	—	—	—	—	—	—
—	—	—	—	—	—	—	—	—	—	—
—	—	—	—	—	—	—	—	—	—	—
—	—	—	—	—	—	—	—	—	—	—
—	—	—	—	—	—	—	—	—	—	—
—	—	—	—	—	—	—	—	—	—	—
—	—	—	—	—	—	—	—	—	—	—
—	—	—	—	—	—	—	—	—	—	—
—	—	—	—	—	—	—	—	—	—	—
—	—	—	—	—	—	1	—	—	—	—
—	—	—	—	1	—	1	—	1	—	—
—	—	—	—	—	—	—	—	—	—	—
—	—	—	—	—	—	—	—	—	—	—
—	—	—	—	—	—	—	—	—	—	—

(*continued*)

Table 17-6 (continued)

	Bismarcks	Solomon Main Chain	Solomon Outliers	Vanuatu	New Caledonia	Fiji	Tonga	Samoa	West Polynesian Outliers	Cook Islands	Society Islands	Marquesas
RAILS												
*Rallina**	1	—	—	—	—	—	—	—	—	—	—	—
Gallirallus	2	3	1	1	1	2	1	1	1	1	1	1
†*Vitirallus*	—	—	—	—	—	1	—	—	—	—	—	—
*Gymnocrex**	1	—	—	—	—	—	—	—	—	—	—	—
Porzana	1	1	1	2	1	1	1	1	1	3	1	1
*Poliolimnas**	1	1	1	1	1	1	—	1	—	—	—	—
*Amaurornis**	1	1	—	—	—	—	—	—	—	—	—	—
Gallinula	—	—	—	—	1	—	—	—	—	—	—	—
Pareudiastes	—	1	—	—	—	—	—	1	—	—	—	—
Porphyrio	2	2	1	1	2	1	1	1	1	1	—	1
KAGUS												
Rhynochetos	—	—	—	—	2	—	—	—	—	—	—	—
JACANAS												
*Irediparra**	1	—	—	—	—	—	—	—	—	—	—	—
STILTS												
*Himantopus**	1	—	—	—	—	—	—	—	—	—	—	
THICK-KNEES												
*Esacus**	1	1	—	—	1	—	—	—	—	—	—	—
PLOVERS												
*Charadrius**	1	—	—	—	—	—	—	—	—	—	—	—
SANDPIPERS												
Coenocorypha?	—	—	—	—	1	—	—	—	—	—	—	—
Prosobonia	—	—	—	—	—	—	—	—	—	1	1	1
COLUMBIDS												
Columba	2	2	—	1	1	1	—	1	—	—	—	—
Macropygia	3	1	1	1	—	—	—	—	—	—	1	1
Reinwardtoena	1	1	—	—	—	—	—	—	—	—	—	—
Ptilinopus	4	3	1	2	1	3	2	2	1	1	1	2
Drepanoptila	—	—	—	—	1	—	—	—	—	—	—	—
Ducula	5	3	1	2	1	3	2	1	2	2	2	1
Gymnophaps	1	1	—	—	—	—	—	—	—	—	—	—
†Undescribed C	—	—	—	—	—	—	1	—	—	—	—	—
Chalcophaps	1	1	1	1	1	—	—	—	—	—	—	—
Henicophaps	1	—	—	—	—	—	—	—	—	—	—	—
Caloenas	1	1	1	—	1	—	1	—	—	—	—	—
Didunculus	—	—	—	—	—	—	1	1	—	—	—	—
Gallicolumba	2	2	1	1	1	1	1	1	1	2	2	2
Microgoura	—	1	—	—	—	—	—	—	—	—	—	—
†Undescribed A	—	1	—	—	—	—	—	—	—	—	—	—

Tuamotu	Tubuai	Pitcairn	Easter Island	Palau	Yap	Marianas	Chuuk	Pohnpei	Kosrae	Remote Central Pacific Islands
—	—	—	—	1	—	—	—	—	—	—
—	—	—	1?	1	—	1	—	—	—	1
—	—	—	—	—	—	—	—	—	—	—
—	—	—	—	—	—	—	—	—	—	—
1	1	1	1	—	—	1	—	—	1	—
—	—	—	—	1	1	1	1	1	—	1
—	—	—	—	—	—	—	—	—	—	—
—	—	—	—	1	—	1	—	—	—	—
—	—	—	—	1	—	1	—	—	—	—
—	—	—	—	—	—	—	—	—	—	—
—	—	—	—	—	—	—	—	—	—	—
—	—	—	—	—	—	—	—	—	—	—
—	—	—	—	—	—	—	—	—	—	—
—	—	—	—	—	—	—	—	—	—	—
—	—	—	—	—	—	—	—	—	—	—
—	—	—	—	—	—	—	—	—	—	—
1	—	1	—	—	—	—	—	—	—	1
—	—	—	—	—	—	—	—	—	—	—
—	—	—	—	—	—	—	—	—	—	—
—	—	—	—	—	—	—	—	—	—	—
1	1	1	—	1	—	1	1	1	1	1
—	—	—	—	—	—	—	—	—	—	—
1	—	2	—	1	1	1	1	1	1	1
—	—	—	—	—	—	—	—	—	—	—
—	—	—	—	—	—	—	—	—	—	—
—	—	—	—	—	—	—	—	—	—	—
—	—	—	—	—	—	—	—	—	—	—
—	—	—	—	1	—	—	—	—	—	—
—	—	—	—	—	—	—	—	—	—	—
2	—	1	—	1	1	2	1	1	—	—
—	—	—	—	—	—	—	—	—	—	—
—	—	—	—	—	—	—	—	—	—	—

(continued)

Table 17-6 (continued)

	Bismarcks	Solomon Main Chain	Solomon Outliers	Vanuatu	New Caledonia	Fiji	Tonga	Samoa	West Polynesian Outliers	Cook Islands	Society Islands	Marquesas
†Undescribed B	—	1	—	—	—	—	—	—	—	—	—	—
†*Natunaornis*	—	—	—	—	1	—	—	—	—	—	—	—
PARROTS												
Cacatua	1	1	—	—	1	—	—	—	—	—	—	—
Chalcopsitta	1	1	—	—	—	—	—	—	—	—	—	—
Trichoglossus	1	1	1	1	1	—	—	—	—	—	—	—
Lorius	2	1	1	—	—	—	—	—	—	—	—	—
Vini	—	—	—	—	—	2	2	1	1	3	3	3
Charmosyna	2	3	1	1	1	1	—	—	—	—	—	—
Micropsitta	2	2	1	—	—	—	—	—	—	—	—	—
Prosopeia	—	—	—	—	—	1	—	—	—	—	—	—
Eunymphicus	—	—	—	—	1	—	—	—	—	—	—	—
Cyanoramphus	—	—	—	—	1	—	—	—	—	—	1	—
Geoffroyus	1	1	1	—	—	—	—	—	—	—	—	—
Eclectus	1	1	—	1	—	—	1	—	—	—	—	—
Loriculus	1	—	—	—	—	—	—	—	—	—	—	—
Genus uncertain	—	—	—	—	—	—	—	—	—	—	—	—
CUCKOOS												
*Cacomantis**	1	1	—	1	1	1	1	—	—	—	—	—
*Chrysococcyx**	—	1	1	1	1	—	—	—	—	—	—	—
*Eudynamys**	1	1	—	—	—	—	—	—	—	—	—	—
Centropus	1	1	—	—	—	—	—	—	—	—	—	—
BARN-OWLS												
Tyto	2	1	1	1	2	1	1	1	1	—	—	—
OWLS												
Pyrroglaux	—	—	—	—	—	—	—	—	—	—	—	—
Ninox	1	1	—	—	1	—	—	—	—	—		
*Asio**	—	—	—	—	—	—	—	—	—	—	—	—
Nesasio	—	1	—	—	—	—	—	—	—	—	—	—
NIGHTJARS												
*Eurostopodus**	—	1	—	—	1	—	—	—	—	—	—	—
*Caprimulgus**	1	1	—	—	—	—	—	—	—	—	—	—
OWLET-NIGHTJARS												
Aegotheles	—	—	—	—	1	—	—	—	—	—	—	—
FROGMOUTHS												
New genus	—	1	—	—	—	—	—	—	—	—	—	—
CRESTED SWIFTS												
*Hemiprocne**	1	1	1	—	—	—	—	—	—	—	—	—
SWIFTS												
Collocalia	4	4	2	2	2	1	1	1	1	1	1	1

Tuamotu	Tubuai	Pitcairn	Easter Island	Palau	Yap	Marianas	Chuuk	Pohnpei	Kosrae	Remote Central Pacific Islands
—	—	—	—	—	—	—	—	—	—	—
—	—	—	—	—	—	—	—	—	—	—
—	—	—	—	—	—	—	—	—	—	—
—	—	—	—	—	—	—	—	—	—	—
—	—	—	—	—	—	—	—	1	—	—
—	—	—	—	—	—	—	—	—	—	—
1	1	1	—	—	—	—	—	—	—	—
—	—	—	—	—	—	—	—	—	—	—
—	—	—	—	—	—	—	—	—	—	—
—	—	—	—	—	—	—	—	—	—	—
—	—	—	—	—	—	—	—	—	—	—
—	—	—	—	—	—	—	—	—	—	—
—	—	—	—	—	—	—	—	—	—	—
—	—	—	—	—	—	—	—	—	—	—
—	—	—	—	—	—	—	—	—	—	—
—	—	—	1	—	—	1	—	—	—	—
—	—	—	—	—	—	—	—	—	—	—
—	—	—	—	—	—	—	—	—	—	—
—	—	—	—	—	—	—	—	—	—	—
—	—	—	—	—	—	—	—	—	—	—
—	—	—	—	—	—	—	—	—	—	—
—	—	—	—	1	—	—	—	—	—	—
—	—	—	—	—	—	—	—	—	—	—
—	—	—	—	—	—	—	—	1	—	—
—	—	—	—	—	—	—	—	—	—	—
—	—	—	—	—	—	—	—	—	—	—
—	—	—	—	1	—	—	—	—	—	—
—	—	—	—	—	—	—	—	—	—	—
—	—	—	—	—	—	—	—	—	—	—
—	—	—	—	—	—	—	—	—	—	—
—	—	—	—	1	1	1	1	1	1	—

(*continued*)

Table 17-6 (continued)

	Bismarcks	Solomon Main Chain	Solomon Outliers	Vanuatu	New Caledonia	Fiji	Tonga	Samoa	West Polynesian Outliers	Cook Islands	Society Islands	Marquesas
KINGFISHERS												
*Alcedo**	1	1	—	—	—	—	—	—	—	—	—	—
Ceyx	2	2	—	—	—	—	—	—	—	—	—	—
Halcyon	3	3	2	2	1	1	1	1	1	1	2	1
Actenoides	—	1	—	—	—	—	—	—	—	—	—	—
*Tanysiptera**	1	—	—	—	—	—	—	—	—	—	—	—
BEE-EATERS												
*Merops**	1	—	—	—	—	—	—	—	—	—	—	—
ROLLERS												
*Eurystomus**	1	1	—	—	—	—	—	—	—	—	—	—
HORNBILLS												
Aceros	1	1	—	—	1	—	—	—	—	—	—	—
TOTAL												
Families	28	21	16	14	23	13	12	10	9	8	8	7
Genera	64	59	31	28	46	27	22	18	14	14	14	12
Species	95	81	35	35	55	34	28	19	15	20	19	16
KEAST TOTALS												
Families	23	23	—	13	19	11	7	9	—	—	7	6
Genera	58	58	—	28	42	28	14	17	—	—	7	8
Species	89	81	—	32	45	32	15	19	—	—	10	9
Number of identified fossils	226	77	73	410	6453	400+	2482	56	452	1653+	107	2427

evaluate it independently. Second, to beat a dead horse, my values incorporate fossil data. This yields major increases in taxonomic diversity in any island group with a substantial fossil record, such as New Caledonia, Tonga, or the Society Islands. Thus my data are closer to reality than those of Keast (1996), although they still underestimate the full range and diversity of many taxa and most if not all island groups. As more fossils are studied, the totals in these tables will increase further. I must note as well that the fossil record is biased toward nonpasserines (see Chapters 4, 14). Outside of New Zealand and the Hawaiian Islands, passerine fossils have been studied extensively in only four Pacific island groups (Table 17-7). In Tonga, they have increased the passerine fauna from five to 13 species (Table 6-10), whereas in the other three groups the changes have been nil or just one species.

The third major difference is that I compare the avifaunas of different island groups by using species richness values that are the highest for that genus for any one island in the group, thereby filtering out interarchipelago variation in the number of islands, which inflate Keast's species richness values for allospecies. My method, therefore, yields lower species values for most genera with evidence of speciation within an island group. For example, although nine species of white-eyes (*Zosterops* spp.) occur in the Solomon Main Chain (Table 5-7), *Zosterops* has a value of three in Table 17-7 because that is the most species recorded on any one island. An archipelago with 20 islands might have had 20 endemic flightless species of *Gallirallus* (one per island at first human contact), whereas a group of just five islands probably would have had only five such species (again, one per island). In either case, *Gallirallus* would have a value of 1 in Table 17-6. Considering only sympatric congeners also eliminates the difficulty of deciding whether distinctive allo-populations should be classified as multiple species, as multiple subspecies of a single species, or a blend thereof.

Tuamotu	Tubuai	Pitcairn	Easter Island	Palau	Yap	Marianas	Chuuk	Pohnpei	Kosrae	Remote Central Pacific Islands
—	—	—	—	—	—	—	—	—	—	—
—	—	—	—	—	—	—	—	—	—	—
1	—	—	—	2	—	1	1	1	—	—
—	—	—	—	—	—	—	—	—	—	—
—	—	—	—	—	—	—	—	—	—	—
—	—	—	—	—	—	—	—	—	—	—
—	—	—	—	—	—	—	—	—	—	—
—	—	—	—	—	—	—	—	—	—	—
6	5	4	3	10	4	9	6	8	4	5
8	5	6	4	19	6	16	10	10	5	7
9	5	7	4	20	6	17	10	10	5	7
—	—	3	—	—	—	—	—	—	—	—
—	—	3	—	—	—	—	—	—	—	—
—	—	3	—	—	—	—	—	—	—	—
4	1	1367	5	5	4	1648+	0	3	0	0

Near Oceania (Bismarcks and Solomon Main Chain) is taxonomically more diverse than anywhere in Remote Oceania (Tables 17-6, 17-7). New Caledonia is Remote Oceania's richest archipelago at any taxonomic level, reflecting its old age and glacial nearness to Australia via former islands in the Chesterfield Reef region (Chapter 1). I would caution, however, that some of this richness also is because New Caledonia has a relatively good nonpasserine fossil record (Chapter 5). Island groups with conspicuously low values in Tables 17-6 and 17-7, such as the Solomon Outliers and Tubuai, will need many more fossils to estimate how depauperate their avifaunas really were at human contact. Of course I would expect substantial increases in taxonomic representation in both cases.

A quick look at Tonga and the Cook Islands demonstrates the importance of prehistory in evaluating faunal attenuation. Keast (1996) listed 20 species of landbirds for Tonga and 33 for nearby Samoa, which has two islands (Upolu, Savai'i) that each have more land area than all Tongan islands combined. Tonga's fossil record increases its avifaunal total to 38 or 39 species, while essentially fossil-free Samoa remains at 33 species. My higher value for Tonga at least approaches reality, although any comparison with the Samoan avifauna remains compromised by the near absence of Samoan fossils. The same may be said for the Cook Islands vs. Tubuai, two adjacent island groups with a shared geological history (Chapter 1). Mangaian fossils have boosted the Cook Island landbird fauna from 10 to 22 or 23 species, whereas that of Tubuai, in a near absence of fossils, remains at six species (Tables 17-6, 17-7). Assuming that the islands of Tubuai experienced comparable levels of anthropogenic extinction as Mangaia, then the Tubuai and Cook Island avifaunas probably were very similar at human arrival.

A last way to look at faunal attenuation is to evaluate the birds of each hemisphere of Oceania as a percentage of the richest, most diverse island group, these being the Bismarcks for the southern hemisphere (Melanesia + Polynesia) and

Table 17-7 Number of resident species in genera of passerines in Oceania. "Bismarcks" includes the Admiralty Islands. "Remote Central Pacific Islands" = the Marshall Islands and Nauru eastward through the Line Islands (see Chapter 8). *, no endemic species known in Oceania. Based on data in Chapters 5–12. These values represent the maximum number of congeneric species found on a single island in each island group. See text for additional explanation.

	Bismarcks	Solomon Main Chain	Solomon Outliers	Vanuatu	New Caledonia	Fiji	Tonga	Samoa	West Polynesian Outliers	Cook Islands	Society Islands
PITTAS											
Pitta	1	1	—	—	—	—	—	—	—	—	—
SWALLOWS											
*Hirundo**	1	1	1	1	1	1	1	—	—	—	1
THRUSHES											
Zoothera	1	1	—	—	—	—	—	—	—	—	—
*Turdus**	1	1	1	1	1	1	1	1	—	—	—
OLD WORLD FLYCATCHERS											
*Saxicola**	1	—	—	—	—	—	—	—	—	—	—
CISTICOLAS											
*Cisticola**	1	1	—	—	—	—	—	—	—	—	—
OLD WORLD WARBLERS											
Cettia	—	1	—	—	—	1	—	—	—	—	—
Acrocephalus	1	1	—	—	—	—	—	—	—	1	1
Phylloscopus	1	1	—	—	—	—	—	—	—	—	—
Megalurus	1	—	—	—	—	—	—	—	—	—	—
Cichlornis	1	1	—	1	1	—	—	—	—	—	—
Trichocichla	—	—	—	—	—	1	—	—	—	—	—
SUNBIRDS											
Nectarinia	2	1	—	—	—	—	—	—	—	—	—
FLOWERPECKERS											
Dicaeum	1	1	—	—	—	—	—	—	—	—	—
WHITE-EYES											
Zosterops	2	3	1	2	2	2	—	1	—	—	—
Rukia	—	—	—	—	—	—	—	—	—	—	—
Cleptornis	—	—	—	—	—	—	—	—	—	—	—
Woodfordia	—	—	1	—	—	—	—	—	—	—	—
Megazosterops	—	—	—	—	—	—	—	—	—	—	—
†Zosteropidae undescribed genus	—	—	—	—	—	—	1	—	—	—	—
STARLINGS											
Aplonis	2	4	2	2	1	1	1	2	1	1	1
Mino	2	1	—	—	—	—	—	—	—	—	—
ESTRILDID FINCHES											
Erythrura	1	1	—	2	1	2	—	1	—	—	—
*Lonchura**	3	1	—	—	—	—	—	—	—	—	—
CUCKOO-SHRIKES											
Coracina	3	5	1	1	2	—	—	—	—	—	—
Lalage	—	1	1	2	1	1	1	2	1	—	—

Marquesas	Tuamotu	Tubuai	Pitcairn	Easter Island	Palau	Yap	Marianas	Chuuk	Pohnpei	Kosrae	Remote Central Pacific Islands
—	—	—	—	—	—	—	—	—	—	—	—
—	—	—	1	—	—	—	—	—	—	—	—
—	—	—	—	—	—	—	—	—	—	—	—
—	—	—	—	—	—	—	—	—	—	—	—
—	—	—	—	—	—	—	—	—	—	—	—
—	—	—	—	—	—	—	—	—	—	—	—
—	—	—	—	—	—	—	—	—	—	—	—
—	—	—	—	—	1	—	—	—	—	—	—
1	1	1	1	—	—	—	1	1	1	1	1
—	—	—	—	—	—	—	—	—	—	—	—
	—	—	—	—	—	—	—	—	—	—	—
—	—	—	—	—	—	—	—	—	—	—	—
—	—	—	—	—	—	—	—	—	—	—	—
—	—	—	—	—	—	—	—	—	—	—	—
—	—	—	—	—	—	—	—	—	—	—	—
—											
—	—	—	—	—	2	1	1	1	2	1	—
—	—	—	—	—	—	1	—	1	1	—	—
—	—	—	—	—	—	—	1	—	—	—	—
—	—	—	—	—	—	—	—	—	—	—	—
—	—	—	—	—	1	—	—	—	—	—	—
—	—	—	—	—	—	—	—	—	—	—	—
—	—	—	—	—	1	1	1	1	2	2	—
—	—	—	—	—	—	—	—	—	—	—	—
—	—	—	—	—	1	—	1	1	1	1	—
—	—	—	—	—	—	—	—	—	—	—	—
—	—	—	—	—	1	1	—	—	1	—	—
—	—	—	—	—	—	—	—	—	—	—	—

(*continued*)

Table 17-7 (continued)

	Bismarcks	Solomon Main Chain	Solomon Outliers	Vanuatu	New Caledonia	Fiji	Tonga	Samoa	West Polynesian Outliers	Cook Islands	Society Islands
GERYGONES											
Gerygone	—	—	1	1	1	—	—	—	—	—	—
AUSTRALIAN ROBINS											
*Monachella**	1	—	—	—	—	—	—	—	—	—	—
Petroica	—	1	—	1	—	1	—	1	—	—	—
Eopsaltria	—	—	—	—	1	—	1	—	—	—	—
WHISTLERS											
Pachycephala	2	2	1	1	2	1	1	1	—	—	—
Colluricincla	—	—	—	—	—	—	—	—	—	—	—
FANTAILS											
Rhipidura	3	4	1	2	2	1	—	1	—	—	—
MONARCHS											
Pomarea	—	—	—	—	—	—	—	—	—	1	1
Mayrornis	—	—	1	—	—	1	—	—	—	—	—
Neolalage	—	—	—	1	—	—	—	—	—	—	—
Clytorhynchus	—	—	1	1	1	2	1	1	1	—	—
Metabolus	—	—	—	—	—	—	—	—	—	—	—
Monarcha	3	2	1	—	—	—	—	—	—	—	—
Myiagra	3	1	1	1	1	2	1	1	—	—	—
Lamprolia	—	—	—	—	—	1	—	—	—	—	—
HONEYEATERS											
Stresemannia	—	1	—	—	—	—	—	—	—	—	—
Guadalcanaria	—	1	—	—	—	—	—	—	—	—	—
Melidectes	1	1	—	—	—	—	—	—	—	—	—
Lichmera	—	—	—	1	1	—	—	—	—	—	—
Myzomela	3	2	1	1	1	1	1	1	1	—	—
Foulehaio	—	—	—	—	—	1	1	1	1	—	—
Philemon	1	—	—	—	1	—	—	—	—	—	—
Gymnomyza	—	—	—	—	1	1	—	1	—	—	—
Phylidonyris	—	—	—	1	1	—	—	—	—	—	—
DRONGOS											
Dicrurus	1	1	—	—	—	—	—	—	—	—	—
WOOD-SWALLOWS											
*Artamus**	1	—	1	1	1	1	—	—	—	—	—
CROWS											
Corvus	2	1	—	—	1	—	—	—	—	—	—
TOTAL											
Families	20	18	11	14	15	13	9	10	4	3	4
Genera	29	29	16	19	21	19	11	13	5	3	4
Species	47	44	17	24	25	23	11	15	5	3	4
KEAST TOTALS											
Families	17	17	—	14	15	12	5	10	—	—	2
Genera	31	26	—	19	21	22	5	13	—	—	2
Species	43	57	—	25	26	26	5	14	—	—	2
Number of identified fossils	29	0	2	83	0	0	822	7	34	105	3

Marquesas	Tuamotu	Tubuai	Pitcairn	Easter Island	Palau	Yap	Marianas	Chuuk	Pohnpei	Kosrae	Remote Central Pacific Islands
—	—	—	—	—	—	—	—	—	—	—	—
—	—	—	—	—	—	—	—	—	—	—	—
—	—	—	—	—	—	—	—	—	—	—	—
—	—	—	—	—	—	—	—	—	—	—	—
—	—	—	—	—	—	—	—	—	—	—	—
—	—	—	—	—	1	—	—	—	—	—	—
—	—	—	—	—	1	1	1	—	1	—	—
1	—	—	—	—	—	—	—	—	—	—	—
—	—	—	—	—	—	—	—	—	—	—	—
—	—	—	—	—	—	—	—	—	—	—	—
—	—	—	—	—	—	—	—	—	—	—	—
—	—	—	—	—	—	—	—	1	—	—	—
—	—	—	—	—	—	1	1	—	—	—	—
—	—	—	—	—	1	—	1	1	1	—	—
—	—	—	—	—	—	—	—	—	—	—	—
—	—	—	—	—	—	—	—	—	—	—	—
—	—	—	—	—	—	—	—	—	—	—	—
—	—	—	—	—	—	—	—	—	—	—	—
—	—	—	—	—	—	—	—	—	—	—	—
—	—	—	—	—	1	1	1	1	1	1	—
—	—	—	—	—	—	—	—	—	—	—	—
—	—	—	—	—	—	—	—	—	—	—	—
—	—	—	—	—	—	—	—	—	—	—	—
—	—	—	—	—	—	—	—	—	—	—	—
—	—	—	—	—	—	—	—	—	—	—	—
—	—	—	—	—	1	—	—	—	—	—	—
—	—	—	—	—	—	—	1	—	—	—	—
2	1	1	2	0	10	6	8	6	8	5	1
2	1	1	2	0	11	7	10	8	9	5	1
2	1	1	2	0	12	7	10	8	11	6	1
2	—	—	1	—	—	—	—	—	—	—	—
2	—	—	1	—	—	—	—	—	—	—	—
2	—	—	1	—	—	—	—	—	—	—	—
3	0	0	160	0	0	3	380+	0	0	0	2

Table 17-8 Taxonomic representation of landbirds in southern Oceania (Melanesia + Polynesia) as a percentage of the Bismarcks avifauna. Based on data in Tables 17-6 and 17-7. *, island groups with >500 landbird fossils. Abbreviations for island groups refer to Figure 17-2.

	Families			Genera			Species		
	Nonpasserine	Passerine	Total	Nonpasserine	Passerine	Total	Nonpasserine	Passerine	Total
Bismarcks* (BI)	100	100	100	100	100	100	100	100	100
Solomon Main Chain* (SM)	75	90	81	95	97	95	86	98	88
Solomon Outliers* (SO)	57	55	54	48	50	51	37	36	37
Vanuatu (VA)	50	70	58	44	63	51	37	53	42
New Caledonia (NC)	82	75	79	72	70	72	58	56	56
Fiji (FI)	46	65	54	44	63	49	36	51	40
Tonga (TO)	43	45	44	36	37	35	29	24	27
Samoa* (SA)	36	50	42	26	43	33	19	33	24
West Polynesian Outliers* (WP)	32	20	27	22	17	20	16	11	14
Cook Islands (CI)	29	15	23	22	10	17	21	7	15
Society Islands* (SI)	29	20	25	22	13	19	20	9	16
Marquesas (MA)	25	10	19	19	7	15	17	4	13
Tuamotu* (TM)	22	5	15	12	3	10	9	2	7
Tubuai* (TB)	18	5	13	8	3	6	5	2	4
Pitcairn (PI)	14	10	13	9	7	9	7	4	6
Easter Island* (EI)	11	0	6	6	0	4	4	0	3

Table 17-9 Taxonomic representation of nonpasserine landbirds in Micronesia as a percentage of the Palau avifauna. Based on data in Tables 17-6 and 17-7. Abbreviations for island groups refer to Figure 17-3.

	Families			Genera			Species		
	Nonpasserine	Passerine	Total	Nonpasserine	Passerine	Total	Nonpasserine	Passerine	Total
Palau (PA)	100	100	100	100	100	100	100	100	100
Yap (YA)	40	60	50	32	64	43	30	58	41
Marianas (MA)	90	80	85	84	91	87	85	83	84
Chuuk (CH)	60	60	60	53	73	60	50	67	56
Pohnpei (PO)	80	80	80	53	82	63	50	92	66
Kosrae (KO)	40	50	45	26	45	33	25	50	34
Remote Central Pacific Islands (RC)	50	10	30	37	9	27	35	8	25

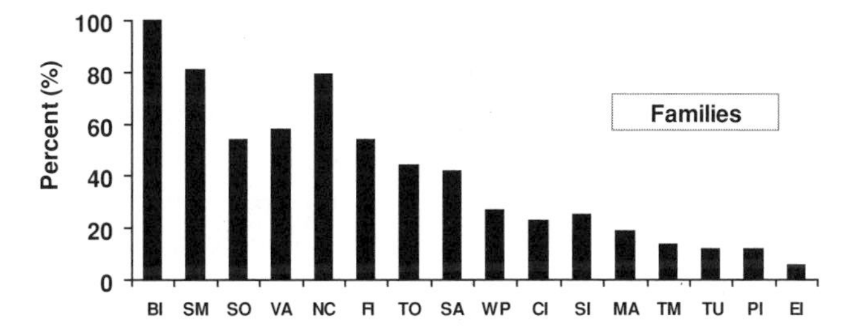

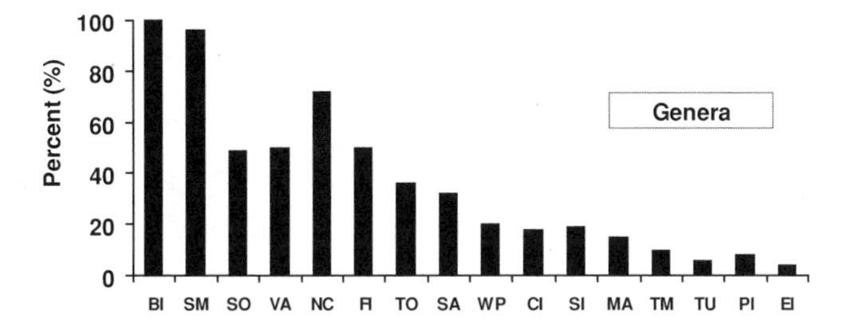

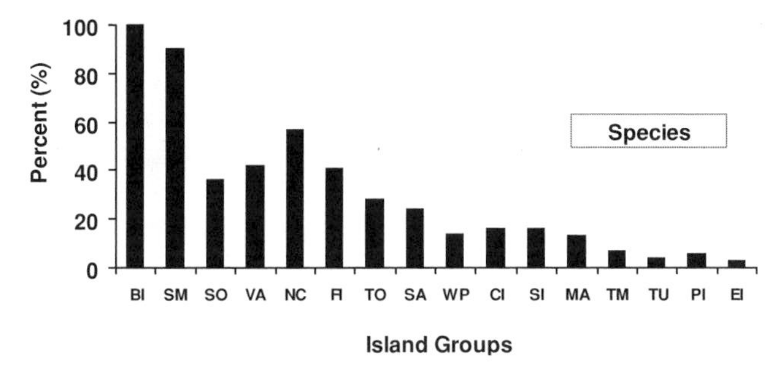

Figure 17-2 Representation of landbirds in southern Oceania (Melanesia & Polynesia) as a percentage of the Bismarcks avifauna, at three taxonomic levels. Abbreviations defined in Table 17-8.

Palau for the northern hemisphere (Micronesia). These analyses (Tables 17-8, 17-9) do not assume that either the Bismarcks or Palau are the source areas for the rest of their regions. Also keep in mind that the percentages in Tables 17-8 and 17-9 do not reflect perfect subsets, i.e., these are only partially nested faunas at any taxonomic level, as detailed in Tables 17-6 and 17-7. Furthermore, the data probably are grossly incomplete (i.e., <80% of the genera or species that existed at first human contact) for Solomon Outliers, Vanuatu, Samoa, West Polynesia Outliers, Tuamotu, Tubuai, and Easter Island (Tables 17-6 to 17-8, Figure 17-2) and for Palau, Yap, Chuuk, Pohnpei, Kosrae, and Remote Central Pacific Islands (Tables 17-6, 17-7, 17-9, Figure 17-3).

Landbird faunas are much more diverse in Near Oceania than Remote Oceania. Avifaunal attenuation within Near Oceania (Bismarcks vs. Solomon Main Chain) is more striking among families than genera or species (Figure 17-2). This reflects an absence in the Solomons of many families that have only one genus and species in the Bismarcks. For the 14 island groups of Remote Oceania, the level of avifaunal attenuation increases as taxonomic resolution goes from family to genus to species (Table 17-8, Figure 17-2). This is because losing a species does not necessarily mean losing a genus, nor does losing a genus always mean losing a family. This same trend is present but less striking in Micronesia (Table 17-9, Figure 17-3), where the data suggest that a substantial fossil record would greatly enhance the avifaunas of Yap or Chuuk, even though neither is as extensive a group of high islands as Palau or the Marianas.

In closing, isolation remains the major force in faunal attenuation, even when fossils are considered. Molecular genetics eventually can estimate the gene flow among extant populations that vary in geographic isolation. Such data should be interpreted in the context of direct observations of dispersal, as well as traditional species/subspecies-level systematics and fossil information. The observational data would be compromised in most island groups by how little fieldwork has taken place. In some island groups, ancient

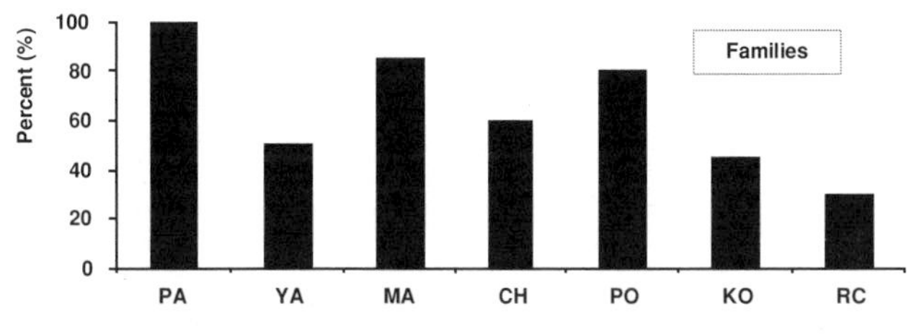

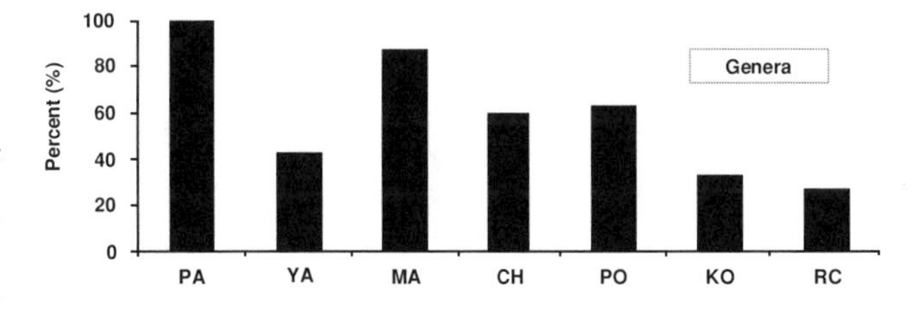

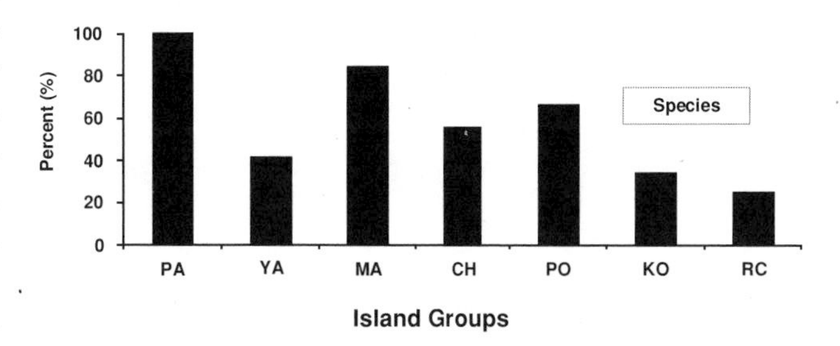

Figure 17-3 Representation of nonpasserine landbirds in Micronesia as a percentage of the Palau avifauna, at three taxonomic levels. Abbreviations defined in Table 17-9.

DNA could complement these studies by comparing extirpated populations and species with those that still exist, and by estimating genetic change on century or millennial timescales.

The date of initial dispersal is not well known for most avian taxa in Oceania and is a rich area for future research. Fossils from New Zealand, the Hawaiian Islands, and Tonga suggest that most of the indigenous species of landbirds (and seabirds) that existed at first human contact in the late Holocene had resided there for more than 100,000 years. What stimulated their original interarchipelago dispersal, and when it occurred, remains a mystery.

Chapter 18 Equilibrium and Turnover

THE EQUILIBRIUM MODEL of island biogeography proposes that the immigration and extinction of species are continuous and linked, resulting in "turnover" in the species composition of some broad taxon (birds, lizards, rodents, bats, all mammals, ants, vascular plants, etc.) on an island. The turnover is fueled primarily by interspecific interactions, such as competition, in which the colonizing species directly or indirectly eliminates one that already had been there. The result is an "equilibrium" value of species richness that remains fairly constant through time, even though species come and go through immigration and extinction.

The equilibrium model looks great on paper (Figures 18-1, 18-2) and has some empirical support from values for immigration, extinction, and species richness over short timescales (decadal or less) in undifferentiated biotas near rich (continental) source areas. As a generalization or even as a possibility for biotas on true oceanic islands, however, I do not believe in the connected concepts of turnover and equilibrium. My disbelief applies especially to landbirds. Without human influence, colonization of oceanic islands by landbirds is rare. Even though most species are volant, there is no Brownian motion of avian propagules hovering over the Pacific Ocean, randomly colonizing islands. This is biology, not physics. Also rare under natural conditions is the extinction of insular birds, which shows no evidence of being linked to colonization. This applies on any oceanic island (e.g., Jehl & Parkes 1983).

No studies of modern turnover in landbirds have been conducted in areas free of human influence because no sets of islands and source areas meet this criterion. Thus it is tempting to play hardball and simply ignore studies of modern turnover (and equilibrium). I will resist that temptation, but only partially, for to review all research on turnover, avian or otherwise, would be a huge task loaded with redundancy. Instead I will compare some representative examples of published turnover data and concepts with what I have learned about landbirds on Pacific islands.

In Remote Oceania and perhaps also in Near Oceania, I am unaware of the loss of an island population of nonaquatic landbird where the species subsequently recolonized under its own power. Since human arrival, we are dealing almost exclusively with loss, not loss and replacement. In other words, there is essentially no turnover today in native landbirds. With rare exceptions, faunal change is in one direction, and that is downward. This artificial decline also influences concepts of species-area relationships (Chapter 19) and community ecology (Chapter 20). During any given time interval, the rate at which indigenous species are lost depends on the level of human impact and how many species are left; once most of them are gone, the rate of extinction (number of species lost per unit time) decreases because fewer candidates are available.

This chapter will expand on the three themes already broached: equilibrium and turnover are flawed concepts for faunas on oceanic islands; colonization and extinction of tropical Pacific islands by landbirds are rare and not linked under natural conditions; and extinction rates and species-richness values are suspect once people show up because of drastic changes in bird communities and the ecosystems that

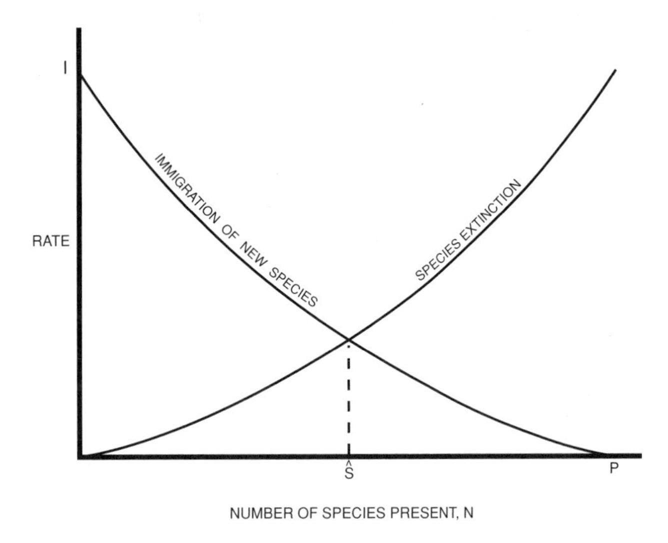

Figure 18-1 Equilibrium model of a fauna on a simple island. I, immigration; P, number of species in mainland source poll; Ŝ, equilibrium number of species. Redrawn from MacArthur and Wilson (1963: Figure 4).

sustain them. I will explain these themes in the context of the conceptual development of equilibrium and turnover, beginning with MacArthur and Wilson (1963, 1967) although aware of earlier efforts (see Brown & Lomolino 1989). I also will ask what would prevent tropical insular biotas, through immigration and speciation, from getting gradually richer through time, i.e., never reaching an equilibrium as long as an island's geological structure is adequate to sustain upland vegetation.

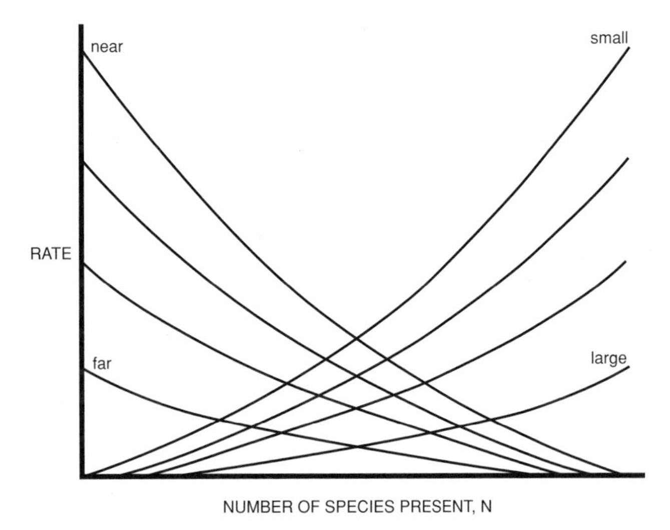

Figure 18-2 Equilibrium model of faunas of several islands that vary in isolation ("near" to "far") and size ("small" to "large"). The intersection of any immigration curve and any extinction curve yields the equilibrium number of species for that island. Redrawn from MacArthur and Wilson (1963: Figure 5).

Problems with Equilibrium and Turnover

Definitions, Inherent Limitations

MacArthur and Wilson (1967:187, 191) defined equilibrium as "the state in which . . . rate of extinction of species in a biota equals rate of immigration of new species" and turnover as "the process of extinction of some species and their replacement by other species." Newer definitions are much the same, e.g., "a condition of balance between opposing forces, such as . . . immigration and extinction rates" for equilibrium and "the rate of replacement of species in a particular area as some taxa become extinct but others immigrate from outside" for turnover (Brown & Lomolino 1998:629, 636).

These definitions would seem to be straightforward, but there is sloppiness in the meaning of "extinction." Biogeographers studying turnover and equilibrium often regard any absence, temporary or not, as an extinction. To me, extinction means the permanent (or essentially permanent, as far as we can determine) loss of a discrete subspecies, species, genus, etc. Extirpation of population(s) is the process through which extinction operates (Chapter 16). Extinction is not the temporary absence (coming and going through time, whether stochastic or not) of undifferentiated populations or individuals. Genuine extinction is why we have a biodiversity crisis that can be traced back, depending on the location, hundreds, thousands, or tens of thousands of years. That a tiny subpopulation (typically several pairs) of a species of British bird fails to breed on some offshore island or some patch of woods for a few years or even decades represents an absence, not an extinction. MacArthur and Wilson (1967:64) recognized this weakness in their model, but most subsequent researchers have not.

When MacArthur and Wilson proposed the equilibrium theory, the only supporting data were for Indo-Pacific landbirds. They closed by writing (MacArthur & Wilson 1963:386), "The main purpose of the paper is to express the criteria and implications of the equilibrium condition, without extending them for the present beyond the Indo-Australian bird faunas." They seem to have been more concerned with the initial buildup of species richness (on young islands?) toward "saturation" than with long-term maintenance of equilibrial faunas. In other words, their main interest was in the unspecified period when an island's fauna is growing, i.e., when the colonization rate exceeds the extinction rate.

Four years later, MacArthur and Wilson did extend the model to a few other places and taxonomic groups, but mainly used hypothetical examples to make claims such as

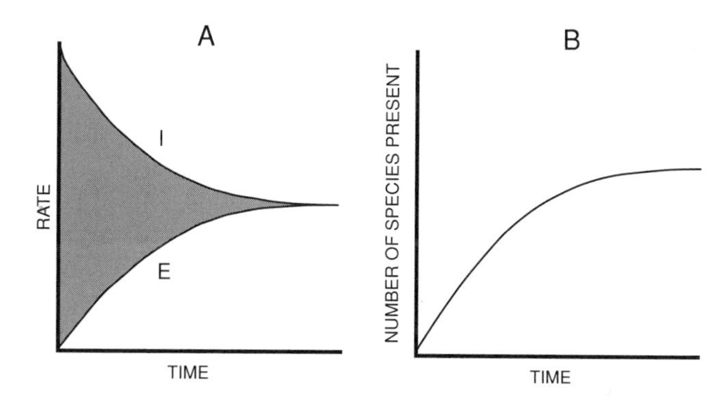

Figure 18-3 A: Integrating the rates of extinction (E) and immigration (I) through time on an island. B: The resulting growth in species richness to equilibrium. Redrawn from MacArthur and Wilson (1967: Figure 20).

(MacArthur & Wilson 1967:32), "The colonization of an island is a dynamic process, accompanied at all stages by turnover in species. During the buildup of species of a given taxon toward its equilibrial (saturation) number, the immigration and extinction rates probably vary as a function of the number of species present. It is reasonable to infer from this consideration alone that the variance of the number of species on different islands of a given size and degree of isolation will also vary with the number of species present."

For any island, MacArthur and Wilson (1967) believed that immigration rates (initially high, but decreasing with time) and "extinction" rates (initially low, then increasing) eventually cancel each other, yielding an equilibrium number for species richness (Figure 18-3). As with Figures 18-1 and 18-2, this is elegant on paper, although Figure 18-3 becomes problematic when you realize that the timescale has no real values and, as with Figures 18-1 and 18-2, neither the immigration nor the extinction rate is known for any oceanic island fauna.

MacArthur and Wilson made the circular claim that species richness can be calculated if we know the immigration and "extinction" curves. In fact, the only data they had about faunas from oceanic islands were species richness values, which themselves were flawed (Chapter 19). They had no measurements of immigration or extinction rates because their model was not calibrated with time-series data for any island except Krakatau, a continental rather than oceanic island (see next section). MacArthur and Wilson admitted (1967:66) that "the measurement of immigration and extinction rates in nature will be difficult to accomplish." For oceanic islands, most of which have had, by orders of magnitude, more time for faunal development than Krakatau, this has proven to be an understatement.

Five years later, MacArthur (1972) cited time-series studies by Diamond (1969, 1972c) and Simberloff and Wilson (1969, 1970) as evidence that some insular faunas really do undergo turnover and reach an equilibrium. None of these studies, however, involved faunas on remote islands. It seems that islands close to continents, not oceanic islands, were indeed what MacArthur (1972:97) had in mind: "As the number of species present on the island increases, fewer of the immigrants are new, so the immigration rate of new species falls, reaching zero when all *mainland* species are present" (emphasis added). No biologist ever would expect all "mainland" species to reach any oceanic island. The circularity continued, in a slightly modified form, when MacArthur (1972:99) stated: "Now, a curve falling to zero must intersect another rising from zero, and at the intersection, immigration and extinction are equal. If the number of species is larger than equilibrium, extinction exceeds immigration, causing a decline; if the number of species is less than equilibrium, immigration exceeds extinction and the number of species increases." The graphics, as in Figures 18-1 to 18-3, not the data, are driving the concept.

This leads to another problem. Nobody ever defined how much change through time (i.e., turnover) was permissible for a fauna to be regarded as in "equilibrium." Thus one could interpret any data set as evidence for or against equilibrium (Gilbert 1980). As noted by Simberloff (1983), a former supporter of equilibrium theory (Simberloff 1974), equilibrium without limits is meaningless. How much of an upward or downward trend in species richness, regardless of composition, should be allowed and still meet the conditions for equilibrium? Is a community equilibrial if the species composition varies, for example, by $\leq$10% between censuses and the species richness varies by $\leq$5%? What about $\leq$20% and $\leq$10%? How many censuses, at what time interval, would be appropriate to determine if the overall trend in species richness is upward, downward, or flat? Any answer to these questions would be arbitrary.

Another arbitrary part of the equilibrium model, noted as early as Sauer (1969), is that all species are equal, whether large or small, common or rare, herbivore or carnivore, canopy or ground dweller, etc. A contemporary version of this need to strip species of their virtues is the proposal by Hubbell (2001) to equate all individuals in a community, not just all species. Losing the properties that define them and allow them to live renders the species (or individuals) easy to manipulate on paper. Thus it becomes mathematically and graphically elegant, even if biologically naïve (Chapter 20), to suppose that the arrival of one species would set the stage for the loss of another, whether the hypothetical process driving to the loss is direct (one-on-one competition or perhaps predation) or diffuse (competition or perhaps predation from multiple species).

MacArthur and Wilson (1967:65) recognized the danger of assuming "that the extinction and immigration curves have a fairly regular shape for different faunas and different islands and for different times on the same island." They also noted that "when a new set of curves must be derived for a new situation, the model loses much of its virtue." An example of "a new set of curves" would be those describing faunas with a "permanent core of resident species" whose presence did not fluctuate through the unspecified time period (Diamond & May 1976). This modification may help to explain European landbirds on continental islands, but is of little use with landbirds on oceanic islands.

Among the practical phenomena that continue to haunt studies of equilibrium and turnover are "cryptoturnover" and "pseudoturnover," both defined by Simberloff (1976). The former is biotic change that has occurred but has gone unobserved by scientists. The latter is the opposite, where an observer believes that a colonization or "extinction" has taken place but it is really just an artifact of inadequate survey. The extent that cryptoturnover and pseudoturnover influence a data set depends on the skill of the fieldworker, the number of surveys, the amount of effort made during each survey, logistical difficulties of the area being surveyed, and how detectable the organisms are. Birds are widely (and perhaps rightly) regarded as relatively easy subjects for census, but studies of insular birds certainly are subject to both cryptoturnover and pseudoturnover. Just as fossil sampling rarely is thorough enough to record each species that was on an island in the past (Chapter 4), even the most skilled eyes and ears often fail to detect each species of landbird living on an island today, especially on rugged, malarial, nearly roadless, infrequently visited islands in Melanesia.

Spatial Scales, Timescales

The most influential research on turnover and equilibrium has been conducted on continental islands with undifferentiated species near rich source areas. Examples (many of them reviewed in Brown & Lomolino 1998:384–390) include insects on mangrove islets in the Florida Keys (Simberloff & Wilson 1969, 1970), birds on the Channel Islands, California (Diamond 1969, Lynch & Johnson 1974), birds on islands off Wales and England or in a "mainland" British woodlot (Lack 1969, Diamond & May 1977, Williamson 1981, Simberloff 1983, Pimm et al. 1988), and small mammals on small islands in the Great Lakes region (Lomolino 1993a).

The physical and biological settings for these analyses are fundamentally different from those of oceanic islands, where much longer dispersal distances are involved, where endemic taxa have evolved, and where (even to a considerable extent in Near Oceania; Salomonsen 1976, Mayr & Diamond 2001) the source areas for most species are more likely to be another island than a continent. Multiple source areas (and therefore variable colonization distances) are not accommodated in Figure 18-2. I must also point out that, at either the level of archipelago or of individual island, the source areas for nearly the entire landbird fauna of Oceania are unknown. The same could be said for most other groups of animals or plants on Pacific islands. Thus the spatial scale for colonists in Figure 18-2, which might be estimated more reasonably for islands lying very close to continents, is fanciful for oceanic islands, especially with Quaternary changes in sea level.

The *y*-axes (the vertical scales) in Figures 18-1 to 18-3 (rates, which are something divided by time) also are fanciful because the timescales of precultural immigration and extinction are generally unknown. Related to the uncalibrated *y*-axis is the failure of equilibrium to account for evolution, which may be rapid for insular birds (although we really do not know) but not instantaneous. In not accommodating speciation, the equilibrium model resembles the species-area model (Chapter 19). The idea that "speciation rates should be significant for only the largest and most isolated islands" (Brown & Lomolino 1998:405) is not true. Recent claims to account for speciation in species-area relationships of West Indian anoles (Losos & Schluter 2000) are riddled with invalid assumptions and simply ignore endemic species on small islands (Chapter 22).

Turnover rates, just like any sort of rate, must have a time component. How can one-time survey data possibly be used as evidence for turnover and its offshoot, equilibrium? This is exactly what MacArthur and Wilson (1963, 1967) did, however, in interpreting distributional data on Polynesian ants (many of them nonnative) and Indo-Pacific birds. As noted above, the time series (multiple survey) data that became available shortly thereafter involved continental rather than oceanic islands.

Just as the perception of macro-ecological pattern depends on the spatial scale of analysis (Rahbek & Graves 2000), so the scale of analysis (spatial or temporal) affects the perceived rate of biological turnover (Whitmore 1989, MacDonald 2002), whether on a continent, continental island, or oceanic island. An example would be the small mammal (rodent and shrew) communities in North America, where little change has occurred over the past 20,000 years on a continental scale (Graham et al. 1996), but on a local spatial scale such as 10 km^2 or 100 km^2, the species assemblages have changed drastically over the same time interval (Stafford et al. 1999).

For undifferentiated species on continental islands, perceived turnover rates depend in part on census intervals, with cryptoturnover acting to depress rates as time intervals between censuses get longer. On oceanic islands in the Pacific, calculations of turnover rates also would depend on census interval if there were any evidence of natural turnover at all. In the Galápagos Islands, for example, the rate of extinction among terrestrial reptiles, birds, and mammals was found to be at least 100 times greater after human arrival (1535 AD) than during the previous 8000 years (Steadman et al. 1991; Table 18-1 herein). Of 33 vertebrate populations lost during this period, all but three vanished during the five centuries of human presence.

Typically, neither prehistoric nor modern data sets for insular faunas are complete. Nevertheless, the evidence in hand suggests that faunal change before human arrival was minimal, whereas that after human arrival has been much greater, but in neither case has been at equilibrium. Rather, the change in species richness before human arrival generally was positive, and that afterward has been dominated by extinction, leading to impoverished communities.

Of course natural colonization of oceanic islands by terrestrial vertebrates does occur through time, as does extinction. The trouble is that colonization by reptiles, birds, and mammals, in the absence of human influence, seems to take place on a geological timescale (i.e., <1 colonization per 10,000 years) on any oceanic island with a long fossil record and fairly well-known geological history (see "Pacific Fossils vs. Equilibrium and Turnover," below in this chapter). Thus we cannot measure colonization rates using ecological time frames, which also do not account for glacial-interglacial changes in the configurations of islands and archipelagos (Chapters 1, 2). Furthermore, barring cataclysmic events such as a large volcanic eruption, or major sea level changes or tsunamis on very low islands, extinction rates on oceanic islands are at least two orders of magnitude lower before human arrival than after.

Succession, Source Areas, Dispersal Distances

While some might cite the extensive data on floral and faunal colonization of volcanically obliterated Krakatau as good support for turnover and perhaps equilibrium, many of Krakatau's biotic changes over the past 120+ years are successional, with pioneer species being replaced by those characteristic of less ephemeral communities (Thornton 1996:209-249). Furthermore, Krakatau lies on Southeast Asia's Sunda Shelf, in a continental setting west of Wallace's Line. No island in Oceania has a nearby source area as rich as Java or Sumatra.

The study of Krakatau does help us to understand how and which plants and animals might colonize a young oceanic island recently devastated by volcanism. Again, however, the sources of Krakatau's propagules are much closer and richer than those for oceanic islands. The most comparable situation would be young volcanic islands very close to New Guinea's northern coast, such as Karkar (Diamond 1971), although the volcanic destruction of Karkar's biota is not as well documented as that of Krakatau, nor has Karkar been surveyed nearly as extensively for any group of organisms.

I agree with MacArthur and Wilson (1963, 1967) that colonization rates decrease as isolation increases. This is a straightforward result of the variable limits of dispersal among species. I do not believe, however, that extinction rates are related to isolation for oceanic islands. By itself, a lower rate of colonization can account for reduced biotic richness with isolation. It is important here to distinguish between single islands and archipelagos. From a standpoint of pure species richness but not taxonomic diversity, the limiting effects of great isolation may be lost or reduced where speciation offsets a low immigration rate. In the Hawaiian Islands, for example, only 20 colonizations account for the 100+ species of landbirds that existed at first human contact (James 1995: Figure 8.1). No other island group in Oceania, however, can claim nearly as much in situ speciation. The 49 species of landbirds known thus far in Tonga (Chapter 6), for example, are based on about 43 colonizations. The 20 species of landbirds on Mangaia probably are based on 20 separate colonizations of the Cook Islands, even if three of the colonizations may have been by the same species (*Porzana tabuensis*) but separated in time. Similarly, there is truly or nearly a one-to-one correspondence between species richness and colonizations in many of the island groups in Chapters 5–8.

Pacific Fossils vs. Equilibrium and Turnover

Fossils are the best evidence for the composition of island faunas before human arrival, although precultural vertebrate fossils are known in Oceania only from several island groups (see below). For lizards and snakes, which have very low species richness values in Remote Oceania, the precultural fossils are especially important in establishing which species are indigenous as opposed to those that arrived with human assistance (Pregill 1993, 1998). For bats, the only diverse precultural fossil assemblage (from 'Eua, Tonga) demonstrates that five species are indigenous (Koopman & Steadman 1995). Only two of these species survived on 'Eua into historic times, although all five bats were present when people first arrived ca. 2800 years ago.

Before focusing on birds from precultural fossil sites, the limitations of fossils must be recognized. There is an apples vs. oranges aspect to comparing prehistoric and modern

Table 18-1 Summary of Holocene vertebrate extinction in the Galápagos Islands. This table excludes isolated records of stray cuckoos, flycatchers, mockingbirds, and Darwin's finches outside of their normal range unless these records strongly suggest the former presence of a resident breeding population. The fossil record is confined to five islands (San Cristóbal, Floreana, Santa Cruz, Rábida, and Isabela). "Modern" ^{14}C dates are those that overlap AD 1950 at 2σ. NF, no fossil record from this island; †, extinct species; *, possibility that a small population may survive; [a]age determined by stratigraphy and faunal associates (particularly presence vs. absence of *Rattus*) rather than ^{14}C dating. From Steadman et al. (1991), with updates for rodents from Dowler et al. (2000).

	Recorded from life (AD 1835 or later)	Fossil associated with ≥ 1 reliable "modern" radiocarbon dates	Fossil associated with prehuman Holocene radiocarbon dates (8500 to 500 yr BP)
REPTILES			
Geochelone elephantopus			
Floreana	X	X	X
Santa Fé	X	NF	NF
Rábida	X	—	X
*Fernandina	X	NF	NF
Phyllodactylus sp.			
*Rábida	—	—	X
Conolophus subcristatus			
Baltra	X	NF	NF
Santiago	X	NF	NF
Rábida	—	—	X
Alsophis biserialis			
Floreana	X	X	X
MAMMALS			
Lasiurus borealis			
*Floreana	—	X	X
†*Oryzomys galapagoensis*			
San Cristóbal	X	X[a]	X[a]
†*Nesoryzomys indefessus*			
Santa Cruz	X	X	X
Baltra	X	NF	NF
†*Nesoryzomys* undesc. sp. 1			
Rábida	—	—	X
†*Nesoryzomys* undesc. sp. 2			
Isabela	—	X	X
†*Nesoryzomys darwini*			
Santa Cruz	X	X	X
†*Nesoryzomys* undesc. sp. 3			
Isabela	—	X	X
†*Megaoryzomys curioi*			
Santa Cruz	—	X	X
†*Megaoryzomys* undesc. sp.			
Isabela	—	X	X

Table 18-1 (continued)

	Recorded from life (AD 1835 or later)	Fossil associated with ≥ 1 reliable "modern" radiocarbon dates	Fossil associated with prehuman Holocene radiocarbon dates (8500 to 500 yr BP)
BIRDS			
Buteo galapagoensis			
San Cristóbal	X	—	—
Floreana	X	—	—
Baltra	X	NF	NF
North Seymour	X	NF	NF
Daphne	X	NF	NF
Tyto punctatissima			
Floreana	—	X[a]	X
Mimus trifasciatus			
Floreana	X	X	X
Geospiza nebulosa			
San Cristóbal	X	—	—
Floreana	X	X	X
Santa Cruz	X	X	X
*Isabela	X	—	—
*Fernandina	X	NF	NF
Geospiza magnirostris			
San Cristóbal	X	X[a]	—
Floreana	X	X	X
TOTAL SPECIES			
X (all islands)	24	16	19
X (five islands with fossils)	15	16	19

data on turnover and equilibrium; each type of information has strengths and weaknesses (James 1995). Both paleontological and neontological data have issues of species identification. If I misidentify a bone in my lab, or a living bird in the field, low-quality data will ensue. Each type of data is time-averaged and therefore unlikely to pick up the short-term absences that characterize many modern studies of turnover. Thus cryptoturnover is potentially a major problem with prehistoric data and would underestimate turnover rates. This is an extreme form of the census interval issue, where increasing length of census interval decreases the perceived turnover rate, a point first made by Diamond and May (1977). In Remote Oceania, however, I do not regard time averaging to be an important problem because these islands do not sustain a variety of ephemeral breeding populations of landbirds like islands near rich continental areas.

Working with continental small mammals, Hadly (1999) noted that while time averaging of prehistoric faunas through paleontological sampling can be a weakness, it also can be a strength because time-averaged data are more likely to sample rare taxa than one-time or short-term data. Intensity of sampling also is important (Chapter 4), with smaller samples of identified fossils increasing the perceived turnover rate through the specter of pseudoturnover, as is clear in the Tongan example below.

Because Remote Oceania, unlike continental islands, has essentially no migrant landbirds except the cuckoo *Eudynamys taitensis*, one does not need to be too concerned whether the species of birds represented by fossils were residents or migrants. Even in Near Oceania, migrant landbirds are a minor part of the avifauna. Regardless, we should question the complete discounting of migrants. During six or more months each year, *E. taitensis* is eating insects, lizards, and perhaps small birds in Remote Oceania and is no less a part of these insular avifaunas than any other uncommon species.

On O'ahu in the Hawaiian Islands, 15 or 16 of the 17 species of landbirds that were present at the 'Ulupau Head fossil site >120 ka still occurred on O'ahu when people arrived <2 ka (James 1987, 1995). (Geologist Paul Hearty now believes that this site may be as much as 500 ka; Helen James pers. comm. 2002.) These 15 or 16 species survived glacial-interglacial climate cycles but not the arrival of people. In this case, as in the others that follow, the fossil data cannot prove that each of these 15 or 16 species resided on O'ahu continuously for the entire 120,000+ years, although I believe that they probably did. Within any oceanic island group, the presence of endemic subspecies and species is evidence for barriers to interisland gene flow in many landbirds, suggesting long, continuous residency times.

Whether evaluated by archipelago or single island, the avifauna of Tonga has a lower level of endemism than that of the Hawaiian Islands. Theory therefore would predict higher turnover rates in Tonga because endemism presumably cannot develop or be sustained unless populations are relatively long-lived and isolated (Diamond & Jones 1980). This can be examined at a cave called 'Anatu ('Eua, Tonga) by comparing the avifaunas of stratum III (>80–60 ka), stratum II (<80–60 but >2.8 ka), and stratum I (<2.8 ka). Only six of the 27 species recorded from strata II and/or III still occur on 'Eua, although 11 of the 21 extinct or extirpated species did survive on 'Eua into the period of human occupation (Tables 6-20, 6-21). At least six of the remaining 10 species persisted until human arrival on other Tongan islands, and presumably on 'Eua as well.

Of 27 species of landbirds from strata II and III, 19 occur in both strata. The eight unshared species may be sampling artifacts rather than colonizations or extinctions. These eight species are represented by only 1–5 bones each (mean, 2.5), compared with 2–88 bones (mean, 20.2) for each of the other 19 species. Of the four extinct/extirpated species from strata II and III not recorded from stratum I or otherwise known to have survived into the human period, three are represented by ≤5 bones. With larger bone samples, I predict that these species also would be found to have survived into human times. More bones also would increase the value for prehuman species richness by recording rare species unsampled thus far. That 10 of the 21 lost species on 'Eua are represented by ≤5 bones suggests that more such species await discovery.

From Mangaia (Cook Islands), Tangatatau Rockshelter (site MAN-44) and Ana Manuku (site MAN-84) have bone-rich cultural strata underlain by precultural Holocene sediments also rich in bird bones (Chapter 7). Of the 17 species of landbirds recorded before the peopling of Mangaia, 16 survived into the human period although 14 of them now are gone (Table 18-2). Most of these losses occurred from 0.8 to 0.5 ka (Chapter 16). The one species absent from cultural deposits is an extinct night-heron (*Nycticorax* undescribed sp. C), known from only six fossils at

Table 18-2 Summary of precultural (paleontological) vs. cultural (archaeological) records of landbirds from Mangaia, Cook Islands. †, extinct species; [e], extirpated species (extant elsewhere, but no longer occurs on Mangaia); i, introduced species (not included in total). From data in Chapter 7.

	Precultural record	Cultural record	Number of bones
HERONS			
Egretta sacra	—	X	1
†*Nycticorax* undescribed sp. C	X	—	6
DUCKS			
Anas superciliosa	—	X	34
CHICKENS			
Gallus gallus (i)	—	X	121
RAILS			
†*Gallirallus ripleyi*	X	X	119
Porzana tabuensis	—	X	54
†*Porzana rua*	X	X	321
†*Porzana* undescribed sp. C	X	X	14
†*Porphyrio* undescribed sp. C	X	X	2
SANDPIPERS			
†*Prosobonia* undescribed sp. A	X	X	5
PIGEONS, DOVES			
[e]*Ptilinopus rarotongensis*	X	X	28
[e]*Ducula aurorae*	X	X	11
[e]*Ducula galeata*	X	X	18
[e]*Gallicolumba erythroptera*	X	X	461
†*Gallicolumba nui*	X	X	23
PARROTS			
†*Vini sinotoi*	X	X	3
†*Vini vidivici*	X	X	68
[e]*Vini kuhlii*	X	X	99
SWIFTS			
†*Collocalia manuoi*	X	X	6
KINGFISHERS			
Halcyon mangaia	X	X	38
WARBLERS			
Acrocephalus kerearako	X	X	105
TOTAL SPECIES/BONES	17	19	1537

Table 18-3 Species of landbirds recorded from precultural (paleontological) contexts on Aguiguan (Layer V of Pisonia Rockshelter; forty-nine identified specimens) and Tinian (Layer III of Railhunter Rockshelter; eight identified specimens), Mariana Islands. †, extinct species. From data in Chapter 8.

	Aguiguan		Tinian	
	Precultural record	Cultural record	Precultural record	Cultural record
RAILS				
†*Gallirallus* undescribed sp. M	X	X	—	—
†-*Gallirallus* undescribed sp. N	—	—	X	X
†*Porzana* undescribed sp. I	X	X	—	—
†*Porzana* undescribed sp. J	—	—	X	X
PIGEONS, DOVES				
Ptilinopus roseicapilla	—	—	X	X
Gallicolumba xanthonura	X	X		
WHITE-EYES				
Zosterops conspicillata	X	X	X	X
STARLINGS				
Aplonis opaca	X	X	—	—
TOTAL SPECIES	5	5	4	4

MAN-84, and thus perhaps a sampling artifact. Although five of the species lost from Mangaia still survive elsewhere, none has recolonized the island in recent centuries. The only native Mangaian landbirds missing from the precultural record are trampy, very widespread, and aquatic (the heron *Egretta sacra*, duck *Anas superciliosa*, and rail *Porzana tabuensis*). The last may be a posthuman colonizer (Chapter 10); the other two may have a longer (although undocumented) history on Mangaia and elsewhere in Remote Oceania (Chapter 13).

At prehistoric sites on Tinian (Railhunter Rockshelter) and Aguiguan (Pisonia Rockshelter) in the Mariana Islands (Steadman 1999a; Chapter 8 herein), each of the eight species of landbirds known to have been present before human arrival at ca. 3.5 ka was also found in the archaeological bone record (Table 18-3). Whether volant or flightless, none of the extinct or extirpated species has recolonized Tinian or Aguiguan since their loss some centuries or millennia ago. As on Mangaia, ʻEua, or Oʻahu, these prehistoric losses were permanent.

The case studies from the Hawaiian Islands, Tonga, Cook Islands, and Marianas are supported by the rich avian fossil record from New Zealand, where Holdaway et al. (2001:120) concluded, "For at least the past 100,000 years, until 2000 years ago, the fauna appears to have been very stable in composition, despite strong cyclic fluctuations in climate and vegetation." Although glacial-interglacial climatic changes led to range shifts among species of landbirds in the late Pleistocene (ca. 25 to 10 ka) vs. the Holocene (<10 ka), no extinctions occurred in New Zealand until after people arrived in the late Holocene (Worthy & Holdaway 1993, 2002).

To summarize, where the oldest fossils of birds (or reptiles and mammals) have been found in Remote Oceania, the message is consistent—the fauna that greeted the first human settlers, only one to several millennia ago, already had been established by the time of the oldest available fossil deposits, which can exceed 100,000 years old. The precultural evidence argues for long-term faunal stability, not turnover. A similar proposal has been made recently for North American forest biodiversity during the Holocene (Clark & McLachlan 2003, 2004).

Colonization rates for terrestrial reptiles, birds, and mammals seem to have been on a geological timescale (i.e., <1 colonization per 10,000 years) on any Pacific island with a long fossil record and fairly well-known geological history. This would include the islands in Oceania just mentioned as well as the Galápagos (Steadman et al. 1991), which are Neotropical. Possible exceptions in Oceania, i.e., species with evidence of colonizations within the past several thousand years, are primarily aquatic or semiaquatic species such as cormorants (*Phalacrocorax carbo*; Chapter 13), herons (*Egretta novaehollandiae*; Chapter 13), or rails (*Gallirallus philippensis*, *Porzana tabuensis*, *Porphyrio porphyrio*; Chapter 10). For any of these species, it is possible that human-induced environmental changes stimulated the recent episodes of colonization, as Jehl and Parkes (1983) showed for Socorro Island off Mexico's Pacific coast.

Even with the time averaging limitations of prehistoric data, I believe that most forest birds are rare colonizers of oceanic islands in the Pacific or elsewhere. Given the long residency times of most insular species of birds, it is unrealistic to expect biologists, during the short time frames covered in their data, to be able to gather much information on natural colonization of oceanic islands by landbirds.

Possible evidence of extinction also is rare in the precultural records of birds from Pacific islands. On islands that have been emergent and without catastrophic volcanism for a million years or more (examples from the Hawaiian Islands, Tonga, and New Zealand; see above), the avifauna that existed when people arrived up to 3000 years ago was pretty much the same one that had been in place roughly 100,000 years ago. Turnover rates were extremely low on oceanic islands before human arrival, even during glacial-interglacial changes in climate and sea level. Schoener (1983) summarized turnover rates (a colonization offset by an extinction) for insular birds; his estimates for the past 10,000 years (the Holocene), as extrapolated by James (1995), would be 500 to 14,000 turnover events per island. By inference, the past 100,000 years would see 5,000 to 140,000 turnover events. Without human influence, on real oceanic islands, I believe that these numbers are overestimates by three to five orders of magnitude.

We also should keep in mind that turnover is incompatible with endemism, unless one believes that endemic species typically evolve and then die out over very short time frames. The fossil record shows that living (or very recently extinct) endemic species already were in existence 100,000 years ago. Until older fossil deposits are found and studied, we simply do not know how much older these species may be.

An Alternative to Turnover and Equilibrium

Qualifications

If we assume that a poor dispersal ability increases vulnerability to human-caused extinction, then the species that have survived would be, on average, better dispersers than those that are gone. Among extant species, the rare ones tend to be poorer dispersers than those that are common. Nevertheless, an extant poor disperser and forest obligate, such as the Tongan whistler *Pachycephala jacquinoti* (Steadman & Freifeld 1998; Figures 14-14, 14-15 herein), may in fact be a better disperser than any of Tonga's extinct species of landbirds.

In clusters of nearby (typically intervisible) oceanic islands, such as the Vava'u and Ha'apai Groups of Tonga, interisland gene flow undoubtedly still occurs to some extent among most extant species. This is not turnover because entire populations are not being lost and replaced on each island. History is important here. In both Vava'u and Ha'apai, the clusters of islands were conjoined into one or several large oceanic islands when sea levels were lower during the late Pleistocene. As sea levels rose (18 to 8 ka; Chapter 1), each large island was subdivided, a process that would be imperceptibly slow to a bird. Thus, in a tight island cluster like Vava'u or Ha'apai, some species might adapt to increasing insularization; those with the best dispersal abilities probably continued to interact among newly isolated populations. This would not be the case during times of major sea level rise in most of Remote Oceania, where tens to hundreds of km of deep ocean separate most individual islands, interisland differentiation typically is higher, and interisland dispersal must be very rare.

As mentioned earlier in this chapter, I do not believe that colonization of oceanic islands by landbirds is a random process. The species differ considerably in morphological, ecological, behavioral, and physiological features that influence their ability or willingness to disperse and establish new populations (McNab 1994a, 1994b, Mayr & Diamond 2001). For example, the birds-of-paradise (Paradiseidae) and bowerbirds (Ptilinorhynchidae) are volant passerines that are well represented in New Guinea and tropical Australia, but occur nowhere in Oceania. Does this mean that no species in either family ever will colonize the Bismarcks or Solomons or perhaps, during glacial intervals, even New Caledonia? Of course not. In fact, as the fossil record of passerines (Chapter 14) improves, I would not be shocked to discover an extinct species of bird-of-paradise or bowerbird somewhere in Melanesia. From evidence in hand, however, both of these families qualify as poor colonizers of islands.

At the other extreme are widespread species that must be excellent colonizers of islands, such as some herons (*Egretta novaehollandiae*, *E. sacra*), ducks (*Anas superciliosa*), rails (*Gallirallus philippensis*, *Porzana tabuensis*, *Porphyrio porphyrio*), pigeons (*Columba vitiensis*, *Ducula pacifica*, *D. oceanica*, *Ptilinopus porphyraceus*), kingfishers (*Halcyon chloris*), and passerines (*Hirundo tahitica*, *Turdus poliocephalus*, *Myzomela cardinalis*). The timing of colonizations and the current level of interisland gene flow must vary among these species because their intraspecific morphological variation ranges from barely detectable to considerable (Chapter 17). It may be among such species that the first long-term colonists of a new island will be found.

On active volcanic islands, populations of landbirds may be lost occasionally to eruptions. On atolls, landbird populations are vulnerable to changes in sea level as well as tsunamis. Thus these two island types are not as well suited for long-term (tens to thousands of millennia) residency of diverse species assemblages of landbirds as eroded volcanic, raised limestone, or composite islands (see Chapter 1). It is on these last three types of islands that my model (if it can be called that; "narrative" may be better) of colonization and extinction applies. This model (Figure 18-4) is not elegant, nor is it very adaptable to bivariate plots with genuine timescales unless we can eliminate more of the uncertainty in the time of colonization. The fossil record is as frustrating as it is informative; with not nearly enough exceptions, it has failed to inform us when colonizations occurred or whether they were clustered in time. These failures render the values on the x-axis in Figure 18-4 as highly speculative. Molecular clocks eventually may help in this regard but typically also will have substantial uncertainty.

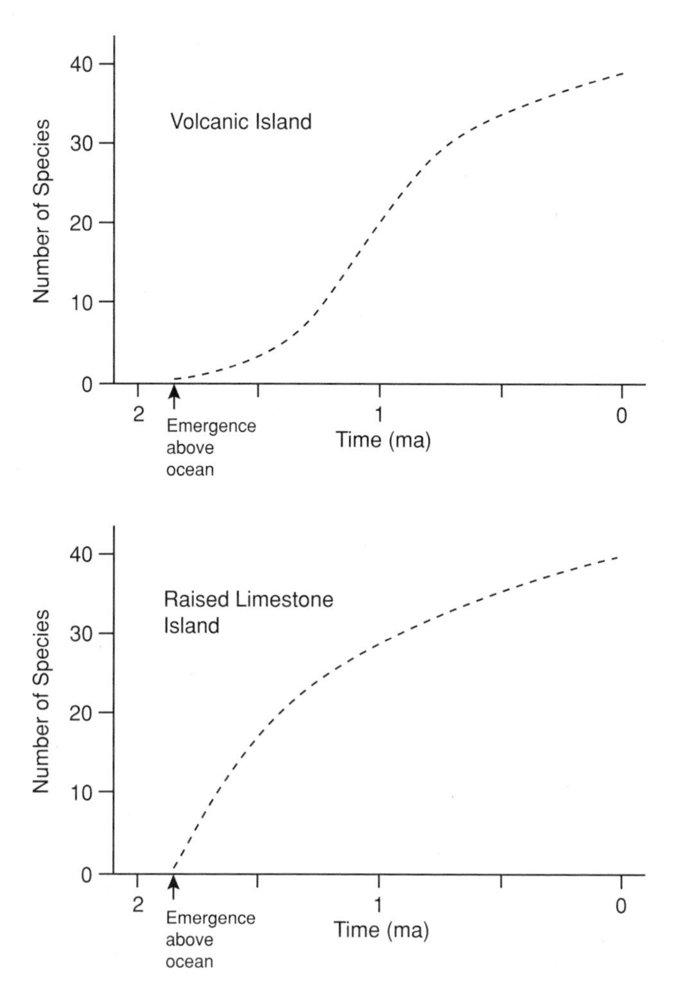

Figure 18-4 A crude model of faunal development on two hypothetical high Pacific islands that, after emergence, have remained high (>50 m elev.) in spite of sea level change. Human impact is not considered. Species richness values are due mainly or entirely to colonization; little if any extinction occurs, except perhaps during the first few hundred thousand years on volcanic islands. On average, species richness values will decrease with distance from New Guinea and with increased isolation within an island group. This example would be in West Polynesia. The timeframe (hypothetical in these examples) also may vary from ca. 0.1 to 10+ million years.

Colonization

Once volcanic activity ceases or becomes confined to a small portion of an island, perhaps 100,000 to 500,000 years after its first appearance above the ocean, long-term maintenance of upland forests becomes more feasible. (Those islands that first emerge from uplift of coralline limestone, or from shearing at transverse plate margins, may be well suited for colonization and long-term establishment of diverse biotas shortly after emergence, depending on the rate and extent of uplift.) Colonization by landbirds (requiring at least one pair), and the forest plants (especially trees and shrubs) needed to sustain most of them, are linked through frugivores such as columbids (especially *Ducula* and *Ptilinopus*) and starlings (*Aplonis*). Trampy frugivores might be among an island's first colonists, capable of dispersing plant propagules even if the habitat still were limited from their own foraging perspective. Insect faunas would build as the flora gets richer, making the island more suitable for insectivorous birds.

More colonization of oceanic islands by landbirds (and other organisms) probably occurs during glacial than interglacial intervals because islands are higher and less isolated during glacial times, which represent ca. 80% of time over the past two million years. After colonization, some species of landbirds may continue to interact with conspecifics on other islands, over variable timescales and at variable rates, whereas others will not. The rate of colonization for an island may decrease with time but never will reach zero. Species richness will increase more rapidly at first and, as long as the island remains substantial, continue to increase but at a slower rate because the most effective colonists already will be established, not because of competitive exclusion. Floral diversification, through dispersal and evolution, would stimulate faunal diversification. Therefore, island age should be an important positive influence on species richness; very old islands, such as New Caledonia's Grande Terre or Fiji's Viti Levu, have high levels of species richness (and endemism) because landbirds have been arriving and evolving here for such a long time. The avifaunas of younger, smaller islands must benefit if they are close to the older, larger, richer islands.

Extinction

As already stated, the precultural fossil records of oceanic islands disclose little or no evidence of extinction. Colonization without extinction is not turnover (see definitions at beginning of chapter) and does not yield an "equilibrium" value for species richness. On substantial islands, biotas simply may get richer with time, although the rate of increase decreases. Some very low rate of extinction may occur, but on average it will be more than offset by new colonization.

Through erosion and subsidence, however, most islands eventually sink beneath the ocean after some millions or tens of millions of years. When this process becomes so advanced that an island's elevation or area dip below a crucial threshold (see Chapter 19), species richness will decline as one after another species of landbird no longer can sustain viable populations. For large, geologically complex islands such as Isabel or Choiseul (Solomons), Grande Terre (New Caledonia), or Viti Levu (Fiji), this final decline in geological structure shows no evidence of having occurred within the past 10,000,000 years, i.e., not within the time frame of the existence of most living genera of birds (Olson 1985b). On these old, stable islands, nonanthropogenic extinction of landbirds probably is exceedingly rare.

Discussion

The concept of turnover-based equilibrium of insular biotas was questioned within a few years of its proposal (Lloyd et al. 1968, Sauer 1969). MacArthur himself (1972:100, 101) noted that "while some nearby islands are doubtless at equilibrium, especially with highly mobile forms like birds and flying insects, remote islands must often still be approaching equilibrium." This was more or less an admission that the equilibrium model might be difficult to apply to nonvolant organisms anywhere, and to any organisms on oceanic islands. Nevertheless, because turnover and equilibrium were proposed in a graphically pleasing way by two biologists as prominent as Robert MacArthur (who, sadly, died prematurely and has been deified ever since) and Edward Wilson (the rightly beloved guru of biodiversity, and admirable if controversial author of *Sociobiology*), these models have persisted for four decades. In spite of occasional criticism, many ecologists continue to accept equilibrium as the normal condition for floras and faunas (Brown & Lomolino 1998, Whittaker 1998), so that the discovery of apparently nonequilibrial situations is regarded as extraordinary (e.g., Bush & Whittaker 1993, Ricklefs & Bermingham 2001). I believe that it is not.

MacArthur and Wilson (1963, 1967) had no time series evidence of equilibrium in native floras and faunas of oceanic islands. Their equilibrium model spawned from Wilson's earlier studies of native and nonnative Polynesian ants (Wilson 1959, 1961) and especially from a flawed species-area plot of landbirds from selected islands (and entire archipelagos) in the Moluccas and across Oceania (MacArthur & Wilson 1963: Figure 2; see Chapter 19 for a critique). When time series evidence began to accumulate, it was for continental rather than oceanic islands. Yearning for generalizations about our complex world, and for "tools" to use in conservation, ecologists rushed to regard almost anything as an island and any absence as an extinction.

Also, in spite of real islands being so well known for endemism, equilibrium theory sees islands as being characteristically inhabited by species (of birds or anything else) with ephemeral populations. In fact a great many island species of birds, and not just flightless ones, are (or were) poor dispersers. Furthermore, the modern avifaunas probably are biased toward species at the upper end of the dispersal scale, because we have eliminated those unable to get out of our way.

The niche concept, which fuels belief in competition, was important in MacArthur and Wilson's vision of building up insular faunas to "saturation" at a hypothetical equilibrium number. "Only a certain fraction of arriving propagules will add a new species to the fauna, however, because except for 'empty' islands at least some ecological positions will be filled" (MacArthur & Wilson 1963:379). Competition-driven assembly and turnover in insular biotas probably was in part an outgrowth of MacArthur's previous studies of "non-overlapping niches" in continental birds (MacArthur 1957, 1958, MacArthur & MacArthur 1961). My hesitance to accept competition as the force that assembles bird communities on Pacific islands will be explained in Chapter 20.

How to measure the stability of a community through short time frames and how to determine which processes control species diversity (richness and relative abundance) continue to be studied and debated (Sankaran & McNaughton 1999). Larger spatial scales increase the apparent stability of communities (Clark & McLachlan 2003). On any timescale, but perhaps mostly in ones measured in 10^4 or 10^5 years (and thus likely to include glacial-interglacial changes), I believe that the species composition of plant or animal communities may be no less stable on high oceanic islands (such as those in Figure 18-4) than for equivalent areas on continents. (Many atolls and volcanic islands would be exceptions because of susceptibility to biotic obliteration from sea level changes or volcanism.) The relative ease of dispersal on continents accounts for my proposal, which might seem illogical at first because we always have been taught that faunal stability increases with land area. Nevertheless, birds (or mammals) can disperse among high-elevation areas in the western United States, for example, merely by crossing over or through the terrestrial habitats that exist at lower elevations. They need not be concerned with drowning in an ocean. Species composition changed more in the past (16 ka) vs. present small mammal communities near Peccary Cave (Arkansas) than in the bird communities that lived on O'ahu (Hawaiian Islands) at >120 ka vs. 2 ka, just before people arrived (James 1987, 1995, Stafford et al. 1999).

Just as incorporating an improved timescale to analyses of West Indian vertebrate faunas refutes the "taxon cycle" model (Pregill & Olson 1981, Pregill & Crother 1999), so the availability of longer timescales for Pacific island avifaunas invalidates the standard dynamic notions of turnover and equilibrium for vertebrate faunas on oceanic islands in the Pacific and probably elsewhere (James 1987, 1995, Steadman et al. 1991, Steadman 1993a, Holdaway et al. 2001). By the way, if a taxon cycle really existed, it would be compatible with equilibrium theory only in its early stages, before any differentiation of populations (Pregill & Crother 1999).

Most paleontologists study phenomena over long timescales that seldom overlap with the short timescales sampled by ecologists. Typically the data collection methods also are quite different. If our civilization allows natural history to be studied to the year 2100 and beyond, future scientists should be able to weave together paleontological and neontological data in much better ways than are possible now. This will be because the timescales of their studies can begin to merge. In some areas, such as the structure of small mammal communities and fisheries productivity, the insightful merging of data sets and timescales already is taking place (Hadly 1999, Finney et al. 2002).

Peter and Rosemary Grant (1996) declared that any field study that goes into a second generation time is "long term." While paleontologists, myself included, might gasp at this notion, we should withhold criticism. The Grants have been studying Darwin's finches in the Galápagos Islands for 30 straight years (Grant & Grant 2002), a phenomenal achievement that has moved their research more than tenfold beyond their own minimum standard for "long term" and into a time frame that allows decadal analyses. We all hope that the research they began continues indefinitely. When a century's worth of neontological data on the ecology and evolution of Darwin's finches becomes available, this should attract as much attention from paleontologists as any other group of scientists. While this may not lead to a utopian "consilience" as proposed by Wilson (1998; see Steadman 1999b), it will be genuine scholarly progress. In as difficult a subject as organismal biology, that is just fine.

Chapter 19 Species-Area Relationships

It seems like hundreds of papers in ecology begin with statements such as: "The species-area (SA) curve is one of the few universally accepted generalizations in community ecology..." (Hanski & Gyllenberg 1997:397); "The great 19th-century scientist Alexander von Humboldt gave ecology its oldest law: Larger areas harbor more species than smaller ones..." (Rosenzweig 1999:276); "The relationship between the number of species and the area sampled is one of the oldest and best-documented patterns in community ecology..." (Crawley & Harral 2001:864). None of these statements explicitly mentions islands, but islands have played an important role in studying the relationship between area and species richness.

As proposed by MacArthur and Wilson (1963, 1967) and developed further in many subsequent studies, the number of species (S) on an island is thought to be related to three physical variables: area (A), elevation, and isolation, of which A has received the most attention. The relationship between S and A, regardless of where or what is being analyzed, is expressed typically as $S = CA^z$, where C is a constant "that depends on the taxon and biogeographic region, and in particular most strongly on the population density determined by these two parameters" and z is another constant "that changes very little among taxa or within a given taxon in different parts of the world" (MacArthur & Wilson 1967:8, 9). Both C and z are fitted to the data available for S and A, although a logical flaw involves C because, contra MacArthur and Wilson (1967), no data on population density are required or even implied by the values for S or A; $S = 1$ for each species regardless of population density. S represents only alpha-diversity (see discussion in Duivenvoorden et al. 2002). For a set of insular floras or faunas, a log-log plot of S (y-axis) vs. A (x-axis) will yield a fitted regression line of supposed predictive abilities where the slope is z and the y-intercept is C. Preston (1962) expressed the species-area equation as $N = KA^z$, but since most authors follow the terminology of MacArthur and Wilson (1967), I will as well.

Like so many scientists I am enticed by the apparent simplicity of a relationship that can be examined by empirically determining just two sets of numbers, only one of which requires fieldwork. As you probably realize by now, I am suspicious of anything so elegant. Anyone studying insular biotas knows that the number of species on an island is due to a number of complex factors, not just land area alone (Pregill & Crother 1999). My suspicion is fueled further because the SA relationship is a product of equilibrium theory, which is based on concepts that I believe do not apply to landbirds on oceanic islands (Chapter 18). As I will describe below, the species-richness data used by MacArthur and Wilson (1963, 1967) for Pacific island landbirds were flawed in multiple ways. The other important data set in the development of species-area concepts was that of Wilson and Taylor (1967a,b) for Polynesian ants, of which most or all species on remote islands were introduced by people rather than arriving through natural means of dispersal. Thus the data used in the conceptual development of the SA relationship were either faulty or not based on natural systems.

Elevation, unlike A, seldom is expressed in a formula with S. This must be because, within any island group I know, the resulting graphics would not be pretty, even if dampened by a log-log scale. Large islands also tend to be high, but there are many exceptions, such as Tonga, where the largest island (Tongatapu, 259 km^2) is 65 m high and some small (<5 km^2) volcanic islands have elevations >300 m. Biogeographers generally have assumed that any positive effect of elevation on *S* is more or less accommodated within *A*. I will show below that neither area or elevation is as influential on *S* as previously believed, at least for landbirds in Remote Oceania.

The third physical variable thought to affect *S* is isolation, greater levels of which decrease species richness. When prehistoric data are considered, increased isolation still has a negative influence on *S* (Chapter 17) although, as I will show later in this chapter, this relationship needs to be evaluated separately between vs. within island groups.

Imagine a not-so-hypothetical island of 3 km^2 in Remote Oceania where the modern landbird fauna ($S = 13$) consists of a heron (body mass 400 g, $N = 10$), a large rail (800 g, $N = 20$), a smaller rail (50 g, $N = 100$), a frugivorous pigeon (400 g, $N = 50$), uncommon and common frugivorous doves (uncommon: 100 g, $N = 20$; common: 100 g, $N = 150$), a barn-owl (350 g, $N = 10$), an insectivorous/saurivorous kingfisher (50 g, $N = 30$), an insectivorous triller (30 g, $N = 50$), an insectivorous white-eye (10 g, $N = 20$), a mostly frugivorous starling (75 g, $N = 60$), and a rare (large) and common (small) nectarivorous honeyeaters (rare: 40 g, $N = 10$; common: 10 g, $N = 200$). How can we possibly evaluate the relative ecological importance of these 13 species of landbirds, each of which has an *S*-value of 1 from an *SA* standpoint? Even if we analyze individuals rather than species, as advocated by Hubbell (2001), how do we compare an 800 g rail with a 10 g honeyeater (or a 5 cm diameter fruitless sapling with an 80 cm diameter fruiting tree) in a way that is biologically meaningful? I do not mean to imply here, by the way, that body mass (or dbh) is an ideal proxy for anything in particular.

Perhaps more than in any other topic in island biogeography, the increase in species richness with increasing land area is one where ecologists, right from the start, blurred the important distinctions among true oceanic islands (as in Oceania), continental islands, and nonislands such as habitat patches or mountaintops. This trend flourishes in the statistical quest for generalization called macroecology. To me, however, these three categories of land are biogeographical apples and oranges in terms of evaluating biotic source areas, isolation, nestedness of land areas and biotas, and evolutionary history. My preference here, therefore, will be to discuss *SA* relationships of landbirds on genuine oceanic islands. I will stray to continents, continental islands, or organisms other than birds only occasionally, to clarify or amplify a point.

So now, with an uneasy feeling that, on our not-so-hypothetical island in Remote Oceania, 20,000 g of fruit-gulping pigeons are the same as 200 g of bug-gleaning white-eyes in an *SA* equation, I will review species-area relationships among landbird faunas in Oceania. I will not analyze the *SA* data for each island group; this information, ever improving as the fossil record grows, is found in Chapters 5-8. Rather, I will cover the principles of *SA* relationships for Pacific island birds, citing specific examples as needed, with an emphasis on island groups with substantial fossil information.

What Is Comparable in *SA* Analyses and What Is Not

As in Chapter 18, I must stress the fundamental differences between oceanic islands (surrounded by ocean, no Neogene connection to a continent) and continental chunks of land, which may be separated from each other today by ocean (land bridge islands), fresh water, different terrestrial habitats, or just geographic scale. For commonsense reasons that are clear to anyone, more species of birds (or of mammals, reptiles, beetles, plants, etc.) live in the United States than in Florida than in Alachua County than on the University of Florida campus than in the courtyard outside my office. These geographic entities are physical subsets of each other; oceanic islands are not. While offshore continental islands may not be subsets of each other or of a continent today, they typically were part of a continental landmass during glacial times. Furthermore, even with today's high sea levels, continental islands lie close to very rich biotic source areas.

Keeping this in mind, let us look at the first species-area plot in MacArthur and Wilson (1963: Figure 1; Figure 19-1 herein). The 23 islands show a fairly steady increase in log-*S* vs. log-*A* over 4+ orders of magnitude of change in *S*. Log-transforming the raw data for *A* and *S* greatly dampens their variation (Gilbert 1980) and makes them much easier to depict on paper, creating the illusion of a tight data set. The more you transform data (logarithmically or otherwise), the better they will look, even if the information is now in a state very different from what originally was measured. Morphometrics also suffers from this trend, even though the data can be more informative when left untransformed (Hayek et al. 2001). MacArthur and Wilson (1963) never presented the raw data for *S* or A for their Figure 1 (Figure 19-1 herein), setting a low standard that still exists (e.g., Murphy et al. 2004).

In Figure 19-1, I underlined the 11 islands that lie on the Sunda Shelf, eight of which were part of mainland Asia during the last glacial interval. Of the 12 other islands, 10 lie east of Wallace's Line and have more Papuan than

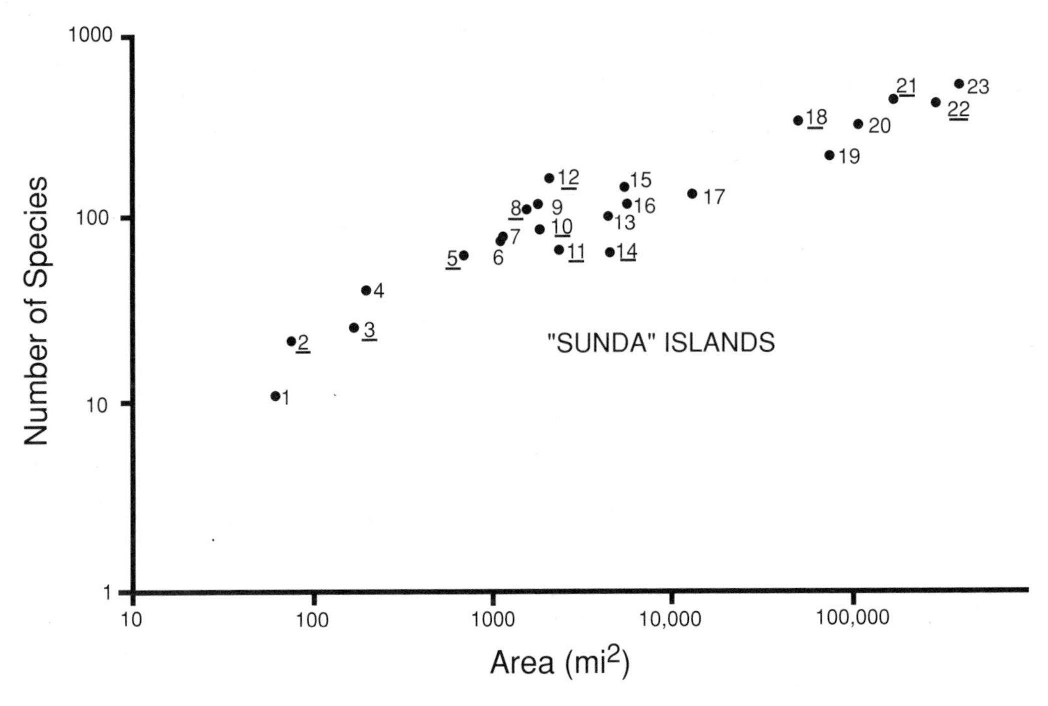

Figure 19-1 Log-log species–area plot of modern landbirds on selected islands from Sumatra to the Philippines and New Guinea. 1, Christmas; 2, Bawean; 3, Engano*; 4, Savu; 5, Simalur*; 6, Alors; 7, Wetar; 8, Nias*; 9, Lombok; 10, Billiton; 11, Mentawei; 12, Bali; 13, Sumba; 14, Bangka; 15, Flores; 16, Sumbawa; 17, Timor; 18, Java; 19, Celebes; 20, Philippines; 21, Sumatra; 22, Borneo; 23, New Guinea. Islands with underlined numbers are on the Sunda Shelf. *, probably not quite connected to mainland Southeast Asia in the Pleistocene. Alternative island names: 3, Enggano; 4, Sawu; 5, Simeulue; 6, Alor; 10, Pulau Belitung; 19, Sulawesi. Redrawn from MacArthur and Wilson (1963: Figure 1).

Asian avifaunas. Christmas Island (#1) is oceanic and isolated. The Philippines (#20) are a special case with a filtered biotic connection with continental Borneo (#22) via Palawan (Reis & Garong 2001). Thus the source areas for landbirds for these 23 islands, which include many of the islands themselves, are highly varied. Another problem is that the Philippines is an archipelago, not a single island, and therefore its *S*-value includes allopatric species. This may also be the case for Mentawei (#11), but cannot be determined from the information presented. Finally, through no fault of MacArthur and Wilson (1963), their *S*-values are underestimates of unknown magnitude because they do not account for the losses of species that undoubtedly took place during the tens to hundreds of millennia that people occupied most of these islands.

The second *SA* plot in MacArthur and Wilson (1963) is redrawn here as Figure 19-2. Again, raw data were not presented. If curious about real values, one is forced to imagine the scales on log-log graph. Even more than Figure 19-1, Figure 19-2 has a diverse set of both oceanic (19) and continental (7) islands, with very different biotic source areas, geological ages, tectonic histories, long-term changes in land area and isolation, and human histories. One would expect analyses of *SA* relationships to be more meaningful for islands within an island group (which are more likely to have shared geological and biological histories) than among archipelagos. Figure 19-2 compares the landbird faunas of 10 single islands with those of 16 island groups, most of which comprise tens if not hundreds of islands. How can this be biologically informative? Furthermore, Figure 19-2 assumes that New Guinea is the sole source of landbirds for the other islands, which is not true (Chapter 17).

While prehistoric data were not available to MacArthur and Wilson (1963, 1967), it is also noteworthy that fossils in the Marquesas (#6), Tonga (#10), and Hawaiian Islands (#20) now have more than doubled the known *S*-values for landbirds (Table 19-1). This obviously would influence, among other things, an assessment of how "saturated" these avifaunas are. Utility of the saturation curve in Figure 19-2 is undermined further by the archipelago-vs.-single-island problem and the invalid assumption that New Guinea is Oceania's exclusive avian source area.

Near Oceania

The *SA* relationship has been examined in some detail among the relatively rich landbird faunas of the Bismarck

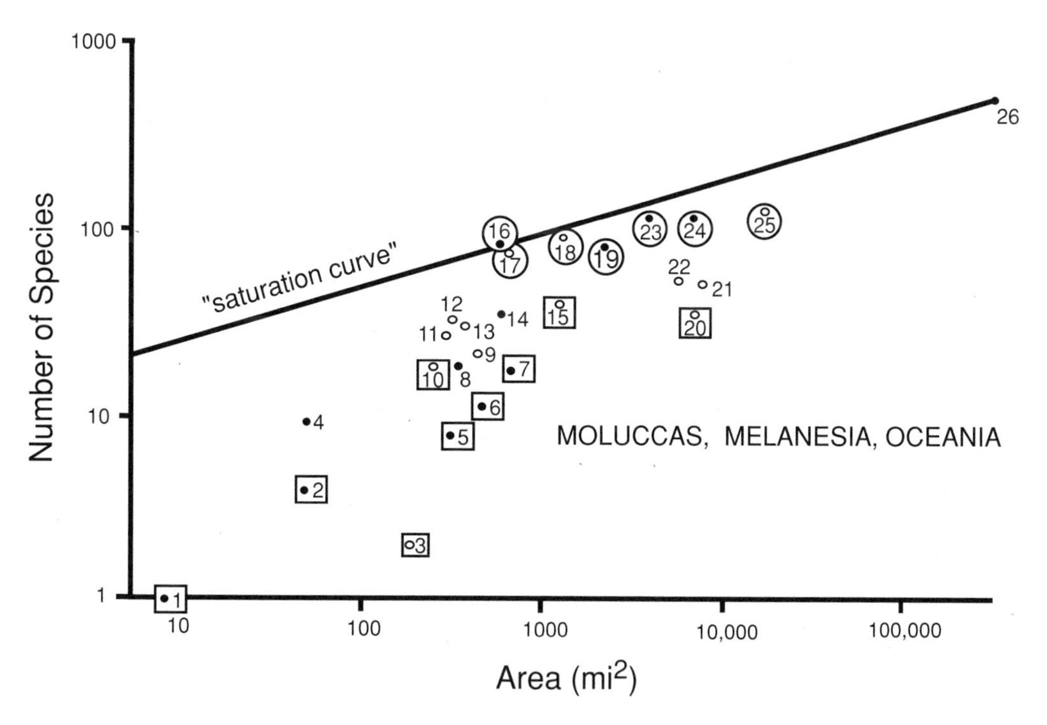

Figure 19-2 Log-log species-area plot of modern landbirds on selected islands in the Moluccas, Melanesia, Micronesia, and Polynesia. 1, Wake; 2, Henderson; 3, Line; 4, Kosrae; 5, Tuamotu; 6, Marquesas; 7, Society; 8, Pohnpei; 9, Marianas; 10, Tonga; 11, Carolines; 12, Palau; 13, Santa Cruz; 14, Rennell; 15, Samoa; 16, Kei; 17, Louisiade; 18, D'Entrecasteaux; 19, Tinambar; 20, Hawaii; 21, Fiji; 22, Vanuatu; 23, Buru; 24, Ceram; 25, Solomons; 26, New Guinea. ●, single islands; ○, island groups; islands <500 miles from New Guinea are enclosed by circles; islands >2000 miles from New Guinea are enclosed by squares. Redrawn from MacArthur and Wilson (1963: Figure 2).

Archipelago and Solomon Islands. Mayr and Diamond (2001) recently summarized species-area data for the Bismarcks. Unlike most authors, they did present some raw data for S and A in their Table 9.1, even if compromised by not providing a complete species vs. island matrix (Kratter 2002).

The species richness of modern landbirds in New Ireland (Bismarcks) was 101 in Diamond (1975b), 106 in Steadman et al. (1999b), 103 in Mayr and Diamond (2001), and 106–110 in my Table 5-3. These differences are small but demonstrate imprecision in S, which always is plotted without confidence intervals. More importantly, bones from archaeological sites (Steadman et al. 1999b; updated in Chapter 5 herein) increase New Ireland's S-value by 10 (see arrow in Figure 19-3). Conservatively, New Ireland has lost ca. 20% of its species of landbirds since human arrival. Among the 40 "extant" taxa from the prehistoric sites, I have been able to identify only 28 to species; it is likely that from one to three additional extinct or extirpated species await detection among the 12 taxa identified only to genus or family. A rough extrapolation would be that 130 to 140 species of landbirds occupied New Ireland at first human contact, compared to the 106 to 110 species believed to live there today. This would move the point for New Ireland in Figure 19-3 to well above the line of best fit, rather than below it (with modern data only) or just above it (with available fossil data). Among comparable islands (those with solid circles), all S-values would increase with decent fossil data. The upward nudge probably would be greater as island area decreases, leading to a decrease in z, the slope of the line of best fit (see below).

SA relationships have received much attention in the Solomon Islands (Diamond et al. 1976, Diamond & Mayr 1976, Gilpin & Diamond 1976, Mayr & Diamond 2001: Table 9.2, Figures 9.2, 9.3). The "classic" *SA* data set for lowland landbirds from the Solomons (Figure 19-4) is suspect, however, because of inadequate modern surveys and grossly insufficient prehistoric data. Our field work on Isabel in 1997 and 1998, for example, added three species of landbirds to its modern avifauna and found that three "montane" species also inhabit the lowlands (Kratter et al. 2001a). The distinction between lowland and highland species of birds is not as clear-cut in Near Oceania as one would surmise from the literature.

Prehistoric data from Buka (geographically part of the Solomons, politically part of Papua New Guinea) are sobering from an *SA* standpoint. A small set of bones from an early archaeological site (Kilu Cave; up to 29 ka) represents

Table 19-1 Maximum known species richness values for islands of different land area categories in major island groups in Oceania. *, includes some prehistoric data; **, includes much prehistoric data. Values with zero or one asterisks are highly suspect and undoubtedly too small; even those with two asterisks are unlikely to represent all species present at first human contact. Based on data in Chapters 5–8.

	Land area (km^2)					
Island group	<1	1–10	10–100	100–1000	1000–10,000	>10,000
Bismarcks	21	20	54	84–86	116–121*	127–132
Solomon Main Chain	35	38	61	82	102	—
Solomon Outliers	1	12*	18	38	—	—
Vanuatu	?	32	36	47*	50	—
New Caledonia	?	?	?	?	?	52*
Fiji	16	25*	37	43	44	52*
Tonga	11	26–27**	37–38**	32**	—	—
Samoa	7	16	16	18	31–32	—
West Polynesian Outliers	—	—	14	14*	—	—
Cook Islands	1	2	20**	—	—	—
Tubuai	0	1	4*	—	—	—
Society Islands	2	5	18**	12	19	—
Tuamotu	2	6	7*	—	—	—
Pitcairn	2	1	10**	—	—	—
Easter Island	0	—	—	5*	—	—
Marquesas	?	3	18**	15*	—	—
Palau	?	30	29	30–31	—	—
Yap	?	7	13	—	—	—
Marianas	2	13*	25**	18	—	—
Chuuk	?	7–8	17	—	—	—
Pohnpei	?	7	—	21*	—	—
Kosrae	—	—	—	11	—	—

17 species of landbirds, 11 of which (65%) no longer live on Buka (Chapter 5). At 611 km^2, Buka is an order of magnitude smaller than New Ireland (7174 km^2), which lost at least 20% of its species (see above). On high islands, all else being equal, levels of anthropogenic extinction tend to be greater on smaller islands in both Near and Remote Oceania (Chapter 16). This artificially increases z-values. For Buka, the 77 landbird bird bones push S from 67–68 (all species, not just "lowland" species) up to 78–79 (Table 5-7; see arrow in Figures 19-4 and 19-5). Extrapolating from a proportional loss of 65%, however, would increase S on Buka from 67–68 to 110-111, which is greater than any modern S-value from the Solomons, the largest of which is 102 for Bougainville, an adjacent but much larger (8591 km^2) and higher (2591 m) island.

With an area of 611 km^2 and an elevation of 402 m, Buka is substantial enough to have sustained viable populations of as many species of landbirds as any island in the Solomons. During glacially lowered sea levels, Buka was joined to Bougainville (and several other major islands) as the immense Greater Bougainville or Greater Bukida (Chapters 1, 5). Of the 11 species known to be lost from Buka, seven (a heron, megapode, three rails, and two pigeons) are extinct and thus presumably were wiped out on Bougainville and other islands as well. One other (a dove) now lives in New Guinea but nowhere in the Solomons. The last three (a rail, pigeon, and owl) still occur on Bougainville, reflecting, I believe, that large and high islands are less susceptible to anthropogenic extinction. Was postglacial "relaxation" (as defined by Diamond 1972b) a

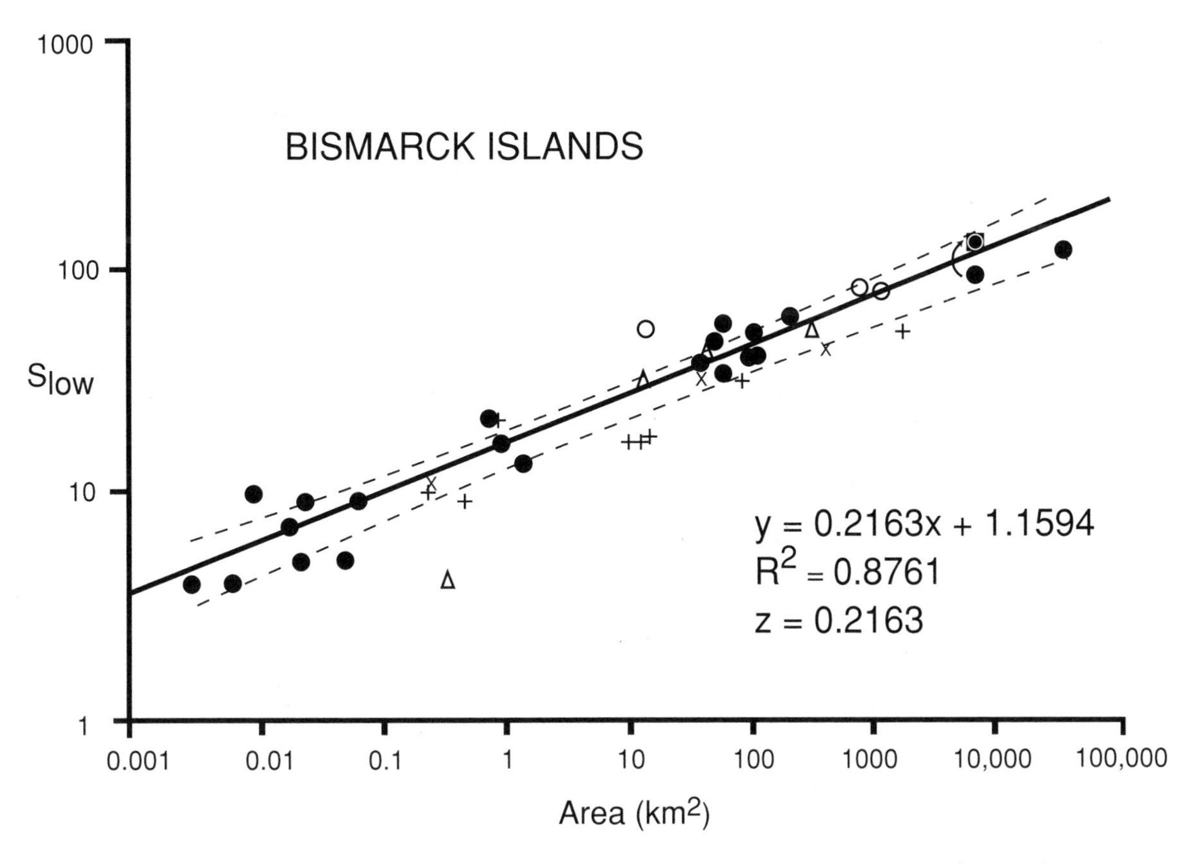

Figure 19-3 Log-log species-area plot of modern lowland landbirds in the Bismarck Archipelago. ●, the Central Bismarck Islands of New Britain, New Ireland, and their satellites; X, the isolated St. Matthias group (Mussau and satellites); +, the very isolated Northwest Bismarcks (Manus and satellites); △, islands recently defaunated by volcanic explosions; ○, islands formerly joined by Pleistocene land bridges to much larger islands. The arrow connects the New Ireland points without (●) and with (◙) prehistoric data. Modified from Mayr and Diamond (2001: Figure 9.1), based on modern data in their Table 9.1.

part of the late Quaternary decline in species richness on Buka? I believe either that it was not, or that its role was minor compared to that of human impact.

The *SA* relationship changes from a power function to an exponential function when $A < 1$ km^2 (dashed line in Figure 19-4), reflecting in part that extremely small islands cannot sustain several or even one pair of many species of landbirds. These same data hover around a straight line in a semilog plot (Figure 19-5). Unlike a log-log plot of *SA* values, a semilog comparison does not visually dampen the interisland variation in *S* (Gilbert 1980), although *A*-values remain dampened. To dampen the depiction of *A*-values by using a logarithmic scale is more or less a visual necessity in geographically diverse archipelagos such as the Solomon Islands. Because I find them to be more instructive, I will present my *SA* data in semilog plots (Figures 19-7 to 19-14), unlike the traditional log-log format (Figures 19-1 to 19-4). For sake of comparison, I will report the equations for both an arithmetic (power function) and a logarithmic (exponential function) line of best fit for *S* vs. *A*, thereby providing the slopes of lines (*z*-values) for all data sets as if they were based on log-log plots.

Remote Oceania

The very limited data from Near Oceania suggest that perhaps 20 to 65% of landbird populations did not survive the 30,000 years of human occupation. In Remote Oceania, where human residency does not exceed 3500 years, major islands have lost from about 20 to 100% of their landbird populations; the Polynesian "heartland" of Tonga, Cook, Society, and Marquesas Islands (Chapters 6, 7) has modern avifaunas so depleted from their condition at human contact (Figure 19-6) as to challenge biologically cogent analyses. No Polynesian island group was spared; the size and ruggedness of an island may temper the speed and extent of extinction but only temporarily (Chapter 16). The isolated corners of the Polynesian triangle (New Zealand, Hawaiian Islands, and Easter Island) were hit particularly

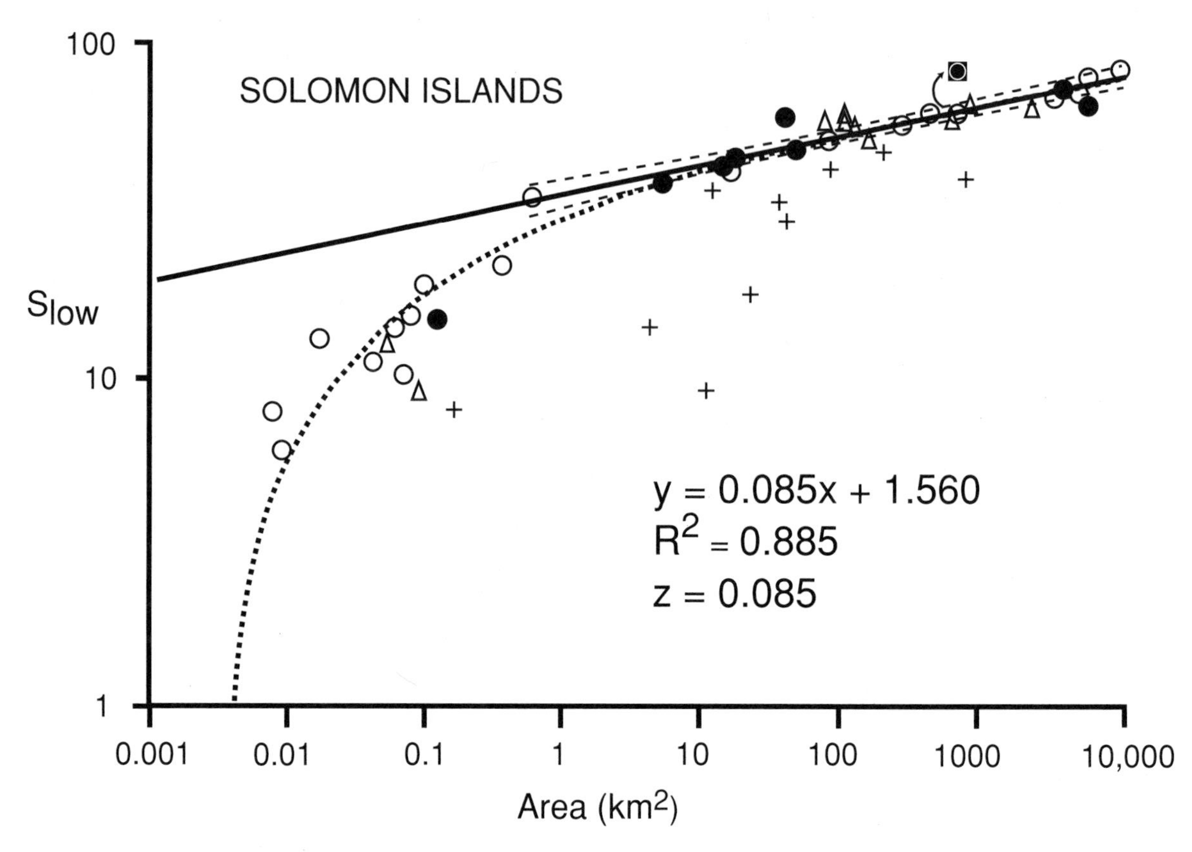

Figure 19-4 Log-log species-area plot of modern lowland landbirds in the Solomon Islands. ●, △, ○, "central" islands, i.e., those with S > 50 species and those within 10 km of such an island; ○, islands derived from the expanded Pleistocene island of Greater Bukida; △, islands similarly derived from Pleistocene Greater Gatumbangra, Greater Vellonga, and Greater Redipari; ●, islands with no recent connections; +, isolated islands, defined as those that support fewer than fifty species and that lie more than 10 km from the nearest island with fifty or more species. The arrow connects the Buka points without (○) and with (◙) prehistoric data. "+" islands are excluded from calculating the line of best fit. Solid line (central islands >0.4 km²) = power function; curved dashed line (all central islands) = exponential function. Modified from Mayr and Diamond (2001: Figure 9.2), based on modern data in their Table 9.2. The dashed lines on either side of the solid line of best fit are the 95% confidence intervals.

hard. Greater isolation may lead to greater vulnerability of native birds to anthropogenic extinction because of the absence or near absence of indigenous predators, pathogens, or parasites, such as the lack of native mosquitoes to serve as vectors of *Plasmodium* and other blood-borne pathogens (van Riper 1986, 2002, Steadman et al. 1990a; Chapter 16 herein). The delays in extinction (decades, centuries, millennia) might be long on ecological timescales but are short in an evolutionary or geological sense.

Within an island group in Remote Oceania, I see little if any pattern of higher species numbers when island elevation or area increases, as long as a minimum land area is reached (see below), isolation is fairly uniform, and prehistoric data are considered. Lower *z*-values are expected for relatively remote island groups because interisland distances within the group are less than interarchipelagal distances, so that only good dispersers colonize the archipelago in the first place (Diamond & Mayr 1976). In the absence of human influence, I believe that intra-archipelago *z*-values would be extremely low across Remote Oceania, regardless of how isolated the island group may be.

Of all island groups in Remote Oceania, only Vanuatu has human malaria, which is a deterrent to human population and thus to avian extinction as well (Chapter 16). Vanuatu also is the only archipelago in Remote Oceania where the *SA* relationship has been examined extensively before now (Diamond & Marshall 1977). The *SA* data from Vanuatu (Table 5-11) may signal the effect of isolation when one examines which islands lie well above or below the line of best fit (Figure 19-7). Those with high outlying *S*-values (Malo, Emae, Emau, and Nguna) lie near large, rich islands such as Santo, Epi, or Efate. The opposite is true for islands lying well below this line, such as Aniwa (8 km², 20 species) or Futuna (11 km², 20 species), which also are distinctive in being nonmalarial. Greater isolation may limit species richness in Vanuatu in both the north (Torres Group, Banks Group, the latter including Mota Lava) and south (Erromango, Tanna, Aneityum, Aniwa, and Futuna; Figure 5-8).

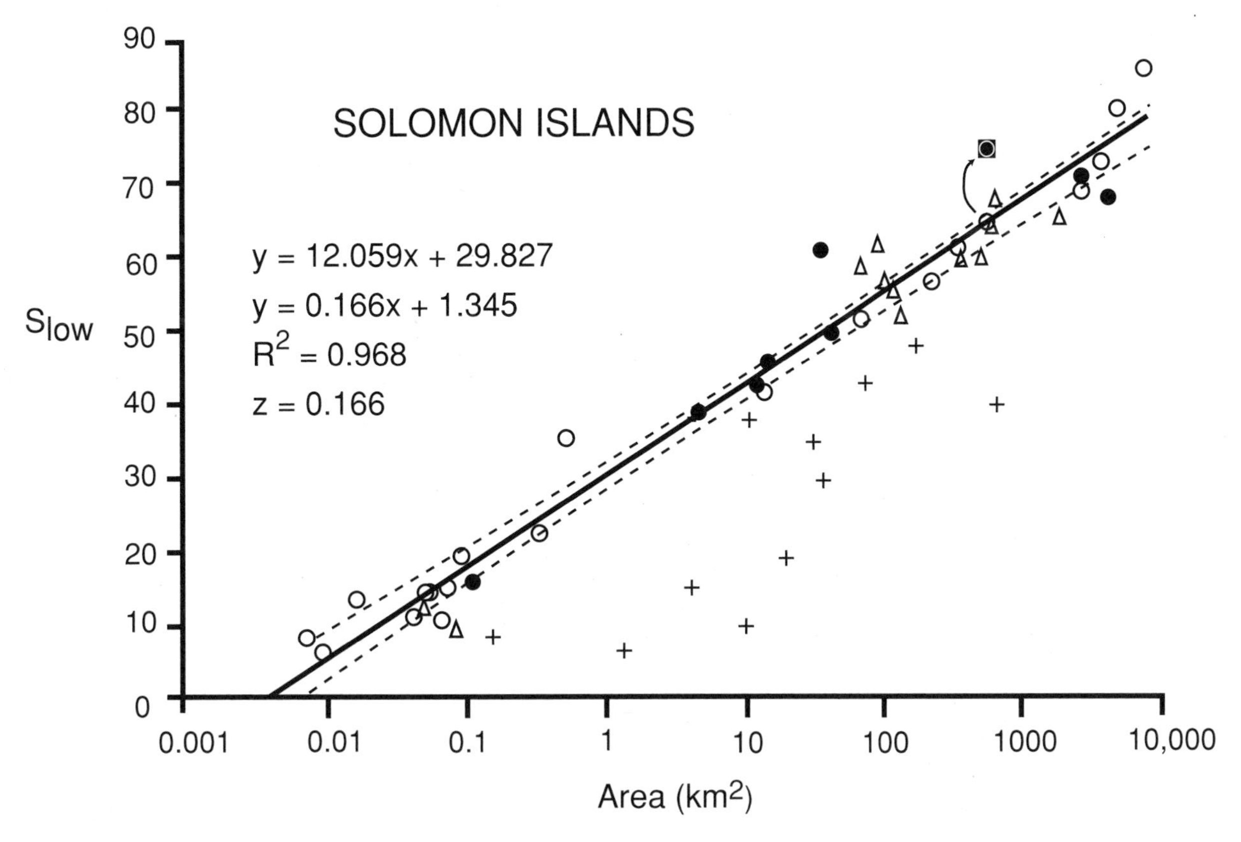

Figure 19-5 Semilog species-area plot of modern lowland landbirds in the Solomon Islands. Symbols as in Figure 19-4. Modified from Mayr and Diamond (2001: Figure 9.3), based on modern data in their Table 9.2. "+" islands are excluded from calculating the line of best fit. The dashed lines on either side of the solid line of best fit are the 95% confidence intervals. The top equation describes the semi-log line of best fit. The top equation describes the semilog line of best fit. The second equation, as well as the R^2 and z values, is derived from a log-log plot.

S-values are based on presence rather than absence. So-called checkerboard distributions of birds in Vanuatu (and elsewhere) are based on both. Concern about whether checkerboard patterns are due to chance, competition, or any other cause (Diamond & Marshall 1976, 1977, Sanderson et al. 1998) is moot until we better estimate which of the absences (and perhaps even presences) are anthropogenic. As in the Bismarcks and Solomons, "incidence functions" (Diamond 1975b, Diamond & Marshall 1977) are tenuous until tested with prehistoric data. Bones have augmented the *S*-values (by only 1–3 species) on just four islands in Vanuatu, where I and others (Spriggs 1997, 1999, Bedford 1999) believe that "first contact" or precontact bone deposits have not been sampled much if at all (Chapter 5). Species-area data will remain difficult to evaluate further in Vanuatu until its fossil record improves.

From several other island groups in Remote Oceania, however, we already have enough prehistoric information to compare modern vs. prehistoric *SA* relationships. In the Marquesas (Figure 3-14), for example, bones from the Hane archaeological site have increased the number of landbirds on Ua Huka (area 78 km^2, elev. 855 m) from 4–5 to 18 (longest arrow in Figure 19-8), which is more than twice the number of species (7) known in modern times from nearby, larger (337 km^2, 1185 m) Nuku Hiva. The three other Marquesan islands with fossils also have substantial jumps in *S*-values, although not as much as Ua Huka, which has 2187 identified landbird bones compared to only 27–146 on Tahuata, Hiva Oa, and Nuku Hiva (Table 7-19). Ua Huka and Nuku Hiva were inhabited prehistorically by identical species or species-groups of landbirds. Most of these are gone on both islands, although forest patches in the steep interior of Nuku Hiva have allowed a few species to persist, such as the pigeon *Ducula galeata*. At human arrival, when the lowlands were forested as well, the distribution of landbirds in the Marquesas and throughout Remote Oceania was more azonal than today, with most or all species occurring in both lowland and upland forest, or just in the former. That some species are confined to montane parts in Remote Oceania may reflect human impact in the lowlands more often than a genuine preference for upland habitats. This is also true to some extent in Near Oceania.

The Marquesas was the only Polynesian or Micronesian island group where MacArthur and Wilson (1967)

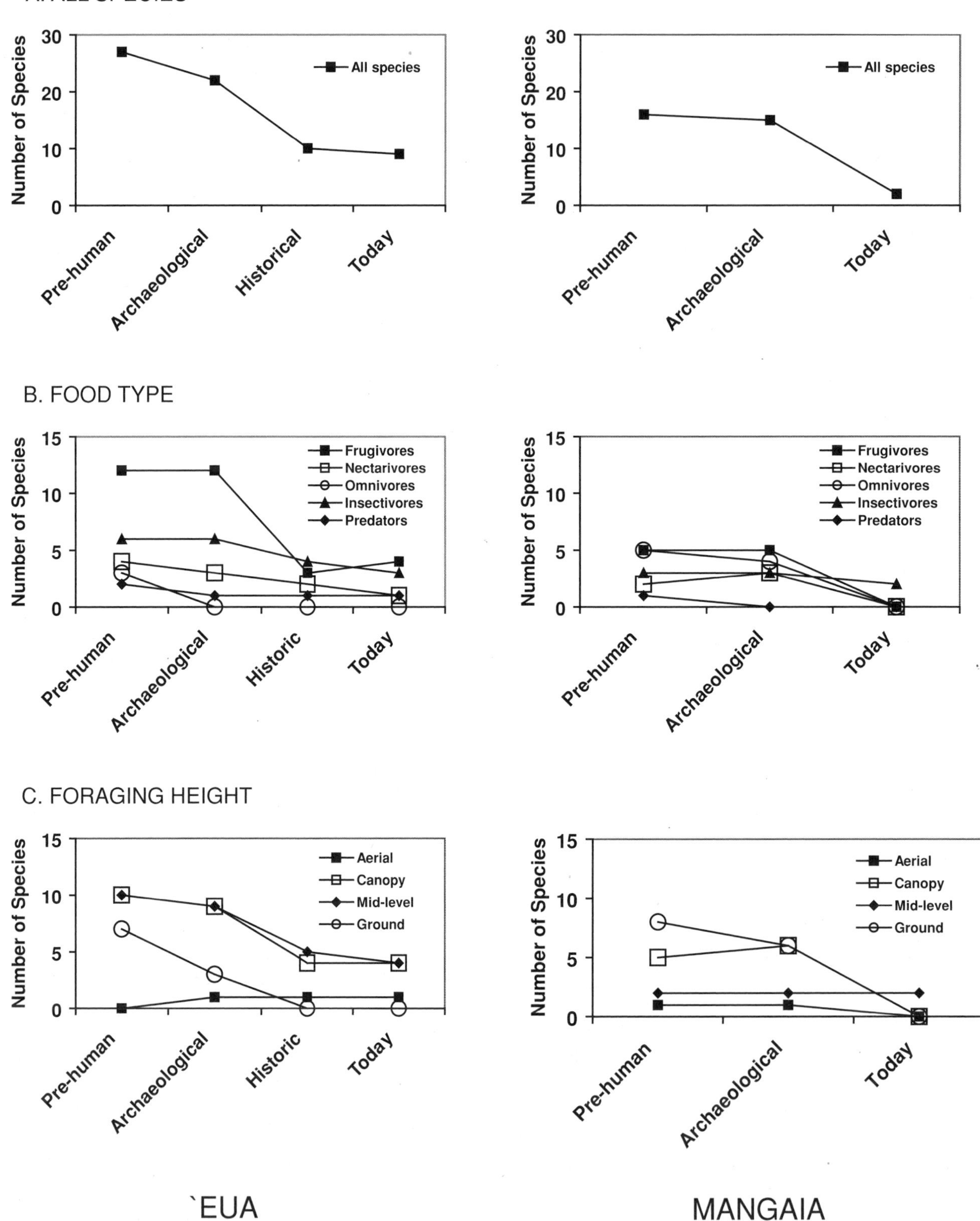

Figure 19-6 Long-term changes in species numbers of landbirds on 'Eua, Tonga (Steadman 1993a, 1995a, updated in Chapter 6) and Mangaia, Cook Islands (Chapter 7).

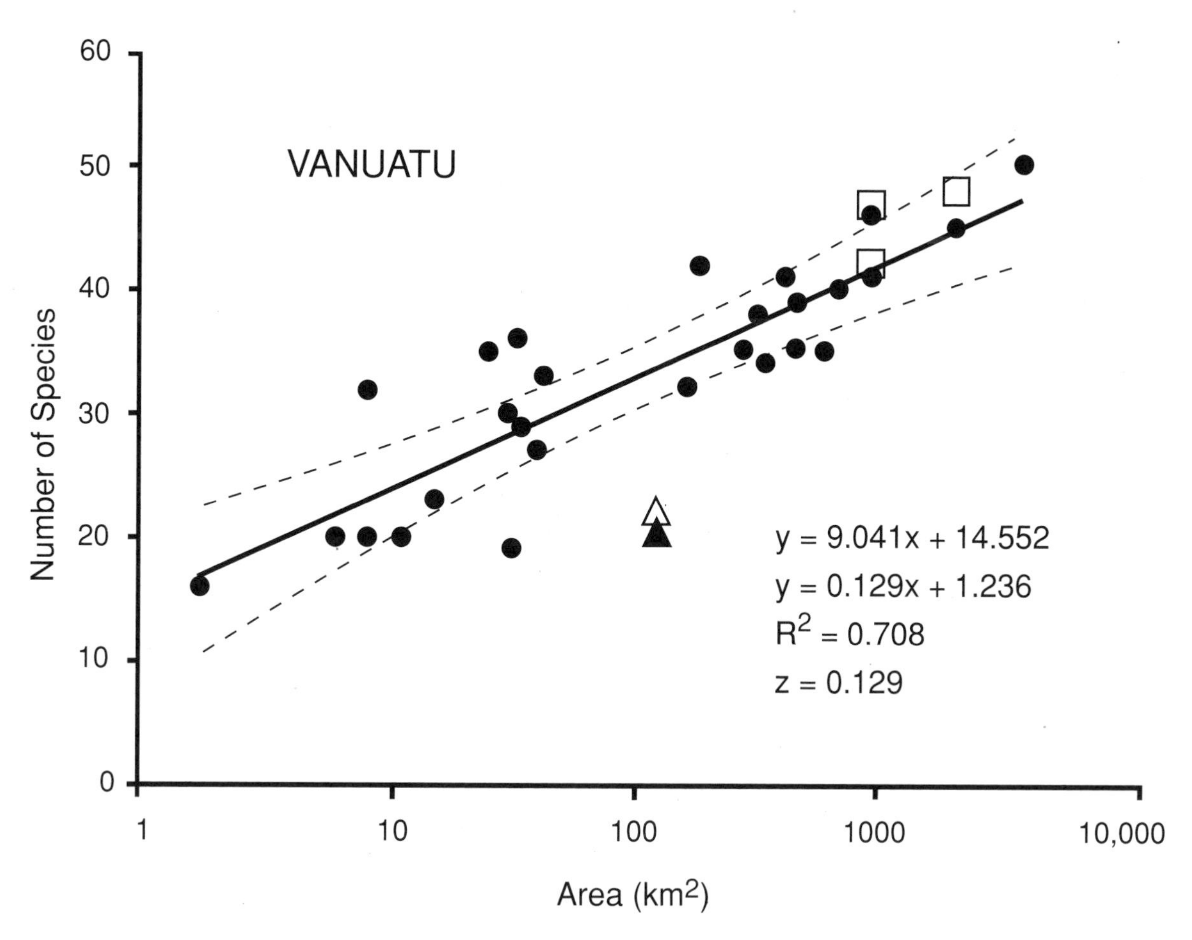

Figure 19-7 Semilog species-area plot of modern (●) and prehistoric (□) landbirds in Vanuatu. All are single islands except for the Torres Group (▲ modern, △ prehistoric); note that □ and △ differ from ● and ■ for only four islands, a reflection of the limited (but growing) fossil record. From data in Table 5-11. The dashed lines on either side of the solid line of best fit are the 95% confidence intervals. The top equation describes the semilog line of best fit. The second equation, as well as the R^2 and z values, is derived from a log-log plot.

presented *SA* landbird data for individual islands (yielding a gently positive slope for the line of best fit; Figure 19-8) rather than the entire archipelago. This may not have been because of a lack of information; following the Whitney South Sea Expedition in the 1920s and 1930s, the modern distribution of landbirds was fairly well known by 1960s standards for all Pacific island groups except the Cook Islands. Perhaps what prevented *SA* analyses in most island groups was that a log-log plot of *S* vs. *A* for landbirds would have yielded a shotgun-scatter of points where drawing a "line of best fit" would have been ridiculous. Such a scatter cannot be interpreted with equilibrium theory (which considers only area and isolation), but is readily explained by considering interisland variation in anthropogenic extinction as well as area, isolation, and quality of survey data.

In the Cook Islands, fossils from Mangaia (52 km^2, 169 m) disclose 20 species of landbirds, 15 of which are extinct or extirpated, compared with eight species known historically (five still surviving) from nearby Rarotonga (67 km^2, 652 m), which lacks fossils (Table 7-2; Figure 19-9). That many more species of landbirds are known from Mangaia than elsewhere in the Cooks is due to Mangaia's much richer fossil record, outlined in Chapters 4 and 7. A similar example would be Rota (Mariana Islands), where a better-than-average fossil record has pushed *S* higher than for any other island in the group (Figure 19-10).

Before continuing to review the implications of prehistoric data on *SA* relationships, let us not forget that modern survey data can be inadequate even in Remote Oceania. Modern *S*-values from the Lau Group of Fiji (Tables 6-4, 6-5) show obvious sampling biases. Three trampy species of landbirds that undoubtedly are much more widespread in Lau than revealed by information gathered during the Whitney South Sea Expedition (which visited Lau in 1924–25) are the volant rails *Gallirallus philippensis* and *Porphyrio porphyrio*, and the nocturnal barn-owl *Tyto alba*. Each can be visually or vocally cryptic, at least during certain seasons (Steadman & Freifeld 1998). Many other absences of

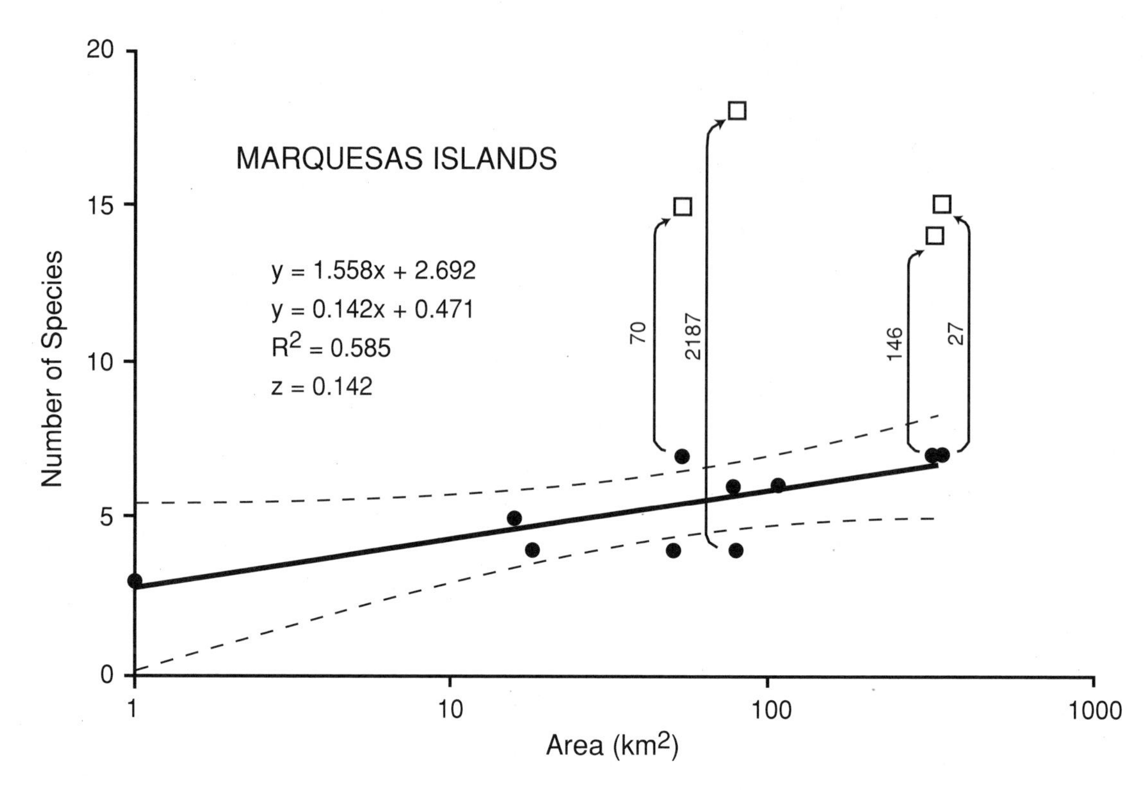

Figure 19-8 Semilog species-area plot of modern (●) and prehistoric (□) landbirds in the Marquesas Islands. Along the arrows pointing to the squares are the numbers of prehistoric landbird bones identified from each island. From data in Table 7-19. The dashed lines on either side of the solid line of best fit are the 95% confidence intervals. The top equation describes the semilog line of best fit. The second equation, as well as the R^2 and z values, is derived from a log-log plot.

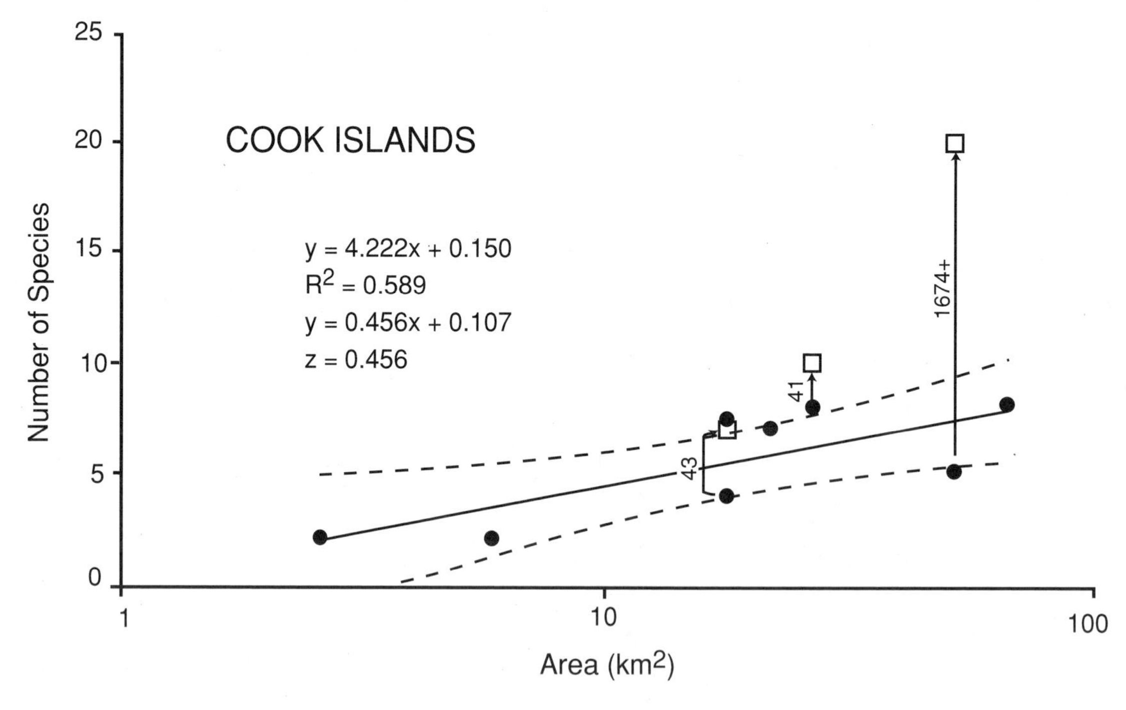

Figure 19-9 Semilog species-area plot of modern (●) and prehistoric (□) landbirds in the Cook Islands. Along the arrows pointing to the squares are the numbers of prehistoric landbird bones identified from each island. From data in Table 7-2. The dashed lines on either side of the solid line of best fit are the 95% confidence intervals. The top equation describes the semilog line of best fit. The second equation, as well as the R^2 and z values, is derived from a log-log plot.

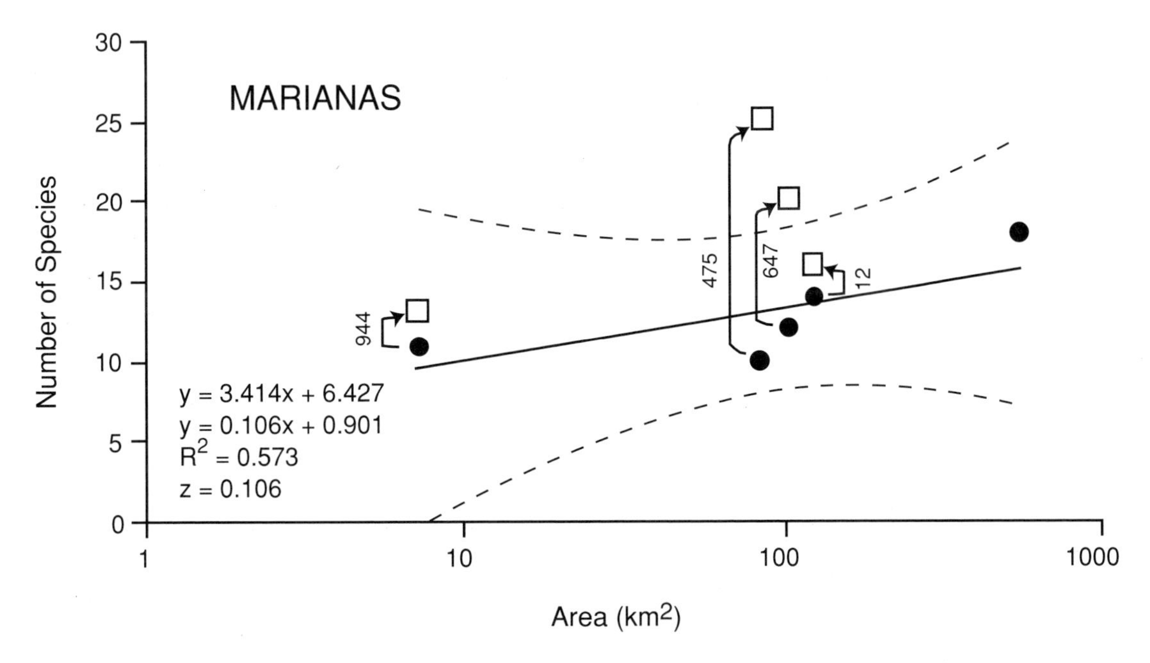

Figure 19-10 Semilog species-area plot of modern (●) and prehistoric (□) landbirds in the Mariana Islands. Along the arrows pointing to the squares are the numbers of prehistoric landbird bones identified from each island. From data in Table 8-4. The dashed lines on either side of the solid line of best fit are the 95% confidence intervals. The top equation describes the semilog line of best fit. The second equation, as well as the R^2 and z values, is derived from a log-log plot.

species on individual islands also are likely to reflect inadequate sampling. I believe that the highly variable species richness values for islands in Lau that are from 1 to 10 km^2 (Figure 19-11) are due to sampling bias as well as differential human impact. In southern Lau, the two islands most intensively surveyed for modern landbirds are Aiwa Levu and Lakeba (Table 6-5); note that the modern SA data for these two islands alone yield a line (long dashes) with a very gentle slope (Figure 19-11). Adding data from their limited fossil records, the tiny (1.21 km^2) but forested, uninhabited Aiwa Levu has more species than Lakeba, which is much larger (55.94 km^2) but is inhabited and mostly deforested.

In the Vava'u Group of Tonga, Steadman and Freifeld (1998) surveyed landbirds on 16 islands in 1995 and 1996 (Tables 6-11 to 6-13). Six of the islands never had been visited before by an ornithologist. The plot of S vs. log-A has a steeper slope for the less comprehensive S-values ($z = 0.21$; Figure 19-12a) than for the presumably more complete S-values ($z = 0.14$; Figure 19-12b). Based on background knowledge about living species and the limitations of our brief surveys, the S-values in Figure 19-12b probably reflect each island's modern avifaunas more accurately than those in Figure 19-12a. The z-values in Figure 19-12a and 19-12b show little evidence of change across four orders of magnitude of area (0.1-100 km^2). If prehistoric data were available, the S-values in Vava'u undoubtedly would be much greater overall, as is true elsewhere in Tonga (Ha'apai,

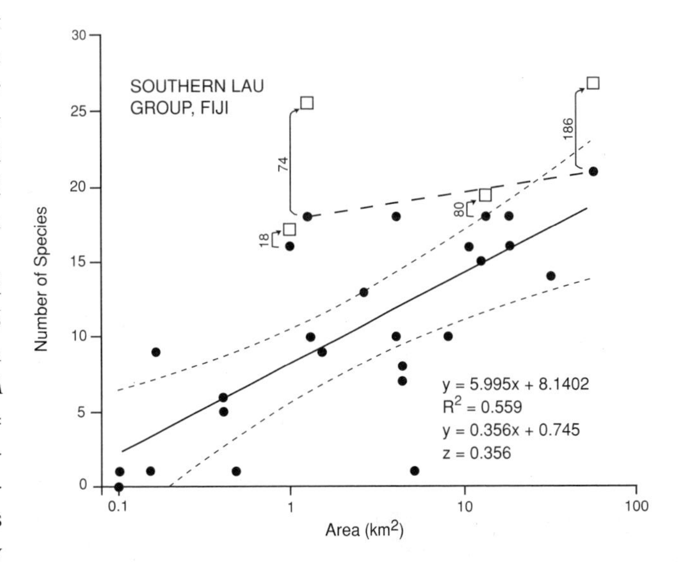

Figure 19-11 Semilog species-area plot of modern (●) and prehistoric (□) landbirds in the Southern Lau Group, Fiji. The heavy dashed line connects only the points for the two islands best surveyed for modern birds, Aiwa Levu (small) and Lakeba (large). Along the arrows pointing to the squares are the numbers of prehistoric landbird bones identified from each island. From data in Tables 6-5, 6-9. The dashed lines on either side of the solid line of best fit are the 95% confidence intervals. The top equation describes the semilog line of best fit. The second equation, as well as the R^2 and z values, is derived from a log-log plot.

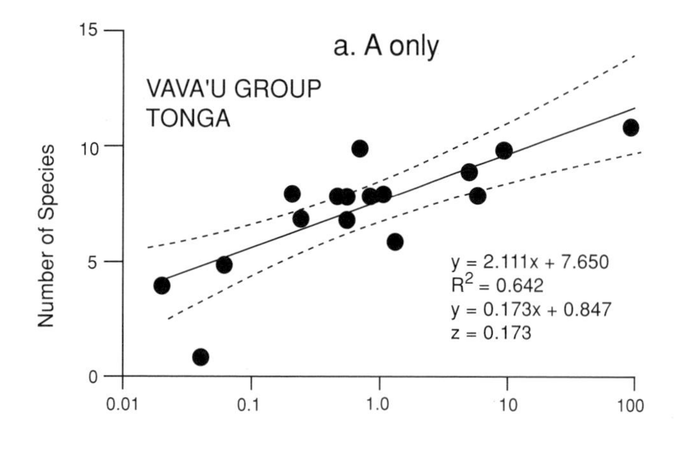

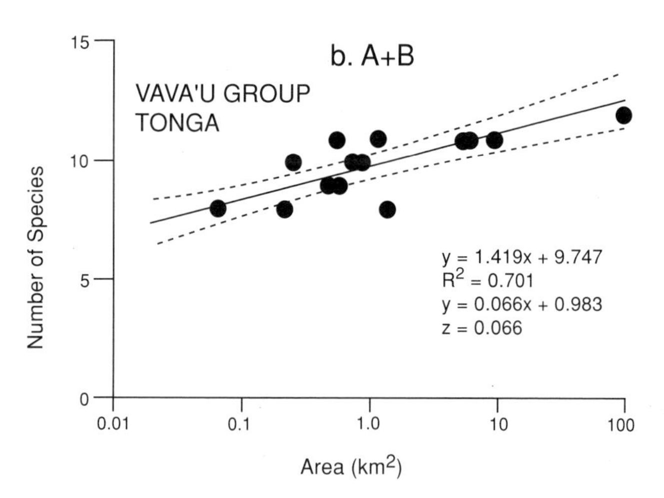

Figure 19-12 Semilog species-area plots of modern landbirds in the Vava`u Group, Tonga. A, species recorded July 1995 or July 1996; B, species not recorded in 1995/1996 but that likely occur on that island. From Steadman and Freifeld (1998) and Table 6-13 herein. The top equation describes the semilog line of best fit. The second equation, as well as the R^2 and z values, is derived from a log-log plot.

'Eua, and Tongatapu) where, after human impact is filtered out, z approaches 0 for islands >1 km^2 (Steadman 1998; Chapter 6 herein; also see below).

Except for the smallest islands (<0.1 km^2), A affects S minimally in Vava'u, where few interisland distances exceed 10 km and most are <5 km. Much of the positive relationship between S and A may be because it has been more difficult for people (through whatever means) to extinguish populations of birds on large islands than on small ones (Steadman 1995a). That habitat loss is important is suggested by examining the points in Figure 19-12b that deviate most from the fitted regression line. Those lying most below the line are the highly disturbed islands of Ofu and Foelifuka. Most above the line are the points for 'Utungake, A'a, and 'Eueiki, three islands that retain much native forest.

As island area approaches 0, one after another resident species of landbird must drop out as survival requirements are not met for even one pair. Morse (1977) discussed this issue for 12 extremely small (0.001-0.04 km^2) islands in Maine, which differ from those in Vava'u in having an immense continent as an obvious, nearby source area. If all islands in Vava'u (or elsewhere in Tonga and Samoa) were still largely forested and lacked native predators, the minimum land area required to sustain the complete modern avifauna may be as small as only 1–2 km^2. Whether this was also the case in the past, when species richness was much greater, has not been tested in Vava'u. In the Ha'apai group of Tonga, however, islands as small and low as Ha'afeva (1.8 km^2, 11 m elev.) seem to have had more or less the full range of species at human contact (Table 6-15, Figure 19-13). In Tonga, the one archipelago with substantial prehistoric data from multiple islands spanning 3+ orders of magnitude of land area (Tables 6-10, 6-19 to 6-21), the increase in S with A is related more to intensity of fossil sampling (number of identified bones) than to A. In Figure 19-14, the four islands with the lowest prehistoric S-values are the same four islands with the fewest identified bones. I note as well that only on 'Eua are there fossils from precultural deposits, which are more effective at sampling the most vulnerable species than cultural sites (Chapters 4, 16).

The prehistoric data from Tonga are inadequate to test the precise effect of isolation on species richness because none of the seven islands with fossil records is very isolated, and biotic source areas within and outside of Tonga are speculative. How much interisland gene flow occurred in

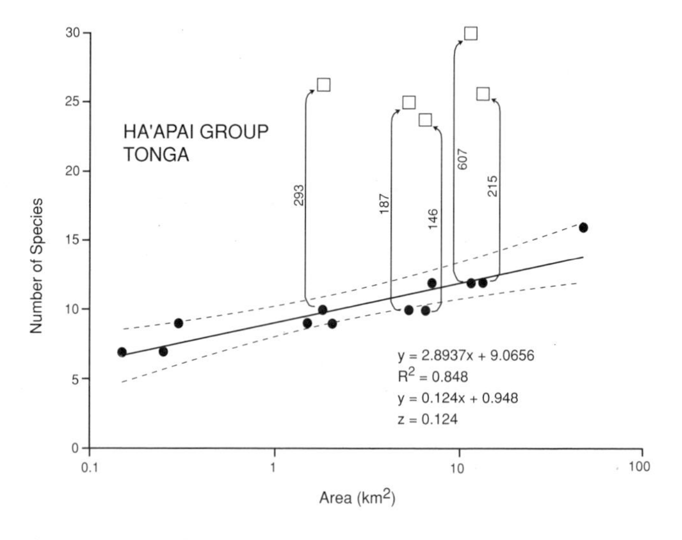

Figure 19-13 Semilog species-area plot of modern landbirds in the Ha`apai Group, Tonga. From data in Steadman (1998) and Table 6-15 herein. The dashed lines on either side of the solid line of best fit are the 95% confidence intervals. The top equation describes the semilog line of best fit. The second equation, as well as the R^2 and z values, is derived from a log-log plot.

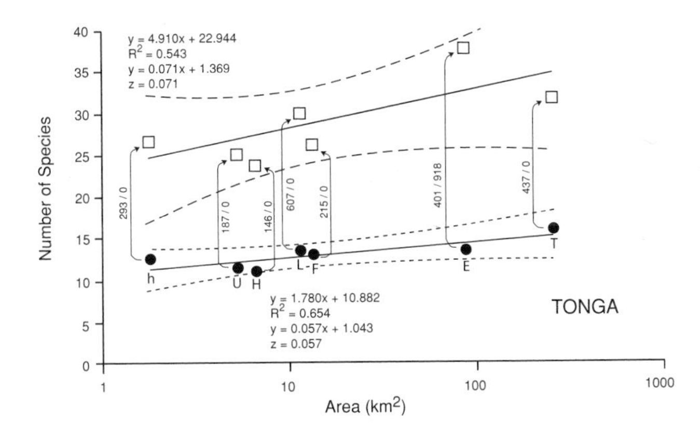

Figure 19-14 Semilog species-area plots of modern (●) and prehistoric (□) landbirds among the seven islands in Tonga with substantial prehistoric records of birds. E, `Eua; F, Foa; H, Ha`afeva; h, Ha`ano; L, Lifuka; T, Tongatapu; U, `Uiha. Along the arrows pointing to the squares are the numbers of prehistoric landbird bones identified from each island. The first of these numbers represents bones from cultural sites; the second number represents bones from precultural sites. From data in Tables 6-10, 6-19. The dashed lines on either side of the solid line of best fit are the 95% confidence intervals. The top equation describes the semilog line of best fit. The second equation, as well as the R^2 and z values, is derived from a log-log plot.

each species of Tongan landbird is a matter of speculation; all we know is that it probably varied considerably. We do not know what the minimum viable population sizes would have been for any Tongan species, whether today or under natural conditions. In relatively tight clusters of islands such as the Vava'u and Ha'apai Groups, it seems likely that most surviving species have interisland gene flow that facilitates persistence on small islands (<0.1 or 1.0 km^2).

z-values

The line of best fit in a log-log plot of S vs. A has a slope called z and a y-intercept called C (McGuinness 1984, Brown & Lomolino 1998:376). In any data set, the value for z can change depending on spatial scale. For British plants in plots of various sizes, z was low (0.1–0.2) at both the smallest and largest spatial scales, and higher (0.4–0.5) at intermediate scales (Crawley & Harral 2001). Hubbell (2001:157–161, 200) reported another sort of triphasic, S-shaped curve as a generalization for Neotropical plants as well as other plants and animals (such as birds worldwide, his Figure 6.2), with a log-log plot being linear only at intermediate A-values and becoming steeper and nonlinear at local and continental/global scales.

For landbirds in Oceania, z seems to get steeper only at very small spatial scales, probably reflecting an inability to establish or maintain long-term populations on tiny islands. This may be more or less the insular equivalent of the observation by Hubbell (2001:200) that "on local spatial scales, species-area curves are sensitive to the local relative abundance of species and are curvilinear on a log-log plot." I find no evidence in Oceania of a steeper slope of a log-log species-area curve for birds at large spatial scales, i.e., the largest islands. Thus, there is no need to explain such steep slopes, which Hubbell (2001:200–201) wrote were because "evolutionary-biogeographic processes on regional scales become increasingly uncoupled and dynamically uncorrelated with one another." This uncoupling may apply to some groups of organisms on continental or intercontinental scales, but does not pertain to interisland, intra-archipelago comparisons of landbirds on Pacific islands.

High-quality S-values for landbirds within Pacific island groups, as derived from both bones and rigorous modern surveys, yield a very low slope (z) for log-A vs. log-S once a minimum island area is reached. This flat or nearly flat slope applies in a strictly mathematical sense and calls into question, as pointed out by Abbott (1983), whether z has much biological meaning in natural systems. On oceanic islands, z for landbirds is nearly flat (approaches 0) over medium to large land areas, whereas in geographically nested continental settings, such as in British plants in tiny plots ranging from 0.01 to 10,000 m^2 [= 0.00000001 to 0.01 km^2] (Crawley & Harral 2001), z mainly reflects sampling. Staying on continents but moving from plants to landbirds, z would again seem to be compromised by sampling at any scale less than 1 km^2 or probably even larger. Therefore, it is tenuous if not illogical to extrapolate any biological inference from tiny continental plots to ones large enough to sustain long-term populations. Moving back to islands, it is similarly inappropriate to infer much from extremely tiny islets (= rocks) such as those studied by Losos et al. (1997) or Schoener et al. (2001), which range from two to six orders of magnitude smaller than 1 km^2, i.e., are more comprehensibly measured in m^2 than km^2.

Isolation

Isolation is important in the distribution of terrestrial life in Oceania. As just noted, isolation is much less absolute on continents than on oceanic islands, where the intervening sea is useless to landbirds and therefore an effective dispersal filter. I do not agree with the suggestion by Brown and Lomolino (1998:405) that speciation contributes significantly to S only on the largest and most isolated islands. Hundreds of exceptions do not prove the rule.

While landbird faunas in Oceania generally do decrease west-to-east in species richness (Chapter 17), and while I present the distances between island groups (Tables 2-1, 2-2) from that perspective, this does not mean that we always should look westward for source areas. Perhaps especially in Remote Oceania, the source areas for nearly all species are hypothetical both within and among island groups. In many cases, island groups to the east are just as likely to be source areas as ones to the west. We have no evidence, for example, whether East Polynesia's two widespread, extinct lorikeets (*Vini vidivici*, *V. sinotoi*) originated in the Cook, Society, or Marquesas Islands, the three places where their bones have been found. The same could be said for columbids formerly widespread in East Polynesia (*Ducula aurorae*, *D. galeata*, *Gallicolumba nui*, *G. erythroptera*). The cumulative result of high levels of multidirectional isolation is that, in simplistic terms, species richness decreases with isolation, which is more important than island area in accounting for the west-to-east faunal attenuation in Oceania.

The negative impact of isolation on species richness is evident across Oceania. In the Bismarcks, the rather large islands of Manus and Mussau have less rich modern avifaunas than comparably sized but less isolated Umboi (Table 5-3, Figures 5-3, 19-3). Landbird fossils are of limited extent on Mussau, unverified on Manus, and unknown on Umboi (Chapter 5), so we cannot yet say to what degree this difference would hold if tested with prehistoric data. I suspect that it would remain.

Within an archipelago in Remote Oceania, most volant species of landbirds were widespread before human impact, yielding highly nested avifaunas. Unlike in Near Oceania, where a few species live only or mainly on small islands (Diamond 1975b, Mayr & Diamond 2001), I have found no evidence of such specialization in Remote Oceania. The reduced species richness on extremely small islands (<1 km^2) that are not highly isolated may be due more often to the failure to maintain viable populations, in spite of occasional arrival, than never having dispersed. Increased susceptibility to environmental disturbance (see Morrison 1997) may be another factor explaining the higher z-values for sets of very small islands, although Franklin et al. (1999) found woody plant communities to be rich in species even on islands <1 km^2 in the Vava'u Group of Tonga, as long as human disturbance was relatively low. The same is true in the Lau Group of Fiji (Janet Franklin pers. comm.). On very small islands that are also highly isolated, pure failure to colonize probably accounts for much of the reduced species richness in plants or birds.

Because I believe that analyses of biotic nestedness in an oceanic island group should not be based on a single hypothesized source area, they are not directly comparable to those involving continental islands (e.g., Simberloff & Martin 1991). Another big difference in evaluating nestedness in insular vs. continental situations is that, for oceanic islands, the plots of land being evaluated are not nested themselves. Surveying the terrestrial biota of each real island requires a separate trip by boat or airplane, with the intervening ocean being a place where S is guaranteed to be zero. The ocean does not provide landbird habitat of even the most marginal quality. This is very different from a continental setting where latitudinally and longitudinally restricted blocks of land can be analyzed (e.g., Jetz & Rahbek 2002). A breeding population of Henslow's sparrow (*Ammodramus henslowi*) in a single field (0.09 km^2) guarantees its presence on my family farm (1 km^2) in McKean Township (86 km^2), Erie County (2077 km^2), Pennsylvania (119,291 km^2), the continental United States (9,612,360 km^2), North America (24,397,000 km^2), the New World (42,968,000 km^2), and all land on earth (510,100,000 km^2), each a spatial subset of the next. In dispersing among weedy hayfields, these sparrows encounter many places where they can rest and probably feed, even if the habitat is not suitable for breeding.

A New Species-Area Model for Birds in Oceania

An alternative default (null) model for the distribution of landbirds in Oceania is that every species or allospecies known to inhabit an archipelago should occur on each high island (non-atoll) of a certain minimum land area (MLA) within that archipelago. The MLA may decrease on average in more isolated island groups. In Remote Oceania, the MLA needed to sustain essentially the entire landbird fauna probably varies from as small as 1–10 km^2 (Tonga, Samoa, East Polynesia, Micronesia) to perhaps 10–30 km^2 (Fiji, Vanuatu). New Caledonia is a special case because Grande Terre and Ile des Pins are much older than the raised limestone Loyalty Group (Chapter 1). In Near Oceania, the MLA is even more speculative but may be as small as 50–100 km^2.

An island group whose landbirds conformed wholly to this model would have a z-value of 0 once the MLA was attained. I do not expect perfect fits for this default model in each island group, but it provides a clear benchmark against which to evaluate species richness on any island. Please keep in mind that species richness is a simplistic measure of taxonomic or ecological diversity on an island; it is uncalibrated by data on relative abundance, habitat association, intra-island distribution, feeding guilds, etc.

Isolation within an island group also affects distributions (see above). In Oceania I believe that interisland distances less than ca. 50-100 km tend to have little long-term influence on species richness. Greater within-archipelago isolation than this probably reduces S-values, as shown in outliers such as Rennell and Bellona in the Solomons, or

Tikopia and Anuta in the Santa Cruz Group, or 'Ata, Niuatoputapu, Niuafo'ou, and Tafahi in Tonga.

An island's geological age influences species richness in a positive way; the longer that an island has existed, the more opportunities it has had for colonization and perhaps speciation (Chapter 18). Two islands where *S*-values might be enhanced by old age are Grande Terre (New Caledonia) and Viti Levu (Fiji) (Chapters 1, 2, 5, 6). On the other hand, islands with very young geological ages should have low *S*-values. This is why atolls do not fit the model; most of them were inundated during the last interglacial (ca. 125,000 years ago) or, even much worse for species richness, during the current interglacial interval (the past 10,000 years; Chapter 1).

My distributional model applies to the majority of high islands in Oceania, which have geological ages of ca. >0.2 but <10 Ma. It can be evaluated and refined with fossils. The strength of the scrutiny will depend on completeness of fossil records, which vary in such attributes as screen size, age of site(s), number of sites, quality of comparative collection, etc. (Chapter 4). Because most fossil data in hand are rather limited in scope, we are just beginning to evaluate this new model.

Feeding the model is information from Tonga and the Marquesas, the two island groups with respectable though highly incomplete fossil records from multiple islands (Tables 6-10, 6-19, 7-19, Figures 19-8, 19-14). In either case, the prehistoric data suggest a much more uniform, widespread distribution of species at human contact than today, leading to less interisland variation in species richness once an MLA is attained (see below). If the Marquesan fossil record were complete, I believe that it would reveal an essentially identical composition of species/allospecies on each of the nine major islands (ranging in area from 337 km^2 down to 16 km^2), with little or no variation in species richness. Perhaps even the 10th-largest Marquesan island (Fatu Huku; 1 km^2) had a full set of species. Under natural conditions, interisland taxonomic variation in the Marquesan landbird faunas would be expressed only in allospecies of flightless rails and in allospecies/allosubspecies of volant landbirds.

The same may have been true or nearly so in Tonga. Please note that, in Figure 19-14, the largest Tongan island (Tongatapu) is well below the line of best fit in spite of having 437 identified prehistoric landbird bones. The second largest island, 'Eua, has the most landbird fossils overall (1319) and is the only one with precultural landbird fossils ($N = 918$), augmenting its high *S*-value by adding seven species not recorded in its cultural deposits (Table 6-20).

Similar analyses could be made from the extensive paleontological data in the Hawaiian Islands or New Zealand (James & Olson 1991, Olson & James 1991, Worthy & Holdaway 2002), although any evaluation of my default model is compromised to some degree by negative evidence. Is the apparent absence of a species on a given island a genuine phenomenon, or is it due to inadequate fossil sampling? We can keep the dragon of negative evidence at bay by continuing to expand the quality and quantity of the fossil record, with an emphasis on recovering fossils of small species and identifying as many passerine fossils as possible. On islands such as 'Eua or Ua Huka, this dragon is not slain but it is bleeding. Even on seemingly saturated Henderson Island (Chapter 7), there may still be a little life in this stubborn beast.

Compatible with the concept of lower *z*-values within more isolated archipelagos (Diamond & May 1976), I believe that *z* was 0 or nearly so at human contact for major islands without great intra-archipelago isolation in East Polynesia and Micronesia. Moving westward through West Polynesia and Melanesia, *z* at human contact may have had very small positive values among major islands within island groups as possible source areas become larger and more numerous. Nowhere in Oceania do I believe that *z*-values among comparable islands exceeded 0.1. Even in Tonga and the Marquesas, the data are insufficient to say confidently that *z* was 0 at human contact. In the Marquesas I suspect that it was; in Tonga it may have been between 0 and 0.05.

Born in paleontological information, my species-area model predicts less interisland variation in occurrence of individual species/allospecies of landbirds than models derived from equilibrium (or neutral?) theory. In this simplicity my model might be seem elegant. Evaluating it, however, is most inelegant, given how much field and lab work is required to develop good fossil records (which never will be had on some islands, regardless of effort), much less complete ones (Chapter 4). Please remember from the preface that my primary goal is to make progress in documenting and understanding the distribution of landbirds in Oceania. I never promised elegance. If improved fossil data argue for *z*-values >0.1 as being normal in Oceania, then my model would lose much of its utility. Finding the MLA concept to be faulty also would undermine my model.

Except for very small or very isolated islands, my default species-area model predicts a uniform interisland distribution of all known species or species-groups within an archipelago (Figure 19-15). This would set a standard called the "fundamental distribution." The "realized distribution" would be the known prehistoric distribution, based on data that improve as the fossil record grows. On islands with few or no fossils the realized distribution would be unknown. A challenge will be to determine whether differences

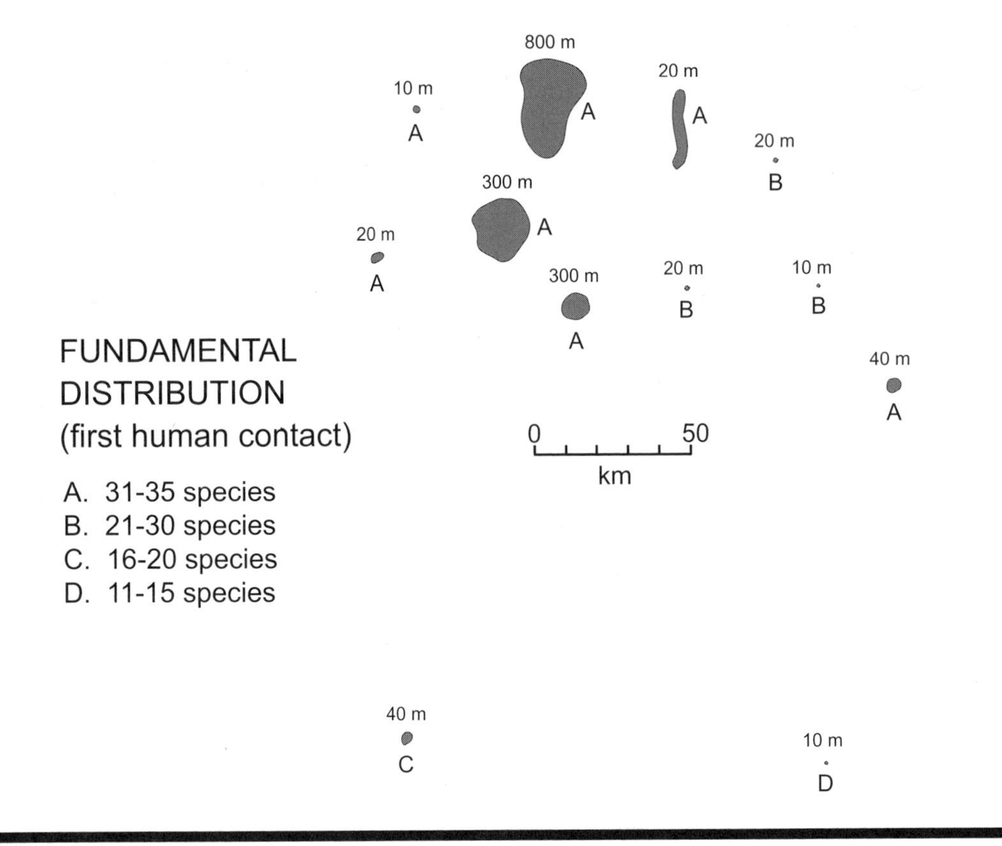

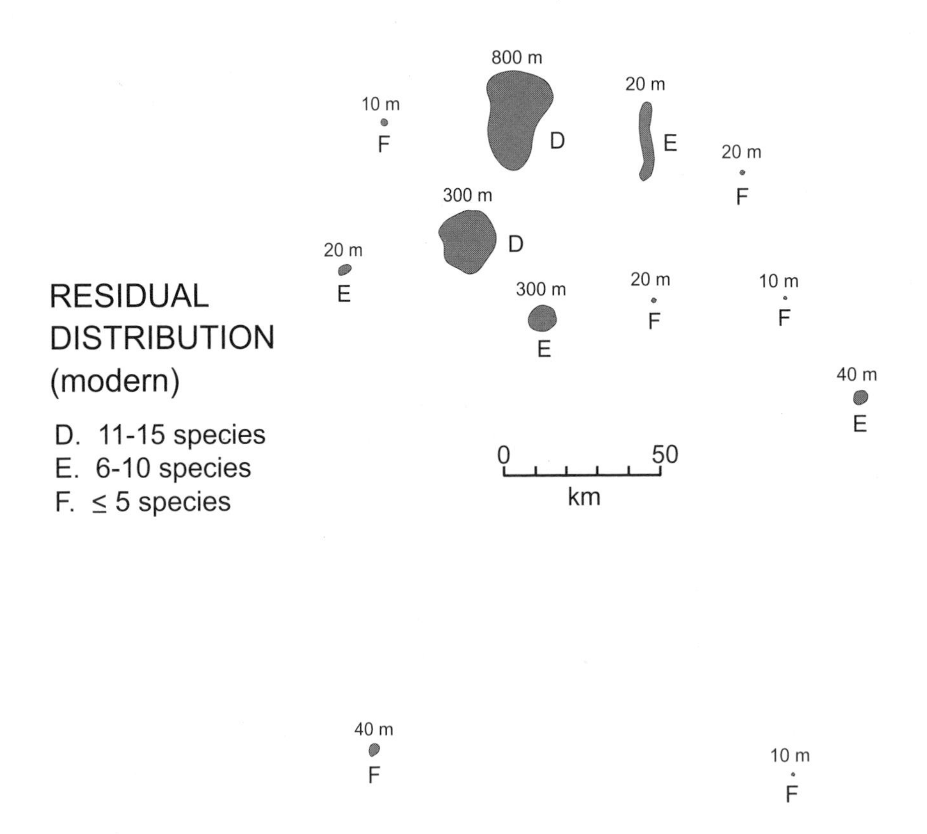

Figure 19-15 Fundamental (first human contact) vs. residual (modern) species richness for a hypothetical island group in Remote Oceania. Elevations in m are given for each island. The approximate isolation and land area of each island can be estimated by the linear (km) scale.

between the fundamental and realized distributions reflect anything other than incomplete fossil sampling. No simple formula can evaluate this with precision. The modern or "residual" distribution consists of the species recorded today or at least in historic times (the past 230 years). It is a subset of the fundamental distribution. On islands with good fossil records, it also is a subset of the realized distribution because, once a certain MLA is attained, much or all of the difference between the known prehistoric (realized) distribution and the modern distribution is due to human impact. Colonization by trampy species since human arrival (Chapters 17, 18) is a minor factor that becomes increasingly insignificant eastward in Oceania. Unlike dynamic, competition-driven models of insular distribution such as that proposed by Lomolino (1999), my species-area model downplays equilibrium and turnover (Chapter 18) and competition (Chapter 20).

Summary and Discussion

I must end this chapter by elaborating on some of the same disclaimers that started it. While the number of species (of birds or any other set of organisms) on an island is some measure of both its taxonomic and ecological diversity, this gauge is crude. For example, an island with 75 endemic, congeneric species of curculionid weevils (see Paulay 1985) certainly would have a high S-value for coleopterans, but may be much less diverse taxonomically than an island with fewer species but more genera and families of beetles. We cannot evaluate ecological diversity until we learn about relative abundances, food habits, and interspecific interactions such as predation, parasitism, competition, seed dispersal, etc. (Chapter 20). It could be that, from a functional standpoint, just as many species (or individuals) of plants are having their juices sucked on an island with one abundant, generalized species of weevil as on an island with 75 specialized species of weevils.

Species-area data based on inadequate sampling (artificially low S-values) are misleading. Many oceanic islands are difficult to survey, even for conspicuous animals such as birds. All survey data, mine included, should be examined critically for completeness, whether birds on oceanic islands (Steadman & Freifeld 1998) or mammals on mountaintops (Grayson & Livingston 1993). Two ways to improve the survey data are to make every effort in the field to ensure that modern S-values reflect all species present, and to determine the extent of anthropogenic change by gathering historic and prehistoric data. In either case, more complete survey data will yield lower z-values, i.e., they will decrease the apparent influence of A on S.

Even the largest S-values for modern landbirds on East Polynesian islands, such as from Tahiti or Nuku Hiva (Chapter 7), are not natural. Instead, these numbers reflect the ability of some species to survive longer in human presence on larger, more mountainous islands. Unlike in Near Oceania, where some species of landbirds may be confined genuinely to montane forest (Mayr & Diamond 1976, 2001; but see Chapters 5, 18, 20), apparent altitudinal differences in Polynesian or Micronesian birdlife probably are artifacts of human impact, such as forest alteration in the lowlands.

The only cases where equilibrium theory might be informative for species richness on oceanic islands would be with sets of undifferentiated organisms on small islands very close to each other but highly isolated from other island groups, such as the plants on individual islets within a single atoll (Woodroffe 1986). These plants are all strand species with good overwater dispersal ability. Even in this case, however, an attribute such as the length of the islet's shoreline (see Buckley & Knedlhans 1986) or nutrient input to soils from seabirds (Anderson & Wait 2001) might be a better predictor of species richness than land area.

Ever since MacArthur and Wilson (1963, 1967) put forth the equilibrium model, biologists have realized that it is based only on immigration vs. "extinction" (= absence; see Chapters 16, 18) and does not account for speciation, relative abundance, or any other biological attribute. Recent attempts to remedy this situation (Losos & Schluter 2000, Hubbell 2001) fail to consider prehistory and therefore are unrealistic. The values used today for species richness and island area remain as Sauer (1969:590) described them, "assemblages of characterless species on various sizes of featureless plains." This biological void is guaranteed by the "neutral" assumptions of Hubbell (2001), which state that every individual in every species in a biological community is identical, and the total abundance of all species is fixed (see critiques by Abrams 2001, de Mazancourt 2001). In nature, nothing could be less true. The neutral prediction that habitat is uniform and that only dispersal and speciation affect how species composition changes with distance is not supported by field data even for tropical trees (Condit et al. 2002).

Losos and Schluter (2000) proposed that *Anolis* has higher S-values on very large (>3000 km^2) West Indian islands because speciation is confined to such islands, with smaller ones sustaining only widespread species. Their proposal does not consider prehistory. Might it not be, for instance, that anthropogenic extinction has removed most endemic species of *Anolis* from the smaller islands? Thus far fossils are silent on this issue because species-level osteological characters in *Anolis* can be difficult to discern (Pregill

1981, Pregill & Crother 1999). Molecular and paleontological data eventually may help to resolve this problem. Nevertheless, we already know that lizards of all sizes have been lost in historic and prehistoric times on West Indian islands of all sizes (Pregill et al. 1994). There is no reason to believe that anoles, especially on small islands, have been immune to human impact.

To make their sweeping generalization, Losos and Schluter (2000) ignored modern endemic species of *Anolis* on the islands they regarded as "small" (which can exceed 1000 km^2 in area, plenty of room for millions of anoles). This is how they calculated a speciation rate of zero for any West Indian island other than the Greater Antillean giants of Puerto Rico, Jamaica, Hispaniola, and Cuba. In the distributional atlas of Schwartz and Henderson (1991), I found 17 species of *Anolis* endemic to single, small West Indian islands and 16 others endemic to multiple small islands. Apparently these 33 species arose through a process other than speciation.

Biogeographers generally have exaggerated the importance of large islands. Evolution and speciation remain incompatible with equilibrium or neutral theory.

Chapter 20 Community Ecology

COMMUNITY ECOLOGY is concerned with how sets of organisms coexist. Here I will examine the species in landbird communities on tropical Pacific islands over different spatial scales. Because our data often are so crude, the species from an entire island will receive more attention than analyses at a finer scale, such as the landbirds from an island's mature forests versus those of its successional habitats. A still finer scale would be feeding guilds within a habitat, such as canopy frugivores or understory insectivores. Related to feeding guilds are taxonomic considerations, such as congeneric or confamilial sets of species.

Most of my attention in this chapter will be given to "assembly rules," a term developed by Diamond (1975b) to explain the composition of landbird communities in the Bismarck Archipelago and other islands off the northern coast of New Guinea. Diamond (1975b) believed that the distribution of landbirds on these islands was not random but was governed by characteristics of the islands (area, and to a lesser extent, elevation and isolation) and intrinsic properties of the birds, especially their highly variable abilities to disperse and to coexist with ecologically similar species. The subsequent research and competition-driven controversies stimulated by Diamond's monograph have been reviewed by authors in Weiher and Keddy (1999). For a brief, even-handed discussion of the composition of biotic communities, especially as it may or may not be influenced by competition, see Gotelli (1999). Much of the information in Diamond (1975b) was either repeated, succinctly summarized, or developed further with additional data in Mayr and Diamond (2001).

Background for Assembly Rules

One reason why it has been difficult to evaluate assembly rules is because Diamond (1975b) did not provide the raw distributional data (a species/island matrix). In spite of nearly three decades of criticism, such a matrix is absent as well in Mayr and Diamond (2001), even though the distributional data are much more extensive (see Kratter 2002). The amount of field effort (person-days on a particular island, when, by whom, where on the island, etc.) was not presented in Diamond's original analyses, but was summarized without detail in Mayr and Diamond (2001). More thorough presentations, such as those for Karkar and Bagabag (Diamond & LeCroy 1979) or Long (Diamond 1981), are needed to assess the quality of distributional data for any island. Such data are essentially lacking, however, for other islands in the Bismarcks. Jared Diamond is an outstanding field ornithologist with ears exceptionally well tuned to the songs and calls of Melanesian birds. The data based on his own fieldwork in New Guinea and the Bismarcks, beginning in the 1960s and ongoing, are undoubtedly of very high quality. Much of the distributional data in Diamond (1975b) and Mayr and Diamond (2001) was collected during or before the Whitney South Sea Expedition (WSSE) of the 1920s and 1930s. Often the WSSE spent only a day or two on an island, and therefore could not record every species of landbird that was present.

The quality of distributional data depends on the effort to document both presences and absences. On large islands, such as Isabel in the Solomons, new fieldwork continues

Table 20-1 Distributional categories (incidence functions) of landbirds in the Bismarck Archipelago. Categories are defined in the text. "Semispecies" is the same as allospecies. S is the species richness value of an island. Lower S_{crit} is the minimum species richness value for the islands where a species occurs. Upper S_{crit} is the maximum species richness value for the islands where a species occurs. From Diamond (1975b:359).

Category	Number of species in category	Number of islands inhabited per species	Endemism level				Lower S_{crit}			Upper S_{crit}		
			None	Subspecies	Semispecies	Species	<10	15–31	>32	≤16	45–83	None
High-S	52	1–5	9	14	21	8	0	0	52	0	0	52
A-tramp	26	3–9	8	14	3	1	1	1	24	0	0	26
B-tramp	17	10–14	4	12	1	0	3	5	9	0	0	17
C-tramp	19	15–19	8	10	1	0	6	9	4	0	0	19
D-tramp	14	20–35	9	5	0	0	11	3	0	0	0	14
Supertramp	13	2–33	2	8	2	1	10	2	1	4	9	0

to augment species richness values for forest birds (Kratter et al. 2001a). Even on very small islands (<10 km^2), a few days or less of fieldwork often are not enough to be sure that a certain species, especially those that are rare or cryptic, is present or absent. In the southern Lau Group of Fiji, for example, inadequate sampling almost surely explains some of the low species-richness values for landbirds on individual islands (Table 6-5, Figure 19-11). Among islands >1 km^2 in southern Lau, those surveyed for only two person-days have lower average values than those surveyed more extensively. Only the two surveyed for >10 person-days by my field parties in 1999–2000 have species richness values that are likely to include >90% of the currently resident species of landbirds (Steadman & Franklin 2000, unpub. data).

To interpret the observed variation in dispersal ability and presence/absence on islands that span a great range of area and species richness, Diamond (1975b) classified Melanesian landbirds into six categories of "incidence functions"–high-S species, A-tramp, B-tramp, C-tramp, D-tramp, and supertramp (Table 20-1). He interpreted the incidence functions of selected species (especially columbids and passerines) "in terms of island area plus a species' habitat requirements, dispersal ability, birth and death schedule, exploitation strategy, and competitive relations" (1975b:346). These complex aspects of natural history often are poorly known, inadequately considered, or difficult to quantify (Graves & Gotelli 1983, Gotelli & Graves 1996).

Figure 20-1 depicts the incidence function for a species in each of the six categories. A high-S species is one that occurs on no low-S islands (≤31 species) or intermediate-S islands (32–83 species), and on one or both of the high-S islands, which are New Ireland ($S = 101$) and New Britain ($S = 127$). The A-tramp species differ from high-S species in that they occur rarely on low-S islands and more commonly on intermediate-S islands. B-tramps occur on more islands than high-S or A-tramp species, including a variable number of low-S islands.

These trends continue for C-tramps and D-tramps. The D-tramp dove *Chalcophaps stephani*, for example, occurs on all islands once S exceeds a minimum value of ca. 20 (Figure 20-1). Supertramps never occur on the two high-S islands; they are the only category of incidence function where S has an upper limit. The dove *Macropygia mackinlayi*, for example, is believed to occur on no island where $S \geq 80$ and is sporadically distributed on the less rich islands (Figure 20-1). "Insular distribution functions" are a form of incidence function that relies more exclusively on an island's area and isolation (Lomolino 1999: Figure 10.1); perhaps they are most applicable to relatively depauperate communities, such as those of small mammals on continental islands.

Diamond (1975b) evaluated incidence functions in terms of resources (= food) available vs. resources used by species on islands that are assumed to differ in their resource base because of variation in area, elevation, and isolation. The resulting "resource utilization functions" were depicted to help explain why certain combinations of species co-occurred but others did not. Resource utilization functions are visually pleasing (Figure 20-2), although nobody ever has quantified the seasonal or annual production vs. avian consumption of any food (fruits, seeds, nectar, insects, etc.) on a Melanesian island, a daunting if not impossible task.

Finally, Diamond (1975b) put forth the assembly rules to explain the observed distribution of landbirds. Sympatric species pairs were called "allowed combinations" or "permitted combinations" because of their hypothesized low levels of competition. Species pairs that generally are allopatric were called "forbidden combinations" as

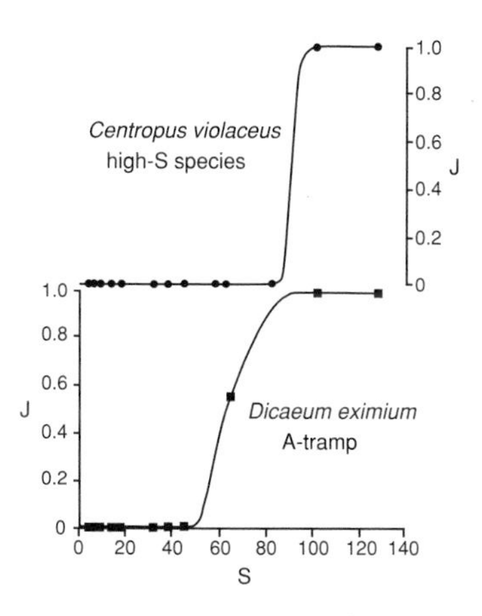

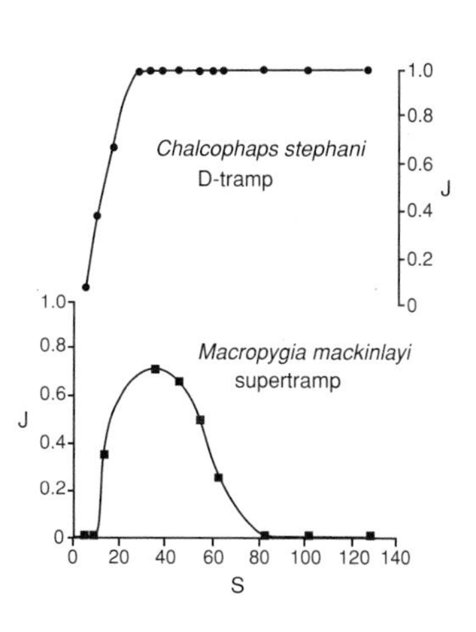

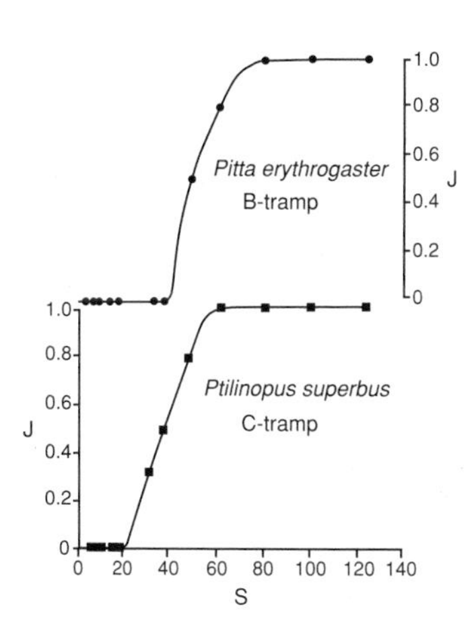

Figure 20-1 Examples of incidence functions in the Bismarck Archipelago, featuring high-S cuckoo *Centropus violaceus*, A-tramp flowerpecker *Dicaeum eximium*, B-tramp pitta *Pitta erythrogaster*, C-tramp dove *Ptilinopus superbus*, D-tramp dove *Chalcophaps stephani*, and supertramp dove *Macropygia mackinlayi*. Islands of the Bismarck Archipelago were divided by Diamond (1975b) into groups such that the value of the total species number (S) of all islands in a given group fell within a narrow range ΔS. ΔS was generally <3 for $S < 20$, 2–10 for $20 < S \leq 65$, and 2–20 for $S > 65$. The ordinate J is the incidence of the given species (i.e., the fraction of the islands in the group in which the species occurs), and the abscissa S is the average species number for the islands of the group. Thus, $J = 1.0$ or $J = 0$ means that a species occurs on all islands or on no island, respectively, that has approximately the indicated species number. Each point is usually based on 3–13 islands, except that the two rightmost points represent one island each ($S = 101$ and 127). Redrawn from Diamond (1975b: Figures 4–6), who provided more detailed captions.

dictated by major differences in incidence functions and/or by hypothesized high levels of competition, thus disobeying "compatibility rules" (p. 395).

Diamond (1975b:423) summarized the assembly rules as follows.

Community assembly involves the following patterns:

- If one considers all the combinations that can be formed from a group of related species, only certain ones of these combinations exist in nature.
- Permissible combinations resist invaders that would transform them into forbidden combinations.
- A combination that is stable on a large or species-rich island may be unstable on a small or species-poor island.
- On a small or species-poor island, a combination may resist invaders that would be incorporated on a larger or more species-rich island.
- Some pairs of species never coexist, either by themselves or as part of a larger combination.
- Some pairs of species that form an unstable combination by themselves may form part of a stable larger combination.
- Conversely, some combinations that are composed entirely of stable subcombinations are themselves unstable.

The "rules" are fueled by interspecific competition, in the absence of which any combinations of species, related or unrelated, should be able to coexist. Diamond (1975b:345) described this as "competition for resources and . . . harvesting of resources by permitted combinations so as to minimize the unutilized resources available to support potential invaders."

I question the importance of competition in structuring insular bird communities. At some level competition among birds must exist (such as between two individual birds, regardless of species, that want to eat the same fruit) and at some level it either does not exist or is trivial (one bird easily can move to eat a different fruit). On 3 March 2000, on the uninhabited, forested island of Aiwa Lailai, in the southern Lau Group of Fiji (12 km SW of Lakeba; Figure 6-2), I spent 20 minutes watching one pigeon (*Ducula pacifica*), 14 fruit-doves (11 *Ptilinopus perousii*, three *P. porphyraceus*), two trillers (*Lalage maculosa*), and five starlings (*Aplonis tabuensis*) feeding voraciously in a fruiting fig tree (*Ficus obliqua*). The fruit-doves and starlings were more aggressive toward individuals of their own species than to any others. Regardless, there were plenty of figs for everyone, and nearby were many other fruiting trees, both fig and non-fig. Multiple species of bird-adapted fruits are available at any season in Fiji, and on Melanesian islands to the west. If competition (leading to "resource partitioning") determines the species composition of the canopy frugivore community (see Diamond 1975b:406–411), how can the more passive *P. porphyraceus* coexist with the similarly sized but more aggressive *P. perousii* in a fig tree with a canopy 9 m wide on an island of only 1 km^2? Furthermore, given the metabolic efficiency of avian flight (Berger & Hart 1974, Walsberg 1983, Norberg 1996), I doubt that it

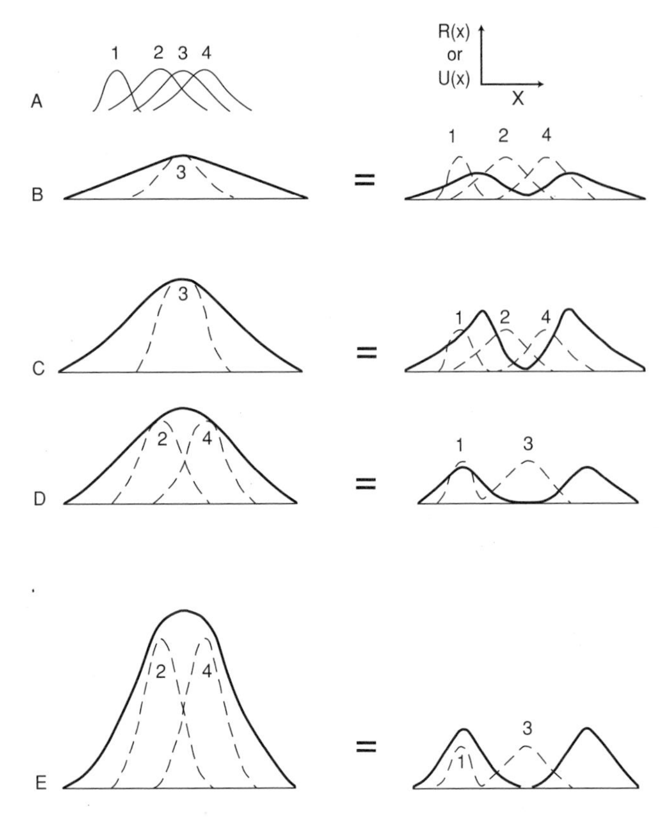

Figure 20-2 Examples of hypothetical relationships between utilization functions $U(x)$ and resource production curves $R(x)$. Resource production is assumed to be distributed along a single dimension (e.g., fruit size or altitude) according to curves such as the solid curves on the left of Figures B–E. Each of four species (1, 2, 3, 4) has a characteristic utilization function or frequency distribution of resources consumed, illustrated by the four curves of Figure A. The areas under these curves represent the resource production rate required for maintaining the smallest population of each species that could survive for a reasonable length of time. The left sides of Figures B–E all depict curves that fit entirely under the illustrated $R(x)$ and thus are sufficient for the indicated one or two species to maintain a stable population on the island. The right sides of Figures B–E represent examples of species that do not fit under the $U(x)$ curve and hence cannot maintain stable populations on the island in coexistence with the species depicted on the right side. In Figure B (left), only species 3 fits under $R_b(x)$. On the right, none of species 1, 2, or 4 could maintain itself alone on this island or share the island with species 3. Figure C represents an island whose area is double that of the island in Figure B, so that $R_c(x)$ is twice as high as $R_b(x)$, and twice as large a population of species 3 can be maintained. On the right, species 1 might be marginally capable of sharing this island with species 3, but species 2 or 4 could not. Figure D depicts the same island as Figure C, so that species 2 and 4 fit under $R_d(x)$, alone or summed, allowing these two species to coexist on this island. On the right, species 2 and species 4 together leave fewer unutilized resources on this island than species 1 and are more likely to occur. Figure E represents an island four times as large as that of Figure B. This island can simultaneously support populations of species 2 and 4 that are 3.8 times the minimum size (dashed curves on left). On the right, species 1 but not species 3 could share this island with species 2 and 4. Thus, in this guild one would usually find species 3 on a small island, species 2 + 4 or occasionally 1 + 3 on a larger island, and species 1 + 2 + 4 on a still larger island. From Diamond (1975b: Figure 44), who provided an even more detailed caption.

really matters very much in terms of metabolic costs how often a bird flies while foraging among trees that are all within a 0.7 km radius. I might also note that Aiwa Lailai's rich landbird community exists in the presence of three resident avian predators (the hawk *Circus approximans*, falcon *Falco peregrinus*, and barn-owl *Tyto alba*).

In attempting to understand more about the composition of communities, I will not stray far from tropical Pacific landbirds. Where the data are adequate, I will analyze assembly rules in a unique way—by comparing modern distributions (upon which the rules are based) with past distributions. I will argue that assembly rules based on modern distributions, as well as the incidence functions on which they depend, can be deceptive if not erroneous because they do not consider anthropogenic extinctions. Modern species assemblages of landbirds from Near Oceania have been affected by human activity (Chapter 5), although direct comparisons of the modern distributional data in Diamond (1975b) with prehistoric distributions are not nearly as extensive as I would like because a substantial set of landbird bones is known from only one island (New Ireland) in his study area, the Bismarck Archipelago. Much more prehistoric data are needed to evaluate the extent of extinction among Melanesian landbirds. In the nearby Solomon Islands, however, the data from Buka give pause, with 11 of 18 prehistorically recorded species of landbirds being extinct or extirpated (Chapter 5).

Furthermore, Remote Oceania has abundant evidence for "forbidden combinations" of species that were widely sympatric before people arrived. If the contemporary distributions of doves (*Macropygia*, *Ptilinopus*) and whistlers (*Pachycephala*) in the Bismarck region (Diamond 1975b:388, 390; Figure 20-3 herein) are partly the result of human-caused extinctions, then the assembly rules for these species may be suspect. It also would be of interest to examine the distributions in the Bismarcks today of other congeneric species pairs or triplets, such as hawks (*Accipiter*), pigeons (*Columba*), ground-doves (*Gallicolumba*), pygmy-parrots (*Micropsitta*), cuckoos (*Centropus*), owls (*Ninox*), kingfishers (*Halcyon*), and a number of passerines. If analyzed in the same detail as those of *Macropygia*, *Ptilinopus*, and *Pachycephala*, how would these distributions be interpreted in terms of assembly rules?

Assembly rules are set in an equilibrium framework where mobile species regularly colonize and go "extinct" on islands. This hypothetical turnover (driven in part by competition) leads to "checkerboard" (more or less mutually exclusive) distributions of related species that are the result of "forbidden combinations" of species believed by Diamond (1975b:344) not to "exist in nature because they would transgress . . . incidence functions." Except for supertramps, I see the landbird communities on Pacific islands as

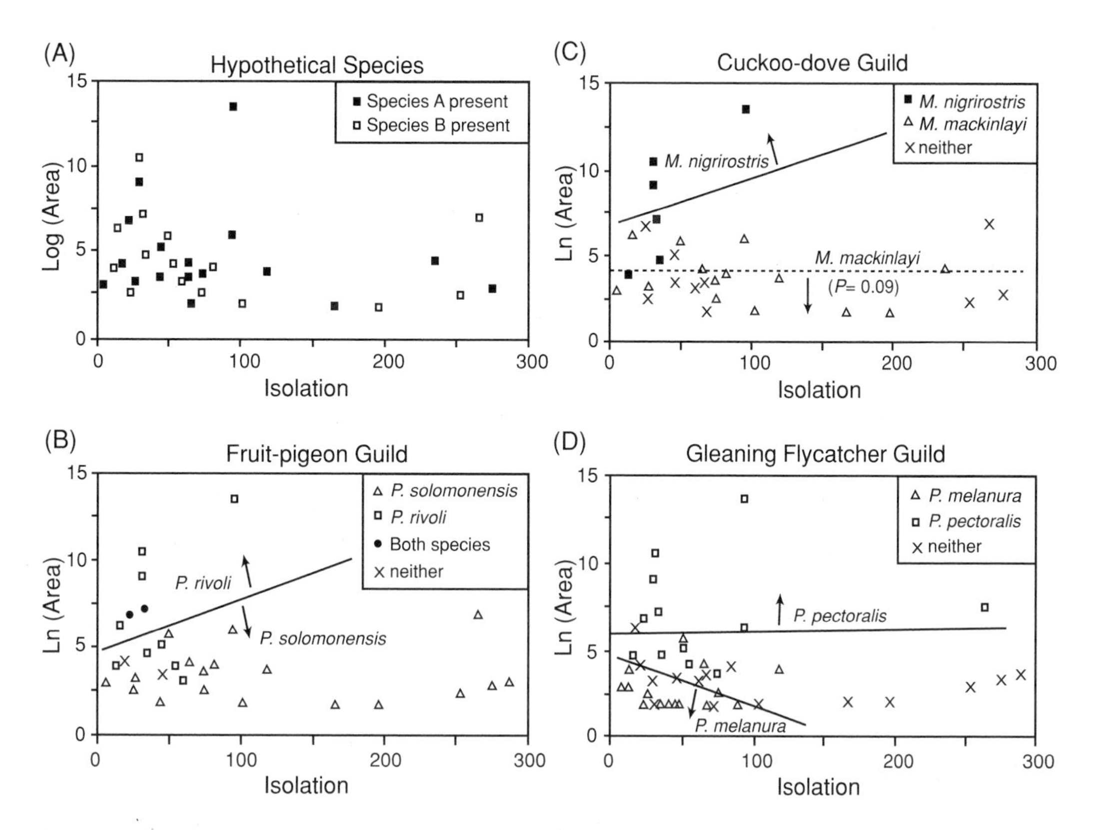

Figure 20-3 (A) Spatial checkerboard distribution of two hypothetical species. In this case, both species are uniformly distributed across an archipelago (i.e., their distributions are independent of isolation and area), but they never occupy the same island. This pattern would be expected where competition is intense but species are essentially equivalent with respect to immigration abilities and resource requirements. In contrast to the hypothetical pattern A, analysis of insular distribution patterns three of Diamond's (1975b) avian guilds reveals that exclusive distributions (B–D) are often achieved by species segregating their realized distributions over different ranges of isolation and area. Regression analysis to estimate linear insular distribution functions of these species found insular distributions significantly ($p < .05$) associated with island area and isolation for *Ptilinopus rivoli, P. solomonensis, Macropygia nigrirostris*, and *Pachycephala pectoralis* (Lomolino 1986). Insular distributions were significantly associated with just area for *P. melanura*, and marginally ($p = .09$) associated with area for *M. mackinlayi*. All plots from Lomolino (1999: Figure 10.18). Caption modified from Lomolino (1999).

much less dynamic. Under natural conditions, colonization and extinction are very rare in Remote Oceania, and even in Near Oceania are much rarer than inferred from competition-based studies (see Chapters 16-18). Whether on large islands close to New Guinea or on smaller, more remote islands, anthropogenic extinctions are more likely than competitive exclusion to explain "forbidden combinations."

Area-isolation plots help to evaluate checkerboard distributions. In Figure 20-3, plot A is a hypothetical "perfect checkerboard" where intense competition leads to the mutually exclusive distribution of two species. There are no "empty squares" (islands with neither species) or islands with both species. I am unaware of any pair of landbirds on Pacific islands with a modern or prehistoric distribution that fits the "perfect checkerboard" pattern. Plots B, C, and D depict field-derived modern distributions from Diamond (1975b). All that I would infer from plot B is that *Ptilinopus solomonensis* may be a better disperser than *P. rivoli*, although both are found on small and isolated islands (Table 5-3). The lines imply a more stringent separation of the species than the points.

The lines are deceptive again in plot C. While the data might suggest that *Macropygia mackinlayi* is a better disperser than *M. nigrirostris*, it also could be that *M. nigrirostris* is more prone to anthropogenic extinction on small islands, which include most of the more remote ones. The two species of *Macropygia* have mutually exclusive distributions in plot C (and in Diamond 1975b: Figure 20) but in fact are sympatric on New Britain, Karkar, and Watom (Diamond & LeCroy 1979, Coates 1985). Mayr and Diamond (2001:164) admit Karkar but regard *M. mackinlayi* as a nonbreeding migrant on the other islands of sympatry. Fieldwork on none of these islands has had enough seasonal or spatial coverage to substantiate this claim.

In the same spirit as in plot C, the islands lacking both species of *Pachycephala* [plot D] may have had relatively more severe environmental degradation, or have not received much survey attention. Although plot D (Diamond 1975b: Figure 21) depicts a mutually exclusive distribution, the two species of *Pachycephala* are in fact sympatric on many islands, even if usually segregated by habitat (Coates 1985).

Please notice that the discussion thus far is limited to presence vs. absence, without considering abundance, which can vary considerably from island to island for any species of landbird (Engbring et al. 1986, Engbring & Ramsey 1989, Steadman & Freifeld 1998). If competition is important in maintaining populations or in repelling newly colonizing species, then population densities should be an important factor. Population data have not been considered, however, in the competition-driven assembly of Pacific landbird communities except in an abstract sense (Case et al. 1979). A rare presence appears as the same point on a graph of incidence functions as an abundant presence.

Before proceeding with the prehistoric data, I should note some differences between the landbird faunas of Near Oceania and Remote Oceania. The most important is that Near Oceania lies just off New Guinea. Thus, unlike Remote Oceania, where most newly colonizing species probably originated on other oceanic islands, the islands off New Guinea are more likely to receive new species straight from New Guinea's rich, continental source pool. (See the list of "non-breeding visitor" landbirds for Near Oceania in Mayr and Diamond 2001:402–405; other than for plovers and sandpipers that breed at high latitudes, no such extensive list of strays can be compiled for Remote Oceania except in western Micronesia.) Also, I see no evidence in Remote Oceania of small island specialists, which do occur in the Bismarcks and Solomons. The one supertramp forest species that inhabits both Near Oceania and Remote Oceania is the pigeon *Ducula pacifica*, which occurs only on small islands in Near Oceania but is found on islands from <1 km^2 to $>10{,}000$ km^2 in Remote Oceania.

Prehistoric Versus Modern Distributions

Near Oceania

This region (Bismarcks, Solomons) is the primary study area of Diamond (1975b) and Mayr & Diamond (2001). The few substantial sets of landbird fossils in Near Oceania (from New Ireland, Mussau, and Buka; Chapter 5) influence what we know about distributions and therefore the incidence functions and assembly rules for certain species. For example, the cockatoo *Cacatua galerita* and flightless rail *Rallus* (= *Gallirallus*) *insignis* share an incidence function that includes only New Britain, the richest of the two high-*S* islands in the Bismarcks (Diamond 1975b: Figure 7; Figure 20-1 herein). A form of *Cacatua* and a flightless species of *Gallirallus* occurred as fossils on New Ireland, the other high-*S* island. A single bone of cf. *Cacatua* has been recovered as well on Mussau, showing that cockatoos once were widespread in the Bismarcks. Given that other flightless species of *Gallirallus* used to live on Buka (611 km^2) and on numerous islands in Remote Oceania with land areas and species richness values as small as Aiwa Levu (1.2 km^2, $S \geq 26$–27; Chapter 6) or Wake (7.4 km^2, $S \geq 1$; Chapter 8), similar rails probably once occurred on small and medium-sized islands in the Bismarcks.

The swamphen *Porphyrio porphyrio* was regarded by Diamond (1975b: Figure 8 as an A-tramp, with an incidence function shown in the top image of Figure 20-4. Mayr & Diamond (2001:425) assigned *P. porphyrio* a dispersal index of 2 in the Bismarcks, meaning either a single record of overwater dispersal in the 20th century or residency on an island with Holocene volcanism. Because *P. porphyrio*

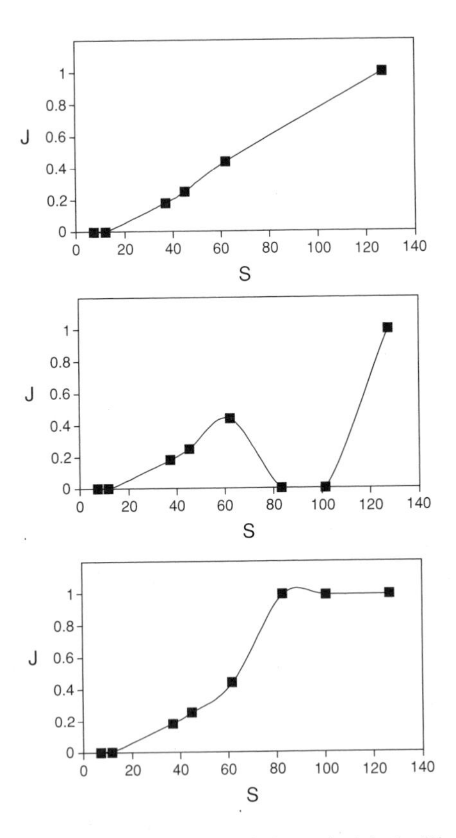

Figure 20-4 Incidence functions for *Porphyrio porphyrio* in the Bismarck Archipelago. Top: as depicted in Diamond (1975b: Figure 8). Middle: as would have been depicted in Diamond (1975b) had the apparent absences on New Ireland and elsewhere been plotted. Bottom: as indicated when including the prehistoric record in Steadman et al. (1999b) and the modern distribution in Coates (1985).

had not yet been recorded on New Ireland, and was of unreported status on islands where $S \sim 80$, a more accurate depiction of its distribution, based on data in Diamond (1975b), would have been the second graph in Figure 20-4. The only record of *P. porphyrio* from New Ireland is a prehistoric bone from the Matenkupkum site (Steadman et al. 1999b; Chapter 5 herein; overlooked in Mayr & Diamond 2001). Including all records of *P. porphyrio* from the Bismarcks would result in the incidence function illustrated at the bottom of Figure 20-4. I would add further that, if survey data were complete, the values for *J* (the fraction of the islands on which the species occurs) on intermediate-*S* islands (~40 to 60) species probably would be higher than in any of the incidence functions in Figure 20-4. *Porphyrio porphyrio* prefers successional habitats and wetlands, may vocalize infrequently, and can be easy to overlook during brief surveys (Steadman 1998). Similar phenomena apply to another widespread rail, *Gallirallus philippensis* (see Diamond 1975b: Figure 10).

Corvus orru, the only crow then known from the Bismarcks, was regarded by Diamond (1975b: Figure 10) as a B-tramp. Fossils of a large species of *Corvus* from New Ireland yield a congeneric species pair that probably once inhabited other islands as well. Today, multiple species of *Corvus* coexist in New Guinea and Australia but nowhere in Oceania. Similarly, the megapode *Megapodius [freycinet] eremita*, regarded by Diamond (1975b: Figure 13) as a D-tramp, had a sympatric congeneric species on New Ireland and presumably elsewhere. That these two sets of congeneric species pairs are known only from New Ireland, the second largest island in the Bismarcks, is almost certainly an artifact of incomplete fossil coverage. Multiple sympatric species of megapodes, unknown in Oceania today, once existed in the Bismarcks, Solomons, Vanuatu, New Caledonia, Fiji, and Tonga. At the family's eastern limit in Tonga, three sympatric species of *Megapodius* occurred on islands as small as 5.3 km^2 (see below). The same may have been true in the other island groups as well.

Fossils also influence incidence functions in showing that modern *S*-values are lower than they would be in the absence of human impact. Because anthropogenic extinction often affects a larger percentage of species on smaller islands (Chapter 16), many islands regarded as low-*S* by Diamond (1975b) probably were medium-*S* or perhaps even high-*S* islands before human arrival. Additional survey of modern landbirds also increases the species richness values of landbirds. New Ireland, for example, is considered to have 106–110 species today (Table 5-3) rather than the 101 used by Diamond (1975b). While not a huge change, this increase would certainly affect presence/absence data, and thus the incidence functions, for the five to nine species added.

Remote Oceania

Now I will develop examples of sympatric congeneric species of landbirds in Remote Oceania. These pairs and triplets would represent "forbidden combinations" by modern standards. Most of these examples, however, involve one or more prehistorically extinct species, which of course could not have been recorded during the past two centuries. My goal here, as throughout the book, is to blend modern ecology, traditional ornithology, paleontology, and zooarchaeology in a way that benefits these closely related, but often independently pursued, fields of study.

As already mentioned, many sets of sympatric congeneric species of landbirds (pairs, occasionally triplets) still exist in Near Oceania (Diamond 1973, 1975a; Chapter 5 herein). They are much rarer in Remote Oceania, where the most are found in Vanuatu (7 pairs, 1 triplet) and Fiji (5 pairs, 1 triplet). In all of East Polynesia, only three sympatric congeneric species pairs (and no triplets) of landbirds are known to have survived into historic times (the past ~230 years). These are the pigeons *Ducula aurorae* and *D. galeata* and the kingfishers *Halcyon tuta* and *H. venerata* in the Society Islands, and the fruit-doves *Ptilinopus dupetithouarsii* and *P. mercierii* in the Marquesas (Diamond 1977, Holyoak & Thibault 1978a, Steadman & Pahlavan 1992; Chapter 7 herein). The species of *Ducula* coexisted only on Tahiti (Olson & Steadman 1987), the fruit-doves on Nuku Hiva, Hiva Oa, and perhaps Fatu Hiva and Tahuata (Steadman 1989a), and the kingfishers on Tahiti, Mo'orea, and Bora Bora (Steadman & Pahlavan 1992). Tahiti being the largest and highest island in East Polynesia, the presence there of two sets of sympatric congeners would not be shocking in terms of competition-driven assembly rules. By the way, none of the three congeneric species pairs may still exist; the only one possibly intact would be the kingfishers on Tahiti, although *H. tuta* has been seen there just twice in the past century, the last time in 1974 (Monnet et al. 1993b).

Before the period of European influence, however, East Polynesian congeneric pairs or triplets included rails (*Porzana*), ground-doves (*Gallicolumba*), parrots (*Vini*), and more instances of *Ducula* and *Ptilinopus*. These sympatric congeners were not confined to just the largest islands. In West Polynesia, congeneric species pairs or triplets are or were widespread in *Megapodius*, *Ducula*, *Ptilinopus*, *Vini*, trillers (*Lalage*), monarch flycatchers (*Myiagra*), and starlings (*Aplonis*). Sympatric congeners have been found in Remote Oceania only in genera that are very widespread in Oceania, probably resulting from variations on the model (Lack 1947) of allopatric speciation followed by dispersal and sympatry.

Table 20-2 Past vs. present distribution of ground-doves (*Gallicolumba*), pigeons (*Ducula*), and lorikeets (*Vini*) on three islands in East Polynesia with extensive data on prehistoric birds. The species are listed in ascending order of body size. M, modern record; P, prehistoric record; † extinct species. Based on data in Chapters 5, 11, 12.

	Mangaia, Cook Is.	Huahine, Society Is.	Ua Huka, Marquesas
GROUND-DOVES			
G. erythroptera	P	P	—
G. rubescens	—	—	P
† *G. nui*	P	P	P
Total (M)	0	0	0
Total (M + P)	2	2	2
PIGEONS			
D. aurorae	P	P	—
D. galeata	P	P	P
Total (M)	0	0	0
Total (M + P)	2	2	1
Columbid bones	543	30	1821
LORIKEETS			
† *V. sinotoi*	P	P	P
† *V. vidivici*	P	P	P
V. kuhlii	P	—	—
V. peruviana	—	M	—
V. ultramarina	—	—	MP
Total (M)	0	1	1
Total (M + P)	3	3	3
Lorikeet bones	189	5	271
Area (km^2)	52	77	78
Elevation (m)	169	669	855

Sympatric species of ground-doves (*Gallicolumba*) occur in Oceania today only in the Bismarcks. Prehistorically, two species of *Gallicolumba* inhabited individual islands in several island groups in East Polynesia (Table 20-2). Considerable extinction of ground-dwelling columbids has taken place in Oceania at least as far west as the Solomon Main Chain (Buka), involving species of *Gallicolumba*, *Microgoura*, *Didunculus*, *Caloenas*, *Natunaornis*, and three undescribed genera (Chapter 11). Thus, there is no reason to believe that these species assemblages are unaffected by human impact, even in the Bismarcks.

The situation with frugivorous canopy pigeons is similar. *Ducula aurorae* and *D. galeata* inhabited Huahine and Mangaia prehistorically (Table 20-2). Both species still existed on Tahiti at European contact, although *D. galeata* has not been recorded there since Captain Cook's visit in the 1770s (Olson & Steadman 1987). Fiji (*D. lakeba, D. latrans*) and Tonga (*D.* undescribed sp.,*D. latrans*) also had at least two large species of *Ducula* at human arrival (Chapters 6, 11). Bones of *D. pacifica* occur with the other two on Tongan islands in Lapita contexts but not in precultural strata. In Fiji today, *D. pacifica* and *D. latrans* are sympatric on some islands (Holyoak & Thibault 1978a, DWS pers. obs.), although details of habitat preference and relative abundance are unknown. Multiple sympatric species of *Ducula* are widespread in the Bismarcks and Solomons, where some species are separated somewhat by habitat (Diamond 1975b: 427–433, Kratter et al. 2001a). The East or West Polynesian islands with sympatric species of *Ducula* have little or no altitudinal zonation of noncoastal forest types. Habitat separation cannot be invoked to explain coexistence.

Endemic to Polynesia, *Vini* comprises small to medium-sized nectar-feeding parrots known as lorikeets. (The species assigned to *Vini* in Diamond (1975b) and Diamond & LeCroy (1979) are referred herein to *Charmosyna*; Chapters 5, 12). *Vini* has a very spotty distribution today, with no more than one species on any island (Steadman & Zarriello 1987). In East Polynesia, three species of *Vini* (Figure 12-3) used to inhabit single islands in the Cook, Marquesas, and Society Islands (Chapter 7). A sympatric species pair (*V. australis*, *V. solitarius*) also lived in Tonga (Chapter 6). These last two lorikeets have current ranges that abut but do not overlap in the southern Lau Group, Fiji (Watling 1985; Table 6-4 herein). In the parlance of assembly rules, *V. australis* and *V. solitarius* would be a forbidden combination.

Among kingfishers (*Halcyon*), modern sympatric species pairs occur in Remote Oceania only in the Society Islands (*H. tuta*, *H. venerata*; see above), Vanuatu (Kratter et al. in press), and Palau (*H. chloris*, *H. cinnamomina*; Chapter 8). Two species of *Halcyon* are found in Samoa today but not on the same islands; the small *H. recurvirostris* inhabits Independent (Western) Samoa, whereas *H. chloris* occurs in American Samoa. This east-west difference does not represent a checkerboard; suspected prehistoric sympatry is untested by fossils.

The East and West Polynesian pairs or triplets of *Ptilinopus* and *Ducula* are frugivores that disperse native plants (Chapters 11, 21). The species of *Vini* are nectarivores undoubtedly involved in pollination. Polynesian islands were more completely forested and richer in woody species of plants before people arrived (Kirch et al. 1992, 1995, Lepofsky et al. 1992, Ellison 1994a), making it easier to visualize the coexistence of closely related species of birds, especially if you believe that competition affects the composition of landbird communities. Regardless, the massive amount of anthropogenic extinction in Remote Oceania has led to an erroneous concept of the potential number of species of landbirds that an island can support. If so many congeneric species pairs, especially of columbids, were able to coexist on Polynesian islands, then it seems odd that competition need be involved to explain the presence or absence of congeneric species on the biotically much richer islands of Near Oceania.

Megapodes and Rails

In Oceania outside of Hawaii and New Zealand, these two families of ground-dwelling birds have lost a larger percentage of species than any other landbird family. Sympatric species of megapodes are unknown in Oceania today, but were very widespread prehistorically (Chapter 9). Multiple prehistoric species of *Megapodius* have been discovered in every island group from the Bismarcks to Tonga (Tables 20-3, 20-4). In currently megapode-free Fiji, at least three species once lived (Worthy 2000). In Tonga, the only megapode to survive is *M. pritchardii* on Niuafo'ou, which nests in volcanically warmed soil (Göth & Vogel 1995, 1997). Prehistorically, *M. pritchardii* also lived on the limestone islands of 'Eua, Foa, Lifuka, and 'Uiha (Table 20-4), where nesting must have been on beaches or in soil warmed by decomposing plant material. At least four species of *Megapodius* inhabited Tonga at human arrival, with three of them known from a single island as small (5.3 km^2) and low (11 m) as 'Uiha (Table 20-4). What sort of niche theory or resource utilization model can account for this congeneric triplet on tiny, flat 'Uiha, when today there is not a single case of truly sympatric multiple species of *Megapodius*, even on huge, rugged, New Ireland, New Britain, or New Guinea? How narrow must their "niches" have been (see Maurer 1999:190-191) or, perhaps, how broad must their niches have been (see Novotny et al. 2002), in order to have coexisted on little specks of land in Tonga? What sort of biologically naïve "scaling law" (i.e., Marquet 2000) should we invoke to account for their differences in body size?

Sympatric congeners among rails (Chapter 10) are also unknown in Oceania today except where the volant

Table 20-3 Past vs. present distribution of megapodes (Megapodiidae) on selected islands. The species are listed in ascending order of body size. M, modern record; P, prehistoric record; † extinct species. Bis, Bismarcks; SC, Santa Cruz Group; SMC, Solomon Main Chain; Van, Vanuatu. Based on data in Chapters 6–9.

	Near Oceania		Remote Oceania						
	Bis	SMC	Van	New Caledonia	Fiji			SC	Carolines
	New Ireland	Buka	Efate	Grande Terre	Viti Levu	Lakeba	Mago	Tikopia	Pohnpei
Megapodius eremita	MP	MP	—	—	—	—	—	—	—
M. layardi	—	—	MP	—	—	—	—	P	—
M. laperouse	—	—	—	—	—	—	—	—	P
†*M. alimentum*	—	—	P	—	—	P	P	—	—
†*M. amissus*	—	—	—	—	P	—	—	—	—
†*M. molistructor*	—	—	—	P	—	—	—	—	—
†*M.* undescribed sp. A	P	—	—	—	—	—	—	—	—
†*M.* undescribed sp. B	—	P	—	—	—	—	—	—	—
†*Megavitiornis altirostrum*	—	—	—	—	P	—	—	—	—
†*Sylviornis neocaledoniae*	—	—	—	P	—	—	—	—	—
TOTAL									
M	1	1	1	0	0	0	0	0	0
M + P	2	2	2	2	2	1	1	1	1
Area (km^2)	11,145	441	915	16,446	10,387	56	21.9	4.6	337
Elevation (m)	2399	498	647	1628	1309	210	204	360	772
Megapode bones	9	2	2	4996+	220	49	1	10	1

Table 20-4 Past vs. present distribution of megapodes (*Megapodius*) and large pigeons (*Ducula*) on selected islands in Tonga. The species are listed in ascending order of body size. M, modern record; P, prehistoric record; †, extinct species; * the total area of Niuafo`ou is 52.3 km², but ca. 29 km² of this is a freshwater lake. Based on data in Chapters 6, 9, 11.

	Tongatapu	`Eua	Niuafo`ou	Foa	Lifuka	Ha`ano	`Uiha	Ha`afeva
MEGAPODES								
M. pritchardii	—	P	M	P	P	—	P	—
† *M. alimentum*	P	P	—	P	P	P	P	P
† *M. molistructor*	P	—	—	P	P	P	P	P
† *M.* undescribed sp. F	—	P	—	—	—	—	—	—
Total (M)	0	0	1	0	0	0	0	0
Total (M + P)	2	3	1	3	3	2	3	2
Megapode bones	70	48	0	14	248	9	39	115
PIGEONS								
D. pacifica	MP	MP	M	M?, P	MP	M?, P	M?, P	M?, P
D. latrans	P	P	—	P	P	P	P	P
† *D.* undescribed sp.	P	P	—	P	P	P	P	P
Total (M)	1	1	1	0–1	1	0–1	0–1	0–1
Total (M + P)	3	3	1	3	3	3	3	3
Columbid bones	167	151	0	57	160	40	65	135
Area (km²)	259	87	33*	13.3	11.4	6.6	5.3	1.8
Elevation (m)	67	325	205	20	16	12	11	10

Gallirallus philippensis co-occurs with a flightless species in the Moluccas, Bismarcks, and Solomons. Prehistorically, most individual islands had multiple species of rails (often flightless) but typically, as far as we know, in different genera. Exceptions would include two species each of *Gallirallus* and *Porphyrio* on New Ireland (Bismarcks) and two species of *Gallirallus* on Buka, Solomons (Chapter 5). Another exception would be the three species of *Porzana* on Mangaia, Cook Islands (*P. tabuensis*, *P. rua*, *P.*undescribed sp. C). They co-occur in the lower strata of Tangatatau Rockshelter and two of the three are found in the same strata at Ana Manuku (Table 7-7).

Other Community-Level Implications of Extinction

Most of the examples just given involve congeneric sets of species. At the family level, a dramatic case of modern versus prehistoric species assemblages would be columbids on Mangaia, where bones represent five species on an island lacking pigeons and doves today. Mangaia's size and forest cover are sufficient that one would expect a species or two of columbid to have survived there, as on the nearby islands of Rarotonga, Atiu, Mitiaro, and Ma'uke (Table 7-2). No ecological rule, only evidence learned from bones, can explain why Mangaia has no columbids today, or predict that the number of species of pigeons and doves once living there was at least five.

Six species of columbids lived 800 years ago on Huahine (Society Islands), where only one species survives (Table 7-14). Marquesan islands each had six species of columbids prehistorically, one to four species in historic times, and one or two species today. Similarly in Tonga, where islands sustain one to three species of columbids today, nine species inhabited individual islands at human arrival. In all cases, the losses include multiple species of both canopy frugivores and understory/mid-level granivores/frugivores (Chapters 6, 7, 11).

Essentially all feeding guilds were more tightly "packed" with species prehistorically. On 'Eua (Tonga), the heaviest losses were among frugivores, nectarivores, and omnivores, whereas the loses were even more devastating on Mangaia, with mid-level insectivores being the only guild of forest birds to survive (Figure 19-6).

Fossils also help us to understand the colonization and extinction potential of individual species. The thrush *Turdus poliocephalus* was regarded by Diamond (1975b:377, 420, 421) as a supertramp "capable of invading" only islands or montane areas with <23–36 species of landbirds even though it occurs on islands in the Bismarcks of highly variable area and elevation. Given its striking geographic variation, with 53 subspecies recognized from Taiwan, the Philippines, and the Greater Sundas to Samoa (Peters 1964:192–199, Ripley 1977, Mayr & Diamond 2001), I would suggest that most or all "invasions" are ancient. There is no evidence of modern interisland dispersal in *T. poliocephalus*, whether in the Bismarcks or in more remote groups such as Vanuatu, where Diamond (1975b:377, 387) regarded it as a C-tramp. Even though this thrush was first discovered on New Ireland only in 1976, this montane population represents an endemic subspecies that therefore is a presumed long-term resident, not a recent colonist (Ripley 1977, Beehler 1978, Diamond 1989). This supposition is strengthened by discovering bones of *T. poliocephalus* in a prehistoric coastal site on New Ireland (Table 5-3), thus showing that it once was part of a richer lowland bird community. The apparent absence of *T. poliocephalus* on most islands in the Solomon Main Chain (see Mayr & Diamond 2001:341) almost certainly reflects extirpations after the arrival of people (Kratter et al. 2001a).

In Tonga, where *T. poliocephalus* is absent today, its fossils range in age from >60-80 ka to after human arrival at ca. 2.8 ka (Steadman 1993a; Chapter 6 herein). The Tongan fossils have been found on islands from as large as Tongatapu (259 km^2, 67 m elev.) and 'Eua (87 km^2, 325 m) to as small as Lifuka (11.4 km^2, 16 m). The failure to record bones of *T. poliocephalus* from Tongan islands even smaller than Lifuka may be only an artifact of the small samples of passerine bones ($N = 21$–26) recovered on Ha'ano, 'Uiha, and Ha'afeva (Table 6-21). *Turdus poliocephalus* is absent today on most of the 50 or so islands >11.4 km^2 between the Solomons and Tonga. I believe that these absences are more likely due to human impact (habitat loss plus predation from people, rats, cats, etc.) on a resident, understory, tasty frugivore/omnivore than to competitive interactions with species trying to thwart an invader. Mayr & Diamond (2001:84) state that "as the number of competing species decreases, the incidence of this thrush increases to include more islands and small islands." I would counter that, before human impact, neither island area, elevation, or species richness had much to do with the distribution of *T. poliocephalus*, a hyper-widespread forest bird.

Discussion

As expected in a field so infused with physics envy (e.g., Maurer 1999:23), much of the debate about understanding insular distributions is statistical rather than biological. Analyses of co-occurring vs. non-co-occurring species pairs (checkerboard distributions) focus on whether these situations are due to anything other than chance. Gilpin and Diamond (1984) summarized the straightforward biological reasons for why the distributions of landbirds in the Bismarcks are not random. The statistical debate begun 25 years ago continues (Connor & Simberloff 1979, Sanderson et al. 1998, Gotelli & Entsminger 2001); the biological issues receive less attention, perhaps because so many organismal biologists are glued to their computers rather than doing fieldwork.

As developed by Jared Diamond 30 years ago in a book dedicated to Robert MacArthur, assembly rules were a logical and creative advancement of equilibrium theory. Given the ecological climate of the time, Diamond's ideas were nothing short of brilliant; whatever praise he has received through the years is richly deserved. I also believe that biologists should be inspired and humbled by the fact that Diamond's lack of participation in a recent volume dedicated to his work (Weiher & Keddy 1999) was because he was doing fieldwork in New Guinea, just as he has done for nearly 40 years. This is not a minor issue. While Diamond was working in one of the world's most unhealthy and logistically difficult places, his younger colleagues were discussing the Humpty-Dumpty effect, the Icarus effect, the Jack Horner effect, the J. P. Morgan effect, and the Narcissus effect (see various chapters in Weiher & Keddy 1999).

My criticisms of assembly rule concepts, made only in a spirit of scientific inquiry, focus on three broad issues—inadequate data on modern distributions, a failure to consider human-caused extinctions, and an understandable inability to document production vs. utilization of resources. To assume that modern surveys of birds in the Bismarcks and other islands off New Guinea were complete, and to assume further that these distributions had not been influenced by 30,000+ years of human activity in this region, were leaps of faith. In 1975, prehistoric extinctions of birds had not been documented in the Pacific outside of the Hawaiian Islands and New Zealand, although biologists did realize that a disproportionate amount of extinction in recent centuries had taken place on oceanic islands. Leaps of faith much less excusable than Diamond's are still being made by those who analyze distributions of tropical Pacific landbirds as if they were natural, such as Adler (1994) and Sanderson et al. (1998) have done for Fiji and Vanuatu. Mayr and Diamond (2001) at least recognized that prehistoric extinctions occurred, even if these losses had little influence on their analyses.

Given the growing body of data on prehistoric extinction of landbirds in Oceania, what might be an appropriate "null" (default) model to test distributional patterns?

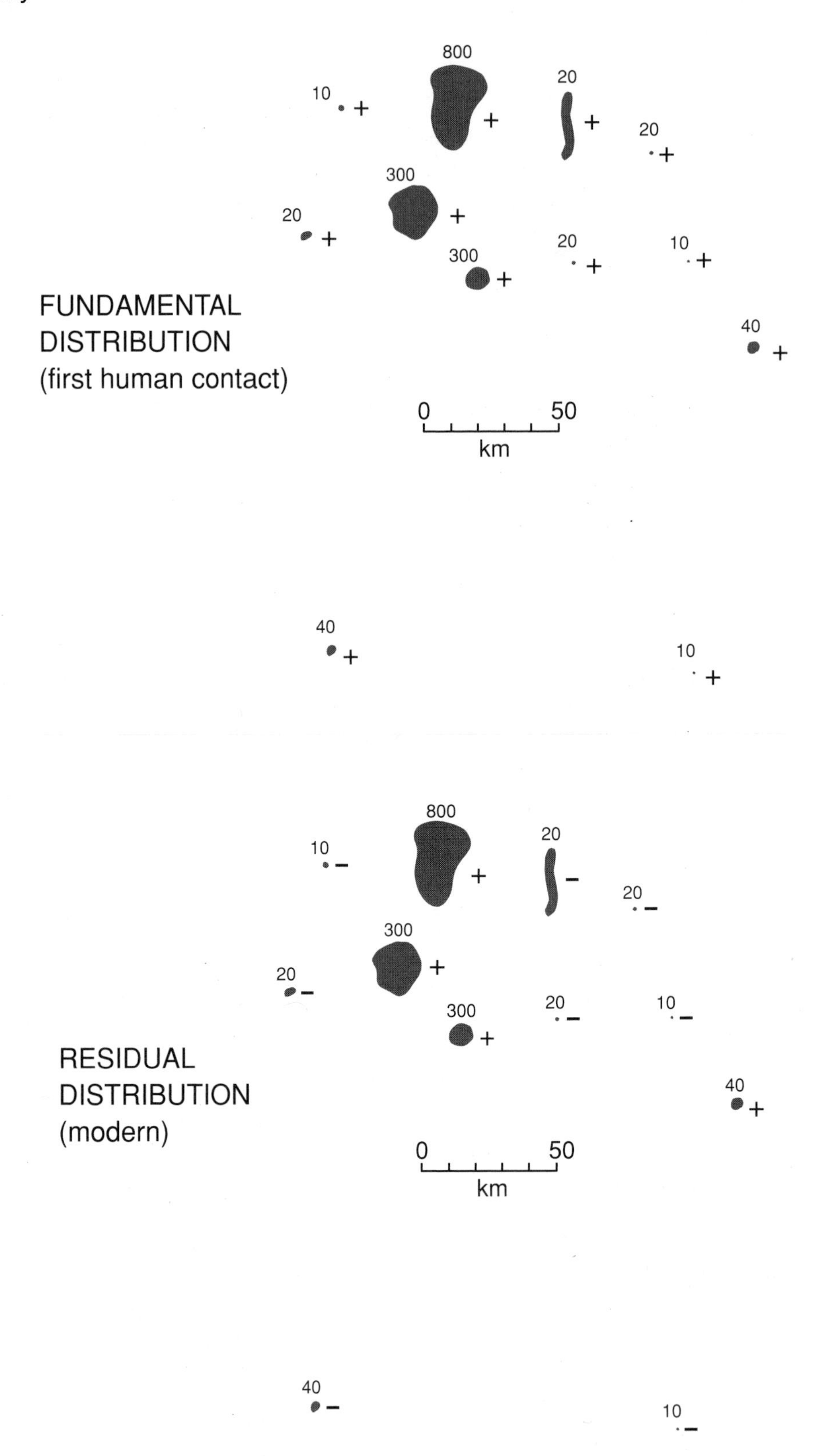

Figure 20-5 Fundamental (first human contact) versus residual (modern) distribution of a well-dispersed canopy species in the same hypothetical island group in Remote Oceania as in Figure 19-15. +, present; –, absent; the numbers are elevations in meters.

Monte Carlo or other computer simulations of random sampling (Connor & Simberloff 1979, 1983, 1984, Roberts & Stone 1990, Sanderson et al. 1998, Simberloff et al. 1999) may not be appropriate because geographical and biological common sense dictates that most colonization of oceanic islands by landbirds is not random (Diamond 1975b, 1981, Gilpin & Diamond 1984:310–312). Random colonization, with or without subsequent competition-driven exclusion, could produce innumerable varieties of checkerboard patterns. As the fossil record grows, however, I believe that many or most of the "empty squares" in any island group will prove to be artifacts of human intervention. Random colonization fits well into the thinking of neutral macroecology (i.e., Brown 1995, Bell 2001, Hubbell 2001) but I see this form of analysis as stripping a biological situation of its biological properties and therefore having little explanatory worth.

My alternative default model for the distribution of landbirds in Oceania is that each species (or allospecies) known to inhabit an archipelago would occur on each island within that group with an area equal to or greater than a minimum land area (MLA; Chapter 19). After initial colonization, interisland gene flow would vary greatly among species. The MLA probably decreases eastward from perhaps ca. 50 to 100 km^2 in Near Oceania, to 10–20 km^2 Vanuatu and Fiji, to perhaps as small as 1–5 km^2 in Tonga, Samoa, and East Polynesia. I do not expect a perfect fit for this default model but it establishes a benchmark for evaluating sets of species on individual islands within a group. My default model predicts that apparent absences of species on large, intervening islands, such as the kingfisher *Actenoides bougainvillei* being on Bougainville and Guadalcanal but not Choiseul, New Georgia, or Isabel (see Mayr & Diamond 2001:316) are probably due to human impact.

Among the species/allospecies known to occur in Remote Oceania, I believe that interisland distances less than ca. 100 km have had a minimal effect on species richness. This is based on data from the five island groups with substantial landbird fossils from multiple islands (Vanuatu, Tonga, Cooks, Marquesas, and Marianas; Chapters 6–8). My model predicts a more uniform interisland, intra-archipelago distribution for each species/allospecies of landbird than those taken from equilibrium theory or random sampling (Chapter 19).

Very small (<MLA) or isolated islands do not fit my model, in which the "fundamental distribution" is the one that a species occupied in the absence of human influence, and the "realized distribution" is where prehistoric bones have been found. The modern or "residual" distribution typically is a subset of the fundamental distribution because of human impact, whether individual species (Figures 20-5, 20-6) or the entire landbird communities (Figure 19-15) are evaluated. Certainly in Remote Oceania, and even to a great extent in Near Oceania, consideration of past human activity removes much of the intrigue about "the puzzling tropical phenomenon of patchy distributions" (Diamond 1975b:344).

Unlike dynamic, competition-driven models of insular distribution such as in Lomolino (1999), my distributional model downplays competition (called "interspecific interactions" by Lomolino). Insular species of landbirds do not evolve or exist in a biotic vacuum; the presence of other species on a newly colonized island, even if confamilials or congeners, is unlikely to be a situation that the arriving species has not faced before. Chances are that the same species were sympatric in the source area, itself an island for most landbirds in Oceania. Thus my default model is a simple one where, within a Pacific island group, differences between the fundamental and realized distributions of landbirds are expected to be slight if substantial fossil data are available. A much larger difference would be that between the fundamental and the residual distributions, because of anthropogenic extinctions. Most of the effect of island area involves minimum required areas, which were small before the prehistoric decline in species richness (Chapter 19). Except for extreme outliers, isolation operates more strongly among island groups than within them. A third factor, geological age, may be important mainly for very old islands (with rich faunas) or very young ones with poor faunas, such as atolls inundated or volcanoes exploded in the late Quaternary.

Regular and random colonization of islands is plausible in equilibrium theory. Abundant evidence, from both the past and the present distributions of landbirds on Pacific islands, argues that very few of these species regularly come and go on any given island through time. I am unaware of an instance in Oceania outside of the Bismarcks (near New Guinea) where a species of landbird has been lost from an island only to recolonize under its own power. Rather, nearly all species of landbirds (except a few trampy ones) establish long-term, resident populations with limited (if any, in the case of flightless species) gene flow after colonization (Chapter 18). This is especially true in Remote Oceania but also applies to most species in Near Oceania. For landbirds on tropical Pacific islands, it is time to abandon equilibrium theory. I would suggest the same for landbirds on other oceanic islands as well.

Equilibrium theory might be useful for studying undifferentiated assemblages of species on continental islands over very short timescales, such as insects on mangroves in Florida (Simberloff & Wilson 1970, Simberloff 1974, 1976), plants on islands in freshwater lakes in Sweden

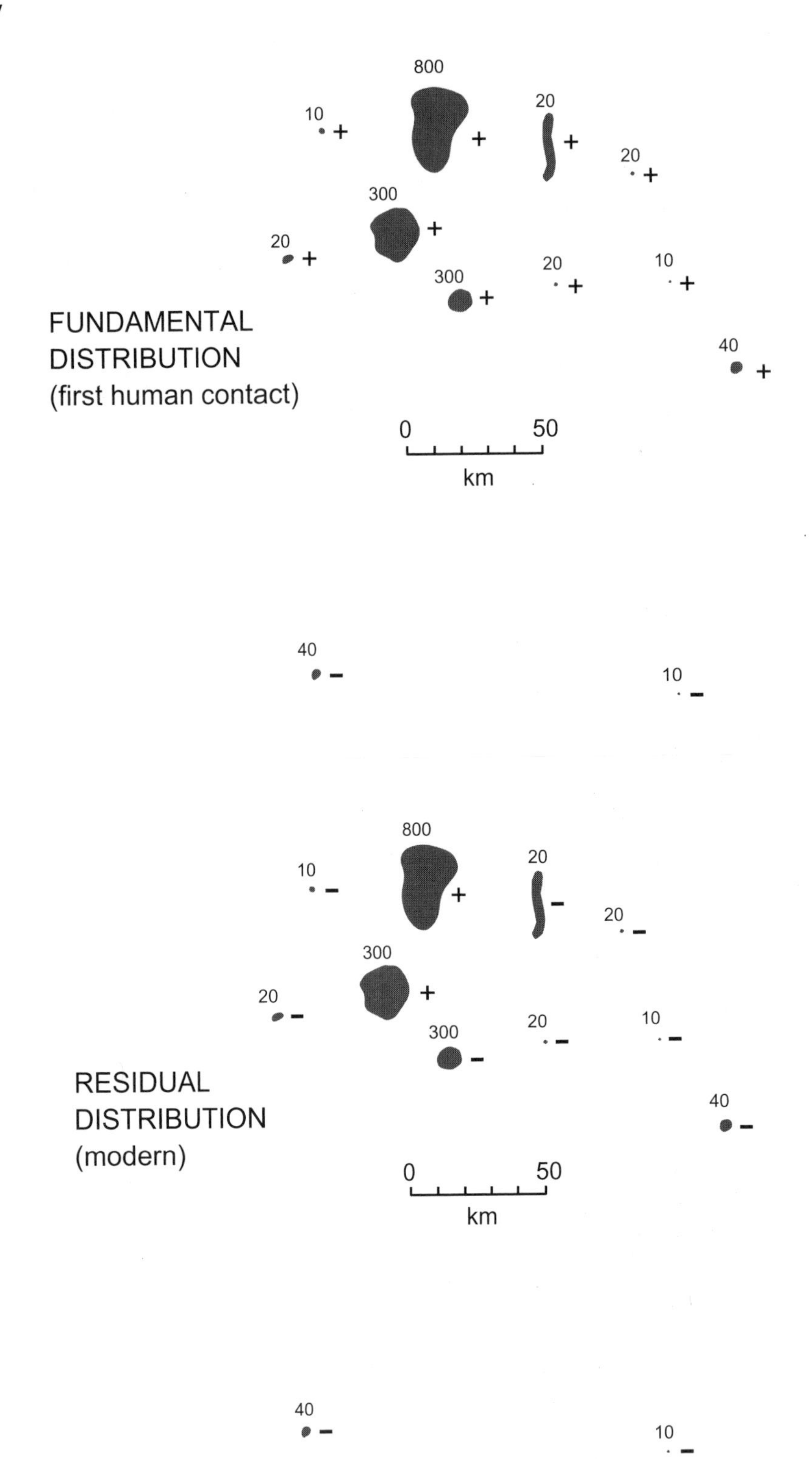

Figure 20-6 Fundamental (first human contact) versus residual (modern) distribution of a poorly dispersed understory species in the same hypothetical island group in Remote Oceania as in Figure 19-15. +, present; −, absent; the numbers are elevations in meters.

(Nilsson & Nilsson 1985), plants on islets within an atoll (Woodroffe 1986), small mammals on ice-connected islands in the St. Lawrence River (Lomolino 1986, 1993a, 1994, 1999), or small mammals in terrestrial habitat patches in Australia (Fox 1999). In regarding almost anything as an island, equilibrium theory has been diluted to the point where some form of it must apply somewhere, at least over short periods of time. Perhaps as good a candidate as any would be local assemblages of continental rodents, all living on or beneath the ground, and with low species-richness values on both their "islands" and source areas. Given how drastically different the lives of birds are from those of plants, protozoans, roundworms, insects, fishes, lizards, rodents, or anything else nonavian, I do not seek universal truths.

Equilibrium theory is least applicable, if at all, on truly oceanic islands. Biologists have gravitated toward islands, whether true islands or ones made up, in searching for general models. I see the quest for generalization as a potential hindrance to scientific progress when it brushes aside interesting variation to seek a vague big picture. One outcome of this quest is to regard carefully collected data as less of a contribution to scientific progress than some speculative stab at an explain-all theory. Scientists often are more interested in the statistical "line of best fit" drawn (without confidence intervals) through a scatter of points, rather than the points themselves, which often already represent transformed instead of raw data.

Both words in the phrase "dynamic equilibrium" are inappropriate for landbird communities in Oceania. Most interisland distributions are not very dynamic under natural conditions. This is especially true in Remote Oceania. "Dynamic" and its derivatives are popular, overused (e.g., Volkov et al. 2004), and misused in many fields today. These words convey energy, a force, even excitement, but especially something that is moving or changing. When describing biological processes on chunks of land surrounded by the roar of the sea and the spirits of Darwin and Wallace, a dynamic theory must be more appropriate than one that is static. I believe, however, that both colonizations and extinctions by volant landbirds are rare events under natural conditions on oceanic islands. A few trampy species aside, the interisland distribution of most indigenous species of landbirds is just not very dynamic, even in time frames measured in centuries or millennia.

I have shared with you my disdain for "dynamic"; what about "equilibrium," the word that it modifies? As is already apparent (Chapter 18), I am not likely to carry the flag of equilibrium into battle either. Rather than believing that extinctions are destined eventually to cancel out colonizations over time, I believe that, all else being equal, the landbird faunas on most oceanic islands were not "turning over" when humans arrived but very slowly had been getting richer through time. There is no endpoint called "equilibrium." The primary type of prehuman extinction in Oceania was catastrophic (inundated atolls, explosive volcanoes) rather than competition driven. Human impact has been the main cause of extinction of landbirds on oceanic islands ever since we arrived. This extinction has occurred on timescales horribly more rapid than that of background extinction or of new colonizations. It has left us with modern distributions of species in Remote Oceania so fragmented as to defy analysis, and in Near Oceania with enough blemishes to saddle any analyses with considerable uncertainty.

My new default model of landbird distribution in Oceania is not problem free. It does not explain how or when colonization occurs. It cannot be tested with rigor on the many islands that never will yield fossils. It does, however, move the ecology of insular landbirds away from the implicit Brownian motion of equilibrium theory.

Chapter 21 Conservation Biology

AVIAN CONSERVATION in Oceania is an uphill battle because of three grim truths. One is that the biology of most species of landbirds is too poorly understood to provide information for conservation any more refined than "don't cut down more forest" or "eliminate the feral cats." (Both are worthy but difficult causes.) The second is that conservation of birdlife is not a high priority for most people or governments on Pacific islands (or anywhere else). The third grim truth is that many if not most species of landbirds in Oceania already are extinct.

Depending on the island group, we are centuries or millennia too late to save the bird communities that existed at human arrival. Extinction really is forever; in spite of what we see on TV or in the movies, attempts to resurrect extinct species through molecular genetic techniques have no chance of success (Höss 2000, Tschentscher 1999). The double-whammy of extinction of native species and introduction of exotic species is homogenizing the biotas of oceanic islands (Wilson 1997). How depressing it is to fly across seven or eight time zones to a Pacific island only to have some ornitho-Euro-trash such as *Columba livia* or *Sturnus vulgaris* be the first species of bird you see.

As native species are replaced by nonnative ones, insular floras and faunas become less interesting as well as less distinctive. Biotas damaged beyond hope of recovery can be found on Easter Island and on most of the Hawaiian Islands, where the replacement of native landbird and plant communities with exotics is complete or nearly so (Scott et al. 2001, van Riper & Scott 2001, Loope et al. 2001). This process is almost done on many other islands as well, such as Tahiti (Meyer 1996, 2004). In the Marianas (Table 21-1), Guam's birdlife is in a more advanced state of decline than that of Saipan or Rota, which may be, however, only a couple of decades behind their more populous neighbor in witnessing the collapse of a landbird community from hyperpredation by the brown tree snake *Boiga irregularis* (see below). That this snake has been found also on Pohnpei (Buden et al. 2001) is frightening.

The grim scene I have painted so far is part of a global extinction crisis that will be difficult to halt (Jenkins 2003). This is especially so because religious conservatives, worldwide, have made it taboo even to talk about, much less carry out, programs of family planning. On a more optimistic note, some extinction that I otherwise would expect to occur in upcoming decades might be delayed or prevented through habitat conservation, predator control, or translocation of birds to islands where they once lived. Even though these programs cannot restore anywhere near the entire avifauna that once existed in Oceania, the surviving populations of landbirds deserve sound conservation efforts, for the sake of the birds and the forests in which they live. Furthermore, not all of the news about landbirds in Oceania is bad. Many islands in Melanesia still have avifaunas uncontaminated by nonnative species (Filardi et al. 1999, Kratter et al. 2001a). Methods for eradicating nonnative mammals such as rats (Figure 2-16) are improving (see below). Secondary succession following agriculture on raised limestone islands in Tonga (and probably many other places in Oceania) leads to forests composed predominantly or exclusively of indigenous trees (Franklin et al. 1999, Franklin 2003). This bodes

Table 21-1 Comparison of four "Christmas Count" surveys of landbirds on Guam, Rota, and Saipan (Mariana Islands) in December 1999 and January 2000. Data compiled from Wiles (2000a,b,c).

	Number of species			Number of individuals		
Locality	Native	Nonnative	% Native	Native	Nonnative	% Native
Dededo, Guam	2	5	29	23	2114	1
Southern Guam	3	6	33	28	777	4
Rota	9	3	75	265	899	23
Saipan	14	2	88	830	364	70

well for the future of forests, as long as they are not cleared to the point where sources of propagules for indigenous trees are too small or remote to be effective (Figures 21-1, 21-2), and as long as feral browsers (Scowcroft & Hobdy 1987) and highly invasive species of plants (see the lists in Meyer 2000) are controlled or eliminated.

In Oceania as elsewhere, conservation is a push-pull of good and bad news, of optimism and pessimism. Conservation is a blend of scientific and nonscientific issues. Lacking the skills to wear both hats, I will write here mainly about the science, although the equally important nonscientific parts will seep in now and then. In either case, my objectivity is compromised by how much I love indigenous island plants and animals and how much I dislike nonnative ones. When possible, my perspective will incorporate paleoecological information to "increase our understanding of the dynamic nature of landscapes and provide a frame of reference for assessing modern patterns and processes" (Swetnam et al. 1999:1189).

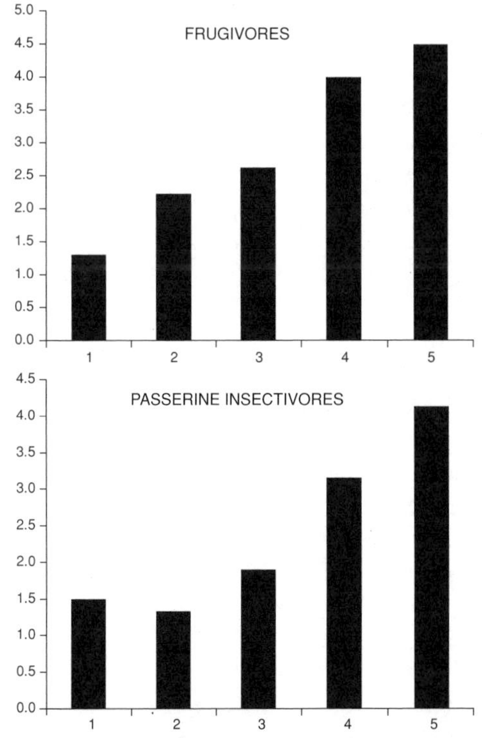

Figure 21-1 Relative abundance (mean, S.E.) of frugivores and passerine insectivores by habitat category, Vava`u Group, Tonga. Frugivores are *Gallicolumba stairi, Ptilinopus porphyraceus, Ducula pacifica*, and *Aplonis tabuensis*. The insectivores are *Lalage maculosa, Pachycephala jacquinoti*, and *Foulehaio carunculata*, none of which is purely insectivorous. 1, village; 2, open plantation; 3, wooded plantation/early successional forest; 4, submature/disturbed mature forest; 5, mature forest. From Steadman & Freifeld (1998: Figure 7).

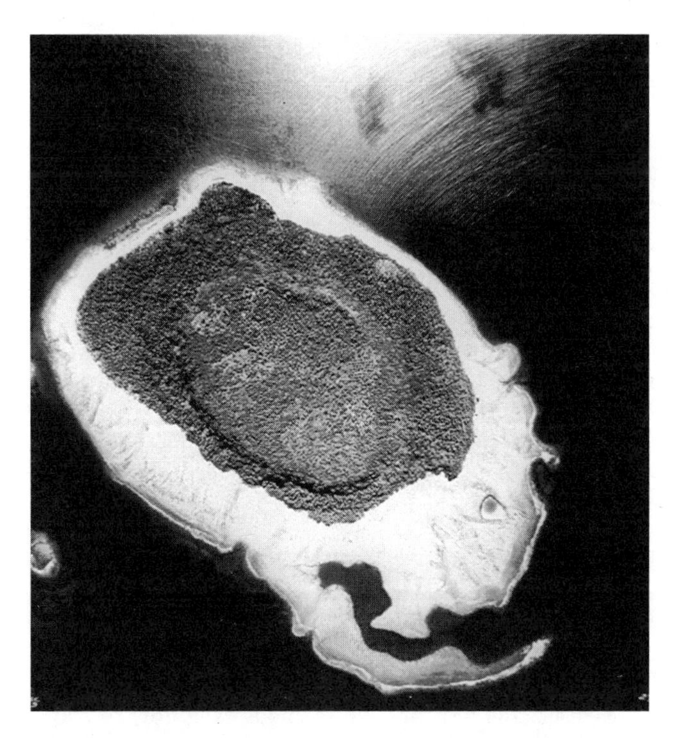

Figure 21-2 Aerial photograph of `Euakata (0.70 km$^2$, elev. 62 m), Vava`u Group, Tonga. No people live on `Euakafa, which lies 2 km WSW of Taunga. Although the central plateau was cleared in recent decades for agriculture, secondary succession is moving in the direction of indigenous rather than nonnative species of plants. Photo by Australian Air Force, 1990.

Table 21-2 Population size, at the end of the breeding season, of the Rarotonga Monarch (*Pomarea dimidiata*) on Rarotonga, Cook Islands. Rat control began in 1989–90. From McCormack & Künzlé (1990a), Robertson et al. (1994), Sanders et al. (1995), and Takitumu Area Conservation Project (unpub. data 2001).

Year	Population size
1988	36
1989	29
1990	38
1991	48
1992	56
1993	60
1994	~88
1995	~108
1996	~139
1997	~156
1998	~172
1999	~188
2000	~202

I hope that more biologists will get interested Pacific island birds. While some recent declines in native bird populations remain unexplained (Fancy & Snetsinger 2001), conservation has improved odds of success when grounded in sound biology. A great example is the monarch *Pomarea dimidiata* (local name *kakerori*), confined to several valleys on Rarotonga, Cook Islands. The population of this understory insectivore increased steadily from just 29 in 1989 to ca. 202 in 2000 (Table 21-2) because of a rigorous program of killing rats (mostly *Rattus rattus*, some *R. exulans*) and putting rat-resistant tree guards on the monarch's nest trees (McCormack & Künzlé 1990a, Robertson et al. 1994, Sanders et al. 1995). This highly commendable success has come at a large investment of time, money, ingenuity, and hard work. The money issue is, of course, relative. What is considered a large amount of money to conservationists is a drop in the bucket to most governments and big corporations.

Sadly, the Rarotonga monarch program is not sustainable without the same effort year after year, because reproductive success depends on annually removing most rats in each nesting territory. No methods exist now to eliminate rats from an island as large as Rarotonga (67 km^2), although rats have been eradicated from islands in New Zealand at least as large as 2.2 km^2 (McFadden & Towns 1991, Taylor & Thomas 1993, Clout & Saunders 1995, Towns et al. 1997). I heartily commend New Zealanders for their herculean efforts in rat removal.

Introduced mammalian predators have caused avain extinction on islands for centuries (Blackburn et al. 2004). Feral cats (*Felis catus*), which are widely distributed in Oceania (pers. obs. 1984–2001), are a poorly studied but likely substantial problem. Birds are a regular part of the diet of feral cats in the Hawaiian Islands (Smucker et al. 2000) and on Mona Island in the Caribbean (Garcia et al. 2001). Mongooses (*Herpestes auropunctatus*) have been introduced to several of the largest islands in Fiji, where most species of ground-dwelling birds are now rare or gone (Watling 2001). Tropical Oceania has been spared the ravages of stoats (*Mustela erminea*), which are abundant and regularly kill birds in New Zealand (King & Moody 1982).

Efforts even as great as those for the Rarotonga monarch may not save species facing massive losses of high-quality habitat, especially when combined with threats from predators or disease. The species of tropical Pacific landbirds most likely to survive in the evolutionary short term or ecological long term (the next few centuries) are those that seem well adapted to patchy forest, secondary forest, or nonforest, and that have relatively good dispersal abilities. In Remote Oceania, likely survivors include the rails *Porzana tabuensis*, *Gallirallus philippensis*, and *Porphyrio porphyrio*, the pigeon *Ducula pacifica*, doves *Ptilinopus porphyraceus* and *P. greyii*, barn-owl *Tyto alba*, swifts *Collocalia spodiopygia* and *C. vanikorensis*, kingfisher *Halcyon chloris*, trillers *Lalage maculosa* and *L. leucopyga*, honeyeaters *Myzomela cardinalis*, *M. jugularis*, *M. rubrata*, *Lichmera incana*, and *Foulehaio carunculata*, and starlings *Aplonis tabuensis* and *A. atrifusca*.

The relative abundance of even these widespread species, however, varies widely from island to island. Their populations can be suppressed or eliminated if habitats are too altered or when predation by nonnative vertebrates becomes too pervasive (Steadman 1998, Steadman & Freifeld 1998, Steadman & Franklin 2000). Nonnative birds are a growing threat, such as the mynas *Acridotheres tristis* and *A. fuscus* that now dominate the birdlife in much of Fiji, Samoa, and the Cook Islands (Holyoak 1980, Watling 1982, Case 1996, Gill 1999; DWS pers. obs. 1984-2001). Another threat is the spread of aggressive species of nonnative insects, especially ants (Wetterer 1998, 2002, Wetterer et al. 1998, Lester & Tavite 2004), which can have major effects on insect communities even on continents (Gotelli & Arnett 2000, Deyrup et al. 2000, Pennisi 2000). The negative effects of nonnative ants could involve nutrient recycling and thus climb right up the food web either in that way or by eliminating or reducing the populations of native insects consumed by birds and lizards. From personal experience in the Solomons, Fiji,

Marianas, and Galápagos, I know that population explosions of nonnative biting ants and stinging wasps can render some islands nearly uninhabitable, at least in the rainy season.

The Need for Surveys

We know little about the precise distributions (interisland and especially intra-island), population sizes and trends, habitat preferences, feeding ecology, and nesting ecology of most species of landbirds in Oceania. Such information helps to carry out good conservation programs (Margules & Pressey 2000). Repeatable, quantified surveys of landbirds are lacking except on certain islands in Samoa (Amerson et al. 1982a, 1982b, Engbring & Ramsey 1989, Freifeld 1999, Freifeld et al. 2001), Tonga (Steadman 1998, Steadman & Freifeld 1998), Fiji (Steadman & Franklin 2000), the Societies (Monnet et al. 1993b), the Marianas (Engbring & Ramsey 1984, Engbring et al. 1986), and Yap, Chuuk, Pohnpei, and Kosrae (Engbring et al. 1990, Buden 2000). Most of these surveys have been one-time or two-time affairs; only on Tutuila in American Samoa have such surveys been done throughout the year (to account for seasonality) and in multiple years (Freifeld 1999, Freifeld et al. 2004).

Partially quantified assessments of relative abundance and habitat preference are available for Isabel (Kratter et al. 2001a) and Rennell (Diamond 1984, Filardi et al. 1999) in the Solomons. For the most part, however, we have no such survey data for landbirds in Near Oceania, a region traditionally protected by difficulty of travel, disease, and rugged landscapes, but that has been subjected to extensive commercial logging over the past several decades (Whitmore 1989).

Accurate, repeatable surveys of landbirds on Pacific islands require good identification skills (sight and sound), adherence to standard methodologies, and a willingness to travel and live in remote areas. This is challenging, uncomfortable work, but it is not rocket science. Given the huge interest worldwide in birds, which has blurred the distinction between ornithologist and birdwatcher, it is puzzling that hundreds of islands in Oceania remain unsurveyed. Where surveys have occurred, especially when combined with vegetation analyses, we have learned a lot about how the distribution and relative abundance of landbirds are related to habitat types, as illustrated by the following example.

Based on fieldwork in 1995 and 1996, my colleagues and I assessed the distribution, relative abundance, and habitat preference of forest plants and birds (and, to some extent, of lizards and mammals) on 17 islands in the Vava'u Group, Tonga (Steadman et al. 1999a). These islands vary in habitat composition, land area (0.02-96 km²), elevation (20–215 m), and distance (0–10.1 km) from the largest island of 'Uta Vava'u (Table 6-11). The most mature category of lowland rain forest persists mainly in areas too steep to cultivate and covers ca. 10% of the land area in Vava'u (Franklin et al. 1999). The interisland and interplot variation in species richness and relative abundance of indigenous plants and vertebrates in Vava'u have been affected more by human deforestation than by any of the classic physical variables of island biogeography–area, elevation, or isolation.

Among landbirds, 11 species are widespread in Vava'u and at least locally common, one (the ground-dove *Gallicolumba stairi*) is extremely rare, and three others have been lost in the past century (Steadman & Freifeld 1998; Tables 6-12, 6-13 herein). Considering all islands surveyed, the mean total abundance and species richness of landbirds increase steadily with increasing forest maturity (Figure 6-7). This trend is especially strong among frugivores and passerine insectivores (Figure 21-1) but does not apply to all species. Two trampy rails, for example, are rare or absent in mature forest. Other species (*Halcyon chloris*, *Lalage maculosa*, and *Aplonis tabuensis*) occur in all habitat categories without clear trends. Similar patterns have been noted in the Lau Group of Fiji (Steadman & Franklin 2000), Western Samoa (Evans et al. 1992), and Cook Islands (Franklin & Steadman 1991).

Our research in Vava'u shows that small islands (<1 km²) are important for conservation, especially given that succession following agricultural abandonment leads to indigenous forests on these raised limestone islands. 'Euakafa (Figure 21-2) is a fine example of a small island (0.70 km²) with much conservation potential. The relative abundance of landbirds on this uninhabited, well-forested island is much greater than on nearby, similarly sized (0.56 km²) Taunga, where 30 to 50 people live and no mature or even submature/disturbed mature forest remains (Table 21-3, Figure 21-3). Small islands off Guam can accommodate species of lizards that are extirpated or endangered on Guam itself because of the snake *Boiga irregularis* (Perry et al. 1998). Most of these 22 islets may not sustain lizard populations over the long-term, however, because of extremely small size (0.0005 to 0.034 km² or 0.05–3.4 ha); only one (Cocos; 0.38 km² or 38 ha) is large enough to have any ecological long-term (centuries or longer) potential to sustain populations of lizards.

The Vava'u data also point to the importance of preventing the establishment of hyperinvasive species of plants. Secondary succession in Vava'u leads to forests of indigenous trees (Franklin 2003) because highly invasive, nonnative, weedy plants are not established here, such as *Miconia calvescens* (Melastomataceae; a Neotropical

Table 21-3 Mean number of birds per station for the nearby islands of `Euakafa (uninhabited, largely forested) versus nearby Taunga (inhabited, largely deforested), Vava`u Group, Tonga. Totals are calculated without the visual but not vocal *Collocalia spodiopygia*. Rounded to the nearest 0.1 except for values <0.06. Data from Steadman & Freifeld (1998: Table 3).

	`Euakafa	Taunga
RAILS		
Porphyrio porphyrio	0.1	0
PIGEONS, DOVES		
Ptilinopus porphyraceus	0.5	0.3
Ducula pacifica	0.6	0.04
SWIFTS		
Collocalia spodiopygia	0.1	0.4
KINGFISHERS		
Halcyon chloris	0.6	0.04
STARLINGS		
Aplonis tabuensis	0.8	1.2
CUCKOO-SHRIKES		
Lalage maculosa	0.4	1.2
WHISTLERS		
Pachycephala jacquinoti	1.2	0
HONEYEATERS		
Foulehaio carunculata	2.2	0.2
TOTAL	6.4	2.9

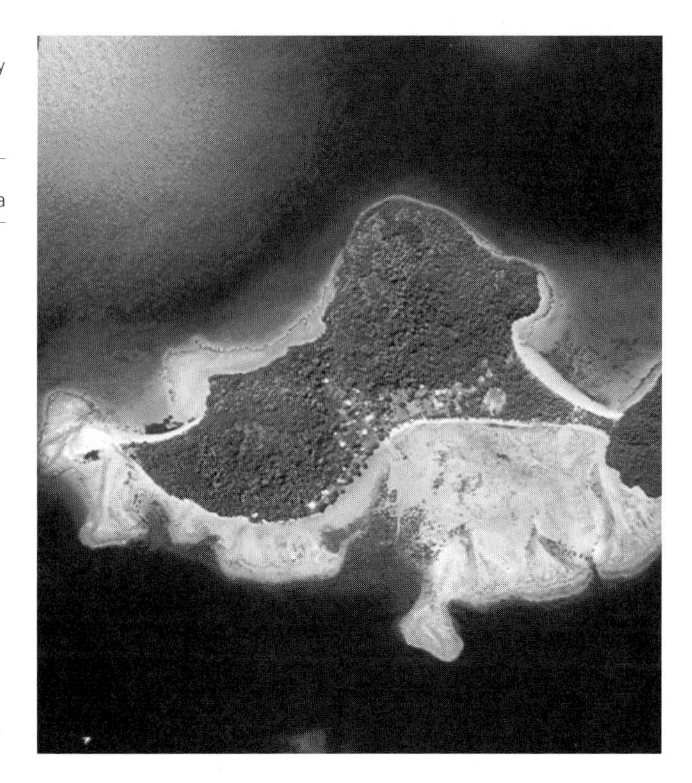

Figure 21-3 Aerial photograph of Taunga (0.56 km^2), elev. 40 m), Vava'u Group, Tonga. Approximately thirty to fifty people live in the village. No late-successional or mature forest remains. Human activity that keeps the vegetation in early successional stages also increases the vulnerability to establishment of nonnative species of plants. Photo by Australian Air Force, 1990.

species displacing native vegetation on Tahiti and Mo'orea; Meyer 1996, 2000) or *Funtumia elastica* (Apocynaceae), an African tree running amuck in Samoa (Whistler 1992, DWS pers. obs. April–May 1999).

Human Perceptions and Involvement

For a moment, let us forget that so many species of Pacific island birds were wiped out hundreds or thousands of years ago. As biologists we want to know as much as we can about each living species. As conservationists, we hope to use this knowledge to help the species to survive. This point becomes moot, of course, unless some respectable vestige of quasi-natural habitat remains and nonnative predators or disease are not out of control.

Global assessments of biodiversity do little to help species in Oceania. Although "Polynesia/Micronesia" is considered one of earth's 25 "hotspots" (Mittermeier et al. 2000, Cincotta et al. 2000), this region is not a "spot" at all, but consists of 1000+ islands in 16 island groups, with a multitude of governments, cultures, and socioeconomic issues. The claim that 49% of the land area in Polynesia/Micronesia is protected (Myers et al. 2000) is grossly inflated. Outside of the Hawaiian Islands and New Zealand, the Pacific region receives relatively little attention from conservation organizations. Amazonia, with its charismatic eagles, macaws, toucans, monkeys, and jaguars, is much more likely to open people's wallets than an atoll in the Tuamotus that sustains three or four indigenous species of landbirds and no mammals. Faunal poverty is also a drawback for "ecotourism" in Oceania, where list-oriented birdwatchers must take expensive flights, perhaps followed by rough voyages in tiny boats, to tick off relatively few new species. (I must note here, however, that the Cook Islands has excellent infrastructure for ecotourism and actively promotes hikes to see the endangered monarch *Pomarea dimidiata*).

Of course, conservation requires more than money alone. About $9,451,664 was spent by various government agencies for avian research and management in the Hawaiian Islands from 1987 to 1997, with very little success at saving species (Steiner 2001). These islands just are too badly damaged. The tragic extinction of Guam's birds in the 1970s and 1980s (Savidge 1987, Rodda et al. 1997, Fritts & Rodda 1998) was either unnoticed or ignored until it was too late by the U.S. Fish and Wildlife Service, an organization that is rich compared to parallel agencies in Pacific island countries. What happens on the ground may be unrelated to the stated goals of conservation programs or government policies. If people need to clear some forest to feed their family, they will do it. If someone decides to release dogs or cats on a formerly carnivore-free island, they will. If a forester from Australia or New Zealand gives some island men seedlings of pines or eucalyptus, the islanders probably will plant these nonnative trees, which may help to control erosion and eventually be marketable, but are sterile for native birds.

Now to flip the coin, we must be careful not to cry wolf about the birds of Oceania. Many species on many islands are abundant. Conservationists need to recognize this. The Tinian monarch (*Monarcha takatsukasae*), for example, has been known to be common on Tinian for 25 years (Pratt et al. 1979), yet continues to be classified federally as "endangered" or "threatened" (Balis-Larsen & Sutterfield 1997). At least 10 species of landbirds in Fiji-Tonga-Samoa are widespread and often common (Steadman & Freifeld 1998, Steadman & Franklin 2000, Freifeld et al. 2001). Dense populations of endemic birds are routine in Near Oceania, where rugged field conditions make it more difficult to assess the status of individual species. On Isabel in the Solomon Islands, for example, we found the flightless rail *Gallirallus woodfordi* (Figure 10-7) and various psittacids, including the cockatoo *Cacatua ducorpsi*, to be common in forests along the Garanga River in 1997 and 1998 (Kratter et al. 2001a,b), even though "global" conservation assessments list these species as rare if not endangered. The rail even had been declared extinct (Collar & Andrew 1988, Johnson & Stattersfield 1990) at the same time that villagers on Isabel considered it a common garden pest, a status that it still has today.

Endemism and rarity do not always go together. The most abundant bird that we found on Rennell in June 1997 was an endemic white-eye *Woodfordia superciliosa* (Filardi et al. 1999; Figure 14-6 herein), one of two species in a genus endemic to Rennell and the Santa Cruz Islands (Murphy 1929). On Mangaia (Cook Islands), the endemic warbler *Acrocephalus kerearako* is widespread and common, even in villages (pers. obs. 1984-2001). There are many other examples of common endemic species across Oceania.

The concept of endemism brings up another form of crying wolf, namely taxonomic inflation to promote conservation, i.e., regarding each population that is different in any way as a separate species. The motivation for this misguided practice is that an endemic species may attract more conservation attention than an endemic subspecies. Because taxonomic rank is one criterion for evaluating conservation priorities, it must be scientifically defensible and therefore based on the best available biological information rather than on political or conservation concerns (Bowen & Karl 1999, Karl & Bowen 1999). Taxonomic inflation tends to be more compatible with the phylogenetic than biological species concept. (For a discussion of species concepts, see Peterson 1998.)

An extreme case of potential taxonomic inflation would be that of the collared kingfisher (*Halcyon chloris*; Figure 13-9), with 47 subspecies (based on plumage or size) from the coast of the Red Sea to West Polynesia (Mayr 1941b, Peters 1945:207–213). To recognize some of these subspecies at the species level would please island-hopping birdwatchers wanting beefier life lists, and might appeal to a few misguided conservationists looking for endemic species wherever they can be found. Such a move, however, would obscure the fascinating if poorly understood evolutionary radiation of the world's most widespread kingfisher. When someone is comparing the resident blue-and-white kingfishers from, for example, Tutuila in Samoa (*H. c. pealei*) and Rennell in the Solomons (*H. c. amoena*) or the Andaman Islands (*H. c. davisoni*), the trinomial suggests that these birds from distant islands are part of the same radiation. A series of binomials has no such implication. The extent of reproductive isolation among the various subspecies of *H. chloris* is unknown, although I believe that pure geographic isolation may restrict interisland gene flow more than any potential biological isolating mechanisms, such as differences in voice or courtship behavior.

Attempts to conserve birdlife in Oceania must consider the importance of involving local people (Hay 1986) including fostering traditional feelings about birds. Aldo Leopold (1991:99, 101) wrote, "Conservationists have, I fear, adopted the pedagogical method of the prophets; we mutter darkly about impending doom if people don't mend their ways. The doom is impending, all right; no one can be an ecologist, even an amateur one, without seeing it. But do people mend their ways for fear of calamity? I doubt it. They are more likely to do it out of pure curiosity and interest." In Oceania, interest in nature is waning as cash economies replace subsistence agriculture. If conservation programs are to succeed, an appreciation of nature must be kept alive. This is difficult in countries with rapid population growth (and concomitant environmental implications; Hern 1993) and an understandable fascination with glitzy Western technology.

Pacific islanders often are portrayed as having minimal impact on their environments (e.g., Geatz 2000). This inaccurate notion is a disservice to conservation (by assuming that indigenous peoples are inherently environmentalists) and to island peoples (by simplifying their complex cultures). Like any other people, Pacific islanders are trying to make a living, which includes eating well and being safe and comfortable. Homes and crops require deforestation, which has been occurring across Oceania since people first arrived. Nobody, myself included, has the right to call this wrong. It just is. Nobody in Oceania, and nobody reading this book, could survive in a purely forested tropical environment. Deforestation, whether by stone adze and fire or chain saw and bulldozer, reduces populations of forest-dwelling birds. Nonnative predators, whether rats 3000 years ago or cats in recent decades, eat native birds. Extinction happens when some blend of habitat loss and predation, perhaps combined with disease, becomes overwhelming.

I doubt that many species of island birds were wiped out intentionally. Instead, the processes that reduced their populations got out of control and could not be halted. Given the importance of red feathers to Polynesians (Steadman 1997b), it is unlikely that any parrots, for example, were intentionally exterminated. Nevertheless, most populations and at least 50% of the species of East Polynesian parrots have been lost since human arrival. Just as with some pigeons on islands, the North American passenger pigeon (*Ectopistes migratorius*) now exists only as museum specimens (Figure 21-4). If one could have polled the citizens of western Pennsylvania in the 1870s and 1880s, I would bet that, all else being equal, most would have preferred that the passenger pigeon not spiral downward from countless millions to extinction. Some sort of compromise would have been nice. Such a desire, however, had no chance of curbing the logging and hunting booms that fueled the 19th-century economy in rural northeastern United States. Similarly today, tropical deforestation yields clear financial benefits at the national level and thus is difficult to stop (Kremen et al. 2000). Nobody knowingly killed the last passenger pigeon in Pennsylvania or the last tooth-billed pigeon (*Didunculus* undescribed sp.) in Tonga. If any of the four critically endangered species of palms in New Caledonia (see Pintaud et al. 1999) are lost in coming decades, it probably will happen without the explicit knowledge of whoever kills the last tree.

Figure 21-4 A specimen of the extinct Passenger Pigeon (*Ectopistes migratorius*) in the New York State Museum. Photo by Christopher Supkis, 1988.

As a boy in western Pennsylvania in the 1950s, I frequently walked up our dirt road to visit my most cherished neighbor, William C. Weaver, who was born in 1886 and had seen passenger pigeons on our farm in his youth. I was struck by what a different place our farm had become in only a couple of human generations. Based on the chronology of extinction in Tonga (Tables 3-3, 3-4), I can imagine a Tongan boy 2700 years ago listening to an old man talk about the large pigeons and lizards he had hunted and eaten in his own youth, but that the boy would never see. The good old days were gone forever.

Pathogens

On both continents and islands, emerging infectious diseases are a concern for common, widespread vertebrates as well as those that are rare or localized (Daszak et al. 2000). In most of Oceania, we know little about the distribution and potential impact on native birds of diseases such as avian pox (Reece 1989, Vargas & Snell 1998), avian tuberculosis (Silva-Krott et al. 1998), avian cholera and botulism (Cooper 1989), and West Nile virus (Stokstad 2004). The same applies to diseases caused by blood-borne parasites (hematozoa), whether avian malaria (caused by *Plasmodium* spp.) or others with morbidity and mortality rates even more poorly known, such as *Haemoproteus*, *Leucocytozoon*, and *Trypanosoma* (Peirce 1989). Nonnative birds, especially chickens, can serve as reservoirs for these pathogens (Silva-Krott et al. 1998). Hematozoan infection is transmitted by biting dipterans, these being mosquitoes (Culicidae) in the case of avian malaria. Another family of biting flies, the Simuliidae or blackflies, is native to East Polynesia (Craig & Joy 2000, Craig et al. 2001). Although they are involved in transmission of disease to insular peoples (Chanteau et al. 1993, Cupp & Cupp 1997), their possible role in avian diseases on islands is unstudied.

Because avian malaria causes considerable mortality of native birds in the highly isolated Hawaiian Islands (Warner 1968, van Riper et al. 1986, Jenkins et al. 1989, Nishida & Evenhuis 2000, Yorinks & Atkinson 2000, Jarvi et al.

2001), determining the distribution of *Plasmodium* (and other hematozoa) across Oceania is a starting point to evaluate disease in native birds. The only Micronesian data on avian hematozoa are from Guam (Marianas), where none was found in 260 blood smears made in the 1980s (Savidge et al. 1992). No hematozoa were found either in 79 avian blood smears from the southern Cook Islands (Steadman et al. 1990a). While this absence would seem to be good news from a conservation standpoint, it may also indicate that native species of birds in the Marianas and Cooks, like those of the Hawaiian Islands, would have little resistance to blood-borne parasites should they be introduced. The Northern Cook Islands lacked mosquitoes in pre-European times (Gill 1885:199-200), strengthening the supposition that avian malaria is not endemic on these atolls. Whether this was true as well on the higher islands in the southern Cooks (and elsewhere in East Polynesia) is unknown, but seems likely. Hematozoa may be absent as well in Tonga (none found in a small sample of 30 blood smears; Greiner & Steadman, unpublished data), which is 1200 km west of the Cook Islands. I must note here that much of the future of hematozoan studies will involve molecular rather than traditional (blood smears) techniques (Fallon et al. 2003).

Even 2000 km to the west of Tonga, the apparent absence of *Plasmodium* (and therefore, avian malaria) in 100+ blood smears from birds on Isabel and Rennell in the Solomons (Frank et al. in press) is fascinating given the high infection rate of human malaria in this island group. A future introduction of avian *Plasmodium* might, therefore, have negative effects in Near Oceania as well as Remote Oceania. On the other hand, birds from both Isabel and Rennell have an infection rate for *Haemoproteus* comparable to that of continental birds (Frank et al. in press). If this has been the case for some time, it would suggest that the deleterious effects of *Haemoproteus* in Near Oceania may be no greater than it is on continents, yet another trait in which the birds of Near Oceania are less "insular" than those of Remote Oceania.

The hematozoan situation is unknown in the island groups between the Solomons and Tonga, i.e., Vanuatu, New Caledonia, and Fiji. Vanuatu has human malaria, whereas New Caledonia and Fiji do not. All three of these island groups have native biting flies (Evenhuis 1989), so vectors for hematozoan infection were present even in precultural times.

Frugivory, Seed Dispersal, and Conservation

Frugivorous birds, especially pigeons and doves (Columbidae) and starlings (Sturnidae), consume and disperse seeds and fruits across the Pacific region from New Guinea (Terborgh & Diamond 1970, Frith et al. 1976, Coates 1985, Pratt & Stiles 1985), the Solomons (Hadden 1981, Filardi et al. 1999, Kratter et al. 2001a), and Vanuatu (Bregulla 1992) eastward to West Polynesia (Steadman 1998, Webb et al. 1999, McConkey & Drake 2002, Meehan et al. 2002) and East Polynesia (Franklin & Steadman 1991, Brooke & Jones 1995, Steadman & Freifeld 1999). Nectarivorous birds, especially lories and lorikeets (Psittacidae: Loriinae) and honeyeaters (Meliphagidae) are also abundant and diverse in Oceania, with some part of their attenuated diversity in Remote Oceania due to human impact (Chapters 6–8, 12, 14). Flying foxes or fruit bats (Chiroptera: Pteropodidae; Figure 2-15) also are important pollinators and seed dispersers as far east as the Cook Islands (Cox et al. 1991, Fujita & Tuttle 1991, Rainey & Pierson 1992, Flannery 1995a, McConkey & Drake 2002). Like birds, flying foxes were hunted and eaten prehistorically, reducing their geographic ranges (Flannery et al. 1988, Steadman & Kirch 1990, Koopman & Steadman 1995).

Although deforestation has been a major factor in reducing or eliminating populations of nectarivores and frugivores, the reverse may also be true; the losses of nectarivores and frugivores in Oceania probably have affected the forests by limiting pollination and the inter- and intra-island dispersal of plant propagules, especially of large-fruited species (Steadman & Freifeld 1999, McConkey & Drake 2002). Similar downward swirls of birds and trees have been described for tropical forests in northeast Brazil and Southeast Asia (Silva & Tabarelli 2000, Brook et al. 2003). The exact nature of these effects in Oceania is speculative, and also must consider effects of nonnative seed predators such as *Rattus* spp. and *Mus musculus* (McConkey et al. 2003). Just as on continents (Kloor 2000, van Gemerden et al. 2003), it is uncertain what an undisturbed forest would be like anywhere in Oceania. We lack data on the reproductive biology and dispersal abilities of most species of plants in Oceania, and on seedling recruitment with versus without dispersers. For most islands, and even some entire island groups, we also lack long-term surveys of birds and forest vegetation, prehistoric pollen records, detailed feeding studies of the birds, and information on the distribution, abundance, and phenology of flowering and fruiting.

Mangaia is one island where at least some of these data data have been gathered. Five species of columbids and two species of trees (*Weinmannia* sp.–Cunoniaceae; *Pritchardia* sp.–Arecaceae) have been lost on Mangaia since human arrival (Ellison 1994a, Kirch et al. 1995; see Chapters 2, 7). Deforestation began about 2500 yr BP and intensified at 1600 and 1100 yr BP (Figure 2-9), reducing the amount of high-quality habitat for columbids. Dry-season burning probably was an important part of the deforestation. As Mangaia's forests became more fragmented, it may

Figure 21-5 Coconut trees (*Cocos nucifera*) in a recently abandoned agricultural plot on Ha'ano, Ha'apai Group, Tonga. Photo by DWS, 12 July 1996.

have become easier for people to burn the remaining forest patches (as in Amazonian forests; Cochrane & Laurance 2002) and to hunt pigeons, doves, and other birds. Rats, dogs, and pigs also preyed upon both seeds and birds. As columbid populations decreased further, they became less effective at dispersing their food plants. Negative reinforcement hastened the decline of columbids and forests on Mangaia, roughly homologous to current deforestation problems in Borneo (Curran et al. 1999).

Mangaia also lost three nectarivores (the lorikeets *Vini sinotoi*, *V. vidivici*, and *V. kuhlii*) about 600 years ago (Chapters 7, 12). Today in Remote Oceania, most species of lorikeets seem to prefer the nectar-rich flowers of coconut trees (*Cocos nucifera*; Figure 21-5). Prehistorically, when most East Polynesian islands had two or three species of lorikeets, perhaps flowers of other species of trees also were important sources of nectar.

In Tonga, the modern forest composition is well documented (Drake et al. 1996, Franklin et al. 1999, Steadman et al. 1999a, Wiser et al. 2002). We also have good estimates of the nectarivorous and frugivorous birds and bats that once lived on these islands (Steadman 1993a, 1995a, Koopman & Steadman 1995). At least 12 species of Tongan forest trees have fruits that may be too large to be consumed by its largest surviving avian frugivore, the pigeon *Ducula pacifica* (McConkey et al. 2004a; Table 21-4 herein). These large fruits probably were eaten by two much larger pigeons lost since human arrival, namely *Ducula* undescribed sp. and Undescribed genus C (Chapters 6, 11). Other examples of oversized native fruits can be found in Fiji and Samoa (J. Franklin pers. comm.).

Translocations

Unknown but substantial numbers of the surviving populations and species of landbirds are at risk of extinction on Pacific islands. Translocation may be the only way to save certain species with small, localized populations (Franklin & Steadman 1991, Witteman et al. 1991). Prehistoric bird bones reveal the natural distributions of species and thus provide a sound geographic framework for recovery programs (Steadman 1999a). In evaluating islands as sites for translocation, consideration should be given to the quality and quantity of forests, the presence or absence of various nonnative vertebrates (especially rats, cats, mongoose, and brown tree snakes; see Atkinson & Atkinson 2000, Fancy et al. 2001), and the current and future human activities, which may not be favorable for newly established avian populations (Griffith et al. 1989, Armstrong & McLean 1995). Whether the island to receive transplanted birds lies near the periphery or center of the former range may not matter because there is no overriding theoretical or practical reason why the periphery of a range should be especially good or bad (Channell & Lomolino 2000). Furthermore, because the range contraction was anthropogenic, and we can assume that all islands within the former range have been influenced to some extent by people, we should be interested only in targeting islands with the best mix of factors that retard extinction (Chapter 16).

In spite of a theoretical propensity for high extinction rates (MacArthur & Wilson 1967, Hubbell 2001), small islands should not be overlooked as potential sites for translocated populations. Many small (0.1–10 km^2) islands have limited agricultural potential or very steep coastlines or both, and therefore little appeal for long-term human habitation, resulting in relatively low levels of deforestation. Sizeable landbird populations persist on small but forested islands in Fiji (Chapter 6), Tonga (Steadman & Freifeld 1998), Samoa (Freifeld et al. 2001), the Bismarcks and

Table 21-4 Selected indigenous trees in Tonga with fruits certainly or possibly too large to be swallowed by Tonga's largest extant avian frugivore, *Ducula pacifica*. Data from McConkey & Drake (2002), Meehan et al. (2002), and J. Franklin (pers. comm.).

Family	Species	Fruit shape and size
Apocynaceae	*Cerbera odollam*	ovoid, 50–60 mm
Apocynaceae	*Neisosperma oppositifolium*	ovoid, 62–81 × 45–66 mm
Burseraceae	*Canarium harveyi*	round, up to 40 × 25 × 20 mm
Clusiaceae	*Calophyllum inophyllum*	spherical, 27–40 mm
Ebenaceae	*Diospyros major*	ovoid, 60 × 30 mm
Meliaceae	*Syzygium quadrangulatum*	spherical, 25–35 mm
Meliaceae	*Syzygium richii*	spherical, 30 mm
Sapindaceae	*Pometia pinnata*	spherical, to 75 mm
Sapotaceae	*Burckella richii*	round, 49 × 44 mm
Sapotaceae	*Planchonella garberi*	nearly spherical, 20–40 × 22–40 mm
Sapotaceae	*Planchonella grayana*	nearly spherical, 26–46 × 30–42 mm
Sapotaceae	*Planchonella membranacea*	nearly spherical, up to 45 × 30 mm

Solomons (Mayr & Diamond 2001), and presumably elsewhere in Oceania. Especially in cases where access by boat is difficult because of cliffs, or where protection can be enforced, small islands may be our best opportunity to save some of the rare birds that survive on Pacific islands.

In temperate New Zealand, where translocations on the two main islands have met with little success (Armstrong & McLean 1995, Bramley & Veltman 1998), many small islands have proven to have great value in translocations, habitat protection, and predator control (Clout & Saunders 1995, Towns 1996, Towns et al. 1997). New Zealanders lead the way in rodent control on islands, with complete eradication of rats being implemented on islands of 2.2 km^2 or even larger (Clout & Saunders 1995). This bodes well for small (up to several km^2), uninhabited, forested islands in tropical Oceania, if the effort were made.

Few translocations of landbirds have been tried in tropical Oceania, although several have shown initial success. The Marquesas lorikeet (*Vini ultramarina*) now lives on Ua Huka and Fatu Hiva, based on birds taken from Ua Pou (Kuehler et al. 1997). In Tonga, the prehistorically widespread megapode *Megapodius pritchardii* has been taken from its final outpost on Niuafo'ou to two other volcanic islands (Late, Fonualei), where it now is breeding (Jones et al. 1995:148, Göth & Vogel 1999, C. Matavalea, pers. comm. 1999, 2001). The ground-dove *Gallicolumba stairi* was taken from Late to 'Ata in Tonga in 1991 by D. Rinke and L. Matavalea. It still was living on 'Ata in August 2001 (pers. obs.). I believe that *M. pritchardii* and *G. stairi* could be released on any Tongan island that is >0.1 km^2, forested, uninhabited, and free of nonnative mammals (or has only *Rattus exulans*).

A more biogeographically impure translocation involves Kuhl's lorikeet (*Vini kuhlii*), native to the Cook and Tubuai (Austral) Islands (Chapter 7). Surviving only on Rimatara (Tubuai) in its original range, *V. kuhlii* was introduced to the northern Line Islands (Kiribati) more than 200 years ago and still lives on Tabuaeran (Fanning) and Teraina (Washington) islands (Kirby 1925, Watling 1995). The blue lorikeet (*V. peruviana*) survives only on Rangiroa, Scilly, and Bellinghausen (Society Islands) in its native range, but has lived for the past century on Aitutaki, Cook Islands (Steadman 1991b; Chapter 7 herein).

One translocation to consider would be to release the Guam rail (*Gallirallus owstoni*; Figure 10-9) on Aguiguan (Figure 21-6), a cliffy, uninhabited, mostly forested island (7.2 km^2, 157 m elev.) that is 150 km NNE of Guam but only 9 km SW of Tinian. This rail became extinct in the wild in the 1980s but thrives in captivity (Witteman et al. 1991, Haig & Ballou 1995). Five islands in the Marianas from Guam through Saipan once were inhabited by flightless species of *Gallirallus* very much like the Guam rail (Chapters 8, 10). Since 1989, at least five labor-intensive releases, involving more than 1000 captive-bred birds, have failed to establish a wild population of Guam rails on nearby Rota (Brock & Beauprez 2000, Suzanne Medina pers. comm. 2005); within weeks or months of being released on Rota, nearly all rails die from one cause or another, perhaps especially from being killed by cats and black rats. The environmental conditions on Aguiguan are much better now than

Figure 21-6 Aguiguan, Northern Mariana Islands. Photo from small boat by DWS, 14 July 1994.

on any other limestone island in the Marianas (Craig 1992c, Craig et al. 1992a). Aguiguan differs from the larger Rota (81 km^2, 459 m elev.) and Guam (540 km^2, 262 m elev.) in lacking snakes, black rats, cats, dogs, cattle, pigs, people, highways, vehicles, a wharf, and an anchorage, which are causes of mortality in rails and other native birds.

Aguiguan is unique among islands south of Saipan in sustaining only the Pacific rat *Rattus exulans* rather than the black rat *R. rattus* or both species. Although neither rat is desirable, *R. exulans* probably is less harmful to native birds (Kepler 1967, Atkinson 1985, 1989, Brown 1997). (The extinction of flightless rails on several islands in the Marianas, however, has a suspicious chronological correlation with the prehistoric arrival of *R. exulans*; Chapters 8, 16). Application of New Zealand's rat eradication techniques might eliminate *R. exulans* on Aguiguan but not on Rota, which is 11+ times larger.

Aguiguan also lacks the nonnative brown tree snake (*Boiga irregularis*), which has decimated Guam's avifauna in recent decades (Savidge 1987, Fritts & Rodda 1996, 1998). Sadly, it is probably just a matter of time before the three inhabited islands just north of Guam (Rota, Tinian, Saipan) also sustain large populations of this snake, with a likelihood of devastating their birdlife as on Guam. In fact, *B. irregularis* already has been recorded on all three islands (Rodda & Fritts 1993), including more than 40 sightings on Saipan from 1991 to 1996 (Rodda et al. 1997), a strong indication that a snake population is being established. Aguiguan, on the other hand, lacks shipping facilities and is accessible (with difficulty) only by helicopter or small boat, with no place to dock or anchor. Therefore, Aguiguan is unlikely to be colonized by snakes, not to mention black rats or cats.

Bones of flightless "Guam-like" rails (*Gallirallus* undescribed spp. L–O, similar in morphology to *G. owstoni*) have been found on Rota, Aguiguan, Tinian, and Saipan. The behavior and ecology of *G. owstoni* (see Jenkins 1979, 1983) probably are not very different from those of the extinct species. At Pisonia Rockshelter, an archaeological site on Aguiguan dating from ca. 1800 to 500 cal BP, the most common species of bird recovered was *Gallirallus* undescribed sp. M (Steadman 1999a; Chapter 8 herein).

To summarize, Guam rails would have an excellent chance of surviving on Aguiguan, a largely forested island (Chandran et al. 1992, Craig 1992a) that still supports healthy populations of many other landbirds (Engbring et al. 1986, Craig 1992b, DWS pers. obs. June 1994). The feral goats and chickens on Aguiguan could be eliminated right away. Eventually, *Rattus exulans* should be eradicated as well.

The Micronesian pigeon (*Ducula oceanica*) is another candidate for translocation to Aguiguan. Bones of *D. oceanica* occur prehistorically on Rota and Tinian, the islands on either side of Aguiguan. Therefore this pigeon, which is capable of living even on atolls, must have inhabited Aguiguan as well. The same logic could apply to the white-browed crake (*Poliolimnas cinereus*), although this seldom-recorded and little-studied species is of unknown conservation concern in Oceania or elsewhere.

Other possible translocations would involve Mangaia (52 km^2, 169 m elev.), Cook Islands, where five species of columbids have been lost (Chapter 7). Of these, *Gallicolumba nui* is extinct, *Ducula aurorae*, *D. galeata*, and *Gallicolumba erythroptera* are very rare and survive today only on one to several islands outside of the Cook Islands, and *Ptilinopus rarotongensis* survives on other islands in the Cooks, as does *Ducula pacifica*, which is more or less the ecological equivalent of *D. aurorae* (Steadman 1997b; Table 7-2 herein). If protected initially from hunting, *D. pacifica* and *P. rarotongensis*, which are canopy-dwelling frugivores, would stand a reasonable chance of reestablishment. Much of Mangaia's rugged limestone terrain sustains forest dominated by native trees that could provide food for columbids, such as *Ficus*, *Elaeocarpus*, *Guettarda*, *Pipturus*, *Canthium*, and *Homalium* (Merlin 1991).

In 1997 and 2001, several Mangaians told me that they wanted to establish and protect populations of native pigeons and doves. With local policy determined by a council that represents each of Mangaia's six districts, it should be feasible to achieve island-wide support for stocking and protecting *Ducula pacifica* and *Ptilinopus rarotongensis*. Mangaia's large populations of black rats, cats, dogs, and pigs preclude consideration of ground-dwelling or understory species for translocation. Rat guards on tree trunks might be needed to protect eggs and nestlings of columbids. Similar protection from rats probably would be required if an attempt were made to inoculate Mangaia once again with the lorikeet *Vini kuhlii*, a nectar feeder found commonly in prehistoric sites.

Translocation may not be able to help some of Oceania's most endangered species of birds. For example, both remaining populations of the canopy frugivore *Ducula aurorae* (Chapters 7, 11) may be so small that it would be difficult to catch enough individuals of both sexes to establish a new population.

Summary and Prospects

On average, birds have been more resilient to human-caused extirpation in Near Oceania than in Remote Oceania, where there is little reason for optimism about the long-term future of many species of landbirds, especially in East Polynesia and Micronesia. Human populations are growing. Forests continue to be cut. Nonnative plants continue to proliferate. Rats, cats, dogs, and pigs flourish. The spread of nonnative ants is ominous (Jourdan 1997, Wetterer 1998, Wetterer et al. 1998, Gotelli & Arnett 2000, O'Dowd et al. 2003). Soil biotas may play an important role in promoting or stifling invasion by exotic plants (Callaway et al. 2004) but almost nothing is known about soil biotas in Oceania. In places such as the Hawaiian Islands, pessimism is just reality, given the pervasive conservation problems (van Riper & Scott 2001, Loope et al. 2001). Many of the surviving species in Remote Oceania prefer open or wetland habitats (a few widespread species of heron, duck, volant rails, and migrant shorebirds) or tolerate some forest clearance (barn-owl and certain species of fruit-doves, pigeons, kingfishers, starlings, honeyeaters, and warblers).

Of course we should try to halt deforestation, control or eliminate established populations of nonnative plants, mammals, birds, snakes, and invertebrates (especially ants and biting dipterans), and keep these introduced producers, herbivores, predators, and sanguivores from spreading further. Rarely, however, are these measures possible, although natural features such as cliffs, knife-edge erosion, and pinnacle karst have curbed deforestation on parts of many islands, and extreme remoteness or difficulty getting ashore has done a fair job of preventing colonization by exotic species of plants and animals on some islands.

Selected species could be translocated to islands where they or a closely related extinct species previously lived, after it is shown that the habitat is reasonably suitable and that current human activities probably are compatible. Uninhabited islands are best qualified, although inhabited islands also can be considered, especially for canopy-dwelling species. Sadly, some uninhabited islands (especially in Fiji and Tonga) are being developed for tourism, which seems to be no less damaging to ecosystems than any other sort of habitation. The initial success in translocating megapodes in Tonga has been low-budget and low-technology, implemented by Tongans working with foreign biologists. No models of population dynamics were consulted; smart people worked hard, armed with common sense, good field skills, and dedication.

This is not to say that exercises such as population viability analysis (PVA) cannot be useful in conservation. For example, Brook et al. (1997) used a PVA to suggest that the flightless rail *Gallirallus sylvestris* on Lord Howe Island would still be highly susceptible to extinction from any number of population perturbations even after the population had "recovered" from 20-30 to 200 individuals.

The chance for success in translocation is enhanced, of course, by good background knowledge of the species involved. The persons who translocated megapode eggs in Tonga had considerable field experience with this precocial species, which also had been studied more thoroughly than any other Tongan bird (Weir 1973, Todd 1983, Rinke 1986, Göth & Vogel 1995, 1999). For most species of landbirds in Oceania, we know much less about current population sizes and trends, habitat preferences, and potential vulnerabilities. Also lacking for most landbirds are good descriptive studies of nesting ecology and food habits. Where such studies have been carried out, such as for the endangered Rarotonga monarch, an effective conservation program was implemented (see above). Many similar programs could be undertaken if more scientists, and agencies that fund research, became interested in the basic biology of Pacific island birds outside of Hawaii and New Zealand.

Our understanding of the molecular genetics of island birds is in its infancy, although the few data available are fascinating. Comparing random amplified polymorphic DNA markers, Zwartjes (1999) found a higher proportion of polymorphic bands in the North American white-eyed vireo (*Vireo griseus*) than in three species of *Vireo* endemic to West Indian islands (*V. latimeri, V. modestus, V. osburni*). Haig & Ballou (1995) found very little genetic variation in the Micronesian kingfisher from Guam (*Halcyon c. cinnamomina*; bottleneck of 29 birds in 1986) but high levels of genetic variation (comparable to that in continental species of rails) in the Guam rail (*Gallirallus owstoni*; bottleneck of 22 birds in 1983). An endangered continental species, the whooping crane *Grus americana*, went through an even smaller bottleneck (only 14 adults alive in 1938) that clearly led to the loss of variation in mitrochondrial DNA (Glenn et al. 1999). The Mauritius kestrel (*Falco punctatus*) underwent the ultimate bottleneck of a single wild breeding pair in 1974, but has increased to 400 to 500 individuals today (Groombridge et al. 2000). Examining 12 microsatellite DNA loci in pre-bottleneck museum skins and modern birds, these authors found, however, that allelic diversity had fallen by 55% and heterozygosity by 57%. Similarly, the black robin (*Petroica traversi*), endemic to New Zealand's Chatham Islands, rebounded to ca. 110 individuals in 1993–94 following a bottleneck of five birds in 1980, but has extremely low genetic variation as revealed by minisatellite DNA markers (Ardern & Lambert 1997, Lambert & Millar 1995).

It is encouraging that some populations of birds can increase by an order of magnitude or more within decades of such low population sizes and accompanying loss of genetic variation. I would caution, however, that two or three decades of population growth is a great start but represents very little time on an evolutionary/geological timescale. The long-term prospects for survival remain unknown for *Halcyon c. cinnamomina*, *Gallirallus owstoni*, *Grus americana*, *Falco punctatus*, and *Petroica traversi*. For now, we can only speculate what these analyses of tiny genomic segments might really mean for long-term survival. I should note as well that some surviving insular endemic species may have gone through anthropogenic genetic bottlenecks in prehistoric rather than historic times (Paxinos et al. 2002).

The forests of Oceania also merit additional research and monitoring. Most species of Pacific landbirds now considered to be rare require forests, which face threats from logging, seed predation, overbrowsing, and encroachment by nonnative plants (Wiles et al. 1996). Detailed descriptions of forest vegetation do not exist for most islands. The ecology of pollination, seed dispersal, seedling establishment, and long-term growth is unknown for most indigenous species of forest trees. During the past few thousand years, the loss of so many populations of frugivores and nectarivores (pigeons, doves, lorikeets, thrushes, starlings, honeyeaters, etc.) must have affected the reproduction and dispersal of trees. Probably no forest in Remote Oceania resembles its prehuman condition. The same may even be true in Near Oceania, although the story here is complicated by human occupation extending into the late Pleistocene with its substantial changes in climate and sea level. Badly needed is more research on the complex interactions among plant communities, nonnative mammals, and native birds, such as King (1983) and Wiser et al. (2002) have done in New Zealand and Tonga.

Many tropical Pacific islands are overpopulated, which stresses natural resources such as soils, watersheds, forests, and wildlife. As overall human populations grow across Oceania, this situation will only get worse. The religious conservatives who control much of global politics have made it difficult or impossible to access information and materials for family planning in developing countries. Sooner or later, some organization has to step up to the plate and fix this shameful situation. I challenge conservation organizations to tackle this issue, rather than dancing around it like just about everyone else.

While conservation programs might help more species per dollar on continents than on islands, there is something undeniably special about insular biotas. Efforts to conserve island birds are important for the sake of the birds themselves, for the well-being of island forests, for preserving a part of the human heritage of Oceania, and for the future of science. It is no coincidence that Charles Darwin and Alfred Russel Wallace came up with many of their most brilliant insights while studying island faunas, as did many other great names of evolutionary biology and biogeography, such as Jared Diamond, David Lack, Robert MacArthur, Ernst Mayr, Storrs Olson, and Edward Wilson. It is also no coincidence that birds, colorful and conspicuous as they are, played a leading role in these discoveries. Just imagine how much better our biological concepts would be had these gifted men been able to study plant and animal communities unaffected by people.

Chapter 22 Conclusions, and Suggestions for Future Research

HUMAN-CAUSED EXTINCTION in Oceania, ongoing but mostly prehistoric, has left us with badly depleted modern avifaunas. This revolutionizes avian biogeography on Pacific islands. When outlining the major evolutionary and biogeographic features of Polynesian birds six decades ago, Ernst Mayr (1941c, 1942) had no prehistoric data to consider. When Robert MacArthur and Edward Wilson (1963, 1967) developed their biogeographic theory for islands, or when Jared Diamond (1975b, 1977) analyzed the distributions of congeneric and confamilial sets of Melanesian and Polynesian birds, they lacked prehistoric data to calibrate their thought-provoking models. They did not realize how much the distributions of the living species they studied had been changed by humans, or how many species already had gone extinct. Regardless of criticism, I commend these biogeographic pioneers.

I am less sympathetic to some current practitioners of island biogeography. For two decades now, abundant evidence has been gathered for prehistoric anthropogenic extinction of insular vertebrates. Unwilling to consider timescales longer than decades, most biogeographers still analyze modern distributions as if they were natural. It is disheartening to see pertinent prehistoric information ignored time after time (e.g., Johnson et al. 2000), whether in the Pacific or any other set of islands. It is just as disappointing when authors do cite a paper or two on prehistoric extinction but then ignore them in their analyses (e.g., Murphy et al. 2004).

Physics envy, simplistic hypotheses, and passion for universal explanation prevail over careful field and laboratory observations, solid data, humility about your own abilities, and appreciation for information from other disciplines. I suspect that detachment from fieldwork and thus from the organisms themselves, as well as an obsession with computers and statistics, is partly to blame. With enough manipulation, some sort of "pattern" can emerge from nearly any data set, especially if you did not collect it yourself and are unaware of its limitations.

"Whatever the reason that nearly all of us hold so steadfastly to the time-honored concept that field relations are the ultimate test of any hypothesis, we should bless them, for we do study the Earth" (Dickinson 1999b:9). These strong words of a geologist apply just as well to organismal biology, where so often the theoretical cart is placed in front of the empirical horse, not behind it.

I am not advocating that we pitch our computers over a cliff and return to slide rules, manual typewriters, and pure description. I am arguing that the balance between empiricism and theory has suffered an ill-advised shift in favor of the latter. This is why theorists gasp when Smith (2000) points out that empirical studies can provide solutions to conservation problems regardless of their theoretical underpinning. The plea that theory and statistics rescue ecology from "a dead end of blind empiricism" (Maurer 2000:768) is shortsighted and polarizing. A glance at any ecology journal should reassure us that the rescue can be postponed. Theory and data need each other. Pure theory will lead us to a dead end as well.

Perhaps my ranting can be summarized this way. If you ever have measured the femur (= thigh) length of a live, squirming anole (*Anolis* spp.), you never would make a claim of "adaptive differentiation" as Losos et al. (1997) did for their Bahamian anoles, based on average differences of 0.01 mm in tiny samples of animals with indeterminate growth on "islands" (= rocks) much too small to sustain any form of evolution. Reality almost always gets in the way of elegance.

Conclusions

Most ideas in this concluding chapter will involve only birds on Pacific islands. Occasionally, as you have just seen, I will venture out onto nonavian, non-Pacific thin ice, thankful that I'm not horribly overweight but still hoping that the water isn't over my head. I am not seeking to understand the distribution of all life on Earth. Within the more narrow topic of birds on Pacific islands, I seek progress, not the final word. Following the lead of Sauer (1988:235), I am not disappointed that some important issues in avian biogeography have little prospect of simple generalization.

Scientists had been studying the modern distribution of birds in Oceania for 200 years when MacArthur and Wilson (1963, 1967) proposed a theoretical framework. The prehistoric distribution has been studied for only 20 years, and by only a few workers. I am trying my best to interpret the growing body of prehistoric data, but the next 20 years of study almost certainly will alter some if not many of my conclusions.

Every aspect of biogeography is influenced by prehistory. Human-caused extirpations (population-level losses) and extinctions (species-level losses) have altered the avifaunas of every island group in Oceania. Thus any sort of research on modern Pacific island birds should be interpreted with the realization that the single or multiple species being studied are part of a human-affected bird community functioning in habitats that also have been changed by people. This statement does not mean that studies of modern birdlife are meaningless. It acknowledges, however, that the species of landbirds that are alive today in Oceania evolved within plant and animal communities that no longer exist. Studies of turnover, species richness, potential competitive interactions, and checkerboard distributions are particularly suspect when based only on modern distributions.

Geographic Perspective

The Oceania I study excludes the Hawaiian Islands, isolated in the north-central Pacific. In spite of outstanding research on avian prehistory by Storrs Olson and Helen James, the independently derived and highly endemic Hawaiian avifauna lacks, for example, four of Oceania's most widespread families of landbirds (megapodes, pigeons, parrots, starlings). I also exclude New Zealand, a south-temperate island group with a highly endemic avifauna of largely independent origin from that of tropical Oceania. As skillfully developed by Richard Holdaway, Phillip Millener, Trevor Worthy, and others, the Quaternary fossil record of New Zealand portrays an extraordinary assemblage of passerine and nonpasserine birds. In both the Hawaiian Islands and New Zealand, avian extinction does resemble that elsewhere in Remote Oceania in having little or no evidence of loss before human arrival, after which the avifauna begins to fall apart (James et al. 1987, Millener 1990, James & Olson 1991, Olson & James 1991, Holdaway et al. 2001, James 2001, Worthy & Holdaway 2002).

All evidence, modern or prehistoric, points to Old World (Papuan) affinities for the landbirds everywhere in Oceania except the Hawaiian Islands. The rich Neotropical avifaunas have had no influence on Pacific islands west of Juan Fernandez, Galápagos, Cocos, and the various Mexican islands, all of which lie relatively close to the American mainland. Given the general east-to-west direction of winds and ocean currents in Oceania, much colonization by birds (and most other groups of organisms) has been against the prevailing wind and current. This applies even to weakly flying species such as rails.

The Papuan relationships of birds in Oceania testify to the importance of pure distance. The island groups in Oceania closest to the Neotropics are 5400 km west of northern Chile (Pitcairn) and 5700 km west of central Peru (Marquesas). These distances ignore Easter Island, which is highly isolated both to the east and west, and where the geographic affinities of landbirds are unresolved. By contrast, birds today potentially could reach the eastern end of Solomon Main Chain from New Guinea's Huon Peninsula through the Bismarcks, without crossing more than 174 km of ocean (Table 2-1). During Pleistocene glacial intervals, the minimum interisland distance would have been only slightly less, perhaps 157 km. Isolation values in Remote Oceania are much greater. For example, getting from the Solomons to Viti Levu in Fiji, via the Santa Cruz Group and Vanuatu, would involve a major ocean crossing of 840 km today. In glacial times, this would have been reduced to 530 km (Table 2-1) or, via currently inundated islands east of Santa Cruz, perhaps only ca. 300 km (Gibbons & Clunie 1986; Figure 1-43 herein).

With maximum efficiency in island hopping, one could reach all but the most remote outposts of West Polynesia from Viti Levu (or vice versa) without crossing more than 300 km of ocean, a distance little changed in glacial times once beyond Fiji. A substantial ocean gap of 1100 km separates East Polynesia (Cook Islands) from West Polynesia (Niue). To reach Tahiti from the Cook Islands would require a crossing of at least 520 km. From the Society Islands and Tuamotus, the Marquesas are at least 400 km away. In the most remote parts of southeastern Polynesia, ocean crossings as great as 400 km are needed to island-hop from Mangareva to Henderson Island.

The avifaunas of Melanesia (Bismarcks, Solomons, Vanuatu, New Caledonia; Chapter 5) are the richest ones in Oceania, a clear result of old age and nearness to New Guinea and Australia. Species-level endemism is high at the archipelago or regional levels, whereas generic level endemism is low among Melanesian landbirds (11 of 128; Table 5-1). Flightless species of rails evolved in *Gallirallus*, *Porzana*, *Pareudiastes*, and *Porphyrio*. Studies of prehistoric extinction in Melanesia are in their infancy. In Near Melanesia, where people first arrived 33,000+ years ago, considerable losses have been found on New Ireland (Bismarcks) and Buka (Solomons), with the greater proportion of extinct species on Buka (65%, vs. 20% on New Ireland) perhaps due to its smaller size and less rugged terrain. The New Ireland data demonstrate that major prehistoric losses of birds occurred on large, high, mostly forested islands close to New Guinea. This would suggest further that even immense New Guinea, which was occupied by people longer than Near Oceania (Groube et al. 1986, Spriggs 1997), has suffered as yet undetected losses of birds since human arrival. In Remote Melanesia, the prehistoric extinction of birds was, as far as is known, extensive in New Caledonia but less so in Vanuatu. The diverse losses in Melanesia include herons, hawks, megapodes, quail, kagu, rails, snipes, pigeons, parrots, barn-owls, strigid owls, owlet-nightjars, hornbills, starlings, and crows.

At any taxonomic level, the landbird faunas of West Polynesia (Fiji, Tonga, Samoa, Rotuma, Wallis & Futuna, Niue; Chapter 6) are less rich than in Melanesia but richer than in East Polynesia. Species-level endemism again is high within or among island groups. Of 54 landbird genera, 10 are endemic to West Polynesia (the megapode *Megavitiornis*, rail *Vitirallus*, pigeons *Natunaornis*, *Didunculus*, and undescribed genus C, parrot *Prosopeia*, undescribed white-eye, monarch *Lamprolia*, warbler *Trichocichla*, and honeyeater *Foulehaio*; Table 6-1). Flightless rails were widespread only in *Gallirallus* but also included species from four other genera. Extinction and extirpation in West Polynesia, which are best documented in Tonga, begin with the arrival of Lapita peoples at ca. 2850 cal BP. These losses involved herons, hawks, ospreys, megapodes, rails, pigeons, parrots, cuckoos, and passerines. A typical Tongan island sustained 30+ species of landbirds at human contact, compared to 10-14 species today.

The isolation of East Polynesia (Cook Islands, Tubuai, Society Islands, Tuamotus, Marquesas, Pitcairn Group, Easter Island; Chapter 7) is reflected in the absence of many avian genera and families found in West Polynesia. Species-level endemism at the archipelago or regional level remains high in East Polynesia, although only one (the monarch *Pomarea*) of its 23 genera of landbirds may be endemic to the region; an extinct pigeon from Henderson (Pitcairn Group) and one or two extinct parrots from Easter Island may eventually be shown to represent endemic genera as well (Table 7-1). Modern East Polynesian avifaunas are, on average, the most anthropogenically depleted ones in Oceania. Rails, pigeons, and parrots suffered the worst, although herons, ducks, sandpipers, swifts, kingfishers, and passerines also were lost. Most of the extinction and extirpation occurred only 1000 to 500 years ago. Other species and populations survived into the past two centuries but now are gone or nearly so. Flightless species of rails, endemic to single islands, were widespread in *Porzana* and *Gallirallus*.

Micronesia and the Remote Central Pacific Island Groups (Palau, Yap, Marianas, Chuuk, Pohnpei, Kosrae, Marshall Islands, Kiribati, Tuvalu, Tokelau; Chapter 8) have substantial species-level endemism at the archipelago or regional level. Four or five of the 44 genera are endemic (the owl *Pyrroglaux*, the white-eyes *Rukia*, *Megazosterops*, and *Cleptornis*, and possibly the monarch *Metabolus*). Many Micronesian island groups are dominated by atolls and have highly depauperate modern landbird faunas. The human-caused loss of birdlife on atolls is uncertain but may have been considerable. Avian losses in Micronesia include falcons, megapodes, rails, pigeons, parrots, swifts, and passerines. Flightless rails were widespread in *Porzana* and *Gallirallus*. At least in the Mariana Islands, which has the region's best prehistoric record, most avian extinction did not rapidly follow first human contact (ca. 3500 cal BP), but instead occurred only after rats were introduced ca. 2500 years later.

Taxonomic Perspective

Megapodes (Chapter 9) have a decent fossil history in Oceania because these large, ground-dwelling birds are easy to hunt, especially near their subsurface nesting grounds. One genus (*Megapodius*) with four allopatric species inhabits Oceania today, in a very discontinuous range from the Bismarcks to Tonga, skipping New Caledonia, Fiji, and all of Micronesia except Palau and the Marianas. Three genera (two extinct, one extant), 16 species (12 extinct, four extant), and 41 extirpated island populations of megapodes have been recorded by bones in Oceania. If the data were complete, these values would probably be in the range of

20 to 30 extinct species and hundreds of lost populations. The extinct megapodes ranged from a tiny, quail-sized species in Tonga to the giant, flightless *Sylviornis neocaledoniae* in New Caledonia, standing 1.2 to 1.4 m high and weighing 30 to 40 kg. Another large (ca. 6-8 kg), flightless megapode (*Megavitiornis altirostris*) once lived in Fiji. Before people showed up, megapodes lived across Melanesia and Western Polynesia as far east as Niue, and in Micronesia at least as far as east as Pohnpei. No East Polynesian island, even Mangaia, which is relatively close to Niue and has many fossils, has yielded evidence of megapodes. Islands or island groups now lacking megapodes but inhabited by them in the past include New Caledonia, the Reef Islands, Fiji, Samoa, Niue, Pohnpei, and the Ha'apai Group, Tongatapu, and 'Eua in Tonga. When people arrived, at least three species of megapodes were sympatric on New Caledonia and on many islands in Tonga and probably elsewhere in West Polynesia and Melanesia. Nowhere in Oceania does more than one species of megapode occur today. Low, flat Tongan islands at least as small as 1.8 km^2 were able to sustain three species of *Megapodius* before human impact.

Because they had evolved endemic flightless species on islands of any reasonable size, rails (Chapter 10) have lost more species to human impact than any other family of birds in Oceania. The extinct flightless rails (in *Porzana*, *Gallirallus*, *Vitirallus*, *Pareudiastes*, and *Porphyrio*) ranged from tiny crakes (*Porzana* spp.) weighing only 20 g to giant swamphens (*Porphyrio* spp.) that stood 60 cm high and weighed 4 kg. The most widespread genus was *Gallirallus*, with flightless species endemic to single islands from the largest in Near Oceania (New Britain) to remote atolls such as Wake, and high islands as far east as the Marquesas. The genus *Porzana* had flightless species from the Marianas and Vanuatu to remote Easter Island, although fossil evidence suggests a less continuous distribution of allopatric, endemic, flightless species than in *Gallirallus*. I crudely estimate that from 444 to 1579 species of flightless rails once existed in tropical Oceania. Much of the variation in this estimate is due to a poor fossil record, especially on atolls. Only five to seven of these flightless species survive (on New Britain, several of the Solomon Islands, Guam, and Henderson). The flightless rails in Near Oceania evolved alongside indigenous rodents, whereas no mammals except bats ever existed in Remote Oceania. The presence or absence of avian predators (herons, hawks, falcons, owls) seems not to have influenced the loss of flight in rails. There is, however, a mutually exclusive distribution of flightless rails and indigenous placental carnivores (order Carnivora).

The modern species richness of pigeons and doves (Chapter 11) exceeds that of any other landbird family in Oceania. Since human arrival, many additional species of columbids have become extinct, including ground-dwellers (in *Caloenas*, *Microgoura*, *Gallicolumba*, *Natunaornis*, and two undescribed genera), canopy species (in *Ducula* and *Ptilinopus*), and mid-level foragers such as *Didunculus*, *Macropygia*, and perhaps the extinct undescribed genus C. Columbid diversity and species richness were and still are greatest in Near Oceania, with up to 10 genera and 20 species on single islands in the Bismarcks and nine genera and 15 species in the Solomons. These values drop to six and eight in Vanuatu and New Caledonia. Individual West Polynesian islands typically sustain from two to four genera and two to five species of columbids today, but once had from four to six genera and six to nine species. In East Polynesia, these values are from zero to three genera and zero to three species today versus three or four genera and five or six species at human contact. Little extinction of columbids has been found in Micronesia outside of the Marianas, where *Ducula oceanica* and an extinct species of *Gallicolumba* used to occur. Extinct species of *Caloenas*, *Gallicolumba*, *Didunculus*, *Macropygia*, *Ducula*, and *Ptilinopus* extend the geographic range of these genera to island groups without living representatives. For *Gallicolumba*, *Ducula*, and *Ptilinopus*, congeneric species pairs once lived on many islands in Remote Oceania where zero or one species is present today. Because columbids are important dispersers of plant propagules, the loss of so many species and populations undoubtedly has affected the intra- and interisland distribution of trees and shrubs.

Parrots (Chapter 12) taste good, have colorful feathers, are entertaining, and are easy to find in their cavity nests. Thus parrots have not fared well on Pacific islands since human arrival. In Remote Oceania, it almost seems a miracle that any parrots survive. None of the six extant species of Polynesian lorikeets (*Vini* spp.) is sympatric today, but in the past as many as three species of *Vini* (including two extinct species larger than any extant forms) coexisted on single East Polynesian islands. Cockatoos (*Cacatua* spp.) and eclectus parrots (*Eclectus* spp.), confined to Indonesia, New Guinea, and Near Oceania today, both lived in Remote Oceania prehistorically. Fossils have added no information on the former distribution of *Cyanoramphus*, which consists of small granivorous parrots with an enigmatic modern range in the New Zealand region, New Caledonia, and the Society Islands.

Among other families of nonpasserine landbirds in Oceania (Chapter 13), those most affected by people are herons (*Butorides*, *Nycticorax*), ducks (*Dendrocygna*, *Anas*), hawks (*Accipiter*), ospreys (*Pandion haliaetus*), and barn-owls (*Tyto*), although lost species or populations have been discovered in nearly every nonpasserine family with a fossil record, including grebes (*Tachybaptus*), falcons (*Falco*), quail (*Coturnix*), kagu (*Rhynochetos*), sandpipers (*Prosobonia*, *Coenocorypha*), strigid owls (*Ninox*, *Nesasio*), owlet-nightjars (*Aegotheles*), swifts (*Collocalia*), hornbills (*Aceros*), and kingfishers (*Halcyon*). A radiation of endemic species of *Nycticorax*, all now extinct, has been discovered in the Solomon Islands, Tonga, Niue, and Cook Islands. These night-herons may have been important terrestrial predators on Pacific islands in prehuman times. Although fossils have increased the distribution and diversity of *Accipiter* in Melanesia and West Polynesia, there is no evidence of resident hawks in East Polynesia or Micronesia. Residing today in Near Oceania and New Caledonia, the discovery that *Pandion* once inhabited Vanuatu and Tonga suggests that osprey bones eventually will be found in Fiji. *Tyto alba*, a widespread barn-owl in Melanesia and West Polynesia today, is absent from precultural fossil deposits in New Caledonia and Tonga and may not have colonized Remote Oceania until people brought rats to these islands. New Caledonia's fossil record includes extinct species of cf. *Coenocorypha*, *Ninox*, *Aegotheles*, and *Aceros*, genera that do not occur there today.

Passerines or songbirds (Chapter 14) also have been depleted by human activities in Oceania. Understory species, such as thrushes (*Turdus*), shrikebills (*Clytorhynchus*), and monarchs (*Pomarea*), have done particularly poorly alongside people, with nonnative predators and habitat alteration probably being mostly to blame. Other passerine losses involve tasty frugivores of the canopy, such as starlings (*Aplonis*) and crows (*Corvus*). On fossil-rich 'Eua (Tonga), at least 10 species of passerines once existed, compared to just three or four today. The lost species include insectivores, omnivores, frugivores, and nectarivores with conspecifics or nearest relatives in the Solomons, Vanuatu, New Caledonia, Fiji, or Samoa. Even considering the extinct forms, passerines make up a smaller percentage of species in the landbird communities of Oceania than on any continent. Most Papuan and Australian species of passerines are poor dispersers to oceanic islands. Of northeastern New Guinea's 31 families of songbirds, only 19 have any representation in Oceania. Just five families (warblers, white-eyes, monarchs, honeyeaters, and starlings) accommodate more than 50% of the passerine species in Oceania.

All families of tropical seabirds (Chapter 15) also have undergone substantial anthropogenic losses. Shearwaters and petrels have been the most adversely affected, with 100 lost populations already documented in Oceania. Compared to landbirds, the prehistoric depletion of seabirds has involved proportionally more extirpation of island populations than extinction of species. Perhaps no species of seabirds in Oceania today has a geographic range or population size as large as it was at human contact. Seabird bones are abundant in most prehistoric sites in Remote Oceania but are much scarcer in Near Oceania, where native rodents and monitor lizards, as well as a more diverse snake fauna, may have restricted nesting. Lying at the southeastern apex of Polynesia, Easter Island sustained more species of seabirds (25+) than any other island in Oceania or perhaps anywhere. Only one species of seabird still nests on Easter Island itself, while several others occur on offshore islets. Seabirds once nested in profusion on Pacific islands of all sizes; our concept of steep, offshore islets being more suitable for seabird breeding is an artifact of human impact.

Conceptual Perspective

To me, extinction (Chapter 16) is the permanent loss of a species. The permanent loss of a subspecies or distinct population is extirpation, a process that leads to extinction. Biogeographers often use "extinction" to mean the temporary absence or "disappearance" of a vagile species from an island or habitat patch. This is an inappropriate use of the word, especially in these times of biodiversity crisis. Genuine extinction (and extirpation) affects every aspect of avian biogeography in Oceania because probably no island in the tropical Pacific has been spared anthropogenic loss of species. The causes of these losses, whether modern or prehistoric, seem obvious (habitat alteration, predation from humans, rats, dogs, cats, etc., and disease) but are based more often on speculative inference from common sense than direct evidence. Finding the burned, broken bones of an extinct species of bird in an archaeological site may be evidence that it was hunted, butchered, and eaten, but does not show that hunting was the primary or even an important cause of the loss. The intensity of avian extinction and extirpation varies widely depending on the abiotic, biotic, and cultural traits of an island. Relatively low, flat islands have experienced more losses than higher islands, where areas of rugged relief hinder hunting and deforestation.

Two of the least understood biological processes in birds or other organisms in Oceania are colonization and evolution (Chapter 17). Colonization is influenced by the ability to disperse across the ocean. Isolation needs to be evaluated between continents and islands, among island groups, and among islands within an archipelago. Each type of isolation is involved in the west-to-east decline in taxonomic diversity and species richness of birds in Oceania.

The great distances from Asia and America might explain, for example, why neither cardueline or emberizine finches have radiated in the main part of Oceania as in the Hawaiian Islands, Galápagos Islands, and West Indies (Bowman 1961, Steadman 1982, Grant & Grant 1986, 1989, Steadman & Morgan 1985, James & Olson 1991,

James in press). In fact, nowhere in Oceania do we find interisland differences among closely related species of passerines that involve major changes in bill morphology, as in the spectacular radiations of finches in the island groups just mentioned.

Many aspects of evolution in island birds are unknown, such as when most species colonized, whether subsequent morphological and genetic changes occurred rapidly or slowly, and what sort of founder effects and selective forces may drive the changes. Fossils have yet to record morphological or molecular changes within a single lineage of landbirds in Oceania. Studies of ancient DNA have some promise in this regard (e.g., Slikas et al. 2002). Issues of speciation are complicated further by differences between the biological and the phylogenetic species concepts, neither of which explains all inter- or intraspecific variation in island birds, even if the former concept is more functional.

The equilibrium theory of island biogeography remains incompatible with evolution. Constant comings and goings of species on an island do not promote speciation, which presumably requires genetically isolated populations. That it is so near to rich continental source areas (Java, Sumatra) is why the famous Krakatau natural experiment, ably reviewed by Thornton (1996), has limited pertinence in Remote Oceania.

Turnover and equilibrium (Chapter 18) are closely related topics that concern rates of colonization and extinction over time. Equilibrium is the central theme of biogeographic theory (MacArthur & Wilson 1963, 1967), based on the idea that comings (colonizations) and goings (often called extinctions, but see above) of species on an island will balance out over time to yield a fairly uniform number of species over the long run. In essence, species composition changes through time but species richness does not. Interspecific competition is believed to be a primary cause of the goings. David Lack (1947) invoked such competition as fueling the adaptive radiation of Darwin's finches (*Geospiza* s.l.) in the Galápagos Islands. Ironically, MacArthur and Wilson played down adaptive radiations on islands; their equilibrium model seemed instead to be based on timescales (unspecified) that did not consider evolutionary changes. The "unified neutral theory of biodiversity and biogeography" (Hubbell 2001, Volkov et al. 2003) attempts to incorporate speciation into an equilibrium model that has yet to be tested on true oceanic islands. Regardless, I am not optimistic that much progress will result from a "neutral" theory where individuals of all species are equal, featureless entities stripped of their biological attributes.

The fossil record of birds in Oceania provides very little evidence of nonanthropogenic turnover. Instead, we see relatively stable avifaunas until the arrival of people causes extinction. This argues for turnover rates that are several orders of magnitude lower than those proposed by ecologists. High islands (other than active volcanoes) had relatively stable avifaunas until people caused true extinction (loss of a species) and extirpation (loss of a subspecies or population) without the offsetting colonization by other species required for turnover.

Species-area relationships (Chapter 19) have been one of the most studied aspects of island biogeography because the data involved seem easy to compile and compare. In reality, we lack rigorous values even for modern species richness in most groups of organisms on most Pacific islands. Turn off your computer and help to change this appalling state of affairs, whether it's mosses, ferns, angiosperms, flatworms, spiders, beetles, microlepidopterans, thrips, snails, lizards, or birds that tickle your fancy. Although only an ornithologist, I have added species to the known herpetofauna on every Pacific island where I've spent time truly searching for lizards. Beyond modern surveys, we now have the sobering and growing evidence for extinction that took place (at least for landsnails, reptiles, bats, and birds) before scientists explored Oceania. Thus the raw data that have, or could have, been used to assess species-area relationships of these groups are, without exception, underestimates. For landbirds, modern species-richness values probably are 10 to 50% too low for large islands (>100 km^2) in Near Oceania. For most major islands (>10 km^2) in East or West Polynesia, modern values underestimate the true species richness by 50 to 100%.

Because extinction has been more severe on small islands (which often are low and flat) than larger islands, the slopes (z-values) of log-log species-area plots based on modern data are artificially too steep. Log-log plots are graphically pleasing, but the log of a species-richness value or the log of an island's area is, in itself, biologically meaningless. Within an island group in Oceania, the increase in species richness of birds with increasing land area is not, as is widely believed to be true, either a steady, linear progression in a log-log plot or an *S*-shaped curvilinear arrangement. Instead, this line has a steep slope among very small islands and soon levels off (i.e., z approaches 0) after a minimum land area (MLA) is attained. MLA values undoubtedly will vary among island groups, with an overall trend of decreasing MLA with increasing distance from New Guinea. In Tonga, islands as small as 1.8 km^2 may have had nearly all available species of landbirds at first human contact. In Near Oceania, perhaps 100 km^2 would be a reasonable though highly speculative estimate for MLA. An island of this size is more than two orders of magnitude smaller than New Britain, the largest island in Near Oceania. Once the MLA is reached, variation in species richness within an island group is more likely to be due to isolation than island area. Determining the MLA for birds in various archipelagos would be useful in planning translocations (Chapter 21).

One aspect of community ecology (Chapter 20) is the development of guidelines or "assembly rules" to explain how species are able to coexist. For birds on oceanic islands, these guidelines have been compromised by imperfectly known past and present distributions. Many cases of "forbidden combinations" or "checkerboard distributions" are invalidated when prehistory is considered. Assembly rules usually assume that interspecific competition can be intense, and that species regularly colonize new islands as well as disappear (over unknown periods of time), as predicted by equilibrium theory. For birds, I believe that interspecific competition generally is trivial and that turnover rates on oceanic islands have been exaggerated. Furthermore, any attempt to formulate universal "rules" for community structure breaks down when applied to situations that are taxonomically or geographically diverse. Sympatric congeneric species of landbirds in Remote Oceania seem to be restricted to very widespread genera, perhaps a result of the Lack model of allopatric speciation followed by dispersal leading to sympatry.

Conservation biology in Oceania (Chapter 21) suffers from inadequate data (on inter- and intra-island distribution, relative abundance, feeding and nesting ecology, pathogens, etc.), low funding, public and private apathy or antagonism, and variable efforts by biologists and agencies who may be reluctant to work on uninhabited islands, hesitant to implement programs with little precedent, and unwilling to address the high rates of human population growth in most island groups. Some conservationists are more likely to tackle bogus issues such as scientific collecting (which actually benefits species) rather than the more difficult problems that do deplete bird populations, such as human population growth. The future of many species of birds in Near Oceania may not be bleak because of rugged topography (which hinders deforestation) and malaria (which brutally controls human population density), although widespread logging in the lowlands assaults the richest part of an island's avifauna. The situation is even worse in Remote Oceania, where I suspect that much of the native birdlife that still exists will be lost over the next century or two. Some of this extinction can be offset by translocation, predator control, habitat restoration, and prohibiting human habitation on certain islands.

Suggestions for Future Research

Now I will attempt to entice the exhausted reader with some ideas for new research on the biogeography of Pacific island birds. I make these suggestions in light of the major empirical and conceptual gaps that remain. Of course the two types of gaps are intimately related, although perfect data never will solve all of the conceptual mysteries surrounding the highly varied birds that inhabit(ed) tropical Pacific islands, nor will conceptual progress ever be such that we can abandon data. Many secrets of island birds will remain safe, probably forever. For the extinct species, we have only bones to study. Bones can instruct us about morphology, systematics, past distributions, and time of occurrence. Bone-based inferences about past relative abundance, nesting biology, food habits, vocalizations, plumages, molt cycles, social behavior, interspecific interactions, etc. will remain speculative.

First and foremost, let me say that we simply need more high-quality data on tropical Pacific birds, whether paleontological or neontological. This simplistic statement is not a cop-out. How will we ever know which species of birds once lived on Tahiti until some prehistoric bird bones are found and identified on this famous island? How can we say which species of birds still live on Fonualei, an active volcano in northern Tonga, as long as no biologist has ever landed there, taken notes on birds, and published them. How can we make an intelligent plan to save the ground-dove *Gallicolumba erythroptera* from extinction when its feeding and breeding ecology are so poorly understood? I could ask a hundred other similar questions.

The fossil record is fascinating, important, and inadequate. Many of the extinct species of herons, megapodes, rails, pigeons, parrots, and passerines now known from Oceania are undescribed. I am guilty of creating much of this backlog. I plan to spend a lot of time in the lab over the coming years so that these species (and genera) become properly described, which also is a step toward developing phylogenies for landbirds in Oceania. Such phylogenies are needed to address evolutionary questions; they should incorporate both molecular and morphological data whenever possible.

Tonga has more islands (seven) with respectable samples of identified landbird bones than any other island group, most of which have fossils from two or fewer islands. Fossils are scarce or lacking in Palau, Samoa, Tuvalu, Tokelau, Kiribati, Tuamotus, and the 10 largest islands in the Solomons. Four of these groups (Tuvalu through Tuamotus) consist solely or primarily of atolls. This leads to one of the great mysteries in Pacific biogeography–what was the nature of landbird and other terrestrial communities on atolls before the arrival of people?

Being small, flat, and low, atolls offer little resistance to the human activities that reduce bird populations. The most vulnerable seabirds and landbirds on atolls probably were exterminated shortly after human colonization. The vegetation on most or all Pacific atolls is anthropogenic to some extent. Indigenous coastal (littoral) plants provide poor nesting and feeding habitat for most extant species of landbirds in Oceania (Steadman & Freifeld 1998). We do not know, however, to what extent upland species of trees and shrubs may have lived on atolls when people arrived. Another possibility is that most species of birds that were adapted to coastal forests, on atolls or otherwise, have been wiped out since human colonization. While it seems unlikely that the original forests and avifaunas of atolls would have been as rich as those on volcanic or raised limestone islands, we simply lack the plant and bird fossils to reconstruct atoll biotas. This is a major challenge for future research.

Most atolls were inundated by an interglacial high sea level stand ca. 130 to 120 ka (Chapter 1). Through most of their subsequent history, however, the islands called atolls today were raised limestone islands with elevations as high as 120 m. They became atolls only with the return to interglacial conditions. During this last major rise in sea level from 18 to 10 ka, we do not know the extent of floral and faunal depletion as raised limestone islands were converted to much lower, smaller atolls. Furthermore, most atolls south of the equator probably were obliterated or nearly so during the mid-Holocene (ca. 6 to 3 ka) because vertical growth of corals did not keep pace with the rising sea (Dickinson et al. 1994, Dickinson 1999a, 2003, 2004a). Perhaps most atolls inundated in the mid-Holocene sustained only seabirds and the most trampy landbirds when people arrived within the next several millennia.

Because of lithospheric flexure, some atolls in the northern hemisphere would not have drowned in the Holocene. These would include Wake and Kirimati (Christmas), where endemic landbirds survived even into historic times (Chapter 8). Given the evidence for rapid past changes in sea level, the increased CO_2 in the atmosphere today raises fear about the tragic possibility of a meter-scale rise in sea level in the near future (Kindler & Hearty 2000, Miller & Douglas 2004, Thomas et al. 2004, Meehl et al. 2005), threatening terrestrial life on many atolls. We American gas guzzlers are largely to blame.

Extinction of birds on most atolls may have been so rapid that few bones were left behind. Evidence of this comes from the Ha'apai Group of Tonga, where most islands are not true atolls but are low (mostly <20 m elev.) raised limestone islands within a barrier reef. Just like atolls, currently flat islands in Ha'apai would have been higher, much larger limestone islands in glacial times. Archaeological sites with Lapita-style pottery are known from five islands in Ha'apai, including Lifuka (11.4 km^2, elev. 16 m), where the terrestrial vertebrate fauna changed drastically after people arrived ca. 2850 years ago (Steadman et al. 2002a). The lower strata contain decorated Lapita-style pottery and bones of extinct iguanas and birds. The upper strata lack these materials and instead feature Polynesian plainware pottery and bones of extant species of vertebrates. Radiocarbon dates on individual bones of an extinct iguana and megapode, and the nonnative chicken, suggest that the loss of the first two species and the introduction of the last occurred within a century or so. The terrestrial environments of atolls probably were even more vulnerable than those on Lifuka. Most species of birds may have been eliminated on atolls within years or decades rather than centuries of human arrival.

Reasonable species lists of modern landbirds exist for most major Pacific islands. The data on species richness could be improved by visiting almost any island lacking an airstrip, or by visiting remote areas of large islands such as New Britain, New Ireland, Bougainville, Choiseul, Isabel, Guadalcanal, or Malakula. In the spirit of going beyond lists of species, an enormous amount of work needs to be done to determine relative abundances, habitat associations, and, with repeat surveys, changes in population size and distribution. In the Vava'u Group of Tonga, for example, in spite of being a hangout for international yachties, 10 of the 16 islands that we visited in 1995 never had been mentioned before in the ornithological literature (Steadman & Freifeld 1998). Surveys had not been conducted previously on any of these 16 islands, nor on any of the 13 islands in the Ha'apai Group of Tonga that I visited in 1995–96 (Steadman 1998). The good news (if you are willing to study birds on remote islands) or bad news (if you would like more complete data the distribution of Tongan birds) is that 63 islands in Vava'u and 49 islands in Ha'apai remain unsurveyed. They will stay that way as long as you are surfing the internet rather than wading through the surf to go ashore on some remote island. The invitation by Jonathan Sauer (1969:593) remains enticing and true: "At the moment there are still plenty of easily recognized species and many beautiful islands that no biogeographer has claimed."

Species-specific information helps to round out the picture, involving such topics as foraging and nesting ecology, social behavior, and spot mapping of territories. This is especially important for rare species. Just as for endangered human languages (Wuethrich 2000), time is running out to study some species of birds with badly diminished populations.

Why, for instance, does the Tongan whistler (*Pachycephala jacquinoti*) thrives only in submature and mature forests in Vava'u (Steadman & Freifeld 1998; Figures 6-6, 14-14, 14-15 herein)? As forest is degraded, are the whistlers affected by a scarcity of food, a lack of nest sites or display/singing perches, an increase in nonnative predators, a simple aversion to bright sunlight, or some other factor? We will not know until someone learns more about this bird than its distribution, relative abundance, and habitat preference.

Some Other Major Mysteries

When did any particular species colonize a certain island or island group? To answer this question we must find older fossil deposits. "Molecular clocks" may help as well, although meaningful calibration of such clocks might be difficult in many island groups. Given the geological variety of islands and taxonomic diversity of birds in Oceania, not to mention the uncertainty of source areas, it is doubtful that any one or even several calibrations could be widely applicable. Furthermore, if some form of founder effect regularly operates subsequent to colonization, a molecular clock might be a poor way to estimate the time of colonization. For example, suppose that most genetic (and morphological?) differentiation occurs within the first 1000 years after colonization, followed by long periods of little change. If mutation rates were low, you potentially could discover similar amounts of genetic (and morphological?) difference among populations that arrived on their respective islands 100,000 years ago, 10,000 years ago, and only 1000 years ago. These vastly different ages may be difficult to resolve with molecular data.

Oceania has some outstanding avian examples of evolutionary radiations that could be studied with molecules and morphology. Such analyses have yet to take place, although the work on monarchs (*Monarcha* spp.) by Filardi and Smith (2005) is a step in the right direction. When possible, extinct taxa should be included because "phylogenetic biogeographic studies on extant organisms that do not include fossil taxa can often be artificially incongruent and inaccurate" (Lieberman 2002:39). Kingfishers (*Halcyon* spp.) and whistlers (*Pachycephala* spp.) are two genera where enough taxa survive (Mayr 1932a, 1932b, 1940a, 1942; Chapters 13, 14 herein) that the data would not be horribly compromised by missing populations. A multifaceted look at evolution in these widespread birds almost surely would yield fascinating results, especially if grounded in good life history data (reproductive and feeding biology) and conducted by researchers not polarized within the "morphology vs. molecules" debate (see Gura 2000). Molecular data also hold great promise in identifying prehistoric bones that lack osteological characters (Sorenson et al. 1999, Fleischer et al. 2000, Fleischer & McIntosh 2001, Huynen et al. 2003, Nicholls et al. 2003, Willerslev et al. 2003).

We know essentially nothing about how natural selection affects interisland variation of landbirds in Oceania at the species or subspecies level. Why, for example, is the adult male of the fruit-dove *Ptilinopus victor* orange, whereas its presumed close relative *P. luteovirens* is bright yellow? Why is the starling *Aplonis tabuensis* dark overall with a yellow iris in northern Tonga (*A. t. nesiotes*, *A. t. tenebrosus*) but much lighter overall with an orangish brown iris on large islands in Fiji (*A. t. vitiensis*)? Why does the song of the cuckoo-shrike *Coracina tenuirostris* vary so much among populations in the Solomons? In Fiji as well as the Solomons, why does the whistler *Pachycephala pectoralis* have a white throat on some islands but a yellow one on others (Mayr 1932a,b)? Why does the kingfisher *Halcyon tuta* have a mostly white crown on Atiu (*H. t. atiu*) yet a mostly blue crown on nearby and similarly sized Ma'uke (*H. t. mauke*), another raised limestone island in the southern Cooks? Are any of these differences adaptive, or are they due only to founder effects or genetic drift? Nobody knows; this is difficult to study.

What about species-packing? Given that no island in Oceania supports more than one species of megapode today, how can we account for three species of *Megapodius* coexisting on low, flat islands as small as 1.8 km^2 in Tonga? How could three or four species of flightless rails coexist on Mangaia, or eight or nine species of columbids on single islands in Tonga, or three species of lorikeets (*Vini* spp.) on single East Polynesian islands? Did the megapodes and rails all forage over the same ground, consuming highly overlapping diets of invertebrates and fallen fruits? Did the nectar-feeding lorikeets all congregate on whichever trees and shrubs were producing nectar? Auffenberg and Auffenberg (1988) found broadly overlapping diets in 11 sympatric species of skinks in the Philippines, suggesting little dietary partitioning. Whether megapode or rail, lorikeet or lizard, was it advantageous on a Pacific island to be a generalist or a specialist? Either path could lead to coexistence.

Does competition in birds even exist? On a trivial level, it must, because two individual birds, regardless of species, cannot eat the same insect or seed, nor can they suck nectar from the same flower at the same time, nor can two pairs of birds nest in an identical place (brood parasites excepted). Is interspecific competition truly important, however, in structuring landbird communities on oceanic islands? Competition is untested among birds in Oceania. The indirect test of plotting modern distributions lacks historic perspective. I believe that the prehuman distribution of landbirds in Oceania was influenced much more by isolation than by competition.

A Final Word

The trend in science away from natural history and toward specialization and theory is out of control (Wilcove & Eisner 2000). Graduate students and upper-level undergraduates are funneled into theory-revising long before they know enough about nature even to evaluate whether what they're trying to support or refute has any biological merit in the first place. Our academic infrastructure has created legions of young biologists who scoff at description even though they cannot describe a natural situation themselves. Their professors are to blame. This has been going on for decades. It may be impossible to change.

How ironic that biologists with increasingly specialized backgrounds are searching for universal truths. Before writing me off as a weirdo, please note that I am not alone. "Attempting to shoe-horn explanation into a one-size-fits-all form represents nothing less than the dumbing-down of explanation" (Taylor 2001:419). I agree with Ernst Mayr (2004) that biololgy suffers from too much reductionism. The search for a grand unit in nature (a "theory of everything") has led to "macroecology," where organisms that operate differently from a highly massaged statistical norm are seen as exceptions that prove the rule. The data sets might involve range size, body size, population density, metabolic rates, or whatever (e.g., Carbone & Gittleman 2002, Olff et al. 2002, Brown 2001, Gillooly et al. 2001, Whitfield 2001). The underlying assumptions are enormous, and historical perspectives usually are absent or superficial. Perhaps in today's frenetic atmosphere of information overload, people need simple answers to complex problems. I agree, however, with Marquet (2001:482) that "macroecology is becoming stuck in contingent explanations for many of the large-scale patterns it addresses."

While I do not share the admirable view of Edward Wilson (1998) that a comprehensive theory eventually will unite the various academic disciplines (see Steadman 1999b), I enthusiastically believe that any field of study can benefit from any other, no matter how seemingly disparate. In this light I encourage biology students to dabble in geology or archaeology. Similarly, I nudge paleontology or archaeology students toward courses in biology and chemistry. I encourage students whose education is being funneled into theoretical issues of the behavioral ecology of birds to obtain a good background in, let's say, avian anatomy and systematics, or vice versa. Finding a school that even teaches anything rigorous about avian anatomy and systematics is not easy. Most students of systematics are more concerned with learning about cookbook molecular techniques and whimsical software packages than the ecology, behavior, physiology, paleontology, or anatomy of the species whose DNA is being analyzed.

Paleobotany now plays an essential role in addressing questions of angiosperm phylogeny and evolution (Dilcher 2000). I hope that fossils can become just as important for birds. Michael Rosenzweig (1995: xv) asked ecologists to "understand and integrate the information that comes from fossils." There is no longer any reason why paleontology and zooarchaeology should be regarded as anything but an integral, mainstream part of biogeography, whether on islands or continents (Jablonski et al. 2003). This will not happen until more scientists respond to Rosenzweig's plea and incorporate prehistoric information into their interpretation of neontological data. Although prehistory is my bread and butter, long ago I realized that I could not interpret it effectively without studying living birds as well.

Rather than look down our noses at scientists whose interests, methods, or philosophies don't coincide with our own, it's time that we all unplug, sit around the same campfire, pass the bottle, and swap lies. We may not solve the world's problems but we'll all learn a lot, and after the fire has died out and we're back home staring at computers, our research programs will improve. Molecules are OK, but so are whole organisms or macroscopic bits thereof. Most empiricists are nice people, as are most theorists. When all is said and done, few of us are purely one or the other. We're all trying to understand nature. That is a privilege to treasure and preserve.

Appendix

Ordinal and familial classification of all species of birds mentioned in the text, with scientific names and common names. The higher-level classification of birds is in extreme flux; the treatment presented here is conservative and fairly traditional. For living species (except columbids), the sequence generally follows that of Clements (2000), with nomenclature variously following Pratt et al. (1987), Clements (2000), and my own preferences. A huge amount of research on scientific nomenclature of birds from Oceania remains undone. The sequence of passerine families begins with non-oscines (Pittidae, Acanthisittidae) followed by the suborders Passerida (Hirundinidae through Icteridae) and Corvida (Campephagidae through Vireonidae). For most extinct species (†), the scientific nomenclature is mine, the common names are unimportant and made up, and the sequence of families, genera, and species is tentative at best, given how little we know about avian phylogenies, with or without extinct species. Commonly used alternative names are given in parentheses. Species names in brackets designate possible species groups, sometimes referred to as superspecies. *, landbird that is not indigenous to tropical Oceania now or prehistorically; m, probable or certain migrant (no breeding population known from tropical Oceania, now or in the past).

ORDER CASUARIIFORMES

Family Casuariidae—Cassowaries

Casuarius casuarius—southern cassowary
Casuarius bennetti—dwarf cassowary
**Casuarius unappendiculatus*—northern cassowary

ORDER PODICIPEDIFORMES

Family Podicipedidae—Grebes

Tachybaptus ruficollis—little grebe
Tachybaptus novaehollandiae—Australian grebe

ORDER PROCELLARIIFORMES

Family Diomedeidae—Albatrosses

Diomedea exulans—wandering albatross (m)
Diomedea epomophora—royal albatross (m)
Diomedea amsterdamensis—Amsterdam Island albatross (m)
Diomedea albatrus—short-tailed albatross
Diomedea leptorhyncha (irrorata)—waved albatross (m)
Diomedea immutabilis—Laysan albatross
Diomedea nigripes—black-footed albatross
Diomedea cauta—shy albatross (m)

Family Procellariidae—Petrels & Shearwaters

Macronectes sp.—giant petrels (m?)
Fulmarus glacialoides—southern fulmar (m?)
Pseudobulweria rostrata—Tahiti petrel
Pseudobulweria becki—Beck's petrel
Pterodroma macroptera—great-winged petrel (m?)
Pterodroma lessonii—white-headed petrel (m?)
Pterodroma alba—phoenix petrel
Pterodroma inexpectata—mottled petrel (m?)
Pterodroma solandri—Providence petrel (m?)
Pterodroma ultima—Murphy's petrel
†*Pterodroma* undescribed sp.—'Eua petrel
Pterodroma neglecta—Kermadec petrel
Pterodroma heraldica—herald petrel
Pterodroma phaeopygia—dark-rumped petrel
Pterodroma externa—Juan Fernandez petrel (m?)
Pterodroma cookii—Cook's petrel (m?)
Pterodrorma nigripennis—black-winged petrel
†Undescribed genus and species—Easter Island petrel
Pachyptila vittata—broad-billed prion
Pachyptila belcheri—slender-billed prion
Bulweria bulwerii—Bulwer's petrel
Bulweria fallax—Jouanin's petrel
†*Bulweria bifax*—St. Helena petrel
Procellaria sp.—large petrel
Puffinus carneipes—flesh-footed shearwater
Puffinus pacificus—wedge-tailed shearwater
Puffinus bulleri—Buller's shearwater
†*Puffinus pacificoides*—St. Helena shearwater
Puffinus griseus—sooty shearwater
Puffinus tenuirostris—short-tailed shearwater
Puffinus nativitatis—Christmas shearwater
Puffinus huttoni—Hutton's shearwater
Puffinus gavia—fluttering shearwater
†*Puffinus spelaeus*—cave shearwater
†*Puffinus* undescribed sp.—'Eua shearwater
Puffinus assimilis—little shearwater
Puffinus lherminieri—Audubon's shearwater

Family Hydrobatidae—Storm-Petrels

Fregetta tropica—black-bellied storm-petrel
Fregetta grallaria—white-bellied storm-petrel
Nesofregetta fuliginosa—Polynesian storm-petrel

ORDER PELECANIFORMES

Family Phaethontidae—Tropicbirds

Phaethon rubricauda—red-tailed tropicbird
Phaethon lepturus—white-tailed tropicbird

Family Sulidae—Boobies

Papasula abbotti abbotti—Abbott's booby
†*Papasula abbotti costelloi*—Abbott and Costello's booby
Sula dactylatra—masked booby
Sula sula—red-footed booby
Sula leucogaster—brown booby

Family Phalacrocoracidae—Cormorants

Phalacrocorax carbo—great cormorant
Phalacrocorax melanoleucos—little pied cormorant

Family Fregatidae—Frigatebirds

Fregata minor—great frigatebird
Fregata ariel—lesser frigatebird

Family Pelecanidae—Pelicans
Pelecanus conspicillatus—Australian pelican

ORDER CICONIIFORMES
Family Ardeidae—Herons
Ardea (Casmerodius) alba—great egret
Egretta intermedia—intermediate egret
Egretta novaehollandiae—white-faced heron
Egretta sacra—Pacific reef-heron
Butorides striatus—striated heron
Nycticorax caledonicus—rufous night-heron
†*Nycticorax* undescribed sp. A—Solomon Islands night-heron
†*Nycticorax* undescribed sp. B—Tongan night-heron
†*Nycticorax kalavikai*—Niue night-heron
†*Nycticorax* undescribed sp. C—Cook Islands night-heron
Ixobrychus sinensis—yellow bittern
Ixobrychus (Dupetor) flavicollis—black bittern
Botaurus poiciloptilus—Australasian bittern
Family Threskiornithidae—Ibises
Threskiornis molucca—Australian ibis
Platalea regia—royal spoonbill (m?)

ORDER ANSERIFORMES
Family Anatidae—Ducks
Dendrocygna guttata—spotted whistling-duck
Dendrocygna arcuata—wandering whistling-duck
**Dendrocygna autumnalis*—black-bellied whistling-duck
†cf. *Dendrocygna* undescribed sp.—Polynesian whistling-duck
Anas gibberifrons—Sunda teal
Anas gracilis—gray teal
Anas platyrhynchos—mallard (m)
Anas superciliosa—Pacific black duck
†*Anas oustaleti*—Marianas mallard
Anas acuta—northern pintail (m)
†*Anas [strepera] couesi*—Coues's gadwall
†cf. *Anatidae* undescribed sp.—Marianas flightless duck
Aythya australis—white-eyed duck

ORDER FALCONIFORMES
Family Pandionidae—Ospreys
Pandion haliaetus—osprey
Family Accipitridae—Hawks & Eagles
Aviceda subcristata—Pacific baza
Henicopernis infuscata—black honey-buzzard
Haliastur sphenurus—whistling kite
Haliastur indus—Brahminy kite
Haliaeetus leucogaster—white-bellied sea-eagle
Haliaeetus sanfordi—Solomon sea-eagle
**Harpyopsis novaeguinae*—New Guinea harpy eagle
Circus approximans—swamp harrier
Accipiter novaehollandiae—gray goshawk
Accipiter fasciatus—brown goshawk
Accipiter albogularis—pied goshawk
Accipiter haplochrous—New Caledonia goshawk
†*Accipiter quartus*—fourth goshawk
Accipiter rufitorques—Fiji goshawk
Accipiter luteoschistaceus—slaty-mantled goshawk
Accipiter imitator—imitator goshawk
Accipiter princeps—New Britain goshawk
Accipiter brachyurus—New Britain sparrowhawk
Accipiter meyerianus—Meyer's goshawk
†*Accipiter efficax*—powerful goshawk
**Buteo galapagoensis*—Galápagos hawk
Family Falconidae—Falcons
**Falco punctatus*—Mauritius kestrel
Falco severus—Oriental hobby
Falco berigora—brown falcon
Falco peregrinus—peregrine falcon

ORDER GALLIFORMES
Family †Quercymegapodiidae—Proto-Megapodes
†**Quercymegapodius depereti*—Deperet's proto-megapode
†**Quercymegapodius brodkorbi*—Brodkorb's proto-megapode
†**Ameripodius silvasantosi*—Brazilian proto-megapode
†**Ameripodius alexis*—French proto-megapode
Family Megapodiidae—Megapodes
†**Ngawupodius minya*—Lake Pinpa pygmy megapode
*†*Progura gallinacea*—Australian Pleistocene megapode
**Alectura lathami*—Australian brush-turkey
**Aepypodius arfakianus*—wattled brush-turkey
**Talegalla cuvieri*—red-billed talegalla
**Talegalla jobiensis*—brown-collared talegalla
**Leipoa ocellata*—malleefowl
**Macrocephalon maleo*—maleo
**Eulipoa wallacei*—Moluccan megapode
**Megapodius cumingii*—Philippine megapode (scrubfowl)
**Megapodius decollatus*—New Guinea megapode
**Megapodius reinwardt*—orange-footed megapode
Megapodius eremita—Melanesian megapode (scrubfowl)
Megapodius layardi—Vanuatu megapode (scrubfowl)
Megapodius pritchardii—Tongan megapode (scrubfowl)
Megapodius laperouse—Micronesian megapode (scrubfowl)
†*Megapodius alimentum*—consumed megapode
†*Megapodius amissus*—lost megapode
†*Megapodius molistructor*—pile-builder megapode
†*Megapodius* undescribed sp. A—large Bismarcks megapode
†*Megapodius* undescribed sp. B—large Solomon Islands megapode
†*Megapodius* undescribed sp. C—large Vanuatu megapode
†*Megapodius* undescribed sp. D—New Caledonia megapode
†*Megapodius* undescribed sp. E—Loyalty megapode
†*Megapodius* undescribed sp. F—small-footed megapode
†Genus uncertain, undescribed sp.—stout Tongan megapode
†*Megavitiornis altirostris*—deep-billed megapode
†*Sylviornis neocaledoniae*—giant flightless megapode
Family Phasianidae—Quails, Chickens, Turkeys
Coturnix ypsilophorus—brown quail
Coturnix chinensis—blue-breasted quail
**Gallus gallus*—red junglefowl (chicken)
**Meleagris ocellata*—ocellated turkey

ORDER GRUIFORMES

Family Turnicidae—Buttonquail

Turnix maculosa—red-backed buttonquail
Turnix varia—painted buttonquail
Turnix novaecaledoniàe—New Caledonia buttonquail

Family Gruidae—Cranes

**Grus americana*—whooping crane

Family Rallidae—Rails

Rallina tricolor—red-necked crake
Rallina eurizonoides—slaty-legged crake
**Rallus limicola*—Virginia rail
**Gallirallus pectoralis*—Lewin's rail
**Gallirallus mirificus*—brown-banded rail
**Gallirallus striatus*—slaty-breasted rail
**Gallirallus torquatus*—barred rail
Gallirallus philippensis—buff-banded rail
Gallirallus (Nesoclopeus) [woodfordi] woodfordi—Woodford's rail
Gallirallus (Nesoclopeus) [woodfordi] immaculatus—Isabel rail
Gallirallus (Nesoclopeus) [woodfordi] tertius—Bougainville rail
†*Gallirallus (Nesoclopeus) poecilopterus*—bar-winged rail
Gallirallus rovianae—Roviana rail
**Gallirallus australis*—weka
*†*Gallirallus dieffenbachii*—Dieffenbach's rail
†*Gallirallus lafresnayanus*—New Caledonian rail
**Gallirallus sylvestris*—Lord Howe rail
**Gallirallus okinawae*—Okinawa rail
Gallirallus (Habroptila) wallacii—invisible rail
Gallirallus insignis—New Britain rail
Gallirallus owstoni—Guam rail
†*Gallirallus wakensis*—Wake rail
†*Gallirallus vekamatolu*—'Eua *rail*
†*Gallirallus huiatua*—Niue rail
†*Gallirallus ripleyi*—Ripley's rail
†*Gallirallus pacificus*—Tahiti rail
†*Gallirallus* undescribed sp. A—New Ireland rail
†*Gallirallus* undescribed sp. B—Buka rail
†*Gallirallus* undescribed sp. C—Toga rail
†*Gallirallus* undescribed sp. D—Lakeba rail
†*Gallirallus* undescribed sp. E—Ha'afeva rail
†*Gallirallus* undescribed sp. F—Tongatapu rail
†*Gallirallus* undescribed sp. G—Huahine rail
†*Gallirallus* undescribed sp. H—Nuku Hiva rail
†*Gallirallus* undescribed sp. I—Hiva Oa rail
†*Gallirallus* undescribed sp. J—Ua Huka rail
†*Gallirallus* undescribed sp. K—Tahuata rail
†*Gallirallus* undescribed sp. L—Rota rail
†*Gallirallus* undescribed sp. M—Aguiguan rail
†*Gallirallus* undescribed sp. N—Tinian rail
†*Gallirallus* undescribed sp. O—Saipan rail
*†*Cabalus modestus*—Chatham Island rail
*†*Capellirallus karamu*—snipe rail
*†*Diaphorapteryx hawkinsi*—Hawkin's rail
†*Vitirallus watlingi*—Viti Levu rail
Gymnocrex plumbeiventris—bare-eyed rail
Porzana tabuensis—spotless crake
†*Porzana monasa*—Kosrae crake
†*Porzana rua*—Mangaian crake
†*Porzana nigra*—Tahiti crake
Porzana atra—Henderson Island crake
†*Porzana* undescribed sp. A—Malakula crake
†*Porzana* undescribed sp. B—Aiwa crake
†*Porzana* undescribed sp. C—small Mangaian crake
†*Porzana* undescribed sp. D—Nuku Hiva crake
†*Porzana* undescribed sp. E—Ua Huka crake 1
†*Porzana* undescribed sp. F—Ua Huka crake 2
†**Porzana* undescribed sp. G—Easter Island crake
†**Porzana* undescribed sp. H—Rota crake
†**Porzana* undescribed sp. I—Aguiguan crake
†*Porzana* undescribed sp. J—Tinian crake
**Porzana pusilla*—Baillon's crake
†*Porzana keplerorum*—Kepler's crake
†*Porzana severnsi*—Severns' Crake
†*Porzana sandwichensis*—Hawaiian crake
†*Porzana palmeri*—Laysan crake
Poliolimnas (Porzana) cinereus—white-browed crake
**Amaurornis olivaceus*—plain bush-hen
Amaurornis moluccanus—rufous-tailed bush-hen
**Amaurornis phoenicurus*—white-breasted water-hen
Megacrex ineptus—New Guinea flightless rail
Gallinula chloropus—common moorhen
Gallinula tenebrosa—dusky moorhen
**Gallinula mortierii*—Tasmanian native-hen
*†*Gallinula hodgeni*—Hodgen's moorhen
**Gallinula nesiotis*—Tristan da Cunha moorhen
**Gallinula comeri*—Gough moorhen
(†?)*Pareudiastes sylvestris*—San Cristobal gallinule
†*Pareudiastes* undescribed sp.—Buka gallinule
†*Pareudiastes pacificus*—Samoan gallinule
Porphyrio porphyrio (poliocephalus)—purple swamphen
*†*Porphyrio mantelli*—North Island takahe
**Porphyrio hochstetteri*—South Island takahe
†*Porphyrio kukwiedei*—divine swamphen
†*Porphyrio* undescribed sp. A—New Ireland swamphen
†*Porphyrio* undescribed sp. B—Buka swamphen
†*Porphyrio* undescribed sp. C—Mangaia swamphen
†*Porphyrio* undescribed sp. D—Huahine swamphen
†*Porphyrio* undescribed sp. E—Mariana swamphen
†*Porphyrio paepae*—Marquesan swamphen
*†*Porphyrio albus*—Lord Howe swamphen
Fulica [americana] alai—[American] Hawaiian coot
*†*Fulica chathamensis*—Chatham Island coot
*†*Fulica prisca*—flightless New Zealand coot

Family Rhynochetidae—Kagus

Rhynochetos jubatus—inland kagu
†*Rhynochetos orarius*—coastal kagu

ORDER CHARADRIIFORMES

Family Jacanidae—Jacanas

Irediparra gallinacea—comb-crested jacana

Family Recurvirostridae—Stilts

Himantopus himantopus—black-winged stilt

Family Burhinidae—Thick-knees

Esacus magnirostris—beach thick-knee

Family Charadriidae—Plovers

Charadrius dubius—little ringed plover
Charadrius mongolus—Mongolian plover (m)
Charadrius leschenaultii—greater sandplover (m)

Pluvialis fulva—Pacific golden plover (m)

Family Scolopacidae—Sandpipers

†*Coenocorypha mirapacifica*—Fijian snipe
†*Coenocorypha?* undescribed sp.—New Caledonia snipe
Limosa lapponica—bar-tailed godwit (m)
Numenius phaeopus—whimbrel (m)
Numenius tahitiensis—bristle-thighed curlew (m)
Numenius arquata—Eurasian curlew (m)
Tringa erythropus—spotted redshank (m)
Heteroscelus incanus—wandering tattler (m)
Prosobonia parvirostris [cancellata]—Tuamotu sandpiper
†*Prosobonia cancellata*—Christmas sandpiper
†*Prosobonia leucoptera*—white-winged sandpiper
†*Prosobonia ellisi*—Ellis's sandpiper
†*Prosobonia* undescribed sp. A—Cook Islands sandpiper
†*Prosobonia* undescribed sp. B—Pitcairn sandpiper
Arenaria interpres—ruddy turnstone (m)
Calidris alba—sanderling (m)

Family Laridae—Gulls

Larus novaehollandiae—silver gull
†*Larus utunui*—Society Island gull
**Larus cirrocephalus*—grey-hooded gull
**Larus atricilla*—laughing gull
**Larus maculipennis*—brown-hooded gull
**Larus ridibundus*—black-headed gull

Family Sternidae—Terns

Sterna bergii—great crested tern
Sterna sumatrana—black-naped tern
Sterna hirundo—common tern (m)
Sterna paradisaea—Arctic tern (m)
Sterna lunata—gray-backed tern
Sterna anaethetus—bridled tern
Sterna fuscata—sooty tern
Anous minutus—black noddy
Anous stolidus—brown noddy
Procelsterna cerulea—blue noddy
Gygis candida—white [fairy] Tern
Gygis microrhyncha—least white [fairy] Tern

ORDER COLUMBIFORMES

Family Columbidae—Pigeons & Doves

*†*Ectopistes migratorius*—passenger pigeon
Columba livia—rock dove, rock pigeon (*)
Columba vitiensis—white-throated pigeon
Columba pallidiceps—yellow-legged pigeon
Macropygia amboinensis—slender-billed cuckoo-dove
Macropygia nigrirostris—black-billed cuckoo-dove
Macropygia mackinlayi—Mackinlay's cuckoo-dove
†*Macropygia arevarevauupa*—Society Island cuckoo-dove
†*Macropygia heana*—Marquesan cuckoo-dove
Reinwardtoena browni—pied cuckoo-dove
Reinwardtoena crassirostris—crested cuckoo-dove
Ptilinopus tannensis—Tanna fruit-dove
Ptilinopus superbus—superb fruit-dove
Ptilinopus perousii—many-colored fruit-dove
Ptilinopus porphyraceus—crimson-crowned fruit-dove
Ptilinopus pelewensis—Palau fruit-dove
Ptilinopus rarotongensis—Cook Islands fruit-dove
Ptilinopus roseicapilla—Mariana fruit-dove
Ptilinopus richardsii—silver-capped fruit-dove
Ptilinopus purpuratus—gray-green fruit-dove
Ptilinopus chalcurus—Makatea fruit-dove
Ptilinopus coralensis—atoll fruit-dove
Ptilinopus greyii—red-bellied fruit-dove
Ptilinopus huttoni—Rapa fruit-dove
Ptilinopus dupetithouarsii—white-capped fruit-dove
†*Ptilinopus mercierii*—red-moustached fruit-dove
†*Ptilinopus* undescribed sp.—Tubuai fruit-dove
Ptilinopus insularis—Henderson Island fruit-dove
Ptilinopus rivoli—white-breasted fruit-dove
Ptilinopus solomonensis—yellow-bibbed fruit-dove
Ptilinopus viridis—claret-breasted fruit-dove
Ptilinopus insolitus—knob-billed fruit-dove
Ptilinopus (Chrysoenas) victor—orange dove
Ptilinopus (Chrysoenas) luteovirens—golden dove
Ptilinopus (Chrysoenas) layardi—velvet dove
Drepanoptila holosericea—cloven-feathered dove
Ducula pacifica—Pacific imperial-pigeon
Ducula rubricera—red-knobbed imperial-pigeon
Ducula oceanica—Micronesian imperial-pigeon
Ducula aurorae—Polynesian imperial-pigeon
Ducula galeata—Marquesas imperial-pigeon
†*Ducula david*—David's imperial-pigeon
†*Ducula lakeba*—Lau imperial-pigeon
†*Ducula* undescribed sp.—Tongan imperial-pigeon
Ducula finschii—Finsch's imperial-pigeon
Ducula pistrinaria—island imperial-pigeon
Ducula latrans—Peale's imperial-pigeon
Ducula brenchleyi—chestnut-bellied imperial-pigeon
Ducula bakeri—Baker's imperial-pigeon
Ducula goliath—New Caledonian imperial-pigeon
Ducula melanochroa—Bismarck imperial-pigeon
Ducula spilorrhoa—Torresian imperial-pigeon
**Hemiphaga novaeseelandiae*—New Zealand pigeon
Gymnophaps albertisii—Papuan mountain-pigeon
Gymnophaps solomonensis—pale mountain-pigeon
†Undescribed genus and species C—royal Tongan pigeon
Chalcophaps indica—emerald dove
Chalcophaps stephani—Stephan's dove
Henicophaps foersteri—New Britain bronzewing
Caloenas nicobarica—Nicobar pigeon
†*Caloenas canacorum*—Kanaka pigeon
Didunculus strigirostris—tooth-billed pigeon
†*Didunculus* undescribed sp.—Tongan tooth-billed pigeon
Gallicolumba rufigula—cinnamon ground-dove
Gallicolumba jobiensis—white-bibbed ground-dove
Gallicolumba kubaryi—Caroline Islands ground-dove
Gallicolumba erythroptera—Polynesian ground-dove
†*Gallicolumba leonpascoi*—Henderson ground-dove
†*Gallicolumba* undescribed sp.—large Marianas ground-dove
Gallicolumba xanthonura—white-throated ground-dove
Gallicolumba stairi—friendly ground-dove
Gallicolumba sanctaecrucis—Santa Cruz ground-dove
†*Gallicolumba ferruginea*—Tanna ground-dove
*†*Gallicolumba norfolciensis*—Norfolk Island ground-dove
Gallicolumba salamonis—thick-billed ground-dove
Gallicolumba rubescens—Marquesas ground-dove
Gallicolumba beccarii—bronze ground-dove

Gallicolumba canifrons—Palau ground-dove
†*Gallicolumba nui*—immense ground-dove
†*Gallicolumba longitarsus*—long-tarsus ground-dove
(†?)*Microgoura meeki*—Meek's ground-pigeon
**Trugon terrestris*—thick-billed ground-pigeon
**Otidiphaps nobilis*—pheasant pigeon
†Undescribed genus and species A—small-winged ground-pigeon
†Undescribed genus and species B—Kilu ground-pigeon
**Goura victoria*—Victoria crowned-pigeon
†*Natunaornis gigoura*—Fiji flightless pigeon
*†*Raphus cucullatus*—dodo
*†*Pezophaps solitaria*—Rodriguez solitaire

ORDER PSITTACIFORMES
Family Psittacidae—Parrots
SUBFAMILY CACATUINAE—COCKATOOS
**Cacatua tenuirostris*—long-billed corella
**Cacatua sanguinea*—little corella
**Cacatua haematuropygia*—Philippine cockatoo
**Cacatua sulphurea*—yellow-crested cockatoo
Cacatua ducorpsii—Ducorps' cockatoo
**Cacatua leadbeateri*—pink cockatoo
**Cacatua [galerita] triton*—sulphur-crested cockatoo
Cacatua [galerita] ophthalmica—blue-eyed cockatoo
†*Cacatua [galerita]* undescribed sp./subsp. A—New Ireland cockatoo
Cacatua moluccensis—salmon-crested cockatoo
**Cacatua alba*—white cockatoo
†*Cacatua* undescribed sp. B—New Caledonia cockatoo
SUBFAMILY LORIINAE—LORIES
Chalcopsitta cardinalis—cardinal lory
Trichoglossus haematodus—rainbow lory
Trichoglossus rubiginosus—Pohnpei lory
Lorius hypoinochrous—purple-bellied lory
Lorius albidinuchus—white-naped lory
Lorius chlorocercus—yellow-bibbed lory
†*Vini sinotoi*—Sinoto's lorikeet
†*Vini vidivici*—conquered lorikeet
Vini (Phigys) solitarius—collared lory
Vini australis—blue-crowned lorikeet
Vini kuhlii—Kuhl's lorikeet
Vini stepheni—Stephen's lorikeet
Vini peruviana—blue lorikeet
Vini ultramarina—ultramarine lorikeet
Charmosyna palmarum—palm lorikeet
Charmosyna rubrigularis—red-chinned lorikeet
Charmosyna meeki—Meek's lorikeet
Charmosyna placentis—red-flanked lorikeet
Charmosyna amabilis—red-throated lorikeet
Charmosyna margarethae—duchess lorikeet
†*Charmosyna diadema*—New Caledonian lorikeet
SUBFAMILY PSITTACINAE—PARROTS
**Strigops habroptilus*—kakapo
Micropsitta pusio—buff-faced pygmy-parrot
Micropsitta bruijnii—red-breasted pygmy-parrot
Micropsitta meeki—Meek's pygmy-parrot
Micropsitta finschii—Finsch's pygmy-parrot
Prosopeia tabuensis—red shining-parrot
Prosopeia personata—masked shining-parrot
Eunymphicus cornutus—horned parakeet
**Cyanoramphus unicolor*—Antipodes parakeet
Cyanoramphus novaezelandiae—red-fronted parakeet
**Cyanoramphus cookii*—Norfolk Island parakeet
**Cyanoramphus auriceps*—yellow-fronted parakeet
†*Cyanoramphus zealandicus*—black-fronted parakeet
†*Cyanoramphus ulietanus*—Ra'iatea parakeet
Geoffroyus heteroclitus—singing parrot
Eclectus roratus—eclectus parrot
†*Eclectus infectus*—Oceanic eclectus parrot
Loriculus (aurantiifrons) tener—green-fronted hanging-parrot
Genus uncertain—Easter Island parrot

ORDER CUCULIFORMES
Family Cuculidae—Cuckoos
Cacomantis variolosus—brush cuckoo
Cacomantis [pyrrhophanus] flabelliformis—fan-tailed cuckoo
Chrysococcyx lucidus—shining bronze-cuckoo
Eudynamys scolopacea—Asian koel
Eudynamys taitensis—long-tailed koel (m)
Scythrops novaehollandiae—channel-billed cuckoo (m)
Centropus milo—buff-headed coucal
Centropus violaceus—violaceous coucal
Centropus ateralbus—pied coucal

ORDER STRIGIFORMES
Family Tytonidae—Barn-Owls
**Tyto tenebricosa*—greater sooty-owl
(*?)*Tyto novaehollandiae*—Australian masked-owl
Tyto [novaehollandiae] manusi—Manus owl
Tyto aurantia—New Britain masked-owl
Tyto longimembris—Australasian grass-owl
**Tyto capensis*—African grass-owl
Tyto alba—barn-owl
**Tyto punctatissima*—Galápagos barn-owl
†*Tyto? letocarti*—Letocart's owl
Family Strigidae—Owls
Pyrroglaux podarginas—Palau owl
**Bubo virginianus*—great horned owl
**Ninox novaeseelandiae*—morepork
Ninox meeki—Manus hawk-owl
Ninox variegata—Bismarck hawk-owl
Ninox odiosa—New Britain hawk-owl
Ninox jacquinoti—Solomon Islands hawk-owl
Asio flammeus—short-eared owl
Nesasio solomonensis—fearful owl

ORDER CAPRIMULGIFORMES
Family Aegothelidae—Owlet-Nightjars
†*Aegotheles savesi*—New Caledonian owlet-nightjar
†*Aegotheles novaezealandiae*—New Zealand owlet-nightjar
Family Podargidae—Frogmouths
Podargus ocellatus—marbled frogmouth
Undescribed genus, *inexpectatus*—Solomon Islands frogmouth
Family Caprimulgidae—Nightjars
Eurostopodus mysticalis—white-throated nightjar

Caprimulgus indicus—jungle nightjar
Caprimulgus macrurus—large-tailed nightjar

ORDER APODIFORMES

Family Hemiprocnidae—Treeswifts
Hemiprocne mystacea—moustached treeswift

Family Apodidae—Swifts
Collocalia esculenta—glossy swiftlet
Collocalia spodiopygia—white-rumped swiftlet
Collocalia orientalis—Mayr's swiftlet
Collocalia vanikorensis—uniform swiftlet
Collocalia whiteheadi—Whitehead's swiftlet
Collocalia sawtelli—Atiu swiftlet
†*Collocalia manuoi*—Mangaia swiftlet
Collocalia leucophaea—Polynesian swiftlet
Collocalia ocista—Marquesan swiftlet

ORDER CORACIIFORMES

Family Alcedinidae—Kingfishers
Alcedo atthis—common kingfisher
Ceyx (Alcedo) websteri—Bismarck kingfisher
Ceyx (Alcedo) pusillus—little kingfisher
Ceyx (Alcedo) lepidus—variable kingfisher
Halcyon albonotata—New Britain kingfisher
Halcyon leucopygia—ultramarine kingfisher
Halcyon farquhari—chestnut-bellied kingfisher
Halcyon recurvirostris—flat-billed kingfisher
Halcyon cinnamomina—Micronesian kingfisher
Halcyon chloris—collared kingfisher
Halcyon saurophaga—beach kingfisher
Halcyon sancta—sacred kingfisher
Halcyon venerata—Tahiti kingfisher
Halcyon mangaia—Mangaia kingfisher
Halcyon tuta—chattering kingfisher
Halcyon godeffroyi—Marquesas kingfisher
Halcyon gertrudae—Niau kingfisher
†*Halcyon gambieri*—Mangareva kingfisher
Actenoides bougainvillei—moustached kingfisher
Tanysiptera [sylvia] nigriceps—black-headed paradise-kingfisher

Family Meropidae—Bee-Eaters
Merops [superciliosus] philippinus—blue-tailed bee-eater
Merops ornatus—rainbow bee-eater (m)

Family Coraciidae—Rollers
Eurystomus orientalis—dollarbird

Family Bucerotidae—Hornbills
Aceros plicatus—Blyth's hornbill
†*Aceros* undescribed sp.—New Caledonian hornbill

ORDER PASSERIFORMES

Family Pittidae—Pittas
Pitta sordida—hooded pitta
Pitta superba—superb pitta
Pitta erythrogaster—red-bellied pitta
Pitta anerythra—black-faced pitta

Family Acanthisittidae—New Zealand Wrens
*†*Traversia lyalli*—Stephens Island wren
*†*Pachyplichas yaldwyni*—Yaldwyn's wren
*†*Dendroscansor decurvirostris*—long-billed wren

SUBORDER PASSERIDA

Family Hirundinidae—Swallows
Hirundo tahitica—Pacific swallow
Hirundo nigricans—tree martin (m)

Family Pycnonotidae—Bulbuls
**Pycnonotus cafer*—red-vented bulbul

Family Turdidae—Thrushes
Zoothera heinei—russet thrush
Zoothera talaseae—New Britain thrush
Zoothera margaretae—San Cristobal thrush
Turdus poliocephalus—island thrush

Family Mimidae—Mockingbirds
**Mimus trifasciatus*—Floreana mockingbird

Family Cisticolidae—Cisticolas
Cisticola exilis—Golden-headed cisticola

Family Sylviidae—Old World Warblers
Cettia annae—palau bush-warbler
Cettia parens—shade warbler
Cettia ruficapilla—Fiji bush-warbler
†cf. *Cettia* sp.—Tongan warbler
Acrocephalus [australis] stentoreus—Australian reed-warbler
Acrocephalus luscinia—nightingale reed-warbler
Acrocephalus syrinx—Caroline reed-warbler
Acrocephalus rehsei—Nauru reed-warbler
Acrocephalus aequinoctialis—Christmas Island reed-warbler
†*Acrocephalus* undescribed sp.—Majuro warbler
Acrocephalus familiaris—millerbird
Acrocephalus caffer—Tahiti reed-warbler
Acrocephalus atyphus—Tuamotu reed-warbler
Acrocephalus mendanae—Marquesan reed-warbler
Acrocephalus [vaughani] kerearako—Mangaian reed-warbler
Acrocephalus [vaughani] kaoko—Mitiaro reed-warbler
Acrocephalus [vaughani] rimatarae—Rimatara reed-warbler
Acrocephalus [vaughani] vaughani—Pitcairn reed-warbler
Acrocephalus [vaughani] taiti—Henderson Island reed-warbler
Phylloscopus [trivirgatus] poliocephalus—island leaf-warbler
Phylloscopus [trivirgatus] makirensis—San Cristobal leaf-warbler
Phylloscopus [trivirgatus] amoenus—Kulambangara leaf-warbler
Megalurus timoriensis—tawny grassbird
Cichlornis mariei—New Caledonian thicketbird
Cichlornis grosvenori—Bismarck thicketbird
Cichlornis llaneae—Bougainville thicketbird
Cichlornis whitneyi—Guadalcanal thicketbird
Cichlornis rubiginosus—rusty thicketbird
Trichocichla rufa—long-legged warbler

Family Muscicapidae—Old World Flycatchers
Saxicola caprata—pied bushchat

Family Nectariniidae—Sunbirds
Nectarinia [sericea] aspasia—black sunbird
Nectarinia jugularis—olive-backed sunbird

Family Dicaeidae—Flowerpeckers
Dicaeum eximium—red-banded flowerpecker

Dicaeum aeneum—midget flowerpecker
Dicaeum tristrami—mottled flowerpecker
Family Zosteropidae—White-Eyes
Zosterops conspicillatus—bridled white-eye
Zosterops semperi—Caroline Island white-eye
Zosterops hypolais—plain white-eye
Zosterops hypoxanthus—black-headed white-eye
Zosterops griseotinctus—Louisiade white-eye
Zosterops rennellianus—Rennell white-eye
Zosterops vellalavella—banded white-eye
Zosterops splendidus—Ranongga white-eye[1]
Zosterops luteirostris—splendid white-eye
Zosterops kulambangrae—Solomon Islands white-eye
Zosterops murphyi—Kulambangara white-eye
Zosterops metcalfii—yellow-throated white-eye
Zosterops rendovae—gray-throated white-eye
Zosterops ugiensis—bicolored white-eye
Zosterops stresemanni—Malaita white-eye
Zosterops santaecrucis—Santa Cruz white-eye
Zosterops inornatus—large Lifou white-eye
Zosterops xanthochroa—green-backed white-eye
Zosterops minutus—small Lifou white-eye
Zosterops explorator—Layard's white-eye
Zosterops lateralis—silvereye
Zosterops flavifrons—yellow-fronted white-eye
Zosterops samoensis—Samoan white-eye
Zosterops finschii—dusky white-eye
Zosterops cinereus—gray white-eye
Rukia oleaginea—Yap white-eye
Rukia ruki—Truk white-eye
Rukia longirostra—long-billed white-eye
Cleptornis marchei—golden white-eye
Woodfordia superciliosa—bare-eyed white-eye
Woodfordia lacertosa—Sanford's white-eye
Megazosterops palauensis—giant white-eye
†Genus and species undescribed—'Eua white-eye
Family Sturnidae—Starlings
Aplonis metallica—metallic starling
Aplonis cantoroides—singing starling
Aplonis feadensis—atoll starling
Aplonis insularis—Rennell starling
Aplonis brunneicapilla—white-eyed starling
Aplonis grandis—brown-winged starling
Aplonis dichroa—San Cristobal starling
Aplonis zelandica—rusty-winged starling
Aplonis striata—striated starling
Aplonis santovestris—mountain starling
Aplonis opaca—Micronesian starling
Aplonis pelzelni—Pohnpei starling
†*Aplonis corvina*—Kosrae starling
Aplonis tabuensis—Polynesian starling
Aplonis atrifusca—Samoan starling
Aplonis cinerascens—Rarotonga starling
†*Aplonis mavornata*—Ma'uke starling
†*Aplonis diluvialis*—Huahine starling

[1]The common names and scientific names of this species of white-eyes and the following three do not seem to agree; following Clements (2000), "splendid white-eye" is the common name of *Zosterops luteirostris* rather than of *Z. splendidus*, and "Kulambangara white-eye" is the common name of *Z. murphyi* rather than of *Z. kulambangarae*.

†*Aplonis* undescribed sp.—Erromango starling
Mino dumontii—yellow-faced myna
**Acridotheres fuscus*—jungle myna
**Acridotheres tristis*—common myna
**Sturnus vulgaris*—common starling
Family Passeridae—Weaver Finches
**Passer domesticus*—house sparrow
Family Estrildidae—Estrildid Finches
Erythrura trichroa—blue-faced parrotfinch
Erythrura psittacea—red-throated parrotfinch
Erythrura pealii—Fiji parrotfinch
Erythrura cyaneovirens—red-headed parrotfinch
Erythrura kleinschmidti—pink-billed parrotfinch
†*Erythrura* undescribed sp.—Marianas parrotfinch
Lonchura spectabilis—hooded munia
Lonchura hunsteini—mottled munia
Lonchura forbesi—New Ireland munia
Lonchura melaena—Bismarck munia
Family Emberizidae—Emberizine Finches
*†*Emberiza alcoveri*—Canary Island flightless finch
**Geospiza nebulosa*—sharp-beaked ground-finch
**Geospiza magnirostris*—large ground-finch
**Ammodramus henslowi*—Henslow's sparrow
Family Icteridae—Blackbirds
**Icterus oberi*—Montserrat oriole
SUBORDER CORVIDA
Family Campephagidae—Cuckoo-Shrikes
Coracina caledonica—Melanesian cuckoo-shrike
Coracina novaehollandiae—black-faced cuckoo-shrike
Coracina lineata—yellow-eyed cuckoo-shrike
Coracina papuensis—white-bellied cuckoo-shrike
Coracina analis—New Caledonian cuckoo-shrike
Coracina tenuirostris—cicadabird
Coracina holopolia—Solomon Islands cuckoo-shrike
Lalage leucomela—varied triller
Lalage sharpei—Samoan triller
Lalage maculosa—Polynesian triller
Lalage leucopyga—long-tailed triller
†cf. *Lalage* undescribed sp.—Tongan triller
Family Acanthizidae—Thornbills & Gerygones
Gerygone flavolateralis—fan-tailed gerygone
Family Pachycephalidae—Australian Robins, Whistlers
SUBFAMILY PETROICINAE—AUSTRALIAN ROBINS
Monachella muelleriana—torrent flycatcher
Petroica multicolor—scarlet robin
**Petroica traversi*—New Zealand black robin
Eopsaltria flaviventris—yellow-bellied robin
SUBFAMILY PACHYCEPHALINAE—WHISTLERS
Pachycephala pectoralis—golden whistler
Pachycephala melanura—black-tailed whistler
Pachycephala caledonica—New Caledonian Whistler
Pachycephala flavifrons—Samoan whistler
Pachycephala jacquinoti—Tongan whistler
Pachycephala implicata—hooded whistler
Pachycephala rufiventris—rufous whistler
Colluricincla (Pitohui) tenebrosa—morningbird
Family Monarchidae—Fantails, Monarchs
SUBFAMILY RHIPIDURINAE—FANTAILS
Rhipidura leucophrys—willy wagtail

Rhipidura rufiventris—northern fantail
Rhipidura cockerelli—white-winged fantail
Rhipidura drownei—brown fantail
Rhipidura tenebrosa—dusky fantail
Rhipidura rennelliana—Rennell fantail
Rhipidura fuliginosa—gray fantail
Rhipidura spilodera—streaked fantail
Rhipidura personata—Kandavu fantail
Rhipidura nebulosa—Samoan fantail
Rhipidura lepida—Palau fantail
Rhipidura matthiae—Matthias fantail
Rhipidura dahli—Bismarck fantail
Rhipidura malaitae—Malaita fantail
Rhipidura rufifrons—rufous fantail
Rhipidura semirubra—Manus fantail
Rhipidura kubaryi—Pohnpei fantail

SUBFAMILY MONARCHINAE—MONARCHS

Pomarea[2] *dimidiata*—Rarotonga monarch
Pomarea[2] *nigra*—Tahiti monarch
†*Pomarea*[2] *pomarea*—Maupiti monarch
Pomarea[2] *iphis*—iphis monarch
Pomarea[2] *mendozae*—Marquesas monarch
Pomarea[2] *whitneyi*—Fatuhiva monarch
Mayrornis[2] *versicolor*—Ogea monarch
Mayrornis[2] *lessoni*—slaty monarch
Mayrornis[2] *schistaceus*—Vanikoro monarch
Neolalage[2] *banksiana*—buff-bellied monarch
Clytorhynchus[2] *pachycephaloides*—southern shrikebill
Clytorhynchus[2] *hamlini*—Rennell shrikebill
Clytorhynchus[2] *vitiensis*—Fiji shrikebill
Clytorhynchus[2] *nigrogularis*—black-throated shrikebill
Metabolus[2] *rugensis*—Truk monarch
Monarcha cinerascens—island monarch
Monarcha castaneiventris—chestnut-bellied monarch
Monarcha richardsii—white-capped monarch
Monarcha infelix—Manus monarch
Monarcha menckei—white-breasted monarch
Monarcha verticalis—black-tailed monarch
Monarcha ateralba—Dyaul monarch
Monarcha browni—Kulambangara monarch
Monarcha viduus—white-collared monarch
Monarcha barbatus—black-and-white monarch
Monarcha godeffroyi—Yap monarch
Monarcha takatsukasae—Tinian monarch
Monarcha chrysomela—golden monarch
Myiagra freycineti—Guam monarch
Myiagra erythrops—Palau monarch
Myiagra pluto—Pohnpei monarch
Myiagra oceanica—Oceanic monarch
Myiagra ferrocyanea—steel-blue monarch
Myiagra cervinicauda—ochre-headed monarch
Myiagra caledonica—Melanesian monarch
Myiagra vanikorensis—Vanikoro monarch
Myiagra albiventris—Samoan monarch
Myiagra azureocapilla—blue-crested monarch
Myiagra cyanoleuca—satin flycatcher
Myiagra alecto—shining flycatcher

[2]Possibly congeneric with *Monarcha*; see Filardi & Smith (2005).

Myiagra hebetior—dull flycatcher
Lamprolia victoriae—silktail

Family Meliphagidae—Honeyeaters

Stresemannia (Lichmera, Melilestes, Meliphaga?) bougainvillei—Bougainville honeyeater
Guadalcanaria (Meliphaga?) inexpectata—Guadalcanal honeyeater
Melidectes (Vosea?) whitemanensis—Bismarck melidectes
Meliarchus (Melidectes?) sclateri—San Cristobal honeyeater
Lichmera incana—dark-brown honeyeater
Myzomela cineracea—ashy myzomela
Myzomela cruentata—red myzomela
Myzomela nigrita—black myzomela
Myzomela pulchella—New Ireland myzomela
Myzomela caledonica—New Caledonian myzomela
Myzomela rubrata—Micronesian myzomela
Myzomela cardinalis—cardinal myzomela
Myzomela chermisina—Rotuma myzomela
Myzomela sclateri—scarlet-bibbed myzomela
Myzomela pammelaena—ebony myzomela
Myzomela lafargei—scarlet-naped myzomela
Myzomela eichhorni—yellow-vented myzomela
Myzomela malaitae—red-bellied myzomela
Myzomela melanocephala—black-headed myzomela
Myzomela tristrami—sooty myzomela
Myzomela jugularis—orange-breasted myzomela
Myzomela erythromelas—black-bellied myzomela
Foulehaio carunculata—wattled honeyeater
Foulehaio provocator—Kandavu honeyeater
Philemon albitorques—white-naped friarbird
Philemon cockerelli—New Britain friarbird
Philemon eichhorni—New Ireland friarbird
Philemon diemenensis—New Caledonian friarbird
Gymnomyza viridis—giant honeyeater
Gymnomyza samoensis—mao
Gymnomyza aubryana—crow honeyeater
Phylidonyris undulata—barred honeyeater
Phylidonyris notabilis—Vanuatu honeyeater

Family Dicruridae—Drongos

Dicrurus megarhynchus—ribbon-tailed drongo
Dicrurus bracteatus—spangled drongo

Family Artamidae—Woodswallows

Artamus leucorhynchus—white-breasted woodswallow
Artamus insignis—Bismarck woodswallow

Family Corvidae—Crows

Corvus moneduloides—New Caledonian crow
Corvus kubaryi—Mariana crow
Corvus woodfordi—Guadalcanal crow
Corvus meeki—Bougainville crow
**Corvus tristis*—gray crow
**Corvus macrorhynchos*—large-billed crow
Corvus orru—Torresian crow
**Corvus bennettii*—little crow
**Corvus coronoides*—Australian raven

Family Vireonidae—Vireos

**Vireo griseus*—white-eyed vireo
**Vireo modestus*—Jamaican vireo
**Vireo latimeri*—Puerto Rican vireo
**Vireo osburni*—blue mountain vireo

Literature Cited

Abbott, I. 1983. The meaning of z in species/area regressions and the study of species turnover in island biogeography. Oikos 41:385–390.

Abrams, P. A. 2001. A world without competition. Nature 412:858–859.

Adams, M., P. R. Baverstock, D. A. Saunders, R. Schodde & G. T. Smith. 1984. Biochemical systematics of the Australian cockatoos (Psittaciformes: Cacatuinae). Australian Journal of Zoology 32:363–377.

Adamson, A. M. 1939. Review of the fauna of the Marquesas Islands and discussion of its origin. Bernice P. Bishop Museum Bulletin 159:1–93.

Adler, G. H. 1994. Avifaunal diversity and endemism on tropical Indian Ocean islands. Journal of Biogeography 21:85–95.

Aharon, P., R. A. Socki & L. Chan. 1987. Dolomitization of atolls by sea water convection flow: test of a hypothesis at Niue, South Pacific. Journal of Geology 95:187–203.

Ainley, D. G. & R. J. Boekelheide. 1983. An ecological comparison of oceanic seabird communities of the South Pacific Ocean. Studies in Avian Biology 8:2–23.

Aitutaki. 1983. NZMS 272/8/3 (map). New Zealand Department of Lands & Survey, Wellington.

Alcover, J. A., B. Seguí & P. Bover. 1999. Extinctions and local disappearances of vertebrates in the western Mediterranean islands. Pp. 165–188 *in* R. D. E. MacPhee (ed.). Extinctions in Near Time: Causes, Contexts, and Consequences. Kluwer Academic/Plenum Publishers, New York.

Allaby, A. & M. Allaby (eds.). 1999. A Dictionary of Earth Sciences, 2nd ed. Oxford University Press, Oxford.

Allen, J. & C. Gosden (eds.). 1991. Report of the Lapita Homeland Project. Department of Prehistory, Australian National University, Canberra.

Allen, J., C. Gosden, R. Jones & J. P. White. 1988. Pleistocene dates for the human occupation of New Ireland, northern Melanesia. Nature 331:707–709.

Allen, J., C. Gosden & J. P. White. 1989. Human Pleistocene adaptations in the tropical island Pacific: recent evidence from New Ireland, a greater Australian outlier. Antiquity 63:548–561.

Allen, M. S. 1992. Temporal variation in Polynesian fishing strategies: the southern Cook Islands in regional perspective. Asian Perspectives 31:183–204.

Allen, M. S. 1997. Coastal morphogenesis, climatic trends, and Cook Islands prehistory. Pp. 124–146 *in* P. V. Kirch & T. L. Hunt (eds.). Historical Ecology in the Pacific Islands. Yale University Press, New Haven, CT.

Allen, M. S. 2002. Resolving long-term change in Polynesian marine fisheries. Asian Perspectives 41:195–212.

Allen, M. S. 2003. Evaluating human impact on Pacific marine fisheries. Pp. 337–345 *in* C. Sands (ed.). Proceedings of the International Conference for the 50th Anniversary of the First Lapita Excavation (July 1952), August 1–7, 2002. Nouméa, New Caledonia.

Allen, M. S., E. Matisoo-Smith & A. Horsburgh. 2001a. Pacific "Babes": issues in the origins and dispersal of Pacific pigs and the potential of mitochondrial DNA analysis. International Journal of Osteoarchaeology 11:4–13.

Allen, M. S., T. N. Ladefoged & J. J. Wall. 2001b. Traditional Rotuman fishing in temporal and regional context. International Journal of Osteoarchaeology 11:56–71.

Allen, M. S. & S. E. Schubel. 1990. Recent archaeological research on Aitutaki, southern Cooks: the Moturakau Rockshelter. Journal of the Polynesian Society 99:265–295.

Allen, M. S. & D. W. Steadman. 1990. Excavations at the Ureia site, Aitutaki, southern Cook Islands: preliminary results. Archaeology in Oceania 25:24–37.

Allison, A. 1996. Zoogeography of amphibians and reptiles of New Guinea and the Pacific region. Pp. 407–436 *in* A. Keast & S. E. Miller (eds.). The Origin and Evolution of Pacific Island Biotas, New Guinea to Eastern Polynesia: Patterns and Processes. SPB Academic Publishing, Amsterdam.

Allison, A. & I. Bigilale. 2001. The herpetofauna of southern New Ireland. Pp. 50–60 *in* B. M. Beehler & L. E. Alonso (eds.). Southern New Ireland, Papua New Guinea: A Biodiversity Assessment. RAP Bulletin of Biological Assessment 21. Bishop Museum of Natural History, Honolulu.

Alroy, J. 1999. Putting North America's end-Pleistocene megafaunal extinction in context: large-scale analyses of spatial patterns, extinction rates, and size distributions. Pp. 105–143 *in* R. D. E. MacPhee (ed.). Extinctions in Near Time: Causes, Contexts, and Consequences. Kluwer Academic/Plenum Publishers, New York.

Amadon, D. 1942a. Birds collected during the Whitney South Sea Expedition. XLIX. Notes on some non-passerine genera, 1. American Museum Novitiates no. 1175.

Amadon, D. 1942b. Birds collected during the Whitney South Sea Expedition. L. Notes on some non-passerine genera, 2. American Museum Novitates no. 1176.

Amadon, D. 1943. Birds collected during the Whitney South Sea Expedition. LII. Notes on some non-passerine genera. American Museum Novitates no. 1237.

Ambrose, W. R. 1997. Contradictions in Lapita pottery, a composite clone. Antiquity 71:525–538.

Amerson, A. B. 1969. Ornithology of the Marshall and Gilbert Islands. Atoll Research Bulletin no. 127.

Amerson, A. B., Jr. & P. C. Shelton. 1976. The natural history of Johnston Atoll, central Pacific Ocean. Atoll Research Bulletin no. 192.

Amerson, A. B., Jr., W. A. Whistler & T. D. Schwaner. 1982a. Wildlife and Wildlife Habitat of American Samoa. I. Environment and Ecology. U.S. Department of Interior, Fish and Wildlife Service, Washington.

Amerson, A. B., Jr., W. A. Whistler & T. D. Schwaner. 1982b. Wildlife and Wildlife Habitat of American Samoa. II. Accounts of Flora and Fauna. U.S. Department of Interior, Fish and Wildlife Service, Washington.

Amesbury, J. R., D. R. Moore & R. L. Hunter-Anderson. 1996. Cultural adaptations and Late Holocene sea level change in the Marianas: recent excavations at Chalan Piao, Saipan, Micronesia. Bulletin of the Indo-Pacific Prehistory Association 15:53–69.

Anderson, A. 1989. Prodigious Birds: Moas and Moa-hunting in Prehistoric New Zealand. Cambridge University Press, Cambridge, UK.

Anderson, A. 1995. Current approaches in East Polynesian colonization research. Journal of the Polynesian Society 104:110–132.

Anderson, A., S. Bedford, G. Clark, I. Lilley, C. Sand, G. Summerhayes & R. Torrence. 2001. An inventory of Lapita sites containing dentate-stamped pottery. Terra Australis 17:1–13.

Anderson, A. & G. Clark. 1999. The age of Lapita settlement in Fiji. Archaeology in Oceania 34:31–39.

Anderson, A., H. Leach, I. Smith & R. Walter. 1994. Reconsideration of the Marquesan sequence in East Polynesian prehistory with particular reference to Hane (MUH 1). Archaeology in Oceania 29:29–52.

Anderson, W. B. & G. A. Polis. 1998. Marine subsidies of island communities in the Gulf of California: evidence from stable carbon and nitrogen isotopes. Oikos 81:75–80.

Anderson, W. B. & G. A. Polis. 1999. Nutrient fluxes from water to land: seabirds affect plant nutrient status on Gulf of California islands. Oecologia 118:324–332.

Anderson, W. B. & D. A. Wait. 2001. Subsidized Island Biogeography Hypothesis: another new twist on an old theory. Ecology Letters 4:289–291.

Anonymous. 1985. Observations of birds on Mussau. Papua New Guinea Bird Society Newsletter 212:7, 11, 12.

Anonymous. 1989. Tuamotus yield new data. World Birdwatch 11:4.

Anonymous. 2002. Rail awaits discovery in the Solomon Islands. World Birdwatch 24:4.

Anthony, N. 2001. Small mammals: an annotated checklist. Pp. 34–45 *in* D. Byrnes (ed.). New Britain Biological Survey. University of Wisconsin New Britain Biological Survey, Madison.

Antón, S. C. & D. W. Steadman. 2003. Mortuary patterns in burial caves on Mangaia, Cook Islands. International Journal of Osteoarchaeology 13:132–146.

Archipel de la Société. 1988. (Map.) Institut Géographique National, Paris, France.

Archipel des Nouvelles-Hébrides. 1976. South Pacific Maps. Distributed by HEMA Maps, Queensland, Australia.

Arendt, A. A., K. A. Echelmeyer, W. D. Harrison, C. S. Lingle & V. B. Valentine. 2002. Rapid wastage of Alaska glaciers and their contribution to rising sea level. Science 297:382–386.

Ardern, S. L. & D. M. Lambert. 1997. Is the black robin in genetic peril? Molecular Ecology 6:21–28.

Armstrong, D. P. & I. G. McLean. 1995. New Zealand translocations: theory and practice. Pacific Conservation Biology 2:39–54.

Arnow, T. 1954. The hydrology of the northern Marshall Islands. Atoll Research Bulletin no. 30.

Ash, J. 1988. Stunted cloud forest in Taveuni, Fiji. Pacific Science 41:191–199.

Ash, J. 1992. Vegetation ecology of Fiji: past, present, and future perspectives. Pacific Science 46:111–127.

Ashmole, N. P. 1963. The regulation of numbers of tropical oceanic birds. Ibis 103:458–473.

Asia West Pacific. 1998. South Pacific Maps. Distributed by HEMA Maps, Queensland, Australia.

Athens, J. S. 1997. Hawaiian native lowland vegetation in prehistory. Pp. 248–270 *in* P. V. Kirch & T. L. Hunt (eds.). Historical Ecology in the Pacific Islands. Yale University Press, New Haven, CT.

Athens, J. S. & J. V. Ward. 1998. Paleoenvironment and Prehistoric Landscape Change: A Sediment Core Record from Lake Hagoi, Tinian, CNMI. International Archaeological Research Institute, Honolulu.

Athens, J. S., H. D. Tuggle, J. V. Ward & D. J. Welch. 2002. Avifaunal extinctions, vegetation change, and Polynesian impacts in prehistoric Hawai‘i. Archaeology in Oceania 37:57–78.

Atkinson, I. A. E. 1985. The spread of commensal species of *Rattus* to oceanic islands and their effects on island avifaunas. International Council for Bird Preservation Technical Publication 3:35–81.

Atkinson, I. A. E. 1989. Introduced animals and extinctions. Pp. 54–79 *in* D. Western & M. C. Pearl (eds.). Conservation for the Twenty-first Century. Oxford University Press, New York.

Atkinson, I. A. E. & T. J. Atkinson. 2000. Land vertebrates as invasive species on the islands in the south Pacific regional environment program. Pp. 19–84 *in* G. Sherley (ed.). Invasive Species in the Pacific: A Technical Review and Draft Regional Strategy. South Pacific Regional Environment Programme, Apia, Samoa.

Atkinson, P. & D. Gibbons. 1998. Status of the Montserrat oriole *Icterus oberi* since the start of a volcanic eruption in July 1995. Bird Conservation International 8:309.

Auffenberg, W. & T. Auffenberg. 1988. Resource partitioning in a community of Philippine skinks (Sauria: Scincidae). Bulletin of the Florida State Museum (Biological Sciences) 32:151–219.

Augstein, E. 1984. The atmospheric boundary layer over the tropical oceans. Pp. 73–103 *in* D. B. Shaw (ed.). Meteorology over the Tropical Oceans. Royal Meteorological Society, Berkshire, UK.

Avise, J. C., D. Walker & G. C. Johns. 1998. Speciation durations and Pleistocene effects on vertebrate phylogeography. Proceedings of the Royal Society of London B 265:1707–1712.

Ayres, W. S. 1979. Easter Island fishing. Asian Perspectives 22:61–92.

Ayres, W. S. 1985. Easter Island subsistence. Journal de la Société des Océanistes 80:103–124.

Ayres, W. S. 1990. Mystery islets of Micronesia. Archaeology 43:58–63.

Ayres, W. S. & A. E. Haun. 1990. Prehistoric food production in Micronesia. Pp. 211–227 *in* D. E. Yen & J. M. J. Mummery (eds.). Pacific Production Systems: Approaches to Economic Prehistory. Australian National University Occasional Papers in Prehistory no. 18.

Bahn, J. & Flenley, J. 1992. Easter Island, Earth Island. Thames and Hudson, New York.

Baird, R. F. 1984. The Pleistocene distribution of the Tasmanian native-hen *Gallinula mortierii mortierii*. Emu 84:119–123.

Baird, R. F. 1986. Tasmanian native-hen *Gallinula mortierii mortierii*: the first late Pleistocene record from Queensland. Emu 86:121–122.

Baird, R. F. 1991. Avian fossils from the Quaternary of Australia. Pp. 809–870 *in* P. V. Rich, J. M. Monaghan, R. F. Baird & T. H. Rich (eds.). Vertebrate Paleontology of Australasia. Monash University, Melbourne, Australia.

Baird, R. F. & M. J. Rowley. 1990. Preservation of avian collagen in Australian Quaternary cave deposits. Palaeontology 33:447–451.

Baker, P. E. 1967. Preliminary account of recent geological investigations on Easter Island. Geological Magazine 104:116–122.

Baker, P. E., F. Buckley & J. G. Holland. 1974. Petrology and geochemistry of Easter Island. Contributions to Mineralogy and Petrology 44:85–100.

Baker, R. H. 1948. Report on collections of birds made by United Stages Naval Medical Research Unit no. 2 in the Pacific war area. Smithsonian Miscellaneous Collections no. 107.

Baker, R. H. 1951. The avifauna of Micronesia, its origin, evolution, and distribution. University of Kansas Publication, Museum of Natural History 3:1–359.

Baldwin, P. H. 1947. The life history of the Laysan rail. Condor 49:14–21.

Balis-Larsen, M. & T. Sutterfield. 1997. Navy protects island monarch. Endangered Species Bulletin 22:10–11.

Ballance, P. F., A. G. Ablaev, I. K. Puschin, S. P. Pletnev, M. G. Birylina, T. Itaya, H. A. Follas & G. W. Gibson. 1999. Morphology and history of the Kermadec trench-arc-backarc basin-remnant arc system at 30 to 32°S: geophysical profile, microfossil and K-Ar data. Marine Geology 159:35–62.

Balouet, J. C. 1984. Les étranges fossiles de Nouvelle-Calédonie. La Recherche 15:390–392.

Balouet, J. C. 1987. Extinctions des vertébrés terrestres de Nouvelle-Calédonie. Mémoires de la Société Géologique de France 150:177–183.

Balouet, J. C. 1991. The fossil vertebrate record of New Caledonia. Pp. 1383–1409 *in* P. Vickers-Rich, J. M. Monaghan, R. F. Baird & T. H. Rich (eds.). Vertebrate Palaeontology of Australasia. Monash University, Melbourne, Australia.

Balouet, J. C. & E. Buffetaut. 1987. *Mekosuchus inexpectatus*, n. g., n. sp., crocodilien nouveau de l'Holocene de Nouvelle Calédonie. Comptes Rendus de l'Académie des Sciences Paris (Série 2) 304:853–856.

Balouet, J. C. & S. L. Olson. 1987. A new extinct species of giant pigeon (Columbidae: *Ducula*) from archaeological deposits on Wallis (Uvea) Island, South Pacific. Proceedings of the Biological Society of Washington 100:769–775.

Balouet, J. C. & S. L. Olson. 1989. Fossil birds from late Quaternary deposits in New Caledonia. Smithsonian Contributions in Zoology no. 469:1–38.

Bandy, M. C. 1937. Geology and petrology of Easter Island. Bulletin of the Geological Society of America 48:1589–1610.

Banks, R. C. 1984. Bird specimens from American Samoa. Pacific Science 38:150–169.

Barash, M. S. & V. M. Kuptsov. 1997. Late Quaternary palaeoceanography of the western Woodlark Basin (Solomon Sea) and Manus Basin (Bismarck Sea), Papua New Guinea, from planktonic foraminifera and radiocarbon dating. Marine Geology 142:171–187.

Bard, E. 2001. Extending the calibrated radiocarbon record. Science 292:2443–2444.

Bard, E., F. Rostek & G. Menot-Combes. 2004. A better radiocarbon clock. Science 303:178–179.

Barnes, I., P. E. Matheus, B. Shapiro, D. Jensen & A. Cooper. 2002. Dynamics of mammal population extinctions in Eastern Beringia during the last glaciation. Science 295:2267–2270.

Barre, N. & A. Barau. 1982. Oiseaux de la Réunion. Privately printed, St. Denis, Île de la Réunion.

Bath, J. E. 1984. Sapwtakai: archaeological survey and testing. Micronesian Archeological Survey Report no. 14.

Bauer, A. M. 1999. The terrestrial reptiles of New Caledonia: the origin and evolution of a highly endemic herpetofauna. Pp. 3–25 *in* H. Ota (ed.). Tropical Island Herpetofauna: Origin, Current Diversity, and Conservation. Elsevier Press, Amsterdam.

Bauer, A. M. & A. P. Russell. 1986. *Hoplodactylus delcourti* n. sp. (Reptilia: Gekkonidae), the largest known gecko. New Zealand Journal of Zoology 13:141–148.

Bauer, A. M. & R. A. Sadlier. 1993. Systematics, biogeography, and conservation of the lizards of New Caledonia. Biodiversity Letters 1:107–122.

Bauer, A. M. & R. A. Sadlier. 2000. The Herpetofauna of New Caledonia. Society for the Study of Amphibians and Reptiles, Ithaca, New York. 310p.

Bauer, A. M. & J. V. Vindum. 1990. A checklist and key to the herpetofauna of New Caledonia, with remarks on biogeography. Proceedings of the California Academy of Sciences 47:17–45.

Bauman, S. 1996. Diversity and decline of land snails on Rota, Mariana Islands. American Malacological Bulletin 12:13–27.

Bayliss-Smith, T. P. 1972. The birds of Ontong Java and Sikaiana, Solomon Islands. Bulletin of the British Ornithologists' Club 92:1–10.

Bayliss-Smith, T. P. 1988. The role of hurricanes in the development of reef islands, Ontong Java Atoll, Solomon Islands. Geographical Journal 154:377–391.

Beaglehole, E. & P. Beaglehole. 1938. Ethnology of Pukapuka. Bernice P. Bishop Museum Bulletin no. 150.

Beardsley, F. R., W. S. Ayres & G. G. Goles. 1991. Characterization of Easter Island obsidian sources. Pp. 179–187 *in* P. Bellwood (ed.). Bulletin of the Indo-Pacific Prehistory Association no. 11.

Beaufort, L., T. de Garidel-Thoron, A. C. Mix & N. G. Pisias. 2001. ENSO-like forcing on oceanic primary production during the late Pleistocene. Science 293:2440–2444.

Beck, J. W., J. Récy, F. Taylor, R. L. Edwards & G. Cabioch. 1997. Abrupt changes in early Holocene tropical sea surface temperature derived from coral records. Nature 385:705–707.

Beck, J. W., D. A. Richards, R. L. Edwards, B. W. Silverman, P. L. Smart, D. J. Donahue, S. Hererra-Osterheld, G. S. Burr, L. Calsoyas, A. J. T. Jull & D. Biddulph. 2001. Extremely large variations of atmospheric ^{14}C concentration during the last glacial period. Science 292:2453–2458.

Becker, J. J. & B. M. Butler. 1988. Nonfish vertebrate remains. Pp. 473–475 *in* B. M. Butler (ed.). Archaeological Investigations on the North Coast of Rota, Mariana Islands. Micronesian Archaeological Survey Report no. 23. Southern

Illinois University at Carbondale, Center for Archaeological Investigations, Occasional Paper no. 8.

Beckon, W. N. 1980. Gizzard structure of the Pacific pigeon, *Ducula pacifica*. Notornis 27:302–303.

Bedford, S. 1999. Lapita and post-Lapita ceramic sequences from Erromango, southern Vanuatu. Pp. 127–137 *in* J. C. Galipaud & I. Lilley (eds.). The Pacific from 5000 to 2000 BP: Colonisation and Transformations. IRD, Paris.

Bedford, S. 2001. Ceramics from Malekula, northern Vanuatu: the two ends of a potential 3000-year sequence. Terra Australis 17:105–114.

Beehler, B. M. 1978. Notes on the mountain birds of New Ireland. Emu 78:65–70.

Beehler, B. M., J. P. Angle, D. Gibbs, M. Hedemark & D. Kuro. 2001. A field survey of the resident birds of southern New Ireland. Pp. 61–66 *in* B. M. Beehler & L. E. Alonso (eds.). Southern New Ireland, Papua New Guinea: A Biodiversity Assessment. RAP Bulletin of Biological Assessment 21. B. P. Bishop Museum, Honolulu.

Beehler, B. M., T. K. Pratt & D. A. Zimmerman. 1986. Birds of New Guinea. Princeton University Press, Princeton, NJ.

Behrenfeld, M. J. & Z. S. Kolber. 1999. Widespread iron limitation of phytoplankton in the South Pacific Ocean. Science 283:840–843.

Beichle, U. 1987. Lebensraum, Bestand und Nahrungsaufnahme der Zahntaube, *Didunculus strigirostris*. Journal für Ornithologie 128:75–89.

Beichle, U. 1991. Status and acoustical demarcation of pigeons of Western Samoa. Notornis 38:81–86.

Bell, G. 2001. Neutral macroecology. Science 293:2413–2418.

Bell, S. C. & H. G. Siegrist, Jr. 1991. Facies development model for emergent Holocene reef limestone: southern Marianas region. Micronesica 24:137–157.

Bellingham, M. & A. Davis. 1988. Forest bird communities in Western Samoa. Notornis 35:117–128.

Bellwood, P. 1979. Man's Conquest of the Pacific: The Prehistory of Southeast Asia and Oceania. Oxford University Press, New York.

Bellwood, P. 1991. The Austronesian dispersal and the origin of languages. Scientific American 265:88–93.

Bellwood, P., J. J. Fox & D. Tryon. 1995. The Austronesians: Historical and Comparative Perspectives. Australian National University, Canberra.

Bengtsson, L. 2001. Hurricane threats. Science 293:440–441.

Benton, T. G. & T. Spencer (eds.). 1995. The Pitcairn Islands: Biogeography, Ecology and Prehistory. Academic Press, London.

Berger, A. J. 1987. Avifauna of Enewetak Atoll. Pp. 215–220 *in* D. M. Devaney, E. S. Reese, B. L. Burch & P. Helfrich (eds.). The Natural History of Enewetak Atoll, vol. I: The Ecosystem: Environments, Biotas, and Processes. U.S. Department of Energy, Office of Scientific and Technical Information, Oak Ridge, TN.

Berger, J., J. E. Swenson & I.-L. Persson. 2001. Recolonizing carnivores and naive prey: conservation lessons from Pleistocene extinctions. Science 291:1036–1039.

Berger, M. & J. S. Hart. 1974. Physiology and energetics of flight. Pp. 416–477 *in* D. S. Farner and J. R. King (eds.). Avian Biology 7. Academic Press, London.

Berggren, W. A., D. V. Kent, C. C. Swisher III & M.-P. Aubry. 1995. A revised Cenozoic geochronology and chronostratigraphy. Pp. 129– 212 *in* W. A. Berggren, D. V. Kent, M.-P. Aubry & J. Hardenbol (eds.). Geochronology Time Scales and Global Stratigraphic Correlations. SEPM Special Publication no. 54.

Bermingham, E. & C. Dick. 2001. The Inga: newcomer or museum antiquity? Science 293:2214–2216.

Best, S. B. 1984. Lakeba: The prehistory of a Fijian island. Ph.D. thesis, University of Auckland, New Zealand.

Best, S. B. 2002. Lapita: A view from the East. New Zealand Archaeological Association Monograph no. 24.

Bier, J. A. 1980. Islands of Samoa. University of Hawai'i Press, Honolulu.

Bier, J. A. 1995. Reference Map of Oceania. University of Hawai'i Press, Honolulu.

Birks, S. M. & S. V. Edwards. 2002. A phylogeny of the megapodes (Aves: Megapodiidae) based on nuclear and mitochondrial DNA sequences. Molecular Phylogenetics and Evolution 23:408–421.

Bismarck Archipelago and Solomon Islands. 1995. Map 82010. United States Defense Mapping Agency, Washington.

Blaber, S. J. M. 1990. A checklist and notes on the current status of the birds of New Georgia, Western Province, Solomon Islands. Emu 90:205–214.

Blaber, S. J. M. & D. A. Milton. 1991. A note on habitat decline and the status of the spotted button-quail *Turnix maculosa salomonis* on Guadalcanal, Solomon Islands. Emu 91:64–65.

Blackburn, T. M., P. Cassey, R. P. Duncan, K. L. Evans & K. J. Gaston. 2004. Avian extinction and mammalian introductions on Oceanic islands. Science 305:1955–1958.

Blake, S. G. 1995. Late Quaternary history of Henderson Island, Pitcairn Group. Biological Journal of the Linnean Society 56:43–62.

Blanvillain, C., C. Florent & V. Thenot. 2002. Land birds of Tuamotu Archipelago, Polynesia: relative abundance and changes during the 20th century with particular reference to the critically endangered Polynesian ground-dove (*Gallicolumba erythroptera*). Biological Conservation 103:139–149.

Blondel, J., P. C. Dias, P. Perret, M. Maistre & M. M. Lambrechts. 1999. Selection-based biodiversity at a small spatial scale in a low-dispersing insular bird. Science 285:1399–1402.

Blust, R. 1999. Subgrouping, circularity and extinction: some issues in Austronesian comparative linguistics. Pp. 31–94 *in* E. Zeitoun & P. J. Li. Selected Papers from the Eighth. International Conference on Austronesian Linguistics. Symposium Series of the Institute of Linguistics (Preparatory Office), Academia Sinica, Taipei, Taiwan.

Blumenstock, D. I. & D. F. Rex. 1960. Microclimatic observations at Eniwetok. Atoll Research Bulletin no. 71.

Bocheński, Z. & P. Kuśnierczyk. 2003. Nesting of the *Acrocephalus* warblers. Acta Zoologica Cracoviensia 46:97–195.

Bogert, C. 1937. The distribution and the migration of the long-tailed cuckoo (*Urodynamis taitensis* Sparrman). Birds collected during the Whitney South Sea Expedition. XXXIV. American Museum Novitates no. 933.

Bolen, E. G. & M. K. Rylander. 1983. Whistling-ducks: zoogeography, ecology, anatomy. Special Publications of the Museum of Texas Tech University. Texas Tech Press, Lubbock.
Boles, W. E. 1993. A new cockatoo (Psittaciformes: Cacatuidae) from the Tertiary of Riversleigh, northwestern Queensland, and an evaluation of rostral characters in the systematics of parrots. Ibis 135:8–18.
Boles, W. E. 2001. A swiftlet (Apodidae: Collocaliini) from the Oligo-Miocene of Riversleigh, northwestern Queensland. Association of Australasian Palaeontologists Memoir 25:45–52.
Boles, W. E. & T. J. Ivison. 1999. A new genus of dwarf megapode (Galliformes: Megapodiidae) from the Late Oligocene of Central Australia. Pp. 199–206 *in* S. L. Olson (ed.). Avian Paleontology at the Close of the 20th Century: Proceedings of the 4th International Meeting of the Society of Avian Paleontology and Evolution, Washington, 4–7 June 1996. Smithsonian Contributions to Paleobiology no. 89.
Boles, W. E. & B. Mackness. 1994. Birds from the Bluff Downs local fauna, Allingham Formation, Queensland. Records of the South Australian Museum 27:139–149.
Bonaccorso, F. J. 1998. Bats of Papua New Guinea. Conservation International Tropical Field Guide Series 2. Conservation International, Washington.
Bouchet, P. & A. Abdou. 2001. Recent extinct land snails (Euconulidae) from the Gambier Islands with remarkable apertural barriers. Pacific Science 55:121–127.
Bouchet, P., P. Lozouet, P. Maestrati & V. Heros. 2002. Assessing the magnitude of species richness in tropical marine environments: exceptionally high numbers of molluscs at a New Caledonia site. Biological Journal of the Linnean Society 75:421–436.
Bourdon, B., S. Turner & C. Allégre. 1999. Melting dynamics beneth the Tonga-Kermadec Island Arc inferred from ^{231}Pa-^{235}U systematics. Science 286:2491–2493.
Bourne, W. R. P. 1976. On subfossil bones of Abbott's booby *Sula abbotti* from the Mascarene Islands, with a note of the proportions and distribution of the Sulidae. Ibis 118:119–123.
Bowen, J. 1996. Notes on the Vanuatu megapode *Megapodius layardi* on Ambrym, Vanuatu. Bird Conservation International 6:401–408.
Bowen, B. W. & S. A. Karl. 1999. In war, truth is the first casualty. Conservation Biology 13:1013–1016.
Bowman, R. I. 1961. Morphological differentiation and adaptation in the Galápagos finches. University of California Publications in Zoology 58:1–302.
Bradley, R. S. 2001. Many citations support global warming trend. Science 292:2011.
Bramley, G. N. & C. J. Veltman. 1998. Failure of translocated, captive-bred North Island weka *Gallirallus australis greyi* to establish a new population. Bird Conservation International 8:195–204.
Bregulla, H. L. 1992. Birds of Vanuatu. Anthony Nelson, Oswestry, UK.
Bretagnolle, V., C. Attie & E. Pasquet. 1998. Cytochrome-B evidence for validity and phylogenetic relationships of *Pseudobulweria* and *Bulweria* (Procellariidae). Auk 115:188–195.
Brock, M. K. & G. M. Beauprez. 2000. The rail road to recovery. Endangered Species Bulletin 25:6–7.
Brodkorb, P. 1963. Catalogue of fossil birds, part 1. Bulletin of the Florida State Museum (Biological Sciences) 7:179–293.
Brodkorb, P. 1971. Origin and evolution of birds. Avian Biology 1, pp. 19–55.
Brook, B. W., L. Lim, R. Harden & R. Frankham. 1997. How secure is the Lord Howe Island Woodhen? A population viability analysis using VORTEX. Pacific Conservation Biology 3:125–133.
Brook, B. W., N. S. Sodhi & P. K. L. Ng. 2003. Catastrophic extinctions follow deforestation in Singapore. Nature 424:420–423.
Brooke, M. de L. 1995. The modern avifauna of the Pitcairn Islands. Biological Journal of the Linnean Society 56:199–212.
Brooke, M. de L. 2002. Seabird systematics and distribution: a review of current knowledge. Pp. 57–85 *in* E. A. Schreiber & J. Burger (eds.). Biology of Marine Birds. CRC Press, Boca Raton, FL.
Brooke, M. de L. & P. J. Jones. 1995. The diet of the Henderson fruit-dove *Ptilinopus insularis*. I. Field observations of fruit choice. Biological Journal of the Linnean Society 56:149–165.
Broome, L. S., K. D. Bishop & D. R. Anderson. 1984. Population density and habitat use by *Megapodius freycinet eremita* in West New Britain. Australian Wildlife Research 1:161–171.
Brown, D. M. & C. A. Toft. 1999. Molecular systematics and biogeography of the cockatoos (Psittaciformes: Cacatuidae). Auk 116:141–157.
Brown, J. H. 1971. Mammals on mountaintops: nonequilibrium insular biogeography. American Naturalist 105:467–478.
Brown, J. H. 1995. Macroecology. University of Chicago Press, Chicago.
Brown, J. H. & M. Lomolino. 1989. Independent discovery of the equilibrium theory of island biogeography. Ecology 70:1954–1957.
Brown, J. H. & M. V. Lomolino. 1998. Biogeography. Sinauer Associates, Sunderland, MA.
Brown, K. 2001. Physiology—all fired up: a universal metabolic rate. Science 5538:2191.
Brown, K. P. 1997. Predation at nests of two New Zealand endemic passerines: implications for bird community restoration. Pacific Conservation Biology 3:91–98.
Browning, M. R. 1993. Species limits of the cave swiftlets (*Collocalia*) in Micronesia. Avocetta 17:101–106.
Bruna, E. M., R. N. Fisher & T. J. Case. 1995. Cryptic species of Pacific skinks (*Emoia*): further support from mitochondrial DNA sequences. Copeia 1995:981–983.
Bruna, E. M., R. N. Fisher & T. J. Case. 1996. Morphological and genetic evolution appear decoupled in Pacific skinks (Squamata: Scincidae: *Emoia*). Proceedings of the Royal Society of London B 263:681–688.
Bruner, P. L. 1972. Field Guide to the Birds of French Polynesia. Pacific Scientific Information Center, Bernice P. Bishop Museum, Honolulu.
Bryan, E. H., Jr. 1926. Introduction. Pp. 1–16 *in* E. H. Bryan, Jr. & collaborators. Insects of Hawaii, Johnston Island, and Wake Island. Bernice P. Bishop Museum Bulletin no. 31.
Bryan, E. H., Jr. 1953. Check list of atolls. Atoll Research Bulletin no. 19.

Bryan, E. H., Jr. 1959. Notes on the geography and natural history of Wake Island. Atoll Research Bulletin no. 66.

Buckley, R. C. & S. B. Knedlhans. 1986. Beachcomber biogeography: interception of dispersing propagules by islands. Journal of Biogeography 13:69–70.

Buden, D. W. 1996a. Reptiles, birds, and mammals of Ant Atoll, Eastern Caroline Islands. Micronesica 29:21–36.

Buden, D. W. 1996b. Reptiles, birds, and mammals of Pakin Atoll, Eastern Caroline Islands. Micronesica 29:37–48.

Buden, D. W. 1996c. Rediscovery of the Pohnpei mountain starling (*Aplonis pelzelni*). Auk 113:229–230.

Buden, D. W. 1998. The reptiles of Kapingamarangi Atoll, Micronesia. Atoll Research Bulletin no. 453.

Buden, D. W. 1999. Reptiles, birds, and mammals of Oroluk Atoll, Eastern Caroline Islands. Micronesica 31:289–300.

Buden, D. W. 2000. A comparison of 1983 and 1994 bird surveys of Pohnpei, Federated States of Micronesia. Wilson Bulletin 112:403–410.

Buden, D. W., D. B. Lynch & G. R. Zug. 2001. Recent records of exotic reptiles on Pohnpei, Eastern Caroline Islands, Micronesia. Pacific Science 55:65–70.

Buffetaut, E. 1983. Sur la persistance tardive d'un crocodilien archaïque dans le Pleistocene d'Ile des Pins (Nouvelle-Calédonie) et sa signification biogeographique. Comptes Rendus de l'Académie des Sciences Paris (Série 2) 297:89–92.

Bull, P. 1983. Chemical sedimentation in caves. Pp. 301–319 *in* A. S. Goudie & K. Pye (eds.). Chemical Sediments and Geomorphology: Precipitates and Residua in the Near-surface Environment. Academic Press, London.

Bulmer, S. 1975. Settlement and economy in prehistoric Papua New Guinea: a review of the archaeological evidence. Journal de la Société des Océanistes 31:7–75.

Bunge, F. M. & M. W. Cooke (eds.). 1984. Oceania: A Regional Study. U.S. Government Printing Office, Washington.

Burgess, S. M. 1987. The Climate and Weather of Western Kiribati. New Zealand Meteorological Service, Wellington.

Burland, J. C. 1964. Some notes on the bird life of Palmerston Atoll. Notornis 11:145–154.

Burley, D. V. 1996. Sport, status, and field monuments in the Polynesian chiefdom of Tonga: the pigeon snaring mounds of northern Ha'apai. Journal of Field Archaeology 23:421–435.

Burley, D. V., W. R. Dickinson, A. Barton & R. Shutler, Jr. 2001. Lapita on the periphery: new data on old problems in the Kingdom of Tonga. Archaeology in Oceania 34:59–72.

Burley, D. V., E. Nelson & R. Shutler, Jr. 1995. Rethinking Tongan Lapita chronology in Ha'apai. Archaeology in Oceania 30:132–134.

Burley, D. V., E. Nelson & R. Shutler, Jr. 1999. A radiocarbon chronology for the eastern Lapita frontier in Tonga. Archaeology in Oceania 34:59–70.

Burney, D. A., R. V. DeCandido, L. P. Burney, F. N. Kostel-Hughes, T. W. Stafford, Jr. & H. F. James. 1995. A Holocene record of climate change, fire ecology, and human activity from montane Flat Top Bog, Maui. Journal of Paleolimnology 13:209–217.

Burney, D. A., H. F. James, L. P. Burney, S. L. Olson, W. Kikuchi, W. L. Wagner, M. Burney, D. McCloskey, D. Kikuchi, F. V. Grady, R. Gage, II & R. Nishek. 2001. Fossil evidence for a diverse biota from Kaua'i and its transformation since human arrival. Ecological Monographs 71:615–641.

Burney, D. A., G. S. Robinson & L. P. Burney. 2003. *Sporormiella* and the late Holocene extinctions in Madagascar. Proceedings of the National Academy of Sciences USA 100:10800–10805.

Burney, D. A., D. W. Steadman & P. S. Martin. 2002. Evolution's second chance. Wild Earth 12:12–15.

Burney, L. D. & D. A. Burney. 2003. Charcoal stratigraphies for Kaua'i and the timing of human arrival. Pacific Science 57:211–226.

Bush, M. B. & R. J. Whittaker. 1993. Non-equilibration in island theory of Krakatau. Journal of Biogeography 20:453–457.

Butler, B. M. 1994. Early prehistoric settlement in the Mariana Islands: new evidence from Saipan. Man and Culture in Oceania 10:15–38.

Butler, B. M. [1991 or later]. An archaeological survey of Aguiguan (Aguijan) Northern Mariana Islands. Micronesian Archaeological Survey Report no. 29, Saipan.

Butler, V. L. 1994. Fish feeding behaviour and fish capture: the case for variation in Lapita fishing strategies. Archaeology in Oceania 29:81–90.

Butler, V. L. 2001. Changing fish use on Mangaia, Southern Cook Islands: resource depression and the prey choice model. International Journal of Osteoarchaeology 11:88–100.

Butler, V. L. & N. J. Bowers. 1998. Ancient DNA from salmon bone: a preliminary study. Ancient Biomolecules 2:17–26.

Cabanes, C., A. Cazenave & C. Le Provost. 2001. Sea level rise during past 40 years determined from satellite and in situ observations. Science 294:840–842.

Cabioch, G. & L. K. Ayliffe. 2001. Raised coral terraces at Malakula, Vanuatu, southwest Pacific, indicate high sea level during marine isotope stage 3. Quaternary Research 56:357–365.

Cabioch, G., K. A. Banks-Cutler, W. J. Beck, G. S. Burr, T. Corrège, R. L. Edwards & F. W. Taylor. 2003. Continuous reef growth during the last 23 cal kyr BP in a tectonically active zone (Vanuatu, south west Pacific). Quaternary Science Reviews 22:1771–1786.

Cabioch, G., L. F. Montaggioni, G. Faure & A. Ribaud-Laurenti. 1999. Reef coralgal assemblages as recorders of paleobathymetry and sea level changes in the Indo-Pacific province. Quaternary Science Reviews 18:1681–1695.

Cain, A. J. 1954. Affinities of the fruit pigeon *Ptilinopus perousii* Peale. Ibis 96:104–110.

Cain, A. J. 1955. A revision of *Trichoglossus haematodus* and of the Australian platycercine parrots. Ibis 97:432–479.

Cain, A. J. 1957. Range-changes and differential selection in fruit-pigeons of the *Ptilinopus purpuratus* species-group. Proceedings of the 8th Pacific Science Congress 3A:1393–1412.

Cain, A. J. & I. C. J. Galbraith. 1956. Field notes on the birds of the eastern Solomon Islands. Ibis 98:100–134, 262–295.

Cairns, D. K. 1992. Bridging the gap between ornithology and fisheries science: use of seabird data in stock assessment models. Condor 94:811–824.

Callaway, R. M., G. C. Thelen, A. Rodriguez & W. E. Holben. 2004. Soil biota and exotic plant invasion. Nature 427:731–733.

Calmant, S. & A. Cazenave. 1987. Anomalous elastic thickness of the oceanic lithosphere in the south-central Pacific. Nature 328:236–238.

Cane, M. A. & P. Molnar. 2001. Closing of the Indonesian seaway as a precursor to east African aridification around 3–4 million years ago. Nature 411:157–162.

Cann, R. L. 2001. Genetic clues to dispersal in human populations: retracing the past from the present. Science 291:1742–1748.

Carbone, C. & J. L. Gittleman. 2002. A common rule for the scaling of carnivore density. Science 295:2273–2276.

Carboneras, C. 1992a. Family Diomedeidae (Albatrosses). Pp. 198–215 *in* J. del Hoyo, A. Elliott & J. Sargatal (eds.). Handbook of the Birds of the World, vol. 1, Ostrich to Ducks. Lynx, Barcelona.

Carboneras, C. 1992b. Family Anatidae (Ducks, Geese, and Swans). Pp. 536–628 *in* J. del Hoyo, A. Elliott & J. Sargatal (eds.). Handbook of the Birds of the World, vol. 1, Ostrich to Ducks. Lynx, Barcelona.

Caress, D. W., M. K. McNutt, R. S. Detrick & J. C. Mutter. 1995. Seismic imaging of hotspot-related underplating beneath the Marquesas Islands. Nature 373:600–603.

Carlquist, S. J. 1965. Island Life: A Natural History of the Islands of the World. Natural History Press, Garden City, NJ.

Carlquist, S. J. 1974. Island Biology. Columbia University Press, New York.

Carney, J. N., A. McFarlane & D. I. J. Mallick. 1985. The Vanuatu island arc: an outline of the stratigraphy, structure, and petrology. Pp. 683–718 *in* A. E. M. Nairn, F. G. Stehli & S. Uyeda (eds.). The Ocean Basins and Margins, vol. 7A, The Pacific Ocean. Plenum Press, New York.

Carter, J. (ed.). 1984. Pacific Islands Year Book. Pacific Publishers, Sydney, Australia.

Case, T. J. 1996. Global patterns in the establishment and distribution of exotic birds. Biological Conservation 78:69–96.

Case, T. J., M. E. Gilpin & J. M. Diamond. 1979. Overexploitation, interference, competition, and excess density compensation in insular faunas. American Naturalist 113:843–854.

Cassells, R. 1984. The role of prehistoric man in the faunal extinctions of New Zealand and other Pacific islands. Pp. 741–767 *in* P. S. Martin and R. G. Klein (eds.). Quaternary Extinctions. University of Arizona Press, Tucson.

Chandran, R., R. J. Craig, Z. Keys, C. Sheu & J. Dubrall. 1992. The structure and tree species composition of Aguiguan forests. Pp. 51–56 *in* R. J. Craig (ed.). The Aguiguan Expedition. Proceedings: Marianas Research Symposium, vol. 1. Northern Marianas College, Saipan.

Chang, J. H. 1972. Atmospheric Circulation Systems and Climates. Oriental Press, Honolulu.

Channell, R. & M. V. Lomolino. 2000. Dynamic biogeography and conservation of endangered species. Nature 403:84–86.

Chanteau, S., Y. Sechan, J. P. Moulia-Pelat, P. Luquiaud, A. Spiegel, J. P. Boutin & J. F. Roux. 1993. The blackfly *Simulium buissoni* and infection by hepatitis B virus on a holoendemic island of the Marquesas archipelago in French Polynesia. American Journal of Tropical Medicine and Hygiene 48:763–770.

Chantler, P. & G. Driessens. 1995. Swifts. A Guide to the Swifts and Treeswifts of the World. Yale University Press, New Haven, CT.

Chappell, J. 1998. Jive talking. Nature 394:130–131.

Chaproniere, G. C. H. 1994a. Late Eocene to Pleistocene foraminiferal biostratigraphy and paleobathymetry of dredge samples from the southern Tonga platform (Cruise L3-84-SP). Pp. 45–65 *in* A. J. Stevenson, R. H. Herzer & P. F. Balance (eds.). Geology and Submarine Resources of the Tonga-Lau-Fiji Region. SOPAC Technical Bulletin no. 8.

Chaproniere, G. C. H. 1994b. Middle and late Eocene, Neogene, and Quaternary foraminiferal faunas from 'Eua and Vava'u Islands, Tonga Group. Pp. 21–44 *in* A. J. Stevenson, R. H. Herzer & P. F. Balance (eds.). Geology and Submarine Resources of the Tonga-Lau-Fiji Region. SOPAC Technical Bulletin no. 8.

Chaproniere, G. C. H. 1994c. Late Oligocene, late Miocene, and Pleistocene foraminiferal biostratigraphy and paleobathymetry of dredge samples from the Lau ridge. Pp. 185–195 *in* A. J. Stevenson, R. H. Herzer & P. F. Balance (eds.). Geology and Submarine Resources of the Tonga-Lau-Fiji Region. SOPAC Technical Bulletin no. 8.

Chauvel, C., W. McDonough, G. Guille, R. Maury & R. Duncan. 1997. Contrasting old and young volcanism in Rurutu Island, Austral chain. Chemical Geology 139:125–143.

Chavez, F. P., P. G. Stutton, G. E. Friederich, R. A. Feely, G. C. Feldman, D. G. Foley & M. J. McPhaden. 1999. Biological and chemical response of the equatorial Pacific Ocean to the 1997–98 El Niño. Science 286:2126–2131.

Chen, W.-P. & M. R. Brudzinski. 2001. Evidence for a large-scale remnant of subducted lithosphere beneath Fiji. Science 292:2475– 2479.

Cheneval, J., L. Ginsburg, C. Mourer-Chauviré & B. Ratanasthien. 1991. The Miocene avifauna of the Li Mae Long locality, Thailand: systematics and paleoecology. Journal of Southeast Asian Earth Sciences 6:117–126.

Child, P. 1960. Birds of the Gilbert and Ellice (Tuvalu) Islands colony. Atoll Research Bulletin no. 74.

Christensen, C. C. & P. V. Kirch. 1986. Nonmarine molluscs and ecological change at Barbers Point, O'ahu, Hawai'i. Bernice P. Bishop Museum Occasional Papers 26:52–80.

Christidis, L., R. Schodde, D. D. Shaw & S. F. Maynes. 1991. Relationship among the Australo-Papuan parrots, lorikeets, and cockatoos (Aves: Psittaciformes): protein evidence. Condor 93:302–317.

Christopher, B. 1994. The Kingdom of Tonga: A Geography Resource for Teachers. Friendly Island Book Shop, Nuku'alofa, Kingdom of Tonga.

Church, J. A. 2001. How fast are sea levels rising? Science 294:802–803.

Cibois, A., J.-C. Thibault & E. Pasquet. 2004. Biogeography of Eastern Polynesian monarchs (*Pomarea*): an endemic genus close to extinction. Condor 106:837–851.

Cincotta, R. P., J. Wisnewski & R. Engelman. 2000. Human population in the biodiversity hotspots. Nature 404:990–992.

Clapp, R. B. 1977. Notes on the vertebrate fauna of Tongareva Atoll. Atoll Research Bulletin 198:1–8.

Clark, G. A., Jr. 1964. Life histories and the evolution of megapodes. Living Bird 3:149–167.

Clark, G. 2003. Shards of meaning: archaeology and the Melanesia-Polynesia divide. The Journal of Pacific History 38:197–215.

Clark, G., A. Anderson & S. Matararaba. 2001. The Lapita site at Votua, northern Lau Islands, Fiji. Archaeology in Oceania 36:134–145.

Clark, J. G. & J. Dymond. 1977. Geochronology and petrochemistry of Easter and Sala y Gomez Islands: implications for the origin of the Sala y Gomez Ridge. Journal of Volcanology and Geothermal Research 2:29–48.

Clark, J. S. & J. S. McLachlan. 2003. Stability of forest biodiversity. Nature 423:635–638.

Clark, J. S. & J. S. McLachlan. 2004. Reply to Volkov et al. Nature 427:696–697.

Clark, J. T. & K. M. Kelley. 1993. Human genetics, paleoenvironments, and malaria: relationships and implications for the settlements of Oceania. American Anthropologist 95:612–630.

Clark, P. U., R. B. Alley & D. Pollard. 1999. Northern hemisphere ice-sheet influences on global climate change. Science 286:1104–1111.

Clark, P. U., J. X. Mitrovica, G. A. Milne & M. E. Tamisiea. 2002. Sea-level fingerprinting as a direct test for the source of global meltwater pulse IA. Science 295:2438–2441.

Clark, P. U. & A. C. Mix. 2000. Ice sheets by volume. Nature 406:689–690.

Clark, P. U. & A. C. Mix. 2002. Ice sheets and sea level of the last glacial maximum. Quaternary Science Reviews 21:1–7.

Clark, R. 1982. Proto-Polynesian birds. Transactions of the Finnish Anthropological Society 11:121–143.

Claustre, H. & S. Maritorena. 2003. The many shades of ocean blue. Science 302:1514–1515.

Cleere, N. 1998. Nightjars: A Guide to the Nighthawks, Nightjars, and Their Relatives. Yale University Press, New Haven, CT.

Cleere, N. & D. Nurney. 1998. Nightjars. Pica Press, Sussex, UK.

Cleere, N., A. W. Kratter, D. W. Steadman, M. J. Braun, C. J. Huddleston, C. E. Filardi & G. Dutson. MS submitted. A new genus of frogmouth (Podargidae) from the Solomon Islands: Results from a taxonomic review of *Podargus ocellatus inexpectatus* Hartert 1901. Ibis.

Clements, J. F. 2000. Birds of the World: A Checklist. Ibis Publishing, Vista, CA.

Clift, P. D. & ODP Leg 135 Scientific Party. 1995. Volcanism and sedimentation in a rifting island-arc terrain: an example from Tonga, SW Pacific. Pp. 29–51 *in* J. L. Smellie (ed.). Volcanism Associated with Extension at Consuming Plate Margins. Geological Society of London Special Publication no. 81.

Clout, M. N. & A. J. Saunders. 1995. Conservation and ecological restoration in New Zealand. Pacific Conservation Biology 2:91–98.

Clunie, F. 1983. The extinction of the wandering whistling-duck in Fiji. Domodomo 1:132–142.

Clunie, F. 1984. Birds of the Fiji Bush. Fiji Museum, Suva.

Clunie, F. 1985. Notes on the bats and birds of Rotuma. Domodomo 3:153–160.

Coates, B. J. 1973. Observations of birds on Emira (or Squally) Island, New Ireland District. From 22.3.73 to 17.4.73. New Guinea Bird Society Newsletter 87:2.

Coates, B. J. 1985. The Birds of Papua New Guinea, vol. I. Non-passerines. Dove Publications, Alderley, Australia.

Coates, B. J. 1990. The Birds of Papua New Guinea, vol. II. Passerines. Dove Publications, Alderley, Australia.

Coates, A. G, J. B. Jackson, L. S. Collins, T. M. Cronin, H. J. Dowsett, L. M. Bybell, P. Jung & J. A. Obando. 1992. Closure of the Isthmus of Panama: the near-shore marine record of Costa Rica and western Panama. Geological Society of America Bulletin 104:814–828.

Cochrane, M. A. & W. F. Laurance. 2002. Fire as a large-scale edge effect in Amazonian forests. Journal of Tropical Ecology 18:311–325.

Cooper, A., J. Rhymer, H. F. James, S. L. Olson, C. E. McIntosh, M. D. Sorenson & R. C. Fleischer. 1996. Ancient DNA and island endemics. Nature 381:484.

Coello, J. J., C. Castillo & E. M. González. 1999. Stratigraphy, chronology, and paleoenvironmental reconstruction of the Quaternary sedimentary infilling of a volcanic tube in Fuerteventura, Canary Islands. Quaternary Research 52:360–368.

Cohen, A. L., K. E. Owens, G. D. Layne & N. Shimizu. 2002. The effect of algal symbionts on the accuracy of Sr/Ca paleotemperatures from coral. Science 296:331–333.

Collar, N. J. & P. Andrew. 1988. Birds to Watch: The ICBP World Checklist of Threatened Birds. ICBP Technical Publication no. 8. International Council for Bird Protection, Cambridge, UK.

Collerson, K. D., S. Hapugoda, B. S. Kamber & Q. Williams. 2000. Rocks from the mantle transition zone: majorite-bearing xenoliths from Malaita, southwest Pacific. Science 288:1215–1223.

Condit, R., P. S. Ashton, P. Baker, S. Bunyavejchewin, S. Gunatilleke, N. Gunatilleke, S. P. Hubbell, R. B. Foster, A. Itoh, J. V. LaFrankie, H. S. Lee, E. Losos, N. Manokaran, R. Sukumar & T. Yamakura. 2000. Spatial patterns in the distribution of tropical tree species. Science 288:1414–1418.

Condit, R., N. Pitman, E. G. Leigh, Jr., J. Chave, J. Terborgh, R. B. Foster, P. Núñez, S. Aguilar, R. Valencia, G. Villa, H. C. Muller-Landau, E. Losos & S. P. Hubbell. 2002. Beta-diversity in tropical forest trees. Science 295:666–669.

Connor, E. F. & D. Simberloff. 1979. The assembly of species communities: chance or competition? Ecology 60:1132–1140.

Connor, E. F. & D. Simberloff. 1983. Interspecific competition and species co-occurrence patterns on islands: null models and the evaluation of evidence. Oikos 41:455–465.

Connor, E. F. & D. Simberloff. 1984. Neutral models of species co-occurrence patterns. Pp. 316–331 *in* D. R. Strong, D. Simberloff, L. G. Abele & A. B. Thistle (eds.) Ecological Communities: Conceptual Issues and the Evidence. Princeton University Press, Princeton, NJ.

Conte, E. & A. Anderson. 2003. Radiocarbon ages for two sites on Ua Huka, Marquesas. Asian Perspectives 42:155–160.

Cook, R. R. & I. Hanski. 1995. On expected lifetimes of small-bodied and large-bodied species of birds on islands. American Naturalist 145:307–315.

Cooper, A., 1997. Ancient DNA and avian systematics: from Jurassic Park to modern island extinctions. Pp. 345–373 *in* D. Mindell (ed.). Avian Molecular Evolution and Molecular Systematics. Academic Press, New York.

Cooper, A., C. Lalueza-Fox, S. Anderson, A. Rambaut, J. Austin & R. Ward. 2001. Complete mitochondrial genome sequences of two extinct moas clarify ratite evolution. Nature 409:704–707.

Cooper, A., J. D. Rhymer, H. F. James, S. L. Olson, C. E. McIntosh, M. D. Sorenson & R. C. Fleischer. 1996. Ancient DNA and island endemics. Nature 381:484.

Cooper, J. E. 1989. The role of pathogens in threatened populations: an historical review. Pp. 51–61 *in* J. E. Cooper (ed.). Disease and threatened birds. International Council for Bird Preservation Technical Publications no. 10.

Coote, T., E. Loeve, J.-Y. Meyer & D. Clarke. 1999. Extant populations of endemic partulids on Tahiti, French Polynesia. Oryx 33:215–222.

Corner, E. J. H. 1963. *Ficus* in the Pacific region. Pp. 233–245 *in* J. L. Gressitt (ed.). Pacific Basin Biogeography. Bernice P. Bishop Museum Press, Honolulu.

Corner, E. J. H. 1967. *Ficus* in the Solomon Islands and its bearing on the post-Jurassic history of Melanesia. Philosophical Transactions of the Royal Society of London B 253:23–159.

Coulson, F. I. & J. G. Vedder. 1986. Geology of the central and western Solomon Islands. Pp. 59–87 *in* J. G. Vedder, K. S. Pound & S. Q. Boundy (eds.). Geology and Offshore Resources of Pacific Island Arcs: Central and Western Solomon Islands. Circum-Pacific Council for Energy and Mineral Resources, Earth Science Series, vol. 4, Houston.

Cowie, J. D. 1980. Soils from andesitic tephra and their variability, Tongatapu, Kingdom of Tonga. Australian Journal of Soil Research 18:273–284.

Cowie, R. H. 1992. Evolution and extinction of Partulidae, endemic Pacific island land snails. Philosophical Transactions of the Royal Society of London B 335:167–191.

Cowie, R. H. & R. J. Rundell. 2002. The land snails of a small tropical pacific island, Aunu'u, American Samoa. Pacific Science 56:143–147.

Cox, P. A., T. Elmqvist, E. D. Pierson & W. E. Rainey. 1991. Flying foxes as strong interactors in south Pacific island ecosystems: a conservation hypothesis. Conservation Biology 5:448–454.

Craib, J. L. 1993. Early occupation at Unai Chulu, Tinian, Commonwealth of the Northern Mariana Islands. Bulletin of the Indo-Pacific Prehistory Association 13:116–134.

Craib, J. 1999. Colonisation of the Mariana Islands: new evidence and implications for human movements in the western Pacific. Pp. 477– 485 *in* J. C. Galipaud & I. Lilley (eds.). The Pacific from 5000 to 2000 BP: Colonisation and Transformations. IRD, Paris.

Craig, D. A., D. C. Currie & D. A. Joy. 2001. Geographical history of the central-western Pacific black fly subgenus *Inseliellum* (Diptera: Simuliidae: *Simulium*) based on a reconstructed phylogeny of the species, hot-spot archipelagoes, and hydrological considerations. Journal of Biogeography 28:1101–1127.

Craig, D. A. & D. A. Joy. 2000. New species and redescriptions in the central-western Pacific subgenus *Inseliellum* (Diptera: Simuliidae). Annals of the Entomological Society of America 93:1236–1262.

Craig, R. J. 1989. Observations on the foraging ecology and social behavior of the bridled white-eye. Condor 91:187–192.

Craig, R. J. 1990. Foraging behavior and microhabitat use of two species of white-eyes (Zosteropidae) on Saipan, Micronesia. Auk 107:500–505.

Craig, R. J. 1992a. Ecological characteristics of a native limestone forest on Saipan, Mariana Islands. Micronesica 25:85–97.

Craig, R. J. 1992b. Territoriality, habitat use, and ecological distinctness of an endangered Pacific island reed-warbler. Journal of Field Ornithology 63:436–444.

Craig, R. J. (ed.). 1992c. The Aguiguan Expedition. Proceedings: Marianas Research Symposium, vol. 1. Northern Marianas College, Saipan.

Craig, R. J. 1996. Seasonal population surveys and natural history of a Micronesian bird community. Wilson Bulletin 108:246–267.

Craig, R. J., R. Chandran & A. Ellis. 1992a. Bird populations of Aguiguan: a ten year update. Pp. 2–15 *in* R. J. Craig (ed.). The Aguiguan Expedition. Proceedings: Marianas Research Symposium, vol. 1. Northern Marianas College, Saipan.

Craig, R. J., R. Kaipat, B. A. Lussier & H. Sabino. 1992b. Foraging differences between small passerines on Aguiguan and Saipan. Pp. 16–22 *in* R. J. Craig (ed.). The Aguiguan Expedition. Proceedings: Marianas Research Symposium, vol. 1. Northern Marianas College, Saipan.

Cranwell, L. M. 1962. Endemism and isolation in the Three Kings Islands, New Zealand: with notes on pollen and spore types of the endemics. Records of the Auckland Institute and Museum 5:215–232.

Cranwell, L. M. (ed.). 1964. Ancient Pacific Floras: The Pollen Story. University of Hawai'i Press, Honolulu.

Crawley, M. J. & J. E. Harral. 2001. Scale dependence in plant biodiversity. Science 291:864–868.

Crombie, R. I. & G. K. Pregill. 1999. A checklist of the herpetofauna of the Palau Islands (Republic of Belau), Oceania. Herpetological Monographs 13:29–80.

Crombie, R. I. & D. W. Steadman. 1986. The lizards of Rarotonga and Mangaia, Cook Island Group, Oceania. Pacific Science 40:44–57.

Crowley, T. J. 2002. Cycles, cycles everywhere. Science 295:1473–1474.

Croxall, J. P. 1987. Conclusions. Pp. 369–381 *in* J. P. Croxall (ed.). Seabirds: Feeding Ecology and Role in Marine Ecosystems. Cambridge University Press, Cambridge, UK.

Cuffey, K. M. & S. J. Marshall. 2000. Substantial contribution to sea-level rise during the last interglacial from the Greenland ice sheet. Nature 404:591–594.

Cullen, D. J. & W. C. Burnett. 1987. "Insular" phosphorite on submerged atolls in the tropical southwest Pacific. Search 18:311–313.

Cunningham, J. K. & K. J. Anscombe. 1985. Geology of 'Eua and other islands, Kingdom of Tonga. Pp. 221–257 *in* D. W. Scholl & T. L. Vallier (eds.). Geology and Offshore Resources of Pacific Island Arcs: Tonga Region. Circum-Pacific Council for Energy and Mineral Resources Earth Sciences Series, vol. 2, Houston.

Cupp, E. W. & M. S. Cupp. 1997. Black fly (Diptera: Simuliidae) salivary secretions: importance in vector competence and disease. Journal of Medical Entomology 34:87–89.

Curnutt, J. & S. L. Pimm. 2001. How many bird species in Hawai'i and the central Pacific before first contact? Pp. 15–30 *in* J. M. Scott, S. Conant & C. van Riper III (eds.). Evolution, Ecology, Conservation, and Management of Hawaiian Birds: A Vanishing Avifauna. Studies in Avian Biology 22.

Curran, L. M., I. Caniago, G. D. Paoli, D. Astianti, M. Kusneti, M. Leighton, C. E. Nirarita & H. Haeruman. 1999. Impact of El Niño and logging on canopy tree recruitment in Borneo. Science 286:2184–2188.

Dahl, A. L. 1979. Marine ecosystems and biotic provinces in the South Pacific area. Pp. 541–546 *in* Proceedings of the International Symposium on Marine Biogeography in the Southern Hemisphere. New Zealand DSIR Information Series no. 137.

Dahl, A. L. 1980. Regional ecosystems survey of the South Pacific area. South Pacific Commission Technical Paper 179:1–99.

Dalrymple, G. B., R. D. Jarrard & D. A. Clague. 1975. K-Ar ages of some volcanic rocks from the Cook and Austral Islands. Geological Society of America Bulletin 86:1463–1467.

Daszak, P., A. A. Cunningham & A. D. Hyatt. 2000. Emerging infectious diseases of wildlife: threats to biodiversity and human health. Science 287:443–449.

Dawson, E. Y. 1959. Changes in Palmyra Atoll and its vegetation through the activities of man, 1913–1958. Pacific Naturalist 1:1–51.

de Garidel-Thoron, T., Y. Rosenthal, F. Bassinot & L. Beaufort. 2005. Stable sea surface temperatures in the western Pacific warm pool over the past 1.75 million years. Nature 433:294–298.

de Laubenfels, D. J. 1988. Coniferales. Flora Malesiana, ser. I: 337– 453.

de Laubenfels, D. J. 1996. Gondwanan conifers on the Pacific rim. Pp. 261–265 *in* A. Keast and S. E. Miller (eds.). The Origin and Evolution of Pacific Island Biotas, New Guinea to Eastern Polynesia: Patterns and Processes. SPB Academic Publishing, Amsterdam.

de Mazancourt, C. 2001. Consequences of community drift. Science 293:1772.

Debus, S. J. S. 1994. Australasian species accounts of Accipitridae and Falconidae *in* J. M. Thiollay. Family Accipitridae (Hawks and Eagles). Pp. 52–205 *in* J. del Hoyo, A. Elliott & J. Sargatal (eds.). Handbook of the Birds of the World, vol. 2, New World Vultures to Guineafowl. Lynx, Barcelona.

Decker, B. G. 1973. Unique dry-island biota under official protection in northwestern Marquesas Islands (Iles Marquises). Biological Conservation 5:66–67.

Dekker, R. W. R. J. 1989. Predation and western limits of megapode distribution (Megapodiidae: Aves). Journal of Biogeography 16:317–321.

Delacour, J. 1966. Guide des oiseaux de Nouvelle Calédonie. Delachaux et Niestlé, Neufchatel, France.

Denevan, W. M. 1992. The pristine myth: the landscape of the Americas in 1492. Annals of the Association of American Geographers 82:369–385.

Denham, T. P., S. G. Haberle, C. Lentfer, R. Fullagar, J. Field, M. Therin, N. Porch & B. Winsborough. 2003. Origins of agriculture at Kuk Swamp in the highlands of New Guinea. Science 301:189–193.

Dening, G. M. 1963. The geographic knowledge of the Polynesians and the nature of inter-island contact. Pp. 102–153 *in* J. Golson (ed.). Polynesian Navigation. Polynesian Society Memoir no. 34.

Derrick, R. A. (revised by C. A. A. Hughes and R. B. Riddell). 1965. The Fiji Islands: A Geographical Handboook. Fiji Government Printing Department, Suva.

Devaney, D. M., E. S. Reese, B. L. Burch & P. Helfrich (eds.). 1987. The Natural History of Enewetak Atoll, vols. I and II. U.S. Department of Energy, Office of Scientific and Technical Information, Oak Ridge, TN.

Deyrup, M., L. Davis & S. Cover. 2000. Exotic ants in Florida. Transactions of the American Entomological Society 126:293–326.

Diamond, J. M. 1969. Avifaunal equilibria and species turnover rates on the Channel Islands of California. Proceedings of the National Academy of Sciences USA 64:57–63.

Diamond, J. M. 1970a. Ecological consequences of island colonization by southwest Pacific birds, I. Types of niche shifts. Proceedings of the National Academy of Sciences USA 67:529–536.

Diamond, J. M. 1970b. Ecological consequences of island colonization by southwest Pacific birds, II. The effect of species diversity on total population density. Proceedings of the National Academy of Sciences USA 67:1715–1721.

Diamond, J. M. 1971. Bird records from west New Britain. Condor 73:481–493.

Diamond, J. M. 1972a. Avifauna of the eastern highlands of New Guinea. Publications of the Nuttall Ornithological Club no. 12.

Diamond, J. M. 1972b. Biogeographic kinetics: estimation of relaxation times for avifaunas of southwest Pacific islands. Proceedings of the National Academy of Sciences USA 69:3199–3203.

Diamond, J. M. 1972c. Comparison of faunal equilibrium turnover rates on a tropical and a temperate island. Proceedings of the National Academy of Sciences USA 68:2742–2745.

Diamond, J. M. 1973. Distributional ecology of New Guinea birds. Science 79:759–769.

Diamond, J. M. 1974. Colonization of exploded volcanic islands by birds: the supertramp strategy. Science 184:803–806.

Diamond, J. M. 1975a. Distributional ecology and habits of some Bougainville birds (Solomon Islands). Condor 77:14–23.

Diamond, J. M. 1975b. Assembly of species communities. Pp. 342–444 *in* M. L. Cody & J. M. Diamond (eds.). Ecology and Evolution of Communities. Belknap Press, Cambridge, MA.

Diamond, J. M. 1976a. Preliminary results of an ornithological exploration of the islands of Vitiaz and Dampier Straits, Papua New Guinea. Emu 76:107.

Diamond, J. M. 1976b. Relaxation and differential extinction on land-bridge islands: applications to natural preserves. Pp. 618–628 *in* Proceedings of the 16th International Ornithological Congress.

Diamond, J. M. 1977. Continental and insular speciation in Pacific island birds. Systematic Zoology 26:263–268.

Diamond, J. M. 1981. Reconstitution of bird community structure on Long Island, New Guinea, after a volcanic explosion. National Geographic Society Research Reports 13:191–204.

Diamond, J. M. 1982. Effect of species pool size on species occurrence frequencies: musical chairs on islands. Proceedings of the National Academy of Sciences USA 79:2420–2424.

Diamond, J. M. 1984. The avifauna of Rennell and Bellona islands. Natural History of Rennell Island, British Solomon Islands 8:127–168.

Diamond, J. M. 1985. Population processes in island birds: immigration, extinction and fluctuation. International Council for Bird Preservation Technical Publication 3:17–21.

Diamond, J. M. 1987. Extant unless proven extinct? Or, extinct unless proven extant? Conservation Biology 1:77–79.

Diamond, J. M. 1989. A new subspecies of the island thrush *Turdus poliocephalus* from Tolokiwa Island in the Bismarck Archipelago. Emu 89:58–60.

Diamond, J. M. 1991. A new species of rail from the Solomon Islands and convergent evolution of insular flightlessness. Auk 108:461–470.

Diamond, J. M. 2000. Taiwan's gift to the world. Nature 403:709–710.

Diamond, J. M. 2001. Unwritten knowledge: preliterate societies depend on the wise words of the older generations. Nature 410:521.

Diamond, J. M. 2002. Dispersal, mimicry, and geographic variation in northern Melanesian birds. Pacific Science 56:1–22.

Diamond, J. M. & H. L. Jones. 1980. Breeding landbirds of the Channel Islands. Pp. 597–612 *in* D. M. Power (ed.). The California Islands. Santa Barbara Museum of Natural History, Santa Barbara, CA.

Diamond, J. M. & M. LeCroy. 1979. Birds of Karkar and Bagabag Islands, New Guinea. Bulletin of the American Museum of Natural History 164:467–531.

Diamond, J. M. & A. G. Marshall 1976. Origin of the New Hebridean avifauna. Emu 76:187–200.

Diamond, J. M. & A. G. Marshall. 1977. Distributional ecology of New Hebridean birds: a species kaleidoscope. Journal of Animal Ecology 46:703–727.

Diamond, J. M. & R. M. May. 1976. Island biogeography and the design of natural reserves. Pp. 228–252 *in* R. M. May (ed.). Theoretical Ecology: Principles and Applications. Blackwell Scientific Publications, Oxford.

Diamond, J. M. & R. M. May. 1977. Species turnover on islands: dependence on census interval. Science 197:266–270.

Diamond, J. M. & E. Mayr. 1976. Species-area relation for birds of the Solomon Archipelago. Proceedings of the National Academy of Sciences USA 73:262–266.

Diamond, J. M., M. E. Gilpin & E. Mayr. 1976. Species-distance relation for birds of the Solomon Archipelago, and the paradox of the great speciators. Proceedings of the National Academy of Sciences USA 73:2160–2164.

Dickinson, W. R. 1998a. Petrography of sand tempers in prehistoric Watom sherds in comparison with other temper suites of the Bismarck Archipelago. New Zealand Journal of Archaeology 20:161–182.

Dickinson, W. R. 1998b. Geomorphology and geodynamics of the Cook-Austral-Island-Seamount Chain in the South Pacific Ocean: implications for hotspots and plumes. International Geology Review 40:1039–1075.

Dickinson, W. R. 1999a. Holocene sea-level record on Funafuti and potential impact of global warming on central Pacific atolls. Quaternary Research 51:124–132.

Dickinson, W. R. 1999b. A century of Cordilleran research. Geological Society of America, Special Paper 338:7–13.

Dickinson, W. R. 2000a. Changing times: the Holocene legacy. Environmental History 5:483–502.

Dickinson, W. R. 2000b. Hydro-isostatic and tectonic influences on emergent Holocene paleoshorelines in the Mariana Islands, Western Pacific Ocean. Journal of Coastal Research 16:735 –746.

Dickinson, W. R. 2001. Petrography and geologic provenance of sand tempers in prehistoric potsherds from Fiji and Vanuatu, South Pacific. Geoarchaeology 16:275–322.

Dickinson, W. R. 2002. Petrologic character and geologic sources of sand tempers in prehistoric New Caledonian pottery. Pp.165–172 *in* S. Bedford, C. Sand & D. Burley (eds.). Fifty Years in the Field: Essays in Honour and Celebration of Richard Shutler Jr.'s Archaeological Career. New Zealand Archaeological Association Monograph 25.

Dickinson, W. R. 2003. Impact of mid-Holocene hydro-isostatic highstand in regional sea level on habitability of islands in Pacific Oceania. Journal of Coastal Research 19:489–502.

Dickinson, W. R. 2004a. Impacts of eustasy and hydro-isostasy on the evolution and landforms of Pacific atolls. Palaeogeography, Palaeoclimatology, Palaeoecology 213:251–269.

Dickinson, W. R. 2004b. Picture essay of Pacific island coasts. Journal of Coastal Research 20:1012–1034.

Dickinson, W. R. & R. C. Green. 1998. Geoarchaeological context of Holocene subsidence at the ferry berth Lapita site, Mulifanua, Upolu, Samoa. Geoarchaeology 13:239–263.

Dickinson, W. R. & R. Shutler, Jr. 2000. Implications of petrographic temper analysis for Oceanian prehistory. Journal of World Prehistory 14:203–266.

Dickinson, W. R., D. V. Burley & R. Shutler, Jr. 1994. Impact of hydro-isostatic Holocene sea-level change on the geologic context of island archaeological sites, northern Ha'apai Group, Kingdom of Tonga. Geoarchaeology 9:85–111.

Dickinson, W. R., D. V. Burley & R. Shutler, Jr. 1999. Holocene paleoshoreline record in Tonga: geomorphic features and archaeological implications. Journal of Coastal Research 15:682–700.

Dickinson, W. R., B. M. Butler, D. R. Moore & M. Swift. 2001. Geologic sources and geographic distribution of sand tempers in prehistoric potsherds from the Mariana Islands. Geoarchaeology 16:827–854.

Dilcher, D. L. 2000. Toward a new synthesis: major evolutionary trends in the angiosperm fossil record. Proceedings of the National Academy of Sciences USA 97:7030–7036.

Dodson, J. R. & M. Intoh. 1999. Prehistory and palaeoecology of Yap, Federated States of Micronesia. Quaternary International 59:17–26.

Dodson, J. R., R. Fullagar & H. Head. 1988. The Naive Lands. Longman-Cheshire, Melbourne, Australia.

Dostal, J., B. Cousens & C. Dupuy. 1998. The incompatible element characteristics of ancient subducted sedimentary component in ocean island basalts from French Polynesia. Journal of Petrology 39:937–952.

Doughty, C., N. Day & A. Plant. 1999. Birds of the Solomons, Vanuatu & New Caledonia. Christopher Helm, London.

Douglas, D. H., III. 1998. Changes in the distribution and abundance of waved albatrosses at Isla Española, Galápagos Islands, Ecuador. Condor 100:737–740.

Douglas, G. 1969. Draft check list of Pacific Ocean islands. Micronesica 5:327–463.

Dove, C. J. 1998. Feather evidence helps clarify locality of anthropological artifacts in the Museum of Mankind. Pacific Studies 21:73–85.

Dowler, R. C., D. S. Carroll & C. W. Edwards. 2000. Rediscovery of rodents (Genus *Nesoryzomys*) considered extinct in the Galápagos Islands. Oryx 34:109–117.

Downie, J. E. & J. P. White. 1978. Balof Shelter, New Ireland: report on a small excavation. Records of the Australian Museum 31:762–802.

Drake, D. R., W. A. Whistler, T. J. Motley & C. T. Imada. 1996. Rain forest vegetation of 'Eua Island, Kingdom of Tonga. New Zealand Journal of Botany 34:65–77.

Dransfield, J., J. R. Flenley, S. M. King, D. D. Harkness & S. Rapu. 1984. A recently extinct palm from Easter Island. Nature 312:750–752.

Driver, M. G. 1989. The account of Fray Juan Pobre's residence in the Marianas, 1602. Micronesian Area Research Center, University of Guam, Mangilao.

Drummond, K. J., J. Corvalán D., E. Inoue, W. D. Palfreyman, H. F. Doutch, C. Craddock, T. Sato, F. W. McCoy, G. W. Moore, P. W. Richards, T. R. Swint-Iki & A. L. Gartner. 2000a. Geologic map of the circum-Pacific region: Pacific Basin sheet. Circum-Pacific Council for Energy and Mineral Resources, U.S. Geological Survey.

Drummond, K. J., G. P. Salas, M. R. Yrigoyen, T. Sumii, H. Natori, M. Kato, W. D. Palfreyman, K. Fujii, E. Inoue, M. Sogabe, G. H. Wood, P. W. Richards, W. V. Bour III, T. R. Swint-Iki, O. Matsubayashi, K. Wakita, J. Corvalán D. & H. F. Doutch. 2000b. Energy-resources map of the circum-Pacific region: Pacific Basin sheet. Circum-Pacific Council for Energy and Mineral Resources, U.S. Geological Survey.

Duivenvoorden, J. F., J.-C. Svenning & S. J. Wright. 2002. Beta diversity in tropical forests. Science 295:636–638.

Duncan, R. A. 1985. Radiometric ages from volcanic rocks along the New Hebrides–Samoa lineament. Pp. 67–76 *in* T. M. Brocher (ed.). Investigations of the Northern Melanesian Borderland: Circum-Pacific Council for Energy and Mineral Resources Earth Sciences Series, vol. 3, Houston.

Duncan, R. A., M. R. Fisk, W. M. White & R. L. Nielsen. 1994. Tahiti: geochemical evolution of a French Polynesian volcano. Journal of Geophysical Research 99:24341–24357.

Duncan, R. A., I. McDougall, R. M. Carter & D. S. Coombs. 1974. Pitcairn Island: another Pacific hot spot? Nature 251:679–682.

Duncan, R. A., T. L. Vallier & D. A. Falvey. 1985. Volcanic episodes at 'Eua, Tonga Islands. Pp. 281–290 *in* D. W. Scholl & T. L. Vallier (eds.). Geology and Offshore Resources of Pacific Island Arcs: Tonga Region. Circum-Pacific Council for Energy and Mineral Resources Earth Science Series, vol. 2, Houston.

DuPon, J. F. 1986. The Effects of Mining on the Environment of High Islands: A Case Study of Nickel Mining in New Caledonia. South Pacific Commission. SPREP Environmental Case Studies no. 1.

duPont, J. E. 1976. South Pacific birds. Delaware Museum of Natural History Monograph no. 3.

Dutson, G. C. L. & J. L. Newman. 1991. Observations on the superb pitta *Pitta superba* and other Manus endemics. Bird Conservation International 1:215–222.

Dwyer, P. D. 1981. Two species of megapode laying in the same mound. Emu 81:173–174.

Dye, T. 1990. The causes and consequences of a decline in the prehistoric Marquesan fishing industry. Pp. 70–84 *in* D. E. Yen & J. M. J. Mummery (eds.). Pacific Production Systems: Approaches to Economic Prehistory. Australian National University Occasional Papers in Prehistory no. 18.

Dye, T. 1996. Assemblage definition, analytic methods, and sources of variability in the interpretation of Marquesan subsistence change. Asian Perspectives 35:73–88.

Dye, T. & D. W. Steadman. 1990. Polynesian ancestors and their animal world. American Scientist 78:209–217.

Eakle, W. L. 1997. Observations of raptors in the Republic of Vanuatu. Journal of Raptor Research 31:303–307.

Ehrhardt, J. P. 1980. L'avifaune de Tubuai. Cahiers du l'Indo Pacifique 2:271–288.

Elkibbi, M. & J. A. Rial. 2002. An outsider's review of the astronomical theory of the climate: is the eccentricity-driven insolation the main driver of the ice ages? Earth-Science Reviews 56:161–177.

Eldredge, L. G., R. T. Tsuda, P. Moore, M. Chernin & S. Neudecker. 1977. A natural history of Maug, Northern Mariana Islands. University of Guam Marine Laboratory Technical Report no. 43.

Eller, G. J. 1989. Grey Ternlets in the Andaman Sea. Notornis 36:159–160.

Ellison, J. C. 1991. The Pacific palaeogeography of *Rhizophora mangle* L. (Rhizophoraceae). Botanical Journal of the Linnean Society 105:271–284.

Ellison, J. C. 1994a. Palaeo-lake and swamp stratigraphic records of Holocene vegetation and sea-level changes, Mangaia, Cook Islands. Pacific Science 48:1–15.

Ellison, J. C. 1994b. Caves and speleogenesis of Mangaia, Cook Islands. Atoll Research Bulletin no. 417.

Elmqvist, T., W. E. Rainey, E. D. Pierson & P. A. Cox. 1994. Effects of tropical cyclones Ofa and Val on the structure of a Samoan lowland rain forest. Biotropica 26:384–391.

Elmqvist, T., M. Wall, A. L. Berggren, L. Blix, Å. Fritioff & U. Rinman. 2001. Tropical forest reorganization after cyclone and fire disturbance in Samoa: remnant trees as biological legacies. Conservation Ecology 5:10.

Emery, K. O., J. I. Tracey & H. S. Ladd. 1954. Geology of Bikini and nearby atolls. U.S. Geological Survey Professional Paper 260-A.

Emory, K. P. 1927. The curved club from a Rurutu cave. Bulletin de la Société des estudies Oceaniennes, Papeete 21:304–306.

Emory, K. P. 1932. The curved club from a Rurutu cave, additional note. Bulletin de la Société des estudies Oceaniennes, Papeete 42:12–14.

Emory, K. P. & Y. H. Sinoto. 1965. Preliminary Report on the Archaeological Excavations in Polynesia. Bernice P. Bishop Museum Library, Honolulu.

Empson, L., J. Flenley & P. Sheppard. 2002. A dated pollen record of vegetation change on Mayor Island (Tuhua) throughout the last 3000 years. Global and Planetary Change 33:329–337.

Emslie, S. D. 1987. Age and diet of fossil California condors in Grand Canyon, Arizona. Science 273:768–770.

Engbring, J. 1983a. Avifauna of the Southwest Islands of Palau. Atoll Research Bulletin no. 267.

Engbring, J. 1983b. First Ponape record of a dollarbird, with a summary of the species' occurrence in Micronesia. 'Elepaio 44:35–36.

Engbring, J. 1988. Field Guide to the Birds of Palau. Conservation Office, Koror, Palau.

Engbring, J. 1992. A 1991 Survey of the Forest Birds of the Republic of Palau. U.S. Fish and Wildlife Service, Honolulu.

Engbring, J. & H. D. Pratt. 1985. Endangered birds in Micronesia: their history, status, and future prospects. Pp. 71–105 *in* S. A. Temple (ed.). Bird Conservation, 2. International Council for Bird Preservation. University of Wisconsin Press, Madison.

Engbring, J. & F. L. Ramsey. 1984. Distribution and abundance of the forest birds of Guam: results of a 1981 survey. U.S. Fish and Wildlife Service FWS/OBS 84/20.

Engbring, J. & F. L. Ramsey. 1989. A 1986 Survey of the Forest Birds of American Samoa. U.S. Fish and Wildlife Service Administrative Report. 145 pp.

Engbring, J., F. L. Ramsey & V. J. Wildman. 1986. Micronesian Forest Bird Survey, 1982: Saipan, Tinian, Aguiguan, and Rota. U.S. Department of the Interior, Fish and Wildlife Service, Honolulu.

Engbring, J., F. L. Ramsey & V. J. Wildman. 1990. Micronesian Forest Bird Surveys, the Federated States: Pohnpei, Kosrae, Chuuk, and Yap. U.S. Fish and Wildlife Service. Department of the Interior, Washington.

Erwin, D. 2000. Life's downs and ups. Nature 404:129–130.

Evans, S. M., F. J. C. Fletcher, P. J. Loader & F. G. Rooksby. 1992. Habitat exploitation by landbirds in the changing Western Samoan environment. Bird Conservation International 2:123–129.

Evenhuis, N. L. (ed.). 1989. Catalog of the Diptera of the Australasian and Oceanian regions. Bernice P. Bishop Museum Special Publication 86, Bishop Museum Press, Honolulu.

Exon, N. F. & M. S. Marlow. 1988. Geology and offshore resource potential of the New Ireland: Manus region. A synthesis. Pp. 241–262 *in* M. S. Marlow, S. V. Dadisman & N. F. Exon (eds.) Geology and Offshore Resources of Pacific Island Arcs: New Ireland and Manus Region, Papua New Guinea. Circum-Pacific Council for Energy and Mineral Resources Earth Science Series, vol. 9, Houston.

Falanruw, M. V. C. 1975. Distribution of the Micronesian megapode *Megapodius laperouse* in the Northern Mariana Islands. Micronesica 11:149–150.

Falanruw, M. V. C. 1994. Food production and ecosystem management on Yap. ISLA 2:5–22.

Falanruw, M. V. C, C. Whitesell, T. Cole, C. MacLean & A. Ambacher. 1987. Vegetation Survey of Yap, Federated States of Micronesia. USDA Forest Service, Pacific Southwest Forest and Range Experiment Station, Resource Bulletin PSW-21.

Falla, R. A. 1942. Review of the smaller Pacific forms of *Pterodroma* and *Cookilaria*. Emu 42:111–118.

Fallon, S. M., E. Bermingham & R. E. Ricklefs. 2003. Island and taxon effects in parasitism revisited: avian malaria in the Lesser Antilles. Evolution 57:606–615.

Falvey, D. A., J. B. Colwell, P. J. Coleman, H. G. Greene, J. G. Vedder & T. R. Bruns. 1991. Petroleum prospectivity of Pacific island arcs: Solomon Islands and Vanuatu. Australian Petroleum Exploration Association Journal 1:191–212.

Fancy, S. G. & T. J. Snetsinger. 2001. What caused the population decline of the bridled white-eye on Rota, Mariana Islands? Studies in Avian Biology 22:274–280.

Fancy, S. G., J. T. Nelson, P. Harrity, J. Kuhn, M. Kuhn, C. Kuhler & J. G. Giffin. 2001. Reintroduction and translocation of a Hawaiian solitaire: a comparison of methods. Studies in Avian Biology 22:347–353.

Fancy, S. G., R. J. Craig & C. W. Kessler. 1999. Forest bird and fruit bat populations on Sarigan, Mariana Islands. Micronesica 31:247–254.

Fariña, J. M., S. Salazar, K. P. Wallem, J. D. Witman & J. C. Ellis. 2003. Nutrient exchanges between marine and terrestrial ecosystems: the case of the Galapagos sea lion *Zalophus wollebaecki*. Journal of Animal Ecology 72:873–887.

Farrand, W. R. 1985. Rockshelter and cave sediments. Pp. 21–39 *in* J. K. Stein & W. R. Farrand (eds.). Archaeological Sediments in Context. Center for the Study of Early Man, Institute for Quaternary Studies, University of Maine, Orono.

Feduccia, A. 1998. Reappraisal of the benefits inferred from cladistic methods: new data proves that alvarezsaurids are birds. Ibis 320:201–205.

Feinberg, R. 1981. Anuta: Social Structure of a Polynesian Island. Institute for Polynesian Studies, Brigham Young University Press, Provo, UT.

Feinberg, R. 1988. Polynesian Seafaring and Navigation: Ocean Travel in Anutan Culture and Society. Kent State University Press, Kent, OH.

Fiji Islands. 1980. NZMS 242 (map). New Zealand Department of Lands & Survey, Wellington.

Filardi, C. E. & C. E. Smith. 2005. Molecular phylogenetics of monarch flycatchers (genus *Monarcha*) with emphasis on Solomon Island endemics. Molecular Phylogenetics and Evolution 37:776–788.

Filardi, C. E., C. E. Smith, A. W. Kratter, D. W. Steadman & H. P. Webb. 1999. New behavioral, ecological, and biogeographic data on the avifauna of Rennell, Solomon Islands. Pacific Science 53:319–340

Finney, B. R. 1979. Voyaging. Pp. 323–351 *in* J. D. Jennings (ed.). The Prehistory of Polynesia. Harvard University Press, Cambridge, MA.

Finney, B. R. 2001. Voyage to Polynesia's land's end. Antiquity 75:172–181.

Finney, B. R., I. Gregory-Eaves, M. S. V. Douglas & J. P. Smol. 2002. Fisheries productivity in the northeastern Pacific Ocean over the past 2,200 years. Nature 416:729–733.

Finsch, O. 1880a. Ornithological letters from the Pacific I. Ibis 4:218–220.

Finsch, O. 1880b. Ornithological letters from the Pacific II. Ibis 4:329–333.

Finsch, O. 1881. Ornithological letters from the Pacific. Ibis 1881:102–114.

Firth, R. 1936. We, the Tikopia: A Sociological Study of Kinship in Primitive Polynesia. American Book Company, New York.

Fisher, A. K. & A. Wetmore. 1931. Report on birds recorded by the Pinchot expedition of 1929 to the Caribbean and Pacific. Proceedings of the United States National Museum 79:1–66.

Fisher, H. I. 1950. The birds of Yap, Western Caroline Islands. Pacific Science 4:55–62.

Fitchett, K. 1987. The physical effects of Hurricane Bebe upon Funafuti Atoll, Tuvalu. Australian Geographer: 18:1–7.

Flannery, T. 1995a. Mammals of the South-West Pacific & Moluccan Islands. Cornell University Press, Ithaca, NY.
Flannery, T. F. 1995b. Mammals of New Guinea. Robert Brown and Associates, Queensland, Australia.
Flannery, T. F. & R. G. Roberts. 1999. Late Quaternary extinctions in Australasia: an overview. Pp. 239–255 *in* R. D. E. MacPhee (ed.). Extinctions in Near Time: Causes, Contexts, and Consequences. Kluwer Academic/Plenum Publishers, New York.
Flannery, T. F. & J. P. White. 1991. Animal translocation: zoogeography of New Ireland mammals. National Geographic Research and Exploration 7:96–113.
Flannery, T. F. & S. Wickler. 1990. Quaternary murids (Rodentia: Muridae) from Buka Island, Papua New Guinea, with descriptions of two new species. Australian Mammalogy 13:127–139.
Flannery, T. F., P. V. Kirch, J. Specht & M. Spriggs. 1988. Holocene mammal faunas from archaeological sites in island Melanesia. Archaeology in Oceania 23:89–94.
Fleischer, R. C. & C. E. McIntosh. 2001. Molecular systematics and biogeography of the Hawaiian avifauna. Studies in Avian Biology 22:51–60.
Fleischer, R. C., S. L. Olson, H. F. James & A. C. Cooper. 2000. Identification of the extinct Hawaiian eagle (*Haliaeetus*) by mtDNA sequence analysis. Auk 117:1051–1056.
Fleischer, R. C., C. L. Tarr, H. F. James, B. Slikas & C. E. McIntosh. 2001. Phylogenetic placement of the Po'ouli, *Melamprosops phaeosoma*, based on mitochondroal DNA sequence data and osteological characters. Studies in Avian Biology 22:98–103.
Flenley, J. R. & S. M. King. 1984. Late Quaternary pollen records from Easter Island. Nature 307:47–50.
Flenley, J. R., S. M. King, J. Jackson, C. Chew, J. T. Teller & M. E. Prentice. 1991. The late Quaternary vegetational and climatic history of Easter Island. Journal of Quaternary Science 6:85–115.
Flood, J. 1983. Archaeology of the Dreamtime: The Story of Prehistoric Australia and Its People. Yale University Press, New Haven, CT.
Florence, J. & D. H. Lorence. 1997. Introduction to the flora and vegetation of the Marquesas Islands. Allertonia 7:226–237.
Florence, J., S. Waldren & A. J. Chepstow-Lusty. 1995. The flora of the Pitcairn Islands: a review. Biological Journal of the Linnean Society 56:79–119.
Foote, M., J. P. Hunter, C. M. Janis & J. J. Sepkoski Jr. 1999. Evolutionary and preservational constraints on origins of biologic groups: divergence times of eutherian mammals. Science 283:1310–1314.
Forbis, R. G. 1978. Some facets of communal hunting. Plains Anthropologist Memoir 14:3–8.
Fosberg, F. R. 1953. Vegetation of Central Pacific atolls. Atoll Research Bulletin 23:1–25.
Fosberg, F. R. 1960. The vegetation of Micronesia. Part I. General descriptions, the vegetation of the Mariana Islands, and a detailed consideration of the vegetation of Guam. Bulletin of the American Museum of Natural History 119:1–75.
Fosberg, F. R. 1966. Northern Marshall Islands land biota: birds. Atoll Research Bulletin no. 114.
Fosberg, F. R. 1990. A review of the natural history of the Marshall Islands. Atoll Research Bulletin no. 330.
Fosberg, F. R. & M. Evans. 1969. A collection of plants from Fais, Caroline Islands. Atoll Research Bulletin no. 133.
Fosberg, F. R., M.-H. Sachet & D. R. Stoddart. 1983. Henderson Island (south-eastern Polynesia): summary of current knowledge. Atoll Research Bulletin 272:1–47.
Fosberg, F. R., J. W. Wells, M. S. Doty & R. Todd. 1965. Other features. Pp. 23–30 *in* F. R. Fosberg & D. Carroll (eds.). Terrestrial sediments and soils of the northern Marshall Islands. Atoll Research Bulletin no. 113.
Foster, T. 1999. Update on the Vanuatu megapode *Megapodius layardi* on Ambrym, Vanuatu. Bird Conservation International 9:63–71.
Fox, B. 1999. The genesis and development of guild assembly rules. Pp. 23–57 *in* E. Weiher & P. Keddy (eds.). Ecological Assembly Rules: Perspectives, Advances, Retreats. Cambridge University Press, Cambridge, UK.
Frank, K. A., D. W. Steadman & E. C. Greiner. In press. Avian hematozoa from Isabel and Rennell, Solomon Islands. Journal of Wildlife Diseases.
Franklin, J. 2003. Regeneration and growth of pioneer and shade-tolerant rain forest trees in Tonga. New Zealand Journal of Botany 41:669–684.
Franklin, J. & M. Merlin. 1992. Species-environment patterns of forest vegetation on the uplifted reef limestone of Atiu, Mitiaro, and Ma'uke, Cook Islands. Journal of Vegetation Science 3:3–14.
Franklin, J. & D. W. Steadman. 1991. The potential for conservation of Polynesian birds through habitat mapping and species translocation. Conservation Biology 5:506–521.
Franklin, J., D. R. Drake, L. A. Bolick, D. S. Smith & T. J. Motley. 1999. Rain forest composition and patterns of secondary succession in the Vava'u Island Group, Tonga. Journal of Vegetation Science 10:51–64.
Franklin, J., D. R. Drake, K. R. McConkey, F. Tonga & L. B. Smith. 2004. The effects of Cyclone Waka on the structure of lowland tropical rain forest in Vava'u, Tonga. Journal of Tropical Ecology 20:409–420.
Frederickson, C. M. Spriggs & W. R. Ambrose. 1993. Pamwak Rockshelter: a Pleistocene site at Manus Island, Papua New Guinea. Pp. 144–152 *in* A. Smith, M. Spriggs & B. Fankhauser (eds.). Sahul in Review: Pleistocene Archaeology in Australia, New Guinea and Island Melanesia. Occasional Papers in Prehistory 24. Australian National University, Canberra.
Freeman, J. D. 1944. The Falemaunga caves. Journal of Polynesian Science 53:86–106.
Freifeld, H. B. 1999. Habitat relationships of forest birds on Tutuila Island, American Samoa. Journal of Biogeography 26:1191–1213.
Freifeld, H. B., C. Solek & A. Tualaulelei. 2004. Temporal variation in forest bird survey data from Tutuila Island, American Samoa. Pacific Science 58:99–117.
Freifeld, H. B., D. W. Steadman & J. K. Sailer. 2001. Landbirds on offshore islands in Samoa. Journal of Field Ornithology 72:72–85.
Frimigacci, D., J. P. Siorat & B. Vienne. 1987. Preliminary report on archaeological sites on Wallis (Uvea) Island yielding bones of pigeons (*Ducula*). P. 775 *in* J. C. Balouet & S. L. Olson (eds.). Proceedings of the Biological Society of Washington 100:769–775.

Frith, H. J. 1956. Breeding habits of the family Megapodiidae. Ibis 98:620–640.

Frith, H. J. 1982. Pigeons and Doves of Australia. Rigby Publishers, Adelaide, Australia.

Frith, H. J., F. H. J. Crome & T. O. Wolfe. 1976. Food of fruit-pigeons in New Guinea. Emu 76:49–58.

Fritts, T. H. 1993. The common wolf snake, *Lycodon aulicus capucinus*, a recent colonist of Christmas Island in the Indian Ocean. Wildlife Research 20:261–266.

Fritts, T. H. & G. H. Rodda. 1996. Trouble in paradise: the brown tree snake in the western Pacific. Aquatic Nuisance Species Digest 1:26–27.

Fritts, T. H. & G. H. Rodda. 1998. The role of introduced species in the degradation of island ecosystems: a case history of Guam. Annual Review of Ecology and Systematics 29:113–140.

Frost, E. L. 1979. Fiji. Pp. 61–81 *in* J. D. Jennings (ed.). The Prehistory of Polynesia. Harvard University Press, Cambridge, MA.

Fry, C. H., K. Fry & A. Harris. 1992. Kingfishers, Bee-eaters, and Rollers: A Handbook. Princeton University Press, Princeton, NJ.

Fryer, P. 1995. Geology of the Mariana Trough. Pp. 237–279 *in* B. Taylor (ed.). Back-arc Basins: Tectonics and Magmatism. Plenum Press, New York.

Fujita, M. S. & M. D. Tuttle. 1991. Flying foxes (Chiroptera: Pteropodidae): threatened animals of key ecological importance. Conservation Biology 5:455–463.

Fullagar, P. J., H. J. De S. Disney & R. De Naurois. 1982. Additional specimens of two rare rails and comments on the genus *Tricholimnas* of New Caledonia and Lord Howe Island. Emu 82:131–136.

Fullard, J. H., R. M. R. Barclay & D. W. Thomas. 1993. Echolocation in free-flying Atiu swiftlets (*Aerodramus sawtelli*). Biotropica 25:334–339.

Fuller, E. 1987. Extinct Birds. Facts on File Publications, New York.

Gadow, H. F. 1898. A Classification of Vertebrata, Recent and Extinct. Black, London.

Gaffney, E. S. 1981. A review of the fossil turtles of Australia. American Museum Novitates 2720:1–38.

Gaffney, E. S., J. C. Balouet & F. de Froin. 1984. New occurrences of extinct meiolaniid turtles in New Caledonia. American Museum Novitates 2800:1–6.

Gagan, M. K., L. K. Ayliffe, J. W. Beck, J. E. Cole, E. R. M. Druffel, R. B. Dunbar & D. P. Schrag. 2000. New views of tropical paleoclimates from corals. Quaternary Science Reviews 19:45–64.

Gagan, M. K., L. K. Ayliffe, D. Hopley, J. A. Cali, G. E. Mortimer, J. Chappell, M. T. McCulloch & M. J. Head. 1998. Temperature and surface-ocean water balance of the mid-Holocene tropical western Pacific. Science 279:1014–1020.

Galbraith, I. O. J. & E. H. Galbraith. 1962. Landbirds of Guadalcanal and the San Cristoval Group, eastern Solomon Islands. Bulletin of the British Museum (Natural History), Zoology 9:1–86.

Gallagher, M. D., D. A. Scott, R. F. G. Ormond, R. J. Connor & M. C. Jennings. 1984. The distribution and conservation of seabirds breeding on the coasts and islands of Iran and Arabia. Pp. 421–456 *in* J. P. Croxall, P. G. H. Evans & R. W. Schreiber (eds.). Status and Conservation of the World's Seabirds. International Council for Bird preservation Technical Publication no. 2, Cambridge, UK.

Galipaud, J.-C. & I. Lilley (eds.). 1999. The South Pacific, 5000 to 2000 B.P.: Colonizations and Transformations. Éditions de Institut de Recherche pour le Développement, Paris.

Garcia, M. A., C. E. Diez & A. O. Alvarez. 2001. The impact of feral cats on Mona Island wildlife and recommendations for their control. Caribbean Journal of Science 37:107–108.

Garnett, M. C. 1984. Conservation of seabirds in the south Pacific region: a review. Pp. 547–558 *in* J. P. Croxall, P. G. H. Evans & R. W. Schreiber (eds.). Status and Conservation of the World's Seabirds. International Council for Bird Preservation Technical Publication no. 2, Cambridge, UK.

Geatz, R. 2000. Roots of tradition. Nature Conservancy 50:20–28.

Gibbons, J. R. H. & F. Clunie. 1986. Sea level changes and Pacific prehistory: new insight into early human settlement of Oceania. Journal of Pacific History 21:58–82.

Gibbons, J. R. H. & I. F. Watkins. 1982. Behavior, ecology, and conservation of South Pacific banded iguanas, *Brachylophus*, including a newly discovered species. Pp. 418–441 *in* G. M. Burghardt & A. S. Rand (eds.). Iguanas of the World: Their Behavior, Ecology, and Conservation. Noyes Publications, Park Ridge, NJ.

Gifford, E. W. & D. S. Gifford. 1959. Archaeological excavations in Yap. University of California Anthropological Records 18:149–224.

Gifford, E. W. & D. Shutler, Jr. 1956. Archaeological excavations in New Caledonia. University of California Anthropological Records 18:1–148.

Gilbert, F. S. 1980. The equilibrium theory of island biogeography: fact or fiction? Journal of Biogeography 7:209–235.

Gilbert Islands to Tuvalu Islands. 1990. Map 83005. United States Defense Mapping Agency, Washington.

Gill, B. J. 1995. Notes on the birds of Wallis and Futuna, southwest Pacific. Notornis 42:17–22.

Gill, B. J. 1996. Notes on certain Cook Island birds. Notornis 43:154–158.

Gill, B. J. 1999. A myna increase: notes on introduced mynas (*Acridotheres*) and bulbuls (*Pycnonotus*) in Western Samoa. Notornis 46:268–269.

Gill, B. J. 2003. Osteometry and systematics of the extinct New Zealand ravens (Aves: Corvidae: *Corvus*). Journal of Systematic Palaeontology 1:43–58.

Gill, B. J. & W. R. Sykes. 1996. T. F. Cheesman's diary of a botanical visit to Rarotonga, Cook Islands, 1899. Records of the Auckland Institute and Museum 33:53–78.

Gill, J. B. 1987. Early geochemical evolution of an oceanic island arc and backarc: Fiji and the South Fiji Basin. Journal of Geology 95:589–615.

Gill, W. W. 1885. Jottings from the Pacific. American Trust Society, New York.

Gill, W. W. 1894. From Darkness to Light in Polynesia. London Missionary Society, London.

Gillie, R. D. 1997a. Coastal processes and causes of coastal erosion on Pacific Islands. Pp. 11–23 *in* A. M. Sherwood (comp.). Coastal and Environmental Geoscience Studies of the Southwest Pacific Islands. SOPAC Technical Bulletin no. 9.

Gillie, R. D. 1997b. Distinctive physical features of Pacific island coastal regions. Pp. 1–10 *in* A. M. Sherwood (comp.). Coastal and Environmental Geoscience Studies of The Southwest Pacific Islands. SOPAC Technical Bulletin no. 9.

Gillieson, D. & M.-J. Mountain. 1983. Environmental history of Nombe rockshelter, Papua New Guinea Highlands. Archaeology in Oceania 18:45–53.

Gillooly, J. F., J. H. Brown, G. B. West, V. M. Savage & E. L. Charnov. 2001. Effects of size and temperature on metabolic rate. Science 293:2248–2251.

Gillot, F. Y., Y. Cornette & G. Guille. 1992. Age (K-Ar) et conditions d'édification du soubassement volcanique de l'atoll de Mururoa (Pacifique Sud). Comptes Rendus de l'Académie des Science (Série 2) 314:393–399.

Gilpin, M. E. & J. M. Diamond. 1976. Calculation of immigration and extinction curves from the species-area-distance relation. Proceedings of the National Academy of Sciences USA 73:4130–4134.

Gilpin, M. E. & J. M. Diamond. 1984. Are species co-occurrences on islands non-random, and are null hypotheses useful in community ecology? Pp. 297–341 *in* D. R. Strong, D. Simberloff, L. G. Abele & A. B. Thistle (eds.). Ecological Communities: Conceptual Issues and the Evidence. Princeton University Press, Princeton, NJ.

Gittleman, J. L. & M. E. Gompper. 2001. The risk of extinction: what you don't know will hurt you. Science 291:997–999.

Gladwin, T. 1970. East Is a Big Bird: Navigation and Logic on Puluwat Atoll. Harvard University Press, Cambridge, Massachusetts.

Glenn, T. C., W. Stephan, M. J. Braun. 1999. Effects of a population bottleneck on whooping crane mitochondrial DNA variation. Conservation Biology 13:1097–1107.

Gobalet, K. W. 2001. A critique of faunal analysis: inconsistency among experts in blind tests. Journal of Archaeological Science 28:377–386.

Good, R. 1974. The Geography of the Flowering Plants. Longman, Greens & Co., London.

Goodacre, S. L. & C. M. Wade. 2001. Molecular evolutionary relationships between partulid land snails of the Pacific. Proceedings of the Royal Society of London B7 268:1–7.

Goodwin, D. 1960. Taxonomy of the genus *Ducula*. Ibis 102:526–535.

Gordon, A. L., R. D. Susanto & K. Vranes. 2003. Cool Indonesian throughflow as a consequence of restricted surface layer flow. Nature 425:824–828.

Gosden, C. 1991. Towards an understanding of the regional archaeological record from the Arawe Islands, West New Britain, Papua New Guinea. Pp. 205–216 *in* J. Allen & C. Gosden (eds.). Report of the Lapita Homeland Project. Department of Prehistory, Australian National University, Canberra.

Gosden, C. & N. Robertson. 1991. Models for Matenkupkum: interpreting a late Pleistocene site from southern New Ireland, Papua New Guinea. Pp. 20–45 *in* J. Allen & C. Gosden (eds.). Report of the Lapita Homeland Project. Department of Prehistory, Australian National University, Canberra.

Gosden, C. & J. Webb. 1994. The creation of a Papua New Guinean landscape: archaeological and geomorphological evidence. Journal of Field Archaeology 21:29–51.

Gosden, C., J. Allen, W. Ambrose, D. Anson, J. Golson, R. Green, P. Kirch, P. I. Lilley, J. Specht & M. Spriggs. 1989. Lapita sites of the Bismarck Archipelago. Antiquity 63:561–586.

Gotelli, N. J. 1999. How do communities come together? Science 286:1684–1685.

Gotelli, N. J. & A. E. Arnett. 2000. Biogeographic effects of red fire ant invasion. Ecology Letters 3:257–261.

Gotelli, N. J. & G. L. Entsminger. 2001. Swap and fill algorithms in null model analysis: rethinking the knight's tour. Oecologia 129:281–291.

Gotelli, N. J. & G. R. Graves. 1996. Null Models in Ecology. Smithsonian Institution Press, Washington.

Göth, A. & U. Vogel. 1995. Status of the Polynesian megapode *Megapodius pritchardii* on Niuafo'ou (Tonga). Bird Conservation International 5:117–128.

Göth, A. & U. Vogel. 1997. Egg laying and incubation of the Polynesian megapode. Annual Review of the World Pheasant Association 1996/97:43–54.

Göth, A. & U. Vogel. 1999. Notes on breeding and conservation of birds on Niuafo'ou Island, Kingdom of Tonga. Pacific Conservation International 5:103–114.

Graham, R. W., E. L. Lundelius, Jr., M. A. Graham, E. K. Schroeder, R. S. Toomey III, E. Anderson, A. D. Barnosky, J. A. Burns, C. S. Churcher, D. K. Grayson, R. D. Guthrie, C. R. Harrington, G. T. Jefferson, L. D. Martin, H. G. McDonald, R. E. Morlan, H. A. Semken Jr., S. D. Webb, L. Werdelin & M. C. Wilson. 1996. Spatial response of mammals to late Quaternary environmental fluctuations. Science 272:1601–1606.

Grant, B. R. & P. R. Grant. 1979. Darwin's finches: population variation and sympatric speciation. Proceedings of the National Academy of Sciences USA 76:2359–2363.

Grant, P. R. & B. R. Grant. 1986. Ecology and Evolution of Darwin's Finches. Princeton University Press, Princeton, NJ.

Grant, B. R. & P. R. Grant. 1989. Evolutionary Dynamics of a Natural Population: The Large Cactus Finch of the Galapagos. University of Chicago Press, Chicago.

Grant, P. R. 1998. Speciation. Pp. 83–101 *in* P. R. Grant (ed.). Evolution on Islands. Oxford University Press, New York.

Grant, P. R. & B. R. Grant. 2002. Unpredictable evolution in a 30-year study of Darwin's finches. Science 296:707–711.

Grant, P. R. & B. R. Grant. 1996. Finch communities in a fluctuating environment. Pp. 343–390 *in* M. L. Cody and J. A. Smallwood (eds.). Long-term Studies of Vertebrate Communities. Academic Press, New York.

Grant, P. R. & B. R. Grant. 1998. Speciation and hybridization of birds on islands. Pp. 142–162 *in* P. R. Grant (ed.). Evolution on Islands. Oxford University Press, New York.

Graves, G. R. 1992. The endemic land birds of Henderson Island, southeastern Polynesia: notes on natural history and conservation. Wilson Bulletin 104:32–43.

Graves, G. R. & N. J. Gotelli. 1983. Neotropical land-bridge avifaunas: new approaches to null hypotheses in biogeography. Oikos 41:322–333.

Gray, R. D. & F. M. Jordan. 2000. Language trees support the express-train sequence of Austronesian expansion. Nature 405:1052–1055.

Gray, S. C., J. R. Hein., R. Hausmann & U. Radtke. 1992. Geochronology and subsurface stratigraphy of Pukapuka and Rakahanga Atolls, Cook Islands: Late Quaternary reef growth and sea level history. Palaeogeography, Palaeoclimatology, Palaeoecology 91:377–394.

Gray, W. M. 1984. Hurricanes: their formation, structure and likely role in tropical circulation. Pp. 155–218 *in* D. B. Shaw (ed.). Meteorology over the Tropical Oceans. Royal Meteorological Society, Berkshire, UK.

Grayson, D. K. & F. Delpech. 1998. Changing diet breadth in the early upper Palaeolithic of southwestern France. Journal of Archaeological Science 25:1119–1129.

Grayson, D. K. & S. D. Livingston. 1993. Missing mammals on Great Basin Mountains: Holocene extinctions and inadequate knowledge. Conservation Biology 7:527–532.

Grayson, D. K. & D. B. Madsen. 2000. Biogeographic implications of recent low-elevation recolonization by *Neotoma cinerea* in the Great Basin. Journal of Mammalogy 81:1100–1105.

Green, H. W., II. 2001. A graveyard for buoyant slabs? Science 292:2445–2446.

Green, R. C. 1976. Lapita sites in the Santa Cruz Group. Pp. 245–265 *in* R. C. Green & M. M. Cresswell (eds.). Southeast Solomon Islands Cultural History: A Preliminary Survey. Bulletin of the Royal Society of New Zealand 11.

Green, R. C. 1988. Those mysterious mounds are for the birds. Archaeology in New Zealand 31:153–158.

Green, R. C. 1991. Near and Remote Oceania: disestablishing "Melanesia" in cultural history. Pp. 491–502 *in* A. Pawley (ed.). Man and a Half: Essays in Pacific Anthropology and Ethnobiology in Honour of Ralph Bulmer. Polynesian Society, Auckland, New Zealand.

Green, R. C. 1999. Integrating historical linguistics with archaeology: insights from research in remote Oceania. Indo-Pacific Prehistory Association Bulletin 18:3–16.

Green, R. C. 2003. The Lapita horizon and traditions: signature for one set of oceanic migrations. Le Cahiers de l'Archéologie en Nouvelle-Calédonie 15:95–120.

Green, R. C. & D. Anson. 1991. The Reber-Rakival Lapita site on Watom. Implications of the 1985 excavations at the SAC and SDI localities. Pp. 170–181 *in* J. Allen & C. Gosden (eds.). Report of the Lapita Homeland Project. Department of Prehistory, Australian National University, Canberra, Australia.

Green, R. C. & J. Davidson (eds.). 1969. Archaeology in Western Samoa, vol. I. Bulletin of the Auckland Institute and Museum 6, Auckland.

Green, R. C. & J. Davidson (eds.). 1974. Archaeology in Western Samoa, vol. II. Bulletin of the Auckland Institute and Museum 7, Auckland.

Green, R. C. & J. S. Mitchell. 1983. New Caledonian cultural history: a review of the archaeological sequence. New Zealand Journal of Archaeology 5:19–67.

Green, R. C., K. Green, R. A. Rappaport, A. Rappaport & J. M. Davidson. 1967. Archeology on the island of Mo'orea, French Polynesia. Anthropological Papers of the American Museum of Natural History no. 51.

Greenway, J. C, Jr. 1958. Extinct and Vanishing Birds of the World. American Committee for International Wild Life Protection, New York.

Gressitt, J. L. 1956. Some distribution patterns of Pacific island faunas. Systematic Zoology 5:11–32.

Gressitt, J. L. 1961. Problems in the zoogeography of Pacific and Antarctic insects. Pacific Insects Monographs 2:1–94.

Griffith, B., J. M. Scott, J. W. Carpenter & C. Reed. 1989. Translocation as a species conservation tool: status and strategy. Science 245:477–480.

Groombridge, J. J., C. G. Jones, M. W. Bruford & R. A. Nichols. 2000. "Ghost" alleles of the Mauritius kestrel. Nature 403:616.

Grossman, E. E., C. H. Fletcher & B. M. Richmond. 1998. The Holocene sea-level highstand in the equatorial Pacific: analysis of the insular paleosea-level database. Coral Reefs 17:309–327.

Groube, L. 1989. The taming of the rain forests: a model for late Pleistocene forest exploitation in New Guinea. Pp. 292–304 *in* D. R. Harris & G. C. Hillman (eds.). Foraging and Farming: The Evolution of Plant Exploitation. Unwin Hyman, London.

Groube, L., J. Chappell, J. Muke & D. Price. 1986. A 40,000-year-old human occupation site at Huon Peninsula, Papua New Guinea. Nature 324:453–455.

Grover, J. C. 1960. The geology of Rennel and Bellona. Pp. 103–119 *in* T. Wolff (ed.). The Natural History of Rennell Island, British Solomon Islands, vol. 3 (Botany and Geology): Scientific results of the Danish Rennell Expedition, 1951 and the British Museum (Natural History) Expedition, 1953. Danish Science Press, Copenhagen.

Guillou, H., R. Brousse, P. Y. Gillot & G. Guille. 1993. Geological reconstruction of Fangataufa Atoll, South Pacific. Marine Geology 110:377–391.

Gulick, L. H. 1858. The climate and productions of Ponape or Ascension Island, one of the Carolines, in the Pacific Ocean. American Journal of Science and Art 26:34–49.

Gura, T. 2000. "Bones, molecules, . . . or both?" Nature 406:230–233.

Haberle, S. G. 2003. Late Quaternary vegetation dynamics and human impact on Alexander Selkirk Island, Chile. Journal of Biogeography 30:239–255.

Hadden, D. 1981. Birds of the North Solomons. Wau Ecology Institute Handbook no. 8.

Hadden, D. 2002. Woodford's rail (*Nesoclopeus woodfordi*) on Bougainville Island, Papua New Guinea. Notornis 49:115–121.

Hadfield, M. G., S. E. Miller & A. H. Carwile. 1993. The decimation of endemic Hawaiian tree snails by alien predators. American Zoologist 33:610–622.

Hadly, E. A. 1999. Fidelity of terrestrial vertebrate fossils to a modern ecosystem. Palaeogeography, Palaeoclimatology, Palaeoecology 149:389–409.

Hagelberg, E. 2001. Genetic affinities of the principal human lineages in the Pacific. Terra Australis 17:167–176.

Haig, S. M. & J. D. Ballou. 1995. Genetic diversity in two avian species formerly endemic to Guam. Auk 112:445–455.

Hall, P. S. & C. Kincaid. 2001. Diapiric flow at subduction zones: a recipe for rapid transport. Science 292:2472–2475.

Halle, F. 1978. Architectural variation at the specific level in tropical trees. Pp. 209–221 *in* P. B. Tomlinson & M. H. Zimmermann (eds.). Tropical Trees as Living Systems. Cambridge University Press, Cambridge, UK.

Hanan, B. B. & J. G. Schilling. 1989. Easter microplate evolution: Pb isotope evidence. Journal of Geophysical Research 94:7432–7448.

Handschumacher, D. W., R. H. Pilger Jr., J. A. Foreman & J. F. Campbell. 1981. Structure and evolution of the Easter plate. Pp. 63–76 *in* L. D. Kulm, J. Dymond, E. J. Dasch & D. M. Hussong (eds.). Nazca Plate: Crustal Formation and Andean Convergence. Geological Society of America Memoir 154.

Hanebuth, T., K. Stattegger & P. M. Grootes. 2000. Rapid flooding of the Sunda Shelf: a late-glacial sea-level record. Science 288:1033–1035.

Hannecart, F. & Y. Letocart. 1980. Oiseaux de Nouvelle Calédonie et des Loyautés, vol. I. Les Editions Cardinalis, Nouméa, New Caledonia.

Hannecart, F. & Y. Letocart. 1983. Oiseaux de Nouvelle Calédonie et des Loyautés, vol. II. Les Editions Cardinalis, Nouméa, New Caledonia.

Hannon, G. E. & R. H. W. Bradshaw. 2000. Impacts and timing of the first human settlement on vegetation of the Faroe Islands. Quaternary Research 54:404–413.

Hanski, I. & M. Gyllenberg. 1997. Uniting two general patterns in the distribution of species. Science 275:397–400.

Harlow, P. S. & P. N. Biciloa. 2001. Abundance of the Fijian crested iguana (*Brachylophus vitiensis*) on two islands. Biological Conservation 98:223–231.

Harrison, C. S. & M. P. Seki. 1987. Trophic relationships among tropical seabirds at the Hawaiian Islands. Pp. 305–326 *in* J. P. Croxall (ed.). Seabirds: Feeding Ecology and Role in Marine Ecosystems. Cambridge University Press, Cambridge, UK.

Harrison, P. & J. R. Jehl, Jr. 1988. Notes on the seabirds of Sala y Gomez. Condor 90:259–261.

Hartert, E. 1924a. The birds of New Hanover. Novitates Zoologicae 31:194–213.

Hartert, E. 1924b. The birds of St. Matthias Island. Novitates Zoologicae 31:261–275.

Hartert, E. 1924c. The birds of Squally or Storm Island. Novitates Zoologicae 31:276–278.

Hartert, E. 1925. A collection of birds from New Ireland. Novitates Zoologicae 32:115–136.

Hartert, E. 1926. On the birds of the District of Talasea in New Britain. Novitates Zoologicae 33:122–145.

Hartlaub, G. & O. Finsch. 1871. On a collection of birds from Savai and Rarotonga Islands in the Pacific. Proceedings of the Zoological Society of London 1871:21–32.

Hastings, P. A. 1990. Southern Oscillation influences on tropical cyclone activity in the Australian/southwest Pacific region. International Journal of Climatology 10:291–298.

Hathway, B. & H. Colley. 1994. Eocene to Miocene geology of southwest Viti Levu, Fiji. Pp. 153–169 *in* A. J. Stevenson, R. H. Herzer & P. F. Balance (eds.). Geology and Submarine Resources of the Tonga-Lau-Fiji Region. SOPAC Technical Bulletin no. 8.

Hau'ofa, E. 1977. Our Crowded Islands. Institute of Pacific Studies, University of the South Pacific, Suva, Fiji.

Hay, R. 1986. Bird conservation in the Pacific Islands. International Council for Bird Preservation Study Report 7.

Hay, R. & R. Powlesland. 1998. Guide to the Birds of Niue. South Pacific Regional Environment Programme, Apia, Samoa.

Hayek, L. C., W. R. Heyer & C. Gascon. 2001. Frog morphometrics: a cautionary tale. Alytes 18:153–177.

Haynes, G. & B. S. Eiselt. 1999. The power of Pleistocene hunter-gatherers: forward and backward searching for evidence about mammoth extinction. Pp. 71–93 *in* R. D. E. MacPhee (ed.). Extinctions in Near Time: Causes, Contexts, and Consequences. Kluwer Academic/Plenum Publishers, New York.

Hearty, P. J., S. L. Olson, D. S. Kaufman, R. L. Edwards & H. Cheng. 2004. Stratigraphy and geochronology of pitfall accumulations in caves and fissures, Bermuda. Quaternary Science Reviews 23:1151–1171.

Heather, B. D. & H. A. Robertson. 1997. The Field Guide to the Birds of New Zealand. Oxford University Press, Oxford.

Heinroth, O. 1902. Ornithologische Ergebnisse der "I. Deutschen Südsee Expedition von Br. Mencke." Journal für Ornithologie 50:390–457.

Heinsohn, R., S. Legge & S. Barry. 1997. Extreme bias in sex allocation in *Eclectus* parrots. Proceedings of the Royal Society of London B 264:1325–1329.

Hekinian, R., D. Bideau, P. Stoffers, J. L. Cheminee, R. Muhe, D. Puteanus & N. Binard. 1991. Submarine intraplate volcanism in the South Pacific: geological setting and petrology of the Society and the Austral regions. Journal of Geophysical Research 96:2409–2438.

Helffrich, G. & B. Wood. 2001. The Earth's mantle. Nature 412:501–507.

Henderson-Sellers, A. & P. J. Robinson. 1986. Contemporary Climatology. John Wiley & Sons, New York.

Hern, W. M. 1993. Is human culture carcinogenic for uncontrolled population growth and ecological destruction? Bioscience 43:768–773.

Herron, E. M. 1972. Two small crustal plates in the South Pacific near Easter Island. Nature: Physical Science 240:35–37.

Herzer, R. H. & N. F. Exon. 1985. Structure and basin analysis of the southern Tonga forearc. Pp. 55–73 *in* D. W. Scholl & T. L. Vallier (eds.). Geology and Offshore Resources of Pacific Island Arcs: Tonga Region. Circum-Pacific Council for Energy and Mineral Resources Earth Sciences Series, vol. 2, Houston.

Heyerdahl, T. 1961. Surface artifacts. Pp. 397–489 *in* T. Heyerdahl & E. N. Ferdon, Jr. (eds.). Archaeology of Easter Island, vol. 1. Monograph of the School of American Research and the Museum of New Mexico no. 24.

Hieronymus, C. F. & D. Bercovici. 1999. Discrete alternating hotspot islands formed by interaction of magma transport and lithospheric flexure. Nature 397:604–607.

Hilder, B. 1963. Primitive navigation in the Pacific-II. Pp. 81–97 *in* J. Golson (ed.). Polynesian Navigation. Polynesian Society Memoir no. 34.

Hill, K. D. 1996. Cycads in the Pacific. Pp. 267–274 *in* A. Keast and S. E. Miller (eds.). The Origin and Evolution of Pacific Island Biotas, New Guinea to Eastern Polynesia: Patterns and Processes. SPB Academic Publishing, Amsterdam.

Hill, R. S. 1996. The riddle of unique southern hemisphere *Nothofagus* on southwest Pacific Islands: its challenge to biogeographers. Pp. 247–260 *in* A. Keast & S. E. Miller (eds.). The Origin and Evolution of Pacific Island Biotas, New Guinea to Eastern Polynesia: Patterns and Processes. SPB Academic Publishing, Amsterdam.

Hiroa, T. R. (P. H. Buck]. 1934. Mangaian society. Bernice P. Bishop Museum Bulletin no. 122.

Hiroa, T. R. (P. H. Buck]. 1944. Arts and crafts of the Cook Islands. Bernice P. Bishop Museum Bulletin no. 179.

Hjerpe, J., H. Hedenås & T. Elmqvist. 2001. Tropical rain forest recovery from cyclone damage and fire in Samoa. Biotropica 33:249–259.

Hoch, H. & M. Asche. 1988. A new troglobitic meenoplid from a lava tube in Western Samoa (*Homoptera, Fulgoroidea, Meenoplidae*). Journal of Natural History 22:1489–1494.

Hoffman, K. A. 1991. Long-lived transitional states of the geomagnetic field and the two dynamo families. Nature 354:273–277.

Hoffmeister, J. E. 1932. Geology of Eua, Tonga. Bernice P. Bishop Museum Bulletin no. 96.

Holdaway, R. N. 1999. Introduced predators and avifaunal extinction in New Zealand. Pp. 189–238 *in* R. D. E. MacPhee (ed.). Extinctions in Near Time: Causes, Contexts, and Consequences. Kluwer Academic/Plenum Publishers, New York.

Holdaway, R. N. & C. Jacomb. 2000. Rapid extinction of the moas (Aves: Dinornithiformes): model, test, and implications. Science 287:2250–2254.

Holdaway, R. N. & T. H. Worthy. 1994. A new fossil species of shearwater *Puffinus* from the late Quaternary of the South Island, New Zealand, and notes on the biogeography and evolution of the *Puffinus gavia* superspecies. Emu 94:201–215.

Holdaway, R. N., T. H. Worthy & A. J. D. Tennyson. 2001. A working list of breeding bird species of the New Zealand region at first human contact. New Zealand Journal of Zoology 28:119–187.

Holden, C. 2000. Pop-up island. Science 288:1735.

Holdridge, L. R. 1967. Life Zone Ecology. Tropical Science Center, San José, Costa Rica.

Holyoak, D. T. 1974. Undescribed land birds from the Cook Islands. Bulletin of the British Ornithologists' Club 94:145–150.

Holyoak, D. T. 1979. Notes on the birds of Viti Levu and Taveuni, Fiji. Emu 79:7–18.

Holyoak, D. T. 1980. Guide to Cook Island Birds. Privately published.

Holyoak, D. T. & J.-C. Thibault. 1977. Habitats, morphologie et interactions ecologiques des oiseaux insectivores de Polynésie orientale. L'Oiseau et Revue Française du Ornithologie 47:115–147.

Holyoak, D. T. & J.-C. Thibault. 1978a. Notes on the phylogeny, distribution, and ecology of frugivorous pigeons in Polynesia. Emu 78:201–206.

Holyoak, D. T. & J.-C. Thibault. 1978b. Undescribed *Acrocephalus* warblers from Pacific Ocean islands. Bulletin of the British Ornithologist's Club 98:122–127.

Holyoak, D. T. & J.-C. Thibault. 1984. Contribution a l'étude des oiseaux de Polynésie Orientale. Memoires du Museum National D'histoire Naturelle, Nouvelle Série, Série A, Zoologie 127:1–209.

Hope, G. 1998. Early fire and forest change in the Baliem Valley, Irian Jaya, Indonesia. Journal of Biogeography 25:453–461.

Hope, G. S. 1996. History of *Nothofagus* in New Guinea and New Caledonia. Pp. 257–270 *in* T. T. Veblen, R. S. Hill & J. Read (eds.). The Ecology and Biogeography of *Nothofagus* Forests. Yale University Press, New Haven, CT.

Hope, G. S. & J. Golson. 1995. Late Quaternary change in the mountains of New Guinea. Antiquity 69 (special no. 265): 818–830.

Hope, G. S. & J. Tulip. 1994. A long vegetation history from lowland Irian Jaya, Indonesia. Palaeogeography, Palaeoclimatology, Palaeoecology 109:385–398.

Hope, G. S., D. O'Dea & W. Southern. 1999. Holocene vegetation histories in the Western Pacific: alternative records of human impact. Pp. 387–404 *in* I. Lilley & J.-C. Galipaud (eds.). Le Pacifique de 5000 à 2000 Avant le Présent: Suppléments à L'histoire d'une Colonisation. IRD, Paris.

Hopper, D. R. & B. D. Smith. 1992. Status of tree snails (Gastropoda: Partulidae) on Guam, with a resurvey of sites studied by H. E. Crampton in 1920. Pacific Science 46:77–85.

Höss, M. 2000. Neanderthal population genetics. Nature 404:453–454.

Hotchkiss, S. C. & J. O. Juvik. 1999. A Late-Quaternary pollen record from Ka'au Crater, O'ahu, Hawai'i. Quaternary Research 52:115–128.

Hough, F. 1990. The Assault on Peleliu. USMC Historical Division (1950), Washington. Reprinted by The Battery Press, Nashville, TN.

Houghton, P. 1996. People of the Great Ocean: Aspects of Human Biology of the Early Pacific. Cambridge University Press, Cambridge, UK.

Howells, W. W. 1970. Anthropometric grouping analysis of Pacific peoples. Archaeology and Physical Anthropology in Oceania 5:192–217.

Hubbell, S. P. 2001. The Unified Neutral Theory of Biodiversity and Biogeography. Princeton University Press, Princeton, NJ.

Hughen, K., S. Lehman, J. Southon, J. Overpeck, O. Marchal, C. Herring & J. Turnbull. 2004. ^{14}C activity and global carbon cycle changes over the past 50,000 years. Science 303:202–207.

Humphrey, P. S. & G. A. Clark, Jr. 1964. The anatomy of waterfowl. Pp. 167–232 *in* J. Delacour (ed.). Waterfowl of the World IV. Country Life, London.

Humphrey, P. S., B. C. Livezey & D. Siegel-Causey. 1987. Tameness of birds of the Falkland Islands: an index and preliminary results. Bird Behaviour 7:67–72.

Hunt, G. R. 1996. Family Rhynochetidae (Kagu). Pp. 218–225 *in* J. del Hoyo, A. Elliott & J. Sargatal (eds.). Handbook of the Birds of the World, vol. 1, Ostrich to Ducks. Lynx, Barcelona.

Hunt, G. R., R. Hay & C. J. Veltman. 1996. Multiple kagu *Rhynochetos jubatus* deaths caused by dog attacks at a high-altitude study site on Pic Ningua, New Caledonia. Bird Conservation International 6:295–306.

Hunt, T. L., K. A. Aronson, E. E. Cochrane, J. S. Field, L. Humphrey & T. M. Rieth. 1999. A preliminary report on archaeological research in the Yasawa Islands, Fiji. Domodomo: Fiji Museum Quarterly 12:5–43.

Hunter-Anderson, R. L. & B. M. Butler. 1995. An overview of Northern Marianas prehistory. Micronesian Archaeological Survey Report no. 31.

Hutchings, J. A. 2000. Collapse and recovery of marine fishes. Nature 406:882–885.

Huynen, L., C. D. Millar, R. P. Scofield & D. M. Lambert. 2003. Nuclear DNA sequences detect species limits in ancient moa. Nature 425:175–178.

Hvidberg, C. S. 2000. When Greenland ice melts. Nature 404:551–552.

Île Mangareva. 1995. Map 83252. United States Defense Mapping Agency, Washington.

Îles Marquises. 1976. Map 83020. United States Defense Mapping Agency, Washington.

Ineich, I. & G. R. Zug. 1996. *Tachygia*, the giant Tongan skink: extinct or extant? Cryptozoology 12:30–35.

Innis, G. J. 1989. Feeding ecology of fruit pigeons in subtropical rainforests of south-eastern Queensland. Australian Wildlife Research 16:365–394.

Institut Géographique National. 1976. Archipel des Nouvelles-Hébrides. Institut Géographique National, Paris.

Institut Géographique National. 1988. Archipel de la Société. Institut Géographique National, Paris.

Intoh, M. 1991. Archaeological research on Fais Island: preliminary report. Historic Preservation Office, Yap State, Federated States of Micronesia.

Intoh, M. 1997. Human dispersal in Micronesia. Anthropological Science 105:15–28.

Irwin, G. 1992. The Prehistoric Exploration and Colonisation of the Pacific. Cambridge University Press, Cambridge, UK.

Isaacson, L. B. & D. F. Heinrichs. 1976. Paleomagnetism and secular variation of Easter Island basalts. Journal of Geophysical Research 81:1476–1482.

Isacks, B., L. R. Sykes & J. Oliver. 1969. Focal mechanisms of deep and shallow earthquakes in the Tonga-Kermadec region and the tectonics of island arcs. Geological Society of America Bulletin 80:1443–1470.

Island of Pohnpei. 2001a. Northeast (map). NIMA 5842 III NE—Series W856. United States Geological Survey, Denver, CO.

Island of Pohnpei. 2001b. Northwest (map). NIMA 5842 II NW—Series W856. United States Geological Survey, Denver, CO.

Island of Pohnpei. 2001c. Southeast (map). NIMA 5842 IV SE—Series W856. United States Geological Survey, Denver, CO.

Island of Pohnpei. 2001d. Southwest (map). NIMA 5842 I SW—Series W856. United States Geological Survey, Denver, CO.

Itoh, S. 1991. Geographical variation of the plumage polymorphism in the eastern reef heron (*Egretta sacra*). Condor 93:383–389.

Jablonski, D. 2002. Survival without recovery after mass extinctions. Proceedings of the National Academy of Sciences USA 99:8139–8144.

Jablonski, D. 2003. The interplay of physical and biotic factors in microevolution. Pp. 235–252 *in* L. Rothschild & A. Lister (eds.). Evolution on Planet Earth. Elsevier Press, Amsterdam.

Jablonski, D. & K. Roy. 2003. Geographical range and speciation in fossil and living molluscs. Proceedings of the Royal Society of London B 270:401–406.

Jablonski, D., K. Roy & J. W. Valentine. 2003. Evolutionary macroecology and the fossil record. Pp. 368–390 *in* T. B. Blackburn & K. J. Gaston (eds.). Macroecology: Concepts and Consequences. Blackwell Scientific, Oxford.

Jaffe, M. 1994. And No Birds Sing. Simon and Schuster, New York.

Jaffre, T. & J. M. Veillon. 1990. Etude floristique et structurale de deux forets denses humides sur roches ultrabasique en Nouvelle Calédonie. Science de la Vie, Série Botanique 3:1–41.

James, H. F. 1987. A Late Pleistocene avifauna from the island of Oahu, Hawaiian Islands. Documents des Laboratories de Geologie de la Faculté des Sciences de Lyon no. 99, Pp. 221–230.

James, H. F. 1995. Prehistoric extinctions and ecological changes on oceanic islands. Ecological Studies 115:88–102.

James, H. F. 2001. Systematics: introduction. Studies in Avian Biology 22:48–50.

James, H. F. In press. Hawaiian finches. Biological Journal of the Linnean Society.

James, H. F. & D. A. Burney. 1997. The diet and ecology of Hawaii's extinct flightless waterfowl: evidence from coprolites. Biological Journal of the Linnean Society 62:279–297.

James, H. F. & S. L. Olson. 1991. Descriptions of thirty-two species of birds from the Hawaiian Islands. II. Passeriformes. Ornithological Monographs no. 46.

James, H. F. & S. L. Olson. 2003. A giant new species of nukupuu (Fringillidae: Drepanidini: *Hemignathus*) from the island of Hawaii. Auk 120:970–981.

James, H. F., T. W. Stafford, Jr., D. W. Steadman, S. L. Olson, P. S. Martin, A. J. T. Jull & P. C. McCoy. 1987. Radiocarbon dates on bones of extinct birds from Hawaii. Proceedings of the National Academy of Sciences USA 84:2350–2354.

Jarvi, S. I., C. T. Atkinson & R. C. Fleischer. 2001. Immunogenetics and resistance to avian malaria in Hawaiian honeycreepers (Drepanidinae). Studies in Avian Biology 22:254–263.

Jehl, J. R., Jr. 1997. Fat loads and flightlessness in Wilson's phalaropes. Condor 99:538–543.

Jehl, J. R., Jr. & K. C. Parkes. 1983. "Replacements" of landbird species on Socorro Island, Mexico. Auk 100:551–559.

Jenkins, C. D., S. A. Temple, C. van Riper & W. R. Hansen. 1989. Disease-related aspects of conserving the endangered Hawaiian crow. Pp. 77–87 *in* J. E. Cooper (ed.). Disease and Threatened Birds. International Council for Bird Preservation Technical Publications no. 10.

Jenkins, J. M. 1979. Natural history of the Guam rail. Condor 81:404–408.

Jenkins, J. M. 1983. The native forest birds of Guam. Ornithological Monographs 31.

Jenkins, M. 2003. Prospects for biodiversity. Science 302:1175–1177.

Jennings, J. D., R. N. Holmer, J. C. Janetslei & H. L. Smith. 1976. Excavations on Upolu, Western Samoa. Pacific Anthropological Records 25:1–113.

Jetz, W. & C. Rahbek. 2002. Geographic range size and determinants of avian species richness. Science 297:1548–1551.

Johnson, A. W., W. R. Millie & G. Moffett. 1970. Notes on the birds of Easter Island. Ibis 112:532–538.

Johnson, D. 1996. Palms: Their Conservation and Sustained Utilization. IUCN/SSC Palm Specialist Group, IUCN, Gland and Cambridge.

Johnson, K. P. & D. H. Clayton. 2000. Nuclear and mitochondrial genes contain similar phylogenetic signal for pigeons and doves (Aves: Columbiformes). Molecular Phylogenetics and Evolution 14:141–151.

Johnson, K. P., F. R. Adler & J. L. Cherry. 2000. Genetic and phylogenetic consequences of island biogeography. Evolution 54:387–396.

Johnson, O. W. & R. J. Kienholz. 1975. New avifaunal records for Enewetok. Auk 92:592–594.
Johnston, R. F. 1962. The taxonomy of pigeons. Condor 64:69–74.
Johnson, T. H. & A. J. Stattersfield. 1990. A global review of island endemic birds. Ibis 132:167–180.
Jones, D. N. 1989. Modern megapode research: a post-Frith review. Corella 13:145–154.
Jones, D. N., R. Dekker & C. S. Roselaar. 1995. The Megapodes. Oxford University Press, Oxford.
Jourdan, H. 1997. Threats on Pacific Islands: the spread of the tramp ant *Wasmannia auropunctata* (Hymenoptera: Formicidae). Pacific Conservation Biology 3:61–64.
Juniper, T. & M. Parr. 1998. Parrots: A Guide to Parrots of the World. Yale University Press, New Haven, CT.
Kaneko, H. & T. Abe. 1979. Fish bones. Pp. 343–347 *in* D. Osborne (ed.). Archaeological test excavations, Palau Islands, 1968–1969. Micronesica Supplement 1:1–353.
Kanton Island. 1998. Map 83103. National Imagery and Mapping Agency, Bethesda, MD.
Kataoka, O. 1990. Man and Culture in Oceania 7:71–106.
Karl, S. A. & B. W. Bowen. 1999. Evolutionary significant units versus geopolitical taxonomy: molecular systematics of an endangered sea turtle (genus *Chelonia*). Conservation Biology 13:990–999.
Karolle, B. 1993. Atlas of Micronesia. Bess Press, Honolulu.
Kaufmann, G. W. 1987. Growth and development of sora and Virginia rail chicks. Wilson Bulletin 99:432–440.
Kaufmann, G. W. 1988. The usefulness of taped spotless crake calls as a census technique. Wilson Bulletin 100:682–686.
Kawahata, H., R. Maeda & H. Ohshima. 2002. Fluctuations in terrestrial-marine environments in the western equatorial Pacific during the Late Pleistocene. Quaternary Research 57:71–81.
Kawahata, H., A. Suzuki & N. Ahagon. 1998. Biogenic sediments in the west Caroline Basin, the western equatorial Pacific during the last 330,000 years. Marine Geology 149:155–176.
Kay, C. E. 1997. The ultimate tragedy of the commons. Conservation Biology 11:1447–1448.
Kay, E. A. 1999. Biogeography. Pp. 76–92 *in* M. Rapaport (ed.). The Pacific Islands: Environment and Society. Bess Press, Honolulu.
Kayanne, H., H. Yamano & R. H. Randall. 2002. Holocene sea-level changes and barrier reef formation on an oceanic island, Palau Islands, western Pacific. Sedimentary Geology 150:47–60.
Kear, D. & B. L. Wood. 1959. The geology and hydrology of Western Samoa. New Zealand Geological Survey Bulletin no. 63.
Keast, A. 1996. Avian geography: New Guinea to the eastern Pacific. Pp. 373–398 *in* A. Keast & S. E. Miller (eds.). The Origin and Evolution of Pacific Island Biotas, New Guinea to Eastern Polynesia: Patterns and Processes. SPB Academic Publishing, Amsterdam.
Keast, A. & S. E. Miller (eds.). 1996. The Origin and Evolution of Pacific Island Biotas, New Guinea to Eastern Polynesia: Patterns and Processes. SPB Academic Publishing, Amsterdam.
Keating, B. 1985. Paleomagnetic studies of the Samoan Islands: results from the islands of Tutuila and Savai'i. Pp. 187–199 *in* T. M. Brocher (ed.). Geological Investigations of the Northern Melanesian Borderland: Circum-Pacific Council for Energy and Mineral Resources Earth Science Series, vol. 3, Houston.
Keating, B. H. 1992. The geology of the Samoan Islands. Pp. 127–178 *in* B. H. Keating & B. R. Bolton (eds.). Geology and Mineral Resources of the Central Pacific Basin: Circum-Pacific Council for Energy and Mineral Resources, Earth Science Series, vol. 14. Springer-Verlag, New York.
Keating, B. H., D. P. Mattey, J. Naughton & C. E. Helsley. 1984. Age and origin of Truk atoll, eastern Caroline Islands: geochemical, radiometric age and paleomagnetic evidence. Geological Society of America Bulletin 95:350–356.
Kemp, A. 1995. The Hornbills, Bucerotiformes. Oxford University Press, Oxford.
Kennedy, R. S., P. C. Gonzales, E. C. Dickinson, H. C. Miranda, Jr. & T. H. Fisher. 2000. A Guide to the Birds of the Phillipines. Oxford University Press, New York.
Kennedy, T. F. 1966. A Descriptive Atlas of the Pacific Islands. A. H. Reed & A. W. Reed, Wellington, New Zealand.
Kent, A., T. Webber & D. W. Steadman. 1999. Distribution, relative abundance, and prehistory of birds on the Taraco Peninsula, Bolivian Altiplano. Ornitologia Neotropical 10:151–178.
Kepler, C. B. 1967. Polynesian rat predation on nesting Laysan albatrosses and other Pacific seabirds. Auk 84:426–430.
Keppel, G. 2005. Botanical studies within the PABITRA Wet-Zone Transect, Viti Levu, Fiji. Pacific Science 59:165–174.
Keppel, G., J. C. Navuso, A. Naikatini, N. T. Thomas, I. A. Rounds, T. A. Osborne, N. Batinamu & E. Senivasa. 2005. Botanical diversity at Savura, a lowland rain forest site along the PABITRA Gateway Transect, Viti Levu, Fiji. Pacific Science 59:175–191.
Kerr, I. S. 1976. Tropical storms and hurricanes in the southwest Pacific. New Zealand Meteorological Service Miscellaneous Publication 148, Wellington.
Kessler, C. C. 1999. New cuckoo record and vagrant bird sightings for the Mariana Islands (1995–1998). Micronesica 31:283–287.
Kindler, P. & P. J. Hearty. 2000. Elevated marine terraces from Eleuthera (Bahamas) and Bermuda: sedimentological, petrographic and geochronological evidence for important deglaciation events during the middle Pleistocene. Global and Planetary Change 24:41–58.
King, C. M. 1983. The relationships between beech (*Nothofagus* sp.) seedfall and populations of mice (*Mus musculus*), and the demographic and dietary responses of stoats (*Mustela erminea*), in three New Zealand forests. Journal of Animal Ecology 52:141–166.
King, C. M. & J. E. Moody. 1982. The biology of the stoat (*Mustela erminea*) in the national parks of New Zealand. New Zealand Journal of Zoology 9:57–80.
King, F. W. & R. L. Burke (eds.). 1997. Crocodilian, tuatura, and turtle species of the world: an online taxonomic and geographic reference. Association of Systematics Collections, Washington.
King, J. E. 1958. Some observations on the birds of Tahiti and the Marquesas Islands. 'Elepaio 19:14–17.
King, T. F. & P. L. Parker. 1984. Archeology in the Tonaachaw Historic District, Moen Island. Micronesian Archaeological Survey Report 18. Southern Illinois University at Carbondale,

Center for Archaeological Investigations, Occasional Paper no. 3.
Kingdom of Tonga. 1969. (Map). Lands & Survey Department, Nuku'alofa, Tonga.
Kingdom of Tonga. 1975a. Vava'u Group (Sheet 5) (map). Tonga Government, Nuku'alofa.
Kingdom of Tonga. 1975b. Ha'apai Group—Tofua & Kao (Sheet 6) (map). Tonga Government, Nuku'alofa.
Kingdom of Tonga. 1976. Ha'apai Group—Lifuka (Sheet 10) (map). Tonga Government, Nuku'alofa.
Kingdom of Tonga. No date. Vava'u Island Group (map). The Moorings, Clearwater, FL.
Kinsky, F. C. & Yaldwyn, J. C. 1981. The bird fauna of Niue Island, south-west Pacific, with special notes on the white-tailed tropic bird and golden plover. National Museum of New Zealand, Miscellaneous Series 2:1–49.
Kirby, H., Jr. 1925. The birds of Fanning Island, Central Pacific Ocean. Condor 27:185–196.
Kirch, P. V. 1973. Prehistoric subsistence patterns in the northern Marquesas Islands, French Polynesia. Archaeology and Physical Anthropology in Oceania 8:24–40.
Kirch, P. V. 1978. The Lapitoid Period in West Polynesia: excavations and survey in Niuatoputapu, Tonga. Journal of Field Archaeology 5:1–13.
Kirch, P. V. 1982. The impact of prehistoric Polynesians on the Hawaiian ecosystem. Pacific Science 36:1–14.
Kirch, P. V. 1983. Man's role in modifying tropical and subtropical Polynesian ecosystems. Archaeology in Oceania 18:26–31.
Kirch, P. V. 1984a. The Evolution of the Polynesian Chiefdoms. Cambridge University Press, Cambridge.
Kirch, P. V. 1984b. The Polynesian outliers. Journal of Pacific History 19:224–238.
Kirch, P. V. 1986. Rethinking East Polynesian prehistory. Journal of the Polynesian Society 95:9–40.
Kirch, P. V. 1988. The Talepakemalai Lapita site and oceanic prehistory. National Geographic Research 4:328–342.
Kirch, P. V. 1989. Second millenium B.C. arboriculture in Melanesia: archaeological evidence from the Mussau Islands. Economic Botany 43:225–240.
Kirch, P. V. 1990. The evolution of sociopolitical complexity in prehistoric Hawaii: an assessment of the archaeological evidence. Journal of World Prehistory 4:311–345.
Kirch, P. V. 1994. The Wet and the Dry: Irrigation and Agricultural Intensification in Polynesia. University of Chicago Press, Chicago.
Kirch, P. V. 1996. Late Holocene human-induced modifications to a central Polynesian island ecosystem. Proceedings of the National Academy of Sciences USA 93:5296–5300.
Kirch, P. V. 1997. The Lapita Peoples: Ancestors of the Oceanic World. Blackwell Publishers, Oxford.
Kirch, P. V. 2000. On the Road of the Winds: An Archaeological History of the Pacific Islands before European Contact. University of California Press, Berkeley.
Kirch, P. V. (ed.). 2001. Lapita and Its Transformations in Near Oceania: Archaeological Investigations in the Mussau Islands, Papua New Guinea, 1985–88. Volume 1: Introduction, Excavations, Chronology. Archaeological Research Facility Contribution no. 59, University of California, Berkeley.
Kirch, P. V. & J. Ellison. 1994. Palaeoenvironmental evidence for human colonization of remote Oceanic islands. Antiquity 68:310–321.
Kirch, P. V. & T. L. Hunt. 1988. Radiocarbon dates from the Mussau Islands and the Lapita colonization of the southwestern Pacific. Radiocarbon 30:161–169.
Kirch, P. V. & T. L. Hunt. 1993. The To'aga site: three millennia of Polynesian occupation in the Manu'a Islands, American Samoa. Contributions of the University of California Archaeological Research Facility, Berkeley no. 51.
Kirch, P. V. & T. L. Hunt (eds.). 1997. Historical Ecology in the Pacific Islands. Yale University Press, New Haven, CT.
Kirch, P. V. & D. Lepofsky. 1993. Polynesian irrigation: archaeological and linguistic evidence for origins and development. Asian Perspectives 32:183–204.
Kirch, P. V. & S. J. O'Day. 2002. New archaeological insights into food and status: a case study from pre-contact Hawaii. World Archaeology 34:484–497.
Kirch, P. V. & P. Rosendahl. 1973. Archaeological investigation of Anuta. Pp. 25–108 *in* E. Yen & J. Gordon (eds.). Anuta: A Polynesian Outlier in the Solomon Islands. Pacific Anthropological Records 21. Bernice P. Bishop Museum, Honolulu.
Kirch, P. V. & P. Rosendahl. 1976. Early Anutan settlement and the position of Anuta in the prehistory of the southwest Pacific. Pp. 223–244 *in* R. C. Green & M. Cresswell (eds.). Southeast Solomon Islands Cultural History: A Preliminary Survey. Royal Society of New Zealand Bulletin 11.
Kirch, P. V. & D. E. Yen. 1982. Tikopia: the prehistory and ecology of a Polynesian outlier. Bernice P. Bishop Museum Bulletin 238:1–396.
Kirch, P. V., W. R. Dickinson & T. L. Hunt. 1988. Polynesian plainware sherds from Hivaoa and their implications for early Marquesan prehistory. New Zealand Journal of Archaeology 10:101–107.
Kirch, P. V., J. R. Flenley & D. W. Steadman. 1991. A radiocarbon chronology for human-induced environmental change on Mangaia, southern Cook Islands, Polynesia. Radiocarbon 33:317–328.
Kirch, P. V., D. W. Steadman, V. L. Butler, J. Hather & M. Weisler. 1995. Prehistory and human ecology in Eastern Polynesia: excavations at Tangatatau Rockshelter, Mangaia, Cook Islands. Archaeology in Oceania 30:47–65.
Kirch, P. V., D. R. Swindler & C. G. Turner II. 1989. Human skeletal and dental remains from Lapita sites (1600–500 B.C.) in the Mussau Islands, Melanesia. American Journal of Physical Anthropology 79:63–76.
Kirch, P. V., M. I. Weisler & E. Casella (eds.). 1997. Towards a Prehistory of the Koné Region, New Caledonia: A Reanalysis of the Pioneering Archaeological Excavations of E. W. Gifford. Kroeber Anthropological Society Papers, vol. 82. University of California Press, Berkeley.
Kirch, P. V., J. R. Flenley, D. W. Steadman, F. Lamont & S. Dawson. 1992. Prehistoric human impacts on an island ecosystem: Mangaia, Central Polynesia. National Geographic Research and Exploration 8:166–179.
Kirch, P. V., A. S. Hartshorn, O. A. Chadwick, P. M. Vitousek, D. R. Sherrod, J. Coil, L. Holm & W. D. Sharp. 2004. Environment, agriculture, and settlement patterns in a

marginal Polynesian landscape. Proceedings of the National Academy of Sciences USA 101:9936–9941.

Kirchner, J. W. & A. Weil. 2000. Delayed biological recovery from extinctions throughout the fossil record. Nature 404:177–180.

Kloor, K. 2000. Returning America's forests to their "natural" roots. Science 287:573–575.

Knox, A. G. & M. P. Walters. 1994. Extinct and Endangered Birds in the Collections of the Natural History Museum. British Ornithologists' Club Occasional Publications 1, Tring, UK.

König, C., F. Weick & J.-H. Becking. 1999. Owls: A Guide to the Owls of the World. Yale University Press, New Haven, CT.

Koopman, K. E. 1957. Evolution in the genus *Myzomela* (Aves: Meliphagidae). Auk 74:49–72.

Koopman, K. F. & D. W. Steadman. 1995. Extinction and biogeography of bats on 'Eua, Kingdom of Tonga. American Museum Novitates no. 3125.

Kosrae Island to Ngatik Atoll. 1990. Map 81002. United States Defense Mapping Agency, Washington.

Koutavas, A., J. Lynch-Stieglitz, T. M. Marchitto, Jr. & J. P. Sachs. 2002. El Niño-like pattern in ice age tropical Pacific sea surface temperature. Science 297:226–230.

Kratter, A. W. 2002. [Book review] Birds of North Melanesia: Speciation, Ecology, and Biogeography by E. Mayr and J. M. Diamond. Auk 119:883–888.

Kratter, A. W. & D. W. Steadman. 2003. First Atlantic Ocean and Gulf of Mexico specimen of a short-tailed shearwater. North American Birds 57:277–279.

Kratter, A. W., D. W. Steadman, C. E. Smith, C. E. Filardi & H. P. Webb. 2001a. Avifauna of a lowland forest site on Isabel, Solomon Islands. Auk 118:472–483.

Kratter, A. W., D. W. Steadman, C. E. Smith & C. E. Filardi. 2001b. Reproductive condition, molt, and body mass of birds from Isabel, Solomon Islands. Bulletin of the British Ornithologists' Club 121:128–144.

Kratter, A. W., J. J. Kirchman & D. W. Steadman. In press. Upland bird communities on Santo, Vanuatu, southwest Pacific. Wilson Bulletin.

Kremen, C., J. O. Niles, M. G. Dalton, G. C. Daily, P. R. Ehrlich, J. P. Fay, D. Grewal & R. P. Guillery. 2000. Economic incentives for rain forest conservation across scales. Science 288:1828–1832.

Kroenke, L. W. 1996. Plate tectonic development of the western and southwestern Pacific: Mesozoic to the present. Pp. 19–34 *in* A. Keast & S. E. Miller (eds.). The Origin and Evolution of Pacific Island Biotas, New Guinea to Eastern Polynesia: Patterns and Processes. SPB Academic, Amsterdam.

Kromer, B., S. W. Manning, P. I. Kuniholm, M. W. Newton, M. Spurk & I. Levin. 2001. Regional $^{14}CO_2$ offsets in the troposphere: magnitude, mechanisms, and consequences. Science 294:2529–2532.

Kuehler, C., A. Lieberman, A. Varney, P. Unitt, R. M. Sulpice, J. Azua & B. Tehevini. 1997. Translocation of ultramarine lories *Vini ultramarina* in the Marquesas Islands: Ua Huka to Fatu Hiva. Bird Conservation International 7:69–79.

Kukla, G. J. 2000. The last interglacial. Science 287:987–988.

Kump, L. R. 2001. Chill taken out of the tropics. Nature 413:470–471.

Lacan, F. & J.-L. Mougin. 1974. Les oiseaux des îles Gambier et de quelques atolls orientaux de l'archipel des Tuamotu (Océan Pacifique). L'Oiseau et Revue Française d'Ornithologie 44:192–280.

Lack, D. 1947. Darwin's Finches: An Essay on the General Biological Theory of Evolution. Cambridge University Press, Cambridge, UK.

Lack, D. 1969. The numbers of bird species on islands. Bird Study 16:193–209.

Ladd, H. S. 1965. Tertiary fresh-water mollusks from Pacific islands. Malacologia 2:189–197.

Lal, K. & B. V. Fortune (eds.). 2000. The Pacific Islands: An Encyclopedia. University of Hawai'i Press, Honolulu.

Lambeck, K., T. M. Esat & E.-K. Potter. 2002. Links between climate and sea levels for the past three million years. Nature 419:199–206.

Lambert, D. M. & C. D. Millar. 1995. DNA science and conservation. Pacific Conservation Biology 2:21–38.

Lambert, D. M., P. A. Ritchie, C. D. Millar, B. Holland, A. J. Drummond & C. Baroni. 2002. Rates of evolution in ancient DNA from Adelie penguins. Science 295:2270–2273.

Lambert, F. R. 1989. Some field observations of the endemic Sulawesi rails. Kukila 4:34–36.

Lambert, F. R. 1998a. A new species of *Gymnocrex* from the Taluad Islands, Indonesia. Forktail 13:1–6.

Lambert, F. R. 1998b. A new species of *Amaurornis* from the Taluad Islands, Indonesia, and a review of taxonomy of bush hens occurring from the Philippines to Australasia. Bulletin of the British Ornithologists' Club 118:67–82.

Lane, I. E. 1960. Vegetation. Pp. 15–19 *in* D. I. Blumenstock & D. F. Rex. Microclimatic observations at Eniwetok. Atoll Research Bulletin no. 71.

Langley, J. 1992. The incidence of dog bites in New Zealand. New Zealand Medical Journal 105:33–35.

Larson, E. E. & R. L. Reynolds, M. Ozima, Y. Aoki, H. Kinoshita, S. Zasshu, N. Kawai, T. Nakajima, K. Hirooka, R. Merril & S. Levi. 1975. Paleomagnetism of Miocene volcanic rocks of Guam and the curvature of the southern Mariana Island arc. Geological Society of America Bulletin 86:346–350.

Layard, E. L. 1875. Ornithological notes from Fiji, with descriptions of supposed new species of birds. Proceedings of the Zoological Society of London 1875:27–30.

Lea, D. W. 2002. The glacial-tropical Pacific: not just a west side story. Science 297:202–203.

Lea, D. W., D. K. Pak & H. J. Spero. 2000. Climate impact of late Quaternary equatorial Pacific sea surface temperature variations. Science 289:1719–1724.

Leach, H. M. 1982. Cooking without pots: aspects of prehistoric and traditional Polynesian cooking. New Zealand Journal of Archaeology 4:149–156.

Leavesley, M. & J. Allen. 1998. Dates, disturbance, and artefact distributions: another analysis of Buang Merabak, a Pleistocene site on New Ireland, Papua New Guinea. Archaeology in Oceania 33:63–82.

LeCroy, M., A. W. Kratter, D. W. Steadman & H. P. Webb. 2001. *Accipiter imitator* on Isabel Island, Solomon Islands. Emu 101:151–155.

Lee, G. 1986. The birdman motif of Easter Island. Journal of New World Archaeology 7:39–49.

Leopold, A. 1949. A Sand County Almanac. Oxford University Press, London.

Leopold, A. 1991. Round River: From the Journals of Aldo Leopold. Northword Press, Minocqua, WI.

Leopold, E. B. 1969. Miocene pollen and spore flora of Eniwetok Atoll, Marshall Islands. U.S. Geological Survey Professional Paper 260-11.

Lepofsky, D. 1989. Eating eggs on Eloaua: the initiation of a mutualistic relationship. Journal of Ethnobiology 9:229–231.

Lepofsky, D. 1995. A radiocarbon chronology for prehistoric agriculture in the Society Islands, French Polynesia. Radiocarbon 37:917–930.

Lepofsky, D., H. C. Harries & M. Kellum. 1992. Early coconuts on Mo'orea Island, French Polynesia. Journal of the Polynesian Society 101:299–308.

Lepofsky, D., P. V. Kirch & K. P. Lertzman. 1996. Stratigraphic and paleobotanical evidence for prehistoric human-induced environmental disturbance on Mo'orea, French Polynesia. Pacific Science 50:253–273.

Lessa, W. A. 1962. The social effects of Typhoon Ophelia (1960) on Ulithi. Micronesica 1:1–47.

Lessa, W. A. 1966. Ulithi: A Micronesian Design for Living. Holt, Rinehart and Winston, New York.

Lester, P. J. & A. Tavite. 2004. Long-legged ants, *Anoplolepsis gracilipes* (Hymenoptera: Formicidae), have invaded Tokelau, changing composition and dynamics of ant and invertebrate communities. Pacific Science 58:391–401.

Lewis, D. 1964. Polynesian navigational methods. Journal of the Polynesian Society 73:364–374.

Lewis, D. 1972. We, the Navigators. University of Hawai'i Press, Honolulu.

Lewis, K. B., R. B. Smith & J. S. Pow. 1997. Future sand supplies for Tongatapu, Kingdom of Tonga. Pp. 175–191 *in* A. M. Sherwood (comp.). Coastal and Environmental Geoscience Studies of the Southwest Pacific Islands. SOPAC Technical Bulletin no. 9.

Lewis, S., T. N. Sherratt, K. C. Hamer & S. Wanless. 2001. Evidence of intra-specific competition for food in a pelagic seabird. Nature 412:816–818.

Lieberman, B. S. 2002. Phylogenetic biogeography with and without the fossil record: gauging the effects of extinction and paleontological incompleteness. Palaeogeography, Palaeoclimatology, Palaeoecology 178:39–52.

Lister, J. J. 1891. Notes on the geology of the Tonga Islands. Geological Society of London Quarterly Journal 47:590–617.

Liston, J., H. D. Tuggle, T. M. Mangieri, M. W. Kaschko & M. Desilets. 1998a. Archaeological Data Recovery for the Compact Road, Babeldaob Island, Republic of Palau. Historic Preservation Investigations Phase II. Volume I: Fieldwork Reports. International Archaeological Research Institute, Honolulu.

Liston, J., T. M. Mangieri, D. Grant, M. W. Kaschko & H. D. Tuggle. 1998b. Archaeological data recovery for the compact road, Babeldaob Island, Republic of Palau. Historic Preservation Investigations Phase II. Volume II. Fieldwork Reports. International Archaeological Research Institute, Honolulu.

Livesey, B. C. 2003. Evolution of flightlessness in Rails (Gruiformes: Rallidae): phylogenetic, ecomorphological, and ontogenetic perspectives. Ornithological Monographs no. 53.

Livingston, S. D. 1989. The taphonomic interpretation of avian skeletal part frequencies. Journal of Archaeological Science 16:537–547.

Lloyd, M., R. F. Inger & F. W. King. 1968. On the diversity of reptile and amphibian species in a Bornean rain forest. American Naturalist 102:497–515.

Lobban, C. S. & M. Schefter. 1997. Tropical Pacific Island Environments. University of Guam Press, Mangilao.

Lockwood, J. L., M. P. Moulton & S. K. Anderson. 1999. Morphological assortment and the assembly of communities of introduced Passeriformes on oceanic islands: Tahiti vs. Oahu. American Naturalist 141:398–408.

Lomolino, M. V. 1986. Mammalian community structure on islands: the importance of immigration, extinction, and interactive effects. Biological Journal of the Linnean Society 28:1–21.

Lomolino, M. V. 1993a. Winter filtering, immigrant selection, and species composition of insular mammals of Lake Huron. Ecography 16:24–30.

Lomolino, M. V. 1993b. Immigrations and distribution patterns of insular mammals: studying fundamental processes in island biogeography. Ecography 16:376–379.

Lomolino, M. V. 1994. Species richness of mammals inhabiting nearshore archipelagoes: area, isolation, and immigration filters. Journal of Mammalogy 75:39–49.

Lomolino, M. V. 1999. A species-based, hierarchical model of island biogeography. Pp. 272–310 *in* E. A. Weiher & P. A. Keddy (eds.). The Search for Assembly Rules in Ecological Communities. Cambridge University Press, New York.

Long, J., M. Archer, T. Flannery & S. Hand. 2002. Prehistoric Mammals of Australia and New Guinea. Johns Hopkins University Press, Baltimore, MD.

Long Island to the Tami Islands. 1996. Map 73650. United States Defense Mapping Agency, Washington.

Longhurst, A. 1998. Ecological Geography of the Sea. Academic Press, New York.

Loope, L. L., F. G. Howarth, F. Kraus & T. K. Pratt. 2001. Newly emergent and future threats of alien species to Pacific birds and ecosystems. Studies in Avian Biology 22:291–304.

Losos, J. B. & D. Schluter. 2000. Analysis of an evolutionary species-area relationship. Nature 408:847–850.

Losos, J. B., K. I. Warheit & T. W. Schoener. 1997. Adaptive differentiation following experimental island colonization in *Anolis* lizards. Nature 387:70–73.

Lourandos, H. 1997. Continent of Hunter-gatherers: New Perspectives in Australian Prehistory. Cambridge University Press, Cambridge, UK.

Lovejoy, T. E. 1997. Biodiversity: what is it? Pp. 7–14 *in* M. L. Reaka-Kudla, D. E. Wilson & E. O. Wilson (eds.). Biodiversity II: Understanding and Protecting Our Biological Resources. Joseph Henry Press, Washington.

Lowe, P. R. 1928. A description of *Atlantisia rogersi*, the diminutive and flightless rail of Inaccessible Island (southern Atlantic), with some notes on flightless rails. Ibis 99–131.

Lynch, J. 1998. Pacific Languages: An Introduction. University of Hawai'i Press, Honolulu.

Lynch, J. F. & N. K. Johnson. 1974. Turnover and equilibria in insular avifaunas, with special reference to the California Channel Islands. Condor 76:370–384.

Lysaght, A. M. 1959. Some eighteenth century bird paintings in the library of Sir Joseph Banks (1743–1820). Bulletin of the British Museum (Natural History), Historical Series 1.

Lyver, P. O'B., H. Moller & C. J. R. Robertson. 2000. Predation at sooty shearwater *Puffinus griseus* colonies on the New Zealand mainland: is there safety in numbers? Pacific Conservation Biology 5:347–357.

MacArthur, R. H. 1957. On the relative abundance of bird species. Proceedings of the National Academy of Sciences USA 43:293–295.

MacArthur, R. H. 1958. Population biology of some warblers of northeastern coniferous forests. Ecology 39:599–619.

MacArthur, R. H. 1972. Geographical Ecology. Harper & Row, New York.

MacArthur, R. H. & J. W. MacArthur. 1961. On bird species diversity. Ecology 42:594–598.

MacArthur, R. H. & E. O. Wilson. 1963. An equilibrium theory of insular zoogeography. Evolution 17:373–387.

MacArthur, R. H. & E. O. Wilson. 1967. The Theory of Island Biogeography. Princeton University Press, Princeton, NJ.

MacDonald, G. M. 2002. Biogeography: Introduction to Space, Time, and Life. John Wiley & Sons, New York.

MacFarlane, A., J. N. Carney, A. J. Crawford & H. G. Greene. 1988. Vanuatu: a review of onshore geology. Pp. 45–91 *in* H. G. Greene & F. L. Wong (eds.). Geology and Offshore Resources of Pacific Island Arcs: Vanuatu Region. Circum-Pacific Council for Energy and Mineral Resources Earth Science Series, vol. 8, Houston.

MacKinnon, J. & K. Phillips. 1993. A Field Guide to the Birds of Borneo, Sumatra, Java, and Bali. Oxford University Press, Oxford.

MacPhee, R. D. E. & C. Flemming. 1999. *Requiem Æternam:* the last five hundred years of mammalian species extinctions. Pp. 333–371 *in* R. D. E. MacPhee (ed.). Extinctions in Near Time: Causes, Contexts, and Consequences. Kluwer Academic/Plenum Publishers, New York.

Magurran, A. E. & P. A. Henderson. 2003. Explaining the excess of rare species in natural species abundance distributions. Nature 422:714–716.

Mammerickx, J., S. M. Smith, I. L. Taylor & T. E. Chase. 1973. Bathymetry of the South Pacific (Chart 13) (map). La Jolla, Scripps Institute of Oceanography, University of California, IMR Technical Report 46A.

Mangaia. 1986. NZMS 272/8/2 (map). New Zealand Department of Lands & Survey, Wellington.

Manner, H. I., D. Mueller-Dombois & M. Rapaport. 1999. Terrestrial Ecosystems. Pp. 93–108 *in* M. Rapaport (ed.). The Pacific Islands: Environment and Society. Bess Press, Honolulu.

Manner, H. I., R. R. Thaman & D. H. Hassall. 1984. Phosphate-mining induced vegetation changes on Nauru Island. Ecology 65:1454–1465.

Manner, H. I., R. R. Thaman & D. H. Hassall. 1985. Plant succession after phosphate mining on Nauru. Australian Geographer 16:185–195.

Manning, S. W., B. Kromer, P. I. Kunihom & M. W. Newton. 2001. Anatolian tree rings and a new chronology for the East Mediterranean Bronze-Iron Ages. Science 294:2532–2535.

Manus Island & Approaches. 1996. Map 82060. United States Defense Mapping Agency, Washington.

Maragos, J. E., G. B. K. Baines & P. J. Beveridge. 1973. Tropical Cyclone Bebe creates new land on Funafuti Atoll. Science 181:1161–1164.

Marche, A.-A. (translated: S. E. Cheng; edited: R. D. Craig). 1982. The Mariana Islands. Micronesian Area Research Center, Mangilao, Guam.

Margules, C. R. & R. L. Pressey. 2000. Systematic conservation planning. Nature 405:243–253.

Mariana Islands. 1975. Map 81005. United States Defense Mapping Agency, Washington.

Markgraf, V., J. R. Dodson, A. P. Kershaw, M. S. McGlone & N. Nicholls. 1992. Evolution of late Pleistocene and Holocene climates in the circum-South Pacific land areas. Climate Dynamics 6:193–211.

Marks, J. S. 1993. Molt of bristle-thighed curlews in the northwestern Hawaiian Islands. Auk 110:573–587.

Marquet, P. A. 2000. Invariants, scaling laws, and ecological complexity. Science 289:1487–1488.

Marquet, P. A. 2001. Ecology goes macro: taking a bird's-eye view reveals the hidden order of ecosystems. Nature 412:481–482.

Marshall, B. & J. Allen. 1991. Excavations at Panakiwuk Cave, New Ireland. Pp. 59–91 *in* J. Allen & C. Gosden (eds.). Report of the Lapita Homeland Project. Department of Prehistory, Australian National University, Canberra.

Marshall, J. F. & G. Jacobsen. 1985. Holocene growth of a mid-Pacific atoll: Tarawa, Kiribati. Coral Reefs 4:11–17.

Marshall, J. T., Jr. 1949. The endemic avifauna of Saipan, Tinian, Guam, and Palau. Condor 51:200–221.

Marshall, M. 1977. Notes on birds from Namoluk Atoll. Micronesica 7:234–236.

Martens, G. H. 1922. Vögel der Südsee Expedition der Hamburger Wissenschaftl. Stiftung, 1908–1909. Archiv für Naturgeschichte 88A:44–54.

Martin, P. S. 1958. Biogeography of reptiles and amphibians in the Gomez Farías region, Tamaulipas, México. Miscellaneous Publications of the Museum of Zoology, University of Michigan 101:1–102.

Martin, P. S. 1967. Prehistoric overkill. Pp. 75–120 *in* P. S. Martin & H. E. Wright (eds.). Pleistocene Extinctions: The Search for a Cause. Yale University Press, New Haven, CT.

Martin, P. S. 1984. Prehistoric overkill: the global model. Pp. 354–403 *in* P. S. Martin & R. G. Klein (eds.). Quaternary Extinctions: A Prehistoric Evolution. University of Arizona Press, Tucson.

Martin, P. S. 1990. 40,000 years of extinctions of the "planet of doom." Palaeogeography, Palaeoclimatology, Palaeoecology 82:187–201.

Martin, P. S. & D. W. Steadman. 1999. Prehistoric extinctions on islands and continents. Pp. 17–55 *in* R. D. E. MacPhee (ed.). Extinctions in Near Time: Causes, Contexts, and Consequences. Kluwer Academic/Plenum Publishers, New York.

Martin, P. S. & A. J. Stuart. 1995. Mammoth extinction: two continents and Wrangell Island. Radiocarbon 37:7–10.

Martinez, F. & B. Taylor. 1996. Backarc spreading, rifting, and microplate rotation between transform faults in the Manus Basin. Marine Geophysical Researches 18:203–224.

Martinez, F. & B. Taylor. 2002. Mantle wedge control on back-arc crustal accretion. Nature 416:417–420.

Martínez-Vilalta, A. & A. Motis. 1992. Family Ardeidae (Herons). Pp. 376–429 *in* J. del Hoyo, A. Elliott & J. Sargatal (eds.). Handbook of the Birds of the World, vol. 1, Ostrich to Ducks. Lynx, Barcelona.

Martinsson-Wallin, H. & S. J. Crockford. 2002. Early settlement of Rapa Nui (Easter Island). Asian Perspectives 40:244–278.

Massey, J. A., S. Matsui, T. Suzuki, E. P. Swift, A. Hibi, N. Ichida, Y. Tsukamoto & K. Sonobe. 1982. A Field Guide to the Birds of Japan. Kodansha International, Tokyo.

Mather, J. & W. J. Fielding. 2001. The threat posed to personal health by dogs in New Providence. Bahamas Journal of Science 8:39–45.

Matisoo-Smith, E. & J. S. Allen. 2001. Name that rat: molecular and morphological identification of Pacific rodent remains. International Journal of Osteoarchaeology 11:34–42.

Matisoo-Smith, E. & J. H. Robins. 2004. Origins and dispersals of Pacific peoples: evidence from mtDNA phylogenies of the Pacific rat. Proceedings of the National Academy of Sciences USA 101:9167–9172.

Matisoo-Smith, E., R. M. Roberts, G. J. Irwin, J. S. Allen, D. Penny & D. M. Lambert. 1998. Patterns of prehistoric human mobility in Polynesia indicated by mtDNA from the Pacific rat. Proceedings of the National Academy of Sciences USA 95:15145–15150.

Mattey, D. P. 1982. The minor and trace element geochemistry of volcanic rocks from Truk, Ponape, and Kusaie, eastern Caroline Islands: the evolution of a young hot spot trace across old Pacific Ocean crust. Contributions to Mineralogy and Petrology 80:1–13.

Maurer, B. A. 1999. Untangling Ecological Complexity: The Macroscopic Perspective. University of Chicago Press, Chicago.

Maurer, B. A. 2000. Ecology needs theory as well as practice. Nature 408:768.

Mayes, C. L., L. A. Lawver & D. T. Sandwell. 1990. Tectonic history and new isochron chart of the South Pacific. Journal of Geophysical Research 95:8543–8567.

Mayr, E. 1931. A systematic list of the birds of Rennell Island with description of new species and subspecies. American Museum Novitates no. 486.

Mayr, E. 1932a. Birds collected during the Whitney South Sea Expedition. XX. American Museum Novitates no. 522.

Mayr, E. 1932b. Birds collected during the Whitney South Sea Expedition. XXI. American Museum Novitates no. 531.

Mayr, E. 1933. Birds collected during the Whitney South Sea Expedition. XXIII. American Museum Novitates no. 609.

Mayr, E. 1935. Descriptions of twenty-five new species and subspecies. Birds collected during the Whitney South Sea Expedition. XXX. American Museum Novitates no. 820.

Mayr, E. 1936. Descriptions of twenty-five species and subspecies. Birds collected during the Whitney South Sea Expedition. XXXI. American Museum Novitates no. 828.

Mayr, E. 1938. Birds collected during the Whitney South Sea Expedition. XXXIX. Notes on New Guinea birds. IV. American Museum Novitates no. 1006.

Mayr, E. 1940a. Birds collected during the Whitney South Sea Expedition. XLI. Notes on New Guinea Birds. VI. American Museum Novitates no. 1056.

Mayr, E. 1940b. Birds collected during the Whitney South Sea Expedition. XLII. On the birds of the Loyalty Islands. American Museum Novitates no. 1057.

Mayr, E. 1940c. Speciation phenomena in birds. American Naturalist 74:249–278.

Mayr, E. 1941a. Birds collected during the Whitney South Sea Expedition. XLV. Notes on New Guinea Birds. VIII. American Museum Novitates no. 1133.

Mayr, E. 1941b. Birds collected during the Whitney South Sea Expedition. XLVII. American Museum Novitates no. 1152.

Mayr, E. 1941c. The origin and the history of the bird fauna of Polynesia. Proceedings of the Sixth Pacific Science Congress 4:197–216.

Mayr, E. 1942. Systematics and the Origin of Species. Columbia University Press, New York.

Mayr, E. 1945. Birds of the Southwest Pacific. Macmillan, New York.

Mayr, E. 1949a. The species concept: semantics versus semantics. Evolution 3:371–372.

Mayr, E. 1949b. Notes on the birds of Northern Melanesia. American Museum Novitates no. 1417.

Mayr, E. 1976. The origin and history of the Polynesian bird fauna. Pp. 601–617 in Evolution and the Diversity of Life: Selected Essays. Harvard University Press, Cambridge, MA.

Mayr, E. 2004. What Makes Biology Unique? Considerations on the Autonomy of a Scientific Discipline. Cambridge University Press, New York.

Mayr, E. & D. Amadon. 1941. Birds collected during the Whitney South Sea Expedition. XLVI. Geographical variation in *Demigretta sacra* (Gmelin). American Museum Novitates no. 1144.

Mayr, E. & J. M. Diamond. 1976. Birds on islands in the sky: origin of the montane avifauna of northern Melanesia. Proceedings of the National Academy of Sciences USA 73:1765–1769.

Mayr, E. & J. Diamond. 2001. The Birds of Northern Melanesia: Speciation, Ecology, and Biogeography. Oxford University Press, New York.

McConkey, K. R. & D. R. Drake. 2002. Extinct pigeons and declining bat populations: are large seeds still being dispersed in the tropical Pacific? Pp. 381–395 *in* D. J. Levey, W. R. Silva & M. Galetti (eds.). Seed Dispersal and Frugivory: Ecology, Evolution, and Conservation. CAB International Publishing, Wallingford, UK.

McConkey, K. R., D. R. Drake, H. J. Meehan & N. Parsons. 2003. Husking stations provide evidence of seed predation by introduced rodents in Tongan rain forests. Biological Conservation 109:221–225.

McConkey, K. R., H. J. Meehan & D. R. Drake. 2004a. Seed dispersal by Pacific pigeons (*Ducula pacifica*) in Tonga, Western Polynesia. Emu 104:1–8.

McConkey, K. R., D. R. Drake, J. Franklin & F. Tonga. 2004b. Effects of Cyclone Waka on flying foxes (*Pteropus tonganus*) in the Vava'u Islands of Tonga. Journal of Tropical Ecology 20:555–561.

McCormack, G. & J. Künzlé. 1990a. Kakerori: Rarotonga's Endangered Flycatcher. Cook Island Conservation Service, Rarotonga.

McCormack, G. & J. Künzlé. 1990b. Rarotonga's Cloud Forest. Cook Island Conservation Service, Rarotonga.

McCormack, G. & J. Künzlé. 1995. Rarotonga's Mountain Tracks and Plants: A Field Guide. Cook Islands Natural Heritage Project, Rarotonga.

McCormack, G. & J. Künzlé. 1996. The 'Ura or Rimatara lorikeet *Vini kuhlii:* its former range, present status, and conservation priorities. Bird Conservation International 6:325–334.

McCoy, M. 1980. Reptiles of the Solomon Islands. Wau Ecology Handbook 7. Wau Ecology Institute, Wau, Papua New Guinea.

McCoy, P. C. 1978. The place of near-shore islets in Easter Island prehistory. Journal of the Polynesian Society 87:193–214.

McCulloch, B. 1994. A likely association of the New Zealand owlet-nightjar (*Megaegotheles novaezealandiae* Scarlett) with early humans in New Zealand. Notornis 41:209–210.

McCulloch, M. T., A. W. Tudhope, T. M. Esat, G. E. Mortimer, J. Chappell, B. Pillans, A. R. Civas & A. Omura. 1999. Coral record of equatorial sea-surface temperatures during the penultimate deglaciation at Huon Peninsula. Science 283:202–204.

McDonald, D. B., W. K. Potts, J. W. Fitzpatrick & G. E. Woolfenden. 1999. Contrasting genetic structures in sister species of North American scrub-jays. Proceedings of the Royal Society of London B 266:1117–1125.

McDougall, I. 1985. Age and evolution of the volcanoes of Tutuila, American Samoa. Pacific Science 39:311–320.

McFadden, I. & D. R. Towns. 1991. Eradication campaigns against kiore (*Rattus exulans*) on Rurima Rocks and Korapuki Island, northern New Zealand. Scientific Research Internal Report no. 97. New Zealand Department of Conservation, Wellington.

McGlone, M. S. 1988. New Zealand. Pp. 557–602 *in* B. Huntley & T. Webb III (eds.). Vegetation History. Kluwers Academic, Dordrecht, The Netherlands.

McGuinness, K. A. 1984. Equations and explanations in the study of species-area curves. Biological Reviews 59:423–440.

McKinney, M. L., J. L. Lockwood & N. R. Frederick. 1997. Does ecosystem and evolutionary stability include rare species? Palaeogeography, Palaeoclimatology, Palaeoecology 127:191–207.

McNab, B. K. 1994a. Energy conservation and the evolution of flightlessness in birds. American Naturalist 144:628–642.

McNab, B. K. 1994b. Resource use and the survival of land and freshwater vertebrates on oceanic islands. American Naturalist 144:643–660.

McNab, B. K. 2000. The influence of body mass, climate, and distribution on the energetics of South Pacific pigeons. Comparative Biochemistry and Physiology Part A 127:309–329.

McNab, B. K. 2003. The energetics of New Zealand's ducks. Comparative Biochemistry and Physiology Part A 135:229–247.

McNab, B. K. & C. A. Salisbury. 1995. Energetics of New Zealand's temperate parrots. New Zealand Journal of Zoology 22:339–349.

McQuarrie, P. 1991. The banded rail: a new bird record from Tuvalu. South Pacific Journal of Natural Science 11:36–39.

Mead, J. I., D. W. Steadman, S. H. Bedford, C. J. Bell & M. Spriggs. 2002. New extinct mekosuchine crocodile from Vanuatu, South Pacific. Copeia 2002:632–641.

Mead, J. I., P. S. Martin, R. C. Euler & A. Long. 1986. Extinction of Harrington's mountain goat. Proceedings of the National Academy of Sciences USA 83:836–839.

Medway, D. G. 1979. Some ornithological results of Cook's third voyage. Journal of the Society for the Bibliography of Natural History 9:315–351.

Medway, L. & A. G. Marshall. 1975. Terrestrial vertebrates of the New Hebrides: origin and distribution. Philosophical Transactions of the Royal Society of London B 272:423–465.

Meehl, G. A. 1987. The annual cycle and interannual variability in the tropical Pacific and Indian Ocean regions. Monthly Weather Review 115:27–50.

Meehl, G. A., W. M. Washington, W. D. Collins, J. M. Arblaster, A. Hu, L. E. Buja, W. G. Strand & H. Teng. 2005. How much more global warming and sea level rise? Science 307:1769–1772.

Meehan, H. J., K. R. McConkey & D. R. Drake. 2002. Potential disruptions to seed dispersal mutualisms in Tonga, western Polynesia. Journal of Biogeography 29:695–712.

Menard, H. W. 1986. Islands. W. H. Freeman, New York.

Merlin, M. D. 1985. Woody vegetation in the upland region of Rarotonga, Cook Islands. Pacific Science 39:81–99.

Merlin, M. D. 1991. Woody vegetation on the raised coral limestone of Mangaia, southern Cook Islands. Pacific Science 45:131–151.

Merlin, M. D. & J. O. Juvik. 1995. Montane cloud forest in the tropical Pacific: some aspects of their floristics, biogeography, ecology, and conservation. Pp. 234–253 *in* L. S. Hamilton, J. O. Juvik & F. N. Scatena (eds.). Tropical Montane Cloud Forests. Springer-Verlag, New York.

Merlin, M., D. Jano, W. Raynor, T. Keene, J. Juvik & B. Sebastian. 1992. Plants of Pohnpei. East-West Center, Honolulu.

Merlin, M., R. Taulung & J. Juvik. 1993. Plants and Environments of Kosrae. East-West Center, Honolulu.

Merlin, M., A. Capelle, T. Keene, J. Juvik & J. Maragos. 1994. Plants and Environments of the Marshall Islands. East-West Center, Honolulu.

Metraux, A. 1940. Ethnology of Easter Island. Bernice P. Bishop Museum Bulletin no. 160.

Meyer, J.-Y. 1996. Status of *Miconia calvescens* (Melastomataceae), a dominant invasive tree in the Society Islands (French Polynesia). Pacific Science 50:66–76.

Meyer, J.-Y. 2000. Preliminary review of invasive plants in the Pacific islands (SPREP member countries). Pp. 85–114 *in* G. Sherley (ed.). Invasive Species in the Pacific: A Technical Review and Draft Regional Strategy. South Pacific Regional Environment Programme, Apia, Samoa.

Meyer, J.-Y. 2004. Threat of invasive alien plants to native flora and forest vegetation of Eastern Polynesia. Pacific Science 58:357–375.

Mildenhall, D. C. 1980. New Zealand Late Cretaceous and Cenozoic plant biogeography: a contribution. Palaeogeography, Palaeoclimatology, Palaeoecology 31:197–233.

Miles, J. A. R. 1964. Notes on the status of certain birds in Fiji. Emu 63:422.

Millener, P. R. 1981. The Quaternary avifauna of North Island, New Zealand. Ph.D. thesis, University of Auckland, New Zealand.

Millener, P. R. 1988. Contributions to New Zealand's late Quaternary avifauna. 1: *Pachyplichas*, a new genus of wren (Aves: Acanthisittidae), with two new species. Journal of the Royal Society of New Zealand 18:383–406.

Millener, P. R. 1990. Evolution, extinction, and the subfossil record of New Zealand's avifauna. Pp. 93–100 *in* B. J. Gill & B. D. Heather (eds.). A flying start: commemorating fifty years of the Ornithological Society of New Zealand 1940–1990. Notornis 37 (supplement).

Millener, P. R. & T. H. Worthy. 1991. Contribution to New Zealand's late Quaternary avifauna. II. *Dendroscansor decurvirostris*, a new genus and species of wren (Aves: Acanthisittidae). Journal of the Royal Society of New Zealand 21:179–200.

Miller, A. H. 1937. Structural modifications in the Hawaiian goose (*Nesochen sandvicensis*): a study in adaptive evolution. University of California Publications in Zoology 42:1–80.

Miller, A. H. 1962. The history and significance of the fossil *Casuarius lydekkeri*. Records of the Australian Museum 25:235–238.

Miller, S. E. 1996. Biogeography of Pacific insects and other terrestrial invertebrates: a status report. Pp. 463–475 *in* A. Keast & S. E. Miller (eds.). The Origin and Evolution of Pacific Island Biotas, New Guinea to Eastern Polynesia: Patterns and Processes. SPB Academic Publishing, Amsterdam.

Miller, G. H., J. W. Magee, B. J. Johnson, M. L. Fogel, N. A. Spooner, M. T. McCulloch & L. K. Ayliffe. 1999. Pleistocene extinction of *Genyornis newtoni*: human impact on Australian megafauna. Science 283:205–208.

Miller, J. M. 1989. The archaic flowering plant family Degeneriaceae: its bearing on an old enigma. National Geographic Research 5:218–231.

Miller, L. & B. C. Douglas. 2004. Mass and volume contributions to twentieth-century global sea level rise. Nature 428:406–409.

Milne, G. A., J. L. Davies, J. X. Mitrovica, H.-G. Scherneck, J. M. Johansson, M. Vermeer & H. Koivula. 2001. Space-geodetic constraints on glacial isostatic adjustment in Fennoscandia. Science 291: 2381–2385.

Mindanao to Palau Islands. 1989. Map 81001. United States Defense Mapping Agency, Washington.

Mittermeier, R. A., N. Myers & C. G. Mittermeier. 2000. Hotspots: Earth's Biologically Richest and Most Endangered Terrestrial Ecoregions. University of Chicago Press, Chicago.

Mizutani, H. & E. Wada. 1988. Isotope ratios in seabird rookeries and the ecological implications. Ecology 69:340–349.

Molnar, R. E. & M. Pole. 1997. A Miocene crocodilian from New Zealand. Alcheringa 21:65–70.

Monnet, C., L. Sanford, P., Siu, J.-C. Thibault & A. Varney. 1993a. Polynesian ground dove (*Gallicolumba erythroptera*) discovered at Rangiroa Atoll, Tuamotu Islands (Polynesia). Notornis 40:128–130.

Monnet, C., J.-C. Thibault & A. Varney. 1993b. Stability and changes during the twentieth century in the breeding landbirds of Tahiti (Polynesia). Bird Conservation International 3:261–280.

Montaggioni, L. F., G. Richard, F. Bourrouilh-Le Jan, C. Gabrie, L. Humbert, M. Monteforte, O. Naim, C. Payri & B. Salvat. 1985. Geology and marine biology of Makatea, an uplifted atoll, Tuamotu Archipelago, central Pacific Ocean. Journal of Coastal Research 1:165–171.

Montgomery, S. L., W. C. Gagne & B. H. Gagne. 1980. Notes on birdlife and nature conservation in the Marquesas and Society Islands. 'Elepaio 40:152–156.

Moors, P. J. & I. A. E. Atkinson. 1984. Predation on seabirds by introduced animals, and factors affecting its severity. Pp. 667–690 *in* J. P. Croxall, P. G. H. Evans & R. W. Schreiber (eds.). Status and Conservation of the World's Seabirds. International Council for Bird Preservation Technical Publication no. 2.

Morat, Ph. 1993. Our knowledge of the flora of New Caledonia: endemism and diversity in relation to vegetation types and substrates. Biodiversity Letters 1:72–81.

Morat, Ph., J.-M. Veillon & H. S. MacKee. 1984. Floristic relationships of New Caledonian rain forest phanerogams. Pp. 71–128 *in* F. J. Radovsky, P. H. Raven & S. H. Sohmer (eds.). Biogeography of the Tropical Pacific: Proceedings of a Symposium. Association of Systematics Collections (Lawrence, KS) and the Bernice P. Bishop Museum (Honolulu).

Morgan, W. N. 1988. Prehistoric Architecture in Micronesia. University of Texas Press, Austin.

Morley, R. J. 2000. Origin and Evolution of Tropical Rain Forests. John Wiley & Sons, New York.

Morrison, C., A. Naikatini, N. Thomas, I. Rounds, B. Thaman & J. Niukula. 2004. Rediscovery of an endangered frog *Platymantis vitianus*, on mainland Fiji: implications for conservation and management. Pacific Conservation Biology 10:237–240.

Morrison, L. W. 1997. The insular biogeography of small Bahamian cays. Journal of Ecology 85:441–454.

Morse, D. H. 1977. The occupation of small islands by passerine birds. Condor 79:399–412.

Mosimann, J. E. & P. S. Martin. 1975. Simulating overkill by Paleoindians. American Scientist 63:304–313.

Motteler, L. S. 1986. Pacific island names: a map and name guide to the new Pacific. Bernice P. Bishop Museum Miscellaneous Publication no. 34.

Mougin, J.-L. & R. de Naurois. 1981. Le noddi bleu des îles Gambier *Procelsterna cerulea murphyi* ssp. nov. L'Oiseau et Revue Française d'Ornithologie 51:201–204.

Mourer-Chauviré, C. 1982. Les oiseaux fossiles des Phosphorites du Quercy (Eocène supérieur à Oligocène supérieur): implications paléobiogéographiques. Pp. 413–426 *in* E. Buffetaut, P. Janvier, J. C. Rage & P. Tassy (eds.). Phylogénie et Paléobiogéographie. Livre jubilaire en l'honneur de Robert Hoffstetter. Geobios, Memoire Spécial 6.

Mourer-Chauviré, C. 1992. The Galliformes (Aves) of Phosphorites du Quercy (France): systematics and biostratigraphy. Pp. 67–95 *in* K. E. Campbell (ed.). Papers in Avian Paleontology Honoring Pierce Brodkorb. Los Angeles County Museum of Natural History, Science Series 36.

Mourer-Chauviré, C. 1995. Dynamics of the avifauna during the Paleogene and the Early Neogene of France. Acta Zoologica Cracov 38:325–342.

Mourer-Chauviré, C. 1999. Influence de l'homine préhistorique sur la répartition de certains oiseaux marins: l'example du grand pingouin *Pinguinus impennis*. Alauda 67:273–279.

Mourer-Chauviré, C. 2000. A new species of *Ameripodius* (Aves: Galliformes: Quercymegapodidae) from the lower Miocene of France. Palaeontology 43:481–493.

Mourer-Chauviré, C., R. Bour, S. Ribes & F. Moutou. 1999. The avifauna of Reunion Island (Mascarene Islands) at the time of the arrival of the first Europeans. Smithsonian Contributions to Paleobiology no. 89:1–38.

Mueller-Dombois, D. & F. R. Fosberg. 1998. Vegetation of the Tropical Pacific Islands. Springer-Verlag, New York.

Munschy, M., C. Antoine & A. Gachon. 1996. Évolution tectonique de la région des Tuamotu, Océan Pacifique Central. Comptes Rendus de l'Académie des Sciences de Paris (Série 2a) 323:941–948.

Murphy, M. T., J. Zysik & A. Pierce. 2004. Biogeography of the birds of the Bahamas with special reference to the island of San Salvador. Journal of Field Ornithology 75:18–30.

Murphy, R. C. 1924a. Birds collected during the Whitney South Sea Expedition. I. American Museum Novitates no. 115.

Murphy, R. C. 1924b. Birds collected during the Whitney South Sea Expedition. II. American Museum Novitates no. 124.

Murphy, R. C. 1929. Birds collected during the Whitney South Sea Expedition. VI. (Sylviidae). American Museum Novitates 350.

Murphy, R. C. 1936. Oceanic Birds of South America. Macmillan, New York; American Museum of Natural History, New York.

Murphy, R. C. & J. M. Pennoyer. 1952. Larger petrels of the genus *Pterodroma*. American Museum Novitates no. 1580.

Murray, J. W., R. T. Barber, M. R. Roman, M. P. Bacon & R. A. Feely. 1994. Physical and biological controls on carbon cycling in the equatorial Pacific. Science 266:58–65.

Murray, P. F. & D. Megirian. 1998. The skull of dromornithid birds: anatomical evidence for their relationship to Anseriformes. Records of the South Australia Museum 31:51–97.

Musgrave, R. J. & J. V. Firth. 1999. Magnitude and timing of New Hebrides arc rotation: paleomagnetic evidence from Nendo, Solomon Islands. Journal of Geophysical Research 104:2841–2853.

Myers, M. J., C. P. Meyer & V. H. Resh. 2000. Neritid and thiarid gastropods from French Polynesian streams: how reproduction (sexual, parthenogenetic) and dispersal (active, passive) affect population structure. Freshwater Biology 44:535–545.

Myers, N., R. A. Mittermeier, C. G. Mittermeier, G. A. B. da Fonseca & J. Kent. 2000. Biodiversity hotspots for conservation priorities. Nature 403:853–858.

Nagaoka, L. 1988. Lapita subsistence: the evidence of non-fish archaeofaunal remains. Pp. 117–133 *in* P. V. Kirch & T. L. Hunt (eds.). Archaeology of the Lapita Cultural Complex: A Critical Review. Burke Memorial Washington State Museum Research Report no. 5.

Natland, J. H. & D. L. Turner. 1985. Age progression and petrological development of Samoan shield volcanoes: evidence from K-Ar ages, lava compositions, and mineral studies. Pp. 139–171 *in* T. M. Brocher (ed.). Geological Investigations of the Northern Melanesian Borderland: Circum-Pacific Council for Energy and Mineral Resources Earth Science Series, vol. 3, Houston.

Neal, C. R., J. J. Mahoney, L. W. Kroenke, R. A. Duncan & M. G. Petterson. 1997. The Ontong Java Plateau. Pp. 183–216 *in* J. J. Mahoney & M. Coffin (eds.). Large Igneous Provinces: Continental, Oceanic, and Planetary Flood Volcanism. Geophysical Monograph 100.

Nee, S. & R. M. May. 1997. Extinction and the loss of evolutionary history. Science 278:692.

Nelson, J. B. 1974. The distribution of Abbott's booby *Sula abbotti*. Ibis 116:368–369.

Nelson, J. B. 1976. Breeding biology of frigatebirds: a comparative review. Living Bird 14:113–156.

Neumann, A. C. & I. MacIntyre. 1985. Reef response to sea level rise: keep-up, catch-up, or give-up. Proceedings of the 5th International Coral Reef Congress 3:105–110.

Newhouse, J. 1980. What is an atoll? South Pacific Bulletin 30:4–8.

New Zealand Department of Lands & Survey. 1986. Atlas of the South Pacific. Government Printing Office, Wellington.

Nicholls, A., E. Matisso-Smith & M. S. Allan. 2003. A novel application of molecular techniques to Pacific archaeofish remains. Archaeometry 45:133–147.

Nicholson, A. J. & D. W. Warner. 1953. The rodents of New Caledonia. Journal of Mammalogy 34:168–179.

Nicholson, E. M. 1969. Draft check list of Pacific Ocean islands. Micronesica 5:327–463.

Niering, W. A. 1963. Terrestrial ecology of Kapingamarangi Atoll, Caroline Islands. Ecological Monographs 33:131–160.

Nilsson, I. N. & S. G. Nilsson. 1985. Experimental estimates of census efficiency and pseudoturnover in islands: error trend and between-observer variation when recording vascular plants. Journal of Ecology 73:65–70.

Nishida, G. M. & N. L. Evenhuis. 2000. Arthropod pests of significance in the Pacific: a preliminary assessment of selected groups. Pp. 115–142 *in* G. Sherley (ed.). Invasive Species in the Pacific: A Technical Review and Draft Regional Strategy. South Pacific Regional Environment Programme, Apia, Samoa.

Niue. 1985. NZMS 250 (map). New Zealand Department of Lands & Survey, Wellington.

Norberg, U. 1996. Energetics of flight. Pp. 199–249 *in* C. Carey (ed.). Avian Energetics and Nutritional Ecology. Chapman and Hall, New York.

North, A. J. 1898. On a species of pigeon frequenting the atolls of the Ellice Group. Records of the Australian Museum 3:85–87.

Nott, J. & M. Hayne. 2001. High frequency of "super-cyclones" along the Great Barrier Reef over the past 5,000 years. Nature 413:508–512.

Novotny, V., Y. Basset, S. E. Miller, G. D. Weiblen, B. Bremer, L. Cizek & P. Drozd. 2002. Low host specificity of herbivorous insects in a tropical forest. Nature 416:841–844.

Nunn, P. D. 1994. Oceanic Islands. Blackwell Publishers, Oxford.

Nürnberg, D. 2000. Taking the temperature of past ocean surfaces. Science 289:1698–1699.

O'Day, S. J., P. O'Day & D. W. Steadman. 2004. Defining the Lau context: recent findings on Nayau, Lau Islands, Fiji. New Zealand Journal of Archaeology 25:31–56.

O'Dowd, D. J., P. T. Green & P. S. Lake. 2003. Invasional "meltdown" on an oceanic island. Ecology Letters 6:812–817.

Odum, E. P. 2002. The Southeastern Region: a biodiversity haven for naturalists and ecologists. Southeastern Naturalist 1:1–2.

Okal, E. A. & R. Batiza. 1987. Hotspots: the first 25 years. Pp. 1– 11 *in* B. H. Keating, P. Fryer, R. Batiza & G. W. Boehlert (eds.). Seamounts, Islands, and Atolls. American Geophysics Union Monograph 43.

Olff, H., M. E. Ritchie & H. H. T. Prins. 2002. Global environmental controls of diversity in large herbivores. Nature 415:901–904.

Ollier, C. D. & P. Zarriello. 1979. Pe'ape'a lava cave, Western Samoa. Transactions of the British Cave Research Association 6:133–142.

Olsen, P. D. 1999. Fearful owl *Nesasio solomonensis*. P. 242 *in* J. del Hoyo, A. Elliott & J. Sargatal (eds.). Handbook of the Birds of the World, vol 5. Lynx, Barcelona.

Olson, S. L. 1970. The relationships of *Porzana flaviviventer*. Auk 87:805–808.

Olson, S. L. 1973a. Evolution of the rails of the South Atlantic islands (Aves: Rallidae). Smithsonian Contributions to Zoology no. 152: 1–53.

Olson, S. L. 1973b. A classification of the Rallidae. Wilson Bulletin 85:381–416.

Olson, S. L. 1975a. The South Pacific gallinules of the genus *Pareudiastes*. Wilson Bulletin 87:1–5.

Olson, S. L. 1975b. The fossil rails of C. W. DeVis, being mainly an extinct form of *Tribonyx mortierii* from Queensland. Emu 75:49–54.

Olson, S. L. 1975c. Paleornithology of St. Helena Island, South Atlantic Ocean. Smithsonian Contributions to Paleobiology no. 23:1–49.

Olson, S. L. 1977. A synopsis of the fossil Rallidae. Pp. 339–373 *in* S. D. Ripley (ed.). Rails of the World. David R. Godine, Boston.

Olson, S. L. 1980. The significance of the distribution of the Megapodiidae. Emu 80:21–24.

Olson, S. L. 1985a. The Italian specimen of *Bulweria fallax* (Procellariidae). Bulletin of the British Ornithologists' Club 105:29–30.

Olson, S. L. 1985b. The fossil record of birds. Pp. 80–238 *in* D. S. Farner, J. R. King & K. C. Parkes (eds.). Avian Biology 8. Academic Press, New York.

Olson, S. L. 1986a. *Aplonis movornata*. Notornis 33:197–208.

Olson, S. L. 1986b. *Gallirallus sharpei* (Butterkofer), nov. comb, a valid species of rail (Rallidae) of unknown origin. Le Gerfaut 76:263–269.

Olson, S. L. 1989. Extinction on islands: man as a catastrophe. Pp. 50–53 *in* D. Western & M. C. Pearl (eds.). Conservation for the Twenty-first Century. Oxford University Press, London.

Olson, S. L. 1990. [Book review] The Birds of New South Wales: A Working List by A. W. McAllen & M. D. Bruce. Auk 107:458–459.

Olson, S. L. 1992. Requiescat for *Tricholimnas conditicius*, a rail that never was. Bulletin of the British Ornithologists' Club 112:174–179.

Olson, S. L. 1994. The endemic vireo of Fernando de Noronha (*Vireo gracilirostris*). Wilson Bulletin 106:1–17.

Olson, S. L. 2000. A new genus for the Kerguelen petrel. Bulletin of the British Ornithologists' Club 120:59–62.

Olson, S. L. & H. F. James. 1982. Fossil birds from the Hawaiian Islands: evidence for wholesale extinction by man before Western contact. Science 217:633–635.

Olson, S. L. & H. F. James. 1984. The role of Polynesians in the extinction of the avifauna of the Hawaiian Islands. Pp. 768–787 *in* P. S. Martin & R. G. Klein (eds.). Quaternary Extinctions: A Prehistoric Revolution. University of Arizona Press, Tuscon.

Olson, S. L. & H. F. James. 1991. Descriptions of thirty-two species of birds from the Hawaiian Islands: Part I. Non-passeriformes. Ornithological Monographs 45:1–88.

Olson, S. L. & P. Jouventin. 1996. A new species of small flightless duck from Amsterdam Island, southern Indian Ocean (Anatidae: *Anas*). Condor 98:1–9.

Olson, S. L. & P. C. Rasmussen. 2001. Miocene and Pliocene birds from the Lee Creek Mine, North Carolina. Pp. 233–365 *in* C. E. Ray & D. J. Bohaska (eds.). Geology and Paleontology of the Lee Creek Mine, North Carolina, II. Smithsonian Contributions to Paleobiology no. 90. Smithsonian Institution Press, Washington.

Olson, S. L. & D. W. Steadman. 1977. New genus of flightless ibis (Threskiornithidae) and other fossil birds from cave deposits in Jamaica. Proceedings of the Biological Society of Washington 90:447–457.

Olson, S. L. & D. W. Steadman. 1987. Comments on the proposed suppression of *Rallus nigra* Miller, 1784 and *Columba R. Forsteri* Wagler, 1829 (Aves). Bulletin of Zoological Nomenclature 44:126–127.

Olson, S. L. & K. I. Warheit. 1988. A new genus for *Sula abbotti*. Bulletin of the British Ornithologists' Club 108:9–12.

Olson, S. L. & D. B. Wingate. 2000. Two new species of flightless rails (Aves: Rallidae) from the middle Pleistocene "crane fauna" of Bermuda. Proceedings of the Biological Society of Washington 113:356–368.

Olson, S. L. & A. C. Ziegler. 1995. Remains of land birds from Lisianski Island, with observations on the terrestrial avifauna of the northwestern Hawaiian Islands. Pacific Science 49:111–125.

Olson, S. L., J. C. Balouet & C. T. Fisher. 1987. The owlet-nightjar of New Caledonia, *Aegotheles savesi*, with comments on the systematics of the Aegothelidae. Gerfaut 77:341–352.

Onley, D. 1982. The nomenclature of the spotless crake (*Porzana tabuensis*). Notornis 29:75–79.

Oppenheimer, S. J. & M. Richards. 2001. Slow boat to Melanesia. Nature 410:166–167.

Orenstein, R. I. 1976. Birds of the Plesyumi area, central New Britain. Condor 78:370–373.

ORSTROM. 1976. New Hebrides Archipelago Atlas of Soils and of Natural Environment Data. Office de la Recherche Scientifique et Technique outre-mer, Bondy, France.

Osborne, D. 1966. The archaeology of the Palau Islands. Bernice P. Bishop Museum Bulletin no. 230.

Osborne, D. 1979. Archaeological test excavations, Palau Islands, 1968–1969. Micronesica Supplement 1.

Owen-Smith, N. 1999. The interaction of humans, megaherbivores, and habitats in the late Pleistocene extinction event. Pp. 57–69 *in* R. D. E. MacPhee (ed.). Extinctions in Near Time: Causes, Contexts, and Consequences. Kluwer Academic/Plenum Publishers, New York.

Oxcal, version 3.3. 1999. University of Oxford Radiocarbon Accelerator Unit, Oxford.

Palau Islands. 1996. Map 81141. United States Defense Mapping Agency, Washington.

Palmer, M. W. 1988. The vegetation and anthropogenic disturbance of Toloa Forest, Tongatapu Island, South Pacific. Micronesica 21:279–281.

Pandolfi, J. M. 1995. Geomorphology of the uplifted Pleistocene atoll at Henderson Island, Pitcairn Group. Biological Journal of the Linnean Society 56:63–77.

Papua New Guinea. No date. (Map). Runaway Publications, New South Wales, Australia.

Paris, J. P. 1981. Géologie de la Nouvelle-Calédonie. Territoire de la Nouvelle-Calédonie Bureau de Recherches Géologiques et Miniéres, Nouméa.

Parker, S. 1967. New information on the Solomon Islands crown pigeon *Microgoura meeki* Rothschild. Bulletin of the British Ornithologists' Club 87:86–89.

Parker, S. A. 1972. An unsuccessful search for the Solomon Islands crowned pigeon. Emu 72:24–26.

Parkes, A. 1997. Environmental change and the impact of Polynesian colonization: sedimentary records from central Polynesia. Pp. 166–199 *in* P. V. Kirch & T. L. Hunt (eds.). Historical Ecology in the Pacific Islands. Yale University Press, New Haven, CT.

Parkes, A., J. T. Teller & J. R. Flenley. 1992. Environmental history of the Lake Vaihiria drainage-basin, Tahiti, French Polynesia. Journal of Biogeography 19:431–447.

Paulay, G. 1985. Adaptive radiation on an isolated oceanic island: the Cryptorhynchinae of Rapa revisited. Biological Journal of the Linnean Society 26:95–187.

Paulay, G. 1991. Henderson Island: biogeography and evolution at the edge of the Pacific plate. Pp. 304–313 *in* E. C. Dudley (ed.). The Unity of Evolutionary Biology. The Proceedings of the Fourth International Congress of Systematic and Evolutionary Biology, vol. 1. Dioscorides Press, Portland, OR.

Paulay, G. 1994. Biodiversity on oceanic islands: its origin and extinction. American Zoologist 34:134–144.

Paulay, G. 1997. Productivity plays a major role in determining atoll life and form: Tarawa, Kiribati. Proceedings of the 8th International Coral Reef Symposium 1:483–488.

Paulay, G. (ed.). 2003. The marine biodiversity of Guam and the Marianas. Micronesica 35–36:1–682.

Paulay, G. & C. Meyer. 2003. Diversification in the tropical Pacific: comparisons between marine and terrestrial systems and the importance of founder speciation. Integrative and Comparative Biology 42:922–934.

Paulay, G. & T. Spencer. 1989. Vegetation of Henderson Island. Atoll Research Bulletin 328:1–13.

Paulay, G. & T. Spencer. 1992. Niue Island: geologic and faunatic history of a Pliocene atoll. Pacific Science Association Information Bulletin 44:21–23.

Pavlides, C. & C. Gosden. 1994. 35,000–year-old sites in the rainforest of West New Britain, Papua New Guinea. Antiquity 68:604–610.

Pawley, A. K. & M. Ross. 1995. The prehistory of Oceanic languages: a current view. Pp. 39–74 *in* P. Bellwood, J. J. Fox & D. Tryon (eds.). The Austronesians: Historical and Comparative Perspectives. Australian National University, Canberra.

Paxinos, E. E., H. F. James, S. L. Olson, J. D. Ballou, J. A. Leonard & R. C. Fleischer. 2002. Prehistoric decline of genetic diversity in the nene. Science 296 :1827.

Payne, R. B. 1997. Family Cuculidae (cuckoos). Pp. 508–607 *in* J. del Hoyo, A. Elliott & J. Sargatal (eds.). Handbook of the Birds of the World, vol. 4, Sandgrouse to Cuckoos. Lynx, Barcelona.

Paytan, A., M. Kastner & F. P. Chavez. 1996. Glacial to interglacial fluctuations in productivity in the equatorial Pacific as indicated by marine barite. Science 274:1355–1357.

Pearson, D. L. & J. W. Knudsen. 1967. Avifaunal records from the Eniwetok Atoll, Marshall Islands. Condor 69:201–203.

Pearson, P. N. & M. R. Palmer. 2000. Atmospheric carbon dioxide concentrations over the past 60 million years. Nature 406:695–699.

Pearson, P. N., P. W. Ditchfield, J. Singano, K. G. Harcourt-Brown, C. J. Nicholas, R. K. Olsson, N. J. Shackleton & M. A. Hall. 2001. Warm tropical sea surface temperatures in the late Cretaceous and Eocene epochs. Nature 413:481–487.

Peirce, M. A. 1989. The significance of avian haematozoa in conservation strategies. Pp. 69–76 *in* J. E. Cooper (ed.). Disease and Threatened Birds. International Council for Bird Preservation Technical Publications no. 10.

Pelletier, B. & R. Louat. 1989. Seismotectonics and present-day relative plate motions in the Tonga-Lau and Kermadec-Havre region. Tectonophysics 165:237–250.

Peltier, W. R. 1994. Ice Age paleotopography. Science 265:195–201.

Pennisi, E. 2000. When fire ants move in, others leave. Science 289:231.

Perrault, G. G. 1992. Endemism and biogeography among Tahitian *Mecyclothorax* species (Coleoptera, Carabidae, Psydrini). Pp. 201–215 *in* G. R. Noonan, G. E. Ball & N. E. Stork (eds.). The Biogeography of Ground Beetles of Mountains and Islands. Intercept, Andover, England.

Perry, G., G. H. Rodda, T. H. Fritts & T. R. Sharp. 1998. The lizard fauna of Guam's fringing islets: island biogeography, phylogenetic history, and conservation implications. Global Ecology and Biogeography Letters 7:353–365.

Peters, D. S. 1996. *Monarcha takatsukasae* (Yamashina 1931): ein Nachweis von Saipan. Senckenbergiana Biologica 76:15–17.

Peters, J. L. 1932. A new genus for *Rallus poeciloptera*. Auk 49:347–348.

Peters, J. L. 1937. Checklist of the Birds of the World, vol. 12. Harvard University Press, Cambridge, MA.

Peters, J. L. 1945. Checklist of the Birds of the World, vol. 5. Harvard University Press, Cambridge, MA.

Peters, J. L. 1964. Checklist of the Birds of the World, vol. 10. Harvard University Press, Cambridge, MA.

Peters, J. L. & L. Griscom. 1928. A new rail and a new dove from Micronesia. Proceedings of the New England Zoological Club 5:99–106.

Petitot, C. & F. Petitot. 1975. Observations ornithologiques dans L'Atoll de Manihi (Archipel des Tuamotu) et dans L'ile de Tubuai (Australes). L'Oiseau et Revue Française du Ornithologie 45:83–88.

Peters, S. E. & M. Foote. 2002. Determinants of extinction in the fossil record. Nature 416:420–424.

Peterson, A. T. 1998. New species and new species limits in birds. Auk 115:555–558.

Petterson, M. G., C. R. Neal, J. J. Mahoney, L. W. Kroenke, A. D. Saunders, T. L. Babbs, R. A. Duncan, D. Tolia & B. McGrail. 1997. Structure and deformation of north and central Malaita, Solomon Islands: tectonic implications for the Ontong Java Plateau-Solomon arc collision, and for the fate of oceanic plateaus. Tectonophysics 283:1–33.

Pigeot, N. 1985. Eléments de typologie et technologie d'un matériel en nacre du site de Fa'ahia, Huahine-Polynésie Française. Centre Polynésien des Sciences Humaines, Te Anavaharau. Punaauia, Tahiti.

Pigeot, N. 1986. Nouvelle recherches sur le site de Fa'ahia, Huahine-Polynésie Française: les fouilles 1983. Centre Polynésien des Sciences Humaines, Te Anavaharau. Punaauia, Tahiti.

Pimm, S. L., H. L. Jones & J. Diamond. 1988. On the risk of extinction. American Naturalist 132:757–785.

Pimm, S. L., M. P. Moulton & L. J. Justice. 1994. Bird extinctions in the central Pacific. Philosophical Transactions of the Royal Society of London B 344:27–33.

Pimm, S. L., M. P. Moulton & L. J. Justice. 1995. Bird extinctions in the central Pacific. Pp. 75–87 *in* J. H. Lawton & R. M. May (eds.). Extinction Rates. Oxford University Press, New York.

Pintaud, J.-C., T. Jaffré & J.-M. Veillon. 1999. Conservation status of New Caledonian palms. Pacific Conservation Biology 5:9–15.

Pirazzoli, P. A. & L. F. Montaggioni. 1988. Holocene sea-level changes in French Polynesia. Palaeogeography, Palaeoclimatology, Palaeoecology 68:153–175.

Pirazzoli, P. A. & B. Salvat. 1992. Ancient shorelines and Quaternary vertical movements on Rurutu and Tubai (Austral Isles, French Polynesia). Zeitschrift für Geomorphologie 36:431–451.

Pole, M. 1993. Early Miocene flora of the Manuherikia Group, New Zealand. 7. Myrtaceae, including *Eucalyptus*. Journal of the Royal Society of New Zealand 23:313–328.

Polhemus, D. A. 1996. Island arcs and their influence on Indo-Pacific biogeography. Pp. 51–66 *in* A. Keast & S. E. Miller (eds.). The Origin and Evolution of Pacific Island Biotas, New Guinea to Eastern Polynesia: Patterns and Processes. SPB Academic Publishing, Amsterdam.

Polis, G. A & S. D. Hurd. 1996. Linking marine and terrestrial food webs: allochthonous input from the ocean supports high secondary productivity on small islands and coastal land communities. American Naturalist 147:396–423.

Poplin, F. 1980. *Sylviornis neocaledoniae* n. g., n. sp. (Aves), Ratite éteint de la Nouvelle-Calédonie. Comptes Rendus de l'Académie des Sciences Paris (Série D) 290:691–694.

Poplin, F. & C. Mourer-Chauviré. 1985. *Sylviornis neocaledoniae* (Aves, Galliformes, Megapodiidae), oiseau géant éteint de l'Ile des Pins (Nouvelle-Calédonie). Gobios 18:73–97.

Poplin, F., C. Mourer-Chauviré & J. Evin. 1983. Position systematique et datation de *Sylviornis neocaledoniae*, Mégapode géant (Aves, Galliformes, Megapodiidae) éteint de la Nouvelle-Calédonie. Comptes Rendus Hebdomadaires des Séances de l'Académie des Sciences, Paris, Ser. III:99–102.

Poulsen, J. I. 1987. Early Tongan Prehistory. Terra Australis 12. Australian National University, Canberra.

Powlesland, R. G. & J. R. Hay. 1998. The Status of Birds, Peka, and Rodents on Niue. Status Report 1994–1995. South Pacific Regional Environment Programme, Apia, Samoa.

Pratt, T. K. & E. W. Stiles. 1985. The influence of fruit size and structure on composition of frugivore assemblages in New Guinea. Biotropica 17:314–321.

Pratt, H. D., P. L. Bruner & D. G. Berrett. 1977. Ornithological observations on Yap, Western Caroline Islands. Micronesica 13:49–56.

Pratt, H. D., P. L. Bruner & D. G. Berrett. 1979. America's unknown avifauna: the birds of the Mariana Islands. American Birds 3:227–235.

Pratt, H. D., P. L. Bruner & D. G. Berrett. 1987. A Field Guide to the Birds of Hawaii and the Tropical Pacific. Princeton University Press, Princeton, NJ.

Pratt, H. D., J. Engbring, P. L. Bruner & D. G. Berrett. 1980. Notes on the taxonomy, natural history, and status of the resident birds of Palau. Condor 82:117–131.

Pregill, G. K. 1981. Late Pleistocene herpetofaunas from Puerto Rico. University of Kansas Museum of Natural History Miscellaneous Publication no. 71.

Pregill, G. K. 1993. Fossil lizards from the Late Quaternary of 'Eua, Tonga. Pacific Science 47:101–114.

Pregill, G. K. 1998. Squamate reptiles from prehistoric sites in the Mariana Islands, Micronesia. Copeia 1998:64–75.

Pregill, G. K. 2001. Review: the herpetofauna of New Caledonia. Copeia 2001:883–884.

Pregill, G. K. & B. I. Crother. 1999. Ecological and historical biogeography of the Caribbean. Pp. 335–356 *in* B. I. Crother (ed.). Caribbean Amphibians and Reptiles. Academic Press, San Diego, CA.

Pregill, G. K. & T. Dye. 1989. Prehistoric extinction of giant iguanas in Tonga. Copeia 1989:505–508.

Pregill, G. K. & S. L. Olson. 1981. Zoogeography of West Indian vertebrates in relation to Pleistocene climatic cycles. Annual Review of Ecology and Systematics 1981:75–98.

Pregill, G. K. & D. W. Steadman. 2000. Fossil vertebrates from Palau, Micronesia: a resource assessment. Micronesica 33:137–152.

Pregill, G. K. & D. W. Steadman. 2004. Human-altered diversity and biogeography of South Pacific iguanas. Journal of Herpetology 38:15–21.

Pregill, G. K. & T. H. Worthy. 2003. A new iguanid lizard (Squamata, Iguanidae) from the late Quaternary of Fiji, southwest Pacific. Herpetologica 59:57–67.

Pregill, G. K., D. W. Steadman & D. R. Watters. 1994. Late Quaternary vertebrate faunas of the Lesser Antilles: historical components of Caribbean biogeography. Bulletin of the Carnegie Museum of Natural History 30:1–51.

Preston, F. W. 1962. The canonical distribution of commonness and rarity. Ecology 43:185–215, 410–432.

Price, J. J., K. P. Johnson & D. H. Clayton. 2004. The evolution of echolocation in swiftlets. Journal of Avian Biology 35:135–143.

Price, J. P. & D. Elliott-Fisk. 2004. Topographic history of the Maui Nui Complex, Hawai'i, and its implications for biogeography. Pacific Science 58:27–45.

Prŷs-Jones, R. P., M. S. Prŷs-Jones & J. C. Lawley. 1981. The birds of Assumption Island, Indian Ocean: past and future. Atoll Research Bulletin no. 248.

Purvis, A., P.-M. Agapow, J. L. Gittleman & G. M. Mace. 2000. Nonrandom extinction and the loss of evolutionary history. Science 288:328–330.

Putnam, M. S. & B. Iova. 2001. Birds of the Nakanai Mountains and Willaumez Peninsula, West New Britain Province, Papua New Guinea. Pp. 64–133 *in* D. Byrnes (ed.). New Britain Biological Survey. University of Wisconsin New Britain Biological Survey, Madison.

Pyle, P. & J. Engbring. 1985. Checklist of the birds of Micronesia. ʻElepaio 46:57–68.

Pyle, P. & J. Engbring. 1988. Nest of Truk greater white-eye. Micronesica 21:281–283.

Radtkey, R. R., B. Becker, R. D. Miller, R. Riblet & T. J. Case. 1996. Variation and evolution of Class I Mhc in sexual and parthenogenetic geckos. Proceedings of the Royal Society of London B 263:1023–1032.

Rahbek, C. & G. R. Graves. 2000. Detection of macro-ecological patterns in South American hummingbirds is affected by spatial scale. Proceedings of the Royal Society of London B 267:2259–2265.

Rainey, W. E. & E. D. Pierson. 1992. Distribution of Pacific island flying foxes: implications for conservation. U.S. Fish & Wildlife Service Biological Report 90: 111–122.

Rando, J. C., M. López & B. Seguí. 1999. A new species of extinct flightless passerine (Emberizidae: *Emberiza)* from the Canary Islands. Condor 101:1–13.

Rapaport, M. (ed.). 1999. The Pacific Islands: Environment and Society. Bess Press, Honolulu.

Raxworthy, C. J., M. R. J. Forstner & R. A. Nussbaum. 2002. Chameleon radiation by oceanic dispersal. Nature 415:784–788.

Raynal, M. 1980–1981. "Koau", l'oiseau insaisissable des îles Marquises. Bulletin de la Société d'Etude des Sciences naturelles de Béziers, n.s. 8:20–26.

Raynal, M. 2002. Une représentation picturale de l'oiseau mystérieux d'Hiva-Oa. Cryptozoologia 47:3–10.

Raynal, M. & M. Dethier. 1990. Lézards géants des Maoris et oiseaux énigmatiques des Marquisiens: la vérité derriére la légende. Bulletin Mensuel de la Société Linnéenne de Lyon 59:85–91.

Reagan, M. K. & A. Meijer. 1984. Geology and geochemistry of early arc-volcanic rocks from Guam. Geological Society of America Bulletin 95:701–713.

Reece, R. L. 1989. Avian pathogens: their biology and methods of spread. Pp. 1–23 *in* J. E. Cooper (ed.). Disease and Threatened Birds. International Council for Bird Preservation Technical Publications no. 10.

Reichel, J. D. 1991. Status and conservation of seabirds in the Mariana Islands. Pp. 248–262 *in* J. P. Croxall (ed.). Seabird Status and Conservation: A Supplement. International Council for Bird Preservation Technical Bulletin no. 11. Cambridge, UK.

Reichel, J. D. & P. O. Glass. 1991. Checklist of the birds of the Mariana Islands. ʻElepaio 51:3–10.

Reichel, J. D., G. J. Wiles & P. O. Glass. 1992. Island extinctions: the case of the endangered nightingale reed-warbler. Wilson Bulletin 104:44–54.

Reimer, P. J. 2001. A new twist in the radiocarbon tale. Science 294:2494–2495.

Reis, K. R. & A. M. Garong. 2001. Late Quaternary terrestrial vertebrates from Palawan Island, Philippines. Palaeogeography, Palaeoclimatology, Palaeoecology 171:409–421.

Reis, K. R. & D. W. Steadman. 1999. Archaeology of Trants, Montserrat. Part 5. Prehistoric avifauna. Annals of Carnegie Museum 68:275–287.

Remsen, J. V., Jr. & T. A. Parker III. 1990. Seasonal distribution of the azure gallinule *(Porphyrula flavirostris)*, with comments on vagrancy in rails and gallinules. Wilson Bulletin 102:380–399.

Reville, B. J., J. D. Tranter & H. D. Yorkston. 1990. Impact of forest clearing on the endangered seabird *Sula abbotti.* Biological Conservation 51:23–38.

Reynolds, M. H. & G. L. Ritchotte. 1997. Evidence of Newells' shearwater breeding in Puna District, Hawaii. Journal of Field Ornithology 68:26–32.

Reynolds, B. C., M. Frank & R. K. O'Nions. 1999. Nd- and Ph-isotope time series from Atlantic ferromanganese crusts: implications for changes in provenance and paleocirculation over the last 8 Myr. Earth and Planetary Science Letters 173:381–396.

Rhymer, J. M. 2001. Evolutionary relationships and conservation of the Hawaiian anatids. Studies in Avian Biology 22:61–67.

Rhymer, J. M., M. J. Williams & R. T. Kingsford. 2004. Implications of phylogeography and population genetics for subspecies taxonomy of grey (Pacific black) duck *Anas superciliosa* and its conservation in New Zealand. Pacific Conservation Biology 10:57–66.

Rich, P. V. & R. J. Scarlett. 1977. Another look at *Megaeotheles*, a large owlet-nightjar from New Zealand. Emu 77:1–8.

Rich, P. V. & G. F. van Tets (eds.). 1990. Kadimakara: Extinct Vertebrates of Australia, 2nd ed. Pioneer Design Studios, Lilydale, Australia.

Rich, P. V., A. R. McEvey & R. Walkley. 1978. A probable masked owl *Tyto novaehollandiae* from Pleistocene deposits of Cooper Creek, South Australia. Emu 78:88–90.

Richmond, B. M. 1997a. Reconnaissance geology of the atoll islets of the Gilbert Islands group, Kiribati. Pp. 25–50 *in* A. M. Sherwood (comp.). Coastal and Environmental Geoscience Studies of the Southwest Pacific Islands. SOPAC Technical Bulletin no. 9.

Richmond, B. M. 1997b. Review of Holocene reef growth, South Pacific islands. Pp. 241–263 *in* A. M. Sherwood (comp.). Coastal and Environmental Geoscience Studies of the Southwest Pacific Islands. SOPAC Technical Bulletin no. 9.

Ricklefs, R. E. and E. Bermingham. 2001. Nonequilibrium diversity dynamics of the Lesser Antillean avifauna. Science 294:1522–1524.

Ridgell, R. 1995. Pacific Nations and Territories: The Islands of Micronesia, Melanesia, and Polynesia. Bess Press, Honolulu.

Ridgway, R. 1893. Descriptions of some new birds collected on the islands of Aldabra and Assumption, northwest of Madagascar, by Dr. W. L. Abbott. Proceedings of the United States National Museum 16:597–600.

Rieppel, O. 2002. Biogeography: a case of dispersing chameleons. Nature 415:744–745.

Rinke, D. 1986. Notes on the avifauna of Niuafoʻou Islands, Kingdom of Tonga. Emu 86:82–86.

Rinke, D. 1987. The avifauna of 'Eua and its offshore islet Kalau, Kingdom of Tonga. Emu 87:26–34.
Rinke, D. 1989. The relationship and taxonomy of the Fijian parrot genus *Prosopeia*. Bulletin of the British Ornithologists' Club 109:185–195.
Rinke, D. 1991. Birds of 'Ata and Late, and additional notes on the avifauna of Niuafo'ou, Kingdom of Tonga. Notornis 38:131–151.
Rinke, D., L. H. Soakai & A. Usback. 1993. Koe Malau: life and future of the Malau. Brehm Fund for International Bird Conservation, Bonn.
Ripley, S. D. 1960. Distribution and niche differentiation in species of megapodes in the Moluccas and Western Papuan area. Proceedings of the 17th International Ornithological Congress:631–640.
Ripley, S. D. 1964. A systematic and ecological study of birds of New Guinea. Peabody Museum of Natural History, Yale University, Bulletin 19:1–87.
Ripley, S. D. 1977. A new subspecies of island thrush *Turdus poliocephalus*, from New Ireland. Auk 94:772–773.
Ripley, S. D. & H. Birckhead. 1942. Birds collected during the Whitney South Sea Expedition. LI. On the fruit pigeons of the *Ptilinopus purpuratus* group. American Museum Novitates no. 1192.
Roberts, A. & L. Stone. 1990. Island-sharing by archipelago species. Oecologia 83:560–567.
Roberts, D. L. & A. R. Solow. 2003. When did the dodo become extinct? Nature 426:245.
Roberts, R. G., T. F. Flannery, L. K. Ayliffe, H. Yoshida, J. M. Olley, G. J. Prideaux, G. M. Laslett, A. Baynes, M. A. Smith, R. Jones, B. L. Smith. 2001. New ages for the last Australian megafauna: continent-wide extinction about 46,000 years ago. Science 292:1888–1892.
Robertson, H. A., J. R. Hay, E. K. Saul & G. V. McCormack. 1994. Recovery of the Kakerori: an endangered forest bird of the Cook Islands. Conservation Biology 8:1078–1086.
Robin, C., M. Monzier & J.-P. Eissen. 1994. Formation of the mid-fifteenth century Kuwae Caldera (Vanuatu) by an initial hydroclastic and subsequent ignimbritic eruption. Bulletin of Volcanology 56:170–183.
Robinet, O. & M. Salas. 1999. Reproductive biology of the endangered Ouvea parakeet *Eunymphicus cornutus uvaeensis*. Ibis 141:660–669.
Rodda, G. H. & T. H. Fritts. 1993. The brown tree snake on Pacific islands: 1993 status. Pacific Science Association Information Bulletin 45:1–3.
Rodda, G. H., T. H. Fritts & D. Chiszar. 1997. The disappearance of Guam's wildlife: new insights for herpetology, evolutionary ecology, and conservation. BioScience 47:565–574.
Rodda, P. 1994. Geology of Fiji. Pp. 131–151 *in* A. J. Stevenson, R. H. Herzer & P. F. Balance (eds.). Geology and Submarine Resources of the Tonga-Lau-Fiji Region. SOPAC Technical Bulletin no. 8.
Rodgers, J. 1948. Phosphate deposits of the former Japanese islands in the Pacific: a reconnaissance report. Economic Geology 43:400–407.
Rogers, B. W. & C. J. Legge. 1992. Karst features of the Palau Islands: a survey of caves and karst features in the Palau Archipelago. Pacific Basin Speleological Survey (National Speleological Society) Bulletin no. 3.
Rogers, G. (ed.). 1986. The Fire Has Jumped: Eyewitness Accounts of the Eruption and Evacuation of Niuafo'ou, Tonga. Institute of Pacific Studies at the University of the South Pacific, Suva, Fiji.
Rohling, E. J., M. Fenton, F. J. Jorissen, P. Bertrand, G. Ganssen & J. P. Caulet. 1998. Magnitudes of sea-level lowstands of the past 500,000 years. Nature 394:162–165.
Rohwer, S. 1990. Foraging differences between white and dark morphs of the Pacific reef heron *Egretta sacra*. Ibis 132:21–26.
Rolett, B. V. 1993. Marquesan prehistory and the origins of East Polynesian culture. Journal de la Société des Océanistes 96:29–47.
Rolett, B. V. 1998. Hanamiai: Prehistoric Colonization and Cultural Change in the Marquesas Islands (East Polynesia). Yale University Publications in Anthropology 81, New Haven, CT.
Rolett, B. V. & E. Conte. 1995. Renewed investigation of the Ha'atuatua dune (Nukuhiva, Marquesas Islands): a key site in Polynesian prehistory. Journal of the Polynesian Society 104:195–228.
Roselaar, C. S. 1994. Systematic notes on Megapodiidae (Aves, Galliformes), including the description of five new subspecies. Bulletin of the Zoological Museum, University of Amsterdam 14:9–36.
Rosenfeld, A. 1997. Excavation at Buang Merabak, central New Ireland. Bulletin of the Pacific Prehistory Association 16:213–224.
Rosenzweig, M. L. 1995. Species Diversity in Space and Time. Cambridge University Press, Cambridge, UK.
Rosenzweig, M. L. 1999. Heeding the warning in biodiversity's basic law. Science 284:276–277.
Ross, C. A. 1988. Weights of some New Caledonian birds. Bulletin of the British Ornithologists' Club 108:91–93.
Ross, B. 1991. Peleliu: Tragic Triumph. Random House, New York.
Ross, J. P. (ed.). 1998. Crocodiles: Status Survey and Conservation Action Plan. 2nd ed. IUCN/SSC Crocodile Specialist Group. IUCN, Gland, Switzerland, and Cambridge, UK.
Rothschild, W. 1904. [Description of a new pigeon]. Bulletin of the British Ornithologists' Club 14:77–78.
Rothschild, W. & E. Hartert. 1924. New species of birds from St. Matthias Island. Bulletin of the British Ornithologists' Club 44:50–53.
Rowe, S. & R. Empson. 1996. Distribution and abundance of the tanga'eo or Mangaian kingfisher (*Halcyon tuta ruficollaris*). Notornis 43:35–42.
Roy, P. S. 1997a. The morphology and surface geology of the islands of Tongatapu and Vava'u, Kingdom of Tonga. Pp. 153–173 *in* A. M. Sherwood (comp.). Coastal and Environmental Geoscience Studies of the Southwest Pacific Islands. SOPAC Technical Bulletin no. 9.
Roy, P. S. 1997b. Quaternary geology of the Guadalcanal coastal plain and adjacent seabed, Solomon Islands: a basis for resources assessment and planning economic developments. Pp. 207–240 *in* A. M. Sherwood (comp.). Coastal and Environmental Geoscience Studies of the Southwest Pacific Islands. SOPAC Technical Bulletin no. 9.

Rutherford, S., S. D'Hondt & W. Prell. 1999. Environmental controls on the geographic distribution of zooplankton diversity. Nature 400:749–753.

Ryan, P. G., B. P. Watkins & W. R. Siegfried. 1989. Morphometrics, metabolic rate, and body temperature of the smallest flightless bird: the inaccessible island rail. Condor 91:465–467.

Rylander, M. K. & E. G. Bolen. 1974. Analysis and comparison of gaits in whistling ducks (*Dendrocygna*). Wilson Bulletin 86:237–245.

Sabadini, R. 2002. Ice sheet collapse and sea level change. Science 295:2376–2377.

Salomonsen, F. 1964. Some remarkable new birds from Dyaul Island, Bismarck Archipelago, with zoogeographical notes. Biologiske Skrifter det Kongelige Danske Videnskabernes Selskab 14:1–37.

Salomonsen, F. 1972. New pigeons from the Bismarck Archipelago (Aves, Columbidae). Steenstrupia (Zoological Museum, University of Copenhagen) 2:183–189.

Salomonsen, F. 1976. The main problems concerning avian evolution on islands. Pp. 585–602 *in* Proceedings of the 16th International Ornithological Congress.

Salvat, F. & B. Salvat. 1991. Nukutipipi Atoll, Tuamotu Archipelago: geomorpology, land and marine flora and fauna, and interrelationships. Atoll Research Bulletin 357:1–43.

Salvat, B., F. Salvat & J.- C. Thibault. 1993. Les oiseaux de Nukutipipi (archipel des Tuamotu, Polynésie). Journal de la Société des Océanistes 20:183–186.

Sand, C. 1999. The beginning of southern Melanesian prehistory: the St. Maurice-Vatcha Lapita site, New Caledonia. Journal of Field Archaeology 26:307–323.

Sand, C. 2001a. Evolutions in the Lapita Cultural Complex: a view from the southern Lapita province. Archaeology in Oceania 36:65–76.

Sand, C. 2001b. Changes in non-ceramic artefacts during the prehistory of New Caledonia. Terra Australis 17:75–92.

Sand, C. (ed.). 2001c. Tiouandé: Archéologie d'un Massif de Karst du Nord-Est de la Grande Terre (Nouvelle-Calédonie). Service des Musées et du Patrimoine, Nouméa, New Caledonia.

Sand, C. 2002. Walpole: ha colo, une île de l'extrême archéologies et histoires. Les Cahiers de l'Archéologie en Nouvelle Calédonie 14:1–122.

Sand, C., J. Bolé & A. Ouetcho. 2002. Site LPO023 of Kurin: Characteristics of a Lapita settlement in the Loyalty Islands (New Caledonia). Asian Perspectives 41:129–147.

Sand, C., A. Ouetcho, J. Bole & D. Baret. 2001. Evaluating the "Lapita Smoke Screen": site SGO015 of Gora, an early Austronesian settlement on the south-east coast of New Caledonia's Grand Terre. New Zealand Journal of Archaeology 22:91–111.

Sanders, K. H., E. O. Minot & R. A. Fordham. 1995. Juvenile dispersion and use of habitat by the endangered kakerori *Pomarea dimidiata* (Monarchinae) on Rarotonga, Cook Islands. Pacific Conservation Biology 2:167–176.

Sanderson, J. G., M. P. Moulton & R. G. Selfridge. 1998. Null matrices and the analysis of species co-occurrences. Oecologia 116:275–283.

Sankaran, M. & S. J. McNaughton. 1999. Determinants of biodiversity regulate compositional stability of communities. Nature 401:691–693.

Santa Cruz Islands. 1993. Map 82449. United States Defense Mapping Agency, Washington.

Sauer, J. D. 1969. Oceanic islands and biogeographical theory. Geographical Review 59:582–593.

Sauer, J. D. 1988. Plant Migration: The Dynamics of Geographical Patterning in Seed Plant Species. University of California Press, Berkeley.

Savidge, J. A. 1987. Extinction of an island forest avifauna by an introduced snake. Ecology 68:660–668.

Savidge, J. A., L. Sileo & L. M. Siegfried. 1992. Was disease involved in the decimation of Guam's avifauna? Journal of Wildlife Diseases 28:206–214.

Schauenberg, P. 1970. Note sur une perruche éteinte de Tahiti *Cyanoramphus zealandicus* (Latham 1781) conservée au Muséum de Genève. Archives des Sciences (Société de physique et d'histoire naturelle de Genève) 22:645–649.

Schluter, D. 2000. The Ecology of Adaptive Radiation. Oxford University Press, Oxford.

Schodde, R. 1977. Contributions to Papuasian Ornithology. VI. Survey of the Birds of Southern Bougainville Island, Papua New Guinea. Division of Wildlife Research Technical Paper no. 34. Commonwealth Scientific Industrial Research Organization, Melbourne, Australia.

Schodde, R. & R. de Naurois. 1982. Patterns of variation and dispersal in the buff-banded rail (*Gallirallus philippensis*) in the south-west Pacific, with description of a new subspecies. Notornis 29:131–142.

Schodde, R., I. J. Mason, M. L. Dudzinski & J. L. McKean. 1980. Variation in the striated heron *Butorides striatus* in Australia. Emu 80:203–212.

Schoener, T. W. 1983. Rate of species turnover decreases from lower to higher organisms: a review of the data. Oikos 41:372–377.

Schoener, T. W., D. A. Spiller & J. B. Losos. 2001. Natural restoration of the species-area relation for a lizard after a hurricane. Science 294:1525–1528.

Schofield, J. C. & C. S. Nelson. 1978. Dolomitisation and Quaternary climate of Niue Island, Pacific Ocean. Pacific Geology 13:37–48.

Scholl, D. W., T. L. Vallier & T. U. Maung. 1985. Introduction. Pp. 3–15 *in* D. W. Scholl & T. L. Vallier (eds.). Geology and Offshore Resources of Pacific Island Arcs: Tonga Region. Circum-Pacific Council for Energy and Mineral Resources Earth Sciences Series, vol. 2, Houston.

Schreiber, E. A. 2002. Climate and weather effects on seabirds. Pp. 179–215 *in* E. A. Schreiber & J. Burger (eds.). Biology of Marine Birds. CRC Press, Boca Raton, FL.

Schreiber, R. W. & E. A. Schreiber. 1984. Central Pacific seabirds and the El Niño southern oscillation. Science 225:713–716.

Schubel, S. E. & D. W. Steadman. 1989. More bird bones from Polynesian archaeological sites on Henderson Island, Pitcairn Group, South Pacific. Atoll Research Bulletin no. 325.

Schwartz, A. & R. W. Henderson. 1991. Amphibians and Reptiles of the West Indies: Descriptions, Distributions, and Natural History. University of Florida Press, Gainesville.

Scoffin, T. P. 1993. The geological effects of hurricanes on coral reefs and the interpretation of storm deposits. Coral Reefs 12:203–221.

Scott, J. M., S. Conant & C. van Riper III (eds.). 2001. Evolution, Ecology, Conservation, and Management of Hawaiian Birds: A Vanishing Avifauna. Studies in Avian Biology 22.

Scowcroft, P. G. & R. Hobdy. 1987. Recovery of goat-damaged vegetation in an insular tropical montane forest. Biotropica 19:208–215.

Seitre, R. & J. Seitre. 1991. Causes de disparition des oiseaux terrestres de Polynésie française. SPREP Occasional Papers Séries 8, South Pacific Commission, Nouméa, New Caledonia.

Seitre, R. & J. Seitre. 1992. Causes of land-bird extinctions in French Polynesia. Oryx 26:215–222.

Seltzer, G. O., D. T. Rodbell, P. A. Baker, S. C. Fritz, P. M. Tapia, H. D. Rowe & R. B. Dunbar. 2002. Early warming of tropical South America at the last glacial-interglacial transition. Science 296:1685–1686.

Serjeantson, D. 2001. The great auk and the gannet: a prehistoric perspective on the extinction of the great auk. International Journal of Osteoarchaeology 11:43–55.

Serjeantson, S. W. & Gao X. 1995. *Homo sapiens* is an evolving species: origins of the Austronesians. Pp. 165–180 *in* P. Bellwood, J. J. Fox & D. Tryon (eds.). The Austronesians: Historical and Comparative Perspectives. Australian National University, Canberra.

Setchell, W. A. 1926. Phytogeographical notes on Tahiti. I. Land vegetation. University of California Publications in Botany 12:241–290.

Shapiro, B., D. Sibthorpe, A. Rambaut, J. Austin, G. M. Wragg, O. R. P. Bininda-Edmonds, P. L. M. Lee & A. Cooper. 2002. Flight of the dodo. Science 295:1683.

Sharp, A. 1964. Ancient Voyagers in Polynesia. University of California Press, Berkeley.

Shealer, D. A. 2002. Foraging behavior and food of seabirds. Pp. 137–177 *in* E. A. Schreiber & J. Burger (eds.). Biology of Marine Birds. CRC Press, Boca Raton, FL.

Sibley, C. G. 1951. Notes on the birds of New Georgia, central Solomon Islands. Condor 53:81–92.

Sibley, C. G. & J. A. Ahlquist. 1990. Phylogeny and Classification of Birds: A Study in Molecular Evolution. Yale University Press, New Haven, CT.

Siegrist, H. G., Jr. & R. H. Randall. 1992. Carbonate Geology of Guam: Summary and Field Trip Guide. Water and Energy Research Institute of the Western Pacific and Marine Laboratory, University of Guam, Mangilao.

Signor, P. W. III & J. H. Lipps. 1982. Sampling bias, gradual extinction patterns and catastrophes in the fossil record. Geological Society of America, Special Paper 190:291–296.

Sillitoe, P. 1998. An Introduction to the Anthropology of Melanesia: Culture and Tradition. Cambridge University Press, Cambridge, UK.

Silva, J. M. C. & M. Tabarelli. 2000. Tree species impoverishment and the future flora of the Atlantic forest of northeast Brazil. Nature 404:72–74.

Silva, K. 1975. Observations of birds in the Bismarck Archipelago. New Guinea Bird Society Newsletter 112:4–6.

Silva-Krott, I., M. K. Brock & R. E. Junge. 1998. Determination of the presence of *Mycobacterium avium* on Guam as a precursor to reintroduction of indigenous bird species. Pacific Conservation Biology 4:227–231.

Simberloff, D. S. 1974. Equilibrium theory of island biogeography and ecology. Annual Review of Ecology and Systematics 5:161–182.

Simberloff, D. S. 1976. Species turnover and equilibrium island biogeography. Science 194:572–578.

Simberloff, D. S. 1983. When is an island community in equilibrium? Science 220:1275–1277.

Simberloff, D. S. & J.-L. Martin. 1991. Nestedness of insular avifaunas: simple summary statistics masking complex species patterns. Ornis Fennica 68:178–192.

Simberloff, D. S. & E. O. Wilson. 1969. Experimental zoogeography of islands: the colonization of empty islands. Ecology 50:278–296.

Simberloff, D. S. & E. O. Wilson. 1970. Experimental zoogeography of islands. A two-year record of colonization. Ecology 51:934–937.

Simberloff, D., L. Stone & T. Dayan. 1999. Ruling out an assembly rule: the method of favored states. Pp. 58–74 *in* E. Weiher & P. A. Keddy (eds.). The Search for Assembly Rules in Ecological Communities. Harvard University Press, Cambridge, MA.

Simkin, T., L. Siebert, L. McClelland, D. Bridge, C. Newhall & J. H. Latter. 1981. Volcanoes of the World. Hutchinson Ross Publishing, Stroudsburg, PA.

Sinclair, J. R. 2002. Selection of incubation mound sites by three sympatric megapodes in Papua New Guinea. Condor 104:395–406.

Sinoto, Y. H. 1966. Tentative prehistoric cultural sequence in northern Marquesas Islands, French Polynesia. Journal of the Polynesian Society 75:287–303.

Sinoto, Y. H. 1970. An archaeologically based assessment of the Marquesas Islands as a dispersal center in east Polynesia. Pp. 105–132 *in* R. C. Green & M. Kelly (eds.). Studies in Oceanic Culture History. Pacific Anthropological Records 11. Bernice P. Bishop Museum, Honolulu.

Sinoto, Y. H. 1979. The Marquesas. Pp. 110–134 *in* J. D. Jennings (ed.). The Prehistory of Polynesia. Harvard University Press, Cambridge, MA.

Sinoto, Y. S. 1983. Analysis of Polynesian migrations based on archaeological assessments. Journal de la Société des Océanistes 39:57–67.

Sinoto, Y. 1988. A waterlogged site on Huahine Island, French Polynesia. Pp. 113–130 *in* B. Purdy (ed.). Wet Site Archaeology. Telford Press, Caldwell, NJ.

Sinoto, Y. H. & P. C. McCoy. 1975. Report on the preliminary excavation of an early habitation site on Huahine, Society Islands. Journal de la Société des Océanistes 31:143–86.

Skottsberg, C. 1956. Derivation of the flora and fauna of Juan Fernandez and Easter Island. Pp. 193–468 *in* C. Skottsberg (ed.). The Natural History of Juan Fernandez and Easter Islands. Almquist & Wiksells, Uppsala, Sweden.

Slikas, B. 2003. Hawaiian birds: lessons from a rediscovered avifauna. Auk 120:953–960.

Slikas, B., I. B. Jones, S. R. Derrickson & R. C. Fleischer. 2000. Phylogenetic relationships of Micronesian white-eyes based on mitochondrial sequence data. Auk 117:355–365.

Slikas, B., S. L. Olson & R. C. Fleischer. 2002. Rapid, independent evolution of flightlessness in four species of pacific Island rails (Rallidae): an analysis based on mitochondrial sequence data. Journal of Avian Biology 33:5–14.

Smart, P. L. & D. A. Richards. 1992. Age estimates for the late Quaternary high sea-stands. Quaternary Science Reviews 11:687–696.

Smith, A. C. 1963. Summary discussion on plant distribution patterns in the tropical Pacific. Pp. 247–253 *in* J. L. Gressitt (ed.). Pacific Basin Biogeography. Bernice P. Bishop Museum Press, Honolulu.

Smith, A. C. 1953. Studies of Pacific island plants. XV. The genus *Elaeocarpus* in the New Hebrides, Fiji, Samoa, and Tonga. Contributions from the U.S. National Herbarium.

Smith, A. C. 1970. The Pacific as a key to flowering plant history. University of Hawaii Harold L. Lyon Arboretum Lecture no. 1.

Smith, G. A. 1975. Systematics of parrots. Ibis 117:18–68.

Smith, G. C., B. J. Hamley & N. Lees. 1998. An estimate of the plumed frogmouth *Podargus ocellatus plumiferus* population size in the Conondale Ranges. Pacific Conservation Biology 4:215–226.

Smith, G. P., D. A. Wiens, K. M. Fischer, L. M. Dorman, S. C. Webb & J. A. Hildebrand. 2001. A complex pattern of mantle flow in the Lau backarc. Science 292:713–716.

Smith, J. 2000. Nice work—but is it science? Nature 408:293.

Smucker, T. D., G. D. Lindsey & S. M. Mosher. 2000. Home range and diet of feral cats in Hawaii forests. Pacific Conservation Biology 6:229–237.

Solem, A. 1976. Endodontoid land snails from Pacific Islands (Mollusca: Pulmonata: Sigmurethra). Part I. Family Endodontidae. Field Museum of Natural History, Chicago.

Solem, A. 1979. Biogeographic significance of land snails, Paleozoic to recent. Pp. 277–287 *in* J. Gray & A. J. Boucot (eds.). Historical Biogeography, Plate Tectonics, and the Changing Environment. Oregon State University Press, Corvallis.

Solem, A. 1983. Endodontoid land snails from Pacific Islands (Mollusca: Pulmonata: Sigmurethra). Part II. Families Punctidae and Charopidae: Zoogeography. Field Museum of Natural History, Chicago.

Solomon Islands. 1976. Structure and Lithology. Map 1–1. DOS 3240A. Directorate of Overseas Surveys, Surrey, England.

Solomon Islands. 1993. South Pacific Maps. Distributed by HEMA Maps, Queensland, Australia.

Solomon Islands. 1996. Map 82015. United States Defense Mapping Agency, Washington.

Soloviev, S. L., Ch. N. Go & Kh. S. Kim. 1992. Catalog of tsunamis in the Pacific 1969–1982. Academy of Sciences of the USSR, Moscow.

Soltis, P. S., D. E. Soltis & M. W. Chase. 1999. Angiosperm phylogeny inferred from multiple genes as a tool for comparative biology. Nature 402:402–404.

Soltis, P. S., D. E. Soltis, M. J. Zanis & S. Kim. 2000. Basal lineages of angiosperms: relationships and implications for floral evolution. International Journal of Plant Science 161:S97–S107.

Sorenson, M. D., A. Cooper, E. E. Paxinos, T. W. Quinn, H. F. James, S. L. Olson & R. C. Fleischer. 1999. Relationships of the extinct moa-nalos, flightless Hawaiian waterfowl, based on ancient DNA. Proceedings of the Royal Society of London B 266:2187–2193.

Spear, L. B. & D. G. Ainley. 1999. Migration routes of sooty shearwaters in the Pacific Ocean. Condor 101:205–218.

Specht, J. 1968. Preliminary report of excavations on Watom Island. Journal of the Polynesian Society 77:117–134.

Spencer, T. 1995. The Pitcairn Islands, South Pacific Ocean: plate tectonics and climatic contexts. Biological Journal of the Linnean Society 56:13–42.

Spencer, T., D. R. Stoddart & C. D. Woodroffe. 1987. Island uplift and lithospheric flexure: observations and cautions from the South Pacific. Zeitschrift für Geomorphologie 63:87–102.

Spennemann, D. H. R. 1989. ‘ata ‘a Tonga mo ‘ata ‘o Tonga: early and later prehistory of the Tongan Islands. Ph.D. thesis, Australian National University, Canberra.

Spennemann, D. H. R. 1997. A Holocene sea-level history from Tongatapu, Kingdom of Tonga. Pp. 115–153 *in* A. M. Sherwood (comp.). Coastal and Environmental Geoscience Studies of the Southwest Pacific Islands. SOPAC Technical Bulletin 9.

Spennemann, D. H. R. 1999. Exploitation of bird plumages in the German Mariana Islands. Micronesica 31:309–318.

Spotila, J. R., R. D. Reina, A. C. Steyermark, P. T. Plotkin & F. V. Paladino. 2000. Pacific leatherback turtles face extinction. Nature 405:529–530.

Spriggs, M. 1984. Early coconut remains from the South Pacific. Journal of the Polynesian Society 93:71–76.

Spriggs, M. 1997. The Island Melanesians. Blackwell Publishers, Oxford.

Spriggs, M. 1999. The stratigraphy of the Ponamla site, northwest Erromango, Vanuatu: evidence for 2700 year old stone structures. Pp. 323–331 *in* J.-C. Galipaud & I. Lilley (eds.). The South Pacific, 5000 to 2000 B.P.: Colonizations and Transformations. ORSTOM, Nouméa, New Caledonia.

Spriggs, M. & A. Anderson. 1993. Late colonization of East Polynesia. Antiquity 67:200–217.

Spriggs, M. & S. Bedford. 2001. Arapus: a Lapita site at Mangaasi in central Vanuatu? Terra Australis 17:93–104.

Stafford, T. W., K. Brendel & R. C. Duhamel. 1988. Radiocarbon ^{13}C and ^{15}N analysis of fossil bone: removal of humates with XAD-2 resin. Geochimica et Cosmochimica Acta 52:2257–2267.

Stafford, T. W., Jr., P. E. Hare, L. Currie, A. J. T. Jull & D. J. Donahue. 1990. Accelerator radiocarbon dating at the molecular level. Journal of Archaeological Science 18:35–72.

Stafford, T. W., Jr., H. A. Semken Jr., R. W. Graham, W. F. Klippel, A. Markova, N. G. Smirnov & J. Southon. 1999. First accelerator mass spectrometry ^{14}C dates documenting contemporaneity of nonanalog species in late Pleistocene mammal communities. Geology 27:903–906.

Stair, J. B. 1897. Old Samoa or Flotsam and Jetsam from the Pacific Ocean. R. McMillan, Papakura, New Zealand.

Stanley, D. 1989. South Pacific Handbook. Moon Publications, Emeryville, CA.

Stapp, P., G. A. Polis & F. S. Pinero. 1999. Stable isotopes reveal strong marine and El Niño effects on island food webs. Nature 401:467–469.

Steadman, D. W. 1980. A review of the osteology and paleontology of turkeys (Aves: Meleagridinae). Contributions to Science, Natural History Museum of Los Angeles County 330:131–207.

Steadman, D. W. 1982. The origin of Darwin's finches (Fringillidae: Passeriformes). Transactions of the San Diego Society of Natural History 19:279–296.

Steadman, D. W. 1985. Fossil birds from Mangaia, southern Cook Islands. Bulletin of the British Ornithologists' Club 105:58–66.
Steadman, D. W. 1986. Holocene vertebrate fossils from Isla Floreana, Galápagos. Smithsonian Contributions in Zoology no. 413:1–103.
Steadman, D. W. 1987. Two new species of rails (Aves: Rallidae) from Mangaia, southern Cook Islands. Pacific Science 40:27–43.
Steadman, D. W. 1988. A new species of *Porphyrio* (Aves: Rallidae) from archeological sites in the Marquesas Islands. Proceedings of the Biological Society of Washington 101:162–170.
Steadman, D. W. 1989a. A new species of starling (Sturnidae, *Aplonis*) from an archaeological site on Huahine, Society Islands. Notornis 36:161–169.
Steadman, D. W. 1989b. Extinction of birds in eastern Polynesia: a review of the record, and comparisons with other Pacific island groups. Journal of Archaeological Science 16:177–205.
Steadman, D. W. 1989c. New species and records of birds (Aves: Megapodiidae, Columbidae) from an archaeological site on Lifuka, Tonga. Proceedings of the Biological Society of Washington 102:537–552.
Steadman, D. W. 1990. Archaeological bird bones from Ofu: extirpations of shearwaters and petrels. Pp. 14–15 *in* P. V. Kirch, T. L. Hunt, L. Nagaoka & J. Tyler. An ancestral, Polynesian occupation site at To'aga, Ofu Island, America Samoa. Archaeology in Oceania 25:1–15.
Steadman, D. W. 1991a. Extinction of species: past, present, and future. Pp. 156–169 *in* R. L. Wyman (ed.). Global Climate Change and Life on Earth. Routledge, Chapman and Hall, New York.
Steadman, D. W. 1991b. Extinct and extirpated birds from Aitutaki and Atiu, southern Cook Islands. Pacific Science 45:325–347.
Steadman, D. W. 1991c. The identity and taxonomic status of *Megapodius stairi* and *M. burnabyi* (Aves: Megapodiidae). Proceedings of the Biological Society of Washington 104:870–877.
Steadman, D. W. 1992a. Extinct and extirpated birds from Rota, Mariana Islands. Micronesica 25:71–84.
Steadman, D. W. 1992b. New species of *Gallicolumba* and *Macropygia* (Aves: Columbidae) from archaeological sites in Polynesia. Los Angeles County Museum of Natural History, Science Series 36:329–349.
Steadman, D. W. 1993a. Biogeography of Tongan birds before and after human impact. Proceedings of the National Academy of Sciences USA 90:818–822.
Steadman, D. W. 1993b. Bird bones from the To'aga site, Ofu, American Samoa: prehistoric loss of seabirds and megapodes. University of California Archaeological Research Facility Contribution 51:217–228.
Steadman, D. W. 1995a. Prehistoric extinctions of Pacific island birds: biodiversity meets zooarchaeology. Science 267:1123–1131.
Steadman, D. W. 1995b. Extinction of birds on tropical Pacific islands. Pp. 33–49 *in* D. W. Steadman and J. I. Mead (eds.). Late Quaternary Environments and Deep History: A Tribute to Paul S. Martin. Mammoth Site, Hot Springs, SD.
Steadman, D. W. 1997a. A re-examination of the bird bones excavated on New Caledonia by E. W. Gifford in 1952. Kroeber Anthropological Society Papers 82:38–48.
Steadman, D. W. 1997b. Prehistoric extinctions of Polynesian birds: reciprocal impacts of birds and people. Pp. 51–79 *in* P. V. Kirch and T. L. Hunt (eds.). Historical Ecology in the Pacific Islands. Yale University Press, New Haven, CT.
Steadman, D. W. 1997c. The historic biogeography and community ecology of Polynesian pigeons and doves. Journal of Biogeography 24:157–173.
Steadman, D. W. 1998. Status of land birds on selected islands in the Ha'apai Group, Kingdom of Tonga. Pacific Science 52:14–34.
Steadman, D. W. 1999a. The prehistory of vertebrates, especially birds, on Tinian, Auguiguan, and Rota, Northern Mariana Islands. Micronesica 31:59–85.
Steadman, D. W. 1999b. [Book review] Consilience: The Unity of Knowledge by E. O. Wilson. Professional Geographer 51:325–326.
Steadman, D. W. 1999c. The Lapita extinction of Pacific island birds: catastrophic versus attritional. Pp. 375–386 *in* J.-C. Galipaud & I. Lilley (eds.). The South Pacific, 5000 to 2000 B.P.: Colonizations and Transformations. ORSTOM, Noumea, New Caledonia.
Steadman, D. W. 1999d. The biogeography and extinction of megapodes in Oceania. Zoologische Verhandelingen 327:7–21.
Steadman, D. W. 2002a. A new species of gull (Laridae: *Larus*) from an archaeological site on Huahine, Society Islands. Proceedings of the Biological Society of Washington 115:1–17.
Steadman, D. W. 2002b. A new species of swiftlet (Aves: Apodidae) from the late Quaternary of Mangaia, Cook Islands, Oceania. Journal of Vertebrate Paleontology 22:326–331.
Steadman, D. W. 2002c. Everything you want to know and moa [review]. Science 298:2136–2137.
Steadman, D. W. in press a. A new species of extinct parrot (Psittacidae: *Eclectus*) from Tonga and Vanuatu, South Pacific. Pacific Science.
Steadman, D. W. in press b. A new species of extinct tooth-billed pigeon *(Didunculus)* from the Kingdom of Tonga, and the concept of endemism in insular landbirds. Journal of Zoology.
Steadman, D. W. & V. E. Burke. 1999. The first highly stratified prehistoric vertebrate sequence from the Galápagos Islands, Ecuador. Pacific Science 53:129–143.
Steadman, D. W. & J. Franklin. 2000. A preliminary survey of landbirds on Lakeba, Lau Group, Fiji. Emu 100:227–235.
Steadman, D. W. & H. B. Freifeld. 1998. Distribution, relative abundance, and habitat relationships of landbirds in the Vava'u Group, Kingdom of Tonga. Condor 100:609–628.
Steadman, D. W. & H. B. Freifeld. 1999. The food habits of Polynesian pigeons and doves: a systematic and biogeographic review. Ecotropica 5:13–33.
Steadman, D. W. & M. Intoh. 1994. Biogeography and prehistoric exploitation of birds on Fais Island, Yap, Federated States of Micronesia. Pacific Science 48:116–135.
Steadman, D. W. & L. J. Justice. 1998. Prehistoric exploitation of birds on Mangareva, Gambier Islands, French Polynesia. Man and Culture in Oceania 14:81–98.

Steadman, D. W. & P. V. Kirch. 1990. Prehistoric extinction of birds on Mangaia, Cook Islands, Polynesia. Proceedings of the National Academy of Sciences USA 87:9605–9609.

Steadman, D. W. & P. V. Kirch. 1998. Biogeography and prehistoric exploitation of birds in the Mussau Islands, Papua New Guinea. Emu 98:13–21.

Steadman, D. W. & P. S. Martin. 2003. The late Quaternary extinction and future resurrection of birds on Pacific islands. Earth-Science Reviews 61:133–147.

Steadman, D. W. & G. S. Morgan. 1985. A new species of bullfinch (Aves, Emberizinae) from a late Quaternary cave deposit on Cayman Brac, West Indies. Proceedings of the Biological Society of Washington 98:544–553.

Steadman, D. W. & S. L. Olson. 1985. Bird remains from an archaeological site on Henderson Island, South Pacific: man-caused extinctions on an "uninhabited" island. Proceedings of the National Academy of Sciences USA 82:6191–6195.

Steadman, D. W. & D. S. Pahlavan. 1992. Prehistoric exploitation and extinction of birds on Huahine, Society Islands, French Polynesia. Geoarchaeology 7:449–483.

Steadman, D. W. & G. K. Pregill. 2004. A prehistoric vertebrate assemblage from Tutuila, American Samoa. Pacific Science 58: 615–624.

Steadman, D. W. & B. Rolett. 1996. A chronostratigraphic analysis of landbird extinction on Tahuata, Marquesas Islands. Journal of Archaeological Science 23:81–94.

Steadman, D. W. & M. C. Zarriello. 1987. Two new species of parrots (Aves: Psittacidae) from archaeological sites in the Marquesas Islands. Proceedings of the Biological Society of Washington 100:518–528.

Steadman, D. W. & S. Zousmer. 1988. Galapagos: Discovery on Darwin's Islands. Smithsonian Institution Press, Washington.

Steadman, D. W., J. A. Stull & S. W. Eaton. 1979. Natural history of the ocellated turkey. World Pheasant Association Journal 4:15–37.

Steadman, D. W., E. C. Greiner & C. S. Wood. 1990a. Absence of blood parasites in indigenous and introduced birds from the Cook Islands, South Pacific. Conservation Biology 4:398–404.

Steadman, D. W., D. S. Pahlavan & P. V. Kirch. 1990b. Extinction, biogeography, and human exploitation of birds on Tikopia and Anuta, Polynesian outliers in the Solomon Islands. Bernice P. Bishop Museum Occasional Papers 30:118–153.

Steadman, D. W., A. Plourde & D. V. Burley. 2002a. Prehistoric butchery and consumption of birds in the Kingdom of Tonga. Journal of Archaeological Science 29:571–584.

Steadman, D. W., G. K. Pregill & D. V. Burley. 2002b. Rapid prehistoric extinction of iguanas and birds in Polynesia. Proceedings of the National Academy of Sciences USA 99:3673–3677.

Steadman, D. W., S. E. Schubel & D. Pahlavan. 1988. A new subspecies and new records of *Papasula abbotti* (Aves: Sulidae) from archaeological sites in the tropical Pacific. Proceedings of the Biological Society of Washington 101:487–495.

Steadman, D. W., T. W. Stafford, Jr., D. J. Donahue & A. J. T. Jull. 1991. Chronology of Holocene vertebrate extinction in the Galápagos Islands. Quaternary Research 36:126–133.

Steadman, D. W., P. Vargas & C. Cristino. 1994. Stratigraphy, chronology, and cultural context of an early faunal assemblage from Easter Island. Asian Perspectives 33:79–96.

Steadman, D. W., J. Franklin, D. R. Drake, H. B. Freifeld, L. A. Bolick, D. S. Smith & T. J. Motley. 1999a. Conservation status of forests and vertebrate communities in the Vava'u Island Group, Tonga. Pacific Conservation Biology 5:191–207.

Steadman, D. W., P. J. White & J. Allen. 1999b. Prehistoric birds from New Ireland, Papua New Guinea: extinction on a large Melanesian island. Proceedings of the Natural Academy of Sciences USA 96:2563–2568.

Steadman, D. W., S. C. Antón & P. V. Kirch. 2000a. Ana Manuku: a prehistoric ritualistic site on Mangaia, Cook Islands. Antiquity 74:873–883.

Steadman, D. W., T. H. Worthy, A. J. Anderson & R. Walter. 2000b. New species and records of birds from prehistoric sites on Niue, Southwest Pacific. Wilson Bulletin 112:165–186.

Stearns, H. T. 1971. Geological setting of an Eocene fossil deposit on Eua island, Tonga. Geological Society of America Bulletin 82:2541–2551.

Steele, D. G. & B. W. Baker. 1993. Multiple predation: a definitive human hunting strategy. Pp. 9–37 *in* J. Hudson (ed.). From Bones to Behavior: Ethnoarchaeological and Experimental Contributions to the Interpretation of Faunal Remains. Center for Archaeological Investigations, Southern Illinois University at Carbondale, Occasional Paper no. 21.

Steig, E. J. 2001. No two latitudes alike. Science 293:2015–2016.

Steiner, W. W. M. 2001. Evaluating the cost of saving native Hawaiian birds. Studies in Avian Biology 22:377–383.

Stevenson, J. 1999. Human impact from the palaeoenvironmental record on New Caledonia. Pp. 251–258 *in* J.-C. Galipaud & I. Lilley (eds.). The South Pacific, 5000 to 2000 B.P.: Colonizations and Transformations. ORSTOM, Nouméa, New Caledonia.

Stinson, D. W. 1994. Birds and mammals recorded from the Marianas Islands. Natural History Research, Special Issue 1:333–344.

Stinson, D. W., M. W. Ritter & J. D. Reichel. 1991. The Mariana common moorhen: decline of an island endemic. Condor 93:38–43.

Stinson, D. W., G. J. Wiles & J. D. Reichel. 1997. Migrant land birds and water birds in the Mariana Islands. Pacific Science 51:314–327.

Stock, J., J. Coil & P. V. Kirch. 2003. Paleohydrology of arid southeastern Maui, Hawaiian Islands, and its implications for prehistoric human settlement. Quaternary Research 59:12–24.

Stoddart, D. R. 1972. Reef islands of Rarotonga. Atoll Research Bulletin no. 160.

Stoddart, D. R. 1975a. Almost-atoll of Aitutaki: geomorphology of reefs and islands. Pp. 31–57 *in* D. R. Stoddart & P. E. Gibbs (eds.). Almost-atoll of Aitutaki: Reef Studies in the Cook Islands, South Pacific. Atoll Research Bulletin no. 190.

Stoddart, D. R. 1975b. Mainland vegetation of Aitutaki. Pp. 117–122 *in* D. R. Stoddart & P. E. Gibbs (eds.). Almost-atoll of Aitutaki: Reef Studies in the Cook Islands, South Pacific. Atoll Research Bulletin no. 190.

Stoddart, D. R. 1981. Abbott's booby on Assumption. Atoll Research Bulletin 255:27–32.

Stoddart, D. R. 1992. Biogeography of the tropical Pacific. Pacific Science 46:276–293.

Stoddart, D. R. & T. P. Scoffin. 1983. Phosphate rock on coral reef islands. Pp. 369–400 *in* A. S. Goudie & K. Pye (eds.). Chemical Sediments and Geomorphology: Precipitates and Residua in the Near-surface Environment. Academic Press, New York.

Stoddart, D. R. & T. Spencer. 1987. Rurutu reconsidered: the development of makatea topography in the Austral Islands. Atoll Research Bulletin no. 297.

Stoddart, D. R., T. Spencer & T. P. Scoffin. 1985. Reef growth and karst erosion on Mangaia, Cook Islands: a reinterpretation. Zeitschrift für Geomorphologie 57:121–140.

Stoddart, D. R., C. D. Woodroffe & T. Spencer. 1990. Mauke, Mitiaro and Atiu: geomorphology of makatea islands in the southern Cooks. Atoll Research Bulletin 341:1–61.

Stokstad, E. 2004. Hawaii girds itself for arrival of West Nile virus. Science 306:603.

Stone, E. L., Jr. 1953. Summary of information on atoll soils. Atoll Research Bulletin no. 22.

Storr, G. M. 1973. List of Queensland Birds. Government Printer, Perth, Australia.

Stott, L., C. Poulsen, S. Lund & R. Thunell. 2002. Super ENSO and global climate oscillations at millennial time scales. Science 297:222–226.

Stratford, J. M. C. & P. Rodda. 2000. Late Miocene to Pliocene palaeogeography of Viti Levu, Fiji Islands. Palaeogeography, Palaeoclimatology, Palaeoecology 162:137–153.

Strauss, E. 1999. Can mitochondrial clocks keep time? Science 283:1435–1438.

Streets, T. H. 1876. Description of a new duck from Washington Island. Bulletin of the Nuttall Ornithological Club 1:46–47.

Stuart, A. J. 1999. Late Pleistocene megafaunal extinctions: a European perspective. Pp. 257–269 *in* R. D. E. MacPhee (ed.). Extinctions in Near Time: Causes, Contexts, and Consequences. Kluwer Academic/Plenum Publishers, New York.

Stuessy, T. F., R. W. Sanders & M. Silva. 1984. Phytogeography and evolution of the flora of the Juan Fernandez Islands: a progress report. Pp. 55–69 *in* F. J. Radovsky, P. H. Raven & S. H. Sohmer (eds.). Biogeography of the Tropical Pacific. Proceedings of a Symposium. Bernice P. Bishop Museum Special Publication, Honolulu.

Stuiver, M., P. J. Reimer & T. F. Braziunas. 1998. High-precision radiocarbon age calibration for terrestrial and marine samples. Radiocarbon 40:1127–1151.

Sturman, A. P. & H. A. McGowan. 1999. Climate. Pp. 3–18 *in* M. Rapaport (ed.). The Pacific Islands: Environment and Society. Bess Press, Honolulu.

Suggs, R. C. 1961. The archaeology of Nuku Hiva, Marquesas Islands, French Polynesia. Anthropological Papers of the American Museum of Natural History 49, Part 1.

Summerhayes, C. P. 1967. Baythmetry and topographic lineation in the Cook Islands. New Zealand Journal of Geology and Geophysics 10:1382–1399.

Summerhayes, G. R. 2001a. Far western, western, and eastern Lapita: a re-evaluation. Asian Perspectives 39:109–138.

Summerhayes, G. R. 2001b. Defining the chronology of Lapita in the Bismarck Archipelago. Terra Australis 17:25–38.

Sutherland, L. 2000. Pumice puzzles. Nature Australia 26/9:66–69.

Swarth, H. S. 1931. The avifauna of the Galápagos Islands. Occasional Papers of the California Academy of Sciences 18:5–299.

Swetnam, T. W., C. D. Allen & J. L. Betancourt. 1999. Applied historical ecology: using the past to manage for the future. Ecological Applications 9:1189–1206.

Sykes, W. R. 1981. The vegetation of Late, Tonga. Allertonia 2:323–353.

Tahiti. 1977. (Map). Institut Géographique National, Paris, France.

Takano, L. L. & S. M. Haig. 2004. Seasonal movement and home range of the Mariana common moorhen. Condor 106:652–663.

Takayama, J. 1982. A brief report on archaeological investigations of the southern part of Yap Island and nearby Ngulu Atoll. Pp. 77–104 *in* M. Aoyagi (ed.). Islanders and Their Outside Worlds. St. Paul's (Rikkyo) University, Tokyo.

Takayama, J. & M. Intoh. 1978. Archaeological excavation at Chukienu shell midden on Tol, Truk. Reports of Pacific Archaeological Survey no. 5, Tezukayama University, Nara City, Japan.

Takayama, J. & T. Seki. 1973. Preliminary archaeological investigations on the island of Tol in Truk. Reports of Pacific Archaeological Survey no. 2, Tezukayama University, Nara City, Japan.

Taomia, J. M. E. 2000. Household units in the analysis of prehistoric social complexity, southern Cook Islands. Asian Perspectives 39:139–164.

Tappin, D. R. 1993. The Tonga frontal-arc basin. Pp. 157–176 *in* P. F. Ballance (ed.). South Pacific Sedimentary Basins: Sedimentary Basins of the World, vol. 2. Elsevier Press, Amsterdam.

Tappin, D. R. & P. F. Ballance. 1994. Contributions to the sedimentary geology of 'Eua island, Kingdom of Tonga: reworking in an oceanic forearc. Pp. 1–20 *in* A. J. Stevenson, R. H. Herzer & P. F. Ballance (eds.). Geology and Submarine Resources of the Tonga-Lau-Fiji Region. SOPAC Technical Bulletin no. 8.

Tarburton, M. K. 1990. Breeding biology of the Atiu swiftlet. Emu 90:175–179.

Tarling, D. H. 1965. The palaeomagnetism of the Samoan and Tongan Islands. Geophysical Journal of the Royal Astronomical Society 10:497–513.

Tarr, C. L. & R. C. Fleischer. 1995. Evolutionary relationships of the Hawaiian Honeycreepers (Aves, Drepanidinae). Pp. 147–159 *in* W. L. Wagner & V. A. Funk (eds.). Hawaiian Biogeography: Evolution on a Hot Spot Archipelago. Smithsonian Institution Press, Washington.

Tayama, R. 1939. Brief report on the geology and ore resources of Babelthuap Island (Palau main island). Tropical Industry Institutional Bulletin 3:1–19.

Taylor, B. 1992. Rifting and the volcanic-tectonic evolution of the Izu-Bonin-Mariana arc. Proceedings of the Ocean Drilling Program, Scientific Results 126:627–651.

Taylor, F. W. & A. L. Bloom. 1977. Coral reefs on tectonic blocks, Tonga island arc. Third International Coral Reef Symposium Proceedings 2:275–281.

Taylor, F. W., R. L. Edwards & G. J. Wasserburg. 1990. Seismic recurrence intervals and timing of aseismic subduction inferred from emerged corals and reefs of the central Vanuatu (New Hebrides) frontal arc. Journal of Geophysical Research 95:393–408.

Taylor, F. W., C. Jouannin & A. L. Bloom. 1985. Quaternary uplift of the Torres islands, northern New Hebrides frontal arc: comparison with Santo and Malekula islands, central New Hebrides frontal arc. Journal of Geology 93:419–438.

Taylor, J. M., J. H. Calaby & H. M. van Deusen. 1982. A revision of the genus *Rattus* (Rodentia, Muridae) in the New Guinea region. Bulletin of the American Museum of Natural History 173:177–336.

Taylor, P. B. 1996. Family Rallidae (rails, gallinules and coots). Pp. 108–209 *in* J. del Hoyo, A. Elliott & J. Sargatal (eds.). Handbook of the Birds of the World, vol. 3, Hoatzin to Auks. Lynx, Barcelona.

Taylor, P. B. 1998. Rails: A Guide to the Rails, Crakes, Gallinules, and Coots of the World. Yale University Press, New Haven, CT.

Taylor, R. C. 1973. An atlas of the Pacific Islands rainfall. Hawai'i Institute of Geophysics Data Report no. 25, HIG-73-9. University of Hawai'i, Honolulu.

Taylor, R. H. 1985. Status, habits and conservation of *Cyanoramphus* parakeets in the New Zealand region. International Council for Bird Preservation Technical Publication 3:195–211.

Taylor, R. H. & B. W. Thomas. 1993. Rats exterminated from rugged Breaksea Island (170 ha), Fiordland, New Zealand. Biological Conservation 65:191–198.

Taylor, T. 2001. Explanatory tyranny. Nature 411:419.

Teimaore, T. (translated by M. O. Walker). 1927. Histoire de la grotte secréte de Rurutu. Bulletin de la Société des Estudies Oceaniennes, Papeete 22:315–317.

Teotónio, H. & M. R. Rose. 2000. Variation in the reversibility of evolution. Nature 408:463–466.

Terborgh, J. & J. M. Diamond. 1970. Niche overlap in feeding assemblages of New Guinea birds. Wilson Bulletin 82:29–52.

Terry, J. P. & R. Raj. 1999. Island environment and landscape responses to 1997 tropical cyclones in Fiji. Pacific Science 53:257–272.

Terry, J. P., R. Raj & R. A. Kostaschuk. 2001. Links between the Southern Oscillation Index and hydrological hazards on a tropical Pacific island. Pacific Science 55:275–283.

Thaman, R. R. 1990. Kiribati Agroforestry: Trees, People, and the Atoll Environment. Atoll Research Bulletin 333. Smithsonian Institution, Washington.

Thaman, R. R. 1992. Vegetation of Nauru and the Gilbert Islands: case studies of poverty, degradation, disturbance, and displacement. Pacific Science 46:128–158.

Thaman, R. R., F. R. Fosberg, H. I. Manner & D. C. Hassall. 1994. The Flora of Nauru. Atoll Research Bulletin 392. Smithsonian Institution, Washington.

Thaman, R. R., G. Keppel, D. Watling, B. Thaman, T. Gaunavinaka, A. Naikatini, B. Thaman, N. Bolaqace, E. Sekinoco & M. Masere. 2005. Nasoata mangrove island, the PABITRA coastal study site for Viti Levu, Fiji Islands. Pacific Science 59:193–204.

Thibault, J.-C. 1974. Les periodes de reproduction des oiseaux de mer dans L'Archipel de la Société (Polynésie française). Alauda 42:437–450.

Thibault, J.-C. 1976. L'avifaune de Tetiaroa (archipel de la Société Polynésie Française). L'Oiseau et Revue Française du Ornithologie 46:29–45.

Thibault, J.-C. & I. Guyot. 1987. Recent changes in the avifauna of Makatea Island (Tuamotus, central Pacific). Atoll Research Bulletin 300:1–10.

Thibault, B. & J.-C. Thibault. 1973. Liste préliminaire des oiseaux de Polynésie orientale. L'Oiseau et Revue Française du Ornithologie 43:55–74.

Thibault, J.-C & A. Varney. 1991a. Breeding seabirds of Rapa (Polynesia): numbers and changes during the 20th century. Bulletin of the British Ornithologists' Club 111:70–77.

Thibault, J.-C. & A. Varney. 1991b. Numbers and habitat of the Rapa fruit-dove *Ptilinopus huttoni*. Bird Conservation International 1:75–81.

Thomas, R., E. Rignot, G. Casassa, P. Kanagaratnam, C. Acuña, T. Akins, H. Brecher, E. Frederick, P. Gogineni, W. Krabill, S. Manizade, H. Ramamoorthy, A. Rivera, R. Russell, J. Sonntag, R. Swift, J. Yungel & J. Zwally. 2004. Accelerated sea-level rise from West Antarctica. Science 306:255–258.

Thomson, J. A. 1921. The geology of Western Samoa. New Zealand Journal of Science and Technology 4:49–66.

Thompson, C. S. 1986a. The climate and weather of the southern Cook Islands. New Zealand Meteorological Service Miscellaneous Publication no. 188.

Thompson, C. S. 1986b. The climate and weather of Tonga. New Zealand Meteorological Service Miscellaneous Publication no. 188.

Thompson, G. M., J. Malpas & I. E. M. Smith. 1998. Volcanic geology of Rarotonga, southern Pacific Ocean. New Zealand Journal of Geology and Geophysics 41:95–104.

Thorne, R. F. 1963. Biotic distribution patterns in the tropical Pacific. Pp. 311–350 *in* J. L. Gressitt (ed.). Pacific Basin Biogeography. Bernice P. Bishop Museum Press, Honolulu.

Thorne, A. & R. Raymond. 1989. Man on the Rim: The Peopling of the Pacific. Angus and Robertson Publishers, Auckland, New Zealand.

Thornton, I. W. B., R. A. Zann & S. van Balen. 1993. Colonization of Rakata (Krakatau Is.) by non-migrant land birds from 1883 to 1992 and implications for the value of island equilibrium theory. Journal of Biogeography 20:441–452.

Thornton, I. 1996. Krakatau: The Destruction and Reassembly of an Island Ecosystem. Harvard University Press, Cambridge, MA. 368 pp.

Tibi, R., D. A. Wiens & H. Inoue. 2003. Remote triggering of deep earthquakes in the 2002 Tonga sequences. Nature 424:921–925.

Tishkoff, S. A., R. Varkonyi, N. Cahinhinan, S. Abbes, G. Argyropoulos, G. Destro-Bisol, A. Drousiotou, B. Dangerfield, G. Lefranc, J. Loiselet, A. Piro, M. Stoneking, A. Tagarelli, G. Tagarelli, E. H. Touma, S. M. Williams & A. G. Clark. 2001. Haplotype diversity and linkage disequilibrium at human G6PD: recent origin of alleles that confer malarial resistance. Science 293:455–462.

Todd, D. M. 1983. Pritchard's megapode on Nuiafo'ou Island, Kingdom of Tonga. Journal of the World Pheasant Association 8:69–88.

Tokelau Islands (General Map of). 1969. NZMS 254. New Zealand Department of Lands & Survey, Wellington.

Tol. 1983. State of Truk (Chuk) (map). United States Geological Survey, Denver.

Tonga Islands. 1990. Map 83560. United States Defense Mapping Agency, Washington.

Towns, D. R. 1996. Changes in habitat use by lizards on a New Zealand island following removal of the introduced Pacific rat *Rattus exulans*. Pacific Conservation Biology 2:286–292.

Towns, D. R., D. Simberloff & I. A. E. Atkinson. 1997. Restoration of New Zealand islands: redressing the effects of introduced species. Pacific Conservation Biology 3:99–124.

Townsend, C. H. & A. Wetmore. 1919. Reports on the scientific results of the expedition to the tropical Pacific in charge of Alexander Agassiz, on the U.S. Fish Commission steamer "Albatross," from August, 1899, to March, 1900, Commander Jefferson F. Moser, U.S.N., Commanding. XXI. The Birds. Bulletin of the Museum of Comparative Zoology, Harvard University 63:151–225.

Tracey, J. I., Jr., S. O. Schlanger, J. T. Stark, D. B. Doan & H. G. May. 1964. General Geology of Guam. U.S. Geological Survey Professional Paper 403–A.

Trewick, S. A. 1996. Morphology and evolution of two takahe: flightless rails of New Zealand. Journal of Zoological Society of London 238:221–237.

Trewick, S. A. 1997. Flightlessness and phylogeny amongst endemic rails (Aves: Rallidae) of the New Zealand region. Philosophical Transactions of the Royal Society of London B 352:429–446.

Triggs, S. J. & C. H. Daugherty. 1996. Conservation and genetics of New Zealand parakeets. Bird Conservation International 6:89–101.

Tryon, R. 1970. Development and evolution of fern floras of oceanic islands. Biotropica 2:76–84.

Truk Islands–Eastern Part. 1996. Map 81327. United States Defense Mapping Agency, Washington.

Trust Territory of the Pacific Islands. (Map). 1985. United States Geological Survey, Reston, VA.

Tschentscher, F. 1999. Too mammoth an undertaking. Science 286:2084.

Tudhope, A. W., C. P. Chilcott, M. T. McCulloch, E. R. Cook, J. Chappell, R. M. Ellam, D. W. Lea, J. M. Lough & G. B. Shimmield. 2001. Variability in the El Niño-southern oscillation through a glacial-interglacial cyle. Science 291:1511–1517.

Turner, C. G., II. 1986. Dentochronological separation estimates for Pacific rim populations. Science 232:140–142.

Turner, D. L. & R. D. Jarrard. 1982. K-Ar dating of the Cook-Austral Island chain: a test of the hot-spot hypothesis. Journal of Volcanology and Geothermal Research 12:187–220.

Turner, S. & C. Hawkesworth. 1997. Constraints on flux rates and mantle dynamics beneath island arcs from Tonga-Kermadec lava geochemistry. Nature 359:568–573.

Udvardy, M. D. F. 1975. A classification of the biogeographical provinces of the world. IUCN Occasional Paper no. 18.

Udvardy, M. D. F. & A. Engilis, Jr. 2001. Migration of northern pintail across the Pacific with reference to the Hawaiian Islands. Studies in Avian Biology 22:124–132.

UNEP/IUCN. 1988. Coral Reefs of the World. Volume 3: Central and Western Pacific. UNEP Regional Seas Directories and Bibliographies. UNEP, Nairobi; IUCN, Gland, Switzerland, and Cambridge, UK.

Urwin, G. J. W. 1997. Facing Fearful Odds: The Siege of Wake Island. University of Nebraska Press, Lincoln.

Valdebenito, H. A., T. F. Stuessy & D. J. Crawford. 1990. A new biogeographic connection between islands in the Atlantic and Pacific Oceans. Nature 347:549–550.

Valeri, V. 2000. The Forest of Taboos. University of Wisconsin Press, Madison.

van Balgooy, M. M. J. 1971. Plant-geography of the Pacific as based on a census of phanerogam genera. Blumea, Supplement 6:1–222.

van Balgooy, M. M. J. 1976. Phytogeography. Pp. 1–22 *in* K. Paijmans (ed.). New Guinea Vegetation. Australian National University Press, Canberra.

van Balgooy, M. M. J., P. H. Hovenkamp & P. C. van Welzen. 1996. Phytogeography of the Pacific: floristic and historical distribution patterns in plants. Pp. 191–213 *in* A. Keast and S. E. Miller (eds.). The Origin and Evolution of Pacific Island Biotas, New Guinea to Eastern Polynesia: Patterns and Processes. SPB Academic Publishing, Amsterdam.

van Gemerden, B. S., H. Olff, M. P. E. Parren & F. Bongers. 2003. The pristine rain forest? Remnants of historical human impacts on current tree species composition and diversity. Journal of Biogeography 30:1381–1390.

van Gils, J. & P. Wiersma. 1996. Scolopacidae. Pp. 444–533 *in* J. del Hoyo, A. Elliott & J. Sargatal (eds.). Handbook of the Birds of the World, vol. 3, Hoatzin to Auks. Lynx, Barcelona.

van Riper, C., III & J. M. Scott. 2001. Limiting factors affecting Hawaiian native birds. Studies in Avian Biology 22:221–233.

van Riper, C., III, S. G. van Riper, M. L. Goff & M. Laird. 1986. The epizootiology and ecological significance of malaria in Hawaiian land birds. Ecological Monographs 56:327–344.

van Riper, C., III, S. G. van Riper & W. Hansen. 2002. The epizootiology and ecological significance of avian pox in Hawaii. Auk 119:929–942.

Van Tilburg, J. A. 1992. HMS Topaze on Easter Island. British Museum Occasional Paper 73.

Van Tilburg, J. A. 1994. Easter Island: Archaeology, Ecology, and Culture. British Museum Press, London.

Vanuatu. 1994. South Pacific Maps. Distributed by HEMA Maps, Queensland, Australia.

Vargas, H. & H. M. Snell. 1998. Marek's disease on the Galápagos Islands. Bird Conservation International 8:312–313.

Vermeer, K. & L. Rankin. 1984. Influence of habitat destruction and disturbance on nesting seabirds. Pp. 723–736 *in* J. P. Croxall, P. G. H. Evans & R. W. Schreiber (eds.). Status and Conservation of the World's Seabirds. International Council for Bird Preservation Technical Publication no. 2.

Vice, D. S. & D. L. Vice. 2004. Prey items of migratory peregrine falcon (*Falco peregrinus*) and Eurasian kestrel (*Falco tinnunculus*) on Guam. Micronesica 37:33–36.

Vimeux, F., V. Masson, J. Jouzel, M. Stievenard & J. R. Petit. 1999. Glacial-interglacial changes in ocean surface conditions in the southern hemisphere. Nature 398:410–413.

Visher, S. S. 1925. Tropical cyclones of the Pacific. Bernice P. Bishop Museum Bulletin 20:1–163.

Visser, K., R. Thunell & L. Stott. 2003. Magnitude and timing of temperature change in the Indo-Pacific warm pool during deglaciation. Nature 421:152–156.

Vitousek, P. M., T. N. Ladefoged, P. V. Kirch, A. S. Hartshorn, M. W. Graves, S. C. Hotchkiss, S. Tuljapurkar & O. A. Chadwick. 2004. Soils, agriculture, and society in precontact Hawai'i. Science 304:1665–1669.

Voelker, G. 1999. Dispersal, vicariance, and clocks: historical biogeography and speciation in a cosmopolitan passerine genus (*Anthus:* Motacillidae). Evolution 53:1536–1552.

Voight, B., R. S. J. Sparks, A. D. Miller, R. C. Stewart, R. P. Hoblitt, A. Clarke, J. Ewart, W. P. Aspinall, B. Baptie, E. S. Calder, P. Cole, T. H. Druitt, C. Hartford, R. A. Herd, P. Jackson, A. M. Lejeune, A. B. Lockhart, S. C. Loughlin, R. Luckett, L. Lynch, G. E. Norton, R. Robertson, I. M. Watson, R. Watts & S. R. Young. 1999. Magma flow instability and cyclic activity at Soufriere Hills volcano, Montserrat, British West Indies. Science 283:1138–1142.

Voisin, C., J.-F. Voisin & D. Mary. 1995. A fifth specimen of the Tahiti parakeet. Bulletin of the British Ornithologists' Club 115:262–263.

Volkov, I., J. R. Banavar, S. P. Hubbell & A. Maritan. 2003. Neutral theory and relative species abundance in ecology. Nature 424:1035–1037.

Volkov, I., J. R. Banavar, A. Maritan & S. Hubbell. 2004. The stability of forest biodiversity. Nature 427:696.

Wagner, W. L. 1991. Evolution of waif floras: a comparison of the Hawaiian and Marquesan archipelagos. Pages 267–284 *in* E. Dudley (ed.). The Unity of Evolutionary Biology: The Proceedings of the 4th International Congress of Systematics and Evolutionary Biology. Dioscorides Press, Portland, OR.

Wagner, W. L. & V. A. Funk (eds.). 1995. Hawaiian Biogeography: Evolution on a Hot Spot Archipelago. Smithsonian Institution Press, Washington.

Waldren, S., J. Florence & A. J. Chepstow-Lusty. 1995. Rare and endemic vascular plants of the Pitcairn Islands, south-central Pacific Ocean: a conservation appraisal. Biological Conservation 74:83–98.

Walsberg, G. E. 1983. Avian ecological energetics. Pp. 161–200 *in* D. S. Farner, J. R. King & K. C. Parkes (eds.). Avian Biology 7. Academic Press, New York.

Walter, H., E. Harnickell & D. Mueller-Dombois. 1975. Climate-Diagram Maps of the Individual Continents and the Ecological Climatic Regions of the Earth. Springer-Verlag, Berlin.

Walters, M. P. 1988. Probable validity of *Rallus nigra* Miller, an extinct species from Tahiti. Notornis 35:265–269.

Walters, M. P. 1989. Comment on the proposed suppression of *Rallus nigra* Miller, 1784 (Aves). Bulletin of Zoological Nomenclature 46:50–52.

Walters, M. P. 1991. *Prosobonia ellisi*, an extinct species of sandpiper from Moorea, Society Islands. Bollettino del Museo Regionale di Scienze Naturali-Torina 9:217–226.

Walters, M. P. 1993. On the status of the Christmas Island sandpiper, *Aechmorhynchus cancellatus*. Bulletin of the British Ornithologists' Club 113:97–102.

Ward, W. T., P. J. Ross & D. J. Colquhoun. 1971. Interglacial high sea levels: an absolute chronology derived from shoreline elevations. Palaeogeography, Palaeoclimatology, Palaeoecology 9:77–99.

Warheit, K. I. 2002. The seabird fossil record and the role of paleontology in understanding seabird community structure. Pp. 17–55 *in* E. A. Schreiber & J. Burger (eds.). Biology of Marine Birds. CRC Press, Boca Raton, FL.

Warner, R. E. 1968. The role of introduced diseases in the extinction of the endemic Hawaiian avifauna. Condor 70:101–120.

Watling, D. 1978. The Cambridge collection of Fijian and Tongan landbirds. Bulletin of the British Ornithologists' Club 98:95–98.

Watling, D. 1982. Birds of Fiji, Tonga, and Samoa. Millwood Press, Wellington, New Zealand.

Watling, D. 1985. The distribution of Fijian land and freshwater birds, based on the collections and observations of the Whitney South Sea Expedition. Domodomo 3:130–152.

Watling, D. 1989. Notes of the fauna of Laucala and Matagi Islands, Fiji. Domodomo 1–4.

Watling, D. 1995. Notes on the status of Kuhl's lorikeet *Vini kuhlii* in the northern Line Islands, Kiribati. Bird Conservation International 5:481–489.

Watling, D. 2001. A Guide to the Birds of Fiji and Western Polynesia including American Samoa, Niue, Samoa, Tokelau, Tonga, Tuvalu, and Wallis & Futuna. Environmental Consultants, Suva, Fiji.

Webb, E. L., B. J. Stanfield & M. L. Jensen. 1999. Effects of topography on rainforest tree community structure and diversity in American Samoa, and implications for frugivore and nectarivore populations. Journal of Biogeography 26:887–897.

Webb, H. P. 1992. Field observations of the birds of Santa Isabel, Solomon Islands. Emu 92:52–56.

Webb, S. D. 1989. The fourth dimension in North American terrestrial mammal communities. Pp. 181–203 *in* D. W. Morris, Z. Abramsky, B. J. Fox & M. R. Willig (eds.). Patterns in the Structure of Mammalian Communities. Special Publications of the Museum, Texas Tech University no. 28, Lubbock.

Webb, S. D. & N. D. Opdyke. 1995. Global climatic influence on Cenozoic land mammal faunas. Pp. 184–208 *in* J. P. Kennett & S. M. Stanley (eds.). Effects of Past Global Change on Life. National Academy Press, Washington.

Weiher, E. & P. Keddy (eds.). 1999. Ecological Assembly Rules: Perspectives, Advances, Retreats. Cambridge University Press, Cambridge, UK.

Weir, A. A. S., J. Chappell & A. Kacelnik. 2002. Shaping of hooks in New Caledonian crows. Science 297:981.

Weir, D. G. 1973. Status and habits of *Megapodius pritchardii*. Wilson Bulletin 85:79–82.

Weisler, M. I. 1994. The settlement of marginal Polynesia: new evidence from Henderson Island. Journal of Field Archaeology 21:83–102.

Weisler, M. I. 1995. Henderson Island prehistory: colonization and extinction on a remote Polynesian island. Biological Journal of the Linnean Society 56:377–404.

Weisler, M. I. 2000. Burial artifacts from the Marshall Islands: description, dating, and evidence for extra-archipelago contacts. Micronesica 33:111–136.

Weller, M. W. 1980. The Island Waterfowl. Iowa State University Press, Ames.

Wells, M. L., G. K. Vallis & E. A. Silver. 1999. Tectonic processes in Papua New Guinea and past productivity in the eastern equatorial Pacific Ocean. Nature 398:601–604.

Western Samoa. 1996. South Pacific Maps. Distributed by HEMA Maps, Queensland, Australia.

Weston, P. H. & M. D. Crisp. 1996. Trans-Pacific biogeographic patterns in the Proteaceae. Pp. 215–232 *in* A. Keast and S. E. Miller (eds.). The Origin and Evolution of Pacific Island

Biotas, New Guinea to Eastern Polynesia: Patterns and Processes. SPB Academic Publishing, Amsterdam.

Wetmore, A. 1925. The Coues Gadwall extinct. Condor 27:36.

Wetterer, J. K. 1998. Nonindigenous ants associated with geothermal and human disturbance in Hawai'i Volcanoes National Park. Pacific Science 52:40–50.

Wetterer, J. K. 2002. Ants of Tonga. Pacific Science 56:125–135.

Wetterer, J. K., P. C. Banko, L. P. Laniawe, J. W. Slotterback & G. J. Brenner. 1998. Nonindigenous ants at high elevations on Mauna Kea, Hawai'i. Pacific Science 52:228–236.

Wetterer, J. K. & D. L. Vargo. 2003. Ants (Hymenoptera: Formicidae) of Samoa. Pacific Science 57:409–419.

Whatley, R. & R. Jones. 1999. The marine podocopid Ostracoda of Easter Island: a paradox in zoogeography and evolution. Marine Micropaleontology 37:327–343.

Wheeler, C. W. & P. Aharon. 1991. Midoceanic carbonate platforms as oceanic dipsticks: examples from the Pacific. Coral Reefs 10:101–114.

Whistler, W. A. 1980. The vegetation of eastern Samoa. Allertonia 2:45–190.

Whistler, W. A. 1983. Vegetation and flora of the Aleipata Islands, Western Samoa. Pacific Science 37:227–250.

Whistler, W. A. 1992. Vegetation of Samoa and Tonga. Pacific Science 46:159–178.

White, C. M. N. 1976. The problem of the cassowary in New Britain. Bulletin of the British Ornithologists' Club 96:66–68.

White, C. M., D. J. Brimm & F. Clunie. 1988. A study of peregrines in the Fiji Islands, South Pacific Ocean. Pp. 275–287 *in* T. J. Cade, J. H. Enderson, C. G. Thelander & C. M. White (eds.). Peregrine Falcon Populations: Their Management and Recovery. Peregrine Fund, Boise, ID.

White, C. M., N. J. Clum, T. J. Cade & W. G. Hunt. 2002. Peregrine falcon *(Falco peregrinus). In* A. Poole & F. Gill (eds.). The Birds of North America, no. 481. The Birds of North America, Philadelphia.

White, J. P., G. Clark & S. Bedford. 2000. Distribution, present and past, of *Rattus praetor* in the Pacific and its implications. Pacific Science 54:105–117.

White, J. P., T. F. Flannery, R. O'Brien, R. V. Hancock & L. Pavlish. 1991. The Balof Shelters, New Ireland. Pp. 46–58 *in* J. Allen & C. Gosden (eds.). Report on the Lapita Homeland Project. Department of Prehistory, Australian National University, Canberra.

Whitfield, J. 2001. All creatures great and small. Nature 413:342–344.

Whitmee, S. J. 1874. Letter from Rev. S. J. Whitmee. Proceedings of the Zoological Society of London 1874:183–186.

Whitmore, T. C. 1969. The vegetation of the Solomon Islands. Philosophical Transactions of the Royal Society of London B 255:259–270.

Whitmore, T. C. 1974. Change with time and the role of cyclones in tropical rainforest on Kolombangara, Solomon Islands. Commonwealth Forestry Institute Paper 46.

Whitmore, T. C. 1989. Changes over twenty-one years in the Kolombangara rain forests. Journal of Ecology 77:469–483.

Whittaker, R. J. 1998. Island Biogeography: Ecology, Evolution, and Conservation. Oxford University Press, Oxford.

Whittier, H. O. 1976. Mosses of the Society Islands. University Presses of Florida, Gainesville.

Wickler, S. 1990. Prehistoric Melanesian exchange and interaction: recent evidence from the northern Solomon Islands. Asian Perspectives 29:135–154.

Wickler, S. 2001. The prehistory of Buka: a stepping stone island in the northern Solomons. Terra Australis 16. Australian National University, Canberra.

Wickler, S. & M. Spriggs. 1988. Pleistocene occupation of the Solomon Islands, Melanesia. Antiquity 62:703–706.

Wiens, H. J. 1962. Atoll Environment and Ecology. Yale University Press, New Haven, CT.

Wiersma, P. 1996. Species accounts (Charadriidae). Pp. 410–442 *in* J. del Hoyo, A. Elliott & J. Sargatal (eds.). Handbook of the Birds of the World, vol. 3, Hoatzin to Auks. Lynx, Barcelona.

Wigley, T. M. L. & S. C. B. Raper. 2001. Interpretation of high projections for global-mean warming. Science 293:451–454.

Wijpkema, J. & T. Wijpkema. 1997. Tuamotu sandpiper. Dutch Birding 19:76–80.

Wilcove, D. S. & T. Eisner. 2000. The impending extinction of natural history. Chronicle of Higher Education, September 15, B24.

Wilder, G. P. 1931. Flora of Rarotonga. Bernice P. Bishop Museum Bulletin no. 86.

Wiles, G. J. 1998. Records of communal roosting in Mariana crows. Wilson Bulletin 110:126–128.

Wiles, G. J. 2000a. Dededo, Guam. American Birds (100th Christmas Bird Count), pp. 589–590.

Wiles, G. J. 2000b. Southern Guam, Guam. American Birds (100th Christmas Bird Count), p. 590.

Wiles, G. J. 2000c. Saipan, C.N.M.I. American Birds (100th Christmas Bird Count), p. 590.

Wiles, G. J. & P. J. Conry. 1990. Terrestrial vertebrates of the Ngerukewid Islands Wildlife Preserve, Palau Islands. Micronesica 23:41–66.

Wiles, G. J., R. E. Beck, C. F. Aguon & K. D. Orcutt. 1993. Recent bird records for the southern Mariana Islands, with notes on a colony of black noddies on Cocos Island, Guam. Micronesica 26:199–215.

Wiles, G. J., N. C. Johnson, J. B. de Cruz, G. Dutson, V. A. Camacho, A. K. Kepler, D. S. Vice, K. L. Garrett, C. C. Kessler & H. D. Pratt. 2004. New and noteworthy bird records for Micronesia. Micronesica 37:69–96.

Wiles, G. J., I. H. Schreiner, D. Nafus, L. K. Jurgensen & J. C. Manglona. 1996. The status, biology, and conservation of *Serianthes nelsonii* (Fabaceae), an endangered Micronesian tree. Biological Conservation 76:229–239.

Willerslev, E., A. J. Hansen, J. Binladen, T. B. Brand, M. T. P. Gilbert, B. Shapiro, M. Bunce, C. Wiuf, D. A. Gilichinsky & A. Cooper. 2003. Diverse plant and animal genetic records from Holocene and Pleistocene sediments. Science 300:791–795.

Williams, G. R. 1960. The birds of the Pitcairn Islands, central South Pacific Ocean. Ibis 102:58–70.

Williams, C. E. 1998. Zooarchaeology of the Pamwak Site, Manus Island, PNG. Ph.D. thesis, Monash University, Melbourne, Australia.

Williams, M. I. & D. W. Steadman. 2001. The historic and prehistoric distribution of parrots (Psittaciformes, Psittacidae) in the West Indies. Pp. 175–187 *in* C. A. Woods & F. E. Sergile (eds.). Biogeography of the West Indies: Patterns and Perspectives. CRC Press, Boca Raton, FL.

Williamson, M. 1981. Island Populations. Oxford University Press, Oxford.
Willis, P. M. A. 1997. A review of fossil crocodilians from Australasia. Australian Zoologist 30:287–298.
Wilson, A. D. & F. G. Beecroft. 1983. Soils of the Ha'apai Group, Kingdom of Tonga. New Zealand Soil Survey Report no. 67.
Wilson, E. O. 1959. Adaptive shift and dispersal in a tropical ant fauna. Evolution 13:122–144.
Wilson, E. O. 1961. The nature of the taxon cycle in the Melanesian ant fauna. American Naturalist 95:169–193.
Wilson, E. O. 1998. Consilience: The Unity of Knowledge. Alfred A. Knopf, New York.
Wilson, E. O. & R. W. Taylor. 1967a. An estimate of the potential evolutionary increase in species density in the Polynesian ant fauna. Evolution 21:1–10.
Wilson, E. O. & R. W. Taylor. 1967b. The ants of Polynesia (Hymenoptera, Formicidae). Pacific Insects Monographs 14.
Wilson, K.-J. 1997. Extinct and introduced vertebrate species in New Zealand: a loss of biodistinctiveness and gain in biodiversity. Pacific Conservation Biology 3:301–305.
Wilson, P. G. 1996. Myrtaceae in the Pacific, with special reference to *Metrosideros*. Pp. 233–245 *in* A. Keast & S. E. Miller (eds.). The Origin and Evolution of Pacific Island Biotas, New Guinea to Eastern Polynesia: Patterns and Processes. SPB Academic Publishing, Amsterdam.
Wilson, S. B. 1907. Notes on birds of Tahiti and the Society Group. Ibis 3:373–379.
Winterhalder, B. & F. Lu. 1997. A forager-resource population ecology model and implications for indigenous conservation. Conservation Biology 11:1354–1364.
Wiser, S. K., D. R. Drake, L. E. Burrows & W. R. Sykes. 2002. The potential for long-term persistence of forest fragments on Tongatapu, a large island in western Polynesia. Journal of Biogeography 29:767–787.
Witteman, G. J., R. E. Beck, Jr., S. L. Pimm & S. R. Derrickson. 1991. The decline and restoration of the Guam rail, *Rallus owstoni*. Endangered Species Update 8:36–39.
Wodzicki, K. 1971. The birds of Niue Island: an annotated checklist. Notornis 18:291–304.
Wodzicki, K. & M. Laird. 1970. Birds and bird lore in the Tokelau Islands. Notornis 17:247–276.
Wolfe, C. J., M. K. McNutt & R. S. Detrick. 1994. The Marquesas archipelagic apron: seismic stratigraphy and implications for volcano growth, mass wasting, and crustal underplanting. Journal of Geophysical Research 99:13591–13608.
Wolff, T. 1958. The Natural History of Rennell Island, British Solomon Islands. Vol. 1 (Vertebrates): Scientific Results of the Danish Rennell Expedition, 1951 and the British Museum (Natural History) Expedition, 1953. Danish Science Press, Copenhagen.
Wood, B. L. 1967. Geology of the Cook Islands. New Zealand Journal of Geology and Geophysics 10:1429–1445.
Wood, B. L. & R. F. Hay. 1970. Geology of the Cook Islands. New Zealand Geological Survey Bulletin no. 82.
Wood, D. S. & G. D. Schnell. 1986. Revised world inventory of avian skeletal specimens, 1986. American Ornithologists' Union, Norman; Oklahoma Biological Survey, Norman.
Woodford, C. M. 1916. On some little-known Polynesian settlements in the neighbourhood of the Solomon Islands. Geographical Journal 48:26–54.
Woodroffe, C. D. 1985. Vegetation and flora of Nui Atoll, Tuvalu. Atoll Research Bulletin 283:1–18.
Woodroffe, C. D. 1986. Vascular plant species-area relationships on Nui Atoll, Tuvalu, central Pacific: a reassessment of the small island effect. Australian Journal of Ecology 11:21–31.
Woodroffe, C. D. 1987. Pacific island mangroves: distribution and environmental settings. Pacific Science 4:166–185.
Woodroffe, C. D., S. A. Short, D. R. Stoddart, T. Spencer & R. S. Harmon. 1991. Stratigraphy and chronology of late Pleistocene reefs in the southern Cook Islands, South Pacific. Quaternary Research 35:246–263.
Woodroffe, C. D., D. R. Stoddart, T. Spencer, T. P. Scoffin & A. W. Tudhope. 1990. Holocene emergence in the Cook Islands, South Pacific. Coral Reefs 9:31–39.
Woollard, G. P. & L. D. Kulm. 1981. History of the Nazca Plate project. Pp. 3–24 *in* L. D. Kulm, J. Dymond, E. J. Dasch & D. M. Hussong (eds.). Nazca Plate: Crustal Formation and Andean Convergence. Geological Society of America Memoir 154.
Worthington, D. J. 1998. Inter-island dispersal of the Mariana common moorhen: a recolonization by an endangered species. Wilson Bulletin 110:414–417.
Worthy, T. H. 1997a. A mid-Pleistocene rail from New Zealand. Alcheringa 21:71–78.
Worthy, T. H. 1997b. Quaternary fossil fauna of South Canterbury, South Island, New Zealand. Journal of the Royal Society of New Zealand 27:67–162.
Worthy, T. H. 2000. The fossil megapodes (Aves: Megapodiidae) of Fiji with descriptions of a new genus and two new species. Journal of the Royal Society of New Zealand 30:337–364.
Worthy, T. H. 2001a. A new species of *Platymantis* (Anura: Ranidae) from Quaternary deposits on Viti Levu, Fiji. Paleontology 44:665–680.
Worthy, T. H. 2001b. A giant flightless pigeon gen. et sp. nov. and a new species of *Ducula* (Aves: Columbidae), from Quaternary deposits in Fiji. Journal of the Royal Society of New Zealand 31:763–794.
Worthy, T. H. 2003. A new extinct species of snipe *Coenocorypha* from Viti Levu, Fiji. Bulletin of the British Ornithologists' Club 123:90–103.
Worthy, T. H. 2004. The fossil rails (Aves: Rallidae) of Fiji with descriptions of a new genus and species. Journal of the Royal Society of New Zealand 34:295–314.
Worthy, T. H. & R. N. Holdaway. 1993. Quaternary fossil faunas from caves in the Punakaiki area, west coast, South Island, New Zealand. Journal of the Royal Society of New Zealand 23:147–254.
Worthy, T. H. & R. N. Holdaway. 1994. Quaternary fossil faunas from caves in Takaka Valley and on Takaka Hill, northwest Nelson, South Island, New Zealand. Journal of the Royal Society of New Zealand 24:297–391.
Worthy, T. H. & R. N. Holdaway. 2002. The Lost World of the Moa. Indiana University Press, Bloomington.
Worthy, T. H. & G. M. Wragg. 2003. A new species of *Gallicolumba:* Columbidae from Henderson Island, Pitcairn Group. Journal of the Royal Society of New Zealand 33:769–793.

Worthy, T. H., A. J. Anderson & R. E. Molnar. 1999. Megafaunal expression in a land without mammals: the first fossil faunas from terrestrial deposits in Fiji. Senckenbergiana Biologica 79:237–242.

Worthy, T. H., R. Walter & A. J. Anderson. 1998. Fossil and archaeological avifauna of Niue Island, Pacific Ocean. Notornis 45:177–190.

Worthy, T. H., C. M. Miskelly & B. A. Ching. 2002a. Taxonomy of North and South Island snipe (Aves: Scolopacidae: *Coenocorypha*), with analysis of a remarkable collection of snipe bones from Greymouth, New Zealand. New Zealand Journal of Zoology 29:231–244.

Worthy, T. H., R. N. Holdaway, B. V. Alloway, J. Jones, J. Winn & D. Turner. 2002b. A rich Pleistocene-Holocene avifaunal sequence from Te Waka #1: terrestrial fossil vertebrate faunas from inland Hawke's Bay, North Island, New Zealand. Part 2. Tuhinga 13:1–38.

Wragg, G. M. 1995. The fossil birds of Henderson Island, Pitcairn Group: natural turnover and human impact, a synopsis. Biological Journal of the Linnean Society 56:405–414.

Wragg, G. M. & M. I. Weisler. 1994. Extinctions and new records of birds from Henderson Island, Pitcairn Group, South Pacific Ocean. Notornis 41:61–70.

Wright, A. C. S. 1963. Soils and land use of Western Samoa. New Zealand Soil Bureau Bulletin no. 22.

Wright, J. D. 2001. The Indonesian valve. Nature 411:142–143.

Wright, S. D., C. G. Young, J. W. Dawson, D. J. Whittaker & R. C. Gardner. 2000. Riding the ice age El Nino? Pacific biogeography and evolution of *Metrosideros* subg. *Metrosideros* (Myrtaceae) inferred from nuclear ribosomal DNA. Proceedings of the National Academy of Sciences USA 97:4118–4123.

Wuethrich, B. 2000. Learning the world's languages: before they vanish. Science 288:1156–1159.

Wuvulu Island to Kaniet Islands. 1995. Map 82050. United States Defense Mapping Agency, Washington.

Wynne-Edwards, V. C. 1955. Low reproductive rates in birds, especially sea-birds. Acta XI International Ornithological Congress, Basel 1954:540–547.

Wyrtki, K. & G. Meyers. 1976. Trade wind field over Pacific Ocean. Journal of Applied Meterology 15:698–704.

Yamashina, Y. 1948. Notes on the Marianas mallard. Pacific Science 2:121–124.

Yan, C. Y. & L. W. Kroenke. 1993. A plate tectonic reconstruction of the southwest Pacific, 0–100 Ma. Proceedings Ocean Drilling Project, Scientific Results 130:697–709.

Yap Islands. 1996. Map 81187. United States Defense Mapping Agency, Washington.

Yokoyama, Y., K. Lambeck, P. D. Deckker, P. Johnston & L. K. Fifield. 2000. Timing of the Last Glacial Maximum from observed sea-level minima. Nature 406:713–716.

Yonekura, N., Y. Saito, Y. Maeda, Y. Matsushima, E. Matsumoto & H. Kayanne. 1988. Holocene fringing reefs and sea-level change in Mangaia Island, southern Cook Islands. Palaeogeography, Palaeoclimatology, Palaeoecology 68:177–188.

Yorinks, N. & C. T. Atkinson. 2000. Effects of malaria on activity budgets of experimentally infected juvenile apapane (*Himatione sanguinea*). Auk 117:731–738.

Ysabel Channel. 1995. Map 82095. United States Defense Mapping Agency, Washington.

Zachos, J., M. Pagani, L. Sloan, E. Thomas & K. Billups. 2001. Trends, rhythms, and aberrations in global climate 56 Ma to present. Science 292:686–693.

Zhang, J., G. Harbottle, C. Wang & Z. Kong. 1999. Oldest playable music instruments found at Jiahu early Neolithic site in China. Nature 401:366–368.

Zink, R. M. & J. B. Slowinski. 1995. Evidence from molecular systematics for decreased avian diversification in the Pleistocene Epoch. Proceedings of the National Academy of Sciences USA 92:5832–5835.

Zug, G. R. 1991. The lizards of Fiji: natural history and systematics. Bernice P. Bishop Museum Bulletins in Zoology 2:1–136.

Zug, G. R. & B. R. Moon. 1995. Systematics of the Pacific slender-toed geckos, *Nactus pelagicus* complex: Oceania, Vanuatu, and Solomon Islands populations. Herpetologica 51:77–90.

Zug, G. R., D. Watling, T. Alefaio, S. Alefaio & C. Ludescher. 2003. A new gecko (Reptilia: Squamata: Genus *Lepidodactylus*) from Tuvalu, south-central Pacific. Proceedings of the Biological Society of Washington 116:38–46.

Zusi, R. L. & J. R. Jehl. 1970. Systematic relationships of *Aechmorhynchus*, *Prosobonia*, and *Phegornis* (Charadriiformes: Charadrii). Auk 87:760–780.

Zwartjes, P. W. 1999. Genetic variability in the endemic vireos of Puerto Rico and Jamaica contrasted with the continental white-eyed vireo. Auk 116:964–975.

Systematic Index

Page numbers in italics refer to figures; page numbers followed by t *refer to tables.*

General Index

Page numbers in italics refer to figures; page numbers followed by t *refer to tables.*